INTRODUCTORY QUANTUM MECHANICS

2ND EDITION

Richard L. Liboff
Cornell University

Addison-Wesley Publishing Company
Reading, Massachusetts • Menlo Park, California • New York
Don Mills, Ontario • Wokingham, England • Amsterdam • Bonn • Sydney
Singapore • Tokyo • Madrid • San Juan • Milan • Paris

Library of Congress Cataloging-in-Publication Data

Liboff, Richard L., 1931-
    Introductory quantum mechanics / Richard L. Liboff. -- 2nd ed.
        p.   cm.
    Includes bibliographical references and index.
    ISBN 0-201-54715-5
    1. Quantum theory. I. Title.
QC174.12.L52 1991
530.1'2–dc20   91-2448
                                                                   CIP

ISBN 0-201–54715-5
2345678910-DO-9594939291

To Myra

"She openeth her mouth with wisdom; and in her tongue is the law of kindness. . . ."

PREFACE

Since the publication of the first edition of this work, physics has progressed both in esoteric and pragmatic directions. The reader will note that this new edition parallels this trend.

Thus, for example, in Chapter 7, a discussion of Feynman's path integral appears, preceded by review of Hamilton's classical principal of least action. A new section on WKB wavefunctions is also included in this chapter.

Two new sections appear in Chapter 8. In the first of these, an approximation technique important to solid-state physics and the modeling of molecular states is presented, commonly known as the linear combination of atomic orbitals method (LCAO). In the second section, expressions for the density of states in various dimensions are derived, including that of the "quantum well." This is a one-dimensional well embedded in 3-space that finds wide application in modern-day microtechnology.

In Chapter 10, a section has been inserted on hybridization with application made to the CH_4 molecule. The section on charged-particle motion in a magnetic field was extended to include a derivation of the degeneracy of Landau levels. In the concluding section of the chapter, the reader is introduced to an approximation technique important to atomic physics known as the Thomas–Fermi model. The Thomas–Fermi equation for the atomic potential is derived and the implied variation of effective atomic size as a function of atomic number is noted.

The Schrödinger, Heisenberg, and interaction pictures are described in Chapter 11. In the concluding section of the chapter, photon polarization states are introduced. These are then applied to an optical experiment related to the Einstein–Podolsky–Rosen paradox. Bell's inequality is derived and discussed in relation to the results of this experiment.

Groupings of atoms in the periodic chart as described in Section 12.4 are found to be consistent with two new graphs in which ionization energies and atomic radii are plotted against atomic number. A discussion is included in this same section on the construction of atomic states for two-electron atoms in accord with the Pauli principle.

Application of this principle is further developed in the closing two sections of Chapter 12. In the first of these, the Fermi–Dirac distribution comes into play in describing conductivity in extrinsic semiconductors. The atomic make up of n- and p-type semiconductors are discussed in relation to the location of respective atoms in the periodic chart. Temperature dependence of the Fermi energy, as well as that of electron and hole concentrations, is noted for both types of materials. This section complements the discussion of intrinsic semiconductors presented previously in Section 8.4.

A component of nuclear physics is introduced in the closing section of this chapter, which addresses the ground state of the deuteron. This state is constructed in accord with the Pauli principle, together with values of magnetic dipole and electric quadrupole moments of the deuteron as well as the principle of conservation of parity for the strong nuclear forces. Furthermore, here the reader is introduced to the concepts of isotopic spin, the nucleon, and noncentral forces.

In Chapter 13, which addresses perturbation theory, a section is included describing an atom interacting with a radiation field. In this analysis, the reader encounters the notion of oscillator strengths, which, in turn, are shown to obey the Thomas–Reiche–Kuhn sum rule. Application is made in calculation of the lifetime of an excited state of an atom.

Two sections were added to the concluding chapter on scattering theory. In the first of these, continuing with the formalism developed in the previous chapter, an expression for the cross section of photon scattering from atoms is obtained in which the notion of the Lorentzian line-shape factor comes into play. A review of formal scattering theory is presented in the concluding section in which the interaction picture, previously developed in Section 11.12, is employed in deriving the Lippmann–Schwinger equation.

A number of self-contained problems, closely allied to the textual material, have been inserted, many of which carry solutions. These problems are listed after the Contents under the heading of "Topical Problems."

Many individuals have been helpful in the development of this new edition. I am indebted to P. Bruce Pipes and Stavros Fallieros for their constructive suggestions for improvement in sections of the first edition. I am equally indebted to Norman Ramsey for calling my attention to an interesting property which enters the Stern–Gerlach experiment (see Problem 11.85). I am grateful to Steven Seidman and George George who were helpful in the preparation of added sections to Chapter 8. Other individuals who contributed to this new edition and whom I would like here to thank are Gregory Schenter, Brian Jones, Kenneth Gardner, Sidney Leibovich, Norman Rostoker, Manfried Kleber, S. Jayaraman, H. Chang, William Case, Donald Yennie, Lloyd Hillman, Jon Jarrett, Terrence Fine, Donald Scarl, Stanly Bashkin, and Vaclav Kostroun.

The wider list of subjects discussed in the present edition should provide instructors with a broader base from which to choose topics for their particular course.

In closing, I would like to express my appreciation to the many individuals who have taught from this text and the many who have learned from it. I sincerely hope that these individuals find this new edition equally valuable.

Ithaca, 1990   R. L. LIBOFF

.תושלב״ע

"I do not know what I may appear to the world; but to myself I seem to have been only like a boy playing on the seashore, and diverting myself in now and then finding a smoother pebble or a prettier shell than ordinary, whilst the great ocean of truth lay all undiscovered before me."

This statement, said by Isaac Newton shortly before his death in 1727, eloquently reflects the sentiments of all mature scientists from the ancient past to the present.

CONTENTS

TOPICAL PROBLEMS

PART I
ELEMENTARY PRINCIPLES AND APPLICATIONS TO PROBLEMS IN ONE DIMENSION

REVIEW OF CONCEPTS
OF CLASSICAL MECHANICS

This is a preparatory chapter in which we review fundamental concepts of classical mechanics important to the development and understanding of quantum mechanics. Hamilton's equations are introduced and the relevance of cyclic coordinates and constants of the motion is noted. In discussing the state of a system, we briefly encounter our first distinction between classical and quantum descriptions. The notions of forbidden domains and turning points relevant to classical motion, which find application in quantum mechanics as well, are also described. The experimental motivation and historical background of quantum mechanics are described in Chapter 2.

1.1 GENERALIZED OR "GOOD" COORDINATES

Our discussion begins with the concept of *generalized* or *good* coordinates.

A bead (idealized to a point particle) constrained to move on a straight rigid wire has *one degree of freedom* (Fig. 1.1). This means that only one variable (or parameter) is needed to uniquely specify the location of the bead in space. For the problem under discussion, the variable may be displacement from an arbitrary but specified origin along the wire.

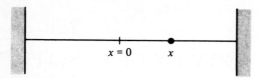

FIGURE 1.1   A bead constrained to move on a straight wire has one degree of freedom.

A particle constrained to move on a flat plane has two degrees of freedom. Two independent variables suffice to uniquely determine the location of the particle in space. With respect to an arbitrary, but specified origin in the plane, such variables might be the Cartesian coordinates (x, y) or the polar coordinates (r, θ) of the particle (Fig. 1.2).

Two beads constrained to move on the same straight rigid wire have two degrees of freedom. A set of appropriate coordinates are the displacements of the individual particles (x_1, x_2) (Fig. 1.3).

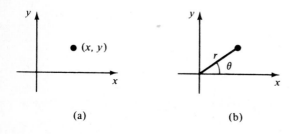

FIGURE 1.2   A particle constrained to move in a plane has two degrees of freedom. Examples of coordinates are (x, y) or (r, θ).

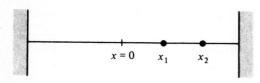

FIGURE 1.3   Two beads on a wire have two degrees of freedom. The coordinates x_1 and x_2 denote displacements of particles 1 and 2, respectively.

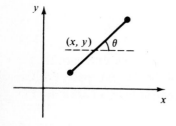

FIGURE 1.4   A rigid dumbbell in a plane has three degrees of freedom. A good set of coordinates are: (x, y), the location of the center, and θ, the inclination of the rod with the horizontal.

A rigid rod (or dumbbell) constrained to move in a plane has three degrees of freedom. Appropriate coordinates are: the location of its center (x, y) and the angular displacement of the rod from the horizontal, θ (Fig. 1.4).

Independent coordinates that serve to uniquely determine the orientation and location of a system in physical space are called *generalized* or *canonical* or *good* coordinates. *A system with N generalized coordinates has N degrees of freedom.* The orientation and location of a system with, say, three degrees of freedom are not specified until all three generalized coordinates are specified. The fact that *good* coordinates may be specified independently of one another means that given the values of all but one of the coordinates, the last coordinate remains arbitrary. Having specified (x, y) for a point particle in 3-space, one is still free to choose z independently of the assigned values of x and y.

PROBLEMS

1.1 For each of the following systems, specify the number of degrees of freedom and a set of good coordinates.

    (a)   A bead constrained to move on a closed circular hoop that is fixed in space.

    (b)   A bead constrained to move on a helix of constant pitch and constant radius.

    (c)   A particle on a right circular cylinder.

    (d)   A pair of scissors on a plane.

    (e)   A rigid rod in 3-space.

    (f)   A rigid cross in 3-space.

    (g)   A linear spring in 3-space.

    (h)   Any rigid body with one point fixed.

    (i)   A hydrogen atom.

    (j)   A lithium atom.

    (k)   A compound pendulum (two pendulums attached end to end).

1.2 Show that a particle constrained to move on a curve of any shape has one degree of freedom.

Answer

A curve is a one-dimensional locus and may be generated by the parameterized equations

$$x = x(\eta), \qquad y = y(\eta), \qquad z = z(\eta)$$

Once the independent variable η (e.g., length along the curve) is given, x, y, and z are specified.

1.3 Show that a particle constrained to move on a surface of arbitrary shape has two degrees of freedom.

Answer

A surface is a two-dimensional locus. It is generated by the equation

$$u(x, y, z) = 0$$

Any two of the three variables x, y, z determine the third. For instance, we may solve for z in the equation above to obtain the more familiar equation for a surface (height z at the point x, y),

$$z = z(x, y)$$

In this case. x and y may serve as generalized coordinates.

1.4 How many degrees of freedom does a classical gas composed of 10^{23} point particles have?

1.2 ENERGY, THE HAMILTONIAN, AND ANGULAR MOMENTUM

These three elements of classical mechanics have been singled out because they have direct counterparts in quantum mechanics. Furthermore, as in classical mechanics, their role in quantum mechanics is very important.

Consider that a particle of mass m in the potential field $V(x, y, z)$ moves on the trajectory

(1.1)
$$x = x(t)$$
$$y = y(t)$$
$$z = z(t)$$

At any instant t, the energy of the particle is

(1.2) $E = \frac{1}{2}mv^2 + V(x, y, z) = \frac{1}{2}m(\dot{x}^2 + \dot{y}^2 + \dot{z}^2) + V(x, y, z)$

The velocity of the particle is **v**. Dots denote time derivatives. The force on the particle **F** is the negative gradient of the potential.

(1.3) $$\mathbf{F} = -\nabla V = -\left(\mathbf{e}_x \frac{\partial}{\partial x} V + \mathbf{e}_y \frac{\partial}{\partial y} V + \mathbf{e}_z \frac{\partial}{\partial z} V\right)$$

The three unit vectors $(\mathbf{e}_x, \mathbf{e}_y, \mathbf{e}_z)$ lie along the three Cartesian axes.

Here are two examples of potential. The energy of a particle in the gravitational force field,

$$\mathbf{F} = -\mathbf{e}_z mg = -\nabla mgz$$

is

(1.4) $$E = \frac{1}{2}m(\dot{x}^2 + \dot{y}^2 + \dot{z}^2) + mgz$$

The particle is at the height z above sea level. For this example,

$$V = mgz$$

An electron of charge q and mass m, between capacitor plates that are maintained at the potential difference Φ_0 and separated by the distance d (Fig. 1.5), has potential

$$V = \frac{q\Phi_0}{d} z$$

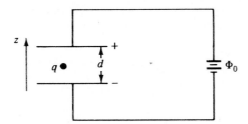

FIGURE 1.5 Electron in a uniform capacitor field.

The displacement of the electron from the bottom plate is z. The electron's energy is

$$(1.5) \qquad E = \tfrac{1}{2}m(\dot{x}^2 + \dot{y}^2 + \dot{z}^2) + \frac{q\Phi_0}{d}\, z$$

In both examples above, the system (particle) has three degrees of freedom. The Cartesian coordinates (x, y, z) of the particle are by no means the only "good" coordinates for these cases. For instance, in the last example, we may express the energy of the electron in spherical coordinates (Fig. 1.6):

$$(1.6) \qquad E = \tfrac{1}{2}m(\dot{r}^2 + r^2\dot{\theta}^2 + r^2\dot{\phi}^2 \sin^2 \theta) + \frac{q\Phi_0}{d}\, r \cos \theta$$

In cylindrical coordinates (Fig. 1.7) the energy is

$$(1.7) \qquad E = \tfrac{1}{2}m(\dot{\rho}^2 + \rho^2\dot{\phi}^2 + \dot{z}^2) + \frac{q\Phi_0}{d}\, z$$

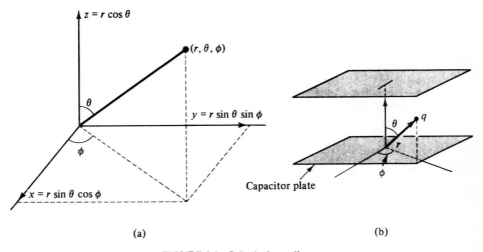

(a)

(b)

FIGURE 1.6 Spherical coordinates.

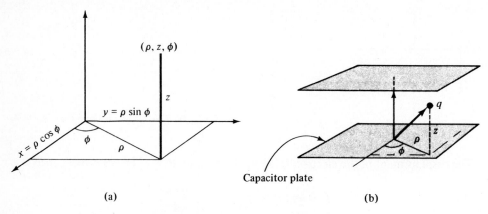

(a) (b)

FIGURE 1.7 Cylindrical coordinates.

The hydrogen atom has six degrees of freedom. If (x_1, y_1, z_1) are the coordinates of the proton and (x_2, y_2, z_2) are the coordinates of the electron, the energy of the hydrogen atom appears as

(1.8)
$$E = \tfrac{1}{2}M(\dot{x}_1{}^2 + \dot{y}_1{}^2 + \dot{z}_1{}^2) + \tfrac{1}{2}m(\dot{x}_2{}^2 + \dot{y}_2{}^2 + \dot{z}_2{}^2)$$
$$- \frac{q^2}{\sqrt{(x_1 - x_2)^2 + (y_1 - y_2)^2 + (z_1 - z_2)^2}}$$

(Fig. 1.8). The mass of the proton is M and that of the electron is m. In all the cases above, the energy is a *constant of the motion*. A constant of the motion is a dynamical function that is constant as the system unfolds in time. For each of these cases,

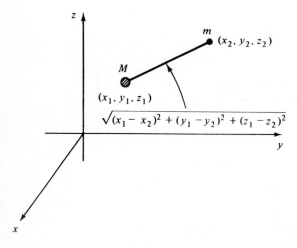

FIGURE 1.8 The hydrogen atom has six degrees of freedom. The Cartesian coordinates of the proton and electron serve as good generalized coordinates.

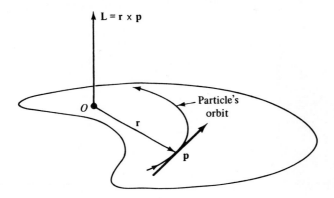

FIGURE 1.9 Angular momentum of a particle with momentum p about the origin O.

whatever E is initially, it maintains that value, no matter how complicated the subsequent motion is. Constants of the motion are extremely useful in classical mechanics and often serve to facilitate calculation of the trajectory.

A system that in no way interacts with any other object in the universe is called an *isolated system*. The total energy, linear momentum, and angular momentum of an isolated system are constant. Let us recall the definition of linear and angular momentum for a particle. A particle of mass m moving with velocity $\mathbf{v}$ has linear momentum

$$(1.9) \qquad\qquad \mathbf{p} = m\mathbf{v}$$

The angular momentum of this particle, measured about a specific origin, is

$$(1.10) \qquad\qquad \mathbf{L} = \mathbf{r} \times \mathbf{p}$$

where $\mathbf{r}$ is the radius vector from the origin to the particle (Fig. 1.9).

If there is no component of force on a particle in a given (constant) direction, the component of momentum in that direction is constant. For example, for a particle in a gravitational field that is in the z direction, p_x and p_y are constant.

If there is no component of torque $\mathbf{N}$ in a given direction, the component of angular momentum in that direction is constant. This follows directly from Newton's second law for angular momentum,

$$(1.11) \qquad\qquad \mathbf{N} = \frac{d\mathbf{L}}{dt}$$

For a particle in a gravitational field that is in the minus z direction, the torque on the particle is

$$\mathbf{N} = \mathbf{r} \times \mathbf{F} = -\mathbf{r} \times \mathbf{e}_z mg$$

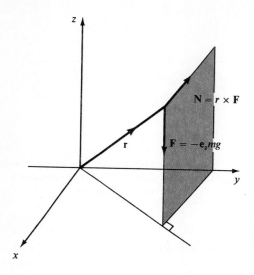

FIGURE 1.10   **The torque r × F has no component in the z direction.**

The radius vector from the origin to the particle is $\mathbf{r}$ (Fig. 1.10). Since $\mathbf{e}_z \times \mathbf{r}$ has no component in the z direction ($\mathbf{e}_z \cdot \mathbf{e}_z \times \mathbf{r} = 0$), it follows that

$$(1.12) \qquad\qquad L_z = xp_y - yp_x = \text{constant}$$

Since p_x and p_y are also constants, this equation tells us that the projected orbit in the xy plane is a straight line (Fig. 1.11).

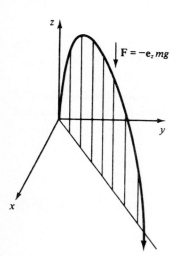

FIGURE 1.11   **The projected motion in the xy plane is a straight line. Its equation is given by the constant z component of angular momentum: $L_z = xp_y - yp_x$.**

Hamilton's Equations

The constants of motion for more complicated systems are not so easily found. However, there is a formalism that treats this problem directly. It is Hamiltonian mechanics. Consider the energy expression for an electron between capacitor plates (1.5). Rewriting this expression in terms of the linear momentum $\mathbf{p}$ (as opposed to velocity) gives

$$(1.13) \quad E(x, y, z, \dot{x}, \dot{y}, \dot{z}) \rightarrow H(x, y, z, p_x, p_y, p_z) = \frac{1}{2m}(p_x^2 + p_y^2 + p_z^2) + \frac{q\Phi_0}{d}z$$

The energy, written in this manner, as a function of coordinates and momenta is called the *Hamiltonian*, H. One speaks of p_x as being the momentum *conjugate* to x; p_y is the momentum conjugate to y; and so on.

The equations of motion (i.e., the equations that replace Newton's second law) in Hamiltonian theory are (for a point particle moving in three-dimensional space)

$$(1.14)$$

$$\frac{\partial H}{\partial x} = -\dot{p}_x \qquad \frac{\partial H}{\partial p_x} = \dot{x}$$

$$\frac{\partial H}{\partial y} = -\dot{p}_y \qquad \frac{\partial H}{\partial p_y} = \dot{y}$$

$$\frac{\partial H}{\partial z} = -\dot{p}_z \qquad \frac{\partial H}{\partial p_z} = \dot{z}$$

Cyclic Coordinates

For the Hamiltonian (1.13) corresponding to an electron between capacitor plates, one obtains

$$(1.15) \qquad \frac{\partial H}{\partial x} = \frac{\partial H}{\partial y} = 0$$

The Hamiltonian does not contain x or y. When coordinates are missing from the Hamiltonian, they are called *cyclic* or *ignorable*. The momentum conjugate to a cyclic coordinate is a constant of the motion. This important property follows directly from Hamilton's equations, (1.14). For example, for the case at hand, we see that $\partial H/\partial x = 0$ implies that $\dot{p}_x = 0$, so p_x is constant; similarly for p_y. (Note that there is no component of force in the x or y directions.) The remaining four Hamilton's equations give

$$(1.16) \qquad \dot{p}_z = -\frac{q\Phi_0}{d}, \qquad p_x = m\dot{x}, \qquad p_y = m\dot{y}, \qquad p_z = m\dot{z}$$

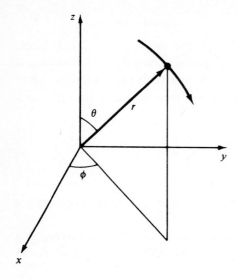

FIGURE 1.12   Motion of a particle in spherical co-ordinates with r and ϕ fixed: $v_\theta = r\dot\theta$, $p_\theta = rmv_\theta = mr^2\,\dot\theta$. The moment arm is r.

The last three equations return the definitions of momenta in terms of velocities. The first equation is the z component of Newton's second law. (For an electron, $q = -|q|$. It is attracted to the positive plate.)

Consider next the Hamiltonian for this same electron but expressed in terms of spherical coordinates. We must transform E as given by (1.5) to an expression involving r, θ, ϕ, and the momenta conjugate to these coordinates. The momentum conjugate to r is the component of linear momentum in the direction of $\mathbf{r}$. If $\mathbf{e}_r$ is a unit vector in the $\mathbf{r}$ direction, then

$$(1.17) \qquad p_r = \frac{\mathbf{r}\cdot\mathbf{p}}{r} = \mathbf{e}_r\cdot\mathbf{p} = m\mathbf{e}_r\cdot\mathbf{v} = m\dot r$$

The momentum conjugate to the angular displacement θ is the component of angular momentum corresponding to a displacement in θ (with r and ϕ fixed). The moment arm for this motion is r. The velocity is $r\dot\theta$. It follows that

$$(1.18) \qquad p_\theta = mr(r\dot\theta) = mr^2\dot\theta$$

(Fig. 1.12).

The momentum conjugate to ϕ is the angular momentum corresponding to a displacement in ϕ (with r and θ fixed). The moment arm for this motion is $r \sin\theta$. The velocity is $r\dot\phi \sin\theta$ (Fig. 1.13). The angular momentum of this motion is

$$(1.19) \qquad p_\phi = mr^2\dot\phi \sin^2\theta$$

Since such motion is confined to a plane normal to the z axis, p_ϕ is the z component of angular momentum. This was previously denoted as L_z in (1.12).

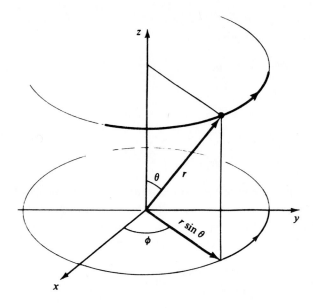

FIGURE 1.13 Motion of a particle with r and θ fixed: $v_\phi = r \sin \theta \, \dot{\phi}$. The moment arm is $r \sin \theta$, $p_\phi = (r \sin \theta) m v_\phi = m r^2 \, \dot{\phi} \sin^2 \theta$.

In terms of these coordinates and momenta, the energy expression (1.6) becomes

$$(1.20) \qquad H(r, \theta, \phi, p_r, p_\theta, p_\phi) = \frac{p_r^2}{2m} + \frac{p_\theta^2}{2mr^2} + \frac{p_\phi^2}{2mr^2 \sin^2 \theta} + \frac{q\Phi_0}{d} r \cos \theta$$

Hamilton's equations for a point particle, in spherical coordinates, become

$$\frac{\partial H}{\partial \theta} = -\dot{p}_\theta \qquad \frac{\partial H}{\partial p_\theta} = \dot{\theta}$$

$$(1.21) \qquad \frac{\partial H}{\partial \phi} = -\dot{p}_\phi \qquad \frac{\partial H}{\partial p_\phi} = \dot{\phi}$$

$$\frac{\partial H}{\partial r} = -\dot{p}_r \qquad \frac{\partial H}{\partial p_r} = \dot{r}$$

From the form of the Hamiltonian (1.20) we see that ϕ is a cyclic coordinate. That is,

$$(1.22) \qquad \frac{\partial H}{\partial \phi} = 0 = -\dot{p}_\phi$$

It follows that p_ϕ, as given by (1.19), is constant. Thus, the component of angular momentum in the z direction is conserved. The torque on the particle has no component in this direction.

Again the momentum derivatives of H in (1.20) return the definitions of momenta in terms of velocities. For example, from (1.20),

$$(1.23) \qquad \frac{\partial H}{\partial p_\theta} = \dot\theta = \frac{p_\theta}{mr^2}$$

which is (1.18). Hamilton's equation for $\dot p_r$ is

$$(1.24) \qquad -\frac{\partial H}{\partial r} = \dot p_r = \frac{p_\theta^2}{mr^3} + \frac{p_\phi^2}{mr^3 \sin^2\theta} - \frac{q\Phi_0}{d}\cos\theta$$

The first two terms on the right-hand side of this equation are the components of centripetal force in the radial direction, due to θ and ϕ displacements, respectively. The last term is the component of electric force $-\mathbf{e}_z q\Phi_0/d$ in the radial direction. Hamilton's equation for $\dot p_\theta$ is

$$(1.25) \qquad -\frac{\partial H}{\partial \theta} = \dot p_\theta = \frac{p_\phi^2 \cos\theta}{mr^2 \sin^3\theta} + \frac{q\Phi_0}{d} r\sin\theta$$

The right-hand side is a component of torque. It contains the centripetal force factor due to the ϕ motion $(p_\phi^2/mr^3 \sin^3\theta)$ and a moment arm factor, $r\cos\theta$. At any instant of time this component of torque is normal to the plane swept out by r due to θ motion alone.

A very instructive example concerns the motion of a free particle. A free particle is one that does not interact with any other particle or field. It is free of all interactions and is an isolated system. A particle moving by itself in an otherwise empty universe is a free particle. In Cartesian coordinates the Hamiltonian for a free particle is

$$(1.26) \qquad H = \frac{1}{2m} p^2 = \frac{1}{2m}(p_x^2 + p_y^2 + p_z^2)$$

All coordinates (x, y, z) are cyclic. Therefore, the three components of momenta are constant and may be equated to their respective initial values at time $t = 0$.

$$(1.27) \qquad \begin{aligned} p_x &= p_x(0) \\ p_y &= p_y(0) \\ p_z &= p_z(0) \end{aligned}$$

Combining these with the remaining three Hamilton's equations gives

$$(1.28) \qquad \begin{aligned} m\dot x &= p_x(0) \\ m\dot y &= p_y(0) \\ m\dot z &= p_z(0) \end{aligned}$$

These are simply integrated to obtain

$$x(t) = \frac{p_x(0)}{m} t + x(0)$$

(1.29)
$$y(t) = \frac{p_y(0)}{m} t + y(0)$$

$$z(t) = \frac{p_z(0)}{m} t + z(0)$$

which are parametric equations for a straight line.

Let us calculate the y component of angular momentum of the (free) particle.

(1.30) $$L_y = zp_x - xp_z = \left[z(0) + \frac{p_z(0)}{m} t \right] p_x(0) - \left[x(0) + \frac{p_x(0)}{m} t \right] p_z(0)$$

Canceling terms, we obtain

(1.31) $$L_y = z(0)p_x(0) - x(0)p_z(0) = L_y(0)$$

and similarly for L_x and L_z. It follows that

(1.32) $$\mathbf{L} = (L_x, L_y, L_z) = \text{constant}$$

for a free particle.

Investigating the dynamics of a free particle in Cartesian coordinates has given us immediate and extensive results. We know that $\mathbf{p}$ and $\mathbf{L}$ are both constant. The orbit is rectilinear.

We may also, consider the dynamics of a free particle in spherical coordinates. The Hamiltonian is

(1.33) $$H = \frac{p_r^2}{2m} + \frac{p_\theta^2}{2mr^2} + \frac{p_\phi^2}{2mr^2 \sin^2 \theta}$$

Only ϕ is cyclic, and we immediately conclude that p_ϕ (or equivalently, L_z) is constant. However, p_r and p_θ are not constant. From Hamilton's equations, we obtain

$$\dot{p}_r = \frac{p_\theta^2}{mr^3} + \frac{p_\phi^2}{mr^3 \sin^2 \theta}$$

(1.34)

$$\dot{p}_\theta = \frac{p_\phi^2 \cos \theta}{mr^2 \sin^3 \theta}$$

These centripetal terms were interpreted above. In this manner we find that the recti-linear, constant-velocity motion of a free particle, when cast in a spherical coordinate frame, involves accelerations in the r and θ components of motion. These accelerations

TABLE 1.1 Hamiltonian of a free particle in three coordinate frames

Frames	Cartesian Coordinates	Spherical Coordinates	Cylindrical Coordinates
	(x, y, z)	(r, θ, ϕ)	(ρ, z, ϕ)
Hamiltonian	$H(x, y, z, p_x, p_y, p_z)$ $= \dfrac{1}{2m}(p_x^{\,2} + p_y^{\,2} + p_z^{\,2})$	$H(r, \theta, \phi, p_r, p_\theta, p_\phi)$ $= \dfrac{1}{2m}\left[p_r^{\,2} + \dfrac{1}{r^2}\left(p_\theta^{\,2} + \dfrac{p_\phi^{\,2}}{\sin^2 \theta} \right) \right]$ $= \dfrac{1}{2m}\left(p_r^{\,2} + \dfrac{L^2}{r^2} \right)$	$H(\rho, z, \phi, p_\rho, p_z, p_\phi)$ $= \dfrac{1}{2m}\left(p_\rho^{\,2} + p_z^{\,2} + \dfrac{p_\phi^{\,2}}{\rho^2} \right)$
Momenta	$p_x = m\dot{x}$ $p_y = m\dot{y}$ $p_z = m\dot{z}$	$p_r = m\dot{r}$ $p_\theta = mr^2\dot{\theta}$ $p_\phi = mr^2\dot{\phi}\sin^2\theta$	$p_\rho = m\dot{\rho}$ $p_z = m\dot{z}$ $p_\phi = mp^2\dot{\phi}$
Cyclic coordinates	x, y, z	ϕ	z, ϕ
Constant momenta	p_x, p_y, p_z	$p_\phi = L_z$	p_z, p_ϕ

16

arise from an inappropriate choice of coordinates. In simple language: Fitting a straight line to spherical coordinates gives peculiar results.

A comparison of the Hamiltonian for a free particle in Cartesian, spherical, and cylindrical coordinates is shown in Table 1.1.

Canonical Coordinates and Momenta

While the reader may feel some familiarity with the components of linear momentum (p_x, p_y, p_z) and angular momentum (p_θ, p_ϕ), it is clear that these intuitive notions are exhausted for a system with, say, 17 degrees of freedom. If we call the seventeenth coordinate q_{17}, what is the momentum p_{17} conjugate to q_{17}? There is a formal procedure for determining the momentum conjugate to a given generalized coordinate. For example, it gives $p_\theta = mr^2\dot{\theta}$ as the momentum conjugate to θ for a particle in spherical coordinates. This procedure is described in any book in graduate mechanics.[1]

The coordinates of a system with N degrees of freedom, $(q_1, q_2, q_3, \ldots, q_N)$, and conjugate momenta $(p_1, p_2, p_3, \ldots, p_N)$ are also called *canonical* coordinates and momenta. A set of coordinates and momenta are canonical if with the Hamiltonian, $H(q_1, \ldots, q_N, p_1, \ldots, p_N, t)$, Hamilton's equations

$$(1.35) \qquad \frac{\partial H}{\partial q_l} = -\dot{p}_l, \qquad \frac{\partial H}{\partial p_l} = \dot{q}_l \qquad (l = 1, \ldots, N)$$

are entirely consistent with Newton's laws of motion. We have seen this to be the case for all the problems considered above. (Time-dependent Hamiltonians are considered in Chapter 13.)

Other important functions and concepts of classical mechanics include the *Lagrangian, action integral,* and *Hamilton's principle.* These topics are discussed in Section 7.11, which addresses the Feynman path integral.

PROBLEMS

1.5 Show that the z component of angular momentum for a point particle

$$L_z = xp_y - yp_x$$

when expressed in spherical coordinates, becomes

$$L_z = p_\phi = mr^2\dot{\phi}\sin^2\theta$$

---

[1] See, for example, H. Goldstein, *Classical Mechanics,* Addison-Wesley, Reading, Mass., 1951.

(*Hint*: Recall the transformation equations

$$z = r \cos \theta$$
$$y = r \sin \theta \sin \phi$$
$$x = r \sin \theta \cos \phi.)$$

1.6 (a) Calculate $\dot{p}_r$, $\dot{p}_\theta$, and $\dot{p}_\phi$ as explicit functions of time for the following motion of a particle.

$$y = y_0, \qquad z = z_0, \qquad x = v_0 t$$

(b) For what type of free-particle orbit are the following conditions obeyed?

(1) $\dot{p}_r = 0$
(2) $\dot{p}_\theta = 0$
(3) $\dot{p}_\phi = 0$
(4) $\dot{p}_r = \dot{p}_\theta = \dot{p}_\phi = 0$

(c) Describe an experiment to measure p_r, at a given instant, for the motion of part (a).

1.7 Show that the energy of a free particle may be written

$$H = \frac{p_r^2}{2m} + \frac{L^2}{2mr^2}$$

where $\mathbf{L} = \mathbf{r} \times \mathbf{p}$. [*Hint*: Use the vector relation

$$L^2 = (\mathbf{r} \times \mathbf{p})^2 = r^2 p^2 - (\mathbf{r} \cdot \mathbf{p})^2$$

together with the definition $p_r = (\mathbf{r} \cdot \mathbf{p})/r$.]

1.8 Show that angular momentum of a free particle obeys the relation

$$L^2 = L_x^2 + L_y^2 + L_z^2 = p_\theta^2 + \frac{p_\phi^2}{\sin^2 \theta}$$

(*Hint*: Employ the results of Problem 1.7.)

1.9 A particle of mass m is in the environment of a force field with components

$$F_z = -Kz, \qquad F_x = 0, \qquad F_y = 0$$

with K constant.

(a) Write down the Hamiltonian of the particle in Cartesian coordinates. What are the constants of motion?

(b) Use the fact that the Hamiltonian itself is also constant to obtain the orbit.

(c) What is the Hamiltonian in cylindrical coordinates? What are the constants of motion?

1.10 Suppose that one calculates the Hamiltonian for a given system and finds a coordinate missing. What can be said about the symmetry of the system?

1.11 A particle of mass m is attracted to the origin by the force

$$\mathbf{F} = -K\mathbf{r}$$

Write the Hamiltonian for this system in spherical and Cartesian coordinates. What are the cyclic coordinates in each of these frames? [*Hint:* The potential for this force, $V(r)$, is given by $\mathbf{F} = -K\mathbf{r} = -\nabla V(r)$.]

1.12 A "spherical pendulum" consists of a particle of mass m attached to one end of a weightless rod of length a. The other end of the rod is fixed in space (the origin). The rod is free to rotate about this point. If at any instant the angular velocity of the particle about the origin is ω, its energy is

$$E = \tfrac{1}{2}ma^2\omega^2 = \tfrac{1}{2}I\omega^2$$

The moment of inertia is I. What is the Hamiltonian of this system in spherical coordinates? (*Hint:* Recall the relation $L = I\omega$.)

1.3 THE STATE OF A SYSTEM

To know the values of the generalized coordinates of a system at a given instant is to know the location and orientation of the system at that instant. In classical physics we can ask for more information about the system at any given instant. We may ask for its motion as well. The location, orientation, and motion of the system at a given instant specify the state of the system at that instant. For a point particle in 3-space, the classical state Γ is given by the six quantities (Fig. 1.14)

(1.36) $$\Gamma = (x, y, z, \dot{x}, \dot{y}, \dot{z})$$

In terms of momenta,

(1.37) $$\Gamma = (x, y, z, p_x, p_y, p_z)$$

More generally, the state of a system is a minimal aggregate of information about the system which is maximally informative. A set of good coordinates and their corresponding time derivatives (generalized velocities) or corresponding momenta (canonical momenta) always serves as such a minimal aggregate which is maximally informative and serves to specify the state of a system in classical physics.

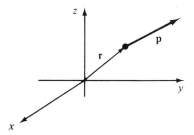

FIGURE 1.14 **The classical state of a free particle is given by six scalar quantities** (x, y, z, p_x, p_y, p_z).

The state of the system composed of two point particles moving in a plane is given by the eight parameters

(1.38) $$\Gamma = (x_1, y_1, x_2, y_2, p_{x_1}, p_{y_1}, p_{x_2}, p_{y_2})$$

Just as the set of generalized coordinates one assigns to a given system is not unique, neither is the description of the state Γ. For instance, the state of a point particle moving in a plane in Cartesian representation is

(1.39) $$\Gamma = (x, y, p_x, p_y)$$

In polar representation it is

(1.40) $$\Gamma = (r, \theta, p_r, p_\theta)$$

All representations of the state of a given system in classical mechanics contain an equal number of variables. If we think of Γ as a vector, then for a system with N degrees of freedom, Γ is $2N$-dimensional. In classical mechanics change of representation is effected by a change from one set of canonical coordinates and momenta (q, p) to another valid set of canonical coordinates and momenta (q', p').

$$\Gamma(q_1, \ldots, q_N, p_1, \ldots, p_N) \to \Gamma(q_1', \ldots, q_N', p_1', \ldots, p_N')$$

One form of canonical transformation results simply from a change in coordinates. For example, the transformation from Cartesian to polar coordinates for a particle moving in a plane effects the following change in representation:

$$\Gamma(x, y, p_x, p_y) \to \Gamma(r, \theta, p_r, p_\theta)$$

Representations in Quantum Mechanics

Next, we turn briefly to the form these concepts take in quantum mechanics. The specification of parameters that determines the state of a system in quantum mechanics is more subtle than in classical mechanics. As will emerge in the course of development of this text, in quantum mechanics one is not free to simultaneously specify certain sets of variables relating to a system. For example, while the classical state of a free particle moving in the x direction is given by the values of its position x, and momentum p_x, in quantum mechanics such simultaneous specification cannot be made. Thus, if the position x of the particle is measured at a given instant, the particle is left in a state wherein the particle's momentum is maximally uncertain. If on the other hand the momentum p_x is measured, the particle is left in a state in which its position is maximally uncertain. Suppose it is known that the particle has a specific value of momentum. One may then ask if there are any other variables whose values may be ascertained without destroying the established value of momentum. For a free particle one may further specify the energy E; that is, in quantum mechanics it is possible for the particle to be in a state such that measurement of momentum definitely finds the value p_x and measurement of energy definitely finds the value E. Suppose there are no further observable properties of the free particle that may be specified simultaneously with those two variables. Consequently, values of p_x and E comprise the most informative statement one can make about the particle and these values may be taken to comprise the state of the system of the particle

$$\Gamma = \Gamma(p_x, E)$$

As remarked above, if the particle is in this state, it is certain that measurement of momentum finds p_x and measurement of energy finds E. Such values of p_x and E are sometimes called *good quantum numbers*. As with their classical counterpart, good quantum numbers are an independent set of parameters which may be simultaneously specified and which are maximally informative.

For some problems in quantum mechanics it will prove convenient to give the state in terms of the Cartesian components of angular momentum: L_x, L_y, and L_z. We will find that specifying the value of L_z, say, induces an uncertainty in the accompanying components of L_x and L_y, so that, for example, it is impossible to simultaneously specify L_z and L_x for a given system. One may, however, simultaneously specify L_z together with the square of the magnitude of the total momentum, L^2. For a particle moving in a spherically symmetric environment, one may also simultaneously specify the energy of the particle. This is the most informative[1] statement one can make about such a particle, and the values of energy, L^2 and L_z, comprise a quantum state of the system.

(1.41) $$\Gamma = (E, L^2, L_z)$$

The values of E, L^2, and L_z are then good quantum numbers. That is, they are an independent set of parameters which may be simultaneously specified and which are maximally informative.

Just as change in representation, as discussed above, plays an important role in classical physics, so does its counterpart in quantum mechanics. A representation in quantum mechanics relates to the observables that one can precisely specify in a given state. In transforming to a new representation, new observables are specified in the state. For a free point particle moving in 3-space, in one representation the three components of linear momentum p_x, p_y, and p_z are specified while in another representation the energy $p^2/2m$, the square of the angular momentum L^2, and

---

[1] More precisely, Γ includes the *parity* of the system. This is a purely quantum mechanical notion and will be discussed more fully in Chapter 6.

any component of angular momentum, say L_z, are specified. In this change of representation,

(1.42)
$$\Gamma(p_x, p_y, p_z) \rightarrow \Gamma(E, L^2, L_z)$$

When treating the problem of the angular momentum of two particles ($\mathbf{L}_1$ and $\mathbf{L}_2$, respectively) in one representation, $(L_1{}^2, L_2{}^2, L_{z_1}, L_{z_2})$ are specified while in another representation, $(L_1{}^2, L_2{}^2, L^2, L_z)$ are specified. Here we are writing $\mathbf{L}$ for the *total angular momentum of the system* $\mathbf{L} = \mathbf{L}_1 + \mathbf{L}_2$. In this change of representation,

(1.43)
$$\Gamma(L_1{}^2, L_2{}^2, L_{z_1}, L_{z_2}) \rightarrow \Gamma(L_1{}^2, L_2{}^2, L^2, L_z{}^2)$$

Finally, in this very brief introductory description, we turn to the concept of the change of the quantum state in time. In classical mechanics, Newton's laws of motion determine the change of the state of the system in time. In quantum mechanics, the evolution in time of the state of the system is incorporated in the *wave (or state) function* and its equation of motion, the *Schrödinger equation*. Through the wave-function, one may calculate (expected) values of observable properties of the system, including the time development of the state of the system.

These concepts of the quantum state—its evolution in time and change in representation—comprise principal themes in quantum mechanics. Their understanding and application are important and are fully developed later in the text.

PROBLEMS

1.13 Write down a set of variables that may be used to prescribe the classical state for each of the 11 systems listed in Problem 1.1.

Answer (partial)

(e)   A rigid rod in 3-space: Since the system has five degrees of freedom, the classical state of the system is given by 10 parameters. For example,

$$\Gamma = \{x, y, z, \theta, \phi, \dot{x}, \dot{y}, \dot{z}, \dot{\theta}, \dot{\phi}\}$$

[*Note:* The quantum state is less informative. For example, such a state is prescribed by five variables (x, y, z, θ, ϕ). Another specification of the quantum state is given by five momenta $(p_x, p_y, p_z, p_\theta, p_\phi)$. However, simultaneous specification of, say, x and p_x is not possible in quantum mechanics.]

1.14   (a)   Use Hamilton's equations for a system with N degrees of freedom to show that H is constant in time if H does not contain the time explicitly. [*Hint:* Write

$$\frac{dH}{dt} = \frac{\partial H}{\partial t} + \sum_{l=1}^{N} \left(\frac{\partial H}{\partial q_l} \dot{q}_l + \frac{\partial H}{\partial p_l} \dot{p}_l \right).\right]$$

(b)   Construct a simple system for which H is an explicit function of the time.

1.15   For a system with N degrees of freedom, the Poisson bracket of two dynamical functions A and B is defined as

$$\{A, B\} \equiv \sum_{l=1}^{N} \left(\frac{\partial A}{\partial q_l} \frac{\partial B}{\partial p_l} - \frac{\partial B}{\partial q_l} \frac{\partial A}{\partial p_l} \right)$$

(a)   Use Hamilton's equations to show that the total time rate of change of a dynamical function A may be written

$$\frac{dA}{dt} = \frac{\partial A}{\partial t} + \{A, H\}$$

where H is the Hamiltonian of the system.

(b)   Prove the following: (1) If $A(q, p)$ does not contain the time explicitly and $\{A, H\} = 0$, then A is a constant of the motion. (2) If A does contain the time explicitly, it is constant if $\partial A/\partial t = \{H, A\}$.

(c)   For a free particle moving in one dimension, show that

$$A = x - \frac{pt}{m}$$

satisfies the equation

$$\frac{\partial A}{\partial t} = -\{A, H\}$$

so that it is a constant of the motion. What does this constant correspond to physically?

1.16   How many degrees of freedom does the compound pendulum depicted in Fig. 1.15 have? Choose a set of generalized coordinates (be certain they are independent). What is the Hamiltonian for this system in terms of the coordinates you have chosen? What are the immediate constants of motion?

1.17   How many constants of the motion does a system with N degrees of freedom have?

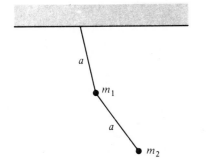

$F = -\mathbf{e}_z mg$

FIGURE 1.15   Compound pendulum composed of two masses connected by weightless rods of length a. The motion is in the plane of the paper. (See Problem 1.16.)

Answer

Each of the coordinates $\{q_i\}$ and momenta $\{p_i\}$ satisfies a first-order differential equation in time (i.e., Hamilton's equations). Every such equation has one constant of integration. These comprise $2N$ constants of the motion.

1.4 PROPERTIES OF THE ONE-DIMENSIONAL POTENTIAL FUNCTION

Consider a particle that is constrained to move in one dimension, x. The particle is in the potential field $V(x)$ depicted in Fig. 1.16. What is the direction of force at the point $x = A$? We can calculate the gradient (in the x direction) and conclude that the direction of force at A is in the $+x$ direction. There is a simpler technique. Imagine that the curve drawn is the contour of a range of mountain peaks. If a ball is placed at A, it rolls down the hill. The force is in the $+x$ direction. If placed at B (or C), it remains there. If placed at D, it rolls back toward the origin; the force is in the $-x$ direction. This technique always works (even for three-dimensional potential surfaces) because the gravity potential is proportional to height z, so the potential surface for a particle constrained to move on the surface of a mountain is that same surface.

The one-dimensional spring potential, $V = Kx^2/2$, is depicted in Fig. 1.17. If the particle is started from rest at $x = A$, it oscillates back and forth in the potential well between $x = +A$ and $x = -A$.

Motion described by a potential function is said to be *conservative*. For such motion, the energy

$$(1.44) \qquad\qquad E = T + V$$

is constant. In terms of the kinetic energy T,

$$(1.45) \qquad\qquad T = \frac{mv^2}{2} = E - V$$

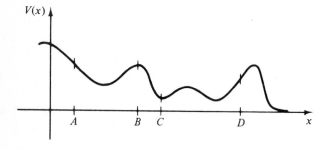

FIGURE 1.16 Arbitrary potential function.

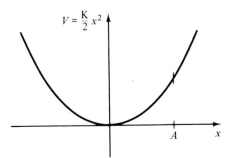

FIGURE 1.17 Spring potential.

Forbidden Domains

From (1.45) we see that if $V > E$, then $T < 0$ and the velocity becomes imaginary. In classical physics, particles are excluded from such domains. They are called *forbidden regions*. Again consider a one-dimensional problem with potential $V(x)$ shown in Fig. 1.18. The constant energy E is superimposed on this diagram. Segments AB and CD are forbidden regions. Points A, B, C, and D are *stationary* or *turning* points. Since $E = V$ at these points, $T = 0$ and $\dot{x} = 0$. Suppose that a particle is started from rest from the point C. What is the subsequent motion? The particle is trapped in the potential well between B and C. It accelerates down the hill, slows down in climbing the middle peak, then slows down further in climbing to B, where it comes to rest and turns around. This periodic motion continues without end.

The one-dimensional potential depicted in Fig. 1.18 can be effected by appropriately charging and spacing a linear array of plates with holes bored along the axis. The potential depicted in Fig. 1.18 is seen by an electron constrained to move along this (x) axis.

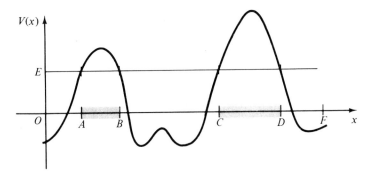

FIGURE 1.18 Forbidden domains at energy E.

PROBLEMS

1.18 A particle constrained to move in one dimension (x) is in the potential field

$$V(x) = \frac{V_0(x - a)(x - b)}{(x - c)^2} \qquad (0 < a < b < c < \infty)$$

(a) Make a sketch of V.

(b) Discuss the possible motions, forbidden domains, and turning points. Specifically, if the particle is known to be at $x = -\infty$ with

$$E = \frac{3V_0}{c - b}(b - 4a + 3c)$$

at which value of x does it reflect?

1.19 A particle of mass m moves in a "central potential," $V(r)$, where r denotes the radial displacement of the particle from a fixed origin.

(a) What is the (vector) force on the particle? Recall here the components of the $\mathbf{V}$ operator in spherical coordinates.

(b) Show that the angular momentum $\mathbf{L}$ of the particle about the origin is constant. (*Hint:* Calculate the time derivative of $\mathbf{L} = \mathbf{r} \times \mathbf{p}$ and recall that $\mathbf{p} = m\dot{\mathbf{r}}$.)

(c) Show that the energy of the particle may be written

$$E = \frac{p_r^{\,2}}{2m} + \frac{L^2}{2mr^2} + V(r)$$

(d) From Hamilton's equations obtain a "one-dimensional" equation for $\dot{p}_r$, in the form

$$\dot{p}_r = -\frac{\partial}{\partial r} V_{\text{eff}}(r)$$

where V_{eff} denotes an "effective" potential that is a function of r only.

(e) For the case of gravitational attraction between two masses (M, m), $V = -GmM/r$, where G is the gravitational constant. Make a sketch of V_{eff} versus r for this case. Use this sketch to establish the conditions for circular motion (assume that M is fixed in space) for a given value of L^2.

1.20 Complex variables play an important role in quantum mechanics. The following two problems are intended as a short review.

If

$$\psi = |\psi| \exp(i\alpha_1)$$
$$\chi = |\chi| \exp(i\alpha_2)$$

show that

$$|\psi + \chi|^2 = |\psi|^2 + |\chi|^2 + 2|\psi\chi| \cos(\alpha_1 - \alpha_2)$$

26

1.21   Use the expansion

$$e^{i\theta} = \cos\theta + i\sin\theta$$

to derive the following relations.

(a) $\cos(\theta_1 + \theta_2) = \cos\theta_1 \cos\theta_2 - \sin\theta_1 \sin\theta_2$

(b) $\sin(\theta_1 + \theta_2) = \cos\theta_1 \sin\theta_2 + \sin\theta_1 \cos\theta_2$

(c) $2\sin\theta_1 \cos\theta_2 = \sin(\theta_1 - \theta_2) + \sin(\theta_1 + \theta_2)$

(d) $2\cos\theta_1 \cos\theta_2 = \cos(\theta_1 + \theta_2) + \cos(\theta_1 - \theta_2)$

(e) $2\cos^2\theta = 1 + \cos 2\theta$

(f) $2\sin^2\theta = 1 - \cos 2\theta$

(g) $e^{i\theta} - 1 = 2ie^{i\theta/2}\sin(\theta/2)$

(h) $\frac{1}{2}|e^{i\theta_1} + e^{i\theta_2}|^2 = \frac{1}{2}(e^{i\theta_1} + e^{i\theta_2})(e^{i\theta_1} + e^{i\theta_2})^* = 1 + \cos(\theta_1 - \theta_2)$

(i) $2\,\mathrm{Re}\,z = z + z^*$

(j) $2i\,\mathrm{Im}\,z = z - z^*$

(k) $(\exp z)^* = \exp z^*$

(l) $|\exp z|^2 = \exp(2\,\mathrm{Re}\,z)$

CHAPTER 2

HISTORICAL REVIEW:
EXPERIMENTS AND THEORIES

The following sections summarize experiments and theories formulated during the early decades of the century. These observations and theories comprise the genesis of quantum mechanics. The important concept of the wavefunction is introduced and the Born interpretation of this function in terms of probability density is described. A more formal presentation of the postulates of quantum mechanics appears in Chapter 3.

2.1 DATES

Physics at the turn of the century was in a state of turmoil. There was a Pandora's Box of experimental observations which, on the grounds of otherwise firmly established classical theory, was totally inexplicable. One by one all these perplexing questions were answered—with the drama and flair of a story told by a masterful raconteur. Out of the turmoil came a new philosophy of science. A new way of thinking was called for. At the very core of natural law lay subjective probability—not objective determinism.

What were some of these perplexing observations? Light exhibits interference and therefore may be assumed to be a wave phenomenon. However, if we try to explain the photoelectric effect (light hitting a metal surface ejects electrons) on the basis of the wave nature of light, we obtain erroneous results. It is found that the energy of an emitted electron is dependent only on the frequency of the incident radiation, not on the intensity as might be expected from the classical theory of light.

In 1911 it was established by Rutherford that an atom has a positive central core and satellite electrons. Hydrogen, for instance, has a proton at its center and one outer electron. But such a circulating (and therefore accelerating) electron radiates and soon should collapse into the nucleus. So why do we not see a burst of ultraviolet radiation emitted as the electron spirals into the nucleus? Why is the frequency spectrum of light emitted from an atom a discrete line spectrum and not a continuous spectrum?

Another dilemma lay in the observations of the spectrum of radiant energy in a cavity whose walls are maintained at a fixed temperature. Theory (on the basis of the wave nature of light) was unable to account for the observed frequency distribution of radiant energy.

The very rapid development of events that occurred in the first three decades of this century, which removed the enigmas posed by these experiments, were as follows:

1901	Planck	Blackbody radiation
1905	Einstein	Photoelectric effect
1913	Bohr	Quantum theory of spectra
1922	Compton	Scattering photons off electrons
1924	Pauli	Exclusion principle
1925	de Broglie	Matter waves
1926	Schrödinger	Wave equation
1927	Heisenberg	Uncertainty principle
1927	Davisson and Germer	Experiment on wave properties of electrons
1927	Born	Interpretation of the wavefunction

In the remainder of this chapter we will outline these topics in more detail, except for the work of Schrödinger, which is formally presented in Chapter 3, and the work of Pauli, which is presented in Chapter 12. The Compton effect is discussed in Problem 2.28.

2.2   THE WORK OF PLANCK. BLACKBODY RADIATION

Place a closed, evacuated container (with a small window in the wall) in an oven of uniform temperature. Wait until all components of the experiment reach the same temperature (thermal equilibrium). At a sufficiently high temperature, visible light emerges from the window of the container cavity. The cavity contains radiant energy, which is in thermal equilibrium with the cavity walls. Suppose that the total radiant energy per unit volume in the cavity (at any instant) is U. How much of this energy is in electromagnetic waves with frequency between v and $v + dv$? Let us call the answer $u(v)\, dv$. The function $u(v)$ then gives the energy per frequency interval per unit volume. The total energy per unit volume in the radiation field in the cavity is

$$(2.1) \qquad\qquad U = \int_0^\infty u(v)\, dv$$

The radiation is called *blackbody radiation* because it is assumed that any light falling on the window is totally absorbed. The window acts as a perfect radiator and a perfect absorber. This property is characteristic of ideal black surfaces. At any given temperature, no object emits or absorbs radiation more efficiently than does an (ideal) blackbody.

The experimentally observed curve of $u(v)$ is shown in Fig. 2.1. Classical electro-dynamic and thermodynamic theory give two properties of the spectral distribution of a radiation field in equilibrium at the temperature T. The Rayleigh–Jeans (1900) approximation

$$u_{RJ}(v) = \frac{8\pi v^2}{c^3} k_B T$$

is appropriate for low frequencies. In this formula k_B is Boltzmann's constant,

$$k_B = 1.381 \times 10^{-16} \text{ erg/K}$$

and c is the speed of light. While this approximation is valid at low frequencies, it is seen to diverge at larger frequencies, where as shown in Fig. 2.1, the correct spectral distribution falls off to zero. Wien's law (1893) specifies that u, as a function of wavelength $\lambda = c/v$, is of the form

$$u_W(\lambda) = \frac{W(\lambda T)}{\lambda^5}$$

where W is an arbitrary function of the product of wavelength λ and temperature T. Although this formula is valid over the whole spectrum of wavelengths, it is incomplete in that $W(\lambda T)$ is undetermined. The complete explicit form for the spectral distribution u cannot be obtained from classical physics. A quantum hypothesis must be invoked. Such was the assumption made by Planck to obtain a uniformly valid formula for $u(v)$. It implied that energy of radiation with frequency v exists only in multiples of hv, where h is a constant of nature (Planck's constant). A quantum of radiation of energy hv is called a *photon*.

$$(2.2) \qquad\qquad E = hv$$

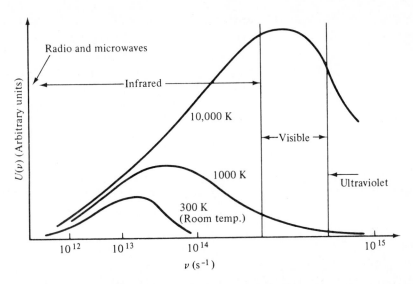

FIGURE 2.1   Spectrum of blackbody radiation. The curves have been distorted to bring out some important features. In reality the curve at 10,000 K is about 37,000 times higher than the curve at 300 K. Also, the radio and microwave domain is only about 1/30,000 of the v axis depicted.

The correct formula for $u(v)$ which results is (see Problems 2.36 and 2.37)

(2.3)
$$u(v) = \frac{8\pi h v^3}{c^3} \frac{1}{e^{hv/k_B T} - 1}$$

$$h = 6.626 \times 10^{-27} \text{ erg-s}$$

This expression precisely matches the experimental curves shown in Fig. 2.1.

PROBLEMS

2.1   (a)   Show that for photons of frequency v and wavelength λ:

(1) $dv = -c\, d\lambda/\lambda^2$

(2) $u(\lambda)\, d\lambda = -u(v)\, dv$

(3) $u(\lambda)\, d\lambda = u(v)c\, d\lambda/\lambda^2$

   (b)   Show that the Rayleigh–Jeans spectral distribution of blackbody radiation, $u_{RJ}(v)$, is of the form required by Wien's law,

$$u_W(\lambda) = \frac{W(\lambda T)}{\lambda^5}$$

   (c)   Obtain the correct form of Wien's undetermined function $W(\lambda T)$ from Planck' formula.

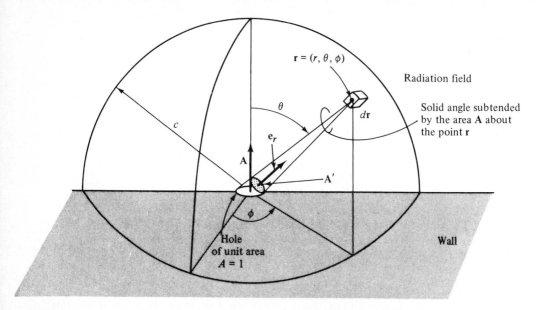

FIGURE 2.2 **The power radiated by an electromagnetic field in equilibrium at temperature** T **is due to photons that lie in a hemisphere of radius** c**, centered at the hole. (See Problem 2.2.)**

2.2 A spherical enclosure is in equilibrium at the temperature T with a radiation field that it contains. Show that the power emitted through a hole of unit area in the wall of enclosure is

$$P = \tfrac{1}{4}cU$$

Answer

Let the cavity be very large, so that its walls can be considered to be flat. The energy that flows through a hole in the wall, of unit area, in 1 s is the power radiated. This energy is due to photons that lie in a hemisphere of radius c, centered at the hole (Fig. 2.2). The energy in the volume element $d\mathbf{r}$ about the point $\mathbf{r}$ is $U\, d\mathbf{r}$. Owing to the isotropy of the radiation field, the amount of this energy that passes through the hole is $U\, d\mathbf{r}$ times the ratio of solid angle Ω subtended by the area of the hole about the point $\mathbf{r}$, to 4π, the total solid angle about the point $\mathbf{r}$.

$$dP = \frac{\Omega}{4\pi}\, U\, d\mathbf{r} = \frac{A'}{4\pi r^2}\, U\, d\mathbf{r} = \frac{\mathbf{e}_r \cdot \mathbf{A} U\, d\mathbf{r}}{4\pi r^2} = \frac{U\cos\theta}{4\pi r^2}\, d\mathbf{r}$$

$$= -\frac{U\, d\phi\, \cos\theta\, d\cos\theta\, r^2\, dr}{4\pi r^2}$$

The radiation energy that passes through the hole in 1 s from all volume elements in the hemisphere is the total power radiated per unit area.

$$P = \int_{\text{hemisphere}} dP = \frac{U}{4\pi}\int_0^{2\pi} d\phi \int_0^1 \cos\theta\, d\cos\theta \int_0^c dr = \tfrac{1}{4}cU$$

2.3 Show that the energy density $U(T)$ of a radiation field in equilibrium at the temperature T is directly proportional to T^4. The corresponding expression for the emitted power is

$$P = \sigma T^4$$

where σ is the Stefan–Boltzmann constant

$$\sigma = \frac{\pi^2}{60} \frac{k_B{}^4}{\hbar^3 c^2} = 0.567 \times 10^{-4} \text{ erg/s-cm}^2\text{-K}^4$$

[*Hint:* Nondimensionalize the integration over (2.3) through the variable $x \equiv h\nu/k_B T$.]

2.4 Use (2.3) to prove Wien's displacement law

$$\lambda_{\max} T = \text{constant} = 0.290 \text{ cm K}$$

The wavelength $\lambda_{\max}$ is such that $u(\lambda_{\max})$ is maximum. [*Hint:* Differentiate $u(\lambda)$ with respect to the variable $x \equiv hc/kT\lambda$ and set equal to zero.]

2.5 From the sketch of u versus ν given in Fig. 2.1, make a sketch of u versus λ, where $\nu\lambda = c$.

2.6 What is the photon flux (photons/cm^2 s) at a distance of 1 km from a light source emitting 50 W of radiation in the visible domain, with wavelength 6000 Å?

2.7 The average energy in a unit volume in the ν frequency mode of a blackbody radiation field is

$$\langle U \rangle = \frac{h\nu}{e^{h\nu/k_B T} - 1}$$

What does $\langle U \rangle$ reduce to in the limit (a) $\nu \to 0$? (b) $T \to \infty$?

2.8 As discussed above, the radiation field interior to a closed cavity whose walls are in thermal equilibrium (i.e., at the same temperature) with the radiation field is called blackbody radiation. Prove that blackbody radiation has the following properties by showing that if any of these properties are not true, a device can be constructed which violates the second law of thermodynamics.

    (a)   The flux of radiation is the same in all directions. (The radiation field is *isotropic.*)

    (b)   The energy density is the same at all points inside the cavity. (The radiation field is *homogeneous.*)

    (c)   The energy density interior to the cavity is the same (function of frequency) at a given temperature, regardless of the material of the cavity wall.

2.9 Prove that the radiation emitted by the surface of an ideal blackbody at the temperature T is the same as that which travels in one direction inside a closed isothermal cavity at the same temperature.

Answer

Immerse an ideal black cube inside the isothermal container. The radiation that falls on any face of the cube is completely absorbed. For equilibrium to be maintained, the radiation emitted must be balanced by that absorbed, so that the radiation emitted is precisely that which flows into the face.

If, on the other hand, the cube is not ideally black, equilibrium is maintained by balancing the absorbed radiation by the reflected plus emitted radiation. Since energy density in the cavity is the same as in the case above (both experiments are at the same temperature), the radiation emitted by the nonblack surface is less intense than that emitted by the ideally black surface.

2.10 One of the theories of the origin of the universe is that it was contained in a primeval fireball which began its expansion about 10^{10} years ago. As it expanded, it cooled. Measurements of the energy spectrum of cosmic photons suggest a (blackbody) temperature of 3 K. At what frequency is maximum energy observed?

2.11 Suppose that you are inside a blackbody radiation cavity which is at temperature T. Your job is to measure the energy in the radiation field in the frequency interval 10^{14} to 89×10^{14} Hz. You have a detector that will do the job. For best results, should the temperature of the detector T' be $T' > T$, $T' = T$, $T' < T$, or $T' = 0$; or is the temperature of the detector irrelevant to the measurement?

2.3 THE WORK OF EINSTEIN. THE PHOTOELECTRIC EFFECT

The experimental setup that exhibits the photoelectric effect is depicted in Fig. 2.3. The observation is as follows. A metal plate (e.g., copper) is irradiated with light of a given frequency. Electrons are ejected from the photo cathode and current is registered in the ammeter A. As the potential on the collecting plate is made more negative, the current diminishes, until finally at the potential V_{stop}, current ceases. The energy that an electron must have in order to climb the potential hill imposed by the negative bias

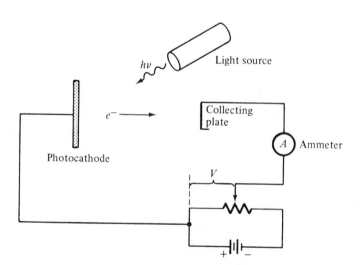

FIGURE 2.3 Photoelectric experiment.

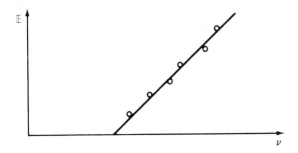

FIGURE 2.4 Typical data showing energy of most energetic electrons as a function of frequency ν in the photoelectric experiment.

V is eV. Only the most energetic electrons reach the plate near V_{stop}. At V_{stop} the electrons with maximum kinetic energy $\mathbb{E}$ have been repelled. Then

$$(2.4) \qquad \mathbb{E} = eV_{stop}$$

At a given frequency ν, one makes a measurement of $\mathbb{E}$ and plots a point on an $\mathbb{E}$ versus ν graph (Fig. 2.4). If the intensity of light is increased while ν is held fixed, $\mathbb{E}$ remains constant. On the other hand, when ν is increased, $\mathbb{E}$ increases. A typical collection of data is shown in Fig. 2.4.

To explain this effect Einstein hypothesized that light is composed of localized bundles of electromagnetic energy called photons. At frequency ν, the energy of a photon is $h\nu$. When striking the metal surface the photon interacts with an electron and ejects it from the metal. Let us consider the *Sommerfeld model* of a conductor (Fig. 2.5). The conductor is composed of fixed positive sites (e.g., Cu^{2+} ions in copper) and free electrons. The positive ions generate a potential well in which the electrons are trapped. The electrons have energy from 0 to E_F, the *Fermi energy*. The minimum work required to remove an electron from the metal is $W - E_F$, which is called the *work function*, Φ. The depth of the well is W.

Electrons distribute themselves in accordance with the *Pauli exclusion principle*. This principle precludes more than one electron existing in the same quantum state. For example, the distribution of electron energies shown in Fig. 2.5 is maintained at

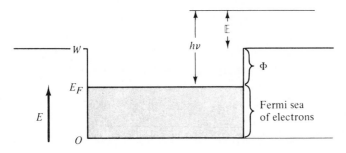

FIGURE 2.5 Sommerfeld model for energy distribution of electrons in a metal.

0 K. At this temperature electrons fall to lowest allowable energies. They cannot all fall to the single lowest level, owing to the Pauli principle. Once this level is occupied, the next electron must seek the next higher level. The maximum value of energy so reached is the Fermi energy E_F.

Suppose that a photon of energy $h\nu$ hits an electron and ejects it with kinetic energy $\mathbb{E}$. The most energetic electrons come from the top of the *Fermi sea*. The energy $\mathbb{E}$ of such an electron ejected by a photon of energy $h\nu$ is given by

(2.5)
$$\mathbb{E} = h\nu - \Phi$$

If we plot $\mathbb{E}$ versus ν from this equation, we obtain the curve shown in Fig. 2.4. Note that the slope of the curve is Planck's constant h, and the ν intercept gives the work function (of the photocathode metal). If $\Phi \equiv h\nu_{\text{th}}$, ν_{th} is called the *threshold frequency*. A few typical values are:

Metal	ν_{th} (Hz)	E_F (eV)
Silver	1.14×10^{15}	5.5
Potassium	0.51×10^{15}	2.1
Sodium	0.56×10^{15}	3.1

Millikan in 1916 used the photoelectric experiment to obtain a value of Planck's constant, h [see (2.3)].

Contact Potential

The preceding description may also be used to explain the phenomenon of *contact potential*, the finite potential that develops between two dissimilar metals which are brought into contact with each other. To describe this effect we consider a parallel-plate capacitor with one plate made of metal A and the other made of metal B. When the plates are isolated and displaced far from each other, the common zero in potential of both metals corresponds to zero free-particle kinetic energy (Fig. 2.6a).

Now let the metals be brought into contact with each other. Electrons then "fall" from the Fermi level of metal A, which has the smaller work function, to the deeper-lying Fermi level of metal B, until the tops of the two electron energy distributions are equalized. Having lost electrons, metal A is left electropositive with respect to metal B and a potential difference exists between the plates (Fig. 2.6b).

This description leads to the conclusion that the contact potential difference

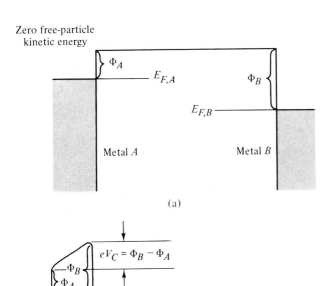

(a)

(b)

FIGURE 2.6 The difference in work functions causes electrons to fall to the lower Fermi level thereby creating a contact potential. (a) The metals far removed from each other. (b) The metals in contact with each other. The sloping curve represents the potential seen by an electron.

V_C between two metals should be well approximated by the difference in work functions:

$$eV_C = \Phi_B - \Phi_A$$

The validity of this relation is borne out by experiment.

PROBLEMS

2.12 (a) A monochromatic point source of light radiates 25 W at a wavelength of 5000 Å. A plate of metal is placed 100 cm from the source. Atoms in the metal have a radius of 1 Å. Assume that the atom can continually absorb light. The work function of the metal is 4 eV. How long is it before an electron is emitted from the metal?

(b) Is there sufficient energy in a single photon in the radiation field to eject an electron from the metal?

2.13 The photoelectric threshold of tungsten is 2300 Å. Determine the energy of the electrons ejected from the surface by ultraviolet light of wavelength 1900 Å.

2.14 The work function of zinc is 3.6 eV. What is the energy of the most energetic photoelectron emitted by ultraviolet light of wavelength 2500 Å?

2.15 Photoelectrons emitted from a cesium plate illuminated with ultraviolet light of wavelength 2000 Å are stopped by a potential of 4.21 V. What is the work function of cesium?

2.4 THE WORK OF BOHR. A QUANTUM THEORY OF ATOMIC STATES

Consider a discharge tube filled with hydrogen gas. At sufficient voltage the gas glows. If the light is examined in a spectroscope, it is seen that only a discrete set of frequencies—a line spectrum—is emitted. Bohr was able to account for the discrete emission spectra in an analysis based on two postulates:

(1) Hydrogen exists in discrete energy states. These states are characterized by discrete values of the angular momentum as given by the relation

$$(2.6) \qquad \oint p_\theta \, d\theta = nh$$

with n an integer greater than zero. In these states the atom does not radiate. The line integral follows the electron in one complete orbit about the nucleus.

(2) When an atom undergoes a change in energy from E_n to E_m, electromagnetic radiation (a photon) is emitted at a frequency v given by

$$(2.7) \qquad hv = E_n - E_m$$

Let us recall how condition (2.6) leads to a discrete set of energies $\{E_n\}$. The energy of a (stationary) hydrogen atom whose electron is moving in circular motion is

$$(2.8) \qquad E = \tfrac{1}{2}mv^2 - \frac{e^2}{r} = \frac{p_\theta^2}{2mr^2} - \frac{e^2}{r}$$

The radius r obeys the centripetal condition

$$(2.9) \qquad \frac{mv^2}{r} = \frac{p_\theta^2}{mr^3} = \frac{e^2}{r^2}$$

so that, with (2.6),

$$(2.10) \qquad \frac{e^2}{r} = \frac{p_\theta^2}{mr^2} = \frac{n^2\hbar^2}{mr^2} \qquad \left(\hbar \equiv \frac{h}{2\pi} \right)$$

$$(2.11) \qquad r_n = \frac{n^2\hbar^2}{me^2}$$

These are the quantized values of r at which the electron persists without radiating. The values of the energy at these radii are

$$(2.12) \qquad E_n = - \frac{p_\theta^2}{2mr^2} = - \frac{n^2\hbar^2}{2m} \left(\frac{me^2}{n^2\hbar^2} \right)^2$$

$$= - \frac{\mathbb{R}}{n^2}$$

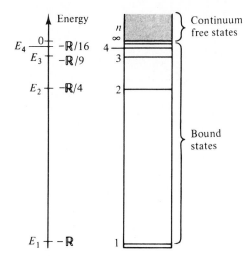

FIGURE 2.7   Bohr spectrum.

where $\mathbb{R}$ is the Rydberg constant:

$$(2.13) \qquad \mathbb{R} = \frac{me^4}{2\hbar^2} = 2.18 \times 10^{-11} \text{ erg} = 13.6 \text{ eV}$$

The negative quality of the energy reflects the fact that we are dealing with *bound states*. When $n = 1$, the atom is in the *ground state* and has energy, $-\mathbb{R}$. To ionize the atom when it is in this state takes $+\mathbb{R}$ ergs of energy. The value of r when the atom is in the ground state is

$$(2.14) \qquad r_1 \equiv a_0 = \frac{\hbar^2}{me^2} = 5.29 \times 10^{-9} \text{ cm}$$

This is a fundamental length in physics. It is called the *Bohr radius*.

   When the electron and proton are infinitely far removed and at rest, $r_n = \infty$. From (2.11) we see that this corresponds to $n = \infty$. In this state $E_n = 0$; there is no kinetic energy and no potential energy. If the electron is given a tap, it becomes a free particle. The composite system of proton plus electron then has positive energy (kinetic only), with all (unquantized) positive values of energy allowed (Fig. 2.7).

   The quality of the emission spectra of hydrogen is generated by the values for E_n (2.12) and the second postulate (2.7). The frequencies so generated (with some minor refinements, e.g., accounting for the motion of the proton) agree to a high degree of accuracy with the data. Characteristically, the spectrum divides into various series

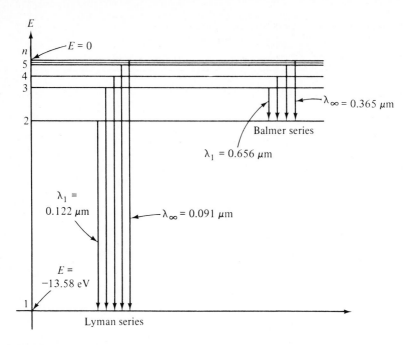

FIGURE 2.8 First two series of emission spectrum for hydrogen. Wavelengths of radiation are given in units of microns (10^{-4} cm).

of lines: *Lyman, Balmer, Paschen,* and so on. The Lyman series is comprised of frequencies generated by transitions to the ground state:

(2.15) $$h\nu_L = E_n - E_1 \qquad (n > 1)$$

The Balmer series is generated by transitions to the second excited state:

(2.16) $$h\nu_B = E_n - E_2 \qquad (n > 2)$$

and so forth (Fig. 2.8).

PROBLEMS

2.16 (a) Consider the spherical pendulum described in Problem 1.12. Use the Bohr formula (2.6) to obtain the quantum energies of this system. Note the identity $p_\theta = L$.

(b) Suppose that the pendulum is comprised of a proton attached to a weightless rod of length $a = 2$ Å. What is the ground rotational state of this system (in eV)? (See Problem 10.39.)

2.17 (a) What is the formula for the frequency ν of radiation emitted when the hydrogen atom decays from state n to state n'? Give your answer in terms of $\mathbb{R}$, n, n', and h only.

(b) What is the corresponding formula for the wavelength λ emitted in the same transition? Now your formula will contain the additional constant c, the speed of light.

2.18 The angular momentum of an isolated system is constant (when referred to any origin). Derive an expression for the angular momentum p_θ carried away by a photon emitted when a hydrogen atom decays from the state n to the state n' (in the Bohr model).

2.19 In classical electromagnetic theory an accelerating charge e radiates energy at the rate

$$W = \frac{2}{3}\frac{a^2 e^2}{c^3} \quad \text{ergs/s}$$

The acceleration is a and c is the speed of light. At time $t = 0$, a hydrogen atom has a radius 1 Å. Assuming classical circular motion:

(a) What is a initially?

(b) What is the initial frequency of radiation that the atom emits?

(c) How long does it take for the radius to collapse from 1 Å to 0.5 Å? (Assume that a is constant.)

(d) What is the frequency of radiation at the radius 0.5 Å?

2.20 The dimensionless number

$$\alpha \equiv \frac{e^2}{\hbar c} = \frac{1}{137.037}$$

is called the *fine-structure constant*.

(a) Show that the Rydberg constant may be written $\mathbb{R} = \frac{1}{2}\alpha^2 mc^2$.

(b) If the rest-mass energy of the electron is $mc^2 = 0.511$ MeV, calculate $\mathbb{R}$ in eV.

(c) Obtain an expression for the Bohr energies E_n in terms of α and mc^2.

2.5 WAVES VERSUS PARTICLES

Suppose that a disturbance propagates from one point in space to another point in space. What is propagating, waves or particles? A principle distinguishing characteristic is that waves exhibit interference, particles do not.

Consider the two-slit experiment shown in Fig. 2.9. A continuous spray of particles is fired from the source S. They strike the wall or pass through the two slits A and B. An intensity I_1 (number/unit area · second) emerges from A and an intensity I_2 emerges from B. When striking the screen, the two streams of particles superimpose and the net intensity measured is

(2.17) $$I = I_1 + I_2$$

This is nothing more than the statement that numbers of particles add.

Now consider the same experimental setup, but instead of a source of particles, let S represent a source of waves, say water waves (Fig. 2.10). Waves are characterized by an amplitude function ψ such that the absolute square of this function gives the intensity I:

(2.18) $$I = |\psi|^2$$

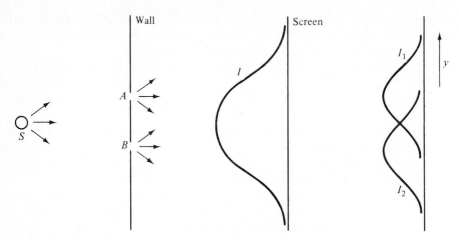

FIGURE 2.9 Particle double-slit experiment. Particle intensities add.

(Absolute values are taken for complex amplitudes.) Let the two propagating wave disturbances have (complex) scalar (as opposed to vector) amplitudes $\psi_1(\mathbf{r}, t)$ and $\psi_2(\mathbf{r}, t)$, respectively. These functions have the representations

$$(2.19) \qquad \psi_1 = |\psi_1| e^{i\alpha_1}, \qquad \psi_2 = |\psi_2| e^{i\alpha_2}$$

where α is the phase of the wave, which in general is also a function of $(\mathbf{r}, t)$. The intensities of these waves are

$$(2.20) \qquad I_1 = |\psi_1|^2, \qquad I_2 = |\psi_2|^2$$

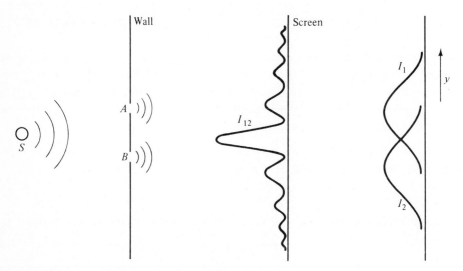

FIGURE 2.10 Wave double-slit experiment. Amplitudes add.

At a common value of $\mathbf{r}$ and t, the two wave amplitudes superimpose to give the resultant amplitude:

(2.21)
$$\psi = \psi_1 + \psi_2$$

The corresponding resultant intensity is

(2.22)
$$\begin{aligned} I = |\psi|^2 = |\psi_1 + \psi_2|^2 &= (\psi_1 + \psi_2)(\psi_1 + \psi_2)^* \\ &= |\psi_1|^2 + |\psi_2|^2 + |\psi_1\psi_2|[e^{i(\alpha_1 - \alpha_2)} + e^{-i(\alpha_1 - \alpha_2)}] \\ &= I_1 + I_2 + 2\sqrt{I_1 I_2} \cos(\alpha_1 - \alpha_2) \end{aligned}$$

Comparing this with the resultant intensity I for the particle case (2.17), we note that the wave intensity carries the additional term

(2.23)
$$\Delta \equiv 2\sqrt{I_1 I_2} \cos(\alpha_1 - \alpha_2)$$

This is an interference term. As the y component of r traverses the screen in Fig. 2.10, Δ oscillates and gives a pattern of the form depicted.

Hence, we have uncovered an operational, distinguishing characteristic between particles and waves. Waves exhibit interference, particles do not. Consider the example of a propagating electric field $\mathscr{E}(\mathbf{r}, t)$. The intensity of the wave (energy flux) is proportional to the time average of $|\mathscr{E}|^2$. If two electric waves $\mathscr{E}_1$ and $\mathscr{E}_2$ are superimposed, the new value of the electric field becomes

(2.24)
$$\mathscr{E} = \mathscr{E}_1 + \mathscr{E}_2$$

The intensity is proportional to the time average of squared amplitude, $|\mathscr{E}_1 + \mathscr{E}_2|^2$.

So we have the following important rule: *when two noninteracting beams of particles combine in the same region of space, intensities add; when waves interact, amplitudes add.* The intensity is then proportional to the time average of the absolute square of the resultant amplitude.

PROBLEMS

2.21 In a given wave double-slit experiment, a detector traces across a screen along a straight line whose coordinate we label y. If one slit is closed, the amplitude

$$\psi_1 = \sqrt{\tfrac{1}{2}} e^{-y^2/2} e^{i(\omega t - ay)}$$

is measured. If the other slit is closed, the amplitude

$$\psi_2 = \sqrt{\tfrac{1}{2}} e^{-y^2/2} e^{i(\omega t - ay - by)}$$

is measured. What is the intensity pattern along the y axis if both slits are open?

2.6 THE DE BROGLIE HYPOTHESIS AND THE DAVISSON–GERMER EXPERIMENT

In preceding sections we have seen that for a consistent explanation of certain experiments it is necessary to ascribe particle (photon) behavior to light. The energy of such a photon of frequency v is $E = hv$. Its momentum is

$$(2.25) \qquad\qquad p = \frac{E}{c} = \frac{hv}{c}$$

This formula can also be written in terms of wavelength λ. The relation between λ and v for light is particularly simple. It is

$$(2.26) \qquad\qquad \lambda v = c$$

In terms of wavenumber k (cm^{-1}) and angular frequency ω,

$$(2.27) \qquad\qquad \omega = 2\pi v, \qquad k = \frac{2\pi}{\lambda}$$

Equations (2.25) appear as

$$(2.28) \qquad E = \hbar\omega, \qquad p = \hbar k, \qquad \omega = ck \qquad \left(\hbar \equiv \frac{h}{2\pi}\right)$$

The last of these three equations is called a *dispersion relation*. It reveals a linear dependence between ω and k. The significance of this is that the phase velocity (ω/k) of a monochromatic wave of frequency ω is independent of ω or k. It is the constant c (speed of light). If a *wave packet* composed of a collection of waves of different wavelengths (or, equivalently, different wavenumbers) is constructed, it propagates with no distortion (dispersion). All component waves have the same speed, c.

   The first two equations of (2.28) reveal that photons, which are in essence particles, are identified by two wave parameters: wavenumber k and frequency ω. Now in what sense is a photon different from other more familiar particles (e.g., electrons, protons, etc.)? A photon is special in that it has zero rest mass and travels only at the speed of light. The more familiar particles with finite rest mass also have wave properties. For a (nonrelativistic) particle of kinetic energy

$$(2.29) \qquad\qquad E = \frac{p^2}{2m}$$

the wavelength for the corresponding ("matter") wave is

$$(2.30) \qquad\qquad \lambda = \frac{h}{p} \quad \text{or} \quad p = \hbar k$$

which we see from (2.25) and (2.26) is equally relevant to photons. Equation (2.30) is, in essence, the *de Broglie hypothesis*. It ascribes a wave property to particles. While the Planck hypothesis, which assigned a particle quality to electromagnetic waves, had strong experimental motivation, the de Broglie hypothesis, when first introduced in 1925, had little. Such motivation lay to a large degree in the mystery that surrounded the Bohr recipe for the hydrogen atom. What was the physical basis of the first rule for stationary orbits (2.6)? For circular orbits of radius r, with electron momentum p, this rule gives

(2.31) $$2\pi r p = nh$$

In terms of the de Broglie wavelength λ, the last equation reads

(2.32) $$2\pi r = n\lambda$$

The stationary orbits in the Bohr model have an integral number of wavelengths precisely fitting the circumference (Fig. 2.11). This is the classical criterion for the existence of (standing) waves on a circle.

Thus, we see that the de Broglie hypothesis returns the stationary orbit radii of the Bohr theory. This result lends support to the idea that the electron has something "wavy" associated with it, this property being characterized by the de Broglie wavelength (2.30). It was not until two years later (1927) that M. Born suggested what is believed today to be the correct interpretation of this wave property (see Section 2.8).

If electrons (in some respect) propagate as waves, they should exhibit interference. This is the essence of the *Davisson-Germer experiment* (1927). Reflect a beam of electrons with well-defined momentum (therefore, wavelength) off a crystal

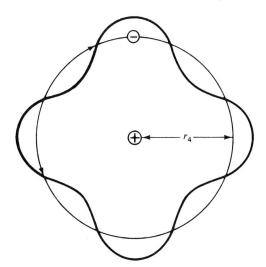

FIGURE 2.11   De Broglie wavelength λ and the $n = 4$ Bohr orbit of the hydrogen atom.

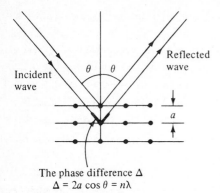

The phase difference Δ
$$\Delta = 2a \cos \theta = n\lambda$$

FIGURE 2.12 Reflection of plane waves from a lattice. Conditions stated are for constructive interference, with n an integer.

surface whose ion sites are separated by a distance a (the lattice constant) which is of the order of the de Broglie wavelength of the electrons. In the actual experiment, low-energy (~ 200-eV) electrons were reflected from the face of a nickel crystal ($a = 3.52 \times 10^{-8}$ cm). An interference pattern was observed which could most consistently be interpreted as the diffraction of plane waves (with de Broglie wavelength) by the regularly spaced atoms of the crystal (Fig. 2.12).

To bring out the full physical interpretation of these results we will consider a simpler experiment in which the same principles are involved. We revert to the two-slit configuration depicted in Fig. 2.10. The source S is able to eject single electrons with well-defined momenta[1] $\mathbf{p} = \hbar\mathbf{k}$. This is a vector normal to the diffracting wall. The distance between S and the diffracting wall is large compared to the distance between slits. The screen is composed of scintillation material. When an electron hits it, there is a localized flash at the point of impact.

In any single run of this experiment, one sees a single localized flash on the screen. There is no interference pattern. If we record the number and location (idealized to one dimension, y) of these flashes, the results of 5 runs are shown in Fig. 2.13a; 10 runs in Fig. 2.13b; 50 runs in Fig. 2.13c; 10,000 runs in Fig. 2.13d. The solid curve is the theoretically calculated diffraction pattern obtained with the de Broglie wavelength.

The electrons begin to distribute themselves in an interference pattern. It follows that if we change the source to eject a *current pulse* containing many electrons, the scintillation plate will show an interference pattern.

Similarly, for a source of light we can use a detection plate made of many photomultipliers. If the source emits a single photon, a single pulse from one of the photomultiplier tubes is registered. There is no diffraction pattern. A single particle under any circumstance always gives a single localized "flash." Wherein lies the wave quality of particles? Clearly, it is centered in a *statistical* interpretation of data.

---

[1] The consistency of this arrangement with the *uncertainty principle* is discussed in the next section.

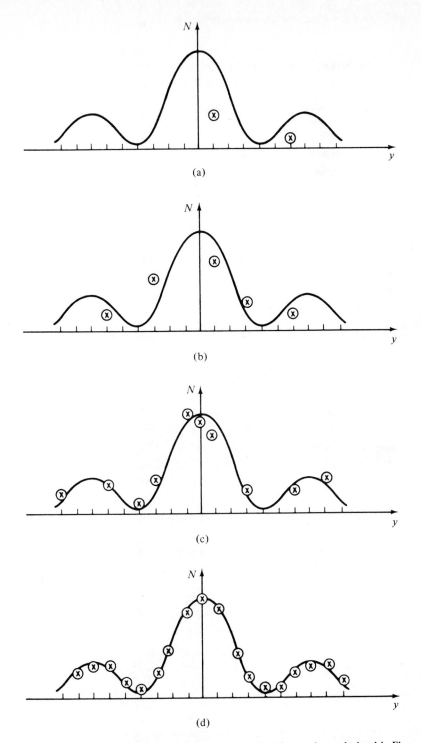

FIGURE 2.13 Number and location of flashes in electron double-slit experiment depicted in Figure 2.10. Each point is an average of flashes over a unit interval. The solid curve is the theoretical interference pattern corresponding to the de Broglie wavelength.

Such description was first presented by Born in 1927. Before turning to this analysis, we will give a brief account of a discovery of Heisenberg, which was to throw the well-established philosophical dogma of the seventeenth to nineteenth centuries into disarray.

PROBLEMS

2.22   A photon of energy $h\nu$ collides with a stationary electron of rest mass m. Show that it is not physically possible for the photon to impart all its energy to the electron.

Answer

We must do this problem relativistically. Let us assume that the photon does give up all its energy to the electron. Conservation of energy and momentum then give

$$h\nu + mc^2 = m\gamma c^2$$

$$\frac{h\nu}{c} = m\gamma\beta c$$

where

$$\beta \equiv \frac{v}{c}, \qquad \gamma \equiv (1 - \beta^2)^{-1/2}, \qquad m = \text{rest mass}$$

The speed of the electron after collision is v. Eliminating $h\nu$ from the conservation equations gives

$$\gamma\beta = \gamma - 1$$

whose only (real) solution is $\beta = 0$ ($\gamma = 1$). This is a contradiction.

2.23   Show that the de Broglie wavelength of an electron of kinetic energy $E(eV)$ is

$$\lambda_e = \frac{12.3 \times 10^{-8}}{E^{1/2}} \text{ cm}$$

and that of a proton is

$$\lambda_p = \frac{0.29 \times 10^{-8}}{E^{1/2}} \text{ cm}$$

2.24   At what speed is the de Broglie wavelength of an α particle equal to that of a 10 keV photon?

2.25   Show that in order to associate a de Broglie wave with the propagation of photons (electromagnetic radiation), photons must travel with the speed of light c and their rest mass must be zero. (Do relativistically.)

Answer

For a de Broglie wave associated with a particle of rest mass m,

$$\lambda = \frac{h}{p} = \frac{h}{m\gamma v}$$

For a photon with rest mass m,

$$\lambda = \frac{c}{v} = \frac{hc}{hv} = \frac{hc}{m\gamma c^2}$$

Equating these relations gives

$$v = c$$

which gives a noninfinite mass, γm, only for $m = 0$.

2.26   The relativistic kinetic energy T of a particle of rest mass m and momentum $p = \gamma mv$ is

$$T = \sqrt{p^2 c^2 + m^2 c^4} - mc^2$$

(a)   Show that

$$T = mc^2(\gamma - 1)$$

(b)   Show that in the limit $\beta \ll 1$,

$$T = \tfrac{1}{2}mv^2 + O(\beta^4)$$

(c)   Show that the relativistic expression for T above gives the correct energy–momentum relation for a photon if $m = 0$.

(d)   What is the total relativistic energy, E (i.e., including rest-mass energy) of a particle of mass m?

(e)   What is the total relativistic energy of a particle moving in a potential field $V(x)$? What is the corresponding Hamiltonian, $H(p, x)$?

Answers (partial)

(d) $E = \gamma mc^2 = \sqrt{p^2 c^2 + m^2 c^4}$.

(e) $E = \gamma mc^2 + V(x); H = \sqrt{p^2 c^2 + m^2 c^4} + V(x)$.

2.27   Assuming the sun to be a blackbody with a surface temperature of 6000 K, (a) calculate the rate at which energy is radiated from it. (b) Determine the loss in solar mass per day due to this radiation.

2.28   In 1922, A. H. Compton applied the photon concept of electromagnetic radiation to explain the scattering of x rays from electrons. In the analysis it is assumed that a photon of energy hv and momentum $hv/c = h/\lambda$ is incident on a stationary but otherwise free electron of rest mass m. The photon scatters from the electron. Its new momentum, h/λ', makes an angle θ with the incident (old) momentum. The momentum of the recoiling electron makes an angle ϕ with the incident momentum (Fig. 2.14). If the system of electron and photon is an isolated system, its energy and total momentum are constant. Conservation of energy reads

$$hv + mc^2 = hv' + m\gamma c^2$$

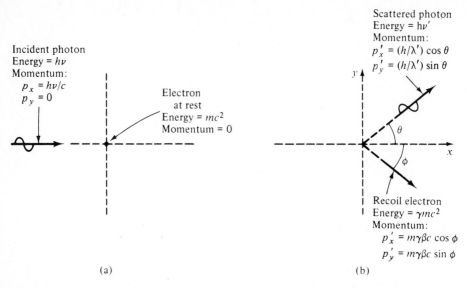

(a)   (b)

FIGURE 2.14 **Angles θ and ϕ in the Compton scattering of photons from electrons. (See Problem 2.28.)**

Conservation of momentum (the whole collision occurs in a plane) gives

$$\frac{h}{\lambda} = m\gamma\beta c \cos\phi + \left(\frac{h}{\lambda'}\right)\cos\theta$$

$$0 = -m\gamma\beta c \sin\phi + \left(\frac{h}{\lambda'}\right)\sin\theta$$

Using these three conservation equations, derive the *Compton effect equation* for the difference in wavelengths:

$$\lambda' - \lambda = \lambda_C(1 - \cos\theta)$$

$$\lambda_C = the\ Compton\ wavelength = \frac{h}{mc} = 2.43 \times 10^{-10}\ cm$$

2.29 The "classical radius of the electron," r_0, is obtained by setting the potential e^2/r_0 equal to the rest-mass energy of the electron, mc^2.

$$\frac{e^2}{r_0} = mc^2$$

$$r_0 = \frac{e^2}{mc^2}$$

Show that successive powers of the fine-structure constant

$$\alpha \equiv \frac{e^2}{\hbar c}$$

are measures of the Bohr radius to the Compton wavelength; and of the Compton wavelength to the classical radius of the electron. That is, show that

$$a_0 : \lambda_C : r_0 = 1 : \alpha : \alpha^2$$

2.7 THE WORK OF HEISENBERG. UNCERTAINTY AS A CORNERSTONE OF NATURAL LAW

It is an essential feature of Newton's second law that given the initial coordinates and velocity of a particle, $\mathbf{r}(0)$ and $\dot{\mathbf{r}}(0)$, respectively, and knowing all the forces on the particle, the orbit $\mathbf{r}(t)$ is uniquely determined. The same holds true for a system of particles. This is the essence of *determinism*. Laplace, in the eighteenth century, took the implications of the latter statements to their extreme: the entire universe consists of bodies moving through space and obeying Newton's laws. Once the interaction between these bodies is precisely known and the position and velocities of all the bodies at any given instant are known, these coordinates and velocities are determined (through Newton's second law) for all time.

Quantum mechanics was to bring down the walls of this deterministic philosophy. The instrument of destruction was the *Heisenberg uncertainty principle*. What Heisenberg put forth in 1927 implied the following: if the momentum of a particle is known precisely, it follows that the position (location) of that same particle is completely unknown. Quantitatively, if an identical experiment involving an electron is performed many times, and in each run of the experiment the position (x) of the electron is measured, then although the experimental setup is identical (same electron momentum) in each run, measurement of the position of the electron does not give the same result. Let the average of these measurements be $\langle x \rangle$. Then we can form the mean-square deviation

(2.33) $$(\Delta x)^2 \equiv \langle (x - \langle x \rangle)^2 \rangle$$

The standard deviation is labeled Δx. If Δx is small compared to some typical length in the experiment, one is more certain to find the value $x = \langle x \rangle$ in any given run. If Δx is large, it is not certain what the measurement of x will yield (Fig. 2.15). For this reason Δx is also called the *uncertainty in x*.

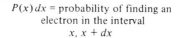

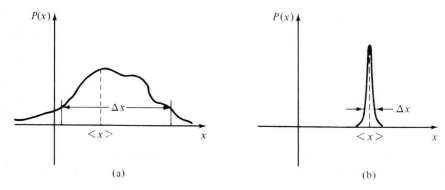

FIGURE 2.15   (a) Large uncertainty in x: $(\Delta x)^2 = \langle x^2 \rangle - \langle x \rangle^2$. (b) Small uncertainty in x.

Similarly, one may speak of an uncertainty in any physically observable quantity: magnetic field $\mathscr{B}$, energy E, momentum $\mathbf{p}$, and so forth.

$$(2.34) \qquad \Delta\mathscr{B}_x = \sqrt{\langle(\mathscr{B}_x - \langle\mathscr{B}_x\rangle)^2\rangle}, \qquad \Delta\mathscr{B}_y = \cdots$$

$$\Delta E = \sqrt{\langle(E - \langle E\rangle)^2\rangle}$$

$$\Delta p_x = \sqrt{\langle(p_x - \langle p_x\rangle)^2\rangle}, \qquad \Delta p_y = \cdots$$

Heisenberg's uncertainty relation for momentum p_x and position x (parallel components) appears as

$$(2.35) \qquad \Delta x \Delta p_x \gtrsim \hbar$$

If it can be said with certainty what the position of a particle is ($\Delta x = 0$), then there is total uncertainty regarding the momentum of the particle ($\Delta p_x = \infty$). Observable parameters obeying a relation such as (2.35) are called *complementary variables*. Examples include: (1) coordinates and momenta (x, p_x); (2) energy and time (E, t); and (3) any two Cartesian components of angular momentum (L_x, L_y). Later in the text a formal technique is presented to determine if two observables are complementary (as opposed to *compatible*).

When an electron (or photon) exists within a well-defined locality of space (momentum is ill defined), it acts very much like a particle. This is the case since in a double-slit experiment such a localized disturbance would only go through one slit; and we can therefore follow it in time, so it is very much like a true particle. When the electron does not exist in a well-defined locality of space, its momentum can be defined more precisely. Under such circumstances the wave character of the electron manifests itself. We cannot follow *it*. A whole wave is propagating. Nevertheless, it should be borne in mind that when a scintillation screen is put across its path one gets a single flash—although one has little idea of when or where the event will occur.

PROBLEMS

2.30 Consider a particle with energy $E = p^2/2m$ moving in one dimension (x). The uncertainty in its location is Δx. Show that if $\Delta x \, \Delta p > h$, then $\Delta E \, \Delta t > h$, where $(p/m) \, \Delta t = \Delta x$.

2.31 The size of an atom is approximately 10^{-8} cm. To locate an electron within the atom, one should use electromagnetic radiation of wavelength not longer than, say, 10^{-9} cm. (a) What is the energy of a photon with such a wavelength (in eV)? (b) What is the uncertainty in the electron's momentum if we are uncertain about its position by 10^{-9} cm?

2.8 THE WORK OF BORN. PROBABILITY WAVES

When discussing the double-slit wave diffraction experiment and the Davisson–Germer experiment, we found it appropriate to introduce an amplitude function ψ, the square of whose modulus, $|\psi|^2$, was set equal to the intensity (2.18) of the wave.

Born suggested in 1927 that, when referred to the propagation of particles, $|\psi|^2$ is more appropriately termed a *probability density*. The function ψ is called the *wavefunction* (also the *state function* or *state vector*) of the particle. Quantitatively, the Born postulate states the following (in Cartesian space). The wavefunction for a particle $\psi(x, y, z, t)$ is such that

$$(2.36) \qquad\qquad |\psi|^2 \, dx \, dy \, dz = P \, dx \, dy \, dz$$

where $P \, dx \, dy \, dz$ is the probability that measurement of the particle's position at the time t finds it in the volume element $dx \, dy \, dz$ about the point x, y, z.

This statement is quite consistent with the discussions above relating to the interference of photons or electrons. In all cases an interference pattern exhibits itself when an abundance of particles is present. The wavefunction ψ generates the interference pattern. Where $|\psi|^2$ is large, the probability that a particle is found there is large. When enough particles are present, they distribute themselves in the probability pattern outlined by the density function $|\psi|^2$.

The rules of quantum mechanics (Chapter 3) give a technique for calculating the wavefunction ψ to within an arbitrary multiplicative constant. The equation one solves to find ψ is called the *Schrödinger equation*. This is a homogeneous linear equation. Suppose that we solve it and obtain a function ψ. Then $A\psi$ is also a solution, where A is any constant. The Born postulate specifies[1] A. For problems where it can be said with certainty that the particle is somewhere in a given volume V,

$$(2.37) \qquad \int_V |\psi|^2 \, dx \, dy \, dz = 1$$

This is a standard property that probability densities satisfy under most conditions. It is the mathematical expression of the certainty that the particle is in the volume V.

As an example, consider the following one-dimensional problem. A particle that is known to be somewhere on the x axis has the wavefunction

$$(2.38) \qquad \psi = A e^{i\omega t} e^{-x^2/2a^2}$$

The frequency ω and length a are known constants. The (real) constant A is to be determined. Since it is certain that the particle is somewhere in the interval $-\infty < x < +\infty$, it follows that

$$(2.39) \qquad 1 = \int_{-\infty}^{\infty} \psi^* \psi \, dx = A^2 \int_{-\infty}^{\infty} e^{-i\omega t} e^{+i\omega t} e^{-x^2/a^2} \, dx$$

$$= A^2 a \int_{-\infty}^{\infty} e^{-\eta^2} \, d\eta = A^2 a \sqrt{\pi}$$

The nondimensional variable $\eta \equiv x/a$. This calculation gives

$$(2.40) \qquad A = \frac{1}{a^{1/2} \pi^{1/4}}$$

The normalized wavefunction is therefore

$$(2.41) \qquad \psi = \frac{1}{a^{1/2} \pi^{1/4}} e^{i\omega t} e^{-x^2/2a^2}$$

For the stated problem, $|\psi|^2$ as obtained from (2.41) is the correct probability density.

PROBLEMS

2.32   The wavefunction for a particle in one dimension is given by

$$|\psi_1| = A_1 e^{-y^2/4}$$

Another state that the particle may be in is

$$|\psi_2| = A_2 y e^{-y^2/8}$$

A third state the particle may be in is

$$|\psi_3| = A_3 (e^{-y^2/4} + y e^{-y^2/8})$$

[1] If A is complex it may be determined only to within an arbitrary phase factor, $e^{i\alpha}$, where α is a real number.

Normalize all three states in the interval $-\infty < y < +\infty$ (i.e., find A_1, A_2, and A_3). Is the probability of finding the particle in the interval $0 < y < 1$ when the particle is in the state ψ_3 the same as the sum of the separate probabilities for the states ψ_1 and ψ_2? Answer the same question for the interval $-1 < y < +1$.

2.33   The energy density (ergs/cm^3) of electromagnetic radiation is proportional to $\mathscr{E}^2$, where $\mathscr{E}$ is the electric field. Present an argument to demonstrate that $\mathscr{E}^2 \, d\mathbf{r}$ is a measure of the probability of finding a photon in the volume element $d\mathbf{r}$. Assume a monochromatic radiation field.

2.34   Suppose that in a sample of 1000 electrons, each has a wavefunction

$$\psi = e^{-|x|} e^{-i\omega t} \cos \pi x$$

Measurements are made (at a specific time, $t = t'$) to determine the locations of electrons in the sample. Approximately how many electrons will be found in the interval $-\frac{1}{2} \le x \le \frac{1}{2}$? A graphical approximation is adequate.

2.35   A beam of monochromatic electromagnetic radiation incident normally on a totally absorbing surface exerts a pressure on the surface of

$$P = U = \frac{\mathscr{E}^2}{8\pi}$$

where $\mathscr{E}$ is the amplitude of the electric field vector. If $P = 3 \times 10^{-6}$ dyne/cm^2 and the wavelength of radiation is $\lambda = 8000$ Å, what is the photon flux (cm^{-2}/s) striking the surface?

2.9   SEMIPHILOSOPHICAL EPILOGUE TO CHAPTER 2

The wavefunction ψ affords information related to experiments on, say, an electron. Consider once again the double-slit experiment of Fig. 2.10. Again, we suppose that the source is able to fire single electrons with well-defined momenta. There is a corresponding (propagating) wavefunction $\psi(\mathbf{r}, t)$ which is diffracted by the slits. When measurements are made, the scintillation screen gives a single flash (for a single electron). If we calculate $|\psi|^2$ at the screen, we find an interference pattern. What is the significance of this pattern? Suppose that the electron is a bullet. We can play a quantum mechanical Russian roulette. The game is to stand at the screen so that the bullet misses you. The first thing to do is solve the Schrödinger equation and calculate $|\psi|^2$ at the screen. Stand where it is minimum. But this, of course, does not guarantee that the bullet does not find its mark. The laws of nature do not provide a more definite knowledge of the electron's trajectory.

Now when a pulse of electrons (assume that they are all independent of one another) is fired at the slits, the scintillation screen registers an interference pattern. Eventually (i.e., when a sufficient number are fired), the electrons begin to follow the dictates of $|\psi|^2$ and fall into place (Fig. 2.13). It is interesting that all the information in ψ cannot be extracted from an experiment on one electron. To get this information one has to do many experiments, each of which involves many more than one electron.

At this point the reader may well ask: If electrons are particles, why not follow their trajectories through the slits (an electron can only go through one slit at a time) and onto the screen? One could then add the intensities of particles stemming from each individual slit and obtain the (noninterference) pattern depicted in Fig. 2.9. Well, we can do exactly that, and the interference pattern does vanish. In the process of "watching" the electrons, the interference is destroyed.

This is seen as follows. Let us see if we can discern which slit the electron goes through. The uncertainty in measurement of its y coordinate must obey the inequality

$$(2.42) \qquad \Delta y \ll \frac{d}{2}$$

If the interference pattern is not to be destroyed, the uncertainty in an electron's y momentum Δp_y, induced by encounter with a photon, must be substantially smaller than that which would displace the electron from a maximum in the interference pattern to a neighboring minimum. With the aid of Fig. 2.16 this condition is

$$(2.43) \qquad \Delta p_y \ll \frac{\theta}{2} p_x = \frac{h}{2d}$$

In the latter equality we have recalled the de Broglie relation (2.30).

The first inequality for Δy, (2.42), enables one to observe which slit the electron goes through. The second inequality for Δp_y, (2.43), guarantees the preservation of the interference pattern. Combining these two inequalities gives the relation

$$(2.44) \qquad \Delta y \Delta p_y \ll \frac{h}{4}$$

which is in contradiction to the Heisenberg uncertainty principle.

We conclude that if it is possible to observe which slit electrons go through, their interference pattern is destroyed. In observing the positions of the electrons, their wave quality (e.g., interference-producing mechanism) diminishes. When the light (whose photons are illuminating the electrons' path) is switched off, the interference pattern reappears.

In general, we may note the fundamental rule that quantum mechanics does not delineate the trajectory of a single particle. One may calculate the probability that an electron is in some region of space, but this is again a probability and not a guarantee that the electron will be found there. To realize this probability, one must in principle observe many experiments on the same system with identical initial conditions obeyed

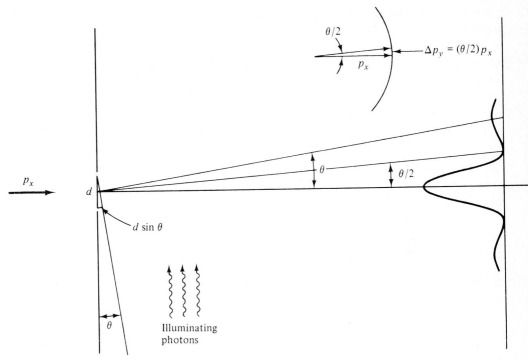

FIGURE 2.16   First maximum at $\theta = 0$; second maximum at $\sin \theta \simeq \theta = \lambda/d$. **The angle between the first minimum and the second maximum** $= \theta/2 = \lambda/2d$.

in each experiment. Average results then fall to the dictates of quantum mechanics. In this regard Einstein[1] has remarked that quantum mechanics is incapable of describing the behavior of a single system (such as an electron).

Hidden Variables

There is another, somewhat philosophical school of thought (due primarily to Bohm and de Broglie[2]) which holds that the impossibility of quantum mechanics to predict with certainty the outcome of a given measurement on an individual system stems from one's inability to know the exact values of certain *hidden variables* relating to the system. In this description, the wavefunction is viewed as a mathematical object which contains all the information one possesses regarding an incompletely known system. Quantum formulas should emerge as averages over the hidden parameters in much

[1] P. A. Schlipp, ed., *Albert Einstein: Philosopher–Scientist*, Harper & Row, New York, 1959.
[2] D. Bohm, *Phys. Rev.* **85**, 166, 180 (1952); *L. de Broglie, Physicien et Penseur*, Albin Michel, Paris, 1953. For further discussion and reference, see J. S. Bell, *Rev. Mod. Phys.* **38**, 447 (1966); F. J. Belinfante, *A Survey of Hidden-Variable Theories*, Pergamon Press, New York, 1973.

the same way as the laws of classical physics do in fact follow in averaging over the quantum equations.[1]

Bethe[2] has argued that the existence of such hidden variables for an electron would imply electronic degrees of freedom which are not specified in atomic physics. However, the success of the present theory in formulation of the periodic table indicates that this is not the case (i.e., no further degrees of freedom exist).[3]

In Chapter 3 we discuss the postulates of quantum mechanics. These are clear-cut formal statements whose mastery enables the student to treat many problems in the quantum mechanical domain. In addition, a deeper understanding is gained of some of the questions raised in this semiphilosophical epilogue. The notion of hidden variables is returned to in Section 11.3 in discussion of the Einstein-Padolsky-Rosen paradox.

PROBLEMS

2.36 In deriving the Planck radiation formula (2.3), one first sets

$$u(v) = hvn(v)$$

where $n(v) \, dv$ is the density of photons in the frequency interval $v, v + dv$. An expression for $n(v)$ is obtained through the relation

$$n(v) = g(v)f_{BE}(v)$$

where $g(v) \, dv$ is the density of modes (i.e., vibrational states) in the said frequency interval and f_{BE}, the Bose–Einstein factor, gives the average number of photons per mode at the frequency v. Use Planck's hypothesis to obtain the expression

$$f_{BE}(v) = \frac{1}{e^{hv/k_B T} - 1}$$

[The calculation of $g(v)$ is considered in Problem 2.37.]

Answer

We seek the average number of photons per mode at the frequency v. At this frequency the modes of excitation of the radiation field have energies $hv, 2hv, 3hv, \ldots$. Let us assume that the probability that the Nth energy mode is excited is given by the *Boltzmann distribution*

$$p(N) = e^{-Nx} \left/ \sum_{N=0}^{\infty} e^{-Nx} \right.$$

$$x \equiv \frac{hv}{k_B T}$$

---

[1] This relation between quantum and classical mechanics is called *Ehrenfest's principle* and is discussed fully in Chapter 6.
[2] H. A. Bethe and R. W. Jackiw, *Intermediate Quantum Mechanics*, 2nd ed., W. A. Benjamin, New York, 1968.
[3] Experimental evidence obtained by S. Freedman and J. Clauser, *Phys. Rev. Lett.* **28**, 938 (1972), also appears to point against a hidden-variable theory.

There are N photons of frequency v in the Nth mode. Averaging over N gives

$$f_{BE} = \langle N \rangle = \sum_N N P(N)$$

$$= \sum_N N e^{-Nx} \bigg/ \sum_N e^{-Nx} = -\frac{\partial}{\partial x} \ln \sum_N e^{-Nx}$$

$$= -\frac{\partial}{\partial x} \ln \sum_N (e^{-x})^N$$

$$= -\frac{\partial}{\partial x} \ln \frac{1}{1 - e^{-x}} = \frac{\partial}{\partial x} \ln (1 - e^{-x})$$

$$= \frac{1}{e^x - 1}$$

2.37 In Problem 2.36 we noted that the number of photons per unit volume in the frequency interval $v, v + dv$, is given by

$$n(v) = g(v) f_{BE}\left(\frac{hv}{k_B T}\right)$$

(a) Calculate the density of states $g(v)$, assuming that the blackbody radiation field consists of standing waves in a cubical box with perfectly reflecting walls.

(b) Obtain the Rayleigh–Jeans law for the radiant energy density $u_{RJ}(v)$, assuming the classical *equipartition hypothesis* for the electromagnetic field: that is, each mode of vibration contains $k_B T$ ergs of energy.

(c) Make a sketch of $u_{RJ}(v)$ and compare it to the Planck formula for $u(v)$. In what frequency domain do the two theories agree?

(d) What property of the vibrational energy levels of the radiation field (at a given frequency) allows the classical description (i.e., u_{RJ}) to be valid?

Answers (partial)

(a) The spatial components of a standing electric field in a cubical box of volume $V = L^3$, with perfectly reflecting walls, are

$$\mathscr{E}_x = A \cos k_x x \sin k_y y \sin k_z z$$
$$\mathscr{E}_y = B \sin k_x x \cos k_y y \sin k_z z$$
$$\mathscr{E}_z = C \sin k_x x \sin k_y y \cos k_z z$$

These fields have the required property that the tangential component of $\mathscr{E}$ vanishes at all six walls provided that

$$k_x L = n_x \pi, \qquad k_y L = n_y \pi, \qquad k_z L = n_z \pi$$

where n_x, n_y, and n_z assume positive integer values. There is a mode of vibration for each triplet of values (n_x, n_y, n_z). We seek the number of such modes in the frequency interval v, $v + dv$. First note that for each mode the square sum

$$n^2 = n_x{}^2 + n_y{}^2 + n_z{}^2 = \left(\frac{L}{\pi}\right)^2 (k_x{}^2 + k_y{}^2 + k_z{}^2)$$

$$= \left(\frac{L}{\pi}\right)^2 k^2 = \left(\frac{2L}{c}\right)^2 v^2$$

is proportional to the square of v, the frequency of vibration ($2\pi v = ck$). Next consider Cartesian n space with axes n_x, n_y, and n_z. Each point in this space corresponds to a mode of vibration. It is clear that all points which fall on a spherical surface of radius $(2Lv/c)$ correspond to modes at the frequency v. It follows that the number of modes in the frequency interval v, $v + dv$, is given by the volume in n space of a spherical shell of thickness dn and radius n (Fig. 2.17):

$$Vg(v)\, dv = 2 \times \tfrac{1}{8} \times 4\pi n^2\, dn = \pi n^2\, dn$$

The factor 2 enters because of the two possible polarizations of an electric field in a given mode. The factor $\tfrac{1}{8}$ is due to the fact we wish only to consider positive frequencies so that only that

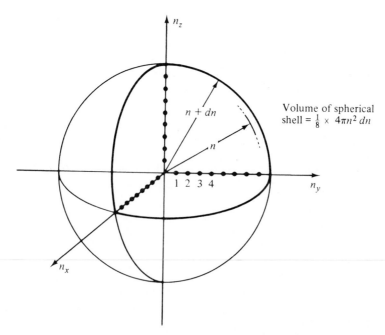

FIGURE 2.17 Cartesian n space for the enumeration of standing electromagnetic wave states in a box of edge length L. All points that fall in the shell of thickness dn and radius $(2Lv/c)$ correspond to modes with frequencies in the interval v, $v + dv$. (See Problem 2.37.)

portion of the shell in the first octant is counted. This gives the desired result:

$$g(v) = \frac{8\pi v^2}{c^3}$$

(b)   If there is $k_B T$ energy per mode, the spectral energy density is

$$u_{RJ} = \frac{8\pi v^2}{c^3} k_B T$$

2.38   For a gas of N noninteracting particles (an *ideal gas*) in thermodynamic equilibrium at temperature T and confined to volume V, the pressure P is given by

$$PV = N\langle p_x v_x \rangle$$

The momentum of a particle is **p** and **v** is its velocity. The average is taken over all particles in the gas. Show that for a *gas* of photons, this relation gives

$$PV = \tfrac{1}{3}E$$

while for a gas of mass points it gives

$$PV = \tfrac{2}{3}E$$

The total energy of the gas is E. (*Note:* The principle of *equipartition of energy* ascribes equal portions of energy, on the average, to each degree of freedom of a particle. Thus, if the average energy of a mass particle is ε, then $\langle \tfrac{1}{2}mv_x^2 \rangle = \varepsilon/3$.)

2.39   The energy density $U = E/V$ of a blackbody radiation field is a function only of temperature, $U = U(T)$. Using this fact together with the result of the last problem, $E = 3PV$, show that Stefan's law, $U = (4\sigma/c)T^4$, follows from purely thermodynamic arguments, thereby establishing the law as a classical result.

Answer

The first two laws of thermodynamics give

$$T\, dS = dE + P\, dV$$

The second law defines the entropy S, while the first law gives the conservation of energy statement in the form: heat added = increase in internal energy + work done. Using the given relations permits this equation to be rewritten

$$dS = \frac{V}{T} dU + \frac{4}{3}\frac{U}{T} dV$$

We recognize this equation to be in the form

$$dS = \frac{\partial S}{\partial U}\bigg|_V dU + \frac{\partial S}{\partial V}\bigg|_U dV$$

It follows that

$$\left.\frac{\partial V/T}{\partial V}\right|_{U(T)} = \frac{4}{3}\left.\frac{\partial U(T)/T}{\partial U}\right|_{V}$$

which integrates to the desired result.

2.40 An *adiabatic process* is one in which a system exchanges no heat with its environment.

(a) What form does the first law of thermodynamics assume for an adiabatic process?

(b) Using the form obtained in part (a), show that for an adiabatic expansion of a black-body radiation field,

$$VT^3 = \text{constant}$$

(c) Consider that the primeval fireball described in Problem 2.10 contains mass $M = \rho V$ and radiation energy $E = UV$, where V is volume. Show that in an expanding universe, the radiation density U decreases faster than the mass density ρ. Thus, although it is believed that radiation density of the primeval fireball far exceeded its mass density, the fact that in our present universe mass density dominates over radiation density is seen to be consistent with an adiabatically expanding universe in which Stefan's law holds.

Answers (partial)

(a) Since $dS = 0$ for an adiabatic process, the first law becomes

$$dE + P\,dV = 0$$

(b) *Hint:* To find $P = P(T)$, use $E = 3PV$ in conjunction with Stefan's law.

(c) *Hint:* Compare $\rho(V)$ at constant M with $U(V)_{adb}$.

2.41 Two plates of an ideal parallel-plate capacitor are made of platinum and silver, respectively. When the plates are brought to the displacement 10^{-3} cm, electrons "tunnel" through[1] the potential barrier, thereby creating a contact potential. What is the electric field (V/cm) between the plates? Which metal is left positive?

2.42 As described in Section 2.3, in a sample of metal at absolute zero, electrons completely fill the lowest levels such that no more than one electron occupies each state. All levels are filled from zero energy to E_F, the Fermi energy (see Fig. 2.5). Let $g(E)\,dE$ represent the number of energy states that are available for occupation in the interval $E, E + dE$, per unit volume. Since each state from 0 to E_F is occupied, the number of free electrons, n, per unit volume is given by the number of available states per unit volume in this interval. That is,

$$n = \int_0^{E_F} g(E)\,dE$$

For free electrons, *the density of states* $g(E)$ (see Section 8.8) is given by

$$g = \frac{8\sqrt{2}\pi m^{3/2}E^{1/2}}{h^3}$$

---

[1] The mechanism of quantum mechanical tunneling is described in Section 7.7.

(a)   Using the expressions above, obtain an explicit formula for E_F for a metal with n free electrons per unit volume.

(b)   Given that $E_F(Cu) = 7.0$ eV and $E_F(Na) = 3.1$ eV, use your formula to obtain the density (cm^{-3}) of Cu and Na nuclei, respectively, in samples at absolute zero. (The periodic chart is given in Table 12.2.)

Answer (partial)

(a) $E_F = (h^2/2m)(3n/8\pi)^{2/3}$.

2.43   (a)   Show that when expressed in terms of *angular* frequency ω, the radiant energy density relevant to a blackbody radiation field is given by

$$u(\omega) = \frac{\hbar\omega^3/\pi^2 c^3}{e^{\hbar\omega/k_B T} - 1}$$

(b)   Show that the corresponding density of states is given by

$$g(\omega) = \frac{\omega^2}{\pi^2 c^3}$$

(c)   Answers to the preceding questions may be found by setting

$$u(\omega)\, d\omega = u(\nu)\, d\nu$$

What is the physical meaning of this equality? (See also Problem 2.1.)

2.44   (a)   What is the wavelength of an electromagnetic wave of frequency ν propagating in vacuum?

(b)   What is the de Broglie wavelength of a photon of frequency ν?

(c)   How are the wavelengths of (a) and (b) related?

2.45   An idealized model of a *plasma* (i.e., a fluid of electrons and ions) is given by a so-called one-component plasma (OCP). In this configuration electrons move in a uniformly distributed charge-neutralizing background. This background is sometimes referred to as "jellium." It was originally suggested by E. P. Wigner that a high-density OCP will crystallize at some critical value of electron density as this parameter is relaxed from its extreme high value.

(a)   What is the potential energy of an electron situated at the center of a sphere of radius a which is uniformly filled with positive neutralizing charge?

(b)   What is the uncertainty in momentum of an electron confined to a region of dimension a? What is the corresponding kinetic energy E_k?

(c)   Assuming that crystallization will occur when potential energy per particle exceeds uncertainty energy, obtain a criterion for crystallization in terms of the distance between electrons a and the Bohr radius a_0.

Answer

(a)   The charge density of the neutralizing medium is

$$\rho = \frac{e}{\frac{4}{3}\pi a^3}$$

The potential of interaction between electron and positive jellium in a shell of thickness dr at r is

$$dV = \frac{e\rho 4\pi r^2\, dr}{r}$$

Thus

$$V = \int_0^a \frac{e\rho 4\pi r^2\, dr}{r} = \frac{3e^2}{2a}$$

(b)   From the uncertainty relation (2.35) we have $\Delta p \simeq \hbar/a$ so that

$$E_k \simeq \frac{(\Delta p)^2}{2m} = \frac{\hbar^2}{2ma^2}$$

(c)   The relation $V > E_k$ gives the order-of-magnitude criterion,

$$a > a_0$$

where a_0 is the Bohr radius. Thus, Wigner crystallization can be expected to occur when interparticle spacing exceeds the Bohr radius.

2.46   In a nuclear reactor, uranium atoms break apart and release several high-energy neutrons. These neutrons have an initial energy of 2 MeV, but after many collisions with other atoms they slow down to an average energy of 0.025 eV. Thus there will be a continuous distribution of neutrons with energies ranging from 2 MeV to *less* than 0.025 eV.

(a)   What are the de Broglie wavelengths of the 2 MeV neutrons and the 0.025 eV neutrons? Give your answer in Å.

(b)   A "low-pass" energy filter can be constructed by passing the neutrons through a long piece of polycrystalline graphite. Explain briefly why low-energy "particles" can travel straight through the block but high-energy "particles" are *reflected* out the sides.

(c)   What is the shortest de Broglie wavelength of the neutrons that are totally transmitted through the crystal? Give your answer in Å. Spacing between lattice planes in graphite is $a \simeq 2$ Å.

Answer (partial)

(c)   The condition for constructive interference from adjacent planes (Fig. 2.12) is

$$n\lambda = 2a \cos \theta \leq 2a$$

$$\lambda \leq \frac{2a}{n} \leq 2a$$

Thus for reflection, $\lambda \leq 2a = \lambda_{\text{max}}$.

2.47   De Broglie waves are incident on a crystal with lattice planes separated by a. Show that constructive interference of waves reflected from planes separated by $2a$ is ruled out by parallel reflection from planes in between.

2.48   As described in Section 7.10, a criterion which discerns if a given configuration is classical or quantum mechanical may be stated in terms of the de Broglie wavelength λ. Namely, if L is a scale length characteristic to the configuration at hand, then one has the following criteria:

$$\lambda \ll L: \quad \text{Classical}$$

$$\lambda \gtrsim L: \quad \text{Quantum mechanical}$$

Use these criteria to describe which physics is relevant to the following configurations:

(a)   An atomic electron. For the typical length choose the Bohr radius. For typical energy choose the Rydberg.

(b)   A proton in a nucleus. For nuclear size choose $\simeq 10^{-13}$ cm (1 fermi). For energy choose $\simeq 10$ MeV.

(c)   An electron in a vacuum tube operating at 10 kV.

(d)   An electron gas of density 10^{20} electrons cm^{-3} and temperature 300 K.

2.49   A particle constrained to move on the x axis has a probability of $\frac{1}{5}$ for being in the interval $(-d - a, -d + a)$ and $\frac{4}{5}$ for being in the interval $(d - a, d + a)$, where $d \gg a$.

(a)   Sketch the wavefunction that describes this situation.

(b)   Call the normalized wavefunction for the "left" interval $\varphi_-(x)$ and that for the "right" interval $\varphi_+(x)$. What is the normalized wavefunction $\varphi(x)$ for the particle?

(c)   Again, with $d \gg a$, what is the probability density $P(x)$ for the particle? What does the integral over all x of $P(x)$ give?

Answer (partial)

(b) $\varphi_\pm = \sqrt{\frac{1}{5}}\,\varphi_- \pm \sqrt{\frac{4}{5}}\,\varphi_+$

(c) $P(x) = |\varphi_\pm|^2 = \sqrt{\frac{1}{5}}|\varphi_-|^2 + \sqrt{\frac{4}{5}}|\varphi_+|^2$

Note that for $d \gg a$, $\varphi_+\varphi_- = 0$.

$$\int_{-\infty}^{\infty} P(x)\, dx = \tfrac{1}{5} + \tfrac{4}{5} = 1$$

Note: For mutually exclusive events a and b, the summational probability $P_1(a) + P_2(b)$ gives the probability of a or b occurring; whereas for independent events, the product probability $P_1(a)P_2(b)$ gives the probability of a and b occurring (and is called the *joint probability*). Note further that mutually exclusive events are not independent.

2.50   Explain the meaning of the following statement: Interference between photons does not occur. A photon can only interfere with itself.[1]

---

[1] For further discussion see P. A. M. Dirac, *The Principles of Quantum Mechanics* 4th ed., Section 3, Oxford University Press, New York, 1958

CHAPTER 3

THE POSTULATES OF
QUANTUM MECHANICS. OPERATORS,
EIGENFUNCTIONS, AND EIGENVALUES

In this chapter we consider four basic postulates of quantum mechanics, which when taken with the Born postulate described in Section 2.8, serve to formalize the rules of quantum mechanics. Mathematical concepts material to these postulates are developed along with the physics. The postulates are applied over and over again throughout the text. We choose the simplest problems first to exhibit their significance and method of application—that is, problems in one dimension.

3.1 OBSERVABLES AND OPERATORS

Postulate I[1]

This postulate states the following: To any self-consistently and well-defined observable in physics (call it A), such as linear momentum, energy, mass, angular momentum, or number of particles, there corresponds an operator (call it $\hat{A}$) such that

---

[1] The order in which these postulates appear is by no means conventional.

measurement of A yields values (call these measured values a) which are eigenvalues of $\hat{A}$. That is, the values, a, are those values for which the equation

(3.1)
$$\hat{A}\varphi = a\varphi \qquad \boxed{\text{an eigenvalue equation}}$$

has a solution φ. The function φ is called the *eigenfunction* of $\hat{A}$ corresponding to the eigenvalue a.

Examples of mathematical operators, which are not necessarily connected to physics, are offered in Table 3.1. (Labels such as D, G, and M are of no special significance.) An operator operates on a function and makes it something else (except for the identity operator $\hat{I}$).

Let us now turn to operators that correspond to physical observables. Two very important such observables are the momentum and the energy.

The Momentum Operator $\hat{\mathbf{p}}$

The operator that corresponds to the observable linear momentum is

(3.2)
$$\hat{\mathbf{p}} = -i\hbar\nabla$$

What are the eigenfunctions and eigenvalues of the momentum operator? Consider that the particle (whose momentum is in question) is constrained to move in one dimension (x). Then the momentum has only one nonvanishing component, p_x. The corresponding operator is

(3.3)
$$\hat{p}_x = -i\hbar\frac{\partial}{\partial x}$$

TABLE 3.1   **Examples of operators**

$\hat{D} = \partial/\partial x$	$\hat{D}\varphi(x) = \partial\varphi(x)/\partial x$
$\hat{\Delta} = -\partial^2/\partial x^2 = -\hat{D}^2$	$\hat{\Delta}\varphi(x) = -\partial^2\varphi(x)/\partial x^2$
$\hat{M} = \partial^2/\partial x\,\partial y$	$\hat{M}\varphi(x, y) = \partial^2\varphi(x, y)/\partial x\,\partial y$
$\hat{I} = $ operation that leaves φ unchanged	$\hat{I}\varphi = \varphi$
$\hat{Q} = \int_0^1 dx'$	$\hat{Q}\varphi(x) = \int_0^1 dx'\varphi(x')$
$\hat{F} = $ multiplication by $F(x)$	$\hat{F}\varphi(x) = F(x)\varphi(x)$
$\hat{B} = $ division by the number 3	$\hat{B}\varphi(x) = \frac{1}{3}\varphi(x)$
$\hat{\Theta} = $ operator that annihilates φ	$\hat{\Theta}\varphi = 0$
$\hat{P} = $ operator that changes φ to a specific polynomial of φ	$\hat{P}\varphi = \varphi^3 - 3\varphi^2 - 4$
$\hat{G} = $ operator that changes φ to the number 8	$\hat{G}\varphi = 8$

The eigenvalue equation for this operator is

(3.4)
$$-i\hbar \frac{\partial}{\partial x} \varphi = p_x \varphi$$

The values p_x represent the possible values that measurement of the x component of momentum will yield. The eigenfunction $\varphi(x)$ corresponding to a specific value of momentum (p_x) is such that $|\varphi|^2 \, dx$ is the probability of finding the particle (with momentum p_x) in the interval $x, x + dx$. Suppose we stipulate that the particle is a *free* particle. It is unconfined (along the x axis). For this case there is no boundary condition on φ and the solution to (3.4) is

(3.5)
$$\varphi = A \exp\left(\frac{ip_x x}{\hbar}\right) = A e^{ikx}$$

where we have labeled the wavenumber k and have deleted the subscript x.

(3.6)
$$k = \frac{p}{\hbar}$$

The eigenfunction given by (3.5) is a periodic function (in x). To find its wavelength λ, we set

(3.7)
$$e^{ikx} = e^{ik(x + \lambda)}$$

$$1 = e^{ik\lambda} = \cos k\lambda + i \sin k\lambda$$

which is satisfied if

(3.8)
$$\cos k\lambda = 1$$
$$\sin k\lambda = 0$$

The first nonvanishing solution to these equations is

(3.9)
$$k\lambda = 2\pi$$

which (with 3.6) is equivalent to the de Broglie relation

(3.10)
$$p = \frac{h}{\lambda}$$

We conclude that the eigenfunction of the momentum operator corresponding to the eigenvalue p has a wavelength that is the de Broglie wavelength h/p.

In quantum mechanics it is convenient to speak in terms of wavenumber k instead of momentum p. In this notation one says that the eigenfunctions and eigenvalues of the momentum operator are

(3.11) $$\varphi_k = Ae^{ikx}, \qquad p = \hbar k$$

The subscript k on φ_k denotes that there is a continuum of eigenfunctions and eigenvalues, $\hbar k$, which yield nontrivial solutions to the eigenvalue equation, (3.4).

The Energy Operator $\hat{H}$

The operator corresponding to the energy is the Hamiltonian $\hat{H}$, with momentum **p** replaced by its operator counterpart, $\hat{\mathbf{p}}$. For a single particle of mass m, in a potential field $V(\mathbf{r})$,

(3.12) $$\hat{H} = \frac{\hat{p}^2}{2m} + V(\mathbf{r}) = -\frac{\hbar^2}{2m}\nabla^2 + V(\mathbf{r})$$

The eigenvalue equation for $\hat{H}$,

(3.13) $$\hat{H}\varphi(\mathbf{r}) = E\varphi(\mathbf{r})$$

is called the *time-independent Schrödinger equation*. It yields the possible energies E which the particle may have. Again consider the free particle. The energy of a free particle is purely kinetic, so

(3.14) $$\hat{H} = \frac{\hat{p}^2}{2m} = -\frac{\hbar^2}{2m}\nabla^2$$

Constraining the particle to move in one dimension, the time-independent Schrödinger equation becomes

(3.15) $$-\frac{\hbar^2}{2m}\frac{\partial^2}{\partial x^2}\varphi = E\varphi$$

In terms of the wave vector

(3.16) $$k^2 = \frac{2mE}{\hbar^2}$$

(3.15) appears as

(3.17) $$\varphi_{xx} + k^2\varphi = 0$$

The subscript x denotes differentiation. For a free particle there are no boundary conditions and we obtain[1]

(3.18) $$\varphi = Ae^{ikx} + Be^{-ikx}$$

---

[1] The solution to (3.17) with boundary conditions imposed is discussed in Section 4.1.

This is the eigenfunction of $\hat{H}$ which corresponds to the energy eigenvalue

$$(3.19) \qquad E = \frac{\hbar^2 k^2}{2m}$$

We have found above (3.11) that the momentum of a free particle is $\hbar k$. This is clearly the same $\hbar k$ that appears in (3.19), since for a free particle

$$(3.20) \qquad E = \frac{p^2}{2m} = \frac{\hbar^2 k^2}{2m}$$

Note also that the eigenfunction of $\hat{H}$ (3.18), with $B = 0$, is also an eigenfunction of $\hat{p}$ (3.11). That $\hat{H}$ and $\hat{p}$ for a free particle have common eigenfunctions is a special case of a more general theorem to be discussed later.[1] The following simple argument demonstrates this fact. Let

$$(3.21) \qquad \hat{p}\varphi = \hbar k \varphi$$

Let us see if φ is also an eigenfunction of $\hat{H}$ (for a free particle).

$$(3.22) \qquad \hat{H}\varphi = \frac{\hat{p}}{2m}(\hat{p}\varphi) = \frac{\hat{p}(\hbar k\varphi)}{2m} = \frac{\hbar k}{2m}\hat{p}\varphi$$

$$= \frac{(\hbar k)^2}{2m}\varphi$$

It follows that φ is also an eigenfunction of $\hat{H}$.

Both the energy and momentum eigenvalues for the free particle comprise a continuum of values:

$$(3.23) \qquad \boxed{E = \frac{\hbar^2 k^2}{2m} \qquad p = \hbar k}$$

That is, these are valid eigenvalues for *any* wavenumber k. The eigenfunction (of both $\hat{H}$ and $\hat{p}$) corresponding to these eigenvalues is

$$(3.24) \qquad \varphi_k = Ae^{ikx}$$

If the free particle is in this state, measurement of its momentum will definitely yield $\hbar k$, and measurement of its energy will definitely yield $(\hbar^2 k^2/2m)$.

Suppose that we measure its position x; what do we find? Well, where is the particle most likely to be? Again we call on the Born postulate. If the particle is in the state φ_k, the probability density relating to the probability of finding the particle in the interval $x, x + dx$, is

$$(3.25) \qquad |\varphi_k|^2 = |A|^2 = \text{constant}$$

---

[1] The commutator theorem, Chapter 5.

The probability density is the same constant value for all x. That means we would be equally likely to find the particle at any point from $x = -\infty$ to $x = +\infty$. This is a statement of maximum uncertainty which is in agreement with the Heisenberg uncertainty principle. In the state φ_k, it is known with absolute certainty that measurement of momentum yields $\hbar k$. Therefore, for the state φ_k, $\Delta p = 0$, whence $\Delta x = \infty$.

We mentioned in Section 2.7 that E and t are complementary variables; that is, they obey the relation $\Delta E \, \Delta t \geq \hbar$. Specifically, this means that if the energy is uncertain by amount ΔE, *the time it takes to measure E is uncertain by* $\Delta t \geq \hbar/\Delta E$. Now for the problem at hand, in the state φ_k, it is certain that measurement of E yields $\hbar^2 k^2/2m$. Therefore, $\Delta E = 0$. To measure E we have to let the particle interact with some sort of energy-measuring apparatus, say a plate with a spring attached to measure the momentum imparted to the plate when the particle hits it head on. Well, if the plate with attached spring is placed in the path of the particle, how long must we wait before we detect something? We can wait 10^{-8} s—or we can wait 10^{10} yr. The uncertainty Δt is infinite in the present case, since there is an infinite uncertainty in Δx.

PROBLEMS

3.1 For each of the operators listed in Table 3.1 ($\hat{D}$, $\hat{\Delta}$, $\hat{M}$, etc.), construct the square, that is, $\hat{D}^2, \hat{\Delta}^2, \ldots$.

Answer (partial)

$$\hat{I}^2\varphi = \hat{I}\varphi = \varphi$$

$$\hat{Q}^2\varphi = \hat{Q}\int_0^1 dx'\varphi(x') = \int_0^1 dx'' \int_0^1 dx'\varphi(x')$$

$$\hat{F}^2\varphi = F^2\varphi$$

$$\hat{B}^2\varphi = \tfrac{1}{9}\varphi$$

$$\hat{P}^2\varphi = \hat{P}(\hat{P}\varphi) = (\varphi^3 - 3\varphi^2 - 4)^3 - 3(\varphi^3 - 3\varphi^2 - 4)^2 - 4$$

3.2 The inverse of an operator $\hat{A}$ is written $\hat{A}^{-1}$. It is such that

$$\hat{A}^{-1}\hat{A}\varphi = \hat{I}\varphi = \varphi$$

Construct the inverses of $\hat{D}$, $\hat{I}$, $\hat{F}$, $\hat{B}$, $\hat{\Theta}$, $\hat{G}$, provided that such inverses exist.

3.3 An operator $\hat{O}$ is *linear* if

$$\hat{O}(a\varphi_1 + b\varphi_2) = a\hat{O}\varphi_1 + b\hat{O}\varphi_2$$

where a and b are arbitrary constants. Which of the operators in Table 3.1 are linear and which are nonlinear?

3.4 The displacement operator $\hat{\mathscr{D}}$ is defined by the equation

$$\hat{\mathscr{D}}f(x) = f(x + \zeta)$$

Show that the eigenfunctions of $\hat{\mathscr{D}}$ are of the form

$$\varphi_\beta = e^{\beta x}g(x)$$

where

$$g(x + \zeta) = g(x)$$

and β is any complex number. What is the eigenvalue corresponding to φ_β?

3.5 An electron moves in the x direction with de Broglie wavelength 10^{-8} cm.
    (a)   What is the energy of the electron (in eV)?
    (b)   What is the time-independent wavefunction of the electron?

3.2 MEASUREMENT IN QUANTUM MECHANICS

Postulate II

The second postulate[1] of quantum mechanics is: measurement of the observable A that yields the value a leaves the system in the state φ_a, where φ_a is the eigenfunction of $\hat{A}$ that corresponds to the eigenvalue a.

    As an example, suppose that a free particle is moving in one dimension. We do not know which state the particle is in. At a given instant we measure the particle's momentum and find the value $p = \hbar k$ (with k a specific value, say 1.3×10^{10} cm^{-1}). This measurement[2] leaves the particle in the state φ_k, so immediate subsequent measurement of p is certain to yield $\hbar k$.

    Suppose that one measures the position of a free particle and the position $x = x'$ is measured. The first two postulates tell us the following. (1) There is an operator corresponding to the measurement of position, call it $\hat{x}$. (2) Measurement of x that yields the value x' leaves the particle in the eigenfunction of $\hat{x}$ corresponding to the eigenvalue x'.

    The operator equation appears as

(3.26) $$\hat{x}\delta(x - x') = x'\delta(x - x')$$

---

[1] This postulate has been the source of some discussion among physicists. For further reference, see B. S. DeWitt, *Phys. Today* **23**, 30 (September 1970).

[2] Measurement is taken in the idealized sense. More formal discussions on the theory of measurement may be found in K. Gottfried, *Quantum Mechanics*, W. A. Benjamin, New York, 1966; J. Jauch, *Foundations of Quantum Mechanics*, Addison-Wesley, Reading, Mass., 1968, and E. C. Kemble, *The Fundamental Principles of Quantum Mechanics with Elementary Applications*, Dover, New York, 1958.

Dirac Delta Function

The eigenfunction of $\hat{x}$ has been written[1] $\delta(x - x')$ and is called the *Dirac delta function*. It is defined in terms of the following two properties. The first are the integral properties

$$\int_{-\infty}^{\infty} f(x')\delta(x - x')\, dx' = f(x)$$

(3.27)

$$\int_{-\infty}^{\infty} \delta(x - x')\, dx' = 1$$

or equivalently, in terms of the single variable y

$$\int_{-\infty}^{\infty} f(y)\delta(y)\, dy = f(0)$$

(3.28)

$$\int_{-\infty}^{\infty} \delta(y)\, dy = 1$$

The second defining property is the value

(3.29) $$\delta(y) = 0 \qquad \text{(for } y \neq 0)$$

A sketch of $\delta(y)$ is given in Fig. 3.1. Properties of $\delta(y)$ are usually proved with the aid of the defining integral (3.27). For instance, consider the relation

(3.30) $$y\delta'(y) = -\delta(y)$$

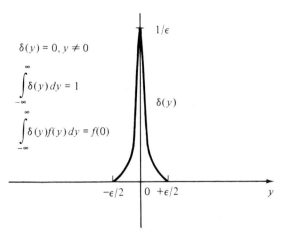

$\delta(y) = 0,\ y \neq 0$

$$\int_{-\infty}^{\infty} \delta(y)\, dy = 1$$

$$\int_{-\infty}^{\infty} \delta(y)f(y)\, dy = f(0)$$

FIGURE 3.1 Dirac delta function $\delta(y)$. The curve is distorted to bring out essential features. A more accurate picture is obtained in the limit $\epsilon \to 0$.

---

[1] More accurately one says that $\delta(x - x')$ is an eigenfunction of $\hat{x}$ in the coordinate representation. This topic is returned to in Section 7.4 and in Appendix A.

To establish this relation we employ a *test function* $f(y)$ and perform the following integration by parts.

(3.31)
$$\int_{-\infty}^{\infty} f(y)y\delta'(y)\,dy = \int_{-\infty}^{\infty} \frac{d}{dy}(fy\delta)\,dy - \int_{-\infty}^{\infty} \delta \frac{d}{dy}(yf)\,dy$$

$$= -\int_{-\infty}^{\infty} \delta(y)\left(y\frac{df}{dy} + f\right)dy$$

$$= -\int_{-\infty}^{\infty} \delta(y)f(y)\,dy$$

which establishes (3.30).

The student should not lose sight of the fact that $\hat{x}$, when operating on a function $f(x)$, merely represents multiplication by x. For example, $\hat{x}f(x) = xf(x)$. These topics will be returned to in Chapter 11 and discussed further in Appendix A.

PROBLEMS

3.6   Establish the following properties of $\delta(y)$.

(a) $\delta(y) = \delta(-y)$

(b) $\delta'(y) = -\delta'(-y)$

(c) $y\delta(y) = 0$

(d) $\delta(ay) = |a|^{-1}\delta(y)$

(e) $\delta(y^2 - a^2) = |2a|^{-1}[\delta(y-a) + \delta(y+a)]$

(f) $\int_{-\infty}^{\infty} \delta(a-y)\delta(y-b)\,dy = \delta(a-b)$

(g) $f(y)\delta(y-a) = f(a)\delta(y-a)$

(h) $y\,\delta'(y) = -\delta(y)$

(i) $\int g(y)\,\delta[f(y) - a]\,dy = \dfrac{g(y)}{|df/dy|}\bigg|_{\substack{y=y_0 \\ f(y_0)=a}}$

3.7   Show that the following are valid representations of $\delta(y)$:

(a) $2\pi\delta(y) = \int_{-\infty}^{\infty} e^{iky}\,dk$

(b) $\pi\delta(y) = \lim_{\eta \to \infty} \dfrac{\sin \eta y}{y}$

Note: In mathematics, an object such as $\delta(y)$, which is defined in terms of its integral properties, is called a *distribution*. Consider all $\chi(y)$ defined on the interval $(-\infty, \infty)$ for which

$$\int_{-\infty}^{\infty} |\chi(y)|^2 \, dy < \infty$$

Then two distributions, δ_1 and δ_2, are *equivalent* if for all $\chi(y)$,

$$\int_{-\infty}^{\infty} \chi \delta_1 \, dy = \int_{-\infty}^{\infty} \chi \delta_2 \, dy$$

When one establishes that a mathematical form such as $\int_{-\infty}^{\infty} \exp\,(iky)\, dy$ is a representation of $\delta(y)$, one is in effect demonstrating that these two objects are *equivalent as distributions*.

3.8   Show that the continuous set of eigenfunctions $\{\delta(x - x')\}$ obeys the "orthonormality" condition

$$\int_{-\infty}^{\infty} \delta(x - x') \, \delta(x - x'') \, dx = \delta(x' - x'')$$

3.9   (a)   Show that $\delta(\sqrt{x}) = 0$.

   (b)   Evaluate $\delta(\sqrt{x^2 - a^2})$.

3.3   THE STATE FUNCTION AND EXPECTATION VALUES

Postulate III

The third postulate of quantum mechanics establishes the existence of the state function and its relevance to the properties of a system: The state of a system at any instant of time may be represented by a state or wave function ψ which is continuous and differentiable. All information regarding the state of the system is contained in the wavefunction. Specifically, if a system is in the state $\psi(\mathbf{r}, t)$, the average of any physical observable C relevant to that system at time t is

(3.32)
$$\langle C \rangle = \int \psi^* \hat{C} \psi \, d\mathbf{r}$$

(The differential of volume is written $d\mathbf{r}$.) The average, $\langle C \rangle$, is called the *expectation value* of C.

The physical meaning of the average of an observable C involves the following type of (conceptual) measurements. The observable C is measured in a specific experiment, X. One prepares a very large number (N) of identical replicas of X. The initial states $\psi(\mathbf{r}, 0)$ in each such replica are all identical. At the time t, one measures C in all these replica experiments and obtains the set of values $C_1, C_2, \ldots, C_N$. The average of C is then given by the rule

$$(3.33) \qquad \langle C \rangle = \frac{1}{N} \sum_{i=1}^{N} C_i \qquad (N \gg 1)$$

The postulate stated above claims that this experimentally calculated average (3.33) is the same as that given by the integral in (3.32). Another way of defining $\langle C \rangle$ is in terms of the probability $P(C_i)$. This function gives the probability that measurement of C finds the value C_i. For $\langle C \rangle$, we then have

$$(3.34) \qquad \langle C \rangle = \sum_{\text{all } C} C_i P(C_i)$$

This is a consistent formula if all the values C may assume comprise a discrete set (e.g., the number of marbles in a box). In the event that the values that C may assume comprise a continuous set (e.g., the values of momentum of a free particle), $\langle C \rangle$ becomes

$$(3.35) \qquad \langle C \rangle = \int C P(C) \, dC$$

The integration is over all values of C. Here $P(C)$ is the probability of finding C in the interval $C, C + dC$.

The quantity $\langle C \rangle$ is also called the *expectation value* of C because it is representative of the value one expects to obtain in any given measurement of C. This will be especially true if the deviation of values of C from the mean value $\langle C \rangle$ is not large. As discussed in Section 2.7, a measure of this spread of values about the value $\langle C \rangle$ is given by the mean-square deviation ΔC, defined through

$$(3.36) \qquad (\Delta C)^2 = \langle (C - \langle C \rangle)^2 \rangle = \langle C^2 \rangle - \langle C \rangle^2$$

In order to become familiar with the operational use of postulate III, we work out the following one-dimensional problem. A particle is known to be in the state

$$(3.37) \qquad \psi(x, t) = A \exp \left[\frac{-(x - x_0)^2}{4a^2} \right] \exp \left(\frac{i p_0 x}{\hbar} \right) \exp (i \omega_0 t)$$

The lengths x_0 and a are constants, as are the momentum p_0 and frequency ω_0. The (real) constant A is determined through normalization. This then ensures that $\psi^* \psi$ is a numerically correct probability density.

$$\int_{-\infty}^{\infty} |\psi|^2 \, dx = A^2 a \int_{-\infty}^{\infty} e^{-\eta^2/2} \, d\eta = \sqrt{2\pi} \, A^2 a = 1$$

$$(3.38)$$

$$A^2 = \frac{1}{a\sqrt{2\pi}}$$

The nondimensional "dummy" variable η and constant η_0 are such that

$$\eta = \frac{x - x_0}{a}$$

(3.39)
$$x = a(\eta + \eta_0)$$

$$\eta_0 = \frac{x_0}{a}$$

Having obtained A, we may now calculate the expectation of x:

$$\langle x \rangle = \int_{-\infty}^{\infty} \psi^* \hat{x} \psi \, dx = \int_{-\infty}^{\infty} \psi^* x \psi \, dx$$

(3.40)
$$= A^2 a^2 \int_{-\infty}^{\infty} e^{-\eta^2/2}(\eta + \eta_0) \, d\eta = a\eta_0 \left(aA^2 \int_{-\infty}^{\infty} e^{-\eta^2/2} \, d\eta \right)$$

which, with the normalization condition (3.38), gives

(3.41)
$$\langle x \rangle = a\eta_0 = x_0$$

[Note that integration of the odd integrand $\eta \exp(-\eta^2)$ in (3.40) vanishes.] That x_0 is the proper value for $\langle x \rangle$ is evident from the sketch of $|\psi|^2$ shown in Fig. 3.2.

If we call

(3.42)
$$|\psi|^2 \, dx = P(x) \, dx$$

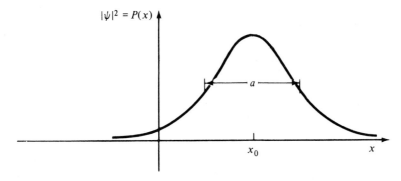

FIGURE 3.2   Gaussian probability density with variance a^2. The variance measures the spread of $P(x)$ about the mean, $\langle x \rangle = x_0$. In quantum mechanics the square root of variance is called the uncertainty in x and is denoted as Δx, so for the case under discussion,

$$a = \Delta x = \sqrt{\langle x^2 \rangle - \langle x \rangle^2}$$

the probability of finding the particle in the interval $x, x + dx$, then

$$(3.43) \qquad \langle x \rangle = \int_{-\infty}^{\infty} x P(x) \, dx$$

This is consistent with definition (3.35).

The probability density

$$P(x) = \frac{1}{a\sqrt{2\pi}} \exp \left[\frac{-(x - x_0)^2}{2a^2} \right]$$

is called the *Gaussian* or *normal distribution*, and a^2 is called the *variance* of x. It is a measure of the spread of $P(x)$ about the mean value

$$\langle x \rangle = x_0$$

As shown in Problem 3.10, the variance of x is the same as the mean-square deviation, $(\Delta x)^2$.

$$(\Delta x)^2 = \langle x^2 \rangle - \langle x \rangle^2 = a^2 + x_0^2 - x_0^2 = a^2$$

If it is known that a particle is in the state $\psi(x)$ at a given instant of time, and that in this state, $\langle x \rangle = x_0$, one may then ask: With what certainty will measurement of x find the value x_0? A measure of the relative *uncertainty* is given by the square root of the variance, Δx. If this value is large (compared to $\langle x \rangle$), one may say with little certainty that measurement will find the particle at x_0. If, on the other hand, Δx is small, one is more certain that measurement will find the particle at $x = x_0$. In quantum mechanics Δx is called the *uncertainty in x*, introduced previously in Section 2.7.

Next, we calculate the expectation of the momentum for a particle in the state ψ, (3.37).

$$(3.44) \qquad \langle p \rangle = \int_{-\infty}^{\infty} \psi^* \hat{p} \psi \, dx = \int_{-\infty}^{\infty} \psi^* \left(-i\hbar \frac{\partial}{\partial x} \right) \psi \, dx$$

$$= A^2 a \int_{-\infty}^{\infty} \left(p_0 + \frac{i\hbar}{2a} \eta \right) e^{-\eta^2/2} \, d\eta = p_0 \left(A^2 a \int_{-\infty}^{\infty} e^{-\eta^2/2} \, d\eta \right)$$

$$= p_0$$

It follows that the parameter p_0 which appears in the state function ψ is the average value of p. In any given measurement of p, any of a continuum of values can be obtained. Only in the event that ψ is an eigenfunction of $\hat{p}$ would measurement of p yield one definite value (i.e., the eigenvalue corresponding to the said eigenfunction).

PROBLEMS

3.10 For the state ψ, given by (3.37), show that

$$(\Delta x)^2 = a^2$$

Argue the consistency of this conclusion with the change in shape that $|\psi|^2$ suffers with a change in the parameter a.

3.11 Calculate the uncertainty Δp for a particle in the state ψ given by (3.37). Do you find your answer to be consistent with the uncertainty principle? (In this problem one must calculate $\langle \hat{p}^2 \rangle$. The operator $\hat{p}^2 = -\hbar^2 \, \partial^2/\partial x^2$.)

3.12 Let s be the number of spots shown by a die thrown at random.
    (a) Calculate $\langle s \rangle$.
    (b) Calculate Δs.

3.13 The number of hairs (N_l) on a certain rare species can only be the number 2^l ($l = 0, 1, 2, \ldots$). The probability of finding such an animal with 2^l hairs is $e^{-1}/l!$. What is the expectation, $\langle N \rangle$? What is ΔN?

3.4 TIME DEVELOPMENT OF THE STATE FUNCTION

Postulate IV

The fourth postulate of quantum mechanics specifies the time development of the state function $\psi(\mathbf{r}, t)$: the state function for a system (e.g., a single particle) develops in time according to the equation

(3.45)
$$i\hbar \frac{\partial}{\partial t} \psi(\mathbf{r}, t) = \hat{H} \psi(\mathbf{r}, t)$$

This equation is called the *time-dependent Schrödinger* equation.[1] The operator $\hat{H}$ is the Hamiltonian operator. For a single particle of mass m, in a potential field $V(\mathbf{r})$, it is given by (3.12). If $\hat{H}$ is assumed to be independent of time, we may write

(3.46)
$$\hat{H} = \hat{H}(\mathbf{r})$$

---

[1] A formulation of the Schrödinger equation that has its origin in the classical principle of least action has been offered by R. P. Feynman, *Rev. Mod. Phys.* **60**, 367 (1948). An elementary description of this derivation may be found in S. Borowitz, *Quantum Mechanics*, W. A. Benjamin, New York, 1967.

Under these circumstances, one is able to construct a solution to the time-dependent Schrödinger equation through the technique of separation of variables. We assume a solution of the form

(3.47)
$$\psi(\mathbf{r}, t) = \varphi(\mathbf{r})T(t)$$

Substitution into (3.45) gives

(3.48)
$$i\hbar \frac{T_t}{T} = \frac{\hat{H}\varphi}{\varphi}$$

The subscript t denotes differentiation with respect to t. Equation (3.48) is such that the left-hand side is a function of t only, while the right-hand side is a function of $\mathbf{r}$ only. Such an equation can be satisfied only if both sides are equal to the same constant, call it E (we do not yet know that E is the energy).

(3.49)
$$\hat{H}\varphi(\mathbf{r}) = E\varphi(\mathbf{r})$$

(3.50)
$$\left(\frac{\partial}{\partial t} + \frac{iE}{\hbar}\right)T(t) = 0$$

The first of these equations is the time-independent Schrödinger equation (3.13). This identification serves to label E, in (3.49), the energy of the system. That is, E, as it appears in this equation, is an eigenvalue of $\hat{H}$. But the eigenvalues of $\hat{H}$ are the allowed energies a system may assume, and we again conclude that E is the energy of the system.

The second equation (3.50) is simply solved to give the oscillating form

(3.51)
$$T(t) = A \exp\left(-\frac{iEt}{\hbar}\right)$$

Suppose that we solve the time-independent Schrödinger equation and obtain the eigenfunctions and eigenvalues

(3.52)
$$\hat{H}\varphi_n = E_n\varphi_n$$

For each such eigensolution, there is a corresponding eigensolution to the time-dependent Schrödinger equation

(3.53)
$$\psi_n(\mathbf{r}, t) = A\varphi_n(\mathbf{r}) \exp\left(-\frac{iE_n t}{\hbar}\right)$$

In equations (3.52) and (3.53) the index n denotes the set of integers $n = 1, 2, \ldots$. This notation is appropriate to the case where solution to the time-independent Schrödinger equation gives a discrete set of eigenfunctions, $\{\varphi_n\}$. Such is the case for problems that pertain to a finite system, such as a particle confined to a finite domain of space. We will encounter this property in Chapter 4 when we solve the problem of a bead constrained to move on a straight wire strung between two impenetrable walls.

In the one-dimensional free-particle case treated in Section 3.2, one obtains a continuum of eigenfunctions $\varphi_k(x)$ and, correspondingly, a continuum of eigenvalues, E_k. To repeat, these values are

$$(3.54) \qquad \hat{H}\varphi_k = E_k\varphi_k$$

$$(3.55) \qquad \varphi_k = A \exp{(ikx)}, \qquad E_k = \frac{\hbar^2 k^2}{2m}$$

For each such time-independent solution, there is a solution to the time-dependent Schrödinger equation

$$(3.56) \qquad \psi_k(x, t) = A e^{i(kx - \omega t)}$$

where we have labeled

$$(3.57) \qquad \hbar\omega = E_k$$

The structure of the solution (3.56) is characteristic of a propagating wave. More generally, any function of x and t of the form

$$(3.58) \qquad f(x, t) = f(x - vt)$$

represents a wave propagating in the positive x direction with velocity v. To see this, we note the following property of f:

$$(3.59) \qquad f(x + v\,\Delta t, t + \Delta t) = f(x, t)$$

At any given instant t, one may plot the x dependence of f (Fig. 3.3). If t increases to $t + \Delta t$, this curve is displaced to the right (as a rigid body) by the amount $v\,\Delta t$. We conclude from these arguments that the disturbance f (3.58) propagates with the wave speed v.

Now let us return to the free-particle eigenstate, (3.56), and rewrite it in the form

$$(3.60) \qquad \psi_k(x, t) = A \exp\left[ik\left(x - \frac{\omega}{k}t\right)\right]$$

Comparison with the waveform (3.58) indicates that (1) ψ_k is a propagating wave (moving to the right), and (2) the speed of this wave is

$$(3.61) \qquad v = \frac{\omega}{k} = \frac{\hbar\omega}{\hbar k} = \frac{p^2/2m}{p} = \frac{p}{2m} = \frac{v_{CL}}{2}$$

The velocity v_{CL} represents the classical velocity of a particle of mass m and momentum p. Thus we find that the wave speed of the state function of a particle with well-defined momentum, $p = \hbar k$, is half the classical speed, $v_{CL} = p/m$.

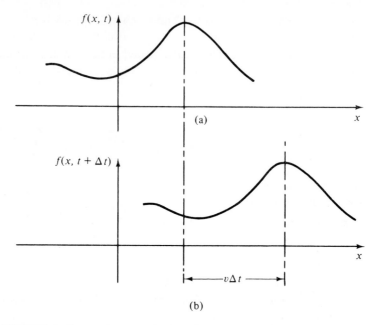

FIGURE 3.3   Propagating wave, $f(x, t) = f(x - vt)$: **(a)** at time t; **(b)** at time $t + \Delta t$.

This discrepancy is due to the following fact. Suppose that we calculate the probability density corresponding to the state given in (3.56). We obtain the result that it is uniformly probable to find the particle anywhere along the x axis. This is not a typical classical property of a particle. The state function that better represents a classical (localized) particle is a wave packet. The shape of such a function is sketched in Fig. 3.4. Such a state may be constructed as a sum of eigenstates of the form given in (3.56) (a Fourier series). The velocity with which the packet moves is called the *group velocity*,[1]

$$(3.62) \qquad\qquad v_g = \frac{\partial \omega}{\partial k}$$

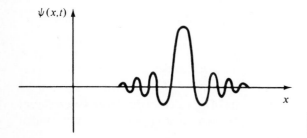

FIGURE 3.4   Wave packet at a given instant of time t.

[1] The concepts of phase and group velocities are returned to in Section 6.1.

For a wave packet composed of free-particle eigenstates, v_g takes the value

(3.63)
$$v_g = \frac{\partial \hbar \omega}{\hbar \, \partial k} = \frac{\partial(\hbar^2 k^2/2m)}{\hbar \, \partial k} = \frac{\hbar k}{m} = \frac{p}{m}$$

$$= v_{CL}$$

The value of k that enters the formula for v_g is the value about which there is a superabundance of ψ_k component waves. These topics will be more fully developed in Chapter 4. For the moment we are concerned only with the identification given in (3.63).

PROBLEMS

3.14 Describe the evolution in time of the following wavefunctions:

$$\psi_1 = A \sin \omega t \cos k(x + ct)$$
$$\psi_2 = A \sin (10^{-5} \, kx) \cos k(x - ct)$$
$$\psi_3 = A \cos k(x - ct) \sin [10^{-5} k(x - ct)]$$

3.15 What is the expectation of momentum $\langle p \rangle$ for a particle in the state

$$\psi(x, t) = A e^{-(x/a)^2} e^{-i\omega t} \sin kx?$$

3.5 SOLUTION TO THE INITIAL-VALUE PROBLEM IN QUANTUM MECHANICS

Functions of Operators

The time-dependent Schrödinger equation permits solution of the initial-value problem: given the initial value of the state function $\psi(\mathbf{r}, 0)$, determine $\psi(\mathbf{r}, t)$. We will formulate the solution to the problem for a time-independent Hamiltonian. The more general case is given as an exercise (Problem 3.18).

First we rewrite (3.45) in the form

(3.64)
$$\frac{\partial}{\partial t} \psi(\mathbf{r}, t) + \frac{i\hat{H}}{\hbar} \psi(\mathbf{r}, t) = 0$$

Next, we multiply this equation (from the left) by the integrating factor $\hat{U}^{-1}$

(3.65)
$$\hat{U}^{-1} = \exp\left(\frac{it\hat{H}}{\hbar}\right)$$

which is the inverse of

(3.66)
$$\hat{U} \equiv \exp\left(-\frac{it\hat{H}}{\hbar}\right)$$

This function of the operator, $\hat{H}$, is itself an operator. It is defined in terms of its Taylor series expansion.

(3.67)
$$\hat{U}^{-1} = \exp\left(\frac{it\hat{H}}{\hbar}\right) = 1 + \frac{it\hat{H}}{\hbar} + \frac{1}{2!}\left(\frac{it\hat{H}}{\hbar}\right)^2 + \cdots$$

More generally for any operator $\hat{A}$, the function operator $f(\hat{A})$ is defined in terms of a series in powers of $\hat{A}$. A few examples are provided in the problems.

Let us return to the problem under discussion. Multiplying the time-dependent Schrödinger equation through by the integrating factor (3.65), one obtains the equation

(3.68)
$$\frac{\partial}{\partial t}\left[\exp\left(\frac{it\hat{H}}{\hbar}\right)\psi(\mathbf{r}, t)\right] = 0$$

Integrating over the time interval $(0, t)$ gives

(3.69)
$$\exp\left(\frac{it\hat{H}}{\hbar}\right)\psi(\mathbf{r}, t) - \psi(\mathbf{r}, 0) = 0$$

Multiplying this equation through by $\hat{U}$ gives the desired result:

(3.70)
$$\psi(\mathbf{r}, t) = \exp\left(-\frac{it\hat{H}}{\hbar}\right)\psi(\mathbf{r}, 0) = \hat{U}\psi(\mathbf{r}, 0)$$

Here we have used the fact that

(3.71)
$$\hat{U}\hat{U}^{-1} = \exp\left(-\frac{it\hat{H}}{\hbar}\right)\exp\left(\frac{it\hat{H}}{\hbar}\right) = \hat{I}$$

where $\hat{I}$ is the identity operator.

Suppose that in solution (3.70) we choose the initial state to be an eigenstate of $\hat{H}$. Call it φ_n, so that

(3.72)
$$\psi_n(\mathbf{r}, 0) = \varphi_n(\mathbf{r})$$
$$\hat{H}\varphi_n = E_n\varphi_n$$

By virtue of the theorem presented in Problem 3.16,

$$\psi_n(\mathbf{r}, t) = \exp\left(-\frac{it\hat{H}}{\hbar}\right)\varphi_n = \exp\left(-\frac{iE_n t}{\hbar}\right)\varphi_n$$

(3.73)
$$= e^{-i\omega_n t}\varphi_n(\mathbf{r})$$

$$\hbar\omega_n = E_n$$

This is the solution of the time-dependent Schrödinger equation, derived in Section 3.4 by the technique of separation of variables. The solution given in (3.70) is more general. It exhibits the development of an arbitrary initial state $\psi(\mathbf{r}, 0)$ in time. It will be used extensively in the chapters to follow, where the student will gain a more workable understanding of the equation.

As a final topic of discussion in this chapter we note the following. Suppose that a system is in an eigenstate of the Hamiltonian at $t = 0$, described by (3.72). At this (initial) time the expectation of an observable A is

(3.74)
$$\langle A \rangle_{t=0} = \int \psi^*(\mathbf{r}, 0)\hat{A}\psi(\mathbf{r}, 0)\, d\mathbf{r} = \int \varphi_n{}^* \hat{A}\varphi_n\, d\mathbf{r}$$

What is $\langle A \rangle$ at a later time, $t > 0$? The state of the system at $t > 0$ is given by (3.73):

(3.75)
$$\psi_n(\mathbf{r}, t) = e^{-i\omega_n t}\varphi_n(\mathbf{r})$$

so that at $t > 0$ (assuming that $\partial\hat{A}/\partial t = 0$),

$$\langle A \rangle_t = \int \psi^*(\mathbf{r}, t)\hat{A}\psi(\mathbf{r}, t)\, d\mathbf{r} = e^{+i\omega_n t}e^{-i\omega_n t}\int \varphi_n{}^* \hat{A}\varphi_n\, d\mathbf{r}$$

(3.76)
$$= \int \varphi_n{}^* \hat{A}\varphi_n\, d\mathbf{r} = \langle A \rangle_{t=0}$$

$$\boxed{\langle A \rangle_{t>0} = \langle A \rangle_{t=0} \qquad \text{in a stationary state}}$$

The expectation of *any* observable is constant in time, if at any instant in time the system is in an eigenstate of the Hamiltonian. For this reason eigenstates of the Hamiltonian are called *stationary states*.

(3.77)
$$\boxed{\psi_n(\mathbf{r}, t) = e^{-i\omega_n t}\varphi_n(\mathbf{r}) \qquad \text{a stationary state}}$$

In the first three sections of this chapter we encountered functions relevant to a system which are eigenfunctions of operators corresponding to observable properties of that same system. In what sense are these eigenfunctions related to the *state function* of the system? From postulate II we know that ideal measurement of A leaves the system in the eigenstate of $\hat{A}$ corresponding to the value of A that was found in measurement. Thus, the state function of the system immediately after measurement is this same eigenstate of $\hat{A}$. The state function then evolves in time according to (3.70).

PROBLEMS

3.16   Let the eigenfunctions and eigenvalues of an operator $\hat{A}$ be $\{\varphi_n\}$ and $\{a_n\}$, respectively, so that

$$\hat{A}\varphi_n = a_n\varphi_n$$

Let the function $f(x)$ have the expansion

$$f(x) = \sum_{l=0}^{\infty} b_l x^l$$

Show that φ_n is an eigenfunction of $f(\hat{A})$ with eigenvalue $f(a_n)$. That is,

$$f(\hat{A})\varphi_n = f(a_n)\varphi_n$$

3.17   If $\hat{p}$ is the momentum operator in the x direction, and $f(x)$ is an arbitrary "well-behaved" function, show that

$$\exp\left(\frac{i\zeta\hat{p}}{\hbar}\right) f(x) = f(x + \zeta)$$

The constant ζ represents a small displacement. In this problem the student must demonstrate that the left-hand side of the equation above is the Taylor series expansion of the right-hand side about $\zeta = 0$.

3.18   If $\hat{H}$ is an explicit function of time, show that the solution to the initial-value problem (by direct differentiation) is

$$\psi(\mathbf{r}, t) = \exp\left[-\frac{i}{\hbar}\int_0^t dt'\hat{H}(t')\right]\psi(\mathbf{r}, 0)$$

You may assume that $\hat{H}(t)\hat{H}(t') = \hat{H}(t')\hat{H}(t)$.

3.19.   What is the effect of operating on an arbitrary function $f(x)$ with the following two operators?

(a) $\hat{O}_1 \equiv (\partial^2/\partial x^2) - 1 + \sin^2(\partial^3/\partial x^3) + \cos^2(\partial^3/\partial x^3)$.

(b) $\hat{O}_2 \equiv \cos(2\partial/\partial x) + 2\sin^2(\partial/\partial x) + \int_a^b dx$.

3.20 (a) The time-dependent Schrödinger equation is of the form

$$a \frac{\partial \psi}{\partial t} = \hat{H} \psi$$

Consider that a is an unspecified constant. Show that this equation has the following property. Let $\hat{H}$ be the Hamiltonian of a system composed of two independent parts, so that

$$\hat{H}(x_1, x_2) = \hat{H}_1(x_1) + \hat{H}_2(x_2)$$

and let the stationary states of system 1 be $\psi_1(x_1, t)$ and those of system 2 be $\psi_2(x_2, t)$. Then the stationary states of the composite system are

$$\psi(x_1, x_2) = \psi_1(x_1, t)\psi_2(x_2, t)$$

That is, show that this product form is a solution to the preceding equation for the given composite Hamiltonian.

Such a system might be two beads that are invisible to each other and move on the same straight wire. The coordinate of bead 1 is x_1 and the coordinate of bead 2 is x_2.

(b) Show that this property is not obeyed by a wave equation that is second order in time, such as

$$a^2 \frac{\partial^2 \psi}{\partial t^2} = \hat{H} \psi$$

(c) Arguing from the Born postulate, show that the wavefunction for a system composed of two independent components must be in the preceding product form, thereby disqualifying the wave equation in part (b) as a valid equation of motion for the wavefunction ψ.

Answer (partial)

(c) If the two components are independent of each other, the joint probability density describing the state of the system is given by

$$P_{12} = P_1 P_2$$

This, in turn, guarantees that the probability density associated with component 1,

$$P_1(x_1) = \int P_{12}(x_1, x_2)\, dx_2$$

is independent of the form of $P_2(x_2)$ (and vice versa). The product form for P_{12} is guaranteed by the product structure for the wavefunction $\psi(x_1, x_2)$.

3.21 It is established in Problem 3.20 that for the joint probability for two independent systems to be consistently described by the time-dependent Schrödinger equation, this equation must be of the form

$$a \frac{\partial \psi}{\partial t} = \hat{H} \psi$$

where a is some number. Show that for this equation to imply wave motion, a must be complex. You may assume that $\hat{H}$ has only real eigenvalues.

Answer

Following development of the general solution (3.70), we find that the given equation implies the solution

$$\psi(\mathbf{r}, t) = \exp\left(\frac{t\hat{H}}{a}\right)\psi(\mathbf{r}, 0)$$

Since $\hat{H}$ has only real eigenvalues, the time dependence of $\psi(\mathbf{r}, t)$ is nonoscillating. It modulates $\psi(\mathbf{r}, 0)$ in time and does not give propagation. Thus, if a is real, ψ cannot represent a propagating wave. (*Note:* The fact that a is complex implies that ψ is complex. These last two problems illustrate the necessity of complex wavefunctions in quantum mechanics.)

3.22 Consider the wavefunction

$$\psi = Ae^{i(kx + \omega t)}$$

where k is real and $\omega > 0$ and is real. Is this wavefunction an admissible quantum state for a free particle? Justify your answer. If your answer is no, in what manner would you change the given function to describe a free particle moving in the $-x$ direction?

CHAPTER 4

PREPARATORY CONCEPTS. FUNCTION SPACES AND HERMITIAN OPERATORS

In this and the following two chapters, we continue development of physical principles and mathematical groundwork important to quantum mechanical descriptions. Included in the present chapter are the notions of Hilbert space and Hermitian operators. First, we obtain wavefunctions relevant to a particle in a one-dimensional box. These, together with previously derived free-particle wavefunctions, then serve as simply understood references for subsequent descriptions of Hilbert space and Hermitian operators.

4.1 PARTICLE IN A BOX AND FURTHER REMARKS ON NORMALIZATION

In chapter 3 we solved the quantum mechanical free-particle problem. We recall that the free-particle Hamiltonian generates a continuous spectrum of eigenvalues, $\hbar^2 k^2/2m$, and eigenfunctions, $\varphi_k = A \exp(ikx)$, as given in (3.55).

The second one-dimensional problem we wish to treat is that of a point mass m, constrained to move on an infinitely thin, frictionless wire which is strung tightly between two impenetrable walls a distance L apart (see Fig. 4.1). The corresponding

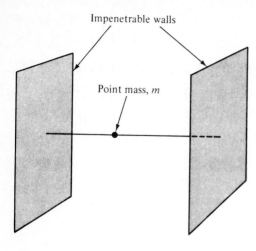

Impenetrable walls

Point mass, m

FIGURE 4.1   One-dimensional "box."

potential has the values

(4.1)
$$V(x) = \infty \qquad (x \le 0, \quad x \ge L) \qquad \text{(domain 1)}$$
$$V(x) = 0 \qquad (0 < x < L) \qquad \text{(domain 2)}$$

and is depicted in Fig. 4.2. This configuration is known as the *one-dimensional box.*[1]

The Hamiltonian for this problem is the following operator:

(4.2)
$$\hat{H}_1 = \frac{\hat{p}^2}{2m} + \infty = \infty \qquad (x \le 0, \quad x \ge L) \qquad \text{(domain 1)}$$

(4.3)
$$\hat{H}_2 = \frac{\hat{p}^2}{2m} \qquad (0 < x < L) \qquad \text{(domain 2)}$$

In domain 1 the time-independent Schrödinger equation gives $\varphi = 0$. For any finite eigenenergy E, in this domain the time-independent Schrödinger equation reads

(4.4)
$$\hat{H}_1 \varphi = E\varphi$$

Since φ and E are finite, the right-hand side is finite. Therefore, the left-hand side is finite and φ must vanish in this domain.

The fact that $\varphi = 0$ in domain 1 implies that there is zero probability that the particle is found there ($|\varphi|^2 = 0$). This is in agreement with the discussion in Chapter 1 on "forbidden domains." These, we recall, are domains where $E < V$. Certainly, this is the case in domain 1 for any finite energy E.

---

[1] A mathematically more accurate description of the one-dimensional box is: an infinitesimally thin, flat sheet of infinite extent and finite mass which moves between two walls of infinite extent. The two walls and sheet are all parallel and the velocity of the sheet is normal to the walls. Every point in space is then characterized by one coordinate, the normal displacement of the sheet from either of the walls.

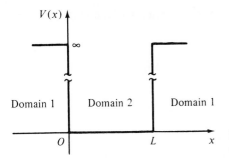

FIGURE 4.2   Potential corresponding to the one-dimensional box.

In domain 2 the time-independent Schrödinger equation is

(4.5)
$$-\frac{\hbar^2}{2m}\frac{\partial^2}{\partial x^2}\varphi_n = E_n \varphi_n$$

The subscript n is in anticipation of a discrete spectrum of energies E_n and eigenfunctions φ_n.

Since φ_n is a continuous function, it must have the values

(4.6)
$$\varphi_n(0) = \varphi_n(L) = 0$$

First we rewrite (4.5) in the form

(4.7)
$$\frac{\partial^2 \varphi_n}{\partial x^2} + k_n{}^2 \varphi_n = 0$$

(4.8)
$$k_n{}^2 = \frac{2mE_n}{\hbar^2}$$

This is merely a change of variables from energy E_n to wavenumber k_n. The solution to (4.7) appears as

(4.9)
$$\varphi_n = A \sin k_n x + B \cos k_n x$$

The boundary conditions (4.6) give

(4.10)
$$B = 0$$

(4.11)
$$A \sin k_n L = 0$$

The second of these equations serves to determine the eigenvalues k_n.

(4.12)
$$k_n L = n\pi, \qquad n = 0, 1, 2, \ldots$$

This is seen to be equivalent to the requirement that an integral number of half-wavelengths, $n(\lambda/2)$, fit into the width L.

The spectrum of eigenvalues and eigenfunctions is discrete. To find the constant A in (4.11), we normalize φ_n.

(4.13)
$$\int_0^L \varphi_n^2 \, dx = 1 = A^2 \int_0^L \sin^2\left(\frac{n\pi x}{L}\right) dx$$

$$1 = \frac{A^2 L}{n\pi} \int_0^{n\pi} \sin^2 \theta \, d\theta = \frac{A^2 L}{2}$$

The dummy variable $\theta = n\pi x/L$.

It follows that the eigenenergies E_n and normalized eigenfunctions φ_n for the one-dimensional box problem are

(4.14)
$$\boxed{E_n = n^2 E_1 \qquad E_1 = \frac{\hbar^2 k_1^2}{2m} = \frac{\hbar^2 \pi^2}{2mL^2}}$$

(4.15)
$$\boxed{\varphi_n = \sqrt{\frac{2}{L}} \sin\left(\frac{n\pi x}{L}\right)}$$

The eigenstate corresponding to $n = 0$ is $\varphi = 0$. This, together with the solution in domain 1, gives $\varphi = 0$ over the whole x axis. There is zero probability of finding the particle anywhere. This is equivalent to the statement that the particle does not exist in the $n = 0$ state. Another argument that disallows the $n = 0$ state follows from the uncertainty principle. The energy corresponding to $n = 0$ is $E = 0$. Since the energy in domain 2 is entirely kinetic, this, in turn, implies that the particle is in a state of absolute rest ($\Delta p = 0$), an illegitimate state of affairs for a particle constrained to move in a finite domain.

The eigenenergies and eigenfunctions given by (4.14) and (4.15), together with the corresponding probability densities $|\varphi_n|^2$, are sketched in Fig. 4.3.

The Arbitrary Phase Factor

In concluding this section we note the following important fact. As described in Section 3.3, the wavefunction ψ gives information about a system through calculation of averages of observable properties of that system, according to the rule

$$\langle C \rangle = \int \psi^* \hat{C} \psi \, dx$$

This equation, as well as the normalization condition

$$\int \psi^* \psi \, dx = 1$$

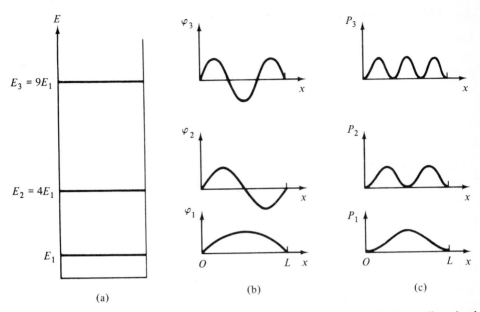

FIGURE 4.3   (a) Eigenenergies for the one-dimensional box problem. (b) Eigenstates for the one-dimensional box problem:

$$\varphi_n = \sqrt{\frac{2}{L}} \sin\left(\frac{n\pi x}{L}\right)$$

(c) Probability densities for the one-dimensional box problem:

$$P_n \equiv |\varphi_n|^2 = \frac{2}{L} \sin^2\left(\frac{n\pi x}{L}\right)$$

are invariant under the transformation $\psi \rightarrow e^{i\alpha}\psi$, where α is any real number. That is, a wavefunction is determined only to within a constant *phase factor* of the form $e^{i\alpha}$. Although associated with all wavefunctions, this arbitrary quality has no effect upon any physical results.[1]

PROBLEMS

4.1   What are the energy eigenfunctions and eigenvalues for the one-dimensional box problem described above if the ends of the box are at $-L/2$ and $+L/2$? [Check your answer with (6.100).]

4.2   For what values of the real angle θ will the constant $C = \frac{1}{2}(e^{i\theta} - 1)$ have no effect in calculations involving the modulus $|C\psi|$?

---

[1] On the other hand, component phase factors for a composite wavefunction such as that discussed in Section 2.5 do contribute to measurable effects, such as interference.

4.2 THE BOHR CORRESPONDENCE PRINCIPLE

Let us now consider the *classical* motion of a particle in a one-dimensional box. As described previously, this configuration is effected by a bead sliding with no friction on a taut wire strung between two impenetrable walls a distance L apart. If the particle is given a velocity v, its motion (between collisions with the wall) is

$$x = x_0 + vt$$

Now suppose that the initial position x_0 is completely unknown. What is the probability $P\,dx$ of finding the particle in the interval x, $x + dx$, at a subsequent time? The answer is: the fraction of time dt/T it spends in this interval.

(4.16)
$$P\,dx = \frac{dt}{T} = \frac{v\,dt}{L} = \frac{dx}{L}$$

so that

(4.17)
$$P = \frac{1}{L} = \text{constant}$$

It is uniformly probable to find the particle at any position on the wire. If we make a large number of replicas of this one-dimensional system, measurement (at random times) of the coordinate x of the bead will find all values ($0 \le x \le L$) occurring equally often (Fig. 4.4).

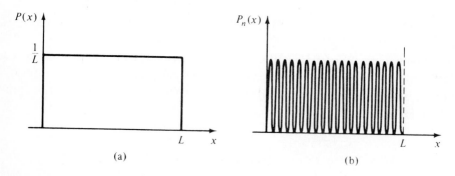

(a)
(b)

FIGURE 4.4 (a) Classical probability density for the one-dimensional box. (b) Quantum mechanical probability density

$$P_n = |\varphi_n|^2$$

for the one-dimensional box problem, for the case $n \gg 1$. The probability P_n vanishes $n + 1$ times in the interval $(0, L)$.

On the other hand, in the quantum mechanical case, if the particle is in the state φ_3, say, the probability density P is peaked at $x = (L/6, L/2, 5L/6)$; see Fig. 4.3. In this case, measurement on an abundant number of replica systems finds the particle spending much of its time in the neighborhood of these three values of x. This situation is quite different from the classical case described above. Suppose we move to higher quantum states. At what values of x is the probability density P peaked? The solution is left to Problem 4.3, where one obtains that $|\varphi_n|^2$ is peaked at the values

(4.18)
$$x_j = \frac{2j + 1}{2n} L, \qquad j = 0, 1, 2, \ldots$$

As n becomes very large, the probability density oscillates with so large a frequency that it begins to assume a uniform quality. For any n, one can divide the interval $(0, L)$ into n strips of equal width Δx such that in each strip the probability $|\varphi_n|^2 \, \Delta x$ of finding the particle is equal. For the classical case the number of such strips is infinite. In the quantum mechanical case, the same situation is approached in the limit $n \to \infty$.

One encounters this transition to classical physics from the quantum mechanical domain in many problems. Bohr was the first to analyze this transition and offered the general rule that a quantum mechanical result must reduce to its classical counterpart in the classical domain. Since classical formulas do not contain $\hbar$, such a transition should be realized in the limit that $\hbar$ becomes small. For many problems this limit is attained in passing to high quantum numbers ($n \to \infty$). This rule is called the *Bohr correspondence principle*.

Classical physics includes the dynamics of macroscopic bodies. An aggregate of particles (e.g., a gas) obeys classical laws when the de Broglie wavelength, λ, of a typical particle is small compared to all relevant lengths. For example, if the density of particles (number/cm^3) is n, the gas obeys classical statistics if $\lambda \ll n^{-1/3}$ (the mean distance between particles is $n^{-1/3}$). In the classical limit, fluctuations about the average become small and the probabilities indigenous to quantum mechanics reduce to certainties.

A rule of thumb in this area is that any quantum mechanical result that does not contain $\hbar$ is in essence a classical result. The first (fortuitous) example of this rule was Rutherford's classical calculation of the Coulomb cross section, relevant to the scattering of charged particles. The correct quantum mechanical calculation of this parameter is found not to contain $\hbar$. Rutherford's classical calculation yields the same result.

More examples of the correspondence principle will arise in the course of development of the text. Coulomb scattering is further described in Section 14.4.

4.3 For the one-dimensional box problem, show that $P = |\varphi_n|^2$ is maximum at the values $x = x_j$ given by

$$x_j = \frac{2j + 1}{2n} L, \qquad j = 0, 1, 2, \ldots, n - 1$$

4.3 DIRAC NOTATION

In this section we introduce a notation that proves to be an invaluable tool in calculation, called the *Dirac notation*. It gives a monogram to the integral of the product of two state functions, $\psi(x)$ and $\varphi(x)$, which appears as

(4.19)
$$\langle \psi | \varphi \rangle = \int_{-\infty}^{\infty} \psi^*(x)\varphi(x)\, dx$$

In Dirac notation, the integral on the right is written in the form shown on the left.

More generally, the integral operation $\langle \psi | \varphi \rangle$ denotes: (1) take the complex conjugate of the object in the first slot ($\psi \to \psi^*$) and then, (2) integrate the product ($\psi^*\varphi$). This operation has the following simple properties. If a is any complex number and the functions ψ and φ are such that

(4.20)
$$\int_{-\infty}^{\infty} \psi^*\varphi\, dx < \infty$$

the following rules hold:

(4.21)
$$\langle \psi | a\varphi \rangle = a\langle \psi | \varphi \rangle$$

(4.22)
$$\langle a\psi | \varphi \rangle = a^*\langle \psi | \varphi \rangle$$

(4.23)
$$\langle \psi | \varphi \rangle^* = \langle \varphi | \psi \rangle$$

(4.24)
$$\langle \varphi + \psi | = \langle \psi | + \langle \varphi |$$

(4.25)
$$\int (\psi_1 + \psi_2)^*(\varphi_1 + \varphi_2)\, dx$$

$$= \langle \psi_1 + \psi_2 | \varphi_1 + \varphi_2 \rangle = (\langle \psi_1 | + \langle \psi_2 |)(|\varphi_1\rangle + |\varphi_2\rangle)$$

$$= \langle \psi_1 | \varphi_1 \rangle + \langle \psi_1 | \varphi_2 \rangle + \langle \psi_2 | \varphi_1 \rangle + \langle \psi_2 | \varphi_2 \rangle$$

The object $\langle \psi |$ (called a "bra vector"). It joins in a product form with a ("ket vector")$|\varphi\rangle$, to form the "bra-ket," $\langle \psi | \varphi \rangle$.

A more fundamental description of Dirac notation in quantum mechanics is given in Appendix A.

4.4 Write the following equations for the state vectors f, g, and so on, in Dirac notation.

(a) $f(x) = g(x)$.

(b) $c = \int g^*(x')h(x')\,dx'$.

(c) $f(x) = \sum_n \varphi_n(x) \int \varphi_n^*(x')f(x')\,dx'$.

(d) $\hat{O} \equiv \psi(x) \int dx'\varphi^*(x')$.

(e) $\dfrac{\partial}{\partial x} f(x) = h(x) \int h^*(x')g(x')\,dx'$.

4.5 Consider the operator $\hat{O} = |\varphi\rangle\langle\psi|$ and the arbitrary state function $f(x)$. Describe the following forms.

(a) $\langle f|\hat{O}$.
(b) $\hat{O}|f\rangle$.
(c) $\langle f|\hat{O}|f\rangle$.
(d) $\langle f|\hat{O}|\psi\rangle$.

Answer (partial)

(a) $\langle f|\hat{O}$ is the bra vector $C\langle\psi|$, where the constant $C \equiv \langle f|\varphi\rangle = \int_{-\infty}^{\infty} f^*\varphi\,dx$.

4.4 HILBERT SPACE

In this section we introduce the concept of a space of functions. Specifically we will deal with a Hilbert space. This serves the purpose of giving a geometrical quality to some of the abstract concepts of quantum mechanics.

We recall that in Cartesian 3-space a vector **V** is a set of three numbers, called components (V_x, V_y, V_z). Any vector in this space can be expanded in terms of the three unit vectors $\mathbf{e}_x, \mathbf{e}_y, \mathbf{e}_z$ (Fig. 4.5). Under such conditions one terms the triad $\mathbf{e}_x, \mathbf{e}_y, \mathbf{e}_z$, a *basis*.

(4.26) $$\mathbf{V} = \mathbf{e}_x V_x + \mathbf{e}_y V_y + \mathbf{e}_z V_z$$

The vectors $\mathbf{e}_x, \mathbf{e}_y, \mathbf{e}_z$ are said to *span* the vector space.

The inner ("dot") product of two vectors (**U** and **V**) in the space is defined as

(4.27) $$\mathbf{V} \cdot \mathbf{U} = V_x U_x + V_y U_y + V_z U_z$$

The length of the vector **V** is $\sqrt{\mathbf{V} \cdot \mathbf{V}}$.

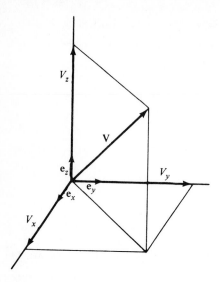

FIGURE 4.5   Vector V in Cartesian 3-space and its components (V_x, V_y, V_z). The orthogonal triad (e_x, e_y, e_z) spans the space.

A Hilbert space is much the same type of object. Its elements are functions instead of three-dimensional vectors. The similarity is so close that the functions are sometimes called vectors. A Hilbert space $\mathfrak{H}$ has the following properties.

1. The space is linear. A function space is linear under the following two conditions: (a) If a is a constant and φ is any element of the space, then $a\varphi$ is also an element of the space. (b) If φ and ψ are any two elements of the space, then $\varphi + \psi$ is also an element of the space.

2. There is an *inner product*, $\langle \psi | \varphi \rangle$, for any two elements in the space. For functions defined in the interval $a \le x \le b$ (in one dimension), we may take

   (4.28)
   $$\langle \varphi | \psi \rangle = \int_a^b \varphi^* \psi \, dx$$

3. Any element of $\mathfrak{H}$ has a norm ("length") that is related to the inner product as follows:

   (4.29)
   $$(\text{norm of } \varphi)^2 = \|\varphi\|^2 = \langle \varphi | \varphi \rangle$$

4. $\mathfrak{H}$ is complete. Every Cauchy sequence of functions in $\mathfrak{H}$ converges to an element of $\mathfrak{H}$. A Cauchy sequence $\{\varphi_n\}$ is such that $\|\varphi_n - \varphi_l\| \to 0$ as n and l approach infinity. (See Problem 4.24.) Loosely speaking, a Hilbert space contains all its limit points.

An example of a Hilbert space is given by the set of functions defined on the interval $(0 \le x \le L)$ with finite norm

(4.30)
$$\|\varphi\|^2 = \int_0^L \varphi^*\varphi \, dx < \infty \qquad \mathfrak{H}_1$$

Another example is the space of functions commonly referred to by mathematicians as "L^2 space." This is the set of square-integrable functions defined on the whole x interval.

(4.31)
$$\|\varphi\|^2 = \int_{-\infty}^{\infty} \varphi^*\varphi \, dx < \infty \qquad \mathfrak{H}_2$$

Let us see how the preceding concept of inner product (4.28) is similar to the definition of the inner product between two finite-dimensional vectors (4.27). To see this we interpret the function $\varphi(x)$ as a vector with infinitely many components. These components are the values that φ assumes at each distinct value of its independent variable x. Just as the inner product between $\mathbf{U}$ and $\mathbf{V}$ is a sum over the products of parallel components, so is the inner product between φ and ψ a sum over parallel components. This sum is nothing but the integral of the product of φ and ψ. The reason we complex-conjugate the first "vector" is to ensure that the "length" (square root of the inner product between a "vector" φ and itself) of a vector φ is real.

Thus we see that Hilbert space is closely akin to a vector space. Mathematicians[1] call it that—an infinite-dimensional vector space (also: a complete, normed, linear vector space). Elements of this space have length and one can form an inner product between any two elements. The vector quality of Hilbert space can be pushed a bit further. We recall that if two vectors $\mathbf{U}$ and $\mathbf{V}$ in three-dimensional vector space are orthogonal to each other, their inner product vanishes. In a similar vein two vectors in Hilbert space, φ and ψ, are said to be orthogonal if

(4.32)
$$\langle \varphi | \psi \rangle = 0$$

Furthermore, we recall that the three unit vectors $\mathbf{e}_x$, $\mathbf{e}_y$, and $\mathbf{e}_z$ "span" 3-space. Similarly, there is a set of vectors that "spans" Hilbert space. For instance, the Hilbert space whose elements all have the property given by (4.30) is spanned by the sequence of functions $\{\varphi_n\}$, which are the eigenfunctions of the Hamiltonian relevant

[1] A more mathematically accurate presentation of function spaces may be found in C. Goffman and G. Pedrick, *First Course in Functional Analysis*, Prentice-Hall, Englewood Cliffs, N.J., 1965. Another book in this area, but more directly related to quantum mechanics, is T. F. Jordan, *Linear Operators for Quantum Mechanics*, Wiley, New York, 1969.

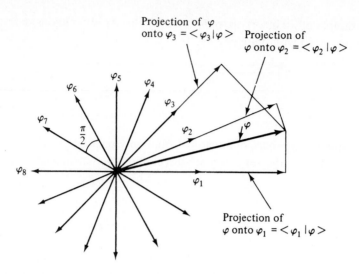

FIGURE 4.6 Projection of φ onto an orthonormal set of eigenfunctions in Hilbert space.

to the one-dimensional box problem (4.15). This means that any function φ in this Hilbert space may be expanded in a series of the sequence $\{\varphi_n\}$.

$$(4.33) \qquad \varphi(x) = \sum_{n=1}^{\infty} a_n \varphi_n(x)$$

The geometrical interpretation of this relation is depicted in Fig. 4.6. The coefficient a_n is the projection of φ onto the vector φ_n. To see this, first we state a fact to be illustrated in the next section. The *basis* vectors $\{\varphi_n\}$ comprise an orthogonal set. That is,

$$(4.34) \qquad \langle \varphi_n | \varphi_{n'} \rangle = 0 \qquad (n \neq n')$$

Furthermore, φ_n is a unit vector; that is, it has unit "length"

$$(4.35) \qquad \langle \varphi_n | \varphi_n \rangle = \| \varphi_n \|^2 = 1$$

These latter two statements may be combined into the single equation

$$(4.36) \qquad \langle \varphi_n | \varphi_{n'} \rangle = \delta_{n, n'}$$

The symbol $\delta_{n, n'}$ is called the *Kronecker delta* and is defined by

$$(4.37) \qquad \delta_{n, n'} = 0 \quad \text{for } n \neq n', \qquad \delta_{n, n'} = 1 \quad \text{for } n = n'$$

Any sequence of functions that obeys (4.36) is called an *orthonormal set*.

To show that a_n is the projection of φ into φ_n, we first rewrite (4.33) in Dirac notation.

$$(4.38) \qquad |\varphi\rangle = \sum_n |a_n \varphi_n\rangle$$

Then we multiply from the left by $\langle \varphi_{n'}|$ and use the relation (4.36).

$$\langle \varphi_{n'}|\varphi\rangle = \sum_n \langle \varphi_{n'}|a_n \varphi_n\rangle$$

$$(4.39) \qquad = \sum_n a_n \langle \varphi_{n'}|\varphi_n\rangle = \sum_n a_n \delta_{n,\,n'} = a_{n'}$$

$$a_{n'} = \langle \varphi_{n'}|\varphi\rangle$$

The coefficient $a_{n'}$ is the inner product between the basis vector $\varphi_{n'}$ and the vector φ. Since $\varphi_{n'}$ is a "unit" vector, $a_{n'}$ is the projection of φ onto $\varphi_{n'}$ (Fig. 4.6). The student should recognize (4.33) to be a discrete Fourier series representation of φ, in terms of the trigonometric sequence (4.15).

Delta-Function Orthogonality

We will continue with the use of the labels $\mathfrak{H}_1$ and $\mathfrak{H}_2$ to denote the two Hilbert spaces defined by (4.30) and (4.31), respectively. As stated previously, the sequence $\{\varphi_n\}$ given by (4.15) "spans" $\mathfrak{H}_1$. The sequence $\{\varphi_n\}$ is a basis of $\mathfrak{H}_1$. What are the vectors which span $\mathfrak{H}_2$? The answer is: the eigenfunctions of the momentum operator $\hat{p}$,

$$(4.40) \qquad \varphi_k(x) = \frac{1}{\sqrt{2\pi}} e^{ikx}$$

Let us see if this (continuous) set of functions is an orthogonal set. Toward these ends we form the inner product

$$(4.41) \qquad \langle \varphi_k|\varphi_{k'}\rangle = \frac{1}{2\pi} \int_{-\infty}^{\infty} e^{ix(k'-k)}\, dx = \delta(k'-k)$$

It follows that the inner product between any two distinct eigenvectors of the operator $\hat{p}$ vanishes.

Any function in $\mathfrak{H}_2$ may be expanded in terms of the eigenvectors $\{\varphi_k\}$. Since this sequence comprises a continuous set, the expansion is not a discrete sum as in (4.33), but an integral. If $\varphi(x)$ is any element of $\mathfrak{H}_2$, then since $\{\varphi_k\}$ spans this space, one may write

$$(4.42) \qquad \varphi(x) = \int_{-\infty}^{\infty} b(k)\varphi_k(x)\, dk$$

This is the Fourier integral representation of $\varphi(x)$. Again, the coefficient of expansion $b(k)$ is the projection of $\varphi(x)$ onto φ_k. To exhibit this fact, we first rewrite the last integral in the form

$$(4.43) \qquad |\varphi\rangle = \int_{-\infty}^{\infty} dk \,|b(k)\varphi_k\rangle$$

Again, if this equation is compared to (4.38), we see how the sum over discrete a_n values is replaced by an integration over the continuum of $b(k)$ values. If we now multiply (4.43) from the left with $\langle\varphi_{k'}|$, there results

$$(4.44) \qquad \langle\varphi_{k'}|\varphi\rangle = \int_{-\infty}^{\infty} dk\langle\varphi_{k'}|b(k)\varphi_k\rangle = \int_{-\infty}^{\infty} dk b(k)\langle\varphi_{k'}|\varphi_k\rangle$$

$$= \int_{-\infty}^{\infty} dk b(k)\delta(k'-k) = b(k')$$

The coefficient of expansion $b(k')$ is the inner product between $\varphi_{k'}$ and φ, hence it may be termed a projection of φ onto $\varphi_{k'}$. But $\varphi_{k'}$ does not appear to be a "unit" vector. Indeed, the vector φ_k is infinitely long.

$$(4.45) \qquad \|\varphi_k\|^2 = \langle\varphi_k|\varphi_k\rangle = \delta(0) = \frac{1}{2\pi}\int_{-\infty}^{\infty} dx = \infty$$

Although this disqualifies the set $\{\varphi_k\}$ for membership in $\mathfrak{H}_2$, they nevertheless span the space. They comprise a valid set of basis vectors and the projection of any function in $\mathfrak{H}_2$ onto any member of the basis $\{\varphi_k\}$ gives a finite result. If φ is any function in $\mathfrak{H}_2$, then

$$(4.46) \qquad \langle\varphi_k|\varphi\rangle < \infty$$

The functions $\{\varphi_k\}$ may, through proper renormalization, be cast in a form which allows them to be members of $\mathfrak{H}_2$. (See Problem 4.6.)

PROBLEMS

4.6 Consider the functions

$$\varphi_k = \frac{1}{\sqrt{L}}e^{ikx}$$

defined over the interval $(-L/2, +L/2)$.

 (a) Show that these functions are all normalized to unity and maintain this normalization in the limit $L \to \infty$.

 (b) Show that these functions comprise an orthogonal set in the limit $L \to \infty$.

4.7 State to which space each of the functions listed belongs, $\mathfrak{H}_1$ or $\mathfrak{H}_2$.

(a) $f_1 = (x^5 - x^4 - Lx^4 + Lx^3)/(x - 2L)$

(b) $f_2 = (\sin x)e^{-x^2}$

(c) $f_3 = \sqrt{\ln[x(x - L) + 1]}$

(d) $f_4 = \sin 2n\pi[x(x - L) + 1]$, $\quad n = 0, 1, 2, \ldots$

(e) $f_5 = e^{i\alpha x}(x^2 + a^2)^{-1}$

(f) $f_6 = x^{10}e^{-x^2}$

(g) $f_7 = 1/\sin kx$

4.8 The function

$$g(x) = x(x - L)e^{ikx}$$

is in $\mathfrak{H}_1$. Calculate the coefficients of expansion, a_n, of this function, in the series representation (4.33), in terms of the constants L and k. Use the basis functions (4.15).

4.9 Two vectors ψ and φ in a Hilbert space are orthogonal. Show that their lengths obey the Pythagorean theorem,

$$\|\psi + \varphi\|^2 = \|\psi\|^2 + \|\varphi\|^2$$

4.10 Consider a free particle moving in one dimension. The state functions for this particle are all elements of $\mathfrak{H}_2$. Show that the expectation of the momentum $\langle p_x \rangle$ vanishes in any state that is purely real ($\psi^* = \psi$). Does this property hold for $\langle H \rangle$? Does it hold for $\langle x \rangle$?

4.5 HERMITIAN OPERATORS

The average of an observable A for a system in the state $\psi(x, t)$ is given by (3.32). In Dirac notation this equation appears as (in one dimension)

$$(4.47) \qquad \langle A \rangle = \int \psi^*(x, t)\hat{A}\psi(x, t)\, dx = \langle \psi | \hat{A}\psi \rangle$$

Since t is a fixed parameter in this equation, we may conclude that the formula gives the expectation of A at the time t. Now one may ask: What are the possible state functions for a particle moving in one dimension at a given instant of time? The answer is: any function in $\mathfrak{H}_2$. For example, the particle could be in any of the following states at some specified time:

$$(4.48) \qquad \psi_1 = Be^{-x^2/a^2}, \qquad \psi_2 = \frac{Ce^{ikx}}{x}, \qquad \psi_3 = \frac{iD}{\sqrt{x^2 + a^2}}$$

where B, C, and D are normalization constants. Again consider the observable A. If the average of this observable is calculated in any of these states (that is, any member of $\mathfrak{H}_2$), the result must be a real number. This is a property that we demand an operator have if it is to qualify as the operator corresponding to a physical observable. The object $\langle \psi | \hat{A}\psi \rangle$ must be real for all ψ in $\mathfrak{H}_2$. When working with the one-dimensional box problem, $\langle \psi | \hat{A}\psi \rangle$ must be real for all ψ in $\mathfrak{H}_1$. For example, if $\hat{H}$ is the operator corresponding to energy, then

$$(4.49) \qquad \langle E \rangle = \langle \psi | \hat{H}\psi \rangle = - \int_0^L \frac{\psi^* \hbar^2}{2m} \frac{\partial^2}{\partial x^2} \psi \, dx$$

must be real for any state function ψ in $\mathfrak{H}_1$.

These observations give rise to the following rule: In quantum mechanics one requires that the eigenvalues of an operator corresponding to a physical observable be real numbers. In this section we discuss the class of operators that have this property. They are called *Hermitian operators* and are a cornerstone in the theory of quantum mechanics.

The Hermitian Adjoint

To understand what a Hermitian operator is, we must first understand what the *Hermitian adjoint* of an operator is. Consider the operator $\hat{A}$. The Hermitian adjoint of $\hat{A}$ is written $\hat{A}^\dagger$. Under most circumstances, it is an entirely different operator from $\hat{A}$. For instance, the Hermitian adjoint of the complex number c is the complex conjugate of c. That is,

$$(4.50) \qquad c^\dagger = c^*$$

How is the Hermitian adjoint defined? First, let us agree that an operator is defined over a specific Hilbert space, $\mathfrak{H}$. Also if $\hat{A}$ is the operator and ψ is any element of $\mathfrak{H}$, then $\hat{A}\psi$ is also in $\mathfrak{H}$. For any two elements of this space, say ψ_l and ψ_n, we can form the inner product

$$(4.51) \qquad \langle \psi_l | \hat{A}\psi_n \rangle$$

Suppose there is another operator, $\hat{A}^\dagger$, also defined over $\mathfrak{H}$, for which

$$(4.52) \qquad \langle \hat{A}^\dagger \psi_l | \psi_n \rangle = \langle \psi_l | \hat{A}\psi_n \rangle$$

Suppose further that this equality holds for *all* ψ_l and ψ_n in $\mathfrak{H}$. Then $\hat{A}^\dagger$ is called the *Hermitian adjoint* of $\hat{A}$. To find the Hermitian adjoint of an operator $\hat{A}$, we have to find the object $\hat{A}^\dagger$ that fits (4.52) for all ψ_l and ψ_n. Consider $\hat{A} = a$, a complex number. Then

$$(4.53) \qquad \langle a^\dagger \psi_l | \psi_n \rangle = \langle \psi_l | a\psi_n \rangle = a \langle \psi_l | \psi_n \rangle = \langle a^* \psi_l | \psi_n \rangle$$

Equating the first and the last terms, we see that $a^\dagger = a^*$. As a second example, consider the operator

(4.54)
$$\hat{D} = \frac{\partial}{\partial x}$$

defined in $\mathfrak{H}_2$. Then

(4.55) $\langle \psi_l | \hat{D}\psi_n \rangle = \int_{-\infty}^{\infty} dx\, \psi_l^* \frac{\partial}{\partial x} \psi_n = [\psi_l^* \psi_n]_{-\infty}^{+\infty} - \int_{-\infty}^{\infty} dx \left(\frac{\partial}{\partial x} \psi_l^* \right) \psi_n$

$$= \langle -\hat{D}\psi_l | \psi_n \rangle$$

The "surface" term is zero since ψ_l and ψ_n are elements of $\mathfrak{H}_2$. Thus we find

(4.56)
$$\hat{D}^\dagger = -\hat{D}$$

For some cases we will find that the Hermitian adjoint of an operator is the operator itself. For such an operator $\hat{A}$, we may write

(4.57)
$$\hat{A}^\dagger = \hat{A}$$

In terms of the defining equation (4.52), this implies that for all ψ_l and ψ_n in $\mathfrak{H}$ (over which $\hat{A}$ is defined),

(4.58)
$$\langle \psi_l | \hat{A}\psi_n \rangle = \langle \hat{A}\psi_l | \psi_n \rangle$$

Operators that have this property are called *Hermitian operators*. The simplest example of a Hermitian operator is any real number a, since

(4.59)
$$\langle \psi_l | a\psi_n \rangle = \langle a\psi_l | \psi_n \rangle$$

If $\hat{A}$ and $\hat{B}$ are two Hermitian operators, is the product operator $\hat{A}\hat{B}$ Hermitian? This is most simply answered with the aid of Problem 4.11(b), according to which

(4.60)
$$(\hat{A}\hat{B})^\dagger = \hat{B}^\dagger \hat{A}^\dagger$$

If $\hat{A}$ and $\hat{B}$ are Hermitian, then

(4.61)
$$(\hat{A}\hat{B})^\dagger = \hat{B}\hat{A}$$

and $\hat{A}\hat{B}$ is not (necessarily) Hermitian. What about $\hat{A}\hat{B} + \hat{B}\hat{A}$?

(4.62)
$$(\hat{A}\hat{B} + \hat{B}\hat{A})^\dagger = \hat{B}^\dagger \hat{A}^\dagger + \hat{A}^\dagger \hat{B}^\dagger = \hat{B}\hat{A} + \hat{A}\hat{B}$$
$$= \hat{A}\hat{B} + \hat{B}\hat{A}$$

It follows that if $\hat{A}$ and $\hat{B}$ are both Hermitian, so is the bilinear form $(\hat{A}\hat{B} + \hat{B}\hat{A})$.
    Is the square of a Hermitian operator Hermitian?

(4.63)
$$(\hat{A}^2)^\dagger = (\hat{A}\hat{A})^\dagger = \hat{A}^\dagger \hat{A}^\dagger = \hat{A}\hat{A} = (\hat{A})^2$$

The answer is yes. Another way of doing this problem is as follows. Look at the inner product,

$$(4.64) \qquad \langle \psi_l | \hat{A}\hat{A}\psi_n \rangle = \langle \hat{A}\psi_l | \hat{A}\psi_n \rangle = \langle \hat{A}\hat{A}\psi_l | \psi_n \rangle$$

The first equality follows because $\hat{A}\psi_n$ is in $\mathfrak{H}$ and $\hat{A}$ is Hermitian, while the second equality follows simply because $\hat{A}$ is Hermitian. Comparing the first and third terms shows that $\hat{A}^2$ is Hermitian.

The Momentum and Energy Operators

Let us test the momentum operator $\hat{p}$ and see if it is Hermitian. For the free-particle case, $\hat{p}$ is Hermitian if for all ψ_l and ψ_n in $\mathfrak{H}_2$,

$$(4.65) \qquad \langle \psi_l | \hat{p}\psi_n \rangle = \langle \hat{p}\psi_l | \psi_n \rangle$$

Developing the left-hand side, we have

$$(4.66) \qquad \int_{-\infty}^{\infty} \psi_l^* \left(-i\hbar \frac{\partial}{\partial x} \psi_n \right) dx = -i\hbar [\psi_l^* \psi_n]_{-\infty}^{\infty} + i\hbar \int_{-\infty}^{\infty} \left(\frac{\partial}{\partial x} \psi_l^* \right) \psi_n \, dx$$

$$= \int_{-\infty}^{\infty} \left(-i\hbar \frac{\partial}{\partial x} \psi_l \right)^* \psi_n \, dx = \langle \hat{p}\psi_l | \psi_n \rangle$$

This technique is, by and large, the principal method by which a specific operator is shown to be Hermitian.

Having shown that $\hat{p}$ is Hermitian, it follows that the free-particle Hamiltonian, $\hat{H}$, is Hermitian.

$$(4.67) \qquad \hat{H} = \frac{\hat{p}^2}{2m}$$

$$(4.68) \qquad \hat{H}^\dagger = \left(\frac{\hat{p}^2}{2m} \right)^\dagger = \frac{\hat{p}^2}{2m} = \hat{H}$$

[Recall (4.63).] For a particle in a potential field $V(x)$,

$$(4.69) \qquad \hat{H} = \frac{\hat{p}^2}{2m} + V(x)$$

Since $V(x)$ is a real function that merely multiplies (say in $\mathfrak{H}_2$), it is Hermitian.

$$(4.70) \qquad \langle \psi_l | V\psi_n \rangle = \int_{-\infty}^{\infty} \psi_l^* V\psi_n \, dx = \int_{-\infty}^{\infty} V\psi_l^* \psi_n \, dx$$

$$= \int (V\psi_l)^* \psi_n \, dx = \langle V\psi_l | \psi_n \rangle$$

It follows that $\hat{H}$ as given by (4.69) is Hermitian.

PROBLEMS

4.11 (a) Show that $(a\hat{A} + b\hat{B})^\dagger = a^*\hat{A}^\dagger + b^*\hat{B}^\dagger$.

(b) Show that $(\hat{A}\hat{B})^\dagger = \hat{B}^\dagger\hat{A}^\dagger$.

(c) What is the Hermitian adjoint of the real number a?

(d) What is the Hermitian adjoint of $\hat{D}^2$? [See (4.54).]

(e) What is the Hermitian adjoint of $(\hat{A}\hat{B} - \hat{B}\hat{A})$?

(f) What is the Hermitian adjoint of $(\hat{A}\hat{B} + \hat{B}\hat{A})$?

(g) What is the Hermitian ajoint of $i(\hat{A}\hat{B} - \hat{B}\hat{A})$?

(h) What is $(\hat{A}^\dagger)^\dagger$?

(i) What is $(\hat{A}^\dagger\hat{A})^\dagger$?

4.12 If $\hat{A}$ and $\hat{B}$ are both Hermitian, which of the following three operators are Hermitian?

(a) $i(\hat{A}\hat{B} - \hat{B}\hat{A})$.

(b) $(\hat{A}\hat{B} - \hat{B}\hat{A})$.

(c) $\left(\dfrac{\hat{A}\hat{B} + \hat{B}\hat{A}}{2}\right)$.

(d) If $\hat{A}$ is not Hermitian, is the product $\hat{A}^\dagger\hat{A}$ Hermitian?

(e) If $\hat{A}$ corresponds to the observable A, and $\hat{B}$ corresponds to B, what is a "good" (i.e., Hermitian) operator that corresponds to the physically observable product AB?

4.13 If $\hat{A}$ is Hermitian, show that

$$\langle\hat{A}^2\rangle \geq 0$$

Answer (in $\mathfrak{H}_2$)

$$\langle\hat{A}^2\rangle = \int_{-\infty}^{\infty} \psi^*\hat{A}^2\psi \, dx = \int_{-\infty}^{\infty} (\hat{A}\psi)^*\hat{A}\psi \, dx$$

$$= \int_{-\infty}^{\infty} |\hat{A}\psi|^2 \, dx \geq 0$$

4.14 If $\hat{A}$ is Hermitian, show that $\langle A\rangle$ is real; that is, show that $\langle A\rangle^* = \langle A\rangle$.

4.15 For a particle moving in one dimension, show that the operator $\hat{x}\hat{p}$ is not Hermitian. Construct an operator which corresponds to this physically observable product that is Hermitian.

4.6 PROPERTIES OF HERMITIAN OPERATORS

The first property of Hermitian operators we wish to establish is that their eigenvalues are real. Let $\hat{A}$ be a Hermitian operator. Let $\{\varphi_n\}$ and $\{a_n\}$ represent, respectively, the eigenfunctions and eigenvalues of the operator $\hat{A}$.

(4.71) $$\hat{A}\varphi_n = a_n\varphi_n$$

In Dirac notation

(4.72) $$|\hat{A}\varphi_n\rangle = |a_n\varphi_n\rangle \quad \text{or equivalently} \quad \hat{A}|\varphi_n\rangle = a_n|\varphi_n\rangle$$

Multiplying from the left with $\langle \varphi_n |$ gives

(4.73) $$\langle \varphi_n | \hat{A}\varphi_n \rangle = \langle \varphi_n | a_n \varphi_n \rangle = a_n \langle \varphi_n | \varphi_n \rangle$$

Since $\hat{A}$ is Hermitian, we can write the left-hand side as

(4.74) $$\langle \hat{A}\varphi_n | \varphi_n \rangle = \langle a_n \varphi_n | \varphi_n \rangle = a_n^* \langle \varphi_n | \varphi_n \rangle$$

Equating the last terms in the latter two equations gives

(4.75) $$a_n^* = a_n$$

and a_n is real.

The second property of Hermitian operators we wish to establish is that *their eigenfunctions are orthogonal*. Again consider (4.72). Now multiply from the left with another eigenvector of $\hat{A}$, $\langle \varphi_l |$. There results

(4.76) $$\langle \varphi_l | \hat{A}\varphi_n \rangle = a_n \langle \varphi_l | \varphi_n \rangle$$

Since $\hat{A}$ is Hermitian, the left-hand side of this equation can be rewritten

(4.77) $$\langle \hat{A}\varphi_l | \varphi_n \rangle = a_l^* \langle \varphi_l | \varphi_n \rangle = a_l \langle \varphi_l | \varphi_n \rangle$$

The eigenvalue a_l is real because it is an eigenvalue of a Hermitian operator (i.e., $\hat{A}$). Subtracting the two equations above gives

(4.78) $$(a_l - a_n)\langle \varphi_l | \varphi_n \rangle = 0$$

If $a_l \neq a_n$, this equation says that

(4.79) $$\langle \varphi_l | \varphi_n \rangle = 0$$

which is the expression of the orthogonality of the set of functions $\{\varphi_n\}$. If these functions are all normalized, then (4.79) may be generalized to read

(4.80) $$\langle \varphi_l | \varphi_n \rangle = \delta_{ln}$$

Thus, the eigenvalues of a Hermitian operator are real, and its eigenfunctions are orthogonal.

PROBLEMS

4.16 Show that if an operator $\hat{B}$ has an eigenvalue $b_1 \neq b_1^*$, then $\hat{B}$ is not Hermitian.

4.17 Consider the operator $\hat{C}$,

$$\hat{C}\varphi(x) = \varphi^*(x)$$

    (a) Is $\hat{C}$ Hermitian?
    (b) What are the eigenfunctions of $\hat{C}$?
    (c) What are the eigenvalues of $\hat{C}$?

4.18 Given that the operator $\hat{O}$ annihilates the ket vector $|f\rangle$, that is, $\hat{O}|f\rangle = 0$, what is the value of the bra vector $\langle f|\hat{O}^{\dagger}$? Interpret the meaning of your answer.

4.19 The parallelogram law of geometry states that: the sum of the squares of the diagonals of a parallelogram equals twice the sum of the squares of the sides. Show that this is also true in Hilbert space; that is, if ψ and φ are any two elements of a Hilbert space, then

$$\|\psi + \varphi\|^2 + \|\psi - \varphi\|^2 = 2\|\psi\|^2 + 2\|\varphi\|^2$$

4.20 Show that the standard properties of $\cos\theta$, together with the definition of the inner product between two vectors φ and ψ, in $\mathfrak{H}$, with respective lengths, $\|\varphi\|$ and $\|\psi\|$, imply the Cauchy–Schwartz inequality

$$|\langle\varphi|\psi\rangle| \le \|\varphi\|\,\|\psi\|$$

4.21 Use the Cauchy–Schwartz inequality to prove the triangle inequality

$$\|\varphi + \psi\|^2 \le (\|\varphi\| + \|\psi\|)^2$$

4.22 Construct the squared length of $(\psi - \varphi)$ to show that

$$\|\psi\|^2 + \|\varphi\|^2 \ge 2\,\mathrm{Re}\,\langle\psi|\varphi\rangle$$

4.23 Let the sequence $\{\varphi_n\}$ be an orthonormal basis in $\mathfrak{H}$. Let the sequence $\{\cos\theta_n\}$ represent the angles between the vectors $\{\varphi_n\}$ and an arbitrary element ψ in $\mathfrak{H}$. Using Bessel's inequality,

$$\sum_{n=1}^{\infty} |\langle\varphi_n|\psi\rangle|^2 \le \|\psi\|^2$$

show that

$$\sum_{n=1}^{\infty} \cos^2\theta_n \le 1$$

Under what circumstances does the equality hold?

4.24 Every convergent sequence is also a *Cauchy sequence*. A sequence $\{\varphi_n(x)\}$ is a Cauchy sequence if

$$\lim_{\substack{n\to\infty \\ l\to\infty}} \|\varphi_n - \varphi_l\| = 0$$

A function space $\mathfrak{H}$ is a *complete space* if every Cauchy sequence in $\mathfrak{H}$ converges to an element of $\mathfrak{H}$. This is a requirement that a function space must satisfy in order that it be termed a Hilbert space. (See property 4 after Eq. 4.27.) Show that the space of functions on the unit interval with the property $\varphi(0) = \varphi(1) = 0$ is not a Hilbert space.

4.25 In addition to a complete space, one also defines a *complete sequence*. An orthonormal sequence $\{\varphi_n\}$ is complete in $\mathfrak{H}$ if there is no vector ψ, in $\mathfrak{H}$ of nonzero length ($\|\psi\| > 0$), which is perpendicular to all the elements in the sequence $\{\varphi_n\}$. Show that if $\{\varphi_n\}$ is an orthonormal basis of $\mathfrak{H}$, it is complete in $\mathfrak{H}$.

Answer

Let $\{\varphi_n\}$ be an orthonormal basis of $\mathfrak{H}$. Let ψ be an element of $\mathfrak{H}$ with nonzero length, which is normal to all the elements of $\{\varphi_n\}$. If $\{\varphi_n\}$ is a basis, then we may expand ψ,

$$\psi = \sum a_n \varphi_n = \sum \langle \varphi_n | \psi \rangle \varphi_n$$

But ψ is normal to all φ_n. Therefore, $\langle \varphi_n | \psi \rangle = 0$, which gives $\psi = 0$, so the hypothesis leads to a contradiction, hence the hypothesis is an incorrect statement and there is no such ψ in $\mathfrak{H}$.

4.26   Show that any operator $\hat{A}$ may be expressed as the linear combination of a Hermitian and an anti-Hermitian ($\hat{B}^\dagger = -\hat{B}$) operator.

Answer

$$\hat{A} = \left(\frac{\hat{A} + \hat{A}^\dagger}{2} \right) + i \left(\frac{\hat{A} - \hat{A}^\dagger}{2i} \right)$$

[*Note:* $\hat{A} + \hat{A}^\dagger$ and $i(\hat{A} - \hat{A}^\dagger)$ are both Hermitian.]

4.27   Show that the wavefunctions for a particle in a one-dimensional box with walls at $x = 0$ and L satisfy the equality

$$\int_0^L \psi^* \psi_{xx} \, dx = - \int_0^L |\psi_x|^2 \, dx$$

The subscript x denotes differentiation.

4.28   Use the equality proved in Problem 4.27 to establish the following *variational principle*. If the expectation $\int \psi^* \hat{H} \psi \, dx$ is minimum, the normalized wavefunction ψ is the ground state. Specifically, establish the theorem for a particle in a one-dimensional box, assuming real wavefunctions.

Answer

Apart from a constant factor and with the results of Problem 4.27, we may write

$$\langle H \rangle = - \int_0^L \psi^* \psi_{xx} \, dx = \int_0^L \psi_x^2 \, dx$$

Let ψ minimize $\langle H \rangle$. Then infinitesimal variation of ψ causes no change in $\langle H \rangle$. Let $\psi \to \psi + \delta\psi$. The variation $\delta\psi$ is an arbitrary infinitesimal function of x that vanishes at $x = 0$ and L. Then

$$\langle H \rangle = \int \psi_x^2 \, dx \to \int (\psi_x + \delta\psi_x)^2 \, dx = \langle H \rangle + \delta\langle H \rangle$$

$$\delta\langle H \rangle = 2 \int \psi_x \, \delta\psi_x \, dx = 2 \int \psi_x \frac{d}{dx} \delta\psi \, dx = 0$$

Integrating the last term by parts and dropping the "surface" terms gives

$$\int \psi_{xx}\, \delta\psi\, dx = 0$$

Variation of the normalization statement (both ψ and $\psi + \delta\psi$ are normalized) gives

$$\lambda \int \psi\delta\psi\, dx = 0$$

where λ is an arbitrary undetermined multiplier. Combining the last two equations yields

$$\int_0^L \delta\psi(\psi_{xx} - \lambda\psi)\, dx = 0$$

If this equation is to be satisfied for arbitrary variation of ψ about the minimizing value, we may conclude

$$\psi_{xx} = \lambda\psi$$

It follows that ψ is an eigenstate of $\hat{H}$, in which case $\langle H\rangle$ is an energy eigenvalue which has minimum value for the ground state.

4.29   Let

$$A_{nl} \equiv \langle \varphi_n | \hat{A}\varphi_l \rangle$$

Show that

$$(\hat{A}^\dagger)_{ln} = (A_{nl})^*$$

Answer

$$A_{nl} = \langle \varphi_n | \hat{A}\varphi_l \rangle$$
$$= \langle \hat{A}\varphi_l | \varphi_n \rangle^* = \langle \varphi_l | \hat{A}^\dagger \varphi_n \rangle^*$$

Taking the complex conjugate of the last and first terms in this equality gives the desired result.

4.30   Generalize the derivation of the variational principle given in Problem 4.28 to the case of a particle in a one-dimensional potential $V(x)$. What properties of the potential are assumed in your derivation?

Answer (partial)
We must show that the variational statement

$$\delta\langle H\rangle = 0$$

where

$$H = \frac{p_x^2}{2m} + V(x)$$

returns the Schrödinger equation. Repeating steps of Problem 4.28 we find

$$\delta\langle H \rangle = \int \left[\left(-\frac{\hbar^2}{2m} \varphi_{xx}\, \delta\varphi^* + V\varphi\, \delta\varphi^* - \lambda\varphi\, \delta\varphi^* \right) + (cc) \right] dx = 0$$

where cc denotes complex conjugate. This latter equation may be rewritten as

$$2 \operatorname{Re} \int \left(-\frac{\hbar^2}{2m} \varphi_{xx} + V\varphi - \lambda\varphi \right) \delta\varphi^*\, dx = 0$$

Since $\delta\varphi^*$ is arbitrary we may take it to be purely imaginary. The preceding equation then yields the Schrödinger equation for the imaginary part of φ. Repeating this procedure for $\delta\varphi^*$ purely real, and adding the two results, gives the Schrödinger equation for the full wavefunction.

SUPERPOSITION AND COMPATIBLE OBSERVABLES

In this chapter we encounter the superposition principle, which is considered by many to be one of the more fundamental concepts of quantum mechanics. This principle represents one of the basic differences between classical and quantum mechanics and also provides a deeper understanding of the uncertainty principle. Closely related to the superposition principle are the commutator theorem and the notions of compatible observables and simultaneous eigenfunctions.

5.1 THE SUPERPOSITION PRINCIPLE

Ensemble Average

Consider again a particle in a one-dimensional box. Let us imagine a large number of identical replicas of the system (called an *ensemble* in statistical mechanics), such as described in Section 3.3. If each such box is in the *same* initial state $\psi(x, 0)$, after an interval of time t, each box will again be in a common state $\psi(x, t)$, as shown in Fig. 5.1. Suppose that we ask what the energy of the particle is in each box, at the time t. The laws of nature are such that the energy measured in each of the identical

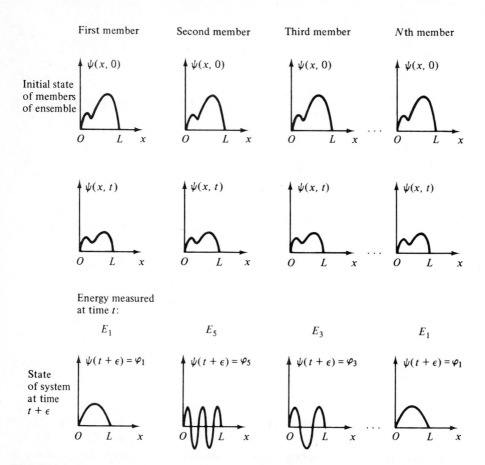

FIGURE 5.1 Measurement of energy of N identical one-dimensional boxes which comprise an "ensemble." All boxes are in the same state at $t = 0$.

boxes, which are all in the identically same state $\psi(x, t)$, are not the same [save for the case that $\psi(x, 0)$ is an eigenstate of $\hat{H}$].

How does one answer the question above: What will the energy be? Since the energy measured at the time t in each box of the ensemble will most likely not be the same, more appropriate questions are: (1) What is the average of the energies measured in all the boxes of the ensemble? (2) If we measure the energy in one box, with what probability will the value, say, E_3 be found? To answer these questions, we first recall that if the probability of finding the value E_n in a given measurement of energy is $P(E_n)$, then the average energy over measurements of all members of the ensemble in the limit as this number becomes large is given by the expression

$$(5.1) \qquad \langle E \rangle = \sum_{\text{all } E_n} P(E_n) E_n$$

(Recall Eq. 3.34.) This formula holds for all physical observables. For example, the average particle position is given by

$$\langle x \rangle = \int_0^L x P(x)\, dx \tag{5.2}$$

In this case the integral is a sum over the continuum of values x may assume.

The quantum mechanical prescription for calculating the average of a dynamical observable in the state ψ is given by the third postulate of quantum mechanics (Section 3.3, Eq. 3.32). Specifically, for the energy we have (in Dirac notation)

$$\langle E \rangle = \langle \psi | \hat{H} \psi \rangle \tag{5.3}$$

Let us expand the state ψ in the eigenstates of $\hat{H}$. These eigenstates obey the equation

$$\hat{H} \varphi_n = E_n \varphi_n \tag{5.4}$$

For the box problem they are explicitly (Eq. 4.15)

$$\varphi_n = \sqrt{\frac{2}{L}} \sin\left(\frac{n\pi x}{L}\right) \tag{5.5}$$

The expansion of ψ in these eigenstates appears as

$$\psi(x, t) = \sum_{n=1}^{\infty} b_n(t) \varphi_n(x) \tag{5.6}$$

The state ψ is that of the system at the time t, so that it is, in general, a function of x and t. Since φ_n is a function of x only, the coefficients of expansion b_n may, in general, be functions of time.

In Dirac notation, (5.6) appears as

$$|\psi\rangle = \sum_{n=1}^{\infty} |b_n \varphi_n\rangle \tag{5.7}$$

Substituting this series into (5.3) gives

$$
\begin{aligned}
\langle E \rangle &= \left\langle \sum_n b_n \varphi_n \middle| \hat{H} \sum_l b_l \varphi_l \right\rangle \\
&= \sum_n \sum_l b_n{}^* b_l \langle \varphi_n | \hat{H} \varphi_l \rangle \\
&= \sum_n \sum_l b_n{}^* b_l E_l \langle \varphi_n | \varphi_l \rangle \\
&= \sum_n \sum_l b_n{}^* b_l E_l \delta_{nl} \\
&= \sum_{n=1}^{\infty} |b_n|^2 E_n
\end{aligned}
\tag{5.8}
$$

Equating this average to that given by (5.1) gives

(5.9)
$$\sum_n |b_n|^2 E_n = \sum_n P(E_n)E_n$$

This equation dictates the following interpretation of the square of the modulus of b_n. It is the probability that at the time t, measurement of the energy of the particle which is in the state $\psi(x, t)$ yields the value E_n.

(5.10)
$$P(E_n) = |b_n|^2$$

These coefficients have the correct normalization, provided that the states ψ and φ_n are normalized. In this case we have

(5.11)
$$1 = \langle \psi | \psi \rangle = \left\langle \sum_n b_n \varphi_n \Big| \sum_l b_l \varphi_l \right\rangle$$

$$= \sum_n \sum_l b_n^* b_l \langle \varphi_n | \varphi_l \rangle$$

$$= \sum_n \sum_l b_n^* b_l \delta_{nl}$$

$$= \sum_n |b_n|^2 = 1$$

When this is the case the coefficient $|b_n|^2$ is an *absolute* probability. If not, the correct expression for the probability that measurement finds E_n is

(5.12)
$$P(E_n) = \frac{|b_n|^2 |C_n|^2}{\sum |b_n|^2 |C_n|^2} = \frac{|b_n|^2 |C_n|^2}{\langle \psi | \psi \rangle}$$

where

$$|C_n|^2 = \langle \varphi_n | \varphi_n \rangle$$

Let us return to the expansion (5.7). The coefficients b_n are calculated in the following manner. Multiply this equation from the left with the bra vector $\langle \varphi_{n'} |$. Owing to the orthonormality of the set $\{\varphi_n\}$, one obtains

(5.13)
$$b_n = \langle \varphi_n | \psi \rangle$$

The coefficient b_n is the projection of ψ onto the eigenvector φ_n. The physical interpretation of b_n is that $|b_n|^2$ is the probability that measuring E finds the value E_n when the system is in the state ψ. This prescription is true for *any* dynamical observable. Consider the symbolic operator $\hat{F}$

(5.14)
$$\hat{F} \varphi_n = f_n \varphi_n$$

At a given time t, the system is in the state $\psi(x, t)$. What is the probability that measurement of F at this time finds the value f_3? The state ψ is a superposition state. It is

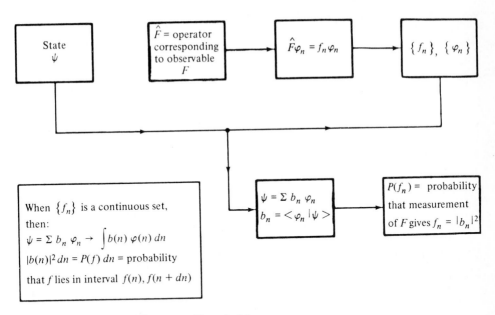

FIGURE 5.2 Elements of the superposition principle.

composed of the eigenstates of $\hat{F}$. Here we are assuming that the eigenstates of $\hat{F}$ are a basis for the Hilbert space that ψ is in. So we may write

(5.15) $$\psi = \sum b_n \varphi_n$$

(5.16) $$b_n = \langle \varphi_n | \psi \rangle$$

This assumption that an arbitrary state ψ may be represented as a superposition of the eigenstates of a physical observable is the essence of the superposition principle. With $\{\varphi_n\}$ and ψ normalized to unity, the probability that measurement finds the value f_3 is $|b_3|^2$. This procedure is depicted in Fig. 5.2.

Hilbert-Space Interpretation

When we look in Hilbert space, $\{\varphi_n\}$ is one set of vectors and ψ is another vector. The system is in the state ψ. Measurement of F causes the state ψ to fall to one of the φ_n vectors. Chances are that it goes to the φ_n vector to which it is most inclined (in the geometrical sense; see Fig. 5.3).

Consider the following illustrative example. A particle of mass m is in a one-dimensional box of width L. At $t = 0$ the particle is in the state

(5.17) $$\psi(x, 0) = \frac{3\varphi_2 + 4\varphi_9}{\sqrt{25}}$$

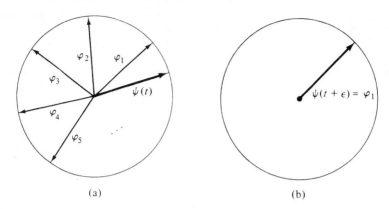

(a)   (b)

FIGURE 5.3 (a) State of the system before measurement at t, superimposed on the basis $\{\varphi_n\}$, which are the eigenvectors of the operator $\hat{F}$. The probability that measurement of F finds the value f_n is proportional to the projection of ψ on φ_n. (b) State of the system immediately after measurement has found the value f_1. Measurement acts as a "wave filter." It filters out all components of the superposition $\psi(x, t) = \sum b_n(t)\varphi_n(x)$, passing only the φ_1 wave.

The φ_n functions are the orthonormal eigenstates of $\hat{H}$:

(5.18)
$$\varphi_n = \sqrt{\frac{2}{L}}\sin\left(\frac{n\pi x}{L}\right)$$

What will measurements of E yield at $t = 0$ and what is the probability of finding this value? First let us see if ψ is normalized. In Dirac notation we have, for the state (5.17),

(5.19)
$$|\psi\rangle = \frac{3|\varphi_2\rangle + 4|\varphi_9\rangle}{\sqrt{25}}$$

so that

(5.20) $\langle\psi|\psi\rangle = \frac{1}{25}\{(3\langle\varphi_2| + 4\langle\varphi_9|)(3|\varphi_2\rangle + 4|\varphi_9\rangle)\}$
$\qquad = \frac{1}{25}\{9\langle\varphi_2|\varphi_2\rangle + 12\langle\varphi_2|\varphi_9\rangle + 12\langle\varphi_9|\varphi_2\rangle + 16\langle\varphi_9|\varphi_9\rangle\}$
$\qquad = 1$

and ψ is normalized. The inner products $\langle\varphi_2|\varphi_2\rangle = \langle\varphi_9|\varphi_9\rangle = 1$ while the other two are zero, owing to the orthogonality of the set $\{\varphi_n\}$.

The superposition principle stipulates the following. If we want the probability that measurement finds the value E_n, we must expand ψ in the eigenstates of $\hat{H}$. The square of the magnitude of the coefficient of φ_n is the said probability.

(5.21)
$$\psi = \sum b_n\varphi_n = \frac{3\varphi_2 + 4\varphi_9}{\sqrt{25}}$$

In this simplified problem, by inspection we find that

$$b_2 = \frac{3}{\sqrt{25}}$$

(5.22)

$$b_9 = \frac{4}{\sqrt{25}}$$

$$b_n = 0 \qquad (n \neq 2 \text{ or } 9)$$

Therefore, the probability $P(E_n)$ that measurement of E at $t = 0$ finds the value E_n is

$$P(E_2) = \frac{9}{25}$$

(5.23)

$$P(E_9) = \frac{16}{25}$$

$$P(E_n) = 0 \qquad (n \neq 2 \text{ or } 9)$$

In an ensemble of 2500 identical one-dimensional boxes, each containing an identical particle in the same state $\psi(x, 0)$ given by (5.17), measurement of E at $t = 0$ finds about 900 particles to have energy $E_2 = 4E_1$ and about 1600 particles to have energy $E_9 = 81E_1$.

Is there a chance that in an ensemble of 10^{17} boxes, measurement of E finds E_2 in all 10^{17} boxes? Yes. This remarkable response carries the philosophical impact of the superposition principle. Although the state $\psi(x, 0)$ is a precise superposition of well-defined eigenstates of the observables being measured, one is not *certain* what measurement will yield. There is nothing in classical physics that is similar to this concept. Any uncertainty in classical physics arises from uncertain initial data. In quantum mechanics, although the initial state $\psi(x, 0)$ is prescribed with perfect accuracy, one is never certain in which eigenstate, φ_n, measurement will leave the system.

However, once E is measured and, say, the value E_9 is found, then one knows with absolute certainty that the state of the system immediately after this measurement is φ_9.

The Initial Square Wave

As a second illustrative example, we consider the following free-particle problem in one dimension. Suppose that at $t = 0$ the system is in the state (Fig. 5.4)

(5.24)

$$\psi(x, 0) \begin{cases} \sqrt{\dfrac{1}{L}} & |x| < \dfrac{L}{2} \\ \\ 0 & \text{elsewhere} \end{cases}$$

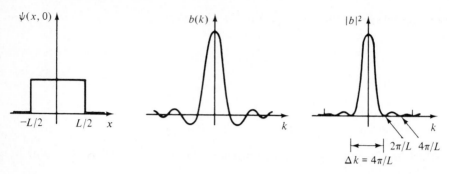

FIGURE 5.4 Square wave packet at $t = 0$ and corresponding momentum eigenstate amplitudes $b(k)$. The interval over which momentum values are most likely to be found is $\Delta p = h\,\Delta k = 4\pi h/L$.

If at this same instant, the momentum of the particle is measured, what are the possible values that will be found, and with what probability will these values occur?

To answer these questions we must first expand $\psi(x, 0)$ in a superposition of the eigenstates of $\hat{p}$:

$$(5.25) \qquad \varphi_k = \frac{1}{\sqrt{2\pi}} e^{ikx}$$

Since these states comprise a continuum, the corresponding superposition of eigenstates of $\hat{p}$ is an integral.

$$(5.26) \qquad \psi(x, 0) = \int_{-\infty}^{\infty} b(k)\varphi_k \, dk$$

Inverting this equation (see Eq. 4.42 et seq.) gives the coefficient $b(k)$.

$$(5.27) \qquad b(k) = \int_{-\infty}^{\infty} \psi(x, 0)\varphi_k{}^* \, dx = \frac{1}{\sqrt{2\pi}} \int_{-\infty}^{\infty} \psi(x, 0)e^{-ikx} \, dx$$

$$= \frac{1}{\sqrt{2\pi L}} \int_{-L/2}^{+L/2} e^{-ikx} \, dx = \frac{1}{\sqrt{2\pi L}} \frac{2}{k} \left(\frac{e^{ikL/2} - e^{-ikL/2}}{2i} \right)$$

$$= \sqrt{\frac{2}{\pi L}} \frac{\sin(kL/2)}{k}$$

Again, this coefficient is the projection of the state $\psi(x, 0)$ onto the eigenstate φ_k. Its square times the differential dk is the probability that measurement of momentum yields $p = \hbar k$, in the interval $\hbar k$, $\hbar(k + dk)$. The corresponding probability density (in momentum space) is

$$(5.28) \qquad |b|^2 = \frac{2}{\pi L} \frac{\sin^2(kL/2)}{k^2}$$

This function has its maximum at $k = 0$. It drops to zero at

(5.29)
$$\frac{kL}{2} = \pi$$

or equivalently at

(5.30)
$$p = \hbar k = \frac{2\pi \hbar}{L}$$

It is most probable that measurement of momentum finds the value $p = 0$. The momentum values $(\pm n2\pi \hbar / L)$ with n an integer greater than 1 are never found, for at these values, $b(k) = 0$.

Referring to Fig. 5.4, we see that the interval of momentum values that measurements are most likely to uncover has the approximate width

(5.31)
$$\Delta k = \frac{4\pi}{L}$$

$$\Delta p = \hbar \, \Delta k = \frac{4\pi \hbar}{L}$$

On the other hand, from (5.24), it is uniformly probable that measurement of x finds the particle anywhere in the interval $(-L/2, +L/2)$, of width

(5.32)
$$\Delta x = L$$

Combining these two latter uncertainties (5.31 and 5.32) gives

(5.33)
$$\Delta x \, \Delta p \simeq \hbar$$

The approximation sign is used because of the qualitative manner in which Δp was calculated. The result (5.33) is another example of the Heisenberg uncertainty principle at work.

The Chopped Beam

To further exhibit the significance of the probability density $|b(k)|^2$, we consider the following problem. Suppose that the free-particle system above is composed of N noninteracting electrons. Every electron is in the state $\psi(x, 0)$ given by (5.24). The density ρ (number/length) is related to ψ through

(5.34)
$$\text{number of particles in } dx = \rho \, dx = N |\psi|^2 \, dx$$

The total number in the whole "beam" is

(5.35)
$$N = \int_{-\infty}^{\infty} \rho(x) \, dx = N \int_{-L/2}^{L/2} |\psi|^2 \, dx = N$$

Suppose that we now ask how many electrons have momentum in the interval $(-2\pi\hbar/L, +2\pi\hbar/L)$, or equivalently, how many have wavenumber in the interval $(-2\pi/L, +2\pi/L)$. For a single electron, the probability of finding an electron with momentum in the interval $\hbar k$ to $\hbar k + \hbar dk$ is

(5.36)
$$P(k)\, dk = |b(k)|^2\, dk$$

This is a correct statement provided that

(5.37)
$$\int_{-\infty}^{\infty} |b(k)|^2\, dk = 1$$

If this is not the case, one must divide $|b(k)|^2$ in (5.36) by the last integral.

For a totality of N electrons in the beam, the number of them that have momentum in the interval $\hbar k, \hbar k + \hbar\, dk$ is

(5.38)
$$\rho(k)\, dk = N|b(k)|^2\, dk$$

The total number in the whole beam is

(5.39)
$$N = \int_{-\infty}^{\infty} \rho(k)\, dk = N \int_{-\infty}^{\infty} |b(k)|^2\, dk$$

For the example at hand

(5.40)
$$\int_{-\infty}^{\infty} |b(k)|^2\, dk = \frac{2}{\pi L} \int_{-\infty}^{\infty} \frac{\sin^2 (kL/2)}{k^2}\, dk$$

$$= \frac{1}{\pi} \int_{-\infty}^{\infty} \frac{\sin^2 \eta\, d\eta}{\eta^2} = 1$$

The dummy variable $\eta \equiv kL/2$. To return to the original question, the number of electrons ΔN in the beam with momentum in the interval $(-2\pi\hbar/L, +2\pi\hbar/L)$ is given by the integral

(5.41)
$$\Delta N = N \int_{-2\pi/L}^{+2\pi/L} \frac{2}{\pi L} \frac{\sin^2 (kL/2)}{k^2}\, dk$$

$$= \frac{N}{\pi} \int_{-\pi}^{+\pi} \frac{\sin^2 \eta}{\eta^2}\, d\eta = 0.903N$$

Thus, we find a majority of the electrons in this momentum interval.

Superposition and Uncertainty

Let us return to the case of a single electron in the state $\psi(x, 0)$ given by (5.24). Suppose at this time, $t = 0$, we measure the electron's momentum. What value do we find? The answer is: (a) the values $p = \pm n2\pi\hbar/L$ are never found; (b) any other value may occur with corresponding probability density $|b(k)|^2$. Let the measurement find the electron to have the momentum

$$(5.42) \qquad p = \frac{\pi h}{L}$$

Immediately after this measurement, what is the state of the particle? The answer is

$$(5.43) \qquad \psi = \frac{1}{\sqrt{2\pi}} \exp\left(\frac{i\pi x}{L}\right)$$

The electron is now in the state (5.43). Suppose that we measure the energy of the particle. What value is found? Since this state is also an eigenstate of $\hat{H}$, it is a certainty that measurement yields

$$(5.44) \qquad E = \frac{(\pi h/L)^2}{2m}$$

The system is still left in the eigenstate (5.43). Suppose that we now measure the position of the particle. What values may occur? The probability density is

$$(5.45) \qquad P = |\psi|^2 = \frac{1}{2\pi}$$

which is a constant. It is uniformly probable to find the electron anywhere along the whole x axis. The uncertainty in x is $\Delta x = \infty$. For this same state it is certain that measurement of momentum finds the value $\pi h/L$, so that $\Delta p = 0$. Again we find corroborating evidence for the Heisenberg uncertainty principle.

Now we place a uniform array of scintillation detectors along the x axis. One of them scintillates at $x = x'$. What is the state of the electron immediately after measurement? The answer is the eigenstate of the position operator corresponding to the eigenvalue x' (Fig. 5.5).

$$(5.46) \qquad \psi = \delta(x - x')$$

Now we measure momentum again. What values can be found? To answer this question, we again call on the superposition recipe: expand ψ in the eigenstates of $\hat{p}$.

$$(5.47) \qquad \delta(x - x') = \frac{1}{\sqrt{2\pi}} \int_{-\infty}^{\infty} b(k)e^{ikx}\, dk$$

$$b(k) = \frac{1}{\sqrt{2\pi}} \int \delta(x - x')e^{-ikx}\, dx = \frac{1}{\sqrt{2\pi}} e^{-ikx'}$$

The corresponding momentum probability density is

$$(5.48) \qquad P(k) = |b(k)|^2 = \frac{1}{2\pi}$$

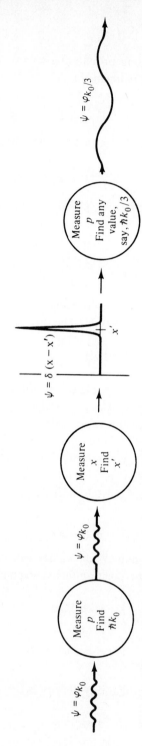

FIGURE 5.5 Measuring x destroys the momentum eigenstate φ_{k_0}.

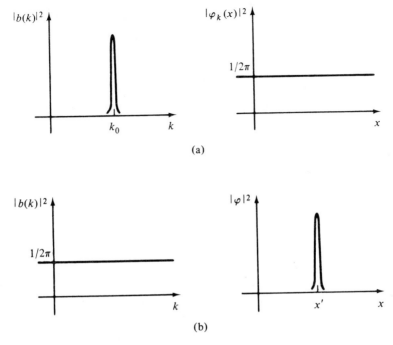

FIGURE 5.6 (a) In the state $\psi = \varphi_{k_0} = (1/\sqrt{2\pi})e^{ik_0x}$, $\Delta p = 0$ and $\Delta x = \infty$. (b) In the state $\psi = \delta(x - x')$, $\Delta x = 0$ and $\Delta p = \infty$.

It is uniformly probable to find the electron with any momentum along the whole k axis. The uncertainty in momentum is $\Delta p = \infty$ for the state (5.46), for which $\Delta x = 0$, and the uncertainty principle holds firm (Fig. 5.6).

We have been using the phrase superposition principle, but have not given a concise statement of this principle. P. A. M. Dirac, one of the early investigators of quantum mechanics, was first to grasp the full significance of this principle. His description[1] is perhaps the most succinct. The superposition principle "requires us to assume that between . . . states there exist peculiar relationships such that whenever the system is definitely in one state we can consider it as being partly in each of two or more other states. The original state must be regarded as the result of a kind of superposition of the two or more new states, in a way that cannot be conceived on classical ideas."

The superposition principle is a cornerstone of quantum mechanics. We have used it previously in some elementary one-dimensional problems. We will return to it in the remainder of the text in relation to more extensive one-dimensional

[1] P. A. M. Dirac, *The Principles of Quantum Mechanics*, 4th ed., Oxford University Press, New York, 1958.

problems as well as more practical problems in two and three dimensions. A sound understanding of this principle is prerequisite to a working knowledge of quantum mechanics.

PROBLEMS

5.1   If an arbitrary initial state function for a particle in a one-dimensional box is expanded in the discrete series of eigenstates of the Hamiltonian relevant to the box configuration, one obtains (5.6)

$$\psi(x, 0) = \sum_{n=1}^{\infty} b_n(0)\varphi_n(x)$$

On the other hand, if the particle is free, its Hamiltonian has a continuous spectrum of eigen-energies and the superposition of an arbitrary initial state in the eigenstates φ_k of $\hat{H}$ becomes an integral (5.26):

$$\psi(x, 0) = \int_{-\infty}^{\infty} b(k)\varphi_k \, dk$$

(a)   What are the dimensions of $|b_n|^2$ and $|b(k)|^2$, respectively?
(b)   What is the source of the difference in dimensionality?
(c)   What are the dimensions and physical interpretation of the integral

$$\int_{-\infty}^{\infty} |b(k)|^2 \, dk?$$

Answer (partial)

    (b)   The term $|b_n|^2$ represents a probability, whereas $|b(k)|^2$ represents a probability density.

5.2   One thousand neutrons are in a one-dimensional box, with walls at $x = 0, x = L$. At $t = 0$, the state of each particle is

$$\psi(x, 0) = Ax(x - L)$$

(a)   Normalize ψ and find the value of the constant A.
(b)   How many particles are in the interval $(0, L/2)$ at $t = 0$?
(c)   How many particles have energy E_5 at $t = 0$?
(d)   What is $\langle E \rangle$ at $t = 0$?

5.3   Using the expressions for φ_k and ψ given by (5.25) and (5.26), respectively, show that

$$\langle \psi | \psi \rangle = 1 \rightarrow \int_{-\infty}^{\infty} |b(k)|^2 \, dk = 1$$

5.4   A pulse 1 m long contains 1000 α particles. At $t = 0$, each α particle is in the state

$$\psi(x, 0) = \begin{cases} \frac{1}{10}e^{ik_0 x}, & |x| \le 50 \text{ cm}, \; k_0 = \pi/50 \\ 0 & \text{elsewhere} \end{cases}$$

(a)   At $t = 0$, how many α particles have momentum in the interval $(0 < \hbar k < \hbar k_0)$?
(b)   At which values of momentum will α particles not be found at $t = 0$?

(c) Describe an experiment to "prepare" such a state.

(d) Construct Δx and Δp for this state, formally. What is $\Delta x \, \Delta p$? [*Hint:* To calculate Δp, use $|b(k)|^2$.]

5.5 At $t = 0$ it is known that of 1000 neutrons in a one-dimensional box of width 10^{-5} cm, 100 have energy $4E_1$, and 900 have energy $225E_1$.

(a) Construct a state function that has these properties.

(b) Use the state you have constructed to calculate the density $\rho(x)$ of neutrons per unit length.

(c) How many neutrons are in the left half of the "box"?

5.6 Over a very long interval of the x axis, a uniform distribution of 10,000 electrons is moving to the right with velocity 10^8 cm/s and 10,000 electrons are moving to the left with velocity 10^8 cm/s. Assuming that the electrons do not interact with one another, construct a state function that yields the preceding properties for the combined beam. Calculate $\langle p \rangle$ for this state.

5.7 Give an argument in support of the conjecture that one cannot measure the momentum of a particle in a one-dimensional box, with absolute accuracy. Support the theoretical argument with an argument involving an experiment.

5.8 A one-dimensional box containing an electron suffers an infinitesimal perturbation and emits a photon of frequency

$$h\nu = 3E_1$$

where E_1 denotes the ground state of the particle. A student concludes that the electron was in the state φ_2 prior to perturbation. Is he correct?

Answer

What the student has in mind is that the photon corresponds to the decay

$$h\nu = E_2 - E_1 = 3E_1$$

However, suppose that the electron was in the superposition state $(3\varphi_2 + 8\varphi_6)/\sqrt{73}$. Then it is still possible that a photon of frequency $h\nu = 3E_1$ is emitted. So the student is incorrect.

5.9 Measurement of the position of a particle in a one-dimensional box with walls at $x = 0$ and $x = L$ finds the value $x = L/2$.

(a) Show that in subsequent measurement, it is equally probable to find the particle in any odd-energy eigenstate.

(b) Show that the probability of finding the particle in any even-eigenstate is zero. (An eigenstate φ_n is even if n is even and odd if n is odd.)

5.10 It is known that at time $t = 0$, a particle in a box (described in Problem 5.9) is not in the right half of the box. The particle is in one of an infinite number of states. Six such states are depicted in Fig. 5.7.

(a) Write down an approximate wavefunction for each of these states.

(b) Calculate $\langle E \rangle$ for each of these states.

(c) Argue that the state depicted in Fig. 5.7a is the state of minimum $\langle E \rangle$ (assuming that $\varphi = A \sin 2\pi x/L, \; x < L/2$).

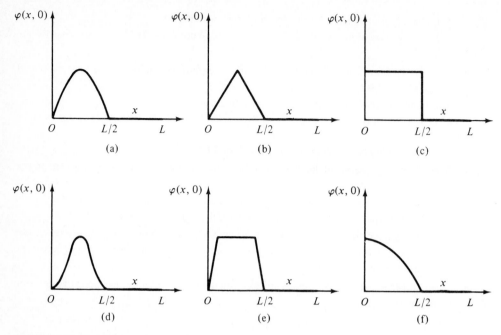

FIGURE 5.7   Six initial states for a particle in a one-dimensional box, with the property that $|\psi|^2 = 0$ in the right half of the box. (See Problem 5.10.)

5.11   A particle in the one-dimensional box described in Problem 5.9 is in the ground state. One of the walls of the box is moved to the position $x = 2L$, in a time short compared to the natural period $2\pi/\omega_1$, where $\hbar\omega_1 = E_1$. If the energy of the particle is measured soon after this expansion, what value of energy is most likely to be found? How does this energy compare to the particle's initial energy (E_1)?

5.2   COMMUTATOR RELATIONS IN QUANTUM MECHANICS

An important operation in quantum mechanics is the *commutator* between two operators, $\hat{A}$ and $\hat{B}$. It is written $[\hat{A}, \hat{B}]$ and is defined as

(5.49)
$$[\hat{A}, \hat{B}] = \hat{A}\hat{B} - \hat{B}\hat{A}$$

An immediate property of the commutator is that

(5.50)
$$[\hat{A}, \hat{B}] = -[\hat{B}, \hat{A}]$$

If

(5.51)
$$[\hat{A}, \hat{B}] = 0$$

the two operators are said to *commute* (A and B are *compatible*) with each other. That is,

$$(5.52) \qquad \hat{A}\hat{B} = \hat{B}\hat{A}$$

Any operator $\hat{A}$ commutes with any constant a.

$$(5.53) \qquad [\hat{A}, a] = 0$$

$$(5.54) \qquad [\hat{A}, a\hat{B}] = [a\hat{A}, \hat{B}] = a[\hat{A}, \hat{B}]$$

Any operator $\hat{A}$ commutes with its own square, $\hat{A}^2$.

$$(5.55) \qquad [\hat{A}, \hat{A}^2] = (\hat{A}\hat{A}^2 - \hat{A}^2\hat{A}) = (\hat{A}\hat{A}\hat{A} - \hat{A}\hat{A}\hat{A}) = 0$$

The meaning of this relation is that, no matter what $\hat{A}$ is, when $[\hat{A}, \hat{A}^2]$ operates on any function $g(x)$, one gets zero,

$$(5.56) \qquad [\hat{A}, \hat{A}^2]g(x) = 0$$

More generally, $\hat{A}$ commutes with any function of $\hat{A}$, $f(\hat{A})$.

$$(5.57) \qquad [f(\hat{A}), \hat{A}] = 0$$

As an example of this rule, consider the following commutator involving the momentum operator $\hat{p}$.

$$(5.58) \qquad [e^{\hat{p}}, \hat{p}] = \left[\sum_{n=0}^{\infty} \frac{\hat{p}^n}{n!}, \hat{p} \right]$$

$$= \sum \frac{1}{n!} [\hat{p}^n, \hat{p}]$$

$$= [1, \hat{p}] + [\hat{p}, \hat{p}] + \frac{1}{2!} [\hat{p}^2, \hat{p}] + \cdots = 0$$

It follows that

$$(5.59) \qquad [e^{\hat{p}}, \hat{p}]g(x) = \left[\exp\left(-\frac{i\hbar\partial}{\partial x} \right), -\frac{i\hbar\partial}{\partial x} \right]g(x) = 0$$

where $g(x)$ represents any function of x.

One of the most important commutators in physics is that between the coordinate, $\hat{x}$, and the momentum, $\hat{p}$. Let us calculate it.

$$(5.60) \qquad [\hat{x}, \hat{p}]g(x) = i\hbar\left(-x\frac{\partial}{\partial x} + \frac{\partial}{\partial x} x \right)g(x)$$

$$= i\hbar\left(-x\frac{\partial g}{\partial x} + x\frac{\partial g}{\partial x} + g \right) = i\hbar g(x)$$

It follows that

(5.61)
$$\boxed{[\hat{x}, \hat{p}] = i\hbar}$$

In other words, the operator $[\hat{x}, \hat{p}]$ has the sole effect of a simple multiplication by the constant $i\hbar$. As an immediate consequence (using Problem 5.12)

(5.62)
$$[\hat{x}, \hat{p}^2] = [\hat{x}, \hat{p}]\hat{p} + \hat{p}[\hat{x}, \hat{p}]$$
$$= 2i\hbar\hat{p}$$

so that

(5.63)
$$[\hat{x}, \hat{p}^2]g(x) = 2\hbar^2 \frac{\partial g}{\partial x}$$

In a similar vein,

(5.64)
$$[\hat{x}^2, \hat{p}] = \hat{x}[\hat{x}, \hat{p}] + [\hat{x}, \hat{p}]\hat{x}$$
$$= 2i\hbar\hat{x} = 2i\hbar x$$

The operator $[\hat{x}^2, \hat{p}]$ multiplies by $2i\hbar x$.

We now prove an important theorem in quantum mechanics which is related to the commutator between two operators. It states: if $\hat{A}$ and $\hat{B}$ commute

(5.65)
$$[\hat{A}, \hat{B}] = 0$$

then $\hat{A}$ and $\hat{B}$ have a set of nontrivial (i.e., other than a constant) common eigenfunctions. The proof is as follows.

Let φ_a be the eigenfunction of $\hat{A}$ that corresponds to the eigenvalue a.

(5.66)
$$\hat{A}\varphi_a = a\varphi_a$$

Then

(5.67)
$$\hat{B}\hat{A}\varphi_a = a\hat{B}\varphi_a$$

Since $\hat{A}$ and $\hat{B}$ commute, the left-hand side of this last equation may be rewritten

(5.68)
$$\hat{A}(\hat{B}\varphi_a) = a(\hat{B}\varphi_a)$$

Inspection of this equation reveals that $\hat{B}\varphi_a$ is also an eigenfunction of $\hat{A}$ corresponding to the eigenvalue a. If φ_a is the *only* linearly independent (defined below) eigenfunction of $\hat{A}$ that corresponds to the eigenvalue a, the function $\hat{B}\varphi_a$ can differ from φ_a by, at most, a multiplicative constant μ. That is,

(5.69)
$$\hat{B}\varphi_a = \mu\varphi_a$$

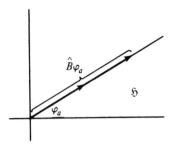

FIGURE 5.8 If $\hat{B}\varphi_a$ is eigenvector of $\hat{A}$ corresponding to the eigenvalue a, $\hat{B}\varphi_a$ and φ_a are in the same direction in Hilbert space $\mathfrak{H}$.

($\hat{B}\varphi_a$ and $\mu\varphi_a$ are in the same *direction* in Hilbert space; see Fig. 5.8.) But this is the eigenvalue equation for the operator $\hat{B}$. It follows that φ_a is also an eigenfunction of $\hat{B}$.

　　We have already encountered the implication of this theorem for the problem of the free particle moving in one dimension. For this case

(5.70)
$$[\hat{p}, \hat{H}] = 0$$

It follows by the theorem above that $\hat{p}$ and $\hat{H}$ have common eigenfunctions. They do. We recall that

(5.71)
$$\hat{p}e^{ikx} = \hbar k e^{ikx}$$

$$\hat{H}e^{ikx} = \frac{\hbar^2 k^2}{2m} e^{ikx}$$

Before pursuing the case when φ_a is not the only linearly independent eigenfunction of $\hat{A}$ corresponding to the eigenvalue a, we consider the definition of linearly independent functions.

Linearly Independent Functions

When is a set of functions a linearly independent set? The N functions of the set $\{\varphi_n\}$ are linearly independent if the linear combination

(5.72)
$$\sum_{n=1}^{N} \lambda_n \varphi_n = 0$$

for all x is *only* satisfied when

(5.73)
$$\lambda_1 = \lambda_2 = \cdots = \lambda_n = 0$$

For example, the two functions e^x and $\sin x$ are linearly independent since

(5.74)
$$\lambda_1 e^x + \lambda_2 \sin x = 0$$

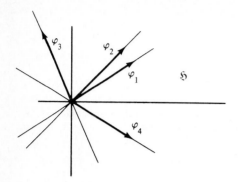

FIGURE 5.9 If $\{\varphi_1, \varphi_2, \varphi_3, \varphi_4\}$ **are a linearly independent set, no two lie along the same axis in Hilbert space $\mathfrak{H}$.**

for all x is only satisfied by

(5.75)
$$\lambda_1 = \lambda_2 = 0$$

The two functions e^x and $3e^x$ are not linearly independent since

(5.76)
$$\lambda_1 e^x + 3\lambda_2 e^x = 0$$

is true for all x if

(5.77)
$$\lambda_1 = -3\lambda_2 \neq 0$$

The concept of linearly independent functions has an interesting geometrical interpretation in Hilbert space. If two "vectors" φ_1 and φ_2 in a Hilbert space $\mathfrak{H}$ are linearly independent, they do not lie along the same axis (line) in $\mathfrak{H}$ (Fig. 5.9). Similarly, if the set of N vectors $\{\varphi_n\}$ are linearly independent, no two elements of this set lie on the same axis. If φ_1 and φ_2 are linearly independent, one must "rotate" φ_1 to align it with φ_2.

If φ_a is the only linearly independent eigenfunction of $\hat{A}$ corresponding to the eigenvalue a, all eigenfunctions of $\hat{A}$ corresponding to a must be of the form $\mu\varphi_a$. The functions φ_a and $\mu\varphi_a$ are two linearly dependent eigenfunctions of $\hat{A}$ corresponding to the eigenvalue a.

(5.78)
$$\hat{A}(\mu\varphi_a) = \mu\hat{A}\varphi_a = \mu a\varphi_a = a(\mu\varphi_a)$$

How many such vectors are there? Since μ can be any constant, there is a continuum of such linearly dependent eigenfunctions of $\hat{A}$ corresponding to the eigenvalue a. In any given problem only one of these states is relevant. For bound states $(|\psi|^2 \rightarrow 0, |x| \rightarrow \infty)$, ψ is fixed (and therefore μ) by normalization. For an unbound state $(|\psi|^2 \nrightarrow 0, |x| \rightarrow \infty)$, ψ is fixed through an appropriate boundary condition. The latter case is appropriate to beam or scattering problems, where the boundary conditions usually involve stipulations on particle current or number density at $|x| = \infty$. These concepts are discussed in greater detail in Section 7.5, which concerns one-dimensional barrier problems.

PROBLEMS

5.12 If $\hat{A}$, $\hat{B}$, and $\hat{C}$ are three distinct operators, show that:
(a) $[\hat{A} + \hat{B}, \hat{C}] = [\hat{A}, \hat{C}] + [\hat{B}, \hat{C}]$
(b) $[\hat{A}\hat{B}, \hat{C}] = \hat{A}[\hat{B}, \hat{C}] + [\hat{A}, \hat{C}]\hat{B}$

5.13 If $\hat{A}$ and $\hat{B}$ are both Hermitian, show that $\hat{A}\hat{B}$ is Hermitian if $[\hat{A}, \hat{B}] = 0$.

5.14 Show that the solution to the time-dependent Schrödinger equation given in Problem (3.18), that is,

$$\psi(\mathbf{r}, t) = \exp\left[-\frac{i}{\hbar}\int_0^t dt' \hat{H}(t')\right]\psi(\mathbf{r}, 0)$$

is correct, provided that

$$[\hat{H}(t), \hat{H}(t')] = 0 \qquad (t \neq t')$$

Answer
For $\psi(\mathbf{r}, t)$ as given above to be a solution, the expansion

$$\frac{\partial}{\partial t} e^{\hat{W}}\psi = e^{\hat{W}}\frac{\partial\psi}{\partial t} + e^{\hat{W}}\hat{H}\psi$$

must be valid in order to obtain the Schrödinger equation (with $i/\hbar = 1$). For this to be so, $e^{\hat{W}}$ in the second term must precede $\hat{H}$. Here we have

$$\hat{W} \equiv \int_0^t \hat{H}(t')\,dt'$$

We must show that

$$\frac{\partial}{\partial t} e^{\hat{W}} = e^{\hat{W}}\frac{\partial\hat{W}}{\partial t}$$

In general

$$\frac{\partial}{\partial t} e^{\hat{W}} = \frac{\partial}{\partial t}\left(1 + \hat{W} + \frac{1}{2}\hat{W}^2 + \frac{1}{6}\hat{W}^3 + \cdots\right)$$

$$= \frac{\partial\hat{W}}{\partial t} + \frac{1}{2}\left(\hat{W}\frac{\partial\hat{W}}{\partial t} + \frac{\partial\hat{W}}{\partial t}\hat{W}\right) + \cdots$$

Thus the equality above holds if we are able to set

$$\left[\hat{W}, \frac{\partial\hat{W}}{\partial t}\right] = 0$$

In this case

$$\frac{\partial}{\partial t} e^{\hat{W}} = \left(\frac{\partial\hat{W}}{\partial t} + \hat{W}\frac{\partial\hat{W}}{\partial t} + \frac{1}{2}\hat{W}^2\frac{\partial\hat{W}}{\partial t} + \cdots\right)$$

$$= e^{\hat{W}}\frac{\partial\hat{W}}{\partial t}$$

In terms of the integral definition of $\hat{\mathscr{W}}$, the commutation criterion above becomes

$$\hat{H}(t) \int_0^t \hat{H}(t')\,dt' = \left(\int_0^t \hat{H}(t')\,dt' \right) \hat{H}(t)$$

which is guaranteed if $[\hat{H}(t).\ \hat{H}(t')] = 0$.

5.15   Discuss the linear independence of the following sets of functions.
  (a) $\{x,\ 3x,\ e^x\}$
  (b) $\{e^{ix},\ \sin x,\ \cos x\}$
  (c) $\{x^2,\ x^3,\ x^5\}$
  (d) $\{x,\ 3,\ \sin^2 x,\ 4\cos^2 x,\ \ln x\}$

5.16   If μ is an arbitrary constant, the two vectors φ and $\mu\varphi$ in $\mathfrak{H}$ are linearly dependent. Show that the cosine of the angle between these two vectors has modulus 1.

$$|\cos \theta| = 1$$

5.17   From Problem 5.16 we conclude that φ and $\mu\varphi$ lie along the same axis in $\mathfrak{H}$. Show also that $\mu\varphi$ is $|\mu|$ times longer than φ, that is, that (see Fig. 5.10)

$$\|\mu\varphi\| = |\mu|\|\varphi\|$$

5.18   Show that if $\hat{A}\varphi_n = a_n\varphi_n$ and $\hat{B}\varphi_n = b_n\varphi_n$ for all eigenvalues $\{a_n\}$ and $\{b_n\}$ of $\hat{A}$ and $\hat{B}$, respectively (i.e., $\hat{A}$ and $\hat{B}$ have completely common eigenstates), then $[A, B] = 0$ on the space of functions spanned by the basis $\{\varphi_n\}$. (*Hint*: Any element of this space may be written

$$\psi = \sum c_n\varphi_n$$

and one need merely show that

$$[\hat{A},\hat{B}] \sum c_n\varphi_n = 0.)$$

Note: In a more general vein one may say the following: let the eigenstates common to $\hat{A}$ and $\hat{B}$ span a subspace $\mathscr{G}$ of a Hilbert space $\mathfrak{H}$. Then $[\hat{A},\hat{B}]\psi = 0$, where ψ is any element of $\mathscr{G}$.

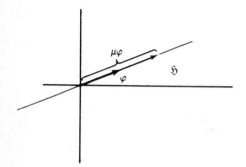

FIGURE 5.10   The vectors φ and $\mu\varphi$ in Hilbert space $\mathfrak{H}$ lie along the same axis and $\|\mu\varphi\| = |\mu|\|\varphi\|$. (See Problem 5.17.)

5.3 MORE ON THE COMMUTATOR THEOREM

The Concept of Degeneracy

Suppose there are two (and *only* two) linearly independent eigenfunctions of the operator $\hat{A}$ which both correspond to the eigenvalue a. Call them φ_1 and φ_2.

$$
(5.79) \qquad \begin{aligned} \hat{A}\varphi_1 &= a\varphi_1 \\ \hat{A}\varphi_2 &= a\varphi_2 \end{aligned}
$$

Under such circumstances one says that *the eigenvalue a is doubly degenerate*. The eigenfunctions φ_1 and φ_2 are degenerate. Now we ask, what is the most general eigenfunction of $\hat{A}$ that corresponds to the eigenvalue a? The answer is, any function of the form

$$
(5.80) \qquad \varphi_a = \alpha\varphi_1 + \beta\varphi_2
$$

with α and β arbitrary constants. Let us test that this is the case.

$$
(5.81) \qquad \begin{aligned} \hat{A}\varphi_a = \hat{A}(\alpha\varphi_1 + \beta\varphi_2) &= \alpha a \varphi_1 + \beta a \varphi_2 \\ &= a(\alpha\varphi_1 + \beta\varphi_2) \end{aligned}
$$

In Hilbert space the two functions φ_1 and φ_2 span a plane (two-dimensional subspace). Equation (5.80) indicates that any vector φ_a in this plane is an eigenfunction of $\hat{A}$ corresponding to the eigenvalue a (Fig. 5.11).

Let us return to the commutator theorem discussed in Section 5.2. The operators $\hat{A}$ and $\hat{B}$ commute. If we operate on the first of Eqs. (5.79) with $\hat{B}$ and use the commuting property of $\hat{A}$ and $\hat{B}$, there results

$$
(5.82) \qquad \hat{B}\hat{A}\varphi_1 = a(\hat{B}\varphi_1) = \hat{A}(\hat{B}\varphi_1)
$$

We conclude that $\hat{B}\varphi_1$ is an eigenstate of $\hat{A}$ that corresponds to the eigenvalue a. But there is a continuum of such eigenstates, all of the form (5.80). All we can say is that there are some α and β such that

$$
(5.83) \qquad \hat{B}\varphi_1 = \mu(\alpha\varphi_1 + \beta\varphi_2)
$$

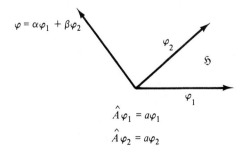

$$\varphi = \alpha\varphi_1 + \beta\varphi_2$$

$$\varphi_2$$

$$\mathfrak{H}$$

$$\varphi_1$$

$$\hat{A}\varphi_1 = a\varphi_1$$

$$\hat{A}\varphi_2 = a\varphi_2$$

FIGURE 5.11 If φ_1 and φ_2 are two linearly independent degenerate eigenvectors of $\hat{A}$, they span a "plane" (two-dimensional subspace) in Hilbert space $\mathfrak{H}$. Any vector in this plane is an eigenvector of $\hat{A}$ corresponding to the eigenvalue a.

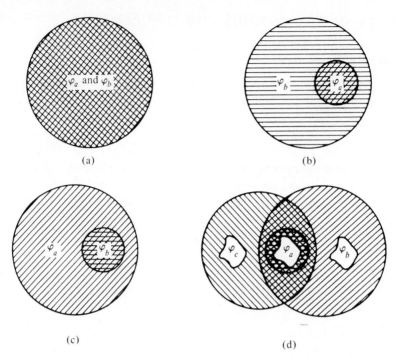

FIGURE 5.12 Various cases pertaining to the sets of eigenfunctions of two compatible operators, $\hat{A}$ and $\hat{B}$. $[\hat{A}, \hat{B}] = 0$. (a) eigenfunctions of $\hat{A}$ = all eigenfunctions of $\hat{B}$. (b) $\hat{A}$ has only nondegenerate eigenfunctions. (c) $\hat{B}$ has only nondegenerate eigenfunctions. (d) $[\hat{A}, \hat{B}] = [\hat{B}, \hat{C}] = [\hat{A}, \hat{C}] = 0$. $\hat{A}$ has only nondegenerate eigenfunctions.

Inspection of this equation [compare with (5.69)] reveals that φ_1 need not be an eigenfunction of $\hat{B}$.

So we have the following rule: If $[\hat{A}, \hat{B}] = 0$, and a is a degenerate eigenvalue of $\hat{A}$, the corresponding eigenfunctions of $\hat{A}$ (which all have the same eigenvalue, a) are not necessarily eigenfunctions of $\hat{B}$. Loosely speaking, degenerate operators have "more" eigenstates than nondegenerate operators. This concept may be illustrated in terms of the Venn diagrams depicted in Fig. 5.12.

A very simple physical example of this situation is provided by the problem of the free particle moving in one dimension. The eigenvalue

(5.84)
$$E_k = \frac{\hbar^2 k^2}{2m}$$

of the Hamiltonian [see (3.17)]

(5.85)
$$\hat{H} = \frac{\hat{p}^2}{2m}$$

is doubly degenerate. All of the following functions are eigenfunctions of $\hat{H}$ corresponding to this eigenvalue.

(5.86) $$(\varphi_1, \varphi_2, \varphi_3) = \{(\cos kx, \sin kx, \exp(ikx)\}$$

This is not a linearly independent set. However, any two are, so that for the free particle the eigenvalue (5.84) is doubly degenerate. For example, the two linearly independent functions, say

(5.87) $$\{\varphi_1, \varphi_2\} = \{\cos kx, \sin kx\}$$

both have the eigenvalue $\hbar^2 k^2/2m$. Although $[\hat{p}, \hat{H}] = 0$, for the free particle, the set of functions (5.87), being degenerate eigenfunctions of energy, need not be eigenfunctions of $\hat{p}$. In fact, they are not.

Another linearly independent set of degenerate eigenstates corresponding to the eigenenergy $\hbar^2 k^2/2m$ is $\{\varphi_2, \varphi_3\}$. Of these, φ_3 is an eigenstate of $\hat{p}$ and φ_2 is not. Of the set $\{\varphi_1, \varphi_3\}$, φ_1 is not an eigenstate of $\hat{p}$, and again φ_3 is.

When there are n (and only n) linearly independent eigenstates of an operator $\hat{A}$ that all correspond to the same eigenvalue, the eigenvalue is *n-fold degenerate*. Suppose that $[\hat{A}, \hat{B}] = 0$. What can then be said is that from these n degenerate eigenstates of $\hat{A}$, one can form n linear combinations which are n linearly independent eigenstates of both $\hat{A}$ and $\hat{B}$.

For instance, from the two degenerate eigenstates (5.87) in the free-particle problem above, we can form

(5.88) $$\varphi_+ = \varphi_1 + i\varphi_2 = \cos kx + i \sin kx = e^{ikx}$$

(5.89) $$\varphi_- = \varphi_1 - i\varphi_2 = \cos kx - i \sin kx = e^{-ikx}$$

These two functions are common eigenstates of $\hat{H}$ and $\hat{p}$. They remain degenerate eigenstates of $\hat{H}$ but are nondegenerate eigenstates of $\hat{p}$.

PROBLEMS

5.19 Construct two linearly independent linear combinations of φ_2 and φ_3 given in (5.86) which are common eigenfunctions of $\hat{H}$ and $\hat{p}$.

5.20 Given that $\hat{x}$ and $\hat{p}$ operate on functions in $\mathfrak{H}_2$ and the relation $[\hat{x}, \hat{p}] = i\hbar$, show that if $\hat{x} = x$ (i.e., multiplication by x), $\hat{p}$ has the representation

$$\hat{p} = -i\hbar \frac{\partial}{\partial x} + f(x)$$

where $f(x)$ is an arbitrary function of x.

[*Note:* Dirac[1] has shown that through proper choice of *phase factor* (Section 4.2), the arbitrary function f may always be made to vanish. Thus, the basic commutator relation between $\hat{x}$ and $\hat{p}$

---

[1] Dirac, *The Principles of Quantum Mechanics.*

is equivalent to explicit operator forms for these variables. Listing of such commutator relations may serve in place of postulate I (Section 3.1).]

5.21   The operators $\hat{A}$ and $\hat{B}$ both have a denumerable number of eigenstates. Of these, the single eigenstate φ is known to be common to both. That is,

$$\hat{A}\varphi = a\varphi, \qquad \hat{B}\varphi = b\varphi.$$

(a)   What can be said about the commutability of $\hat{A}$ and $\hat{B}$?

(b)   Suppose that it is further known that all the eigenstates of $\hat{A}$ and $\hat{B}$ are degenerate. Does this additional information in any way change your answer to part (a)?

5.4   COMMUTATOR RELATIONS AND THE UNCERTAINTY PRINCIPLE

As we have seen above, owing to the fact that for a free particle, $\hat{p}$ and $\hat{H}$ commute, they have a set of simultaneous eigenfunctions. Namely, any function of the form

$$(5.90) \qquad\qquad \varphi = Ae^{ikx}$$

is a common eigenstate of both $\hat{p}$ and $\hat{H}$. If the system (particle) is in this state, it is certain that measurement of p gives $\hbar k$ and measurement of energy gives $\hbar^2 k^2 / 2m$. Since (5.90) is a common eigenstate of $\hat{p}$ and $\hat{H}$, measurement of p, which (absolutely) gives $\hbar k$, leaves the particle in the state (5.90). Subsequent measurement of E gives $\hbar^2 k^2 / 2m$ and also leaves the particle in the state (5.90).[1] The operators $\hat{H}$ and $\hat{p}$ are *compatible*; that is, they commute. Quantum mechanics allows p and E to be simultaneously specified (for a free particle). Furthermore, there is only one (unique) state which gives these two values, the state (5.90).

Although there exists a state in which both energy and momentum may be specified simultaneously, the same is not true for the observables $\hat{x}$ and $\hat{p}$. There is no state in which measurement is certain to yield definite values of x and p. Measurement of p leaves the system in an eigenstate of $\hat{p}$ (5.90). Subsequent measurement of x is infinitely uncertain. The state (5.90) is not an eigenstate of $\hat{x}$. Conversely, measurement of x that finds x' leaves the system in the eigenstate of $\hat{x}$,

$$(5.91) \qquad\qquad \psi = \delta(x - x')$$

---

[1] Here we mean an *ideal* measurement. This is a measurement which *least* perturbs the system. Any real measurement causes the system to suffer a greater perturbation. After the energy of the particle in the state (5.90) is measured, *ideal* measurement maintains that state. However, it is also possible that after finding $\hbar^2 k^2 / 2m$, the particle is in any linear combination of the independent degenerate energy eigenfunctions of $\hat{H}$ which correspond to this eigenvalue (e.g., $\alpha \cos kx + \beta \sin kx$). However, measurement that leaves the particle in this state must have interfered with the momentum, since this state is a superposition of momentum eigenstates. Measurement that leaves the system in the original state (5.90) does not perturb the momentum. It is the *ideal* measurement of energy.

When the particle is in this state, measurement of momentum is infinitely uncertain.

For the free particle, there are states in which the uncertainty in energy and momentum obeys the relation

$$\Delta E \, \Delta p = 0 \tag{5.92}$$

On the other hand, in any state, the uncertainties in observation of p and x are such that the product $\Delta p \, \Delta x$ is always greater than a fixed magnitude.

$$\Delta x \, \Delta p \geq \frac{\hbar}{2} \tag{5.93}$$

It is quite clear at this point that these uncertainty relations have their origin in the compatibility properties (5.51) of the operators that correspond to the observables being measured.

Suppose that two observables $\hat{A}$ and $\hat{B}$ are not compatible:

$$[\hat{A}, \hat{B}] = \hat{C} \neq 0 \tag{5.94}$$

For example, such is the case for displacement and kinetic energy. Then one can show the following[1]: If measurement of A, in the state ψ, is uncertain by the amount ΔA, then measurement of B is uncertain by the amount ΔB, such that[2]

$$\Delta A \, \Delta B \geq \tfrac{1}{2} |\langle C \rangle| \tag{5.95}$$

We recall (Section 3.3) that the uncertainty of an observable A in the state ψ is the root mean square of the deviation of A away from the mean $\langle A \rangle$.

$$(\Delta A)^2 \equiv \langle (A - \langle A \rangle)^2 \rangle = \langle A^2 \rangle - \langle A \rangle^2$$

Expectation values in (5.95) are calculated in the state ψ. For example,

$$\langle C \rangle = \langle \psi | \hat{C} \psi \rangle, \qquad (\Delta A)^2 = \langle (\hat{A} - \langle A \rangle) \psi | (\hat{A} - \langle A \rangle) \psi \rangle \tag{5.96}$$

The mechanism at work behind these uncertainty relations is as follows. If $\hat{A}$ and $\hat{B}$ do not commute, then the eigenstate φ_a of $\hat{A}$ which the system goes into on measurement of A is not necessarily an eigenstate of $\hat{B}$. Subsequent measurement of B will give any of the spectrum of eigenvalues of $\hat{B}$ with a corresponding probability distribution $P(b)$. This probability distribution is obtained from the coefficients in the expansion of φ_a in the eigenstates φ_b of $\hat{B}$.

$$P(b) = |\langle \varphi_b | \varphi_a \rangle|^2 \tag{5.97}$$

(with $\{\varphi_a\}$ and $\{\varphi_b\}$ normalized). Remeasurement of A is then in no way certain of finding the system in the state φ_a.

---

[1] See Problem 5.42.

[2] This generalization of the uncertainty principle is sometimes called the Robertson–Schrödinger relation [H. P. Robertson, *Phys. Rev.* **35**, 667A (1930); E. Schrödinger, *Sitzungsber. Preuss. Akad. Wiss.* (1930), p. 296].

We note that the commutator-uncertainty relation, (5.94) and (5.95), is among the more fundamental relations in quantum mechanics. In addition to its important practical significance, it stands as an immutable barrier separating quantum and classical physics.

PROBLEMS

5.22   How do the states for a free particle

$$\varphi_1 = A e^{ikx}$$

$$\varphi_2 = B \cos kx$$

differ with regard to measurements of momentum and energy?

5.23   For a particle in a one-dimensional potential field $V(x)$, show that

$$\Delta E \, \Delta x \geq \frac{\hbar}{2m} \langle p_x \rangle$$

5.24   Consider three observables, $\hat{A}$, $\hat{B}$, and $\hat{C}$. If it is known that

$$[\hat{B}, \hat{C}] = \hat{A}$$

$$[\hat{A}, \hat{C}] = \hat{B}$$

show that

$$\Delta(AB) \, \Delta C \geq \tfrac{1}{2} \langle A^2 + B^2 \rangle$$

5.25   Obtain uncertainty relations for the following products
  (a) $\Delta x \, \Delta E$
  (b) $\Delta p_x \, \Delta E$
  (c) $\Delta x \, \Delta T$
  (d) $\Delta p_x \, \Delta T$
relevant to a particle whose kinetic energy is T and whose total energy is E. (A closely related example is discussed in Problem 2.30.)

5.26   If $g(x)$ is an arbitrary function of x, show that

$$[\hat{p}_x, g] = -i\hbar \frac{dg}{dx}$$

5.27   If $g(x)$ and $f(x)$ are both analytic functions, show that

$$g(\hat{A}) f(\varphi) = g(a) f(\varphi), \qquad \text{where } \hat{A}\varphi = a\varphi$$

5.28   The time-dependent Schrödinger equation permits the identification

$$\hat{E} = i\hbar \frac{\partial}{\partial t}$$

Using this identification together with the rule (5.95), give a formal derivation of the uncertainty relation

$$\Delta E \, \Delta t \geq \tfrac{1}{2} \hbar$$

Note that in a stationary state (eigenstate of $\hat{H}$), $\Delta E = 0$. The implication for this case is that a stationary state may last indefinitely.

5.29 Can the total energy and linear momentum of a particle moving in one dimension in a constant potential field be measured consecutively with no uncertainty in the values obtained?

5.30 If

$$[\hat{A}, \hat{B}] = i\hat{C}$$

and $\hat{A}$ and $\hat{B}$ are both Hermitian, show that $\hat{C}$ is also Hermitian.

5.31 Prove that if $\hat{A}$ and $\hat{B}$ are Hermitian, $[\hat{A}, \hat{B}]$ is Hermitian if and only if $[\hat{A}, \hat{B}] = 0$.

Answer (partial)

Set $[\hat{A}, \hat{B}] \equiv \hat{K}$. Then $\hat{K} = -\hat{K}^\dagger$. But $\hat{K}^\dagger = \hat{K}$, hence $\hat{K} = -\hat{K}$.

5.32 (a) Obtain an uncertainty relation for mass and time from the relativistic mass-energy equivalency formula.

(b) A free neutron has a mean lifetime of $\simeq 10^3$ s. Apply the uncertainty relation found in part (a) to find the uncertainty in the neutron's mass.

Answer (partial)

(b) $\Delta m \simeq 10^{-27}$ amu ($M_p \simeq 1$ amu)

5.5 "COMPLETE" SETS OF COMMUTING OBSERVABLES

We have already seen that for the free particle in one dimension, the eigenvalues of $\hat{H}$ are doubly degenerate. The two eigenfunctions of $\hat{H}$ corresponding to the eigenvalue $\hbar^2 k^2 / 2m$ are $\exp(+ikx)$ and $\exp(-ikx)$. However, once we specify what p is (say $+\hbar k$), in addition to E, then one can say that the system is in one and only one state, $\exp(+ikx)$ (to within a multiplicative constant). Merely prescribing the energy of the particle does not uniquely determine the state of the particle. Further specifying the momentum removes this ambiguity and the state of the particle is uniquely determined.

Suppose that an operator $\hat{A}$ has degenerate eigenvalues. If a is one of these values, specifying a does not uniquely determine which state the system is in. Let $\hat{B}$ be another operator which is compatible with $\hat{A}$. Consider all the eigenstates $\{\varphi_{ab}\}$ which are common to $\hat{A}$ and $\hat{B}$. Of the degenerate eigenstates of $\hat{A}$, only a subset of these are also eigenfunctions of $\hat{B}$. Under such conditions, if we specify the eigenvalue b and the eigenvalue a, then the state that the system can be in is a smaller set than that determined by specification of a alone. Suppose further that there is only one other operator $\hat{C}$ which is compatible with both $\hat{A}$ and $\hat{B}$. Then they all share a set of common eigenstates. Call these states φ_{abc}. Then

(5.98)
$$\hat{A}\varphi_{abc} = a\varphi_{abc}$$
$$\hat{B}\varphi_{abc} = b\varphi_{abc}$$
$$\hat{C}\varphi_{abc} = c\varphi_{abc}$$

These functions are still a smaller set than the set $\{\varphi_a\}$ or $\{\varphi_{ab}\}$. Indeed, let us consider that φ_{abc} is *uniquely* determined by the values a, b, and c. This means that having measured a, b, and c: (1) Since φ_{abc} is a common eigenstate of $\hat{A}$, $\hat{B}$ and $\hat{C}$, simultaneous measurement (or a succession of three immediately repeated "ideal" measurements) of A, B, and C will definitely find the values a, b, and c. (2) The state φ_{abc} cannot be further resolved by more measurement. This state contains a maximum of information which is permitted by the laws of quantum mechanics. (3) There are no other operators independent of $\hat{A}$, $\hat{B}$, and $\hat{C}$ which are compatible with these. If there were, the state φ_{abc} could be further resolved. An exhaustive set (in the sense that there are no other independent operators compatible with $\hat{A}$, $\hat{B}$, and $\hat{C}$) of commuting operators such as $\hat{A}$, $\hat{B}$, and $\hat{C}$ above, whose common eigenstates are uniquely determined by the eigenvalues a, b, and c and are a basis of Hilbert space, is called a *complete set of commuting operators*.

Maximally Informative States

The values a, b, and c, which may be so specified in the state φ_{abc}, are sometimes referred to as *good quantum numbers*. These are analogous to the generalized coordinates whose values determine the state of a system classically. As discussed in Section 1.1, such classical coordinates are also labeled *good variables*.

Suppose that there are, in all, five independent operators that specify the properties of a system: $\hat{A}$, $\hat{B}$, $\hat{C}$, $\hat{D}$, and $\hat{F}$. Of these, $\hat{A}$, $\hat{B}$, and $\hat{C}$ are compatible with one another and $\hat{D}$ and $\hat{F}$ are compatible. However, these two sets are incompatible with one another, so that, for example,

(5.99)
$$[\hat{A}, \hat{D}] \neq 0$$

One can simultaneously specify either the eigenvalues a, b, and c or the eigenvalues d and f. One cannot, for instance, say that the system is in a state for which measurement of A definitely gives a and measurement of D definitely gives d. For this case there are two sets of states that are maximally informative: $\{\varphi_{abc}\}$ and $\{\varphi_{df}\}$.

Suppose that $\hat{A}$ has degenerate eigenvalue a. What is the state of the system after one has measured and found a? The state lies in a subspace of Hilbert space which is spanned by the degenerate eigenfunctions that correspond to a. This subspace $\mathfrak{H}_a$ has dimensionality $\mathcal{N}_a$ (a is an $\mathcal{N}_a$-fold degenerate eigenvalue). After measurement of $\hat{B}$, the state of the system lies in the space $\mathfrak{H}_{ab}$, which is a subspace of $\mathfrak{H}_a$ and is spanned by the eigenfunction common to $\hat{A}$ and $\hat{B}$. This subspace has dimensionality $\mathcal{N}_{ab}$, which is not greater than $\mathcal{N}_a$.

(5.100)
$$\mathcal{N}_{ab} \leq \mathcal{N}_a$$

Subsequent measurement of $\hat{C}$ (mutually compatible with $\hat{A}$ and $\hat{B}$) leaves the state of the system in a space $\mathfrak{H}_{abc}$ that is a subspace of $\mathfrak{H}_{ab}$ and whose dimensionality does not exceed that of $\mathfrak{H}_{ab}$.

(5.101)
$$\mathcal{N}_{abc} \leq \mathcal{N}_{ab}$$

In this manner we can proceed to measure more and more mutually compatible observables. At each step of the way the eigenstate is forced into subspaces of lesser and lesser dimensionality, until finally after the successive measurement of $A, B, C, D, \ldots$ the state of the system is forced into a subspace of dimensionality $N = 1$. This is a space spanned by only one function. It is the eigenstate common to the complete set of observables $(\hat{A}, \hat{B}, \hat{C}, \hat{D}, \ldots)$: namely, $\varphi_{abcd} \ldots$. This state cannot be further resolved by additional measurements. Measurement of any of the observables $(A, B, C, D, \ldots)$ in this state is certain to find the respective values $(a, b, c, d, \ldots)$.

PROBLEMS

5.33 (a) Show that for a particle in a one-dimensional box, in an arbitrary state $\psi(x, 0)$,

$$\langle E \rangle \geq E_1$$

(b) Under what conditions does the equality maintain?

5.34 A free particle at a given instant of time is in the state

$$\psi = \frac{A}{(xk_0)^2 + 4}$$

At this same instant, (ideal) measurement of the energy finds that

$$E = \frac{\hbar^2 k_0^2}{2m}$$

The measurement leaves the momentum uncertain. Under such circumstances, what is the state $\tilde{\psi}$ of the particle immediately after measurement?

Answer

Since the momentum is uncertain after measurement, we know that the state is not one of the eigenstates of momentum $\varphi_{\pm k_0}$. Instead, one may say that the state vector lies in a subspace of $\mathfrak{H}$ spanned by the vectors $\cos k_0 x$ and $\sin k_0 x$.

$$\tilde{\psi} = \alpha \cos k_0 x + \beta \sin k_0 x$$

The coefficients α and β are proportional to the projections of ψ on $\cos k_0 x$ and $\sin k_0 x$, respectively. Since

$$\langle \psi | \sin k_0 x \rangle = 0$$

it follows that after measurement, the state of the particle is

$$\tilde{\psi} = \alpha \cos k_0 x$$

5.35   Show that

$$e^{\hat{A}} \hat{B} e^{-\hat{A}} = \hat{B} + [\hat{A}, \hat{B}] + \frac{1}{2!} [\hat{A}, [\hat{A}, \hat{B}]] + \frac{1}{3!} [\hat{A}, [\hat{A}, [\hat{A}, \hat{B}]]] + \cdots$$

(*Hint:* Taylor-series expand $\tilde{f}(\eta) \equiv e^{\eta \hat{A}} \hat{B} e^{-\eta \hat{A}}$ about $\eta = 0$. Also note the derivative property of $\hat{f}(\eta)$: $d\hat{f}/d\eta = [\hat{A}, \hat{f}]$.)

5.36   Show that [Baker–Hausdorf lemma]

$$e^{\hat{A}} e^{\hat{B}} = e^{\hat{A} + \hat{B}} e^{(1/2)[\hat{A}, \hat{B}]}$$

given that $\hat{A}$ and $\hat{B}$ each commutes with $[\hat{A}, \hat{B}]$. (*Hint:* First show that $[e^{\eta \hat{A}}, \hat{B}] = \eta e^{\eta \hat{A}} [\hat{A}, \hat{B}]$. Then establish that the derivative of

$$\hat{g}(\eta) \equiv e^{\eta \hat{A}} e^{\eta \hat{B}} e^{-\eta(\hat{A} + \hat{B})}$$

is

$$\frac{d\hat{g}}{d\eta} = \eta [\hat{A}, \hat{B}] \hat{g}$$

and integrate.) Note that for $\beta \ll 1$, one may always write

$$e^{\beta(\hat{A} + \hat{B})} \simeq e^{\beta \hat{A}} e^{\beta \hat{B}} e^{-(1/2)\beta^2 [\hat{A}, \hat{B}]}$$

This relation is important in statistical mechanics, where β plays the role of inverse temperature and $\hat{A} + \hat{B}$ is the Hamiltonian.

5.37   The operator $\hat{A}$ has only nondegenerate eigenvectors and eigenvalues, $\{\varphi_n\}$ and $a_n\}$. What are the eigenvectors and eigenvalues of the inverse operator, $\hat{A}^{-1}$? Is your answer consistent with the commutator theorem?

5.38   (a)   Construct a one-dimensional wave packet that has zero probability density outside a domain of length L at time $t = 0$ and which has average momentum $\langle p \rangle = +\hbar k_0$. That is, it is propagating to the right.

(b)   The wave packet collides with a mass m. Estimate the probability that the mass is deflected to the left with momentum $\hbar k_0/10 \pm \hbar k_0/100$. Take $k_0 L = 10\pi$. (Assume that complete momentum exchange occurs simultaneously at $t = 0$. The mass is located at the origin.)

5.39   Show that the expectation of an observable A of a system that is in the superposition state

$$\psi(x, t) = \sum_n b_n \varphi_n e^{i\omega_n t}$$

may be written in the form

$$\langle A \rangle = 2 \sum_{n>l} \sum b_n^* b_l \langle n|A|l \rangle \cos(\omega_n - \omega_l)t + \sum_n |b_n|^2 \langle n|A|n \rangle$$

for $\{b_n^* b_l\}$ and $\langle n|A|l \rangle$ real. The states $\varphi_n \exp(i\omega_n t)$ are eigenstates of the Hamiltonian of the system. Here we are writing $|n\rangle$ for φ_n.

5.40 What is the average $\langle x \rangle$ and square root of variance Δx for the following probability densities?

(a) $P(x) = A[1 + (x - x_0)^2]^{-1}$

(b) $P(x) = Ax^2 e^{-x^2/2a^2}$

(c) $P(x) = A \sin^2 \left(\dfrac{x - x_0}{\sqrt{2a}} - 8\pi \right) \exp \left\{ - \left[\dfrac{(x - x_0)^2}{2a^2} \right] \right\}$

5.41 (a) Show that for a particle in a one-dimensional box with walls at $(-L/2, L/2)$

$$\Delta p_{\min} = \sqrt{\langle p^2 \rangle_{\min}} = \frac{h}{2L}$$

(b) Show for this same configuration that

$$\Delta x_{\max} = \sqrt{\langle x^2 \rangle_{\max}} = \frac{L}{2\sqrt{3}}$$

(c) In which states are $\Delta p_{\min}$ and $\Delta x_{\max}$ realized?

(d) From part (a) obtain the following momentum uncertainty relation for this configuration:

$$L \, \Delta p \geq \frac{h}{2}$$

5.42 Given that $\hat{A}$ and $\hat{B}$ are Hermitian operators and that

$$[\hat{A}, \hat{B}] = i\hat{C}$$

show that

$$\Delta A \, \Delta B \geq \tfrac{1}{2} |\langle C \rangle|$$

Answer

The uncertainties in $\hat{A}$ and $\hat{B}$, when written in terms of the operators

$$\hat{\delta}_A \equiv \hat{A} - \langle A \rangle$$
$$\hat{\delta}_B \equiv \hat{B} - \langle B \rangle$$

appear as

$$(\Delta A)^2 = \langle \hat{\delta}_A \psi | \hat{\delta}_A \psi \rangle = \| \hat{\delta}_A \psi \|^2$$

$$(\Delta B)^2 = \| \hat{\delta}_B \psi \|^2$$

These expressions may be incorporated into the Schwartz inequality (Problem 4.20). There results

$$\|\hat{\delta}_A\psi\|^2\|\hat{\delta}_B\psi\|^2 \geq |\langle\hat{\delta}_A\psi|\hat{\delta}_B\psi\rangle|^2$$
$$(\Delta A)^2(\Delta B)^2 \geq |\langle\hat{\delta}_A\psi|\hat{\delta}_B\psi\rangle|^2 = |\langle\psi|\hat{\delta}_A\hat{\delta}_B\psi\rangle|^2$$

The latter equality is due to the Hermiticity of $\hat{\delta}_A$. We now recall that any operator can be written as a linear combination of two Hermitian operators:

$$\hat{\delta}_A\hat{\delta}_B = \frac{1}{2}(\hat{\delta}_A\hat{\delta}_B + \hat{\delta}_B\hat{\delta}_A) + \frac{1}{2}[\hat{\delta}_A, \hat{\delta}_B] \equiv \hat{G} + \frac{i}{2}\hat{C}$$

Here we have used the fact that $[\hat{\delta}_A, \hat{\delta}_B] = [\hat{A}, \hat{B}]$. Substituting the expression above into the preceding inequality gives

$$(\Delta A)^2(\Delta B)^2 \geq \left|\left\langle\psi\left|\left(\hat{G} + \frac{i}{2}\hat{C}\right)\psi\right\rangle\right|^2 = \left|\langle G\rangle + \frac{i}{2}\langle C\rangle\right|^2$$

Owing to the Hermiticity of $\hat{G}$ and $\hat{C}$, their expectation values are both real. It follows that

$$(\Delta A)^2(\Delta B)^2 \geq |\langle G\rangle|^2 + \tfrac{1}{4}|\langle C\rangle|^2 \geq \tfrac{1}{4}|\langle C\rangle|^2$$

5.43 The linear independence of two functions $u(x)$ and $v(x)$ may be specified in terms of their *Wronskian*,

$$W(u, v) = \begin{vmatrix} u & v \\ u' & v' \end{vmatrix}$$

Thus, if u and v are solutions to a linear, second-order differential equation, and $W(u, v) \neq 0$ in some interval, then u and v are independent solutions in this interval. Employing this criterion, establish that the two functions given by (5.87) are independent over the entire x axis. What value of W do you find for this case?

5.44 Determine an expression for the Bohr radius a_0 from the following crude approximation. The electron moves to the nucleus to lower its potential energy,

$$V(r) = -\frac{e^2}{r}$$

If the electron is in the domain $0 \leq r \leq \bar{r}$, then we may write $\Delta p \simeq \hbar/\bar{r}$, with corresponding kinetic energy $\hbar^2/2m\bar{r}^2$. With this information estimate a_0 by minimizing the total energy. How does your answer compare with the actual expression for a_0 given by (2.14)?

5.45 Show that in three dimensions, the coordinate-momentum Commutation relation (5.61) may be written

$$[\mathbf{r}, \mathbf{p}] = i\hbar\hat{I}$$

In this expression, the dyadic commutator has nine components and may be written as a 3×3 matrix and $\hat{I}$ is the identity operator.

5.46 Show that in three dimensions, the coordinate-momentum uncertainty relation (5.93) may be written

$$(\Delta r)^2 (\Delta p)^2 \geq \tfrac{3}{4}\hbar^2$$

where

$$(\Delta r)^2 = \langle (\mathbf{r} - \langle \mathbf{r} \rangle)^2 \rangle$$
$$(\Delta p)^2 = \langle (\mathbf{p} - \langle \mathbf{p} \rangle)^2 \rangle$$

Answer

Expanding the given uncertainty relation gives

$$(\Delta r)^2 (\Delta p)^2 = (\langle r^2 \rangle - \langle \mathbf{r} \rangle^2)(\langle p^2 \rangle - \langle \mathbf{p} \rangle^2)$$

Next note that

$$\langle \mathbf{r} \rangle = \langle \mathbf{e}_x x + \mathbf{e}_y y + \mathbf{e}_z z \rangle = \mathbf{e}_x \langle x \rangle + \mathbf{e}_y \langle y \rangle + \mathbf{e}_z \langle z \rangle$$

where $\mathbf{e}_x$ is a unit vector in the x direction. Thus we may write

$$\langle \mathbf{r} \rangle^2 = \langle \mathbf{r} \rangle \cdot \langle \mathbf{r} \rangle = \langle x \rangle^2 + \langle y \rangle^2 + \langle z \rangle^2$$

Furthermore

$$\langle r^2 \rangle = \langle x^2 \rangle + \langle y^2 \rangle + \langle z^2 \rangle$$

It follows that the given expansion contains 36 Cartesian products. However, contributions such as $(\Delta x)^2 (\Delta p_y)^2$ are zero. Each such zero contribution contains 4 Cartesian products. In all, there are 6 such contributions, which gives 24 terms that drop out of the expansion, leaving 12 terms. These remaining terms comprise the relation

$$(\Delta x)^2 (\Delta p_x)^2 + (\Delta y)^2 (\Delta p_y)^2 + (\Delta z)^2 (\Delta p_z)^2 \geq \frac{3\hbar^2}{4}$$

The validity of this relation follows from summing the squares of the uncertainty relations for each of the Cartesian components.

5.47 (a) A particle of mass m moves in one dimension (x). It is known that the momentum of the particle is $p_x = \hbar k_0$, where k_0 is a known constant. What is the *time-independent* (unnormalized) wavefunction of this particle, $\psi_a(x)$?

(b) The particle interacts with a system. After interaction it is known that the probability of measuring the momentum of the particle is $\frac{1}{5}$ for $p_x = 2\hbar k_0$ and $\frac{4}{5}$ for $p_x = 8\hbar k_0$. What is the time-independent (unnormalized) wavefunction of the particle in this state, $\psi_b(x)$?

(c) What is the average momentum for the particle, $\langle p_x \rangle$, in the state $\psi_b(x)$?

(d) What is the particle's average kinetic energy $\langle T \rangle$ in the state $\psi_b(x)$? Express your answer in terms of the constant $E_0 \equiv \hbar^2 k_0{}^2 / 2m$.

5.48 Consider a situation where it is equally likely that an electron has momentum $\pm \mathbf{p}_0$. Measurement at a given instant of time finds the value $+\mathbf{p}_0$. A student concludes that the electron must have had this value of momentum prior to measurement. Is the student correct?

Answer

The given information indicates that the electron was in a superposition state prior to measurement. In quantum mechanics one cannot rely on the premise of inference. The student is incorrect.

5.49   (a)   Show that in one dimension, the energy spectrum of bound states is always *nondegenerate*. That is, to any eigenenergy there corresponds only one linearly independent eigenstate.

(b)   In what step in your derivation does the Wronskian (Problem 5.43) come into play?

(c)   In what manner does your proof depend on the given bound-state property?

(d)   What is the nature of the potential you have included in your proof?

[*Hint:* See Problem 10.67. (The notion of degeneracy is discussed in detail in Chapter 8 et seq.)]

CHAPTER 6

TIME DEVELOPMENT, CONSERVATION THEOREMS, AND PARITY

In this chapter we pursue the study of time development of the state function in greater generality than we did in our previous discussion in Chapter 3. This description leads naturally to the concept of constants of the motion in quantum mechanics and again to the notion of stationary states. The distortion of a wave packet in time is obtained with the aid of the free-particle propagator. Classical motion of the packet is obtained in the limit $\hbar \to 0$. The significance to physics of constants of the motion was described in Chapter 1. We now find that such constants stem from related fundamental symmetries in nature. In the two chapters to follow, the principles and mathematical formalism developed to this point are applied to some practical one-dimensional problems.

6.1 TIME DEVELOPMENT OF STATE FUNCTIONS

The Discrete Case

Let us recall the recipe for solution to the *initial-value problem* in quantum mechanics (Section 3.5). The initial-value problem poses the question: Given the state $\psi(x, 0)$, at time $t = 0$, what is the state at $t > 0$, $\psi(x, t)$? The answer is: Eq. 3.70.

(6.1)
$$\psi(x, t) = \exp\left(\frac{-i\hat{H}t}{\hbar}\right)\psi(x, 0)$$

We recall that the exponential operation is written for its series representation,

$$(6.2) \qquad \exp\left(\frac{-i\hat{H}t}{\hbar}\right) = 1 - \frac{i\hat{H}t}{\hbar} - \frac{\hat{H}^2t^2}{2!\hbar^2} + \cdots$$

Suppose that this exponential operator operates on an eigenfunction φ_n of $\hat{H}$. Then $\hat{H}$ as it appears in the exponential is simply replaced by E_n; that is,

$$(6.3) \qquad \exp\left(\frac{-i\hat{H}t}{\hbar}\right)\varphi_n = \exp\left(\frac{-iE_nt}{\hbar}\right)\varphi_n$$

As an application of this property we consider the problem of a particle in a one-dimensional box with walls at $(0, L)$, which is initially in an eigenstate of the Hamiltonian of this system.

$$(6.4) \qquad \psi_n(x, 0) = \varphi_n(x)$$

Then the state at time t is

$$\psi_n(x, t) = \exp\left(\frac{-i\hat{H}t}{\hbar}\right)\varphi_n(x) = e^{-i\omega_n t}\varphi_n(x)$$

$$(6.5) \qquad \begin{array}{c} \psi_n(x, t) = e^{-i\omega_n t}\varphi_n(x) \\ \hbar\omega_n = E_n = n^2 E_1 \end{array}$$

As described in Section 3.6, the time-dependent eigenstates, $\psi_n(x, t)$ of $\hat{H}$, are called *stationary states*. We recall a very important property of a stationary state (3.76)—that the expectation of any operator (which does not contain the time explicitly) is constant in a stationary state. As an example of a stationary state, consider the $n = 5$ eigenstate of the problem at hand,

$$(6.6) \qquad \psi_5(x, t) = e^{-i25E_1t/\hbar}\sqrt{\frac{2}{L}}\sin\left(\frac{5\pi x}{L}\right)$$

The eigenstate ψ_5 oscillates with the frequency $25E_1/\hbar$. Both real and imaginary parts of $\psi_5(x, t)$ are *standing waves*. The expectation of energy in this state is constant and equal to $25E_1$.

Suppose, on the other hand, that $\psi(x, 0)$ is not an eigenstate of $\hat{H}$. Under such circumstances, to determine the time development of $\psi(x, 0)$ one calls on the superposition principle and writes $\psi(x, 0)$ as a linear superposition of the eigenstates of $\hat{H}$.

$$(6.7) \qquad \begin{array}{c} \psi(x, 0) = \sum b_n \varphi_n(x) \\ b_n = \langle \varphi_n | \psi(x, 0) \rangle \end{array}$$

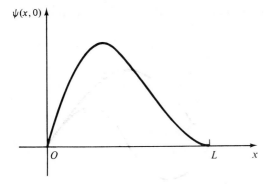

FIGURE 6.1   Initial state

$$\psi(x, 0) = \sqrt{\frac{2}{L}} \left(\frac{\sin (2\pi x/L) + 2 \sin (\pi x/L)}{\sqrt{5}} \right)$$

If we now invoke (6.1), the calculation of $\psi(x, t)$ becomes tractable.

$$\psi(x, t) = \exp \left(\frac{-i\hat{H}t}{\hbar} \right) \sum b_n \varphi_n(x)$$

(6.8)
$$= \sum b_n \exp \left(\frac{-i\hat{H}t}{\hbar} \right) \varphi_n(x)$$

$$= \sum b_n e^{-i\omega_n t} \varphi_n(x)$$

$$\hbar \omega_n = E_n = n^2 E_1$$

This solution indicates that each component amplitude $b_n \varphi_n$ oscillates with the corresponding angular eigenfrequency ω_n.

Consider the specific example in which the initial state is

(6.9)
$$\psi(x, 0) = \sqrt{\frac{2}{L}} \frac{\sin (2\pi x/L) + 2 \sin (\pi x/L)}{\sqrt{5}}$$

This state is depicted in Fig. 6.1 and is simply the superposition of the two eigenstates φ_2 and φ_1. That is, in the expansion (6.7), one obtains

(6.10)
$$b_1 = \frac{2}{\sqrt{5}}, \qquad b_2 = \frac{1}{\sqrt{5}}$$

$$b_n = 0 \qquad \text{(for all other } n)$$

The state of the system at $t > 0$ is given by (6.8).

(6.11)
$$\psi(x, t) = \sqrt{\frac{2}{L}} \left(\frac{e^{-i\omega_2 t} \sin (2\pi x/L) + 2e^{-i\omega_1 t} \sin (\pi x/L)}{\sqrt{5}} \right)$$

How are these time-dependent solutions related to experimental observations? Let us rewrite (6.8) in the form

$$(6.12) \qquad \psi(x, t) = \sum \bar{b}_n(t) \varphi_n(x)$$

so that $\bar{b}_n(t)$ now includes the exponential time factor

$$(6.13) \qquad \bar{b}_n(t) \equiv e^{-i\omega_n t} b_n$$

Suppose that the energy is measured at $t > 0$. What values will result, and with what probabilities will these values occur? As in Section 5.1, calculation of the expectation of E yields

$$(6.14) \qquad \langle E \rangle = \sum |\bar{b}_n(t)|^2 E_n$$

Again, we find that the square of the coefficient of expansion $b_n(t)$ gives the probability that measurement of E at the time t finds the value E_n.

$$(6.15) \qquad P(E_n) = |\bar{b}_n(t)|^2$$

For the state (6.9) this probability distribution is

$$(6.16) \qquad \begin{aligned} P(E_1) &= \tfrac{4}{5} \\ P(E_2) &= \tfrac{1}{5} \\ P(E_n) &= 0 \qquad \text{(for all other } n) \end{aligned}$$

For the initial state (6.9), at *any time* $t > 0$, the probability that measurement of energy finds the value E_1 is $\tfrac{4}{5}$. Similarly, the probability that measurement finds the value $4E_1$ is $\tfrac{1}{5}$.

What is the expectation of E at $t > 0$ for the initial state (6.9)?

$$(6.17) \quad \langle E \rangle_{t>0} = \frac{(\langle e^{-i\omega_2 t} \varphi_2| + \langle 2 e^{-i\omega_1 t} \varphi_1|)(\hat{H}|e^{-i\omega_2 t}\varphi_2\rangle + \hat{H}|2e^{-i\omega_1 t}\varphi_1\rangle)}{5}$$

$$= \frac{E_2 + 4E_1}{5} = \langle E \rangle_{t=0} = \frac{8}{5} E_1$$

The "cross terms" vanish due to orthogonality of the eigenstates of $\hat{H}$, and one finds that the expectation of energy is constant in time. More generally, for any isolated system, in any initial state: (1) the probability of finding a specific energy E_n is constant in time; (2) the expectation $\langle E \rangle$ is constant in time.

These rules follow directly from (6.13)–(6.15).

$$P(E_n) = |\bar{b}_n(t)|^2 = e^{+i\omega_n t} e^{-i\omega_n t} b_n{}^* b_n$$

$$(6.18) \qquad P(E_n) = |b_n|^2 = \text{constant in time}$$

$$(6.19) \qquad \langle E \rangle = \sum |b_n(t)|^2 E_n = \sum |b_n|^2 E_n = \text{constant in time}$$

The Continuous Case. Wave Packets

Next, consider the problem of a free particle moving in one dimension. Let the particle be initially in a localized state $\psi(x, 0)$ such as that depicted in Fig. 6.2.

Since the eigenstates of the Hamiltonian for a free particle comprise a continuum, the representation of $\psi(x, 0)$ as a superposition of energy eigenstates is an integral (see Eqs. 5.26 et seq.).

$$(6.20) \qquad \psi(x, 0) = \frac{1}{\sqrt{2\pi}} \int_{-\infty}^{\infty} b(k)e^{ikx}\, dk$$

$$b(k) = \frac{1}{\sqrt{2\pi}} \int_{-\infty}^{\infty} \psi(x, 0)e^{-ikx}\, dx$$

The state of the particle at $t > 0$ follows from (6.1).

$$(6.21) \qquad \psi(x, t) = \exp\left(\frac{-i\hat{H}t}{\hbar}\right) \frac{1}{\sqrt{2\pi}} \int_{-\infty}^{\infty} b(k)e^{ikx}\, dk$$

$$\psi(x, t) = \frac{1}{\sqrt{2\pi}} \int_{-\infty}^{\infty} b(k)e^{i(kx - \omega t)}\, dk$$

$$(6.22)$$

$$\hbar\omega = \frac{\hbar^2 k^2}{2m} = E_k$$

While the component amplitudes of the state function of a particle in a box oscillate as standing waves, the k-component amplitudes of the free-particle state function propagate. For each value of k, the integrand of (6.22) appears as

$$(6.23) \qquad b(k) \exp\left[ik\left(x - \frac{\omega}{k}t\right)\right]$$

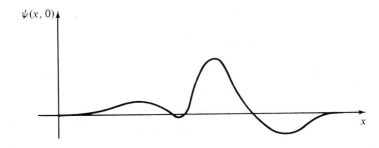

$\psi(x, 0)$

x

FIGURE 6.2 Initial state for a free particle.

The phase of this component, $[x - (\omega/k)t]$, is constant on the propagating "surface,"

(6.24)
$$x = \frac{\omega}{k} t$$

This is a surface of constant phase. It propagates with the phase velocity

(6.25)
$$v = \frac{\omega}{k} = \frac{\hbar k}{2m}$$

The components with larger wavenumbers (shorter wavelengths) propagate with larger speeds. The long-wavelength components propagate more slowly.

Suppose that at $t = 0$, the state $\psi(x, 0)$ is a tight bundle of eigenstates of $\hat{H}$. When the clocks begin to move, each k-component propagates with a distinct phase velocity. The initial state begins to distort. It may be that the initial state remains somewhat intact and moves. In this case one speaks of a *propagating wave packet*. To have a wave packet propagate, it is necessary that the average momentum of the particle in the initial state does not vanish.

(6.26)
$$\langle p \rangle_{t=0} = \langle \psi(x, 0) | \hat{p} \psi(x, 0) \rangle \neq 0$$

Furthermore, since the packet is localized in space,

(6.27)
$$|\psi(x, 0)|^2 \neq 0 \qquad \text{only over a small domain}$$

The velocity with which such a packet moves is called the *group velocity*.

(6.28)
$$v_g = \frac{\partial \omega}{\partial k}\bigg|_{k_{max}}$$

The meaning of k_{max} is that the amplitude $|b(k)|^2$ is maximum at $k = k_{max}$.

(6.29)
$$\hbar k_{max} \simeq \langle p \rangle = \int_{-\infty}^{\infty} |b(k)|^2 \hbar k \, dk$$

This approximation becomes more accurate the more peaked is[1] $|b(k)|^2$.

Combining (6.28) and (6.29) gives

(6.30)
$$v_g = \frac{\partial \omega}{\partial k}\bigg|_{k_{max}} = \frac{\partial \hbar \omega}{\hbar \, \partial k}\bigg|_{k_{max}} = \frac{\partial (\hbar^2 k^2 / 2m)}{\hbar \, \partial k}\bigg|_{k_{max}}$$

$$= \frac{\hbar k_{max}}{m} = \frac{\langle p \rangle}{m} = v_{CL}$$

The packet moves with the classical velocity $\langle p \rangle / m$.

---

[1] However, if $|b(k)|^2$ becomes too peaked, condition (6.27) is violated; that is, $\psi(x)$ spreads out too much.

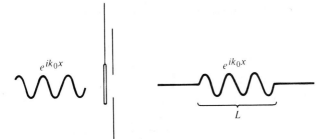

FIGURE 6.3   Chopped wave of length L.

As an example of these concepts, consider a beam of neutrons each of which has momentum $\hbar k_0$. The beam is "chopped," producing a pulse L cm long and containing N neutrons (Fig. 6.3). The state function for each neutron at the instant after the pulse is produced is

$$(6.31) \qquad \psi(x, 0) = \begin{cases} \dfrac{1}{\sqrt{L}} e^{ik_0 x} & -\dfrac{L}{2} \le x \le +\dfrac{L}{2} \\ 0 & \text{elsewhere} \end{cases}$$

If the momentum of any one of the neutrons is measured at $t > 0$, what values may be found and with what probability do these values occur? To answer this question, we need calculate only the expansion coefficients $b(k)$ of (6.20).

$$(6.32) \qquad b(k) = \frac{1}{\sqrt{2\pi L}} \int_{-L/2}^{+L/2} e^{ik_0 x} e^{-ikx}\, dx$$

$$= \sqrt{\frac{2}{\pi L}} \frac{\sin\left[(k - k_0)L/2\right]}{k - k_0}$$

The state at time $t > 0$ is

$$(6.33) \qquad \psi(x, t) = \frac{1}{\pi\sqrt{L}} \int_{-\infty}^{\infty} \frac{\sin\left[(k - k_0)L/2\right]}{k - k_0} e^{i(kx - \omega t)}\, dk$$

with

$$(6.34) \qquad \hbar\omega = \frac{\hbar^2 k^2}{2m}$$

The amplitude $b(k)$ is sketched in Fig. 6.4.

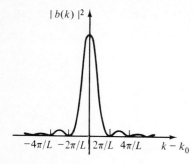

FIGURE 6.4 Momentum probability density corresponding to the pulsed wave of Fig. 6.3.

The momentum probability density $P(k)$ gives the probability that measurement of momentum of any of the neutrons yields a value in the interval $\hbar k$ to $\hbar(k + dk)$. It is given by

(6.35)
$$P(k) = \frac{|b(k)|^2}{\int_{-\infty}^{\infty} |b(k)|^2 \, dk} = |b(k)|^2$$

$$= \frac{2}{\pi L} \frac{\sin^2 [(k - k_0)L/2]}{(k - k_0)^2}$$

This probability density is constant in time.[1] At any time $t > 0$, it is most likely that measurement of momentum of any particle in the pulse finds the value

(6.36)
$$p = \hbar k_{max} = \hbar k_0$$

Recall that this was the only momentum the neutrons had before the beam was chopped.

At any time $t > 0$, the momentum values

(6.37)
$$\hbar k = \hbar k_0 + \frac{2n\pi\hbar}{L} \qquad (n = 1, 2, 3, \ldots)$$

have zero probability of being found. These momentum eigenstates do not enter into the superposition construction of $\psi(x, 0)$.

How many neutrons will be found with momentum in the interval $\hbar(k - k_0) - \hbar k_0$ to $\hbar(k - k_0) + \hbar k_0$? The answer is

(6.38)
$$\Delta N = N \int_{(k - 2k_0)}^{k} |b(k)|^2 \, dk$$

This number is also constant in time.

[1] This property of the free-particle momentum probability density is more fully developed in Section 7.4.

Consider next the Fourier decomposition of a square wave packet as depicted in Figure 5.4. There we see that the largest k component corresponds to $k = 0$ with

$$(6.39) \qquad \varphi_0 = \frac{1}{\sqrt{2\pi}}$$

This is a "flat" wave. The other k components in the superposition of the square wave serve to taper the sides of the pulse. Since $p = 0$ for this packet, it does not propagate— it only *diffuses*.

The Gaussian Wave Packet

A more rewarding problem both from the pedagogical and physical points of view is that of the diffusion and propagation of a *Gaussian wave packet*, discussed previously in Section 3.3. The initial state is

$$(6.40) \qquad \psi(x, 0) = \frac{1}{a^{1/2}(2\pi)^{1/4}} e^{ik_0 x} e^{-x^2/4a^2}$$

The corresponding initial probability density

$$(6.41) \qquad P(x, 0) = \psi^*\psi = \frac{1}{a\sqrt{2\pi}} e^{-x^2/2a^2}$$

is properly normalized as

$$\int_{-\infty}^{\infty} P \, dx = 1$$

The initial uncertainty in position of a particle in the state (6.40) is the square root of the variance

$$(6.42) \qquad \Delta x = a$$

The complex modulation $\exp(ik_0 x)$ in the state (6.40) serves to give the particle the average momentum

$$(6.43) \qquad \langle p \rangle = \hbar k_0$$

It follows that the initial Gaussian state function (6.40) represents a particle localized within a spread of a about the origin and moving with an average momentum $\hbar k_0$.

The momentum amplitude corresponding to this initial state is

$$(6.44) \qquad b(k) = \frac{1}{a^{1/2}(2\pi)^{3/4}} \int_{-\infty}^{\infty} e^{-x'^2/4a^2} e^{ix'(k_0 - k)} \, dx'$$

$$= \sqrt{\frac{2a}{\sqrt{2\pi}}} e^{-a^2(k_0 - k)^2}$$

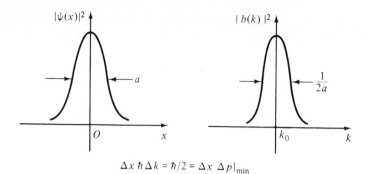

$$\Delta x\, \hbar\, \Delta k = \hbar/2 = \Delta x\, \Delta p|_{min}$$

FIGURE 6.5 The momentum probability density $|b|^2$ corresponding to a Gaussian position probability density, $|\psi|^2$, is Gaussian. In this state $\Delta p\, \Delta x$ has its minimum value at $\hbar/2$.

The Fourier transform of a Gaussian is itself Gaussian (see Fig. 6.5). The initial momentum probability density

(6.45)
$$|b(k)|^2 = \frac{2a}{\sqrt{2\pi}}\, e^{-2a^2(k_0 - k)^2}$$

is normalized, centered about the value $k = k_0$, and has a spread $\Delta k = (2a)^{-1}$. It follows that in the initial Gaussian state,

(6.46)
$$\Delta x\, \Delta p\bigg|_{Gauss} = \Delta x\hbar\, \Delta k = \frac{\hbar}{2} = \Delta x\, \Delta p\bigg|_{min}$$

The product of uncertainties has its minimum value in a Gaussian packet.

Free-Particle Propagator

Next we turn to the construction of $\psi(x, t)$ from the initial state (6.40). The value of this function may be obtained from (6.21 et seq.)

(6.47)
$$\psi(x, t) = \frac{1}{2\pi} \int_{-\infty}^{\infty} \int_{-\infty}^{\infty} dx'\, dk e^{-ikx'}\psi(x', 0)e^{i(kx - \omega t)}$$

$$= \frac{1}{a^{1/2}}\frac{1}{(2\pi)^{5/4}} \int_{-\infty}^{\infty} dx'\, \exp\left(ik_0 x' - \frac{x'^2}{4a^2}\right)$$

$$\times \int_{-\infty}^{\infty} dk\, \exp\left\{i\left[k(x - x') - \frac{k^2 a^2 t}{\tau}\right]\right\}$$

where the time constant τ is defined as

(6.48)
$$\tau\omega = k^2 a^2 \qquad \omega = \frac{\hbar k^2}{2m}$$

Let us take advantage of our construction of $\psi(x, t)$ at this point of the analysis to introduce the free-particle propagator, $K(x', x; t)$. This function provides a formal solution to the free-particle, initial-value problem through the prescription

$$(6.49) \qquad \psi(x, t) = \int_{-\infty}^{\infty} dx' \psi(x', 0) K(x', x; t)$$

The explicit form of $K(x', x; t)$ is inferred from (6.47).

$$(6.50) \qquad K(x', x; t) = \frac{1}{2\pi} \int_{-\infty}^{\infty} dk \, \exp \left\{ i \left[k(x - x') - \frac{k^2 a^2 t}{\tau} \right] \right\}$$

With the aid of the integral (see Problem 6.5)

$$(6.51) \qquad \int_{-\infty}^{\infty} e^{-uy^2} e^{vy} \, dy = \sqrt{\frac{\pi}{u}} \, e^{v^2/4u} \qquad (\text{Re } u > 0)$$

there results[1]

$$(6.52) \qquad K(x', x; t) = \sqrt{\frac{\tau}{i4\pi a^2 t}} \, \exp \left[\frac{i(x - x')^2 \tau}{4a^2 t} \right]$$

$$= \sqrt{\frac{m}{2\pi i h t}} \, \exp \left[\frac{im(x - x')^2}{2ht} \right]$$

Having found this explicit form for the free-particle propagator, let us return to (6.49) and see its meaning. The wavefunction $\psi(x, t)$ gives the probability amplitude related to finding the particle at x at the instant t. If the particle was at x' at $t = 0$, then the probability that it is found at x at $t > 0$ depends on the probability that the particle propagated from x' to x in the interval t. This is what (6.49) says. The probability amplitude that the particle is at x at time t is equal to the initial amplitude that the particle is at x' multiplied by the probability amplitude of propagation from x' to x in the interval t, summed over all x'. Thus we may interpret $K(x, x'; t)$ as the probability amplitude that a particle initially at x' propagates to x in the interval t. It should be noted that the explicit form (6.52) is appropriate only for free-particle propagation. For more general problems involving interaction, the form of (6.49) still maintains, although the propagator function is more complicated (see Problem 6.26).

---

[1] To obtain a convergent integral, first replace i by $\alpha \equiv i + \epsilon$, where ϵ is a small real positive number. After integrating, let $\epsilon \to 0$.

Distortion of the Gaussian State in Time

Let us return to the calculation of $\psi(x, t)$, given the initial Gaussian distribution (6.40). To complete the calculation one need merely complete the x' integration in (6.49).

$$\psi(x, t) = \frac{1}{a^{1/2}(2\pi)^{1/4}} \int_{-\infty}^{\infty} dx' \left[\exp\left(i k_0 x' - \frac{x'^2}{4a^2} \right) \right] K(x', x; t)$$

Employing the explicit form (6.52) for K and once again utilizing the integral formula (6.51) gives the desired result.

(6.53)

$$\psi(x, t) = \frac{1}{a^{1/2}(2\pi)^{1/4}(1 + it/\tau)^{1/2}} \exp\left[i\frac{\tau}{t}\left(\frac{x}{2a}\right)^2 \right] \exp\left[-\frac{(i\tau/4a^2 t)(x - \hbar k_0 t/m)^2}{1 + it/\tau} \right]$$

The corresponding probability density is

$$(6.54) \quad P(x, t) = |\psi(x, t)|^2 = \frac{1}{a\sqrt{2\pi}(1 + t^2/\tau^2)^{1/2}} \exp\left[-\frac{(x - \hbar k_0 t/m)^2}{2a^2(1 + t^2/\tau^2)} \right]$$

If we compare this form with the initial probability density we see that the generic shape of $P(x, 0)$ (i.e., that of a bell) has remained intact with three modifications. It has become wider,

$$a \to a(1 + t^2/\tau^2)^{1/2}$$

Second, the center of symmetry of the packet is now at

$$x = v_0 t$$

where we have labeled

$$v_0 \equiv \frac{\hbar k_0}{m}$$

It follows that the probability density of a Gaussian wave packet propagates with a velocity that is directly related to the expectation of momentum of the particle in the Gaussian state. Finally, the height of the density function has diminished.

$$\frac{1}{a\sqrt{2\pi}} \to \frac{1}{a\sqrt{2\pi}(1 + t^2/\tau^2)^{1/2}}$$

The area under the curve P, at any time, remains unity.

A sequence of packet contours is shown in Fig. 6.6. It is quite clear that the packet begins to distort significantly after a time interval τ. If we represent a piece of chalk by a wave packet, $a \simeq 1$ cm, $m \simeq 1$ g, there results

$$\tau \simeq 10^{27} \text{ s} \simeq 10^{20} \text{ yr}$$

But the universe is only $\sim 10^{10}$ yr old. That is why classical objects are never observed to suffer a quantum mechanical spreading.

160

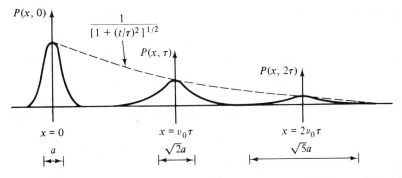

FIGURE 6.6 Shrinkage and spreading of the probability distribution corresponding to a Gaussian wave packet. At any time t,

$$\int_{-\infty}^{\infty} P(x, t)\, dx = 1$$

Flattening of the δ Function

There are two limits that can be taken on the probability density $P(x, t)$ related to the Gaussian wave packet which are very revealing. The first evolves from the initial state

(6.55) $$P(x, 0) = |\psi(x, 0)|^2 = \delta(x)$$

A valid representation of the delta function is given by the limit

(6.56) $$\delta(x) = \lim_{a \to 0} \frac{1}{a(2\pi)^{1/2}} e^{-x^2/2a^2}$$

This function has the correct delta function properties (3.27 et seq.).

Measurement of the position of a particle which finds the value $x = 0$ leaves the particle in the state

(6.57) $$\psi = \delta(x)$$

This state is not normalizable. The state given by (6.55) is a little less sharply peaked than (6.57) and is normalizable.

To obtain the probability density $P(x, t)$ which follows from the initial value (6.55), we merely examine (6.54) in the limit, $a \to 0$. There results

(6.58) $$\lim P(x, t) = \lim \frac{2ma}{t\hbar\sqrt{2\pi}} \exp\left[-\frac{2a^2(x - \hbar k_0 t/m)^2}{t^2 \hbar^2/m^2} \right]$$

$$= \lim \frac{2ma}{t\hbar\sqrt{2\pi}} [1 + O(a^2)]$$

The notation $O(a^2)$ denotes "order" of a^2. It stands for a group of terms, the sum of which goes to zero like a^2, with decreasing a.

From expression (6.58) we see that for all $t > 0$, P vanishes uniformly for all x, in the limit $a \to 0$. This instantaneous flattening of an infinitely peaked state (6.55) is due to the following circumstance. The momentum probability density $|b(k)|^2$ corresponding to such a state, depicted in Fig. 5.6b, is flat. This means that it is equally probable to find any k value, no matter how large k is, in this state. At any instant $t > 0$, at any point x, the components φ_k with k values which obey the inequality, $(\hbar k/m)t \geq x$, have overtaken that point. The initial infinitely peaked distribution assumes an (almost) instantaneous flattening.[1]

The Classical Particle

The second limit we wish to consider relative to the probability density $P(x, t)$, (6.54), changes $P(x, t)$ to the classical probability relating to a point particle of mass m moving with velocity $\hbar k_0/m$. This is accomplished by setting $\hbar \to 0$ in $P(x, t)$ (except where $\hbar$ appears in $p_0 = \hbar k_0$).

(6.59)
$$\lim_{\hbar \to 0} P(x, t) = \frac{1}{a\sqrt{2\pi}} \exp\left[-\frac{(x - p_0 t/m)^2}{2a^2} \right]$$

$$\hbar k_0 = p_0 = \text{constant}$$

For this probability to relate to a "point particle" we impose the additional constraint, $a \to 0$. This gives

(6.60)
$$\lim P(x, t) = \delta\left(x - \frac{p_0 t}{m} \right) = P_{CL}(x, t)$$

The probability of finding the particle at t is zero everywhere except on the classical trajectory

(6.61)
$$x = \frac{p_0 t}{m}$$

This is another example of the correspondence principle at work. In essence, the "leading term" (i.e., the term not containing $\hbar$) in the expansion of $P(x, t)$ about $\hbar = 0$ gives the classical result.

---

[1] Of course, these conclusions become erroneous for $x/t \geq c$. To obtain a completely physically valid solution for the infinitely peaked initial state, it is necessary to solve the relativistic form of the Schrödinger equation. See related discussions on the *Dirac equation* in A. Messiah, *Quantum Mechanics*, Wiley, New York, 1966.

PROBLEMS

6.1 (a) Find $\psi(x, t)$ and $P(E_n)$ at $t > 0$, relevant to a particle in a one-dimensional box with walls at $(0, L)$, for each of the following initial states.

(b) If measurement of E finds that $E = 4E_1$ at 6 s, what is $\psi(x, t)$ at $t > 6$ s for each of these initial states?

(1) $\psi(x, 0) = A_1 \sin\left(\dfrac{3\pi x}{L}\right) \cos\left(\dfrac{\pi x}{L}\right)$

(2)[1] $\psi(x, 0) = A_2 x^2 (x - L)^2$

(3) $\psi(x, 0) = A_3 [e^{i\pi(x - L)/L} - 1]$

6.2 Consider the following three dispersion relations.

(1) $\omega^2 = gk$

(2) $\omega^2 = \dfrac{c^2 k^2}{1 - (\omega_0/\omega)^2}$

(3) $\omega^2 = \omega_p^2 + 3C^2 k^2$

The first relation obtains to deep-water surface waves (g is the acceleration due to gravity), the second to electromagnetic waves in a waveguide, and the third to longitudinal waves in a "warm" plasma (ω_p is the *plasma frequency* and C is the *thermal speed*). For all three cases find (1) the phase velocity and (2) the group velocity of a wave packet propagating in the respective medium.

6.3 (a) Show that the free-particle propagator (6.52) has the following property and interpret the result physically.

$$K(x', x; 0) = \delta(x' - x)$$

(b) Show that K satisfies the integral equation

$$K(x', x; t - t_0) = \int K(x', x''; t - t_1) K(x'', x; t_1 - t_0) \, dx''$$

and interpret this result physically in terms of the evolution in time of the state $\psi(x, t_0)$, first from t_0 to t_1 and then from t_1 to t.

Answer (partial)

(a) Set $it = \epsilon^2$ and compare with (C6) of Appendix C. The interpretation of this result is that for infinitesimally short time intervals, the probability amplitude for propagation away from the initial point x' is zero, except in a small neighborhood about the initial point.

6.4 At $t = 0$, 10^5 noninteracting protons are known to be on a line segment 10 cm long. It is equally probable to find any proton at any point on this segment. How many protons remain on the segment at $t = 10$ s? [*Hint:* Let the center of the segment be at $x = 0$. Then the formal answer to the problem with $\psi(x, t)$ normalized is

$$\Delta N = 10^5 \int_{-5}^{5} |\psi(x, 10)|^2 \, dx$$

[1] Faulty apparatus. E_2 cannot be measured in this state.

163

To construct $\psi(x, t)$, the initial square pulse must first be written as a superposition of φ_k states. With $b(k)$ calculated,

$$|\psi(x, t)|^2 = \frac{1}{2\pi} \int_{-\infty}^{\infty} \int_{-\infty}^{\infty} dk\, dk'\, b(k) b^*(k') e^{i(\omega' - \omega)t} e^{i(k - k')x}$$

where ω' is written for $\omega(k')$.]

6.5   The integration involved in obtaining (6.51) is of the type

$$S = \int_{-\infty}^{\infty} e^{-uy^2} e^{vy}\, dy$$

where u and v are constants. Evaluate this integral.

Answer

The aim is to transform the exponent $-uy^2 + vy$ to a perfect square. First we set

$$uy^2 - vy \equiv \alpha^2 y^2 - 2\alpha\beta y$$

which gives

$$\alpha^2 = u, \qquad \beta^2 = \frac{v^2}{4u}$$

The exponent may now be written

$$uy^2 - vy = (\alpha y - \beta)^2 - \beta^2$$

and the integral S becomes (with $\eta = \alpha y - \beta$)

$$S = \frac{e^{\beta^2}}{\alpha} \int_{-\infty}^{\infty} e^{-\eta^2}\, d\eta = \frac{\sqrt{\pi} e^{\beta^2}}{\alpha} = \sqrt{\frac{\pi}{u}}\, e^{v^2/4u}$$

6.6   (a)   An electron is in a Gaussian wave packet. If the packet is to remain intact for at least the time it takes light to move across 1 Bohr diameter, $2a_0 = 2\hbar^2/me^2$, what is the minimum width, a, that the Gaussian packet may have (in centimeters)?

   (b)   What is the diffusion time (τ) for an electron in a Gaussian wave packet of width e^2/mc^2 (in seconds)? This is the classical radius of the electron. How far does light travel in this time (in centimeters)?

6.7   A free particle of mass m moving in one dimension is known to be in the initial state

$$\psi(x, 0) = \sin(k_0 x)$$

   (a)   What is $\psi(x, t)$?

   (b)   What value of p will measurement yield at the time t, and with what probabilities will these values occur?

   (c)   Suppose that p is measured at $t = 3$ s and the value $\hbar k_0$ is found. What is $\psi(x, t)$ at $t > 3$ s?

6.8   A particle moving in one dimension has the wavefunction

$$\psi(x, t) = A \exp\left[i(ax - bt)\right]$$

where a and b are constants.

(a)   What is the potential field $V(x)$ in which the particle is moving?

(b)   If the momentum of the particle is measured, what value is found (in terms of a and b)?

(c)   If the energy is measured, what value is found?

6.2   TIME DEVELOPMENT OF EXPECTATION VALUES

The law that covers the time development of the expectation of an observable, $\langle A \rangle$, follows from the time-dependent Schrödinger equation. We wish to calculate $d\langle A \rangle/dt$. Since $\langle A \rangle$ has all its spatial dependence integrated out, it is at most a function of time. We may therefore write

$$(6.62) \qquad \frac{d\langle A \rangle}{dt} = \frac{\partial \langle A \rangle}{\partial t}$$

In the state $\psi(x, t)$, this expression becomes

$$(6.63) \qquad \frac{d\langle \psi | \hat{A}\psi \rangle}{dt} = \int dx \, \frac{\partial}{\partial t}\left(\psi^* \hat{A}\psi\right)$$

The time derivative of the product is

$$(6.64) \qquad \frac{\partial}{\partial t}\left(\psi^* \hat{A}\psi\right) = \left(\frac{\partial \psi^*}{\partial t}\right)\hat{A}\psi + \psi^* \hat{A}\,\frac{\partial \psi}{\partial t} + \psi^* \frac{\partial \hat{A}}{\partial t}\,\psi$$

Employing the time-dependent Schrödinger equation

$$(6.65) \qquad \frac{\partial \psi}{\partial t} = \frac{-i\hat{H}}{h}\,\psi, \qquad \frac{\partial \psi^*}{\partial t} = \frac{i\hat{H}\psi^*}{h}$$

in (6.64) gives

$$(6.66) \qquad \frac{\partial}{\partial t}\left(\psi^* \hat{A}\psi\right) = \frac{i}{h}\left(\hat{H}\psi^* \hat{A}\psi - \psi^* \hat{A}\hat{H}\psi + \frac{h}{i}\,\psi^* \frac{\partial \hat{A}}{\partial t}\,\psi\right)$$

Substituting this expansion in (6.63) gives

$$(6.67) \qquad \frac{d\langle \hat{A} \rangle}{dt} = \frac{i}{h}\left(\langle \hat{H}\psi | \hat{A}\psi \rangle - \langle \psi | \hat{A}\hat{H}\psi \rangle + \frac{h}{i}\left\langle \psi \left| \frac{\partial \hat{A}}{\partial t}\,\psi \right\rangle \right.\right)$$

Since $\hat{H}$ is Hermitian, the first term on the right-hand side of (6.67) may be rewritten to yield the final result,

(6.68)
$$\boxed{\frac{d\langle A\rangle}{dt} = \left\langle \frac{i}{\hbar}[\hat{H}, \hat{A}] + \frac{\partial \hat{A}}{\partial t} \right\rangle}$$

If $\hat{A}$ does not contain the time explicitly, then the last term on the right-hand side vanishes and

(6.69)
$$\frac{d\langle A\rangle}{dt} = \frac{i}{\hbar}\langle [\hat{H}, \hat{A}]\rangle$$

In the event that $\hat{A}$ commutes with $\hat{H}$, the quantity $\langle A\rangle$ is constant in time and A is called a *constant of the motion*. For a free particle, $\hat{p}$ commutes with $\hat{H}$ and $\langle p\rangle$ is constant in time for any state (wave packet). Since $\hat{H}$ commutes with itself, $\langle H\rangle$, the expectation of the energy, is always constant in time.

Let a particle moving in one dimension be in the presence of the potential $V(x)$. The Hamiltonian of the particle is

(6.70)
$$\hat{H} = \frac{\hat{p}^2}{2m} + V(x)$$

How does $\langle x\rangle$ vary in time? Eq. (6.69) gives

(6.71)
$$\frac{d\langle x\rangle}{dt} = \frac{i}{\hbar}\langle [\hat{H}, \hat{x}]\rangle$$

$$= \frac{i}{\hbar}\left\langle \left[\frac{\hat{p}^2}{2m}, \hat{x}\right]\right\rangle = \frac{i}{2m\hbar}\langle \hat{p}[\hat{p}, \hat{x}] + [\hat{p}, \hat{x}]\hat{p}\rangle$$

$$= \frac{i}{2m\hbar}\langle -2i\hbar p\rangle = \left\langle \frac{p}{m}\right\rangle$$

or, equivalently,

(6.72)
$$m\frac{d\langle x\rangle}{dt} = \langle p\rangle$$

This equation bears the same relation between expected values of displacement and momentum as in the classical case. Equation (6.72) cannot hold for the eigenvalues of $\hat{x}$ and $\hat{p}$, since such an equation implies that $x(t)$ and $p(t)$ are simultaneously known.

Ehrenfest's Principle

The reduction of quantum mechanical equations to classical forms when averages are taken, such as demonstrated above, is known as *Ehrenfest's principle*. Newton's second law follows from the commutator $[\hat{H}, \hat{p}]$, which for the Hamiltonian (6.70) is

$$(6.73) \qquad [\hat{H}, \hat{p}] = i\hbar \frac{\partial V}{\partial x}$$

Again using (6.68), one obtains

$$(6.74) \qquad \frac{d\langle p \rangle}{dt} = -\left\langle \frac{\partial V}{\partial x} \right\rangle$$

which is the x component of the vector relation

$$(6.75) \qquad \frac{d\langle \mathbf{p} \rangle}{dt} = -\langle \nabla V(x, y, z) \rangle = \langle \mathbf{F}(x, y, z) \rangle$$

where $\mathbf{F}$ is the force at (x, y, z). In any state $\psi(x, t)$, the time development of the averages of $\hat{x}$ and $\hat{p}$ follow the laws of classical dynamics, with the force at any given point replaced by its expectation in the state $\psi(x, t)$. (See Problem 6.31.)

PROBLEMS

6.9 Show that if $[\hat{H}, \hat{A}] = 0$ and $\partial \hat{A}/\partial t = 0$, then $\langle \Delta A \rangle$ is constant in time.

6.10 Show that

$$\frac{d}{dt} \langle A \rangle = 0$$

in a stationary state, provided that $\partial \hat{A}/\partial t = 0$, using the commutator relation (6.68).

Answer

$$\frac{d\langle A \rangle}{dt} = \frac{i}{\hbar} \langle \varphi_n | [\hat{H}, \hat{A}] \varphi_n \rangle = \frac{i}{\hbar} \langle \varphi_n | (\hat{H}\hat{A} - \hat{A}\hat{H}) \varphi_n \rangle$$

$$= \frac{i}{\hbar} (\langle \hat{H}\varphi_n | \hat{A}\varphi_n \rangle - \langle \varphi_n | \hat{A}\hat{H}\varphi_n \rangle)$$

$$= \frac{i}{\hbar} E_n (\langle \varphi_n | \hat{A}\varphi_n \rangle - \langle \varphi_n | \hat{A}\varphi_n \rangle) = 0$$

6.11 Show that for a wave packet propagating in one dimension,

$$m \frac{d\langle x^2 \rangle}{dt} = \langle xp \rangle + \langle px \rangle$$

6.12   A particle moving in one dimension interacts with a potential $V(x)$. In a stationary state of this system show that

$$\tfrac{1}{2}\left\langle x \frac{\partial}{\partial x} V \right\rangle = \langle T \rangle$$

where $T = p^2/2m$ is the kinetic energy of the particle.

Answer

In a stationary state,

$$\frac{d}{dt} \langle xp \rangle = \frac{i}{\hbar} \langle [\hat{H}, \hat{x}\hat{p}] \rangle = 0$$

Expanding the right-hand side, we obtain

$$0 = \langle \hat{x}[\hat{H}, \hat{p}] + [\hat{H}, \hat{x}]\hat{p} \rangle$$
$$= \langle x[\hat{V}, \hat{p}] + [\hat{T}, \hat{x}]\hat{p} \rangle$$
$$= i\hbar \left\langle x \frac{\partial V}{\partial x} - 2T \right\rangle$$

6.13   Consider an operator $\hat{A}$ whose commutator with the Hamiltonian $\hat{H}$ is the constant c.

$$[\hat{H}, \hat{A}] = c$$

Find $\langle A \rangle$ at $t > 0$, given that the system is in a normalized eigenstate of $\hat{A}$ at $t = 0$, corresponding to the eigenvalue a.

6.14   A system is in a superposition of the two energy eigenstates φ_1 and φ_2. Physical properties of the system characteristically depend on the probability density $\psi^*\psi$. Show that resolution of any such property involves measurements over an interval $\Delta t > \hbar/|E_1 - E_2|$.

Answer

The superposition state is

$$\psi(\mathbf{r}, t) = \varphi_1(\mathbf{r}) \exp\left(\frac{-iE_1 t}{\hbar}\right) + \varphi_2(\mathbf{r}) \exp\left(\frac{-iE_2 t}{\hbar}\right)$$

so that

$$\psi^*\psi = |\varphi_1|^2 + |\varphi_2|^2 + 2 \operatorname{Re} \varphi_1^*\varphi_2 \exp\left[\frac{i(E_1 - E_2)t}{\hbar}\right]$$

This function oscillates between the two extremes $(|\varphi_1| + |\varphi_2|)^2$ and $(|\varphi_1| - |\varphi_2|)^2$ with the period $\hbar/|E_1 - E_2|$. It follows that changes in related properties become discernible only after an interval greater than or of the same order as this period. The situation is similar to the process of tuning an oscillator to a frequency ω_0 by "listening" for beats. The period between beats varies as the inverse frequency $(\omega - \omega_0)^{-1}$. Thus one is certain that $\omega = \omega_0$ only after an infinite interval.

6.3 CONSERVATION OF ENERGY, LINEAR AND ANGULAR MOMENTUM

The principle of conservation of energy in classical physics states that the energy of an *isolated system* or a *conservative system* is constant in time. A conservative system is one whose dynamics are describable in terms of a potential function. A particle in a one-dimensional box is a conservative system. Suppose that at $t = 0$, the state of the particle is

$$(6.76) \qquad \psi(x, 0) = \frac{3\varphi_1 + 4\varphi_5}{\sqrt{25}}$$

What can be said of the energy of the particle at the time $t > 0$? Measurement of the energy has a $\frac{9}{25}$ probability of finding the value E_1 and a $\frac{16}{25}$ probability of finding the value $25E_1$. At $t > 0$ the state (6.76) becomes

$$(6.77) \qquad \psi(x, t) = \frac{3\varphi_1(x)e^{-iE_1t/\hbar} + 4\varphi_5 e^{-E_5t/\hbar}}{\sqrt{25}}$$

The probability that measurement yields E_1 is

$$(6.78) \qquad P(E_1) = \frac{(3e^{-iE_1t/\hbar})^*(3e^{-iE_1t/\hbar})}{25} = \frac{9}{25}$$

A similar calculation of $P(E_5)$ yields the constant value $\frac{16}{25}$. In other words, in the state given, one cannot say with certainty what the energy is at $t \geq 0$. In what sense is energy conserved? The answer is: in the average sense. It follows directly from (6.69) that

$$(6.79) \qquad \langle H \rangle = \langle E \rangle = \text{constant}$$

For the example given, at any instant in time the expectation of the energy is

$$(6.80) \qquad \langle E \rangle = \frac{9E_1 + 16E_5}{25} = 16.36E_1 = \text{constant}$$

For a free particle, $\hat{p}$ also commutes with $\hat{H}$, hence we can conclude from (6.68) that

$$(6.81) \qquad \langle p \rangle = \text{constant}$$

The energy and total momentum of an isolated system are constants of the motion.

Conservation theorems in physics are closely related to symmetry principles. Consider, for example, the fact that the laws of physics do not depend on the time at which they are applied. Newton's second law, Maxwell's equations, and so on, do not change their structure with time. This symmetry of time (i.e., homogeneity) gives rise to the conservation of energy. Let H be the Hamiltonian of the whole universe.

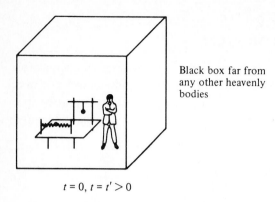

Black box far from any other heavenly bodies

$t = 0, t = t' > 0$

FIGURE 6.7   The laws of physics are the same at t and t' → $\partial H/\partial t = 0$ → $\langle E \rangle$ = constant.

Homogeneity of time implies that H is not an explicit function of time. This together with (6.68) implies that $d\langle E \rangle/dt = 0$. We may reach the same conclusion for any *isolated system* (Fig. 6.7).

Conservation of momentum for an isolated system depends on the homogeneity of space. Go out in space to a point far removed from other objects. Enclose yourself in a box with no windows and opaque walls. Let the box suffer a "virtual" displacement (Fig. 6.8). There is no experiment which will reveal that the box is at a new location. Consequently, for example, the dynamical laws of an isolated system of particles can only depend on the relative orientation of particles, not on the distances from these particles to some arbitrarily chosen origin. Equivalently, the Hamiltonian of the system can always be transformed so that it does not contain these variables (i.e., the coordinates of the center of mass).

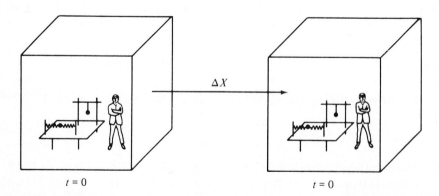

$t = 0$ $t = 0$

FIGURE 6.8   The laws of physics stay the same   → $\partial H/\partial x = 0$ → $\langle p \rangle$ = constant. Note that in this "thought" experiment, translation occurs in zero time. This is called a "virtual" displacement. It is demanded by the details of the argument. To physically effect a virtual displacement one merely imagines two identical, noninterfering boxes a distance Δx apart.

To find the basis of the relation between the homogeneity of space and conservation of linear momentum, we turn back to Problem 3.17, where it was shown that the $\hat{p}$ operator effects the displacement

(6.82)
$$\hat{\mathcal{D}}(\zeta)f(x) = \exp\left(\frac{i\zeta\hat{p}_x}{\hbar}\right)f(x) = f(x + \zeta)$$

In this expression f is any differentiable function of x. For infinitesimal displacement ($\zeta \to 0$), the displacement operator becomes

(6.83)
$$\hat{\mathcal{D}}(\zeta) = \hat{I} + \frac{i\zeta\hat{p}_x}{\hbar}$$

or, equivalently,

$$\hat{p}_x = \frac{\hbar}{i\zeta}[\hat{\mathcal{D}}(\zeta) - \hat{I}]$$

where the identity operator is $\hat{I}$. As observed previously, the Hamiltonian of an isolated system cannot depend on displacement of the system from an origin at an arbitrary point in space. Therefore, the displacement operator $\hat{\mathcal{D}}$ commutes with $\hat{H}$, whence $\hat{p}_x$ does also. Again calling on (6.69), we recapture the constancy of $\langle p_x \rangle$. However, in the present argument we see how this conservation theorem finds its origin in the symmetry of the homogeneity of space.

In three dimensions the displacement operator becomes

(6.84)
$$\hat{\mathcal{D}}f(\mathbf{r}) = \exp\left(\frac{i\boldsymbol{\zeta} \cdot \hat{\mathbf{p}}}{\hbar}\right)f(\mathbf{r}) = f(\mathbf{r} + \boldsymbol{\zeta})$$

Again, for an isolated system, one may conclude that $\hat{H}$ commutes with $\hat{\mathcal{D}}$ and therefore with $\hat{\mathbf{p}}$, the total linear momentum of the system. It follows that the vector $\langle \mathbf{p} \rangle$ is conserved.

Let us return to the experimental "black box" described above. The fact that experiments performed within the box are impervious to the box's location in space or time implies, respectively, conservation of linear momentum and energy. Suppose now that the box undergoes a rotation through the angle $\Delta\phi$ about an arbitrary fixed axis in space. Owing to the isotropy of space, experiments within the box cannot detect such rotational displacement. They are impervious to the box's orientation in space (Fig. 6.9). It follows that the Hamiltonian of the system cannot depend on ϕ, the rotational orientation with respect to some fixed axis, in the same way that it cannot depend on the displacement ζ from an arbitrary point in space. As a consequence of this rotational symmetry, the total angular momentum of the system is conserved.

Suppose that there is a property of the system which is dependent on the system's rotational orientation ϕ about a fixed axis, which we designate the z axis. Let

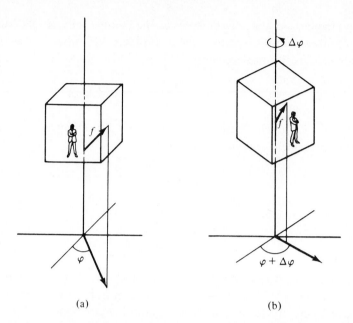

(a)   (b)

FIGURE 6.9 If $f(\phi)$ is a property of an isolated system that depends on the orientation ϕ of the system about an arbitrary fixed axis, rotation of the system through the angle $\Delta\phi$ causes the property to change to $f(\phi + \Delta\phi)$. Isotropy of space precludes the existence of such a property. This invariance with respect to rotation implies conservation of angular momentum.

the measure of this property be $f(\phi)$. After rotation of the system through the angle $\Delta\phi$, $\phi \to \phi + \Delta\phi$ and $f(\phi) \to f(\phi + \Delta\phi)$. This transformation of function is effected by the rotation operator $\hat{R}_{\Delta\phi}$:

$$\hat{R}_{\Delta\phi} f(\phi) = f(\phi + \Delta\phi)$$

(6.85)

$$\hat{R}_{\Delta\phi} = \exp\left(\frac{i\Delta\phi \hat{L}_z}{\hbar}\right)$$

Here $\hat{L}_z$ is the z component of the total angular momentum of the system.[1] Since the Hamiltonian of the (isolated) system cannot depend on ϕ, it is insensitive to the rotation operator, $\hat{R}_{\Delta\phi}$; that is, $\hat{H}$ commutes with $\hat{R}_{\Delta\phi}$, hence it also commutes with $\hat{L}_z$ and we may conclude that $\langle L_z \rangle$ is constant. More generally, rotation through the vector angle $\Delta\boldsymbol{\phi}$ (the direction of $\Delta\boldsymbol{\phi}$ is parallel to the axis of rotation) is effected by the operator

(6.86)

$$\hat{R}_{\Delta\boldsymbol{\phi}} = \exp\left(\frac{i\Delta\boldsymbol{\phi} \cdot \hat{\mathbf{L}}}{\hbar}\right)$$

[1] This relation is derived in Problem 9.17.

The argument demonstrating the constancy of $\hat{L}_z$ carries over to $\hat{\mathbf{L}}$, the total angular momentum of the system.

In summary, with $\mathbf{p}$ and $\mathbf{L}$ denoting, respectively, the total linear and angular momentum of an isolated system whose Hamiltonian is H, the following symmetry-conservation principles hold.

Homogeneity of Space

$$(6.87) \qquad\qquad [\hat{H}, \hat{\mathbf{p}}] = 0 \rightarrow \frac{d}{dt}\langle\mathbf{p}\rangle = 0$$

Isotropy of Space

$$(6.88) \qquad\qquad [\hat{H}, \hat{\mathbf{L}}] = 0 \rightarrow \frac{d}{dt}\langle\mathbf{L}\rangle = 0$$

Homogeneity of Time

$$(6.89) \qquad\qquad \frac{\partial\hat{H}}{\partial t} = 0 \rightarrow \frac{d}{dt}\langle E\rangle = 0$$

PROBLEMS

6.15 Under what conditions is the expectation of an operator $\hat{A}$ (which does not contain the time explicitly) constant in time?

Answer

Under either of the following conditions:
  (a) $[\hat{A}, \hat{H}] = 0$.
  (b) $\langle A\rangle$ is calculated in a stationary state.

6.4 CONSERVATION OF PARITY

Consider an experiment and its mirror image (Fig. 6.10). Such an experiment might be the observation of the orbit of a missile fired in a uniform gravity field, or two particles colliding. These phenomena obey certain physical laws. Suppose that we formulate the laws obeyed by the image orbits in the mirror. They turn out to be the same as the laws that the orbits in the real world obey. This is a symmetry principle. It has no further implication in classical physics. However, in quantum mechanics it is associated with a conservation law, *conservation of parity*.[1]

---

[1] This conservation principle belongs to a class of phenomena called "broken symmetries." Such symmetries are not universally maintained. For example, parity is not conserved in weak-interaction β-decay processes. For further discussion of this topic, see H. Frauenfelder and E. Henley, *Subatomic Physics*, Prentice-Hall, Englewood Cliffs, N.J., 1974.

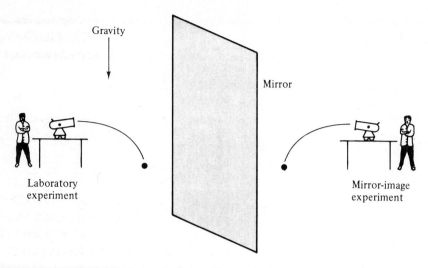

FIGURE 6.10   **The laws of physics are the same for the lab experiment and for the mirror-image experiment. This symmetry statement gives rise to the conservation of parity.**

Parity is a property of a function. A function $f(x)$ has *odd parity* if

$$f(-x) = -f(x)$$

A function has *even parity* if (see Fig. 6.11)

$$f(-x) = f(x)$$

The parity operator $\hat{\mathbb{P}}$ is defined as[1]

(6.90) $$\hat{\mathbb{P}}f(x) = f(-x)$$

What are the eigenvalues of $\hat{\mathbb{P}}$? Let g be an eigenfunction of $\hat{\mathbb{P}}$ with eigenvalue α; then

(6.91) $$\hat{\mathbb{P}}g(x) = g(-x) = \alpha g(x)$$

To find α we operate again with $\hat{\mathbb{P}}$.

(6.92) $$\hat{\mathbb{P}}\hat{\mathbb{P}}g(x) = \hat{\mathbb{P}}g(-x) = g(x) = \alpha^2 g(x).$$

Hence

(6.93) $$\alpha^2 = 1, \qquad \alpha = \pm 1$$

For $\alpha = +1$, from (6.91), we obtain

(6.94) $$g(-x) = g(x)$$

---

[1] In three dimensions $\hat{\mathbb{P}}f(x, y, z) = f(-x, -y, -z)$. See Problem 6.23.

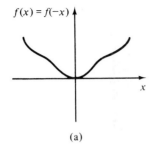

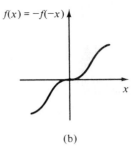

(a)   (b)

FIGURE 6.11 **Any even function is an eigenfunction of $\hat{\mathbb{P}}$ with eigenvalue $+1$:**

$$\hat{\mathbb{P}}f(x) = +1f(x)$$

Any odd function is an eigenfunction of $\hat{\mathbb{P}}$ with eigenvalue -1:

$$\hat{\mathbb{P}}f(x) = -1f(x)$$

Any even function is an eigenfunction of $\hat{\mathbb{P}}$ with eigenvalue $+1$. For $\alpha = -1$,

(6.95) $$g(-x) = -g(x)$$

Any odd function is an eigenfunction of $\hat{\mathbb{P}}$ with eigenvalue -1. The order of degeneracy of $\alpha = \pm 1$ is infinite. There are no other eigenvalues of $\hat{\mathbb{P}}$.

How is this parity property connected with the symmetry principle relating to mirror images mentioned above? Consider that a particle (m) moving in one dimension interacts with another stationary particle (M) which is at the position $x = 0$. The potential of interaction between the particles is $V(x)$. Suppose that the (moving) particle is at a position $x' > 0$. The image of the particle seen in a mirror which intersects the x axis normally at $x = 0$ is at $x = -x' < 0$ (Fig. 6.12). The temporal behavior of the image particle will be the same as that for the laboratory particle if $V(x) = V(-x)$. [The potential "seen" by the image particle is $V(-x)$.]

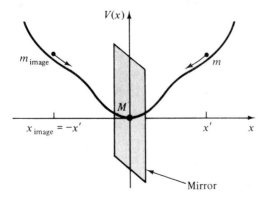

FIGURE 6.12 **A mass m interacting with stationary mass M through potential $V(x)$. If image dynamics is to be the same as lab dynamics, $V(x) = V(-x)$.**

The Hamiltonian for the particle in the laboratory system is

(6.96)
$$\hat{H} = \frac{\hat{p}^2}{2m} + V(x)$$

For $V(x)$ an even function, $\hat{H}$ commutes with $\hat{\mathbb{P}}$. To show that $\hat{\mathbb{P}}$ commutes with $V(x)$, let $g(x)$ be an arbitrary function of x. Then

(6.97)
$$\hat{\mathbb{P}}V(x)g(x) = V(-x)g(-x) = V(x)\hat{\mathbb{P}}g(x)$$

The fact that $\hat{\mathbb{P}}$ commutes with the kinetic-energy part of $\hat{H}$ is shown in Problem 6.16. Since $\hat{\mathbb{P}}$ commutes with both parts of $\hat{H}$, it commutes with $\hat{H}$ itself.

(6.98)
$$[\hat{H}, \hat{\mathbb{P}}] = 0$$

Together with (6.69) this gives the conservation principle

(6.99)
$$\langle \mathbb{P} \rangle = \text{constant}$$

The parity of the state of a system is a constant of the motion.

As an example of this principle, consider a one-dimensional box centered at the origin, so that its walls are at $x = L/2, x = -L/2$ (see Problem 4.1). The eigenstates and eigenenergies of the Hamiltonian for this system are

(6.100)
$$
\left.
\begin{aligned}
\tilde{\varphi}_n &= \sqrt{\frac{2}{L}} \sin\left(\frac{n\pi x}{L}\right) \qquad (n = 2, 4, 6, \ldots, 2j, \ldots) \\
\\
\varphi_n &= \sqrt{\frac{2}{L}} \cos\left(\frac{n\pi x}{L}\right) \qquad (n = 1, 3, 5, \ldots, 2j + 1, \ldots)
\end{aligned}
\right\} \quad E_n = n^2 E_1
$$

The eigenstates $\tilde{\varphi}_n$ are odd while φ_n are even.

(6.101)
$$\hat{\mathbb{P}}\tilde{\varphi}_n = -\tilde{\varphi}_n$$
$$\hat{\mathbb{P}}\varphi_n = \varphi_n$$

Suppose that at $t = 0$ the particle is in the state

(6.102)
$$\psi(x, 0) = \sqrt{\frac{2}{45L}} \left[6 \sin\left(\frac{2\pi x}{L}\right) + 3 \cos\left(\frac{\pi x}{L}\right) \right]$$

$$= \frac{1}{\sqrt{45}} (6\tilde{\varphi}_2 + 3\varphi_1)$$

At $t > 0$,

(6.103)
$$\psi(x, t) = \sqrt{\frac{2}{45L}} \left[6 \sin\left(\frac{2\pi x}{L}\right) e^{-iE_2 t/\hbar} + 3 \cos\left(\frac{\pi x}{L}\right) e^{-iE_1 t/\hbar} \right]$$

The expectation of $\mathbb{P}$ at $t = 0$ is

(6.104)
$$\langle \mathbb{P} \rangle = \langle \psi(x, 0) | \hat{\mathbb{P}} \psi(x, 0) \rangle$$

$$\hat{\mathbb{P}} \psi(x, 0) = \frac{1}{\sqrt{45}} (-6\tilde{\varphi}_2 + 3\varphi_1)$$

$$\langle \mathbb{P} \rangle = \frac{1}{45} \{ -36 \langle \tilde{\varphi}_2 | \tilde{\varphi}_2 \rangle + 9 \langle \varphi_1 | \varphi_1 \rangle \} = -\frac{27}{45}$$

Since $[\hat{\mathbb{P}}, H] = 0$, this is the value of $\langle \mathbb{P} \rangle$ for all time. In that the initial state (6.102) is a superposition of the eigenstates of $\hat{\mathbb{P}}$, the squares of the coefficients of expansion give the probability that measurement finds the system in a state of even or odd parity. Measurements on an ensemble of 4500 boxes all of whose particles are in the initial state (6.102) at the time $t = 0$ would find approximately 3600 of the particles in the odd state $\tilde{\varphi}_2(t)$ and approximately 900 of the particles in the even state $\varphi_1(t)$ at the subsequent time $t > 0$.

PROBLEMS

6.16 (a) Show that $\hat{\mathbb{P}}$ *anticommutes* with the momentum operator $\hat{p}$. That is, show that

$$[\hat{\mathbb{P}}, \hat{p}]_+ \equiv \hat{\mathbb{P}} \hat{p} + \hat{p} \hat{\mathbb{P}} = 0$$

(b) Use your answer to part (a) to show that $\hat{\mathbb{P}}$ commutes with the kinetic-energy operator $\hat{T} = \hat{p}^2/2m$.

6.17 A particle in one dimension is in the energy eigenstate

$$\varphi_{k_0} = A \cos(k_0 x)$$

Ideal measurement of energy finds the value

$$E = \frac{\hbar^2 k_0{}^2}{2m}$$

What is the state of the particle after measurement?

Answer
If we recall postulate II of quantum mechanics (Section 3.3), the system is left in the eigenstate of $\hat{H}$ corresponding to the eigenenergy above. Any state in the two-dimensional subspace of Hilbert space spanned by $\sin(k_0 x)$ and $\cos(k_0 x)$ [or, equivalently, $\exp(ik_0 x)$ and $\exp(-ik_0 x)$] gives the eigenenergy above. However, an *ideal measurement* perturbs the system least. In the state before measurement the probability distribution relating to momentum is $1/2$ for $p = \pm \hbar k_0$. If we guess that the system is left in the state $\exp(ik_0 x)$ after measurement, the momentum distribution of the original state was disturbed. If we guess that the state of the measurement is $\sin(k_0 x)$, then measurement has not disturbed the momentum distribution; however, this measurement has

disturbed the parity. In the original state the parity is $+1$ (with respect to the origin $x = 0$) while that of the hypothesized state after measurement is -1. This is still not the ideal measurement. We can find a measurement that perturbs the system even less. Consider that the particle is left in the state $\cos (k_0 x)$. The corresponding measurement did not perturb the momentum distribution or the parity of the original state. It is the ideal measurement.

6.18 (a) If $f(x)$ is any function, show that

$$f_+ = \frac{f(x) + f(-x)}{2} = \text{even function}$$

$$f_- = \frac{f(x) - f(-x)}{2} = \text{odd function}$$

(b) Show that

$$\hat{P}_+ \equiv \frac{\hat{I} + \hat{P}}{2}$$

is such that

$$\hat{P}_+ f(x) = f_+(x)$$

The identity operator is $\hat{I}$.

(c) Show that

$$\hat{P}_- \equiv \frac{\hat{I} - \hat{P}}{2}$$

is such that

$$\hat{P}_- f(x) = f_-(x)$$

The operator $\hat{P}_+$ "projects" f onto f_+ while $\hat{P}_-$ projects f onto f_-.

(d) Show that the *projection operators* $\hat{P}_+$ and $\hat{P}_-$ satisfy the following properties:

$$\hat{P}_\pm^2 = \hat{P}_\pm$$
$$[\hat{P}_+, \hat{P}_-] = 0$$
$$\hat{P}_+ + \hat{P}_- = \hat{I}$$

6.19 What is $\langle \mathbb{P} \rangle$ for a particle in a one-dimensional box with walls at $(-L/2, +L/2)$ in the initial state

$$\psi(x, 0) = \frac{1}{\sqrt{29}} (3\tilde{\varphi}_2 + 4\tilde{\varphi}_4 + 2\varphi_3)$$

6.20 For the same one-dimensional box as described in Problem 6.19, it is known that the particle is in a state with energy probabilities

$$P(E_1) = \tfrac{1}{3}, \qquad P(E_2) = \tfrac{1}{3}, \qquad P(E_3) = \tfrac{1}{3}$$
$$P(E_n) = 0, \qquad (n \neq 1, 2, 3)$$

Answer (partial)

(a)   The formal solution to the time-dependent Schrödinger equation appears as (see Eq. 3.70)

$$\psi(x, t) = \exp\left(-\frac{it\hat{H}}{\hbar}\right)\psi(x, 0)$$

Expanding the initial state in eigenstates of $\hat{H}$ (see Eq. 5.6) gives

$$\psi(x, 0) = \sum_n b_n \varphi_n(x)$$

$$b_n = \int \psi(x', 0)\varphi_n{}^*(x')\, dx'$$

Substituting in the above gives

$$\psi(x, t) = \sum_n b_n \exp\left(-\frac{iE_n t}{\hbar}\right)\varphi_n(x)$$

$$= \int \psi(x', 0)\left[\sum_n \varphi_n{}^*(x')\varphi_n(x)e^{-iE_n t/\hbar}\right] dx'$$

$$= \int \psi(x', 0)K(x', x; t)\, dx'$$

which serves to identify the propagator

$$K(x', x; t) = \sum_n \varphi_n{}^*(x')\varphi_n(x)e^{-iE_n t/\hbar}$$

In Dirac notation this equation appears as

$$\hat{K}(x', x; t) = \sum_n |\varphi_n(x)\rangle e^{-iE_n t/\hbar}\langle\varphi_n(x')|$$

which allows the solution to be written:

$$|\psi(x, t)\rangle = \hat{K} | \psi(x, 0)\rangle$$

6.27   The wavefunction of a particle of mass m which moves in one dimension is

$$\Psi(x, t) = Ae^{i(kx - \omega t)}$$

where A, k, and ω are constants. Determine the potential $V(x)$ in which the particle moves in terms of m, $\hbar$, k, and ω.

6.28   Consider a particle of mass m moving in one dimension (x). Let $\varphi(x)$, $\bar{\varphi}(x)$ be eigenstates of the particle with corresponding eigenenergies $E > \bar{E}$. Let x_1, x_2 be two successive zeros of $\bar{\varphi}(x)$. [Values of x for which $\bar{\varphi}(x) = 0$ are called "zeros" of $\bar{\varphi}(x)$.]

(a)   Show that $\varphi(x)$ has at least one zero in the interval (x_1, x_2).

(b)   If energies are discrete and E is the next larger energy than $\bar{E}$, then show that $\varphi(x)$ has only one zero in the interval (x_1, x_2). (Zeros are then said to be *interlaced*. See, for example, Fig. 10.3.)

Answer (partial)

(a)   The Schrödinger equation may be written [compare with (7.2)]

$$\varphi''(x) + k^2(x)\varphi(x) = 0$$

$$\bar{\varphi}''(x) + \bar{k}^2(x)\bar{\varphi}(x) = 0$$

$$k^2 > \bar{k}^2$$

where primes denote differentiation. Let x_1, x_2 be two successive zeros of $\bar{\varphi}(x)$. Assume that $\varphi(x) \neq 0$ on (x_1, x_2) and that (without loss of generality) $\varphi(x)$, $\bar{\varphi}(x) > 0$ on (x_1, x_2). Consider the Wronskian of these two solutions:

$$W(\varphi, \bar{\varphi}) = \varphi\bar{\varphi}' - \bar{\varphi}\varphi'$$

$$\frac{dW(\varphi, \bar{\varphi})}{dx} = \varphi\bar{\varphi}'' - \bar{\varphi}\varphi''$$

$$= \varphi\bar{\varphi}(k^2 - \bar{k}^2) > 0$$

In the second equation we employed the preceding Schrödinger equations. Thus $W(x)$ has positive slope, and we may conclude

$$W(x_2) - W(x_1) > 0$$

Since $\bar{\varphi}(x)$ vanishes at x_1, x_2, we obtain (from the first of the preceding equations)

$$W(x_1) = \varphi(x_1)\bar{\varphi}'(x_1) > 0$$

$$W(x_2) = \varphi(x_2)\bar{\varphi}'(x_2) < 0$$

so that

$$W(x_2) - W(x_1) < 0$$

which contradicts our previous result. Thus our assumptions are incorrect and we conclude that $\varphi(x)$ has at least one zero in (x_1, x_2). (*Note:* The preceding result is often referred to as the *Sturm comparison theorem* in the theory of differential equations.)

6.29   (a)   What is the value of the spread a of the classical probability density, (6.60)?

(b)   What is the spreading time τ of this distribution? Are your answers compatible with the picture one has of a classical particle?

6.30   An electron is initially in the Gaussian state, (6.40). If $a \approx \lambda_e$, the de Broglie wavelength, what is the spread time τ of the subsequent time development of the wave packet? Is this a rapid or a slow spread?

6.31   To complete Ehrenfest's correspondence principle, one must convert (6.74) to the classical relation

$$\frac{d\langle p \rangle}{dt} = -\frac{\partial V\langle x \rangle}{\partial \langle x \rangle}$$

For what class of potentials would this relation be valid?

Answer

If $V(x)$ is slowly varying. To show this, write

$$\frac{\partial V}{\partial x} \equiv G(x)$$

Then under the said condition we may expand

$$G(x) = G(\langle x \rangle) + (x - \langle x \rangle)G'(\langle x \rangle) + \frac{(x - \langle x \rangle)^2}{2!} G''(\langle x \rangle) + \cdots$$

Neglecting derivatives of G (second derivatives of V) gives the desired result:

$$\frac{\partial V(x)}{\partial x} \approx \frac{\partial V(\langle x \rangle)}{\partial \langle x \rangle}$$

6.32   Let $\hat{\Gamma}$ represent a symmetry operation such as: translation, rotation, etc. Suppose a given system with Hamiltonian $\hat{H}$ is invariant under an operation represented by $\hat{\Gamma}$.

(a)   What is the value of $[\hat{H}, \hat{\Gamma}]$?

(b)   If the eigenstates of $\hat{H}$ are known, what can you say about the eigenstates of $\hat{\Gamma}$?

(c)   What can you say about the expectation, $\langle \Gamma \rangle$?

(d)   From descriptions given in this chapter, state four examples of $\hat{\Gamma}$ and respective systems on which $\hat{\Gamma}$ operates for which your answers to (a), (b), and (c) are valid.

CHAPTER 7

ADDITIONAL ONE-DIMENSIONAL
PROBLEMS. BOUND AND UNBOUND STATES

In this and the following chapters we examine some practical and fundamental problems in one dimension. Included are the very important examples of the harmonic oscillator and scattering configurations in one dimension. Creation and annihilation operators are introduced in algebraic construction of eigenenergies of the harmonic oscillator. The purely quantum mechanical effect of tunneling is encountered in a study of transmission through a barrier. The chapter continues with a description of the WKB technique of solution appropriate in the near-classical domain. This method of approximation finds application in still more realistic configurations, such as cold emission from a metal surface and α decay from a radioactive nucleus. A review of Hamilton's principal of least action precedes a description of Feynman's path integral formalism.

7.1 GENERAL PROPERTIES OF THE ONE-DIMENSIONAL SCHRÖDINGER EQUATION

The time-independent Schrödinger equation for a particle of mass m moving in one dimension in a potential field $V(x)$ appears as

(7.1)
$$\left[-\frac{\hbar^2}{2m}\frac{\partial^2}{\partial x^2} + V(x) \right]\varphi(x) = E\varphi(x)$$

With subscripts denoting differentiation, this equation may be rewritten

(7.2)
$$\varphi_{xx} = -k^2(x)\varphi$$
$$\frac{\hbar^2 k^2}{2m} = E - V$$

The partition of energy

(7.3)
$$E = T + V$$

permits us to identify $\hbar^2 k^2/2m$ as the kinetic energy of the particle

(7.4)
$$T = \frac{\hbar^2 k^2}{2m}$$

This identification is especially relevant if $E > V$. More generally, there are three distinct possibilities (Fig. 7.1). These are $E > V$, $E < V$, and $E = V$. In the first case the kinetic energy is positive and the corresponding classical motion is permitted. Classical motion is forbidden in the second domain, where the kinetic energy is negative. The points where $E = V$ are the classical *turning points*. (Recall Section 1.4.)

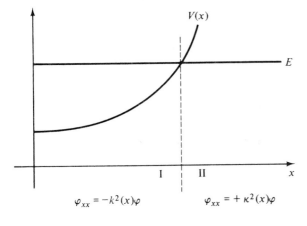

FIGURE 7.1 Domains relevant to a particle of energy E moving in a one-dimensional potential field $V(x)$. I: $E > V$. Kinetic energy is positive. II: $E < V$. Kinetic energy is negative ("forbidden domain"). III: $E = V$. This is a *turning point* of the corresponding classical motion.

In the domain where the kinetic energy is negative, the Schrödinger equation becomes

$$\varphi_{xx} = \kappa^2(x)\varphi$$

(7.5)
$$\frac{\hbar^2 \kappa^2}{2m} = V - E > 0$$

$$\text{kinetic energy} = -\frac{\hbar^2 \kappa^2}{2m} = E - V < 0$$

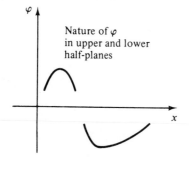

Nature of φ
in upper and lower
half-planes

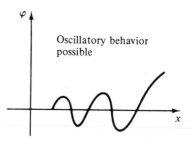

Oscillatory behavior
possible

(a)

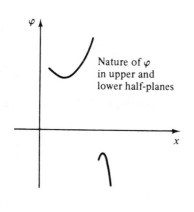

Nature of φ
in upper and
lower half-planes

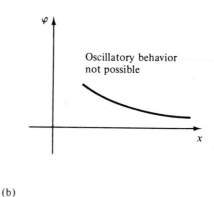

Oscillatory behavior
not possible

(b)

FIGURE 7.2 **(a) Kinetic energy positive:**

$$\varphi_{xx} = -k^2\varphi \qquad k^2 > 0$$

$$\varphi_{xx} < 0 \qquad \textbf{for} \quad \varphi > 0 \textbf{ (upper half-plane)}$$

$$\varphi_{xx} > 0 \qquad \textbf{for} \quad \varphi < 0 \textbf{ (lower half-plane)}$$

(b) Kinetic energy negative:

$$\varphi_{xx} = \kappa^2\varphi \qquad \kappa^2 > 0$$

$$\varphi_{xx} > 0 \qquad \textbf{for } \varphi > 0 \textbf{ (upper half-plane)}$$

$$\varphi_{xx} < 0 \qquad \textbf{for } \varphi < 0 \textbf{ (lower half-plane)}$$

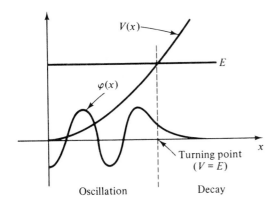

FIGURE 7.3   Characteristic   behavior   of wavefunction corresponding to the configuration shown in Fig. 7.1.

Recall now that in analytic geometry, φ_{xx} is related to the *curvature* of φ (at the point x). If $\varphi_{xx} > 0$, then φ is concave upward. If $\varphi_{xx} < 0$, then φ is concave downward. When the kinetic energy is positive, the Schrödinger equation takes the form (7.2) and φ has the following properties: φ_{xx} is less than zero in the upper half-plane so φ is concave downward; φ_{xx} is greater than zero in the lower half-plane, so φ is concave upward. As shown in Fig. 7.2, these conditions permit *oscillating solutions*. When the kinetic energy is negative, the Schrödinger equation takes the form (7.5) and the following properties pertain: φ_{xx} is greater than zero in the upper half-plane so φ is concave upward; φ_{xx} is less than zero in the lower half-plane and φ is concave downward. Again referring to Fig. 7.2, these conditions are seen to lead to growing or decaying solutions (as opposed to oscillating solutions). At a turning point, $\varphi_{xx} = 0$ and φ has a constant slope.

For the potential shown in Fig. 7.1, one might then expect an eigenfunction of the Hamiltonian to behave as depicted in Fig. 7.3.

PROBLEMS

7.1   (a)   Let a particle of mass m move in a one-dimensional potential field with energy E as sketched in Fig. 1.18. Write down the form of the time-independent Schrödinger equation (i.e., Eq. 7.2 or Eq. 7.5) for the four domains that lie in the interval $0 \leq x \leq D$. In each case identify the wavenumber k or κ.

   (b)   Given $\varphi(0) = \varphi_0 > 0$, make a rough sketch of $\varphi(x)$ in the interval $0 \leq x \leq F$.

7.2   THE HARMONIC OSCILLATOR

The configuration of a harmonic oscillator is depicted in Fig. 7.4. The classical equation of motion of a particle of mass m is given by Hooke's law,

(7.6)
$$m \frac{d^2 x}{dt^2} = -\mathrm{K}x$$

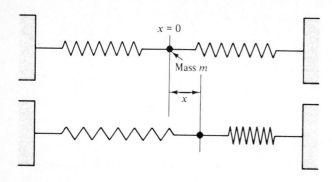

FIGURE 7.4 The one-dimensional harmonic oscillator. Displacement from equilibrium $(x = 0)$ is denoted by x.

The spring constant is K. In terms of the natural frequency ω_0,

(7.7)
$$\omega_0{}^2 = \frac{K}{m}$$

the above equation appears as

(7.8)
$$\frac{d^2x}{dt^2} + \omega_0{}^2x = 0$$

Multiplying this equation by $\dot{x}$ gives

(7.9)
$$\frac{d}{dt}\left[\frac{1}{2}(\dot{x}^2 + \omega_0{}^2x^2)\right] = 0$$

Integrating, one obtains the constant of motion

(7.10)
$$\frac{E}{m} = \frac{1}{2}(\dot{x}^2 + \omega_0{}^2x^2)$$

$$E = \frac{1}{2}m\dot{x}^2 + \frac{K}{2}x^2$$

The potential energy is

(7.11)
$$V = \frac{K}{2}x^2$$

When the particle comes to rest, the energy is entirely potential.

(7.12)
$$E = \frac{K}{2}x_0{}^2$$

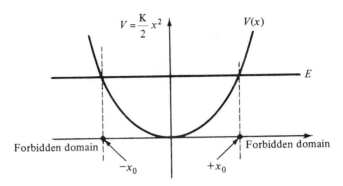

FIGURE 7.5 The turning points of the harmonic oscillator are at $x = \pm x_0$, where

$$\frac{K x_0^2}{2} = E$$

Such points (x_0) are turning points. For $x^2 > x_0^2$, the kinetic energy T is negative, so that classically this is a forbidden domain.

$$T = E - V = \frac{K}{2}(x_0^2 - x^2)$$

(7.13)

$$T < 0 \quad (\text{for } x^2 > x_0^2)$$

See Fig. 7.5.

    With these properties of the classical motion established, we turn next to the quantum mechanical formulation of the harmonic oscillator problem. The Hamiltonian for a particle of mass m in the potential (7.11) is

(7.14)
$$H = \frac{p^2}{2m} + \frac{K}{2}x^2$$

The corresponding Schrödinger equation appears as

(7.15)
$$-\frac{\hbar^2}{2m}\frac{\partial^2 \varphi}{\partial x^2} + \frac{K}{2}x^2\varphi = E\varphi$$

In the classically accessible domain, $E > Kx^2/2$, and this equation may be written

$$\varphi_{xx} = -k^2\varphi$$

(7.16)
$$\frac{\hbar^2 k^2(x)}{2m} = E - \frac{K}{2}x^2 > 0$$

The wavefunction φ is oscillatory in this domain.

In the classically forbidden domain where $x^2 > x_0{}^2$, $E < Kx^2/2$ and the Schrödinger equation becomes

(7.17)
$$\varphi_{xx} = \kappa^2 \varphi$$
$$\frac{\hbar^2 \kappa^2}{2m} = \frac{K}{2} x^2 - E > 0$$

so the wavefunction is nonoscillatory in this domain. In the asymptotic domain $Kx^2/2 \gg E$, the Schrödinger equation becomes

(7.18)
$$\varphi_{xx} = \frac{mK}{\hbar^2} x^2 \varphi \equiv \beta^4 x^2 \varphi$$

where β is the characteristic wavenumber

(7.19)
$$\beta^2 \equiv \frac{m\omega_0}{\hbar}$$

In terms of the nondimensional displacement

(7.20)
$$\xi \equiv \beta x$$

(7.18) appears as

$$\varphi_{\xi\xi} = \xi^2 \varphi$$

In the domain under consideration, $\xi \gg 1$ and the solution to the latter equation appears as

$$\varphi \sim A \exp\left(\pm \frac{\xi^2}{2}\right) = A \exp\left[\pm \frac{(\beta x)^2}{2}\right]$$

The growing solution $(+)$ violates the normalization condition

(7.21)
$$\int_{-\infty}^{\infty} \varphi^* \varphi \, dx < \infty$$

and one is left with the exponentially decaying wavefunction

(7.22)
$$\varphi \sim A \exp\left(- \frac{\xi^2}{2}\right) = A \exp\left[- \frac{(\beta x)^2}{2}\right]$$

The character of the wavefunction changes from oscillatory for $x^2 < x_0{}^2$ to decaying for $x^2 > x_0{}^2$, so the turning points $x = \pm x_0$ are also physically relevant in quantum mechanics. These properties are depicted in Fig. 7.6.

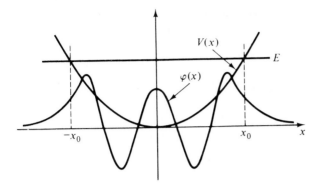

FIGURE 7.6 Typical behavior of energy eigenfunction for the simple harmonic oscillator.

Annihilation and Creation Operators

We turn to a general formulation of the solution to (7.15). The technique of solution we will develop is known as the *algebraic method*. It involves the operators

(7.23)

$$\hat{a} \equiv \frac{\beta}{\sqrt{2}}\left(\hat{x} + \frac{i\hat{p}}{m\omega_0}\right)$$

$$\hat{a}^\dagger = \frac{\beta}{\sqrt{2}}\left(\hat{x} - \frac{i\hat{p}}{m\omega_0}\right)$$

Inasmuch as $\hat{a} \neq \hat{a}^\dagger$, $\hat{a}$ is non-Hermitian. The properties that these operators have are determined through the fundamental commutator relation

(7.24)
$$[\hat{x}, \hat{p}] = i\hbar$$

For instance, it is readily shown that (see Problem 7.5)

(7.25)
$$[\hat{a}, \hat{a}^\dagger] = 1$$
$$\hat{a}\hat{a}^\dagger = 1 + \hat{a}^\dagger\hat{a}$$

With the aid of the inverse of (7.23),

(7.26)
$$\hat{x} = \frac{\hat{a} + \hat{a}^\dagger}{\sqrt{2}\beta}, \qquad \hat{p} = \frac{m\omega_0}{i}\frac{\hat{a} - \hat{a}^\dagger}{\sqrt{2}\beta}$$

the Hamiltonian for the harmonic oscillator becomes

(7.27)
$$\hat{H} = \frac{\hat{p}^2}{2m} + \frac{K\hat{x}^2}{2} = \hbar\omega_0(\hat{a}^\dagger\hat{a} + \tfrac{1}{2})$$

In this manner we see that the problem of finding the eigenvalues of $\hat{H}$ has been transformed to that of finding the eigenvalues of the operator

(7.28)
$$\hat{N} \equiv \hat{a}^{\dagger}\hat{a}$$

Let φ_n be the eigenfunction of $\hat{N}$ corresponding to the eigenvalue n, so that

(7.29)
$$\hat{N}\varphi_n = n\varphi_n$$

(We do not assume that n is an integer at this point. This property is established later.) Consider the effect of operating on $\hat{a}\varphi_n$ with $\hat{N}$.

(7.30)
$$\hat{N}\hat{a}\varphi_n = \hat{a}^{\dagger}\hat{a}\hat{a}\varphi_n = (\hat{a}\hat{a}^{\dagger} - 1)\hat{a}\varphi_n = \hat{a}(\hat{a}^{\dagger}\hat{a} - 1)\varphi_n$$
$$\hat{N}\hat{a}\varphi_n = \hat{a}(\hat{N} - 1)\varphi_n = \hat{a}(n - 1)\varphi_n = (n - 1)\hat{a}\varphi_n$$

It follows that $\hat{a}\varphi_n$ is the eigenfunction of $\hat{N}$ which corresponds to the eigenvalue $n - 1$. That is (apart from normalization factors),

(7.31)
$$\hat{a}\varphi_n = \varphi_{n-1}$$

Similarly,

(7.32)
$$\hat{a}\varphi_{n-1} = \varphi_{n-2}$$

and so forth. Because of this property, $\hat{a}$ is called an *annihilation* or *stepdown* or *demotion* operator.

In similar manner, if we consider the operation $\hat{N}\hat{a}^{\dagger}\varphi_n$, there results

(7.33)
$$\hat{N}\hat{a}^{\dagger}\varphi_n = (n + 1)\hat{a}^{\dagger}\varphi_n$$

This equation implies that $\hat{a}^{\dagger}\varphi_n$ is the eigenfunction of $\hat{N}$ corresponding to the eigenvalue $n + 1$.

(7.34)
$$\hat{a}^{\dagger}\varphi_n = \varphi_{n+1}$$

Similarly,

(7.35)
$$\hat{a}^{\dagger}\varphi_{n+1} = \varphi_{n+2}$$

and so forth. The operator $\hat{a}^{\dagger}$ is called a *creation* or *stepup* or *promotion* operator (Fig. 7.7).

Since the Hamiltonian for the harmonic oscillator is the sum of the squares of two Hermitian operators,

(7.36)
$$\langle H \rangle \geq 0$$

(see Problem 4.13). In the eigenstate φ_n,

(7.37)
$$\hat{H}\varphi_n = \hbar\omega_0(\hat{N} + \tfrac{1}{2})\varphi_n = \hbar\omega_0(n + \tfrac{1}{2})\varphi_n$$
$$\langle \varphi_n | \hat{H}\varphi_n \rangle = \hbar\omega_0(n + \tfrac{1}{2}) \geq 0$$

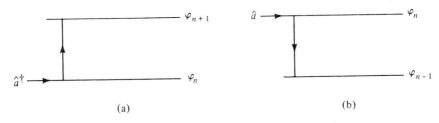

FIGURE 7.7   Schematic representation of the raising and lowering operators $\hat{a}^\dagger$ and $\hat{a}$.

This implies that the eigenvalues n must obey the condition

(7.38)
$$n \geq -\tfrac{1}{2}$$

That is, all eigenstates of $\hat{H}$, or equivalently $\hat{N}$, corresponding to eigenvalues $n < -\tfrac{1}{2}$ must vanish identically. For the harmonic oscillator such states do not exist. This condition is guaranteed if we set

(7.39)
$$\hat{a}\varphi_0 = 0$$

With (7.31) we obtain

(7.40)
$$\hat{a}\varphi_0 = \varphi_{-1} = 0$$
$$\hat{a}(\varphi_{-1}) = \varphi_{-2} = 0$$

As will be shown, (7.39) has a nontrivial (i.e., other than zero) solution for φ_0. Furthermore,

(7.41)
$$\hat{N}\varphi_0 = \hat{a}^\dagger \hat{a}\varphi_0 = 0 = 0\varphi_0$$

and we may conclude that the eigenvalue of $\hat{N}$ corresponding to the eigenfunction φ_0 is zero. It follows that

(7.42)
$$\hat{N}\hat{a}^\dagger \varphi_0 = \hat{a}^\dagger \hat{a}\hat{a}^\dagger \varphi_0 = \hat{a}^\dagger(\hat{a}^\dagger \hat{a} + 1)\varphi_0 = \hat{a}^\dagger \varphi_0$$
$$\hat{N}\hat{a}^\dagger \varphi_0 = 1\hat{a}^\dagger \varphi_0 = \varphi_1$$

The eigenvalue of $\hat{N}$ corresponding to φ_1 is the integer 1. This construction (same as in Eq. 7.34 et seq.) allows one to conclude that the index n, which labels the eigenfunction φ_n, is indeed an integer.
Repeating (7.37),

(7.43)
$$\hat{H}\varphi_n = \hbar\omega_0(n + \tfrac{1}{2})\varphi_n$$

one finds that the energy eigenvalues of the simple harmonic oscillator are

(7.44)
$$E_n = \hbar\omega_0(n + \tfrac{1}{2}) \qquad (n = 0, 1, 2, \ldots)$$

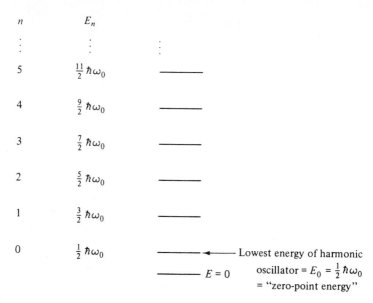

FIGURE 7.8   **The energy levels of the simple harmonic oscillator are equally spaced.**

The energy levels are equally spaced by the interval $\hbar\omega_0$ (Fig. 7.8). If a molecule, for example HCl, which resembles a dumbbell, has vibrational modes of excitation (the arm of the dumbbell acts as a spring), the Bohr frequencies emitted by the molecule fall in the scheme

(7.45)
$$hv = E_{n'} - E_n = \hbar\omega_0(n' + \tfrac{1}{2}) - \hbar\omega_0(n + \tfrac{1}{2})$$
$$= \hbar\omega_0(n' - n) = \hbar\omega_0 s$$
$$v = sv_0, \qquad \omega_0 \equiv 2\pi v_0$$

In the latter sequence of equations, n and n' are integers, so their difference, s, is also an integer. It follows that the frequencies emitted by a vibrational diatomic molecule are integral multiples of the natural frequency of the molecule, v_0 (Fig. 7.9).

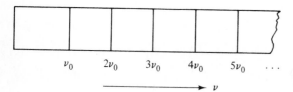

$v_0 \qquad 2v_0 \qquad 3v_0 \qquad 4v_0 \qquad 5v_0 \qquad \cdots$

$\longrightarrow v$

FIGURE 7.9   **Spectrum of a vibrational diatomic molecule. (See Problem 10.58.)**

PROBLEMS

7.2 A harmonic oscillator consists of a mass of 1 g on a spring. Its frequency is 1 Hz and the mass passes through the equilibrium position with a velocity of 10 cm/s. What is the order of magnitude of the quantum number associated with the energy of the system?

7.3 The spacing between vibrational levels of the CO molecule is 2170 cm^{-1}. Taking the mass of C to be 12 amu and O to be 16 amu, compute the effective spring constant K, which is a measure of the bond stiffness between the atoms of the molecule. [*Hint:* The mass that enters is the reduced mass, $mM/m + M$. The spacing between lines is given in terms of wavenumber $k = 2\pi/\lambda$, where $\omega = ck$ ($c =$ the speed of light), so that $\Delta\omega = c\,\Delta k$.]

7.4 The derivation in the text of the eigenvalues of $\hat{N}$ is based on the constraint that there are no states corresponding to the eigenvalues $n < -\frac{1}{2}$. This constraint was guaranteed by setting $\hat{a}\varphi_0 = 0$. It would appear that it can also be guaranteed by setting $\hat{a}\varphi_{1/2} = 0$, for in this case

$$\hat{a}\varphi_{1/2} = \varphi_{-1/2} = 0$$

Show that $\varphi_{1/2}$ as defined is an eigenfunction of $\hat{N}$ with the eigenvalue zero, hence $\varphi_{1/2}$ is more properly termed φ_0.

7.5 Using the fundamental commutator relation

$$[\hat{x}, \hat{p}] = i\hbar$$

show that

$$[\hat{a}, \hat{a}^\dagger] = 1$$

7.3 EIGENFUNCTIONS OF THE HARMONIC OSCILLATOR HAMILTONIAN

When written in terms of the nondimensional displacement ξ (7.20),

$$(7.46) \qquad \xi^2 \equiv \frac{m\omega_0}{\hbar} x^2 \equiv \beta^2 x^2$$

the operators $\hat{a}$ and $\hat{a}^\dagger$ become

$$(7.47) \qquad \hat{a} = \frac{\beta}{\sqrt{2}}\left(\hat{x} + \frac{i\hat{p}}{m\omega_0}\right) = \frac{\beta}{\sqrt{2}}\left(x + \frac{\hbar}{m\omega_0}\frac{\partial}{\partial x}\right) = \frac{1}{\sqrt{2}}\left(\xi + \frac{\partial}{\partial\xi}\right)$$

$$\hat{a}^\dagger = \frac{\beta}{\sqrt{2}}\left(\hat{x} - \frac{i\hat{p}}{m\omega_0}\right) = \frac{\beta}{\sqrt{2}}\left(x - \frac{\hbar}{m\omega_0}\frac{\partial}{\partial x}\right) = \frac{1}{\sqrt{2}}\left(\xi - \frac{\partial}{\partial\xi}\right)$$

The time-independent Schrödinger equation becomes

(7.48)
$$\left(2\hat{a}^\dagger\hat{a} + 1 - \frac{2E}{\hbar\omega_0}\right)\varphi = \varphi_{\xi\xi} + \left(\frac{2E}{\hbar\omega_0} - \xi^2\right)\varphi = 0$$

The ground-state wavefunction φ_0 of the simple harmonic oscillator Hamiltonian obeys (7.39)

$$\hat{a}\varphi_0 = 0$$

or, equivalently,

(7.49)
$$\frac{1}{\sqrt{2}}\left(\xi + \frac{\partial}{\partial\xi}\right)\varphi_0 = 0$$

This has the solution

(7.50)
$$\varphi_0 = A_0 e^{-\xi^2/2}$$

The requirement that $\varphi_0(\xi)$ be normalized implies that

(7.51)
$$1 = \int_{-\infty}^{\infty}|\varphi_0|^2\,d\xi = A_0^2\int_{-\infty}^{\infty}e^{-\xi^2}\,d\xi = \sqrt{\pi}\,A_0^2$$
$$A_0 = \pi^{-1/4}$$

so

(7.52)
$$\varphi_0(\xi) = \pi^{-1/4}e^{-\xi^2/2}$$

In terms of the dimensional displacement x, the normalized ground state is

(7.53)
$$\varphi_0(x) = B_0 e^{-\xi^2/2} = B_0 e^{-(\beta x)^2/2}$$

Normalization gives

(7.54)
$$1 = \int_{-\infty}^{\infty}|\varphi_0(x)|^2\,dx = \frac{B_0^2}{\beta}\int_{-\infty}^{\infty}e^{-\xi^2}\,d\xi = \frac{B_0^2\sqrt{\pi}}{\beta}$$
$$\varphi_0(x) = \left(\frac{\beta^2}{\pi}\right)^{1/4}e^{-(\beta x)^2/2}$$

The ground state φ_0 is a purely exponentially decaying wavefunction. It has no oscillatory component. The higher-energy eigenstates, on the other hand, will be found to oscillate in the classically allowed domain and decay exponentially in the classically forbidden domain.

With φ_0 given by (7.52), the remaining normalized eigenstates of the harmonic oscillator Hamiltonian are generated with the aid of the creation operator $\hat{a}^\dagger$, in the following manner:

$$\varphi_1 = \hat{a}^\dagger \varphi_0$$

(7.55)
$$\varphi_2 = \frac{1}{\sqrt{2}} \hat{a}^\dagger \varphi_1 = \frac{1}{\sqrt{2}} (\hat{a}^\dagger)^2 \varphi_0$$

$$\varphi_n = \frac{1}{\sqrt{n!}} (\hat{a}^\dagger)^n \varphi_0$$

With $\hat{a}^\dagger$ written in terms of ξ, as in (7.47), the equation for φ_1 above becomes

$$\varphi_1 = A_1 \left(\xi - \frac{\partial}{\partial \xi} \right) e^{-\xi^2/2}$$

(7.56)
$$\varphi_1 = A_1 2\xi e^{-\xi^2/2}$$

$$A_1 = (2\sqrt{\pi})^{-1/2}$$

where A_1 is the normalization constant of φ_1. The nth eigenstate is given by the formula

(7.57)
$$\varphi_n = A_n \left(\xi - \frac{\partial}{\partial \xi} \right)^n e^{-\xi^2/2}$$

The nth-order differential operator $(\hat{a}^\dagger)^n$, when acting on the exponential form $\exp(-\xi^2/2)$, reproduces the same exponential factor, multiplied by an nth-order polynomial in ξ.

(7.58)
$$\left(\xi - \frac{\partial}{\partial \xi} \right)^n e^{-\xi^2/2} = \mathscr{H}_n(\xi) e^{-\xi^2/2}$$

Thus the nth eigenstate of the simple harmonic oscillator Hamiltonian may be written together with its eigenvalue as

(7.59)
$$\boxed{\begin{aligned} \varphi_n &= A_n \mathscr{H}_n(\xi) e^{-\xi^2/2} \\ E_n &= \hbar \omega_0 (n + \tfrac{1}{2}) \end{aligned}}$$

The nth-order polynomials $\mathscr{H}_n(\xi)$ are well-known functions in mathematical physics. They are called *Hermite polynomials*. From (7.56) we see that $\mathscr{H}_1 = 2\xi$. The first six Hermite polynomials are listed in Table 7.1.

TABLE 7.1   The first six eigenenergies and eigenstates of the simple harmonic oscillator Hamiltonian

n	E_n	φ_n
0	$\hbar\omega_0/2$	$A_0 e^{-\xi^2/2}$
1	$3\hbar\omega_0/2$	$A_1 2\xi e^{-\xi^2/2}$
2	$5\hbar\omega_0/2$	$A_2(4\xi^2 - 2)e^{-\xi^2/2}$
3	$7\hbar\omega_0/2$	$A_3(8\xi^3 - 12\xi)e^{-\xi^2/2}$
4	$9\hbar\omega_0/2$	$A_4(16\xi^4 - 48\xi^2 + 12)e^{-\xi^2/2}$
5	$11\hbar\omega_0/2$	$A_5(32\xi^5 - 160\xi^3 + 120\xi)\,e^{-\xi^2/2}$

$$A_n = (2^n n! \sqrt{\pi})^{-1/2}$$

The nth-order Hermite polynomial $\mathcal{H}_n$ enters in the eigenfunctions φ_n of the quantum mechanical harmonic oscillator as

$$\varphi_n(\xi) = A_n \mathcal{H}_n(\xi)e^{-\xi^2/2}$$

$\mathcal{H}_n$ is a solution to *Hermite's equation*,

$$\mathcal{H}_n'' - 2\xi\mathcal{H}_n' + 2n\mathcal{H}_n = 0$$

The formulas connecting φ_n and φ_{n+1} (see Problem 7.9) are very useful in many problems relating to the simple harmonic oscillator. In Dirac notation they appear as

(7.60)
$$\hat{a}|\varphi_n\rangle = n^{1/2}|\varphi_{n-1}\rangle$$
$$\hat{a}^\dagger|\varphi_n\rangle = (n + 1)^{1/2}|\varphi_{n+1}\rangle$$

In place of $|\varphi_n\rangle$, let us write the ket vector $|n\rangle$. In this notation the equations above appear as

(7.61)
$$\hat{a}|n\rangle = n^{1/2}|n - 1\rangle$$
$$\hat{a}^\dagger|n\rangle = (n + 1)^{1/2}|n + 1\rangle$$

Let us check that

(7.62)
$$\hat{N}|n\rangle = n|n\rangle$$

With the aid of (7.61), we obtain

(7.63)
$$\hat{a}^\dagger\hat{a}|n\rangle = \hat{a}^\dagger n^{1/2}|n - 1\rangle = n^{1/2}n^{1/2}|n\rangle$$
$$\hat{a}^\dagger\hat{a}|n\rangle = \hat{N}|n\rangle = n|n\rangle$$

Inasmuch as $\{\varphi_n\}$ are normalized and are eigenstates of a Hermitian operator, they comprise an orthonormal sequence.

(7.64) $$\int_{-\infty}^{\infty} \varphi_n{}^* \varphi_l \, d\xi = \langle n|l \rangle = \delta_{nl}$$

To gain familiarity with the manner in which these concepts are used in problems, we will work out a few illustrative examples.

First, consider the question: What is $\langle x \rangle$ in the nth eigenstate φ_n? Here we must calculate

(7.65) $$\langle x \rangle = \langle n|\hat{x}|n \rangle$$

$$= \frac{1}{\sqrt{2\beta}} \langle n|\hat{a} + \hat{a}^\dagger|n \rangle$$

$$= \frac{1}{\sqrt{2\beta}} \{ n^{1/2} \langle n|n - 1 \rangle + (n + 1)^{1/2} \langle n|n + 1 \rangle \}$$

$$= 0$$

The last step follows from the orthogonality relation (7.64). The fact that the average value of x in any eigenstate φ_n vanishes is a consequence of the symmetry of the probability density $P = |\varphi_n|^2$ about the origin (see Fig. 7.10).

The second example we consider is the expectation of momentum p, in the nth eigenstate φ_n.

(7.66) $$\langle p \rangle = \langle n|\hat{p}|n \rangle = \frac{m\omega_0}{\sqrt{2i\beta}} \langle n|\hat{a} - \hat{a}^\dagger|n \rangle$$

$$= \frac{m\omega_0}{\sqrt{2i\beta}} \{ n^{1/2} \langle n|n - 1 \rangle - (n + 1)^{1/2} \langle n|n + 1 \rangle \}$$

$$= 0$$

In any eigenstate φ_n of the Hamiltonian of the simple harmonic oscillator, the probability of finding the particle with momentum $\hbar k$ is equal to that of finding the particle with momentum $-\hbar k$. Were we to express $\varphi_n(x)$ as a superposition of momentum eigenstates, $\exp(ikx)$, we would find the probability amplitude $b(k)$ to be an even (symmetric) function of k [i.e., $b(k) = b(-k)$].

Correspondence Principle

Next, we consider the manner in which the solution to the quantum mechanical harmonic oscillator problem obeys the *correspondence principle*. To these ends let us

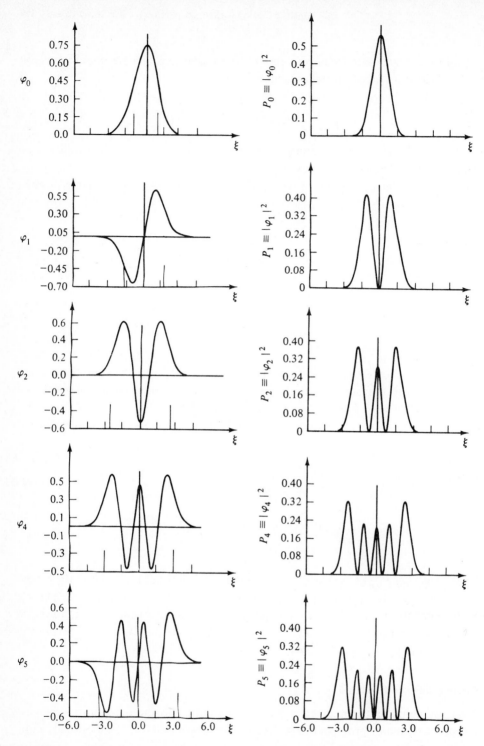

FIGURE 7.10 The first few eigenstates of the simple harmonic oscillator and corresponding probability densities. Turning points, $\xi_0^{(n)} = \sqrt{1 + 2n}$, are denoted by vertical marks.

calculate the classical probability density P, corresponding to a one-dimensional spring with natural frequency ω_0. Let the particle be at the origin at $t = 0$ with velocity $x_0 \omega_0$. The displacement at the time t is then given by

$$x = x_0 \sin (\omega_0 t)$$

(7.67)

$$\dot{x} = x_0 \omega_0 \cos (\omega_0 t)$$

This gives the correct initial data

$$x(0) = 0$$

(7.68)

$$\dot{x}(0) = x_0 \omega_0$$

The product $P(x)\, dx$ is the probability of finding the particle in the interval dx about the point x at any time. If T_0 is the period of oscillation

(7.69)
$$T_0 = \frac{2\pi}{\omega_0}$$

then

(7.70)
$$P\, dx = \frac{dt}{T_0} = \frac{\omega_0\, dt}{2\pi}$$

where

(7.71)
$$dt = \frac{dx}{\dot{x}}$$

Using (7.67), one obtains

(7.72)
$$dt = \frac{dx}{\omega_0 \sqrt{x_0{}^2 - x^2}}$$

so that

(7.73)
$$P\, dx = \frac{\omega_0}{2\pi}\, dt = \frac{dx}{2\pi \sqrt{x_0{}^2 - x^2}}$$

The probability density so found is normalized with respect to the angular displacement $d\theta = \omega_0\, dt$, $0 \leq \theta \leq 2\pi$. The interval in displacement x is one-half as long, so the properly normalized P function, over the interval $-x_0 < x < +x_0$, is

$$P = \frac{1}{\pi \sqrt{x_0{}^2 - x^2}}$$

(7.74)

$$\int_{-x_0}^{+x_0} P(x)\, dx = 1$$

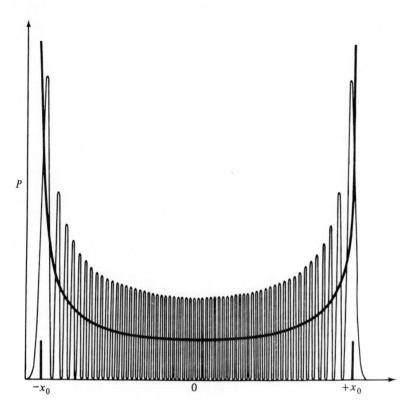

$-x_0$ 0 $+x_0$

FIGURE 7.11   Classical probability density

$$P^{\mathrm{CL}} = \frac{1}{\pi\sqrt{x_0{}^2 - x^2}}$$

superimposed on the quantum mechanical probability density

$$P_n{}^{\mathrm{QM}} = |\varphi_n|^2$$

For the case $n \gg 1$,

$$\lim_{n \to \infty} \langle P_n{}^{\mathrm{QM}} \rangle = P^{\mathrm{CL}}$$

This function is sketched in Fig. 7.11, where it is superimposed on the quantum mechanical probability density corresponding to a state with $n \gg 1$. The singularities in P at the turning points $\pm x_0$ are due to the fact that the particle comes to rest at these points.

    The correspondence which the quantum mechanical formulation displays in the present case is clearly exhibited in Fig. 7.11, where we see that

(7.75)

$$\lim_{n \to \infty} \langle P_n{}^{\mathrm{QM}} \rangle = P^{\mathrm{CL}}$$

$$\langle P_n{}^{\mathrm{QM}} \rangle = \frac{1}{2\epsilon} \int_{x-\epsilon}^{x+\epsilon} \varphi_n{}^*(y) \varphi_n(y)\, dy$$

The superscripts QM and CL denote quantum mechanical and classical, respectively, and ϵ is an arbitrarily small interval. The integral above is called a *local average*. It represents the average of P^{QM} in a small interval centered at x.

The classical configuration that corresponds to the quantum state in which a set of commuting observables are specified is the configuration which includes these same parameters as constant and known. Thus, in the problem of a particle confined to a one-dimensional box, considered in Chapter 4, when one concludes that the classical probability density is uniform, it should be noted that this is the case provided that all one knows about the particle is its energy. The classical state of this system permits a more elaborate description. Unlike the quantum case, for the classical particle one may specify both its energy and position in time, $x(t)$. Given this maximally informative classical description, the configurational probability density becomes $\delta[x(t) - x]$. When one speaks of configurational correspondence in the limit of high quantum numbers, what is usually meant is that in this limit the quantum probability density goes to the classical probability density in which, consistent with the quantum description, not more than the energy is specified.

In our consideration of correspondence for the harmonic oscillator, this rule is again obeyed. The expression (7.74) for the classical probability density is relevant to the case where only the amplitude x_0, or equivalently the energy $E = Kx_0{}^2/2$, is known. The quantum density sketched in Fig. 7.11 is likewise connected to the energy eigenstate φ_n for which measurement of energy finds with certainty the value E_n.

PROBLEMS

7.6 (a) Show that the Hermite polynomials generated in the Taylor series expansion

$$\exp(2\xi t - t^2) = \sum_{n=0}^{\infty} \frac{\mathscr{H}_n(\xi)}{n!} t^n$$

are the same as generated in (7.58).

(b) Show that $\mathscr{H}_n$ as generated by

$$\mathscr{H}_n(\xi) = (-1)^n \left(e^{\xi^2} \frac{\partial^n}{\partial \xi^n} e^{-\xi^2} \right)$$

are equivalent to those given by (7.58).

(c)   Use any of the preceding relations to establish

$$\mathcal{H}_n{}' = 2n\mathcal{H}_{n-1}$$

(d)   and the recursion relation

$$\mathcal{H}_{n+1} = 2\xi\mathcal{H}_n - 2n\mathcal{H}_{n-1}$$

(e)   Use the generating formula of part (b) to find $\mathcal{H}_0(\xi)$, $\mathcal{H}_1(\xi)$, and $\mathcal{H}_2(\xi)$.

7.7   The general formula for the normalization constant of φ_n is

$$A_n = (2^n n!\sqrt{\pi})^{-1/2}$$

Show that this gives correct normalization for φ_4.

7.8   Show directly from the form of φ_n given by (7.57) that

$$\hat{P}\varphi_n = (-)^n\varphi_n$$

where $\hat{P}$ is the parity operator.

7.9   (a)   Show that the normalized nth eigenstate φ_n is generated from the normalized ground state φ_0 through

$$\varphi_n = \frac{1}{\sqrt{n!}}(\hat{a}^\dagger)^n\varphi_0$$

(b)   Show that part (a) implies the following relations.

$$\hat{a}\varphi_n = n^{1/2}\varphi_{n-1}$$
$$\hat{a}^\dagger\varphi_n = (n+1)^{1/2}\varphi_{n+1}$$

where φ_n, φ_{n-1}, and φ_{n+1} are all normalized.

7.10   Show that in the nth eigenstate of the harmonic oscillator, the average kinetic energy $\langle T\rangle$ is equal to the average potential energy $\langle V\rangle$ (the virial theorem). That is,

$$\langle V\rangle = \frac{K}{2}\langle x^2\rangle = \langle T\rangle = \frac{1}{2m}\langle p^2\rangle = \frac{1}{2}\langle E\rangle = \frac{\hbar\omega_0}{2}\left(n + \frac{1}{2}\right)$$

Answer (partial)

$$\frac{K}{2}\langle x^2\rangle = \left(\frac{K}{4\beta^2}\right)\langle n|(\hat{a} + \hat{a}^\dagger)^2|n\rangle$$

$$= \left(\frac{K}{4\beta^2}\right)\langle n|\hat{a}^2 + \hat{a}^{\dagger 2} + (\hat{a}\hat{a}^\dagger + \hat{a}^\dagger\hat{a})|n\rangle$$

$$= \left(\frac{K}{4\beta^2}\right)\{0 + 0 + \langle n|(1 + 2\hat{N})|n\rangle\}$$

$$\langle V\rangle = \frac{\hbar\omega_0}{4}(1 + 2n) = \frac{\hbar\omega_0}{2}\left(n + \frac{1}{2}\right)$$

7.11 In Problem 7.2, what is the average spacing (in centimeters) between zeros of an eigenstate with such a quantum number?

7.12 A harmonic oscillator is in the initial state

$$\psi(x, 0) = \varphi_n(x)$$

that is, an eigenstate of $\hat{H}$. What is $\psi_n(x, t)$?

7.13 For a harmonic oscillator in the superposition state

$$\psi(x, t) = \frac{1}{\sqrt{2}} [\psi_0(x, t) + \psi_1(x, t)]$$

show that

$$\langle x \rangle = C \cos (\omega_0 t)$$

In the notation above,

$$\psi_n(x, t) \equiv \varphi_n(x) \exp \left(-\frac{iE_n t}{\hbar} \right)$$

7.14 Show that in the nth state of the harmonic oscillator,

$$\langle x^2 \rangle = (\Delta x)^2$$
$$\langle p^2 \rangle = (\Delta p)^2$$

7.15 Find $\langle x \rangle$ for a harmonic oscillator in the superposition state

$$\psi(x, t) = \frac{1}{\sqrt{2}} [\psi_0(x, t) + \psi_3(x, t)]$$

The harmonic oscillator has natural frequency ω_0.

7.16 A large dielectric cube with edge length L is uniformly charged throughout its volume so that it carries a total charge Q. It fills the space between condenser plates, which have a potential difference Φ_0 across them. An electron is free to move in a small canal drilled in the dielectric normal to the plates (Fig. 7.12).

The Hamiltonian for the electron is (with x measured from the center of the canal and e written for $-|e|$)

$$\hat{H} = \frac{\hat{p}^2}{2m} + \frac{Kx^2}{2} + \frac{e\Phi_0}{L} x$$

(a) What is the spring constant K in terms of the total charge Q?

(b) What are the eigenenergies and eigenfunctions of $\hat{H}$? [*Hint*: Rewrite the potential energy of the electron as

$$V = \frac{K}{2} (x^2 + 2\gamma x) = \frac{K}{2} [(x + \gamma)^2 - \gamma^2]$$

$$\gamma \equiv \frac{e\Phi_0}{LK}$$

then change variables to $z \equiv x + \gamma$. To evaluate K, use Gauss's law (neglecting "edge effects").]

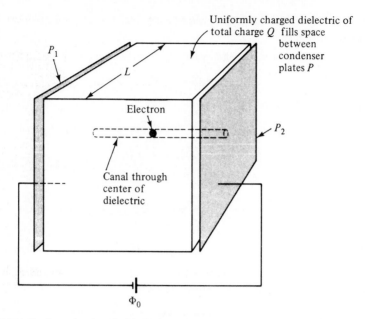

FIGURE 7.12 **Configuration described in Problem 7.16.**

7.17 (a) Show that the time-independent Schrödinger equation for the harmonic oscillator, with the energy eigenvalues (7.44), may be written

$$\varphi_{\xi\xi} + (2n + 1 - \xi^2)\varphi = 0$$

(b) Using the relations of Problem 7.6, show that

$$\varphi_n = \mathscr{H}_n(\xi)e^{-\xi^2/2}$$

is a solution to this equation.

(c) Obtain Hermite's equation

$$\mathscr{H}_n'' - 2\xi\mathscr{H}_n' + 2n\mathscr{H}_n = 0$$

7.18 Use the uncertainty principle between x and p to derive the "zero-point" energy

$$E_0 = \tfrac{1}{2}\hbar\omega_0$$

of a harmonic oscillator with natural frequency ω_0 (see Fig. 7.8).

7.19 Show that

$$\hat{a}^\dagger\hat{a} = \frac{1}{2}\left(\xi^2 - \frac{\partial^2}{\partial\xi^2} - 1\right)$$

in the nondimensional ξ notation.

7.20 Show that the asymptotic exponential behavior of $\varphi_n(\xi)$ agrees with that obtained directly from the Schrödinger equation, in the limit that $\xi \to \infty$.

7.21 Show that

$$\left(\xi + \frac{\partial}{\partial\xi}\right)^\dagger = \xi - \frac{\partial}{\partial\xi}$$

in $\mathfrak{H}_2$ (see Eq. 4.31).

7.22 (a) What is the asymptotic solution φ_n to the Schrödinger equation (as given in Problem 7.17)

$$\varphi_{\xi\xi} + (2n + 1 - \xi^2)\varphi = 0$$

in the domain

$$\xi^2 \ll 1 + 2n \simeq 2n?$$

(b) Show that

$$\lim_{n \gg 1} \langle P_n \rangle = \langle |\tilde{\varphi}_n|^2 \rangle = \text{constant}$$

Answer

In this domain the Schrödinger equation above becomes

$$\tilde{\varphi}_{\xi\xi} + 2n\tilde{\varphi} = 0$$

which has the (even) solution

$$\tilde{\varphi}_n = C \cos\left(\sqrt{2n}\,\xi\right)$$

It follows that the local average of $|\tilde{\varphi}_n|^2$ in this domain is given by

$$\langle |\tilde{\varphi}_n|^2 \rangle = \frac{C^2}{2\epsilon} \int_{\xi-\epsilon}^{\xi+\epsilon} \cos^2\left(\sqrt{2n}\,\xi\right) d\xi$$

$$= \frac{C^2}{2\epsilon}\left\{\epsilon + \frac{1}{2\sqrt{2n}}\left[\sin\left(2\sqrt{2n}\epsilon\right)\cos\left(2\sqrt{2n}\,\xi\right)\right]\right\}$$

$$\lim_{n \to \infty} \langle P_n \rangle = \frac{C^2}{2}$$

This result explains the flatness of $\langle P^{\text{QM}} \rangle$ in the central domain $\xi^2 \ll 2n$, as seen in Fig. 7.11.

7.23 Estimate the length of interval about $x = 0$ which corresponds to the classically allowed domain for the ground state of the simple harmonic oscillator.

Answer

The turning points occur at

$$\xi = \pm 1 \quad \text{or equivalently at} \quad x = \pm\sqrt{\frac{\hbar}{m\omega_0}}$$

At this value, $|\varphi_0|^2$ is e^{-1} times smaller than its value at the origin (Fig. 7.10).

7.24 Show that in the nth stationary state $|n\rangle$ of a harmonic oscillator with fundamental frequency ω_0,

$$\Delta p \, \Delta x = \frac{E_n}{\omega_0} = \hbar(n + \tfrac{1}{2})$$

7.4 THE HARMONIC OSCILLATOR IN MOMENTUM SPACE

Representations in Quantum Mechanics

Let us recall Eqs. (4.42) et seq., which relate the wavefunction $\varphi(x)$ to the momentum coefficient $b(k)$.

(7.76)

$$\varphi(x) = \int_{-\infty}^{\infty} b(k)\varphi_k \, dk$$

$$b(k) = \int_{-\infty}^{\infty} \varphi(x)\varphi_k^* \, dx$$

The eigenfunction of momentum corresponding to the value $p = \hbar k$ is φ_k. The wavefunction $\varphi(x)$ gives the probability density in coordinate space through the Born relation

(7.77) $$P(x) = |\varphi(x)|^2$$

The momentum coefficient $b(k)$ gives the probability density [probability of finding the particle to have momentum in the interval $\hbar k$ to $\hbar(k + dk)$] in momentum (k) space through the relation

(7.78) $$P(k) = |b(k)|^2$$

The integral formulas (7.76) serve to determine $\varphi(x)$ given $b(k)$, and vice versa. It follows that any information contained in $\varphi(x)$ can be obtained from knowledge of $b(k)$ and vice versa. Given the Hamiltonian of a system, $\varphi(x)$ is determined. Let us construct an equation which similarly determines $b(k)$ from the Hamiltonian [i.e., without first finding $\varphi(x)$]. To these ends we first recall the time-independent Schrödinger equation for the harmonic oscillator.

(7.79) $$\left(-\frac{\hbar^2}{2m}\frac{\partial^2}{\partial x^2} + \frac{Kx^2}{2}\right)\varphi(x) = E\varphi(x)$$

Substituting the Fourier decomposition of $\varphi(x)$ above and noting the equality

(7.80)
$$x\varphi_k = +i\frac{\partial\varphi_k}{\partial k}$$

gives

(7.81)
$$\int_{-\infty}^{\infty} dk\, b(k)\left(\frac{\hbar^2 k^2}{2m} - \frac{K}{2}\frac{\partial^2}{\partial k^2}\right)\varphi_k = E\int_{-\infty}^{\infty} dk\, b(k)\varphi_k$$

Integrating the second term on the left-hand side by parts twice and setting

(7.82)
$$b(k)|_{k=\pm\infty} = 0$$

gives

(7.83)
$$\int_{-\infty}^{\infty} dk\,\varphi_k\left[\left(\frac{\hbar^2 k^2}{2m} - \frac{K}{2}\frac{\partial^2}{\partial k^2} - E\right)b(k)\right] = 0$$

It follows that the term in brackets is the Fourier transform of zero, which is zero. We conclude that $b(k)$ (appropriate to the harmonic oscillator) satisfies the *k-dependent Schrödinger equation*

(7.84)
$$\left(\frac{\hbar^2 k^2}{2m} - \frac{K}{2}\frac{\partial^2}{\partial k^2}\right)b(k) = Eb(k)$$

This equation is also called the *Schrödinger equation in momentum representation.*
We note that the Hamiltonian in momentum representation includes the simple multiplicative operator $\hbar k$ in place of p and the differential operator $+i\partial/\partial k$ in place of x. This rule for obtaining the structure of the Hamiltonian in momentum representation always holds providing the potential $V(x)$ is an analytic function[1] of x (i.e., has a well-defined power-series expansion). For such cases the Schrödinger equation in either coordinate or momentum space is obtained through the recipes:

$$\text{In } x\text{-space:}\quad \hat{H}(x, p) \to \hat{H}\left(x, -\frac{i\hbar\,\partial}{\partial x}\right)$$

$$\text{In } p\text{-space:}\quad \hat{H}(x, p) \to \hat{H}\left(+\frac{i\partial}{\partial k}, \hbar k\right)$$

---

[1] In the more general case the Schrödinger equation in momentum space becomes an integral equation. These topics are discussed in greater detail in E. Merzbacher, *Quantum Mechanics*, 2nd ed., Wiley, New York, 1970.

The time-dependent Schrödinger equation in momentum representation appears as

(7.85)
$$i\hbar \frac{\partial}{\partial t} b(k, t) = \hat{H}(k)b(k, t)$$

Paralleling the development of (3.70) permits the solution to (7.85) for the initial-value problem for $b(k, t)$ to be written

(7.86)
$$b(k, t) = \exp\left(-\frac{it\hat{H}}{\hbar}\right)b(k, 0)$$

For free-particle motion with $\hat{H} = \hbar^2 k^2/2m$, the latter relation gives

(7.87)
$$|b(k, t)|^2 = |b(k, 0)|^2$$

The momentum probability density for free-particle motion is constant in time.

Geometrically, the function $b(k)$ is the projection of the state $\varphi(x)$ onto the momentum eigenstate φ_k (recall Eq. 4.44).

(7.88)
$$b(k) = \langle \varphi_k | \varphi \rangle$$

For any given state $\varphi(x)$, the function $b(k)$ represents a distribution of values, corresponding to the projections of $\varphi(x)$ onto the set of basis vectors $\{\varphi_k\}$. The functions $b(k)$ and $\varphi(x)$ are equally informative. In momentum representation a state of the system is represented by its projections onto the basis of Hilbert space $\{\varphi_k\}$. (See Fig. 4.6.)

This is analogous to the statement that a vector **B** in 3-space is represented by its projections onto the three unit vectors $\mathbf{e}_x$, $\mathbf{e}_y$, and $\mathbf{e}_z$, namely B_x, B_y, and B_z. These are not the only basis vectors one can use to represent the vector **B**. For instance, one can employ the basis $\mathbf{e}_x'$, $\mathbf{e}_y'$, and $\mathbf{e}_z'$ given by

(7.89)
$$\mathbf{e}_x' = \frac{1}{\sqrt{2}}(\mathbf{e}_x + \mathbf{e}_y)$$

$$\mathbf{e}_y' = \frac{1}{\sqrt{2}}(\mathbf{e}_x - \mathbf{e}_y)$$

$$\mathbf{e}_z' = \mathbf{e}_z$$

(See Fig. 7.13.) In this basis **B** is represented by the three components

$$\mathbf{B} = (B_x', B_y', B_z') = \frac{1}{\sqrt{2}}(B_x + B_y, B_x - B_y, \sqrt{2}B_z)$$

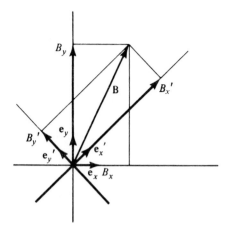

FIGURE 7.13 Projections onto two sets of basis vectors (in the xy plane) of the vector B. The two bases are related through (7.89). The z component is the same in both representations.

There are countless other triads of unit vectors which are valid bases of 3-space (i.e., they all span 3-space). The three components of **B** in any one of these representations completely specify **B**.

Similarly, one may describe the state of a system in quantum mechanics in different representations. In each of these, a distinct set of vectors serves as a basis of Hilbert space. Particularly important in the theory of representations is the concept of common eigenfunctions of some *complete set of commuting operators* relevant to a given system. Suppose that the complete set of commuting operators are $\hat{A}$, $\hat{B}$, and $\hat{C}$. In the state φ_{abc} (common eigenstates of $\hat{A}$, $\hat{B}$, and $\hat{C}$) one may specify the "good" quantum numbers a, b, and c. The state of the system cannot be further resolved. Such states may serve as a basis of Hilbert space. The representation in which all states are referred to the basis $\{\varphi_{abc}\}$ is called the *abc representation*, just as we call the representation in which states are referred to the eigenfunctions of momentum, the *momentum representation*.[1] One also speaks of the *abc* representation as the one in which $\hat{A}$, $\hat{B}$, and $\hat{C}$ are *diagonal*.

PROBLEMS

7.25   Show that $b(k)$ is even for any even potential $V(x) = V(-x)$. What can be concluded about the oddness or evenness of $b(k)$ if $V(x)$ is an odd function, $V(x) = -V(-x)$?

7.26   What are the eigenfunctions $b_n(k)$ of the harmonic oscillator Hamiltonian $\hat{H}(k)$ in momentum space (as given by Eq. 7.84)? [*Hint:* Note the similarity between $\hat{H}(k)$ and $\hat{H}(x)$, and the boundary conditions of $\varphi_n(x)$ and $b_n(k)$.]

---

[1] For further discussion of the coordinate and momentum representations, see Appendix A.

7.27   What is the Schrödinger equation in momentum representation for a free particle moving in one dimension? What are the eigenfunctions $b(k)$ of this equation?

7.28   Consider the Gaussian wave packet whose initial momentum probability density is given by (6.45).

    (a)   What is $|b(k, t)|^2$ at $t > 0$?

    (b)   What is $\Delta x\, \Delta p$ at $t > 0$?

7.29   Consider an arbitrary differentiable function of p, $\varphi(p)$. Show that with $\hat{p} = p$ and $\hat{x} = +i\hbar\, \partial/\partial p$,

$$[\hat{x}, \hat{p}]\varphi(p) = i\hbar\varphi(p)$$

7.30   What is the eigenfunction of the operator $\hat{x}$, in the momentum representation, corresponding to the eigenvalue x? That is, give the solution to the equation

$$\hat{x}\varphi_x(p) = x\varphi_x(p)$$

7.31   Let $|x'\rangle$ denote an eigenvector of the position operator $\hat{x}$ with eigenvalue x' and let $|k'\rangle$ denote an eigenvector of the momentum operator $\hat{p}$ with eigenvalue $\hbar k'$. Show that

    (a) $\langle k|k'\rangle = \delta(k - k')$

    (b) $\langle x|x'\rangle = \delta(x - x')$

    (c) $\langle x|k\rangle = \dfrac{1}{\sqrt{2\pi}}\exp{(ikx)}$

For each case, state in which representation you are working.

7.32   Suppose that the operators $\hat{a}$ and $\hat{a}^\dagger$ in

$$\hat{H} = \hbar\omega_0(\hat{a}^\dagger\hat{a} + \tfrac{1}{2})$$

obey the *anticommutation* relation

$$\{\hat{a}, \hat{a}^\dagger\} \equiv \hat{a}\hat{a}^\dagger + \hat{a}^\dagger\hat{a} = 1$$

    (a)   What are the values of $\hat{a}|n\rangle$ and $\hat{a}^\dagger|n\rangle$ that follow from the anticommutation relation above?

    (b)   Since $\langle H\rangle \geq 0$, for consistency we may again set

$$\hat{a}|0\rangle = 0$$

Combining this fact with your answer to part (a), which are the only nonvanishing states $|n\rangle$?

    (c)   If, in addition to the anticommutation property above, $\hat{a}$ and $\hat{a}^\dagger$ also obey the relations, $\{\hat{a}, \hat{a}\} = \{\hat{a}^\dagger, \hat{a}^\dagger\} = 0$, show that $\hat{N}^2 = \hat{N}$.

Answer (partial)

    (a) $\hat{a}|n\rangle = \sqrt{n}|1 - n\rangle$

        $\hat{a}^\dagger|n\rangle = \sqrt{1 - n}|1 - n\rangle$

    (b)   The only nonvanishing states are $|0\rangle$ and $|1\rangle$. [*Note:* Anticommutation relations between $\hat{a}$ and $\hat{a}^\dagger$ are used to describe particles that obey the *Pauli exclusion principle*.[1] In this context the operator $\hat{N}$ denotes the number of particles in a given state so that (b) implies that there is no more than one particle in any state. The $|0\rangle$ state is called the *vacuum state*. The formalism is known as *second quantization*.[2]]

7.33   What is the lowest value of kinetic energy $\langle T \rangle$ a harmonic oscillator with frequency ω_0 can have?

Answer

In Problem 7.10 we found that in the nth eigenstate of the oscillator

$$\langle V \rangle = \langle T \rangle = \frac{\hbar\omega_0}{2}\left(n + \frac{1}{2}\right) \geq \frac{\hbar\omega_0}{4}$$

Thus, the lowest allowed energy of the oscillator is $\hbar\omega_0/2$. It is impossible to force the oscillator to a lower energy. In a solid, for example, whose nuclei are bound together by harmonic forces, this zero-point energy persists at 0 K. (See Problem 7.18.)

7.5   UNBOUND STATES

If a wavefunction ψ represents a bound state (in one dimension), then

(7.90) $$|\psi|^2 \rightarrow 0, \qquad |x| \rightarrow \infty$$

for all t. A wavefunction that does not obey this condition represents an *unbound state*. The square modulus of a *bound state* gives a finite integral over the infinite interval.

(7.91) $$\int_{-\infty}^{\infty} |\psi|^2 \, dx < \infty$$

The square modulus of an *unbound state* gives a finite integral over *any* finite interval.

(7.92) $$\int_{a}^{b} |\psi|^2 \, dx < \infty, \qquad |b - a| < \infty$$

The eigenstate of the momentum operator

(7.93) $$\varphi_n(x) = \frac{1}{\sqrt{2\pi}} e^{ikx}$$

---

[1] A formal statement of this principle is given in Chapter 12. See also Appendix B.
[2] Second quantization will be encountered again in Chapter 13.

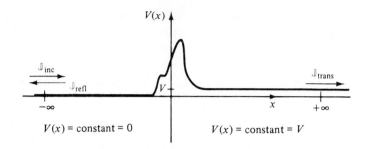

FIGURE 7.14   One-dimensional scattering problem. Incident particle current $\mathbb{J}_{inc}$ initiated at $x = -\infty$ is partially transmitted ($\mathbb{J}_{trans}$) and partially reflected ($\mathbb{J}_{refl}$) by a potential barrier $V(x)$. The potential is constant outside the scattering domain.

represents an unbound state. The eigenfunction of the simple harmonic oscillator Hamiltonian

$$(7.94) \qquad \varphi_n(\xi) = A_n \mathcal{H}_n(\xi)e^{-\xi^2/2}$$

(see Section 7.3) represents a bound state. Unbound states are relevant to scattering problems. Such problems characteristically involve a beam of particles which is incident on a potential barrier (Fig. 7.14).

Since $\int_{-\infty}^{\infty} |\psi|^2 \, dx$ diverges for unbound states, it is convenient to normalize the wavefunction for scattering problems in terms of the particle density ρ. For one-dimensional scattering problems we take

$$|\psi|^2 \, dx = \rho \, dx = dN$$

$$= \text{number of particles in the interval } dx$$

$$(7.95)$$

$$\int_a^b |\psi|^2 \, dx = N$$

$$= \text{number of particles in the interval } (b - a)$$

For a one-dimensional beam of 10^3 neutrons/cm, all moving with momentum $p = \hbar k_0$, the wavefunction is written

$$\psi = 10^{3/2} e^{i(k_0 x - \omega t)}, \qquad |\psi|^2 = 10^3 \text{ cm}^{-1}$$

$$(7.96)$$

$$\frac{\hbar^2 k_0^2}{2m} = \hbar\omega$$

The sole difference between ψ so defined and a wavefunction whose square modulus is probability density is a multiplicative constant. It follows that $|\psi|^2$, when referred to particle density, is proportional to probability density also. For uniform beams, $|\psi|^2$ is constant, which in turn implies that it is uniformly probable to find particles anywhere along the beam. This is consistent with the uncertainty principle. For instance, for the wavefunction (7.96), the momentum of any neutron in the beam is $\hbar k_0$, whence its position is maximally uncertain.

Continuity Equation

One-dimensional barrier problems involve incident, reflected, and transmitted *current densities*, $\mathcal{J}_{\text{inc}}$, $\mathcal{J}_{\text{refl}}$, and $\mathcal{J}_{\text{trans}}$, respectively. In three dimensions the number density and current density **J** are related through the *continuity equation*

$$(7.97) \qquad \frac{\partial \rho}{\partial t} + \nabla \cdot \mathbf{J} = 0$$

To clarify the physical meaning of this equation, we integrate it over a volume V and obtain

$$(7.98) \qquad \frac{\partial N}{\partial t} = - \int_S \mathbf{J} \cdot d\mathbf{S}$$

The total number of particles in the volume V is

$$(7.99) \qquad N = \int_V \rho \, d\mathbf{r}$$

(Gauss's theorem was used to transform the divergence term.) The surface S encloses the volume V (Fig. 7.15). Equation (7.98) says that the number of particles in the volume V changes by virtue of a net flux of particles out of (or into) the volume V. It is a statement of the *conservation of matter* because it says that this is the *only* way the total number of particles in V can change. If particles are born spontaneously in V with no net flux of particles through the surface S, then $\partial N / \partial t > 0$, while $\int \mathbf{J} \cdot d\mathbf{S} = 0$ and (7.98) is violated.

If particles are moving only in the x direction,

$$(7.100) \qquad \mathbf{J} = (\mathcal{J}_x, 0, 0)$$

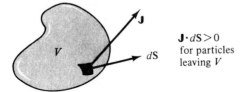

$$\mathbf{J} \cdot d\mathbf{S} > 0$$
for particles
leaving V

FIGURE 7.15 Geometry relevant to integration of the continuity equation.

and the continuity equation becomes

(7.101)
$$\frac{\partial \rho}{\partial t} + \frac{\partial \mathcal{J}_x}{\partial x} = 0$$

We already have identified ρ with $|\psi|^2$. To relate $\mathcal{J}_x$ to ψ, we must construct an equation that looks identical to (7.101) with $|\psi|^2$ in place of ρ. Then the functional of ψ which appears after $\partial/\partial x$ is $\mathcal{J}_x$.

The wavefunction for particles in the beam obeys the Schrödinger equation

(7.102)
$$\frac{\partial \psi}{\partial t} = -\frac{i}{\hbar} \hat{H}\psi, \qquad \frac{\partial \psi^*}{\partial t} = +\frac{i}{\hbar} \hat{H}\psi^*$$

The time derivative of the particle density $\psi^*\psi$ is

(7.103)
$$\frac{\partial \psi^*\psi}{\partial t} = \psi^* \frac{\partial \psi}{\partial t} + \psi \frac{\partial \psi^*}{\partial t} = \psi^* \left(\frac{-i\hat{H}}{\hbar} \psi \right) + \psi \left(\frac{+i\hat{H}}{\hbar} \psi^* \right)$$

For the typical one-dimensional Hamiltonian

(7.104)
$$\hat{H} = \frac{\hat{p}^2}{2m} + V(x)$$

the latter equation becomes

(7.105)
$$\frac{\partial \psi^*\psi}{\partial t} = \frac{i\hbar}{2m} (\psi^*\psi_{xx} - \psi\psi^*_{xx})$$

$$\frac{\partial \psi^*\psi}{\partial t} + \frac{\partial}{\partial x} \left[\frac{\hbar}{2mi} \left(\psi^* \frac{\partial \psi}{\partial x} - \psi \frac{\partial \psi^*}{\partial x} \right) \right] = 0$$

(The subscript x denotes differentiation.) Comparison of this equation with (7.101) permits the identification

(7.106)
$$\mathcal{J}_x = \frac{\hbar}{2mi} \left(\psi^* \frac{\partial \psi}{\partial x} - \psi \frac{\partial \psi^*}{\partial x} \right)$$

Note that the dimensions of $\mathcal{J}_x$ are number per second. In three dimensions the current density is written

(7.107)
$$\mathbf{J} = \frac{\hbar}{2mi} (\psi^* \nabla \psi - \psi \nabla \psi^*)$$

and has dimensions $\text{cm}^{-2}\,\text{s}^{-1}$.

Transmission and Reflection Coefficients

For one-dimensional scattering problems, the particles in the beam are in plane-wave states with definite momentum. Given the wavefunctions relevant to incident, reflected, and transmitted beams, one may calculate the corresponding current densities according to (7.106). The *transmission coefficient* T and *reflection coefficient* R are defined as

(7.108)
$$T \equiv \left| \frac{\mathcal{J}_{trans}}{\mathcal{J}_{inc}} \right|, \qquad R \equiv \left| \frac{\mathcal{J}_{refl}}{\mathcal{J}_{inc}} \right|$$

These one-dimensional barrier problems are closely akin to problems on the transmission and reflection of electromagnetic plane waves through media of varying index of refraction (see Fig. 7.16). In the quantum mechanical case, the scattering is also of waves.

For one-dimensional barrier problems there are three pertinent beams. Particles in the incident beam have momentum

(7.109)
$$p_{inc} = \hbar k_1$$

Particles in the reflected beam have the opposite momentum

(7.110)
$$p_{refl} = -\hbar k_1$$

In the event that the environment (i.e., the potential) in the domain of the transmitted beam ($x = +\infty$) is different from that of the incident beam ($x = -\infty$), the momenta in these two domains will differ. Particles in the transmitted beam will have momentum $\hbar k_2 \neq \hbar k_1$,

(7.111)
$$p_{trans} = \hbar k_2$$

In all cases the potential is constant in the domains of the incident and transmitted beams (see Fig. 7.14), so the wavefunctions in these domains describe free particles, and we may write

(7.112)
$$\psi_{inc} = A e^{i(k_1 x - \omega_1 t)}, \qquad \hbar\omega_1 = E_{inc} = \frac{\hbar^2 k_1{}^2}{2m}$$

$$\psi_{refl} = B e^{-i(k_1 x + \omega_1 t)}, \qquad \hbar\omega_1 = E_{refl} = E_{inc}$$

$$\psi_{trans} = C e^{i(k_2 x - \omega_2 t)}, \qquad \hbar\omega_2 = E_{trans} = \frac{\hbar^2 k_2{}^2}{2m} + V$$

$$= E_{inc} = \hbar\omega_1$$

Energy is conserved across the potential hill so that frequency remains constant ($\omega_1 = \omega_2$). The change in wavenumber k corresponds to changes in momentum and

217

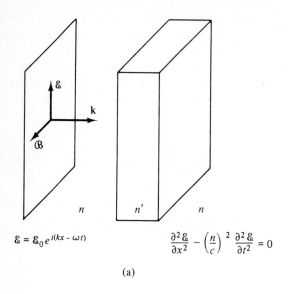

$$\mathcal{E} = \mathcal{E}_0 \, e^{\, i(kx - \omega t)}$$

$$\frac{\partial^2 \mathcal{E}}{\partial x^2} - \left(\frac{n}{c}\right)^2 \frac{\partial^2 \mathcal{E}}{\partial t^2} = 0$$

(a)

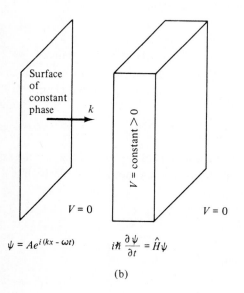

$$\psi = A e^{\, i(kx - \omega t)}$$

$$i\hbar \frac{\partial \psi}{\partial t} = \hat{H}\psi$$

(b)

FIGURE 7.16 (a) Scattering of plane electromagnetic waves through domains of different index of refraction n. (b) Scattering of plane, free-particle wavefunctions through domains of different potential.

kinetic energy. Using (7.106) permits calculation of the currents

$$\mathbb{J}_{inc} = \frac{\hbar}{2mi} 2ik_1 |A|^2$$

(7.113)
$$\mathbb{J}_{trans} = \frac{\hbar}{2mi} 2ik_2 |C|^2$$

$$\mathbb{J}_{refl} = -\frac{\hbar}{2mi} 2ik_1 |B|^2$$

It should be noted that these relations are equivalent to the classical prescription for particle current, $\mathbb{J} = \rho v$, with $\rho = |\psi|^2$ and $v = \hbar k/m$. These formulas, together with (7.108), give the T and R coefficients

(7.114)
$$T = \left|\frac{C}{A}\right|^2 \frac{k_2}{k_1}, \qquad R = \left|\frac{B}{A}\right|^2$$

In the event that the potentials in domains of incident and transmitted beams are equal, $k_1 = k_2$ and $T = |C/A|^2$. More generally, to calculate C/A and B/A as functions of the parameters of the scattering experiment (namely, incident energy, structure of potential barrier), one must solve the Schrödinger equation across the domain of the potential barrier.

PROBLEMS

7.34 Show that the current density **J** may be written

$$\mathbf{J} = \frac{1}{2m} [\psi^* \hat{\mathbf{p}}\psi + (\psi^* \hat{\mathbf{p}}\psi)^*]$$

where $\hat{\mathbf{p}}$ is the momentum operator.

7.35 Show that for a one-dimensional wavefunction of the form

$$\psi(x, t) = A \exp [i\phi(x, t)],$$

$$\mathbb{J} = \frac{\hbar}{m} |A|^2 \frac{\partial \phi}{\partial x}$$

7.36 Show that for a *wave packet* $\psi(x, t)$, one may write

$$\int_{-\infty}^{\infty} \mathbb{J} \, dx = \frac{1}{2m} (\langle p \rangle + \langle p \rangle^*) = \frac{\langle p \rangle}{m}$$

7.37 Show that a complex potential function, $V^*(x) \neq V(x)$, contradicts the continuity equation (7.97).

7.38 (a) Show that if $\psi(x, t)$ is real, then

$$\mathbb{J} = 0$$

for all x.

(b) What type of wave structure does a real state function correspond to?

7.6 ONE-DIMENSIONAL BARRIER PROBLEMS

In a one-dimensional scattering experiment, the intensity and energy of the particles in the incident beam are known in addition to the structure of the potential barrier $V(x)$. Three fundamental scattering configurations are depicted in Fig. 7.17. The energy of the particles in the beam is denoted by E.

The Simple Step

Let us first consider the simple step (Fig. 7.17a) for the case $E > V$. We wish to obtain the space-dependent wavefunction φ for all x. The potential function is zero for $x < 0$ and is the constant V, for $x \geq 0$. The incident beam comes from $x = -\infty$. To construct φ we divide the x axis into two domains: region I and region II, depicted in Fig. 7.18. In region I, $V = 0$, and the time-independent Schrödinger equation appears as

$$(7.115) \qquad -\frac{\hbar^2}{2m}\varphi_{xx} = E\varphi$$

In this domain the energy is entirely kinetic. If we set

$$(7.116) \qquad \frac{\hbar^2 k_1^2}{2m} = E$$

then the latter equation becomes

$$(7.117) \qquad \varphi_{xx} = -k_1^2\varphi \qquad \text{in region I}$$

In region II the potential is the constant V and the time-independent Schrödinger equation appears as

$$(7.118) \qquad -\frac{\hbar^2}{2m}\varphi_{xx} = (E - V)\varphi$$

The kinetic energy decreases by V and is given by

$$(7.119) \qquad \frac{\hbar^2 k_2^2}{2m} = E - V$$

In terms of k_2, (7.118) appears as

$$(7.120) \qquad \varphi_{xx} = -k_2^2\varphi \qquad \text{in region II}$$

Writing φ_I for the solution to (7.117) and φ_{II} for the solution to (7.120), one obtains

$$(7.121) \qquad \begin{aligned} \varphi_I &= Ae^{ik_1 x} + Be^{-ik_1 x} \\ \varphi_{II} &= Ce^{ik_2 x} + De^{-ik_2 x} \end{aligned}$$

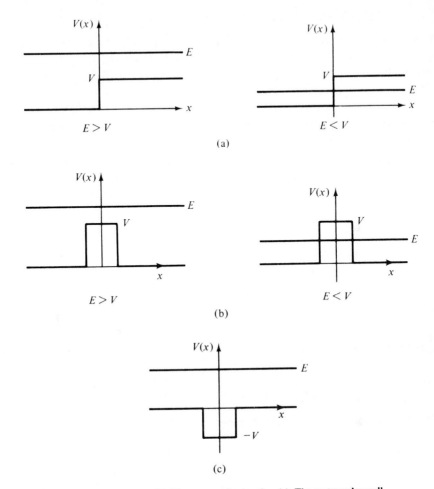

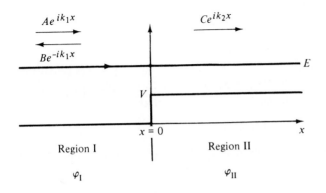

FIGURE 7.17 (a) The simple step. (b) The rectangular barrier. (c) The rectangular well.

FIGURE 7.18 Domains relevant to the simple-step scattering problem for the case $E \geq V$.

Since the term $De^{-ik_2 x}$ (together with the time-dependent factor $e^{-i\omega_2 t}$) represents a wave emanating from the right ($x = +\infty$ in Fig. 7.18), and there is no such wave, we may conclude that $D = 0$. The interpretation of the remaining A, B, and C terms is given in Eq. (7.112). To repeat, $A \exp(ik_1 x)$ represents the incident wave; $B \exp(-ik_1 x)$, the reflected wave; and $C \exp(ik_2 x)$, the transmitted wave.

It is important at this time to realize that φ_I and φ_II (with $D \equiv 0$) represent a single solution to the Schrödinger equation for all x, for the potential curve depicted in Fig. 7.18. Since any wavefunction and its first derivative are continuous (see Section 3.3), at the point $x = 0$ where φ_I and φ_II join it is required that

$$\varphi_\text{I}(0) = \varphi_\text{II}(0)$$

(7.122)
$$\frac{\partial}{\partial x}\varphi_\text{I}(0) = \frac{\partial}{\partial x}\varphi_\text{II}(0)$$

These equalities give the relations

$$A + B = C$$

(7.123)
$$A - B = \frac{k_2}{k_1} C$$

Solving for C/A and B/A, one obtains

(7.124)
$$\frac{C}{A} = \frac{2}{1 + k_2/k_1}, \qquad \frac{B}{A} = \frac{1 - k_2/k_1}{1 + k_2/k_1}$$

Substituting these values into (7.114) gives

(7.125)
$$T = \frac{4k_2/k_1}{[1 + (k_2/k_1)]^2}, \qquad R = \left| \frac{1 - k_2/k_1}{1 + k_2/k_1} \right|^2$$

The ratio k_2/k_1 is obtained from (7.116) and (7.119).

(7.126)
$$\left(\frac{k_2}{k_1}\right)^2 = 1 - \frac{V}{E}$$

In the present case $E \geq V$, so $0 \leq k_2/k_1 \leq 1$. For $E \gg V$, $k_2/k_1 \to 1$ and $T \to 1$, $R \to 0$. There is total transmission. For $E = V$, $k_2/k_1 = 0$ and $T = 0$, $R = 1$. There is total reflection and zero transmission. The T and R curves for the simple-step potential are sketched in Fig. 7.19. For all values of (k_2/k_1) we note that

(7.127)
$$T + R = 1$$

The validitiy of this relation for all one-dimensional barrier problems is proved in Problem 7.39.

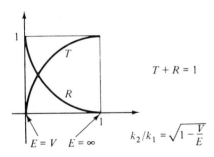

$$k_2/k_1 = \sqrt{1 - \frac{V}{E}}$$

$T + R = 1$

FIGURE 7.19 T and R versus k_2/k_1 for the simple-step scattering problem for $E \geq V$.

In the second configuration for the simple-step barrier, $E < V$ (see Fig. 7.17a). Again the x domain is divided into two regions, as shown in Fig. 7.20. In region I the Schrödinger equation becomes

$$(7.128) \qquad \varphi_{xx} = -k_1^2 \varphi \qquad \text{in region I}$$

where

$$(7.129) \qquad \frac{\hbar^2 k_1^2}{2m} = E$$

In region II the Schrödinger equation is

$$(7.130) \qquad \varphi_{xx} = \kappa^2 \varphi \qquad \text{in region II}$$

where

$$(7.131) \qquad \frac{\hbar^2 \kappa^2}{2m} = V - E > 0$$

The kinetic energy in this domain is negative $(-\hbar^2 \kappa^2/2m)$. In classical physics region II is a "forbidden" domain. In quantum mechanics, however, it is possible for particles to penetrate the barrier.

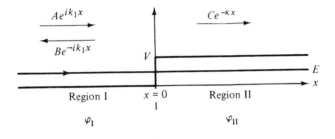

FIGURE 7.20   Domains relevant to the simple-step scattering problem for the case $E \leq V$.

Again calling the solution to (7.128) φ_I and the solution to (7.130) φ_{II}, we obtain

(7.132)
$$\varphi_I = Ae^{ik_1x} + Be^{-ik_1x}$$
$$\varphi_{II} = Ce^{-\kappa x}$$

Continuity of φ and φ_x at $x = 0$ gives

(7.133)
$$1 + \frac{B}{A} = \frac{C}{A}$$
$$1 - \frac{B}{A} = i\frac{\kappa}{k_1}\frac{C}{A}$$

Solving for (C/A) and (B/A) one obtains

(7.134)
$$\frac{C}{A} = \frac{2}{1 + i\kappa/k_1}$$
$$\frac{B}{A} = \frac{1 - i\kappa/k_1}{1 + i\kappa/k_1}$$

The coefficient B/A is of the form z^*/z, where z is a complex number. It follows that $|B/A| = 1$, so

(7.135)
$$R = \left|\frac{B}{A}\right|^2 = 1, \qquad T = 0$$

There is total reflection, hence the transmission must be zero.

To obtain the latter result analytically from our equations above, we must calculate the transmitted current. The function φ_{II} is of the form of a complex amplitude times a real function of x (7.132). Such wavefunctions do not represent propagating waves. They are sometimes called *evanescent waves*. That they carry no current is most simply seen by constructing J_{trans} (7.106).

(7.136)
$$J_{trans} = \frac{\hbar}{2mi}|C|^2\left(e^{-\kappa x}\frac{\partial}{\partial x}e^{-\kappa x} - e^{-\kappa x}\frac{\partial}{\partial x}e^{-\kappa x}\right) = 0$$

We conclude that $T = 0$.

PROBLEMS

7.39 Show that

$$T + R = 1$$

for all one-dimensional barrier problems.

Answer

Since the scattering process is assumed to be steady-state, the continuity equation (7.101) becomes

$$\frac{\partial \mathbb{J}_x}{\partial x} = 0$$

Integrating this equation, one obtains

$$\int_{-\infty}^{\infty} \left(\frac{\partial \mathbb{J}_x}{\partial x}\right) dx = \mathbb{J}_{+\infty} - \mathbb{J}_{-\infty} = 0$$

But

$$\mathbb{J}_{-\infty} = \mathbb{J}_{inc} - \mathbb{J}_{refl}$$

$$\mathbb{J}_{+\infty} = \mathbb{J}_{trans}$$

so that the equation above becomes

$$\mathbb{J}_{trans} + \mathbb{J}_{refl} = \mathbb{J}_{inc}$$

Dividing through by $\mathbb{J}_{inc}$ gives the desired result.

7.40 Electrons in a beam of density $\rho = 10^{15}$ electrons/m are accelerated through a potential of 100 V. The resulting current then impinges on a potential step of height 50 V.

   (a)   What are the incident, reflected, and transmitted currents?

   (b)   Design an electrostatic configuration that gives a simple-step potential.

7.41 Show that the reflection coefficients for the two cases depicted in Fig. 7.21 are equal.

7.42 For the scattering configuration depicted in Fig. 7.20, given that $V = 2E$, at what value of x is the density in region II half the density of particles in the incident beam?

7.43 Equation (7.123) may be written in the matrix form

$$\begin{pmatrix} -1 & 1 \\ 1 & k_2/k_1 \end{pmatrix} \begin{pmatrix} B/A \\ C/A \end{pmatrix} = \begin{pmatrix} 1 \\ 1 \end{pmatrix}$$

Calling the 2×2 matrix $\mathscr{D}$, the left column vector $\mathscr{V}$, and the right column vector $\mathscr{U}$ permits this equation to be more simply written

$$\mathscr{D}\mathscr{V} = \mathscr{U}$$

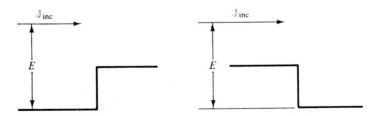

FIGURE 7.21   **Reflection coefficients for these two configurations are equal. (See Problem 7.41.)**

This inhomogeneous matrix equation has the solution

$$\mathscr{V} = \mathscr{D}^{-1}\mathscr{U}$$

where $\mathscr{D}^{-1}$ is the inverse of $\mathscr{D}$, that is,

$$\mathscr{D}^{-1}\mathscr{D} = \begin{pmatrix} 1 & 0 \\ 0 & 1 \end{pmatrix}$$

(a) Find $\mathscr{D}^{-1}$ and then construct $\mathscr{V}$ using the technique above. Check your answer with (7.124).

(b) Do the same for (7.133) and (7.134).

7.7 THE RECTANGULAR BARRIER. TUNNELING

The scattering configuration we now wish to examine is depicted in Fig. 7.17b. The energy of the particles in the beam is greater than the height of the potential barrier, $E > V$. For the case at hand there are three relevant domains (see Fig. 7.22):

Region I: $x < -a$, $V = 0$.

(7.137)   Region II: $-a \leq x \leq +a$, $V > 0$,   and constant.

Region III: $a < x$, $V = 0$.

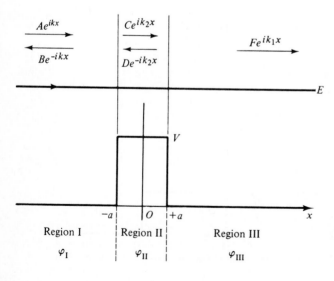

FIGURE 7.22 Domains relevant to the rectangular barrier scattering problem for the case $E \geq V$.

The solutions to the time-independent Schrödinger equation in each of the three domains are:

$$\varphi_I = A e^{ik_1 x} + B e^{-ik_1 x}, \qquad \frac{\hbar^2 k_1^2}{2m} = E$$

$$\varphi_{II} = C e^{ik_2 x} + D e^{-ik_2 x}, \qquad \frac{\hbar^2 k_2^2}{2m} = E - V$$

(7.138)

$$\varphi_{III} = F e^{ik_1 x}, \qquad \frac{\hbar^2 k_1^2}{2m} = E$$

$$(ak_1)^2 - (ak_2)^2 = \frac{2ma^2 V}{\hbar^2} \equiv \frac{g^2}{4}$$

The parameter g contains all the barrier (or well) characteristics. The latter equation (conservation of energy) reveals the simple manner in which ak_1 and ak_2 are related. In Cartesian ak_1, ak_2 space they lie on a hyperbola (Fig. 7.23). The permitted values of k_1 (and therefore E) comprise a positive unbounded continuum. For each such eigen-k_1-value, there is a corresponding eigenstate (φ_I, φ_{II}, φ_{III}) which is determined in terms of the coefficients, $(B/A, C/A, D/A, F/A)$. Knowledge of these coefficients gives the scattering parameters

$$T = \left| \frac{F}{A} \right|^2; \qquad R = \left| \frac{B}{A} \right|^2$$

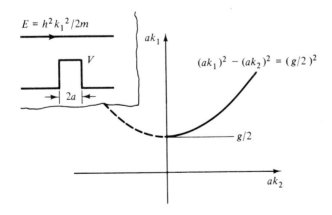

$$E = \hbar^2 k_1^2 / 2m$$

$$(ak_1)^2 - (ak_2)^2 = (g/2)^2$$

FIGURE 7.23 For rectangular-barrier scattering with $E \geq V$, ak_1 and ak_2 lie on a hyperbola.

$$ak_1 \geq ak_2 \geq 0$$

The energy spectrum $\hbar^2 k_1^2 / 2m$ comprises an unbounded continuum.

The coefficients are determined from the boundary conditions at $x = a$ and $x = -a$,

(7.139)

$$e^{-ik_1a} + \left(\frac{B}{A}\right)e^{ik_1a} = \left(\frac{C}{A}\right)e^{-ik_2a} + \left(\frac{D}{A}\right)e^{ik_2a}$$

$$k_1\left[e^{-ik_1a} - \left(\frac{B}{A}\right)e^{ik_1a}\right] = k_2\left[\left(\frac{C}{A}\right)e^{-ik_2a} - \left(\frac{D}{A}\right)e^{ik_2a}\right]$$

$$\left(\frac{C}{A}\right)e^{ik_2a} + \left(\frac{D}{A}\right)e^{-ik_2a} = \left(\frac{F}{A}\right)e^{ik_1a}$$

$$k_2\left[\left(\frac{C}{A}\right)e^{ik_2a} - \left(\frac{D}{A}\right)e^{-ik_2a}\right] = k_1\left(\frac{F}{A}\right)e^{ik_1a}$$

These are four linear, algebraic, inhomogeneous equations for the four unknowns: (B/A), (C/A), (D/A), and (F/A). Solving the last two for (D/A) and (C/A) as functions of (F/A) and substituting into the first two permits one to solve for (B/A) and (F/A). These appear as

(7.140)

$$\frac{F}{A} = e^{-2ik_1a}\left[\cos(2k_2a) - \frac{i}{2}\left(\frac{k_1^2 + k_2^2}{k_1k_2}\right)\sin(2k_2a)\right]^{-1}$$

$$2\left(\frac{B}{A}\right) = i\left(\frac{F}{A}\right)\frac{k_2^2 - k_1^2}{k_1k_2}\sin(2k_2a)$$

The transmission coefficient is most simply obtained from the second of these, together with the relation

(7.141)

$$T + R = \left|\frac{F}{A}\right|^2 + \left|\frac{B}{A}\right|^2 = 1$$

There results

(7.142)

$$\frac{1}{T} = \left|\frac{A}{F}\right|^2 = 1 + \frac{1}{4}\left(\frac{k_1^2 - k_2^2}{k_1k_2}\right)^2\sin^2(2k_2a)$$

Rewriting k_1 and k_2 in terms of E and V as given by (7.138), one obtains

(7.143)

$$\boxed{\frac{1}{T} = 1 + \frac{1}{4}\frac{V^2}{E(E - V)}\sin^2(2k_2a) \qquad E > V}$$

The reflection coefficient is $1 - T$.

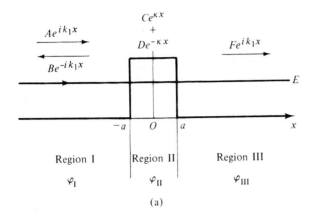

Region I   Region II   Region III

φ_I φ_II φ_III

(a)

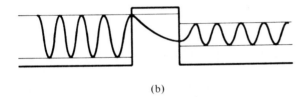

(b)

FIGURE 7.24 **(a) Domains relevant to the rectangular barrier scattering problem, for the case $E \le V$. (b) Real part of φ for the case above, showing the hyperbolic decay in the barrier domain and decrease in amplitude of the transmitted wave.**

For the case $E < V$, as depicted in Fig. 7.24a, we find that the structure of the solutions (7.138) are still appropriate, with the simple modification

(7.144)
$$ik_2 \to \kappa, \qquad \frac{\hbar^2\kappa^2}{2m} = V - E > 0$$

$$(ak_1)^2 + (a\kappa)^2 = \frac{2ma^2V}{\hbar^2} \equiv \frac{g^2}{4}$$

This latter conservation of energy statement indicates that the variables ak_1 and $a\kappa$ lie on a circle of radius $g/2$ (Fig. 7.25). The permitted eigen-k_1-values now comprise a positive, bounded continuum, so that the eigenenergies

$$E = \frac{\hbar^2 k_1{}^2}{2m}$$

also comprise a positive, bounded continuum.

The algebra leading to (7.140) remains unaltered so that the transmission coefficient for this case is obtained by making the substitution of (7.144) into (7.142).

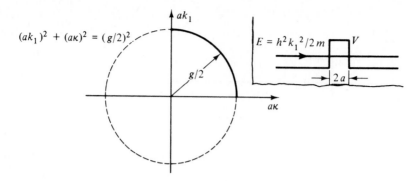

$$(ak_1)^2 + (a\kappa)^2 = (g/2)^2$$

FIGURE 7.25 For rectangular barrier scattering with $E \leq V$, ak_1 and $a\kappa$ lie on a circle $ak_1 \geq 0$, $a\kappa \geq 0$. The energy spectrum $(\hbar^2 k_1{}^2/2m)$ comprises a bounded continuum.

We also recall that $\sin(iz) = i \sinh z$. There results

(7.145)
$$\frac{1}{T} = 1 + \frac{1}{4}\left(\frac{k_1{}^2 + \kappa^2}{k_1\kappa}\right)^2 \sinh^2(2\kappa a)$$

which, with (7.144), gives

(7.146)
$$\frac{1}{T} = 1 + \frac{1}{4}\frac{V^2}{E(V - E)} \sinh^2(2\kappa a)$$

Writing this equation in terms of T,

(7.147)
$$\boxed{T = \frac{1}{1 + \dfrac{1}{4}\dfrac{V^2}{E(V - E)} \sinh^2(2\kappa a)} \qquad E < V}$$

indicates that in the domain $E < V$, $T < 1$. The limit that $E \to V$ deserves special attention. With

$$\frac{V - E}{V} = \frac{\hbar^2 \kappa^2}{2mV} \equiv \iota \to 0$$

one obtains

(7.148)
$$T = \frac{1}{1 + g^2/4} + O(\iota) < 1$$

$$g^2 \equiv \frac{2m(2a)^2 V}{\hbar^2}$$

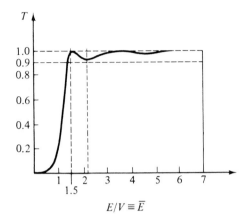

FIGURE 7.26   Transmission coefficient T versus E/V for scattering from a rectangular barrier with $2m(2a)^2 V/\hbar^2 \equiv g^2 = 16$. The additional lines are in references to Problems 7.50 et seq.

The expression $O(\epsilon)$ represents a sum of terms whose value goes to zero with ϵ. We conclude that for scattering from a potential barrier, the transmission is less than unity at $E = V$ (Fig. 7.26).

Returning to the case $E \neq V$, (7.143) indicates that $T = 1$ when $\sin^2(2k_2 a) = 0$, or equivalently when

(7.149) $$2ak_2 = n\pi \qquad (n = 1, 2, \ldots)$$

Setting $k_2 = 2\pi/\lambda$, the latter statement is equivalent to

(7.150) $$2a = n\left(\frac{\lambda}{2}\right)$$

When the barrier width $2a$ is an integral number of half-wavelengths, $n(\lambda/2)$, the barrier becomes transparent to the incident beam; that is, $T = 1$. This is analogous to the case of total transmission of light through thin refracting layers.

Written in terms of E and V, the requirement for perfect transmission, (7.149), becomes

(7.151) $$E - V = n^2\left(\frac{\pi^2 \hbar^2}{8a^2 m}\right) = n^2 E_1$$

where E_1 is the ground-state energy of a one-dimensional box of width $2a$ (see Eq. 4.14).

Equations (7.143) and (7.146) give the transmission coefficient T, as a function of E, V, and the width of the well $2a$. The former of these indicates that $T \to 1$ with increasing energy of the incident beam. The transmission is unity for the values of E given by (7.151). Equation (7.146) gives T for $E \leq V$. The transmission is zero for

$E = 0$ and is less than 1 for $E = V$. A sketch of T versus $E/V \equiv \bar{E}$ for the case $g^2 = 16$ is given in Fig. 7.26.

The fact that T does not vanish for $E < V$ is a purely quantum mechanical result. This phenomenon of particles passing through barriers higher than their own incident energy is known as *tunneling*. It allows emission of α particles from a nucleus and field emission of electrons from a metal surface in the presence of a strong electric field.

PROBLEMS

7.44   In terms of the new variables,

$$\alpha_\pm \equiv \frac{k_1{}^2 \pm k_2{}^2}{2k_1 k_2}, \qquad \beta \equiv 2k_2 a$$

$$\frac{F}{A} = \sqrt{T}\, e^{i\phi_T}, \qquad \frac{B}{A} = \sqrt{R}\, e^{i\phi_R}$$

(7.140) may be rewritten in the simpler form

$$\sqrt{T}\, e^{i\phi_T} = \frac{e^{2iak_1}}{\cos \beta - i\alpha_+ \sin \beta}$$

$$\sqrt{R}\, e^{i\phi_R} = i\alpha_- \sqrt{T}\, e^{i\phi_T} \sin \beta$$

Use these expressions to show:

(a) $T + R = 1$.
(b) $\phi_T = \phi_R - n(\pi/2)$, $n = 1, 2, 3, \ldots$.
(c) $\tan(\phi_T - 2k_1 a) = \alpha_+ \tan \beta$.
(d)   What is ϕ_R for the infinite potential step: $V(x) = \infty$, $x \geq 0$; $V(x) = 0$, $x < 0$?

Answers (partial)

(a)   Solving for $T + R$ from (7.140) gives

$$T + R = \frac{1 + \alpha_-{}^2 \sin^2 \beta}{\cos^2 \beta + \alpha_+{}^2 \sin^2 \beta}$$

Substituting the definitions of $\alpha_\pm$ gives the desired result.

(c)   From the first of the two given equations above, we obtain

$$\sqrt{T}\, e^{i(\phi_T - 2k_1 a)} = \frac{1}{\cos \beta + i\alpha_+ \sin \beta}$$

$$= \frac{e^{-i\phi}}{\sqrt{\cos^2 \beta + \alpha_+{}^2 \sin^2 \beta}}$$

Equating the tangents of the phases of both sides gives the desired result.

7.45   An electron beam is sent through a potential barrier 1 cm long. The transmission coefficient exhibits a third maximum at $E = 100$ eV. What is the height of the barrier?

7.46   An electron beam is incident on a barrier of height $10\,\text{eV}$. At $E = 10\,\text{eV}$, $T = 3.37 \times 10^{-3}$. What is the width of the barrier?

7.47   Use the correspondence principle with (7.147) to show that $T = 0$ for $E < V$, for the classical case of a beam of particles of energy E incident on a potential barrier of height V.

7.8   THE RAMSAUER EFFECT

The configuration for this case is depicted in Fig. 7.17c. The relevant domains are shown in Fig. 7.27. Once again Eqs. 7.138 et seq. apply with the modification

$$(7.152) \qquad \frac{\hbar^2 k_2{}^2}{2m} = E - V = E + |V|$$

The transmission coefficient (7.143) becomes, for $E \geq 0$,

$$(7.153) \qquad \boxed{\frac{1}{T} = 1 + \frac{1}{4}\frac{V^2}{E(E + |V|)}\sin^2(2k_2 a)}$$

Again there is perfect transmission when an integral number of half-wavelengths fit the barrier width.

$$(7.154) \qquad 2ak_2 = n\pi \qquad (n = 1, 2, \ldots)$$

This condition may also be cast in terms of the eigenenergies of a one-dimensional box of width $2a$:

$$(7.155) \qquad E + |V| = n^2 E_1$$

From (7.153) we see that $T \to 1$ with increasing incident energy. At $E = 0$, $T = 0$. Thus we obtain an idea of the shape of T versus E. It is similar to the curve shown in

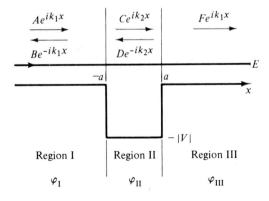

FIGURE 7.27   Domains relevant to the rectangular well scattering problem, $E > 0$.

Fig. 7.26. The transmission is zero for $E = 0$ and rises to the first maximum (unity) at $E = E_1 - |V|$. It has successive maxima of unity at the values given by (7.155), and approaches 1 with growing incident energy E.

The preceding theory of scattering of a beam of particles by a potential well has been used as a model for the scattering of low-energy electrons from atoms. The attractive well represents the field of the nucleus, whose positive charge becomes evident when the scattering electrons penetrate the shell structure of the atomic electrons. The reflection coefficient is a measure of the scattering cross section.[1] Experiments in which this cross section is measured (for rare gas atoms) detect a low-energy minimum which is consistent with the first maximum that T goes through for typical values of well depth and width according to the model above, (7.153). This transparency to low-energy electrons of rare gas atoms is known as the Ramsauer effect.

The student should not lose sight of the following fact. For any of the solutions to the scattering problems considered in these last few sections, we have in essence found the eigenfunctions and eigenenergies for the corresponding Hamiltonian. These Hamiltonians are of the form

(7.156)
$$H = \frac{p^2}{2m} + V(x)$$

with the potential $V(x)$ depicted by any of the configurations of Fig. 7.17. In each case considered, the spectrum of energies is a continuum, $E = \hbar^2 k^2/2m$. For each value of k, a corresponding set of coefficient ratios (B/A, C/A for the simple step and B/A, C/A, D/A, F/A for the rectangular potential) are determined. The coefficient A is fixed by the data on the incident beam. These coefficients then determine the wavefunction, which is an eigenfunction of the Hamiltonian above. All such scattering eigenstates are unbound states. A continuous spectrum is characteristic of unbound states, while a discrete spectrum is characteristic of bound states (e.g., particle in a box, harmonic oscillator).

The transmission coefficients corresponding to the one-dimensional potential configurations considered above are summarized in Table 7.2.

PROBLEMS

7.48   The scattering cross section for the scattering of electrons by a rare gas of krypton atoms exhibits a low-energy minimum at $E \simeq 0.9$ V. Assuming that the diameter of the atomic well seen by the electrons is 1 Bohr radius, calculate its depth.

---

[1] The notion of scattering cross section is discussed in Chapter 14.

TABLE 7.2 Transmission coefficients for three elementary potential barriers

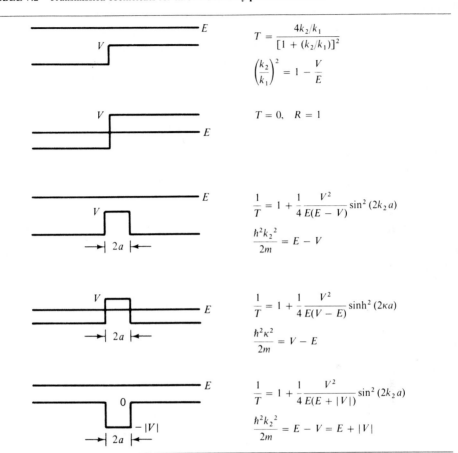

$$T = \frac{4k_2/k_1}{[1 + (k_2/k_1)]^2}$$

$$\left(\frac{k_2}{k_1}\right)^2 = 1 - \frac{V}{E}$$

$$T = 0, \quad R = 1$$

$$\frac{1}{T} = 1 + \frac{1}{4}\frac{V^2}{E(E - V)}\sin^2(2k_2 a)$$

$$\frac{\hbar^2 k_2^2}{2m} = E - V$$

$$\frac{1}{T} = 1 + \frac{1}{4}\frac{V^2}{E(V - E)}\sinh^2(2\kappa a)$$

$$\frac{\hbar^2 \kappa^2}{2m} = V - E$$

$$\frac{1}{T} = 1 + \frac{1}{4}\frac{V^2}{E(E + |V|)}\sin^2(2k_2 a)$$

$$\frac{\hbar^2 k_2^2}{2m} = E - V = E + |V|$$

7.49 Show that the transmission coefficient for the rectangular barrier may be written in the form

$$T = T(g, \bar{E})$$

where

$$g^2 \equiv \frac{2m(2a)^2 V}{\hbar^2}$$

$$\bar{E} \equiv \frac{E}{V}$$

Answer (partial)
For $\bar{E} \geq 1$,

$$T^{-1} = 1 + \frac{1}{4}\frac{1}{\bar{E}(\bar{E} - 1)}\sin^2\sqrt{g^2(\bar{E} - 1)}$$

7.50   Using your answer to Problem 7.49, derive an equation for an approximation to the curve on which minimum values of T fall.

$$T_{min} = T_{min}(\bar{E})$$

Show that the values of T and $\bar{E}$ at the first minimum in the sketch of T versus $\bar{E}$ depicted in Fig. 7.26 ($g^2 = 16$) agree with your equation. [*Hint:* The minima of T fall at the values of $\bar{E}$ where T^{-1} is maximum. From Problem 7.49,

$$T^{-1} \leq 1 + \frac{1}{4} \frac{1}{\bar{E}(\bar{E} - 1)}.\Big]$$

7.51   For the rectangular barrier:

(a)   Write the values of $\bar{E}$ for which $T = 1$ as a function of g.

(b)   Using your answer to part (a) and the two preceding problems, make a sketch of T versus $\bar{E}$ in the two limits $g \gg 1$, $g \ll 1$. Cite two physical situations to which these limits pertain.

(c)   Show that for an electron, $g^2/V \equiv 2m(2a)^2/\hbar^2 = 0.26(2a)^2(eV)^{-1}$, where a is in angstroms.

7.52   For the case depicted in Fig. 7.26, show that the first maximum falls at a value consistent with your answer to part (a) of Problem 7.51.

7.53   Write the transmission coefficient for the rectangular well as a function of g and $\bar{E}$.

Answer

$$T^{-1} = 1 + \frac{1}{4} \frac{1}{\bar{E}(\bar{E} + 1)} \sin^2 \sqrt{g^2(\bar{E} + 1)}$$

7.54   In the limit $g^2 \gg 1$, show that the minima of T for the rectangular well fall on a curve which is well approximated by

$$T_{min} = 4\bar{E}$$

Use this result together with (7.155) for the values of $\bar{E}$ where $T = 1$ to obtain a sketch of T versus $\bar{E}$ for the case $g^2 = 10^5$.

Answer

See Fig. 7.28.

7.55   Show that the spaces between resonances in T for the case of scattering from a potential well grow with decreasing g.

7.56   (a)   Calculate the transmission coefficient T for the double potential step shown in Fig. 7.29a.

(b)   If we call T_1 the transmission coefficient appropriate to the single potential step V_1, and T_2 that appropriate to the single potential step V_2, show that

$$T_2 \leq T_1, \qquad T \geq T_2$$

Offer a physical explanation for these inequalities.

(c)   What are the three sets of conditions under which T is maximized? What do these conditions correspond to physically?

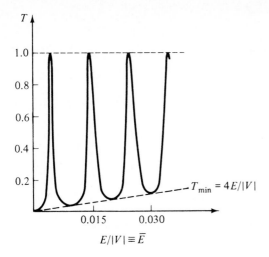

$$E/|V| \equiv \bar{E}$$

FIGURE 7.28   Resonances in the transmission coefficient for scattering by a potential well for $g^2 = 10^5$. (See Problems 7.54 et seq.)

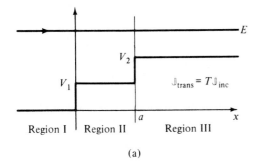

(a)

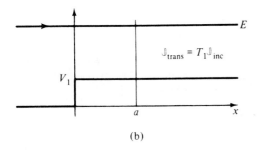

(b)

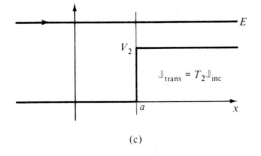

(c)

FIGURE 7.29   (a) Double potential step showing three regions discussed in Problem 7.56. (b) and (c) Two related single potential steps: $T_1 \geq T_2$ and $T_2 \leq T$.

(d) A student argues that T is the product $T_1 T_2$ on the following grounds. The particle current that penetrates the V_1 barrier is $T_1 \mathcal{J}_{inc}$. This current is incident on the V_2 barrier so that $T_2(T_1 \mathcal{J}_{inc})$ is the current transmitted through the second barrier. What is the incorrect assumption in his argument?

Answer (partial)

Applying boundary conditions to the wavefunctions

$$\varphi_{\mathrm{I}} = Ae^{ik_1 x} + Be^{-ik_1 x} \qquad \text{(region I)}$$
$$\varphi_{\mathrm{II}} = Ce^{ik_2 x} + De^{-ik_2 x} \qquad \text{(region II)}$$
$$\varphi_{\mathrm{III}} = Fe^{ik_3 x} \qquad \text{(region III)}$$

at $x = 0$ and $x = a$, respectively, and solving for $T = (k_3/k_1)|F/A|^2$ gives the desired result:

$$T = \frac{4k_1 k_3 k_2^{\,2}}{k_2^{\,2}(k_1 + k_3)^2 + (k_3^{\,2} - k_2^{\,2})(k_1^{\,2} - k_2^{\,2})\sin^2(k_2 a)} \qquad (k_1 \geq k_2 \geq k_3)$$

Note that

$$T = \frac{4k_1 k_3 k_2^{\,2}}{k_2^{\,2}(k_1 + k_3)^2 - \Delta^2} \geq \frac{4k_1 k_3 k_2^{\,2}}{k_2^{\,2}(k_1 + k_3)^2} = T_2$$

where Δ^2 is as implied. With (7.125) we see that $T_1 \geq T_2$.

7.57 Calculate the transmission coefficient for the potential configuration and energy of incident particles depicted in Fig. 7.30. (*Note:* T is easily obtained from the answer given to Problem 7.56.)

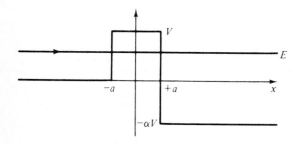

FIGURE 7.30 **Tunneling configuration for Problem 7.57. The constant α is real and greater than zero.**

7.9 KINETIC PROPERTIES OF A WAVE PACKET SCATTERED FROM A POTENTIAL BARRIER

The time-dependent one-dimensional scattering problem addresses itself primarily to the problem of a wave packet incident on a potential barrier. It seeks the shape of the reflected and transmitted pulse. We will restrict our discussion to the kinematic properties of these pulses.

To formulate this problem we first construct a wave packet whose center is at $x = -X$ at $t = 0$. In previous chapters we obtained such wave packets centered at

$x = 0$ at $t = 0$. They are of the form

(7.157)
$$\psi(x, t) = \frac{1}{\sqrt{2\pi}} \int_{-\infty}^{\infty} b(k)e^{i(kx - \omega t)} \, dk$$

For this same packet to be centered at $x = -X$ initially, one merely effects a translation in x so that

(7.158)
$$\psi(x, t) = \frac{1}{\sqrt{2\pi}} \int_{-\infty}^{\infty} b(k)e^{ik(x + X)}e^{-i\omega t} \, dk$$

$$= \frac{1}{\sqrt{2\pi}} \int_{-\infty}^{\infty} b(k)e^{ikX}e^{i(kx - \omega t)} \, dk$$

For example, for a chopped pulse, L cm long, containing particles moving with momentum $\hbar k_0$, $b(k)$ is given by (6.32):

(7.159)
$$b(k) = \sqrt{\frac{2}{\pi L}} \frac{\sin (k - k_0)L/2}{k - k_0}$$

See Fig. 7.31. The group velocity of this packet is $v_0 = \hbar k_0/m$. Let us call the wave packet (7.158), ψ_{inc}. This packet is a superposition of plane-wave states of the form (7.112). Each such incident k-component plane wave is reflected and transmitted. The corresponding reflected and transmitted waves are constructed from the amplitude ratios B/A and F/A given by (7.140), which are functions of k (k_1 in Eq. 7.140). Reassembling all of these waves, one obtains

$$x < -a \quad \psi_{\text{inc}} = \frac{1}{\sqrt{2\pi}} \int_{-\infty}^{\infty} b(k)e^{ikX}e^{i(kx - \omega t)} \, dk$$

(7.160)
$$x < -a \quad \psi_{\text{refl}} = \frac{1}{\sqrt{2\pi}} \int_{-\infty}^{\infty} \sqrt{R}\, e^{i\phi_R} b(k)e^{ikX}e^{-i(kx + \omega t)} \, dk$$

$$x > +a \quad \psi_{\text{trans}} = \frac{1}{\sqrt{2\pi}} \int_{-\infty}^{\infty} \sqrt{T}\, e^{i\phi_T} b(k)e^{ikX}e^{i(kx - \omega t)} \, dk$$

Here we are using the notation of Problem 7.44.

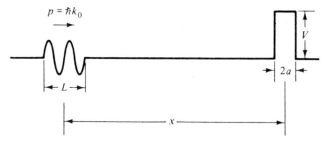

FIGURE 7.31   Wave packet incident on a potential barrier. $x \gg L \simeq 2a$.

To uncover the kinetic properties of these packets, we use the method of *stationary phase*. This relies on the fact that the major contribution in a Fourier integral is due to the k component with stationary phase. If we call this component k_0, then the phase of the Fourier integral for ψ_{refl} vanishes when

$$(7.161) \qquad \frac{\partial}{\partial k}(\phi_R + kX - kx - \omega t) = 0$$

This gives the trajectory of the reflected packet,

$$(7.162) \qquad x = -\frac{\hbar k_0}{m} t + X + \left(\frac{\partial \phi_R}{\partial k}\right)_{k_0} \qquad x < -a$$

In like manner, for the incident and transmitted packets, one obtains

$$(7.163) \qquad x = \frac{\hbar k_0}{m} t - X, \qquad\qquad x < -a$$

$$(7.164) \qquad x = \frac{\hbar k_0}{m} t - X - \left(\frac{\partial \phi_T}{\partial k}\right)_{k_0}, \qquad x > a$$

The latter three equations illustrate the effect of a potential barrier on the trajectory of an incident wave packet. Were there no barrier, the packet would move freely in accordance with (7.163). However, there is a delay for both the transmitted and reflected packets. The transmitted pulse arrives at any plane $x > a$, $(\partial \phi_T/\partial k_0)v_0^{-1}$ seconds after the free pulse. The reflected pulse arrives at any plane $x < -a$, $(\partial \phi_R/\partial k_0)v_0^{-1}$ seconds after the free pulse would be reflected from an impenetrable wall at the $x = 0$ plane.[1]

PROBLEMS

7.58 For a pulse such as described in (7.158) and (7.159), containing 1.5-keV electrons, which scatters from a potential well of width 0.5×10^{-7} cm and of depth 25 keV, what is the delay in the transmitted beam (in seconds) imposed by the well?

7.59 Is there a delay in the scattering of a wave packet from a simple-step potential? Present an argument in support of your answer.

7.60 In the text we mentioned the method of stationary phase for evaluating Fourier integrals.

---

[1] The time development of a wave packet scattering from a potential barrier is graphically depicted in D. A. Saxon, *Elementary Quantum Mechanics*, Holden-Day, San Francisco, 1968.

Use this method to show that

$$\int_{-\infty}^{\infty} f(k)e^{is(k)} dk \simeq \sqrt{\frac{2\pi}{|s''(k_0)|}} f(k_0)e^{i[s(k_0) \pm \pi/4]}$$

$$s'(k_0) = 0$$

The phase factor $+i\pi/4$ applies when $s''(k_0) > 0$ and $-i\pi/4$ applies when $s''(k_0) < 0$. Primes denote k differentiation. [*Hint:* Expand $s(k)$ in a Taylor series about $k = k_0$, keeping $O(k^2)$ terms.]

7.61   There is a tacit assumption in the construction of (7.160) that no interaction occurs between the incident wave packet and the potential barrier in the interval $0 \leq t \leq X/v_0$. Is this a valid assumption?

Answer

All k components in the distribution (7.159) with $k > k_0$ reach the barrier in a time less than X/v_0. The number of such components diminishes in the limit $X \gg 2a \simeq L$.

7.10   THE WKB APPROXIMATION[1]

Correspondence

In Section 7.3 we found that the quantum probability density goes over to the classical probability density in the limit of large quantum numbers. Such quantum states have many zeros and suffer rapid spatial oscillation. Equivalently, we may say that in this classical domain the local quantum (de Broglie) wavelength is small compared to characteristic distances of the problem. For the harmonic oscillator, such a characteristic distance is the maximum displacement or amplitude x_0 (7.12). More generally, this characteristic distance may be taken as the typical length over which the potential changes. Since the de Broglie wavelength changes only by virtue of a change in potential, the latter condition may be incorporated in the criterion (for classical behavior) that the quantum wavelength not change appreciably over the distance of one wavelength. Now the change in wavelength over the distance δx is

$$\delta\lambda = \frac{d\lambda}{dx} \delta x$$

In one wavelength this change is

$$\delta\lambda = \frac{d\lambda}{dx} \lambda$$

---

[1] Named for G. Wentzel, H. A. Kramers, and L. Brillouin.

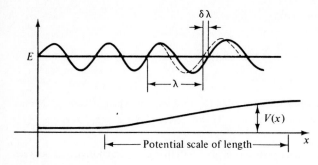

FIGURE 7.32 In the WKB analysis, the fractional change $\delta\lambda/\lambda \ll 1$. The potential scale of length is also large compared to wavelength.

In the classical domain, $\delta\lambda \ll \lambda$ (Fig. 7.32). This gives the criterion

$$(7.165) \qquad \left|\frac{\delta\lambda}{\lambda}\right| = \left|\frac{d\lambda}{dx}\right| \ll 1$$

In terms of the momentum p, we find that

$$\left(\frac{h}{\lambda}\right)^2 = p^2 = 2m(E - V)$$

$$\frac{d(\lambda^2)}{dx} = -\frac{h^2}{p^4}\frac{d(p^2)}{dx} = -\frac{h^2}{p^4}\left(-2m\frac{dV}{dx}\right)$$

or, equivalently,

$$\frac{d\lambda}{dx} = \frac{mh}{p^3}\frac{dV}{dx}$$

Thus, the condition (7.165) for near-classical behavior becomes

$$(7.166) \qquad \left|\frac{d\lambda}{\lambda}\right| = \left|\frac{mh}{p^3}\frac{dV}{dx}\right| \ll 1$$

The WKB Expansion

We seek solutions to the time-independent Schrödinger equation (7.1) which are valid in the near-classical domain (7.166).

If the potential V is slowly varying, one expects the wavefunction to closely approximate the free-particle state

$$\varphi(x) = Ae^{ikx} = Ae^{ipx/\hbar}$$

Thus we will look for solutions in the form

(7.167)
$$\varphi(x) = Ae^{iS(x)/\hbar}$$

Substitution of this function into (7.1) gives

(7.168)
$$-i\hbar \frac{\partial^2 S}{\partial x^2} + \left(\frac{\partial S}{\partial x}\right)^2 = p^2(x)$$

$$p^2 = 2m[E - V(x)]$$

To further bias the solution (7.167) to the classical domain we examine the solutions to the nonlinear equation (7.168) in the limit $\hbar \to 0$. Recall (Section 6.1) that it is in this limit that the Gaussian packet reduces to the classical particle. To these ends we expand $S(x)$ in powers of $\hbar$ as follows:

(7.169)
$$S(x) = S_0(x) + \hbar S_1(x) + \frac{\hbar^2}{2} S_2(x) + \cdots$$

Substituting this expansion into (7.168) gives

(7.170)
$$0 = \left[\left(\frac{\partial S_0}{\partial x}\right)^2 - p^2\right] + 2\hbar\left(\frac{\partial S_0}{\partial x}\frac{\partial S_1}{\partial x} - \frac{i}{2}\frac{\partial^2 S_0}{\partial x^2}\right)$$
$$+ \hbar^2\left[\frac{\partial S_0}{\partial x}\frac{\partial S_2}{\partial x} + \left(\frac{\partial S_1}{\partial x}\right)^2 - i\frac{\partial^2 S_1}{\partial x^2}\right] + O(\hbar^3)$$

Since this equation must be satisfied for small but otherwise arbitrary values of $\hbar$, it is necessary that the coefficient of each power of $\hbar$ vanish separately. In this manner we obtain the following series of coupled equations for the sequence $\{S_n\}$.

(7.171)
$$\left(\frac{\partial S_0}{\partial x}\right)^2 = p^2$$

$$\frac{\partial S_0}{\partial x}\frac{\partial S_1}{\partial x} = \frac{i}{2}\frac{\partial^2 S_0}{\partial x^2}$$

$$\frac{\partial S_0}{\partial x}\frac{\partial S_2}{\partial x} + \left(\frac{\partial S_1}{\partial x}\right)^2 - i\frac{\partial^2 S_1}{\partial x^2} = 0$$

$$\vdots$$

Integrating the first of these equations gives

$$S_0(x) = \pm \int_{x_0}^{x} p(x)\, dx$$

or, equivalently, in terms of wavenumber $k = p/\hbar$,

(7.172)
$$\frac{S_0}{\hbar} = \pm \int_{x_0}^{x} k(x)\, dx$$

Substituting this solution into the second equation in (7.171) and integrating gives

$$S_1 = \frac{i}{2} \ln\left(\frac{\partial S_0}{\partial x}\right) = \frac{i}{2} \ln \hbar k$$

or, equivalently,

(7.173)
$$\exp(iS_1) = \frac{1}{\hbar^{1/2} k^{1/2}}$$

Substituting (7.172) and (7.173) into the third equation in (7.171) and integrating gives

(7.174)
$$S_2 = \frac{1}{2} \frac{m(\partial V/\partial x)}{p^3} - \frac{1}{4} \int \frac{m^2(\partial V/\partial x)^2}{p^5}\, dx$$

In that S_1 is the log of the derivative of S_0, we cannot in general ignore S_1 compared to S_0, and both terms must be retained in the expansion (7.169). However, comparison of S_2 (7.174) with the criterion (7.166) shows that in the near-classical domain, the contribution of the second-order term $\hbar S_2/2$ to the phase of φ is small compared to unity. Higher-order contributions to $S(x)$ are likewise small. Thus, it is consistent to say that near the classical domain, φ is well described by the first two terms in the expansion (7.169). Inserting these solutions into (7.167) gives

(7.175)
$$\varphi(x) = \frac{A}{k^{1/2}} \exp\left(i \int k\, dx\right) + \frac{B}{k^{1/2}} \exp\left(-i \int k\, dx\right)$$

The Near-Classical Domain

In what sense does the solution (7.175) approximate classical behavior? To answer this question we consider the probability density $\varphi^*\varphi$. Specifically, consider that the momentum of the particle is specified so that it is known that the particle is moving to larger values of x. Then the corresponding WKB solution (7.175) reduces to

$$\varphi(x) = \frac{A}{k^{1/2}} \exp\left(i \int k\, dx\right)$$

The probability density for this state is

$$P(x) = \varphi^*\varphi = \frac{|A|^2}{k} = \frac{|A|^2 \hbar/m}{v}$$

where v is written for the classical velocity, $v = p/m$. The probability of finding the particle in the interval dx about x is

$$P\,dx = \left(\frac{|A|^2 \hbar}{m}\right) dt$$

This result, apart from a multiplicative constant, is the same as the classical probability, $P\,dx \sim dt$ (see Eq. 7.70).

To obtain correspondence with classical current, we renormalize φ so that it is relevant to a beam of N particles such as described in (7.95). Calculation of the current (7.106) gives

$$\mathbb{J} = \frac{N\hbar|A|^2}{m} = NP(x)v(x)$$

$$\mathbb{J} = \rho(x)v(x)$$

This is the classical expression for the current across a plane at the point x of a beam of particles with number density $\rho(x)$ moving with velocity $v(x)$.

Thus, the lowest-order WKB solution (7.175) reproduces the classical probability and current.

Application to Bound States

Consider the potential shown in Fig. 7.33. The WKB solution (7.175) is invalid at the classical turning points x_1 and x_2, for at these points $E = V$ and $\hbar k = 0$, thereby violating the criterion (7.166). However, the WKB solution becomes valid in regions far removed from the turning points where $|E - V|$ is sufficiently large.

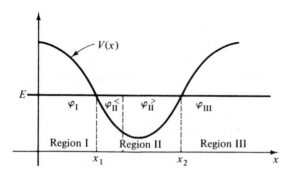

FIGURE 7.33   Domains relevant to the WKB approximation of bound states.

In region I, far to the left of x_1 ($x \to -\infty$), the solution is

(7.176)

$$\varphi_1 = \frac{1}{\sqrt{\kappa}} \exp \int_{x_1}^{x} \kappa \, dx$$

$$\frac{\hbar^2 \kappa^2}{2m} = V - E > 0$$

Far to the right of x_2, the wavefunction also decays exponentially.

(7.177)

$$\varphi_{\text{III}} = \frac{A}{\sqrt{\kappa}} \exp \left(-\int_{x_2}^{x} \kappa \, dx \right)$$

In the classically allowed region II, the WKB solution is oscillatory. It is necessary in the WKB construction of φ to separate this component of the solution into two parts.

(7.178a)

$$\varphi_{\text{II}}^{<}(x) = \frac{C}{\sqrt{k}} \sin \left(\int_{x_1}^{x} k \, dx + \delta \right), \qquad x_1 < x$$

(7.178b)

$$\varphi_{\text{II}}^{>}(x) = \frac{B}{\sqrt{k}} \sin \left(\int_{x}^{x_2} k \, dx + \delta \right), \qquad x < x_2$$

$$\frac{\hbar^2 k^2}{2m} = E - V > 0$$

Through connection formulas obtained below, $\varphi_{\text{II}}^{<}$ is matched to φ_{I} and $\varphi_{\text{II}}^{>}$ is matched to φ_{III}. This connecting process will serve to determine all but one of the constants A, B, C, and δ. The remaining constant is determined in stipulating that $\varphi_{\text{II}}^{<}$ join smoothly to $\varphi_{\text{II}}^{>}$. This continuity condition will also be found to generate energy eigenvalues within the WKB approximation.

Connecting Formulas for Bound States

If φ_{I}, $\varphi_{\text{II}}^{<}$, $\varphi_{\text{II}}^{>}$, and φ_{III} were valid representations of φ throughout their respective domains, the constants of these functions could be obtained by simply matching these component solutions as was done in preceding sections of this chapter. This method clearly cannot be followed in the present analysis since the WKB solutions are invalid at the turning points.

The technique of matching φ_{I} to $\varphi_{\text{II}}^{<}$ and $\varphi_{\text{II}}^{>}$ to φ_{III} in the WKB approximation is as follows. The Schrödinger equation is solved exactly in the regions of the turning points for potentials that approximate $V(x)$ in these domains. The asymptotic forms of these exact solutions are then used to match φ_{I} to $\varphi_{\text{II}}^{<}$ and to match $\varphi_{\text{II}}^{>}$ to φ_{III}.

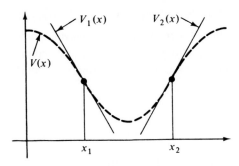

FIGURE 7.34   Approximate linear potentials $V_1(x)$ and $V_2(x)$ valid in the neighborhoods of the turning points x_1 and x_2, respectively.

Following this prescription, we approximate $V(x)$ in the neighborhood of x_1 with the linear potential $V_1(x)$.

(7.179) $$V(x) \simeq V_1(x) = E - F_1(x - x_1)$$

The constant F_1 is the slope of $V(x)$ at x_1. Similarly, in the neighborhood of x_2 we write

(7.180) $$V(x) \simeq V_2(x) \equiv E + F_2(x - x_2)$$

(See Fig. 7.34.) The Schrödinger equation then appears as

(7.181) $$\frac{d^2\varphi}{dx^2} + \frac{2mF_1}{\hbar^2}(x - x_1)\varphi = 0 \qquad x \text{ near } x_1$$

(7.182) $$\frac{d^2\varphi}{dx^2} - \frac{2mF_2}{\hbar^2}(x - x_2)\varphi = 0 \qquad x \text{ near } x_2$$

Further simplification of these equations is accomplished through the change in variable

$$y = -\left(\frac{2mF_1}{\hbar^2}\right)^{1/3}(x - x_1)$$

in (7.181) and

$$y = \left(\frac{2mF_2}{\hbar^2}\right)^{1/3}(x - x_2)$$

in (7.182). Both equations then reduce to the same equation,

$$\frac{d^2\varphi}{dy^2} - y\varphi = 0$$

The solutions to this equation are called[1] *Airy functions* and are denoted by the symbols $Ai(y)$ and $Bi(y)$ (see Table 7.3). For the problem at hand, the wavefunction $\varphi(x)$ must approach zero in the domains $x \ll x_1$ and $x \gg x_2$. Both these regions correspond to large positive values of $|y|$. The function with this property is $Ai(y)$, which has asymptotic forms

(7.183)

$$Ai(y) \sim \frac{1}{2\sqrt{\pi}\,y^{1/4}} \exp\left(-\frac{2}{3}y^{3/2}\right) \qquad (y > 0)$$

$$Ai(y) \sim \frac{1}{\sqrt{\pi}(-y)^{1/4}} \sin\left[\frac{2}{3}(-y)^{3/2} + \frac{\pi}{4}\right] \quad (y < 0)$$

It is exponentially decaying for $y > 0$ and oscillatory for $y < 0$ and strongly resembles the behavior of a harmonic oscillator wavefunction across a turning point, such as shown in Fig. 7.6.

In the neighborhood of x_1, from (7.179), we obtain

$$p^2 = 2m(E - V_1) \simeq 2mF_1(x - x_1) = -(2mF_1\hbar)^{2/3}y$$

$$2mF_1\,dx = -(2mF_1\hbar)^{2/3}\,dy$$

To the left of x_1, $p^2 = -\hbar^2\kappa^2$, so

$$\hbar^2\kappa^2 = (2mF_1\hbar)^{2/3}y$$

and we may write

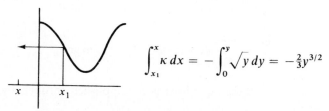

$$\int_{x_1}^{x} \kappa\,dx = -\int_{0}^{y}\sqrt{y}\,dy = -\tfrac{2}{3}y^{3/2}$$

To the right of x_1, in the oscillatory well domain, $p^2 = \hbar^2 k^2$ and

$$\hbar^2 k^2 = -(2mF_1\hbar)^{2/3}y$$

so y is negative in this domain. The integral of k gives

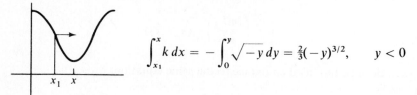

$$\int_{x_1}^{x} k\,dx = -\int_{0}^{y}\sqrt{-y}\,dy = \tfrac{2}{3}(-y)^{3/2}, \qquad y < 0$$

---

[1] Named for an English astronomer, G. B. Airy (1801–1892).

TABLE 7.3 Properties of Airy functions[a]

Differential Equation

$$\frac{\partial^2 \varphi}{\partial x^2} - x\varphi = 0$$

Solutions

    (a)   *Series representation*

$$Ai(x) = af(x) - bg(x)$$

$$Bi(x) = \sqrt{3}[af(x) + bg(x)]$$

where

$$a = 3^{-2/3}/\Gamma(2/3) = 0.3550, \qquad b = 3^{-1/3}/\Gamma(1/3) = 0.2588$$

$$f(x) = 1 + \frac{1}{3!}x^3 + \frac{1\cdot 4}{6!}x^6 + \frac{1\cdot 4\cdot 7}{9!}x^9 + \cdots$$

$$g(x) = x + \frac{2}{4!}x^4 + \frac{2\cdot 5}{7!}x^7 + \frac{2\cdot 5\cdot 8}{10!}x^{10} + \cdots$$

    (b)   *Integral representation*

$$Ai(x) = \frac{1}{\pi}\int_0^\infty \cos\left(\frac{s^3}{3} + sx\right) ds$$

$$Bi(x) = \frac{1}{\pi}\int_0^\infty \left[e^{sx - (1/3)s^3} + \sin\left(\frac{s^3}{3} + sx\right)\right] ds$$

Relations to Bessel functions of fractional order

    With $y \equiv \frac{2}{3}x^{3/2}$, the following relations hold.

$$Ai(x) = \frac{1}{\pi}\sqrt{x/3}\, K_{1/3}(y)$$

$$Ai(-x) = \frac{1}{3}\sqrt{x}[J_{1/3}(y) + J_{-1/3}(y)]$$

$$Bi(x) = \sqrt{x/3}[I_{-1/3}(y) + I_{1/3}(y)]$$

$$Bi(-x) = \sqrt{x/3}[J_{-1/3}(y) - J_{1/3}(y)]$$

    The I and K functions are modified Bessel functions of the first and second kind, respectively.

Asymptotic forms

    For large $|x|$, leading terms in asymptotic series are as follows:

$$Ai(x) \sim \frac{1}{2\sqrt{\pi}x^{1/4}}\exp\left(-\frac{2}{3}x^{3/2}\right), \qquad x > 0$$

$$Ai(x) \sim \frac{1}{\sqrt{\pi}(-x)^{1/4}}\sin\left[\frac{2}{3}(-x)^{3/2} + \frac{\pi}{4}\right], \qquad x < 0$$

$$Bi(x) \sim \frac{1}{\sqrt{\pi}x^{1/4}}\exp\left(\frac{2}{3}x^{3/2}\right), \qquad x > 0$$

$$Bi(x) \sim \frac{1}{\sqrt{\pi}(-x)^{1/4}}\cos\left[\frac{2}{3}(-x)^{3/2} + \frac{\pi}{4}\right], \qquad x < 0$$

---

[a] For further properties of these functions, see *Handbook of Mathematical Functions*, N. Abramowitz and I. A. Stegun, eds., Dover, New York, 1964; H. and B. S. Jeffries, *Methods of Mathematical Physics*, 3rd ed., Cambridge University Press, New York, 1956.

In these same respective domains, the **WKB** functions φ_I (7.176) and $\varphi_{II}^<$ (7.178a), when written in terms of the variable y, appear as

$$\varphi_I = \frac{1}{y^{1/4}} \exp\left(-\tfrac{2}{3}y^{3/2}\right) \qquad\qquad y > 0$$

$$\varphi_{II}^< = \frac{C}{(-y)^{1/4}} \sin\left[\tfrac{2}{3}(-y)^{3/2} + \delta\right] \qquad y < 0$$

These agree with the asymptotic forms (7.183) for the exact Airy function solutions [corresponding to the approximate linear potential (7.179, 180)] provided that we set $C = 2$ and $\delta = \pi/4$.

In this manner we find that the **WKB** approximation in region I,

(7.184) $$\varphi_I(x) = \frac{1}{\sqrt{\kappa}} \exp\left(\int_{x_1}^x \kappa\,dx\right) \qquad (x < x_1)$$

matches (or "connects") with the **WKB** approximation

(7.185) $$\varphi_{II}^<(x) = \frac{2}{\sqrt{k}} \sin\left(\int_{x_1}^x k\,dx + \frac{\pi}{4}\right) \qquad (x_1 < x)$$

in region II.

In like manner we find that the **WKB** approximation in region III

(7.186) $$\varphi_{III} = \frac{A}{\sqrt{\kappa}} \exp\left(-\int_{x_2}^x \kappa\,dx\right) \qquad (x_2 < x)$$

matches with the **WKB** approximation

(7.187) $$\varphi_{II}^> = \frac{2A}{\sqrt{k}} \sin\left(\int_x^{x_2} k\,dx + \frac{\pi}{4}\right) \qquad (x < x_2)$$

in region II. The remaining constant A is determined in matching $\varphi_{II}^<$ to $\varphi_{II}^>$.

The Four Connection Formulas

There are in total four connection formulas which serve to relate **WKB** component wavefunctions across turning points. In the preceding analysis two of these relations were uncovered. Namely, these are given by the manner through which φ_I connects to $\varphi_{II}^<$ (Eqs. 7.184 and 7.185) and that by which $\varphi_{II}^>$ connects to φ_{III} (Eqs. 7.186 and 7.187). Carrying through a parallel analysis and employing the asymptotic forms for the Airy functions $Bi(y)$ gives the remaining two relations. The complete list of four connecting formulas is given on page 242[1] with $x_{1,2}$ denoting either x_1 or x_2.

[1] Here we are assuming that no other linearly independent components of the wavefunction enter the analysis.

		$x < x_{1,2}$		$x_{1,2} < x$

(7.188a) — $\dfrac{2}{\sqrt{k}}\cos\left(\int_x^{x_2} k\,dx - \dfrac{\pi}{4}\right) \rightleftharpoons \dfrac{1}{\sqrt{\kappa}}\exp\left(-\int_{x_2}^x \kappa\,dx\right)$

(7.188b) — $\dfrac{1}{\sqrt{k}}\sin\left(\int_x^{x_2} k\,dx - \dfrac{\pi}{4}\right) \rightleftharpoons -\dfrac{1}{\sqrt{\kappa}}\exp\left(\int_{x_2}^x \kappa\,dx\right)$

(7.189a) — $\dfrac{1}{\sqrt{\kappa}}\exp\left(\int_{x_1}^x \kappa\,dx\right) \rightleftharpoons \dfrac{2}{\sqrt{k}}\cos\left(\int_{x_1}^x k\,dx - \dfrac{\pi}{4}\right)$

(7.189b) — $-\dfrac{1}{\sqrt{\kappa}}\exp\left(-\int_{x_1}^x \kappa\,dx\right) \rightleftharpoons \dfrac{1}{\sqrt{k}}\sin\left(\int_{x_1}^x k\,dx - \dfrac{\pi}{4}\right)$

Bohr–Sommerfeld Quantization Rules

The energy levels of the finite well depicted in Fig. 7.33 may be obtained to within the accuracy of the WKB approximation by joining $\varphi_{\mathrm{II}}^<$ and $\varphi_{\mathrm{II}}^>$ smoothly within the well. This gives

$$\sin\left(\int_{x_1}^x k\,dx + \frac{\pi}{4}\right) = A\sin\left(\int_x^{x_2} k\,dx + \frac{\pi}{4}\right)$$

With

$$\eta \equiv \int_{x_1}^{x_2} k\,dx, \qquad a \equiv \int_x^{x_2} k\,dx + \frac{\pi}{4}$$

the continuity condition above becomes

$$\sin\left(\eta + \frac{\pi}{2} - a\right) = A\sin a$$

or, equivalently,

$$\sin\left(\eta + \frac{\pi}{2}\right)\cos a - \cos\left(\eta + \frac{\pi}{2}\right)\sin a = A\sin a$$

The solution to this equation which gives A, constant and independent of the parameter a, is obtained by setting[1]

(7.190) $\qquad\qquad \eta + \dfrac{\pi}{2} = (n+1)\pi \qquad (n = 0, 1, 2, \ldots)$

---

[1] Writing $n + 1$ instead of n ensures that η is nonnegative.

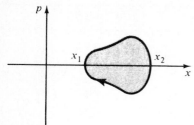

FIGURE 7.35 Classical vibrational motion between the turning points (x_1, x_2) depicted in (x, p) "phase space." The enclosed surface has the area $\oint p\, dx$.

Corresponding values of A are $(-1)^n$. Thus, continuity of φ_{II} implies the condition

$$\eta = \int_{x_1}^{x_2} k\, dx = (n + \tfrac{1}{2})\pi$$

When written in terms of momentum $p = \hbar k$, this criterion appears as

(7.191)
$$\int_{x_1}^{x_2} p\, dx = \left(n + \frac{1}{2}\right)\frac{h}{2}$$

In the corresponding classical motion, the particle oscillates between the turning points x_1 and x_2. In Cartesian x, p space, this "orbit" is a closed loop, as depicted in Fig. 7.35, with area $\oint p\, dx$. With (7.191) we find that

(7.192)
$$\oint p\, dx = (n + \tfrac{1}{2})h$$

This equation is nearly the same as the Bohr–Sommerfeld quantization rule[1] (2.6). As discussed in Section 2.4, this rule prescribes that an integral number of wavelengths fit the orbit perimeter (see Fig. 2.11). When cast in terms of wavelength $\lambda = h/p$, (7.192) becomes

(7.193)
$$\oint \frac{dx}{\lambda} = n + \tfrac{1}{2}$$

The integral represents the number of wavelengths in the orbit perimeter. The distinction between this result and the Bohr–Sommerfeld prescription is that in the present WKB analysis, the wavefunction may extend into the classically forbidden region, or, equivalently, that the wavefunction need not vanish at the turning points. If the wavefunction must vanish at the turning points, such as is the case for a very sharply rising potential, a half-integer number of wavelengths are allowed between turning points, thereby returning the Bohr formula. On the other hand, leakage of the

---

[1] The distinction between the loop integral in (2.6) and that in (7.192) is that $\oint p_\theta d\theta$ is relevant to rotational motion while $\oint p\, dx$ is relevant to vibrational motion. In either case such integrals play a major role in the study of periodic motion and are called *action* integrals. The Bohr–Sommerfeld quantization rule stipulates that these action integrals have only discrete values, nh. That is, in quantization of periodic systems, one quantizes the action variables $\oint p\, dx$.

wavefunction into the classically forbidden domain is evident for a potential with a gradual slope at the turning points, which property is seen to be consistent with the WKB criterion (7.166).

WKB Eigenenergies

The continuity result (7.191) serves to determine eigenenergies within the WKB approximation. Since this analysis becomes more accurate for large energies, values so found will generally give better estimates for large quantum number n. In that this number is also a measure of the number of zeros of the wavefunction between turning points, we see that in this limit the wavelength becomes small compared to the distance between the turning points. As described in the first paragraph of this section, such is the domain of the classical WKB analysis.

As an example of the application of (7.191), let us consider calculation of the energies of the harmonic oscillator. For this configuration the momentum is given by

$$p = \sqrt{2m(E - m\omega_0{}^2 x^2/2)}$$

with turning points given by (7.12). Introducing the variable

$$\cos\theta \equiv x\sqrt{m\omega_0{}^2/2E}$$

permits the condition (7.191) to be written

$$\frac{4E}{\omega_0} \int_0^\pi \sin^2\theta \, d\theta = (n + \tfrac{1}{2})h$$

$$\frac{2\pi E_n}{\omega_0} = (n + \tfrac{1}{2})h$$

These are seen to be, somewhat fortuitously, the exact eigenenergies of the harmonic oscillator (7.44), valid for all n.

An example in the opposite extreme is given by the one-dimensional box potential (4.1). There is no penetration of the particle wavefunction through the sharply rising potential wall, and the validity of the WKB formula (7.191) becomes very questionable, especially at low energies. Using this formula, we readily obtain the eigenenergies

$$E_n^{\text{WKB}} = \left(n + \frac{1}{2}\right)^2 \frac{h^2}{8mL^2} = E_n\left(1 + \frac{4n + 1}{4n^2}\right)$$

Here we have written E_n for the exact eigenenergies (4.14), $E_n = n^2 E_1$. As expected, the estimate gives a large fractional error for small quantum numbers. However, in the high-quantum-number domain, $n \gg 1$, where the walls of the potential

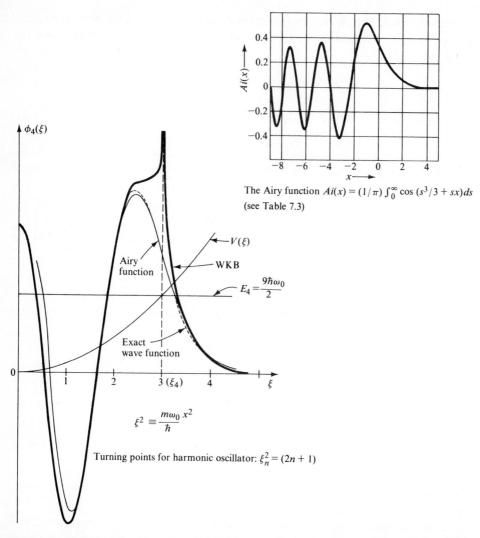

The Airy function $Ai(x) = (1/\pi) \int_0^\infty \cos(s^3/3 + sx)\,ds$
(see Table 7.3)

$$\xi^2 \equiv \frac{m\omega_0 \, x^2}{\hbar}$$

Turning points for harmonic oscillator: $\xi_n^2 = (2n + 1)$

FIGURE 7.36 WKB approximation for the fourth state of the harmonic oscillator, together with a graph of the Airy function. Also shown are the potential $V(\xi)$ and the fourth eigenenergy. Note the divergence of the WKB approximation at the turning point ξ_4. This calculation was performed previously by J. D. Powell and B. Crasemann (*Quantum Mechanics*, Addison-Wesley, Reading, Mass. 1965). For extensive discussion of numerical techniques in the WKB analysis, see C. M. Bender and S. A. Orszag, *Advanced Mathematical Methods for Scientists and Engineers*, McGraw-Hill, New York, 1978.

are many wavelengths apart, we find that the WKB estimate agrees with the exact result

$$E_n^{WKB} \simeq E_n \qquad (n \gg 1)$$

Note, however, that for this singular case where wavefunction penetration does not occur, the simpler Bohr–Sommerfeld rule (2.6) gives exact results.

WKB Wavefunctions

Wavefunctions in the WKB approximation incorporate Airy functions together with matching conditions (7.184) et seq. Results of this analysis to the fourth state of the harmonic oscillator are shown in Fig. 7.36. Here we may observe the dramatic disparity between the exact wavefunction and the WKB approximation in the vicinity of the turning points.

Application to Transmission Problems

In concluding this section we will obtain a very important formula for the transmission coefficient relevant to a potential barrier such as depicted in Fig. 7.37. As described previously in Section 7.5, the transmitted wave in region III has only one momentum component, so that to within the WKB approximation, we may write

(7.194)
$$\varphi_{III} = \frac{A}{\sqrt{k}} \exp\left[i\left(\int_{x_2}^{x} k \, dx - \frac{\pi}{4} \right) \right]$$

The procedure we will follow to obtain the incident component of φ_I is as follows. Rewriting (7.194) as a combination of trigonometric functions permits application of the connection formulas (7.189), which allows calculation of φ_{II}. With φ_{II} so found, we again rewrite it in a manner that permits application of (7.188) to connect φ_{II} to φ_I. Finally, φ_I is decomposed into incident and reflected components. Comparison of the incident component with φ_{III} permits calculation of the transmission coefficient.

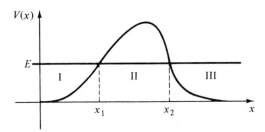

FIGURE 7.37 Domains relevant to the WKB approximation of the transmission through a potential barrier.

Rewriting (7.194) in the form

$$
(7.195) \qquad \varphi_{\text{III}} = \frac{A}{\sqrt{k}} \left[\cos \left(\int_{x_2}^{x} k \, dx - \frac{\pi}{4} \right) + i \sin \left(\int_{x_2}^{x} k \, dx - \frac{\pi}{4} \right) \right]
$$

permits application of (7.189) and we obtain

$$
(7.196) \qquad \varphi_{\text{II}} = \frac{A}{2\sqrt{\kappa}} \exp \left(-\int_{x}^{x_2} \kappa \, dx \right) - \frac{iA}{\sqrt{\kappa}} \exp \left(\int_{x}^{x_2} \kappa \, dx \right)
$$

Let r denote the integral (compare with η, p. 251)

$$
(7.196\text{a}) \qquad r \equiv \exp \left(\int_{x_1}^{x_2} \kappa \, dx \right)
$$

Appropriate division of the interval of integration gives the relations

$$
\exp \left(-\int_{x}^{x_2} \kappa \, dx \right) = r^{-1} \exp \left(\int_{x_1}^{x} \kappa \, dx \right)
$$

$$
\exp \left(\int_{x}^{x_2} \kappa \, dx \right) = r \exp \left(-\int_{x_1}^{x} \kappa \, dx \right)
$$

Substituting these expressions into (7.196) gives

$$
(7.197) \qquad \varphi_{\text{II}} = \frac{A}{2r\sqrt{\kappa}} \exp \left(\int_{x_1}^{x} \kappa \, dx \right) - \frac{iAr}{\sqrt{\kappa}} \exp \left(-\int_{x_1}^{x} \kappa \, dx \right)
$$

which allows application of the connection formulas (7.188). There results

$$
(7.198) \qquad \varphi_{\text{I}} = -\frac{A}{2r\sqrt{k}} \sin \left(\int_{x}^{x_1} k \, dx - \frac{\pi}{4} \right) - \frac{i2Ar}{\sqrt{k}} \cos \left(\int_{x}^{x_1} k \, dx - \frac{\pi}{4} \right)
$$

We are now at the point where we must extract the incident component of φ_1. If we label the argument $\int_{x}^{x_1} k \, dx - \pi/4 \equiv z$ and express both trigonometric terms as exponentials, (7.198) may be rewritten

$$
(7.199) \qquad \varphi_{\text{I}} = -\frac{A}{2r\sqrt{k}} \frac{e^{iz} - e^{-iz}}{2i} - \frac{i2Ar}{\sqrt{k}} \frac{e^{iz} + e^{-iz}}{2}
$$

$$
= i \frac{A}{\sqrt{k}} \left(\frac{1}{4r} - r \right) e^{iz} - i \frac{A}{\sqrt{k}} \left(\frac{1}{4r} + r \right) e^{-iz}
$$

Now if the wavenumber k were constant, z would have the value $-k(x - x_1) - \pi/4$. From this we may infer that the second term in the last equation for φ_1 represents the incident component wavefunction. Employing the expression (7.108) for the transmission coefficient T, with the second term in (7.199) representing the incident wave-function and (7.194) the transmitted wavefunction, we obtain

(7.200)
$$T = \frac{1}{(r + 1/4r)^2} = \frac{1}{r^2 + 1/2 + 1/16r^2}$$

It is consistent with the WKB criterion (7.166) to neglect all but the term r^2 in the denominator of (7.200), thereby obtaining (see Problem 7.89)

(7.201)
$$\boxed{T = r^{-2} = \exp\left(-2 \int_{x_1}^{x_2} \kappa \, dx\right)}$$

The simplest application of this formula is in calculation of the transmission through a square potential barrier. Exact analysis gives the result (7.147). We should find that this expression reduces to the WKB formula in the limit

$$\kappa a = \sqrt{2ma^2(V - E)/\hbar^2} \gg 1.$$

In this limit (7.147) gives the transmission coefficient

$$T \simeq \frac{16E}{V} e^{-4\kappa a}$$

whereas (7.201) gives

$$T = e^{-4\kappa a}$$

which is seen to be in good order-of-magnitude agreement with the limiting form of the exact result given above. Further application of the exceedingly important result (7.201) is left to the problems. A discussion on the Feynman path integral, closely allied to the WKB analysis, is given in Section 7.11.

PROBLEMS

7.62   In the phenomenon of *cold emission*, electrons are drawn from a metal (at room temperature) by an externally supported electric field. The potential well that the metal presents to the free electrons before the electric field is turned on is depicted in Fig. 2.5. After application of the constant electric field $\mathscr{E}$, the potential at the surface slopes down as shown in Fig. 7.38, thereby allowing electrons in the Fermi sea to "tunnel" through the potential barrier. If the surface of the metal is taken as the $x = 0$ plane, the new potential outside the surface is

$$V(x) = \Phi + E_F - e\mathscr{E}x$$

where E_F is the Fermi level and Φ is the work function of the metal.

(a)   Use the WKB approximation to calculate the transmission coefficient for cold emission.

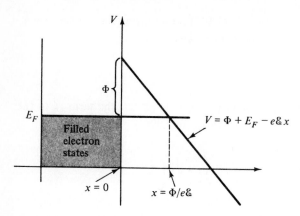

FIGURE 7.38 Potential configuration for the phenomenon of "cold emission." See Problem 7.62.

(b)   Estimate the field strength $\mathscr{E}$, in volt/cm, necessary to draw current density of the order of mA/cm^2 from a potassium surface. For $\mathbb{J}_{inc}$ (see Eq. 7.108) use the expression $\mathbb{J}_{inc} = env$, where n is electron density and v is the speed of electrons at the top of the Fermi sea. The relevant expression for E_F may be found in Problem 2.42. Data for potassium is given in Section 2.3.

Answer (partial)

(a)   Using (7.201), and the form of the potential exterior to the metal given in the statement of the problem, we obtain, for transmission at the Fermi level ($V - E_F = \Phi - e\mathscr{E}x$),

$$T = \exp\left[-\frac{2}{\hbar}\int_0^{\Phi/e\mathscr{E}}\sqrt{2m(\Phi - e\mathscr{E}x)}\,dx\right]$$

$$= \exp\left(-\frac{4}{3}\frac{\sqrt{2m}}{\hbar}\frac{\Phi^{3/2}}{e\mathscr{E}}\right)$$

This equation is referred to as the *Fowler–Nordheim* equation.

7.63   An α particle is the nucleus of a helium atom. It is a tightly bound entity comprised of two protons and two neutrons, for which the energy required to remove one neutron is 20.5 MeV. (This is the rest-mass energy of ~41 electrons.) A primary mode of decay for radioactive nuclei is through the process of α decay. A consistent model for this process envisions the α particle bound to the nucleus by a spherical well potential.[1] Outside the well the α particle is repelled from the residual nucleus by the potential barrier

$$V = \frac{2(Z-2)e^2}{r} \equiv \frac{A}{r}$$

The original radioactive nucleus has charge Ze, while the α particle has charge $2e$ (Fig. 7.39).

---

[1] The rigid spherical well potential is described in Problem 10.16. The effective one-dimentional Hamiltonian for the configuration at hand is given by (10.93) with angular momentum L set equal to zero. This corresponds to assuming in part (b) that the bounce motion of the α-particle is through the origin.

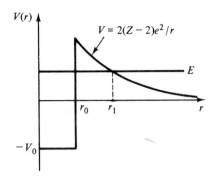

FIGURE 7.39 Nuclear α particle potential model for the process of α decay. See Problem 7.63.

(a) Use the WKB approximation to calculate the transmissivity T of the nuclear barrier to α decay in terms of the velocity $v = \sqrt{2E/m}$ and the dimensionless ratio $\sqrt{(r_0/r_1)} \equiv \cos W$. What form does T assume in the limit $r_0 \to 0$?

(b) Assuming that the α particle "bounces" freely between the walls presented by the spherical well potential with a speed $\sim 10^9$ cm/s and that the radius of the heavy radioactive nucleus (e.g., uranium) is $\sim 10^{-12}$ cm, one obtains that the α particle strikes the nuclear wall at the rate $\sim 10^{21}$ s^{-1}. In each collision the probability that the α particle penetrates the nuclear Coulomb barrier is equal to the transmissivity of the barrier T. It follows that the probability of tunneling through this barrier, per second, is

$$P = 10^{21} T$$

and that the mean lifetime of the nucleus is

$$\tau = \frac{1}{P} = \frac{10^{-21}}{T}$$

Use your answer to part (a) for T and the following expression for the nuclear radius

$$r_0 = 2 \times 10^{-13} Z^{1/3} \text{ cm}$$

to estimate the mean lifetime for uranium α decay.

Answer (partial)

(a)

$$T = \exp\left[-\frac{2}{\hbar}\int_{r_0}^{r_1}\sqrt{2m\left(\frac{A}{r} - E\right)}\,dr\right]$$

$$r_1 = \frac{A}{E}$$

Integrating, one obtains *Gamow's formula*,

$$T = \exp\left[-\frac{2A}{\hbar v}(2W - \sin 2W)\right]$$

As $r_0 \to 0$, $W \to \pi/2$ and $T \sim \exp(-2\pi A/\hbar v)$.

7.64   Use the WKB relation (7.191) to estimate the eigenenergies of the displaced spring potential

$$V = \frac{K}{2}(x^2 + 2\phi x)$$

How do these values compare with the exact values obtained in Problem 7.16?

7.65   An electron with charge $-e$ and mass m, constrained to move in the x direction, interacts with a uniform electric field $\mathscr{E}$, which points in the positive x direction.

(a)   Show that energy eigenstates may be written as Airy functions.

(b)   Do eigenenergies comprise a continuous or a discrete spectrum? *Hint for part (a):* Set $e\mathscr{E}x - E = Kx'$ and then find the value of K that gives Airy's equation. (See Table 7.3.)

7.66   Use the WKB approximation to determine the bound-state energies of the potential well

$$V(x) = \frac{V_0}{a}|x|, \qquad |x| \le a$$

$$V(x) = V_0 = \frac{1}{m}\left(\frac{h}{a}\right)^2, \qquad |x| > a$$

Answer

Eigenenergies appear as

$$E_n = \bar{E}_0(n + \tfrac{1}{2})^{2/3}$$

$$\bar{E}_0^{3/2} = \frac{3V_0 h}{8a\sqrt{2m}}$$

With V_0 as given, we obtain $\bar{E}_0 = 0.41V_0$, so there are four bound states: $E_0 = 0.63\bar{E}_0$, $E_1 = 1.31\bar{E}_0$, $E_2 = 1.85\bar{E}_0$, and $E_3 = 2.31\bar{E}_0$.

7.67   Show that for the singular potential

$$V = aV_0\,\delta(x)$$

boundary conditions on $\varphi(x)$ become

(a) $\varphi(0)_- = \varphi(0)_+$

(b) $\dfrac{h^2}{2m}[\varphi'(0)_+ - \varphi'(0)_-] = aV_0\varphi(0)$

7.68   Use the result of Problem 7.67 to construct the bound state of the potential well, $V = -aV_0\,\delta(x)$. (*Hint:* Set $E = -|E|$ and look for the solution for $x \ne 0$.)

7.69   Find T and R for the potential barrier $V = aV_0\,\delta(x)$.

7.70   The initial state for the harmonic oscillator

$$|\psi(0)\rangle = e^{-g^2/2}\sum_{n=0}^{\infty}\frac{g^n}{\sqrt{n!}}|n\rangle$$

represents a minimum uncertainty wave packet. The parameter g is real.

(a) Show that

$$|\psi(0)\rangle = e^{-g^2/2} \sum \frac{(g\hat{a}^\dagger)^n}{n!} |0\rangle = e^{-g^2/2} e^{g\hat{a}^\dagger} |0\rangle$$

where $|0\rangle$ is the $n = 0$ eigenstate.

(b) Show that for this state

$$\langle x\rangle = g\sqrt{\frac{2\hbar}{\omega_0 m}} \qquad\qquad \langle p\rangle = 0$$

$$\langle x^2\rangle = \frac{\hbar}{2\omega_0 m}(4g^2 + 1) \qquad \langle p^2\rangle = \frac{\hbar\omega_0 m}{2}$$

so that

$$(\Delta x)^2 = \frac{\hbar}{2\omega_0 m} \qquad (\Delta p)^2 = \frac{\hbar\omega_0 m}{2}$$

and

$$\Delta x\, \Delta p = \frac{\hbar}{2}$$

This property establishes the fact that the given superposition state represents a wave packet of minimum uncertainty.

(c) Show that for this state

$$\langle H\rangle = \hbar\omega_0(g^2 + \tfrac{1}{2})$$

which gives a physical interpretation of the parameter g.

(d) Show that $\psi(t)$ is

$$|\psi(t)\rangle = e^{-i\omega_0 t/2} e^{-g^2/2} \exp\left(ge^{-i\omega_0 t}\hat{a}^\dagger\right)|0\rangle$$

[Hint:

$$|\psi(t)\rangle = e^{-i\omega_0 t(\hat{a}^\dagger a + 1/2)}|\psi(0)\rangle$$

Also:

$$\exp\left(-i\omega_0 t\hat{a}^\dagger\hat{a}\right) \sum \frac{g^n}{\sqrt{n!}} |n\rangle = \sum \frac{(ge^{-i\omega_0 t})^n}{\sqrt{n!}} |n\rangle$$

Here we have recalled that $f(\hat{a}^\dagger\hat{a})|n\rangle = f(n)|n\rangle$.]

(e) Show that in the state $\psi(t)$

$$\langle x\rangle = g\sqrt{\frac{2\hbar}{\omega_0 m}}\cos\omega_0 t \qquad \langle p\rangle = -g\sqrt{2\hbar\omega_0 m}\sin\omega_0 t$$

$$\langle x^2\rangle = \langle x\rangle^2 + \frac{\hbar}{2\omega_0 m} \qquad \langle p^2\rangle = \langle p\rangle^2 + \frac{\hbar\omega_0 m}{2}$$

so that

$$\Delta x\, \Delta p = \frac{\hbar}{2}$$

at any time t. The packet remains a packet of minimum uncertainty for all time and oscillates in classical simple harmonic motion. Note also that the probability density $\langle \psi(x, t)|\psi(x, t)\rangle$ is a Gaussian form, centered at $\langle x \rangle$.

7.71   In closed form the wave packet of minimum uncertainty for the harmonic oscillator appears (at time $t = 0$) as

$$\psi(x, 0) = \sqrt{\beta/\pi^{1/2}} \, \exp\left[\frac{ixp_0}{\hbar} - \frac{1}{2}\beta^2(x - x_0)^2 \right]$$

(a)   Show that

$$\langle x \rangle = x_0 \qquad \Delta x = \frac{1}{\beta\sqrt{2}}$$

$$\langle p \rangle = p_0 \qquad \Delta p = \frac{\hbar\beta}{\sqrt{2}}$$

Hence $\Delta x\, \Delta p = \hbar/2$, and we are justified in calling ψ a packet of minimum uncertainty.

(b)   Show that in the initial state above,

$$\langle H \rangle = \frac{1}{2m}(p_0{}^2 + m^2\omega^2 x_0{}^2) + \tfrac{1}{2}\hbar\omega$$

and

$$\langle xp + px \rangle = 2x_0 p_0$$

(c)   In order to establish that $\psi(x)$ remains a packet of minimum uncertainty for all time, one must show that $\Delta x\, \Delta p$ is constant. Recalling the equation of motion for the average of an operator (6.68),

$$\frac{d}{dt}\langle A \rangle = \frac{1}{i\hbar}\langle [\hat{A}, \hat{H}] \rangle + \left\langle \frac{\partial \hat{A}}{\partial t} \right\rangle$$

and introducing the operator

$$\hat{\eta} = \hat{x}\hat{p} + \hat{p}\hat{x} - 2\langle x \rangle\langle p \rangle$$

show that

$$\frac{d}{dt}(\Delta x)^2 = \frac{\langle \eta \rangle}{m}$$

$$\frac{d}{dt}(\Delta p)^2 = -m\omega^2 \langle \eta \rangle$$

$$\frac{d}{dt}\langle \eta \rangle = \frac{2(\Delta p)^2}{m} - 2m\omega^2(\Delta x)^2$$

Using these results, show that for the initial state above, $(\Delta x)^2$ and $(\Delta p)^2$ are both constants in time.

(d)   Show that $\psi(x, 0)$ is an eigenfunction of the annihilation operator $\hat{a}$. What is the eigenvalue of $\hat{a}$ in this state? [*Hint*: Employ the representation for $\psi(x, t)$ given in Problem 7.70.]

Note: In quantum optics the radiation field is viewed as a collection of harmonic oscillators. In this representation, the appropriate generalization of the state of minimum uncertainty is called the *coherent state*.

7.72   The Hamiltonian of a particle is

$$\hat{H} = A\hat{a}^\dagger\hat{a} + B(\hat{a} + \hat{a}^\dagger)$$

where A and B are constants. What are the energy eigenvalues of the particle? (*Hint*: Introduce the operator $\hat{b} = \alpha\hat{a} + \beta$; $\hat{b}^\dagger = \alpha\hat{a}^\dagger + \beta$.)

7.73   What is the form of the potential that gives the Gaussian probability density with variance a^2 in the ground state?

7.74   The reflection coefficient for the smooth potential step

$$V(x) = \frac{V_0}{1 + e^{-\gamma x}}$$

for $E > V_0$ is[1]

$$R = \left(\frac{\sinh\left[\pi(k_1 - k_2)/\gamma\right]}{\sinh\left[\pi(k_1 + k_2)/\gamma\right]}\right)^2$$

The energy of incident particles at $x = -\infty$ is

$$E = \frac{\hbar^2 k_1^2}{2m}$$

while the kinetic energy of transmitted particles at $x = +\infty$ is

$$\frac{\hbar^2 k_2^2}{2m} = E - V_0$$

(a)   Make a sketch of the potential $V(x)$ and indicate roughly the length scale of potential and its relation to the wavenumber γ.

---

[1] The coefficients R and T given in Problems 7.74 and 7.75, respectively, are calculated in L. Landau and E. Lifshitz, *Quantum Mechanics*, 2nd ed., Addison-Wesley, Reading, Mass., 1965.

(b) Show that in the limit that $V(x)$ approaches the simple step (Fig. 7.18), R goes to the value given by (7.125).

(c) Show that the classical value of R emerges for wavelengths small compared to the potential scale of length.

7.75 The transmission coefficient for the symmetric potential hill

$$V(x) = \frac{V_0}{\cosh^2{(\gamma x)}}$$

for $E < V$ is

$$T = \frac{\sinh^2{(\pi k/\gamma)}}{\sinh^2{(\pi k/\gamma)} + \cosh^2[(\pi/2)\sqrt{\rho^2 - 1}]}$$

where

$$\rho^2 \equiv \frac{8mV_0}{\hbar^2\gamma^2} > 1$$

Incident and transmitted particles at $x = -\infty$ and $x = +\infty$, respectively, have energy

$$E = \frac{\hbar^2 k^2}{2m}$$

(a) Sketch the potential and indicate roughly the length scale of potential and its relation to the wavenumber γ.

(b) Show that the classical value of T emerges for values of γ appropriate to the classical domain.

(c) Formulate an expression for the next-order approximation to the entirely classical result (b) for the transmission coefficient using the WKB analysis.

(d) Obtain an explicit expression for the transmission coefficient in the near-classical domain that you have formulated in part (c) by expanding the exact formula for T given in the statement of this problem.

Answer (partial)

(d) The classical limit is attained in the limit $\gamma \to 0$. From the given expression for T, we obtain

$$T \simeq \frac{e^{\pi k/\gamma}}{e^{\pi k/\gamma} + e^{\pi\rho/2}} = \frac{1}{1 + \exp{(\pi/\gamma\hbar)(\hbar\rho\gamma/2 - \hbar k)}}$$

$$\simeq \exp\left[-\frac{\pi}{\gamma\hbar}(\sqrt{2mV_0} - \sqrt{2mE})\right]$$

7.76 A uniform homogeneous beam of electrons is incident on a rectangular potential barrier of height V. Each electron in the beam has energy $E > V$ and unit amplitude wavefunction

$$\varphi_{inc} = e^{ik_1 x}$$

If the transmitted electrons have wavefunction

$$\varphi_{\text{trans}} = \varphi_{\text{III}} = 0.97 e^{ik_1 x}$$

(a)   What is the total wavefunction φ_1, of electrons in region I?
(b)   If $E = 10$ eV and $V = 5$ eV, what is the minimum barrier width compatible with the information given above?

Answers

(a)   In general for unit amplitude incident waves,

$$\varphi_1 = e^{ik_1 x} \pm i\sqrt{R} e^{-ik_1 x}$$

$$\varphi_{\text{III}} = \sqrt{T} e^{ik_1 x}$$

where the $\pm$ signs refer to the sign of $\sin(2k_2 a)$. (See Problem 7.44.) It follows that

$$T = (0.97)^2 = 0.94, \qquad R = 1 - T = 0.06, \quad \text{and} \quad \sqrt{R} = 0.24,$$

so that

$$\varphi_1 = e^{ik_1 x} + i0.24 e^{-ik_1 x}$$

(b)   We then find that $\sin^2(k_2 2a) = 8R/T = 0.51$, $k_2 2a = 0.80 < \pi/2$, and

$$k_2 = \sqrt{\frac{2m(E - V)}{\hbar^2}} = 1.14 \times 10^8 \text{ cm}^{-1}$$

Therefore,

$$2a = 0.70 \text{ Å}$$

7.77   (a)   What are the values of k and κ at the "turning points" of a potential hill or potential barrier?
(b)   What are the values of the WKB wavefunctions, $|\varphi_{\text{I}}|, |\varphi_{\text{II}}|, |\varphi_{\text{III}}|$, at these points (for either bound or unbound states)?
(c)   A student argues the following: We see from part (b) of this problem that WKB wavefunctions blow up at the turning points and are therefore invalid. Such wavefunctions cannot be of any use. Is the student correct? Explain.

Answer

(a) $k = \kappa = 0$ at turning points.
(b) $|\varphi_{\text{I}}| = |\varphi_{\text{II}}| = |\varphi_{\text{III}}| = \infty$ at turning points.
(c)   The WKB wavefunctions are valid in domains removed from the turning points (where they were derived from the Schrödinger equation). For bound-state problems these solutions give an estimate for eigenenergies. For unbound-state problems they give an estimate for transmission and reflection coefficients.

7.11 PRINCIPLE OF LEAST ACTION AND FEYNMAN'S PATH INTEGRAL FORMULATION

Action Integral and the Lagrangian

Classical dynamics may be formulated in terms of *Hamilton's principle of least action*. This principle states the following: the classical trajectory between two fixed points $x_1(t_1)$ and $x_2(t_2)$ renders the integral

$$(7.202) \qquad S = \int_{t_1}^{t_2} L[x(t), \dot{x}(t)] \, dt$$

a minimum. Here we have written

$$(7.203) \qquad L(x, \dot{x}) \equiv T(x, \dot{x}) - V(x)$$

for the *Lagrangian*, where T represents kinetic energy and V potential energy and a dot represents differentiation with respect to time. (Whereas these expressions are written in one dimension, they are easily generalized to three dimensions.)

As noted above, Hamilton's principle states that of all possible paths between the fixed points x_1 and x_2, the path which minimizes the integral (7.202) is the actual physical path between these two points. Thus Hamilton's principle may be restated as follows: the physical path between the points x_1 and x_2 renders the integral (7.202) stationary. That is,

$$(7.204) \qquad \delta \int_{t_1}^{t_2} L(x, \dot{x}) \, dt = 0$$

where δ represents an arbitrary, infinitesimal variation about the true motion of the system. The variable S in (7.202) is called the *action*, and the rule (7.204) is alternatively called Hamilton's principle or the *principle of least action*.

Relation to the Hamiltonian

The relation of the Lagrangian (7.203) to the classical Hamiltonian (1.13) (for the present one-dimensional configuration) is given by

$$(7.205) \qquad H(x, p) = \dot{x} \frac{\partial L(x, \dot{x})}{\partial \dot{x}} - L(x, \dot{x})$$

Here we have written

$$(7.205a) \qquad p \equiv \frac{\partial L(x, \dot{x})}{\partial \dot{x}}$$

for the momentum *conjugate* to x. Employing the relation (7.205) in (7.202) and following a variational calculation similar to that described in Problem (4.28) leads to Hamilton's equations (1.35).[1] Thus Hamilton's principle (7.204) is an alternative description of classical mechanics.

Minimum Action

Let us examine the meaning of Hamilton's principle by way of a specific case. Thus, for example, consider the harmonic oscillator described previously in Section 7.2. Recalling (7.203) the corresponding Lagrangian is given by

$$(7.206) \qquad L = \tfrac{1}{2}m\dot{x}^2 - \tfrac{1}{2}Kx^2$$

the integral (7.202) then becomes

$$(7.207) \qquad S = \int_{t_1}^{t_2} \tfrac{1}{2}(m\dot{x}^2 - Kx^2)\,dt$$

Consider the specific case that at $t = 0$ the particle is at $x = x_0$ with energy E. The motion for this problem is then given by

$$x(t) = x_0 \cos \omega t$$

$$(7.208)$$

$$E = \tfrac{1}{2}Kx_0^2, \qquad \omega^2 = \frac{K}{m}$$

Substituting these values into (7.207) with $t_1 = 0$ and t_2 related to time t, we find

$$(7.209) \qquad S(t) = -\frac{E}{2\omega}\sin 2\omega t$$

Now in what sense is this value of the action minimum? To answer this question, we consider the varied motion

$$(7.210) \qquad x(t) = x_0 \cos \omega t + \varepsilon^2 x_0 \sin \frac{\omega t}{\varepsilon}$$

where $\varepsilon \ll 1$. The preceding function is seen to represent a high-frequency, small-amplitude oscillation, which follows the unperturbed $\cos \omega t$ motion. See Fig. 7.40.

Constructing the Lagrangian for the motion (7.210), we find [keeping terms to $O(\varepsilon^2)$]

$$L = L_0 - 2\varepsilon E\left[\sin \omega t \cos \frac{\omega t}{\varepsilon} + \varepsilon \cos \omega t \sin \frac{\omega t}{\varepsilon} - \frac{\varepsilon}{2}\cos^2 \frac{\omega t}{\varepsilon}\right]$$

[1] See, for example, H. Goldstein, *Classical Mechanics*, 2nd ed., Addison-Wesley, Reading, Mass., 1980. Sec. 8.5.

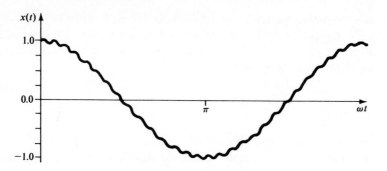

FIGURE 7.40 High-frequency, small-amplitude motion of (7.210).

where L_0 is the Lagrangian corresponding to the unperturbed motion (7.208). Integrating the preceding over time and passing to the limit $\varepsilon \to 0$, we see that contributions from the first two of the bracketed terms "wash out," whereas the third term yields

$$(7.211) \qquad S = S_0 + \frac{E\varepsilon^3}{2\omega}\left(\frac{\omega t}{\varepsilon} + \sin\frac{\omega t}{\varepsilon}\cos\frac{\omega t}{\varepsilon}\right) = S_0 + \tfrac{1}{2}\varepsilon^2 Et + O(\varepsilon^3)$$

where S_0 is the action (7.209). Thus to $O(\varepsilon^2)$ we may write

$$(7.212) \qquad S > S_0$$

We may conclude that the action S, (7.211), corresponding to the varied motion (7.210), is larger than S_0, (7.209), corresponding to the true motion (7.208), thereby corroborating Hamilton's principle.

Feynman Path Integral

In 1948, R. P. Feynman[1] presented a new formulation of quantum mechanics based in large part on the preceding classical concepts.[2]

Our discussion of this formalism begins with the solution of the initial-value problem (3.70) written in Dirac notation.

$$(7.213) \qquad |\psi(t)\rangle = e^{-it\hat{H}/\hbar}|\psi(0)\rangle$$

In coordinate representation (see Appendix A) this equation becomes

$$(7.214) \qquad \langle x|\psi(t)\rangle = \int \langle x|e^{-it\hat{H}/\hbar}|x'\rangle\langle x'|\psi(0)\rangle \, dx'$$

[1] R. P. Feynman, *Rev. Mod. Phys.* **20**, 367 (1948).
[2] For further discussion, see R. P. Feynman and A. R. Hibbs, *Quantum Mechanics and Path Integrals,* McGraw-Hill, New York, 1965.

Note that in this expression $\hat{H} = \hat{H}(\hat{x}, \hat{p})$, whereas in (3.70) $\hat{H} = \hat{H}(x, -i\hbar\partial/\partial x)$. (See Problem 7.86.) We may write (7.214) in its equivalent form

$$(7.215) \qquad \psi(x, t) = \int \langle x|e^{-it\hat{H}/\hbar}|x'\rangle\psi(x',0)\,dx'$$

In the Feynman description one assumes the following form for the propagation term in (7.215):

$$(7.216) \qquad \langle x|e^{-it\hat{H}/\hbar}|x'\rangle = \int_{\substack{all \\ paths}} \exp\left[\frac{i}{\hbar}\int_{t'}^{t} L(x, \dot{x})\,dt\right]\mathcal{D}[x', x]$$

The integration on the right represents a sum of the exponential over all paths $x = x(t)$ between x' and x and $\mathcal{D}[x', x]$ is a differential-like measure related to the sum over paths. Furthermore, $L(x, \dot{x})$ is the classical Lagrangian. Over each path (7.216), $L(x, \dot{x})$ is evaluated from the given $x = x(t)$. Substituting (7.216) into (7.215) gives

$$(7.217) \qquad \psi(x, t) = \iint_{\substack{all \\ paths}} e^{iS/\hbar}\psi(x', 0)\,dx'\mathcal{D}[x', x]$$

which is Feynman's solution to the initial-value problem, where S is the action (7.202). Consider that it is known that the particle is at x_0 at $t = 0$. Then

$$\psi(x', 0) = \delta(x' - x_0)$$

and (7.217) reduces to

$$(7.218) \qquad \psi(x, t) = \int_{\substack{all \\ paths}} \left[\exp\frac{i}{\hbar}\int_{0}^{t} L(x, \dot{x})\,dt\right]\mathcal{D}[x, \dot{x}]$$

This expression represents an extension of the WKB approximation (7.167) to the full quantum domain.

Domains of Large and Small Contribution to the Wavefunction

In the integration over all paths in (7.218), most domains of integration do not contribute for the following reason. Away from the region where action is stationary, a small change in path causes a large change in action compared with $\hbar$. The corresponding rapid fluctuation in the exponential in (7.218) causes cancellation, thereby diminishing contribution to the wavefunction. On the other hand, for paths near the classical orbit, the action is stationary and the exponential term likewise does not vary, resulting in a net contribution to the wavefunction.

We may conclude that the main contribution to the wavefunction (7.218) stems from paths lying near the classical orbit. Thus, to lowest approximation, (7.218) returns the quasiclassical WKB expression (7.167):

(7.219)
$$\psi(x) = A e^{(i/\hbar)S(x)}$$

In this expression the constant x_0 in (7.218) was absorbed in $S(x)$. Furthermore, the action $S(x)$ in (7.219) is evaluated on the classical orbit.

The reader will note that there is a difference between the action integral in (7.204) and that given by the result (7.172). The latter expression may be rewritten

Now note that
$$S_0(x) = \pm \int_0^t 2T\, dt$$

$$L = T - V = 2T - E$$

Thus, with the constraint $\delta E = 0$ in the variation (7.204), the principle of least action becomes

$$\delta \int 2T\, dt = 0$$

This relation contains the action relevant to (7.172).

PROBLEMS

7.78   The *scattering amplitude S*, in one dimension, may be defined by the equation

$$\psi_{\text{trans}} = S\psi_{\text{inc}}$$

For transmission above a potential well, one finds that the bound-state energies of the well are given by the negative real zeros of $S^{-1}(E)$.

(a)   What is the transmission coefficient T in terms of S corresponding to the configuration shown in Fig. 7.14? [See Eqs. (7.112).]

(b)   Consider the transmission past the rectangular potential described in Section 7.8 (See Fig. 7.27.) Construct an expression for S for this configuration.

(c)   Using your answer to (b) and the rule stated above, obtain the energy eigenvalues of the well (as discussed in Section 8.1).

Answers

(a)
$$T = \frac{k_2}{k_1} |S|^2$$

In the event that $V(+\infty) = V(-\infty)$, then

$$T = |S|^2$$

(b)
$$S = \frac{F}{A} = \frac{e^{2ik_2a}}{\cos 2k_2a - (i/2)[(k_1^2 + k_2^2)/k_1k_2] \sin 2k_2a}$$

(c) Setting $k_1^2 \to -\kappa^2 = -2m|E|/\hbar^2$, and $k_2^2 \to k^2$, we find that the poles of S occur at

$$\tan 2ka = \frac{2\kappa k}{k^2 - \kappa^2}, \qquad (ak)^2 + (a\kappa)^2 = \frac{2ma^2V}{\hbar^2}$$

which are in the desired form (8.64).

7.79 (a) Do the two integral expressions given in the connecting formula (7.188a) join smoothly at $x = x_1$?

(b) Explain your answer.

Answers

(a) They do not join smoothly.

(b) These two expressions represent asymptotic forms of the Airy function on either side of the turning point $x = x_1$. The (approximate) wavefunction at the turning point is the continuation of the Airy function from these asymptotic values.

7.80 Electrons in a beam have the wavefunction

$$\varphi_{\text{inc}}(x) = f(x)e^{ik_1x}$$

The beam passes through a potential barrier and electrons emerge with the wavefunction

$$\varphi_{\text{trans}} = g(x)e^{ik_2x}$$

In these expressions $f(x)$ and $g(x)$ are real functions and k_1 and k_2 are real constants. Show that

$$T = \frac{k_2}{k_1} \left| \frac{\varphi_{\text{trans}}}{\varphi_{\text{inc}}} \right|^2 = \frac{k_2}{k_1} \left| \frac{g(x)}{f(x)} \right|^2$$

7.81 A particle of mass m_0 confined to a potential well of width L_0 and depth V_0 is known to have $N_0 \gg 1$ bound states. How many bound states N does the well have if:

(a) $L = 2L_0$ and $V = 25V_0$?

(b) $L = L_0/2$ and $V = 4V_0$?

(c) $m = 0.10m_0$, $V = 30V_0$, and $L = 0.577L_0$?

7.82 In physics, one requires that fundamental equations of motion be reversible in time. In quantum mechanics, the operation of *time reversibility* is given by $\psi(\mathbf{r}, t) \to \psi^*(\mathbf{r}, -t)$. Show that the quantum continuity equation (7.97) [with ρ given by (7.95) and $\mathbf{J}$ given by (7.107)] remains invariant under this operation.

7.83 Consider the plane wave

$$\psi(x, t) = A \exp i(kx - \omega t)$$

In the operation of time reversibility, what is the effect of the successive operations:

(a) $t \to -t$?

(b) $\psi \to \psi^*$?

Answer

(a)   This component of the transformation causes the wave to move in the reverse direction.

(b)   This component of the transformation causes momentum $\hbar k$ to reverse direction. Both of these properties come into play in classical dynamic reversibility.

7.84   (a)   Consider a particle of mass m in the complex potential field

$$\Phi = V(\mathbf{r}) + \frac{i\hbar}{2}\,\omega(\mathbf{r})$$

where $V(\mathbf{r})$ and $\omega(\mathbf{r})$ are real functions. What form does the continuity equation (7.97) assume for this potential?

(b)   Offer an interpretation for the field $\omega(\mathbf{r})$.

(c)   Is the new continuity equation found in (a) time-reversible?

Answer

(a)   Repeating steps leading to (7.105), we obtain

$$\frac{\partial \rho}{\partial t} + \boldsymbol{\nabla} \cdot \mathbf{J} = \omega\rho$$

(b)   The complex potential implies a source of particles (in violation of conservation of matter).

(c)   The equation does not obey time reversibility.

7.85   (a)   Show that the Gaussian wave packet given in Problem 7.71 satisfies the eigenvalue equation

$$\hat{a}\psi(x) = c\psi(x)$$

(b)   Under what conditions will the eigenvalue c be zero?

(c)   If conditions of (b) are satisfied, what function does the Gaussian wave packet reduce to?

7.86   Given that [see (A.7)]

$$\langle x|\hat{p}|x'\rangle = -i\hbar\,\frac{\partial}{\partial x}\,\delta(x - x')$$

show that for free-particle motion

$$\langle x|e^{-it\hat{H}(\hat{x},\,\hat{p})/\hbar}|x'\rangle = \left[\exp-\frac{it}{\hbar}\,H\!\left(x,\,-i\hbar\,\frac{\partial}{\partial x}\right)\right]\delta(x - x')$$

Answer

For free-particle motion we label

$$\frac{it}{\hbar}\,H(\hat{x},\,\hat{p}) \equiv i\alpha\hat{p}^2$$

We may then write

$$\langle x|e^{-i\alpha\hat{p}^2}|x'\rangle = \langle x|[1 - i\alpha\hat{p}^2 + \tfrac{1}{2}(-i\alpha\hat{p}^2)^2] + \cdots|x'\rangle$$

For the second term in the sum we obtain

$$\langle x|\hat{p}^2|x'\rangle = \int dx'' \langle x|\hat{p}|x''\rangle\langle x''|\hat{p}|x'\rangle$$

$$= \int dx'' \left[-i\hbar\frac{\partial}{\partial t}\delta(x - x'')\right]\left[-i\hbar\frac{\partial}{\partial x''}\delta(x'' - x')\right]$$

$$= -i\hbar\frac{\partial}{\partial x}\int dx''\,\delta(x - x'')\left[-i\hbar\frac{\partial}{\partial x''}\delta(x'' - x')\right]$$

$$= \left(-i\hbar\frac{\partial}{\partial x}\right)^2\delta(x - x')$$

Following in this manner we obtain

$$\langle x|e^{-i\alpha\hat{p}^2}|x'\rangle = \exp\left[-i\alpha\left(-i\hbar\frac{\partial}{\partial x}\right)^2\right]\delta(x - x')$$

which was to be shown.

7.87   (a)   What is the value of the classical action $S_0(x_0, x)$ corresponding to free-particle motion of a particle of mass m which is at x_0 at time $t = 0$ and x at $t > 0$?

(b)   Introduce an infinitesimal perturbation about the motion found in (a) and find the new action $S(x_0, x)$. Following the steps (7.210 et seq.), show that $S > S_0$.

(c)   What is the lowest-order wavefunction, $\psi(x, t)$, corresponding to the classical limit for this problem?

7.88   (a)   Do the action functions $S_0, S_1, \ldots$ in the expansion (7.169) have the same dimensions?

(b)   If not, what are their dimensions?

Answers

(a)   They are not the same.

(b)   From (7.167) we see that the dimensions of $S(x)$ are those of $\hbar$. Thus, from the series (7.169) we conclude that $[\hbar^n S_n] = \hbar$ or, equivalently, $[S_n] = \hbar^{1-n}$.

7.89   In developing the WKB expression for the transmission coefficient (7.201), it was assumed that $r \gg 1$, where r is given by (7.196a). Show that this assumption is valid within this approximation.

Answer

For slowly varying $V(x)$ of a potential barrier, one obtains $r \simeq (2a/\hbar)\sqrt{2m(V - E)}$, which, in the limit $V \gg E$, grows large. In this same limit $T \to \bar{0}$ [see (7.201)], which corresponds to the classical value. This, we recall, is the domain of validity of the WKB expansion.

7.90 An electron beam is incident on a rectangular barrier of width 5 Å and height 34.8 eV. The electrons have energy 2.9 eV. If the incident electron current is 10^{-2} cm^{-2} s^{-1}, what is the reflected current, within the WKB approximation? Is this a good approximation for the problem at hand? Why?

7.91 Again consider the configuration of Problem 7.62. However, now let the external potential be given by the following parabolic form:

$$V(x) = \left(\frac{E_F + \Phi}{x_0^{\,2}}\right)(x_0 - x)^2, \qquad 0 \le x \le x_0$$

$$V(x) = 0, \qquad\qquad\qquad\qquad x_0 \le x$$

(a) Draw the appropriate figure for this problem. Identify the Fermi energy.

(b) Calculate the transmission coefficient for emission through this potential barrier in the WKB approximation. Call your answer T_a. How does T_a compare to the transmission coefficient corresponding to the linear potential of Problem 7.26?

CHAPTER **8**

FINITE POTENTIAL WELL, PERIODIC LATTICE, AND SOME SIMPLE PROBLEMS WITH TWO DEGREES OF FREEDOM

In this chapter we meet perhaps the most eminently successful application of quantum mechanics to a one-dimensional configuration. This is the problem of a charged particle in a periodic potential. When coupled with the exclusion principle for electrons, the analysis of this configuration provides a deep understanding of the process of conduction in solids. Some elementary problems in two dimensions are given, together with a discussion of degeneracy in quantum mechanics. The chapter continues with an approximation technique important to molecular and solid-state physics which carries the acronym LCAO. A concluding section describes density of states in various dimensions.

8.1   THE FINITE POTENTIAL WELL

Eigenstates

Scattering from a rectangular potential well was discussed previously in Section 7.8. The configuration is depicted again in Fig. 8.1. The scattering, unbound states

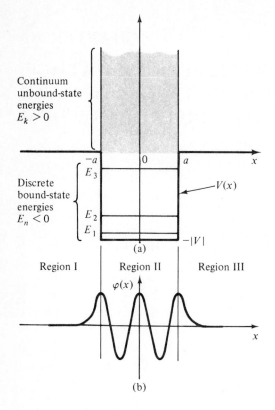

Continuum
unbound-state
energies
$E_k > 0$

Discrete
bound-state
energies
$E_n < 0$

E_3

E_2

E_1

$-V(x)$

$-|V|$

(a)

Region I   Region II   Region III

$\varphi(x)$

(b)

FIGURE 8.1 Finite rectangular potential well. (a) The potential function $V(x)$ and energy spectrum. (b) Typical structure of a bound eigenstate. Function oscillates in region II where kinetic energy is positive and decays in regions I and III, where kinetic energy is negative.

correspond to a continuum of eigenenergies [e.g., (7.129)]:

$$E_k = \frac{\hbar^2 k^2}{2m}, \qquad E_k > 0$$

If we seek solutions to the Schrödinger equation for negative energies, $E < 0$, only a finite, discrete number of eigenstates are found. For the three regions depicted in Fig. 8.1, the Schrödinger equation and corresponding solutions are (for $|E| < |V|$, $E < 0$, $V < 0$):

Region I: $x < -a$

$$-\frac{\hbar^2}{2m}\varphi_{xx} = -|E|\varphi, \qquad \varphi_{xx} = \kappa^2 \varphi$$

(8.1)

$$\varphi_I = A e^{\kappa x}, \qquad \frac{\hbar^2 \kappa^2}{2m} = |E| > 0$$

Region II: $-a \le x \le a$

$$-\frac{\hbar^2}{2m}\varphi_{xx} = (|V| - |E|)\varphi, \qquad \varphi_{xx} = -k^2\varphi$$

(8.2)

$$\varphi_{\mathrm{II}} = Be^{ikx} + Ce^{-ikx}, \qquad \frac{\hbar^2 k^2}{2m} = |V| - |E| > 0$$

Region III: $x > a$

$$-\frac{\hbar^2}{2m}\varphi_{xx} = -|E|\varphi, \qquad \varphi_{xx} = \kappa^2\varphi$$

(8.3)

$$\varphi_{\mathrm{III}} = De^{-\kappa x}, \qquad \frac{\hbar^2 \kappa^2}{2m} = |E| > 0$$

First we note that k and κ obey the constraint

(8.4)
$$k^2 + \kappa^2 = \frac{2m|V|}{\hbar^2}$$

The coefficients A, B, C, and D determine the eigenstate corresponding to the eigen-energy $\hbar^2\kappa^2/2m$. These coefficients are determined by the continuity conditions at $x = a$, $x = -a$. Equating φ and its first derivative at these points gives

$$Ae^{-\kappa a} = Be^{-ika} + Ce^{ika}$$

$$Be^{ika} + Ce^{-ika} = De^{-\kappa a}$$

(8.5)
$$\kappa Ae^{-\kappa a} = ik(Be^{-ika} - Ce^{ika})$$

$$ik(Be^{ika} - Ce^{-ika}) = -\kappa De^{-\kappa a}$$

These are four linear, homogeneous equations for the four unknowns A, B, C, and D. They may be cast in the matrix form (where the right-hand side denotes the null column vector)

(8.6)
$$\mathscr{D}\mathscr{V} \equiv \begin{vmatrix} e^{-\kappa a} & -e^{-ika} & -e^{ika} & 0 \\ 0 & e^{ika} & e^{-ika} & -e^{-\kappa a} \\ \kappa e^{-\kappa a} & -ike^{-ika} & ike^{ika} & 0 \\ 0 & ike^{ika} & -ike^{-ika} & \kappa e^{-\kappa a} \end{vmatrix} \begin{pmatrix} A \\ B \\ C \\ D \end{pmatrix} = 0$$

which serves to define the coefficient matrix $\mathscr{D}$ and the column vector $\mathscr{V}$. Cramer's rule tells us that this system has nontrivial solutions (i.e., other than $A = B = C = D = 0$) only if the determinant of the coefficient matrix vanishes.

(8.7)
$$\det \mathscr{D} = 0$$

After a little manipulation (8.6) is rewritten

(8.8)
$$
\begin{vmatrix}
G^* & G & 0 & 0 \\
G & G^* & 0 & 0 \\
e^{-ika} & -e^{ika} & -\dfrac{\kappa}{ik}e^{-\kappa a} & 0 \\
-e^{ika} & e^{-ika} & 0 & -\dfrac{\kappa}{ik}e^{-\kappa a}
\end{vmatrix}
\begin{pmatrix}
B \\ C \\ A \\ D
\end{pmatrix} = 0
$$

where

(8.9)
$$
G \equiv (\kappa + ik)e^{ika}
$$

(Note the rearrangement of the column vector $\mathscr{V}$.) Expanding about the fourth column, one obtains

(8.10) $\det \mathscr{D} = \det$
$$
\begin{vmatrix}
G^* & G & 0 & 0 \\
G & G^* & 0 & 0 \\
0 & 0 & \dfrac{\kappa}{ik}e^{-\kappa a} & 0 \\
0 & 0 & 0 & -\dfrac{\kappa}{ik}e^{-\kappa a}
\end{vmatrix}
= [G^2 - (G^*)^2]\left(\frac{\kappa}{ik}\right)^2 e^{-2\kappa a}
$$

This is zero when

(8.11)
$$
G^2 = (G^*)^2
$$

or, equivalently, when

(8.12)
$$
G = \pm G^*
$$

Rewriting (8.9) as

$$
G \equiv (\kappa + ik)e^{ika} \equiv |G|e^{i(ka + \phi)}
$$

(8.13)
$$
\tan \phi = \frac{k}{\kappa}
$$

allows the conditions (8.12) to be recast in the form

(8.14)
$$
e^{i(ka + \phi)} = \pm e^{-i(ka + \phi)}
$$

The positive root gives $ka + \phi = 0$, or, equivalently,

(8.15)
$$\tan \phi = \frac{k}{\kappa} = -\tan ka$$

This may be put in the more normal form

(8.16)
$$k \cot ka = -\kappa, \qquad \frac{G}{G*} = 1$$

The negative root gives $ka + \phi = \pi/2$ or, equivalently,

(8.17)
$$\tan \phi = \frac{k}{\kappa} = \tan \left(\frac{\pi}{2} - ka\right) = \cot ka$$

This may also be put in the more normal form

(8.18)
$$k \tan ka = \kappa, \qquad \frac{G}{G*} = -1$$

The values of k that make $\det \mathscr{D} = 0$ fall into two categories. These are the solutions to (8.16) and (8.18), respectively. From our starting matrix equation (8.8) we see that these values of k imply the relations

(8.19)
$$\frac{B}{C} = -\frac{G}{G*} = \pm 1$$

The minus sign corresponds to the roots (8.16). Substituting this value $(B = -C)$ into the last two equations of the set (8.5) gives

(8.20)
$$\frac{C}{B} = -1, \qquad \frac{A}{B} = -\frac{D}{B} = -2i \sin (ka)e^{\kappa a}$$

Substituting these values into (8.2) et seq. gives the eigenstate

(8.21)
$$\left.\begin{aligned} \varphi_{\text{I}} &= -2iB \sin (ka)e^{\kappa(x+a)} \\ \varphi_{\text{II}} &= 2iB \sin kx \\ \varphi_{\text{III}} &= 2iB \sin (ka)e^{-\kappa(x-a)} \end{aligned}\right\} k \cot ka = -\kappa$$

This state has *odd parity*; that is,

(8.22)
$$\varphi(x) = -\varphi(-x)$$

The second class of solutions corresponds to the plus sign in (8.19) and stems from the roots (8.18). Substituting this value $(B = +C)$ into the last two equations of the set (8.5) gives

(8.23)
$$\frac{C}{B} = +1, \qquad \frac{A}{B} = \frac{D}{B} = 2 \cos (ka)e^{\kappa a}$$

The corresponding eigenstate is

(8.24)
$$\left.\begin{array}{l} \varphi_{\mathrm{I}} = 2B \cos (ka)e^{\kappa(x+a)} \\ \varphi_{\mathrm{II}} = 2B \cos kx \\ \varphi_{\mathrm{III}} = 2B \cos (ka)e^{-\kappa(x-a)} \end{array}\right\} k \tan ka = \kappa$$

This state has *even parity*.

Since both eigenstates (8.21) and (8.24) are bound states, we may impose the normalization condition

(8.25)
$$\int_{-\infty}^{\infty} |\varphi|^2 \, dx = 1$$

This determines the remaining constant B.

Next we turn to construction of the eigenenergies corresponding to the eigenstates (8.21) and (8.24). The energy is directly determined from κ. For the even eigenstates, the eigenenergies are determined from (8.4) and (8.24). Written in terms of nondimensional wavenumbers,

(8.26)
$$\xi = ka, \qquad \eta = \kappa a$$

these equations appear as

(8.27)
$$\begin{array}{ll} \xi \tan \xi = \eta & \\ \xi^2 + \eta^2 = \dfrac{2ma^2|V|}{\hbar^2} \equiv \dfrac{g^2}{4} \equiv \rho^2 & \textit{even eigenstates} \end{array}$$

For a given potential width $2a$, depth $|V|$, and particle mass m, (8.27) describes a circle of radius ρ, in Cartesian $\xi\eta$ space. The intersections of this circle (in the first quadrant) with the graph of the first equation of (8.27) determine the eigenenergies corresponding to the even eigenstates (8.24). This graphical technique is sketched in Fig. 8.2 for the case ρ slightly less than π. The sketch tells us that for this value of ρ, the finite potential well has only one bound even eigenstate.

The eigenenergies of the odd eigenstates (8.22) are the intersections of the two curves

(8.28)
$$\begin{array}{ll} \xi \cot \xi = -\eta & \\ \xi^2 + \eta^2 = \rho^2 & \textit{odd eigenstates} \end{array}$$

These curves are sketched in Fig. 8.3 for the case ρ slightly less than π. For this choice of data, we see that there is only one bound odd eigenstate. These two lowest-energy eigenstates, (8.21) and (8.24), are sketched in Fig. 8.4.

At this point we wish to consider again the difference between the unbound scattering states of Chapter 7 and the bound states just encountered. The continuity conditions on the wavefunction φ and its derivative, together with the statement of

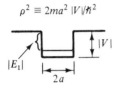

$E_1 = (\eta_1/\rho)^2 V$

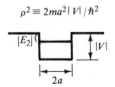

$\rho^2 \equiv 2ma^2 |V|/\hbar^2$

FIGURE 8.2 The curves $\eta = \xi \tan \xi$ and the circle $\xi^2 + \eta^2 = \rho^2$ for the case ρ slightly less than π. Intersections in the first quadrant give bound-state eigenenergies for the potential well Hamiltonian which correspond to even eigenstates.

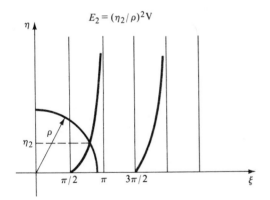

$E_2 = (\eta_2/\rho)^2 V$

$\rho^2 \equiv 2ma^2 |V|/\hbar^2$

FIGURE 8.3 The curves $\eta = -\xi \cot \xi$ and the circle $\xi^2 + \eta^2 = \rho^2$ again for the case ρ slightly less than π. Intersections in the first quadrant give bound-state eigenenergies for the potential well Hamiltonian which correspond to odd eigenstates. Note that E_2 lies higher than the ground state E_1.

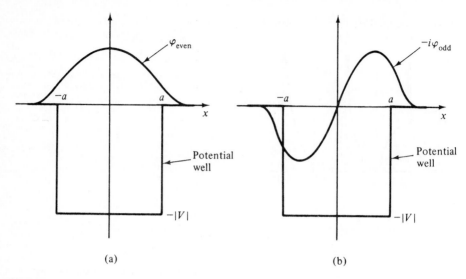

FIGURE 8.4   First two bound eigenstates for the potential well problem. For $2ma^2|V|/\hbar^2 < \pi^2$, these are the only bound states.

conservation of energy, determine eigenenergies and eigenstates. For scattering states, the continuity conditions (7.139) are in the form of an *inhomogeneous* matrix equation,

$$(8.29) \qquad \mathcal{D}(k_1)\mathcal{V} = \mathcal{U}$$

where, for example, the column vector $\mathcal{V}$ is

$$(8.30) \qquad \mathcal{V} = \begin{pmatrix} B/A \\ C/A \\ D/A \\ F/A \end{pmatrix}$$

The solution to (8.29) is

$$(8.31) \qquad \mathcal{V} = \mathcal{D}^{-1}(k_1)\mathcal{U}$$

For these unbound scattering states, conservation of energy serves only to relate wavenumbers connected to distinct potential domains. The eigen-k_1-values comprise a continuum. For each such k_1 value, there corresponds an eigenstate of the form (7.138).

For bound states the continuity conditions (8.5) are in the form of a *homogeneous* matrix equation (8.6),

$$\mathcal{D}(\kappa)\mathcal{V} = 0$$

which has nontrivial solutions ($\mathscr{V} \neq 0$) only if

(8.32) $$\det \mathscr{D}(\kappa) = 0$$

This dispersion relation restricts the eigen-κ-values to values that obey certain transcendental relations [the first equation in (8.27) and in (8.28)]. In addition, κ is further restricted by the conservation-of-energy statement, namely the second equation in (8.27). The intersections of this circle (depicted in Figs. 8.2 and 8.3) with the said transcendental curves generate a discrete spectrum of eigenenergies

$$E_n = -\frac{\hbar^2 \kappa_n^2}{2m}$$

Let us consider the time dependence of the eigenstates corresponding to the finite potential well. The bound time-dependent eigenstates appear as

(8.33) $$\psi_n(x, t) = \varphi_n(x)e^{-iE_n t/\hbar}, \qquad E_n < 0$$

with $\varphi_n(x)$ given by (8.1) et seq. For positive energy, the unbound time-dependent eigenstates form a continuum,

(8.34) $$\psi_{k_1}(x, t) = \varphi_{k_1}(x)e^{-iE_{k_1} t/\hbar}, \qquad E_{k_1} > 0$$

where $\varphi_{k_1}(x)$, for example, is of the form (7.138) with the modification $V \rightarrow -|V|$.

To employ the superposition principle in problems relating to the finite potential well, one must call on the finite number of bound states and infinite continuum of unbound states.[1]

The $E = 0$ Line

As stated above, energies relevant to the finite potential well are directly obtained from κ or, equivalently, η:

$$|E| = \frac{\hbar^2 \kappa^2}{2m} = \frac{\hbar^2 \eta^2}{2ma^2}$$

These energies are measured with respect to the top of the well taken as the $E = 0$ line. It is sometimes convenient to measure energies with respect to the bottom of the well as the zero energy line (as was the case, for example, for the infinitely deep potential

---

[1] Note the continuum of unbound states developed in this chapter excludes states with negative k in region III. The states discussed are appropriate to the superposition of a wave packet incident on a potential barrier from the left. For the superposition of a state with zero average momentum, one must include the negative k waves in region III.

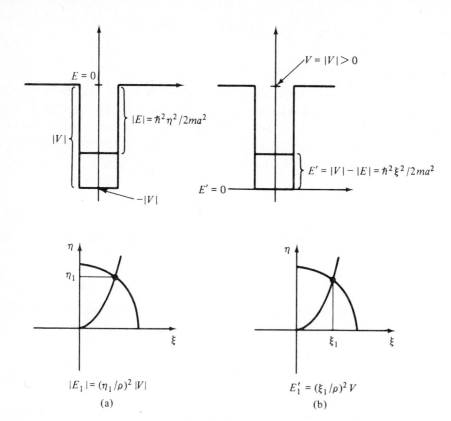

FIGURE 8.5   Relative orientations of bound-state energies for the finite one-dimensional well.

well treated in Chapter 4). The energies, E', measured with respect to the bottom of the well are directly obtained from k or, equivalently, ξ:

$$E' = |V| - |E| = \frac{\hbar^2 k^2}{2m} = \frac{\hbar^2 \xi^2}{2ma^2} > 0$$

See (8.2) and Fig. 8.5.

In practical use, the potential well shown in Fig. 8.5a, with negative-energy bound states, may be employed in one-dimensional modeling of one-electron atoms, for which bound states have negative energy and unbound states (free electrons) have positive energy. (See Chapter 10.) On the other hand, the potential well shown in Fig. 8.5b comes into play, for example, in the modeling of quantum wells in superlattice structures where the $E = 0$ line might, for instance, be set equal to the minimum energy of the conduction band. (See Sections 8.8 and 12.9.)

PROBLEMS

8.1 A deuteron, which is a neutron and a proton bound together, has only one bound state. Assume that the potential of interaction between the two particles may be described as a square well. The effective mass of the system is 0.84×10^{-24} g. The range of nuclear force is approximately 2.3×10^{-13} cm, while the ground state of the deuteron is 2.23 MeV below the zero-energy free-particle state. Assuming that only the odd-parity solutions are permitted for this case, estimate the depth of the potential well, $|V|$, which you may take to be large compared to the binding energy of the system.

8.2 An electron trapped in a potential well 10^{-9} cm wide can be in at most three bound states. The binding energy of the highest state is a factor 10^{-8} smaller than the energy $(\pi h \cdot 10^9)^2/2m$. Estimate the energy of the ground state.

8.3 Show that the graphical solutions of Figs. 8.2 and 8.3 give the eigenenergies of a one-dimensional box, in the limit that the well becomes infinitely deep.

Answer

In the said limit $\rho \to \infty$, the circles of constant ρ cut the tan and cot curves on the vertical asymptotes

$$\xi = \frac{n\pi}{2} \qquad (n = 1, 2, \ldots)$$

(Compare Eq. 4.12.)

8.4 Given that

$$\frac{g^2}{4} = \frac{2ma^2|V|}{h^2} = \left(\frac{7\pi}{4}\right)^2$$

for an electron in a potential well of depth $|V|$ and width $2a = 10^{-7}$ cm, if a 100-keV neutron is scattered by such a system, calculate the possible decrements in energy that the neutron may suffer.

8.5 For the potential well described in Problem 8.4, what is the parity of the eigenstate of maximum energy? How many zeros does this state have?

8.6 Consider a rectangular potential well of depth $|V|$ and width $2a$, such that $2ma^2|V|/h^2 = (8\pi/18)^2$. The lowest-energy normalized bound state, φ_1, has wavenumber $k \simeq \pi/4a$. Let $\tilde{\varphi}$ be a wavefunction that is a square wave of height $1/\sqrt{4a}$ and width $4a$. The centers of $\tilde{\varphi}$ and the rectangular potential well are coincident. At $t = 0$ a particle of mass m is in the state

$$\psi(x, 0) = \frac{3\varphi_1 + 4\tilde{\varphi}}{5}$$

At time $t = 0$:
  (a) What is the expectation of momentum of the particle?
  (b) What is the expectation of energy?
  (c) What is the parity of the state?
  (d), (e), (f) Repeat parts (a), (b), and (c) for $t > 0$.

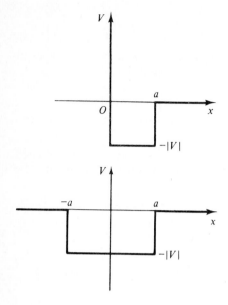

FIGURE 8.6   Semiinfinite potential well and its companion finite potential well. (See Problem 8.7.)

8.7   Consider the semiinfinite potential well

$$V(x) = \begin{cases} \infty, & x < 0 \\ -|V|, & 0 \le x \le a \\ 0, & a < x \end{cases}$$

(see Fig. 8.6).

(a)   Using the solutions to the finite potential well (width $2a$) developed in the text, sketch the first three eigenfunctions of lowest energy for a particle in this well.

(b)   Which ground-state energy is lower—that of the finite potential well (width $2a$) or that of the semiinfinite well (width a)?

(c)   Are the eigenfunctions you have sketched eigenstates of the Hamiltonian appropriate to the finite potential well?

8.8   An electron is trapped in a rectangular potential well of width 3 Å and depth 1 eV. What are the possible frequencies of emission of this system (in hertz)?

8.9   Establish the following criteria for the number of bound states in a finite potential well:

(a) $(n\pi)^2 < \rho^2 < (n+1)^2\pi^2$ ($n + 1$ symmetric states).

(b) $(n - \frac{1}{2})^2\pi^2 < \rho^2 < (n + \frac{1}{2})^2\pi^2$ (n antisymmetric states).

(c)   Total number of bound states = maximum integer (ρ/π).

8.2   PERIODIC LATTICE. ENERGY GAPS

In this section we consider the problem of a particle in a periodic potential. This is of extreme practical importance in the theory of conduction and insulation in solids.

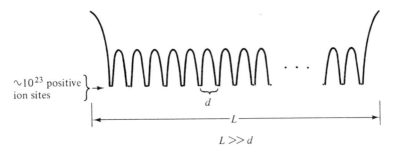

FIGURE 8.7   Periodic potential that an electron sees in a one-dimensional crystalline solid.

Consider the simple model of a solid (more precisely, a metal) in which the positive ions comprise a uniform array of fixed sites. The valence electrons are assumed to be free. They are the conduction electrons. For sodium, for instance, there is one free electron per ion. Each such electron finds itself in a periodic potential supported by the ions. Such a one-dimensional potential configuration is depicted in Fig. 8.7.

If the distance between sites is d, then inside the metal the potential is periodic in the distance d.

$$(8.35) \qquad V(x) = V(x + d)$$

A simple potential function that maintains this periodic quality and all the salient properties of the more realistic potential sketched in Fig. 8.7 is the *Kronig–Penney potential*, depicted in Fig. 8.8. The periodic property of $V(x)$ as given by (8.35) fails at

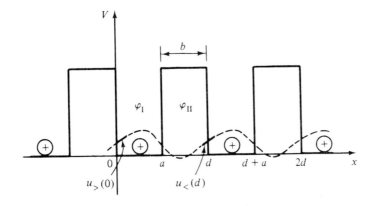

FIGURE 8.8   The Kronig–Penney model for a potential due to fixed ion sites separated by the distance d. The dashed curve represents a hypothetical periodic u component of the Bloch function $\varphi = u(x) \exp(ikx)$. The eigenfunction φ (8.48 et seq.) and corresponding dispersion relation, (8.53) and (8.55), are obtained by matching u and u' at $x = 0 + \varepsilon$ to their respective values at $x = d - \varepsilon$ and matching φ and φ' across the potential barrier at $x = a$.

the ends of the lattice. To remove this difficulty the model is further simplified. This simplification derives from the fact that there are an overwhelmingly large number of ion sites in the length of the sample. The change in the character of the potential at the ends of the sample is therefore relatively unimportant to the transport properties of an interior electron. For this reason we change the ends of the sample to best facilitate analysis. It is assumed that when an electron leaves the end of the sample, it reenters the front of the sample. This idea is best realized if the one-dimensional potential function is assumed to lie on a circle of radius r which is very large compared to the distance between ion sites, d (see Fig. 8.9). The Hamiltonian for an electron in this potential is

$$H = \frac{p^2}{2m} + V(x)$$

(8.36)

$$V(x) = V(x + d)$$

Bloch Wavefunctions

To find the eigenfunctions of this Hamiltonian, we first recall the displacement operator $\hat{\mathcal{D}}$, introduced in Problem 3.4:

(8.37)
$$\hat{\mathcal{D}}f(x) = f(x + d)$$

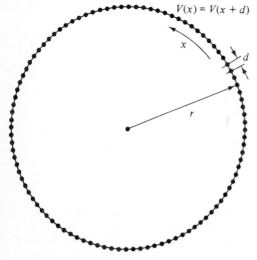

$V(x) = V(x + d)$

x

d

r

FIGURE 8.9   Ring model of a one-dimensional periodic potential. Black dots represent positive ion sites. For N sites in all, and $N \gg 1$, $Nd \simeq 2\pi r$.

The eigenfunctions of this operator are

(8.38)
$$\varphi = e^{ikx}u(x)$$
$$u(x) = u(x + d)$$

with k arbitrary. The eigenvalue of $\hat{\mathcal{D}}$ corresponding to φ is exp (ikd). Although both factors of φ, namely exp (ikx) and $u(x)$, are periodic, φ need not be. The eigenfunction $\varphi(x)$ is periodic if d, the period of u, is commensurate with $2\pi/k$, the period of exp (ikx): that is, if $2\pi/kd$ is a rational number.

Since $\hat{\mathcal{D}}$ commutes with $\hat{H}$

(8.39)
$$[\hat{\mathcal{D}}, \hat{H}] = 0$$

these two operators have common eigenfunctions. We conclude that the eigenfunctions of the Hamiltonian (8.36) are of the form (8.38). These functions are called *Bloch wavefunctions*. The related theorem that the eigenstates of a periodic Hamiltonian such as (8.36) are in the product form (8.38) is called *Bloch's theorem*.[1] We have obtained these functions using the displacement operator $\hat{\mathcal{D}}$. More simply, one may argue that on the average, the density of an electron beam propagating through a crystal with a periodic potential should exhibit the same periodicity as the crystal. That is, one expects that

$$|\varphi(x)|^2 = |\varphi(x + d)|^2$$

This equation admits the solutions

$$\varphi(x) = u(x) \exp [i\alpha(x)]$$

where, again, $u(x)$ is periodic with period d and $\alpha(x)$ is any real function independent of d. In the limit that the periodic potential becomes constant, $V = $ constant, $d \rightarrow \infty$, and the wavefunction $\varphi(x)$ becomes the free-particle wavefunction exp (ikx), with k arbitrary but real. Since $\alpha(x)$ is independent of the period length (or *lattice constant*) d, this value of α (i.e., kx) is its value for all d and we again obtain the Bloch wavefunction

$$\varphi(x) = e^{ikx}u(x)$$

The shape of this wavefunction suggests the manner in which the crystal structure influences the wavefunctions of particles propagating through the crystal. This structure is primarily contained in the periodic factor $u(x)$, which in turn includes the lattice constant d and which modulates the free-particle form, *exp* (ikx).

Another way of writing (8.38) is

(8.40)
$$\varphi(x + d) = e^{ikd}\varphi(x)$$
$$\varphi(x) = e^{ikd}\varphi(x - d)$$

[1] F. Bloch, *Z. Physik* **52** (1928).

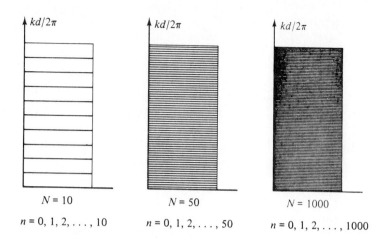

$N = 10$ $N = 50$ $N = 1000$

$n = 0, 1, 2, \ldots, 10$ $n = 0, 1, 2, \ldots, 50$ $n = 0, 1, 2, \ldots, 1000$

FIGURE 8.10   Permitted values of k for the periodic ring model depicted in Fig. 8.9. For $N \gg 1$ the spectrum of permitted k values approximates a continuum.

If the eigenstate φ is known over any cell in the periodic lattice (more generally over any interval of length d), equations (8.40) generate the values of φ in all other cells.

For any value of k, the corresponding function φ, given by (8.38), is an eigenstate of $\mathscr{D}$. When φ is also an eigenstate of $\hat{H}$, the values that k may assume become restricted. For example, the eigenstates of $\hat{H}$, with V defined over a ring, have the property

(8.41) $\varphi(x) = \varphi(x + Nd)$

Substitution into (8.38) gives

(8.42) $e^{ikNd} = 1, \qquad kNd = 2n\pi \qquad (n = 0, \pm 1, \pm 2, \ldots)$

This implies that the allowed values of k form a discrete spectrum $[k_n = n(2\pi/L)]$. However, since N is very large (e.g., $N \simeq 10^{23}$), the difference between successive values of k is very small and the spectrum of the permitted values of k may be taken to comprise a continuum (see Fig. 8.10). With k restricted to the values given by (8.42), the ratio $2\pi/kd = N/n$, a rational number. It follows that for the closed-ring periodic potential, the eigenfunctions of $\hat{H}$ in the Bloch waveform (8.38) are periodic.

The Quasi-momentum

The variable $\hbar k$ is called the *quasi-momentum* of the particle. We list four of the properties of $\hbar k$ which motivate this name.

1.   The eigenstates given in (8.38) resemble the form

(8.43) $\varphi_k = e^{ikx} \times \text{constant}$

This is the momentum eigenfunction of a free particle with momentum $\hbar k$. The momentum of an electron in a periodic lattice is, of course, not constant due to the lattice's space-dependent potential field. Nevertheless, there is a constant value of $\hbar k$ associated with every eigenenergy of the Hamiltonian (8.36).

2. The average velocity of a particle in an eigenstate of $\hat{H}$ is

(8.44)
$$\langle v \rangle = \frac{\partial E(k)}{\partial \hbar k}$$

Here[1] we have labeled the eigenenergy of the said eigenstate, $E(k)$. The relation above follows the classical recipe for obtaining the velocity of a free particle with energy E, provided that we associate $\hbar k$ with its momentum.

3. If a particle in a lattice is acted upon by an outside force $\mathbf{F}$, its acceleration is not $\mathbf{F}/m$, but $\mathbf{F}/m^*$. The "effective mass" m^* may be less than m, greater than m, negative, and even infinite. In one dimension m^* is given by

(8.45)
$$m^* = \frac{\hbar^2}{\partial^2 E/\partial k^2}$$

which is suggestive of the classical relation for a free particle $E = p^2/2m^*$, again with $p = \hbar k$.

4. Eigenenergies $E(k)$ are periodic in k with period $2\pi/d$, so

$$E(k) = E(k + 2\pi n/d),$$

where n is a positive or negative integer. The "central" $E(k)$ curve lies near the parabola $E = \hbar^2 k^2/2m$, which again suggests a free particle with momentum $\hbar k$.

Eigenstates

Next we turn to construction of the eigenstates and eigenenergies of the Kronig–Penney Hamiltonian. We know that eigenstates are in the Bloch form (8.38). The continuity conditions that apply to $\varphi(x)$ clearly apply also to the periodic component $u(x)$. It follows that $u(x)$ and $u'(x)$ must vary continuously from the right side of the point $x = 0$ to the left side of the point $x = d$, which is one periodic length displaced from $x = 0$ (see Fig. 8.8). With $u_>(0)$ denoting $u(x)$ evaluated at $x = 0 + \epsilon$, where ϵ is an infinitesimal, this condition on the periodic continuous quality of u and u' gives

---

[1] These properties of the quasi-momentum $\hbar k$ are derived in L. D. Landau and E. M. Lifshitz, *Quantum Mechanics*, 2nd ed., Addison-Wesley, Reading, Mass., 1965.

the two equations

(8.46a)
$$u_>(0) = u_<(d)$$

(8.46b)
$$u_>'(0) = u_<'(d)$$

Now

$$u = e^{-ikx}\varphi(x)$$

so that

$$u' = \varphi'e^{-ikx} - iku$$

and the continuity of u' (8.46b) across a periodic length becomes

(8.47)
$$\varphi_>'(0) = \varphi_<'(d)e^{-ikd}$$

In the well domain of the potential array

(8.48)
$$\varphi_I(x) = Ae^{ik_1x} + Be^{-ik_1x} \qquad (0 \le x \le a)$$
$$\frac{\hbar^2 k_1{}^2}{2m} = E$$

In the barrier domain (with $E > V$)

(8.49)
$$\varphi_{II}(x) = Ce^{ik_2x} + De^{-ik_2x} \qquad (a \le x \le a + b = d)$$
$$\frac{\hbar^2 k_2{}^2}{2m} = E - V$$

The continuity conditions (8.46, 8.47) on $u(x)$ then become

(8.50)
$$A + B = e^{-ikd}(Ce^{ik_2d} + De^{-ik_2d})$$
$$k_1(A - B) = k_2 e^{-ikd}(Ce^{ik_2d} - De^{-ik_2d})$$

The remaining two equations for the four coefficients (A, B, C, D) are obtained by invoking the continuity of $\varphi(x)$ and $\varphi'(x)$ across the potential barrier at $x = a$. This gives

(8.51)
$$Ae^{ik_1a} + Be^{-ik_1a} = Ce^{ik_2a} + De^{-ik_2a}$$
$$k_1(Ae^{ik_1a} - Be^{-ik_1a}) = k_2(Ce^{ik_2a} - De^{-ik_2a})$$

The latter four equations may be rewritten in the matrix notation

$$\begin{vmatrix} 1 & 1 & -e^{id(k_2-k)} & -e^{-id(k_2+k)} \\ k_1 & -k_1 & -k_2e^{id(k_2-k)} & k_2e^{-id(k_2+k)} \\ e^{ik_1a} & e^{-ik_1a} & -e^{ik_2a} & -e^{-ik_2a} \\ k_1e^{ik_1a} & -k_1e^{-ik_1a} & -k_2e^{ik_2a} & k_2e^{-ik_2a} \end{vmatrix} \begin{vmatrix} A \\ B \\ C \\ D \end{vmatrix} = 0$$

With $\mathscr{D}$ representing the above 4×4 coefficient matrix and $\mathscr{V}$ the four-column vector, the preceding equation may be written

$$\mathscr{D}(k, k_1, k_2)\mathscr{V} = 0$$

This homogeneous equation has nontrivial solutions only if

(8.52) $\det \mathscr{D} = 0$

This is the desired dispersion relation which is seen to involve the propagation constant k and the wavenumbers k_1 and k_2. The latter two variables contain the energy (8.48, 8.49), so for a given value of k, the dispersion relation (8.52) determines the eigenenergy E. As will be shown, this dispersion relation also exhibits the band-gap quality of the energy spectrum attendant to all periodic potentials. The dispersion relation (8.52) is similar to (8.7), which gives the eigenenergies for the bound states of the potential well problem. The states encountered in the present case may also be considered bound states, although the distinction is somewhat academic. The domain of existence of the wavefunctions of $\hat{H}$ is over the finite interval $0 \le x \le Nd$, which makes them normalizable. However, these eigenstates propagate throughout the crystal and in this sense carry the quality of an unbound state. Our main goal is to obtain the energies of these states.

From (8.52) one obtains the dispersion relation (after a bit of algebra)

$E > V$

(8.53a) $\cos k_1 a \cos k_2 b - \dfrac{k_1{}^2 + k_2{}^2}{2k_1 k_2} \sin k_1 a \sin k_2 b = \cos kd$

(8.53b) $k_1{}^2 - k_2{}^2 = \dfrac{2mV}{\hbar^2}$

The related formula for the case $E < V$ is simply obtained from the latter relation through the substitution

(8.54) $ik_2 \to \kappa, \qquad \dfrac{\hbar^2 \kappa^2}{2m} = V - E$

There results

$E < V$

(8.55a) $\cos k_1 a \cosh \kappa b - \dfrac{k_1{}^2 - \kappa^2}{2k_1 \kappa} \sin k_1 a \sinh \kappa b = \cos kd$

(8.55b) $k_1{}^2 + \kappa^2 = \dfrac{2mV}{\hbar^2}$

Equations 8.53 and 8.55 are implicit equations for the eigenenergies E as a function of the propagation constant k, valid for all energies. Owing to the transcendental nature

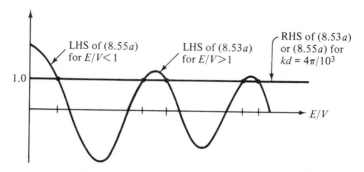

FIGURE 8.11 Graphical evaluation of eigenenergies of the Kronig–Penney Hamiltonian corresponding to $kd = 4\pi/10^3$. Eigenenergies are given by intersections of the horizontal line and the oscillating curve.

of these equations, one turns to a numerical technique for obtaining $E(k)$. For example, consider that $n = 2, N = 1000$. Then the right-hand side of (8.55a) is $\cos(4\pi/10^3) \simeq 1$. One then plots the left side of the same equation as a function of the dimensionless energy E/V. Superimposed on this same curve is the line RHS $= 1$ (Fig. 8.11). The values where these curves cross give the eigenenergies $E(k)$.

Energy Gaps

The fact that values of the right-hand sides of both (8.53a) and (8.55a) lie between $+1$ and -1 ($|\cos kd| \leq 1$) implies that the only solutions to these equations are values of E for which the left-hand sides of these respective equations fall in the same interval, that is, values of E for which

(8.56)　　　　$-1 \leq$ [left-hand sides of (8.53a) and (8.55a)] $\leq +1$

Values of E that violate this condition are excluded from the energy spectrum.

　　　The condition (8.56) gives rise to a "band" structure for the spectrum of eigenenergies. This is again well exhibited with a diagram. In Fig. 8.12, the left-hand sides of (8.53a) and (8.55a) are plotted versus E/V. On the same graph we draw the lines that represent the constant ordinates, $+1$ and -1. The values of E that qualify as eigenenergies are values for which the oscillating curve falls between the two horizontal lines, $+1$ and -1.

　　　This construction illustrates the band property of the energy spectrum of a particle in a periodic potential. This band feature is also illustrated in a plot of E versus k which may be inferred from the graph of Fig. 8.12. At the left of Fig. 8.12, values of $\cos kd$ are marked off. If a horizontal line is drawn from one of these values (e.g., $\cos kd = 1/\sqrt{2}$, $kd = \pm\pi/4$), the intersections of this line with the oscillating curve give all the energies that correspond to the propagation-constant values,

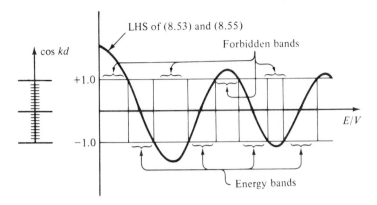

FIGURE 8.12 Band structure of the energy spectrum of the Kronig–Penney Hamiltonian. The only eigenenergies are those values for which the left-hand side of (8.53a) or (8.55a) falls between ±1.

$k = \pm \pi/4d$. There are infinitely many of them. Continuing this process for all values of kd gives the curve of E versus k sketched in Fig. 8.13a.

If we look at any single band, the curve E versus k is periodic in k. This results from the fact that the right-hand sides of (8.53a) and (8.55a) maintain the same values if kd is replaced by $(kd + n2\pi)$, where n is an integer. The value of E that satisfies this equation (i.e., either Eq. 8.53 or 8.55) for a given value of kd satisfies it for $(kd + n2\pi)$. It suffices then to draw all bands in the single interval $-\pi \leq kd \leq \pi$. This gives the *reduced-zone* description (Fig. 8.13b) of eigenstates. These bands consist of very closely packed discrete energies (recall Fig. 8.10) and constitute all the eigen-energies of the Hamiltonian (8.36). This discrete nature of the energy spectrum is a consequence of the boundedness of the system. The quasi-continuous quality (bands of closely packed levels) of the spectrum reflects the propagating nature of the eigenstates.

Superimposed on the E versus k curves in Fig. 8.13a are the free-particle energy curves

$$(8.57) \qquad E = \frac{\hbar^2(k + n2\pi d^{-1})^2}{2m} \qquad (n = \pm 0, 1, 2, \ldots)$$

This corresponds to a free-particle momentum

$$(8.58) \qquad p = \hbar(k + n2\pi d^{-1})$$

From Fig. 8.13 we see that (1) much of the locus of the E versus k curves falls near the free-particle energy curves, and (2) energy gaps occur at the values

$$(8.59) \qquad kd = q\pi$$

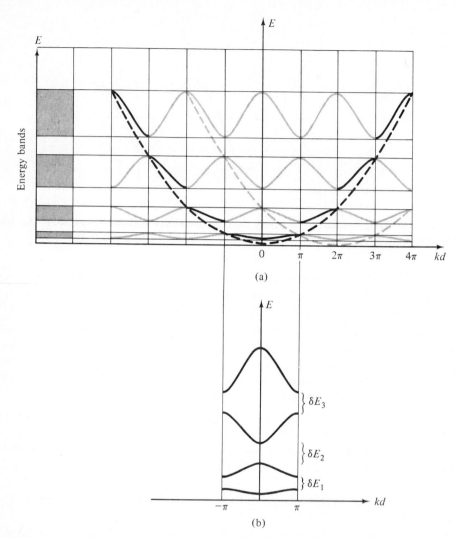

FIGURE 8.13 (a) Typical E versus k curves for the Kronig–Penney potential. The graininess of the curves stems from the fact that the k values in each band are discrete [i.e., $kd = (n/N)2\pi$, $N \gg 1$ or, equivalently, $(\Delta k)_{min} = \pi/L$]. (b) The first four bands in the reduced-zone scheme. Also shown are the first three energy gaps, δE_1, δE_2, and δE_3.

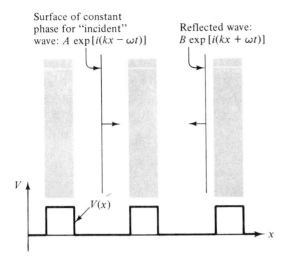

Surface of constant
phase for "incident"
wave: $A \exp[i(kx - \omega t)]$

Reflected wave:
$B \exp[i(kx + \omega t)]$

FIGURE 8.14 Shaded regions depict domains of constant potential. They are slabs that extend out of the paper. Surfaces of constant phase are also normal to the plane of the paper.

where q is a positive or negative integer. At these values of k an integral number of half-wavelengths span the distance d between ions.

Bragg Reflection

To understand the physical origin of energy gaps at these values of k, it is best to recall that the one-dimensional solution we have found is appropriate to propagation of plane waves through slabs of constant potential V, thickness b, and spaced a normal distance d from one another. This situation is depicted in Fig. 8.14, in which two typical wavefronts are also sketched. Suppose that these plane waves are incident at the angle θ, such as is drawn in Fig. 8.15. In the Bragg model of reflection, wavefronts scatter from ion-site lattice planes. We recall that the condition that reflected waves from adjacent planes add constructively is given by Bragg's formula,

$$(8.60) \qquad kd \cos \theta = q\pi$$

where θ is the angle that the incident **k** vector makes with the normal to the lattice plane. In the limit of normal incidence, $\theta \to 0$, and the one-dimensional model of the analysis above becomes relevant. Equation (8.60) reduces to the condition (8.59), which k satisfies at an energy gap. At these values of k, an integral number of wavelengths fit the distance $2d$ and the reflected waves constructively superpose. They are *Bragg-reflected*. Consider that a wave is moving in one direction with a critical k value (8.59). It is soon Bragg-reflected and propagates in the opposite direction. There is a similar reversal of direction of propagation on each reflection until finally the only steady-state solution is that which contains an equal number of waves traveling in

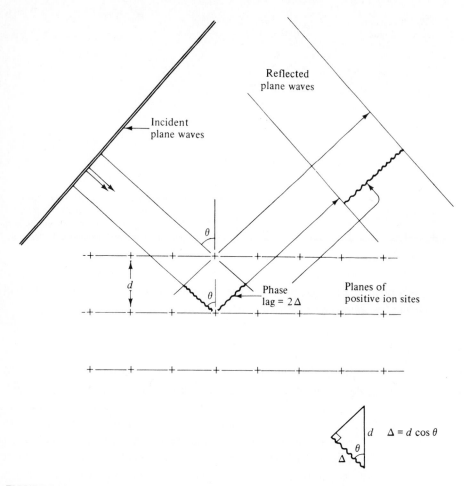

FIGURE 8.15   Constructive interference between reflected waves from different planes occurs when

$$2d \cos \theta = l\lambda = \frac{2\pi l}{k}$$

which gives

$$kd \cos \theta = l\pi$$

For normal incidence, $\theta = 0$, and this condition becomes

$$kd = l\pi$$

either direction. As will be shown in the next section, at these critical values of k, the eigenstates of $\hat{H}$ are composed equally of waves moving to the right and left, so that, for example, in the well domain of the periodic potential, $\varphi \sim \exp(ik_1 x) + \exp(-ik_1 x) \sim \cos(k_1 x_1)$, which is the spatial component of a standing wave. A similar standing-wave structure prevails across the barrier domain. When these solutions are matched at the potential steps, a standing wave ensues over the whole periodic potential. In such states $\langle p \rangle = 0$. Electrons are trapped and lose their free-particle quality. The effective mass, m^*, introduced in (8.45) grows large as k approaches inflection points of $E(k)$ between band edges. Energy curves appropriate to a one-dimensional periodic potential are shown in Fig. 8.13b.

Spreading of the Bound States

Let us now demonstrate that the band-energy spectrum relevant to an electron in a periodic potential collapses to the discrete bound-state spectrum appropriate to a particle in a single finite potential well in the limit that the wells of the periodic potential grow far apart ($b \to \infty$). Toward these ends it suffices to demonstrate that the dispersion relation (8.55), for the states $E < V$ for a periodic potential, in the limit of $b \to \infty$, gives the relations (8.27) and (8.28):

$$(8.61) \qquad \tan \xi = \frac{\eta}{\xi}, \qquad \tan \xi = \frac{-\xi}{\eta}$$

These are the relations for the even and odd states, respectively, of a particle in a finite potential well.

Let us recall that the nondimensional parameters ξ, η, and ρ introduced in (8.26) and (8.27) contain half the well width, which for the Kronig–Penney potential is $a/2$, so for the present case

$$(8.62) \qquad \begin{aligned} \xi &= \frac{k_1 a}{2}, \qquad \eta = \frac{\kappa a}{2} \\[2mm] \xi^2 &+ \eta^2 = \rho^2 = \frac{2m(a/2)^2 V}{\hbar^2} \end{aligned}$$

In terms of these variables, (8.55a) becomes

$$(8.63) \qquad \cos 2\xi \cosh\left(\frac{2b\eta}{a}\right) + \frac{\eta^2 - \xi^2}{2\eta\xi} \sin 2\xi \sinh\left(\frac{2b\eta}{a}\right) = \cos kd$$

This equation must reduce to the two equations (8.61) in the limit $b \to \infty$, $a = $ constant, $V = $ constant. First we note that in this limit

$$\cosh\left(\frac{2b\eta}{a}\right) \simeq \sinh\left(\frac{2b\eta}{a}\right) \simeq \frac{1}{2}\exp\left(\frac{2b\eta}{a}\right)$$

Dividing through by this exponential factor and allowing b to grow infinitely large reduces (8.63) to the form

(8.64) $$\tan 2\xi = \frac{2\xi\eta}{\xi^2 - \eta^2}$$

The double-angle formula for tangents permits this equation to be rewritten as

$$\frac{\tan \xi}{1 - \tan^2 \xi} = \frac{\xi\eta}{\xi^2 - \eta^2}$$

which in turn may be rewritten as

$$\tan^2 \xi + \frac{\xi^2 - \eta^2}{\xi\eta} \tan \xi - 1 = 0$$

This is a quadratic equation for $\tan \xi$. Solving for the two roots gives

$$2 \tan \xi = \frac{\eta^2 - \xi^2}{\xi\eta} \pm \frac{\eta^2 + \xi^2}{\xi\eta}$$

These are the two relations (8.61) that give the discrete bound states of a single isolated finite potential well.

Thus we find that the band structure of the energy spectrum of a particle in a periodic potential collapses to the discrete energy spectrum of a particle in a finite potential well in the limit that the wells of the periodic array became far removed from one another. Consider, for instance, a finite potential well that has two bound states. Such, for example, is the case if $\rho = 5\pi/4$ (see Fig. 8.2). For a periodic array of such potentials, the relation (8.55) applies in the domain $E < V$. If the left-hand side of this equation is plotted versus E such as in Fig. 8.12, two bands will be found to fall in the domain $E < V$. This transition[1] from the discrete states of an isolated well to the band structure of a lattice is illustrated in Fig. 8.16.

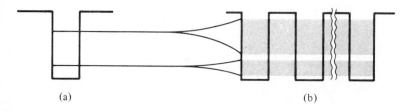

(a)   (b)

FIGURE 8.16   (a) Single isolated finite potential well with two bound states. (b) Corresponding periodic potential with two energy bands. For N wells each band contains N states.

---

[1] Details of a numerical analysis for this transition for the case of a well with four bound states may be found in V. Rojansky, *Introductory Quantum Mechanics*, Prentice-Hall, Englewood Cliffs, N.J., 1938.

8.10 (a) What is the expectation of momentum for an electron propagating in a Bloch wavefunction with spatial component

$$\varphi(x) = e^{ikx}u(x)?$$

(b) Show that if the periodic function $u(x)$ is real, $\langle p \rangle = \hbar k$.

Answer (partial)

(a) $\langle p \rangle = \hbar k + \langle u | \hat{p} | u \rangle$

8.11 What is the period of the Bloch wavefunction under the following conditions?

(a) $kd = 2l\pi$

(b) $kd = (2l + 1)\pi$

(c) $kd = n\pi/q$

Here l, n, and q are integers.

8.12 (a) Use the dispersion relation (8.55) to obtain the dispersion relation for the propagation of electrons through an infinite array of equally spaced delta-function potentials[1] separated by d cm (see Fig. 8.17). Note that the delta-function potential may be effected by constructing a potential barrier whose height is infinite and whose width is infinitesimal such that the area under the potential curve is fixed. This limit is easily constructed with the model at hand by setting

$$\lim_{\substack{\kappa \to \infty \\ b \to 0}} \kappa^2 bd = 2F = \text{constant}$$

(b) Make a plot of your dispersion function for the value $F = 3\pi/2$ and thereby illustrate the persistence of the band structure of the energy spectrum in this delta-function limit.

(c) How is it that electrons are able to propagate through the infinitely high potentials presented by the delta functions?

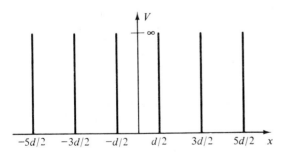

FIGURE 8.17 Periodic delta-function potential. The explicit form of the symmetric periodic delta-function potential is given by (see Problem 3.6e)

$$V(x) = V_0 d \left\{ \sum_{n=0}^{\infty} \delta\left[x - (2n + 1)\left(\frac{d}{2}\right) \right] \right.$$

$$\left. + \sum_{n=0}^{\infty} \delta\left[x + (2n + 1)\left(\frac{d}{2}\right) \right] \right\}$$

$$= V_0 d^2 \sum_{n=0}^{\infty} (2n + 1)\delta\left[x^2 - (2n + 1)^2 \left(\frac{d}{2}\right)^2 \right]$$

See Problem 8.12.

[1] This limiting case was, in fact, the one treated by R. de L. Kronig and W. G. Penney in their original paper, *Proc. Roy. Soc.* **A130**, 499 (1931).

301

(d) Write down a formal expression for the potential you have considered.

(e) What are the eigenstates at the band edges $kd = n\pi$? Show that one of the energies at a band edge is the free-particle energy $\hbar^2 k^2/2m$, while the other energy is larger. In this manner obtain an expression for the width of the energy gap at $kd = n\pi$.

Answer (partial)

(a) In the limit given above it follows that

$$\kappa b \to \frac{2F}{\kappa d} \to 0$$

Hence $\sinh \kappa b \to \kappa b$, $\cosh \kappa b \to 1$, and $d \to a$. The resulting dispersion relation appears as

$$\frac{F \sin k_1 d}{k_1 d} + \cos k_1 d = \cos kd$$

8.13 Show that the $E(k)$ spectrum for the arbitrary finite Kronig–Penney array draws close to the free-particle parabola $E = \hbar^2 k^2/2m$ in the limit $E \gg V$.

Answer

With $V/E = \epsilon \ll 1$, one obtains

$$k_2{}^2 = \frac{2m}{\hbar^2} E(1 - \epsilon) = k_1{}^2 + O(\epsilon)$$

Substitution into the dispersion relation (8.53) gives

$$\cos k_1 a \cos k_1 b - \sin k_1 a \sin k_1 b + O(\epsilon) = \cos k_1(a + b) + O(\epsilon) = \cos kd$$

Neglecting terms of $O(\epsilon)$ gives the spectrum $k_1 = k$ or, equivalently, $E = \hbar^2 k^2/2m$. Recalling further that k is discrete (8.42), one obtains $E_n = n^2(\hbar^2/2mL^2)$.

8.14 The $E(k)$ spectrum for an electron in a periodic lattice, such as illustrated in Fig. 8.13, does not fall to zero at $k = 0$. Estimate this *zero-point* energy using results appropriate to a particle confined to a one-dimensional domain of length L. What value of k is implied by your answer? How does this value compare to the minimum value of k for a crystal of length L?

8.15 (a) Show that the eigenenergies of the one-dimensional box of width a ($k_1 a = n\pi$) lie in the energy gaps of the Kronig–Penney potential of well width a (in the domain $E < V$). [*Hint:* Use (8.55).]

(b) Show that these box energies become the lower energies of the band gaps for the periodic delta-function potential described in Problem 8.12.

8.16 (a) Show that in the limit that the atomic sites of the Kronig–Penney model become far removed from each other ($b \to \infty$), energies of the more strongly bound electrons ($E \ll V$) become the eigenenergies $k_1 a = n\pi$ of a one-dimensional box of width a.

(b) In this limit what do the lower-band $E(k)$ curves shown in Fig. 8.13 become? What is the functional form of $E(k)$ for these bands? (*Note:* The approximation in which one begins with electronic states of isolated atoms is called the *tight-binding approximation*.)

8.17 (a)   Construct an equation for the periodic component $u(k)$ of the Bloch wavefunction $\varphi = u \exp(ikx)$, from the Schrödinger equation with a periodic potential $V(x)$.

(b)   The periodic potential $V(x)$ may be expanded in a Fourier series as follows:

$$V(x) = \sum_{n=-\infty}^{\infty} V_n \exp\left[i2\pi n\left(\frac{x}{d}\right)\right]$$

Expand the periodic component $u(x)$ in a similar series, substitute in the equation obtained in part (a), and derive coupled equations for the coefficients of expansion u_n.

Answers

(a) $\dfrac{\hbar^2}{2m}(u'' + 2iku' - k^2 u) + [E - V(x)]u = 0$

(b) $[E - B_q(k)]u_q = \sum_{l=-\infty}^{\infty} V_{q-l}u_l, \quad 2mB_q(k)/\hbar^2 \equiv (2\pi q/d)^2 + 2k(2\pi q/d) + k^2$

8.18   Show that in the limit $V \to 0$, the equation for the periodic component u obtained in Problem 8.17 gives the free-particle eigenenergy

$$E = \frac{\hbar^2 k^2}{2m}$$

with $u = \text{constant}$.

8.19   What is the number $\mathcal{N}_E$ of discrete energies in any of the energy bands depicted in Fig. 8.13?

Answer

The k values that enter the lowest energy band are given by the sequence

$$k = 0, \pm 1 \times \frac{2\pi}{L}, \pm 2 \times \frac{2\pi}{L}, \ldots, \pm \frac{N}{2}\frac{2\pi}{L}$$

$$L \equiv Nd$$

This series is cut off at $|kd| = \pi$ inasmuch as energy values begin to repeat beyond this value. There is a distinct energy corresponding to each value of $|k|$ in the sequence above. This gives

$$\mathcal{N}_E = \frac{N}{2} + 1 \simeq \frac{N}{2}$$

8.20   What is the number $\mathcal{N}_k$ of independent eigenstates in a band for a one-dimensional crystal comprised of N uniformly spaced ions?

Answer

There is a distinct eigenstate (Eq. 8.46 et seq.) corresponding to each value of k in the series in the example above. However, the state corresponding to $kd = -\pi$ is the same as the one corresponding to $kd = +\pi$, as may be seen from (8.50) and (8.51). We conclude that the eigenstates corresponding to $kd = \pm\pi$ are one and the same. Finally, there is only one eigenstate at $kd = 0$. Thus we obtain

$$\mathcal{N}_k = 2\left(\frac{N}{2} + 1\right) - 1 - 1 = N$$

There are as many eigenstates in a sample as there are ion sites. There are approximately half as many eigenenergies.

This result may also be obtained geometrically. Referring to the reduced-zone energy diagram (Fig. 8.13b), each energy band has width $(\Delta kd)_b = 2\pi$. The minimum interval in each band is $(\Delta kd)_{min} = 2\pi/N$ or, equivalently, $(\Delta k)_{min} = 2\pi/Nd = 2\pi/L$. Thus the number of points (states) in each band is

$$\mathcal{N}_k = \frac{(\Delta kd)_b}{(\Delta kd)_{min}} = N$$

(*Note:* With the two spin orientations taken into account, one obtains $2N$ independent states in each band.[1] The concept of spin is described in Chapter 11.)

8.3   STANDING WAVES AT THE BAND EDGES

Let us return to the nature of the eigenstates of $\hat{H}$ at the band edges, that is, at $kd = n\pi$. We will demonstrate that these eigenstates are standing waves and illustrate the relation between the eigenenergies of these states and the energy gaps at the band edges.

The eigenstates of the Kronig–Penney Hamiltonian established above have components [see Eqs. (8.48) and (8.49)]

(8.65)
$$\varphi_{\mathrm{I}} = Ae^{ik_1x} + Be^{-ik_1x}$$
$$\varphi_{\mathrm{II}} = Ce^{ik_2x} + De^{-ik_2x}$$

In order for these to be components of a standing wave, the magnitude of the amplitudes of the waves moving to the right and left must be equal. That is, at the critical values $kd = n\pi$, one must have

$$\frac{|A|}{|B|} = 1, \qquad \frac{|C|}{|D|} = 1$$

We will establish the first equality and leave the second as a problem. At the values $kd = n\pi$, $\exp(ikd) = (-1)^n$. Consider that n is even so that $exp(ikd) = +1$. With this value substituted into the equations of continuity, (8.50) and (8.51), one quickly obtains the following two equations for the expression $2Ck_2 \exp(ik_2a)$:

$$2Ck_2e^{ik_2a} = e^{-ik_2b}[A(k_1 + k_2) + B(k_2 - k_1)]$$
$$2Ck_2e^{ik_2a} = e^{ik_1a}A(k_1 + k_2) + e^{-ik_1a}B(k_2 - k_1)$$

---

[1] This result maintains in three dimensions, where N represents the number of *primitive cells* in the crystal. For further discussion, see C. Kittel, *Introduction to Solid State Physics*, 5th ed., Wiley, New York, 1976.

Setting these two expressions equal to each other and solving for A/B gives

(8.66)
$$\frac{A}{B} = \frac{k_2 - k_1}{k_2 + k_1} \left(\frac{e^{-ik_1 a} - e^{-ik_2 b}}{e^{-ik_2 b} - e^{ik_1 a}} \right)$$

Forming the square of the modulus $|A/B|^2 = (A/B)(A/B)^*$ gives

(8.67)
$$\left| \frac{A}{B} \right|^2 = \left| \frac{k_2 - k_1}{k_2 + k_1} \right|^2 \frac{2 - 2\cos(k_1 a - k_2 b)}{2 - 2\cos(k_1 a + k_2 b)}$$

$$= \left| \frac{k_2 - k_1}{k_2 + k_1} \right|^2 \frac{1 - \cos k_1 a \cos k_2 b - \sin k_1 a \sin k_2 b}{1 - \cos k_1 a \cos k_2 b + \sin k_1 a \sin k_2 b}$$

We must show that this expression is unity for the allowable values of k_1 and k_2, that is, those values which are obtained from the dispersion relation (8.53). Again with $\exp(ikd) = +1$, this relation reads

$$\cos k_1 a \cos k_2 b = 1 + \frac{k_1{}^2 + k_2{}^2}{2k_1 k_2} \sin k_1 a \sin k_2 b$$

When this formula for the cos product is substituted into (8.67), the desired result, $|A/B| = 1$, follows.

Let us proceed to construct such a standing-wave state. If the potential is placed in a symmetric position about the origin, such as in Fig. 8.18, then the Hamiltonian commutes with the parity operator $\hat{P}$ and these two operators share a set of common eigenfunctions; that is, $\hat{H}$ has even and odd eigenstates. It will be shown that eigenstates at the band edges exist in pairs, with each pair containing an even and an odd eigenstate. Very simply, one expects that this is the case since in the steady-state situation, electron density $|\varphi|^2$ should enjoy the same symmetry as the periodic

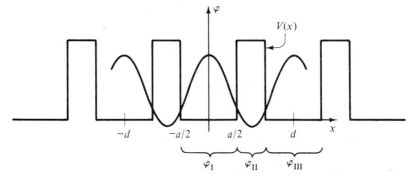

FIGURE 8.18 Standing-wave eigenfunction referred to in the construction $\varphi(x)$ as given by (8.68) to (8.70). Owing to the symmetric form of φ, it suffices to match φ_{I} to φ_{II} at $x = a/2$.

potential $V(x)$; that is, $|\varphi|^2$ is even, so φ is either even or odd. We will find that if kd/π is an even integer, eigenstates have period d. If kd/π is an odd integer, eigenstates have period $2d$. Let us consider the construction of the pair of eigenstates at a band edge corresponding to kd/π an even integer. Consider first the symmetric eigenstate. In the well domain about the origin

$$(8.68) \qquad \varphi_{\mathrm{I}}(x) = \cos k_1 x \qquad (-a/2 \leq x \leq a/2)$$

To obtain φ_{III} in the second well domain, one uses Bloch's theorem (8.40) together with the value $\exp(ikd) = +1$.

$$(8.69) \quad \varphi_{\mathrm{III}}(x) = e^{ikd}\varphi_{\mathrm{I}}(x-d) = \cos k_1(x-d) \qquad (a/2 + b \leq x \leq a/2 + d)$$

The standing wave in the barrier region II which joins these waves is symmetric about the midpoint $d/2$ (see Fig. 8.18).

$$(8.70) \qquad \varphi_{\mathrm{II}}(x) = D \cos k_2\left(\frac{x-d}{2}\right) \qquad (a/2 \leq x \leq a/2 + b)$$

The coefficient D and energy $\hbar^2 k_1{}^2/2m$ are obtained by matching the components of φ and φ' at the potential interface at $x = a/2$. There results for the even eigenstates,

$$(8.71) \qquad \begin{aligned} \cos\left(\frac{ak_1}{2}\right) &= D \cos\left(\frac{k_2 b}{2}\right) \\ k_1 \sin\left(\frac{ak_1}{2}\right) &= -Dk_2 \sin\left(\frac{k_2 b}{2}\right) \end{aligned}$$

Identical equations are obtained by matching φ_{II} to φ_{III} at $a/2 + b$ (see Fig. 8.18). Equations 8.71, together with the energy statement

$$k_1{}^2 - k_2{}^2 = \frac{2mV}{\hbar^2}$$

are three equations for the three unknowns k_1, k_2, and D. They are reminiscent of (8.27) and (8.28) and appropriate to the bound states of a particle in a finite potential well. Here, as there, solution may be effected through a numerical procedure (see Problem 8.24). The companion odd eigenstate $\tilde{\varphi}$ may similarly be constructed with the modification that the standing wave in the barrier region II is odd about the midpoint $d/2$. One obtains

$$\tilde{\varphi}_{\mathrm{I}}(x) = \sin \tilde{k}_1 x \qquad\qquad (-a/2 \leq x \leq a/2)$$

$$(8.72) \qquad \tilde{\varphi}_{\mathrm{II}}(x) = \tilde{D} \sin \tilde{k}_2\left(\frac{d}{2} - x\right) \qquad (a/2 \leq x \leq a/2 + b)$$

$$\tilde{\varphi}_{\mathrm{III}}(x) = \sin \tilde{k}_1(x - d) \qquad (a/2 + b \leq x \leq a/2 + d)$$

Matching conditions at the interface position $a/2$ gives the following relations for the odd eigenstates:

$$\sin\left(\frac{\tilde{k}_1 a}{2}\right) = \tilde{D} \sin\left(\frac{\tilde{k}_2 b}{2}\right)$$

(8.73)
$$\tilde{k}_1 \cos\left(\frac{\tilde{k}_1 a}{2}\right) = -\tilde{k}_2 \tilde{D} \cos\left(\frac{\tilde{k}_2 b}{2}\right)$$

$$\tilde{k}_1{}^2 - \tilde{k}_2{}^2 = \frac{2mV}{\hbar^2}$$

Again numerical procedure yields values for the energy $\hbar^2 \tilde{k}_1{}^2/2m$ and eigenstate parameter $\tilde{D}$.

The most significant result of such calculation is the width of the energy gap δE_n at the band edge $kd = n\pi$. This is the difference in energy between the even and odd standing-wave eigenstates.

$$(\delta E)_n = \frac{\hbar^2}{2m}(k_1{}^2 - \tilde{k}_1{}^2)$$

An analytic evaluation of this energy jump may be obtained in the "nearly free electron" model. This model is described in Section 13.4.

Parity Properties

Next we turn to a discussion of the parity properties of these standing-wave eigenstates at the band edges. These states are either even or odd in x. Again consider the case that kd/π is an even integer. Then the relation

(8.74)
$$\varphi(x + d) = e^{ikd}\varphi(x)$$

which is true for any eigenstate of the Kronig–Penney Hamiltonian, gives

(8.75)
$$\varphi(x + d) = \varphi(x)$$

It follows that for kd an even multiple of π, the period of φ is d. Setting $x = -d/2$ in (8.75) gives

(8.76)
$$\varphi\left(\frac{d}{2}\right) = \varphi\left(\frac{-d}{2}\right)$$

From this equation one concludes that φ can be an odd eigenfunction provided that

(8.77)
$$\varphi\left(\frac{d}{2}\right) = \varphi\left(\frac{-d}{2}\right) = 0$$

This property, taken together with the fact that φ has period d, gives

(8.78)
$$\varphi\left(\pm n\,\frac{d}{2}\right) = 0 \qquad (\varphi \text{ is odd}, \quad kd = 2q\pi)$$

with n an integer. The only stipulation on the even eigenfunctions is that they are of period d.

In this manner we find that the eigenfunctions of the Kronig–Penney Hamiltonian at the band edges $kd = 2q\pi$ exist in pairs. Each such pair contains an even eigenfunction and an odd eigenfunction. A typical pair of these functions is sketched in Fig. 8.19. The eigenenergies that accompany these eigenstates are the close-spaced pairs of values depicted in Fig. 8.13, where the vertical lines $kd = 2q\pi$ intersect the oscillating curves.

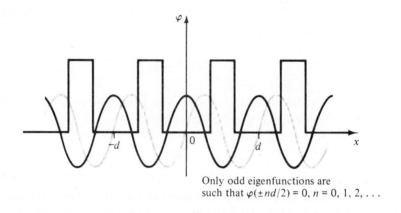

Only odd eigenfunctions are such that $\varphi(\pm nd/2) = 0, n = 0, 1, 2, \ldots$

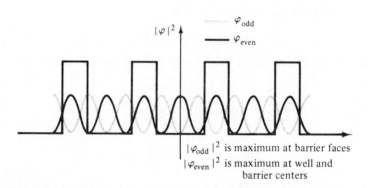

φ_{odd}

φ_{even}

$|\varphi_{odd}|^2$ is maximum at barrier faces

$|\varphi_{even}|^2$ is maximum at well and barrier centers

FIGURE 8.19 Typical pair of eigenfunctions for the Kronig–Penney Hamiltonian at the band edges: $kd = 2q\pi$. **Periodicity of φ is d.**

Having treated the case where kd is an even multiple of π, we next consider the case $kd = (2q + 1)\pi$, again with q an integer. From (8.74) one obtains

(8.79)
$$\varphi(x + d) = -\varphi(x)$$
$$\varphi(x + 2d) = -\varphi(x + d) = +\varphi(x)$$

It follows that for kd an odd multiple of π, the period of φ is $2d$. Setting $x = -d/2$ in the equation above gives

(8.80)
$$\varphi\left(\frac{d}{2}\right) = -\varphi\left(\frac{-d}{2}\right)$$

while $x = -d$ gives

(8.81)
$$\varphi(d) = \varphi(-d)$$

Equation (8.80) indicates that φ can be an even eigenfunction provided that

(8.82)
$$\varphi\left[\pm(1 + 2n)\frac{d}{2}\right] = 0 \qquad [kd = (2q + 1)\pi]$$

with n an integer.

Equation (8.81) indicates that φ can be an odd eigenstate provided that

(8.83)
$$\varphi(\pm nd) = 0 \qquad [kd = (2q + 1)\pi]$$

again with n an integer.

A typical pair of eigenfunctions is sketched in Fig. 8.20, together with accompanying plots of electron density $|\varphi|^2$. From this sketch one notes that each pair of eigenstates, corresponding to kd an odd multiple of π, contains one eigenstate with density $|\varphi|^2$, maximum at the ion sites and minimum at the barrier centers, while the other eigenstate has its extremum values of $|\varphi|^2$ reversed.

Thus we conclude that at the band edges $[kd = 2q\pi$ or $kd = (2q + 1)\pi]$ eigenfunctions appropriate to the Kronig–Penney Hamiltonian are standing waves and that there are two such functions with opposite parity at each edge.

PROBLEMS

8.21   The standing-wave quality of the eigenstate $\varphi(x)$ at the band edges was demonstrated for the component of φ in the valley regions of the potential ($|A/B| = 1$) for the case $\exp(ikd) = +1$. Following this analysis, demonstrate that the component of φ in the barrier domain is also a standing wave (i.e., show that $|C/D| = 1$) for the case $\exp(ikd) = -1$. [See Eq. (8.65) et seq.]

8.22   Show that the expectation of momentum $\langle p \rangle$ vanishes for a particle in a standing-wave eigenstate.

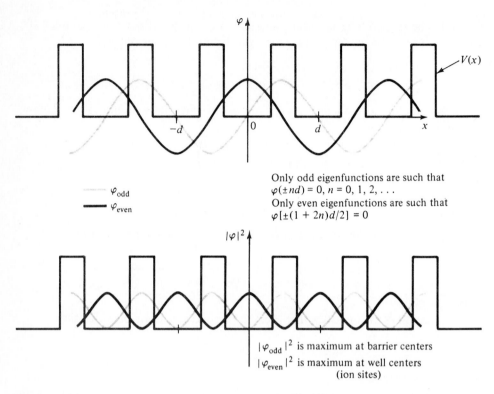

Only odd eigenfunctions are such that
$\varphi(\pm nd) = 0$, $n = 0, 1, 2, \ldots$
Only even eigenfunctions are such that
$\varphi[\pm(1 + 2n)d/2] = 0$

$|\varphi_{odd}|^2$ is maximum at barrier centers
$|\varphi_{even}|^2$ is maximum at well centers
(ion sites)

FIGURE 8.20 **Typical pair of eigenfunctions for the Kronig–Penney Hamiltonian at the band edges:** $kd = (2q + 1)\pi$. **Periodicity of φ is $2d$.**

8.23 Consider a typical pair of eigenstates appropriate to $kd = 2q\pi$ and the adjacent pair appropriate to $kd = (2q + 1)\pi$. By inspection only, conclude which pair of corresponding eigen-energies is of higher value.

8.24 (a) Introducing nondimensional variables

$$\xi \equiv \frac{k_1 a}{2}, \qquad \eta \equiv \frac{k_2 b}{2}, \qquad \rho^2 \equiv \frac{ma^2 V}{2\hbar^2}$$

show that (8.71) and (8.73) relevant to even and odd standing-wave band-edge solutions, re-spectively, give the dispersion relations

$$\xi \tan \xi = -\sqrt{\xi^2 - \rho^2} \tan\left(\frac{b}{a}\sqrt{\xi^2 - \rho^2}\right) \qquad \text{(even)}$$

$$\xi \cot \xi = -\sqrt{\xi^2 - \rho^2} \cot\left(\frac{b}{a}\sqrt{\xi^2 - \rho^2}\right) \qquad \text{(odd)}$$

(b)   Numerical solution of either equation may be effected by plotting the right-hand side and the left-hand side of the equation as functions of ξ (or $\tilde{\xi}$) on the same graph. Intersections then give the eigenenergies $E = (2\hbar^2/ma^2)\xi^2$. Use this procedure to estimate the lowest even and odd eigenstate energy corresponding to the barrier parameters, $\rho = \pi/2$, $(a/b)^2 = 15$.

(c)   Use your answer to part (b) to obtain the width of the energy gap δE at this band edge.

8.25   It was shown in Problem 8.13 that in the high-energy domain $E \gg V$, the $E(k)$ spectrum approaches the free-particle curve $E = \hbar^2 k^2/2m$. Show that the dispersion relation appropriate to the band edge $kd = 2n\pi$, (8.73), yields a free-particle standing wave in this limit.

Answer

With $k_1 \simeq k_2$, (8.73) gives (dropping the tilda notation)

$$D = \frac{\sin(k_1 a/2)}{\sin(k_1 b/2)} = -\frac{\cos(k_1 a/2)}{\cos(k_1 b/2)}$$

The second equality gives

$$\sin\left[\frac{k_1(a+b)}{2}\right] = 0$$

so

$$k_1 d = 2n\pi = kd$$

When substituted back into the expression above, one obtains $D = 1$, which is necessary in order that the eigenstate (8.72) with $k_1 = k_2$ be a free-particle standing wave.

8.4   BRIEF QUALITATIVE DESCRIPTION OF THE THEORY OF CONDUCTION IN SOLIDS

The spectrum of eigenenergies of electrons in an actual three-dimensional crystalline solid closely parallels that of the Kronig-Penney model described previously. In the three-dimensional case one also obtains a band structure for the allowed eigen-energies. The electrons in a solid occupy these bands. The properties of the two bands of highest energy for most practical cases determine whether the solid is an insulator, a conductor, or a semiconductor.

Suppose that the band structure of a solid is such that the band of highest energy is full (see Fig. 8.21). Furthermore, the gap between this filled band and the next completely unoccupied band is reasonably large. For example, for diamond this gap has a width of 6 eV. When an electric field is applied, electrons in the filled bands have no nearby unoccupied states to accelerate to. The sample remains nonconductive. It is an insulator. Furthermore, photons that comprise the visible spectrum do not have

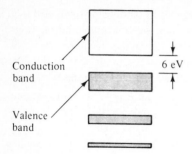

FIGURE 8.21   Energy bands of diamond, a good insulator.

sufficient energy (*hv*) to raise electrons from the *valence band* (last filled band) to the *conduction band*, so that diamond is transparent to light.

These statements are precisely true at absolute zero ($-273°$C). The student will recall that at absolute zero a system of particles falls to its lowest energy state, called the ground state of the system. When the temperature is raised, thermal agitation excites electrons to states of higher energy. For instance, for diamond at room temperature the characteristic energy of thermal agitation is $\simeq 0.03$ eV. The concentration of electrons which are raised to the conduction band is $\simeq 1.1 \times 10^{-34}$ electron/cm³. This gives rise to a conductivity which is lower than can be measured with present-day equipment, and diamond remains an insulator at room temperature.

In some crystalline solids, the conduction band is empty and the energy gap to the valence band is not prohibitively large. For instance, in silicon, this gap is 1.11 eV wide. In germanium it is 0.72 eV wide. At room temperature the concentration of electrons in the conduction band in silicon is 7×10^{10} electrons/cm³. In germanium it is 2.5×10^{13} electrons/cm³ (see Fig. 8.22). These densities give measurable conductivities. Such materials are called *intrinsic semiconductors*. The conduction of an *extrinsic semiconductor* is due to the presence of impurities in the sample.

A semiconductor acts as an insulator at sufficiently low temperatures. It begins to conduct at higher temperatures. In a semiconductor, charge transfer in the valence

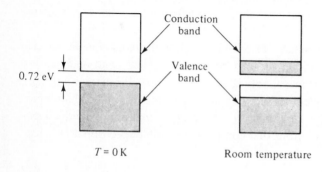

FIGURE 8.22   Valence and conduction bands for germanium, a typical semiconductor, at absolute zero and room temperature.

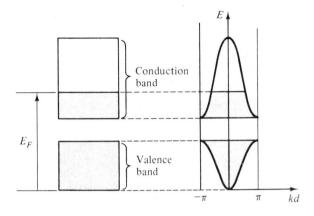

FIGURE 8.23 In a conductor the states in the conduction band are partially filled. The diagram to the right indicates the manner in which electrons fill the corresponding bands in the reduced-zone description for the idealized one-dimensional model. The Fermi energy E_F is also shown.

band may also contribute to conduction. In this case it is simpler for, say, calculation of conductivity, to speak of *hole conduction*. A hole is an unfilled state, found usually in the valence band.

In a metal the band of highest energy is only partially filled and electrons are readily accelerated by an electric field, to states of higher energy (see Fig. 8.23). Photons also fall prey to these electrons, which explains the opacity of metals to light. Note that the Fermi energy, described in Chapter 2, appears in the conduction band.

The description we have presented for the band structure of energy levels in periodic structures is a vast simplification of that which occurs in actual solids. A more accurate description of the formation of such bands with shrinking interatomic distance is shown in Fig. 8.24 for the metal sodium. Note in particular the strong band overlap at very small interatomic distance. This property suggests why materials become electrically conductive under extreme compression. The theory of semiconductors is returned to in Section 12.9.

PROBLEMS

8.26 What is the minimum frequency of radiation to which diamond is opaque? What kind of radiation is this (e.g., x rays, etc.)?

8.27 The mobility μ of an electron in an electric field $\mathbf{E}$ is defined by

$$\mathbf{v} = \mu \mathbf{E}$$

where $\mathbf{v}$ is the drift velocity of the electron. In a given semiconductor the mobility of electrons is μ_n, while the mobility of holes is μ_p. If at a given temperature, the density of conduction electrons is n electrons/cm^3 and the density of holes is p holes/cm^3, obtain an expression for the current flow in the semiconductor if an electric field $\mathbf{E}$ is applied across it.

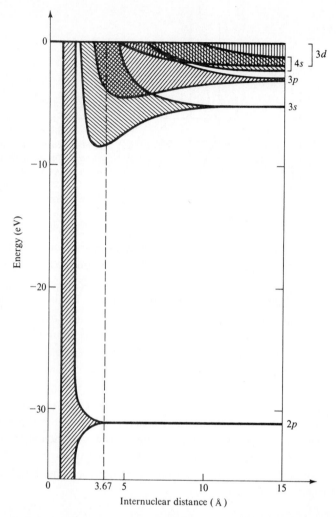

FIGURE 8.24 Energy-level diagram for sodium showing the development of a band structure as internucleon distance decreases. Labeling of states, $2p$, $3s$, etc., is fully described in Section 12.4. In the ground state of sodium, electrons fill up to the $3s$ level, which is half empty. This property accounts for the conductivity of the metal.

8.5 TWO BEADS ON A WIRE AND A PARTICLE IN A TWO-DIMENSIONAL BOX

Exchange Degeneracy

In this and the remaining section we discuss some simple examples of quantum mechanical systems with two degrees of freedom (see Section 1.2). The first such example is that of two beads constrained to move on a straight frictionless wire that is tightly stretched between two perfectly reflecting, rigid walls. The space between walls is L (see Fig. 8.25). We will assume that the particles do not interact with each other (they are "invisible" to each other). The Hamiltonian for this system is

$$(8.84) \qquad \hat{H}(x_1, x_2) = \frac{\hat{p}_1^{\,2}}{2m} + \frac{\hat{p}_2^{\,2}}{2m} + V(x_1) + V(x_2)$$

The two particles have the same mass, m. The potential functions $V(x_1)$ and $V(x_2)$ are relevant to a one-dimensional box. Their properties are given in Section 4.1.

This Hamiltonian may be partitioned into two independent terms,

$$\hat{H}(x_1, x_2) = \hat{H}_1(x_1) + \hat{H}_2(x_2)$$

$$(8.85) \qquad \hat{H}_1(x_1) = \frac{\hat{p}_1^{\,2}}{2m} + V(x_1)$$

$$\hat{H}_2(x_2) = \frac{\hat{p}_2^{\,2}}{2m} + V(x_2)$$

Under such circumstances, solution of the Schrödinger equation

$$(8.86) \qquad \hat{H}\varphi(x_1, x_2) = E\varphi(x_1, x_2)$$

is greatly simplified. It is given by the product

$$(8.87) \qquad \varphi_{n_1 n_2}(x_1, x_2) = \varphi_{n_1}(x_1)\varphi_{n_2}(x_2)$$

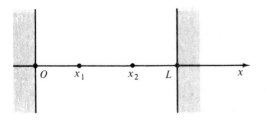

FIGURE 8.25 Coordinates of two beads on a wire stretched between two perfectly reflecting walls separated by the distance L.

where

(8.88)
$$\hat{H}_1 \varphi_{n_1}(x_1) = E_{n_1} \varphi_{n_1}(x_1)$$
$$\hat{H}_2 \varphi_{n_2}(x_2) = E_{n_2} \varphi_{n_2}(x_2)$$

The function φ_{n_1} is the eigenfunction of $\hat{H}_1$ corresponding to the energy E_{n_1}, while φ_{n_2} is the eigenfunction of $\hat{H}_2$ corresponding to the eigenenergy E_{n_2}.

(8.89)
$$\varphi_n = \sqrt{\frac{2}{L}} \sin\left(\frac{n\pi x}{L}\right), \qquad E_n = n^2 E_1$$

where n denotes either n_1 or n_2.

Let us test to see if $\varphi_{n_1 n_2}$, as given by (8.87), is an eigenstate of $\hat{H}(x_1, x_2)$.

(8.90)
$$\hat{H}\varphi_{n_1 n_2}(x_1, x_2) = (\hat{H}_1 + \hat{H}_2)\varphi_{n_1}(x_1)\varphi_{n_2}(x_2)$$
$$= \varphi_{n_2}\hat{H}_1 \varphi_{n_1} + \varphi_{n_1}\hat{H}_2 \varphi_{n_2}$$
$$= E_{n_1}\varphi_{n_1}\varphi_{n_2} + E_{n_2}\varphi_{n_1}\varphi_{n_2}$$
$$\hat{H}\varphi_{n_1 n_2} = (E_{n_1} + E_{n_2})\varphi_{n_1 n_2}$$

Thus we find that $\varphi_{n_1}\varphi_{n_2}$ is an eigenstate of $\hat{H}(x_1, x_2)$, and furthermore that the eigenenergy corresponding to this state is

(8.91)
$$E_{n_1 n_2} = E_{n_1} + E_{n_2} = (n_1{}^2 + n_2{}^2)E_1$$

For example, the eigenstate

(8.92)
$$\varphi_{2,3} = \frac{2}{L} \sin\left(\frac{2\pi x_1}{L}\right) \sin\left(\frac{3\pi x_2}{L}\right)$$

has corresponding eigenenergy

(8.93)
$$E_{2,3} = E_1(4 + 9) = 13E_1$$

This energy is *doubly degenerate* since the eigenstate

(8.94)
$$\varphi_{3,2} = \frac{2}{L} \sin\left(\frac{3\pi x_1}{L}\right) \sin\left(\frac{2\pi x_2}{L}\right)$$

also corresponds to the eigenenergy $E_{2,3}$. One may look upon the difference between $\varphi_{2,3}$ and $\varphi_{3,2}$ as being due to the exchange in the positions of particle 1 and particle 2. Such degeneracy is called *exchange degeneracy*.

Symmetric and Antisymmetric States

If two eigenstates correspond to the same eigenenergy, any linear combination of these eigenstates also corresponds to the same eigenenergy. Of all such linear combinations, two are of particular physical significance. These are of the form

(8.95)

$$\varphi_S = \frac{1}{\sqrt{2}} [\varphi_{n_1}(x_1)\varphi_{n_2}(x_2) + \varphi_{n_1}(x_2)\varphi_{n_2}(x_1)]$$

$$\varphi_A = \frac{1}{\sqrt{2}} [\varphi_{n_1}(x_1)\varphi_{n_2}(x_2) - \varphi_{n_1}(x_2)\varphi_{n_2}(x_1)]$$

The *symmetric state* φ_S has the property that

(8.96)
$$\varphi_S(x_1, x_2) = \varphi_S(x_2, x_1)$$

It is symmetric under the exchange of the particles. The *antisymmetric state* φ_A has the property that

(8.97)
$$\varphi_A(x_1, x_2) = -\varphi_A(x_2, x_1)$$

It is antisymmetric under the exchange of particles.

When referred to systems with two degrees of freedom, such as that of two particles in a one-dimensional box, the probability amplitude related to the system is given by (see Problem 3.20)

(8.98)
$$P_{12} \, dx_1 \, dx_2 = |\varphi(x_1, x_2)|^2 \, dx_1 \, dx_2$$

$P_{12} \, dx_1 \, dx_2$ is the probability of finding particle 1 in the interval dx_1 about the point x_1 *and* particle 2 in the interval dx_2 about the point x_2, in any given measurement.

When the two particles in the one-dimensional box are identical ($m_1 = m_2$), such as in the case considered, we note that for both classes of wavefunctions (symmetric and antisymmetric)

(8.99) $\quad |\varphi_S(x_1, x_2)|^2 = |\varphi_S(x_2, x_1)|^2, \quad |\varphi_A(x_1, x_2)|^2 = |\varphi_A(x_2, x_1)|^2$

Physical properties of the system are not affected by an exchange of the position of the two particles. This is a manifestation of a quantum mechanical property attached to identical particles: that is, in quantum mechanics identical particles are also indistinguishable (they cannot be labeled). In the scattering of electrons off electrons, for example, the scattered beam contains both incident and target electrons. The indistinguishability of these particles must be taken into account in any consistent formulation of the theory of such scattering. It is the indistinguishability of identical particles which selects φ_A or φ_S (8.95) to be the physically relevant linear combination of eigenstates for the two-particle problem.

If the masses of the two particles in our one-dimensional box are different (m_1 and m_2), the Hamiltonian (8.84) becomes

(8.100)
$$\hat{H}(x_1, x_2) = \frac{\hat{p}_1{}^2}{2m_1} + \frac{\hat{p}_2{}^2}{2m_2} + V(x_1) + V(x_2)$$

The particles are now distinguishable and the states of the system do not suffer exchange degeneracy. The eigenstate

(8.101)
$$\varphi_{n_1 n_2} = \varphi_{n_1}(x_1)\varphi_{n_2}(x_2)$$

corresponds to the eigenenergy

(8.102)
$$E_{n_1 n_2} = E_{n_1} + E_{n_2} = \left(\frac{n_1{}^2}{m_1} + \frac{n_2{}^2}{m_2}\right)\frac{\hbar^2\pi^2}{2L^2}$$

The exchange state $\varphi_{n_2 n_1}$ corresponds to the eigenenergy

(8.103)
$$E_{n_2 n_1} = \left(\frac{n_2{}^2}{m_1} + \frac{n_1{}^2}{m_2}\right)\frac{\hbar^2\pi^2}{2L^2} \neq E_{n_1 n_2}$$

Thus the exchange degeneracy associated with systems containing identical particles is removed.

We now turn to the time-dependent Schrödinger equation for systems with two degrees of freedom.

(8.104)
$$i\hbar\frac{\partial\psi}{\partial t} = \hat{H}\psi$$

The solution of this equation is (see Section 3.5)

(8.105)
$$\psi_{n_1 n_2} = \varphi_{n_1 n_2}(x_1, x_2)\exp\left(-\frac{iE_{n_1 n_2}t}{\hbar}\right)$$

Given the arbitrary initial state $\psi(x_1, x_2, 0)$, the state at time $t > 0$ is

$$\psi(x_1, x_2, t) = \exp\left(-\frac{it\hat{H}}{\hbar}\right)\psi(x_1, x_2, 0)$$

Examples of the use of this equation are given in the problems that follow the next subsection.

Symmetry and Accidental Degeneracy

Much of the preceding analysis may be carried over to the problem of a single particle moving in a two-dimensional box (see Fig. 8.26). This is another case of a system with two degrees of freedom. In the example of two beads on a wire, cited above, the

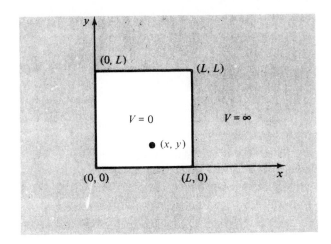

FIGURE 8.26   Particle in a two-dimensional box.

good coordinates are (x_1, x_2). For the single particle in a two-dimensional box, good coordinates are (x, y). The Hamiltonian for this system appears as

$$(8.106) \qquad \hat{H}(x, y) = \frac{\hat{p}_x^{\,2}}{2m} + \frac{\hat{p}_y^{\,2}}{2m} + V(x) + V(y)$$

The potential $V(x)$ is the same as that of a one-dimensional box which lies between $x = 0$ and $x = L$ on the x axis, whereas $V(y)$ is the same as that of a one-dimensional box that lies between $y = 0$ and $y = L$ on the y axis. Eigenfunctions and eigenenergies are

$$(8.107) \qquad \varphi_{n_1 n_2}(x, y) = \frac{2}{L} \sin\left(\frac{n_1 \pi x}{L}\right) \sin\left(\frac{n_2 \pi y}{L}\right)$$

$$E_{n_1 n_2} = E_1(n_1^{\,2} + n_2^{\,2})$$

This eigenenergy also corresponds to the eigenstate

$$\varphi_{n_2 n_1}(x, y) = \frac{2}{L} \sin\left(\frac{n_2 \pi x}{L}\right) \sin\left(\frac{n_1 \pi y}{L}\right)$$

The probability density $|\varphi|^2$ may be plotted as a height above the xy plane. The distinction between $|\varphi_{n_1 n_2}|^2$ and $|\varphi_{n_2 n_1}|^2$ is then as follows. The surface $|\varphi_{n_1 n_2}|^2$ is obtained from the surface $|\varphi_{n_2 n_1}|^2$ by reflecting this surface through the plane $x - y = 0$. The energy corresponding to both these distributions is the same. The degeneracy of these states is sometimes called *symmetry degeneracy* (as opposed to exchange degeneracy). Degeneracy that is neither symmetric nor exchange is often referred to as *accidental degeneracy* (see Problem 8.34).

PROBLEMS

8.28 At time $t = 0$, two particles of mass m_1 and m_2, respectively, in a one-dimensional box of length L, are known to be in the state

$$\psi(x_1, x_2, 0) = \frac{3\varphi_5(x_1)\varphi_4(x_2) + 7\varphi_9(x_1)\varphi_8(x_2)}{\sqrt{58}}$$

(a) If the energy of the system is measured, what values will be found and with what probability will these values occur?

(b) Suppose that measurement finds the value $E_{5,4}$. What is the time-dependent state of the system subsequent to measurement?

(c) What is the probability of finding particle 1 (with mass m_1) in the interval $(0, L/2)$ at $t = 0$?

Answer (partial)

(c) If the state of the two-particle system is $\varphi(x_1, x_2)$, the probability of finding particle 1 in the interval dx_1 (independent of where particle 2 is) is

$$P(x_1)\, dx_1 = dx_1 \int_0^L |\varphi(x_1, x_2)|^2\, dx_2$$

8.29 Show that $\varphi_A(x_1, x_2)$ in (8.95) may be written as a 2×2 determinant.

8.30 Show that φ_A and φ_S in (8.95) correspond to the same eigenenergy relevant to the Hamiltonian (8.84).

8.31 In the event that two particles in a one-dimensional box are identical one must ask: What is the probability of finding a particle in the interval dx about x? In this vein, show that for either $\varphi_S(x_1, x_2)$ or $\varphi_A(x_1, x_2)$, the two integrations

$$\int |\varphi|^2\, dx_1 \quad \text{and} \quad \int |\varphi|^2\, dx_2$$

give the same functional form.

8.32 Consider two identical particles in a one-dimensional box of length L. Calculate the expected value for the square of the interparticle displacement

$$d^2 \equiv (x_1 - x_2)^2$$

in the two states φ_S and φ_A. Show that

$$\langle d^2 \rangle_S \leq \langle d^2 \rangle_A$$

thus establishing that (in a statistical sense) particles in a symmetric state attract one another while particles in an antisymmetric state repel one another. Such attractions and repulsions are classified as exchange phenomena. They are discussed in further detail in Chapter 12.

Answer

With $\varphi_{n_1}(x_1)$ represented by $|n_1\rangle$, $\varphi_{n_1}(x_2)$ by $|\bar{n}_1\rangle$, and $\varphi_{n_1}(x_1)\varphi_{n_2}(x_2)$ by $|n_1\bar{n}_2\rangle$, the symmetric and antisymmetric states appear as

$$|\varphi_{S,A}\rangle = \frac{1}{\sqrt{2}}(|n_1\bar{n}_2\rangle \pm |\bar{n}_1 n_2\rangle)$$

Thus

$$2\langle d^2\rangle_{S,A} = (\langle n_1\bar{n}_2| \pm \langle\bar{n}_1 n_2|)d^2(|n_1\bar{n}_2\rangle \pm |\bar{n}_1 n_2\rangle)$$
$$= \langle n_1\bar{n}_2|d^2|n_1\bar{n}_2\rangle + \langle\bar{n}_1 n_2|d^2|\bar{n}_1 n_2\rangle \pm \langle n_1\bar{n}_2|d^2|\bar{n}_1 n_2\rangle$$
$$\pm\langle\bar{n}_1 n_2|d^2|n_1\bar{n}_2\rangle$$

In the last two $\pm$ contributions with $d^2 = x_1^2 + x_2^2 - 2x_1 x_2$, only the $-2x_1 x_2$ term is found to survive. Consider the term $\langle n_1\bar{n}_2|d^2|\bar{n}_1 n_2\rangle = -2\langle n_1\bar{n}_2|x_1 x_2|\bar{n}_1 n_2\rangle = -2\langle n_1|x_1|n_2\rangle\langle\bar{n}_2|x_2|\bar{n}_1\rangle$. In that $\langle n_1|x_1|n_2\rangle = \langle\bar{n}_1|x_2|\bar{n}_2\rangle \equiv x_{12}$ (write out the integrals and change variables), one obtains

$$\langle n_1\bar{n}_2|x_1 x_2|\bar{n}_1 n_2\rangle = \langle n_1|x_1|n_2\rangle\langle n_2|x_1|n_1\rangle = |x_{12}|^2$$

There results

$$\langle d^2\rangle_S = \langle d^2\rangle_A - 4|x_{12}|^2 \le \langle d^2\rangle_A$$

8.33   For a single particle in a two-dimensional box such as described in the text, one may also construct symmetric and antisymmetric states. The symmetric state φ_S has the property

$$\varphi_S(x, y) = \varphi_S(y, x)$$

while the antisymmetric state has the property

$$\varphi_A(x, y) = -\varphi_A(y, x)$$

What are the eigenstates, φ_S and φ_A, that correspond to the energy $29E_1$? The symmetry of these eigenstates reflects the fact that there is no intrinsic distinction between the diagonal halves of the box depicted in Fig. 8.26.

8.34   Construct the eigenstates and eigenenergies of a particle in a two-dimensional rectangular box of edge lengths L and $2L$. Take the origin to be at a corner of the rectangle. Account geometrically for the removal of most of the degeneracy present in the case of the square, two-dimensional box described previously. The degeneracy present for this configuration (e.g., the energy $5E$ is doubly degenerate) is sometimes called *accidental degeneracy*, in that it is neither exchange- nor symmetry-degenerate.

8.6   TWO-DIMENSIONAL HARMONIC OSCILLATOR

The two-dimensional problem we consider now is that of a point particle of mass m, constrained by a set of four coplanar, orthogonal springs, all with the same spring constant K (see Fig. 8.27).

$$\hat{H}(x, y) = \frac{\hat{p}_x^2}{2m} + \frac{\hat{p}_y^2}{2m} + \frac{K}{2}x^2 + \frac{K}{2}y^2$$

(8.108)

$$\hat{H}(x, y) = \hat{H}(x) + \hat{H}(y)$$

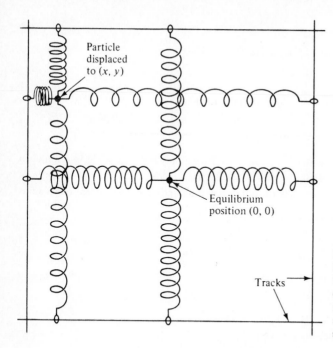

FIGURE 8.27 Two-dimensional harmonic oscillator. Springs are free to move on tracks but are otherwise constrained to displacements parallel to the coordinate axes. All springs have the same spring constant K.

Again, we find that the total Hamiltonian partitions into two independent parts, $\hat{H}(x)$ and $\hat{H}(y)$. These are the Hamiltonians relevant to one-dimensional harmonic oscillation in the x and y directions, respectively (see Sections 7.2 through 7.4). The eigenstates and eigenenergies of these Hamiltonians are

$$\varphi_{n_1}(\xi) = A_{n_1}\mathcal{H}_{n_1}(\xi)e^{-\xi^2/2}$$
$$E_{n_1} = \hbar\omega_0(n_1 + \tfrac{1}{2})$$
(8.109)
$$\varphi_{n_2}(\eta) = A_{n_2}\mathcal{H}_{n_2}(\eta)e^{-\eta^2/2}$$
$$E_{n_2} = \hbar\omega_0(n_2 + \tfrac{1}{2})$$

The nondimensional displacements ξ and η are defined by (7.46)

$$\xi^2 \equiv \frac{m\omega_0 x^2}{\hbar} \equiv \beta^2 x^2$$
(8.110)
$$\eta^2 = \frac{m\omega_0 y^2}{\hbar} \equiv \beta^2 y^2$$

while $\mathcal{H}_n(\xi)$ is the nth-order Hermite polynomial (7.58) and A_n is a normalization constant (Problem 7.7).

Owing to the separability of $\hat{H}(x, y)$, it follows that its eigenstates are the product forms

(8.111)
$$\varphi_{n_1 n_2}(\xi, \eta) = \varphi_{n_1}(\xi)\varphi_{n_2}(\eta)$$
$$\varphi_{n_1 n_2} = A_{n_1 n_2}\mathcal{H}_{n_1}(\xi)\mathcal{H}_{n_2}(\eta)e^{-(\xi^2 + \eta^2)/2}$$

while the eigenenergies of $\hat{H}(x, y)$ are the sums

(8.112)
$$E_{n_1 n_2} = E_{n_1} + E_{n_2}$$
$$E_{n_1 n_2} = \hbar\omega_0(n_1 + \tfrac{1}{2} + n_2 + \tfrac{1}{2}) = \hbar\omega_0(n_1 + n_2 + 1)$$

For example, the ground state of the two-dimensional harmonic oscillator is

$$\varphi_{0,0} = A_0 A_0 \mathscr{H}_0(\xi)\mathscr{H}_0(\eta)e^{-(\xi^2 + \eta^2)/2}$$

(8.113)
$$\varphi_{0,0}(\xi, \eta) = \frac{1}{\sqrt{\pi}} e^{-(\xi^2 + \eta^2)/2} = \frac{1}{\sqrt{\pi}} \exp\left[\frac{-\beta^2(x^2 + y^2)}{2}\right]$$

$$E_{0,0} = \hbar\omega_0$$

This is the only nondegenerate eigenstate of the two-dimensional harmonic oscillator. All the remaining states are degenerate. The order of the degeneracy of the eigenenergy $E_{n_1 n_2}$ is obtained from (8.112), from which we see that any eigenfunction $\varphi_{n_1'}\varphi_{n_2'}$ whose indices n_1', n_2' sum to the value $(n_1 + n_2)$ corresponds to the same eigenenergy, $E_{n_1 n_2}$.

(8.114)
$$\begin{bmatrix} \text{eigenfunctions corresponding to} \\ E_{n_1 n_2} \end{bmatrix} = \begin{bmatrix} \varphi_{n_1'}\varphi_{n_2'}, \text{ such that} \\ n_1' + n_2' = n_1 + n_2 \end{bmatrix}$$

For example, to find the eigenstates that correspond to the eigenenergy

(8.115)
$$E = 5\hbar\omega_0 = (4 + 1)\hbar\omega_0$$

one must find all pairs of integers n_1' and n_2' that sum to 4.

(8.116)
$$n_1' + n_2' = 4$$
$$(n_1', n_2') = (0, 4), (4, 0), (1, 3), (3, 1), (2, 2)$$

It follows that $E = 5\hbar\omega_0$ is a fivefold-degenerate eigenenergy. The five degenerate eigenstates are

(8.117)
$$\left. \begin{array}{l} \varphi_0(\xi)\varphi_4(\eta) = \varphi_{04} \\ \varphi_4(\xi)\varphi_0(\eta) = \varphi_{40} \\ \varphi_1(\xi)\varphi_3(\eta) = \varphi_{13} \\ \varphi_3(\xi)\varphi_1(\eta) = \varphi_{31} \\ \varphi_2(\xi)\varphi_2(\eta) = \varphi_{22} \end{array} \right\} = \text{eigenstates corresponding to } E = 5\hbar\omega_0$$

Of these five states, φ_{04} suffers symmetry degeneracy with φ_{40}, as does φ_{13} with φ_{31}. On the other hand, the three states φ_{04}, φ_{13}, and φ_{22} are accidentally degenerate with each other.

8.35   What is the order of degeneracy of the eigenstate

$$E_s = \hbar\omega_0(s + 1)$$

of the two-dimensional harmonic oscillator?

Answer

The degeneracy equals the number of ways of writing an integer s as the ordered sum of two numbers. There are $(s + 1)$ ways to do this.

8.36   (a)   Write down the Hamiltonians, eigenenergies, and eigenstates for a two-dimensional harmonic oscillator with distinct spring constants K_x and K_y.

(b)   If $K_y = 4K_x$, show that the eigenenergies may be written

$$E_{n_1 n_2} = \hbar\omega_0(n_1 + 2n_2 + \tfrac{3}{2})$$

where n_1 corresponds to x motion and n_2 to y motion.

(c)   For part (b), what is the order of degeneracy of $E_{2,3}$? List the corresponding eigenstates. Account for the absence of symmetry degeneracy among these states.

8.37   A right circular cylinder of infinite height and large, but finite radius is uniformly, positively charged throughout its volume. The charge density is ρ_0 esu/cm³. An electron moves in a plane normal to the cylinder. Its position is close to the central axis of the cylinder (see Fig. 8.28).

(a)   What is the electrostatic potential Φ near the central axis of the cylinder?

(b)   What are the eigenenergies of the electron? [*Hints:* For part (a), use Poisson's equation, $\nabla^2\Phi = -4\pi\rho = -4\pi\rho_0$. The radial operator in ∇^2, in cylindrical coordinates, is $r^{-1}\,\partial/\partial r(r\,\partial/\partial r)$. From symmetry you may assume $\Phi = \Phi(r)$. For part (b), note that the potential energy of the electron is $V(r) = -|e|\Phi(r)$, where $r^2 = x^2 + y^2$.]

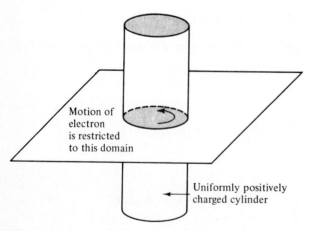

Motion of electron is restricted to this domain

Uniformly positively charged cylinder

FIGURE 8.28   Configuration for Problem 8.37.

8.38   A particle moves in the xy plane in the potential field

$$V = V(x) + V(y)$$

$$V(x) = V_1 = \text{constant}$$

$$V(y) = V_2 = \text{constant}$$

at constant energy E. Give the time-dependent wavefunction $\psi(x, y, t)$ of the particle corresponding to the initial data

$$\psi(x, 0, 0) = \psi(0, y, 0) = 0$$

8.39   A particle of mass m is confined to move on the two-dimensional strip

$$-a < x < a, \qquad -\infty < y < \infty$$

by two impenetrable parallel walls at $x = \pm a$.
   (a)   What is the minimum energy of the particle that measurement can find?
   (b)   Suppose that two additional walls are inserted at $y = \pm a$. Can measurement of the particle's energy find the value $3\pi^2\hbar^2/8ma^2$? Explain your answer.

8.40   Consider that three-dimensional space is divided into two semiinfinite domains of constant potential V_1 and V_2.

$$V = V_1, \qquad z > 0$$

$$V = V_2, \qquad z \leq 0$$

A beam of particles carrying the current $\hbar k_1 |A|^2/m$ particles/cm²-s is incident on the $z = 0$ interface and is in part reflected and transmitted. Particles in the reflected beam have momentum $\hbar k_1'$, while those in the transmitted beam have momentum $\hbar k_2$. The vectors k_1, k_1', k_2 are all parallel to the xz plane. The configuration is shown in Fig. 8.29.
   (a)   What is the wavefunction ψ_1 appropriate to a particle in the upper half-space $z > 0$? What is the wavefunction ψ_2 of a particle in the lower half-space (i.e., that of a particle in the transmitted beam)?
   (b)   Determine the relation between the angles α, α', and α'' through matching ψ_1 to ψ_2 and their derivatives across the $z = 0$ plane.
   (c)   Using the matching equations obtained in part (b) determine the transmission coefficient T and reflection coefficient R. Show that $T + R = 1$.

Answers (partial)

   (a)   The wavefunctions of particles in the upper and lower half-spaces are

$$\psi_1 = Ae^{i\phi_1} + Be^{i\phi_2}$$

$$\psi_2 = Ce^{i\phi_3}$$

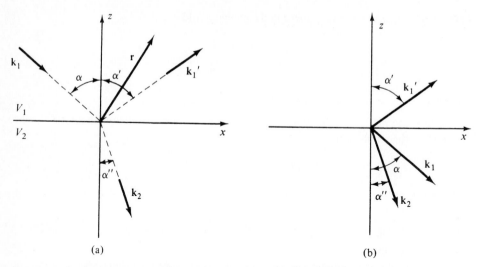

FIGURE 8.29 **Orientation of k vectors for a beam of particles incident on the $z = 0$ plane at the angle α.**
(See Problem 8.40.)

The phases ϕ_1, ϕ_2, and ϕ_3 are

$$\phi_1 = \mathbf{k}_1 \cdot \mathbf{r} - \omega t = k_1 x \sin \alpha - k_1 z \cos \alpha - \omega t$$

$$\phi_2 = \mathbf{k}_1' \cdot \mathbf{r} - \omega t = k_1 x \sin \alpha' + k_1 z \cos \alpha' - \omega t$$

$$\phi_3 = \mathbf{k}_2 \cdot \mathbf{r} - \omega t = k_2 x \sin \alpha'' - k_2 z \cos \alpha'' - \omega t$$

$$\frac{\hbar^2 k_1^2}{2m} = E - V_1, \qquad \frac{\hbar^2 k_2^2}{2m} = E - V_2, \qquad \hbar\omega = E$$

(b) Matching ψ_1 to ψ_2 on $z = 0$ gives

$$A e^{ik_1 x \sin \alpha} + B e^{ik_1 x \sin \alpha'} = C e^{ik_2 x \sin \alpha''}$$

For this equation to be satisfied for all x, it is necessary that the phases be equal.

$$k_1 \sin \alpha = k_1 \sin \alpha' = k_2 \sin \alpha''$$

$$\alpha = \alpha' \quad \text{and} \quad \frac{\sin \alpha}{\sin \alpha''} = \frac{k_2}{k_1} = \sqrt{\frac{E - V_2}{E - V_1}} = \sqrt{\frac{1 - V_2/E}{1 - V_1/E}} = n$$

where n is the relative *index of refraction*. (Compare with Snell's laws of optical refraction.) These results, together with the matching equation $\partial\psi_1/\partial z = \partial\psi_2/\partial z$ on $z = 0$, give

$$A + B = C$$

$$A - B = \frac{k_2 \cos \alpha''}{k_1 \cos \alpha} C$$

which serve to determine T and R: namely,

$$R = \left|\frac{B}{A}\right|^2, \qquad T = \frac{k_2 \cos \alpha''}{k_1 \cos \alpha} \left|\frac{C}{A}\right|^2$$

The transmission coefficient is seen to involve only the normal components of incident and transmitted fluxes.

8.41 Calculate the reflection coefficient of sodium metal for low-energy electrons as a function of electron energy and angle of incidence. For electrons of sufficiently long wavelength, the potential barrier at the metal surface can be treated as discontinuous. Assume that the potential energy of an electron in the metal is -5 eV. Calculate the "index of refraction" of the metal for electrons. (See Problem 8.40.)

8.42 A beam of electrons of energy E in a potential-free region is incident on a potential step of 5 V at an incident angle of 45°. Is there a threshold value of E below which all the electrons will be reflected? If so, what is this value?

Answer

Call the threshold (or "critical") incident energy E_c. Then at E_c, the angle of refraction $\alpha'' = \pi/2$. That is, at E_c the transmitted ray runs along the interface between the two media. If E is increased above E_c (at the same angle of incidence α), electrons penetrate the potential step and $R < 1$. On the other hand, if E is decreased below E_c, there is no transmitted ray at all. The analytic manifestation of this observation is that α'' becomes imaginary for $E < E_c$, while R maintains its value of unity for all such values of E. From Snell's law (for $\alpha = \pi/4$),

$$\sin \alpha = \frac{1}{\sqrt{2}} = n \sin \alpha''$$

where n is the index of refraction,

$$n = \sqrt{1 - \frac{V}{E}}$$

At E_c, $\sin \alpha'' = 1$ and one obtains $E_c = 2V = 10$ V. The reflection coefficient is given by

$$R = \left|\frac{\cos \alpha - n \cos \alpha''}{\cos \alpha + n \cos \alpha''}\right|$$

For the problem under discussion

$$\cos \alpha'' = \sqrt{1 - \sin^2 \alpha''} = \sqrt{1 - \frac{1}{2n^2}}$$

The critical value of $2n^2$ is $2n_c^2 = 1$. If $E < E_c$, then $2n^2 < 1$ and $\cos \alpha''$ becomes imaginary so that for these values of incident energy, R assumes the form $R = |\bar{z}/z|$, where z is a complex number and $\bar{z}$ is its conjugate. It follows that $R = 1$ for $E < E_c$.

8.43 This problem addresses the notion of *supersymmetry*[1] relevant to eigenenergies of a given Hamiltonian.

(a) Show that, apart from the ground state, the two Hamiltonians

$$H_\pm = \frac{p^2}{2m} + [W^2 \pm W']$$

have the same eigenenergies, where a prime denotes differentiation.

(b) What is the symmetric potential $\bar{V}(x)$ corresponding to the harmonic oscillator potential? Hint: Work in nondimensional units and write

$$V(x) = \frac{x^2}{2} + C$$

where C is a constant.

(c) Show that eigenenergies corresponding to $V(x)$ and $\bar{V}(x)$ in the preceding question satisfy the stated theorem.

8.7 LINEAR COMBINATION OF ATOMIC ORBITALS APPROXIMATION (LCAO)

The LCAO approximation is employed to estimate the states of a molecule in terms of a linear combination s of quantum states ("orbitals") of isolated constituent atoms. This formalism finds wide application in solid-state physics in the tight-binding approximation.[2]

In the present analysis, the LCAO method is employed to estimate the ground-state energy and wavefunction of a given molecule.

The Molecular Ion

We will apply the LCAO method to the potential shown in Fig. 8.30, which, in the present discussion, is representative of a one-dimensional model of a homonuclear, diatomic molecular ion, such as $H_2{}^+$. For fixed internucleon spacing this example permits our analysis to remain purely one-dimensional, with the displacement of the valence electron as the only free variable.

The Hamiltonian for our system is given by

$$(8.118) \qquad \hat{H} = \frac{\hat{p}_x{}^2}{2m} + V_1(x) + V_2(x)$$

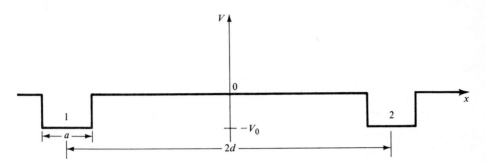

FIGURE 8.30   Configuration of two identical wells (1 and 2) separated by the distance 2d.

where $V_1(x)$ and $V_2(x)$ are identical square wells separated by the distance $2d$, as illustrated in Fig. 8.30.

Let us introduce the normalized ground-state wavefunctions $\varphi_1(x)$ and $\varphi_2(x)$ relevant to well 1 and well 2, respectively, given by the symmetric even wavefunction (8.24). See Fig. 8.31. For wells sufficiently separated, $d \gg a$, it is evident that

(8.119) $$\varphi_1(x)V_2(x) = \varphi_2(x)V_1(x) = 0$$

With these equalities at hand, we may write

(8.120)
$$\hat{H}\varphi_1 = E_0\varphi_1$$
$$\hat{H}\varphi_2 = E_0\varphi_2$$

where E_0 is the ground state of an isolated well. It follows that the ground state of our composite decoupled molecule is doubly degenerate. Furthermore, as is evident from (8.120), this ground-state energy is the same as that of either one of the isolated atoms. See Fig. 8.32.

When atoms are brought together to form a molecular ion, it is assumed that the Hamiltonian maintains the summational form (8.118) but that φ_1 and φ_2 are no longer eigenstates of $\hat{H}$.

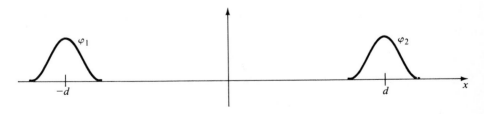

FIGURE 8.31   Ground states of isolated atoms corresponding to the configuration depicted in Fig. 8.30.

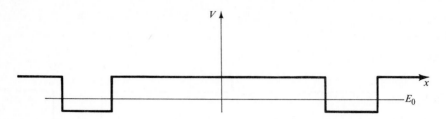

FIGURE 8.32   In the limit that atoms grow uncoupled, the ground-state energy of the composite molecule is the same as that of one of its isolated atoms.

Ground-State Energy and Wavefunction

To estimate the ground state of our molecular ion in the LCAO approximation, one writes

$$(8.121) \qquad \varphi(x) = c_1 \varphi_1(x) + c_2 \varphi_2(x)$$

where $\varphi_1(x)$ and $\varphi_2(x)$ remain centered about respective atomic sites.

To evaluate the coefficients c_1 and c_2, we recall Problems 4.28 and 4.30 according to which the ground state may be obtained through the variational statement

$$(8.122) \qquad \delta \langle E \rangle = 0$$

where

$$(8.123) \qquad \langle E \rangle = \frac{\langle \varphi | H | \varphi \rangle}{\langle \varphi | \varphi \rangle}$$

In developing the variational statement (8.122), we assume c_1, c_2, φ_1, and φ_2 to be real. There results

$$(8.124) \quad \langle E \rangle = \frac{\langle c_1 \varphi_1 + c_2 \varphi_2 | \hat{H} | c_1 \varphi_1 + c_2 \varphi_2 \rangle}{\langle c_1 \varphi_1 + c_2 \varphi_2 | c_1 \varphi_1 + c_2 \varphi_2 \rangle} = \frac{c_1^2 H_{11} + c_2^2 H_{22} + 2 c_1 c_2 H_{12}}{c_1^2 + c_2^2 + 2 c_1 c_2 S}$$

where

$$(8.124a) \qquad S = \langle \varphi_1 | \varphi_2 \rangle < 1$$

represents the overlap integral between φ_1 and φ_2. Due to the symmetry of our configuration,

$$H_{11} = \langle \varphi_1 | \hat{H} \varphi_1 \rangle = H_{22} = \langle \varphi_2 | \hat{H} \varphi_2 \rangle$$

As c_1 and c_2 are the only free variables in (8.124), to minimize $\langle E \rangle$ it suffices to set

$$(8.125) \qquad \frac{\partial \langle E \rangle}{\partial c_1} = \frac{\partial \langle E \rangle}{\partial c_2} = 0$$

There results

(8.126)
$$c_1(H_{11} - \langle E \rangle) + c_2(H_{12} - \langle E \rangle S) = 0$$

$$c_1(H_{12} - \langle E \rangle S) + c_2(H_{22} - \langle E \rangle) = 0$$

with c representing a column vector with components (c_1, c_2). The preceding equations may be written in the matrix form

(8.126a)
$$Rc = 0$$

where R is the implied coefficient matrix. A nontrivial solution to (8.131a) results, providing det $R = 0$. There results

(8.127)
$$\begin{vmatrix} H_{11} - \langle E \rangle & H_{12} - \langle E \rangle S \\ H_{12} - \langle E \rangle S & H_{11} - \langle E \rangle \end{vmatrix} = 0$$

which gives

(8.128)
$$\langle E \rangle_{\pm} = \frac{H_{11} \pm H_{12}}{1 \pm S}$$

Substituting this finding into (8.130) reveals that $\langle E \rangle_-$ corresponds to the odd wavefunction $(c_1 = -c_2)$ and $\langle E \rangle_+$ to the even wavefunction $(c_1 = c_2)$. Consequently, we conclude that the even wavefunction, φ_+ corresponds to $\langle E \rangle_+$ and that the odd wavefunction, φ_- corresponds $\langle E \rangle_-$. These states, with corresponding energies, are listed below (with $\langle E \rangle_{\pm}$ replaced by $E_{\pm}$).

(8.129)
$$E_- = \frac{H_{11} - H_{12}}{1 - S}, \qquad \varphi_- = \frac{A_-}{\sqrt{2}}(\varphi_1 - \varphi_2)$$

$$E_+ = \frac{H_{11} + H_{12}}{1 + S}, \qquad \varphi_+ = \frac{A_+}{\sqrt{2}}(\varphi_1 + \varphi_2)$$

where the normalization constants are given by

(8.129a)
$$A_{\pm} \equiv \frac{1}{\sqrt{1 \pm S}}$$

For the configuration at hand (see Fig. 8.30) we may set

(8.130)
$$H_{11} = -|H_{11}|$$

$$H_{12} = -|H_{12}|$$

Substituting these values into (8.129) reveals that for sufficiently small wavefunction overlap, the ground-state wavefunction and energy for the molecule at hand are

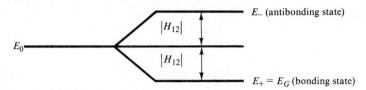

FIGURE 8.33   The unperturbed twofold degeneracy in energy of the uncoupled molecule is removed when atoms interact. For small wavefunction overlap, within the LCAO approximation, the spread in new energies is $2|H_{12}|$. **In the bonding state the valence electron partially occupies the midregion between ions, thereby** tending to bind them and lower the energy.

given by

$$\varphi_G = \varphi_+$$

(8.131)

$$E_G = E_+ = -\frac{(|H_{11}| + |H_{12}|)}{1 + S}$$

The difference between new energies is

(8.132)

$$\Delta E = E_- - E_+ = 2\frac{(|H_{12}| - S|H_{11}|)}{1 - S^2}$$

For small wavefunction overlap, $S \ll 1$, we find

(8.133)

$$\Delta E = 2|H_{12}|$$

Thus we find that the interaction incurred by bringing the atoms in close proximity removes the twofold degeneracy of the uncoupled molecule. [Recall (8.120).] This situation is depicted in Fig. 8.33.

The lower energy of the symmetric state, φ_+, may be understood on the basis of the following. In this state the valence electron has a finite probability of occupying the mid-domain between positive ions, as opposed to the situation in the antisymmetric state, φ_-, for which the electron has zero probability of being found in this domain. Thus, the charge density associated with φ_+ tends to bind the positive ions, thereby lowering the energy of the system. A more realistic description of this situation is given in Section 12.7, which addresses the actual H_2 molecule.

Additional problems in this category of estimating properties of a system are encountered in Chapter 13 under the general topic of *perturbation theory*. In this context the configuration $d \gg a$ in the problem above is viewed as the *unperturbed system*. Perturbation is incurred due to the interaction between atoms, which comes into play with decrease in the separation d. Note in particular that the ground state in the present configuration is doubly degenerate and that the perturbation removes this degeneracy. See Fig. 8.33. That is, with the interaction between atoms "turned on," the twofold degenerate energy E_0 is split into two separate values, $E_\pm$. *Degenerate perturbation theory* is described in detail in Section 13.2.

PROBLEMS

8.44   Consider the following Gaussian form for the wavefunction of atom 1 corresponding to the configuration shown in Fig. 8.30.

$$\varphi_1(x) = A \exp\left[-\left(\frac{x+d}{\sqrt{2a}}\right)^2\right]$$

(a)   What is the value of the normalization constant A, in terms of the separation distance $2d$?

(b)   What is the form of the companion wavefunction $\varphi_2(x)$?

(c)   Obtain an expression for the overlap integral S (8.130a). In what manner does your expression vary with the parameter $\lambda \equiv d/a$?

8.45   Show that E_+, as given by (8.128), corresponds to $(c_1/c_2) = \pm 1$ in (8.121).

8.46   Establish the equality $H_{11} = H_{22}$. [See equation above (8.125).]

8.47   A linear molecule comprised of three identical, singly ionized atoms share a single electron. This molecular ion is modeled by three identical quantum wells, each of width a and depth V_0. The wells are separated from each other by the distance d and are symmetrically displayed about $x = 0$.

Let $\varphi_1, \varphi_2, \varphi_3$ be the ground-state wavefunctions of the three separate wells.

(a)   What is the net charge (in units of e) of the molecular ion?

(b)   Write down the Hamiltonian for the ionized triatomic molecule. What are the criteria for an uncoupled molecule?

(c)   What are the possible states φ of the uncoupled molecule in terms of $\varphi_1, \varphi_2, \varphi_3$ for the case of equal probabilities of occupation among the three wells?

(d)   What is the LCAO *form* of the ground state for the triatomic molecule? Call your constants c_1, c_2, c_3.

(e)   What is the form of $\langle E \rangle$ for this molecule?

(f)   Estimate the eigenenergies of this molecule corresponding to the wavefunction you have constructed, through the null variation of $\langle E \rangle$. Assume $H_{11} = H_{22} = H_{33} \equiv A$, $H_{12} = H_{21} = B$, and $H_{13} = H_{31} = S_{12} = S_{23} = S_{13} = 0$.

(g)   What are the relations among the constants c_1, c_2, c_3 corresponding to the energies you have found in part (f)? Sketch the corresponding wavefunctions and compare with corresponding sketches for part (c).

(h)   What is the ground-state wavefunction φ_G and ground-state energy E_G for this molecule, within the LCAO approximation?

Answers (partial)

   (a) $+2|e|$
   (c) $\varphi = \varphi_1 \pm (\varphi_2 \pm \varphi_3)$
   (d) $\varphi = c_1\varphi_1 + c_2\varphi_2 + c_3\varphi_3$

8.48   A broad frequency band of photons is incident on a rare gas of homonuclear, singly ionized, diatomic molecules. It is noted that photons of frequency $v = 0.75 \times 10^{15}$ Hz are absorbed by the gas. Values of constants in the one-dimensional LCAO model of this molecule are $|H_{11}| = 10$ eV, $|H_{12}| = 3$ eV. What is the value of the overlap integral S? State units of your answer.

8.8   DENSITY OF STATES IN VARIOUS DIMENSIONS

We recall (see Problem 2.37) that the density of states $g(E)$ is defined so that $g(E)\,dE$ gives the number of energy states in the interval $E, E + dE$.

   In this section we wish to obtain expressions for $g(E)$, relevant to a particle confined to boxes in one, two and three dimensions. These results are then applied to a quantum well-defined in "slab geometry" important to electron-device physics.[1]

One Dimension

For a one-dimensional box, eigenenergies are given by (4.14):

$$(8.134) \qquad E_n = n^2 E_1, \qquad E_1 = \frac{h^2}{8mL^2}$$

It follows that there is an energy state for each value of the quantum number n. As is evident from (8.134), in the limit of large L, the separation between energy states diminishes and the energy spectrum grows quasi-continuous.[2] In this limit we may write

$$(8.135) \qquad g(E) = \frac{\Delta n}{\Delta E} \simeq \frac{dn}{dE}$$

which, with (8.134), gives

$$(8.136) \qquad g_1(E) = \frac{1}{2\sqrt{E_1 E}}$$

---

[1] For further discussion, see G. Burns, *Solid State Physics*, Academic Press, New York, 1985.
[2] Since energy increments grow with n, this quasi-continuous spectrum is realized for unbounded L.

where the subscript 1 denotes one dimension. Thus, for a particle confined to a one-dimensional box, the density of states vanishes in the limit of large energy or, equivalently, large quantum number.

Two Dimensions

The two-dimensional square box was discussed in detail in Section 8.5. With the energy eigenvalue equation (8.107) we write

(8.137)
$$E_{n_1 n_2} = E_1(n_1^2 + n_2^2)$$

There is an energy eigenvalue for every pair of quantum numbers (n_1, n_2). It follows that in Cartesian $n_1 n_2$ space, there is an energy eigenstate at every point (n_1, n_2) of this space. (Compare with Fig. 2.17.) Let

$$n = \sqrt{n_1^2 + n_2^2}$$

denote the radius vector in this space. Thus, all points that lie in the annular region $(n, n + dn)$ have the same energy $(E \propto n^2)$. Again, in the extreme of large L, eigenenergies form a quasi-continuum and the number of such points is given by the area of the annular region. There results

(8.138)
$$g(E)\, dE = \tfrac{1}{4} \times 2\pi n\, dn = \frac{\pi}{4E_1}\, dE$$

The factor $\tfrac{1}{4}$ insures that only positive quantum numbers are included. It follows that in two dimensions, the density of states is given by

(8.139)
$$g_2(E) = \frac{\pi}{4E_1}$$

which is constant.

Three Dimensions

The energy eigenvalue equation for a particle confined to a three-dimensional cubical box is given by (see Table 10.2)

(8.140)
$$E_{n_1 n_2 n_3} = E_1(n_1^2 + n_2^2 + n_3^2)$$

In this case there is an energy eigenvalue corresponding to each point in Cartesian $n_1 n_2 n_3$ space. Again we define the radius vector

$$n = \sqrt{n_1^2 + n_2^2 + n_3^2}$$

and note once more that in the limit of large L all points within the spherical shell, $n, n + dn$, have the same energy. There results

$$g(E)\, dE = \tfrac{1}{8} \times 4\pi n^2\, dn = \frac{\pi}{4} \frac{\sqrt{E}}{E_1^{3/2}} dE$$

and

(8.141)
$$g_3(E) = \frac{\pi}{4} \frac{\sqrt{E}}{E_1^{3/2}}$$

where the factor of $\tfrac{1}{8}$ in the preceding equation insures that only positive quantum numbers are included.

Here is a recapitulation of preceding results.

(8.142a)
$$g_1 = \frac{1}{2\sqrt{E_1 E}}$$

(8.142b)
$$g_2 = \frac{\pi}{4E_1}$$

(8.142c)
$$g_3 = \frac{\pi}{4} \frac{\sqrt{E}}{E_1^{3/2}}$$

If the particles we are considering have spin $\tfrac{1}{2}$, then each of the g-values above is multiplied by the factor 2. (The concept of spin is discussed in Chapter 11.)

Density of States per Unit Volume

With (8.135) we write

(8.143)
$$E_1 = \left(\frac{h^2}{8m}\right) \frac{1}{L^2} \equiv \frac{\alpha}{L^2}$$

where α is the inferred constant, which is seen to be independent of the box dimension L. Inserting this value into (8.142) gives the following expression for the density of states per unit volume for a particle confined to boxes in one-, two- and three-dimensional Cartesian space, respectively.

(8.144a)
$$\bar{g}_1 = \frac{g_1}{L} = \frac{1}{2\sqrt{\alpha E}}$$

(8.144b)
$$\bar{g}_2 = \frac{g_2}{L^2} = \frac{\pi}{4\alpha}$$

(8.144c)
$$\bar{g}_3 = \frac{g_3}{L^3} = \frac{\pi}{4} \frac{\sqrt{E}}{\alpha^{3/2}}$$

(See Problem 8.49.)

Slab Geometry

We wish to apply the preceding results to calculate the density of states of a particle of mass m confined to a well in slab geometry defined by the following potential:

(8.145a)
$$\left. \begin{array}{l} 0 \le z \le a \\ 0 \le x \le L \\ 0 \le y \le L \end{array} \right\} V = 0$$

(8.145b)
$$V = \infty, \text{ elsewhere}$$

(8.145c)
$$a \ll L$$

See Fig. 8.34.

The density of states for the particle so confined is an appropriate combination of $g_1{}^{(a)}$ (8.142a) relevant to the one-dimensional z confinement and $g_2{}^{(L)}$ (8.142b) relevant to the two-dimensional xy confinement.

We first note that the energy eigenfunctions of this particle may be written in the product form [compare with (8.111)]:

(8.146)
$$\varphi(x, y, z) = \varphi_a(z)\varphi_L(x, y)$$

where

(8.146a)
$$\varphi_a(z) = \sqrt{\frac{2}{a}} \sin \frac{n_1 \pi z}{a}$$

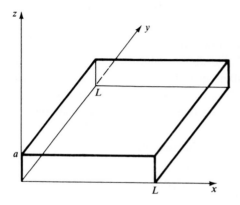

FIGURE 8.34 Sketch of the quantum well corresponding to the potential (8.145), with $L \gg a$.

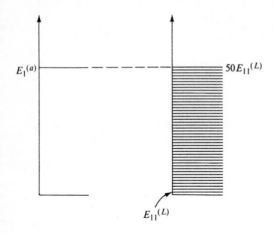

FIGURE 8.35 Sketch of the eigenenergies corresponding to the z and xy motions in the quantum well (8.145) for the case $a = L/10$. States with energy $E < E_1^{(a)}$ do not exist.

is relevant to the one-dimensional z box and

(8.146b)
$$\varphi_L(x, y) = \frac{2}{L} \sin \frac{n_2 \pi x}{L} \sin \frac{n_3 \pi y}{L}$$

is relevant to the two-dimensional xy box.

Eigenenergies are given by the sum

(8.147)
$$E_{n_1 n_2 n_3} = \frac{h^2}{8m} \left[\frac{n_1^2}{a^2} + \left(\frac{n_2^2 + n_3^2}{L^2} \right) \right] \equiv E_{n_1}^{(a)} + E_{n_1 n_2}^{(L)}$$

As $L \gg a$, $E_{1,1}^{(L)} \ll E_1^{(a)}$ [recall (8.134)]. For example, if $a = L/10$, then $E_1^{(a)} = 50E_{1,1}^{(L)}$. See Fig. 8.35. However, as the wavefunction of the confined particle is the product (8.146), no motion exists for energy less than the ground-state energy of the z motion. Thus, $g(E < E_1^{(a)}) = 0$. Furthermore, the density of states for the xy motion is constant. [See (8.142).]

Thus, for the given configuration we have the following picture. The density of states $g(E)$ is zero for $E < E_1^{(a)}$. At $E = E_1^{(a)}$, two-dimensional motion is allowed and $g(E_1^{(a)}) = \pi/4E_1^{(L)}$, where $E_1^{(L)}$ is the constant in (8.134). For larger energy, $g(E)$ maintains this constant value until the second excited state of the z motion is reached. At this value, $g(E_2^{(a)}) = 2g(E_1^{(a)})$, as now there are two allowed (z) modes for the xy motion.

This sequential development of $g(E)$ leads to the formula

(8.148)
$$g(E) = \frac{\pi}{4E_1^{(L)}} \sum_{n=1}^{\infty} \theta(E - E_n^{(a)})$$

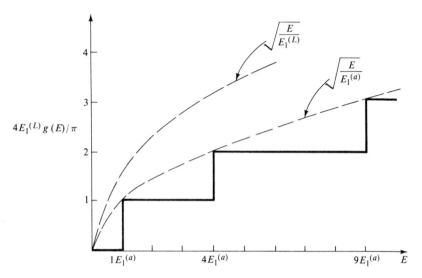

FIGURE 8.36   In the limit $a \to L$, with L large, $E_1^{(a)} \to 0$ and $g(E) \to g_3(E)$, relevant to a large cubical box of edge length L. In this same limit, the dashed curves coalesce.

where $\theta(x)$ is the step function

(8.148a)
$$\theta(x) = 1, \qquad x > 0$$
$$\theta(x) = 0, \qquad x < 0$$

See Fig. 8.36.

As a check on our calculation we should find that

$$g(E) \to g_3(E)$$

in the limit

$$a \to L$$

As is evident from Fig. 8.36,

$$\text{envelope of} \left[\sum_{n=1}^{\infty} \theta(E - E_n^{(a)}) \right] = \sqrt{\frac{E}{E_1^{(a)}}}$$

In the limit that $a = L$, which is large, the increments of the θ function grow closer and θ is well approximated by the envelope function. Thus

$$g(E) \to \frac{\pi}{4E_1^{(L)}} \sqrt{\frac{E}{E_1^{(L)}}} = g_3(E)$$

In device physics, the configuration in slab geometry considered above is called a "quantum well" and is realized by epitaxial growth of layers of semiconductors with different band gaps (see Section 12.9). Interfaces between different semiconductors are called *heterojunctions*. A multilayered structure of alternate semiconductors effects an array of quantum wells and is called a *superlattice*.

PROBLEMS

8.49   (a)   Obtain the density of states per unit volume for a free particle of mass m moving in 3-space. Call your answer $\bar{g}_0(E)$.

(b)   Compare your expression for $g_0(E)$ with $\bar{g}_3(E)$ as given by (8.144c). Explain the difference (or equality) between these results.

Answers

(a)   In six-dimensional Cartesian $\mathbf{x} - \mathbf{p}$ space there is an energy state at each point $(\mathbf{x}, \mathbf{p})$. Due to the uncertainty relation, the minimum volume for a state in phase space[1] is h^3. Thus,

$$g_0(E)\, dE = \frac{d\mathbf{x}\, d\mathbf{p}}{h^3}$$

For a free particle

$$E = \frac{p^2}{2m}$$

so that states in momentum space in the spherical shell $(p, p + dp)$ all have the same energy. As the volume of this shell is $4\pi p^2\, dp$, we obtain

$$\bar{g}_0(E)\, dE = \frac{4\pi p^2\, dp}{h^3}$$

With the preceding expression there results

$$\bar{g}_0(E) = 2\pi \frac{(2m)^{3/2}}{h^3}\sqrt{E} = \frac{\pi}{4}\frac{\sqrt{E}}{\alpha^{3/2}} = \bar{g}_3(E)$$

(b)   We conclude that $\bar{g}_0(E)$ and $\bar{g}_3(E)$ are equal. However, it should be borne in mind that in summing states, $\bar{g}_3(E)$ is summed over the discrete energies of the box, whereas $\bar{g}_0(E)$ is integrated over the continuum of free-particle E-values. We may conclude that in any finite volume and energy interval, there are more free-particle energy states than box energy states.

8.50   Describe qualitatively how Fig. 8.36 changes if the well in the z direction of the quantum well is of finite potential.

---

[1] If the particle we are considering has, say, spin $\frac{1}{2}$, then this minimum volume is decreased by $\frac{1}{2}$. The concept of spin is discussed in detail in Chapter 11.

8.51   A particle of mass m is confined to a one-dimensional box of edge length L. Let the edge length be increased to $L + \delta L$. If $\Delta_0 E$ represents the difference between adjacent energies for the first box and $\Delta_1 E$ the difference of adjacent energies for the expanded box, show that

$$\Delta_1 E < \Delta_0 E$$

That is, show that energies grow closer with increase in L.

8.52   A particle of mass m is confined to move in two dimensions within an impenetrable rectangular barrier with edge-lengths (a, b). Let the rectangle be situated in the first quadrant of Cartesian space with one corner at the origin and adjoining edges aligned with the (x, y) axes.

   (a)   Working in this configuration, obtain the eigenstates, $\varphi_{nn'}(x, y)$, and eigenenergies, $E_{nn'}$ of the particle. Write eigenenergies in terms of E_0, the ground state of a particle in a one-dimensional box of edge-length a.

   (b)   Show that if a/b is a rational number, then states of the system are degenerate.

PART II

FURTHER DEVELOPMENT OF THE THEORY AND APPLICATIONS TO PROBLEMS IN THREE DIMENSIONS

CHAPTER 9

ANGULAR MOMENTUM

Our study of the applications of quantum mechanics to three-dimensional problems begins with a description of the properties of angular momentum. We first consider orbital angular momentum, which is closely akin to angular momentum encountered in classical physics. Angular momentum in quantum mechanics, however, is a more general concept than its classical counterpart. In quantum mechanics, in addition to orbital angular momentum, one also encounters spin angular momentum. Spin angular momentum is an intrinsic, or internal, property of elementary particles such as electrons and photons, and has no classical counterpart. The operators corresponding to the Cartesian components of angular momentum in quantum mechanics obey a set of fixed, fundamental commutator relations. These relations are first derived for orbital angular momentum and then employed as the defining relations for angular momentum in general. Eigenvalues of angular momentum stemming from these commutator relations are obtained, and it is at this point that a distinction between orbital and spin angular momentum first emerges.

9.1 BASIC PROPERTIES

The significance of angular momentum in classical physics is that it is one of the fundamental constants of motion (together with linear momentum and energy) of an isolated system. As we will find, the counterpart of this statement also holds for isolated quantum mechanical systems. This conservation principle for angular momentum stems from the isotropy of space. That is, as described previously in Section 6.3, the physical laws relating to an isolated system are in no way dependent on the orientation of that system with respect to some fixed set of axes in space.

Classically, angular momentum of a particle is a property that depends on the particle's linear momentum $\mathbf{p}$ and its displacement $\mathbf{r}$ from some prescribed origin. It is given by (see Fig. 1.9)

$$\text{(9.1)} \qquad \mathbf{L} = \mathbf{r} \times \mathbf{p}$$

One may also speak of the angular momentum of a system of particles, or of a rigid body. For such extended aggregates, one must add the angular momentum of all particles in the system to obtain the total angular momentum of the system.

Cartesian Components

The classical Cartesian components of the orbital angular momentum $\mathbf{L}$ for a particle with momentum $\mathbf{p} = (p_x, p_y, p_z)$ at the displacement $\mathbf{r} = (x, y, z)$ are

$$\text{(9.2)} \qquad \begin{aligned} L_x &= yp_z - zp_y \\ L_y &= zp_x - xp_z \\ L_z &= xp_y - yp_x \end{aligned}$$

The quantum mechanical operators $\hat{L}_x$, $\hat{L}_y$, and $\hat{L}_z$, corresponding to these observables, derive their definitions directly from the classical expressions above, with $\hat{\mathbf{p}}$ replaced by its corresponding gradient operator. There follows

$$\text{(9.3)} \qquad \begin{aligned} \hat{L}_x &= \hat{y}\hat{p}_z - \hat{z}\hat{p}_y = -i\hbar\left(y\frac{\partial}{\partial z} - z\frac{\partial}{\partial y} \right) \\[2mm] \hat{L}_y &= \hat{z}\hat{p}_x - \hat{x}\hat{p}_z = -i\hbar\left(z\frac{\partial}{\partial x} - x\frac{\partial}{\partial z} \right) \\[2mm] \hat{L}_z &= \hat{x}\hat{p}_y - \hat{y}\hat{p}_x = -i\hbar\left(x\frac{\partial}{\partial y} - y\frac{\partial}{\partial x} \right) \end{aligned}$$

In terms of the three-dimensional vector linear momentum operator

$$\text{(9.4)} \qquad \hat{\mathbf{p}} = (\hat{p}_x, \hat{p}_y, \hat{p}_z) = -i\hbar\left(\frac{\partial}{\partial x}, \frac{\partial}{\partial y}, \frac{\partial}{\partial z} \right) = -i\hbar\nabla$$

the equations above may be written as the single vector equation

$$(9.5) \qquad \mathbf{L} = -i\hbar \mathbf{r} \times \nabla$$

Commutator Relations

Let us examine the commutation properties of these operators. If, for example, $\hat{L}_x$ does not commute with $\hat{L}_y$, then these components of angular momentum cannot be simultaneously specified in a single state, that is, these operators do not have common eigenfunctions.

To examine this specific question, we employ the basic commutator relation

$$(9.6) \qquad [\hat{x}, \hat{p}_x] = i\hbar$$

There follows

$$(9.7) \qquad \begin{aligned} [\hat{L}_x, \hat{L}_y] &= \hat{L}_x \hat{L}_y - \hat{L}_y \hat{L}_x \\ &= (\hat{y}\hat{p}_z - \hat{z}\hat{p}_y)(\hat{z}\hat{p}_x - \hat{x}\hat{p}_z) - (\hat{z}\hat{p}_x - \hat{x}\hat{p}_z)(\hat{y}\hat{p}_z - \hat{z}\hat{p}_y) \\ &= \hat{x}\hat{p}_y(\hat{z}\hat{p}_z - \hat{p}_z\hat{z}) - \hat{y}\hat{p}_x(\hat{z}\hat{p}_z - \hat{p}_z\hat{z}) \\ &= i\hbar(\hat{x}\hat{p}_y - \hat{y}\hat{p}_x) \\ &= i\hbar\hat{L}_z \end{aligned}$$

In similar fashion we obtain

$$(9.8) \qquad \boxed{\begin{aligned} [\hat{L}_y, \hat{L}_z] &= i\hbar\hat{L}_x \\ [\hat{L}_z, \hat{L}_x] &= i\hbar\hat{L}_y \\ [\hat{L}_x, \hat{L}_y] &= i\hbar\hat{L}_z \end{aligned}}$$

These commutator relations are sometimes combined in the single vector equation

$$(9.9) \qquad i\hbar\hat{\mathbf{L}} = \hat{\mathbf{L}} \times \hat{\mathbf{L}}$$

which in determinantal form appears as

$$(9.10) \qquad i\hbar(\mathbf{e}_x \hat{L}_x + \mathbf{e}_y \hat{L}_y + \mathbf{e}_z \hat{L}_z) = \begin{vmatrix} \mathbf{e}_x & \mathbf{e}_y & \mathbf{e}_z \\ \hat{L}_x & \hat{L}_y & \hat{L}_z \\ \hat{L}_x & \hat{L}_y & \hat{L}_z \end{vmatrix}$$

As illustrated in Problem 9.1, only one of the three Cartesian components of angular momentum may be specified in a quantum mechanical state. Suppose, for example, that φ is an eigenstate of $\hat{L}_z$. What will measurement of $\hat{L}_x$ find? To answer this question we must bring the superposition principle into play. Expand φ in the eigenstates of $\hat{L}_x$. The squares of the coefficients of expansion give the distribution of probabilities of finding different values of L_x.

Although no two values of the Cartesian components of angular momentum can be simultaneously specified in a quantum mechanical state, if one component, say the value of L_z, is specified, it is still possible to specify an additional property of angular momentum in that state. This additional property is the value of the square of the total angular momentum, L^2, or, equivalently, the magnitude of $\mathbf{L}$ ($L = \sqrt{\mathbf{L} \cdot \mathbf{L}} = \sqrt{L^2}$).

The total angular momentum operator is the vector operator

$$(9.11) \qquad \hat{\mathbf{L}} = \mathbf{e}_x \hat{L}_x + \mathbf{e}_y \hat{L}_y + \mathbf{e}_z \hat{L}_z$$

from which we may form $\hat{L}^2$.

$$(9.12) \qquad \hat{L}^2 = \hat{L}_x{}^2 + \hat{L}_y{}^2 + \hat{L}_z{}^2$$

To show that there are states of a system in which L_z and L^2 are simultaneously specified, one need merely show that $\hat{L}_z$ and $\hat{L}^2$ commute. Then we know that these operators have simultaneous eigenfunctions. That is, there are states that are eigenfunctions of both $\hat{L}_z$ and $\hat{L}^2$. Let us prove the commutability of $\hat{L}_z$ and $\hat{L}^2$.

$$
\begin{aligned}
[\hat{L}_z, \hat{L}^2] &= [\hat{L}_z, \hat{L}_x{}^2 + \hat{L}_y{}^2 + \hat{L}_z{}^2] \\
&= [L_z, L_x{}^2] + [L_z, L_y{}^2] + 0 \\
&= L_x[L_z, L_x] + [L_z, L_x]L_x + L_y[L_z, L_y] + [L_z, L_y]L_y \\
&= i\hbar[L_x L_y + L_y L_x - L_y L_x - L_x L_y] \\
&= 0
\end{aligned}
$$

In similar manner we find that $\hat{L}_x$ and $\hat{L}_y$ also commute with $\hat{L}^2$. This must be the case because we have in no way given any special significance to the z direction. In general

$$(9.13) \qquad \begin{aligned} [\hat{L}_x, \hat{L}^2] &= [\hat{L}_y, \hat{L}^2] = [\hat{L}_z, \hat{L}^2] = 0 \\ [\hat{\mathbf{L}}, \hat{L}^2] &= 0 \end{aligned}$$

It follows that the Cartesian components of $\hat{\mathbf{L}}$ have simultaneous eigenfunctions with $\hat{L}^2$. However, the individual components of $\hat{\mathbf{L}}$ do not have common eigenstates with one another (except for the special case of zero angular momentum). These properties are depicted in a Venn diagram in Fig. 9.1.

The preceding discussion tells us that $\hat{L}^2$ and $\hat{L}_z$, say, have common eigenfunctions. Let us call these eigenfunctions φ_{lm}. The integral indices l and m are related to the eigenvalues of $\hat{L}^2$ and $\hat{L}_z$ as in the following eigenfunction equations.

$$(9.14) \qquad \begin{aligned} \hat{L}^2 \varphi_{lm} &= \hbar^2 l(l+1)\varphi_{lm} \qquad &(l = 0, 1, 2, \ldots) \\ \hat{L}_z \varphi_{lm} &= \hbar m \varphi_{lm} \qquad &(m = -l, \ldots, +l \text{ in integral steps}) \end{aligned}$$

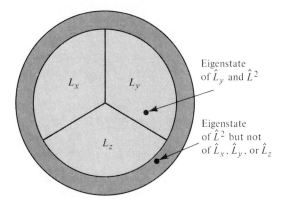

FIGURE 9.1 Venn diagram for the eigenstates of $\hat{L}_x$, $\hat{L}_y$, $\hat{L}_z$, and $\hat{L}^2$. Every point represents an eigenfunction of $\hat{L}^2$. Depending on which sector the point is in, it is also an eigenfunction of $\hat{L}_x$, $\hat{L}_y$, or $\hat{L}_z$. The state at the center is the null eigenvector of $\hat{\mathbf{L}}$ and $\hat{L}^2$. It corresponds to the eigenvalues $L_x = L_y = L_z = 0$. Peripheral points depict states that are eigenstates of $\hat{L}^2$ only. Can you think of one such function? Note that the space of eigenstates of $\hat{L}^2$ is "bigger" than the space containing all the eigenstates of $\hat{L}_x$, $\hat{L}_y$, and $\hat{L}_z$. Compare with Fig. 5.12.

(These equations are derived in the next section.) The form of the first equation indicates the following. Suppose that a system (e.g., a wheel) is rotating somewhere in space, far removed from other objects. We measure the magnitude of its angular momentum. What possible values can be found? The values that experiment finds are only of the form $L = \hbar\sqrt{l(l + 1)}$, where l is some integer. For example, one would never measure the value $L = \hbar\sqrt{7}$, since it is not of the form $L^2 = \hbar^2 l(l + 1)$. There is no integer for which $l(l + 1) = 7$. This is similar to the fact that a particle in a one-dimensional box is never found to have the energy $E = 7E_1$. This value does not fit the energy eigenvalue recipe $E = n^2 E_1$.

Suppose that we measure the magnitude of angular momentum of the wheel and find the value $L^2 = 30\hbar^2$. This corresponds to the l value $l = 5$. Having measured L^2, the system is left in an eigenstate of $\hat{L}^2$. What value does subsequent measurement of L_z yield? The answer is given by the form of the eigenvalues of $\hat{L}_z$ given in (9.14). For the case in point, since $l = 5$, L_z can *only* be found to have one of the eleven values

$$L_z = 5\hbar,\, 4\hbar,\, 3\hbar,\, 2\hbar,\, \hbar,\, 0,\, -\hbar,\, -2\hbar,\, -3\hbar,\, -4\hbar,\, -5\hbar$$

Suppose that measurement finds $L_z = 3\hbar$. Then the wheel is left in the state $\varphi_{5,3}$.

The form of equations (9.14) indicates that the eigenvalues of $\hat{L}^2$ are $(2l + 1)$-fold degenerate. For the problem considered, all the eleven states $\varphi_{5,5}; \varphi_{5,4}; \dots; \varphi_{5,-5}$ correspond to the same value of L^2 (i.e., $L^2 = 30\hbar^2$). (See Fig. 9.2.)

Uncertainty Relations

Angular momentum is a vector. The magnitude of this vector is given by L^2. Having measured L^2, is it possible to measure any of the three Cartesian components of $\mathbf{L}$

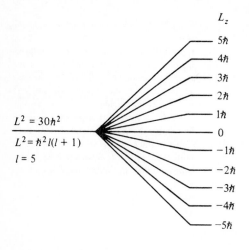

L_z

— $5\hbar$
— $4\hbar$
— $3\hbar$
— $2\hbar$
— $1\hbar$
— 0

$L^2 = 30\hbar^2$

$L^2 = \hbar^2 l(l+1)$

$l = 5$

— $-1\hbar$
— $-2\hbar$
— $-3\hbar$
— $-4\hbar$
— $-5\hbar$

FIGURE 9.2   The eigenvalue $L^2 = \hbar^2 l(l+1)$ is $(2l+1)$-**fold degenerate. For a fixed magnitude,** $L = \hbar\sqrt{l(l+1)}$, **there are only** $2l+1$ **possible projections of L onto a given axis. (See Fig. 9.3c.)**

and leave the system (such as a wheel, a particle, an atom, a rigid rod, etc.) with the same value of L^2 that it had before measurement? Specifically, suppose that we measure L^2 and L_z and find the values $56\hbar^2$ and $3\hbar$, respectively ($l = 7$, $m = 3$). We know that the system is left in a simultaneous eigenstate of L^2 and L_z, namely, $\varphi_{7,3}$.

It is impossible to further resolve the state of the system. We cannot obtain more information on the vector **L** without destroying part of the information already known. Suppose that L_x is measured and the value $5\hbar$ is found. In measuring L_x, the information about L_z previously determined is destroyed.[1] The system is left in a simultaneous eigenstate of $\hat{L}^2$ and $\hat{L}_x$. Since this is not an eigenstate of $\hat{L}_z$, subsequent measurement of L_z is not certain to yield any specific value. Similarly for measurement of L_y. This conclusion is contained in the uncertainty relation

(9.15) $$\Delta L_y\,\Delta L_z \geq \frac{\hbar}{2}|\langle L_x\rangle| = \frac{\hbar L_x}{2} = \frac{5\hbar^2}{2}$$

Consider the case of a wheel whose center is fixed in space. L^2 and L_z are measured. What motion of the wheel will preserve these values but not preserve L_x and L_y? A very worthwhile model for such motion is given by a classical solution in which the angular momentum vector of constant magnitude precesses about the z axis at a constant inclination to that axis (see Fig. 9.3), thereby maintaining L_z. (Such motion is realized by a spinning top, with fixed vertex, in a gravity field.)

In the classical problem **L** is precisely determined as a function of time. At any instant **L** may be observed and completely specified. Not so for the quantum mechanical motion. If the wheel is in an eigenstate of $\hat{L}^2$ and $\hat{L}_z$, it is in a superposition

---

[1] That is, the outcome of subsequent measurement of L_z is rendered more uncertain.

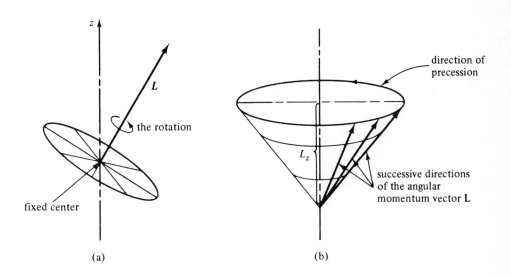

(a)

the rotation

L

fixed center

z

direction of
precession

successive directions
of the angular
momentum vector **L**

L_z

(b)

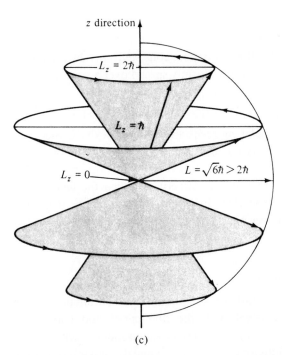

z direction

$L_z = 2\hbar$

$L_z = \hbar$

$L_z = 0$

$L = \sqrt{6}\hbar > 2\hbar$

(c)

FIGURE 9.3 (a) The angular momentum vector **L** of a rotating wheel whose center is fixed in space.
(b) Classical precession of **L** about the z axis with the constant projection L_z. (c) For $l = 2$, $L^2 = 6h^2$. The only
possible orientations of **L** onto the z axis are the five values shown. The precessional motion depicted preserves
L^2 and L_z. $\theta = \cos^{-1} 2/\sqrt{6}$ is the smallest possible angle between **L** and the z axis.

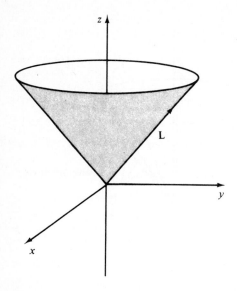

FIGURE 9.4 For the quantum mechanical state in which L^2 and L_z are specified, L may be pictured as being uniformly distributed over the surface of a cone with half-apex-angle $\theta = \cos^{-1} m/\sqrt{l(l+1)}$.

state (i.e., a linear combination of the eigenstates) of $\hat{L}_x$ or $\hat{L}_y$. At best one can only speak of the *probability* of finding a certain value of L_x or L_y. If a system is in such a state with definite l and m values, it is therefore more consistent to view the related configuration as one in which the **L** vector is uniformly spread over a cone about the z axis with half apex angle $\theta = \cos^{-1} m/\sqrt{l(l+1)}$ (see Fig. 9.4).

For a given value of L [i.e., $\hbar\sqrt{l(l+1)}$] the maximum value of L_z is $\hbar l$. But $l < \sqrt{l(l+1)}$. It follows that the angular momentum vector is never aligned with a given axis. Furthermore, there are only a discrete, finite $(2l + 1)$ number of inclinations that **L** makes with any given axis. This extraordinary property (classical physics permits a continuum of inclinations) is sometimes called the *quantization of space*. For reasons that will become clear in the following sections, l is often referred to as the *orbital quantum number* while m is often referred to as the *azimuthal* or *magnetic quantum number*.

Orbital Versus Spin Angular Momentum

The commutator relations (9.8) are the trademark of angular momentum in quantum mechanics. Although they are consistent with the differential and coordinate-momentum operator relations (9.3 et seq.), they may be taken independent of these and assumed to be the defining relations for quantum mechanical angular momentum. When such is the case, angular momentum need not refer to the space coordinates or linear momentum components of a particle, since the relations (9.8) by themselves do not. The first example that incorporates this concept is given in Section 9.2, where the eigenvalues of angular momentum are obtained using only the commutator relations (9.8). As will be shown in Section 9.3, only a subset of these eigenvalues are relevant to *orbital* angular momentum. Orbital angular momentum derives from the space and momentum coordinates of a particle and is akin to classical (**r** × **p**) angular momentum. In contrast, *spin* angular momentum does not relate to a particle's coordinates or momenta, nor are the eigenstates of spin dependent on boundary conditions imposed in coordinate space. Spin, as mentioned previously, is an internal property of a particle, like mass or charge. It is an extra degree of freedom attached to a quantum mechanical particle, and must be prescribed together with the values of all other compatible properties of a particle in order to designate the state of the particle. The properties of spin are developed in detail in Chapter 11.

PROBLEMS

9.1 Show that if a state exists which is a simultaneous eigenstate of $\hat{L}_x$ and $\hat{L}_y$, this state has the eigenvalues $L_x = L_y = L_z = 0$.

Answer

Let φ be the said state. Then

$$0 = [\hat{L}_x, \hat{L}_y]\varphi = i\hbar\hat{L}_z\varphi$$

It follows that φ is an eigenstate of $\hat{L}_z$ corresponding to the eigenvalue $L_z = 0$. (φ is a "null eigenfunction" of $\hat{L}_z$.) From the uncertainty principle, (5.94) and (5.95), and the fact that φ is an eigenstate of $\hat{L}_x$ and $\hat{L}_y$ we find that

$$0 = \Delta L_x \, \Delta L_z \geq \frac{\hbar}{2}|\langle L_y\rangle|$$

Since φ is an eigenstate of $\hat{L}_y$, there is no spread in the values obtained on measurement of L_y in this state. This fact, combined with the preceding equation, gives

$$\langle L_y\rangle = L_y = 0$$

Similarly, $L_x = 0$.

It follows that a state of a system corresponding to finite angular momentum cannot be a simultaneous eigenstate of any two of the Cartesian components of $\hat{\mathbf{L}}$. Furthermore, from the defining equations for $\hat{L}_x$, $\hat{L}_y$, and $\hat{L}_z$, (9.3), it follows that any constant is a simultaneous, null eigenfunction of $\hat{L}_x$, $\hat{L}_y$, and $\hat{L}_z$.

9.2 Show that $\hat{L}_x$ and $\hat{L}^2$ are Hermitian.

Answer (partial)

To prove the Hermiticity of $\hat{L}_x$, we must show that

$$\hat{L}_x = \hat{L}_x^\dagger$$

or, equivalently, that

$$(\hat{y}\hat{p}_z - \hat{z}\hat{p}_y)^\dagger = (\hat{y}\hat{p}_z - \hat{z}\hat{p}_y)$$

Look at the $\hat{y}\hat{p}_z$ term.

$$(\hat{y}\hat{p}_z)^\dagger = \hat{p}_z{}^\dagger\hat{y}^\dagger = \hat{p}_z\hat{y} = \hat{y}\hat{p}_z$$

The last two equalities follow from (a) $\hat{p}_z$ and $\hat{y}$ are Hermitian, and (b) $[\hat{y}, \hat{p}_z] = 0$.

9.3 Measurements are made of the angle θ that **L** makes with the x axis of a collection of non-interacting rotators, all of which are known to have angular momentum $L = \hbar\sqrt{56}$. What is the minimum θ that will be measured?

9.4 If $[\hat{A}, \hat{L}_x] = [\hat{A}, \hat{L}_y] = [\hat{A}, \hat{L}_z] = 0$, what is $[\hat{A}^2, \hat{L}^2]$?

9.2   EIGENVALUES OF THE ANGULAR MOMENTUM OPERATORS

In this section we derive the eigenvalues of angular momentum that follow from the commutator relations (9.8). Eigenvalues relevant to two classes of angular momentum emerge: orbital and spin. In the remainder of the text $\hat{\mathbf{J}}$ will be used to denote angular momentum in general while $\hat{\mathbf{L}}$ will be reserved for orbital angular momentum and $\hat{\mathbf{S}}$ for spin. The operator $\hat{\mathbf{J}}$ may represent $\hat{\mathbf{L}}$, or $\hat{\mathbf{S}}$, or the combination $\hat{\mathbf{L}} + \hat{\mathbf{S}}$. The defining relations for the components of $\hat{\mathbf{J}}$ are:

(9.16)
$$[\hat{J}_x, \hat{J}_y] = i\hbar\hat{J}_z$$
$$[\hat{J}_y, \hat{J}_z] = i\hbar\hat{J}_x$$
$$[\hat{J}_z, \hat{J}_x] = i\hbar\hat{J}_y$$

(9.17)
$$\hat{J}^2 = \hat{J}_x{}^2 + \hat{J}_y{}^2 + \hat{J}_z{}^2$$

The components of $\hat{\mathbf{J}}$ obey all rules obtained above from the commutator relations (9.8). These include:

(9.18)
$$[\hat{J}_x, \hat{J}^2] = [\hat{J}_y, \hat{J}^2] = [\hat{J}_z, \hat{J}^2] = 0$$
$$\Delta J_x \, \Delta J_y \geq \frac{\hbar}{2} |\langle J_z \rangle|$$

Ladder Operators

We seek the eigenvalues of $\hat{J}^2$ and $\hat{J}_z$. To facilitate the derivation we introduce the "ladder operators" $\hat{J}_+$ and $\hat{J}_-$. The reader will find these similar to the annihilation and creation operators $(\hat{a}, \hat{a}^\dagger)$ introduced in Section 7.2. The ladder operators are defined according to

(9.19)
$$\hat{J}_+ = \hat{J}_x + i\hat{J}_y$$
$$\hat{J}_- = \hat{J}_x - i\hat{J}_y = \hat{J}_+{}^\dagger$$

Some immediate properties of these operators are

(9.20)

$$[\hat{J}_z, \hat{J}_+] = \hbar \hat{J}_+$$
$$[\hat{J}_z, \hat{J}_-] = -\hbar \hat{J}_-$$

$$\boxed{[\hat{J}_z, \hat{J}_\pm] = \pm \hbar \hat{J}_\pm}$$

$$[\hat{J}^2, \hat{J}_+] = 0$$
$$[\hat{J}^2, \hat{J}_-] = 0$$

$$\boxed{[\hat{J}^2, \hat{J}_\pm] = 0}$$

The latter two equations follow from (9.8). To establish the first two relations one merely inserts the definitions of $\hat{J}_+$ and $\hat{J}_-$. For example,

(9.21)
$$[\hat{J}_z, \hat{J}_+] = [\hat{J}_z, \hat{J}_x + i\hat{J}_y] = [\hat{J}_z, \hat{J}_x] + i[\hat{J}_z, \hat{J}_y]$$
$$= i\hbar \hat{J}_y - i \cdot i\hbar \hat{J}_x = \hbar(\hat{J}_x + i\hat{J}_y) = \hbar \hat{J}_+$$

Other relations that $\hat{J}_+$ and $\hat{J}_-$ satisfy are

(9.22)
$$\hat{J}^2 = \hat{J}_- \hat{J}_+ + \hat{J}_z^2 + \hbar \hat{J}_z$$
$$= \hat{J}_+ \hat{J}_- + \hat{J}_z^2 - \hbar \hat{J}_z$$

$$\boxed{\hat{J}^2 = \hat{J}_\mp \hat{J}_\pm + \hat{J}_z^2 \pm \hbar \hat{J}_z}$$

$$[\hat{J}_+, \hat{J}_-] = 2\hbar J_z, \qquad 2(\hat{J}^2 - \hat{J}_z^2) = \hat{J}_+ \hat{J}_- + \hat{J}_- \hat{J}_+$$

Consider the relation

(9.23)
$$\hat{J}^2 = (\hat{J}_x - i\hat{J}_y)(\hat{J}_x + i\hat{J}_y) + \hat{J}_z^2 + \hbar \hat{J}_z$$
$$= \hat{J}_x^2 + \hat{J}_y^2 + \hat{J}_z^2 + i(\hat{J}_x \hat{J}_y - \hat{J}_y \hat{J}_x) + \hbar \hat{J}_z$$

With these relations between $\hat{J}^2$, $\hat{J}_z$, $\hat{J}_+$, and $\hat{J}_-$ established we turn to construction of the eigenvalues of $\hat{L}_z$ and $\hat{L}^2$.

Let

(9.24)
$$\hat{J}_z \varphi_m = \hbar m \varphi_m$$

We wish to show that m is either an integer or an odd multiple of one-half. Consider the operation

(9.25)
$$\hat{J}_z \hat{J}_+ \varphi_m = (\hbar \hat{J}_+ + \hat{J}_+ \hat{J}_z)\varphi_m = (\hbar \hat{J}_+ + \hat{J}_+ \hbar m)\varphi_m$$
$$\hat{J}_z(\hat{J}_+ \varphi_m) = \hbar(m + 1)(\hat{J}_+ \varphi_m)$$

where we have employed (9.21). The latter equation (9.25) implies that $\hat{J}_+ \varphi_m$ is an (unnormalized) eigenfunction of $\hat{J}_z$ corresponding to the eigenvalue $\hbar(m + 1)$. That is,

(9.26)
$$\hat{J}_+ \varphi_m = \varphi_{m+1}$$

Applying $\hat{J}_+$ again gives

(9.27)
$$\hat{J}_+(\hat{J}_+ \varphi_m) = \hat{J}_+ \varphi_{m+1} = \varphi_{m+2}$$

In a similar manner, we obtain

(9.28)
$$\hat{J}_- \varphi_m = \varphi_{m-1}, \qquad \hat{J}_- \varphi_{m-1} = \varphi_{m-2}$$

Thus we have found a scheme of generating a sequence of (unnormalized) eigenfunctions of $\hat{J}_z$ from a single eigenfunction φ_m, with successive values of m in the sequence differing by unity.

(9.29)
$$(\dots, \varphi_{m-2}, \varphi_{m-1}, \varphi_m, \varphi_{m+1}, \varphi_{m+2}, \dots)$$

Since $\hat{J}^2$ commutes with $\hat{J}_z$, these operators have common eigenfunctions. Let φ_m be a common eigenfunction with the eigenvalue $\hbar^2 K^2$, that is,

(9.30)
$$\hat{J}^2 \varphi_m = \hbar^2 K^2 \varphi_m$$

Operating on this equation with $\hat{J}_+$ gives [using the third equation in (9.20)]

(9.31)
$$\hat{J}_+ \hat{J}^2 \varphi_m = \hbar^2 K^2 (\hat{J}_+ \varphi_m) = \hat{J}^2 (\hat{J}_+ \varphi_m)$$

The last equality asserts that $\hat{J}_+ \varphi_m = \varphi_{m+1}$ is also an eigenfunction of $\hat{J}^2$ corresponding to the eigenvalue $\hbar^2 K^2$. It follows that the sequence of eigenfunctions of $\hat{J}_z$ found previously (9.29) are all eigenfunctions of $\hat{J}^2$ corresponding to the same eigenvalue $\hbar^2 K^2$. How many such eigenfunctions are there? From (9.30) one obtains

(9.32)
$$\langle J^2 \rangle = \hbar^2 K^2 = \langle J_x{}^2 \rangle + \langle J_y{}^2 \rangle + \langle J_z{}^2 \rangle$$
$$\hbar^2 K^2 = \langle J_x{}^2 \rangle + \langle J_y{}^2 \rangle + \hbar^2 m^2$$

where the average has been taken in the φ_m state. It follows that

(9.33)
$$\hbar^2 K^2 \geq \hbar^2 m^2$$

(recall $\langle J_x{}^2 \rangle \geq 0$; see Problem 4.13) or, equivalently,

(9.34)
$$|K| \geq |m|$$

For a given value of $K > 0$, the possible values of m in the sequence (9.29) fall between $+K$ and $-K$. If m_{max} is the maximum value that m can assume for a given magnitude of angular momentum, $\hbar K$, then

(9.35)
$$\hat{J}_+ \varphi_{m_{max}} = 0$$

Similarly,

(9.36)
$$\hat{J}_- \varphi_{m_{min}} = 0$$

From (9.22) and the last two equations, one obtains

(9.37)
$$\hat{J}^2 \varphi_{m_{max}} = \hbar^2 K^2 \varphi_{m_{max}} = \hat{J}_z{}^2 \varphi_{m_{max}} + \hbar \hat{J}_z \varphi_{m_{max}}$$
$$\hbar^2 K^2 = \hbar^2 m_{max}(m_{max} + 1)$$
$$\hat{J}^2 \varphi_{m_{min}} = \hbar^2 K^2 \varphi_{m_{min}} = \hat{J}_z{}^2 \varphi_{m_{min}} - \hbar \hat{J}_z \varphi_{m_{min}}$$
$$\hbar^2 K^2 = \hbar^2 m_{min}(m_{min} - 1)$$

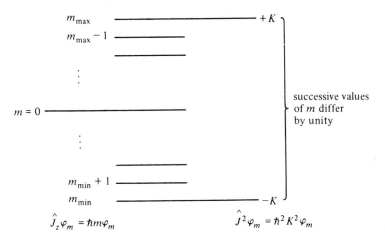

FIGURE 9.5   The possible values that m may assume, for a given value of $J^2 = \hbar^2 K^2$, form a symmetric sequence about $m = 0$.

It follows that

(9.38)
$$m_{max}(m_{max} + 1) = m_{min}(m_{min} - 1)$$

which is satisfied if

(9.39)
$$m_{max} = -m_{min}$$

The possible values that m may assume for a given value of J^2 form a symmetric sequence about $m = 0$ (see Fig. 9.5).

Let us call

(9.40)
$$m_{max} \equiv j$$

Since m runs from $-j$ to $+j$ in unit steps, one obtains

(9.41)

$j =$ an integer   if $m = 0$ is included in the sequence of m values

$j = \frac{1}{2} \times$ an odd integer   if $m = 0$ is not included in the sequence of m values

Furthermore, if j is an integer, the related m values are integers. If j is an odd multiple of one-half, the related m values are odd multiples of one-half (Fig. 9.6).

In either case, inserting $j = m_{max} = -m_{min}$ into (9.37) gives the form of the eigenvalues of $\hat{J}^2$.

(9.42)
$$J^2 = \hbar^2 K^2 = \hbar^2 j(j + 1)$$

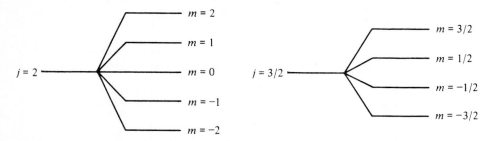

FIGURE 9.6   The angular momentum quantum number j, which enters in the eigenvalue expression $J^2 = \hbar^2 j(j + 1)$, may be either integral or an odd multiple of one-half. In either case, for a given value of j, the azimuthal quantum number, m, runs from $-j$ to $+j$ in unit steps.

Angular Momentum Eigenstates

In this manner we find that the eigenvalues of $\hat{J}^2$ and $\hat{J}_z$ take the form

(9.43)
$$J^2 = \hbar^2 j(j + 1)$$
$$J_z = \hbar m_j \qquad (m_j = -j, \ldots, +j)$$

with j an integer or half an odd integer. The structure of these eigenvalue equations is very significant and is another trademark of quantum mechanical angular momentum. In that they stem directly from the commutation relations (9.16), which in turn are obeyed by all quantum mechanical angular momenta, it follows that such eigenvalue relations are also appropriate to orbital angular momentum, $\hat{\mathbf{L}}$; spin angular momentum, $\hat{\mathbf{S}}$; or their sum, $\hat{\mathbf{L}} + \hat{\mathbf{S}}$. Such, for example, are the eigenvalues of $\hat{L}^2$, $\hat{L}_z$ as given in (9.14).

As will be shown in the following section, boundary conditions imposed on the common eigenstates of $\hat{L}^2$, $\hat{L}_z$ infer that the related eigenvalues (l, m_l) be integral. Thus, of the entire spectrum of quantum angular momentum j values, only a subset $(l = j = \text{integer})$ correspond to orbital angular momentum. The complete j spectrum (integral and half-odd-integral values) will be found to correspond to either spin angular momentum or the combination of spin plus orbital angular momentum. An example of the latter case is given by atomic electrons which have both orbital and spin angular momentum and for which one must write $\hat{\mathbf{J}} = \hat{\mathbf{L}} + \hat{\mathbf{S}}$. For the present we will concentrate on orbital angular momentum.

The eigenvalue equations for the orbital angular momentum operators $\hat{L}^2$ and $\hat{L}_z$ (with m written for m_l), together with the equations for $\hat{L}_\pm$, appear as

(9.44)
$$\hat{L}^2 \varphi_{lm} = \hbar^2 l(l + 1) \varphi_{lm}$$
$$\hat{L}_z \varphi_{lm} = \hbar m \varphi_{lm} \qquad (m = -l, \ldots, +l)$$
$$\hat{L}_+ \varphi_{lm} = \varphi_{l, m+1} \qquad (\hat{L}_+ = \hat{L}_x + i\hat{L}_y)$$
$$\hat{L}_- \varphi_{lm} = \varphi_{l, m-1} \qquad (\hat{L}_- = \hat{L}_x - i\hat{L}_y)$$

Since $m = l$ is the maximum value of m and $m = -l$ is the minimum value of m,

(9.45)
$$L_+ \varphi_{ll} = 0$$
$$L_- \varphi_{l,-l} = 0$$

These equations will be used in the next section for the derivation of the φ_{lm} eigenfunctions.

The Rigid Rotator

As an application of the preceding results relevant to the eigenvalues of $\hat{L}^2$ and $\hat{L}_z$, let us consider the problem of the energy spectrum of a rigid rotator. The rotator has two particles each of mass M separated by a weightless rigid rod of length $2a$. The midpoint of the rotator is fixed in space (Fig. 9.7). The moment of inertia of the rotator, taken about this point, is

$$I = 2Ma^2$$

Let the rotator be far removed from any force fields so that its energy is purely kinetic.

(9.46)
$$E = \frac{L^2}{2I}$$

The quantum mechanical Hamiltonian operator is

(9.47)
$$\hat{H} = \frac{\hat{L}^2}{2I}$$

and the time-independent Schrödinger equation for this system appears as

(9.48)
$$\hat{H}\varphi = \left(\frac{\hat{L}^2}{2I}\right)\varphi = E\varphi$$

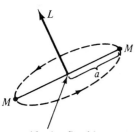

midpoint fixed in
space

FIGURE 9.7   Rigid rotator with fixed midpoint. Moment of inertia about midpoint is $I = 2Ma^2$.

The eigenvalues of $\hat{H}$ are the same as those of the square angular momentum operator $\hat{L}^2$. With the results obtained we may rewrite the equation above with the l, m indices.

$$\left(\frac{\hat{L}^2}{2I}\right)\varphi_{lm} = E_l\varphi_{lm}$$

(9.49)

$$E_l = \frac{\hbar^2 l(l+1)}{2I}$$

This energy is $(2l + 1)$-fold degenerate. For any value of l, there are $(2l + 1)$ eigenfunctions

(9.50) $\varphi_{l,l}, \ldots, \varphi_{l,-l} = \{\varphi_{lm}\}$

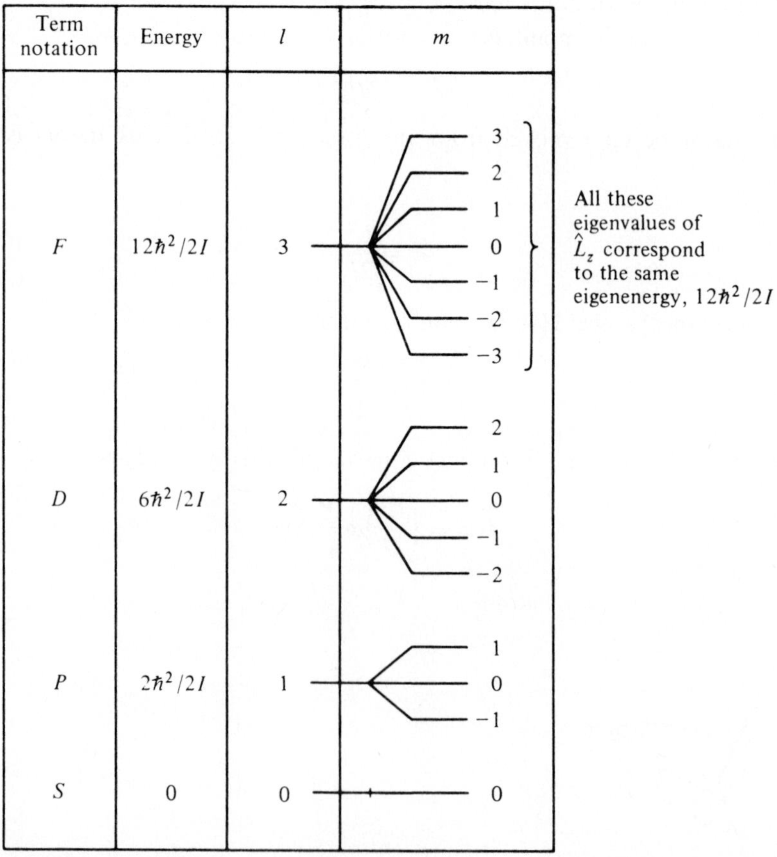

FIGURE 9.8   **Term diagram for the rigid rotator of moment of inertia, I. The lth eigenenergy, $\hbar^2 l(l+1)/2I$, is $(2l+1)$-fold degenerate.**

all corresponding to the same eigenenergy, (9.49). The energy of the rotator does not depend on the projection of **L** into the z axis or onto any other prescribed direction. The energy-level diagram for this system is sketched in Fig. 9.8, together with the "term notation" of levels. This notation is common to atomic spectroscopy and will be used in the next three chapters. When a particle is in a state of definite orbital angular momentum, characterized by the quantum number $l = 0, 1, 2, \ldots$, one speaks of the particle being respectively in an $S, P, D, F, \ldots$ state.

PROBLEMS

9.5 Show that the frequencies of photons due to energy decays between successive levels of a rotator with moment of inertia I are given by

$$\hbar\omega = \left(\frac{\hbar^2}{I}\right)(l + 1), \quad \text{or} \quad \left(\frac{\hbar^2}{I}\right)l$$

9.6 An HCl molecule may rotate as well as vibrate. Discuss the difference in emission frequencies associated with these two modes of excitation. Assume that only $l \to l \pm 1$ transitions between rotational states are allowed. Spring constant and moment of inertia may be inferred from the equivalent temperature values for HCl: $\hbar\omega_0/k_B = 4150$ K; $\hbar^2/2Ik_B = 15.2$ K.

9.7 Show that

(a) $[\hat{L}_x, \hat{x}] = 0$
(b) $[\hat{L}_x, \hat{y}] = i\hbar\hat{z}$
(c) $[\hat{L}_y, \hat{z}] = i\hbar\hat{x}$
(d) $[\hat{L}_z, \hat{x}] = i\hbar\hat{y}$
(e) $[\hat{L}_y, \hat{z}] = [\hat{y}, \hat{L}_z]$

(f) $[\hat{p}_x, \hat{L}_x] = 0$
(g) $[\hat{p}_x, \hat{L}_y] = i\hbar\hat{p}_z$
(h) $[\hat{p}_y, \hat{L}_z] = i\hbar\hat{p}_x$
(i) $[\hat{p}_z, \hat{L}_x] = i\hbar\hat{p}_y$
(j) $[\hat{L}_y, \hat{p}_z] = [\hat{p}_y, \hat{L}_z]$

9.8 Calculate

(a) $\hat{L}_z kr$
(b) $\hat{L}_z \sin kr$
(c) $\hat{L}_z f(kr)$

explicitly in Cartesian coordinates, with $r^2 = x^2 + y^2 + z^2$. The function f is an arbitrary function of r, and k is a constant wavenumber.

9.9 (a) Prove that

$$\hat{\Theta} \equiv \hat{\mathbf{L}} \times \hat{\mathbf{r}} - i\hbar\hat{\mathbf{r}} = i\hbar\hat{\mathbf{r}} - \hat{\mathbf{r}} \times \hat{\mathbf{L}}$$

(b) Show that this operator is Hermitian.

(c) Show that

$$[\hat{L}^2, \hat{\mathbf{r}}] = -2i\hbar\hat{\Theta}$$

9.10 Show that

$$[\hat{L}_x^2, \hat{L}_y^2] = [\hat{L}_y^2, \hat{L}_z^2] = [\hat{L}_z^2, \hat{L}_x^2]$$

9.11 Evaluate

(a) $[\hat{L}^2, \hat{\mathbf{p}}]$ (c) $[\hat{\mathbf{L}}, \hat{p}^2]$

(b) $[\hat{\mathbf{L}}, \hat{\mathbf{p}}]$ (d) $[\hat{\mathbf{L}}, \hat{\mathbf{L}} \times \hat{\mathbf{L}}]$

Note that parts (b) and (d) have nine components. They are called *dyadic operators*.

9.12 Show that

$$[\hat{L}_x, \hat{r}^2] = [\hat{L}_y, \hat{r}^2] = [\hat{L}_z, \hat{r}^2] = 0$$

9.13 Show that the expression

$$\langle J^2 \rangle = \hbar^2 j(j + 1)$$

is implied directly by the two assumptions:

(a) The only possible values that the components of angular momentum can have on any axis are $\hbar(-j, \ldots, +j)$.

(b) All these components are equally probable.

Answer

Because all axes are equivalent,

$$\langle J^2 \rangle = \langle J_x^2 + J_y^2 + J_z^2 \rangle = \langle J_x^2 \rangle + \langle J_y^2 \rangle + \langle J_z^2 \rangle = 3\langle J_x^2 \rangle$$

Since all values of J_x^2 are equally probable,

$$\langle J_x^2 \rangle = \hbar^2 \langle m^2 \rangle = \hbar^2 \frac{\sum_{m=-j}^{j} m^2}{2j + 1} = \frac{2\hbar^2 \sum_{m=1}^{j} m^2}{2j + 1}$$

Substituting the relation

$$\sum_{m=1}^{j} m^2 = \frac{j(j + 1)(2j + 1)}{6}$$

into the above gives

$$\langle J_x^2 \rangle = \frac{\hbar^2 j(j + 1)}{3} = \frac{1}{3}\langle J^2 \rangle$$

$$\langle J^2 \rangle = \hbar^2 j(j + 1)$$

9.3 EIGENFUNCTIONS OF THE ORBITAL ANGULAR MOMENTUM OPERATORS $\hat{L}^2$ AND $\hat{L}_z$

Spherical Harmonics

There are two techniques for obtaining the common eigenfunctions φ_{lm} of the orbital angular momentum operators $\hat{L}^2$ and $\hat{L}_z$. First, one may directly solve the eigenvalue equations

(9.51)

$$\hat{L}^2 \varphi_{lm} = \hbar^2 l(l + 1)\varphi_{lm}$$
$$\hat{L}_z \varphi_{lm} = \hbar m \varphi_{lm}$$

Second, one may seek solution to the equation

(9.52) $$\hat{L}_+ \, \varphi_{ll} = 0$$

Once having found φ_{ll}, the remaining eigenfunctions of $\hat{L}^2$ and $\hat{L}_z$, corresponding to the orbital quantum number l,

(9.53) $$\{\varphi_{lm}\} = (\varphi_{ll}, \varphi_{l,\,l-1}, \ldots, \varphi_{l,\,-l})$$

are obtained by applying $\hat{L}_-$ to φ_{ll}. That is,

(9.54) $$\varphi_{l,\,l-1} = \hat{L}_- \, \varphi_{ll}$$
$$\varphi_{l,\,l-2} = \hat{L}_- \, \varphi_{l,\,l-1}$$

In either technique for obtaining the eigenfunctions φ_{lm}, it proves both convenient and practical to work in spherical coordinates (r, θ, ϕ) (see Fig. 1.6). These coordinates are related to the Cartesian coordinates (x, y, z) through the transformation equations

(9.55) $$x = r \sin \theta \cos \phi$$
$$y = r \sin \theta \sin \phi$$
$$z = r \cos \theta$$

With these equations, the Cartesian components of $\hat{\mathbf{L}}$, (9.3), are transformed to (see Problem 9.14)

(9.56) $$\hat{L}_x = i\hbar \left(\sin \phi \, \frac{\partial}{\partial \theta} + \cot \theta \cos \phi \, \frac{\partial}{\partial \phi} \right)$$
$$\hat{L}_y = i\hbar \left(-\cos \phi \, \frac{\partial}{\partial \theta} + \cot \theta \sin \phi \, \frac{\partial}{\partial \phi} \right)$$
$$\hat{L}_z = -i\hbar \, \frac{\partial}{\partial \phi}$$

Using expressions (9.56) we obtain first the ladder operators

(9.57) $$\hat{L}_+ = \hat{L}_x + i\hat{L}_y = \hbar e^{i\phi} \left(i \cot \theta \, \frac{\partial}{\partial \phi} + \frac{\partial}{\partial \theta} \right)$$
$$\hat{L}_- = \hat{L}_x - i\hat{L}_y = \hbar e^{-i\phi} \left(i \cot \theta \, \frac{\partial}{\partial \phi} - \frac{\partial}{\partial \theta} \right)$$

and second, the operator $\hat{L}^2$

(9.58) $$\hat{L}^2 = -\hbar^2 \left[\frac{1}{\sin \theta} \frac{\partial}{\partial \theta} \left(\sin \theta \, \frac{\partial}{\partial \theta} \right) + \frac{1}{\sin^2 \theta} \frac{\partial^2}{\partial \phi^2} \right]$$

We are now prepared to seek solutions to (9.51). This is the first technique, as mentioned above, for finding the eigenstates φ_{lm}. These solutions are quite common to many branches of physics. They are called *spherical harmonics* and are universally denoted by the symbol Y_l^m. Following this protocol we change notation: $\varphi_{lm} \rightarrow Y_l^m$.

Angular Momentum and Rotation

Before discussing these solutions we note two points. First, all the angular momentum operators, when expressed in spherical coordinates as listed above, are independent of r. They are functions only of the angular variables (θ, ϕ). This means that the eigenfunctions of $\hat{L}^2$ and $\hat{L}_z$ may be chosen independent of r, that is, $Y_l^m = Y_l^m(\theta, \phi)$. This property stems from the fact that angular momentum operators are related to rotation. For instance, the operator

$$(9.59) \qquad \hat{R}_{\delta\phi} = 1 + i\delta\phi \cdot \frac{\mathbf{L}}{\hbar}$$

(described previously in Section 6.3) when acting on $f(\mathbf{r})$ rotates $\mathbf{r}$ through the azimuthal displacement $\delta\phi$, so that

$$(9.60) \qquad \begin{aligned} \hat{R}_{\delta\phi} f(\mathbf{r}) &= f(\mathbf{r} + \delta\mathbf{r}) \\ \delta\mathbf{r} &= \delta\phi \times \mathbf{r} \end{aligned}$$

So the effect of the operation $\delta\phi \cdot \hat{\mathbf{L}}$ on a function $\mathbf{r}$ is to cause a rotational displacement of $\mathbf{r}$. If $\delta\phi$ is parallel to the z axis, $\delta\phi \cdot \hat{\mathbf{L}} = \delta\phi L_z$. This operator induces a rotation of $\mathbf{r}$ about the z axis, without changing the magnitude of $\mathbf{r}$. If we write $f(\mathbf{r}) = f(r, \theta, \phi)$, then $\hat{L}_z$ when operating on f affects only the variable ϕ. When L^2 operates on this function, θ and ϕ are both affected, but not r. So here we have the reason that the eigenstates of $\hat{L}^2$ and $\hat{L}_z$ may be chosen independent of r.

Normalization

The second point we wish to note relates to the normalization of the Y_l^m functions. This normalization is taken over the surface of a unit sphere. The differential element of area dS, on the surface of a sphere of radius a, is conveniently expressed in terms of the element of *solid angle* $d\Omega$.

$$(9.61) \qquad dS = a^2\, d\Omega = a^2 \sin\theta\, d\theta\, d\phi$$

(see Fig. 9.9). The solid angle subtended by dS about the origin is $dS/a^2 = d\Omega$. The solid angle subtended by a sphere (more generally any closed surface) about the origin is

$$(9.62) \qquad \int_{\text{all directions}} d\Omega = \int_0^{2\pi} d\phi \int_0^{\pi} \sin\theta\, d\theta = \int_0^{2\pi} d\phi \int_{-1}^{1} d\cos\theta = 4\pi \text{ steradians}$$

which is the same as the area of a unit sphere (a sphere of unit radius).

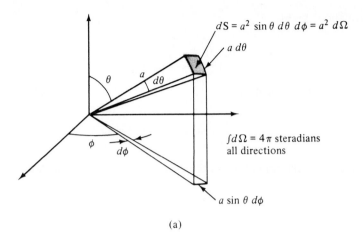

(a)

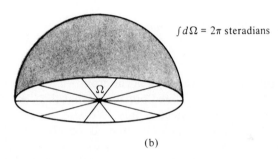

(b)

FIGURE 9.9 (a) Element of the solid angle $d\Omega = dS/a^2$. (b) Solid angle subtended by the hemisphere about the origin O is 2π.

To normalize the eigenfunctions Y_l^m, set

(9.63)
$$\int_{4\pi} |Y_l^m|^2 \, d\Omega = 1$$

which we see is the same as requiring that $|Y_l^m|^2$ integrate to unity over the surface of a unit sphere.

We are now prepared to discuss the solutions to (9.51). The eigenfunction equation for $\hat{L}_z$ gives

(9.64)
$$\frac{\partial}{\partial\phi} Y_l^m = im Y_l^m$$

This equation determines only the ϕ dependence of Y_l^m. If we set

(9.65)
$$Y_l^m(\theta, \phi) = \Phi_m(\phi)\Theta_l^m(\theta)$$

the equation above gives

(9.66)
$$\Phi_m(\phi) = \frac{1}{\sqrt{2\pi}} e^{im\phi}$$

which satisfies the normalization

(9.67)
$$\int_0^{2\pi} d\phi \, |\Phi_m|^2 = 1$$

The index m can be determined from the single valuedness[1] of the wavefunction Φ. That is

(9.68)
$$\Phi(\phi) = \Phi(\phi + 2\pi)$$
$$e^{im\phi} = e^{im(\phi + 2\pi)}$$
$$e^{im2\pi} = 1$$

which is only satisfied for integral values of m:

(9.69)
$$m = 0, 1, 2, \ldots$$

As demonstrated in Section 9.2 the values of m run from $-l$ to $+l$, whence l is also an integer. Thus we obtain the result stated previously that the l, m orbital angular momentum quantum numbers are integers only. We also see how this property follows directly from boundary conditions imposed on the wavefunctions Y_l^m. Spin, being an intrinsic property of a particle, is not so constrained, and the related quantum s numbers may assume half-odd-integral as well as integral values.

Legendre Polynomials

We have found that the eigenfunction Y_l^m has the structure

(9.70)
$$Y_l^m = \frac{1}{\sqrt{2\pi}} e^{im\phi} \Theta_l^m(\theta)$$

Substituting this function into (9.51), together with the explicit expression for $\hat{L}^2$ as given by (9.58), gives the following equation for Θ_l^m (deleting l and m indices, for the moment):

(9.71)
$$\frac{1}{\sin\theta} \frac{d}{d\theta} \left(\sin\theta \frac{d\Theta}{d\theta} \right) + \left[l(l+1) - \frac{m^2}{\sin^2\theta} \right] \Theta = 0$$

---

[1] On physical grounds it is more appropriate to require that $|\Phi|^2$ be single valued. However, this can be shown to be equivalent to the single valuedness of Φ for the case in point. For further discussion, see K. Gottfried, *Quantum Mechanics*, W. A. Benjamin, New York, 1966.

or equivalently, in terms of the variable,

$$\mu \equiv \cos\theta$$

(9.72)
$$\frac{d}{d\mu}\left[(1-\mu^2)\frac{d\Theta}{d\mu}\right] + \left[l(l+1) - \frac{m^2}{1-\mu^2}\right]\Theta = 0$$

$$-1 \leq \mu \leq +1$$

Let us outline the method by which this equation is solved.[1] Setting $m = 0$ and $l(l+1) = \lambda$ in (9.72) gives *Legendre's equation*,

(9.73)
$$\frac{d}{d\mu}\left[(1-\mu^2)\frac{d\Theta_l}{d\mu}\right] + \lambda\Theta_l = 0$$

where we have set $\Theta_l^0 \equiv \Theta_l$. Referring to (9.58), we see that (9.73) is an eigenvalue equation for $\hat{L}^2/\hbar^2$ (corresponding to $L_z = 0$), with eigenvalue λ. A solution to (9.73) may be sought as a series[2] in powers of μ. The requirement that this series solution remain bounded in the interval $-1 \leq \mu \leq +1$ implies that: (1) the eigenvalue λ must be of the form $l(l+1)$, where $l \geq 0$ and is an integer, and (2) the series solution for Θ_l contains, at most, a finite number of terms. The first conclusion returns the form of the eigenvalues of L^2 given previously by (9.14), namely, $L^2 = \hbar^2 l(l+1)$. The second conclusion indicates that Θ_l is a polynomial of order l. These polynomials, called *Legendre polynomials*, are commonly denoted as $P_l(\mu)$, so that apart from a multiplicative constant, $\Theta_l(\mu) = P_l(\mu)$. The series summation for this solution may be expressed in the concise form, called the *formula of Rodrigues*,

(9.74)
$$P_l(\mu) = \frac{1}{2^l l!}\frac{d^l}{d\mu^l}(\mu^2 - 1)^l$$

With this solution to (9.73) at hand, the solution to (9.72) is obtained by first constructing the *associated Legendre polynomials*. These are defined by[3] the following differential operation on $P_l(\mu)$:

(9.75)
$$P_l^m(\mu) = (-1)^m(1-\mu^2)^{m/2}\frac{d^m P_l(\mu)}{d\mu^m}$$

---

[1] For a more detailed description of this method of solution, see E. Merzbacher, *Quantum Mechanics*, 2nd ed., Wiley, New York, 1970. A closely related but more concise technique of solution is described in P. Stehle, *Quantum Mechanics*, Holden-Day, San Francisco, 1966.

[2] This method of series solution is explicitly demonstrated in Chapter 10 in the generation of Laguerre polynomials, which are components of the wavefunctions for the hydrogen atom.

[3] Another popular notation for these polynomials includes the $(-1)^m$ factor explicitly in the Y_l^m functions.

for positive integers $m \leq l$. Differentiating Legendre's equation (9.73) m times with $\lambda = l(l + 1)$, and Θ_l set equal to P_l, and employing the definition (9.75), one readily deduces the equation

$$(9.76) \qquad \frac{d}{d\mu}\left[(1 - \mu^2)\frac{dP_l^m}{d\mu}\right] + \left[l(l + 1) - \frac{m^2}{1 - \mu^2}\right]P_l^m = 0$$

Comparison with (9.72) indicates that $P_l^m(\mu)$ is a solution to this same equation. Furthermore, (9.72) remains unchanged if m is replaced by $-m$, and we may conclude that $P_l^{-m}(\mu)$ is also a solution to this equation, so that apart from a multiplicative constant, P_l^m is equal to P_l^{-m}.

In summary, we have found that the solutions $\Theta_l^m(\mu)$ to (9.72) are given by the associated Legendre polynomials $P_l^m(\mu)$. In addition, we see from the foregoing construction how the quantum conditions (9.14) emerge from the requirements that $Y_l^m(\Theta, \phi)$ remain nonsingular and single valued in the intervals $-1 \leq \mu \leq +1$, $0 \leq \phi \leq 2\pi$.

The precise relation between $\Theta_l^m(\mu)$ and $P_l^m(\mu)$ as defined by (9.75) follows from the normalization condition (9.63).

$$\int_{4\pi} |Y_l^m|^2 \, d\Omega = \int_0^{2\pi} d\phi \left|\frac{e^{im\phi}}{\sqrt{2\pi}}\right|^2 \int_{-1}^{1} d\mu |\Theta_l^m(\mu)|^2 = 1$$

$$\int_{-1}^{1} d\mu |\Theta_l^m(\mu)|^2 = 1$$

There results

$$(9.77) \qquad \Theta_l^m(\mu) = \left[\frac{2l + 1}{2}\frac{(l - m)!}{(l + m)!}\right]^{1/2} P_l^m(\mu)$$

The first few spherical harmonics, Y_l^m, are listed in Table 9.1. Important properties of the Legendre polynomials, P_l, are listed in Table 9.2, while properties of the associated Legendre polynomials, P_l^m, are listed in Table 9.3.

Polar Plots of Y_l^m and Spherical Harmonic Expansions

When a system such as a rigid rotator is in an eigenstate of $\hat{L}^2$ and $\hat{L}_z$, the z axis is said to be *preferred*. Namely, measurement of L_z is certain to find a specific value. However, in this state, it is still true that the x direction is in no way preferred over the y direction. Thus the probability density, $|Y_l^m|^2$, is rotationally symmetric about the z axis or, equivalently (from 9.70), $|Y_l^m|$ is independent of ϕ. The function $|Y_l^m|$ is a surface of revolution about the z axis.

$$(9.78) \qquad |Y_l^m| = \frac{1}{\sqrt{2\pi}}|\Theta_l^m(\cos\theta)| = \left[\frac{2l + 1}{4\pi}\frac{(l - m)!}{(l + m)!}\right]^{1/2}|P_l^m(\cos\theta)|$$

TABLE 9.1 The first few normalized spherical harmonics and corresponding associated Legendre polynomials[a]

$$Y_l^m(\theta, \phi) = \left[\frac{2l+1}{4\pi}\frac{(l-m)!}{(l+m)!}\right]^{1/2} P_l^m(\cos\theta)e^{im\phi}$$

$$\int_{-1}^{1} d\cos\theta \int_{0}^{2\pi} d\phi \; Y_l^m(Y_{l'}^{m'})^* = \delta_{mm'}\delta_{ll'}$$

$P_0 = 1$

$P_1^{\ 1} = -\sin\theta$

$P_1^{\ 0} = \cos\theta$

$P_1^{\ -1} = \frac{1}{2}\sin\theta$

$P_2^{\ 2} = 3\sin^2\theta$

$P_2^{\ 1} = -3\sin\theta\cos\theta$

$P_2^{\ 0} = \frac{1}{2}(3\cos^2\theta - 1)$

$P_2^{\ -1} = \frac{1}{2}\sin\theta\cos\theta$

$P_2^{\ -2} = \frac{1}{8}\sin^2\theta$

$P_3^{\ 3} = -15\sin^3\theta$

$P_3^{\ 2} = 15\sin^2\theta\cos\theta$

$P_3^{\ 1} = -\frac{3}{2}\sin\theta(5\cos^2\theta - 1)$

$P_3^{\ 0} = \frac{1}{2}(5\cos^3\theta - 3\cos\theta)$

$P_3^{\ -1} = \frac{1}{8}\sin\theta(5\cos^2\theta - 1)$

$P_3^{\ -2} = \frac{1}{8}\sin^2\theta\cos\theta$

$P_3^{\ -3} = \frac{1}{48}\sin^3\theta$

$$Y_l^{-l} = \frac{1}{2^l l!}\sqrt{\frac{(2l+1)!}{4\pi}}\sin^l\theta\, e^{-il\phi}$$

$$Y_l^0 = \sqrt{\frac{2l+1}{4\pi}}\, P_l(\cos\theta)$$

$$\sum_{m=-l}^{l} |Y_l^m(\theta,\phi)|^2 = \frac{2l+1}{4\pi}$$

$$Y_l^{-m} = (-1)^m(Y_l^m)^*$$

$$Y_0^0 = \left(\frac{1}{4\pi}\right)^{1/2}$$

$$Y_1^1 = -\frac{1}{2}\left(\frac{3}{2\pi}\right)^{1/2}\sin\theta\, e^{i\phi}$$

$$Y_1^0 = \frac{1}{2}\left(\frac{3}{\pi}\right)^{1/2}\cos\theta$$

$$Y_1^{-1} = \frac{1}{2}\left(\frac{3}{2\pi}\right)^{1/2}\sin\theta\, e^{-i\phi}$$

$$Y_2^2 = \frac{1}{4}\left(\frac{15}{2\pi}\right)^{1/2}\sin^2\theta\, e^{2i\phi}$$

$$Y_2^1 = -\frac{1}{2}\left(\frac{15}{2\pi}\right)^{1/2}\sin\theta\cos\theta\, e^{i\phi}$$

$$Y_2^0 = \frac{1}{4}\left(\frac{5}{\pi}\right)^{1/2}(3\cos^2\theta - 1)$$

$$Y_2^{-1} = \frac{1}{2}\left(\frac{15}{2\pi}\right)^{1/2}\sin\theta\cos\theta\, e^{-i\phi}$$

$$Y_2^{-2} = \frac{1}{4}\left(\frac{15}{2\pi}\right)^{1/2}\sin^2\theta\, e^{-2i\phi}$$

$$Y_3^3 = -\frac{1}{8}\left(\frac{35}{\pi}\right)^{1/2}\sin^3\theta\, e^{3i\phi}$$

$$Y_3^2 = \frac{1}{4}\left(\frac{105}{2\pi}\right)^{1/2}\sin^2\theta\cos\theta\, e^{2i\phi}$$

$$Y_3^1 = -\frac{1}{8}\left(\frac{21}{\pi}\right)^{1/2}\sin\theta(5\cos^2\theta - 1)e^{i\phi}$$

$$Y_3^0 = \frac{1}{4}\left(\frac{7}{\pi}\right)^{1/2}(5\cos^3\theta - 3\cos\theta)$$

$$Y_3^{-1} = \frac{1}{8}\left(\frac{21}{\pi}\right)^{1/2}\sin\theta(5\cos^2\theta - 1)e^{-i\phi}$$

$$Y_3^{-2} = \frac{1}{4}\left(\frac{105}{2\pi}\right)^{1/2}\sin^2\theta\cos\theta\, e^{-2i\phi}$$

$$Y_3^{-3} = \frac{1}{8}\left(\frac{35}{\pi}\right)^{1/2}\sin^3\theta\, e^{-3i\phi}$$

[a] Defining relations for $P_l(\mu)$ and $P_l^{\ m}(\mu)$ are given in Table 9.3. Comparison with other notations for the spherical harmonics and their related functions may be found in D. Park, *Introduction to the Quantum Theory*, 2nd ed., McGraw-Hill, New York, 1974.

TABLE 9.2 Properties of the Legendre polynomials

Generating Formulas

$$(1 - 2\mu s + s^2)^{-1/2} = \sum_{l=0}^{\infty} P_l(\mu) s^l$$

$$P_l(\mu) = \frac{1}{2^l l!} \frac{d^l}{d\mu^l} (\mu^2 - 1)^l \begin{cases} -1 \le \mu \le 1 \\ l = 0, 1, 2, 3, \dots \end{cases}$$

Legendre's Equation

$$(1 - \mu^2) \frac{d^2 P_l(\mu)}{d\mu^2} - 2\mu \frac{dP_l(\mu)}{d\mu} + l(l + 1)P_l(\mu) = 0$$

Recurrence Relations

$$(l + 1)P_{l+1}(\mu) = (2l + 1)\mu P_l(\mu) - lP_{l-1}(\mu)$$

$$(1 - \mu^2) \frac{d}{d\mu} P_l(\mu) = -l\mu P_l(\mu) + lP_{l-1}(\mu)$$

Normalization and Orthogonality

$$\int_{-1}^{1} P_l(\mu) P_m(\mu) \, d\mu = \frac{2}{2l + 1} \qquad (l = m)$$

$$= 0 \qquad (l \neq m)$$

The First Few Polynomials

$$P_0 = 1 \qquad P_2 = \tfrac{1}{2}(3\mu^2 - 1) \qquad P_4 = \tfrac{1}{8}(35\mu^4 - 30\mu^2 + 3)$$
$$P_1 = \mu \qquad P_3 = \tfrac{1}{2}(5\mu^3 - 3\mu) \qquad P_5 = \tfrac{1}{8}(63\mu^5 - 70\mu^3 + 15\mu)$$

Special Values

$$P_l(\mu) = (-1)^l P_l(-\mu) \qquad P_l(1) = 1$$

Polar plots of these functions for $l = 0, 1, 2$, and all accompanying m values, in any plane through the z axis, are sketched in Fig. 9.10.

The functions $Y_l^m(\theta, \phi)$ are a basis of the Hilbert space of square-integrable functions $\varphi(\theta, \phi)$ defined on the unit sphere. Such functions may be normalized as follows.

(9.79)
$$\|\varphi(\theta, \phi)\|^2 = \langle \varphi | \varphi \rangle = \int_0^{2\pi} d\phi \int_{-1}^{1} d\cos\theta \, \varphi^* \varphi = 1$$

The expansion of φ in spherical harmonics is given by

(9.80)
$$\varphi(\theta, \phi) = \sum_{l=0}^{\infty} \sum_{|m| \le l} a_{lm} Y_l^m(\theta, \phi)$$

The coefficient of expansion a_{lm} is given by the inner product,

(9.81)
$$a_{lm} = \langle Y_l^m | \varphi \rangle = \int_0^{2\pi} d\phi \int_{-1}^{1} d\cos\theta \, [Y_l^m(\theta, \phi)]^* \varphi(\theta, \phi)$$

TABLE 9.3   Properties of the associated Legendre polynomials

Definition

$$P_l^m(\mu) = (-1)^m (1 - \mu^2)^{m/2} \frac{d^m}{d\mu^m} P_l(\mu); \quad P_l^0 = P_l$$

$$P_l^{-m}(\mu) = (-1)^m \frac{(l-m)!}{(l+m)!} P_l^m(\mu); \quad P_l^{-l} = \frac{1}{2^l l!} \sin^l \theta$$

For these equations, m is taken as ≥ 0. In the formulas below, however, m may be < 0 also; $l = 0, 1, 2, 3, \ldots$, $|m| \leq l$.

Differential Equation

$$(1 - \mu^2) \frac{d^2 P_l^m(\mu)}{d\mu^2} - 2\mu \frac{dP_l^m(\mu)}{d\mu} + \left[l(l+1) - \frac{m^2}{1 - \mu^2} \right] P_l^m(\mu) = 0$$

Recurrence Relations

$$(2l + 1)\mu P_l^m(\mu) = (l - m + 1)P_{l+1}^m(\mu) + (l + m)P_{l-1}^m(\mu)$$

$$(2l + 1)(1 - \mu^2)^{1/2} P_l^m(\mu) = P_{l-1}^{m+1}(\mu) - P_{l+1}^{m+1}(\mu)$$

$$(1 - \mu^2) \frac{dP_l^m(\mu)}{d\mu} = (l + 1)\mu P_l^m(\mu) - (l - m + 1)P_{l+1}^m(\mu)$$

$$= -l\mu P_l^m(\mu) + (l + m)P_{l-1}^m(\mu)$$

$$(1 - \mu^2)^{1/2} P_l^{m+1}(\mu) = (l - m)\mu P_l^m(\mu) - (l + m)P_{l-1}^m(\mu)$$

$$= -(l + m + 1)P_l^m(\mu) + (l - m + 1)P_{l+1}^m(\mu)$$

Normalization and Orthogonality

$$\int_{-1}^1 P_l^m(\mu)P_k^m(\mu)\, d\mu = \frac{2}{2l + 1} \frac{(l + m)!}{(l - m)!} \quad (l = k)$$

$$= 0 \quad (l \neq k)$$

Suppose that at a given instant a system (e.g., a rigid rotator) is in the state $\varphi(\theta, \phi)$. Then the probability that measurement of L^2 finds the value $\hbar^2 l(l + 1)$ is

(9.82)
$$P[\hbar^2 l(l + 1)] = \sum_{m=-l}^{+l} |a_{lm}|^2$$

while the probability of finding L_z with the value $\hbar m$ is

(9.83)
$$P(\hbar m) = \sum_{l=|m|}^{\infty} |a_{lm}|^2$$

For example, consider that a rotator is in the state

$$\varphi(\theta, \phi) = A \sin^2 \theta \cos 2\phi$$

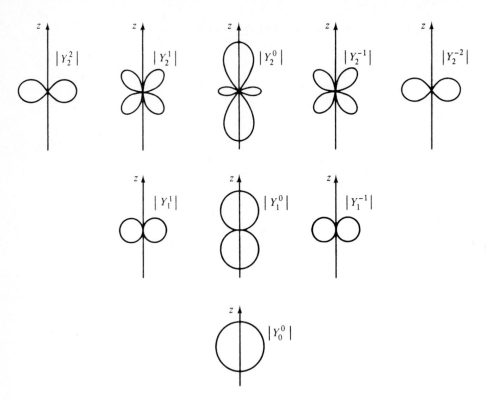

FIGURE 9.10   Polar plots of $|Y_l^m|$ versus θ in any plane through the z axis for $l = 0, 1, 2$. The equality $|Y_l^m| = |Y_l^{-m}|$ is exhibited.

What values of L^2 and L_z will measurement find? To answer this question, in principle we should first evaluate the coefficients a_{lm} given by the operation (9.81). However, for the case at hand, reference to Table 9.1 reveals that φ is the simple superposition

$$\varphi = A'(Y_2{}^2 + Y_2{}^{-2})$$

where A and A' are constants. So the only coefficients that enter the expansion (9.80) are a_{22} and a_{2-2}. We may conclude that measurement will find the value $L^2 = 6\hbar^2$ with probability 1 and the values $L_z = \pm 2\hbar$ with equal probabilities of $\frac{1}{2}$. No other values of L_z and L^2 would be found for a rotator in the state given above.

Second Construction of the Spherical Harmonics

Let us now turn to the second procedure for finding the Y_l^m eigenfunctions, initiated by (9.52). Consider that we have already solved for the eigenfunction of $\hat{L}_z$, so that

φ_{lm} is known to be in the form given by (9.70). Equation (9.52) then becomes

(9.84) $$\hat{L}_+ e^{il\phi}\Theta_l^l(\theta) = \hbar e^{i\phi}\left(\frac{\partial}{\partial\theta} + i\cot\theta\frac{\partial}{\partial\phi}\right)e^{il\phi}\Theta_l^l(\theta) = 0$$

Bringing the exp $(il\phi)$ factor through the differential operator gives (deleting the l-scripts for the moment)

(9.85) $$\frac{\partial}{\partial\theta}\Theta = l\cot\theta\,\Theta$$

Substituting the relation

(9.86) $$l\cot\theta = \frac{\partial}{\partial\theta}\ln\sin^l\theta$$

and then dividing through by Θ gives

(9.87) $$\frac{1}{\Theta}\frac{\partial}{\partial\theta}\Theta = \frac{\partial}{\partial\theta}\ln\Theta = \frac{\partial}{\partial\theta}\ln\sin^l\theta$$

This is simply integrated to yield

(9.88) $$\Theta_l^l = A_{ll}\sin^l\theta$$

where A_{ll} is a normalization constant. It follows that Y_l^l is

(9.89) $$Y_l^l = \frac{A_{ll}}{\sqrt{2\pi}}\sin^l\theta\,e^{il\phi}$$

which agrees with the values given in Table 9.1. The eigenfunction Y_l^{l-1} is obtained from Y_l^l through the operator $\hat{L}_-$.

(9.90) $$\hat{L}_- Y_l^l = Y_l^{l-1}$$

In this manner we obtain

(9.91) $$Y_l^{l-1} = A_{l,l-1}e^{i(l-1)\phi}\sin^{l-1}\theta\cos\theta$$

which is also in agreement with the values given in Table 9.1. The relations between $\hat{L}_+$, $\hat{L}_-$, and the Y_l^m functions with correct normalization factors are given in Table 9.4.

We conclude this section with the following example. Suppose that a rigid rotator is in the eigenstate of $\hat{L}^2$ and $\hat{L}_z$ corresponding to $l = 1$ and $m = 1$ (i.e., Y_1^1). What is the probability that measurement of L_x finds the respective values $m = 0$, ± 1? To answer this question we must expand Y_1^1 in the eigenfunctions of $\hat{L}_x$. These eigenfunctions are solutions to the equation

(9.92) $$\hat{L}_x X(\theta, \phi) = \hbar\alpha X(\theta, \phi)$$

TABLE 9.4   Normalized relations between $\hat{L}_+$, $\hat{L}_-$, $\hat{L}_x$, $\hat{L}_y$ and the states $|lm\rangle$[a]

$\hat{L}_z|lm\rangle = m\hbar|lm\rangle$

$\hat{L}_+|lm\rangle = \hbar[(l-m)(l+m+1)]^{1/2}|l, m+1\rangle$

$\hat{L}_-|lm\rangle = \hbar[(l+m)(l-m+1)]^{1/2}|l, m-1\rangle$

$\hat{L}_x|lm\rangle = \frac{1}{2}\hbar[(l-m)(l+m+1)]^{1/2}|l, m+1\rangle + \frac{1}{2}\hbar[(l+m)(l-m+1)]^{1/2}|l, m-1\rangle$

$\hat{L}_y|lm\rangle = -\frac{1}{2}i\hbar[(l-m)(l+m+1)]^{1/2}|l, m+1\rangle + \frac{1}{2}i\hbar[(l+m)(l-m+1)]^{1/2}|l, m-1\rangle$

$\hat{L}_\pm|lm\rangle = \hbar[l(l+1) - m(m\pm1)]^{1/2}|l, m\pm1\rangle$

[a] These normalization relations also apply to the total angular momentum operators, $\hat{J}_\pm$, $\hat{J}_x$, $\hat{J}_y$, $\hat{J}_z$, and $\hat{J}^2$, where

$$\hat{J}^2|jm_j\rangle = \hbar^2 j(j+1)|jm_j\rangle$$
$$\hat{J}_z|jm_j\rangle = \hbar m_j|jm_j\rangle$$

The student may question why these functions are not simply the spherical harmonics Y_l^m. After all, there is no intrinsic difference between $\hat{L}_x$ and $\hat{L}_z$. The answer is that the eigenfunctions of $\hat{L}_x$ are the Y_l^m functions if we define the x axis as the polar axis, so that θ is angular displacement from the x axis. However, for the problem at hand the z axis is the polar axis and the X functions are a bit more complicated.

Writing $\hat{L}_x$ as

(9.93)
$$\hat{L}_x = \tfrac{1}{2}(\hat{L}_+ + \hat{L}_-)$$

it is clear that $\hat{L}_x Y_l^m$ gives a combination of spherical harmonics with the same l value. Also, since all Y_l^m functions with $|m| \le l$ are eigenfunctions of $\hat{L}^2$ with eigenvalue $\hbar^2 l(l+1)$, any combination of such functions is an eigenfunction of $\hat{L}^2$ with eigenvalue $\hbar^2 l(l+1)$.

With these properties in mind we seek a solution to (9.92) in the form

(9.94)
$$X = aY_1^{\ 1} + bY_1^{\ 0} + cY_1^{\ -1}$$

The problem of finding the eigenfunctions of $\hat{L}_x$ (corresponding to $l = 1$) is then reduced to finding the coefficients a, b, and c in the expression above.

From the properties of $\hat{L}_+$ and $\hat{L}_-$ listed in Table 9.4, we have

(9.95)
$$\hat{L}_+ Y_1^{\ 0} = \sqrt{2}\,\hbar Y_1^{\ 1}$$
$$\hat{L}_+ Y_1^{\ -1} = \sqrt{2}\,\hbar Y_1^{\ 0}$$
$$\hat{L}_- Y_1^{\ 0} = \sqrt{2}\,\hbar Y_1^{\ -1}$$
$$\hat{L}_- Y_1^{\ 1} = \sqrt{2}\,\hbar Y_1^{\ 0}$$

Substituting the expansion (9.94) into the eigenvalue equation (9.92) and using the relations above gives the equation

(9.96) $(aY_1^{\ 0} + bY_1^{\ 1} + bY_1^{\ -1} + cY_1^{\ 0}) = \sqrt{2}\alpha(aY_1^{\ 1} + bY_1^{\ 0} + cY_1^{\ -1})$

Since the Y_l^m functions form a linearly independent sequence, it follows that the only way to guarantee equality for all values of θ and ϕ in the equation above is to set the coefficients of individual Y_l^m functions equal to zero. This gives the following set of three homogeneous algebraic equations:

(9.97)
$$\begin{pmatrix} -\sqrt{2}\alpha & 1 & 0 \\ 1 & -\sqrt{2}\alpha & 1 \\ 0 & 1 & -\sqrt{2}\alpha \end{pmatrix} \begin{pmatrix} a \\ b \\ c \end{pmatrix} = 0$$

A nontrivial solution of these equations occurs only for values of α that make the determinant of the coefficient matrix vanish. Setting the determinant equal to zero, one obtains

(9.98)
$$\alpha(\alpha^2 - 1) = 0$$

which gives the eigenvalues

(9.99)
$$\alpha = 0, \qquad \alpha = 1, \qquad \alpha = -1$$

Substituting these values back into (9.97) gives the (normalized) eigenvectors

$$X_0 = \frac{1}{\sqrt{2}}(Y_1^{\ 1} - Y_1^{\ -1}), \qquad\qquad \alpha = 0$$

(9.100)
$$X_+ = \tfrac{1}{2}(Y_1^{\ 1} + \sqrt{2}\,Y_1^{\ 0} + Y_1^{\ -1}), \qquad \alpha = +1$$

$$X_- = \tfrac{1}{2}(Y_1^{\ 1} - \sqrt{2}\,Y_1^{\ 0} + Y_1^{\ -1}), \qquad \alpha = -1$$

With these eigenfunctions of $\hat{L}_x$ at hand it becomes a matter of inspection to construct the linear combination that gives $Y_1^{\ 1}$. It is given by

(9.101)
$$Y_1^{\ 1} = \tfrac{1}{2}(X_+ + \sqrt{2}\,X_0 + X_-)$$

It follows that if the rotator is in the eigenstate of $\hat{L}^2$ and $\hat{L}_z$ corresponding to $l = 1$, $m = 1$, then the probability that measurement of L_x finds the value $+\hbar$ is $\tfrac{1}{4}$, the probability of finding $-\hbar$ is $\tfrac{1}{4}$, and the probability of finding 0 is $\tfrac{2}{4}$ (Fig. 9.11).

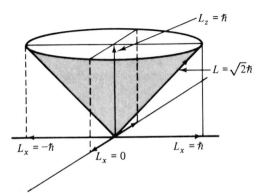

FIGURE 9.11 Given the state $L^2 = 2\hbar^2$, $L_z = \hbar$, what is the probability that measurement of L_x finds the values $\pm\hbar$, 0? Geometrical construction shows that two projections of L give $L_x = 0$, while one projection gives $L_x = +\hbar$ and one projection gives $L_x = -\hbar$.

PROBLEMS

9.14 Use transformation equations (9.55) to obtain the expression

$$\hat{L}_z = -i\hbar \frac{\partial}{\partial \phi}$$

Answer

From (9.55) we obtain the following useful relations.

$$r^2 = x^2 + y^2 + z^2, \qquad \cos \theta = \frac{z}{r}, \qquad \tan \phi = \frac{y}{x}$$

$$\frac{\partial \theta}{\partial x} = \frac{\cos \phi \cos \theta}{r} \qquad\qquad \frac{\partial \phi}{\partial x} = -\frac{y}{x^2} \cos^2 \phi$$

$$\frac{\partial \theta}{\partial y} = \frac{\sin \phi \cos \theta}{r} \qquad\qquad \frac{\partial \phi}{\partial y} = \frac{\cos^2 \phi}{x}$$

$$\frac{\partial \theta}{\partial z} = -\frac{\sin \theta}{r} \qquad\qquad \frac{\partial \phi}{\partial z} = 0$$

$$\frac{\partial r}{\partial x} = \frac{x}{r}, \qquad \frac{\partial r}{\partial y} = \frac{y}{r}, \qquad \frac{\partial r}{\partial z} = \frac{z}{r}$$

For example, from $\cos \theta = z/r$, one obtains

$$-\sin \theta \frac{\partial \theta}{\partial x} = z \frac{\partial}{\partial x} \frac{1}{r} = -\frac{zx}{r^3} = -\frac{(z/r)(x/r)}{r} = -\frac{\cos \theta \sin \theta \cos \phi}{r}$$

Substituting these expressions in the expansion

$$\hat{L}_z = -i\hbar \left(x \frac{\partial}{\partial y} - y \frac{\partial}{\partial x} \right)$$

$$= -i\hbar \left[x \left(\frac{\partial \theta}{\partial y} \frac{\partial}{\partial \theta} + \frac{\partial \phi}{\partial y} \frac{\partial}{\partial \phi} + \frac{\partial r}{\partial y} \frac{\partial}{\partial r} \right) - y \left(\frac{\partial \theta}{\partial x} \frac{\partial}{\partial \theta} + \frac{\partial \phi}{\partial x} \frac{\partial}{\partial \phi} + \frac{\partial r}{\partial x} \frac{\partial}{\partial r} \right) \right]$$

gives the desired result.

9.15 (a) What is $[\hat{\phi}, \hat{L}_z]$?

(b) Calculate the root-mean-square deviation $\Delta\phi$ for a particle in the uniform state $\varphi = 1/\sqrt{2\pi}$. (*Hint:* Perform your integrals over the interval $-\pi, \pi$.)

(c) Write down an uncertainty relation seemingly implied by your answer to part (a) and argue the physical inconsistency of this relation in view of your answer to part (b).

Answers

(a) $[\hat{\phi}, \hat{L}_z] = i\hbar$

(b) $\Delta\phi|_{max} = \pi/\sqrt{3}$

(c) One is tempted to write $\Delta\phi \, \Delta L_z \geq \hbar/2$; however, by virtue of the result in part (b),

376

uncertainty in ϕ greater than $\pi/\sqrt{3}$ has little physical meaning. In the extreme that the system is in an eigenstate (e.g., Y_l^m) of $\hat{L}_z$, $\Delta L_z = 0$ and the uncertainty relation gives $\Delta\phi = \infty$. Thus we may conclude that the assumed uncertainty relation is erroneous. [*Note:* Consider the space of functions $\mathfrak{H}_\phi$ whose elements have finite norm on the finite interval $(0, 2\pi)$ (i.e., $\int_0^{2\pi} \varphi^*\varphi \, d\phi < \infty$). It has been pointed out by D. Judge[1] that $\hat{L}_z$ is not Hermitian on this space. As a consequence, the derivation of the uncertainty relation between ϕ and $\hat{L}_z$ from their commutator relation fails. The non-Hermiticity of $\hat{L}_z$ on $\mathfrak{H}_\phi$ may be seen as follows. It is evident that the Hermiticity condition $\langle \hat{L}_z \varphi_1 | \varphi_2 \rangle = \langle \varphi_1 | \hat{L}_z \varphi_2 \rangle$ is valid only on the subspace $\mathfrak{H}_\phi' \subset \mathfrak{H}_\phi$ whose elements are periodic: $\varphi(0) = \varphi(2\pi)$. Hence $\hat{L}_z$ is non-Hermitian on $\mathfrak{H}_\phi$. Specifically, note that even though $\varphi(\phi)$ is periodic, the product $\phi\varphi(\phi)$ is not periodic and one may not invoke Hermiticity of $\hat{L}_z$ with respect to functions of this type. This is the crux of the breakdown in the proof of the uncertainty relation. See Problem 5.42.]

9.16 In regard to inconsistencies presented by the azimuthal angle ϕ, as discussed in Problem 9.15, it has been pointed out by W. Louisell[2] that more consistent angle variables are $\sin \phi$ and $\cos \phi$.

(a) Show that

$$[\sin \phi, \hat{L}_z] = i\hbar \cos \phi$$
$$[\cos \phi, \hat{L}_z] = -i\hbar \sin \phi$$

(b) Use these commutator formulas to obtain uncertainty relations between $\sin \phi$, L_z and $\cos \phi$, L_z.

Answer (partial)

(b)
$$\Delta L_z \, \Delta \sin \phi \geq \frac{\hbar \langle \cos \phi \rangle}{2}$$

$$\Delta L_z \, \Delta \cos \phi \geq \frac{\hbar \langle \sin \phi \rangle}{2}$$

9.17 (a) Show that the operator

$$\hat{R}_{\Delta\phi} \equiv \exp\left(\frac{i \, \Delta\phi \hat{L}_z}{\hbar}\right)$$

when acting on the function $f(\phi)$ changes f by a rotation of coordinates about the z axis so that the radius through ϕ is rotated to the radius through $\phi + \Delta\phi$. That is, show that

$$\hat{R}_{\Delta\phi} f(\phi) = f(\phi + \Delta\phi)$$

(b) Show that the operator

$$\hat{R}_{\Delta\boldsymbol{\phi}} = \exp\left(\frac{i \, \Delta\boldsymbol{\phi} \cdot \hat{\mathbf{L}}}{\hbar}\right)$$

[1] D. Judge, *Nuovo Cimento* **31**, 332 (1964). For further discussion and reference, see P. Carruthers and N. Nieto, *Rev. Mod. Phys.* **40**, 411 (1968)
[2] W. Louisell, *Phys. Lett.* **7**, 60 (1963).

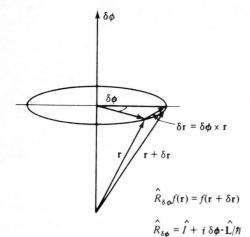

$\hat{R}_{\delta\phi}f(\mathbf{r}) = f(\mathbf{r} + \delta\mathbf{r})$

$\hat{R}_{\delta\phi} = \hat{I} + i\,\delta\boldsymbol{\phi}\cdot\hat{\mathbf{L}}/\hbar$

FIGURE 9.12 The rotation operator $\hat{R}_{\delta\phi}$ changes $f(\mathbf{r})$ by rotating $\mathbf{r}$ through the azimuthal increment $\delta\phi$. See Problem 9.17.

when acting on $f(\mathbf{r})$ changes f by rotating $\mathbf{r}$ to a new value on the surface of the sphere of radius r, but rotated away from $\mathbf{r}$ through the azimuth $\Delta\phi$, so that $\mathbf{r}(\theta, \phi) \to \mathbf{r}' = \mathbf{r}(\theta, \phi + \Delta\phi)$. For infinitesimal displacements $\delta\phi$, we may write

$$\hat{R}_{\delta\phi}f(\mathbf{r}) = f(\mathbf{r} + \delta\mathbf{r})$$

$$\delta\mathbf{r} = \delta\boldsymbol{\phi} \times \mathbf{r}$$

See Fig. 9.12.

Answers

(a) $\hat{R}_{\Delta\phi}f = \left[\exp\left(\Delta\phi\,\dfrac{\partial}{\partial\phi}\right)\right]f$

$= f(\phi) + \Delta\phi\,\dfrac{\partial f}{\partial\phi} + \dfrac{(\Delta\phi)^2}{2}\,\dfrac{\partial^2 f}{\partial\phi^2} + \cdots = f(\phi + \Delta\phi)$

(b) Let $\delta\phi$ be an infinitesimal angle so that $\Delta\phi = n\delta\phi$ in the limit that $n \gg 1$. For the infinitesimal rotation

$$\mathbf{r}' = \mathbf{r} + \delta\mathbf{r} = \mathbf{r} + \delta\boldsymbol{\phi} \times \mathbf{r}$$

so that

$$f(\mathbf{r} + \delta\mathbf{r}) = f(\mathbf{r}) + \delta\boldsymbol{\phi} \times \mathbf{r}\cdot\nabla f(\mathbf{r})$$

$$= f(\mathbf{r}) + \delta\boldsymbol{\phi}\cdot\mathbf{r} \times \nabla f(\mathbf{r})$$

$$= f(\mathbf{r}) + \dfrac{i}{\hbar}\,\delta\boldsymbol{\phi}\cdot\mathbf{r} \times \hat{\mathbf{p}}f(\mathbf{r})$$

$$= f(\mathbf{r}) + \dfrac{i}{\hbar}\,\delta\boldsymbol{\phi}\cdot\hat{\mathbf{L}}f(\mathbf{r})$$

In the Taylor series expansion of $f(\mathbf{r} + \delta\mathbf{r})$ above we have only kept terms of $O(\delta\phi)$. [The expression $\delta\mathbf{r} = \delta\boldsymbol{\phi} \times \mathbf{r}$ is valid only to terms of $O(\delta\phi)$.] In this manner we obtain

$$f(\mathbf{r} + \delta\mathbf{r}) = \left(\hat{I} + \frac{i}{\hbar} \delta\boldsymbol{\phi} \cdot \hat{\mathbf{L}}\right) f(\mathbf{r}) = \hat{R}_{\delta\phi} f(\mathbf{r})$$

For a finite rotational displacement through the angle $\Delta\phi = n\delta\phi$, we apply the operator $\hat{R}_{\delta\phi}$, n times:

$$\hat{R}_{n\delta\phi} = (\hat{R}_{\delta\phi})^n = \left(\hat{I} + \frac{i}{\hbar} \delta\boldsymbol{\phi} \cdot \hat{\mathbf{L}}\right)^n$$

and pass to the limit $n \to \infty$ or, equivalently, $\Delta\phi/\delta\phi \to \infty$.

$$\hat{R}_{\Delta\phi} = \lim_{\Delta\phi/\delta\phi \to \infty} \left(\hat{I} + \frac{i}{\hbar} \delta\boldsymbol{\phi} \cdot \hat{\mathbf{L}}\right)^{\Delta\phi/\delta\phi} = e^{i\Delta\boldsymbol{\phi} \cdot \hat{\mathbf{L}}/\hbar}$$

(*Note:* The operator $\hat{R}_{\delta\phi}$ rotates $\mathbf{r}$ to $\mathbf{r} + \delta\boldsymbol{\phi} \times \mathbf{r}$ with respect to a fixed coordinate frame. If, on the other hand, the coordinate frame is rotated through $\delta\boldsymbol{\phi}$ with $\mathbf{r}$ fixed in space, then in the new coordinate frame this vector has the value $\mathbf{r} - \delta\boldsymbol{\phi} \times \mathbf{r}$. Thus, rotation of coordinates through $\delta\boldsymbol{\phi}$ is generated by the operator $\hat{R}_{-\delta\phi}$.)

9.18 Show that $\hat{L}^2$ may be written as

$$\hat{L}^2 = -\hbar^2 \left(\frac{\partial^2}{\partial\theta^2} + \cot\theta \frac{\partial}{\partial\theta} + \frac{1}{\sin^2\theta} \frac{\partial^2}{\partial\phi^2}\right)$$

9.19 Show by direct operation that

$$\hat{L}^2 Y_2^2 = 6\hbar^2 Y_2^2$$

$$\hat{L}_z Y_2^2 = 2\hbar Y_2^2$$

9.20 First calculate $P_2(\mu)$ using the generating function $(1 - 2\mu s + s^2)^{-1/2}$. Then obtain $P_2^1(\mu)$ using the relation between P_l and P_l^m given in Table 9.3. Having found P_2^1, form Θ_2^1 and then Y_2^1. Check your answers with the values given in Table 9.1.

9.21 Using the explicit form of Y_l^m, show that

$$\langle Y_l^m \mid Y_{l'}^{m'} \rangle = \langle lm \mid l'm' \rangle = 0 \qquad m \neq m'$$

9.22 Operate on Y_l^{l-1} with $\hat{L}_-$ to obtain the angular dependent factor of Y_l^{l-2}.

9.23 Assume that a particle has an orbital angular momentum with z component $\hbar m$ and square magnitude $\hbar^2 l(l + 1)$.

(a) Show that in this state

$$\langle L_x \rangle = \langle L_y \rangle = 0$$

(b) Show that

$$\langle L_x^2 \rangle = \langle L_y^2 \rangle = \frac{\hbar^2 l(l + 1) - m^2\hbar^2}{2}$$

[*Hints:* For (a), use $\hat{L}_+$ and $\hat{L}_-$. For (b), use $\hat{L}^2 = \hat{L}_x^2 + \hat{L}_y^2 + \hat{L}_z^2$.]

9.24   The same conditions hold as in Problem 9.23. What is the expectation of the operator $\frac{1}{2}(\hat{L}_x\hat{L}_y + \hat{L}_y\hat{L}_x)$ in the Y_l^m state?

9.25   A D_2 molecule at 30 K, at $t = 0$, is known to be in the state

$$\psi(\theta, \phi, 0) = \frac{3Y_1{}^1 + 4Y_7{}^3 + Y_7{}^1}{\sqrt{26}}$$

(a)   What values of L and L_z will measurement find and with what probabilities will these values occur?

(b)   What is $\psi(\theta, \phi, t)$?

(c)   What is $\langle E \rangle$ for the molecule (in eV) at $t > 0$?

(*Note:* For the purely rotational states of D_2, assume that $h/4\pi Ic = 30.4$ cm^{-1}.)

9.26   At a given instant of time, a rigid rotator is in the state

$$\varphi(\theta, \phi) = \sqrt{\frac{3}{4\pi}} \sin \phi \sin \theta$$

(a)   What possible values of L_z will measurement find and with what probability will these values occur?

(b)   What is $\langle \hat{L}_x \rangle$ for this state?

(c)   What is $\langle \hat{L}^2 \rangle$ for this state?

9.27   Suppose that a rotator is in the state $Y_1{}^{-1}$. What values will measurement of $\hat{L}_x$ find and with what probability will these values occur? (*Hint:* Most of the analysis in the text [(9.92) et seq.] involving the expansion of the state $Y_1{}^1$ may be used here.)

9.28   A one-particle system is in the angular state Y_l^m. Measurement is made of the component of **L** along the z' axis. The z' axis makes an angle λ with the z axis. What is the expectation of this component? What is the expectation of the square of this component? (See Fig. 9.13.)

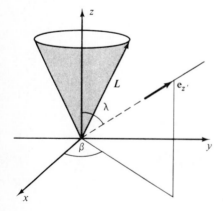

FIGURE 9.13   Configuration relevant to Problem 9.28.

Answer

For the first problem we must calculate $\langle \mathbf{e}_{z'} \cdot \hat{\mathbf{L}} \rangle$, where $\mathbf{e}_{z'}$ is the unit vector in the direction of the z' axis. For the second problem, we must calculate $\langle (\mathbf{e}_{z'} \cdot \hat{\mathbf{L}})^2 \rangle$. The components of $\mathbf{e}_{z'}$ are

$$\mathbf{e}_{z'} = (\sin \lambda \cos \beta, \sin \lambda \sin \beta, \cos \lambda)$$

where β is the azimuthal coordinate of $\mathbf{e}_{z'}$ with respect to the original axes.

$$\langle \mathbf{e}_{z'} \cdot \hat{\mathbf{L}} \rangle = \sin \lambda \cos \beta \langle L_x \rangle + \sin \lambda \sin \beta \langle L_y \rangle + \cos \lambda \langle L_z \rangle = \hbar m \cos \lambda$$

$$\langle (\mathbf{e}_{z'} \cdot \hat{\mathbf{L}})^2 \rangle = \sin^2 \lambda \langle L_x{}^2 \rangle + \cos^2 \lambda \langle L_z{}^2 \rangle$$

9.29 With $\Theta_l(\mu)$ replaced by $P_l(\mu)$ in (9.73), show that single differentiation of this equation gives (9.72) with $\Theta(\mu) = P_l{}^1(\mu)$ and $m = 1$.

9.4 ADDITION OF ANGULAR MOMENTUM

Two Electrons

In this section we examine the relation between the angular momentum of a total system and that of its constituents. This problem is of practical importance in atomic and nuclear physics where one encounters systems of many particles (e.g., electrons, neutrons, protons, etc.). In many cases one is chiefly concerned with the resultant angular momentum of the atom or nucleus.

Consider two systems that are rotating about a common origin. They could be two rotators or two electrons in an atom (Fig. 9.14). We will speak in terms of an atom. If the angular momentum (neglecting spin) of the first electron is $\mathbf{L}_1$ and that

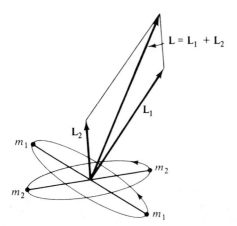

FIGURE 9.14 Classical addition of angular momentum. The two angular momentum vectors $\mathbf{L}_1$ and $\mathbf{L}_2$ add to give the resultant $\mathbf{L}$.

of the second electron is $\mathbf{L}_2$, the magnitude and z component of the total angular momentum of the composite system of the two electrons is

(9.102)
$$\hat{L}^2 = (\hat{\mathbf{L}}_1 + \hat{\mathbf{L}}_2)^2 = \hat{L}_1{}^2 + \hat{L}_2{}^2 + 2\hat{\mathbf{L}}_1 \cdot \hat{\mathbf{L}}_2$$
$$\hat{L}_z = \hat{L}_{1z} + \hat{L}_{2z}$$

Suppose that the total system is in a state with definite values of L_{1z}, L_{2z} (e.g., $|m_1 m_2\rangle$). How much further may this state be resolved? Since there are only two good quantum numbers associated with each electron (i.e., $m_1 l_1$ and $m_2 l_2$), one suspects that the composite system will have no more than four good quantum numbers. As it turns out, the eigenstate $|m_1 m_2\rangle$ may further be resolved to the state $|l_1 l_2 m_1 m_2\rangle$. This state cannot be further resolved. For instance, one might wish to measure L^2. If the atom is in the state $|l_1 l_2 m_1 m_2\rangle$ before measurement, we are not assured that it will be in that state after measurement of L^2. That this is so follows from the fact that $\hat{L}^2$ does not commute with, say, $\hat{L}_{1z}$.

(9.103)
$$[\hat{L}_{1z}, \hat{L}^2] = [\hat{L}_{1z}, \hat{L}_1{}^2 + \hat{L}_2{}^2 + 2\hat{\mathbf{L}}_1 \cdot \hat{\mathbf{L}}_2]$$
$$= 2[\hat{L}_{1z}, \hat{\mathbf{L}}_1 \cdot \hat{\mathbf{L}}_2] = 2i\hbar(\hat{L}_{1y}\hat{L}_{2x} - \hat{L}_{1x}\hat{L}_{2y})$$

In order to establish that the set of eigenvalues (l_1, l_2, m_1, m_2) are *good quantum numbers* (i.e., that these values may be simultaneously specified in an eigenstate $|l_1 l_2 m_1 m_2\rangle$), one must show that the set of four operators $(\hat{L}_{1z}, \hat{L}_{2z}, \hat{L}_1{}^2, \hat{L}_2{}^2)$ are a set of mutually commuting operators. The fact that no other commuting operators (restricting the discussion to the angular momentum properties of the system) can be attached to this set indicates that $(\hat{L}_{1z}, \hat{L}_{2z}, \hat{L}_1{}^2, \hat{L}_2{}^2)$ is a *complete* set of commuting operators.

We wish to show that

(9.104)
$$[\hat{L}_{1z}, \hat{L}_{2z}] = [\hat{L}_{1z}, \hat{L}_1{}^2] = [\hat{L}_{1z}, \hat{L}_2{}^2] = [\hat{L}_{2z}, \hat{L}_1{}^2]$$
$$= [\hat{L}_{2z}, \hat{L}_2{}^2] = [\hat{L}_1{}^2, \hat{L}_2{}^2] = 0$$

The fact that $[\hat{L}_1{}^2, \hat{L}_{1z}] = 0$ was shown in Section 9.1. The commutators $[\hat{L}_{1z}, \hat{L}_{2z}]$ vanish because the coordinates of system 1 are independent of the coordinates of system 2, so that, for example,

$$\left[z_1, \frac{\partial}{\partial z_2}\right] = 0$$

All other terms in (9.104) vanish for similar reasons.

Suppose that we measure L^2 and L_z and establish the state $|lm\rangle$. Can this state be further resolved? Yes. One may subsequently measure $L_1{}^2$ and $L_2{}^2$ and not destroy the eigenvalues of L^2 and L_z already established. After measurement, the system is left in the state $|lml_1l_2\rangle$. To show that l, m, l_1, and l_2 are good quantum numbers, we must establish that the set $(\hat{L}_1{}^2, \hat{L}_2{}^2, \hat{L}^2, \hat{L}_z)$ is a set of commuting operators. The only questionable pairs are of the form

$$[\hat{L}_1{}^2, \hat{L}^2] \quad \text{and} \quad [\hat{L}_1{}^2, \hat{L}_z]$$

Expanding these, we obtain

$$\begin{aligned} (9.105) \quad & [\hat{L}_1{}^2, \hat{L}_1{}^2 + \hat{L}_2{}^2 + 2\hat{\mathbf{L}}_1 \cdot \hat{\mathbf{L}}_2] = 2[\hat{L}_1{}^2, \hat{\mathbf{L}}_1 \cdot \hat{\mathbf{L}}_2] = 2[\hat{L}_1{}^2, \hat{\mathbf{L}}_1] \cdot \hat{\mathbf{L}}_2 = 0 \\ & [\hat{L}_1{}^2, \hat{L}_z] = [\hat{L}_1{}^2, \hat{L}_{1z} + \hat{L}_{2z}] = [\hat{L}_1{}^2, \hat{L}_{1z}] = 0 \end{aligned}$$

Coupled and Uncoupled Representations

Thus we find, in quantum mechanics, that the angular momentum states for a composite system consisting of two subsystems are characterized by either of two sets of good quantum numbers. These correspond, respectively, to the eigenstates $|l_1l_2m_1m_2\rangle$ and $|lml_1l_2\rangle$. The latter states pertain to problems where the total angular momentum of the composite system is important. We will call this representation where L^2 and L_z (together with $L_1{}^2$ and $L_2{}^2$) are specified the *coupled representation* (Fig. 9.15). The representation where the z component and magnitude of angular momentum are specified for all subcomponents (i.e., $L_1{}^2$, L_{1z}, $L_2{}^2$, L_{2z}) will be called the *uncoupled representation*.

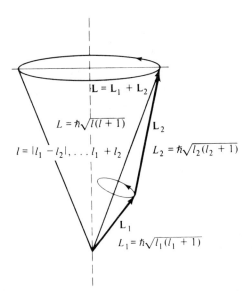

FIGURE 9.15 In the coupled representation, L_1 and L_2 couple to give L, which then exhibits discrete orientations along any prescribed axis. In this vector-model sketch of the state $|lml_1l_2\rangle$, the z components of L_1 and L_2 are not conserved. This corresponds to the fact that most generally, $|lml_1l_2\rangle$ is a superposition state involving all m_1, m_2 values with fixed $m_1 + m_2 = m$.

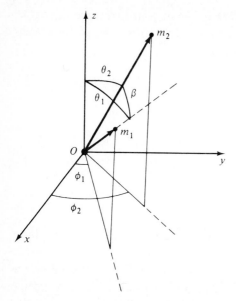

FIGURE 9.16 Angular coordinates for particle m_1 and particle m_2. Two important addition theorems involving the angle β between Om_1 and Om_2 are

(a) $\cos \beta = \cos \theta_1 \cos \theta_2$
$$+ \sin \theta_1 \sin \theta_2 \cos (\phi_1 - \phi_2)$$

(b) $P_l(\cos \beta) = \sum\limits_{m=-l}^{l} \dfrac{(l - m)!}{(l + m)!}$
$$\times P_l^m(\cos \theta_1) P_l^m(\cos \theta_2) e^{im(\phi_1 - \phi_2)}$$
$$= \dfrac{4\pi}{2l + 1}$$
$$\times \sum\limits_{m=-l}^{l} [Y_l^m(\theta_1, \phi_1)]^* Y_l^m(\theta_2, \phi_2)$$

The eigenstates in either representation are constructed from products of the eigenstates $|m_1 l_1\rangle$ and $|m_2 l_2\rangle$. In the uncoupled representation the simultaneous eigenstates of $(\hat{L}_1{}^2, \hat{L}_{1z}, \hat{L}_2{}^2, \hat{L}_{2z})$ are given by the products

(9.106)
$$|l_1 l_2 m_1 m_2\rangle = |l_1 m_1\rangle |l_2 m_2\rangle$$

or, equivalently,

$$Y_{l_1}{}^{m_1}(\theta_1, \phi_1) Y_{l_2}{}^{m_2}(\theta_2, \varphi_2)$$

The spherical coordinates of electron 1 are θ_1, ϕ_1 while θ_2, ϕ_2 are the coordinates of electron 2 (Fig. 9.16). For given values of l_1 and l_2 there are $(2l_1 + 1)(2l_2 + 1)$ linearly independent eigenstates of the composite system each of the form (9.106) and each with specified values of (l_1, l_2, m_1, m_2).

Eigenstates $|lml_1 l_2\rangle$ of the coupled representation are simultaneous eigenstates of the commuting operators

(9.107)
$$\hat{L}^2 = \hat{L}_1{}^2 + \hat{L}_2{}^2 + 2\hat{\mathbf{L}}_1 \cdot \hat{\mathbf{L}}_2$$
$$\hat{L}_z = \hat{L}_{1z} + \hat{L}_{2z}$$
$$\hat{L}_1{}^2, \hat{L}_2{}^2$$

Any such state may be written as a superposition of the eigenstates of the uncoupled representation (9.106). In both representations l_1 and l_2 are good quantum numbers. It follows that in the expansion

(9.108)
$$|lml_1 l_2\rangle = \sum\limits_{m_1 + m_2 = m}\sum |l_1 l_2 m_1 m_2\rangle\langle l_1 l_2 m_1 m_2 | lml_1 l_2\rangle$$

summation can only run over the quantum numbers m_1 and m_2. The constraint $m_1 + m_2 = m$ stems from the middle equation (9.107) and the orthogonality of the states $|l_1 l_2 m_1 m_2\rangle$. Equation (9.108) may be rewritten

$$(9.109) \qquad |lml_1 l_2\rangle = \sum_{\substack{m_1 + m_2 = m}} \sum C_{m_1 m_2} |l_1 l_2 m_1 m_2\rangle$$

$$C_{m_1 m_2} \equiv \langle l_1 l_2 m_1 m_2 | lml_1 l_2\rangle$$

The expansion coefficients $C_{m_1 m_2}$ are called *Clebsch–Gordan* coefficients and their significance is as follows. Let the composite system be two electrons. In the state $|lml_1 l_2\rangle$ it is known that these electrons have respective angular momentum quantum numbers l_1 and l_2, and total angular momentum and z component quantum numbers l and m. The question then arises as to what measurement of L_{1z} and L_{2z} will find in the state $|lml_1 l_2\rangle$. The answer to this question is that

$$(9.110) \qquad |C_{m_1 m_2}|^2 = \text{probability that at fixed } L \text{ and } L_z$$
$$\text{measurement finds one electron with } L_{1z} = m_1 \hbar$$
$$\textit{and the other electron with } L_{2z} = m_2 \hbar$$

As an elementary illustration of the technique employed to construct these coefficients, let us consider expansion of the state

$$|lml_1 l_2\rangle = |1, -1, 1, 1\rangle$$

With $m_1 + m_2 = -1$, the expansion (9.108) becomes

$$|1, -1, 1, 1\rangle = C_{0-1}|1, 0\rangle|1, -1\rangle + C_{-10}|1, -1\rangle|1, 0\rangle$$

The coefficients C_{0-1} and C_{-10} are determined by normalization and propitious application of the $\hat{L}_+$ and $\hat{L}_-$ operators. For the case at hand, we operate on the last equation with

$$\hat{L}_- = \hat{L}_{1-} + \hat{L}_{2-}$$

There results

$$\hat{L}_-|1, -1, 1, 1\rangle = 0$$
$$= (\hat{L}_{1-} + \hat{L}_{2-})(C_{0-1}|1, 0\rangle|1, -1\rangle + C_{-10}|1, -1\rangle|1, 0\rangle)$$
$$= \sqrt{2}(C_{0-1} + C_{-10})|1, -1\rangle|1, -1\rangle$$

We may conclude that

$$C_{0-1} = -C_{-10}$$

Normalization of the state $|1, -1, 1, 1\rangle$ gives $C_{0-1} = 1/\sqrt{2}$. So it is equally probable that measurement finds $(m_1, m_2) = (0, -1)$ or $(-1, 0)$.

Coordinate Representation

We have been writing $|lm\rangle$ for the eigenvectors of $\hat{L}^2, \hat{L}_z$. The coordinate representation of these states is given by the projection

$$\langle \theta, \phi \,|\, lm \rangle = Y_l^m(\theta, \phi)$$

Likewise, the coordinate representation of the composite state $|l_1 l_2 m_1 m_2\rangle$ is given by the projection

$$\langle \theta_1 \phi_1 \theta_2 \phi_2 \,|\, l_1 l_2 m_1 m_2 \rangle = Y_{l_1}^{m_1}(\theta_1 \phi_1) Y_{l_2}^{m_2}(\theta_2, \phi_2)$$

In this manner, the expansion (9.109) gives the coordinate representation

$$\langle \theta_1 \phi_1 \theta_2 \phi_2 \,|\, lm l_1 l_2 \rangle = \sum_{m_1 + m_2 = m} \sum C_{m_1 m_2} \langle \theta_1 \phi_1 \theta_2 \phi_2 \,|\, l_1 l_2 m_1 m_2 \rangle$$

$$= \sum \sum C_{m_1 m_2} Y_{l_1}^{m_1}(\theta_1, \phi_1) Y_{l_2}^{m_2}(\theta_2, \phi_2)$$

Thus we see that the Clebsch–Gordan expansion affords a means of obtaining the coordinate representation of the composite state $|lm l_1 l_2\rangle$.

The theory of representations is discussed further in Section 11.1 and Appendix A.

Values of l for Two Electrons

Next we consider the problem of finding the allowed values of (l, m), given (l_1, l_2). This problem is directly related to "two-electron atoms," such as He and Ca, whose energy levels are l-dependent. Suppose that one electron is a p-electron (i.e., it is in a P state) and the other electron is a d-electron. What values can result for l and m (still neglecting spin)?

Let us consider the general case where the two electrons have respective l values l_1 and l_2. Since

$$L_z = L_{1z} + L_{2z}$$

it follows that the maximum value m can have is

$$m_{\max} = m_{1\max} + m_{2\max}$$

or, equivalently,

(9.111)
$$m_{\max} = l_1 + l_2$$

It is clear that of the various values the total angular momentum quantum number l may assume, the maximum value is equal to $m_{\max}$. With (9.111) we then obtain

(9.112)
$$l_{\max} = l_1 + l_2$$

In that l is an angular momentum quantum number, successive values of l differ from l_{max} in unit steps down to some minimum value. What is this minimum value? To obtain it, we note the following.

As noted previously, in the uncoupled representation there are $(2l_1 + 1)(2l_2 + 1)$ independent, common eigenstates of $\hat{L}_1{}^2$, $\hat{L}_2{}^2$, $\hat{L}_{1z}$, and $\hat{L}_{2z}$ relevant to the two-electron system. These states span a $(2l_1 + 1)(2l_2 + 1)$-dimensional space. A change in representation[1] from this basis to the common eigenstates of $\hat{L}^2$, $\hat{L}_z$, $\hat{L}_1{}^2$, and $\hat{L}_2{}^2$ maintains the dimensionality of this space. This observation affords a method of obtaining l_{min}. That is, keep decreasing l_{max} in unit steps until the total number of independent states equals $(2l_1 + 1)(2l_2 + 1)$.

Now the number of independent eigenstates with a given l number is $2l + 1$. Then the value of l_{min} we seek satisfies the equation

$$\sum_{l=l_{min}}^{l_1 + l_2} (2l + 1) = (2l_1 + 1)(2l_2 + 1)$$

This relation is satisfied if we set

(9.113) $$l_{min} = |l_1 - l_2|$$

In this manner we find that the values of l corresponding to a system comprised of two electrons with respective l values l_1 and l_2 are

(9.114) $$l = |l_1 - l_2|, \ldots, l_1 + l_2$$

For the problem cited above with one p electron ($l_1 = 1$) and one d electron ($l_2 = 2$), the total angular momentum may be any of the values

$$l = 1, 2, 3$$
$$L = \hbar\sqrt{2}, \quad L = \hbar\sqrt{6}, \quad L = \hbar\sqrt{12}$$

There are a totality of

$$N = (2 \times 1 + 1) + (2 \times 2 + 1) + (2 \times 3 + 1) = (2 \times 1 + 1)(2 \times 2 + 1)$$
$$= 15$$

states, corresponding to these three values of l. For the case of two p electrons ($l_1 = l_2 = 1$) there are nine eigenstates. These are listed in Table 9.5.

---

[1] The notion of changes in representation was discussed in Section 7.4 and will be developed further in Chapter 11.

TABLE 9.5   Nine common eigenstates $|lml_1l_2\rangle$ of the opeators: $\hat{L}^2$, $\hat{L}_z$, $\hat{L}_1{}^2$, $\hat{L}_2{}^2$ for two p electrons

Diagramatic Representation[a]	$\|lml_1l_2\rangle$	$= \sum C_{m_1m_2}\|l_1m_1\rangle_1\|l_2m_2\rangle_2$
↑↑	$\|22\ 11\rangle$	$= \|11\rangle_1\|11\rangle_2$
↑· + ·↑	$\|21\ 11\rangle$	$= \sqrt{\tfrac{1}{2}}\|11\rangle_1\|10\rangle_2 + \sqrt{\tfrac{1}{2}}\|10\rangle_1\|11\rangle_2$
↑↓ + ·· + ↓↑	$\|20\ 11\rangle$	$= \sqrt{\tfrac{1}{6}}\|11\rangle_1\|1,-1\rangle_2 + \sqrt{\tfrac{2}{3}}\|10\rangle_1\|10\rangle_2 + \sqrt{\tfrac{1}{6}}\|1,-1\rangle_1\|11\rangle_2$
·↓ + ↓·	$\|2,-1\ 11\rangle$	$= \sqrt{\tfrac{1}{2}}\|10\rangle_1\|1,-1\rangle_2 + \sqrt{\tfrac{1}{2}}\|1,-1\rangle_1\|10\rangle_2$
↓↓	$\|2,-2\ 11\rangle$	$= \|1,-1\rangle_1\|1,-1\rangle_2$
↑· − ·↑	$\|11\ 11\rangle$	$= \sqrt{\tfrac{1}{2}}\|11\rangle_1\|10\rangle_2 - \sqrt{\tfrac{1}{2}}\|10\rangle_1\|11\rangle_2$
↑↓ − ↓↑	$\|10\ 11\rangle$	$= \sqrt{\tfrac{1}{2}}\|11\rangle_1\|1,-1\rangle_2 - \sqrt{\tfrac{1}{2}}\|1,-1\rangle_2\|11\rangle_2$
·↓ − ↓·	$\|1.-1\ 11\rangle$	$= \sqrt{\tfrac{1}{2}}\|10\rangle_1\|1,-1\rangle_2 - \sqrt{\tfrac{1}{2}}\|1,-1\rangle_1\|10\rangle_2$
↑↓ − ·· + ↓↑	$\|00\ 11\rangle$	$= \sqrt{\tfrac{1}{3}}\|11\rangle_1\|1,-1\rangle_2 - \sqrt{\tfrac{1}{3}}\|10\rangle_1\|10\rangle_2 + \sqrt{\tfrac{1}{3}}\|1,-1\rangle_1\|11\rangle_2$

[a] The diagramatic representation of states is such that an up-arrow, a down-arrow, and a dot represent, respectively, $m = 1, -1$, and 0 of individual electrons.

The distinction between the coupled and uncoupled representations is brought out in the following two sets of eigenstate equations.

Coupled Representation

$$\begin{pmatrix}\hat{L}^2 \\ \hat{L}_z \\ \hat{L}_1{}^2 \\ \hat{L}_2{}^2\end{pmatrix}|lml_1l_2\rangle = \hbar^2 \begin{pmatrix} l(l+1) \\ m/\hbar \\ l_1(l_1+1) \\ l_2(l_2+1)\end{pmatrix}|lml_1l_2\rangle$$

Uncoupled Representation

$$\begin{pmatrix}\hat{L}_{1z} \\ \hat{L}_{2z} \\ \hat{L}_1{}^2 \\ \hat{L}_2{}^2\end{pmatrix}|l_1l_2m_1m_2\rangle = \hbar^2 \begin{pmatrix} m_1/\hbar \\ m_2/\hbar \\ l_1(l_1+1) \\ l_2(l_2+1)\end{pmatrix}|l_1l_2m_1m_2\rangle$$

These equations are relevant, for example, to the case of two electrons, given that one is an l_1 electron and the other, an l_2 electron.

For three electrons, in the uncoupled representation the six operators

$$(\hat{L}_1{}^2, \hat{L}_{1z}, \hat{L}_2{}^2, \hat{L}_{2z}, \hat{L}_3{}^2, \hat{L}_{3z})$$

form a complete commuting set. Good quantum numbers associated with these states are $(l_1, m_1, l_2, m_2, l_3, m_3)$. In the more relevant coupled representation, the six operators

$$(\hat{L}^2, \hat{L}_z, \hat{L}_1{}^2, \hat{L}_2{}^2, \hat{L}_3{}^2, \hat{A}_1{}^2)$$

form a complete commuting set. The operator $\hat{A}_1{}^2$ is given by

(9.115) $\qquad \hat{A}_1{}^2 \equiv a_{12}(\hat{\mathbf{L}}_1 + \hat{\mathbf{L}}_2)^2 + a_{13}(\hat{\mathbf{L}}_1 + \hat{\mathbf{L}}_3)^2 + a_{23}(\hat{\mathbf{L}}_2 + \hat{\mathbf{L}}_3)^2$

where the a coefficients are arbitrary. If l' is the eigen-l-value related to $\hat{A}_1{}^2$, six good quantum numbers for the case at hand are $(l, m, l_1, l_2, l_3, l')$.

PROBLEMS

9.30 What are the eigenvalues of the set of operators $(\hat{L}_1{}^2, \hat{L}_{1z}, \hat{L}_2{}^2, \hat{L}_{2z})$ corresponding to the product eigenstate $|m_1 l_1\rangle|m_2 l_2\rangle$?

9.31 Let $\hat{\mathbf{J}}_1$ and $\hat{\mathbf{J}}_2$ be the respective angular momenta of the individual components of a two-component system. The total system has angular momentum $\hat{\mathbf{J}} = \hat{\mathbf{J}}_1 + \hat{\mathbf{J}}_2$. Show that:

(a) $\hat{\mathbf{J}}_1 \cdot \hat{\mathbf{J}}_2 = \frac{1}{2}(\hat{J}_{1+}\hat{J}_{2-} + \hat{J}_{1-}\hat{J}_{2+}) + \hat{J}_{1z}\hat{J}_{2z}$

(b) $\hat{J}^2 = \hat{J}_1{}^2 + \hat{J}_2{}^2 + 2\hat{J}_{1z}\hat{J}_{2z} + (\hat{J}_{1+}\hat{J}_{2-} + \hat{J}_{1-}\hat{J}_{2+})$

9.32 (a) Using the expansions developed in Problem 9.31, operate on the coupled angular momentum eigenstates for two p electrons as listed in Table 9.5 with $\hat{L}^2$ and $\hat{L}_z$, respectively, to verify the lm entries in each of the nine $|lml_1 l_2\rangle$ eigenstates.

(b) What are the Clebsch–Gordan coefficients involved in the expansion of the state $|0011\rangle$?

(c) What is the inner product $\langle 2011|0011\rangle$?

9.33 (a) With respect to the diagrammatic representation of states depicted in Table 9.5, what are the states corresponding to the diagrams

$$\psi_0 = \cdot\cdot, \qquad \psi_1 = \uparrow\downarrow + \downarrow\uparrow, \qquad \psi_3 = \uparrow\downarrow?$$

(b) Expand each of these functions in terms of the nine diagrams listed in Table 9.5.

(c) Are any of these three states eigenstates of $\hat{L}^2$? [*Hint*: Use the expansions obtained in part (b).]

(d) Two electrons are known to be in the coupled state ψ_1. What values of total angular momentum L will measurement find and with what probabilities will these values occur?

9.34 Two p electrons are in the state $|lml_1 l_2\rangle = |1, -111\rangle$. If measurement is made of L_{1z} in this state, what values may be found and with what probability will these values occur?

9.35 Two p electrons are in the coupled angular momentum state $|lml_1 l_2\rangle = |2, -2, 11\rangle$. What is the joint probability of finding the two electrons with $L_{1z} = L_{2z} = -\hbar$?

9.36 How many independent eigenstates are there in the coupled representation for a two-component system, given that $l_1 = 5$ and $l_2 = 1$? Make a table listing the ml values for all these states.

9.37 Show that $\hat{A}_1{}^2$ as given by (9.115) commutes with $\hat{L}^2$ and $\hat{L}_z$.

9.38 The eigenstate corresponding to maximum l for the three-electron case is

$$|l, m, l_1, l_2, l_3, l'\rangle = |l_1, l_1\rangle |l_2, l_2\rangle |l_3, l_3\rangle$$

(a) What are the eigenvalues of $\hat{L}^2$, $\hat{L}_z$ corresponding to this state?

(b) What is the eigenvalue of the operator $\hat{A}_1{}^2$ given by (9.115) corresponding to this state?

9.5 TOTAL ANGULAR MOMENTUM FOR TWO OR MORE ELECTRONS

We are now concerned with the possible values the total angular momentum l numbers may assume for a system of N electrons with respective l_i values: $l_1, l_2, \ldots, l_N$, in the coupled representation. A totality of $(2l_1 + 1)(2l_2 + 1) \cdots (2l_N + 1)$ product states may be formed which are simultaneous eigenstates of the set of operators

$$(\hat{L}^2, \hat{L}_z{}^2, \hat{L}_1{}^2, \hat{L}_2{}^2, \ldots, \hat{L}_N{}^2)$$

We must make sure that our procedure for calculating these l values preserves this number of states. This affords a check that we have found all l values.

The possible values that l can assume may be obtained by one of two techniques. The first technique follows from the rule (9.114) for the addition of the angular momenta of two electrons with respective l values: l_1 and l_2. In this case the combined angular momentum

$$\hat{L}^2 = (\hat{\mathbf{L}}_1 + \hat{\mathbf{L}}_2)^2$$

has eigenvalues, $\hbar^2 l(l + 1)$, where

$$l = |l_1 + l_2|, \ldots, |l_1 - l_2|$$

Consider the case of three electrons. Their total angular momentum is given by

$$\hat{L}^2 = (\hat{\mathbf{L}}_1 + \hat{\mathbf{L}}_2 + \hat{\mathbf{L}}_3)^2$$

This may be rewritten in the form

$$\hat{L}^2 = (\hat{\mathbf{L}}' + \hat{\mathbf{L}}_3)^2$$
$$\hat{\mathbf{L}}' = \hat{\mathbf{L}}_1 + \hat{\mathbf{L}}_2$$

Suppose that one of the l values corresponding to $\hat{L}'^2$ is l'. Then the l values corresponding to the total angular momentum $\hat{L}^2$ are

(9.116) $$l = |l' + l_3|, \ldots, |l' - l_3|$$

This again follows the rule of (9.114). For example, consider the case of three p electrons ($l_1 = l_2 = l_3 = 1$). Then for the first two electrons we have

$$l' = |l_1 + l_2|, \ldots, |l_1 - l_2| = 0, 1, 2$$

Adding the third electron gives [using (9.116)] the l values

$$l' = 0, 1, 2 \qquad l = |l' + l_3|, \ldots, |l' - l_3|$$
$$l' = 0 \longrightarrow l = 1$$
$$l' = 1 \longrightarrow l = 0, 1, 2$$
$$l' = 2 \longrightarrow l = 1, 2, 3$$

Thus we obtain the result

$$\boxed{l = 0, 1, 2, 3 \qquad \text{for three } p \text{ electrons}}$$

There is a distinct eigenstate for each distinct manner in which l may be formed. This gives a total number of

$$(2 \times 0 + 1) + 3(2 \times 1 + 1) + 2(2 \times 2 + 1) + (2 \times 3 + 1)$$
$$= 1 + 9 + 10 + 7 = 27$$

states, which agrees with the product

$$(2l_1 + 1)(2l_2 + 1)(2l_3 + 1) = 3 \times 3 \times 3 = 27$$

For the case of N electrons with respective l values $l_1, l_2, \ldots, l_N$, we follow a similar procedure. First, we add the angular momenta of the first two electrons. This gives

$$l' = |l_1 + l_2|, \ldots, |l_1 - l_2|$$

To these values we add the angular momentum of the third electron to obtain

$$l'' = |l' + l_3|, \ldots, |l' - l_3|$$

There is a separate sequence of l'' values for each value of l'. Adding the angular momentum of the fourth electron gives

$$l''' = |l'' + l_4|, \ldots, |l'' - l_4|$$

We continue in this manner until all individual angular momentum l values are accounted for. The final sequence gives all possible values of l. For three electrons the sequence of l'' gives all the values of l. For four electrons the sequence for l''' gives the values of l.

Addition Rules

The values of total l obtained by sequential addition as described above may more simply be arrived at by the following rule. Consider N electrons with respective angular momentum values: $l_1, l_2, \ldots, l_N$. These values may always be ordered so that

$$l_1 \leq l_2 \leq \cdots \leq l_N$$

Let

$$\Lambda = \sum_{i=1}^{N-1} l_i$$

Then we have the following:

    (a) If $l_N - \Lambda > 0$,

(9.117)
$$l^{min} = l_N - \Lambda$$

    (b) If $l_N - \Lambda \leq 0$,

(9.118)
$$l^{min} = 0$$

    (c) In all cases

(9.119)
$$l^{max} = \sum_{i=1}^{N} l_i$$

    (d) The possible values of l that give the values of total L,

$$L^2 = (\mathbf{L}_1 + \mathbf{L}_2 + \cdots + \mathbf{L}_N)^2 = \hbar^2 l(l+1)$$

are given by

(9.120)
$$l = |l^{max}|, |l^{max} - 1|, \ldots, |l^{min}|$$

As a simple example of this technique, consider the case of two p electrons and one f electron ($l_1 = l_2 = 1, l_3 = 3$). Then

$$\Lambda = 1 + 1 = 2$$
$$l_3 - \Lambda = 3 - 2 = 1 = l^{min}$$
$$l^{max} = 1 + 1 + 3 = 5$$

Therefore,

$$\boxed{\begin{array}{ll} l = 1, 2, 3, 4, 5 & \text{for two } p\text{-electrons and} \\ & \text{one } f\text{-electron} \end{array}}$$

The electron orbital angular momentum notation s, p, d, f, ... stems from atomic physics. The correspondence between these letters and l values of individual electrons follows the scheme

symbol	s	p	d	f	g	h	$\cdots$
l value	0	1	2	3	4	5	$\cdots$

This notation will be used again in Chapter 12.

In Chapter 10 we will see how $\hat{L}^2$ enters the Hamiltonian for one and two-particle systems. The Y_l^m functions will take on further significance. They will emerge as the angular dependent factors of the energy eigenfunctions for these systems.

The topic of the addition of angular momentum is returned to in Chapter 11, where the rules developed above are applied to the addition of spin angular momentum. In Chapter 12 these rules are again applied to the addition of orbital and spin angular momentum as related to one- and two-electron atoms. In general, the rules developed above for the addition of angular momentum, are valid for orbital, **L**, spin, **S**, and total angular momentum, **J**.

PROBLEMS

9.39   What are the possible values of l for:

    (a)   Four p electrons?

    (b)   Three p and one f ($l_4 = 3$) electrons?

9.40   What is the wavefunction (in Dirac notation) for three p electrons in the state with $l = m = 3$?

9.41   Show that the two schemes for obtaining the total l value for three electrons with respective l values l_1, l_2, and l_3, as described in the text, are equivalent.

9.42   (a)   Show that the technique of sequential addition for obtaining total l values in the coupled representation gives

$$(2l_1 + 1)(2l_2 + 1) \cdots (2l_N + 1)$$

eigenstates. (*Hint:* Assume that $l_1 < l_2 < \cdots < l_N$.)

    (b)   How many eigenstates are there for three f electrons?

9.43   In the uncoupled representation, N electrons are described by the simultaneous eigenstates of the $2N$ operators

$$(\hat{L}_1{}^2, \hat{L}_{1z}, \hat{L}_2{}^2, \hat{L}_{2z}, \dots, \hat{L}_N{}^2, \hat{L}_{Nz})$$

In the coupled representation, the $N + 2$ commuting operators

$$(\hat{L}^2, \hat{L}_z, \hat{L}_1{}^2, \hat{L}_2{}^2, \ldots, \hat{L}_N{}^2)$$

are relevant, and there are $N + 2$ good quantum numbers corresponding to these operators. One suspects that $2N - (N + 2) = N - 2$ operators may be added to this sequence, yielding a set of $2N$ commuting operators.

(a) Construct such a set of $N - 2$ operators, $\{\hat{A}_i{}^2\}$.

(b) Show explicitly that the terms in the sum $\hat{A}_2{}^2$ commute with the sequence of $N + 2$ operators given above.

Answer (partial)

(a) The first operator is

$$\hat{A}_1{}^2 = a_{12}(\hat{L}_1 + \hat{L}_2)^2 + a_{13}(\hat{L}_1 + \hat{L}_3)^2 + \cdots = \sum_{i_1 \neq i_2}^{N} a_{i_1 i_2}(\hat{L}_{i_1} + \hat{L}_{i_2})^2$$

The second operator is

$$\hat{A}_2{}^2 = \sum_{i_1 \neq i_2 \neq i_3} a_{i_1 i_2 i_3}(\hat{L}_{i_1} + \hat{L}_{i_2} + \hat{L}_{i_3})^2$$

The $(N - 2)$nd operator is

$$\hat{A}_{N-2}{}^2 = \sum_{i_1 \neq \cdots \neq i_{N-1}}^{N} a_{i_1 \cdots i_{N-1}}(\hat{L}_{i_1} + \cdots \hat{L}_{i_{N-1}})^2$$

9.44 The spherical harmonics $Y_l^m(\theta, \phi)$ are simultaneous eigenstates of $\hat{L}_z$ and $\hat{L}^2$. How must the Cartesian x, y, z axes be aligned with the spherical r, θ, ϕ frame in order for this to be true, or is the validity of this statement independent of the relative orientation of these two frames?

9.45 Suppose that L^2 is measured for a free particle and the value $6\hbar^2$ is found. If L_y is then measured, what possible values can result?

9.46 The parity operator, $\hat{P}$, in three dimensions is defined by the equation $\hat{P}f(r, \theta, \phi) = f(r, \pi - \theta, \pi + \phi)$. Show that $\hat{P} Y_l^m = (-)^l Y_l^m$. That is, the parity of Y_l^m (odd or even) is the same as that of l. (Compare with Problem 6.23.)

9.47 Establish the following equalities.

$$x = -r\sqrt{\frac{2\pi}{3}}(Y_1{}^1 - Y_1{}^{-1}), \qquad y = -\frac{r}{i}\sqrt{\frac{2\pi}{3}}(Y_1{}^1 + Y_1{}^{-1}), \qquad z = r\sqrt{\frac{4\pi}{3}} Y_1{}^0$$

$$xy = \frac{r^2}{i}\sqrt{\frac{2\pi}{15}}(Y_2{}^2 - Y_2{}^{-2}), \qquad yz = -\frac{r^2}{i}\sqrt{\frac{2\pi}{15}}(Y_2{}^1 + Y_2{}^{-1}), \qquad zx = -r^2\sqrt{\frac{2\pi}{15}}(Y_2{}^1 - Y_2{}^{-1})$$

$$x^2 - y^2 = r^2\sqrt{\frac{8\pi}{15}}(Y_2{}^2 + Y_2{}^{-2}), \qquad 2z^2 - x^2 - y^2 = r^2\sqrt{\frac{16\pi}{5}} Y_2{}^0,$$

$$y^2 - z^2 = -r^2\sqrt{\frac{2\pi}{15}}(Y_2{}^2 + \sqrt{6} Y_2{}^0 + Y_2{}^{-2})$$

9.48 Using the preceding relations, argue that x/r is an eigenfunction of $\hat{L}_x$ and y/r is an eigenfunction of $\hat{L}_y$. What are the respective eigenvalues of these functions? (*Note:* Such functions are widely employed in quantum chemistry and are commonly called *orbitals*. These topics are returned to in our discussion of *hybridization* in the following chapter.)

9.49 The Clebsch–Gordan expansion (9.109) affords a coordinate representation of angular momentum states in the coupled scheme. What is the explicit $\theta_1, \phi_1, \theta_2, \phi_2$ representation of the coupled state $|lml_1l_2\rangle = |1, -1, 1, 1\rangle$, relevant to two p electrons?

9.50 (a) Show that the energy eigenvalues relevant to a rigid rotator (9.49) follow from the Bohr–Sommerfeld quantization rule (7.192) in the limit of large quantum numbers.

(b) What conclusion may be inferred from your answer to (a) concerning the domain of relevance of the Bohr–Sommerfeld rules?

Answer

(a) We find

$$\oint L \, d\theta = (l + \tfrac{1}{2})h$$

$$L = (l + \tfrac{1}{2})\hbar$$

$$E = \left\langle \frac{L^2}{2I} \right\rangle = \frac{(2l + 1)^2\hbar^2}{8I}$$

Thus, for $l \gg 1$ (but not neglecting l compared with l^2),

$$E \sim \frac{l(l + 1)\hbar^2}{2I}$$

(b) If we associate large quantum numbers with the classical domain, the preceding example indicates that the Bohr–Sommerfeld rules are relevant to this same region. Note also that the first-order solution which enters the near-classical **WKB** analysis (7.172) is the *action integral*, $\int p \, dx$.

9.51 (a) Show that in the state $|lm\rangle$

$$\langle L_x^2 \rangle = \langle L_y^2 \rangle = \hbar^2[l(l + 1) - m^2] - \tfrac{1}{4}\langle [\hat{L}_+, \hat{L}_-]_+ \rangle$$

Here $[,]_+$ represents the *anticommutator*

$$[\hat{A}, \hat{B}]_+ \equiv \hat{A}\hat{B} + \hat{B}\hat{A}$$

(b) Evaluate $\langle [\hat{L}_+, \hat{L}_-]_+ \rangle$ in terms of a function of l and m.

9.52 Establish the following relations relevant to a system in a state with definite angular momentum $\hbar\sqrt{l(l + 1)}$. The degeneracy factor is $g_l = 2l + 1$.

(a) $\hbar^{-2}\langle L^2 \rangle = \displaystyle\sum_{m=-l}^{l} |m|$

(b) $\hbar^{-4}\langle L^2 \rangle^2 = 2 \displaystyle\sum_{m=-l}^{l} |m|^3$

(c) $\hbar^{-2}g_l\langle L^2\rangle = 3\sum_{m=-l}^{l} m^2$

In addition, show the following:

(d) $\sum_{l=0}^{n-1} g_l = n^2$

(e) $\sum_{l=|l_1-l_2|}^{l_1+l_2} g_l = g_{l_1}g_{l_2}$

9.53   A student argues that the rotational kinetic energy of a classical rigid sphere, spinning about a fixed origin with angular frequency ω, may also be associated with quantum mechanical spin. Is the student correct? Explain your answer.

Answer

The student is incorrect. The kinetic energy of the sphere obeys the relation

$$E = \frac{\omega S}{2}$$

where S is the "spin" angular momentum of the sphere. When evaluating S, one integrates differential elements of orbital angular momentum over the volume of the sphere. So, as with all classical angular momentum, classical "spin" is orbital angular momentum. With the preceding relation, the kinetic energy of the sphere, E, is likewise associated with orbital angular momentum.

PROBLEMS IN
THREE DIMENSIONS

In this chapter we discuss the structure of the Schrödinger equation for a particle moving in three dimensions. General properties are developed through examination of the free-particle problem in Cartesian and spherical coordinates. Separation of variables in spherical coordinates yields product solutions for the free-particle problem comprised of spherical harmonics and spherical Bessel functions. Solution to the corresponding radial wave equation for the hydrogen atom gives Laguerre polynomials. Application is also directed toward the motion of a charged particle in a magnetic field. An elementary description of the theory of radiation from atoms and the formulation of selection rules are given. The chapter concludes with a description of the Thomas–Fermi model important to atomic physics.

10.1 THE FREE PARTICLE IN CARTESIAN COORDINATES

We again recall that the linear momentum operator $\hat{\mathbf{p}}$ is given by

$$(10.1) \qquad\qquad \hat{\mathbf{p}} = -i\hbar\nabla$$

Inserting this form into the Hamiltonian for a free particle of mass m moving in three dimensions gives

(10.2)
$$\hat{H} = \frac{\hat{p}^2}{2m} = -\frac{\hbar^2}{2m}\nabla^2$$

It follows that the time-independent Schrödinger equation for this same particle appears as

(10.3)
$$\hat{H}\varphi = -\frac{\hbar^2}{2m}\left(\frac{\partial^2}{\partial x^2} + \frac{\partial^2}{\partial y^2} + \frac{\partial^2}{\partial z^2}\right)\varphi = E\varphi$$

or, alternatively,

$$\nabla^2\varphi = -k^2\varphi$$

(10.4)
$$E = \frac{\hbar^2 k^2}{2m}$$

Separating variables

(10.5)
$$\varphi \equiv X(x)Y(y)Z(z)$$

permits (10.4) to be rewritten as

(10.6)
$$\left(\frac{X_{xx}}{X} + \frac{Y_{yy}}{Y} + \frac{Z_{zz}}{Z}\right) = -k^2$$
$$-\frac{X_{xx}}{X} = k^2 + \left(\frac{Y_{yy}}{Y} + \frac{Z_{zz}}{Z}\right) \equiv k_x^2$$

In the last equation, the left-hand side is a function only of x, while the middle term is a function only of y and z. The only way for the equality to hold for all (x, y, z) is for both terms to be equal to the same constant. Labeling this constant k_x^2 gives the equation

(10.7)
$$X_{xx} + k_x^2 X = 0$$

which has a solution[1]

(10.8)
$$X = A'e^{ik_x x}$$

In similar manner we obtain

(10.9)
$$Y = B'e^{ik_y y}, \qquad Z = C'e^{ik_z z}$$

where

(10.10)
$$k^2 = k_x^2 + k_y^2 + k_z^2$$

---

[1] Here we consider only the forward propagating wave.

Combining all three factors X, Y, and Z gives the solution

(10.11) $\varphi = A'B'C' \exp\left[i(k_x x + k_y y + k_z z)\right] = \varphi_{\mathbf{k}} = Ae^{i\mathbf{k}\cdot\mathbf{r}}$

The wavevector $\mathbf{k}$ and position vector $\mathbf{r}$ have components

(10.12)
$$\mathbf{k} = (k_x, k_y, k_z)$$
$$\mathbf{r} = (x, y, z)$$

The function $\varphi_{\mathbf{k}}$ so obtained is an eigenfunction of $\hat{H}$ (10.2) with the eigenvalue

(10.13)
$$E_k = \frac{\hbar^2 k^2}{2m}$$

Plane Waves

The corresponding solution to the time-dependent Schrödinger equation (3.52) appears as

(10.14)
$$\psi_{\mathbf{k}}(\mathbf{r}, t) = Ae^{i(\mathbf{k}\cdot\mathbf{r} - \omega t)}$$
$$\hbar\omega = E_k$$

This solution represents a propagating plane wave. At any instant of time, $\psi_{\mathbf{k}}(r, t)$ is constant on the surfaces $\mathbf{k} \cdot \mathbf{r} = $ constant. These are surfaces normal to $\mathbf{k}$. Consider one such surface. The projection of $\mathbf{r}$ onto $\mathbf{k}$

(10.15)
$$r_{\parallel} = \frac{\mathbf{k}\cdot\mathbf{r}}{k}$$

from any point on this surface is constant. This is the normal displacement between the origin and the surface. See Fig. 10.1. Rewriting (10.14) in the form

(10.16) $\psi_{\mathbf{k}}(\mathbf{r}, t) = Ae^{ik[r_{\parallel} - (\omega/k)t]}$

reveals that the rate of increase of $r_{\parallel}$ with respect to a surface of constant $\psi_{\mathbf{k}}$ is the *wave speed*

(10.17)
$$v = \frac{\omega}{k}$$

The normalization constant A may be chosen so that

(10.18) $\langle \psi_{\mathbf{k}} | \psi_{\mathbf{k}'} \rangle = \langle \varphi_{\mathbf{k}} | \varphi_{\mathbf{k}'} \rangle = \langle \mathbf{k} | \mathbf{k}' \rangle = \iiint \varphi_{\mathbf{k}}^* \varphi_{\mathbf{k}'} \, dx \, dy \, dz = \delta(\mathbf{k} - \mathbf{k}')$

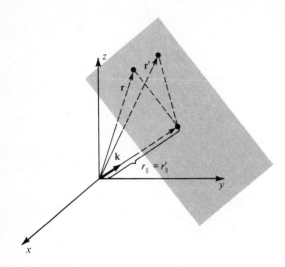

FIGURE 10.1 At any instant of time, the plane wave

$$\psi_{\mathbf{k}}(\mathbf{r}, t) = A \exp\left[i(\mathbf{k} \cdot \mathbf{r} - \omega t)\right]$$

is constant on the surface $\mathbf{k} \cdot \mathbf{r} = $ constant. These are surfaces normal to $\mathbf{k}$. At every point $\mathbf{r}$, on such a surface, the projection $r_\parallel = \mathbf{k} \cdot \mathbf{r}/k$ is constant.

The three-dimensional delta function is defined as the product

(10.19) $$\delta(\mathbf{r} - \mathbf{r}') = \delta(x - x')\delta(y - y')\delta(z - z')$$

and has the representation

(10.20)
$$\delta(\mathbf{r} - \mathbf{r}') = \frac{1}{(2\pi)^3} \iiint e^{i\mathbf{k} \cdot (\mathbf{r} - \mathbf{r}')} \, d\mathbf{k}$$

$$d\mathbf{k} = dk_x \, dk_y \, dk_z$$

Comparison of this representation with (10.18) yields the normalized wavefunction

(10.21) $$\varphi_{\mathbf{k}} = \frac{1}{(2\pi)^{3/2}} e^{i\mathbf{k} \cdot \mathbf{r}}$$

Superposition of Free-Particle States

A free-particle wave packet may be represented by the superposition

(10.22) $$\psi(\mathbf{r}, t) = \frac{1}{(2\pi)^{3/2}} \iiint b(\mathbf{k}, t) e^{i(\mathbf{k} \cdot \mathbf{r} - \omega t)} \, d\mathbf{k}$$

with corresponding inverse

(10.23) $$b(\mathbf{k}, t) = \frac{1}{(2\pi)^{3/2}} \iiint \psi(\mathbf{r}, t) e^{-i(\mathbf{k} \cdot \mathbf{r} - \omega t)} \, d\mathbf{r}$$

$$d\mathbf{r} = dx \, dy \, dz$$

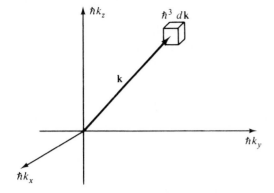

FIGURE 10.2   In the plane-wave decomposition of the free wave packet $\psi(\mathbf{r}, t)$, as given by (10.22), the Fourier amplitude $b(\mathbf{k}, t)$ is such that

$$|b(\mathbf{k}, t)|^2 \, d\mathbf{k}$$

is the probability that measurement finds the particle with momentum in the volume element $\hbar^3 \, d\mathbf{k}$ about the value $\hbar\mathbf{k}$.

As in the one-dimensional case, the coefficient b gives the probability

$$(10.24) \qquad P(\mathbf{k}) \, d\mathbf{k} = |b(\mathbf{k}, t)|^2 \, d\mathbf{k}$$

that measurement at the instant t finds the particle with momentum in the volume element $\hbar^3 \, d\mathbf{k}$ about the value $\hbar k$ (Fig. 10.2).

If the probability amplitude $b(\mathbf{k}, t)$ is peaked about a value of $\mathbf{k}$, say $\mathbf{k}_0$, then the three-dimensional wave packet (10.22) propagates with the *group velocity*

$$(10.25) \qquad \mathbf{v}_g = \mathbf{V}_{\mathbf{k}} \, \omega(k) |_{\mathbf{k} = \mathbf{k}_0}$$

where $\mathbf{V}_{\mathbf{k}}$ is written for the gradient with respect to $\mathbf{k}$. Inasmuch as (10.22) depicts the state of a free particle, for each $\mathbf{k}$-wave component one has

$$(10.26) \qquad E_k = \hbar\omega = \frac{\hbar^2 k^2}{2m}$$

This gives $\omega(k)$, which with (10.25) yields

$$(10.27) \qquad \mathbf{v}_g = \frac{\hbar\mathbf{k}_0}{m} = \mathbf{v}_{\mathrm{CL}}$$

This is the classical velocity of a particle of mass m, moving with momentum $\hbar\mathbf{k}_0$.

For a free particle

$$(10.28) \qquad [\hat{\mathbf{p}}, \hat{H}] = \frac{1}{2m} [\hat{\mathbf{p}}, \hat{p}^2] = 0$$

so that $\hat{\mathbf{p}}$ and $\hat{H}$ have simultaneous eigenstates. These are the functions $\varphi_{\mathbf{k}}(\mathbf{r})$. In the eigenstate $\varphi_{\mathbf{k}}$, the linear momentum $\hbar\mathbf{k}$ and energy $\hbar^2 k^2/2m$ are specified. The state cannot be further resolved. For instance, suppose that we measure the z component of the angular momentum of the particle L_z. This measurement destroys the information in the state before measurement, relating to the linear momentum $\mathbf{p}$. The components of $\mathbf{p}$ and $\mathbf{L}$, in general, do not commute.

What, then, are the states for the free particle, which include specification of L^2 and L_z? To find these states it proves most convenient to express $\hat{H}$ in spherical coordinates. This is discussed in the next section.

PROBLEMS

10.1 If $\psi(\mathbf{r}, t)$ is a free-particle state and $b(\mathbf{k}, t)$ the momentum probability amplitude for this same state, show that

$$\iiint \psi^* \psi \, d\mathbf{r} = \iiint b^* b \, d\mathbf{k}$$

10.2 At time $t = 0$, a free particle is in the superposition state

$$\psi(\mathbf{r}, 0) = \frac{\pi^{-3/2}}{2} \sin 3x \exp\left[i(5y + z)\right]$$

(a) If the energy of the particle is measured at $t = 0$, what value is found?

(b) What possible values of momentum (p_x, p_y, p_z) will measurement find at $t = 0$, and with what probability will these values occur?

(c) Given the above state $\psi(\mathbf{r}, 0)$, what is $\psi(\mathbf{r}, t)$?

(d) If $\mathbf{p}$ is measured at $t = 0$ and the value $\mathbf{p} = \hbar(3\mathbf{e}_x + 5\mathbf{e}_y + \mathbf{e}_z)$ is found, what is $\psi(\mathbf{r}, t)$?

10.3 (a) What is the Hamiltonian for N free, noninteracting particles of mass m?

(b) What is the eigenstate of this Hamiltonian corresponding to the eigenvalue

$$E = \frac{\hbar^2}{2m} \sum_{j=1}^{N} k_j^2$$

(c) Show that the eigenstate found in part (b) is also an eigenstate of the momentum of the center of mass. What is the velocity of the center of mass in this state?

Answers (partial)

(a) $\hat{H} = -\sum_{j=1}^{N} \frac{\hbar^2}{2m} \nabla_j^2, \quad \nabla_j = \left(\frac{\partial}{\partial x_j}, \frac{\partial}{\partial y_j}, \frac{\partial}{\partial z_j}\right)$

(b) $\psi_{\mathbf{k}_1, \mathbf{k}_2, \ldots, \mathbf{k}_N} = A \exp\left(i \sum_{j=1}^{N} \mathbf{k}_j \cdot \hat{\mathbf{r}}_j\right)$

(c) $\mathbf{v}_{CM} = \sum \hbar \mathbf{k}_j / M$

10.2 THE FREE PARTICLE IN SPHERICAL COORDINATES

Hamiltonian

We wish to express the free-particle Hamiltonian (10.2) in spherical coordinates, (r, θ, ϕ) (see Fig. 1.6). We have already found the classical expression for H in spherical coordinates in Chapter 1 [see (1.20)]. Let us again construct this classical form. However, in the present instance we wish H to include the angular momentum term

(10.29)
$$L^2 = (\mathbf{r} \times \mathbf{p})^2 = r^2 p^2 - (\mathbf{r} \cdot \mathbf{p})^2$$

The linear momentum is written $\mathbf{p}$. It follows that

(10.30)
$$H = \frac{p^2}{2m} = \frac{p_r^2}{2m} + \frac{L^2}{2mr^2}$$

where p_r is written for the radial component of the particle's momentum.

(10.31)
$$p_r = \frac{1}{r}(\mathbf{r} \cdot \mathbf{p})$$

If we wish to carry (10.29) over to quantum mechanics, we must make sure that all terms in $\hat{H}$ are Hermitian. The two operators in (10.30) are $r^{-2}\hat{L}^2$ and p_r^2. To examine the Hermiticity of the first operator, we note the following.

Rotation and Angular Momentum

In Section 9.3 we found that the effect of the rotation operator $\hat{R}_{\delta\phi}$ when operating on a function $f(\mathbf{r})$ is to change f by rotating $\mathbf{r}$ to $\mathbf{r} + \delta\phi \times \mathbf{r}$. Suppose that a function f is isotropic[1] in $\mathbf{r}$; that is, f is independent of the direction of $\mathbf{r}$. It depends only on the magnitude of $\mathbf{r}$. Any function of the form $f(r^2)$ is isotropic in $\mathbf{r}$. For example, $f = ar^2 + br^4$, where a and b are constants, is isotropic in $\mathbf{r}$. What is the value of $f(r^2)$ on the surface of a sphere of radius r_0? The answer is, the constant $f(r_0^2)$. An isotropic function is constant on the surface of any given sphere about the origin. Now suppose that we operate on an isotropic function with $\hat{R}_{\delta\phi}$. This causes f to change by rotating $\mathbf{r}$ to the value $\mathbf{r} + \delta\phi \times \mathbf{r}$. The new vector lies on the same sphere on which $\mathbf{r}$ lies.

$$(\mathbf{r} + \delta\phi \times \mathbf{r})^2 = r^2 + 2\mathbf{r} \cdot \delta\phi \times \mathbf{r} + O(\delta\phi^2) = r^2$$

Terms of $O(\delta\phi^2)$ are neglected while the middle term vanishes because $\mathbf{r}$ is normal to $\delta\phi \times \mathbf{r}$. It follows that the operator $\hat{R}_{\delta\phi}$ has no effect on $f(r^2)$.

(10.32)
$$\hat{R}_{\delta\phi} f(r^2) = f(r^2)$$

---

[1] One may also say that f is spherically symmetric.

Since $\hat{R}_{\delta\phi} = (1 + i\delta\phi \cdot \mathbf{L}/\hbar)$, we may conclude that

(10.33)
$$\frac{i\delta\phi \cdot \hat{\mathbf{L}}}{\hbar} f(r^2) = 0$$

In that this statement is true for all axes of rotation about the origin, or equivalently, for all directions of the vector $\delta\phi$, it follows that any isotropic function is a *null eigenstate* of the three components of angular momentum as well as of $\hat{L}^2$.

(10.34)
$$\hat{L}_x f(r^2) = \hat{L}_y f(r^2) = \hat{L}_z f(r^2) = \hat{L}^2 f(r^2) = 0$$

As noted previously in Section 9.2, these spherically symmetric states are called *S* states.

If $g(\mathbf{r})$ is any function of $\mathbf{r}$ (for example, $g = x/r$) and $f(r^2)$ is any isotropic function, then owing to the conclusion immediately above,

$$\hat{L}^2 f(r^2) g(\mathbf{r}) = f(r^2) \hat{L}^2 g(\mathbf{r})$$

$$(\hat{L}^2 f(r^2) - f(r^2) \hat{L}^2) g(\mathbf{r}) = [\hat{L}^2, f(r^2)] g(\mathbf{r}) = 0$$

Since this latter equality holds for all differentiable functions g, we obtain

$$[\hat{L}^2, f(r^2)] = 0$$

Similarly,

(10.35)
$$[\hat{L}_x, f(r^2)] = [\hat{L}_y, f(r^2)] = [\hat{L}_z, f(r^2)] = 0$$

We are now prepared to investigate the Hermiticity of the term $r^{-2}\hat{L}^2$ in the Hamiltonian (10.30). With $\hat{r}^{-2}$ denoting multiplication by r^{-2}, we write

(10.36)
$$(\hat{r}^{-2}\hat{L}^2)^{\dagger} = \hat{L}^{2\dagger}\hat{r}^{-2\dagger} = \hat{L}^2\hat{r}^{-2} = \hat{r}^{-2}\hat{L}^2$$

so that $\hat{r}^{-2}\hat{L}^2$ is Hermitian. In the last equality we used the fact that $\hat{L}^2$ commutes with the isotropic function r^{-2}.

Radial Momentum

Next we consider the operator

(10.37)
$$\hat{p}_r = r^{-1}(\mathbf{r} \cdot \hat{\mathbf{p}}) = r^{-1}(x\hat{p}_x + y\hat{p}_y + z\hat{p}_z)$$

Forming the Hermitian adjoint of $\hat{p}_r$ gives

(10.38)
$$\begin{aligned}(\hat{p}_r)^{\dagger} &= [\hat{r}^{-1}(\hat{x}\hat{p}_x + \hat{y}\hat{p}_y + \hat{z}\hat{p}_z)]^{\dagger} \\ &= (\hat{x}\hat{p}_x)^{\dagger}(\hat{r}^{-1})^{\dagger} + (\hat{y}\hat{p}_y)^{\dagger}(\hat{r}^{-1})^{\dagger} + (\hat{z}\hat{p}_z)^{\dagger}(\hat{r}^{-1})^{\dagger} \\ &= \hat{p}_x\hat{x}\hat{r}^{-1} + \cdots \\ &\neq \hat{r}^{-1}\hat{x}\hat{p}_x + \cdots\end{aligned}$$

The operators $\hat{r}^{-1}$ and $\hat{x}$ cannot be brought through $\hat{p}_x$. Similarly for the other two terms. We conclude that $\not{p}_r$ is not Hermitian.

The more appropriate operator corresponding to radial momentum is given by the symmetric form (see Problems 10.5 and 10.6)

$$(10.39) \qquad \hat{p}_r = \tfrac{1}{2}(\not{p}_r + \not{p}_r{}^\dagger)$$

or, equivalently,

$$(10.40) \qquad \hat{p}_r = \frac{1}{2}\left(\frac{1}{r}\mathbf{r}\cdot\hat{\mathbf{p}} + \hat{\mathbf{p}}\cdot\mathbf{r}\frac{1}{r}\right)$$

The component of $\hat{\mathbf{p}}$ in the direction of $\mathbf{r}$ is given by

$$(10.41) \qquad \frac{1}{r}\mathbf{r}\cdot\hat{\mathbf{p}} = -i\hbar\frac{1}{r}\mathbf{r}\cdot\mathbf{\nabla} = -i\hbar\frac{\partial}{\partial r}$$

while the second term in $\hat{p}_r$ is given by

$$(10.42) \qquad \hat{\mathbf{p}}\cdot\mathbf{r}\frac{1}{r} = -i\hbar\mathbf{\nabla}\cdot\mathbf{e}_r$$

where $\mathbf{e}_r$ is written for the unit radius vector, $\mathbf{r}/r$. Let $f(\mathbf{r})$ be a differentiable function of the radius vector $\mathbf{r}$. Consider the operation

$$(10.43) \qquad \hat{p}_r f(r) = \frac{-i\hbar}{2}\left(\frac{\partial}{\partial r} + \mathbf{\nabla}\cdot\mathbf{e}_r\right)f$$

$$= \frac{-i\hbar}{2}\left(\frac{\partial f}{\partial r} + \mathbf{e}_r\cdot\mathbf{\nabla}f + f\mathbf{\nabla}\cdot\mathbf{e}_r\right)$$

$$= \frac{-i\hbar}{2}\left(\frac{\partial f}{\partial r} + \frac{\partial f}{\partial r} + \frac{2f}{r}\right) = -i\hbar\left(\frac{\partial f}{\partial r} + \frac{f}{r}\right)$$

Equivalently, we may write

$$\hat{p}_r f = -i\hbar\frac{1}{r}\frac{\partial}{\partial r}rf$$

$$(10.44)$$

$$\hat{p}_r = -i\hbar\frac{1}{r}\frac{\partial}{\partial r}r$$

With the above definition of $\hat{p}_r$, we may write the following for the Hamiltonian operator:

$$(10.45) \qquad \hat{H} = \frac{\hat{p}_r{}^2}{2m} + \frac{\hat{L}^2}{2mr^2}$$

The student may well ask the following question at this point. We know that the Hamiltonian $\hat{H}$ has the correct representation

$$\text{(10.46)} \qquad \hat{H} = \frac{\hat{p}^2}{2m} = -\frac{\hbar^2 \nabla^2}{2m}$$

How are we assured that $\hat{H}$, as given by (10.45), with the definition of $\hat{p}_r$ obtained by symmetrization of $\hat{\rho}_r$, is equivalent to this correct form (10.46)? This question is answered by demonstration. The representation of the Laplacian operator ∇^2, in spherical coordinates, is

$$\text{(10.47)} \qquad \nabla^2 = \frac{1}{r}\frac{\partial^2}{\partial r^2} r + \frac{1}{r^2}\left(\frac{1}{\sin\theta}\frac{\partial}{\partial\theta}\sin\theta\frac{\partial}{\partial\theta} + \frac{1}{\sin^2\theta}\frac{\partial^2}{\partial\phi^2}\right)$$

Noting the equality

$$\text{(10.48)} \qquad \left(\frac{1}{r}\frac{\partial}{\partial r}r\right)^2 = \frac{1}{r}\frac{\partial}{\partial r}r\left(\frac{1}{r}\frac{\partial}{\partial r}r\right) = \frac{1}{r}\frac{\partial^2}{\partial r^2}r$$

and recalling the expression for $\hat{L}^2$, as given by (9.58), permits the equation

$$\text{(10.49)} \qquad \hat{H} = -\frac{\hbar^2\nabla^2}{2m} = \frac{\hat{p}_r{}^2}{2m} + \frac{\hat{L}^2}{2mr^2}$$

This is the correct form of $\hat{H}$, in spherical coordinates. In the next section we will examine its eigenfunctions and eigenvalues.

PROBLEMS

10.4   What is the time-independent wavefunction in spherical coordinates of a free particle of mass m, zero angular momentum, and energy E which satisfies the property $|r\varphi| = 0$ at $r = 0$? (*Hint:* Introduce the function $u \equiv r\varphi$.)

10.5   (a)   Show that

$$[\hat{r}, \hat{p}_r] = i\hbar$$

(b)   What properties of φ and ψ insure that

$$\langle \varphi | \hat{p}_r\psi \rangle = \langle \hat{p}_r\varphi | \psi \rangle$$

[*Note:* $\langle \varphi | \psi \rangle = \int d\Omega \int drr^2 \varphi^* \psi$.]

10.6   The current vector **J** associated with a wavefunction $\psi(\mathbf{r}, t)$ is given by (7.107)

$$\mathbf{J} = \frac{\hbar}{2mi}(\psi^*\nabla\psi - \psi\nabla\psi^*)$$

The wavefunction $\psi(\mathbf{r}, t)$ may be termed *source-free*, if $\nabla \cdot \mathbf{J} = 0$ for all values of **r**.
(a)   What is the eigenfunction of $\hat{p}_r$, corresponding to the eigenvalue $\hbar k$?
(b)   Calculate $\nabla \cdot \mathbf{J}$ for this eigenfunction of $\hat{p}_r$.

Answers

(a)   Integration of the eigenvalue equation

$$-i\hbar \frac{1}{r}\frac{\partial}{\partial r} r\tilde{\varphi}_k = \hbar k \tilde{\varphi}_k$$

gives

$$\tilde{\varphi}_k = A \frac{e^{ikr}}{r}$$

The corresponding time-dependent solution is

$$\tilde{\psi}_k = \frac{A}{r} e^{i(kr - \omega t)}$$

This "outgoing wave" is a solution to the time-dependent Schrödinger equation for a free particle with no angular momentum. It is important in the construction of scattering states, which will be discussed in Chapter 14.

(b)   The current vector corresponding to $\tilde{\varphi}_k$ only has an r component.

$$\mathbb{J}_r = \frac{|A|^2(\hbar k/m)}{r^2} \equiv \frac{\Gamma(0)}{4\pi r^2}$$

The divergence of this current is

$$\mathbf{V} \cdot \mathbf{J} = \frac{1}{r^2}\frac{\partial}{\partial r} r^2 \mathbb{J}_r = 0 \qquad (\text{for } r \neq 0)$$

Since $\mathbf{J}$ is radial and a function only of r, we may write

$$\int_V \mathbf{V} \cdot \mathbf{J} \, d\mathbf{r} = \int_{r=R} \mathbf{J} \cdot d\mathbf{S} = \int_{4\pi} \frac{\Gamma(0)}{4\pi r^2} r^2 \, d\Omega = \Gamma(0)$$

The spherical volume V has radius R and is centered at the origin, while $d\Omega$ is an element of solid angle about this same origin. Given these two properties of $\mathbf{V} \cdot \mathbf{J}$, it follows that

$$\mathbf{V} \cdot \mathbf{J} = \Gamma(0)\delta(\mathbf{r})$$

The three-dimensional Dirac delta function is $\delta(\mathbf{r})$ [see (10.19)].

[*Note:* Thus we see that the eigenstates of $\hat{p}_r$ have the unreasonable property of implying that a constant flux of particles, $\Gamma(0)$, emanates from the origin. We may infer from this that the operator $\hat{p}_r$, in spite of its symmetric form (10.39), and proper commutation property with $\hat{r}$ (Problem 10.5), is not a good observable equivalent. Nevertheless, for problems involving a central potential, in quantum mechanics the operator $\hat{p}_r$ proves to be a valuable tool. It is interesting to note that inconsistencies that accompany p_r are also found in classical mechanics. If a free point particle crosses the origin, p_r changes sign instantaneously. This jump in p_r stems from a choice of coordinate frame. It is in no way associated with a force (the particle is free).[1]]

---

[1] For further discussion of this problem, see R. L. Liboff, I. Nebenzahl, and H. A. Fleishmann, *Am. J. Phys.* **41**, 976 (1973).

10.7 Show that the kinetic-energy operator

$$\hat{T} = -\frac{\hbar^2}{2m} \nabla^2$$

is Hermitian for functions in $\mathfrak{H}_2$ (the space of square integrable functions—see Section 4.4). [*Hint:* Use *Green's theorem*

$$\int_V (f\nabla^2 g - g\nabla^2 f)\, d\mathbf{r} = \int_S (f\nabla g - g\nabla f) \cdot d\mathbf{S}$$

The volume V is enclosed by the surface S.]

10.8 Show that

$$\mathbf{\nabla} \cdot \mathbf{J}(\psi) = 0$$

for the superposition state

$$\psi = \psi_1 + \psi_2$$

provided that

$$\mathbf{\nabla} \cdot \mathbf{J}(\psi_1) = \mathbf{\nabla} \cdot \mathbf{J}(\psi_2) = 0$$

and

$$\mathrm{Im}(\psi_1{}^*\nabla^2\psi_2 + \psi_2{}^*\nabla^2\psi_1) = 0$$

10.3   THE FREE-PARTICLE RADIAL WAVEFUNCTION

The time-independent Schrödinger equation for a free particle in spherical coordinates appears as

(10.50) $$\frac{1}{2m}\left(\hat{p}_r{}^2 + \frac{\hat{L}^2}{r^2}\right)\varphi_{klm} = E_{klm}\varphi_{klm}$$

The quantum number k is defined below. The radial kinetic energy operator $\hat{p}_r{}^2/2m$ is inferred from (10.48), while the angular momentum operator $\hat{L}^2$ is given by (9.58). Insofar as $\hat{p}_r{}^2$ is a function only of r, and $\hat{L}^2$ is a function only of the angle variables (θ, ϕ), one may seek solution to (10.50) by separation of variables. Substituting the product form

(10.51) $$\varphi_{klm}(r, \theta, \phi) = R_{kl}(r)Y_l^m(\theta, \phi)$$

into (10.50) gives

(10.52) $$\left[-\left(\frac{1}{r}\frac{d^2}{dr^2}r\right) + \frac{l(l+1)}{r^2}\right]R_{kl}(r) = \frac{2mE}{\hbar^2}R_{kl}(r)$$

In obtaining (10.52) we have recalled the eigenvalue equation for $\hat{L}^2$ (9.51). With the substitution

(10.53)
$$E \equiv \frac{\hbar^2 k^2}{2m}$$

$$x \equiv kr$$

(10.52) becomes the "spherical Bessel differential equation"

(10.54)
$$\frac{d^2}{dx^2} R(x) + \frac{2}{x} \frac{dR(x)}{dx} + \left[1 - \frac{l(l + 1)}{x^2} \right] R(x) = 0$$

Spherical Bessel Functions

This ordinary linear equation for the radial function R has two linearly independent solutions.[1] They are called spherical Bessel and Neumann functions and are denoted conventionally by the symbols $j_l(x)$ and $n_l(x)$, respectively. The first few values of these functions are

$$j_0(x) = \frac{\sin x}{x} \qquad\qquad n_0(x) = -\frac{\cos x}{x}$$

(10.55)
$$j_1(x) = \frac{\sin x}{x^2} - \frac{\cos x}{x} \qquad n_1(x) = -\frac{\cos x}{x^2} - \frac{\sin x}{x}$$

$$j_2(x) = \left(\frac{3}{x^3} - \frac{1}{x}\right) \sin x - \frac{3}{x^2} \cos x \qquad n_2(x) = -\left(\frac{3}{x^3} - \frac{1}{x}\right) \cos x - \frac{3}{x^2} \sin x$$

These functions are sketched in Fig. 10.3, from which it is evident that of the two classes of functions, only the spherical Bessel functions $\{j_l\}$ are regular at the origin. These are the solutions appropriate to the Schrödinger equation (10.50) inasmuch as they are not singular anywhere. Some additional properties of these spherical Bessel and Neumann functions are listed in Table 10.1.

In this manner we find that the eigenstates and eigenenergies of the free-particle Hamiltonian in spherical coordinates are

(10.56)
$$\varphi_{klm}(r, \theta, \phi) = j_l(kr) Y_l^m(\theta, \phi)$$

$$E_k = \frac{\hbar^2 k^2}{2m}$$

[1] One obtains these solutions by the method of series substitution. Details may be found in most books on mathematical physics, e.g., G. Goertzel and N. Tralli, *Some Mathematical Methods in Physics*, McGraw-Hill, New York, 1960.

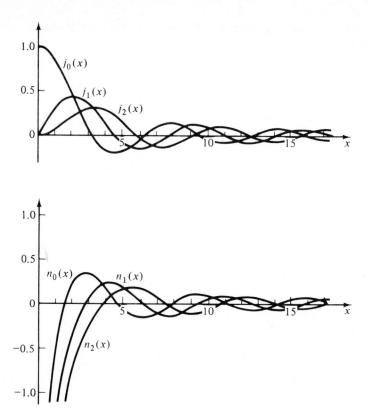

FIGURE 10.3   **Spherical Bessel functions $j_l(x)$ and spherical Neumann functions $n_l(x)$ for $l = 0, 1, 2$. Note that only $j_l(x)$ are regular at the origin.**

The orthonormality of this sequence $\{\varphi_{klm}\}$ is given by the relation

$$(10.57) \quad \langle lmk | l'm'k' \rangle = \int_{4\pi} d\Omega [Y_l^m(\theta, \phi)]^* Y_{l'}^{m'}(\theta, \phi) \int_0^\infty j_l(kr) j_{l'}(k'r) r^2 \, dr$$

$$= \delta_{ll'} \delta_{mm'} \frac{\pi}{2k^2} \delta(k - k')$$

The vector $\mathbf{r}$ has the spherical coordinates (r, θ, ϕ). This orthonormality condition is similar to that corresponding to the free-particle states expressed in Cartesian coordinates (10.18), as well as that corresponding to free-particle motion in one dimension (4.41). In all these cases, the allowed values of momentum, $\hbar k$, comprise a continuum.

Once again we note that the projection

$$\langle r\theta\phi \,|\, lmk \rangle = j_l(kr)Y_l^m(\theta, \phi)$$

gives the coordinate representation of the ket vector $|lmk\rangle$. In similar manner, the coordinate representation of the free-particle ket vector $|\mathbf{k}\rangle$ is given by the projection

$$\langle \mathbf{r} \,|\, \mathbf{k} \rangle = \frac{1}{(2\pi)^{3/2}} e^{i\mathbf{k}\cdot\mathbf{r}}$$

(See Problem 7.31.)

TABLE 10.1   Properties of the spherical Bessel and Neumann functions

Spherical Bessel Functions	*Spherical Neumann Functions*
$j_l(kr) = \left(-\dfrac{r}{k}\right)^l\left(\dfrac{1}{r}\dfrac{d}{dr}\right)^l j_0(kr)$	$n_l(kr) = \left(-\dfrac{r}{k}\right)^l\left(\dfrac{1}{r}\dfrac{d}{dr}\right)^l n_0(kr)$
$j_0(kr) = \dfrac{\sin kr}{kr}$	$n_0(kr) = -\dfrac{\cos kr}{kr}$

Equation

$$f'' + \frac{2}{x}f + \left[1 - \frac{l(l+1)}{x^2}\right]f = 0$$

Asymptotic Values

$x \to 0$

$$j_l(x) \sim \frac{x^l}{1\cdot3\cdot5\cdots(2l+1)} \qquad n_l(x) \sim -\frac{1\cdot3\cdot5\cdots(2l-1)}{x^{l+1}}$$

$x \to \infty$

$$j_l(x) \sim \frac{1}{x}\cos\left[x - \frac{\pi}{2}(l+1)\right] \qquad n_l(x) \sim \frac{1}{x}\sin\left[x - \frac{\pi}{2}(l+1)\right]$$

Recurrence Relations (f is written for j or n)

$$f_{l-1}(x) + f_{l+1}(x) = (2l+1)x^{-1}f_l(x) \qquad \frac{d}{dx}[x^{l+1}j_l(x)] = x^{l+1}j_{l-1}(x)$$

$$lf_{l-1}(x) - (l+1)f_{l+1}(x) = (2l+1)\frac{d}{dx}f_l(x) \qquad \frac{d}{dx}[x^{-l}j_l(x)] = -x^lj_{l+1}(x)$$

Generating Functions

$$\frac{1}{x}\cos\sqrt{x^2 - 2xs} = \sum_0^\infty \frac{s^l}{l!}j_{l-1}(x) \qquad \frac{1}{x}\sin\sqrt{x^2 + 2xs} = \sum_0^\infty \frac{(-s)^l}{l!}n_{l-1}(x)$$

Orthogonality

$$\int_0^\infty j_l(kr)j_l(k'r)r^2\,dr = \frac{\pi}{2k^2}\delta(k - k') \qquad \int j_1(x)\,dx = -j_0(x), \quad \int j_0(x)x^2\,dx = x^2j_1(x)$$

Connection to Bessel and Neumann Functions of Integral Order, J_l and N_l

$$j_l(kr) = \sqrt{\frac{\pi}{2kr}}\,J_{l+1/2}(kr) \qquad n_l(kr) = \sqrt{\frac{\pi}{2kr}}\,N_{l+1/2}(kr)$$

Measurements on a Free Particle

Given that a particle is in the eigenstate φ_{klm}, measurement of:

(10.58)
$$E \quad \text{gives} \quad \hbar^2 k^2/2m$$
$$L^2 \quad \text{gives} \quad \hbar^2 l(l+1)$$
$$L_z \quad \text{gives} \quad \hbar m$$

How do we know that these values may be measured simultaneously? The answer is that φ_{klm} is a simultaneous eigenfunction of $\hat{H}$, $\hat{L}^2$, and $\hat{L}_z$. The existence of such common eigenfunctions follows from the fact that $\hat{H}$, $\hat{L}^2$, and $\hat{L}_z$ are a commuting set of operators. We have already discussed the commutability of $\hat{L}^2$ and $\hat{L}_z$ in Chapter 9. The fact that these operators commute with $\hat{H}$ follows if they commute with $\hat{p}_r^2$. But $\hat{p}_r^2$ is an isotropic operator; $\hat{p}_r^2 f(r)$ is constant on the surface of any given sphere. It follows that $\hat{p}_r^2$ is unaffected by rotations about the origin, hence

(10.59)
$$[\hat{p}_r^2, \hat{L}_z] = [\hat{p}_r^2, \hat{L}^2] = 0$$

and

(10.60)
$$[\hat{H}, \hat{L}_z] = [\hat{H}, \hat{L}^2] = 0$$

The solution $\varphi_{klm}(r, \theta, \phi)$ should be compared to the eigenstate of the free-particle Hamiltonian in Cartesian coordinates (10.11),

(10.61)
$$\varphi_k(\mathbf{r}) = Ae^{i\mathbf{k}\cdot\mathbf{r}}$$

Given that a particle is in this state, measurement of:

(10.62)
$$E \quad \text{gives} \quad \frac{\hbar^2 k^2}{2m}$$
$$p_x \quad \text{gives} \quad \hbar k_x$$
$$p_y \quad \text{gives} \quad \hbar k_y$$
$$p_z \quad \text{gives} \quad \hbar k_z$$

In the spherical representation, (L_z, L^2, E) are specified. In the Cartesian representation, $(\mathbf{p}, E)$ are specified. In the latter representation, E is redundant ($E = p^2/2m$), but in the former representation it is not. It is not determined by L^2 and L_z. Thus we find that in either Cartesian or spherical representations, there are three good quantum numbers [recall (1.41)].

Free Particle S States

The special case $L = 0$ is of interest. For this case the Schrödinger equation (10.50) becomes

(10.63)
$$\left(\frac{\hat{p}_r^2}{2m}\right)\varphi_k = E_k \varphi_k$$

412

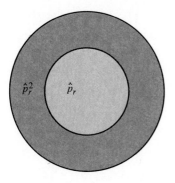

FIGURE 10.4 Central domain represents the eigenstates common to $\hat{p}_r^2$ and $\hat{p}_r$. Peripheral domain represents only eigenstates of $\hat{p}_r^2$, which alone are the physically relevant ones.

The radial kinetic-energy operator $\hat{p}_r^2/2m$ commutes with the radial momentum operator $\hat{p}_r$ and they have common eigenstates. Due to degeneracy, however (the eigenstates E_k are doubly degenerate), eigenfunctions of $\hat{p}_r^2$ are not necessarily eigenfunctions of $\hat{p}_r$ (see Fig. 10.4). Owing to the inadmissibility of the eigenfunctions of $\hat{p}_r$, it is the eigenstates of $\hat{p}_r^2$ alone which are physically relevant. Namely, these functions are

$$(10.64) \qquad \varphi_k = j_0(kr) = \frac{\sin kr}{kr}$$

Rewriting φ_k in the form

$$(10.65) \qquad \varphi_k = \tilde{\varphi}_{+k} + \tilde{\varphi}_{-k} = \frac{1}{2i}\left(\frac{e^{ikr}}{kr} - \frac{e^{-ikr}}{kr}\right)$$

reveals that it is a superposition of the *outgoing* wave, $\tilde{\varphi}_{+k}$, and the *ingoing* wave, $\tilde{\varphi}_{-k}$, which gives zero flux at the origin.

Measurement of L_z and L^2 for a Plane Wave

Next, we consider the following important problem. Suppose that a particle of mass m is "prepared" so that it has momentum $\hbar\mathbf{k}$. Then we know that it is in the plane-wave state (10.11).

$$(10.66) \qquad \varphi_\mathbf{k} = Ae^{i\mathbf{k}\cdot\mathbf{r}}$$

Measurement of E is certain to find $\hbar^2 k^2/2m$. Measurement of momentum is certain to find $\hbar\mathbf{k}$. What will measurement of L_z or L^2 find? And in what states do such measurements leave the particle? To answer this question we must expand the given plane wave in the simultaneous eigenstates of $\hat{H}$, $\hat{L}^2$, and $\hat{L}_z$, that is, φ_{klm}, as given by

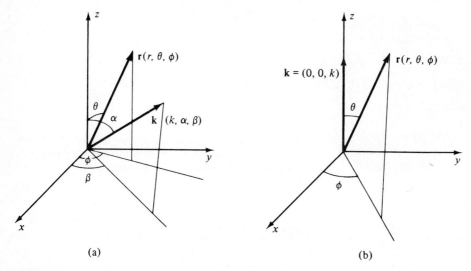

FIGURE 10.5 Coordinates relevant to the expansion of a plane wave in the eigenstates of $\hat{L}^2$ and $\hat{L}_z$. (a) Direction of k is arbitrary. (b) k in the direction of the polar axis.

(10.56). This expansion appears as[1]

(10.67)
$$e^{i\mathbf{k}\cdot\mathbf{r}} = \sum_{l=0}^{\infty} \sum_{m=-l}^{l} a_{lm}(\mathbf{k})\varphi_{klm}$$

where the coefficients of expansion, a_{lm}, are

(10.68)
$$a_{lm} = 4\pi i^l [Y_l^m(\alpha, \beta)]^*$$

and (k, α, β) are the spherical coordinates of **k** (Fig. 10.5).

The probability that measurement of L^2 finds the value $\hbar^2 l(l+1)$ is the partial sum [see (9.82) and (9.83)]

(10.69)
$$P[\hbar^2 l(l+1)] = \sum_{m=-l}^{l} |a_{lm}|^2 = (4\pi)^2 \sum_{m=-l}^{l} |Y_l^m(\alpha, \beta)|^2$$

The probability that measurement of L_z finds the value $\hbar m$ is

(10.70)
$$P[\hbar m] = (4\pi)^2 \sum_{l=|m|}^{\infty} |Y_l^m(\alpha, \beta)|^2$$

---

[1] See Goertzel and Tralli, *Some Mathematical Methods in Physics*. This expansion is also discussed in Problem 10.11.

PROBLEMS

10.9 Calculate the divergence of particle current, $\mathbf{V} \cdot \mathbf{J}$, for a collection of particles that are all in the state

$$\psi(r, t) = j_2(kr)e^{-i\omega t}$$

$$\hbar\omega = \frac{\hbar^2 k^2}{2m}$$

10.10 A spherically propagating shell contains N neutrons, which are all in the state

$$\psi(\mathbf{r}, 0) = 4\pi i \left[\frac{\sin kr}{(kr)^2} - \frac{\cos kr}{kr} \right] \frac{3Y_1{}^0(\theta, \phi) + 5Y_1{}^{-1}(\theta, \phi)}{\sqrt{34}}$$

at $t = 0$.

    (a) What is $\psi(\mathbf{r}, t)$?

    (b) What is the expectation of the energy for this "beam"?

    (c) What possible values of L^2 and L_z will measurement find and how many neutrons will have these values?

    (d) If at $t = 0$, measurement of L^2 finds the value $2\hbar^2$, what is $\psi(\mathbf{r}, t)$?

    (e) If at $t = 0$, measurement of L_z finds the value $-\hbar$, what is $\psi(\mathbf{r}, t)$?

10.11 Use the expansion of a plane wave in spherical harmonics,

$$e^{i\mathbf{k}\cdot\mathbf{r}} = 4\pi \sum_{l=0}^{\infty} \sum_{m=-l}^{l} i^l j_l(kr)[Y_l^m(\alpha, \beta)]^* Y_l^m(\theta, \phi)$$

together with the spherical coordinate representation of $\delta(\mathbf{r} - \mathbf{r}')$,

$$\delta(\mathbf{r} - \mathbf{r}') = \frac{\delta(r - r')\delta(\theta - \theta')\delta(\phi - \phi')}{r^2 \sin \theta} = \left(\frac{1}{2\pi}\right)^3 \int_0^{2\pi} \int_0^{\pi} \int_0^{\infty} e^{i\mathbf{k}\cdot(\mathbf{r}-\mathbf{r}')}k^2 \, dk \sin \alpha \, d\alpha \, d\beta$$

(the spherical coordinates of $\mathbf{k}$ are k, α, β; see Fig. 10.5a) to obtain the orthonormality condition

$$\delta(\mathbf{r} - \mathbf{r}') = \frac{2}{\pi} \sum_l \sum_m [Y_l^m(\theta', \phi')]^* Y_l^m(\theta, \phi) \int_0^{\infty} j_l(kr)j_l(kr')k^2 \, dk$$

The spherical coordinates of $\mathbf{r}$ are (r, θ, ϕ), and those of $\mathbf{r}'$ are (r', θ', ϕ'). [Compare with (C.14) in Appendix C.]

10.12 Use the addition theorem for spherical harmonics (see Fig. 9.16) to reduce the first equation of Problem 10.11 to the expansion

$$e^{ikz} = e^{ikr\cos\theta} = \sum_{l=0}^{\infty} (2l + 1)i^l j_l(kr)P_l(\cos \theta)$$

Note that in this description, $\mathbf{k}$ is aligned with the polar (z) axis, so that $\mathbf{k} \cdot \mathbf{r} = kz$ (see Fig. 10.5b). This expansion is important to the theory of partial wave scattering and will be called upon in Chapter 14.

10.13   The expansion in Problem 10.12 of the plane wave e^{ikz} indicates that the probability of measuring $L^2 = \hbar^2 l(l+1)$ is

$$P[\hbar^2 l(l+1)] \simeq (2l+1)^2$$

Give a semiclassical heuristic argument in support of this conclusion (i.e., that $P \sim l^2$).

Answer

Consider a surface S, of constant phase of the plane wave, $\exp(ikz)$. (See Fig. 10.6). All points in the annular region $dS = 2\pi r_\perp\, dr_\perp = \pi d(r_\perp{}^2)$ correspond to angular momentum $L = r_\perp p_z = r_\perp \hbar k$. It follows that

$$dS = \pi d\left(\frac{L^2}{\hbar^2 k^2}\right)$$

The probability of finding such "points" is proportional to the annular surface dS, so (k^2 is constant)

$$dP \sim dS \sim dL^2$$

In the classical (correspondence) limit, $L^2 \sim \hbar^2 l^2$ and $P \sim l^2$.

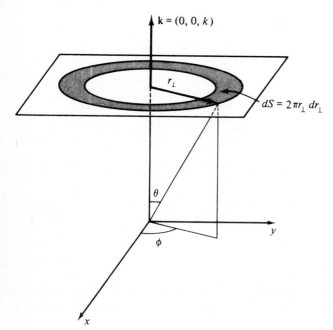

FIGURE 10.6   The probability of finding a particle in a plane-wave state with angular momentum $\sim \hbar l$ increases as l^2 (see Problem 10.13).

10.14   How many independent eigenstates are there corresponding to a free particle moving with energy $E_k = \hbar^2 k^2 / 2m$ in
   (a)   The Cartesian coordinate representation?
   (b)   The spherical coordinate representation?
Give a classical description of the different orbits corresponding to these degenerate states.

Answers (partial)
   (a)   In the Cartesian representation, any state

$$\varphi_k = A e^{i\mathbf{k}\cdot\mathbf{r}}, \qquad k^2 = \frac{2mE}{\hbar^2}$$

is an eigenstate corresponding to the given value of E. These $\mathbf{k}$ vectors describe a sphere of radius $\sqrt{2mE/\hbar^2}$. This continuum of states corresponds to aiming the particle in different directions, while holding its speed, $\hbar k/m$, fixed.
   (b)   In the spherical representation, any state

$$\varphi_{klm} = j_l(kr) Y_l^m(\theta, \phi), \qquad k^2 = \frac{2mE}{\hbar^2}$$

is an eigenstate corresponding to the given value of E. Different states are obtained by choosing different values of l and m. This countable infinity of states corresponds to propitious choice of straight-line trajectories about the origin, all at constant speed, $\hbar k/m$.

10.15   The Laplacian operator ∇^2 in cylindrical coordinates appears as

$$\nabla^2 \varphi = \frac{1}{\rho}\frac{\partial}{\partial\rho}\left(\rho\frac{\partial\varphi}{\partial\rho}\right) + \frac{1}{\rho^2}\frac{\partial^2\varphi}{\partial\phi^2} + \frac{\partial^2\varphi}{\partial z^2}$$

Consider the cylindrical potential well

$$V(\rho) = 0, \qquad \rho < a$$

$$V(\rho) = \infty, \qquad \rho \geq a$$

   (a)   What is the time-independent Schrödinger equation for an arbitrary potential $V(\rho, z, \phi)$, in cylindrical coordinates?
   (b)   Consider the ϕ, z independent wavefunctions $\varphi = R(\rho)$ appropriate to the given potential well. Show that φ obeys *Bessel's equation* (of zero order)

$$\rho\frac{d}{d\rho}\left(\rho\frac{dR}{d\rho}\right) + k^2\rho^2 R = 0, \qquad \hbar^2 k^2 = 2mE$$

   (c)   The class of solutions of this equation which are finite at the origin are the *zeroth-order Bessel functions*, $J_0(x)$ (set $x \equiv k\rho$). The values of x where $J_0(x)$ vanishes are called *the zeros* of J_0. Given that the three lowest values of these zeros are $x_1 = 2.41$, $x_2 = 5.52$, and $x_3 = 8.65$, what are the three lowest eigenenergies and eigenfunctions (as a function of ρ) for the given potential?

(d) If φ is permitted to depend also on the aximuthal angle ϕ, how does this change the three eigenenergies you have just obtained?

10.16 (a) What are the eigenenergies and eigenstates of a particle of mass m in a spherical well

$$V(r) = 0, \quad r < a$$

$$V(r) = \infty, \quad r \geq a$$

Your answers will involve the wavenumbers $k_{l,n}$, which are solutions to the equation

$$j_l(k_{l,n}a) = 0$$

(b) Using the asymptotic form of $j_l(x)$ given in Table 10.1, obtain an explicit expression for the large-order eigenenergies. Consulting Fig. 10.3, obtain a numerical value for the ground energy (eV) of a neutron in a well of radius 10^{-13} cm.

10.4 A CHARGED PARTICLE IN A MAGNETIC FIELD

A closely allied motion to that of a free particle is the motion of a charged particle (e.g., an electron) in a uniform, constant magnetic field $\mathscr{B}$. The Hamiltonian for the electron is given by

$$(10.71) \qquad H = \frac{1}{2m}\left(\mathbf{p} - \frac{e}{c}\mathbf{A}\right)^2$$

The magnetic field is related to the vector potential $\mathbf{A}$ through the relation

$$\mathscr{B} = \nabla \times \mathbf{A}$$

The Cartesian components of $\mathbf{A}$,

$$\mathbf{A} = (-y\mathscr{B}, 0, 0)$$

generate a uniform magnetic field which points in the z direction.

$$\mathscr{B} = (0, 0, \mathscr{B})$$

Substituting this value of $\mathbf{A}$ into the Hamiltonian above gives the time-independent Schrödinger equation

$$(10.72) \qquad \hat{H}\varphi = \left[\frac{1}{2m}\left(\hat{p}_x + \frac{ey\mathscr{B}}{c}\right)^2 + \frac{\hat{p}_y^{\,2}}{2m} + \frac{\hat{p}_z^{\,2}}{2m}\right]\varphi = E\varphi$$

Since the coordinates x and z are missing from the Hamiltonian, it follows that

$$[\hat{p}_x, \hat{H}] = [\hat{p}_z, \hat{H}] = 0$$

and we may conclude that $\hat{p}_x$, $\hat{p}_z$, and $\hat{H}$ have simultaneous eigenstates. The eigenstates of $\hat{p}_x$ and $\hat{p}_z$ appear as

$$\varphi_{k_x k_z} = e^{i(k_x x + k_z z)}$$

so that we may write the common eigenstates of $\hat{H}$, $\hat{p}_x$, and $\hat{p}_z$ in the form

(10.73) $$\varphi = e^{i(k_x x + k_z z)} f(y)$$

Substituting this product into (10.72) gives

(10.74) $$\left[\frac{\hat{p}_y^2}{2m} + \frac{K}{2}(y - y_0)^2 \right] f = \left(E - \frac{\hbar^2 k_z^2}{2m} \right) f$$

where we have set

$$y_0 \equiv -\frac{c\hbar k_x}{e\mathscr{B}}$$

$$\frac{K}{m} \equiv \left(\frac{e\mathscr{B}}{mc} \right)^2 \equiv \Omega^2$$

The frequency Ω is called the *cyclotron frequency*. This is the frequency of rotation corresponding to the classical motion of a charged particle in a uniform magnetic field (see Problem 10.17).

The Schrödinger equation (10.74) is the same as that for a simple harmonic oscillator constrained to move along the y axis, about the point y_0, with natural frequency Ω. From Section 7.2 we recall that the eigenenergies of this equation are

$$\left(E_n - \frac{\hbar^2 k_z^2}{2m} \right) = \hbar\Omega(n + \tfrac{1}{2})$$

which gives the desired result

(10.75) $$\boxed{E_n = \hbar\Omega\left(n + \frac{1}{2} \right) + \frac{\hbar^2 k_z^2}{2m}}$$

The kinetic-energy term $\hbar^2 k_z^2/2m$ corresponds to free, linear motion parallel to the z axis. Classically, such motion is unaffected by a magnetic field in the z direction. The first term in E_n corresponds to the rotational motion normal to the $\mathscr{B}$ field. In the corresponding classical motion the charged particle moves in a helix of constant radius, constant energy, constant rotational frequency, and constant z velocity. The projection of the motion onto the xy plane is a circle with a fixed center (Fig. 10.7). The energy levels (10.75) are commonly referred to as *Landau levels*.

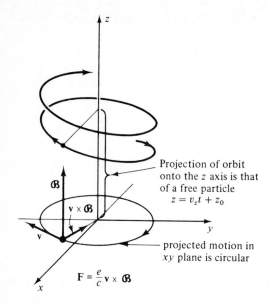

Projection of orbit
onto the z axis is that
of a free particle
$z = v_z t + z_0$

projected motion in
xy plane is circular

$$\mathbf{F} = \frac{e}{c}\mathbf{v} \times \boldsymbol{\mathscr{B}}$$

FIGURE 10.7   Helical motion of a positive charge in a uniform, constant magnetic field that points in the z direction.

The eigenfunction corresponding to the eigenenergy (10.75) is

$$f_n = A_n \mathscr{H}_n\left[\sqrt{\frac{m\Omega}{\hbar}}\,(y - y_0)\right]\exp\left[-\frac{1}{2}\sqrt{\frac{m\Omega}{\hbar}}\,(y - y_0)^2\right]$$

[recall (7.59)]. The nth-order Hermite polynomial is written $\mathscr{H}_n$, while A_n is a normalization constant. Together with (10.73), this form for f_n gives the wavefunction (10.76)

$$\varphi_n = A_n \mathscr{H}_n\left[\sqrt{\frac{m\Omega}{\hbar}}\,(y - y_0)\right]\exp\left[-\frac{1}{2}\sqrt{\frac{m\Omega}{\hbar}}\,(y - y_0)^2 + i(k_x x + k_z z)\right]$$

for a charged particle moving in a uniform magnetic field which points in the z direction.

Degeneracy of Landau Levels

In this section we wish to discover the manner in which the continuous energy spectrum of a free electron changes when the electron is in the presence of a magnetic field. In the course of this discussion we will obtain an expression for the degeneracy of Landau levels.

To examine this problem, we consider the electron to be enclosed in a large cubical box of edge length L. Free-particle wavefunctions and energies are given by (10.14). To account for the finite enclosure, for L sufficiently large, say $L^2 \gg \hbar/m\Omega$, we impose *periodic boundary conditions*:

$$\varphi(x, y, z) = \varphi(x + L, y + L, z + L)$$

As the Landau levels given by (10.75) are evidently degenerate in k_x, we will focus on this wavenumber for the free-particle motion. Substituting the free-particle state (10.14) into the relation above gives

$$k_x = \frac{2\pi n_x}{L}$$

$$n_x = 0, \pm 1, \pm 2, \ldots$$

Thus we discover the following important fact. The continuous energy spectrum of a free, unconfined particle changes to a nearly continuous discrete spectrum when the particle is confined in a large enclosure.

In the presence of a magnetic field, the Shrödinger equation for the electron is given by (10.74). As noted in Problem 10.17, y_0 is associated with the center of the corresponding classical circular motion. Thus, for the present case we may take y_0 to lie between 0 and L.

It follows that the maximum value of this parameter $\bar{y}_0$ is given by

$$\bar{y}_0 = L = \frac{c\hbar\bar{k}_x}{|e|\mathscr{B}} = \frac{c\hbar\bar{n}_x}{|e|\mathscr{B}L}$$

We may conclude that n_x values are positive numbers and have the maximum value[1]

(10.77) $$\bar{n}_x \equiv \bar{g} = \left(\frac{|e|\mathscr{B}}{\hbar c}\right) L^2$$

This is the desired expression for the degeneracy $\bar{g}$ of a Landau level. The presence of L^2 in (10.77) corresponds to the property that a given helical orbit can be displaced anywhere in the xy plane without changing the energy of the electron (see Fig. 10.7).

Thus, we find that the energy spectrum of a confined electron changes from a nearly continuous one for $\mathscr{B} = 0$ to a discrete spectrum for $\mathscr{B} > 0$, with degeneracy given by (10.77).

---

[1] We employ the barred variable $\bar{g}$ to distinguish it from the unbarred variable g used to denote density of states elsewhere in the text.

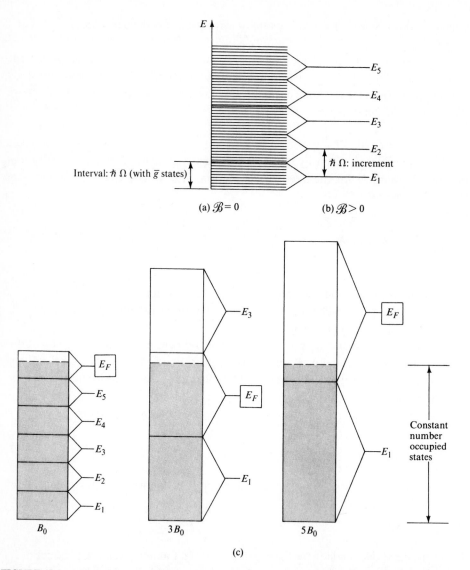

FIGURE 10.8 (a) **Nearly continuous discrete spectrum for a particle confined to a large box with $\mathscr{B} = 0$.** (b) **Equally spaced Landau levels corresponding to $\mathscr{B} > 0$. Each increment of energy, $\hbar\Omega$, corresponds to $\bar{g}$ free-particle states, which, in turn, is the degeneracy of each Landau level.** (c) **Variation of the Fermi energy with change in $\mathscr{B}$.**

From (10.75) we see that the spacing between Landau levels, at fixed k_z, is the constant value

$$\Delta E = \hbar\Omega$$

The degeneracy $\bar{g}$ given by (10.77) gives the number of free-particle states that contribute to the increment ΔE. See Fig. 10.8. Note, in particular, the resemblance between the equally spaced Landau levels and the equally spaced levels of the harmonic oscillator shown in Fig. 7.8. This congruence of spectra stems from the previously described parallel structure of the two respective Hamiltonians. In either event, the density of states, $g(E)$ (Section 8.8), when plotted as a function of energy, is a series of equally spaced delta functions.

Note further that in the classical limit $\hbar \to 0$, the degeneracy $\bar{g}$ grows infinite and the spacing between levels ΔE goes to zero.

Fermi Energy and Landau Levels

Application of the preceding results may be made to the conduction electrons in a two-dimensional metal. In this event, the degenerate states shown in Fig. 10.8a become filled with electrons. The highest energy state occupied (at 0 K) is that of the Fermi energy, E_F, [Sections 2.3 (Fig. 2.3), 8.4, 12.9]. With a magnetic field $\mathscr{B}$ normal to the plane of the sample, we may identify E_F as the highest Landau level occupied. How does E_F change with change in $\mathscr{B}$? To answer this question, we first note that the maximum number of electrons at the Fermi level is given by the degeneracy factor $\bar{g}$ (10.77). Let $n^{oc} \leq \bar{g}$ denote the number of occupied states in this level. The value of E_F remains constant as long as $n^{oc} > 0$. When this Landau level is vacated, E_F changes to the value of the new partially occupied Landau level. See Fig. 10.8c. Such variation of E_F exhibits a periodicity with respect to $\mathscr{B}^{-1}$. This phenomenon is a component of the *de Haas-van Alphen effect.*[1]

PROBLEMS

10.17 The following is a problem in classical physics. The force on a charged particle in a uniform magnetic field $\mathscr{B}$ is

$$\mathbf{F} = \frac{d}{dt}(m\mathbf{v}) = \frac{e}{c}\mathbf{v} \times \mathscr{B}$$

(a) Show that

$$\tfrac{1}{2}mv^2 = \text{constant}$$

with $\mathscr{B} = (0, 0, \mathscr{B})$.

---

[1] For further discussion, see C. Kittel, *Introduction to Solid State Physics*, 6th ed., Wiley, New York, 1986, Chap. 9.

(b) Show that

$$p_z = mv_z = \text{constant}$$

(c) Show that the motion of the particle is that of a helix whose axis is parallel to $\mathscr{B}$ and whose projection onto the xy plane is circular with constant angular frequency Ω.

(d) Show that the center of this circle in the xy plane has coordinates

$$y_0 = \frac{-cp_x}{e\,\mathscr{B}} = \frac{-cmv_x}{e\,\mathscr{B}} + y$$

$$x_0 = \frac{cmv_y}{e\,\mathscr{B}} + x = \frac{cp_y}{e\,\mathscr{B}} + x$$

Note that p_x, canonical momentum, is not equal to mv_x for $\mathbf{A} = (A_x, 0, 0)$. The correct relation follows from (1.14) and (10.71).

10.18 Show that the operator

$$\hat{x}_0 \equiv \hat{x} + \frac{c\hat{p}_y}{e\,\mathscr{B}}$$

commutes with $\hat{H}$ as given in (10.72) but does not commute with

$$\hat{y}_0 = \frac{-c\hat{p}_x}{e\,\mathscr{B}}$$

These operators correspond to the coordinates of the center of the related projected classical motion in the xy plane. In quantum mechanics we see that although x_0 and E, or y_0 and E, may, respectively, be specified simultaneously, x_0 and y_0 may not be simultaneously specified.

10.19 (a) What is the vector potential $\mathbf{A}$ which gives the uniform $\mathscr{B}$ field $(0, 0, \mathscr{B})$ which includes $A_x = A_z = 0$?

(b) What is the form of the wavefunctions φ_n corresponding to this choice of vector potential? How do they compare to the wavefunctions corresponding to $A_y = A_z = 0$ found in the text?

(c) How do the eigenenergies compare to those found in the representation $A_y = A_z = 0$?

10.20 What is the nature of the frequency spectrum emitted by a charged particle moving in a uniform magnetic field? (Assume that the kinetic energy parallel to $\mathscr{B}$ does not change.) For an electron moving in a $\mathscr{B}$ field of 10^4 gauss, what type of radiation is this (x rays, microwaves, etc.)?

10.5 THE TWO-PARTICLE PROBLEM

Coordinates Relative to the Center of Mass

When dealing with systems containing more than one particle (e.g., an atom), it is convenient to separate the motion into that of the center of mass of the system and motion relative to the center of mass. This separation is effected through a partitioning of the Hamiltonian into a part, H_{CM}, involving center of mass coordinates, and a part, H_{rel}, containing coordinates relative to the center of mass.

For example, consider the two-particle Hamiltonian

(10.78)
$$H = \frac{p_1{}^2}{2m_1} + \frac{p_2{}^2}{2m_2} + V(|\mathbf{r}_1 - \mathbf{r}_2|)$$

The potential of interaction $V(|\mathbf{r}_1 - \mathbf{r}_2|)$ is a function only of the radial distance between the particles. For instance, for the hydrogen atom, the interaction V is the Coulomb potential

(10.78a)
$$V = -\frac{e^2}{r}$$

where we have written r for the distance between particles, $|\mathbf{r}_1 - \mathbf{r}_2|$. Such potentials, which are only a function of the scalar distance r, are called *central potentials*.

In the Hamiltonian above, $\mathbf{p}_1$ and $\mathbf{p}_2$ are the linear momenta of particle 1 and particle 2, respectively, while m_1 and m_2 are the respective masses of these particles.

A two-particle system has six degrees of freedom. These are characterized by the parameters $(\mathbf{r}_1, \mathbf{p}_1; \mathbf{r}_2, \mathbf{p}_2)$. The partitioning of the Hamiltonian into $H_{CM} + H_{rel}$ is generated through the transformation of variables

(10.79)
$$(\mathbf{r}_1, \mathbf{p}_1; \mathbf{r}_2, \mathbf{p}_2) \to (\mathbf{r}, \mathbf{p}; \mathcal{R}, \mathcal{P})$$

where

$$\mathbf{r} = \mathbf{r}_2 - \mathbf{r}_1, \qquad \mathcal{P} = \mathbf{p}_1 + \mathbf{p}_2$$

(10.80)
$$\mathbf{p} = \frac{m_2 \mathbf{p}_1 - m_1 \mathbf{p}_2}{m_1 + m_2}, \qquad \mathcal{R} = \frac{m_1 \mathbf{r}_1 + m_2 \mathbf{r}_2}{m_1 + m_2}$$

Using these equations the Hamiltonian (10.77) is transformed to the sum

(10.81)
$$H = \frac{\mathcal{P}^2}{2M} + \left[\frac{p^2}{2\mu} + V(r)\right] \equiv H_{CM} + H_{rel}$$

where the reduced mass μ and the total mass M are

(10.82)
$$\mu = \frac{m_1 m_2}{m_1 + m_2}, \qquad M = m_1 + m_2$$

Equation (10.81) represents the desired separation of H into the Hamiltonian of the center of mass, H_{CM}, and the Hamiltonian of the coordinates relative to the center of mass, H_{rel} (Fig. 10.9). Since $\mathcal{R}$ is absent in H (i.e., $\mathcal{R}$ is a cyclic coordinate; see Section 1.2), the momentum of the center of mass, $\mathcal{P}$, is constant. The center of mass moves in straight rectilinear motion, characteristic of a free particle of mass M. The motion relative to the center of mass is that of a particle of mass μ moving in the central potential $V(r)$.

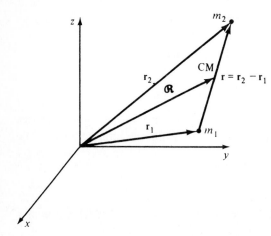

FIGURE 10.9   The relative vector **r** and the center-of-mass vector $\mathscr{R}$. In the classical motion, $\mathscr{P} = M\dot{\mathscr{R}} = \text{constant} = \mathscr{P}(0) =$ the initial value of $\mathscr{P}(t)$. At any time t

$$\mathscr{R}(t) = \mathscr{R}(0) + \frac{\mathscr{P}(0)t}{M}$$

Solving the dynamical equations (viz., Hamilton's equations) using H_{rel} gives $\mathbf{r}(t)$, which when affixed to $\mathscr{R}(t)$ gives the motion in the "lab frame."

The Transformation of $\hat{H}$

For the quantum mechanical case, the transformation of $\hat{H}$ is again effected with the equations (10.80), which are now interpreted as operator relations. Cartesian components of "old" coordinate and momentum operators $(\hat{\mathbf{r}}_1, \hat{\mathbf{p}}_1; \hat{\mathbf{r}}_2, \hat{\mathbf{p}}_2)$ obey the commutation relations

(10.83)
$$[\hat{r}_{1j}, \hat{p}_{1j}] = i\hbar$$
$$[\hat{r}_{2j}, \hat{p}_{2j}] = i\hbar$$
$$j = 1, 2, 3$$

These are the only nonvanishing commutators. With these relations and (10.80), one obtains that the only nonvanishing commutator relations for components of the "new" operators $(\hat{\mathbf{r}}, \hat{\mathbf{p}}; \hat{\mathscr{R}}, \hat{\mathscr{P}})$ are

(10.84)
$$[\hat{r}_j, \hat{p}_j] = i\hbar$$
$$[\hat{\mathscr{R}}_j, \hat{\mathscr{P}}_j] = i\hbar$$
$$j = 1, 2, 3$$

Thus, in obtaining

$$\hat{H} = \hat{H}_{\text{CM}} + \hat{H}_{\text{rel}}$$

(10.85)
$$\hat{H}_{\text{CM}} = \frac{\hat{\mathscr{P}}^2}{2M}$$

$$\hat{H}_{\text{rel}} = \frac{\hat{p}^2}{2\mu} + V(r)$$

the Hamiltonian is separated into two parts involving components that are independent of one another. For such cases, the Schrödinger equation has product eigenfunctions

(10.86)
$$\bar{\varphi} = \varphi_{\text{CM}}(\mathscr{R})\varphi_{\text{rel}}(\mathbf{r})$$

and summational eigenvalues

(10.87)
$$\bar{E} = E_{CM} + E_{rel}$$

where

$$\hat{H}\bar{\varphi} = \bar{E}\bar{\varphi}$$

(10.88)
$$\hat{H}_{CM}\varphi_{CM} = E_{CM}\varphi_{CM}$$

$$\hat{H}_{rel}\varphi_{rel} = E_{rel}\varphi_{rel}$$

The Schrödinger equation for the center of mass appears explicitly as

(10.89)
$$\frac{\hat{\mathscr{P}}^2}{2M}\varphi_{CM} = E_{CM}\varphi_{CM}$$

This is the Schrödinger equation for a free particle of mass M. Its solution was obtained in the previous section. With the linear momentum $\mathscr{P}$ specified, the states are

$$\varphi_{CM} = Ae^{i\mathbf{K}\cdot\mathscr{R}}$$

(10.90)
$$\mathscr{P} = \hbar\mathbf{K}, \qquad E_{CM} = \frac{\hbar^2 K^2}{2M}$$

In the representation where $L_{CM}{}^2$ and L_{CM_z} are specified, the eigenstates are

(10.91)
$$\varphi_{CM} = j_{l_C}(K\mathscr{R})Y_{l_C}{}^{m_C}(\theta_C, \phi_C)$$

The spherical coordinates of $\mathscr{R}$ are $(\mathscr{R}, \theta_C, \phi_C)$ (see Fig. 10.10a).

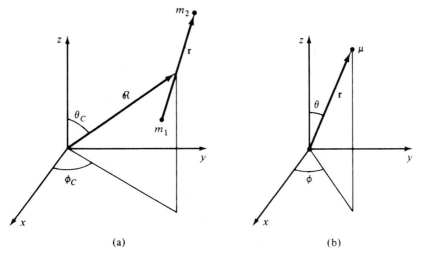

(a)   (b)

FIGURE 10.10 (a) Spherical angle variables for the center-of-mass radius vector $\mathscr{R}$. (b) Coordinates relative to the center of mass.

Radial Equation for a Central Potential

The Schrödinger equation for φ_{rel} appears as (dropping the "rel" subscript)

$$(10.92) \qquad \left[\frac{\hat{p}^2}{2\mu} + V(r)\right]\varphi = E\varphi$$

For central potential functions $V(r)$, it proves most convenient to express the above Hamiltonian in spherical coordinates. The interparticle radius **r** has coordinates (r, θ, ϕ) with the polar axis depicted as lying in the z direction (see Figure 10.10b).

In these coordinates the Schrödinger equation above becomes

$$(10.93) \qquad \hat{H}\varphi = \left[\frac{\hat{p}_r^2}{2\mu} + \frac{\hat{L}^2}{2\mu r^2} + V(r)\right]\varphi = E\varphi$$

First, we note that $\hat{L}^2$ and $\hat{L}_z$ both commute with $\hat{H}$. The remaining components in $\hat{H}$ are all isotropic forms and are therefore unaffected by angular momentum operators. It follows that $\hat{H}$, $\hat{L}^2$, and $\hat{L}_z$ have simultaneous eigenstates. These are given by the product form

$$(10.94) \qquad \varphi = R(r)Y_l^m(\theta, \phi)$$

Substituting this solution into the Schrödinger equation above gives the "radial" equation

$$(10.95) \qquad \left[\frac{\hat{p}_r^2}{2\mu} + \frac{\hbar^2 l(l+1)}{2\mu r^2} + V(r)\right]R(r) = ER(r)$$

This is an ordinary, second-order, linear differential equation for the radial dependent component of the wavefunction $R(r)$. Since only one variable is involved in (10.95), it is suggestive of one-dimensional motion with the effective potential

$$(10.96) \qquad V_{eff} = V(r) + \frac{\hbar^2 l(l+1)}{2\mu r^2}$$

The second term in this expression is called the "angular momentum barrier." It becomes infinitely high as $r \to 0$ and acts as a repulsive core, which for $l > 0$ prevents collapse of the system (see Fig. 10.11).

The normalization of the eigenstates (10.94) is given by the integral

$$(10.97) \qquad \langle R Y_l^m | R Y_l^m \rangle = \int_0^\infty dr\, r^2 \int_{4\pi} d\Omega |R(r)Y_l^m(\theta, \phi)|^2 = 1$$

$$= \int_0^\infty r^2 |R(r)|^2\, dr = 1$$

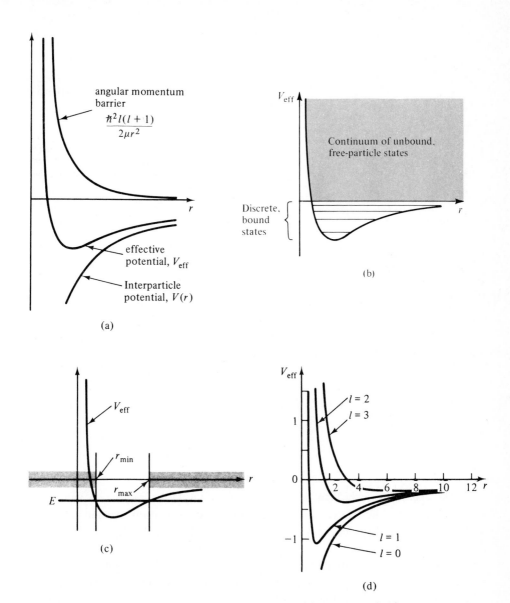

FIGURE 10.11 **(a) The effective potential in relation to the angular momentum barrier**

$$V_{\text{eff}} = V(r) + \frac{\hbar^2 l(l+1)}{2\mu r^2}$$

(b) Nature of the quantum mechanical energy spectrum for central potential problems. (c) The classical motion corresponding to the energy E. Shaded regions define classically forbidden domains. (d) The effective potential energy V_{eff} for hydrogen for several values of the orbital quantum number l. Units of r are angstroms. V_{eff} is in units of 10^{-11} erg.

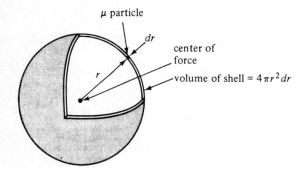

μ particle

dr

center of
force

volume of shell = $4\pi r^2 dr$

r

FIGURE 10.12 Probability of finding
the fictitious μ particle in a spherical shell
between r and $r + dr$ is

$$P_r \, dr = |R(r)r|^2 \, dr$$

This is also the probability of finding m_2
in a spherical shell about m_1 in the configuration shown (or m_1 in a shell about
m_2).

The radial displacement r separates the two particles m_1 and m_2. If we envision
particle m_1 at the origin, then

(10.98) $$Pr^2 \, dr \, d\Omega = |R(r)Y_l^m(\theta, \phi)|^2 r^2 \, dr \, d\Omega$$

is the probability of finding m_2 in the *volume element* $r^2 \, dr \, d\Omega$ about m_1 (an equally
valid statement is obtained with m_1 and m_2 reversed). What is the probability of finding m_2 in a *spherical shell* of radius between r and $r + dr$, about m_1? The answer is
(Fig. 10.12)

(10.99) $$P_r \, dr = \left(\int_{4\pi} Pr^2 \, d\Omega \right) dr = |R(r)|^2 r^2 \, dr \equiv |u(r)|^2 \, dr$$

so that

$$\int_0^\infty |u(r)|^2 \, dr = 1$$

The classically forbidden domains (see Chapter 1) correspond to values of r
for which $E < V$. The related property for a spherical quantum mechanical system
is that the probability density $|u(r)|^2$ becomes small in these domains (see Figs.
10.11c and 10.13).

Having found the radial function $R(r)$, in a specific two-body problem, the wavefunction for the system relative to the laboratory frame (as opposed to the center-of-
mass frame) is either of the forms

(10.100) $$\bar{\varphi} = Ae^{i\mathscr{P} \cdot \mathscr{R}/\hbar}R(r)Y_l^m(\theta, \phi)$$

$$\bar{\varphi} = j_{l_C}(K\mathscr{R})Y_{l_C}^{m_C}(\theta_C, \phi_C)R(r)Y_l^m(\theta, \phi)$$

In the first representation, the six parameters $(\mathscr{P}; L^2, L_z, E)$ are specified. In the second
representation, the six parameters $(E_{CM}, L_{CM}^2, L_{CM_z}; L^2, L_z, E)$ are specified.

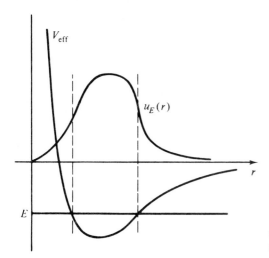

FIGURE 10.13 The radial probability amplitude u_E corresponding to the energy E decays to
zero in the classically forbidden domain.

Continuity and Boundary Conditions

Some general properties of the radial wavefunction are as follows. With $R(r)$ everywhere bounded, we note first that $u(r) \equiv rR(r)$ must vanish at the origin.[1] For
$r > 0$, with energy E and potential energy $V(r)$ bounded, the radial equation (10.95)
indicates that

$$\hat{p}_r{}^2 R(r) = -\hbar^2 \frac{1}{r} \frac{\partial}{\partial r} \left[\frac{\partial}{\partial r} u(r) \right]$$

is likewise bounded. It follows that $\partial u/\partial r$ is continuous. The existence of this derivative
implies that $u(r)$ is continuous. The latter two conditions infer continuity of the
logarithmic derivative

$$\frac{1}{u} \frac{du}{dr} = \frac{d \ln u}{dr}$$

These conditions on the wavefunction in spherical coordinates are employed
in obtaining the ground state of the deuteron (Problem 10.30) and in construction
of the bound states of the hydrogen atom as described in the following section. They
will also come into play in construction of the states for low-energy scattering from a
spherical well (Section 14.2).

---

[1] Dirac obtains this boundary condition from the stipulation that solutions to the Schrödinger equation in spherical
coordinates agree with those obtained in Cartesian coordinates. For further discussion, see P. A. M. Dirac, *The Principles
of Quantum Mechanics*, 4th ed., Oxford University Press, New York, 1958.

PROBLEMS

10.21 Consider a two-particle system. The momenta of the particles are $\mathbf{p}_1$ and $\mathbf{p}_2$, respectively.

(a) What is $[\hat{\mathbf{p}}_1, \hat{\mathbf{p}}_2]$? Explain your answer.

(b) Use the answer to part (a) to show that $[\hat{\mathbf{p}}, \mathscr{P}] = 0$, where $\mathbf{p}$ is the "relative" momentum [defined in (10.80)] and $\mathscr{P}$ is the momentum of the center of mass.

(c) The particles interact under a potential that is a function of the distance between them. As shown in the text, transformation to coordinates relative to the center of mass effects a partitioning of the Hamiltonian, $\hat{H} = \hat{H}_{\mathrm{CM}} + \hat{H}_{\mathrm{rel}}$. What is $[\hat{H}_{\mathrm{CM}}, \hat{H}_{\mathrm{rel}}]$?

10.22 Prove that the following equations are compatible with the transformation equations (10.80).

(a) $\dfrac{p_1^{\,2}}{2m_1} + \dfrac{p_2^{\,2}}{2m_2} = \dfrac{p^2}{2\mu} + \dfrac{\mathscr{P}^2}{2M}$

(b) $m_1 r_1 + m_2 r_2 = \mu r^2 + M\mathscr{R}^2$

(c) $\mathbf{p}_1 \cdot \mathbf{r}_1 + \mathbf{p}_2 \cdot \mathbf{r}_2 = \mathbf{p} \cdot \mathbf{r} + \mathscr{P} \cdot \mathscr{R}$

(d) $\mathbf{L}_1 + \mathbf{L}_2 = \mathbf{L} + \mathbf{L}_{\mathrm{CM}}$

In part (d),

$$\mathbf{L}_1 = \mathbf{r}_1 \times \mathbf{p}_1 \qquad \mathbf{L} = \mathbf{r} \times \mathbf{p}$$

$$\mathbf{L}_2 = \mathbf{r}_2 \times \mathbf{p}_2 \qquad \mathbf{L}_{\mathrm{CM}} = \mathscr{R} \times \mathscr{P}$$

10.23 At a particular time, the wavefunction of a mass m moving in a three-dimensional potential field is

$$\varphi = A(x + y + z)e^{-k_0 r}$$

(a) Calculate the normalization constant A.

(b) What is the probability that measurement of L^2 and L_z finds $2\hbar^2$ and 0, respectively? (See Table 9.1.)

(c) What is the probability of finding the particle in the sphere $r \le k_0^{-1}$?

10.24 For a two-particle system (m_1, m_2), what is the fractional distance to the center of mass from m_1 and m_2, respectively? What are these numbers for hydrogen?

10.25 Let $\mathbf{e}$ be a unit vector in an arbitrary but fixed direction. Show that the commutators between the components of $\hat{\mathbf{r}}$ and $\hat{\mathbf{p}}$, respectively, with the component $\mathbf{e} \cdot \hat{\mathbf{L}}$ obey the relations

$$[\hat{\mathbf{p}}, \mathbf{e} \cdot \hat{\mathbf{L}}] = i\hbar \mathbf{e} \times \hat{\mathbf{p}}$$

$$[\hat{\mathbf{r}}, \mathbf{e} \cdot \hat{\mathbf{L}}] = i\hbar \mathbf{e} \times \hat{\mathbf{r}}$$

10.26 Use the results of Problem 10.25 to show that $\hat{p}^2$, $\hat{r}^2$, and $\hat{\mathbf{r}} \cdot \hat{\mathbf{p}}$ all commute with every component of $\hat{\mathbf{L}}$. Then show that every component of $\hat{\mathbf{L}}$ commutes with any isotropic function $f(r^2)$.

432

Answer (partial)

If the statement is true for arbitrary **e**, it is true for all components of $\hat{\mathbf{L}}$.

$$[\hat{p}^2, \mathbf{e} \cdot \hat{\mathbf{L}}] = \hat{\mathbf{p}}[\hat{\mathbf{p}}, \mathbf{e} \cdot \hat{\mathbf{L}}] + [\hat{\mathbf{p}}, \mathbf{e} \cdot \hat{\mathbf{L}}] \cdot \hat{\mathbf{p}}$$
$$= i\hbar(\hat{\mathbf{p}} \cdot \mathbf{e} \times \hat{\mathbf{p}} + \mathbf{e} \times \hat{\mathbf{p}} \cdot \hat{\mathbf{p}}) = 0$$

10.27   Two free particles of mass m_1 and m_2, respectively, move in 3-space. They do not interact. Write the eigenfunctions and eigenenergies of this system in as many representations as you can. Indicate the number of parameters specified in the eigenstates associated with these representations.

10.28   Write down the time-dependent wavefunction corresponding to eigenstates for a two-particle system in the two representations (10.100).

10.29   Consider two particles that attract each other through the potential

$$V(r) = -\frac{\hbar^2 K^2}{2\mu r^2}$$

The displacement between particles is r, K is a constant, and μ is reduced mass. In states of definite angular momentum, what are the values of the angular momentum quantum number l for which the effective force between particles is repulsive?

10.30   In Problem 8.1 the depth of the potential well appropriate to a deuteron was evaluated using a one-dimensional approximation. A more refined estimate may be obtained using a three-dimensional spherical well with characteristics

$$V(r) = -|V|, \quad r < a \quad \text{region I}$$
$$V(r) = 0 \quad\quad r \geq a \quad \text{region II}$$

(a)   Construct components of the ground-state wavefunction in regions I and II, respectively.

(b)   Show that matching conditions at $r = a$ give the dispersion relation

$$\eta = -\xi \cot \xi$$
$$\rho^2 = \xi^2 + \eta^2$$

where

$$\rho^2 \equiv \frac{2m|V|a^2}{\hbar^2}, \quad \xi \equiv ka, \quad \eta \equiv \kappa a$$

$$|E| = \frac{\hbar^2 \kappa^2}{2m}, \quad |V| - |E| = \frac{\hbar^2 k^2}{2m}$$

(c)   To within the same approximation suggested in Problem 8.1, obtain a numerical value for the depth $|V|$ of the three-dimensional deuteron well. From the ratio $|E|/|V|$ for this bound state, would you say that the deuteron is a strongly or a weakly bound nucleus?

Answers (partial)

(a) Component wavefunctions in the well domain, region I, are the spherical Bessel functions. The ground-state component is therefore

$$\varphi_I = \frac{\sin kr}{kr}$$

In region II the component ground-state wavefunction is exponentially damped.

$$\varphi_{II} = A \frac{e^{-\kappa r}}{\kappa r}$$

TABLE 10.2   Solutions to the three fundamental box problems in quantum mechanics

The Rectangular Box	The Cylindrical Box	The Spherical Box
Edge Lengths a_1, a_2, a_3	Radius a, Height b	Radius a

Hamiltonian

$$\hat{H} = (\hat{p}_x^2 + \hat{p}_y^2 + \hat{p}_z^2)/2M$$

$$\hat{p}_x^2 = \left(-i\hbar \frac{\partial}{\partial x}\right)^2$$

$$\hat{H} = (\hat{p}_\rho^2 + \hat{p}_z^2 + \hat{L}_z^2/\rho^2)/2M$$

$$\hat{p}_\rho^2 = -\hbar^2 \frac{1}{\rho}\frac{\partial}{\partial \rho}\left(\rho \frac{\partial}{\partial \rho}\right)$$

$$\hat{H} = (\hat{p}_r^2 + \hat{L}^2/r^2)/2M$$

$$\hat{p}_r^2 = -\hbar^2\left(\frac{1}{r}\frac{\partial}{\partial r} r\right)^2$$

Eigenfunction

$$\varphi_{qst} = A_{qst} \sin k_q x \sin k_s y \sin k_t z$$
$$(A_{qst})^2 = 8/a_1 a_2 a_3$$
$$\sin k_q a_1 = \sin k_s a_2 = \sin k_t a_3 = 0$$

$$\varphi_{qmn} = A_{qmn} J_m(K_{mn}\rho) \sin k_q z e^{im\phi}$$
$$(A_{qmn})^2 = 2/\pi b[a J_m{}'(K_{mn} a)]^2$$
$$\sin k_q b = J_m(K_{mn} a) = 0$$

$$\varphi_{nlm} = A_{nlm} j_l(k_{ln} r) Y_l{}^m(\theta, \phi)$$
$$(A_{nlm})^2 = 2/a^3[j_l{}'(k_{ln} a)]^2$$
$$j_l(k_{ln} a) = 0$$

Wave Equation

$$\left(\frac{d^2}{dx^2} + k^2\right)\sin kx = 0$$

Bessel's Equation

$$\left[\frac{1}{x^2}\left(x\frac{d}{dx}\right)^2 + 1 - \frac{m^2}{x^2}\right]J_m(x) = 0$$

Spherical Bessel Equation

$$\left[\left(\frac{1}{x}\frac{d}{dx} x\right)^2 + 1 - \frac{l(l+1)}{x^2}\right]j_l(x) = 0$$

Eigenenergy

$$E_{qst} = \hbar^2(k_q^2 + k_s^2 + k_t^2)/2M$$

$$E_{qmn} = \hbar^2(K_{mn}^2 + k_q^2)/2M$$

$$E_{nl} = \hbar^2 k_{ln}^2/2M$$

(b)   Here one must invoke continuity of $d \ln u/dr$.

(c)   You should obtain the answer $|E|/|V| = 0.08$. This value implies that the binding energy is small compared to the depth of well, and we may conclude that the proton and neutron are weakly bound. A sketch of the normalized wavefunction further reveals that there is approximately only one chance in three that the nucleons are closer together than the well radius a.

10.31   Show by explicit calculation that the eigenfunctions and eigenenergies as given in Table 10.2 are correct for each of the three respective "box" configurations shown. (Primes denote differentiation.)

10.6   THE HYDROGEN ATOM

Hamiltonian and Eigenenergies

The (relative) Hamiltonian for the hydrogen atom (more accurately, for a "hydrogenic" atom[1] of atomic number Z) appears as

$$(10.101) \qquad \hat{H} = \frac{\hat{p}_r^{\,2}}{2\mu} + \frac{\hat{L}^2}{2\mu r^2} - \frac{Ze^2}{r}$$

The corresponding Schrödinger equation is

$$(10.102) \qquad \left(\frac{\hat{p}_r^{\,2}}{2\mu} + \frac{\hat{L}^2}{2\mu r^2} - \frac{Ze^2}{r} \right)\varphi = E\varphi = -|E|\varphi$$

We are seeking the *bound states* of hydrogen. These correspond to the negative eigenenergies, $E = -|E|$. Setting $\varphi = R(r)Y_l^m(\theta, \phi)$ in the latter equation gives the radial equation (10.95)

$$(10.103) \qquad \left[\frac{-\hbar^2}{2\mu} \left(\frac{1}{r}\frac{d^2}{dr^2} r \right) + \frac{\hbar^2 l(l+1)}{2\mu r^2} - \frac{Ze^2}{r} + |E| \right] R = 0$$

Changing the dependent variable to

$$u = rR$$

introduced previously in (10.99), gives

$$(10.104) \qquad \left(-\frac{d^2}{dr^2} + \frac{l(l+1)}{r^2} - \frac{2\mu}{\hbar^2}\frac{Ze^2}{r} + \frac{2\mu|E|}{\hbar^2} \right) u = 0$$

---

[1] Hydrogenic atoms are atoms that are ionized with all but one electron bound to the nucleus which carries the charge $+Ze$ (e.g., He^+, Li^{++}, etc.).

Introducing the notation

$$\rho \equiv 2\kappa r, \qquad \frac{\hbar^2 \kappa^2}{2\mu} = |E|$$

(10.105)
$$\lambda^2 = \left(\frac{Z}{\kappa a_0}\right)^2 = \frac{Z^2 \mathbb{R}}{|E|}$$

$$\mathbb{R} = \frac{\hbar^2}{2\mu a_0{}^2}, \qquad a_0 = \frac{\hbar^2}{\mu e^2}$$

where $\mathbb{R}$ is the Rydberg constant[1] (2.13) and a_0 is the Bohr radius (2.14), the radial equation may be further simplified to the form

(10.106)
$$\frac{d^2 u}{d\rho^2} - \frac{l(l+1)}{\rho^2} u + \left(\frac{\lambda}{\rho} - \frac{1}{4}\right) u = 0$$

For large values of ρ this equation reduces to

$$\frac{d^2 u}{d\rho^2} - \frac{u}{4} = 0$$

so that

$$u \sim A e^{-\rho/2} + B e^{\rho/2}$$

In order that u vanish as $\rho \to \infty$, we set $B = 0$, so

$$u \sim e^{-\rho/2} \qquad (\rho \to \infty)$$

In the neighborhood of the origin, (10.106) reduces to

$$\frac{d^2 u}{d\rho^2} - \frac{l(l+1)}{\rho^2} u = 0$$

Substitution of the trial solution $u = \rho^q$ gives

$$u \sim A \rho^{-l} + B \rho^{l+1}$$

In order for u to vanish at the origin, we must set $A = 0$. This gives

$$u \sim \rho^{l+1} \qquad (\rho \to 0)$$

With these two asymptotic forms at hand, we are prepared to solve (10.106) through a polynomial expansion. Solution in the form

$$u(\rho) = e^{-\rho/2} \rho^{l+1} F(\rho)$$

(10.107)
$$F(\rho) = \sum_{i=0}^{\infty} C_i \rho^i$$

---

[1] The Rydberg constant written with m in place of μ (i.e., assuming infinite proton mass) is sometimes written $\mathbb{R}_x$.

with F finite everywhere, gives the proper behavior at $\rho \sim 0$ and $\rho \sim \infty$. Substituting (10.107) for u into (10.106), we obtain

$$(10.108) \qquad \left[\rho \frac{d^2}{d\rho^2} + (2l + 2 - \rho) \frac{d}{d\rho} - (l + 1 - \lambda) \right] F(\rho) = 0$$

Note that for a given value of the orbital quantum number l, this is an eigenvalue equation with eigenvalue λ. The values of λ (or, equivalently, the eigenenergies, $|E|$) are those values which ensure that $F(\rho)$ is finite for all ρ. Substituting the series (10.107) into the latter equation and equating coefficients of equal powers in ρ gives the *recurrence relation*

$$(10.109) \qquad C_{i+1} = \frac{(i + l + 1) - \lambda}{(i + 1)(i + 2l + 2)} C_i \equiv \Gamma_{il} C_i$$

In the limit that $i \to \infty$, this relation becomes

$$C_{i+1} \sim \frac{C_i}{i}$$

which is the same ratio of coefficients obtained in the expansion

$$e^{\rho} = \sum C_i \rho^i = \sum \frac{\rho^i}{i!}$$

$$\frac{C_{i+1}}{C_i} = \frac{i!}{(i + 1)!} = \frac{1}{i + 1} \sim \frac{1}{i}$$

It follows that the form of $u(\rho)$ generated by the series (10.107) behaves as

$$u(\rho) \sim e^{-\rho/2} \rho^{l+1} e^{\rho} = e^{\rho/2} \rho^{l+1}$$

which diverges for large ρ. To obtain a finite wavefunction, the expansion (10.107) for any given value of l must terminate at some finite value of i, which we will call $i_{\max}$. At this value of i, $\Gamma_{il} = 0$. Since all parameters in (10.109) are positive, Γ_{il} can only vanish if

$$i_{\max} + l + 1 = \lambda$$

The function u so generated is a polynomial and, due to the exponential term in the form (10.107), we see that, as demanded, the wavefunction is finite everywhere.

Since i and l are integers, it follows that λ is also an integer, which is called the *principal quantum number, n.*

$$n = i_{\max} + l + 1$$

Thus the above cutoff condition on the series (10.107), which ensures that $u(\rho)$ is finite for all ρ, also serves to determine the eigenenergies λ.

$$\lambda_n{}^2 = n^2 = \frac{Z^2\mathbb{R}}{|E_n|}$$

(10.110)

$$\boxed{E_n = -|E_n| = -\frac{Z^2\mathbb{R}}{n^2}}$$

These are the same values found previously in the simpler Bohr model (Section 2.4).

Laguerre Polynomials

The hydrogen eigenfunction corresponding to the eigenvalue E_n is given by (10.107) with the series over i cut off at the value

(10.111) $$i_{max} = n - l - 1$$

and the recurrence relation for the coefficients $\{C_i\}$ given by (10.109).

(10.112)
$$u_{nl}(\rho) = e^{-\rho/2}\rho^{l+1}F_{nl}(\rho) = A_{nl}e^{-\rho/2}\rho^{l+1}\sum_{i=0}^{n-l-1} C_i\rho^i$$

$$C_{i+1} = \Gamma_{il}C_i, \qquad \rho \equiv 2\kappa_n r, \qquad \kappa_n = \frac{Z}{a_0 n}$$

where A_{nl} is a normalization constant. The polynomials $F_{nl}(\rho)$ (of order $n - l - 1$) so obtained are better known as the *associated Laguerre polynomials*, L_{n-l-1}^{2l+1} (see Table 10.3). The reader should take note of the fact that the scale of length ρ changes

TABLE 10.3 Eigenfunctions of hydrogen in terms of associated Laguerre polynomials

The Normalized Eigenfunctions of Hydrogen ($Z = 1$)

$$\varphi_{nlm}(r, \theta, \phi) = (2\kappa)^{3/2}A_{nl}\rho^l e^{-\rho/2}F_{nl}(\rho)Y_l{}^m(\theta, \phi) = R_{nl}(r)Y_l{}^m(\theta, \phi)$$

$$\rho = 2\kappa r = \frac{2Z}{a_0 n}r \qquad \int_0^\infty |R_{nl}(r)|^2 r^2\, dr = 1$$

$$A_{nl} = \sqrt{\frac{(n-l-1)!}{2n[(n+l)!]^3}} \qquad \boxed{\varphi_{100} = \frac{1}{\sqrt{8\pi}}\left(\frac{2Z}{a_0}\right)^{3/2}e^{-(Z/a_0)r}}$$

$$F_{nl}(\rho) = L_{n-l-1}^{2l+1}(\rho) = L_{i_{max}}^{2l+1}(\rho) = \sum_{i=0}^{n-l-1}\frac{(-1)^i[(n+l)!]^2\rho^i}{i!(n-l-1-i)!(2l+1+i)!}$$

Associated Laguerre Polynomials $L_p^q(\rho)$ and Laguerre Polynomials $L_p(\rho)$

Differential equation:

$$\left[\rho\frac{d^2}{d\rho^2} + (q + 1 - \rho)\frac{d}{d\rho} + p\right]L_p^q(\rho) = 0$$

(Continued)

TABLE 10.3 (*Continued*)

Generating function:

$$\frac{e^{-\rho s/(1-s)}}{(1-s)^{q+1}} = \sum_{p=0}^{\infty} \frac{s^p}{(p+q)!} L_p^q(\rho), \qquad L_0^p(0) = p!$$

Orthonormality:

$$\int_0^{\infty} e^{-\rho} \rho^q L_p^q L_{p'}^q \, d\rho = \frac{[(p+q)!]^3}{p!} \delta_{pp'}$$

Rodrigues' formula:

$$L_p(\rho) \equiv L_p^0(\rho) = e^{\rho} \frac{d^p}{d\rho^p} (\rho^p e^{-\rho}), \qquad L_1(\rho) = 1 - \rho, \qquad L_2(\rho) = 2!\left(1 - 2\rho + \frac{\rho^2}{2}\right)$$

$$L_p^q(\rho) \equiv (-1)^q \frac{d^q}{d\rho^q} [L_{q+p}(\rho)]$$

Recurrence relations:

$$\rho L_p^q(\rho) = (2p + q + 1)L_p^q(\rho) - [(p+1)/(p+q+1)]L_{p+1}^q(\rho) - (p+q)^2 L_{p-1}^q(\rho)$$

$$\left(\rho \frac{d}{d\rho} + q - \rho\right)L_p^q(\rho) = (p+1)L_{p+1}^{q-1}(\rho)$$

$$\frac{d}{d\rho} L_p^q(\rho) = -L_{p-1}^{q+1}(\rho)$$

Relation to Other Notations

Alternative notations for the polynomial L_p^q may be found in other texts. The relation between this notation (B) and our own (A) is given by the following table.[a]

Notation A (e.g., found in Merzbacher, Messiah, and here)	Notation B (e.g., found in Pauling and Wilson, Schiff, and Tomonaga)
$(-)^q L_p^q$	L_{p+q}^q
$\rho(L_p^q)'' + (q + 1 - \rho)(L_p^q)' + pL_p^q = 0$	$\rho(L_{p+q}^q)'' + (q + 1 - \rho)(L_{p+q}^q)' + pL_{p+q}^q = 0$
$R_{nl} = A(2\kappa)^{3/2} e^{-\rho/2} \rho^l L_{n-l-1}^{2l+1}(\rho)$	$R_{nl} = -A(2\kappa)^{3/2} e^{-\rho/2} \rho^l L_{n+l}^{2l+1}(\rho)$
L_p^q is a polynomial of order p	L_{p+q}^q is a polynomial of order p or, equivalently, L_b^q is a polynomial of order $(b - a)$

The first row in this table tells us that L_p^q is written L_{p+q}^q in notation B. The second row indicates that both L functions satisfy the same differential equation. The third row gives the forms of the radial solution R_{nl} in both notations. Still another notation appears in I. S. Gradshteyn and I. M. Ryzhik,[b] where

$$L_p^q(\rho)[\text{here}] = (p + q)! L_p^q(\rho)[\text{G and R}]$$

---

[a] E. Merzbacher, *Quantum Mechanics*, 2nd ed., Wiley, New York, 1970.

A. Messiah, *Quantum Mechanics*, Wiley, New York, 1966.

L. Pauling and E. B. Wilson, *Introduction to Quantum Mechanics*, McGraw-Hill, New York, 1935.

L. Schiff, *Quantum Mechanics*, 3rd ed., McGraw-Hill, New York, 1968.

S. Tomonaga, *Quantum Mechanics*, North-Holland, Amsterdam, 1966.

[b] I. S. Gradshteyn and I. M. Ryzhik, *Tables of Integrals, Series and Products*, Academic Press, New York, 1965.

with different values of n. This is due to the radial displacement r being nondimensionalized through the wavenumber κ_n, which is dependent on n.

Degeneracy

Since $i_{max} \geq 0$, with (10.111) we obtain

$$l \leq n - 1$$

So for a given value of the *principal quantum number n*, the *orbital quantum number l* cannot exceed the value

(10.113) $$l_{max} = n - 1$$

This corresponds to the values $l = 0, 1, 2, \ldots, (n - 1)$. Each of these l values corresponds to different values of i_{max} and therefore different wavefunctions. Inasmuch as the eigenenergy E_n depends only on the principal quantum number n, these n distinct orbital states are degenerate. For instance, there are three distinct radial functions that correspond to the eigenenergy E_3. These are $u_{3,0}$, $u_{3,1}$, and $u_{3,2}$.

The complete eigenstate of the Hamiltonian (10.101) contains the factor $Y_l^m(\theta, \phi)$ [see (10.94)]. For each value of l, there are $2l + 1$ values of m_l: $m_l = -l, \ldots, +l$, which correspond to distinct Y_l^m functions that give the same eigenvalues of $\hat{L}^2$ [i.e., $\hbar^2 l(l + 1)$]. All these $2l + 1$ functions when substituted into (10.102) give the same radial equation, (10.103), which contains only the orbital number l. It follows that for each solution u_{nl} of (10.106), there are $(2l + 1)$ solutions to the Schrödinger equation (10.102) corresponding to the same eigenenergy E_n (see Table 10.4). In this manner we obtain[1]

(10.114) $$\text{degeneracy of } E_n = \sum_{l=0}^{n-1} (2l + 1) = n^2$$

To recapitulate, the allowed values of n, l, and m are (see Fig. 10.14)

$$n = 1, 2, 3, \ldots$$

(10.115) $$l = 0, 1, 2, \ldots, (n - 1)$$

$$m = -l, -l + 1, \ldots, 0, 1, 2, \ldots, +l$$

---

[1] Including spin, degeneracy of states is $2n^2$. This topic is more fully discussed in Section 12.4.

TABLE 10.4 Allowed values of l and m_l for $n = 1, 2, 3$

n	1	2		3		
l	0	0	1	0	1	2
Spectroscopic notation of state	1S	2S	2P	3S	3P	3D
m_l	0	0	$-1, 0, +1$	0	$-1, 0, +1$	$-2, -1, 0, +1, +2$
Degeneracy of state (n^2)	1	4		9		

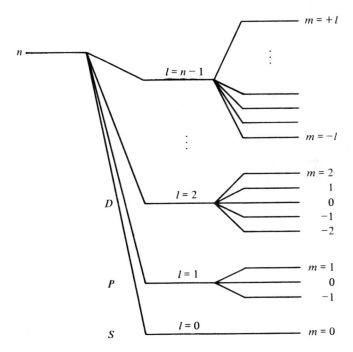

FIGURE 10.14 Term diagram for a hydrogenic atom illustrating all n^2 degenerate states corresponding to the principal quantum number n.

Additional Properties of the Eigenstates

The eigenfunctions and eigenenergies of the hydrogenic Hamiltonian (10.101) are

(10.116)

$$\varphi_{nlm}(r, \theta, \phi) = R_{nl}(r)Y_l^m(\theta, \phi)$$

$$R_{nl} = \frac{A_{nl}u_{nl}}{r}$$

$$E_n = -\frac{Z^2 \mathbb{R}}{n^2} = -\frac{\mu(Ze^2)^2}{2\hbar^2 n^2}$$

A term diagram of these energies is given in Fig. 10.15 (compare Fig. 2.8). The normalization constant A_{nl} (see Table 10.3) is determined by the condition

$$\langle \varphi_{nlm} | \varphi_{nlm} \rangle = \int_{4\pi} d\Omega \int_0^\infty r^2 \, dr \, \varphi_{nlm}{}^* \varphi_{nlm}$$

$$= |A_{nl}|^2 \int_0^\infty |u_{nl}|^2 \, dr = 1$$

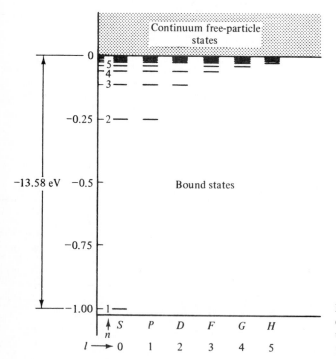

FIGURE 10.15 Energy-level diagram for hydrogen, including the $l \leq 5$ terms. Energy is measured in units of $\mathbb{R}$.

Note also the orthogonality of these functions

$$\langle \varphi_{n'l'm'} | \varphi_{nlm} \rangle = \delta_{nn'} \delta_{ll'} \delta_{mm'}$$

The Ground State

To construct the *ground state* φ_{100} ($n = 1$, $l = 0$, $m = 0$) we must first find u_{10}. From (10.112), with $C_0 = 1$, and inserting the normalization constant[1] A_{10}, one obtains

$$u_{10} = A_{10} e^{-\rho/2} \rho$$

Normalization gives

$$A_{10}{}^2 \frac{1}{2\kappa_1} \int_0^\infty \rho^2 e^{-\rho} \, d\rho = 1$$

$$A_{10}{}^2 = \frac{1}{a_0}$$

This gives the normalized ground-state wavefunction

$$u_{10} = \frac{1}{\sqrt{a_0}} \rho e^{-\rho/2} = \frac{2r}{a_0{}^{3/2}} e^{-r/a_0}$$

$$R_{10} = \frac{u_{10}}{r} = \frac{2}{a_0{}^{3/2}} e^{-r/a_0}$$

$$\varphi_{100} = R_{10} Y_0{}^0 = \frac{2}{(4\pi)^{1/2} a_0{}^{3/2}} e^{-r/a_0}$$

Additional Properties

The first few normalized eigenstates of hydrogen, with corresponding eigenenergies obtained as outlined above, are listed in Table 10.5.

In Fig. 10.16 nondimensionalized radial functions, $\bar{R}_{nl} = a_0{}^{3/2} R_{nl}$, are plotted together with the corresponding nondimensionalized probability density functions, $\bar{P}_{nl} = 4\pi a_0 u_{nl}{}^2 = 4\pi a_0 P_r$. These sketches reveal the shell structure of hydrogen found earlier in the Bohr theory.

The time development of the states of hydrogen follows from (3.70). Consider that

(10.117)
$$\psi(\mathbf{r}, 0) = \varphi_{nlm} = R_{nl} Y_l{}^m$$

[1] Alternatively, we may take $C_0 = A_{10}$.

TABLE 10.5

Spectroscopic Notation	Several Normalized Time-Independent Eigenstates of Hydrogen
1S	$\varphi_{100} = \dfrac{2}{a_0^{3/2}} e^{-r/a_0} Y_0{}^0(\theta, \phi)$
2S	$\varphi_{200} = \dfrac{2}{(2a_0)^{3/2}} (1 - r/2a_0) e^{-r/2a_0} Y_0{}^0(\theta, \phi)$
2P	$\begin{pmatrix} \varphi_{211} \\ \varphi_{210} \\ \varphi_{21-1} \end{pmatrix} = \dfrac{1}{\sqrt{3}(2a_0)^{3/2}} \dfrac{r}{a_0} e^{-r/2a_0} \begin{pmatrix} Y_1{}^1(\theta, \phi) \\ Y_1{}^0(\theta, \phi) \\ Y_1{}^{-1}(\theta, \phi) \end{pmatrix}$
3S	$\varphi_{300} = \dfrac{2}{3(3a_0)^{3/2}} [3 - 2r/a_0 + 2(r/3a_0)^2] e^{-r/3a_0} Y_0{}^0(\theta, \phi)$
3P	$\begin{pmatrix} \varphi_{311} \\ \varphi_{310} \\ \varphi_{31-1} \end{pmatrix} = \dfrac{4\sqrt{2}}{9(3a_0)^{3/2}} \dfrac{r}{a_0} (1 - r/6a_0) e^{-r/3a_0} \begin{pmatrix} Y_1{}^1(\theta, \phi) \\ Y_1{}^0(\theta, \phi) \\ Y_1{}^{-1}(\theta, \phi) \end{pmatrix}$
3D	$\begin{pmatrix} \varphi_{322} \\ \varphi_{321} \\ \varphi_{320} \\ \varphi_{32-1} \\ \phi_{32-2} \end{pmatrix} = \dfrac{2\sqrt{2}}{27\sqrt{5}(3a_0)^{3/2}} \left(\dfrac{r}{a_0}\right)^2 e^{-r/3a_0} \begin{pmatrix} Y_2{}^2(\theta, \phi) \\ Y_2{}^1(\theta, \phi) \\ Y_2{}^0(\theta, \phi) \\ Y_2{}^{-1}(\theta, \phi) \\ Y_2{}^{-2}(\theta, \phi) \end{pmatrix}$

The state at time $t \geq 0$ is then

(10.118)
$$\psi(\mathbf{r}, t) = e^{-i\hat{H}t/\hbar}\psi(\mathbf{r}, 0) = e^{-iE_{nt}/\hbar}\varphi_{nlm}$$

The charge density associated with this state is

(10.119)
$$q(\mathbf{r}, t) = e|\psi_{nlm}(\mathbf{r}, t)|^2 = q(\mathbf{r}) = e|\varphi_{nlm}(\mathbf{r})|^2$$

which is independent of time. The electronic charge is e. Thus the atom suffers no radiation in these states. This topic will be returned to in the next section.

The density configurations, $|\varphi_{nlm}|^2$, corresponding to some of the eigenstates of hydrogen are sketched in Fig. 10.17. Since the angular dependence of $|\varphi_{nlm}|^2$ is entirely contained in the factor $|Y_l^m|^2$, it follows that $|\varphi_{nlm}|^2$ is independent of the azimuthal angle ϕ [see (9.78)]. It is rotationally symmetric about the z axis. Thus we need only present a representation of $|\varphi|^2$ in any plane which includes the z axis, such as is depicted in Fig. 10.17. The value of $|\varphi|^2$ is proportional to the density of whiteness in each of the states depicted.

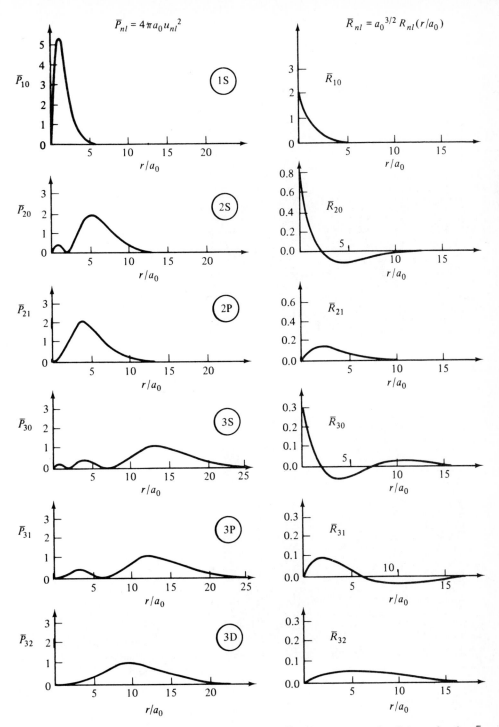

FIGURE 10.16 Nondimensional radial probability density $\bar{P}$ and nondimensional radial wavefunction $\bar{R}$, vs. nondimensional radius r/a_0, for hydrogen. Note that the probability density $\bar{P}$ exhibits the shell structure of the atom.

1S	2P	3D	4F

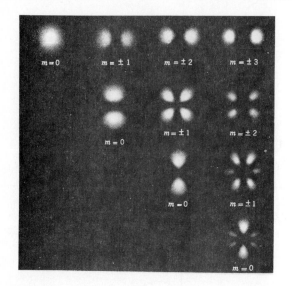

2S	3P	4D	5F

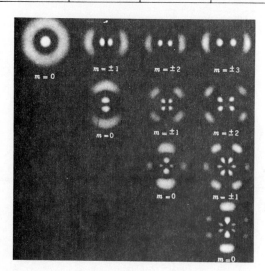

FIGURE 10.17 Probability density, $|\varphi_{nlm}|^2$, for various states of hydrogen. The plane of the paper contains the polar axis, which points from the bottom to the top of the figure. From *Principles of Modern Physics* by R. B. Leighton. Copyright 1959 by McGraw-Hill. Used with permission of the McGraw-Hill Book Company.

Hybridization

Having discovered the wavefunctions of hydrogen, a significant consequence of quantum mechanics emerges important to the formation of molecules. This property concerns the manner in which atomic wavefunctions exhibit geometric orientation in atom–atom binding.

Consider the $2P$ states of hydrogen listed in Table 10.5. Owing to their degeneracy, any linear combination of these states is an eigenstate corresponding to the same eigenenergy, $-\mathbb{R}/4$. Thus, for example, the following set of three orthogonal wavefunctions span the same subdimensional Hilbert space as the original three $2P$ states:

(10.120a)
$$\varphi_{2P_x} = -\frac{1}{\sqrt{2}}(\varphi_{211} - \varphi_{21-1}) = Axe^{-r/2a_0}$$

(10.120b)
$$\varphi_{2P_y} = \frac{1}{\sqrt{2}}(\varphi_{211} + \varphi_{21-1}) = -iAye^{-r/2a_0}$$

(10.120c)
$$\varphi_{2P_z} = \varphi_{210} = Aze^{-r/2a_0}$$

Angular plots (at fixed r) of the corresponding probability densities $|\varphi_{2P_x}|^2$, $|\varphi_{2P_y}|^2$, $|\varphi_{2P_z}|^2$ reveal that they are figure-eight surfaces of revolution about the x, y, and z axes, respectively (similar to those shown in Fig. 9.11).

Linear combinations of the wavefunctions (10.120) come into play in the formation of the methane molecule CH_4. The four outer-shell electrons of carbon are in the $2s^2 2p^2$ configuration (see Table 12.2). Wavefunctions appropriate to these four electrons are formed from linear combinations of the three $2P$ states (10.120) together with the $2S$ state of hydrogen and are given by

(10.121a)
$$\psi_1 = \frac{1}{\sqrt{4}}(\varphi_{2S} + \varphi_{2P_x} + \varphi_{2P_y} + \varphi_{2P_z})$$

(10.121b)
$$\psi_2 = \frac{1}{\sqrt{4}}(\varphi_{2S} + \varphi_{2P_x} - \varphi_{2P_y} + \varphi_{2P_z})$$

(10.121c)
$$\psi_3 = \frac{1}{\sqrt{4}}(\varphi_{2S} + \varphi_{2P_x} - \varphi_{2P_y} - \varphi_{2P_z})$$

(10.121d)
$$\psi_4 = \frac{1}{\sqrt{4}}(\varphi_{2S} - \varphi_{2P_x} + \varphi_{2P_y} - \varphi_{2P_z})$$

With the four outer electrons of carbon in these respective four ψ orbitals, the following picture emerges. Angular plots of the probability densities of these wavefunctions about a common origin reveal maxima along the $(1, 1, 1)$, $(-1, -1, 1)$,

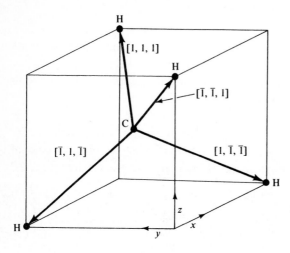

FIGURE 10.18   Orientation of C and H atoms in the CH_4 molecule. Here we have employed crystallographic notation for direction, where a bar over a direction number indicates negative direction.

$(1, -1, -1)$, and $(-1, 1, -1)$ directions, respectively. With the carbon atom at the center of a cube, these orbitals reach out to four tetrahedral corners and covalently bond with hydrogen atoms at these sites to form the CH_4 molecule. See Fig. 10.18.

These orbitals come into play in the chaining of CH_3 molecules. Thus, for example, in the formation of C_2H_6, an uncoupled electronic ψ orbital of CH_3 bonds with a like orbital of an adjacent CH_3 molecule to form the ethane molecule. The concept of atomic binding in the formation of molecules is returned to in Chapter 12.

PROBLEMS

10.32 With $C_0 = 1$ in the recurrence relation (10.109), obtain C_1. Then use (10.112) to show that

$$u_{20} = A_{20}re^{-r\,2a_0}\left(1 - \frac{r}{2a_0}\right)$$

Calculate A_{20} and φ_{200}. Check your answer with the value given in Table 10.5.

10.33 Show that $u_{10}{}^2$ has its maximum at $r = a_0$, the Bohr radius.

10.34 Solve the equation $f_x + f = 0$ by expansion technique and check with the solution $f = Ae^{-x}$.

Answer

Assume that

$$f = \sum_0^\infty C_i x^i$$

to obtain

$$\sum_0^\infty C_i i x^{i-1} + \sum_0^\infty C_i x^i = 0$$

With $s = i - 1$ in the first series, we get

$$\sum_{s=-1}^{\infty} C_{s+1}(s + 1)x^s + \sum_{i=0}^{\infty} C_i x^i = 0$$

Since the first term in the first series is zero, we may write this equation in the form

$$\sum_{i=0}^{\infty} [C_{i+1}(i + 1) + C_i]x^i = 0$$

which is satisfied if and only if

$$C_{i+1} = -\frac{C_i}{i + 1}$$

10.35 The average energy of the hydrogen atom in an arbitrary bound state $\chi(\mathbf{r})$ is given by the integral

$$\langle E \rangle = \langle \chi(\mathbf{r}) | \hat{H} | \chi(\mathbf{r}) \rangle$$

Show that

$$\langle E \rangle \geq \langle 100 | \hat{H} | 100 \rangle = E_1$$

Answer

Since $\chi(\mathbf{r})$ lies in the Hilbert space spanned by the basis $\{\varphi_{nlm}\}$, we may expand

$$|\chi\rangle = \sum_{n=1}^{\infty} \sum_{l=0}^{n-1} \sum_{m=-l}^{l} b_{nlm} |nlm\rangle$$

so that

$$\langle \chi | \hat{H} | \chi \rangle = \sum_n \sum_l \sum_m |b_{nlm}|^2 E_n$$

$$\sum \sum \sum |b|^2 = 1$$

(The ket vector $|nlm\rangle$ represents the state φ_{nlm}.) Owing to the fact that all eigenenergies are negative, $E_n \leq 0$, the statement to be proven is equivalent to the inequality

$$|\langle E \rangle| \leq |E_1|$$

$$|\langle E \rangle| = \sum \sum \sum |b|^2 |E_n|$$

$$\leq |E|_{max} \sum \sum \sum |b|^2 = |E|_{max} = |E_1|$$

10.36 (a) What is the effective Bohr radius and ground-state energy for each of the following two-particle systems?

(1) H^2, a deuteron and an electron (heavy hydrogen).

(2) He^+, a singly ionized helium.

(3) Positronium, a bound positron and electron.

(4) Mesonium, a proton and negative μ meson. The μ meson has mass $207m_e$ and lasts $\sim 10^{-6}$ s.

(5) Two neutrons bound together by their gravitational field.

(b) Calculate the frequencies of the $(n = 2) \to (n = 1)$ transition for each of the systems above.

10.37 At time $t = 0$, a hydrogen atom is in the superposition state

$$\psi(\mathbf{r}, 0) = \frac{4}{(2a_0)^{3/2}} \left[\frac{e^{-r/a_0}}{\sqrt{4\pi}} + A \frac{r}{a_0} e^{-r/2a_0}(-iY_1{}^1 + Y_1{}^{-1} + \sqrt{7}\, Y_1{}^0) \right]$$

(a) Calculate the value of the normalization constant A.

(b) What is the probability that measurement of L^2 finds the value $\hbar^2 l(l + 1)$?

(c) What is the probability density $P_r(r)$ [see (10.99)] that the electron is found in the shell of thickness dr about the proton at the radius r?

(d) At what value of r is $P_r(r)$ maximum?

(e) Given the initial state $\psi(\mathbf{r}, 0)$, what is $\psi(\mathbf{r}, t)$?

(f) What is $\psi(\mathbf{r}, t)$ if at $t = 0$, measurement of L_z finds the value $\hbar$?

(g) What is $\psi(\mathbf{r}, t)$ if at $t = 0$, measurement of L_z finds the value zero?

(h) What is the expectation of the "spherical energy operator," $\langle H_S \rangle$, where

$$\hat{H}_S \equiv \hat{H} - \frac{\hat{p}_r{}^2}{2\mu}$$

at $t = 0$?

(i) What is the lowest value of energy that measurement will find at $t = 0$? (Lowest means the negative value farthest removed from zero.)

10.38 Find the lowest energy and the smallest value for the classical turning radius of the H-atom electron in the state with $l = 6$ (see Fig. 10.11).

10.39 In what sense does the Bohr analysis of the hydrogen atom give erroneous results for the magnitude of angular momentum, L?

Answer

The Bohr analysis that yields the eigenenergies $-\mathbb{R}/n^2$ assumes circular orbits. Circular orbits do not exist in the Schrödinger theory. Quantization of the action, $\oint p_\theta\, d\theta$, in the Bohr theory gives $L = n\hbar$. In the Schrödinger theory, the maximum value of L is $\hbar\sqrt{n(n - 1)}$, which is less than the value that L assumes $(n\hbar)$ in the Bohr theory.

10.40 What is the ionization energy of a hydrogen atom in the $3P$ state?

10.41 Show that $R_{nl}(r)$ has $(n - l - 1)$ zeros (not counting zeros at $r = 0$ and $r = \infty$).

10.42 (a) Show that the expectation of the interaction potential $V(r)$ for hydrogenic atoms is

$$\langle nlm | V(r) | nlm \rangle = -\left\langle \frac{Ze^2}{r} \right\rangle = -\frac{\mu Z^2 e^4}{\hbar^2 n^2} = -\frac{2Z^2 \mathbb{R}}{n^2}$$

(b) Calculate $\langle nlm | T | nlm \rangle$, where the kinetic-energy operator $\hat{T}$ is given

$$\hat{T} = \frac{\hat{p}_r{}^2}{2\mu} + \frac{\hat{L}^2}{2\mu r^2}$$

What relation do $\langle T \rangle$ and $\langle V \rangle$ satisfy? (*The virial theorem.*)

10.43   Obtain an explicit expression for the probability density $P_r(r)$ corresponding to the state whose energy is E_2, for a hydrogenic atom [see (10.99)].

Answer

There are four degenerate eigenstates corresponding to the energy E_2. Since no direction is preferred for a Hamiltonian whose only interaction term is the central potential $V(r)$, all these degenerate states carry the same "weight" (all lm states are equally probable). There results

$$4\pi P(r)r^2\, dr = \int_{4\pi} \frac{1}{4}\left[\varphi_{200}{}^*\varphi_{200} + \varphi_{21-1}{}^*\varphi_{21-1} + \varphi_{210}{}^*\varphi_{210} + \varphi_{211}{}^*\varphi_{211}\right]r^2\, dr\, d\Omega$$

$$= \int_{4\pi} \frac{1}{128\pi}\left(\frac{Z}{a_0}\right)^3 e^{-Zr/a}\left[\left(2 - \frac{Zr}{a_0}\right)^2 + \left(\frac{Zr}{a_0}\right)^2\left(\frac{1}{2}\sin^2\theta + \frac{1}{2}\sin^2\theta + \cos^2\theta\right)\right]$$

$$\times r^2\, dr\, d\Omega$$

$$P(r) = \frac{1}{128\pi}\left(\frac{Z}{a_0}\right)^3 e^{-Zr/a_0}\left[\left(2 - \frac{Zr}{a_0}\right)^2 + \left(\frac{Zr}{a_0}\right)^2\right], \qquad P_r(r) = 4\pi r^2 P(r)$$

10.44   Give a physical argument in support of the conjecture that the sum

$$\Sigma_n \equiv \sum_{l=0}^{n-1}\sum_{m=-l}^{+l}\left[Y_l^m\right]^*\left[Y_l^m\right]\left[R_{nl}\right]^2$$

is independent of θ or ϕ.

10.45   Show that for a hydrogen atom in the state corresponding to maximum orbital angular momentum ($l = n - 1$),

$$\langle n, n - 1|r|n, n - 1\rangle = a_0 n(n + \tfrac{1}{2})$$

$$\langle n, n - 1|r^2|n, n - 1\rangle = a_0{}^2 n^2(n + 1)(n + \tfrac{1}{2})$$

10.46   Use the results of Problem 10.45 to show that for large values of n and l,

$$\sqrt{\langle r^2\rangle} \to a_0 n^2$$

$$\frac{\Delta r}{\langle r\rangle} \to 0$$

$$E_n \to -\frac{1}{2}\frac{e^2}{n^2 a_0}$$

That is, show that for large values of n, the electron is localized near the surface of a sphere of radius $a_0 n^2$ and has energy which is the same as that of a classical electron in a circular orbit of the same radius. Recall: $(\Delta r)^2 = \langle r^2\rangle - \langle r\rangle^2$.

10.47   Calculate $\langle \mathbf{r} \rangle$ in the state φ_{nlm} of hydrogen.

Answer

$$\langle \mathbf{r} \rangle = \mathbf{e}_x \langle x \rangle + \mathbf{e}_y \langle y \rangle + \mathbf{e}_z \langle z \rangle$$

$$\langle x \rangle = \iiint r \cos \phi \sin \theta |Y_l^m|^2 R_{nl}^2 r^2 \, dr \, d \cos \theta \, d\phi$$

$$= 0$$

since

$$\int_0^{2\pi} \cos \phi \, d\phi = 0$$

and $|Y_l^m|^2$ is independent of ϕ. Similarly, $\langle y \rangle = 0$. For $\langle z \rangle$ we must calculate

$$\langle z \rangle = \int_{-1}^{1} |P_l^m|^2 \cos \theta \, d \cos \theta \int \cdots$$

Using the recurrence relations listed in Table 9.3, we find that $\langle z \rangle = 0$. It follows that $\langle \mathbf{r} \rangle = 0$.

10.48   Establish the following properties for hydrogen in the stationary state φ_{nlm}.

(a) $\dfrac{s+1}{n^2} \langle r^s \rangle - (2s+1)a_0 \langle r^{s-1} \rangle + \dfrac{s}{4}[(2l+1)^2 - s^2]a_0^2 \langle r^{s-2} \rangle = 0, \quad s > -2l - 1$

(b) $\langle r \rangle = n^2 \left[1 + \dfrac{1}{2}\left(1 - \dfrac{l(l+1)}{n^2}\right) \right] a_0$

(c) $\left\langle \dfrac{1}{r^2} \right\rangle = \dfrac{2}{(2l+1)n^3 a_0^2}$

(d) $\langle r^2 \rangle = \dfrac{1}{2}[5n^2 + 1 - 3l(l+1)]n^2 a_0^2$

(e) $\left\langle \dfrac{1}{r} \right\rangle = \dfrac{1}{n^2 a_0}$

(f) $\left\langle \dfrac{1}{r^3} \right\rangle = \dfrac{2}{a_0^3 n^3 l(l+1)(2l+1)}$

[*Hint:* Multiply (10.106) by $\{\rho^{s+1} u' + [(s+1)/2]\rho^s u\}$ and integrate by parts several times. Note that for hydrogenic atoms, a_0 is replaced by a_0/Z.]

10.49   Show that the most probable values of r for the $l = n - 1$ states of hydrogen are

$$\tilde{r} = n^2 a_0$$

These are values that satisfy the equation

$$\frac{d}{dr}(u_{nl})^2 = 0$$

10.7 ELEMENTARY THEORY OF RADIATION

In the last section we found that the hydrogen atom does not radiate in its eigen (stationary) states. The charge density (10.119) is fixed in space with configurations such as depicted in Fig. 10.17. In these states the hydrogen atom is stable against radiation. This is opposed to the classical description in which the electron loses kinetic energy to the radiation field and collapses to the nucleus (see Section 2.1).

The student may be perplexed about the absence of radiation from the state ψ_{nlm}. He/she may well ask: Doesn't the electron have a well-defined angular momentum in such a state, and doesn't this correspond to accelerated motion which gives rise to radiation? His/her friend answers: Maybe the orbit of the electron is so peculiar that, on the average, the radiation field washes out. After all, we know that if L^2 is specified, two of the three components of $\mathbf{L}$ remain uncertain.

The best way to see what the electron is doing in quantum mechanics is to calculate $\langle \mathbf{r} \rangle$. Specifically, we must calculate this expectation in the state ψ_{nlm}. Suppose that we find $\langle \mathbf{r} \rangle \sim \mathbf{e}_z \cos \omega t$. Then the electron is suffering linear, simple harmonic oscillation. Such oscillation gives rise to *dipole radiation*. But we have already calculated $\langle \mathbf{r} \rangle$ in Problem 10.47, where we found that $\langle \mathbf{r} \rangle = 0$. Not only is $\langle \mathbf{r} \rangle$ time-independent in the eigenstates of hydrogen, but it is also centered at the origin. Note that we may also reach this conclusion by the much simpler argument: Calculate

$$\langle \mathbf{r} \rangle = \langle \psi_{nlm} | \mathbf{r} | \psi_{nlm} \rangle = \iiint \mathbf{r} | \varphi_{nlm} |^2 \, d\mathbf{r}$$

The average of $\mathbf{r}$ is independent of time, hence it must also be zero since the Hamiltonian, (10.101), is isotropic. It contains no vectors. It in no way implies a "preferred" direction, so $\langle \mathbf{r} \rangle$ cannot be a finite constant vector.

Thus while the stability of the hydrogen atom to radiative collapse is totally inexplicable on classical grounds, our quantum mechanical model renders a denumerably infinite set of states $\{\psi_{nlm}\}$ in which the atom suffers no radiation.

How, then, does the atom radiate? In the Bohr theory of radiation, we recall that a photon is emitted when there is a transition from one eigenenergy state to a lower one. Such a decay might be induced by the collision of the atom with another atom in a gas. It might also be induced by collision with an electron in a discharge tube. It might also be induced by collision with a photon in the interior of a star.[1]

Suppose that at time $t = 0$, the atom is in an excited (stationary) state ψ_n (n denotes the sequence nlm). The atom is perturbed, emits radiation, and decays to the

---

[1] Fundamentally, all these collision processes involve the exchange of photons. For further discussion, see E. G. Harris, *A Pedestrian Approach to Quantum Field Theory*, Wiley, New York, 1972.

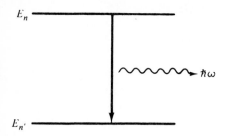

FIGURE 10.19   **Atom emits a photon in decaying from state ψ_n to ψ_n.**

state $\psi_{n'}$ (Fig. 10.19). We may conclude that in the interim, the atom is in the super-position state

(10.122) $$\psi = a\psi_n + b\psi_{n'} \qquad |a|^2 + |b|^2 = 1$$

At any time t, in this interim (by the superposition principle) $|a|^2$ represents the probability that the atom is in the state ψ_n and $|b|^2$ represents the probability that it is in the state $\psi_{n'}$. These coefficients are therefore time-dependent. At $t = 0$, $(|a| = 1, |b| = 0)$. At $t = \infty, (|a| = 0, |b| = 1)$. Let us calculate the expected value of the position of the electron during this collapse.

$$\langle \mathbf{r} \rangle = \langle a\psi_n + b\psi_{n'}|\mathbf{r}|a\psi_n + b\psi_{n'}\rangle$$
$$= |a|^2\langle\psi_n|\mathbf{r}|\psi_n\rangle + |b|^2\langle\psi_{n'}|\mathbf{r}|\psi_{n'}\rangle$$
$$+ a^*b\langle\psi_n|\mathbf{r}|\psi_{n'}\rangle + b^*a\langle\psi_{n'}|\mathbf{r}|\psi_n\rangle$$

The first two terms are time independent and do not contribute to radiation. The last two terms combine to yield

(10.123) $$\langle \mathbf{r}(t)\rangle = a^*be^{i(E_n - E_{n'})t/\hbar}\langle\varphi_n|\mathbf{r}|\varphi_{n'}\rangle + b^*ae^{i(E_{n'} - E_n)t/\hbar}\langle\varphi_{n'}|\mathbf{r}|\varphi_n\rangle$$
$$= 2 \operatorname{Re}\left[a^*b\langle\varphi_n|\mathbf{r}|\varphi_{n'}\rangle e^{i(E_n - E_{n'})t/\hbar}\right]$$
$$= 2|a^*b\langle\varphi_n|\mathbf{r}|\varphi_{n'}\rangle|\cos(\omega_{nn'}t + \delta) = 2|\mathbf{r}_{nn'}|\cos(\omega_{nn'}t + \delta)$$

where $\omega_{nn'}$ is the Bohr frequency

$$\hbar\omega_{nn'} = E_n - E_{n'}$$

δ is a phase factor and $|a^*b|$ is assumed to be slowly varying and of order unity.

Atomic transitions typically occur in an interval of the order of 10^{-9} s. The frequency of emitted radiation, on the other hand, is typically of the order of 10^{15} s^{-1}, so the radiative oscillatory behavior of $\langle \mathbf{r}(t)\rangle$ is due almost exclusively to the cos term, with accompanying Bohr frequency $\omega_{nn'}$ (Fig. 10.20). When the atom is undergoing a transition between the states ψ_n and $\psi_{n'}$, the average position of the electron oscillates with the Bohr frequency corresponding to the energy difference between these states. At the beginning and conclusion of the transition, the atom is in stationary states in which it does not radiate. (These topics are returned to in Section 13.9.)

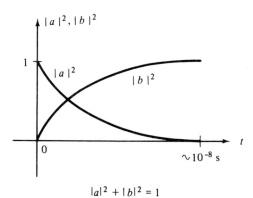

$$|a|^2 + |b|^2 = 1$$

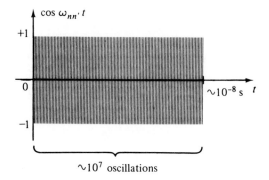

$\sim 10^7$ oscillations

FIGURE 10.20 Radiation decay from the ψ_n state to the $\psi_{n'}$ state involves the superposition state $\psi = a\psi_n + b\psi_{n'}$ with time-varying coefficients a and b. The "beat" frequency $\omega_{nn'}$ between the ψ_n and $\psi_{n'}$ states is much greater than the switchover frequency of ψ.

Selection Rules

Harmonic oscillation of an electron about a proton gives rise to what is commonly referred to as *dipole radiation*[1] (Fig. 10.21). The average radiated power from such an oscillating dipole is[2]

(10.124)
$$P = \frac{1}{3}\frac{\omega^4}{c^3}|\mathbf{d}|^2$$

where $\mathbf{d}$ is the dipole moment

$$\mathbf{d} = e\mathbf{r}_0, \qquad \langle \mathbf{r} \rangle = \mathbf{r}_0 \cos \omega t$$

[1] An atom also radiates in higher multipole channels (e.g., quadrapole). For the most part, dipole radiation is predominant.

[2] See J. D. Jackson, *Classical Electrodynamics*, 2nd ed., Wiley, New York, 1975.

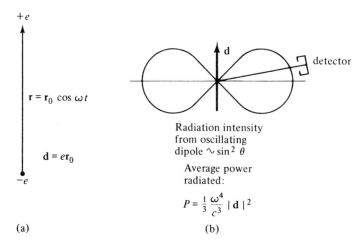

(a)   (b)

FIGURE 10.21 Energy characteristics of an oscillating dipole. (a) Dipole configuration. (b) Radiation profile of a dipole.

We may apply this formula to calculate the power radiated when the hydrogen atom decays from the nth state to the (n')th state. From (10.123) we obtain

$$\mathbf{d} = 2e\mathbf{r}_{nn'}$$

so that, with (10.124),

(10.125)
$$P = \frac{4}{3} \frac{(\omega_{nn'})^4 e^2}{c^3} |\mathbf{r}_{nn'}|^2$$

Calculation (see Problems 10.51 to 10.53) of the squared matrix element $|\mathbf{r}_{nn'}|^2$ with n standing for nlm and n' for $n'l'm'$ gives the following *selection rules*: The only conditions under which $|\mathbf{r}_{nn'}|^2$ (and therefore P) is not zero are

(10.126) $\Delta l = l' - l = \pm 1$ *and* $\Delta m_l = m' - m = 0, \pm 1$

For example, the transition $3S \rightarrow 1S$ ($\Delta l = 0$) is *forbidden*, as is the transition $3D \rightarrow 2S$ ($\Delta l = 2$). Such transitions are not accompanied by any (dipole) radiation and therefore are excluded by conservation of energy. The exclusion of the transitions between S states finds analogy with the classical theorem that spherically symmetric oscillatory charge distributions do not radiate.

The rule $\Delta l = \pm 1$, together with the law of conservation of angular momentum, indicates that for $\Delta l = -1$ the electromagnetic field (i.e., the photon) carries away angular momentum. As it turns out, photons have angular momentum quantum number equal to 1 and are therefore called *bosons*.

There are no restrictions on an atomic transition corresponding to change in the principal quantum number n. This is in agreement with the pre-Schrödinger spectral notation for emission from hydrogen: namely, the Lyman series corresponds to transitions from all n states to the ground state, the Balmer series corresponds to transitions to the $n = 2$ states, etc. (see Fig. 2.8).

PROBLEMS

10.50 What are the allowed transitions from the $5D$ states of hydrogen to lower states? Accompany your answer with a sketch representing these transitions.

10.51 With n representing the triplet nlm and n' the triplet $n'l'm'$, show that the matrix elements of **r** have the following complex representation.

$$|\mathbf{r}_{nn'}|^2 \equiv |\langle n|\mathbf{r}|n'\rangle|^2 = \tfrac{1}{2}\{|\langle n|x + iy|n'\rangle|^2 + |\langle n|x - iy|n'\rangle|^2\} + |\langle n|z|n'\rangle|^2$$

[*Hint:* Call $\langle n|x|n'\rangle \equiv x_{nn'}$, etc. Note also that $|\langle \mathbf{r}\rangle|$ denotes the magnitude of the vector $\langle \mathbf{r}\rangle$, while $|\langle x + iy\rangle|$ denotes the modulus of the complex variable $\langle x + iy\rangle$.]

10.52 Show that the matrix elements of $x \pm iy$ have the following integral representation:

(a) $\langle n|x \pm iy|n'\rangle = \displaystyle\int_{4\pi} [Y_l^m]^* Y_{l'}^{m'} \sin\theta \, e^{\pm i\phi} \, d\Omega \int_0^\infty R_{nl} R_{n'l'} r^3 \, dr$

(b) $\langle n|z|n'\rangle = \displaystyle\int_{4\pi} [Y_l^m]^* Y_{l'}^{m'} \cos\theta \, d\Omega \int_0^\infty R_{nl} R_{n'l'} r^3 \, dr$

10.53 Using the results of the last two problems and Tables 9.1 and 9.3, establish the selection rules for dipole radiation (10.126).

10.54 At the start of Section 10.7 it was noted that the energy eigenstates of an atom are stable against radiative decay, owing to their stationarity in time. However, these states do contain angular momentum, and one may argue that they therefore also contain rotating charge, which does radiate. Although the premise of this argument is correct, why is there still no radiation from the stationary states (from a classical point of view)?

Answer

An aggregate of N uniformly spaced point charges confined to move with fixed speed in a closed loop will radiate as a result of the acceleration of individual charges. However, in the limit that the charges approach a uniformly continuous distribution, $N \to \infty$, $\Delta \to 0$, $q \to 0$ (with total charge Nq, and line charge density q/Δ, constant), the radiation may be shown[1] to vanish. This limiting case closely resembles the state of affairs for the stationary states of an atom. Although there is rotating charge, such charge is continuously distributed and, in accord with the classical prescription, does not radiate.

10.55 The interaction potential of an electron moving in the far field of a dipole **d** is

$$V = -\frac{ed}{r^2} \cos\theta$$

[1] See Jackson, *Classical Electrodynamics*, Chap. 14.

The dipole is at the origin and points in the z direction. The spherical coordinates of the electron are (r, θ, ϕ).

(a) Write down the time-independent Schrödinger equation for this system corresponding to zero total energy.

(b) Show that solutions to this equation are of the form $\varphi = r^s f(\theta, \phi)$. Obtain an equation for $f(\theta, \phi)$.

10.56 Two particles that are isolated from all other objects interact with each other through a central potential. As was established in Problem 10.22(d), the total angular momentum of the system may be written

$$\mathbf{L}_1 + \mathbf{L}_2 = \mathbf{L} + \mathbf{L}_{CM}$$

Show quantum mechanically that this total angular momentum is conserved.

10.57 (a) Prove that the *Runge–Lenz* vector

$$\hat{\mathbf{K}} = \frac{1}{2\mu e^2}[\mathbf{L} \times \hat{\mathbf{p}} - \hat{\mathbf{p}} \times \hat{\mathbf{L}}] + \frac{\mathbf{r}}{r}$$

commutes with the Hamiltonian of the hydrogen atom (10.101).

(b) Show that the operator

$$\hat{\mathbf{A}} = \sqrt{-\mu e^4 / 2E}\ \hat{\mathbf{K}}$$

satisfies the commutation relations

$$[\hat{A}_x, \hat{A}_y] = i\hbar \hat{L}_z, \quad \text{etc.}$$

(c) Use the result above to show that the operators

$$\hat{\mathbf{B}}_+ = \tfrac{1}{2}(\hat{\mathbf{L}} + \hat{\mathbf{A}})$$
$$\hat{\mathbf{B}}_- = \tfrac{1}{2}(\hat{\mathbf{L}} - \hat{\mathbf{A}})$$

obey the angular momentum commutation relations and the equality $\hat{B}_+{}^2 = \hat{B}_-{}^2$.

(d) Derive the relation

$$\hat{B}_+{}^2 + \hat{B}_-{}^2 = -\tfrac{1}{2}\left(\hbar^2 + \frac{\mu e^4}{2E}\right)$$

and use it to obtain the Bohr formula for the energy levels of hydrogen.

10.58 The classical harmonic oscillator with spring constant K and mass m oscillates at the *single* frequency (independent of energy)

$$\omega_0 = \sqrt{K/m}$$

The quantum mechanical oscillator, on the other hand, gives frequencies at all integral multiples of ω_0, as follows directly from the eigenenergies

$$E_n = \hbar\omega_0(n + \tfrac{1}{2}) \qquad (n = 0, 1, 2, \ldots)$$

If one end of the oscillator is charged, dipole radiation is emitted. In the classical domain, this radiation has frequency ω_0. Show that selection rules that follow from calculation of the dipole matrix elements $x_{nn'}$ reduce the quantum mechanical spectrum to the classical one. (The concept of matrix elements of an observable is developed formally in Chapter 11.)

10.59 Consider a gas of noninteracting rigid dumbbell molecules with speeds small compared to the speed of light. The moment of inertia of each molecule is I.

(a) What is the Hamiltonian of a molecule in the gas?

(b) What are the eigenenergies of this Hamiltonian?

(c) Let a molecule undergo spontaneous decay between two rotational states. Owing to the recoil of the center of mass, there is a change in momentum of the center of mass of the molecule as well. Show that the frequency of the photon emitted in this process is

$$\nu = \nu_l\left(1 - \frac{\hbar \mathbf{k} \cdot \mathbf{n}}{Mc}\right)$$

The initial momentum of the center of mass is $\hbar\mathbf{k}$, $\mathbf{n}$ is a unit vector in the direction of the momentum of the emitted photon, and $\{\nu_l\}$ is the rotational line spectrum.

(d) What is the nature of the frequency spectrum emitted by the gas?

Answers

(a) Let $\mathscr{P}$ and M denote the momentum and mass, respectively, of the center of mass. Then

$$\hat{H} = \frac{\mathscr{P}^2}{2M} + \frac{\hat{L}^2}{2I}$$

(b) $E_{k,l} = \dfrac{\hbar^2 k^2}{2M} + \dfrac{\hbar^2 l(l+1)}{2I}$

(c) The frequency of photons emitted by a molecule is given by the change in energy

$$h\nu = \Delta E = \frac{\hbar^2 \mathbf{k} \cdot \Delta \mathbf{k}}{M} + h\nu_l$$

where $h\nu_l$ is written for the change in rotational energy. Since the molecule is a free particle, the momentum of the center of mass $\hbar\mathbf{k}$ can change only by virtue of the momentum carried away by the photon emitted in the transition. As a first approximation we will assume that the momentum carried away is $h\nu_l/c$. If $\hbar\mathbf{k}'$ is the momentum of the center of mass after emission, then by conservation of momentum we have (see Fig. 10.22)

$$\hbar\mathbf{k} = \hbar\mathbf{k}' + \frac{h\nu_l}{c}\mathbf{n}$$

so that

$$\hbar\,\Delta\mathbf{k} = \hbar(\mathbf{k}' - \mathbf{k}) = -\frac{h\nu_l}{c}\,\hat{\mathbf{n}}$$

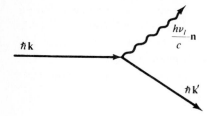

FIGURE 10.22 Change in linear momentum of rigid rotator due to recoil in the emission of a photon (see Problem 10.59).

Substituting this value in the expression above for $h\nu$ gives

$$\nu = \nu_l\left(1 - \frac{\hbar k \cdot n}{Mc}\right)$$

Since $\hbar k \ll Mc$, the assumption that momentum carried away by this radiation field is approximately $h\nu_l/c$ is justified.

(d) The rotational spectrum $\{\nu_l\}$ remains a line spectrum with an infinitesimal broadening of lines.

10.60 Consider two identical rigid spheres of diameter a.

(a) What is the Hamiltonian of this system?

(b) Separate out the center-of-mass motion to obtain an equation for the wavefunction for relative motion, $\varphi(\mathbf{r})$.

(c) What boundary conditions must $\varphi(\mathbf{r})$ satisfy?

(d) What are the eigenstates and eigenenergies for this system?

(e) What are the parities of these eigenstates?

Answers

(a) $\hat{H}(1, 2) = \dfrac{\hat{p}_1^{\,2}}{2m} + \dfrac{\hat{p}_2^{\,2}}{2m} + V(|\mathbf{r}_1 - \mathbf{r}_2|)$

$$V(|\mathbf{r}_1 - \mathbf{r}_2|) \equiv V(r) = \infty, \qquad r \le a$$
$$V(r) = 0, \qquad r > a$$

(b) $\left[\dfrac{\hat{p}_r^{\,2}}{2\mu} + \dfrac{\hat{L}^2}{2\mu r^2} + V(r) - E\right]\varphi(\mathbf{r}) = 0$

(c) $\varphi_{klm}(\mathbf{r}) = 0, \quad r \le a$

(d) $\varphi_{klm}(\mathbf{r}) = \varphi_{lm}(kr)Y_l^m(\theta, \phi), \quad r > a$

$\varphi_{lm}(kr) = A[n_l(ka)j_l(kr) - j_l(ka)n_l(kr)]$

$E_k = \dfrac{\hbar^2 k^2}{2\mu}$

(e) Referring to Problem 9.46, we obtain

$$\hat{P}\varphi_{lm}(kr) = (-)^l\varphi_{lm}(kr)$$

10.8 THOMAS–FERMI MODEL

The Thomas–Fermi model is appropriate to atoms with sufficient number of electrons which permits the system to be treated in a statistical sense. The model views an atom as a spherically symmetric gas of Z electrons which surrounds a nucleus of charge Ze. It is further assumed that the potential in the medium, $\Phi(r)$, varies sufficiently slowly so that it may be taken as constant in small volumes of the electron gas. The description of the Fermi energy, such as given in Section 2.3, then applies. (See Fig. 10.23.) There, we recall the Fermi energy was equated to the maximum kinetic energy of an electron in an aggregate of electrons at 0 K. This description maintains in the present model since at normal temperature $k_B T \ll \Phi(r)$. (The Fermi energy is discussed further in Section 12.9.)

The relation between E_F and electron density $n(r)$ is given in Problem 2.42, and with the preceding description we write

$$(10.127) \qquad E_{max}^{Kin} = E_F = \frac{\hbar^2}{2m} [3\pi^2 n(r)]^{2/3}$$

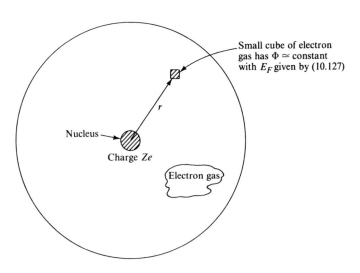

FIGURE 10.23   The Thomas–Fermi statistical atom.

The energy of an electron in the medium at a distance r from the nucleus is

(10.128)
$$E(r) = \frac{p^2}{2m} - e\Phi(r)$$

where we have written $e = |e|$. The preceding equation indicates that for $p^2/2m > e\Phi(r)$, $E > 0$, which implies that the electron escapes the atom. As this does not occur in the present model, we may conclude that maximum kinetic energy may also be set equal to $e\Phi$. With (10.127) we may then write

$$e\Phi(r) = \frac{\hbar^2}{2m} [3\pi^2 n(r)]^{2/3}$$

An additional relation between the potential $\Phi(r)$ and density $n(r)$ is given by Poisson's equation (cgs),

$$\nabla^2 \Phi = 4\pi e n$$

Working in the given spherically symmetric geometry, and eliminating $n(r)$ from the latter two equations, gives

(10.129)
$$\nabla^2 \Phi = \frac{1}{r} \frac{d^2}{dr^2} (r\Phi) = \frac{4\pi e}{3\pi^2} \frac{(2me)^{3/2}}{\hbar^3} \Phi^{3/2}$$

This is the desired single equation for $\Phi(r)$.

The following boundary conditions apply to the potential. Near the origin $\Phi(r)$ is due predominantly to the nucleus, and we write

$$\Phi \simeq \frac{Ze}{r}$$

or, equivalently,

(10.130a)
$$\lim_{r \to 0} r\Phi(r) = Ze$$

At large r we obtain

(10.130b)
$$\lim_{r \to \infty} r\Phi(r) = 0$$

The resulting equation (10.129) may be written in a more concise form in terms of new dimensionless variables (x, χ) defined by

(10.131a)
$$r = ax$$

(10.131b)
$$a \equiv \frac{1}{2} \left(\frac{3\pi}{4} \right)^{2/3} \frac{\hbar^2/me^2}{Z^{1/3}} = \frac{0.885a_0}{Z^{1/3}}$$

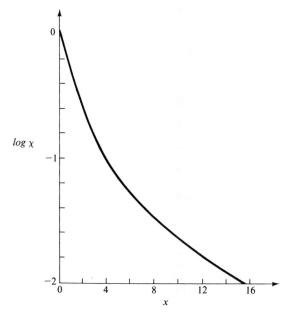

FIGURE 10.24 Semilogarithmic plot of solution to the Thomas–Fermi equation (10.132) subject to the boundary conditions (10.133).

(10.131c)
$$\Phi(r) = \frac{Ze}{r}\,\chi(x)$$

When written in terms of these new variables (see Problem 10.68), (10.129) reduces to

(10.132)
$$x^{1/2}\,\frac{d^2\chi}{dx^2} = \chi^{3/2}$$

This is the standard form of the (highly nonlinear) Thomas–Fermi equation. Boundary conditions (10.130), when written in terms of the new potential $\chi(x)$, appear as

(10.133)
$$\lim_{x\to 0}\chi = 1, \qquad \lim_{x\to\infty}\chi = 0$$

The semilogarithmic plot shown in Fig. 10.24 reveals the rapid decay of the potential χ with increase in x. Thus, for example, for $x > 4$, $\chi < 0.1$.[1]

---

[1] Numerical integration of the Thomas–Fermi equation (10.132) was originally performed by V. Bush and S. H. Caldwell, *Phys. Rev.* **38**, 1898 (1931).

Variation of Atomic Size with Z

An important conclusion of the preceding model is as follows: It is evident from (10.131b) that the parameter a may be viewed as an effective radius of the atom (i.e., the radius of a sphere which contains a fixed fraction of atomic electrons). Thus there is an implied decrease in effective atomic size with increase in Z. However, it should be borne in mind that the shell structure of atoms is lost in the Thomas–Fermi model. Thus one might expect the implied decrease in atomic size to be roughly valid within a given atomic shell of fixed principal quantum number. Observations find this property to be approximately obeyed. (See Fig. 12.11b.)

PROBLEMS

10.61 A hydrogen atom suffers a decay from the $3D$ to the $2P$ state and emits a photon.

    (a) What is the frequency of this photon (in Hz)?

    (b) Given that the decay time of this transition is $\simeq 10^{-10}$ s, what is the power (in watts) emitted in the transition?

10.62 Show that the three φ_{2P} states (10.120) comprise an orthonormal set. What is the value of the constant A? Leave your answer in terms of square roots and a_0.

10.63 Of what angular momentum operators are the respective three φ_{2P} states (10.120) eigenfunctions?

Answer

As described in Chapter 9, when operating on a wavefunction $f(\mathbf{r})$, the operators $\hat{L}_x, \hat{L}_y, \hat{L}_z$ effect a rotation of $\mathbf{r}$ about the x, y, z axes, respectively. Thus φ_{2P_x} is an eigenstate of $\hat{L}_x$; φ_{2P_y} is an eigenstate of $\hat{L}_y$; and φ_{2P_z} is an eigenstate of $\hat{L}_z$. All three functions are eigenstates of $\hat{L}^2$.

10.64 Show that $\langle \psi_1 | \psi_2 \rangle = 0$ for the orbitals (10.121).

10.65 (a) Show that the function φ_{2P_x} as given by (10.120a) is maximum along the x axis.

    (b) With this property at hand, argue that ψ_1 as given by (10.121a) exhibits maxima along the (1, 1, 1) directions.

Answers (partial)

    (a) We may write

$$\varphi_{2P_x} = A \sin\theta \cos\phi\, re^{-r/2a_0}$$

A polar plot of $\sin\theta$ reveals a circle tangent to the z axis at the origin. Thus $\sin\theta$ generates a circular torus about the z axis. A polar plot of $\cos\phi$ reveals a circle in the xy plane tangent to the y axis at the origin and with maximum at $\phi = 0$ (positive x axis). The intersection of these two factors, $\sin\theta \cos\phi$, is evidently maximum along the positive x axis. The remaining r-dependent component merely modulates this result.

10.66 This problem concerns annihilation and creation operators for spherical Bessel functions.[1] Introducing the operators (for $l > 0$)

$$b_l^{\pm} = -\frac{i}{x}\frac{d}{dx}x \pm i\frac{l}{x}$$

(a) Show that

$$\hat{b}_l^+\hat{b}_l^- = \hat{b}_{l+1}^-\hat{b}_{l+1}^+ = -\frac{1}{x}\frac{d^2}{dx^2} + \frac{l(l+1)}{x^2}$$

(b) Show that Bessel's equation (10.54) may be written

$$(\hat{b}_l^+\hat{b}_l^- - 1)R_l(x) = 0$$

(c) Show that

$$\hat{b}_{l+1}^+R_l = R_{l+1}$$

(d) Show that

$$\hat{b}_l^- R_l = R_{l-1}$$

(e) Apart from normalization, employ the preceding formalism to obtain:
   (1) $j_1(x)$ from $j_0(x)$.
   (2) $n_1(x)$ from $n_0(x)$.

10.67 The radial wavefunction satisfies the second-order differential equation (10.95). The student may recall that any such equation has two independent solutions. Show that for bound states there is only one linearly independent solution for given l and E values.

Answer

The equation may be written

$$\left[\hat{D}^2 - \frac{l(l+1)}{r^2} + \frac{2\mu}{\hbar^2}(E-V)\right]u = 0$$

$$\hat{D} \equiv \frac{d}{dr}$$

$$u \equiv rR$$

Let solutions to the preceding equation be written u_1 and u_2. Multiplying the equation for u_2 by u_1 and vice versa and subtracting the resulting equations gives

$$u_1\hat{D}^2u_2 - u_2\hat{D}^2u_1 = 0$$

or, equivalently,

$$\hat{D}[u_1\hat{D}u_2 - u_2\hat{D}u_1] = 0$$

[1] L. Infeld, *Phys. Rev.* **59**, 737 (1941).

Integrating, we find

$$u_1 \hat{D} u_2 - u_2 D u_1 = C$$

where C is a constant. For bound systems one may assume that either $u_1, u_2 \to 0$ or $\hat{D} u_1, \hat{D} u_2 \to 0$ as $r \to \infty$, which, in either case, gives $C = 0$. There results

$$\frac{d \ln u_2}{dr} = \frac{d \ln u_1}{dr}$$

which gives

$$u_2 = a u_1$$

where a is a constant. Thus, u_2 and u_1 are not independent.

10.68   Employing the transformation equations (10.131), show that (10.129) reduces to the Thomas–Fermi equation (10.132).

10.69   This problem addresses motion of an electron in a steady magnetic field as described in Section 10.4.

(a)   Find the mean square radius a^2 of an electron moving in a plane normal to a steady magnetic field.

(b)   Obtain your answer to part (a) once again, but now through use of the correspondence principle.

Answers

(a)   Referring to (10.72) et seq. indicates that

$$a^2 = \langle (y - y_0)^2 \rangle$$

Due to the harmonic oscillator structure of (10.74), we may write

$$\langle (y - y_0)^2 \rangle = \frac{2}{K} \langle V \rangle$$

where V is potential energy. With Problem 7.10 we may then write

$$a^2 = \frac{2}{K} \times \frac{\langle E \rangle}{2} = \frac{1}{K} \langle E \rangle$$

$$a^2 = \frac{\hbar}{m\Omega} \left(n + \frac{1}{2} \right)$$

(b)   The classical kinetic energy of the electron is

$$\tfrac{1}{2} m v^2 = \tfrac{1}{2} m a^2 \Omega^2 = \tfrac{1}{2} E$$

Thus

$$a^2 = \frac{E}{m\Omega^2}$$

Inserting the quantum value for E gives the desired result.

CHAPTER **11**

ELEMENTS OF
MATRIX MECHANICS.
SPIN WAVEFUNCTIONS

In this chapter some mathematical formalism is developed which is necessary for a more complete description of spin angular momentum. This formalism involves the theory of representations described briefly in Chapter 7 and matrix mechanics. Spin angular momentum operators are cast in the form of the Pauli spin matrices. The spinor wavefunction of a propagating spinning electron is constructed and used in problems involving the Stern–Gerlach apparatus. Examples involving the precession of an electron in a magnetic field and magnetic resonance are included as well as a prescription for adding spins. The coupled spin states so obtained are used extensively in the following chapter in conjunction with the Pauli principle in some basic atomic and molecular physics problems. The density matrix relevant to mixed states is introduced. Descriptions are included of the Heisenberg and interaction pictures in quantum mechanics. The chapter continues with a review of polarization states and concludes with a description of Bell's theorem in relation to present-day experiments concerning the notion of hidden variables.

11.1 BASIS AND REPRESENTATIONS

Matrix Mechanics

At very nearly the same time that Schrödinger introduced his wavefunction development of quantum mechanics, an alternative but equivalent description of the same physics was formulated. It is known as *matrix mechanics* and is due to W. Heisenberg, M. Born, and P. Jordan.

We have already encountered the concept of representations in quantum mechanics in Section 7.4. There we noted that in the "A representation," states are referred to a basis comprised of the eigenfunctions of $\hat{A}$. What we will now find is that within any such representation it is always possible to express operators and wavefunctions as matrices. Operator equations become matrix equations. For example, an equation of the form $\psi = \hat{F}\psi'$ may be rewritten as a matrix equation with the wavefunctions ψ and ψ' written as column vectors and the operator $\hat{F}$ written as a square matrix.

Basis

Previously we have found that wavefunctions related to a given quantum mechanical problem must satisfy certain criteria. Examples include:

Configuration	Wavefunctions
(a) Particle in a box	$\psi(0) = \psi(L) = 0, \int |\psi|^2 \, dx < \infty$
(b) One-dimensional harmonic oscillator $V(x) = \dfrac{K}{2} x^2$	$\int |\psi|^2 \, dx < \infty, \psi \to 0 \text{ as } |x| \to \infty$
(c) Particle in a central potential $V = V(r)$	$\iint |\psi|^2 r^2 \, dr \, d\Omega < \infty, |\psi|^2 r^2 \to 0 \text{ as } r \to 0$

For a given problem each such set of conditions implies a related space of functions. Consider a specific problem. Let the space of functions relevant to that configuration be called $\mathfrak{H}$. Let the set of functions

$$(11.1) \qquad \mathfrak{B} = \{\varphi_1, \varphi_2, \ldots\}$$

be a basis of $\mathfrak{H}$. For instance, for a particle in a one-dimensional box, these functions

468

are

$$\mathcal{B} = \left\{ \sqrt{\frac{2}{L}} \sin\left(\frac{n\pi x}{L}\right) \right\}$$

For the hydrogen atom, they are

$$\mathcal{B} = \{R_{nl}(r)Y_l^m(\theta, \phi)\}$$

whereas for a free particle (in spherical coordinates), they are

$$\mathcal{B} = \{j_l(kr)Y_l^m(\theta, \phi)\}$$

Inasmuch as $\mathcal{B}$ is a basis of $\mathfrak{H}$, any function ψ in $\mathfrak{H}$ may be expanded in terms of the basis functions φ_n:

(11.2) $$\psi = \sum_n \varphi_n a_n$$

or, equivalently,

(11.3) $$|\psi\rangle = \sum_n |\varphi_n\rangle\langle\varphi_n|\psi\rangle$$

The coefficients of expansion, a_n, represent ψ in the representation where $\mathcal{B}$ is the basis. These coefficients are projections of ψ onto the basis vectors (see Fig. 4.6). The equivalence of $\{a_n\}$ to the state function ψ is akin to the equivalence between a three-dimensional vector $\mathbf{A}$ and its components (A_x, A_y, A_z).

If $\{a_n\}$ is equivalent to ψ, one should be able to rewrite equations involving ψ as equations involving only $\{a_n\}$. Consider the typical quantum mechanical equation, where $\hat{F}$ is an arbitrary operator

(11.4) $$\psi = \hat{F}\psi'$$
$$|\psi\rangle = \hat{F}|\psi'\rangle$$

Expanding the right-hand side of the latter equation in accordance with (11.3) and multiplying from the left with $\langle\varphi_q|$ gives

(11.5) $$\langle\varphi_q|\psi\rangle = \sum_n \langle\varphi_q|\hat{F}|\varphi_n\rangle\langle\varphi_n|\psi'\rangle$$

or, equivalently,

(11.6) $$a_q = \sum_n F_{qn} a_n'$$

where

(11.7) $$F_{qn} \equiv \langle\varphi_q|\hat{F}|\varphi_n\rangle \equiv \int \varphi_q^* \hat{F}\varphi_n \, d\mathbf{r}$$

is *the matrix representation of the operator* $\hat{F}$ *in the basis* $\mathcal{B}$. The term F_{qn} is also called the *matrix element connecting* φ_q *to* φ_n. Equation (11.6), involving only the expansion coefficients $\{a_q\}$, $\{a_n'\}$, and the matrix elements $\{F_{qn}\}$, is equivalent to (11.4) involving the wavefunctions ψ, ψ' and the operator $\hat{F}$. Equation (11.6) is called a *matrix equation*. It may be written in the form

(11.8)
$$\begin{pmatrix} a_1 \\ a_2 \\ \vdots \end{pmatrix} = \begin{pmatrix} F_{11} & F_{12} & \cdots \\ F_{21} & F_{22} & \cdots \\ \vdots & \vdots & \end{pmatrix} \begin{pmatrix} a_1' \\ a_2' \\ \vdots \end{pmatrix}$$

In this equation the wavefunction ψ is represented by the column vector on the left, and ψ' is represented by the column vector on the right.[1]

(11.9)
$$\psi \rightarrow \begin{pmatrix} a_1 \\ a_2 \\ a_3 \\ \vdots \end{pmatrix} \qquad \psi' \rightarrow \begin{pmatrix} a_1' \\ a_2' \\ a_3' \\ \vdots \end{pmatrix}$$

The operator $\hat{F}$ is represented by the matrix F_{qn}.

(11.10)
$$\hat{F} = \begin{pmatrix} F_{11} & F_{12} & F_{13} & \cdots \\ F_{21} & F_{22} & F_{23} & \cdots \\ F_{31} & F_{32} & F_{33} & \cdots \\ \vdots & \vdots & \end{pmatrix}$$

The infinite dimensionality of these matrix equations is a consequence of the infinite dimensionality of Hilbert space. Finite matrix equations are relevant to vector spaces of finite dimension.

Diagonalization of an Operator

Let the orthogonal basis $\mathfrak{B}$ be comprised of the eigenfunctions of a Hermitian operator $\hat{G}$:

(11.11)
$$\hat{G}\varphi_n = g_n\varphi_n$$

The matrix elements of $\hat{G}$ are

$$\langle \varphi_q | \hat{G} | \varphi_n \rangle = g_n \langle \varphi_q | \varphi_n \rangle = g_n \delta_{qn}$$
$$G_{qn} = g_n \delta_{qn}$$

---

[1] The arrow in these identifications denotes "is represented by."

$$(11.12) \qquad \hat{G} = \begin{pmatrix} g_1 & 0 & 0 & \cdots \\ 0 & g_2 & 0 & \cdots \\ 0 & 0 & g_3 & \cdots \\ \vdots & \vdots & \vdots & \end{pmatrix}$$

Thus the matrix of an operator in a basis of the eigenfunctions of that operator is diagonal. The column vector representations of the eigenfunctions φ_n are the coefficients $\{a_q^{(n)}\}$ in the expansion

$$|\varphi_n\rangle = \sum_q a_q^{(n)} |\varphi_q\rangle$$

Multiplying from the left by $\langle\varphi_p|$ gives

$$\delta_{pn} = \sum_q a_q^{(n)} \delta_{pq} = a_p^{(n)}$$

$$(11.13)$$

$$a_p^{(n)} = \delta_{pn}$$

Thus the matrix representation of the eigenvector φ_n is a column vector with a single nonzero unit entry in the nth slot

$$\varphi_1 \rightarrow \begin{pmatrix} a_1^{(1)} \\ a_2^{(1)} \\ \vdots \end{pmatrix} = \begin{pmatrix} 1 \\ 0 \\ 0 \\ 0 \\ 0 \\ \vdots \end{pmatrix}, \qquad \varphi_2 \rightarrow \begin{pmatrix} a_1^{(2)} \\ a_2^{(2)} \\ \vdots \end{pmatrix} = \begin{pmatrix} 0 \\ 1 \\ 0 \\ 0 \\ 0 \\ \vdots \end{pmatrix},$$

$$(11.14)$$

$$\varphi_3 \rightarrow \begin{pmatrix} 0 \\ 0 \\ 1 \\ 0 \\ 0 \\ 0 \\ \vdots \end{pmatrix}, \qquad \varphi_4 \rightarrow \begin{pmatrix} 0 \\ 0 \\ 0 \\ 1 \\ 0 \\ 0 \\ \vdots \end{pmatrix}$$

The eigenvalue equation (11.11) can be written

$$\sum_q \langle\varphi_p|\hat{G}|\varphi_q\rangle\langle\varphi_q|\varphi_n\rangle = g_n\langle\varphi_p|\varphi_n\rangle$$

$$\sum_q G_{pq} a_q^{(n)} = g_n a_p^{(n)}$$

For $a^{(3)}$ it appears as

(11.15)
$$
\begin{pmatrix}
g_1 & 0 & 0 & 0 & \cdots \\
0 & g_2 & 0 & 0 & \cdots \\
0 & 0 & g_3 & 0 & \cdots \\
& & & & \\
\vdots & \vdots & \vdots & \vdots &
\end{pmatrix}
\begin{pmatrix}
0 \\ 0 \\ 1 \\ 0 \\ 0 \\ \vdots
\end{pmatrix}
= g_3
\begin{pmatrix}
0 \\ 0 \\ 1 \\ 0 \\ 0 \\ \vdots
\end{pmatrix}
$$

The "length" (squared) of a vector ψ is given by [recall (4.30)]

(11.16)
$$
\|\psi\|^2 = \langle \psi | \psi \rangle = \sum_q \langle \psi | \varphi_q \rangle \langle \varphi_q | \psi \rangle
$$
$$
= \sum_q |a_q|^2
$$

The lengths of the orthonormal basis vectors $\{\varphi_n\}$ are

(11.17)
$$
\|\varphi_n\|^2 = \sum_q |a_q^{(n)}|^2 = 1
$$

In matrix representation,

$$
\|\varphi_4\|^2 = \langle \varphi_4 | \varphi_4 \rangle \rightarrow \overbrace{000100}\ \cdots
\begin{pmatrix}
0 \\ 0 \\ 0 \\ 1 \\ 0 \\ 0 \\ \vdots
\end{pmatrix}
= 1
$$

Suppose that $\hat{G}$ is known to be diagonal in the basis $\{\varphi_n\}$. Then

(11.18)
$$
G_{qn} = g_n \delta_{qn}
$$

or, equivalently,

$$
\langle \varphi_q | \hat{G} | \varphi_n \rangle = g_n \langle \varphi_q | \varphi_n \rangle
$$

Multiplying from the left with the sum $\sum_q |\varphi_q\rangle$ gives

(11.19)
$$
\sum_q |\varphi_q\rangle \langle \varphi_q | \hat{G} | \varphi_n \rangle = g_n \sum_q |\varphi_q\rangle \langle \varphi_q | \varphi_n \rangle
$$

Recognizing the sum over φ_q products to be the unity operator $\hat{I}$, (Problem 11.1)

allows this latter equation to be rewritten as

$$\hat{I}(\hat{G}|\varphi_n\rangle - g_n|\varphi_n\rangle) = 0$$

which in turn implies that

(11.20) $$\hat{G}|\varphi_n\rangle = g_n|\varphi_n\rangle$$

Thus we find that if $\hat{G}$ is diagonal in a basis $\mathfrak{B}$, then $\mathfrak{B}$ is comprised of the eigenvectors of $\hat{G}$. One then notes the following important observation. The problem of finding the eigenvalues of an operator is equivalent to finding a basis which diagonalizes that operator.

Complete Sets of Commuting Operators

Suppose that $\hat{A}$, $\hat{B}$, and $\hat{C}$ are a "complete" set of three commuting operators. Let $\mathfrak{B} = \{\varphi_1, \varphi_2, \ldots\}$ be a set of simultaneous eigenstates of these three operators. Then with respect to this basis, $\hat{A}$, $\hat{B}$, and $\hat{C}$ are all diagonal and one speaks of "working in a representation in which $\hat{A}$, $\hat{B}$, and $\hat{C}$ are diagonal." For example, for a free particle moving in 3-space, the representation in which $\hat{H}$, $\hat{L}^2$, and $\hat{L}_z$ are diagonal contains the basis (10.56), while the representation in which $\hat{p}_x$, $\hat{p}_y$, and $\hat{p}_z$ are diagonal contains the basis (10.11).

The Continuous Case

In some cases the indices of a matrix range over a continuum of values. Such, for example, is the Hamiltonian matrix for a free particle in the basis (10.11). In one dimension this basis is comprised of the states (3.24), and the Hamiltonian matrix assumes the continuous form

$$\langle k|\hat{H}|k'\rangle = \left\langle k \left| \frac{\hat{p}_x^2}{2m} \right| k' \right\rangle = \frac{\hbar^2 k'^2}{2m} \delta(k - k')$$

Summing over the index of a continuous matrix is equivalent to integration. For example,

$$\sum_{k'} \langle k|\hat{H}|k'\rangle \rightarrow \int_{-\infty}^{\infty} dk' \frac{\hbar^2 k'^2}{2m} \delta(k - k') = \frac{\hbar^2 k^2}{2m}$$

The matrix representation of a quantum mechanical equation involving a wavefunction ψ is the corresponding equation for the projection coefficients of ψ into the basis $\mathfrak{B}$. If this set of coefficients forms a discrete set, then equations are of the form (11.6), involving summations over a discrete index. For the continuous case, these sums become integrals. In Section 7.4 we considered the case of the simple harmonic oscillator in "momentum space." Since the eigenstates of the momentum operator form a continuous set, the Schrödinger equation for the projection coefficients $\{b(k)\}$ becomes the integral equation (7.81). Owing to the simple form of the harmonic oscillator potential, this in turn is reducible to the differential form

(7.84)
$$\left(\frac{\hbar^2 k^2}{2m} - \frac{K}{2}\frac{\partial^2}{\partial k^2}\right)b(k) = Eb(k)$$

Thus the "matrix" form of the Schrödinger equation in the momentum representation remains a simple differential equation. Its argument is the single component $b(k)$ of the "column vector" $\{b(k)\}$.[1]

PROBLEMS

11.1 Let $\{\varphi_n\}$ be a *complete* orthonormal basis of a Hilbert space, $\mathfrak{H}$. Show that the identity operation $\hat{I}$ has the representation

$$\hat{I} = \sum_n |\varphi_n\rangle\langle\varphi_n|$$

in $\mathfrak{H}$. (This is sometimes called the *spectral resolution of unity*.)

Answer

Forming the matrix elements of $\hat{I}$, as given above, gives

$$I_{pq} = \langle\varphi_p|\hat{I}|\varphi_q\rangle = \sum_n \langle\varphi_p|\varphi_n\rangle\langle\varphi_n|\varphi_q\rangle$$
$$= \sum_n \delta_{pn}\delta_{nq} = \delta_{pq}$$

These are the matrix elements of the identity operator. This is a square matrix with unit entries along the diagonal.

$$\hat{I} = \begin{pmatrix} 1 & & & 0 \\ & 1 & & \\ & & 1 & \\ 0 & & & \ddots \end{pmatrix}$$

(*Note:* As described in Chapter 4, in order that an operator be a valid quantum mechanical representation of an observable, it must be Hermitian. To ensure further consistency of the theory, one also demands that the eigenstates of the operator comprise a complete set.[2])

---

[1] For further remarks on the $\hat{x}$ and $\hat{p}$ representations, see Appendix A.

[2] While completeness of eigenvectors is ensured for Hermitian operators in finite-dimensional spaces, this association is not guaranteed in infinite-dimensional spaces. For further discussion, see P. T. Matthews, *Introduction to Quantum Mechanics*, 2nd ed., McGraw-Hill, New York, 1968.

11.2 Show that if $\psi = 0$, then $a_n = 0$, for all n, where $\psi = \sum_n a_n \varphi_n$ and $\{\varphi_n\}$ is an orthogonal sequence.

11.3 What is the matrix representation of the operator $\hat{p}_x$ in the momentum representation?

11.4 Show that the diagonal elements of $\hat{D} \equiv \partial/\partial x$, in $\mathfrak{H}_1$ (4.30), in any basis are purely imaginary.

11.5 Determine the wavefunctions $b(k)$ in the momentum representation for a particle of mass m in a homogeneous force field $\mathbf{F} = (F_0, 0, 0)$. (Compare with Problem 7.65.)

Answer

The Hamiltonian is

$$\hat{H} = \frac{p^2}{2m} - iF_0 \frac{\partial}{\partial k}$$

and the time-independent Schrödinger equation appears as

$$-iF_0 \frac{\partial b}{\partial k} + \left(\frac{\hbar^2 k^2}{2m} - E\right) b = 0$$

which has the solutions

$$b_E(k) = \frac{1}{\sqrt{2\pi F_0}} \exp\left[\frac{ik}{F_0}\left(E - \frac{\hbar^2 k^2}{6m}\right)\right]$$

These solutions obey the normalization

$$\int_{-\infty}^{\infty} b_E{}^*(k) b_{E'}(k)\, dk = \delta(E' - E)$$

11.2 ELEMENTARY MATRIX PROPERTIES

The following are a series of definitions and properties of matrices and operators relevant to the theory of matrix mechanics.

The Product of Two Matrices

(11.21)
$$(\hat{A}\hat{B})_{nq} = \sum_p A_{np} B_{pq}$$

As an example of matrix multiplication, consider the product of the two (finite) 2×2 matrices:

$$\begin{pmatrix} A_{11} & A_{12} \\ A_{21} & A_{22} \end{pmatrix} \begin{pmatrix} B_{11} & B_{12} \\ B_{21} & B_{22} \end{pmatrix} = \begin{pmatrix} (A_{11}B_{11} + A_{12}B_{21}) & (A_{11}B_{12} + A_{12}B_{22}) \\ (A_{21}B_{11} + A_{22}B_{21}) & (A_{21}B_{12} + A_{22}B_{22}) \end{pmatrix}$$

The Product of Two Wavefunctions

(11.22) $\langle\psi|\psi'\rangle = \sum_n \langle\psi|\varphi_n\rangle\langle\varphi_n|\psi'\rangle = \sum_n a_n^* a_n'$

$$= \overline{a_1^* \quad a_2^* \quad a_3^* \quad \cdots} \begin{vmatrix} a_1' \\ a_2' \\ a_3' \\ \vdots \end{vmatrix} = a_1^* a_1' + a_2^* a_2' + \cdots$$

The Inverse of $\hat{A}$ The inverse of $\hat{A}$ is labeled $\hat{A}^{-1}$. It has the property that

(11.23) $$\hat{A}^{-1}\hat{A} = \hat{A}\hat{A}^{-1} = \hat{I}$$

For instance, if

$$\hat{A} = \begin{pmatrix} a_1 & b \\ c & a_2 \end{pmatrix}$$

then

$$\hat{A}^{-1} = \frac{1}{bc - a_1 a_2} \begin{pmatrix} -a_2 & b \\ c & -a_1 \end{pmatrix}$$

The Transpose of $\hat{A}$ The transpose of $\hat{A}$ is written $\tilde{A}$. The matrix elements of $\tilde{A}$ are obtained by "reflecting" the elements A_{nq} through the major diagonal of the matrix of $\hat{A}$.

(11.24) $$(\tilde{A})_{nq} = A_{qn}$$

$\hat{A}$ Is Symmetric or Antisymmetric   If $\hat{A}$ is symmetric, then

(11.25) $$\tilde{A} = \hat{A}$$

If $\hat{A}$ is antisymmetric, then

(11.26) $$\tilde{A} = -\hat{A}$$

The Trace of $\hat{A}$ The trace of $\hat{A}$ is the sum over its diagonal elements. It is written

(11.27) $$\operatorname{Tr}\hat{A} \equiv \sum_q A_{qq}$$

The Hermitian Adjoint of $\hat{A}$ This operator is written $\hat{A}^\dagger$. To construct $\hat{A}^\dagger$, one first forms the complex conjugate of $\hat{A}$ and then transposes.

(11.28) $$\hat{A}^\dagger = \tilde{A}^*$$

Matrix elements of $\hat{A}^\dagger$ are given by

(11.29)
$$(A^\dagger)_{nq} = (A_{qn})^*$$

or, more explicitly,

$$\langle \varphi_n | \hat{A}^\dagger \varphi_q \rangle = \langle \varphi_q | \hat{A} \varphi_n \rangle^* = \langle \hat{A} \varphi_n | \varphi_q \rangle$$

$\hat{A}$ *Is Hermitian* If $\hat{A}^\dagger = \hat{A}$, then $\hat{A}$ is Hermitian or, equivalently, if

$$(A^\dagger)_{nq} = A_{nq}$$

With (11.29), this definition becomes

(11.30)
$$(A_{qn})^* = A_{nq}$$

or

$$\tilde{A}^* = \hat{A}$$

$\hat{U}$ *Is Unitary*   If the Hermitian adjoint $\hat{U}^\dagger$ of an operator $\hat{U}$ is equal to $\hat{U}^{-1}$, the inverse of $\hat{U}$, i.e.,

(11.31)
$$\hat{U}^\dagger = \hat{U}^{-1}$$

then $\hat{U}$ is said to be unitary. The matrix elements of $\hat{U}$ satisfy the relations

(11.32)
$$(U^\dagger)_{nq} = (U^{-1})_{nq}$$
$$(U_{qn})^* = (U^{-1})_{nq}$$
$$\tilde{U}^* = \hat{U}^{-1}$$

TABLE 11.1   Matrix properties

Matrix	Definition	Matrix Elements
Symmetric	$A = \tilde{A}$	$A_{pq} = A_{qp}$
Antisymmetric	$A = -\tilde{A}$	$A_{pp} = 0; A_{pq} = -A_{qp}$
Orthogonal	$A = \tilde{A}^{-1}$	$(\tilde{A}A)_{pq} = \delta_{pq}$
Real	$A = A^*$	$A_{pq} = A_{pq}^*$
Pure imaginary	$A = -A^*$	$A_{pq} = iB_{pq}; B_{pq}$ real
Hermitian	$A = A^\dagger$	$A_{pq} = A_{qp}^*$
Anti-Hermitian	$A = -A^\dagger$	$A_{pq} = -A_{qp}^*$
Unitary	$A = (A^\dagger)^{-1}$	$(A^\dagger A)_{pq} = \delta_{pq}$
Singular	$\det A = 0$	

If $\hat{U}$ is unitary, then

(11.33)
$$\hat{U}\hat{U}^{\dagger} = \hat{I}$$
$$(\hat{U}\hat{U}^{\dagger})_{nq} = \delta_{nq}$$
$$\sum_{p} U_{np}(U_{qp})^* = \delta_{nq}$$

These matrix properties are summarized in Table 11.1.

PROBLEMS

11.6 Rotation of the xy axes about a fixed z axis through the angle ϕ_1 changes the components (x, y) of the vector $\mathbf{r}$ to (x', y') (see Fig. 11.1). These new components are related to the original components through the rotation matrix $\hat{R}(\phi_1)$ in the following way:

$$\mathbf{r}' = \begin{pmatrix} x' \\ y' \end{pmatrix} = \begin{pmatrix} \cos\phi_1 & \sin\phi_1 \\ -\sin\phi_1 & \cos\phi_1 \end{pmatrix}\begin{pmatrix} x \\ y \end{pmatrix} \equiv \hat{R}(\phi_1)\mathbf{r}$$

A second rotation through ϕ_2 gives

$$\mathbf{r}'' = \hat{R}(\phi_2)\mathbf{r}' = \hat{R}(\phi_2)\hat{R}(\phi_1)\mathbf{r}$$

(a) Show that the rotation matrix $\hat{R}$ has the "group property"

$$\hat{R}(\phi_1)\hat{R}(\phi_2) = \hat{R}(\phi_1 + \phi_2)$$

(b) Show that

$$[\hat{R}(\phi_1), \hat{R}(\phi_2)] = 0$$

(c) Show that $\hat{R}$ is an *orthogonal* matrix (see Table 11.1).

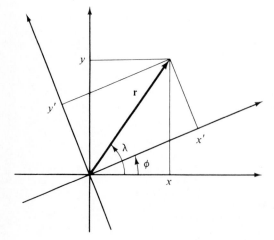

FIGURE 11.1   The rotation operator

$$\hat{R}(\phi) = \begin{pmatrix} \cos\phi & \sin\phi \\ -\sin\phi & \cos\phi \end{pmatrix}$$

is unitary,

$$\hat{R}^{\dagger}\hat{R} = I$$

and obeys the group property

$$\hat{R}(\phi_1)\hat{R}(\phi_2) = \hat{R}(\phi_1 + \phi_2)$$

(See Problem 11.6.)

(d)   In 3-space, rotation about the z axis is effected through the matrix

$$\hat{R} = \begin{pmatrix} \cos\phi & \sin\phi & 0 \\ -\sin\phi & \cos\phi & 0 \\ 0 & 0 & 1 \end{pmatrix}$$

Show that the eigenvalues of $\hat{R}$ have unit magnitude. (The angle λ in Fig. 11.1 remains fixed under rotation.)

11.7   Fill in the missing components of the matrix $\hat{C}$ which make $\hat{C}$ Hermitian.

$$\hat{C} \doteq \begin{pmatrix} 1 & 3i & 4 & 2 \\ - & 2 & 7i & 3 \\ - & - & 4 & 9 \\ - & - & - & 6 \end{pmatrix}$$

11.8   Construct a 2×2 unitary matrix which has at least two imaginary elements.

11.9   What is the inverse of the matrix

$$\hat{A} = \begin{pmatrix} 2 & 3i \\ 4 & 6i \end{pmatrix}?$$

Answer

Calculation shows that $\hat{A}$ has no inverse. Under such circumstances $\hat{A}$ is said to be *singular*.

11.10   Show that if $\hat{U}$ is unitary, then the eigenvalues a_n of $\hat{U}$ are of unit magnitude.

Answer

$$\hat{U}\varphi_n = a_n\varphi_n$$
$$\langle\hat{U}\varphi_n|\hat{U}\varphi_n\rangle = \langle\varphi_n|\hat{I}\varphi_n\rangle = \langle\varphi_n|\varphi_n\rangle = a_n{}^*a_n\langle\varphi_n|\varphi_n\rangle$$
$$a_n{}^*a_n = |a_n|^2 = 1$$

11.3   UNITARY AND SIMILARITY TRANSFORMATIONS IN QUANTUM MECHANICS

The significance of unitary operators in quantum mechanics is due to the following. We are already aware of the fact that a given Hilbert space has many bases. This is similar to the fact that 3-space is spanned by one of a continuum of triad basis vectors. One can obtain a new orthogonal triad basis in 3-space through a rotation of axes about the origin. In the new basis a vector $\mathbf{V}$ has components $V_x{}'$, $V_y{}'$, and $V_z{}'$. The length of $\mathbf{V}$ remains the same ($\mathbf{V} \cdot \mathbf{V} = \mathbf{V}' \cdot \mathbf{V}'$). Furthermore, the angle between any two vectors remains fixed ($\mathbf{V} \cdot \mathbf{F} = \mathbf{V}' \cdot \mathbf{F}'$). The final orientation of the new Cartesian frame with respect to the old may be obtained by a single rotation about a fixed axis (through the origin). The eigenvectors of the rotation matrix lie along this axis. Any vector along this axis remains fixed during the rotation. It follows that the eigenvalues of the rotation matrix are all of unit magnitude.

The related transformation from one basis to another basis in Hilbert space is a unitary transformation. It has all the properties listed above that a rigid rotation in 3-space has. These properties are as follows (for the most part, proofs are left to the problems).

Transformation of Basis

Let the sequence $\{f_n\}$ denote a new basis. These are related to the old basis $\{\varphi_n\}$ through the unitary transformation $\hat{U}$,

(11.34) $$|f_n\rangle = \sum_p |\varphi_p\rangle\langle\varphi_p|f_n\rangle = \sum_p (U_{np})^* |\varphi_p\rangle$$

(11.35) $$(U_{np})^* = \langle\varphi_p|f_n\rangle, \qquad U_{np} = \langle f_n|\varphi_p\rangle$$

These matrix elements are the projections of old basis vectors into the new ones. (The fact that $\hat{U}$ is unitary is established in Problem 11.11.)

Transformation of the State Vector

Let us consider how the components of an arbitrary state vector ψ transform to components in the basis $\{f_n\}$. In the basis $\{f_n\}$, these components (elements of a column vector) are

(11.36) $$\psi_n' = \langle f_n|\psi\rangle$$

Taking the complex conjugate of (11.34) gives

(11.37) $$\langle f_n| = \sum_p U_{np}\langle\varphi_p|$$

Substituting into (11.36) we obtain

(11.38) $$\psi_n' = \sum_p U_{np}\langle\varphi_p|\psi\rangle = \sum_p U_{np}\psi_p$$

This is the matrix representation of the equation

(11.39) $$|\psi'\rangle = \hat{U}|\psi\rangle$$

This equation tells us how an arbitrary state vector transforms under a change of basis.

If $|\varphi\rangle$ and $|\psi\rangle$ are two arbitrary vectors in Hilbert space, then under a transformation of basis ($\hat{U}$), these vectors transform to $|\varphi'\rangle$ and $|\psi'\rangle$ according to (11.39).

$$|\varphi'\rangle = \hat{U}|\varphi\rangle$$
$$|\psi'\rangle = \hat{U}|\psi\rangle$$

Under such transformation the inner product $\langle\psi|\varphi\rangle$ is preserved.

$$\langle\psi'|\varphi'\rangle = \langle\psi|\varphi\rangle$$

Setting $|\psi\rangle = |\varphi\rangle$ gives $\langle\varphi'|\varphi'\rangle = \langle\varphi|\varphi\rangle$. Thus a unitary transformation preserves the length of vectors and the angle between vectors.

The Unitary-Similarity Transformation

Next we consider the manner in which operators transform under a change of basis. A typical quantum mechanical equation appears as

$$\hat{F}|\varphi\rangle = |\psi\rangle$$

In the new basis, the two state vectors transform according to (11.39). Multiplying these equations from the left with $\hat{U}^{-1}$ gives

$$|\varphi\rangle = \hat{U}^{-1}|\varphi'\rangle$$
$$|\psi\rangle = \hat{U}^{-1}|\psi'\rangle$$

Substituting these forms into our typical equation above gives

$$\hat{F}\hat{U}^{-1}|\varphi'\rangle = U^{-1}|\psi'\rangle$$

Multiplying from the left with $\hat{U}$, we obtain the result

$$\hat{F}'|\varphi'\rangle = |\psi'\rangle$$

where

(11.40) $$\hat{F}' = \hat{U}\hat{F}\hat{U}^{-1}$$

This transformation preserves the form of our typical equation. As a special case ($\varphi = \psi$) we see that the eigenvalue equation for $\hat{F}$ is preserved under such a transformation. Equation (11.40), which describes how an operator transforms under a change of basis, is called a *unitary-similarity transformation*. The more general class of transformations, $\hat{A} \rightarrow \hat{A}' = \hat{S}\hat{A}\hat{S}^{-1}$, where $\hat{S}$ is not necessarily unitary, are called *similarity transformations*. However, of these, the unitary-similarity transformations are more relevant to quantum mechanics.

Invariance of Eigenvalues

Since the eigenvalues of an operator corresponding to an observable are physically measurable quantities, one does not expect these values to be effected by a transformation of basis in Hilbert space. The eigenenergies of a harmonic oscillator are $\hbar\omega_0(n + \frac{1}{2})$ in all representations. In a similar vein, the eigenvalues of such Hermitian operators must be real. It follows that (1) the eigenvalues of a Hermitian operator are preserved under a unitary-similarity transformation, and (2) the Hermiticity of an operator is preserved under a unitary-similarity transformation.

PROBLEMS

11.11 Show that $\hat{U}$, with matrix elements

$$U_{np} = \langle f_n | \varphi_p \rangle$$

is unitary. The sequences $\{f_n\}$ and $\{\varphi_p\}$ are complete and orthonormal.

Answer

We must establish the property (11.32) for $\hat{U}$.

$$(\hat{U}^{-1})_{np} = (\hat{U}^\dagger)_{np} = (U_{pn})^*$$

Equivalently, we must show that

$$I_{qp} = \delta_{qp} = \sum_n U_{qn}(U^{-1})_{np} = \sum_n U_{qn}(U_{pn})^*$$

$$= \sum_n \langle f_q | \varphi_n \rangle \langle \varphi_n | f_p \rangle$$

$$= \langle f_q | \left(\sum_n | \varphi_n \rangle \langle \varphi_n | \right) | f_p \rangle = \langle f_q | f_p \rangle = \delta_{qp}$$

11.12 Show that the inner product, $\langle \psi | \varphi \rangle$, is preserved under a unitary transformation.

Answer

$$\langle \psi' | \varphi' \rangle = \langle \hat{U}\psi | \hat{U}\varphi \rangle = \langle \psi | \hat{U}^\dagger \hat{U}\varphi \rangle$$
$$= \langle \psi | \hat{U}^{-1} \hat{U}\varphi \rangle = \langle \psi | \varphi \rangle$$

11.13 What is the matrix representation of the equation

$$|\varphi'\rangle = \hat{U}|\varphi\rangle$$

in the basis $\{f_n\}$? Write this equation explicitly, depicting elements of column and square matrices.

11.14 The matrix elements of $\hat{F}$ in the basis $\{\varphi_n\}$ are

$$F_{nq} = \langle \varphi_n | \hat{F} | \varphi_q \rangle$$

In the basis $\{f_n\}$ they are

$$F_{nq}' = \langle f_n | \hat{F} | f_q \rangle$$

Show that

$$F_{nq}' = (\hat{U}\hat{F}\hat{U}^{-1})_{nq} = \langle \varphi_n | \hat{U}\hat{F}\hat{U}^{-1} | \varphi_q \rangle$$

Answer

$$F_{nq}' = \langle f_n | \hat{F} | f_q \rangle = \sum_r \sum_p \langle f_n | \varphi_r \rangle \langle \varphi_r | \hat{F} | \varphi_p \rangle \langle \varphi_p | f_q \rangle$$

$$= \sum_r \sum_p U_{nr} F_{rp} (U_{qp})^* = \sum_r \sum_p U_{nr} F_{rp} (U^\dagger)_{pq}$$

$$= \sum_r \sum_p U_{nr} F_{rp} (U^{-1})_{pq} = (\hat{U} \hat{F} \hat{U}^{-1})_{nq}$$

11.15 Let $\hat{A}' = \hat{U} \hat{A} \hat{U}^{-1}$, where $\hat{U}$ is unitary. Show that this transformation may be rewritten $\hat{A}' = \hat{T}^{-1} \hat{A} \hat{T}$, where $\hat{T}$ is unitary. (*Note:* It follows that both $\hat{U} \hat{A} \hat{U}^{-1}$ and $\hat{U}^{-1} \hat{A} \hat{U}$ represent unitary-similarity transformations. For demonstrating certain properties of the unitary-similarity transformation, it may prove more convenient to work with the form $\hat{U}^{-1} \hat{A} \hat{U}$.)

11.16 (a) Show that $\hat{A}$ and $\hat{U}^{-1} \hat{A} \hat{U}$ have the same eigenvalues. Must $\hat{U}$ be unitary for this to be true?

(b) If the eigenvectors of $\hat{A}$ are $\{\varphi_n\}$, what are the eigenvectors of $\hat{U}^{-1} \hat{A} \hat{U}$?

11.17 If $\hat{U}$ is unitary and $\hat{A}$ is Hermitian, then show that $\hat{U} \hat{A} \hat{U}^{-1}$ is also Hermitian. That is, show that the Hermitian quality of an operator is preserved under a unitary-similarity transformation.

11.18 Show that the form of the operator equation

$$\hat{G} = \hat{A} \hat{B}$$

is preserved under a similarity transformation.

11.19 Consider the following decomposition of an arbitrary unitary operator $\hat{U}$:

$$\hat{U} = \frac{\hat{U} + \hat{U}^\dagger}{2} + i \frac{\hat{U} - \hat{U}^\dagger}{2i} \equiv \hat{A} + i\hat{B}$$

(a) Show that $\hat{A}$ and $\hat{B}$ are Hermitian.

(b) Show that $[\hat{A}, \hat{B}] = [\hat{A}, \hat{U}] = [\hat{B}, \hat{U}] = 0$.

(c) From part (b) we may conclude that $\hat{A}$, $\hat{B}$, and $\hat{U}$ have common eigenfunctions. Call them $|ab\rangle$. Use these eigenstates to show that the eigenvalues of $\hat{U}$ have unit magnitude.

11.20 (a) Show that diagonal matrices commute.

(b) Let $A_{ik} = a_i \delta_{ik}$, $B_{jl} = b_j \delta_{jl}$, and $C_{nm} = c_n \delta_{nm}$ be three matrices. What are the components of ABC?

(c) Again consider the diagonal matrix $A_{ik} = a_i \delta_{ik}$. What is the matrix representation of $\exp \hat{A}$? What is the matrix representation of $\sin \hat{A}$?

11.21 If $\hat{A}$, $\hat{B}$, and $\hat{C}$ are three $n \times n$ square matrices, show that

$$\text{Tr} (\hat{A} \hat{B} \hat{C}) = \text{Tr} (\hat{C} \hat{A} \hat{B}) = \text{Tr} (\hat{B} \hat{C} \hat{A})$$

11.22 Let $\hat{A}$ and $\hat{B}$ be two $n \times n$ square matrices. Employ the following property of determinants,[1]

$$\det \hat{A} \hat{B} = \det \hat{A} \det \hat{B}$$

---

[1] It is assumed that the student is familiar with the concept of a determinant. However, a definition of determinants may be found in Section 12.5. For further discussion, see G. Birkoff and S. MacLane, *A Survey of Modern Algebra*, Macmillan, New York, 1953.

to show that

$$\det \hat{A}\hat{B} = \det \hat{B}\hat{A}$$

11.23   A property of a matrix $\hat{A}$ which remains the same under a unitary transformation $\hat{A} \to \hat{U}\hat{A}\hat{U}^{-1}$ is called an *invariant*. Show that the trace of $\hat{A}$ is an invariant. That is, show that

$$\text{Tr }\hat{A} = \text{Tr }\hat{U}\hat{A}\hat{U}^{-1}$$

In your proof, establish that

$$\text{Tr }\hat{A} = \sum_n a_n$$

where a_n are the eigenvalues of $\hat{A}$.

Answer (partial)

Let $\hat{U}$ diagonalize $\hat{A}$ so that the diagonal matrix

$$\hat{U}\hat{A}\hat{U}^{-1} = \hat{A}'$$

is comprised of the eigenvalues of $\hat{A}$. With Problem 11.21, we have

$$\sum a_n = \text{Tr }\hat{A}' = \text{Tr }\hat{U}\hat{A}\hat{U}^{-1} = \text{Tr }\hat{A}\hat{U}^{-1}\hat{U} = \text{Tr }\hat{A}$$

11.24   Let $\hat{A}$ be an $n \times n$ square matrix with eigenvalues $a_1, a_2, \ldots, a_n$. Show that

$$\det \hat{A} = a_1 a_2 \cdots a_n$$

(*Hint:* Let $\hat{U}$ diagonalize $\hat{A}$ and refer to Problem 11.22.)

   Note: This problem establishes that $\det \hat{A}$ is another invariant property of $\hat{A}$. Together with the trace, these two fundamental properties appear as

$$\text{Tr }\hat{A} = \sum_i a_i$$

$$\det \hat{A} = \prod_i a_i$$

In general, an $n \times n$ matrix has n invariants, two of which are the trace and determinant. These n invariants are the coefficients of the characteristic equation for the eigenvalues of $\hat{A}$ [i.e., $\det (\hat{A} - \hat{I}a) = 0$], which itself is invariant. Namely, $\det (\hat{A} - \hat{I}a) = \det [\hat{U}(\hat{A} - \hat{I}a)\hat{U}^{-1}] = \det (\hat{U}\hat{A}\hat{U}^{-1} - \hat{I}a)$.

11.25   Let $\hat{A}$ be a Hermitian $n \times n$ matrix. Let the column vectors of the $n \times n$ matrix $\hat{S}$ be comprised of the orthonormalized eigenvectors of $\hat{A}$.

   (a)   Show that $\hat{S}$ is unitary.

   (b)   Show that $\hat{S}^{-1}\hat{A}\hat{S}$ is a diagonal matrix comprised of the eigenvalues of $\hat{A}$.

(*Note:* This establishes that a Hermitian matrix is always diagonalizable by a unitary-similarity transformation.)

11.26   Again, let $\hat{A}$ be a Hermitian $n \times n$ matrix. However, now let the column vectors of the $n \times n$ matrix $\hat{T}$ be comprised of the unnormalized, but still orthogonal eigenvectors of $\hat{A}$.

   (a)   Is $\hat{T}$ unitary?

   (b)   Is $\hat{T}^{-1}\hat{A}\hat{T} = \hat{A}'$ diagonal? If so, what are the elements of $\hat{A}'$?

(c)   Is the inner product between two n-dimensional column vectors preserved under this transformation?

Answers

(a)   We note that although $\hat{T}^{\dagger}\hat{T}$ is diagonal, it is not the unit operator, so that $\hat{T}$ is not unitary.

(b)   Let the eigenvector equation for $\hat{A}$ be written

$$\hat{A}|n\rangle = a_n|n\rangle$$

It follows that the column vectors of $\hat{A}\hat{T}$ are $a_1|1\rangle, a_2|2\rangle, \ldots, a_n|n\rangle$. This matrix may be rewritten

$$\hat{A}\hat{T} = \hat{T}\hat{A}'$$

where the diagonal matrix $\hat{A}'$ is comprised of the eigenvalues of $\hat{A}$. We then have

$$\hat{T}^{-1}\hat{A}\hat{T} = \hat{A}'$$

(*Note:* Although the similarity transformation described in this problem diagonalizes $\hat{A}$ and yields a diagonal matrix comprised of the eigenvalues of $\hat{A}$, it is not a unitary-similarity transformation and therefore is not relevant to quantum mechanics. Changes in representations in quantum mechanics must preserve the inner product between state vectors, which in turn ensures preservation of the Hermiticity of operators. These invariances are maintained in a unitary-similarity transformation.)

11.27   In the Schrödinger description of quantum mechanics, the wavefunction evolves in time according to the equation (3.70)

$$\psi(\mathbf{r}, t) = e^{-i\hat{H}t/\hbar}\psi(\mathbf{r}, 0)$$

(a)   Show that the operator

$$\hat{U} = \exp\left(\frac{-i\hat{H}t}{\hbar}\right)$$

is unitary. (*Hint:* Use the property $\hat{H}^{\dagger} = \hat{H}$.)

(b)   Having shown that $|\psi(t)\rangle = \hat{U}|\psi(0)\rangle$, show that the normalization of ψ, $\langle\psi(t)|\psi(t)\rangle$, is constant.

[*Note:* In this description the state of the system is represented by a vector $|\psi(\mathbf{r}, t)\rangle$, which migrates in Hilbert space according to the unitary transformation above. This behavior is opposed to that of eigenvectors corresponding to observables (e.g., $\hat{L}^2$, $\hat{H}$, $\hat{p}_x$, etc.). These are fixed in the Hilbert space.]

11.28   Show that if $\hat{A}$ is Hermitian, then

$$\hat{U} = (\hat{A} + i\hat{I})(\hat{A} + i\hat{I})^{-1}$$

is unitary. [*Hint:* Multiply from the right with $(\hat{A} + i\hat{I})$.]

11.29   Show that if the unitary operator $\hat{U}$ does not have the eigenvalue 1, then

$$\hat{A} \equiv i(\hat{I} + \hat{U})(\hat{I} - \hat{U})^{-1}$$

is Hermitian.

11.30 Consider two Hermitian operators $\hat{A}$ and $\hat{B}$, which satisfy the commutation relation, $[\hat{A}, \hat{B}] = i\hbar$. Suppose a system is in an eigenstate $|a\rangle$ of $\hat{A}$. What can be said of the probability distribution relating to B (i.e., $|\langle a|b\rangle|^2$)? Does your argument apply to the observables ϕ and L_z? (Recall Problem 9.15.)

11.4 THE ENERGY REPRESENTATION

One-Dimensional Box

In the energy representation, the Hamiltonian is diagonal. This representation includes a basis comprised of the eigenfunctions of the Hamiltonian. For a particle in a one-dimensional box, the basis in which $\hat{H}$ is diagonal is

$$(11.41) \qquad \mathfrak{B} = \sqrt{\frac{2}{L}} \left\{ \sin \frac{\pi x}{L}, \sin \frac{2\pi x}{L}, \sin \frac{3\pi x}{L}, \ldots \right\}$$

The Hamiltonian matrix in this representation is

$$(11.42) \qquad \hat{H} = E_1 \begin{pmatrix} 1 & & & & & \\ & 4 & & & & \\ & & 9 & & & \mathbf{0} \\ & & & 16 & & \\ & \mathbf{0} & & & \ddots & \\ & & & & & n^2 \\ & & & & & & \ddots \end{pmatrix}$$

Simple Harmonic Oscillator

For the one-dimensional harmonic oscillator, the basis that diagonalizes $\hat{H}$ is [recall (7.59)]

$$(11.43) \qquad \begin{aligned} \mathfrak{B} &= e^{-\xi^2/2}\{A_1 \mathcal{H}_1(\xi), A_2 \mathcal{H}_2(\xi), \ldots\} \\ &\equiv \{|1\rangle, |2\rangle, |3\rangle, \ldots\} \\ \xi^2 &\equiv \beta^2 x^2, \qquad \beta^2 \equiv \frac{m\omega_0}{\hbar} \end{aligned}$$

The nth-order Hermite polynomial is written $\mathcal{H}_n(\xi)$. The Hamiltonian matrix in this representation is

(11.44)
$$\hat{H} = \hbar\omega_0 \begin{pmatrix} 1/2 & & & & \\ & 3/2 & & \huge{0} & \\ & & \ddots & & \\ & \huge{0} & & (2n+1)/2 & \\ & & & & \ddots \end{pmatrix}$$

Position and Momentum Operators

Let us calculate the matrix representation of the position operator $\hat{x}$ for the harmonic oscillator in the energy representation. Recalling (7.61) and with k and n representing nonnegative integers, we have

(11.45)
$$\langle n|\hat{x}|k\rangle = \frac{1}{\sqrt{2\beta}} \langle n|\hat{a} + \hat{a}^\dagger |k\rangle$$

$$= \frac{1}{\sqrt{2\beta}} [k^{1/2}\delta_{n,k-1} + (k+1)^{1/2}\delta_{n,k+1}]$$

This gives the matrix

(11.46)
$$\hat{x} = \frac{1}{\sqrt{2\beta}} \begin{pmatrix} 0 & \sqrt{1} & 0 & 0 & 0 \\ \sqrt{1} & 0 & \sqrt{2} & 0 & 0 \\ 0 & \sqrt{2} & 0 & \sqrt{3} & 0 & \cdots \\ 0 & 0 & \sqrt{3} & 0 & \sqrt{4} \\ 0 & 0 & 0 & \sqrt{4} & 0 \\ & & & \vdots \end{pmatrix}$$

For the momentum operator $\hat{p}$, we find that

(11.47)
$$\langle n|\hat{p}|k\rangle = \frac{m\omega_0}{\sqrt{2}i\beta} [k^{1/2}\delta_{n,k-1} - (k+1)^{1/2}\delta_{n,k+1}]$$

which gives the matrix

(11.48)
$$\hat{p} = \frac{m\omega_0}{\sqrt{2}i\beta} \begin{pmatrix} 0 & \sqrt{1} & 0 & 0 & 0 \\ -\sqrt{1} & 0 & \sqrt{2} & 0 & 0 \\ 0 & -\sqrt{2} & 0 & \sqrt{3} & 0 & \cdots \\ 0 & 0 & -\sqrt{3} & 0 & \sqrt{4} \\ & & & \vdots \end{pmatrix}$$

Creation and Annihilation Operators

For the creation and annihilation operators we have [recall (7.61)]

(11.49)
$$a_{nk} = \langle n|\hat{a}|k\rangle = k^{1/2}\langle n|k - 1\rangle = k^{1/2}\delta_{n,k-1}$$
$$a_{nk}{}^{\dagger} = \langle n|\hat{a}^{\dagger}|k\rangle = (k + 1)^{1/2}\langle n|k + 1\rangle = (k + 1)^{1/2}\delta_{n,k+1}$$

which gives the matrices

$$\hat{a} = \begin{pmatrix} 0 & \sqrt{1} & 0 & 0 & 0 \\ 0 & 0 & \sqrt{2} & 0 & 0 & \cdots \\ 0 & 0 & 0 & \sqrt{3} & 0 \\ & & \vdots & & \end{pmatrix}$$

(11.50)

$$\hat{a}^{\dagger} = \begin{pmatrix} 0 & 0 & 0 & 0 & 0 \\ \sqrt{1} & 0 & 0 & 0 & 0 \\ 0 & \sqrt{2} & 0 & 0 & 0 & \cdots \\ 0 & 0 & \sqrt{3} & 0 & 0 \\ & & \vdots & & \end{pmatrix}$$

Let us check that these matrix operators promote and demote according to (7.61). The eigenfunctions $\{|n\rangle\}$ for the harmonic oscillator Hamiltonian are column vectors with the only nonzero entry in the $(n + 1)$st slot.

(11.51)
$$|0\rangle = \begin{pmatrix} 1 \\ 0 \\ 0 \\ 0 \\ 0 \\ 0 \\ \vdots \end{pmatrix}, \quad |1\rangle = \begin{pmatrix} 0 \\ 1 \\ 0 \\ 0 \\ 0 \\ 0 \\ \vdots \end{pmatrix}, \quad |2\rangle = \begin{pmatrix} 0 \\ 0 \\ 1 \\ 0 \\ 0 \\ 0 \\ \vdots \end{pmatrix}, \cdots$$

The time-dependent eigenstates of $\hat{H}$ appear as

$$\psi_0(x, t) = e^{-i\hat{H}t/\hbar}|0\rangle = e^{-i\omega_0 t/2}|0\rangle = e^{-i\omega_0 t/2}\begin{pmatrix} 1 \\ 0 \\ 0 \\ 0 \\ 0 \\ 0 \\ \vdots \end{pmatrix}$$

$$(11.52) \qquad \psi_1(x, t) = e^{-i3\omega_0 t/2} \begin{pmatrix} 0 \\ 1 \\ 0 \\ 0 \\ 0 \\ 0 \\ \vdots \end{pmatrix}, \qquad \psi_2(x, t) = e^{-i5\omega_0 t/2} \begin{pmatrix} 0 \\ 0 \\ 1 \\ 0 \\ 0 \\ 0 \\ \vdots \end{pmatrix}$$

Consider the operations $\hat{a}|2\rangle$ and $\hat{a}^\dagger|2\rangle$.

$$\hat{a}|2\rangle = \begin{pmatrix} 0 & \sqrt{1} & 0 & 0 & 0 & 0 \\ 0 & 0 & \sqrt{2} & 0 & 0 & 0 \\ 0 & 0 & 0 & \sqrt{3} & 0 & 0 \\ 0 & 0 & 0 & 0 & \sqrt{4} & 0 \\ & & & \vdots & & \end{pmatrix} \cdots \begin{pmatrix} 0 \\ 0 \\ 1 \\ 0 \\ 0 \\ \vdots \end{pmatrix} = \sqrt{2} \begin{pmatrix} 0 \\ 1 \\ 0 \\ 0 \\ \vdots \end{pmatrix} = \sqrt{2}|1\rangle$$

(11.53)

$$\hat{a}^\dagger|2\rangle = \begin{pmatrix} 0 & 0 & 0 & 0 & 0 \\ \sqrt{1} & 0 & 0 & 0 & 0 \\ 0 & \sqrt{2} & 0 & 0 & 0 \\ 0 & 0 & \sqrt{3} & 0 & 0 \\ 0 & 0 & 0 & \sqrt{4} & 0 \\ & & \vdots & & \end{pmatrix} \cdots \begin{pmatrix} 0 \\ 0 \\ 1 \\ 0 \\ 0 \\ \vdots \end{pmatrix} = \sqrt{3} \begin{pmatrix} 0 \\ 0 \\ 0 \\ 1 \\ 0 \\ \vdots \end{pmatrix} = \sqrt{3}|3\rangle$$

These equations very simply illustrate the promotion and demotion properties of the $\hat{a}^\dagger$ and $\hat{a}$ operators.

The Number Operator

In addition to the Hamiltonian (7.27)

$$\hat{H} = \hbar\omega_0(\hat{a}^\dagger \hat{a} + \tfrac{1}{2})$$

the number operator (7.28)

$$\hat{N} = \hat{a}^\dagger \hat{a}$$

is also diagonal in the energy representation.

$$\hat{N} = \begin{pmatrix} 0 & 0 & 0 & 0 \\ \sqrt{1} & 0 & 0 & 0 \\ 0 & \sqrt{2} & 0 & 0 \\ 0 & 0 & \sqrt{3} & 0 \\ & & \vdots & \end{pmatrix} \cdots \begin{pmatrix} 0 & \sqrt{1} & 0 & 0 & 0 \\ 0 & 0 & \sqrt{2} & 0 & 0 \\ 0 & 0 & 0 & \sqrt{3} & 0 \\ 0 & 0 & 0 & 0 & \sqrt{4} \\ & & \vdots & & \end{pmatrix} \cdots$$

(11.54)

$$= \begin{pmatrix} 0 & & & & & \\ & 1 & & & & \\ & & 2 & & \mathbf{0} & \\ & & & 3 & & \\ & & & & \ddots & \\ \mathbf{0} & & & & & n \\ & & & & & & \ddots \end{pmatrix}$$

The reader may readily check that the column vectors $\{|n\rangle\}$, as given by (11.51), are eigenvectors of both $\hat{H}$ as given by (11.44) and $\hat{N}$ as given by (11.54), with respective eigenvalues $\{\hbar\omega_0(n + \frac{1}{2})\}$ and $\{n\}$.

PROBLEMS

11.31 In Section 5.5 the importance of complete sets of commuting observables was discussed. The number of such variables (*good quantum numbers*, see Section 1.3) are analogous to the number of *canonical variables* relevant to the description of a classical system. It is important in classical physics that the number of such variables be preserved under a *canonical transformation*. In quantum mechanics it is equally significant that the number of operators comprising complete sets of commuting observables be preserved under a unitary transformation.

(a) Let such a set of compatible operators be $\hat{A}$, $\hat{B}$, $\hat{C}$, and $\hat{D}$. Show that this compatibility is preserved under a unitary transformation.

(b) Let $\hat{F}$ not commute with any element in the set $\hat{A}$, $\hat{B}$, $\hat{C}$, $\hat{D}$. Is this property preserved under a unitary transformation?

11.32 Show that

$$\det(\hat{I} + \epsilon\hat{A}) = 1 + \epsilon \operatorname{Tr} \hat{A} + O(\epsilon^2)$$

where $\hat{A}$ is an $n \times n$ matrix and $\hat{I}$ is the identity matrix in n dimensions.

11.33 Show that

$$\det(\exp \hat{A}) = \exp(\operatorname{Tr} \hat{A})$$

where $\hat{A}$ is an $n \times n$ matrix.

Answer

This equality may be established in two independent ways. In the first method we let $D(\lambda) \equiv \det [\exp (\lambda \hat{A})]$. Then with Problem 11.22 we obtain

$$\frac{dD}{d\lambda} = D \lim_{\epsilon \to 0} \left\{ \frac{\det [\exp (\epsilon \hat{A})] - 1}{\epsilon} \right\}$$

In the limit that ϵ goes to zero,

$$\det [\exp (\epsilon \hat{A})] = \det (1 + \epsilon \hat{A}) = 1 + \epsilon \operatorname{Tr} \hat{A}$$

where we have used the results of Problem 11.32 in establishing the second equality. It follows that

$$\frac{dD}{d\lambda} = D \operatorname{tr} \hat{A}$$

which has the solution

$$D(\lambda) = D(0) \exp (\lambda \operatorname{Tr} \hat{A})$$

But $D(0) = 1$, hence

$$D(1) = \det (\exp \hat{A}) = \exp (\operatorname{Tr} \hat{A})$$

In the second way we first note that if $\hat{U}$ diagonalizes $\hat{A}$, it also diagonalizes $e^{\hat{A}}$. Furthermore, $\det [\hat{U}(\exp \hat{A})\hat{U}^{-1}] = \det \exp \hat{A}$ (see Problem 11.23).

Now the diagonal matrix $U(\exp \hat{A})U^{-1}$ has the explicit form

$$\hat{U}(\exp \hat{A})\hat{U}^{-1} = \begin{pmatrix} e^{a_1} & 0 & 0 \\ 0 & e^{a_2} & 0 \\ 0 & 0 & e^{a_3} \\ & \vdots & \end{pmatrix} \cdots$$

where $\{a_i\}$ are the eigenvalues of A. The determinant of this matrix is

$$\det [U(\exp \hat{A})\hat{U}^{-1}] = e^{a_1}e^{a_2} \cdots = \exp \left(\sum_i a_i \right)$$

This is the value (in all representations) of the left-hand side of the equality to be established. For the right-hand side we recall that the trace is independent of representations (Problem 11.23) so that

$$\exp (\operatorname{Tr} \hat{A}) = \exp (\operatorname{Tr} \hat{A}') = \exp \left(\sum_i a_i \right) \qquad \text{Q.E.D.}$$

The matrix $\hat{A}'$ is the diagonal representation of $\hat{A}$.

11.34 Use the matrix representation (11.46) and (11.48) for $\hat{x}$ and $\hat{p}$ to obtain the matrix representation for the commutator $[\hat{x}, \hat{p}]$ for the harmonic oscillator in the energy representation.

11.35   Calculate the matrix representations for $\hat{x}^2$ and $\hat{p}^2$ for the harmonic oscillator in the energy representation.

11.36   Using the fact that any Hermitian matrix can be diagonalized by a unitary matrix, show that two Hermitian matrices, $\hat{A}$ and $\hat{B}$, can be diagonalized by the same unitary transformation $\hat{U}$ if and only if $[\hat{A}, \hat{B}] = 0$.

11.37   Consider the following equations:

$$\hat{A}^2 = 0, \qquad \hat{A}\hat{A}^\dagger + \hat{A}^\dagger\hat{A} = \hat{I}, \qquad \hat{B} = \hat{A}^\dagger\hat{A}$$

(a)   Show that $\hat{B}^2 = \hat{B}$.
(b)   Obtain explicit (2×2) matrices for $\hat{A}$ and $\hat{B}$.

Answer (*partial*)

(b)
$$\hat{A} = \frac{1}{2}\begin{pmatrix} 1 & i \\ i & -1 \end{pmatrix}$$

11.5   ANGULAR MOMENTUM MATRICES

The $\hat{L}$ Matrices

It was shown above that the matrix representation of an operator $\hat{A}$ in the basis consisting of the eigenvectors of $\hat{A}$, is diagonal. In Chapter 9 we found that the eigenfunctions of the angular momentum operators $\hat{L}^2$ and $\hat{L}_z$ are the spherical harmonics $Y_l{}^m(\theta, \phi)$. It follows that in the basis $\mathfrak{B} = \{Y_l{}^m\}$, the matrices $L^2_{lm, l'm'}$ and the matrices $(L_z)_{lm, l'm'}$ are diagonal. That is,

(11.55)
$$L^2_{lm, l'm'} = \langle lm|\hat{L}^2|l'm'\rangle = \int_{-1}^{1}\int_{0}^{2\pi} d\cos\theta\, d\phi (Y_l{}^m)^* \hat{L}^2 Y_{l'}{}^{m'}$$

$$= \hbar^2 l(l + 1)\delta_{ll'}\delta_{mm'}$$

(11.56)
$$(L_z)_{lm, l'm'} = \langle lm|\hat{L}_z|l'm'\rangle = \hbar m \delta_{ll'}\delta_{mm'}$$

The manner in which these elements are displayed is as follows. The rows and columns of a given matrix are ordered so that for every value of l, m_l runs from $(-l, \ldots, +l)$. For each of these m_l values, l is fixed. The diagonal matrix for $\hat{L}^2$ appears as

(11.57)

$l\downarrow$	$m\downarrow$									
$l' \rightarrow$		0	1	1	1	2	2	2	2	2
$m' \rightarrow$		0	1	0	−1	2	1	0	−1	−2
0	0	0		0				0		
1	1		2	0	0					
1	0	0	0	2	0			0		
1	−1		0	0	2					
2	2					6	0	0	0	0
2	1					0	6	0	0	0
2	0	0		0		0	0	6	0	0
2	−1					0	0	0	6	0
2	−2					0	0	0	0	6

$L^2 = \hbar^2$

In this same scheme, we obtain the following diagonal matrix for $\hat{L}_z$:

$$(11.58) \qquad L_z = \hbar \begin{pmatrix} 0 \\ & 1 \\ & & 0 \\ & & & -1 \\ & & & & 2 \\ & & & & & 1 \\ & & & & & & 0 \\ & & & & & & & -1 \\ & & & & & & & & -2 \\ & & & & & & & & & \ddots \end{pmatrix}$$

To obtain the matrices for $\hat{L}_x$ and $\hat{L}_y$ in the representation in which $\hat{L}^2$ and $\hat{L}_z$ are diagonal, we first construct the matrices for the ladder operators $\hat{L}_+$ and $\hat{L}_-$ (see Section 9.2). Since $\hat{L}_- = \hat{L}_+{}^\dagger$, one merely needs to calculate the $\hat{L}_+$ matrix and then, from its Hermitian adjoint, find $\hat{L}_-$. Once these matrices are known, $\hat{L}_x$ and $\hat{L}_y$ are obtained from

(11.59)
$$\hat{L}_x = \tfrac{1}{2}(\hat{L}_+ + \hat{L}_-)$$
$$\hat{L}_y = -\frac{i}{2}(\hat{L}_+ - \hat{L}_-)$$

Using the relation (see Table 9.4)

(11.60)
$$\hat{L}_- Y_l^m = [(l + m)(l - m + 1)]^{1/2}\hbar Y_l^{m-1}$$

one obtains

(11.61)
$$(L_\pm)_{lm,\, l'm'} = [(l' \mp m')(l' \pm m' + 1)]^{1/2}\hbar\delta_{ll'}\,\delta_{m,\, m' \pm 1}$$

and the matrices (exhibiting only the $l \le 2$ terms)

(11.62)

Adding and subtracting these two matrices according to (11.59) gives (again exhibiting only the $l \leq 2$ terms)

$$
L_x = \frac{\hbar}{2}
\left(
\begin{array}{c|ccc|ccccc}
0 & 0 & & & 0 & & & & \\
\hline
0 & 0 & \sqrt{2} & 0 & & & & & \\
 & \sqrt{2} & 0 & \sqrt{2} & & & 0 & & \\
 & 0 & \sqrt{2} & 0 & & & & & \\
\hline
0 & & 0 & & 0 & 2 & 0 & 0 & 0 \\
 & & & & 2 & 0 & \sqrt{6} & 0 & 0 \\
 & & & & 0 & \sqrt{6} & 0 & \sqrt{6} & 0 \\
 & & & & 0 & 0 & \sqrt{6} & 0 & 2 \\
 & & & & 0 & 0 & 0 & 2 & 0 \\
\end{array}
\right)
$$

(11.63)

$$
L_y = \frac{\hbar}{2i}
\left(
\begin{array}{c|ccc|ccccc}
0 & 0 & & & 0 & & & & \\
\hline
0 & 0 & \sqrt{2} & 0 & & & & & \\
 & -\sqrt{2} & 0 & \sqrt{2} & & & 0 & & \\
 & 0 & -\sqrt{2} & 0 & & & & & \\
\hline
0 & & 0 & & 0 & 2 & 0 & 0 & 0 \\
 & & & & -2 & 0 & \sqrt{6} & 0 & 0 \\
 & & & & 0 & -\sqrt{6} & 0 & \sqrt{6} & 0 \\
 & & & & 0 & 0 & -\sqrt{6} & 0 & 2 \\
 & & & & 0 & 0 & 0 & -2 & 0 \\
\end{array}
\right)
$$

Next we consider the matrix representation of the eigenvectors of $\hat{L}^2$ and $\hat{L}_z$. These are column vectors whose elements are the coefficients of expansion of $Y_l{}^m$ in the basis $\{Y_{l'}{}^{m'}\}$.

(11.64)
$$
Y_l{}^m = \sum_{l'} \sum_{m'=-l'}^{l'} a_{lm,\,l'm'}\, Y_{l'}{}^{m'}
$$

$$
a_{lm,\,l'm'} = \delta_{ll'}\,\delta_{mm'}
$$

The elements of these column vectors have zero entries for all values of l, m except at $l = l'$, $m = m'$, where the entry is unity. For example, the representations of the $l = 1$ eigenstates are [entries in these vectors follow the scheme of (11.57)]

$$(11.65) \qquad Y_1{}^1 \to \begin{pmatrix} 0 \\ 1 \\ 0 \\ 0 \\ 0 \\ \vdots \end{pmatrix}, \qquad Y_1{}^0 \to \begin{pmatrix} 0 \\ 0 \\ 1 \\ 0 \\ 0 \\ \vdots \end{pmatrix}, \qquad Y_1{}^{-1} \to \begin{pmatrix} 0 \\ 0 \\ 0 \\ 1 \\ 0 \\ \vdots \end{pmatrix}$$

When the matrix for $\hat{L}^2$ operates on any of these three column vectors it gives $2\hbar^2$ times the vector. When $\hat{L}_z$ operates on them, it gives the respective values $(\hbar, 0, -\hbar)$.

Sub L Matrices

One often speaks of the submatrices of $\hat{L}^2$, $\hat{L}_z$, ... corresponding to a given value of l. For example, the $\hat{L}^2$ matrix for $l = 1$ is

$$\hat{L}^2 = \hbar^2 \begin{pmatrix} 2 & 0 & 0 \\ 0 & 2 & 0 \\ 0 & 0 & 2 \end{pmatrix}$$

while the $\hat{L}_x$, $\hat{L}_y$, and $\hat{L}_z$ matrices corresponding to $l = 1$ are

$$\hat{L}_x = \frac{\hbar}{\sqrt{2}} \begin{pmatrix} 0 & 1 & 0 \\ 1 & 0 & 1 \\ 0 & 1 & 0 \end{pmatrix}, \quad \hat{L}_y = \frac{\hbar}{\sqrt{2}} \begin{pmatrix} 0 & -i & 0 \\ i & 0 & -i \\ 0 & i & 0 \end{pmatrix}, \quad \hat{L}_z = \hbar \begin{pmatrix} 1 & 0 & 0 \\ 0 & 0 & 0 \\ 0 & 0 & -1 \end{pmatrix}$$

(11.66)

The $\hat{L}_x$, $\hat{L}_y$, and $\hat{L}_z$ matrices corresponding to $l = 2$ are

$$\hat{L}_x = \frac{\hbar}{2} \begin{pmatrix} 0 & 2 & 0 & 0 & 0 \\ 2 & 0 & \sqrt{6} & 0 & 0 \\ 0 & \sqrt{6} & 0 & \sqrt{6} & 0 \\ 0 & 0 & \sqrt{6} & 0 & 2 \\ 0 & 0 & 0 & 2 & 0 \end{pmatrix}, \quad \hat{L}_y = \frac{\hbar}{2} \begin{pmatrix} 0 & -i2 & 0 & 0 & 0 \\ i2 & 0 & -i\sqrt{6} & 0 & 0 \\ 0 & i\sqrt{6} & 0 & -i\sqrt{6} & 0 \\ 0 & 0 & i\sqrt{6} & 0 & -i2 \\ 0 & 0 & 0 & i2 & 0 \end{pmatrix},$$

$$(11.67) \qquad\qquad \hat{L}_z = \hbar \begin{pmatrix} 2 & 0 & 0 & 0 & 0 \\ 0 & 1 & 0 & 0 & 0 \\ 0 & 0 & 0 & 0 & 0 \\ 0 & 0 & 0 & -1 & 0 \\ 0 & 0 & 0 & 0 & -2 \end{pmatrix}$$

We may consider the eigenvectors corresponding to these matrices. These are also subcomponents of the infinitely dimensional column vectors (11.64). For example, the eigenvectors of $\hat{L}_x$ for the case $l = 1$ appear as

$$\xi_x{}^0 = \frac{1}{\sqrt{2}} \begin{pmatrix} 1 \\ 0 \\ -1 \end{pmatrix}, \qquad \hat{L}_x \xi_x{}^0 = 0\hbar\xi_x{}^0$$

(11.68)
$$\xi_x{}^{-1} = \frac{1}{2} \begin{pmatrix} 1 \\ -\sqrt{2} \\ 1 \end{pmatrix}, \qquad \hat{L}_x \xi_x{}^{-1} = -\hbar\xi_x{}^{-1}$$

$$\xi_x{}^1 = \frac{1}{2} \begin{pmatrix} 1 \\ \sqrt{2} \\ 1 \end{pmatrix}, \qquad \hat{L}_x \xi_x{}^1 = +\hbar\xi_x{}^1$$

In the representation where $\hat{L}_x$ is the differential operator [recall (9.56)]

$$\hat{L}_x = i\hbar\left(\sin\phi\,\frac{\partial}{\partial\theta} + \cot\theta\cos\phi\,\frac{\partial}{\partial\phi}\right)$$

the eigenvector $\xi_x{}^1$ corresponds to the linear combination of spherical harmonics

$$\xi_x{}^1 = \tfrac{1}{2}(Y_1{}^1 + \sqrt{2}\,Y_1{}^0 + Y_1{}^{-1})$$

which was previously labeled X_+ in (9.100).

The $\hat{\mathbf{J}}$ Matrices

In the preceding construction of the matrices for the angular momentum operators $\hat{\mathbf{L}}$ and $\hat{L}^2$, it proved convenient to work from the $Y_l{}^m(\theta, \phi)$ eigenstates relevant to the coordinate representation. As we recall from Chapter 9, the more inclusive angular momentum, $\hat{\mathbf{J}}$ (which may represent $\hat{\mathbf{L}}$, $\hat{\mathbf{S}}$, or $\mathbf{L} + \hat{\mathbf{S}}$), is defined in terms of the commutation relations (9.16) and the rule of length appropriate to vectors (9.17).

The procedure for obtaining the matrices for $\hat{\mathbf{J}}$ and $\hat{J}^2$ (in a representation where $\hat{J}^2$ and $\hat{J}_z$ are diagonal) parallels the construction above. In place of (11.60), one writes the operationally identical equations (see Table 9.4)

$$\hat{J}_+ |jm\rangle = \hbar[(j - m)(j + m + 1)]^{1/2}|j, m + 1\rangle$$
$$\hat{J}_- |jm\rangle = \hbar[(j + m)(j - m + 1)]^{1/2}|j, m - 1\rangle$$

Thus the matrices found above for $\hat{\mathbf{L}}$ and $\hat{L}^2$ are also valid for $\hat{\mathbf{J}}$ and $\hat{J}^2$. Such matrices have the correct commutation properties and obey the Pythagorean length rule (see Problem 11.41). However, while $\hat{\mathbf{L}}$ matrices are restricted to integral l values and are therefore of odd $(2l + 1)$ dimension, $\hat{\mathbf{J}}$ matrices also incorporate j values that are half-odd integral. Such matrices are of even dimension.

Since there are $2j + 1$ values of J_z for each value of j, the matrix of $\hat{J}_z$ has $2j + 1$ diagonal elements. For a given j value, the operators $\hat{\mathbf{J}}$ and $\hat{J}^2$ are $(2j + 1) \times (2j + 1)$ square matrices and operate on column vectors $2j + 1$ elements long. That is, $\hat{\mathbf{J}}$ and $\hat{J}^2$ operate on a $(2j + 1)$-dimensional space.

The diagonal matrices for $\hat{J}^2$ and $\hat{J}_z$ (for a given value of j) are simple to construct. The first four $(j = \frac{1}{2}, 1, \frac{3}{2}, 2)$ such pairs appear as

$$j = \tfrac{1}{2}: \quad \hat{J}^2 = \frac{3\hbar^2}{4}\begin{pmatrix} 1 & 0 \\ 0 & 1 \end{pmatrix}, \qquad\qquad \hat{J}_z = \frac{\hbar}{2}\begin{pmatrix} 1 & 0 \\ 0 & -1 \end{pmatrix}$$

$$j = 1: \quad \hat{J}^2 = 2\hbar^2\begin{pmatrix} 1 & 0 & 0 \\ 0 & 1 & 0 \\ 0 & 0 & 1 \end{pmatrix}, \qquad \hat{J}_z = \hbar\begin{pmatrix} 1 & 0 & 0 \\ 0 & 0 & 0 \\ 0 & 0 & -1 \end{pmatrix}$$

$$j = \tfrac{3}{2}: \quad \hat{J}^2 = \frac{15\hbar^2}{4}\begin{pmatrix} 1 & 0 & 0 & 0 \\ 0 & 1 & 0 & 0 \\ 0 & 0 & 1 & 0 \\ 0 & 0 & 0 & 1 \end{pmatrix}, \qquad \hat{J}_z = \frac{\hbar}{2}\begin{pmatrix} 3 & 0 & 0 & 0 \\ 0 & 1 & 0 & 0 \\ 0 & 0 & -1 & 0 \\ 0 & 0 & 0 & -3 \end{pmatrix}$$

$$j = 2: \quad \hat{J}^2 = 6\hbar^2\begin{pmatrix} 1 & 0 & 0 & 0 & 0 \\ 0 & 1 & 0 & 0 & 0 \\ 0 & 0 & 1 & 0 & 0 \\ 0 & 0 & 0 & 1 & 0 \\ 0 & 0 & 0 & 0 & 1 \end{pmatrix}, \qquad \hat{J}_z = \hbar\begin{pmatrix} 2 & 0 & 0 & 0 & 0 \\ 0 & 1 & 0 & 0 & 0 \\ 0 & 0 & 0 & 0 & 0 \\ 0 & 0 & 0 & -1 & 0 \\ 0 & 0 & 0 & 0 & -2 \end{pmatrix}$$

When j is an integer and $\mathbf{J}$ represents orbital angular momentum, $\mathbf{L}$, one may transform from the $|jm\rangle$ column vector representation to the coordinate $Y_l^m(\theta, \phi)$ representation. In this representation the ladder operators $\hat{J}_\pm$ appear as [see (9.57)]

$$\hat{J}_\pm = \hbar e^{\pm i\phi}\left(i\cot\theta\,\frac{\partial}{\partial\phi} \pm \frac{\partial}{\partial\theta}\right)$$

and eigenstates of J^2 and J_z are the spherical harmonics. When $\hat{J}$ represents spin, $\hat{S}$, spherical harmonic eigenstates become inappropriate.

The Rotation Operator

A distinction between angular momenta corresponding to j integral or half-odd integral is found in the rotation operator,[1] described in Section 10.2.

$$\hat{R}_\phi = \exp\left(\frac{i\boldsymbol{\phi}\cdot\hat{\mathbf{J}}}{\hbar}\right)$$

---

[1] A more fundamental distinction involves the theory of group representations. For a concise, self-contained discussion of this topic, see L. I. Schiff, *Quantum Mechanics*, 3rd ed., McGraw-Hill, New York, 1968, Chapter 7.

When $\hat{R}$ operates on a function $f(\mathbf{r})$, it rotates $\mathbf{r}$ through the angle $\boldsymbol{\phi}$. If rotation is solely about the z axis, $\boldsymbol{\phi} = \mathbf{e}_z \phi$, then $\hat{R}$ becomes

$$\hat{R}_\phi = \exp\left(\frac{i\phi \hat{J}_z}{\hbar}\right)$$

Let $|jm\rangle$ denote a common eigenstate of $\hat{J}^2$ and $\hat{J}_z$. Then, in particular,

$$\hat{J}_z |jm\rangle = \hbar m |jm\rangle$$

and

$$\hat{R}_\phi |jm\rangle = e^{i\phi m} |jm\rangle$$

For the case that j, and therefore m, is half-odd integral, $e^{i2\pi m} = -1$, and one obtains the somewhat surprising result

$$(2j + 1 = \text{even no.}) \qquad \hat{R}_{2\pi} |jm\rangle = -1 |jm\rangle$$

That is, the eigenstates of $\hat{J}^2$ and $\hat{J}_z$ corresponding to half-odd integral j values change sign under complete rotation of axes. If, on the other hand, j is integral, one obtains

$$(2j + 1 = \text{odd no.}) \qquad \hat{R}_{2\pi} |jm\rangle = +1 |jm\rangle$$

In the coordinate representation, eigenstates for this case become spherical harmonics. These functions return to their original values under complete rotation

The smallest finite value j may assume is $j = \frac{1}{2}$. This spin quantum value is a profoundly important case and is developed in detail in the next section. Eigenstates corresponding to $j = \frac{1}{2}$ are called *spinors*. We will find (Problem 11.76), in accord with the discussion above, that spinors change sign under complete rotation of axes.

PROBLEMS

11.38 The state column vectors ξ, corresponding to the case $l = 1$, exist in a three-dimensional vector space. Any element ξ of this space is a set of three numbers of the form

$$\xi = \begin{pmatrix} a \\ b \\ c \end{pmatrix}$$

Write the vector

$$\xi = \frac{1}{\sqrt{14}} \begin{pmatrix} 1 \\ 2 \\ 3 \end{pmatrix}$$

as a linear combination of the vectors $(\xi_x{}^0, \xi_x{}^{-1}, \xi_x{}^1)$, as given by (11.68).

Answer

$$\xi = -\frac{1}{\sqrt{7}} \xi_x{}^0 + \frac{1}{\sqrt{7}} (\sqrt{2} - 1)\xi_x{}^{-1} + \frac{1}{\sqrt{7}} (\sqrt{2} + 1)\xi_x{}^1$$

11.39 Use the results of Problem 11.38 to answer the following question. A rigid rotator with moment of inertia I is in the state

$$\psi(t) = \frac{1}{\sqrt{14}} \begin{pmatrix} 1 \\ 2 \\ 3 \end{pmatrix} e^{-iEt/\hbar}, \qquad E = \frac{\hbar^2}{I}$$

(a) What is the probability that measurement of L_x finds the value $-\hbar$?

(b) What is the column vector representation of the time-dependent state of the rotator after measurement finds the value $L_x = -\hbar$?

11.40 What is the column vector representation of the angular momentum state

$$\psi = \frac{1}{\sqrt{24}} (Y_2{}^2 + 3Y_2{}^1 + 2Y_2{}^0 + 3Y_2{}^{-1} + Y_2{}^{-2})$$

in the representation in which $\hat{L}_z$ is diagonal?

11.41 Show that the $l = 2$ angular momentum matrices satisfy the relation

$$\hat{L}^2 = \hat{L}_x{}^2 + \hat{L}_y{}^2 + \hat{L}_z{}^2 = 6\hbar^2 \begin{pmatrix} 1 & 0 & 0 & 0 & 0 \\ 0 & 1 & 0 & 0 & 0 \\ 0 & 0 & 1 & 0 & 0 \\ 0 & 0 & 0 & 1 & 0 \\ 0 & 0 & 0 & 0 & 1 \end{pmatrix}$$

11.42 Show directly, by matrix multiplication, that $\langle \hat{L}_x{}^2 \rangle$ in the state $\xi_x{}^{-1}$ is $\hbar^2$.

11.43 What are the matrix representations of $\hat{L}^2$, $\hat{L}_z$, $\hat{L}_+$, and $\hat{L}_-$ for the case $l = 3$?

11.44 What are the column eigenvectors of $\hat{L}_z$ corresponding to $l = 3$? To what combinations of $Y_l{}^m$ functions do these correspond?

11.6 THE PAULI SPIN MATRICES

Spin Operators

In Section 9.2 it was concluded that there are two classes of angular momentum. These, we recall, stem from the fact that orbital angular momentum l values cover only a subset of the spectrum of j values appropriate to $\hat{J}^2$ and $\hat{J}_z$. The second class includes angular momentum called *spin*. Spin, as described in Section 9.1, is not related to the spatial coordinates of a particle as is orbital angular momentum. It is an intrinsic or internal property. Other intrinsic properties of a particle are charge, mass, dipole moment, moment of inertia, and so forth. Values of such parameters comprise *internal degrees of freedom* for a particle.

Spin angular momentum is denoted by the symbol $\hat{\mathbf{S}}$. The Cartesian components of $\hat{S}$, being angular momentum components, obey the commutation rules

(11.69) $[\hat{S}_x, \hat{S}_y] = i\hbar\hat{S}_z,$ $[\hat{S}_y, \hat{S}_z] = i\hbar\hat{S}_x,$ $[\hat{S}_z, \hat{S}_x] = i\hbar\hat{S}_y$

These are the fundamental relations from which all other properties of spin follow. Similar to the introduction of $\hat{J}_+$ and $\hat{L}_+$ in Chapter 9, one may also introduce ladder operators for spin.

(11.70) $$\hat{S}_\pm = \hat{S}_x \pm i\hat{S}_y$$

Furthermore, since the eigenvalue equations (9.43) for $\hat{J}^2$ and $\hat{J}_z$ were derived from the fundamental angular momentum commutator relations (9.16), we may conclude that a similar structure exists for the eigenvalues of $\hat{S}^2$ and $\hat{S}_z$.

(11.71) $$\hat{S}^2|sm_s\rangle = \hbar^2 s(s+1)|sm_s\rangle, \qquad \hat{S}_z|sm_s\rangle = \hbar m_s|sm_s\rangle$$

For a given value of s, the azimuthal quantum number m_s runs in integral steps from $-s$ to $+s$. The lowest value s can have is zero. Mesons are particles that have zero spin.[1] Photons have unit spin. For $s = 1$ there are three values of m_s: $-1, 0, 1$.

Spin Eigenstates

Electrons, protons, and neutrons have a spin of one-half. There are two values of m_s for $s = \frac{1}{2}$. These are $m_s = +\frac{1}{2}, -\frac{1}{2}$. Let us call the eigenstate corresponding to $(s = \frac{1}{2}, m_s = \frac{1}{2})$ α (also α_z) and the eigenstate corresponding to $(s = \frac{1}{2}, m_s = -\frac{1}{2})$ β (also β_z). These eigenstates obey the eigenvalue equations

(11.72)
$$\hat{S}^2\alpha = \frac{3}{4}\hbar^2\alpha, \quad \hat{S}_z\alpha = \frac{\hbar}{2}\alpha$$

$$\hat{S}^2\beta = \frac{3}{4}\hbar^2\beta, \quad \hat{S}_z\beta = -\frac{\hbar}{2}\beta$$

The raising and lowering operators have the property that[2]

(11.73)
$$\hat{S}_+|s, m_s\rangle = \hbar\sqrt{s(s+1) - m_s(m_s+1)}|s, m_s+1\rangle$$
$$\hat{S}_-|s, m_s\rangle = \hbar\sqrt{s(s+1) - m_s(m_s-1)}|s, m_s-1\rangle$$

It follows that

(11.74)
$$\hat{S}_+\alpha = 0 \qquad \hat{S}_+\beta = \hbar\alpha$$
$$\hat{S}_-\alpha = \hbar\beta \qquad \hat{S}_-\beta = 0$$

---

[1] More precisely, *vector mesons* have spin one; *pseudoscalar* mesons have spin zero.

[2] See Table 9.4.

Matrix Representation

We now wish to construct the matrix elements of $\hat{S}_x$, $\hat{S}_y$, and $\hat{S}_z$ for $s = \frac{1}{2}$ in the α, β basis. In this basis $\hat{S}^2$ and $\hat{S}_z$ are diagonal. The first two equations on the left in (11.74) appear explicitly as

$$(\hat{S}_x + i\hat{S}_y)\alpha = 0$$
$$(\hat{S}_x - i\hat{S}_y)\alpha = \hbar\beta$$

Adding and subtracting these two equations, respectively, gives

(11.75)
$$\hat{S}_x\alpha = \frac{1}{2}\hbar\beta$$

$$\hat{S}_y\alpha = \frac{i}{2}\hbar\beta$$

In similar manner, addition and subtraction of the remaining two equations in (11.74) gives

(11.76)
$$\hat{S}_x\beta = \frac{\hbar}{2}\alpha$$

$$\hat{S}_y\beta = -\frac{i}{2}\hbar\alpha$$

Combining these with the following equations from (11.72),

(11.77)
$$\hat{S}_z\alpha = \frac{\hbar}{2}\alpha$$

$$\hat{S}_z\beta = -\frac{\hbar}{2}\beta$$

establishes all the six equations needed to calculate the matrix elements of $\hat{S}_x$, $\hat{S}_y$, and $\hat{S}_z$. For example, the element $\langle\alpha|\hat{S}_x|\alpha\rangle$ is [using the first equation in (11.75)]

$$\langle\alpha|\hat{S}_x|\alpha\rangle = \tfrac{1}{2}\hbar\langle\alpha|\beta\rangle = 0$$

The vectors α and β are eigenvectors of a Hermitian operator (i.e., $\hat{S}_z$) and are therefore orthogonal. We also take them individually to be normalized. Continuing in this way we find that

(11.78)
$$\hat{S}_x = \begin{pmatrix} \langle\alpha|\hat{S}_x|\alpha\rangle & \langle\alpha|\hat{S}_x|\beta\rangle \\ \langle\beta|\hat{S}_x|\alpha\rangle & \langle\beta|\hat{S}_x|\beta\rangle \end{pmatrix} = \frac{\hbar}{2}\begin{pmatrix} 0 & 1 \\ 1 & 0 \end{pmatrix}$$

$$\hat{S}_y = \frac{i\hbar}{2}\begin{pmatrix} 0 & -1 \\ 1 & 0 \end{pmatrix}, \qquad \hat{S}_z = \frac{\hbar}{2}\begin{pmatrix} 1 & 0 \\ 0 & -1 \end{pmatrix}$$

The matrix representations of the eigenvectors α and β are the two-dimensional column vectors

$$(11.79) \qquad \alpha = \begin{pmatrix} 1 \\ 0 \end{pmatrix}, \qquad \beta = \begin{pmatrix} 0 \\ 1 \end{pmatrix}$$

In this representation the orthonormal relations between α and β appear as

$$\langle \alpha | \beta \rangle = \overline{1 \quad 0} \begin{pmatrix} 0 \\ 1 \end{pmatrix} = 0 + 0 = 0$$

$$\langle \beta | \alpha \rangle = \overline{0 \quad 1} \begin{pmatrix} 1 \\ 0 \end{pmatrix} = 0 + 0 = 0$$

$$(11.80)$$

$$\langle \alpha | \alpha \rangle = \overline{1 \quad 0} \begin{pmatrix} 1 \\ 0 \end{pmatrix} = 1 + 0 = 1$$

$$\langle \beta | \beta \rangle = \overline{0 \quad 1} \begin{pmatrix} 0 \\ 1 \end{pmatrix} = 0 + 1 = 1$$

The operator

$$(11.81) \qquad \hat{\boldsymbol{\sigma}} \equiv \frac{2}{\hbar} \hat{\mathbf{S}}$$

is called the *Pauli spin operator*. The matrix representation of its Cartesian components (in the basis that diagonalizes $\hat{S}^2$ and $\hat{S}_z$) is

$$(11.82) \qquad \hat{\sigma}_x = \begin{pmatrix} 0 & 1 \\ 1 & 0 \end{pmatrix}, \qquad \hat{\sigma}_y = i \begin{pmatrix} 0 & -1 \\ 1 & 0 \end{pmatrix}, \qquad \hat{\sigma}_z = \begin{pmatrix} 1 & 0 \\ 0 & -1 \end{pmatrix}$$

These are called the *Pauli spin matrices*. They will be brought into play shortly when we consider the quantum mechanical motion of a spinning electron in a magnetic field.

PROBLEMS

11.45 (a) For spin corresponding to $s = \frac{1}{2}$, show that the eigenvectors of $\hat{S}_x$ and $\hat{S}_y$ are

$$\alpha_x = \frac{1}{\sqrt{2}} \begin{pmatrix} 1 \\ 1 \end{pmatrix}, \qquad \beta_x = \frac{1}{\sqrt{2}} \begin{pmatrix} 1 \\ -1 \end{pmatrix}$$

$$\alpha_y = \frac{1}{\sqrt{2}} \begin{pmatrix} 1 \\ i \end{pmatrix}, \qquad \beta_y = \frac{1}{\sqrt{2}} \begin{pmatrix} 1 \\ -i \end{pmatrix}$$

(b) What are the eigenvalues corresponding to these eigenvectors?

(c) Show that these eigenvectors comprise two sets of orthonormal vectors.

11.46 Spin-$\frac{1}{2}$ state vectors $\binom{a}{b}$ are called *spinors*. Spinors exist in a two-dimensional, complex space. Any element ξ of this space is a set of two complex numbers. Any two linearly independent spinors span this space. In particular, show that this space is spanned by any of the three pairs of eigenstates $(\alpha_x, \beta_x; \alpha_y, \beta_y; \alpha_z, \beta_z)$. That is, show that any spinor $\binom{a}{b}$ may be expressed as a linear combination of any one of these three pairs of eigenstates.

Answer

$$\binom{a}{b} = a\alpha_z + b\beta_z$$

$$\binom{a}{b} = \frac{1}{\sqrt{2}}[(a + b)\alpha_x + (a - b)\beta_x]$$

$$\binom{a}{b} = \frac{1}{\sqrt{2}}[(a - ib)\alpha_y + (a + ib)\beta_y]$$

11.47 Measurement of the z component of the spin of a neutron finds the value $S_z = \hbar/2$.
(a) What spin state is the particle in after measurement?
(b) Show that in this state

$$\langle S_x^2 \rangle = \langle S_y^2 \rangle = \frac{1}{2}\langle S^2 - S_z^2 \rangle = \frac{\hbar^2}{4} = \langle S_z^2 \rangle$$

by direct calculation.

11.48 An electron is known to be in the spin state α_z. Show that in this state

$$\langle S_x \rangle = \langle S_y \rangle = 0$$

Explain this result geometrically (Fig. 11.2).

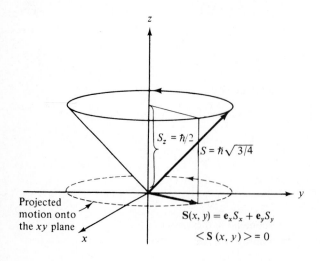

FIGURE 11.2 **Dynamical conception of the spin-$\frac{1}{2}$ state. The angular momentum vector precesses maintaining a constant component about an axis. The projection of S onto a plane normal to the precession axis averages to zero. (See Problem 11.48.)**

11.49 Show that it is impossible for a spin-$\frac{1}{2}$ particle to be in a state $\xi = \binom{a}{b}$ such that

$$\langle S_x \rangle = \langle S_y \rangle = \langle S_z \rangle = 0$$

Answer
Since $\hat{\boldsymbol{\sigma}}$ and $\hat{\mathbf{S}}$ are related through a constant (11.81), we may work with $\hat{\boldsymbol{\sigma}}$ instead of $\hat{\mathbf{S}}$. First we find the relation between a and b which gives $\langle \hat{\sigma}_z \rangle = 0$.

$$\langle \xi | \hat{\sigma}_z | \xi \rangle = \frac{(a^* \; b^*)}{} \begin{pmatrix} 1 & 0 \\ 0 & -1 \end{pmatrix} \begin{pmatrix} a \\ b \end{pmatrix} = |a|^2 - |b|^2 = 0$$

Setting $|a| = 1$, with no loss of generality, this condition implies that ξ is of the form

$$\xi = \begin{pmatrix} e^{i\phi} \\ e^{i\psi} \end{pmatrix}$$

where ϕ and ψ are the phases of a and b, respectively. Substituting this vector into $\langle \hat{\sigma}_x \rangle = 0$ gives

$$\langle \xi | \hat{\sigma}_x | \xi \rangle = (e^{-i\phi} \; e^{-i\psi}) \begin{pmatrix} 0 & 1 \\ 1 & 0 \end{pmatrix} \begin{pmatrix} e^{i\phi} \\ e^{i\psi} \end{pmatrix} = 2 \cos \alpha = 0$$

$$\alpha \equiv \psi - \phi$$

while $\langle \hat{\sigma}_y \rangle = 0$ gives

$$\langle \xi | \hat{\sigma}_y | \xi \rangle = 2 \sin \alpha = 0$$

Since there is no value of α for which $\sin \alpha = \cos \alpha = 0$, we conclude that the stated hypothesis is correct.

11.50 Show that the components of $\hat{\boldsymbol{\sigma}}$ anticommute. For example, show that

$$\hat{\sigma}_x \hat{\sigma}_y + \hat{\sigma}_y \hat{\sigma}_x = 0$$

11.7 FREE-PARTICLE WAVEFUNCTIONS, INCLUDING SPIN

The coordinates of a particle include the spin variables (s, m_s) and the position variables (x, y, z). The operators corresponding to these variables $(\hat{S}^2, \hat{S}_z, \hat{x}, \hat{y}, \hat{z})$ are assumed to commute. Their eigenvalues may therefore be prescribed simultaneously (one may locate a particle without destroying its spin state). Another set of commuting operators for a free particle is $(\hat{S}^2, \hat{S}_z, \hat{p}_x, \hat{p}_y, \hat{p}_z)$.

The Hamiltonian of a free particle is

$$\hat{H} = \frac{\hat{p}^2}{2m}$$

The reason that $\hat{S}$ does not enter in the Hamiltonian of a free particle is that the spin manifests itself only in the presence of an electromagnetic field. It follows that $\hat{H}$ also commutes with $\hat{S}^2$ and $\hat{S}_z$. These operators have simultaneous eigenstates. Let $\varphi_{\mathbf{k}}$ be an eigenstate of $\hat{H}$ corresponding to the eigenvalue $\hbar^2 k^2/2m$ and (α, β) be the eigenstates of $\hat{S}^2, \hat{S}_z$ corresponding to the respective eigenvalues $3\hbar^2/4$ and $\pm\hbar/2$. Then

$$(11.83) \qquad \varphi_+ \equiv \varphi_{\mathbf{k}}(\mathbf{r})\alpha = Ae^{i\mathbf{k}\cdot\mathbf{r}}\begin{pmatrix} 1 \\ 0 \end{pmatrix}$$

gives

$$\hat{H}\varphi_+ = \frac{\hbar^2 k^2}{2m}\varphi_+, \qquad \hat{S}^2\varphi_+ = \frac{3\hbar^2}{4}\varphi_+, \qquad \hat{S}_z\varphi_+ = \frac{\hbar}{2}\varphi_+$$

The eigenstate

$$(11.84) \qquad \varphi_- = \varphi_{\mathbf{k}}(\mathbf{r})\beta = Ae^{i\mathbf{k}\cdot\mathbf{r}}\begin{pmatrix} 0 \\ 1 \end{pmatrix}$$

gives

$$\hat{H}\varphi_- = \frac{\hbar^2 k^2}{2m}\varphi_-, \qquad \hat{S}^2\varphi_- = \frac{3\hbar^2}{4}\varphi_-, \qquad \hat{S}_z\varphi_- = -\frac{\hbar}{2}\varphi_-$$

Consider that $\mathbf{k}$ is unidirectional. Then the state φ_+ gives the same energy $\hbar^2 k^2/2m$ for the two vectors $\pm\mathbf{k}$. The same is true for φ_-. Thus we find that eigenenergies of the spinning free particle propagating in one dimension are fourfold degenerate. This is illustrated below.

$$E_k = \frac{\hbar^2 k^2}{2m} \quad \begin{cases} \varphi_+(+\mathbf{k}) \\ \varphi_+(-\mathbf{k}) \\ \varphi_-(+\mathbf{k}) \\ \varphi_-(-\mathbf{k}) \end{cases}$$

The time-dependent wavefunction for an electron with momentum $\hbar\mathbf{k}$ and with z component of spin $-\hbar/2$ is the column vector

$$\psi_{\mathbf{k}}(\mathbf{r}, t) = Ae^{i(\mathbf{k}\cdot\mathbf{r}-\omega t)}\begin{pmatrix} 0 \\ 1 \end{pmatrix}$$

$$\hbar\omega = \frac{\hbar^2 k^2}{2m}$$

With $A = (2\pi)^{-3/2}$, one obtains

$$\langle \mathbf{k}'|\mathbf{k}\rangle = \iiint \psi_{\mathbf{k}'}{}^*\psi_{\mathbf{k}}\, d\mathbf{r} = \delta(\mathbf{k}' - \mathbf{k})$$

thereby regaining the normalization (10.18) relevant to free-particle motion.

11.51 A beam propagating in the x direction is comprised of 1.5-keV electrons, 20% of which have spin polarized in the $+z$ direction and 80% of which have spin polarized in the $-z$ direction.

    (a)   What is the wavefunction of an electron in the beam?

    (b)   What are the values of the wavenumber k and frequency ω of these electrons?

11.52 A free electron is known to have the following properties:

    1.   Its orbital angular momentum about a prescribed origin is

$$L = \hbar\sqrt{l(l+1)}$$

    2.   Its z component of orbital angular momentum is $\hbar m_l$.

    3.   Its z component of spin is $-\hbar/2$.

    4.   It has kinetic energy

$$E = \frac{\hbar^2 k^2}{2m}$$

    (a)   What is the time-dependent wavefunction, $|\psi\rangle = |k, l, m_l, s, m_s, t\rangle$, for this electron?

    (b)   If in an ideal measurement, the x component of linear momentum is measured and the value

$$p_x = \hbar k_x$$

is found, what time-dependent state is the electron left in?

    (c)   What is the degeneracy of the eigenenergy corresponding to part (b)?

11.8 THE MAGNETIC MOMENT OF AN ELECTRON

Bohr Magneton

The student will recall that a circular loop of wire carrying a current I, which is of cross-sectional area A, produces a magnetic moment[1]

(11.85) $$\mu = IA$$

The magnetic field due to this current loop is sketched in Fig. 11.3. In the limit that $A \rightarrow 0$ and $I \rightarrow \infty$ such that the product $IA = \mu$ remains finite, the magnetic field generated by the loop becomes a dipole field, whose components are also given in Fig. 11.3.

---

[1] In cgs, A is in cm^2 and I is in emu/s. The units of μ are erg/gauss. Also recall that 1 emu = 1 esu/c.

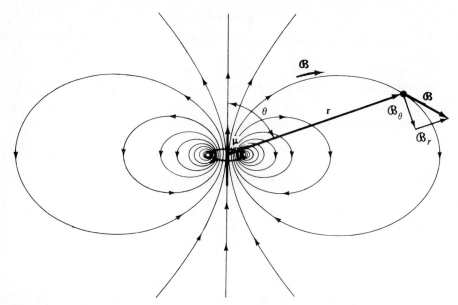

FIGURE 11.3 Magnetic field lines of a current loop. The magnetic dipole of the current loop is μ. The magnitude of μ is IA, where A is the area of the loop. At distances far removed from the origin, the $\mathscr{B}$ field is that due to a magnetic dipole at the origin. In spherical coordinates, with μ parallel to the polar axis, the components of the dipole magnetic field are

$$\mathscr{B}_r = \frac{2\mu \cos \theta}{r^3}, \qquad \mathscr{B}_\theta = \frac{\mu \sin \theta}{r^3}$$

These values are computed from the Biot–Savart law

$$\mathscr{B}(\mathbf{r}) = \frac{1}{c} \int \mathbf{J}(\mathbf{r}') \times \frac{(\mathbf{r} - \mathbf{r}')}{|\mathbf{r} - \mathbf{r}'|^3} \, d\mathbf{r}'$$

which gives the magnetic field at r, due to the current density J distributed over the source points, r'.

For an electron, one finds that the magnetic moment is directly proportional to its spin angular momentum. It is given by[1]

(11.86)
$$\boxed{\boldsymbol{\mu} = \frac{e}{mc}\, \mathbf{S} = \frac{e\hbar}{2mc}\, \boldsymbol{\sigma} = -\mu_b \boldsymbol{\sigma}}$$

[1] A more detailed analysis, including radiative corrections, finds a slightly larger value of electron magnetic moment and (11.86) is more accurately written $\boldsymbol{\mu} = (1.001\mu_b)\boldsymbol{\sigma}$. Thus to within 0.1 % accuracy, the electron magnetic moment has the value of 1 Bohr magneton.

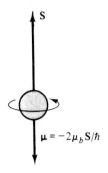

$\mu = -2\mu_b S/\hbar$

FIGURE 11.4   For an electron, the spin and magnetic moment are antialigned.

The quantity $eh/2mc$ is called a *Bohr magneton*. It has the value

(11.87)
$$\mu_b = \frac{|e|\hbar}{2mc} = 0.927 \times 10^{-20} \text{ erg/gauss}$$

Since the charge of the electron is negative, (11.86) may be written

$$\boldsymbol{\mu} = -\mu_b \boldsymbol{\sigma} = -2\left(\frac{\mu_b}{\hbar}\right)\mathbf{S}$$

That is, the spin and magnetic moment of an electron are antialigned (Fig. 11.4).

If a magnetic moment is placed in a uniform, constant magnetic field, a torque is exerted on it about its origin, given by

(11.88)
$$\mathbf{N} = \boldsymbol{\mu} \times \mathcal{B}$$

(Fig. 11.5). It follows that a magnetic moment tends to align itself with a magnetic field in which it is immersed. We may use (11.88) to calculate the work done in rotating the moment from the parallel orientation ($\theta = 0$) to the inclination $\theta > 0$ (Fig. 11.6):

(11.89)
$$V = \int_0^\theta N \, d\theta = -\mu\mathcal{B} \cos \theta + \text{constant}$$
$$V = -\boldsymbol{\mu} \cdot \mathcal{B}$$

If we plot this potential versus θ (Fig. 11.7), it is evident that $\theta = 0$ is the stable orientation of the dipole. While the torque vanishes at $\theta = \pi$ ($\boldsymbol{\mu}$ antiparallel to $\mathcal{B}$), any fluctuation about this orientation will cause the moment to "flip" to the stable position[1] $\theta = 0$. Although there is a torque on a magnetic moment in a uniform $\mathcal{B}$ field, there is no net force on the dipole. However, the expression (11.89) for the

---

[1] This is so if we neglect the angular momentum of the dipole (due to the rotating charge); if the angular momentum of the moment is brought into play, precession results.

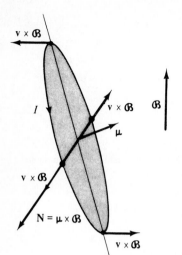

FIGURE 11.5 Torque on a magnetic dipole in a uniform $\mathscr{B}$ field. The force on a point charge moving in a $\mathscr{B}$ field, with velocity v, is

$$\mathbf{F} = \frac{e}{c} \mathbf{v} \times \mathscr{B}$$

Four components of this force along the ring current are shown. These forces tend to align μ with $\mathscr{B}$ so that μ and $\mathscr{B}$ are in the same direction.

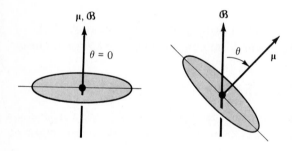

FIGURE 11.6 The energy of interaction between a magnetic dipole μ and a magnetic field $\mathscr{B}$ is a function of the inclination angle θ.

$$V = -\mu \cdot \mathscr{B}$$

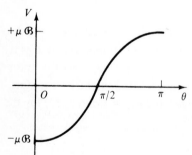

FIGURE 11.7 V versus θ for a magnetic dipole in a uniform, constant $\mathscr{B}$ field.

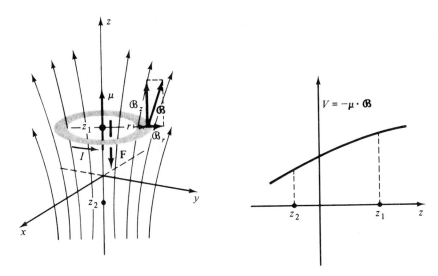

FIGURE 11.8   Force on a magnetic dipole in an inhomogeneous magnetic field. For the configuration shown, it is the r component of magnetic field, $\mathcal{B}_r$, which causes the downward force on the ring. This may be seen by calculating the force $(e/c)\mathbf{v} \times \mathcal{B}$ on the ring, due to the r and z components of $\mathcal{B}$, respectively. In terms of interaction energy, $V = -\boldsymbol{\mu} \cdot \mathcal{B}$, V is larger at z_1 than at z_2. This causes the dipole to "fall" from z_1 to z_2.

interaction energy[1] between $\boldsymbol{\mu}$ and $\mathcal{B}$ suggests that there is a net force on a dipole in a nonuniform $\mathcal{B}$ field. This force causes the dipole, at a given orientation with $\mathcal{B}$, to "fall" to a neighboring point in space where the interaction energy, $-\boldsymbol{\mu} \cdot \mathcal{B}$, is smaller (Fig. 11.8). The net force on the dipole in a nonuniform $\mathcal{B}$ field is given by

$$(11.90) \qquad \mathbf{F} = -\boldsymbol{\nabla} V = -\boldsymbol{\nabla}(-\boldsymbol{\mu} \cdot \mathcal{B}) = \boldsymbol{\nabla}(\boldsymbol{\mu} \cdot \mathcal{B})$$

Stern–Gerlach Experiment

Equation (11.90) reveals the nature of the force which occurs in the Stern–Gerlach (S-G) experiment originally performed in 1922 using a beam of silver atoms. In a prototype of this experiment a beam of electrons[2] with an isotropic distribution of dipole orientations is passed through an inhomogeneous magnetic field, as shown in Fig. 11.9. The predominant component of $\mathcal{B}$ is $\mathcal{B}_z$. Furthermore, $\mathcal{B}_z$ varies most strongly with changes in z so that $\boldsymbol{\nabla}\mathcal{B}_z \simeq \mathbf{e}_z\, \partial\mathcal{B}_z/\partial z$. It follows that the force on electrons as they pass through the pole pieces is

$$(11.91) \qquad \mathbf{F} = \boldsymbol{\nabla}\boldsymbol{\mu} \cdot \mathcal{B} \simeq \mathbf{e}_z \mu_z \frac{\partial \mathcal{B}_z}{\partial z} = \mathbf{e}_z F_z$$

---

[1] This expression for the energy does not take into account the energy supplied by the source that maintains the current in the dipole. It gives correct forces, however, if current and $\mathcal{B}$-field are constant in time. For further discussion on this topic, see R. Feynman, *Lectures on Physics*, Vol. II, Addison-Wesley, Reading, Mass., 1964.

[2] See Problem 11.85.

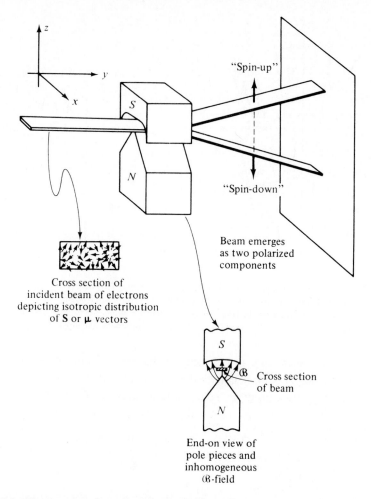

FIGURE 11.9   Elements of the Stern–Gerlach experiment.

The predominant force on the electrons is in the z direction. In addition, the sign of this force is solely determined by the sign of μ_z.

This z component of force causes electrons in the beam to be deflected. They strike the detection screen "off-axis." If we know where an electron strikes the screen, we can[1] calculate F_z. Since $\partial \mathcal{B}_z/\partial z$ is known (it is a property of the apparatus), the equation above then allows one to determine the z component of magnetic moment. Thus the S-G apparatus is a device that measures the z component of magnetic moment. More significantly, since μ is directly proportional to the spin S, the S-G apparatus becomes an instrument that measures the z component of spin. Written in

[1] The kinetic energy of the incident electron is assumed known.

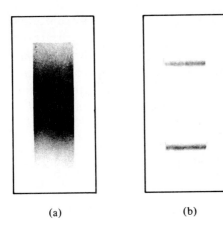

(a) (b)

FIGURE 11.10 Traces of electron beam on detection screen in the Stern–Gerlach experiment. (a) The trace predicted from classical physics,

$$F_z = -\left(\mu_b \frac{\partial \mathscr{B}}{\partial z}\right)\cos\theta$$

Continuous distribution of $\cos\theta$ values implies a continuous trace. (b) The trace observed experimentally. If $s = \frac{1}{2}$ for an electron, then quantum mechanics predicts that two discrete beams emerge from the Stern–Gerlach apparatus.

terms of spin of the electron, the force (11.90) becomes

$$(11.92) \qquad \mathbf{F} = -\frac{2\mu_b}{\hbar}\nabla\mathbf{S}\cdot\mathscr{B} = -\mathbf{e}_z\frac{2\mu_b}{\hbar}S_z\frac{\partial\mathscr{B}_z}{\partial z}$$

At the start of the description of this experiment, we remarked that the incident beam carries an isotropic distribution of directions of magnetic moments or, equivalently, spin vectors (see Fig. 11.9). This means that at any point in the beam it is equally likely to find μ_z with any value from $-\mu$ to $+\mu$. Thus according to (11.91) one expects a uniform distribution of deflections. The pattern that the deflected particles in the beam make on the screen should be a continuous one. However, experiment finds that the beam divides into two discrete components (Fig. 11.10). Thus experiment indicates that μ_z or, equivalently, S_z has only two components. But this is precisely what one expects if $s = \frac{1}{2}$, for in this case S_z has only the two values $\pm\hbar/2$. So according to (11.92), an electron going through the Stern–Gerlach $\mathscr{B}$ field is acted on by only one of two possible values of force:

$$F_z = \pm\frac{\mu_b\partial\mathscr{B}_z}{\partial z}$$

These two oppositely directed components of force divide the beam into two separate components.

Superposition Spin State

Instead of the beam containing an isotropic distribution of spin orientations, let electrons in the incident beam all be polarized with spins in the $+x$ direction. That is, each electron is in the eigenstate α_x of $\hat{S}_x$ and has the wavefunction

$$(11.93) \qquad \psi = Ae^{i(ky-\omega t)}\alpha_x = \frac{A}{\sqrt{2}}e^{i(ky-\omega t)}\begin{pmatrix}1\\1\end{pmatrix}$$

The beam propagates in the $+y$ direction. Let the beam be incident on a S-G apparatus whose $\mathscr{B}$ field is aligned with the z axis. What portion of electrons in the incident beam emerges with spins in the $+z$ and $-z$ directions, respectively? Equivalently, we may ask, what is the probability that measurement of the z component of spin of an electron in the beam finds the respective values of $+\hbar/2$ or $-\hbar/2$? To answer this question, we call on the superposition principle. Expanding the column vector α_x in terms of the eigenvectors of $\hat{S}_z$ gives

$$(11.94) \qquad \alpha_x = \frac{1}{\sqrt{2}}\begin{pmatrix} 1 \\ 1 \end{pmatrix} = \frac{1}{\sqrt{2}}\left[\begin{pmatrix} 1 \\ 0 \end{pmatrix} + \begin{pmatrix} 0 \\ 1 \end{pmatrix}\right] = \frac{1}{\sqrt{2}}(\alpha_z + \beta_z)$$

The two coefficients of expansion are equal. It follows that the probability of measuring $S_z = +\hbar/2$

$$P\left(+\frac{\hbar}{2}\right) = \frac{1}{2} = |\langle \alpha_x | \alpha_z \rangle|^2$$

is equal to that of measuring $S_z = -\hbar/2$

$$P\left(-\frac{\hbar}{2}\right) = \frac{1}{2} = |\langle \alpha_x | \beta_z \rangle|^2$$

Thus if a beam of polarized electrons all in the state α_x enters a S-G apparatus, two equally populated beams emerge. These beams are also polarized, with electrons in spin states α_z and β_z, respectively.

PROBLEMS

11.53 Show that the magnetic moment due to the orbital motion (as opposed to the spin) of the electron in the Bohr-model hydrogen atom is given by

$$\mu = \frac{\mu_b L}{\hbar}$$

The orbital angular momentum of the electron is $\mathbf{L}$. (*Hint:* If the electron moves in a circle with velocity $\mathbf{v}$ and radius r, the related current is

$$I = \frac{ev}{2\pi rc} \quad \text{emu/s)}.$$

11.54 Consider that a polarized beam containing electrons in the α_z state is sent through a S-G analyzer which measures S_x. What values will be found, and with what probabilities will these values occur?

Answer

We write α_z as a linear combination of α_x and β_x.

$$\alpha_z = \begin{pmatrix} 1 \\ 0 \end{pmatrix} = \frac{1}{\sqrt{2}} (\alpha_x + \beta_x) = \frac{1}{2} \left\{ \begin{pmatrix} 1 \\ 1 \end{pmatrix} + \begin{pmatrix} 1 \\ -1 \end{pmatrix} \right\}$$

The coefficients of expansion are equal so that it is equally likely to find the values $S_x = +\hbar/2$ or $S_x = -\hbar/2$.

11.55   A proton is in the spin state α_y. What is the probability that measurement finds each of the following?

(a) $S_y = -\hbar/2$
(b) $S_x = +\hbar/2$
(c) $S_x = -\hbar/2$
(d) $S_z = +\hbar/2$
(e) $S_z = -\hbar/2$

Answers

(a) 0
(b) $\frac{1}{2}$
(c) $\frac{1}{2}$
(d) $\frac{1}{2}$
(e) $\frac{1}{2}$

11.56   A collection of electrons has an isotropic distribution of spin values. For an electron chosen at random, what is the probability of finding it with the following spin components?

(a) $S_x = +\hbar/2, -\hbar/2$
(b) $S_y = +\hbar/2, -\hbar/2$
(c) $S_z = +\hbar/2, -\hbar/2$

Answers

If the x component of spin is measured, only two values can be found. If spin is isotropic, these two values are equally likely, hence there is a probability of $\frac{1}{2}$ that $S_x = \hbar/2$ and a probability of $\frac{1}{2}$ that $S_x = -\hbar/2$; similarly for S_y and S_z.

11.9   PRECESSION OF AN ELECTRON IN A MAGNETIC FIELD

In this section we consider the motion of a spinning, but otherwise fixed electron which is in a constant uniform magnetic field that points in the z direction. Suppose that the electron is initially in the α_x state

(11.95)
$$\xi(0) = \alpha_x = \frac{1}{\sqrt{2}} \begin{pmatrix} 1 \\ 1 \end{pmatrix}$$

What is $\xi(t)$? To answer this question, we write down the time-dependent Schrödinger equation for the state $\xi(t)$.

(11.96)
$$ih \frac{\partial}{\partial t} \xi = \hat{H}\xi$$

The Hamiltonian is the interaction energy

$$\hat{H} = -\hat{\boldsymbol{\mu}} \cdot \boldsymbol{\mathscr{B}} = +\mu_b \hat{\boldsymbol{\sigma}} \cdot \boldsymbol{\mathscr{B}} = \mu_b \mathscr{B}\hat{\sigma}_z$$

In the matrix representation with $\hat{S}_z$ diagonal, $\hat{H}$ becomes

(11.97)
$$\hat{H} = \mu_b \mathscr{B} \begin{pmatrix} 1 & 0 \\ 0 & -1 \end{pmatrix}$$

We seek the solution to (11.96) with this Hamiltonian, for the state vector

$$\xi(t) = \begin{pmatrix} a(t) \\ b(t) \end{pmatrix}$$

Substitution of this form into (11.96) with (11.97) substituted for $\hat{H}$ gives the column vector equation

$$\begin{pmatrix} \dot{a} \\ \dot{b} \end{pmatrix} = -\frac{i\Omega}{2} \begin{pmatrix} a \\ -b \end{pmatrix}$$

where $\Omega/2$ is the *Larmor frequency* and

(11.98)
$$\Omega = \frac{|e|\mathscr{B}}{mc}$$

is commonly referred to as the *cyclotron frequency* (previously encountered in Section 10.4). The column vector equation above is equivalent to the two independent equations

$$\dot{a} = -\frac{i\Omega}{2} a$$

$$\dot{b} = +\frac{i\Omega}{2} b$$

which has the solution

(11.99)
$$\xi(t) = \begin{pmatrix} a \\ b \end{pmatrix} = \frac{1}{\sqrt{2}} \begin{pmatrix} e^{-i(\Omega/2)t} \\ e^{+i(\Omega/2)t} \end{pmatrix}$$

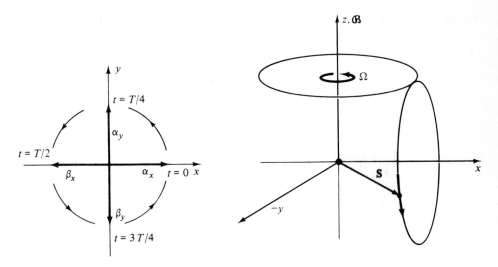

FIGURE 11.11   Precession of a spinning but otherwise fixed electron in a constant, uniform magnetic field. Electron is initially in the α_x state.

This solution is compatible with the initial conditions (11.95). At later times (including $t = 0$) one obtains

$$\xi(0) = \alpha_x$$
$$\xi(T/4) = e^{-i\pi/4}\alpha_y$$
$$\xi(T/2) = e^{-i\pi/2}\beta_x$$
$$\xi(3T/4) = e^{-i3\pi \, 4}\beta_y$$

where we have written T for the period $2\pi/\Omega$. Apart from constant phase factors, we may conclude the following. At $t = 0$, the electron is in an eigenstate of S_x corresponding to the eigenvalue $+\hbar/2$. At $T/4$ it is in an eigenstate of S_y corresponding to the eigenvalue $+\hbar/2$. Proceeding in this manner one finds that the spin of the electron precesses about the z axis with angular frequency Ω (see Fig. 11.11).

Eigenenergies

We now consider the problem of calculating the eigenstates and eigenenergies of this same system, i.e., a spinning but otherwise fixed electron in a constant uniform magnetic field that points in the z direction. To solve this problem we use the time-independent Schrödinger equation. For the case at hand, it appears as

(11.100)
$$\hat{H}\xi = E\xi$$
$$\hat{H} = \mu_b \mathcal{B}\hat{\sigma}_z$$

Setting $\xi = \begin{pmatrix} a \\ b \end{pmatrix}$ gives

$$\mu_b \mathcal{B} \begin{pmatrix} 1 & 0 \\ 0 & -1 \end{pmatrix} \begin{pmatrix} a \\ b \end{pmatrix} = E \begin{pmatrix} a \\ b \end{pmatrix}$$

or, equivalently,

$$(\mu_b \mathcal{B} - E)a = 0$$
$$(\mu_b \mathcal{B} + E)b = 0$$

If $a \neq 0$, $E = +\mu_b \mathcal{B}$ and $b = 0$. If $b \neq 0$, $E = -\mu_b \mathcal{B}$, and $a = 0$. Thus we obtain the (normalized) eigenstates, and eigenenergies

(11.101)
$$\alpha = \begin{pmatrix} 1 \\ 0 \end{pmatrix}, \qquad E = \mu_b \mathcal{B} = \frac{\hbar \Omega}{2}$$

$$\beta = \begin{pmatrix} 0 \\ 1 \end{pmatrix}, \qquad E = -\mu_b \mathcal{B} = -\frac{\hbar \Omega}{2}$$

In the state of higher energy, the spin of the electron is parallel to $\mathcal{B}$, so the magnetic moment is antiparallel to $\mathcal{B}$ and the interaction energy $-\boldsymbol{\mu} \cdot \mathcal{B}$ is maximum. In the state of lower energy, the spin is antiparallel to $\mathcal{B}$, so the magnetic moment is parallel to $\mathcal{B}$ and $-\boldsymbol{\mu} \cdot \mathcal{B}$ is minimum (Fig. 11.12).

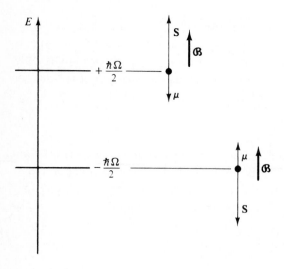

FIGURE 11.12   Energy eigenvalues of a spinning but otherwise fixed electron in a uniform, constant magnetic field. The orientation of μ with respect to $\mathcal{B}$ is also shown for these two states.

Magnetic Resonance

As found above (11.86), the relation between the magnetic moment of an electron and its spin is given by

$$\boldsymbol{\mu} = -2\left(\frac{\mu_b}{\hbar}\right)\mathbf{S}$$

This expression may be written in terms of the *Landé g factor*,

$$\boldsymbol{\mu} = g\left(\frac{\mu_b}{\hbar}\right)\mathbf{S}$$

with $g = -2$. For nuclear particles such as a proton or a neutron, the relevant unit of magnetic moment is the *nuclear magneton*

$$\mu_N = \frac{e\hbar}{2M_p c} = 0.505 \times 10^{-23} \text{ erg/gauss}$$

where M_p is the mass of the proton. The magnetic moment of a nuclear particle is written

$$\boldsymbol{\mu} = g\left(\frac{\mu_N}{\hbar}\right)\mathbf{S}$$

For a proton $g = 2(2.79)$, while for a neutron $g = 2(-1.91)$. We wish now to describe a technique of measuring g or, equivalently, the magnetic moment.

As found in the first part of this section, if a magnetic moment is immersed in a steady $\mathscr{B}$ field in the z direction, the moment will precess about the z axis with the "Larmor" frequency $g\Omega/2$. Measurement of the magnetic moment is made through inducing a spin flip of the magnetic moment between its two energy states (for a spin-$\frac{1}{2}$ particle) parallel and antiparallel to $\mathscr{B}$ (Fig. 11.12). As in the corresponding classical configuration, in order to change the angle that $\boldsymbol{\mu}$ makes with the z axis, it is necessary to impose an additional transverse magnetic field normal to the plane through the z axis and $\boldsymbol{\mu}$ (Fig. 11.13). Since this plane rotates with the Larmor frequency, in the corresponding quantum mechanical motion one may expect a spin flip of the magnetic moment to be induced when the frequency of an imposed rotating transverse magnetic field is equal to the Larmor precessional frequency. Let us examine quantitatively the manner in which this resonance occurs for a spin-$\frac{1}{2}$ nuclear particle with magnetic moment $g(\mu_N/\hbar)\mathbf{S}$.

Since $\mathscr{B}$ has three components, the Schrödinger equation (11.96) takes the form

$$i\hbar \frac{\partial}{\partial t}\begin{pmatrix} a \\ b \end{pmatrix} = -\frac{g\mu_N}{2}\left[\begin{pmatrix} 0 & 1 \\ 1 & 0 \end{pmatrix}\mathscr{B}_x + i\begin{pmatrix} 0 & -1 \\ 1 & 0 \end{pmatrix}\mathscr{B}_y + \begin{pmatrix} 1 & 0 \\ 0 & -1 \end{pmatrix}\mathscr{B}_z\right]\begin{pmatrix} a \\ b \end{pmatrix}$$

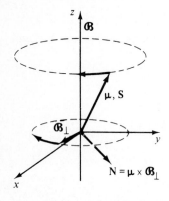

FIGURE 11.13 Transverse field $\mathscr{B}_\perp$ imposes a torque N that causes μ (or, equivalently, S) to change its orientation with respect to the z axis.

Let the transverse magnetic field ($\mathscr{B}_x$, $\mathscr{B}_y$) rotate with frequency ω and let $\mathscr{B}_z$ maintain at the constant value $\mathscr{B}_\parallel$.

$$\mathscr{B}_x = \mathscr{B}_\perp \cos \omega t, \qquad \mathscr{B}_y = -\mathscr{B}_\perp \sin \omega t, \qquad \mathscr{B}_z = \mathscr{B}_\parallel$$

Substituting these values into the preceding equation gives the coupled equations

(11.102)
$$\frac{\partial a}{\partial t} = i(\Omega_\perp b e^{i\omega t} + \Omega_\parallel a)$$

$$\frac{\partial b}{\partial t} = i(\Omega_\perp a e^{-i\omega t} - \Omega_\parallel b)$$

The frequencies $\Omega_\perp$ and $\Omega_\parallel$ are defined through the relations

$$2\hbar\Omega_\perp = g\mu_N \mathscr{B}_\perp, \qquad 2\hbar\Omega_\parallel = g\mu_N \mathscr{B}_\parallel$$

We seek solution to equations (11.102) corresponding to the initial conditions

$$a = 1, \qquad b = 0 \quad \text{at } t = 0$$

Let us look for solutions in the form

(11.103)
$$a = \bar{a}e^{i\omega_a t}$$
$$b = \bar{b}e^{i\omega_b t}$$

The coefficients $\bar{a}$ and $\bar{b}$ are assumed independent of time. Substituting these forms into (11.102) gives the homogeneous matrix equation

(11.104)
$$\begin{pmatrix} (\omega_a - \Omega_\parallel) & -\Omega_\perp e^{-i\phi t} \\ -\Omega_\perp e^{i\phi t} & (\omega_b + \Omega_\parallel) \end{pmatrix} \begin{pmatrix} \bar{a} \\ \bar{b} \end{pmatrix} = 0$$

$$\phi \equiv \omega_a - \omega_b - \omega$$

In order that $\bar{a}$ and $\bar{b}$ be independent of time, we must set $\phi = 0$, which gives

$$\omega_a = \omega_b + \omega$$

Setting the determinant of the coefficient matrix in (11.104) equal to zero gives the frequencies

(11.105)
$$\omega_b = -\frac{\omega}{2} \pm \bar{\omega}, \qquad \omega_a = \frac{\omega}{2} \pm \bar{\omega}$$

$$\bar{\omega}^2 \equiv \left(\Omega_{\parallel} - \frac{\omega}{2}\right)^2 + \Omega_{\perp}^2$$

It is only for these values of ω_a and ω_b that the proposed forms (11.103) are solutions to (11.102). The fact that two frequencies emerge for ω_a and ω_b corresponds to the property that (11.102) represents two independent, second-order differential equations in time. The general solution for, say, $b(t)$ is a linear combination of the two frequency components.

$$b(t) = b_1 \exp\left[-i\left(\frac{\omega}{2} - \bar{\omega}\right)t\right] + b_2 \exp\left[-i\left(\frac{\omega}{2} + \bar{\omega}\right)t\right]$$

To match this solution to the initial condition $b(0) = 0$, we choose $b_1 = -b_2 = C/2i$. There results

$$b(t) = C \sin \bar{\omega}t \, e^{-i(\omega/2)t}$$

To fix the coefficient C, we insert this solution into the second of equations (11.102) and set $t = 0$, $a(0) = 1$, to find

$$C = \frac{i\Omega_{\perp}}{\bar{\omega}}$$

The solution $a(t)$ corresponding to the specific form $b(t)$ obtained above is simply constructed from the second equation of (11.102).

Combining our results we obtain

(11.106)
$$|\xi\rangle = \begin{pmatrix} a(t) \\ b(t) \end{pmatrix} = \frac{\sin \bar{\omega}t}{\bar{\omega}} \begin{pmatrix} e^{i(\omega/2)t}[i(\Omega_{\parallel} - \omega/2) + \bar{\omega} \cot \bar{\omega}t] \\ ie^{-i(\omega/2)t}\Omega_{\perp} \end{pmatrix}$$

Since normalization of the spinor $|\xi\rangle$ is obeyed at $t = 0$,

$$\langle \xi(0)|\xi(0)\rangle = |a(0)|^2 + |b(0)|^2 = 1$$

we should find that it is maintained for all time. From (11.106) we obtain

$$|a|^2 + |b|^2 = \frac{\sin^2 \bar{\omega}t}{\bar{\omega}^2}\left[\left(\Omega_{\parallel} - \frac{\omega}{2}\right)^2 + \Omega_{\perp}^2\right] + \cos^2 \bar{\omega}t$$

Recalling the definition of $\bar{\omega}$ (11.105), we find that

$$\langle \xi(t)|\xi(t)\rangle = \sin^2 \bar{\omega}t + \cos^2 \bar{\omega}t = 1$$

so that normalization is maintained for all time.

Let us see how the solution (11.106) implies the resonant spin-flip behavior when the driving frequency ω of the transverse field $\mathscr{B}_\perp$ is equal to the frequency $2\Omega_{||}$. From the form of $|b|^2$,

(11.107)
$$|b(t)|^2 = \left[\frac{\Omega_\perp^2}{(\Omega_{||} - \omega/2)^2 + \Omega_\perp^2}\right]\sin^2 \bar{\omega}t$$

we may infer the following. In an ensemble of spins that are all pointing in $+z$ direction at $t = 0$, the fraction $|b(t)|^2$ will be found pointing in the $-z$ direction at the time t. From the preceding expression for $|b|^2$, it is clear that the number of such spin flips is maximized at the resonant frequency

(11.108)
$$\omega = 2\Omega_{||} = \frac{g\mu_N \mathscr{B}_{||}}{\hbar}$$

which, as expected, corresponds to the precessional frequency. The structure of the amplitude of $|b(t)|^2$ further indicates that this resonant phenomenon may be made more pronounced by choosing the transverse field small in comparison to the steady parallel field $\mathscr{B}_{||}$.

Experimental Description

The molecules in a sample of water have zero magnetic moment save for that carried by the protons of the hydrogen nuclei. When placed in a steady magnetic field these protons align in one of two possible orientations. In thermal equilibrium there are slightly more protons in the lower-energy orientation with magnetic moment parallel to $\mathscr{B}$ (Fig. 11.12). One may utilize the presence of this net excess of magnetic moment to measure the g factor of protons. As described above, spin flips of the aligned protons may be induced by an additional rotating transverse magnetic field. This effect is maximized at the resonant frequency (11.108)

$$\omega = 2\Omega_{||} = g\frac{e}{2M_p c}\mathscr{B}_{||}$$

With the value of this frequency observed, the relation above may be solved for g, since all other quantities in the equation are known.

In inducing a transition from the lower to the upper energy state, energy in the amount

$$\hbar\Omega = \hbar g\,\frac{e}{2M_p c}\,\mathscr{B}_{\parallel}$$

is absorbed by the transition. This energy comes from the supporting transverse magnetic field coils. Spin flips to the lower parallel orientation expel energy $\hbar\Omega$. Since there are slightly more protons in the lower energy state than in the upper energy state in thermal equilibrium, there will be a net detectable absorption of energy from the transverse coils at the resonant frequency. Measurement of this frequency yields the value $g = 5.58$ for the proton.

In concluding this discussion we note the following convention regarding magnetic moment. Consider a particle with intrinsic spin **S**. Its magnetic moment is given by

$$\boldsymbol{\mu} = g\left(\frac{\mu_N}{\hbar}\right)\mathbf{S}$$

The experimentally quoted value of magnetic moment refers to the expectation of the z component of $\boldsymbol{\mu}$ with **S** inclined maximally toward the z axis, that is, in the state $m_s = s$. With the wavefunction of the particle written $|s, m_s\rangle$ this value is given by

$$\mu = \langle s, s|g\left(\frac{\mu_N}{\hbar}\right)\hat{S}_z|s, s\rangle$$

$$= g\left(\frac{\mu_N}{\hbar}\right)\hbar s = gs\mu_N$$

For a proton

$$\boldsymbol{\mu}_p = 2(2.79)\left(\frac{\mu_N}{\hbar}\right)\mathbf{S}$$

which gives the value

$$\mu_p = 2.79\mu_N$$

For a neutron

$$\boldsymbol{\mu}_n = -2(1.91)\left(\frac{\mu_N}{\hbar}\right)\mathbf{S}$$

to which corresponds the value

$$\mu_n = -1.91\mu_N$$

Similarly, for an electron we find that

$$\mu_e = -\mu_b$$

PROBLEMS

11.57 What frequencies are emitted by an electron gas of low density ($\sim 10^8$ cm^{-3}) which is immersed in a uniform magnetic field of strength 10^4 gauss due to spin flips? What type of radiation is this (x ray, microwaves, etc.)? How does this emission frequency compare to the classical frequency emitted when an electron is executing Larmor rotation (see Section 10.4)?

11.58 In a nuclear magnetic resonance (NMR) experiment with $\mathscr{B}_\parallel$ set at 5000 gauss, resonant energy absorption by a sample of water is detected when the frequency of the transverse components of magnetic field passes through the value 21.2 MHz. What value of g for a proton does this data imply?

11.10 THE ADDITION OF TWO SPINS

In Section 9.4 it was concluded that when adding the angular momenta of two components of a system, one may work in one of two representations: the *coupled representation* or the *uncoupled representation*. This also holds, of course, for the addition of spin angular momenta. As in the orbital case, the construction of wavefunctions in the uncoupled representation proves simpler. In this representation, wavefunctions for the two-electron system are simultaneous eigenstates of the four commuting operators $\hat{S}_1^2, \hat{S}_2^2, \hat{S}_{1z}$, and $\hat{S}_{2z}$. They may be written

(11.109) $$\xi = |s_1 s_2 m_{s_1} m_{s_2}\rangle = |\tfrac{1}{2}, \tfrac{1}{2}, \pm\tfrac{1}{2}, \pm\tfrac{1}{2}\rangle$$

where, for example,

$$\hat{S}_1^2 |s_1 s_2 m_{s_1} m_{s_2}\rangle = \hbar^2 s_1(s_1 + 1)|s_1 s_2 m_{s_1} m_{s_2}\rangle$$
$$= \tfrac{3}{4}\hbar^2 |s_1 s_2 m_{s_1} m_{s_2}\rangle$$

These eigenstates, $(\xi_1, \ldots, \xi_4)$, are simple products of the eigenstates $\alpha(1), \beta(1)$ of $\hat{S}_1^2, \hat{S}_{1z}$ and $\alpha(2), \beta(2)$ of $\hat{S}_2^2, \hat{S}_{2z}$. They are listed in Table 11.2. The column on the left of Table 11.2 contains diagrammatic representations of these states in which the relative orientation of the two spins is suggested.

TABLE 11.2 Spin wave function for two electrons in the uncoupled representation

Spin alignment	Wavefunction $\xi = \|s_1 s_2 m_{s_1} m_{s_2}\rangle$	m_{s_1}	m_{s_2}	s_1	s_2
↑ ↑	$\xi_1 = \alpha(1)\alpha(2)$	$+\frac{1}{2}$	$+\frac{1}{2}$	$\frac{1}{2}$	$\frac{1}{2}$
↓ ↓	$\xi_2 = \beta(1)\beta(2)$	$-\frac{1}{2}$	$-\frac{1}{2}$	$\frac{1}{2}$	$\frac{1}{2}$
↑ ↓	$\xi_3 = \alpha(1)\beta(2)$	$+\frac{1}{2}$	$-\frac{1}{2}$	$\frac{1}{2}$	$\frac{1}{2}$
↓ ↑	$\xi_4 = \beta(1)\alpha(2)$	$-\frac{1}{2}$	$+\frac{1}{2}$	$\frac{1}{2}$	$\frac{1}{2}$

In the coupled representation, one constructs simultaneous eigenstates of the four commuting operators

$$\hat{S}^2 = (\hat{\mathbf{S}}_1 + \hat{\mathbf{S}}_2)^2 = \hat{S}_1^{\,2} + \hat{S}_2^{\,2} + 2\hat{\mathbf{S}}_1 \cdot \hat{\mathbf{S}}_2$$
$$\hat{S}_z = \hat{S}_{1z} + \hat{S}_{2z}$$
$$\hat{S}_1^{\,2}, \hat{S}_2^{\,2}$$

The simultaneous eigenstates of these four operators may be written $|s m_s s_1 s_2\rangle$. Since $s = (0, 1)$, again there are four such independent eigenstates. In constructing these states we show first that two of the states appropriate to the uncoupled representation are also eigenstates of the coupled representation. Toward these ends note that all four uncoupled states are eigenstates of the total z component of spin.

(11.110)
$$\hat{S}_z \xi_1 = \hbar \xi_1$$
$$\hat{S}_z \xi_2 = -\hbar \xi_2$$
$$\hat{S}_z \xi_3 = \hat{S}_z \xi_4 = 0$$

The eigenstate ξ_1 is an eigenvector of $\hat{S}_z$, $\hat{S}_{1z}$, and $\hat{S}_{2z}$. If it is also an eigenstate of $\hat{S}^2$, it is one of the eigenstates appropriate to the coupled representation. To see if this is indeed the case, we employ the relation (see Problem 9.31)

(11.111)
$$\hat{\mathbf{S}}_1 \cdot \hat{\mathbf{S}}_2 = (\hat{S}_{1x}\hat{S}_{2x} + \hat{S}_{1y}\hat{S}_{2y}) + \hat{S}_{1z}\hat{S}_{2z}$$
$$= \tfrac{1}{2}(\hat{S}_{1+}\hat{S}_{2-} + \hat{S}_{1-}\hat{S}_{2+}) + \hat{S}_{1z}\hat{S}_{2z}$$

in the cross term of $\hat{S}^2$ to obtain the expansion

(11.111a) $$\hat{S}^2 = \hat{S}_1^{\,2} + \hat{S}_2^{\,2} + 2\hat{S}_{1z}\hat{S}_{2z} + (\hat{S}_{1+}\hat{S}_{2-} + \hat{S}_{1-}\hat{S}_{2+})$$

The raising operator S_{1+} is defined by[1]

$$\hat{S}_{1+} = \hat{S}_{1x} + i\hat{S}_{1y}$$

[1] See Table 9.4.

with similar definitions carrying over to $\hat{S}_{2+}$, $\hat{S}_{1-}$, and $\hat{S}_{2-}$. Using the above representation of $\hat{\mathbf{S}}_1 \cdot \hat{\mathbf{S}}_2$, we find that

$$\hat{\mathbf{S}}_1 \cdot \hat{\mathbf{S}}_2 \xi_1 = [\tfrac{1}{2}(\hat{S}_{1+}\hat{S}_{2-} + \hat{S}_{1-}\hat{S}_{2+}) + \hat{S}_{1z}\hat{S}_{2z}]\alpha(1)\alpha(2)$$

$$= \left[\frac{1}{2}(0+0) + \frac{\hbar^2}{4}\right]\xi_1 = \frac{\hbar^2}{4}\xi_1$$

so that

$$\hat{S}^2\xi_1 = (\hat{S}_1{}^2 + \hat{S}_2{}^2 + 2\hat{\mathbf{S}}_1 \cdot \hat{\mathbf{S}}_2)\xi_1 = (\tfrac{3}{4}\hbar^2 + \tfrac{3}{4}\hbar^2 + \tfrac{2}{4}\hbar^2)\xi_1$$

$$= 2\hbar^2\xi_1 = \hbar^2 1(1+1)\xi_1$$

We conclude that ξ_1 is also an eigenstate of $\hat{S}^2$, hence it is one of the eigenstates of $\hat{S}^2$ and $\hat{S}_z$ for two spin-$\frac{1}{2}$ particles in the coupled representation. We relabel this eigenstate $\xi_S{}^{(1)}$, for reasons that will become apparent immediately.

$$\xi_1 = \alpha(1)\alpha(2) \equiv \xi_S{}^{(1)}$$

Similarly, we find that ξ_2 is also a common eigenstate of $\hat{S}^2$ and S_z. This eigenstate is relabeled $\xi_S{}^{(-1)}$.

$$\xi_2 \equiv \beta(1)\beta(2) \equiv \xi_S{}^{(-1)}$$

However, ξ_3 and ξ_4 are not eigenstates of $\hat{S}^2$.

The Exchange Operator

To find the remaining two eigenstates (which will be called $\xi_S{}^{(0)}$ and ξ_A) of the coupled representation we introduce the *exchange operator*, $\hat{\mathfrak{X}}$. When $\hat{\mathfrak{X}}$ operates on a function of coordinates (spin or space) of two particles, it exchanges these coordinates. If $\varphi(1, 2)$ is a function of the coordinates of two particles (the spin coordinates of particle 1 are labeled "1," those of particle 2 are labeled "2"), then

(11.112) $$\hat{\mathfrak{X}}\varphi(1, 2) = \varphi(2, 1)$$

From the definition of $\hat{\mathfrak{X}}$, one obtains

$$\hat{\mathfrak{X}}^2\varphi(1, 2) = \hat{\mathfrak{X}}\varphi(2, 1) = \varphi(1, 2)$$

so that the eigenvalues of $\hat{\mathfrak{X}}$ are ± 1. One may construct the corresponding eigenfunctions of $\hat{\mathfrak{X}}$ from any state $\varphi(1, 2)$ as follows:

$$\varphi_S = \varphi(1, 2) + \varphi(2, 1) \qquad \text{(``symmetric'')}$$

clearly has eigenvalue $+1$, whereas

$$\varphi_A = \varphi(1, 2) - \varphi(2, 1) \qquad \text{(``antisymmetric'')}$$

has eigenvalue -1.

Since $\hat{S}^2$ and $\hat{S}_z$ commute with $\hat{\mathfrak{X}}$, it follows that $\hat{S}^2$, $\hat{S}_z$, and $\hat{\mathfrak{X}}$ have common eigenstates. We already know two of them, $\xi_S^{(1)}$ and $\xi_S^{(-1)}$.

$$\hat{S}^2\xi_S^{(1)} = 2\hbar^2\xi_S^{(1)}, \qquad \hat{S}_z\xi_S^{(1)} = \hbar\xi_S^{(1)}, \qquad \hat{\mathfrak{X}}\xi_S^{(1)} = +1\xi_S^{(1)}$$
$$\hat{S}^2\xi_S^{(-1)} = 2\hbar^2\xi_S^{(-1)}, \qquad \hat{S}_z\xi_S^{(-1)} = -\hbar\xi_S^{(-1)}, \qquad \hat{\mathfrak{X}}\xi_S^{(-1)} = +1\xi_S^{(-1)}$$

These equations serve to explain our notation. The superscript on $\xi_S^{(1)}$ (i.e., $+1$) is the eigenvalue of $\hat{S}_z$, while the subscript S denotes "symmetric." The eigenstates $\xi_S^{(1)}$ and $\xi_S^{(-1)}$ are symmetric with respect to particle exchange.

Of the remaining two simultaneous eigenstates of $\hat{S}^2$, $\hat{S}_z$, and $\hat{\mathfrak{X}}$, one is symmetric, $\xi_S^{(0)}$, and one is antisymmetric, ξ_A. Their construction follows simply from that of φ_S and φ_A given above, using the two degenerate eigenstates of $\hat{S}_z$ (i.e., ξ_3 and ξ_4):

(11.113)

$$\xi_S^{(0)} = \frac{1}{\sqrt{2}}(\xi_3 + \xi_4) = \frac{1}{\sqrt{2}}[\alpha(1)\beta(2) + \beta(1)\alpha(2)]$$

$$\xi_A = \frac{1}{\sqrt{2}}(\xi_3 - \xi_4) = \frac{1}{\sqrt{2}}[\alpha(1)\beta(2) - \beta(1)\alpha(2)]$$

Using the expansion (11.111) for $\hat{S}^2$, one finds that

$$\hat{S}^2\xi_S^{(0)} = 2\hbar^2\xi_S^{(0)}$$
$$\hat{S}^2\xi_A = 0\hbar^2\xi_A$$

while

$$\hat{S}_z\xi_S^{(0)} = 0\hbar\xi_S, \qquad \hat{\mathfrak{X}}\xi_S^{(0)} = +1\xi_S(0)$$
$$\hat{S}_z\xi_A = 0\hbar\xi_A, \qquad \hat{\mathfrak{X}}\xi_A = -1\xi_A$$

In this manner we obtain that the four independent eigenstates of $(\hat{S}_1{}^2, \hat{S}_2{}^2, \hat{S}^2, \hat{S}_z)$ are $(\xi_S^{(1)}, \xi_S^{(0)}, \xi_S^{(-1)}, \xi_A)$. Properties of these eigenstates are listed in Table 11.3. The three ξ_S states correspond to $s = 1$, whereas the ξ_A state corresponds to $s = 0$. That is,

$$\hat{S}^2\xi_S = 2\hbar^2\xi_S$$
$$\hat{S}^2\xi_A = 0$$

Spin Values for Two and Three Electrons

In the coupled representation one may speak of the *total* angular momentum of the two-electron system. In Section 9.4 we concluded that the total angular momentum quantum numbers for a two-component system vary in unit steps from $|l_1 + l_2|$ to $|l_1 - l_2|$ [see (9.114)]. In similar manner the total spin angular momentum of a two-electron system has spin quantum numbers varying from $|\frac{1}{2} + \frac{1}{2}|$ to $|\frac{1}{2} - \frac{1}{2}|$ in

TABLE 11.3   Spin wavefunctions for two electrons in the coupled representation

Spin alignment	Wavefunction $\xi = \lvert s_1 s_2 s m_s \rangle$	s	m_s	s_1	s_2
↑ ↑	$\xi_S^{(1)} = \alpha(1)\alpha(2)$	1	1	$\frac{1}{2}$	$\frac{1}{2}$
↑ ↓ + ↓ ↑	$\xi_S^{(0)} = \dfrac{1}{\sqrt{2}}\,[\alpha(1)\beta(2) + \beta(1)\alpha(2)]$	1	0	$\frac{1}{2}$	$\frac{1}{2}$
↓ ↓	$\xi_S^{(-1)} = \beta(1)\beta(2)$	1	−1	$\frac{1}{2}$	$\frac{1}{2}$
↑ ↓ − ↓ ↑	$\xi_A = \dfrac{1}{\sqrt{2}}\,[\alpha(1)\beta(2) - \beta(1)\alpha(2)]$	0	0	$\frac{1}{2}$	$\frac{1}{2}$

unit intervals. This gives the two values $s = 0$ and $s = 1$. For $s = 1$ there are three m_s values: $m_s = 1, 0$, and -1. For $s = 0$ there is one m_s value, $m_s = 0$. This partitioning of states according to total spin number s is depicted in Table 11.3 in the (s, m_s) column.

If we use the angular momentum addition rules of Section 9.4 to calculate the spin quantum number corresponding to the possible total spin values for three electrons, one obtains the series

$$s = \lvert \tfrac{1}{2} + \tfrac{1}{2} + \tfrac{1}{2} \rvert, \ldots, \lvert \tfrac{1}{2} + \tfrac{1}{2} - \tfrac{1}{2} \rvert$$

which gives

$$s = \tfrac{3}{2}, \tfrac{1}{2}$$

There are four m_s values for $s = \frac{3}{2}$ and two for $s = \frac{1}{2}$.

These values of spin quantum number are appropriate to the coupled representation. The corresponding eigenstates are of the form

$$\xi = \lvert s, m_s, s_1, s_2, s_3 \rangle$$

which are simultaneous eigenstates of the five commuting operators

(11.114) $$\{\hat{S}^2, \hat{S}_z, \hat{S}_1{}^2, \hat{S}_2{}^2, \hat{S}_3{}^2\}$$

These concepts of spin angular momentum are very relevant to topics in atomic physics, as will be discussed further in Chapter 12.

PROBLEMS

11.59   Show that the coupled spin states $\xi_s^{(0)}$ and ξ_A given in (11.113) are eigenvectors of $\hat{S}^2$ with eigenvalues $2\hbar^2$ and 0, respectively.

11.60   Obtain the four spin states $|sm_s s_1 s_2\rangle$, listed in Table 11.3, relevant to two electrons in the coupled representation, through a Clebsch–Gordan expansion of the form

$$|sm_s s_1 s_2\rangle = \sum_{m_1 + m_2 = m_s} C_{m_1 m_2} |s_1 m_1\rangle |s_2 m_2\rangle$$

11.61   Consider a spin-1 particle. For integral angular momentum quantum number, the matrices developed in Section 11.5 for orbital angular momentum also apply to spin angular momentum. Using the three-component column eigenvectors of the Cartesian components of $\hat{S}$ [e.g., (11.68)], determine if it is possible for a spin-1 particle to be in a state

$$\xi = \begin{pmatrix} a \\ b \\ c \end{pmatrix}$$

such that

$$\langle \hat{S}_x \rangle = \langle \hat{S}_y \rangle = \langle \hat{S}_z \rangle = 0$$

11.62   What is the form of a spin eigenstate, in Dirac notation, in the *uncoupled* representation for the three-electron case? How many "good" quantum numbers are there?

Answer

$$\xi = |s_1, s_2, s_3, m_{s_1}, m_{s_2}, m_{s_3}\rangle$$

There are six good quantum numbers.

11.63   Using the results of Problem 9.43 (and the discussion preceding), construct an operator, $\hat{A}_1^2$, which commutes with the five commuting operators (11.114) relevant to the addition of three electron spins in the coupled representation.

11.64   For the case of four electrons, in the coupled representation,
   (a)   What are the s eigenvalues?
   (b)   Write down the form of six commuting operators explicitly in terms of the vector operators $\hat{S}_1, \hat{S}_2, \hat{S}_3$, their inner products, and their z components.
   (c)   What is the form of an eigenstate in Dirac notation? How many independent states of this form are there?

Answer (partial)
   (a) $s = |\tfrac{1}{2} + \tfrac{1}{2} + \tfrac{1}{2} + \tfrac{1}{2}|, \ldots, |\tfrac{1}{2} + \tfrac{1}{2} - \tfrac{1}{2} - \tfrac{1}{2}|$
         $= 2, 1, 0$

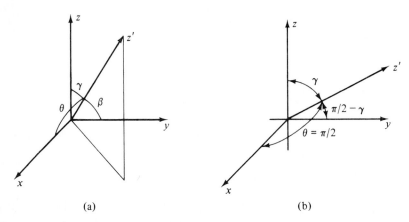

FIGURE 11.14 (a) Direction angles (θ, β, γ) locate the z' axis with respect to the original Cartesian frame. (b) The angles (θ, β, γ) for the case described in Problem 11.65. The z' axis lies in the yz plane.

11.65 The expression for the component of spin along an axis z', which makes respective angles (θ, β, γ) with the (x, y, z) axes, is

$$\hat{\sigma}_{z'} = \hat{\sigma}_x \cos\theta + \hat{\sigma}_y \cos\beta + \hat{\sigma}_z \cos\gamma$$

$$\cos^2\theta + \cos^2\beta + \cos^2\gamma = 1$$

See Fig. 11.14. This relation follows from the vector quality of $\hat{\boldsymbol{\sigma}}$.

    (a)   Write the matrix for $\hat{\sigma}_{z'}$ in terms of θ, β, γ and show that the eigenvalues of $\hat{\sigma}_{z'}$ are the same as those of $\hat{\sigma}_z$.

    (b)   An electron is known to be in the state α_z. What is the probability that measurement of $S_{z'}$ finds the respective values $\pm\hbar/2$ for the angular displacements $\theta = \pi/2$, $\beta = \pi/2 - \gamma$? See Fig. 11.14b.

    (c)   Describe a double S-G experiment in which such measurement may be effected.

Answer (partial)

    (b)   To obtain the answer, one must find the eigenvectors of $\hat{\sigma}_{z'}$ corresponding to the eigenvalues ± 1 and then expand α_z as a superposition of these two states. The squares of the coefficients of expansion give the respective probabilities

$$P\left(+\frac{\hbar}{2}\right) = \frac{1 + \cos\gamma}{2}$$

$$P\left(-\frac{\hbar}{2}\right) = \frac{1 - \cos\gamma}{2}$$

As $\gamma \to 0$, the z' and z axes merge and $P(+\hbar/2) \to 1$, $P(-\hbar/2) \to 0$. For $\gamma = \pi/2$, the z' axis is normal to the z axis and $P(+\hbar/2) = P(-\hbar/2) = \frac{1}{2}$.

11.66   Show, employing explicit matrix representations, that

$$\hat{S}_+\alpha = \hat{S}_-\beta = 0$$
$$\hat{S}_+\beta = \hbar\alpha, \qquad \hat{S}_-\alpha = \hbar\beta$$

11.67   Show that the Pauli spin operators obey the relation

$$\hat{\sigma}_x\hat{\sigma}_y\hat{\sigma}_z = i\hat{I}$$

11.68   Establish the following properties of the Pauli spin matrices:
(a) $\hat{\sigma}_x{}^2 = \hat{\sigma}_y{}^2 = \hat{\sigma}_z{}^2 = \hat{I}$ $(\hat{\sigma}^2 = 3\hat{I})$.
(b) $\{\hat{\sigma}_x, \hat{\sigma}_y\} = 0$, etc., where $\{,\}$ denotes the anticommutator.
(c) $\hat{\sigma}_x\hat{\sigma}_y = i\hat{\sigma}_z$, etc.
(d)   The preceding relations are included in the equality

$$(\hat{\boldsymbol{\sigma}} \cdot \mathbf{a})(\hat{\boldsymbol{\sigma}} \cdot \mathbf{b}) = \mathbf{a} \cdot \mathbf{b}\hat{I} + i\hat{\boldsymbol{\sigma}} \cdot (\mathbf{a} \times \mathbf{b})$$

where $\mathbf{a}$ and $\mathbf{b}$ are arbitrary vectors.
(e) $[\hat{\sigma}_x, \hat{\sigma}_y] = i2\hat{\sigma}_z$, etc.
(f)   If $\mathbf{e}$ is a unit vector in an arbitrary direction, what is $(\hat{\boldsymbol{\sigma}} \cdot \mathbf{e})^2$?

11.69   Measurement is made of the sum of the x and y components of the spin of an electron. What are the possible results of this experiment? After this measurement, the z component of spin is measured. What are the respective probabilities of obtaining the values $\pm \hbar/2$?

11.11   THE DENSITY MATRIX

Pure and Mixed States

In our description of quantum mechanics to this point, we have considered systems which by and large have satisfied the idealization of being isolated and totally uncoupled to any external environment. The free particle, the particle in a box, the harmonic oscillator, and the hydrogen atom are cases in point. Any such isolated system possesses a wavefunction of the coordinates only of the system. This wavefunction determines the state of the system.

Consider, on the other hand, a system that is coupled to an external environment. Such, for example, is the case of a gas of N particles maintained at a constant temperature through contact with a "temperature bath." Very simply, if x denotes coordinates of a system and y coordinates of its environment, then whereas the closed composite of system plus environment has a self-contained Hamiltonian and wavefunction $\psi(x, y)$, this wavefunction does not, in general, fall into a product $\psi_1(x)\psi_2(y)$. Under such circumstances, we may say that the system does not have a wavefunction. A system that does not have a wavefunction is said to be in a *mixed state*. A system that does have a wavefunction is said to be in a *pure state*.

It may also be the case that, owing to certain complexities of the system, less than complete knowledge of the state of the system is available. In quantum mechanics, such maximum information is contained in a wavefunction that simultaneously diagonalizes a complete set of commuting operators relevant to the system, such as that described in Section 5.5. Let a set of such operators be $\hat{A}$, $\hat{B}$, $\hat{C}$ with common wavefunction ψ_{abc}. Now suppose that the system is such that it is virtually impossible to measure A, B, and C in an appropriately small interval of time. Then the wavefunction ψ_{abc} cannot be determined and under such circumstances one also speaks of the system being in a mixed state. As in classical statistical mechanics, this situation arises for systems with a very large number of degrees of freedom, such as, for example, a mole of gas. The quantum state of such a system involves specification of $\sim 10^{23}$ momenta. The study of such complex systems is called *quantum statistical mechanics* and it is in this discipline that the density operator finds its greatest use.

The Density Operator

In dealing with situations where less than maximum information on the state of the system is available, one takes the point of view that a wavefunction for the system exists but that it is not completely determined. In place of the wavefunction, one introduces the density operator $\hat{\rho}$. If A is some property of the system, the density operator determines the expectation of A through the relation

$$(11.115) \qquad \langle A \rangle = \text{Tr}\, \hat{\rho}\hat{A}$$

and

$$(11.116) \qquad \text{Tr}\, \hat{\rho} = 1$$

The trace operation, written Tr, denotes summation over diagonal elements [see (11.27)]. From the density operator we may calculate expectation values of all relevant properties of the system.

Let us calculate the matrix elements of $\hat{\rho}$ for the case of a system whose wavefunction ψ is known. Then

$$\langle A \rangle = \langle \psi | \hat{A}\psi \rangle$$

Let the basis $\{|n\rangle\}$ span the Hilbert space containing ψ. One may expand ψ in this basis to obtain

$$|\psi\rangle = \sum_n |n\rangle\langle n|\psi\rangle$$

Substituting this expansion into the preceding equation gives

$$\langle A \rangle = \sum_q \sum_n \langle \psi | q \rangle \langle q | \hat{A} | n \rangle \langle n | \psi \rangle$$

$$= \sum_q \sum_n \rho_{nq} A_{qn} = \text{Tr } \hat{\rho}\hat{A}$$

Here, we have made the identification

$$\rho_{nq} = \langle q | \psi \rangle^* \langle n | \psi \rangle = a_q^* a_n$$

The coefficient a_n represents the projection of the state ψ onto the basis vector $|n\rangle$. The nth diagonal element of $\hat{\rho}$ is

(11.117) $$\rho_{nn} = |\langle \psi | n \rangle|^2 = a_n^* a_n = P_n$$

which we recognize to be the probability P_n of finding the system in the state $|n\rangle$. Thus, *the diagonal elements of $\hat{\rho}$ are probabilities* and must sum to 1. This is the rationale for the property (11.116), Tr $\hat{\rho} = 1$.

The Mixed State

Now consider a system that is in a mixed state. Let $\hat{N}$ be a measurable property of the system such as its energy. Let $\{|n\rangle\}$ be the eigenstates of $\hat{N}$. Since the system is in a mixed state, we may assume that the projections a_n are not determined quantities. In this case we define the elements ρ_{nq} to be the *ensemble averages* (see Section 5.1),

(11.118) $$\rho_{nq} = \overline{a_q^* a_n}$$

The diagonal element

(11.119) $$\rho_{nn} = \overline{a_n^* a_n}$$

represents the probability that a system chosen at random from the ensemble is found in the nth state.

Equation of Motion

Suppose again that a system is in a pure state and has the wavefunction ψ. Again, let $\hat{N}$ be a measurable property of the system with eigenstates $\{|n\rangle\}$. Expanding ψ in terms of the projections a_n gives

$$|\psi\rangle = \sum_n a_n(t)|n\rangle$$

From the Schrödinger equation for ψ, we obtain

$$i\hbar \sum_n \frac{\partial a_n}{\partial t} |n\rangle = \sum_n a_n \hat{H} |n\rangle$$

Operating on this equation from the left with $\langle l|$ gives

(11.120)
$$i\hbar \frac{\partial a_l}{\partial t} = \sum_n H_{ln} a_n, \qquad -i\hbar \frac{\partial a_l^*}{\partial t} = \sum_n H_{ln}^* a_n^*$$

We may use these relations to obtain an equation of motion for the matrix elements of $\hat{\rho}$.

$$i\hbar \frac{\partial \rho_{nl}}{\partial t} = i\hbar \frac{\partial a_l^* a_n}{\partial t} = i\hbar \left(\frac{\partial a_l^*}{\partial t} a_n + a_l^* \frac{\partial a_n}{\partial t} \right)$$

Substituting the expressions (11.120) for the time derivatives of the projections a_n and setting $H_{lk}^* = H_{kl}$ gives

(11.121)
$$i\hbar \frac{\partial \rho_{nl}}{\partial t} = \sum_k (H_{nk} \rho_{kl} - \rho_{nk} H_{kl})$$

This may be written in operator form as

(11.122)
$$i\hbar \frac{\partial \hat{\rho}}{\partial t} = [\hat{H}, \hat{\rho}]$$

Random Phases

Consider the matrix elements of $\hat{\rho}$ (11.118),

$$\rho_{nm} = \overline{a_m^* a_n}$$

The indeterminacy of the state of the system may be manifest in a corresponding indeterminacy of the phases $\{\phi_n\}$ of the projections $\{a_n\}$. These phases are defined by the relation

$$a_n = c_n e^{i\phi_n}$$

where c_n and ϕ_n are real. What is the consequence of assuming that the phases $\{\phi_n\}$ are random over the sample systems in the ensemble? Consider the matrix element

$$\rho_{nm} = \overline{c_m^* c_n \exp[i(\phi_n - \phi_m)]} = \overline{c_m^* c_n [\cos(\phi_n - \phi_m) + i \sin(\phi_n - \phi_m)]}$$

If phases are random, then in averaging over the ensemble, $\cos(\phi_n - \phi_m)$ will enter with positive value equally often as with negative value. Similarly, for $\sin(\phi_n - \phi_m)$, so that

$$\overline{\cos(\phi_n - \phi_m)} = \overline{\sin(\phi_n - \phi_m)} = 0$$

except when $n = m$. In this case

$$\rho_{nn} = \overline{c_n{}^*c_n}\ \overline{\cos(\phi_n - \phi_n)} = \overline{c_n{}^*c_n}$$

It follows that for the case of random phases, $\hat{\rho}$ is diagonal.

(11.123) $$\rho_{nm} = \rho_{nn}\delta_{nm}$$

Evolution in Time of Diagonal Density Matrix

Suppose that $\hat{\rho}$ is diagonal. How does $\hat{\rho}$ then evolve in time? Specifically, does it remain diagonal? From (11.121) we conclude that the evolution in time of the diagonal elements of $\hat{\rho}$ depends on the off-diagonal elements of $\hat{\rho}$. If these off-diagonal elements begin to grow away from zero, $\{\rho_{ll}\}$ will change. The equation for the off-diagonal elements is obtained from (11.121). If at $t = 0$, $\hat{\rho}$ is diagonal, then

(11.124) $$i\hbar\frac{\partial\rho_{nl}}{\partial t} = H_{nl}(\rho_{ll} - \rho_{nn}) \neq 0 \qquad (t = 0)$$

Thus the diagonal distribution is not, in general, constant in time. However, it is quite clear that the *uniform* distribution

$$\rho_{nl} = \rho_0\delta_{nl}$$

(all states equally populated) is stationary in time.

Density Matrix for a Beam of Electrons

An electron beam generated at a cathode is known to have an isotropic distribution of spins, so that

$$\langle S_x \rangle = \langle S_y \rangle = \langle S_z \rangle = 0$$

Let us calculate a density matrix that gives this property in a representation in which $\hat{S}_z$ is diagonal. The matrices for $\hat{S}_x$, $\hat{S}_y$, and $\hat{S}_z$ are given by (11.78). Since the spins are isotropically oriented, the probabilities of finding S_z with values $\pm\hbar/2$ are both $1/2$. In the said representation, and with the property (11.117), we conclude that the diagonal elements of $\hat{\rho}$ are $\frac{1}{2}$ and $\frac{1}{2}$. It follows that the matrix

(11.125) $$\hat{\rho} = \tfrac{1}{2}\begin{pmatrix} 1 & 0 \\ 0 & 1 \end{pmatrix}$$

gives

$$\langle S_z \rangle = \text{Tr } \hat{\rho}\hat{S}_z = \text{Tr} \frac{\hbar}{4}\begin{pmatrix} 1 & 0 \\ 0 & 1 \end{pmatrix}\begin{pmatrix} 1 & 0 \\ 0 & -1 \end{pmatrix} = 0$$

From (11.78) we quickly conclude that this choice of $\hat{\rho}$ also renders $\langle S_x \rangle = \langle S_y \rangle = 0$.

Projection Representation

As noted above, if the wavefunction of a system is indeterminate, one may describe properties of the system through the use of an ensemble of replica systems. Consider that states ψ of the ensemble systems are distributed with probability P_ψ. An alternative form of the density operator is given by the projection sum over states of the ensemble.

(11.126)
$$\hat{\rho} = \sum_\psi |\psi\rangle P_\psi \langle \psi|$$

In the preceding example of a beam of isotropic spins, we found the density operator to be given by (11.125). This operator is written in a representation in which $\hat{S}_z$ is diagonal, for which the states corresponding to $S_z = \pm\hbar/2$ are $|\alpha\rangle$ and $|\beta\rangle$ given by (11.79). Since the ensemble of systems contains only two values of P_{S_z}, the summation (11.126) for this example may be written

(11.127)
$$\hat{\rho} = |\alpha\rangle\tfrac{1}{2}\langle\alpha| + |\beta\rangle\tfrac{1}{2}\langle\beta|$$

Let us use this form to calculate $\langle S_x \rangle$.

$$\langle S_x \rangle = \text{Tr } \hat{\rho}\hat{S}_x = \langle\alpha|\hat{\rho}\hat{S}_x|\alpha\rangle + \langle\beta|\hat{\rho}\hat{S}_x|\beta\rangle$$

$$= \tfrac{1}{2}\langle\alpha|\hat{S}_x|\alpha\rangle + \tfrac{1}{2}\langle\beta|\hat{S}_x|\beta\rangle = \frac{\hbar}{4}\left[\overbrace{\begin{pmatrix} 1 & 0 \end{pmatrix}}\begin{pmatrix} 0 & 1 \\ 1 & 0 \end{pmatrix}\begin{pmatrix} 1 \\ 0 \end{pmatrix} + \overbrace{\begin{pmatrix} 0 & 1 \end{pmatrix}}\begin{pmatrix} 0 & 1 \\ 1 & 0 \end{pmatrix}\begin{pmatrix} 0 \\ 1 \end{pmatrix} \right] = 0$$

In like manner, we find that (11.127) gives $\langle\hat{S}_y\rangle = 0$.

Let a system be in a mixed state. Although the wavefunction is not determined, it is known that the probability that measurement of energy finds the value E_n is P_n. In this case (11.126) becomes

(11.128)
$$\hat{\rho} = \sum_n |\psi_n\rangle P_n \langle\psi_n|$$

Since $\{|\psi_r\rangle\}$ is an orthonormal sequence, it follows that $\hat{\rho}$ is diagonal in this representation with diagonal elements equal to P_n. If the system is a gas of N particles, the number of particles with energy E_n is NP_n, which for nondegenerate states is the same as the number of particles in the state $|\psi_n\rangle$. Thus the diagonal elements of $\hat{\rho}$ in the case at hand give the occupation numbers for the states of a system.[1]

---

[1] For further discussion and problems on the density matrix, see R. H. Dicke and J. P. Wittke, *Introduction to Quantum Mechanics*, Addison-Wesley, Reading, Mass., 1960.

PROBLEMS

11.70 Is a system that is in a superposition state in a pure or a mixed state?

Answer

A system in a superposition state is in a pure state. The wavefunction of the system is known, and all properties of the system may be determined to the maximum degree that quantum mechanics allows. Let $\hat{A}$, $\hat{B}$ be a complete set of compatible operators for the system. Let the common eigenstates $\{\psi\}$ of $\hat{A}$, $\hat{B}$ span the Hilbert space $\mathfrak{H}$. Then ψ', which is a superposition state with respect to the observables A and B, does not lie along any of the basis vectors $\{\psi\}$. As described earlier in the chapter, ψ', which exists in $\mathfrak{H}$, is related to ψ through a unitary transformation, $\psi' = \hat{U}\psi$. In this manner, one may obtain a new set of states $\{\psi'\}$ which also span $\mathfrak{H}$. In this new basis, the operator $\hat{A}$ has the value $\hat{A}' = \hat{U}\hat{A}\hat{U}^{-1}$. So $\hat{A}'\psi' = \hat{U}\hat{A}\hat{U}^{-1}\hat{U}\psi = \hat{U}a\psi = a\psi'$. Furthermore, $[A', B'] = 0$ if $[A, B] = 0$. Thus we may conclude that ψ' is a common eigenstate of A', B'.

11.71 What is the spin polarization of a beam of electrons described by the density operator

$$\hat{\rho} = \tfrac{1}{2}\begin{pmatrix} 1 & 1 \\ 1 & 1 \end{pmatrix}?$$

11.72 (a) What is the density operator corresponding to an isotropic distribution of deuterons (spin 1) in the representation in which $\hat{S}_z$ is diagonal?

(b) What is the value of $\langle \hat{S}_y \rangle$?

(*Hint:* Your answer should appear as a 3×3 matrix.)

11.73 Consider a particle in a one-dimensional box with walls at $x = 0$ and $x = L$. Eigenenergies are $E_n = n^2 E_1$. It is known that the probability of finding the particle with energy E_1 is $\tfrac{1}{2}$ and that of finding it with energy E_5 is $\tfrac{1}{2}$.

(a) What is the density matrix for this system in the energy representation?

(b) Construct two normalized wavefunctions that give the same probabilities and, therefore, the same density matrix.

11.74 The *canonical form*[1] of the density operator is given by

$$\hat{\rho} = A \exp\left(\frac{-\hat{H}}{k_B T}\right)$$

where k_B is Boltzmann's constant and T denotes temperature. Consider that $\hat{H}$ is the Hamiltonian of a one-dimensional harmonic oscillator with fundamental frequency ω_0. Working in the energy representation:

(a) Find the diagonal elements of $\hat{\rho}$.

(b) Determine the normalization constant A.

(c) Calculate the expectation $\langle E \rangle$ of the oscillator.

(d) Construct the projection representation of $\hat{\rho}$ (11.126).

(*Hint:* For summation of series, see Problem 2.36.)

---

[1] This density matrix is relevant to a system with Hamiltonian $\hat{H}$, which is maintained in equilibrium at the temperature T through contact with a heat reservoir. For a further discussion, see K. Huang, *Statistical Mechanics*, Wiley, New York, 1963.

11.75 Show that

$$(\hat{\sigma}_z)^{2n} = \hat{I}, \qquad (\hat{\sigma}_z)^{2n+1} = \hat{\sigma}_z$$

where n is an integer. (This result also holds for the operator $\mathbf{e} \cdot \hat{\boldsymbol{\sigma}}$, where $\mathbf{e}$ is any fixed unit vector.)

11.76 (a) Using the results of Problem 11.75, show that

$$\hat{R}_\phi = \exp\left(\frac{i\phi \hat{S}_z}{\hbar}\right) = \exp\left(i\frac{\phi}{2}\hat{\sigma}_z\right)$$

$$= \left(\cos\frac{\phi}{2}\right)\hat{I} + i\left(\sin\frac{\phi}{2}\right)\hat{\sigma}_z$$

(Recall Problem 9.17.) This rotation operator tells us how spinors transform under rotation. The transformed spinor is given by $\xi' = \hat{R}_\phi \xi$.

(b) What is the matrix form of $\hat{R}_\phi$?

(c) Show that $\hat{R}_\phi$ preserves the length of ξ. That is, show that $\langle \xi'|\xi'\rangle = \langle \xi|\xi\rangle$.

(d) Show that $\hat{R}_\phi$, at most, changes the phases of the eigenvectors of $\hat{\sigma}_x$, $\hat{\sigma}_y$, and $\hat{\sigma}_z$.

(e) Show that under a complete rotation ($\phi = 2\pi$) about the z axis, $\xi \to \xi' = -\xi$.

11.77 Prove the general expansion

$$\exp(i\mathbf{e} \cdot \hat{\boldsymbol{\sigma}}\phi) = (\cos\phi)\hat{I} + i(\sin\phi)\mathbf{e} \cdot \hat{\boldsymbol{\sigma}}$$

where, again, $\mathbf{e}$ is an arbitrarily oriented unit vector.

11.78 (a) Show that the spin-exchange operator $\hat{x}$ for two electrons has the representation

$$\hbar^2 \hat{x} = 2\hat{\mathbf{S}}_1 \cdot \hat{\mathbf{S}}_2 + \frac{\hbar^2}{2}$$

(b) Show that $\hat{x}$ may also be written

$$\hbar^2 \hat{x} = \hat{S}^2 - \hbar^2$$

[*Hint:* For part (a), let $\hat{x}$ operate on $\alpha(1)\beta(2)$ and $\alpha(2)\beta(1)$, respectively.]

11.79 A beam of neutrons with isotropically distributed spins has the density matrix

$$\hat{\rho} = \frac{1}{2}\begin{pmatrix} 1 & a^* \\ a & 1 \end{pmatrix}$$

in a representation where $\hat{S}^2$ and $\hat{S}_z$ are diagonal. From the condition $\langle \mathbf{S}\rangle = 0$, show that $a = 0$.

11.80 (a) If $\boldsymbol{\mu}$ is the magnetic moment of the electron, in what state will $|\langle \mu_z\rangle| = \mu_b$?

(b) What is the value of $\langle \mu^2\rangle$ in this state?

11.81 Consider a process in which an electron and a positron are emitted collinearly in the $+y$ and $-y$ directions, respectively. Spins are polarized to lie in the $\pm z$ directions. The pair is emitted with zero linear and spin-angular momentum and with total energy $\hbar\omega$. With the electron labeled 1 and the positron labeled 2:

(a) Write down a spin-coordinate, time-dependent product wavefunction for the electron positron pair which contains these properties.

(b)   What is the probability that measurement finds the electron's z component of spin equal to $+\hbar/2$?

(c)   Suppose that measurement of the positron's z component of spin finds the value $-\hbar/2$. What is the wavefunction for the pair immediately after this measurement?

(d)   What will measurement of the electron's z component of spin now find?

Answers

(a)   The appropriate zero spin factor of the wavefunction is found in Table 11.3.

$$\psi(1, 2) = \frac{1}{\sqrt{2(2\pi)}} (\alpha(1)\beta(2) - \alpha(2)\beta(1)) \exp\left[i(ky_1 - ky_2 - \omega t)\right]$$

$$2\left(\frac{\hbar^2 k^2}{2m}\right) = \hbar\omega$$

(b)   It is equally likely to find the values $\pm\hbar/2$ for the electron's z component of spin. The same is true of the positron's z component of spin.

(c) $$\psi_{\text{after}}(1, 2) = \frac{1}{2\pi} \alpha(1)\beta(2) \exp\left[i(k(y_1 - y_2) - \omega t)\right]$$

Here we are assuming that measurement preserves S_z and energy.

(d) $+\hbar/2$.

(*Note:* This problem contains a key tool of the Einstein–Podolsky–Rosen paradox.[1] Consider two observers O_1 and O_2 positioned along the y axis equipped, respectively, with detectors S-G$_1$ and S-G$_2$ oriented for measurement of S_z. Up until the time that O_1 makes his measurement, O_2 is equally likely to measure $\pm\hbar/2$. Once O_1 makes the measurement and measures, say, $+\hbar/2$, O_2 is certain to find the value $-\hbar/2$ upon measurement. This situation, presumably, maintains for O_1 and O_2 sufficiently far from each other with electron and positron beyond each other's range of interaction, thereby violating the *principle of locality*.)

11.82   Consider the antisymmetric spin zero state

$$\xi_A(x) = \frac{1}{\sqrt{2}} [\alpha_x(1)\beta_x(2) - \beta_x(1)\alpha_x(2)]$$

relevant to two spin-1/2 particles labeled 1 and 2, respectively. This state is an eigenstate of $\hat{S}_x$ and $\hat{S}^2$ with eigenvalue zero.

(a)   Show that $\xi_A(x)$ is also an eigenstate of $\hat{S}_z$, thereby establishing that it is a common eigenstate of the two noncommuting operators, $\hat{S}_x$ and $\hat{S}_z$.

(b)   What property renders the result in part (a) compatible with the commutator theorem (Section 5.2)?

[1] A. Einstein, B. Podolsky, and N. Rosen, *Phys. Rev.* **47**, 777 (1935). For further discussion and references on this topic, see M. O. Scully, R. Shea, and J. D. McCullen, *Phys. Repts.* **43**, 501 (1978). See also B. d'Espagnat, "The Quantum Theory and Reality," *Scientific American* (Nov. 1979), p. 158. [*Note:* The configuration described in Problem 11.81 is due to D. Bohm (1952).] These topics are returned to in Section 11.13.

Answers

(a)   From the forms of the spinors α_x, β_x as given in Problem 11.45, we obtain

$$\alpha_x = \frac{1}{\sqrt{2}}(\alpha_z + \beta_z)$$

$$\beta_x = \frac{1}{\sqrt{2}}(\alpha_z - \beta_z)$$

Substituting into $\xi_A(x)$ gives $\xi_A(x) = a\xi_A(z)$, where a is a constant.

(b)   As pointed out previously in Fig. 9.1 and Section 10.2, the null eigenstates of angular momentum are common eigenstates of the individual Cartesian components of angular momentum.

11.83   Employing the rotation operator obtained in Problem 11.76 relevant to spinors, show that the null eigenstate $\xi_A(x)$ of S_x and S^2, introduced in Problem 11.82, is invariant under rotation about the z axis.

Answer

The rotation operator is given by

$$\hat{R}_\phi = \cos\frac{\phi}{2}\,\hat{I} + i\sin\frac{\phi}{2}\,\hat{\sigma}_z$$

When applied to the product state $\alpha_x(1)\beta_x(2)$, we obtain

$$\hat{R}_\phi[\alpha_x(1)\beta_x(2)] = [\hat{R}_\phi\alpha_x(1)][\hat{R}_\phi\beta_x(2)]$$

$$= \left[\left(\cos\frac{\phi}{2}\right)\alpha_x(1) + i\left(\sin\frac{\phi}{2}\right)\beta_x(1)\right]\left[\left(\cos\frac{\phi}{2}\right)\beta_x(2) + i\left(\sin\frac{\phi}{2}\right)\alpha_x(2)\right]$$

Applying a similar rotation to the product $\alpha_x(2)\beta_x(1)$ and carrying out the multiplication gives

$$\hat{R}_\phi[\alpha_x(1)\beta_x(2) - \alpha_x(2)\beta_x(1)] = \left(\sin^2\frac{\phi}{2} + \cos^2\frac{\phi}{2}\right)[\alpha_x(1)\beta_x(2) - \alpha_x(2)\beta_x(1)]$$

$$\hat{R}_\phi\xi_A(x) = \xi_A(x)$$

[*Note:* This problem establishes that zero spin states are invariant under rotation of coordinates and thus transform as a scalar. A like quality is shared by the null orbital angular momentum states which, we recall, are given by any spherically symmetric function $f(r^2)$. See discussion preceding (10.34). For both null spin and null orbital angular momentum states, $\langle J_x \rangle = \langle J_y \rangle = \langle J_z \rangle = 0$. There is no preferred direction for a system in any of these states.]

11.84   One of the puzzles of the early theory of neutron decay, $n \rightarrow p + e$, was the fact that such a process could not conserve spin angular momentum. The neutron n, proton p, and electron e

each have spin 1/2. To answer this objection, Pauli[1] proposed that together with the proton and electron, a massless, chargeless, spin-1/2 particle was emitted, which he called a neutrino.

(a) Explain how the original process cannot conserve angular momentum.

(b) Explain how the corrected process, $n \rightarrow p + e + \bar{\nu}$, can conserve angular momentum.

11.12 OTHER "PICTURES" IN QUANTUM MECHANICS

Schrödinger and Heisenberg Pictures

In addition to representations in quantum mechanics stemming from transformation of bases in Hilbert space, one also speaks of different "pictures" of the theory. These alternative formulations stem from the fact that wavefunctions and operators are not objects of measurement. As we have found previously, the superposition theorem specifies the possible outcome of a measurement in terms of certain projections in Hilbert space. Thus, for example, if $\hat{A}|a\rangle = a|a\rangle$, and the system is in the state ψ, the probability that measurement of A finds the value a is given by the absolute square, $|\langle a|\psi\rangle|^2$.

The so-called *Schrödinger picture* refers to the formulation which is based on the Schrödinger equation (3.45). With the preceding description one requires that any alternative picture satisfies two basic requirements: (1) Eigenvalues of operators corresponding to observables in the new picture must be the same as in the Schrödinger picture; (2) inner products of wavefunctions must maintain their values as well.

In Section 11.3 these properties were found to be obeyed by any unitary transformation. In particular, the so-called *Heisenberg picture* stems from the unitary operator (3.66). If initial time $t = 0$ is labeled $t = t_0$, then this operator becomes

(11.129)
$$\hat{U}(t, t_0) = \exp\left[-\frac{i}{\hbar}(t - t_0)\hat{H} \right]$$

$$\hat{U}^\dagger(t, t_0) = \exp\left[\frac{i}{\hbar}(t - t_0)\hat{H} \right] = U^{-1}(t, t_0)$$

where we have assumed that $\hat{H}$ is not an explicit function of time. With (3.70) we may write

(11.130)
$$\psi(t) = \hat{U}(t, t_0)\psi(t_0)$$

$$\psi(t_0) = \hat{U}^{-1}(t, t_0)\psi(t)$$

---

[1] W. Pauli, *Rapports du Septième Conseil de Physique, Solvay, Brussels, 1933*, Gauthier-Villars, Paris, 1934.

Let ψ' be a wavefunction in the Heisenberg picture. Then

(11.131)
$$\psi' = \hat{U}^{-1}\psi$$
$$\hat{U}\psi' = \psi$$

It follows that

(11.132)
$$\psi'(t) = \hat{U}^{-1}\psi(t) = \hat{U}^{-1}\hat{U}\psi(t_0)$$
$$\psi'(t) = \psi(t_0)$$

That is, in the Heisenberg picture the wavefunction remains constant. On the other hand, operators, which in the Schrödinger picture are constant, vary in time in the Heisenberg picture. This follows from the transformation (11.40).

(11.133)
$$\hat{A}'(t) = \hat{U}^{-1}(t)\hat{A}\hat{U}(t)$$

An important exception to this conclusion is the Hamiltonian. If $\hat{H}$ is constant in the Schrödinger picture, it remains constant in the Heisenberg picture.

An equation of motion for an operator in the Schrödinger picture is given by (6.68). To obtain an equation of motion for an operator in the Heisenberg picture, first with (11.133), we write

(11.134)
$$\frac{d\hat{A}'}{dt} = \frac{d\hat{U}^{-1}}{dt}\hat{A}\hat{U} + \hat{U}^{-1}\hat{A}\frac{d\hat{U}}{dt} + \hat{U}^{-1}\frac{\partial\hat{A}}{\partial t}\hat{U}$$

In obtaining the last term in this equation, we have noted the following. Consider, for example, that $\hat{A} = \hat{A}(\hat{q}, \hat{p}, t)$. Then in the Schrödinger picture, as $\hat{q}$ and $\hat{p}$ are stationary in time,

(11.134a)
$$\frac{d\hat{A}(\hat{q}, \hat{p}, t)}{dt} = \frac{\partial\hat{A}(\hat{q}, \hat{p}, t)}{\partial t}$$

To find a relation for the time derivative of $\hat{U}$, we consider the following sequence of events. Consider a particle which at the time t_0 is in the state $\psi(t_0)$. At time $t_1 > t_0$ it evolves to the state

$$\psi(t_1) = \hat{U}(t_1, t_0)\psi(t_0)$$

At time $t_2 > t_1$ it is in the state

$$\psi(t_2) = \hat{U}(t_2, t_1)\psi(t_1) = \hat{U}(t_2, t_1)\hat{U}(t_1, t_0)\psi(t_0)$$
$$= \hat{U}(t_2, t_0)\psi(t_0)$$

Thus we may write[1]

(11.135)
$$\hat{U}(t_2, t_0) = \hat{U}(t_2, t_1)\hat{U}(t_1, t_0)$$

Recalling the definition of differentiation, we write

(11.136)
$$\frac{d\hat{U}(t, t_0)}{dt} = \lim_{\varepsilon \to 0} \frac{U(t + \varepsilon, t_0) - U(t, t_0)}{\varepsilon}$$

which, with (11.134), gives

(11.137)
$$\hat{U}(t + \varepsilon, t_0) = U(t + \varepsilon, t)U(t, t_0)$$

Substitution into (11.136) gives

$$\frac{d\hat{U}(t, t_0)}{dt} = \lim_{\varepsilon \to 0} \frac{[\hat{U}(t + \varepsilon, t) - 1]\hat{U}(t, t_0)}{\varepsilon}$$

For sufficiently small ε we may write

$$\hat{U}(t + \varepsilon, t) = \exp\left(\frac{-itH}{\hbar}\right) = 1 - \frac{i\varepsilon\hat{H}}{\hbar}$$

When substituted into (11.137), this relation gives

(11.138a)
$$i\hbar \frac{d\hat{U}}{dt} = \hat{H}\hat{U}$$

Taking the Hermitian adjoint gives

(11.138b)
$$-i\hbar \frac{dU^\dagger}{dt} = \hat{U}^\dagger \hat{H}$$

Substitution of these latter two equations into (11.134) gives the desired result:

(11.139a)
$$\boxed{i\hbar \frac{d\hat{A}'}{dt} = [\hat{A}', \hat{H}'] + i\hbar \frac{\partial \hat{A}'}{\partial t}}$$

where we have set

(11.139b)
$$\frac{\partial \hat{A}'}{\partial t} \equiv \hat{U}^{-1} \frac{\partial \hat{A}}{\partial t} \hat{U}$$

---

[1] Steps leading to (11.135) remain valid for $\hat{H}$ an explicit function of time. See Problem 3.18.

TABLE 11.4   Elements of the Schrödinger and Heisenberg pictures

	Wavefunction	Operator
Schrödinger picture	Varying	Constant
Heisenberg picture	Constant	Varying

Note in particular the similarity between (11.139a) and the equation of motion in the Schrödinger picture (6.68). Properties of wavefunctions and operators in these two representations are illustrated in Table 11.4. The right-hand column in this table refers to operators that are time-independent in the Schrödinger picture. It should be noted, however, that operators may be time-dependent in the Schrödinger picture. Such cases are discussed in Section 13.5 as well as in the description below of the interaction representation.

The Heisenberg equation of motion (11.139) often comes into play in discussions concerning the correspondence principle. If one makes the identification

$$(11.140) \qquad \frac{1}{i\hbar}\,[\hat{A},\hat{H}]_{\mathrm{QM}} \to \{A,H\}_{\mathrm{CL}}$$

then the equation of motion for an operator in the Heisenberg picture (11.139a) is seen to reduce to the corresponding classical equation of motion for a dynamical function as given in Problem 1.15.[1]

Interaction Picture

Finally, we turn to an approximation scheme important to time-dependent perturbation theory known as the *interaction picture*. Consider the Hamiltonian

$$\hat{H} = \hat{H}_0 + \lambda\hat{V}(t)$$

where the "unperturbed" Hamiltonian $\hat{H}_0$ is assumed to be time-independent and λ is a nondimensional parameter of smallness.

In analogy to the Heisenberg picture (11.131), wavefunctions in the new picture are given by

$$(11.141) \qquad \psi_I = \hat{U}_0^{-1}\psi$$

[1] For further discussion, see R. L. Liboff, *Found. Phys.* **17**, 981 (1987).

where

(11.141a)
$$\hat{U}_0 \equiv \exp\left[-\frac{i}{\hbar}(t - t_0)\hat{H}_0\right]$$

and, in analogy with (11.133),

(11.142)
$$\hat{A}_I = \hat{U}_0^{-1}\hat{A}\hat{U}_0$$

Taking the time derivative of (11.141), and employing the Schrödinger equation (3.45) and noting that

$$[\hat{U}_0, \hat{H}_0] = 0$$

gives the desired equation of motion,

(11.143)
$$\boxed{i\hbar\frac{\partial}{\partial t}\psi_I = \lambda\hat{V}_I\psi_I}$$

which is Schrödinger-like in form but only involves the interaction potential V_I. Integrating the preceding equation, we obtain

(11.144)
$$\psi_I(t) = \psi_I(t_0) + \frac{\lambda}{i\hbar}\int_{t_0}^{t}\hat{V}_I(t')\psi_I(t')\,dt'$$

This equation leads naturally to a series solution. Namely, substituting $\psi_I(t)$ as given by the RHS of (11.144) for $\psi_I(t')$ in the integrand and continuing this iteration gives the series

(11.145)
$$\psi_I(t) = \psi_I(t_0) + \frac{\lambda}{i\hbar}\int_{t_0}^{t}dt'\,\hat{V}_I(t')\psi_I(t_0)$$

$$+ \left(\frac{\lambda}{i\hbar}\right)^2\int_{t_0}^{t}dt'\int_{t_0}^{t'}dt''\,\hat{V}_I(t')\hat{V}_I(t'')\psi_I(t_0) + \cdots$$

Other techniques in time-dependent perturbation theory are discussed in Section 13.5 et seq. Application of the preceding formalism is given in Problem 11.96.[1]

---

[1] The interaction representation is encountered again in Chapter 13 (Problem 13.57) and in Section 14.5 in derivation of the Lippmann–Schwinger equation.

11.13 POLARIZATION STATES. EPR REVISITED

In this section we introduce eigenstates of photon polarization. This formalism is then applied to the Einstein–Padolsky–Rosen (EPR) paradox (see Problem 11.81) and closely allied notion of *hidden variables* briefly discussed in Chapter 2.

Polarization States

We recall that photons have zero rest mass and are spin-one bosons with spin either aligned or antialigned with photon linear momentum.

The states of polarization of a photon may be written in terms of the two basis eigenstates $|H\rangle$ and $|V\rangle$, representing horizontal and vertical polarization, respectively. With x and y axes taken as horizontal and vertical directions, respectively, related electric fields at fixed z are given by

(11.146)
$$\mathscr{E}_H = \mathscr{E}_0[\cos \omega t, 0, 0] \Rightarrow |H\rangle$$
$$\mathscr{E}_V = \mathscr{E}_0[0, \cos \omega t, 0] \Rightarrow |V\rangle$$

Let us consider the field of the superposition state

(11.147)
$$|R\rangle = \frac{1}{\sqrt{2}}(|H\rangle + i|V\rangle)$$

Since $i|V\rangle = \exp(i\pi/2)|V\rangle$ corresponds to $|V\rangle$ shifted by $\pi/2$ radians, the field corresponding to $|R\rangle$ is

(11.148)
$$\mathscr{E}_R = \mathscr{E}_0[\cos \omega t, \sin \omega t, 0] \Rightarrow |R\rangle$$

which we recognize to be right-handed, circularly polarized radiation—viewing the photon head-on. The wave (11.148) propagates in the $+z$ direction. The closely allied field

(11.149)
$$\mathscr{E}_L = \mathscr{E}_0[\cos \omega t, -\sin \omega t, 0] \Rightarrow |L\rangle$$

represents left-handed, circularly polarized radiation corresponding to the superposition state

(11.150)
$$|L\rangle = \frac{1}{\sqrt{2}}(|H\rangle - i|V\rangle)$$

Note that these polarization states have the orthonormal properties

(11.151)
$$\langle H|H\rangle = \langle V|V\rangle = \langle R|R\rangle = \langle L|L\rangle = 1$$
$$\langle H|V\rangle = \langle R|L\rangle = 0$$

The preceding states of polarization are depicted graphically in Fig. 11.15.

546

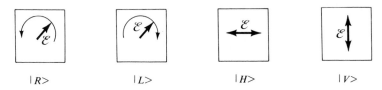

FIGURE 11.15 Diagramatic representation of four photon states of polarization (viewing the photon head-on).

Experimental Setup

With these preliminaries at hand, the question we examine relevant to the EPR paradox is whether quantum theory is consistent or whether it is necessary to bolster the theory with hidden variables. An operational means of answering this question is given by Bell's theorem.[1]

Two processes relevant to this theorem are considered. The first of these is described by the experimental setup depicted in Fig. 11.16 in which an atom undergoes a two-photon decay with no net change in angular momentum. The notation is such that $|R_1\rangle$ is the polarization state of photon 1, etc.

The zero net change in angular momentum of the atomic decay indicates that the photons must be in the symmetric state [analogous to the antisymmetric spin state of Problem 11.81(a)]:

$$(11.152) \qquad |P_{12}\rangle = \frac{1}{\sqrt{2}} [|R_1\rangle|R_2\rangle + |L_1\rangle|L_2\rangle]$$

In this expression $|R_1\rangle$ and $|R_2\rangle$ are antialigned states so that their total angular momentum is zero. The same is true of $|L_1\rangle$ and $|L_2\rangle$.

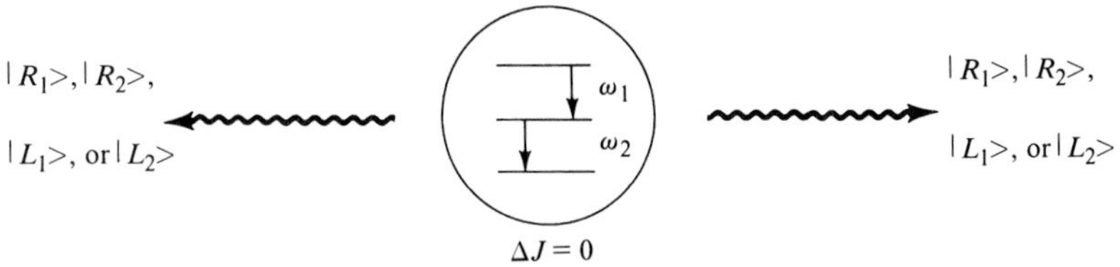

$$|R_1\rangle, |R_2\rangle, \qquad\qquad\qquad\qquad\qquad |R_1\rangle, |R_2\rangle,$$

$$|L_1\rangle, \text{ or } |L_2\rangle \qquad\qquad\qquad\qquad |L_1\rangle, \text{ or } |L_2\rangle$$

$$\omega_1$$
$$\omega_2$$

$$\Delta J = 0$$

FIGURE 11.16 Schematic of two-photon atomic decay and possible circular polarization states of emitted photons.

[1] J. S. Bell, *Physics* **1**, 195 (1964). This paper and other early works of Bell are reprinted in *Quantum Theory of Measurement*, J. A. Wheeler and W. H. Zurek, eds., Princeton University Press, Princeton, N.J., 1983.

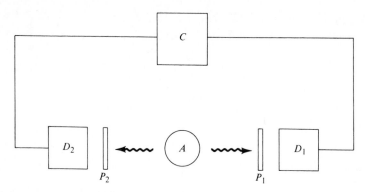

FIGURE 11.17 Photons from source A pass through the polarizers P to detectors D. Signals are then transmitted to coincidence counter C.

Now consider a polarizer which may be set to pass only $|R\rangle, |L\rangle, |H\rangle, |V\rangle$, or $|H_\theta\rangle$. Here we have written $|H_\theta\rangle$ for a linearly polarized state at an angle θ from the horizontal. Two such polarizers are placed in front of photon detectors as shown in Fig. 11.17. Note in particular that the photons emanating from A in Fig. 11.17 are in the superposition state (11.152).

The polarizers are set so that P_1 only passes $|H_{\theta_1}\rangle$ and P_2 only passes $|H_{\bar{\theta}_2}\rangle$. (The motivation for the special notation $\bar{\theta}_2$ is given below.) The counter C responds positively when photons are detected in D_1 and D_2 simultaneously.

We wish to obtain the number of coincidence counts N as a function of the angles θ_1 and $\bar{\theta}_2$. To obtain this relation, first we recall the superpositions (11.147) and (11.150):

$$|R\rangle = \frac{1}{\sqrt{2}}(|H\rangle + i|V\rangle)$$

$$|L\rangle = \frac{1}{\sqrt{2}}(|H\rangle - i|V\rangle)$$

Substituting these transformations into (11.152) gives

(11.153)
$$|P_{12}\rangle = \frac{1}{\sqrt{2}}[|H_1\rangle|H_2\rangle - |V_1\rangle|V_2\rangle]$$

The two-body polarization state so written indicates, for example, that if the polarization of photon 1 is measured to be H_1, then measurement of the polarization of photon 2 is certain to find H_2.

The function $N(\theta_1, \bar{\theta}_2)$ we seek is given by

(11.154)
$$N(\theta_1, \bar{\theta}_2) \propto |\langle H_{\theta_1} H_{\theta_2}|P_{12}\rangle|^2$$

We note that photon polarization states rotate like a spin-one vector.[1] Thus

(11.155) $$|H_\theta\rangle \pm i|V_\theta\rangle = e^{\pm i\theta}(|H\rangle + i|V\rangle)$$

where $|V_\theta\rangle$ represents a polarization state at θ radians from the vertical. There results

(11.156) $$|H_\theta\rangle = \cos\theta|H\rangle - \sin\theta|V\rangle$$

These expressions together with (11.153) give

(11.157) $$\begin{aligned}|\langle H_{\theta_1} H_{\bar\theta_2}|P_{12}\rangle|^2 &= |(\cos\theta_1\langle H_1| - \sin\theta_1\langle V_1|)(\cos\bar\theta_2\langle H_2| - \sin\bar\theta_2\langle V_2|) \\ &\quad \times (|H_1\rangle|H_2\rangle - |V_1\rangle|V_2\rangle)|^2 \\ &= [\cos\theta_1\cos\bar\theta_2 - \sin\theta_1\sin\bar\theta_2]^2 = \cos^2(\theta_1 + \bar\theta_2)\end{aligned}$$

Polarizer Angles

An important point is now made concerning the appropriate angle for polarizer 2 (in the configuration of Fig. 11.17). We have been working in the convention where the field description corresponding to the polarization state of a photon is written with respect to viewing the photon head-on. The Cartesian axes at detectors D_1 and D_2 are shown in Fig. 11.18.

With reference to this diagram we see that the angle polarizer 2 makes with the common axis defined by polarizer 1 is θ_2. From the diagram we also note

$$\bar\theta_2 = \pi - \theta_2$$

Thus (11.157) gives

(11.158) $$N(\theta_1, \theta_2) \propto \cos^2(\theta_1 - \theta_2 + \pi) = \cos^2(\theta_1 - \theta_2)$$

This is our desired relation. It is plotted in Fig. 11.19. The functional dependence of N on $(\theta_1 - \theta_2)$ reflects the fact that the experimental arrangement of Fig. 11.17 is invariant with respect to rotation about the axis D_2AD_1.

For $\chi \equiv \theta_1 - \theta_2 = 0$, polarizers P_1 and P_2 are aligned and there is maximum coincidence. As is evident from (11.153), if measurement of photon 1 finds H_1, then measurement of photon 2 is certain to find H_2. At $\chi = \pi/2$ polarizers are set at right angles and there is no coincidence.

For further reference we set

(11.159) $$N(\chi) = \frac{1}{2}\cos^2\chi$$

See Fig. 11.19. Application of this result is made below.

---

[1] See Section 11.5 concerning the rotation operator.

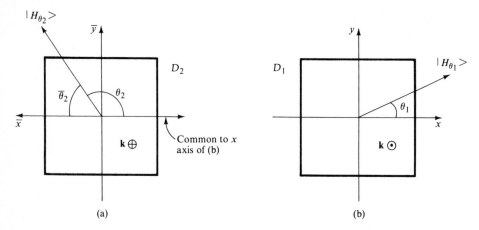

FIGURE 11.18 (a) Coordinate frame at D_2. The propagation vector k is into the page. (b) Coordinate frame at D_1. The vector k is out of the page. Note that looking at (a) from the back of the page gives a right-handed frame with k pointed toward the reader. It is with respect to this frame that $H_{\bar{\theta}_2}$ is defined in (11.157).

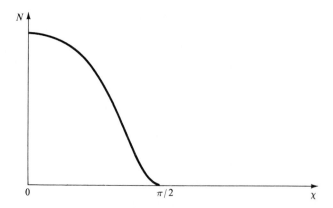

FIGURE 11.19 Coincidence counts N as a function of the angle $\chi \equiv \theta_1 - \theta_2$.

Bell's Theorem

For discussion of Bell's theorem, first we return to the configuration of Problem 11.81 relevant to electron–positron decay. The appropriate antisymmetric spin-zero state is given by

(11.160)
$$\xi_A(1, 2) = \frac{1}{\sqrt{2}} [\alpha(1)\beta(2) - \alpha(2)\beta(1)]$$

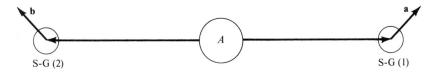

FIGURE 11.20   Two S-G devices situated at equal but opposite distances from the decay process at S.

Consider that two Stern–Gerlach devices are positioned at equal but opposite distances from the source A, as illustrated in Fig. 11.20. The setup is such that S-G(1) detects the projection of particle spin on the unit vector $\mathbf{a}$. Similarly, S-B(2) detects the projection of particle spin on the unit vector $\mathbf{b}$. We consider the expectation

(11.161)
$$P(\mathbf{a}, \mathbf{b}) = \langle \xi_A | (\hat{\sigma}_1 \cdot \mathbf{a})(\hat{\sigma}_2 \cdot \mathbf{b}) | \xi_A \rangle$$

where $\hat{\sigma}$ is the Pauli spin matrix (11.82). Inserting the spin state (11.160) into the preceding gives (see Problem 11.97)

(11.162)
$$P_{\mathrm{QM}}(\mathbf{a}, \mathbf{b}) = -\mathbf{a} \cdot \mathbf{b}$$

Here we have inserted the subscript QM to denote that the expectation is quantum mechanical.

   A formulation of this average stemming from hidden variables proceeds as follows. We introduce a hidden variable λ such that to each state of the two-particle system there corresponds a definite value of λ. Let $\rho(\lambda)$ denote the probability density of λ. Then with A and B denoting the results of measurements of the spin projections onto the directions $\mathbf{a}$ and $\mathbf{b}$ respectively, the expectation $P(\mathbf{a}, \mathbf{b})$ becomes

(11.163)
$$P_h(\mathbf{a}, \mathbf{b}) = \int d\lambda \, \rho(\lambda) A(\mathbf{a}, \lambda) B(\mathbf{b}, \lambda)$$

where h denotes "hidden variables," and

(11.164a)
$$\int \rho(\lambda) \, d\lambda = 1$$

Furthermore, with the above interpretation of A and B we set

(11.164b) $|A(\mathbf{a}', \lambda)| \leq 1, \quad |B(\mathbf{b}', \lambda)| \leq 1, \quad$ for all $\mathbf{a}', \mathbf{b}'$

Note that in writing (11.163), it is tacitly assumed that measurements of the two spin projections do not influence one another. With apparatuses for these measurements spatially removed from one another, such as in the present situation, the preceding assumption is an example of the *principle of locality*, which is a basic premise in physics.

Stemming from (11.163) and the conditions (11.164b), we obtain *Bell's inequality* (see Problem 11.99).

(11.165) $S \equiv |P(\mathbf{a}, \mathbf{b}) - P(\mathbf{a}, \mathbf{b}')| - |P(\mathbf{a}', \mathbf{b}') - P(\mathbf{a}', \mathbf{b})| \leq 2$

where $\mathbf{a}'$, $\mathbf{b}'$ are arbitrary unit vectors. Note in particular that this result is independent of the probability density $\rho(\lambda)$.

We may also form the ratio (11.165) employing $P_{QM}(\mathbf{a}, \mathbf{b})$ as given by (11.162). Thus, with $\cos^{-1}(\mathbf{a}' \cdot \mathbf{a}) = \phi$, $\cos^{-1}(\mathbf{a} \cdot \mathbf{b}) = \theta$, and $\cos^{-1}(\mathbf{b} \cdot \mathbf{b}') = \gamma$, and assuming that these angles are coplanar, the difference S becomes

(11.166) $S_{QM} = |\cos\theta - \cos(\theta + \gamma)| - |\cos(\phi + \theta + \gamma) + \cos(\phi + \theta)| \leq 2$

In what sense is the expression (11.166) quantum mechanical? The answer is that it stems from the quantum mechanical expectation (11.161), which in turn involves the superposition state ξ_A given by (11.160). This state, as described in Section 5.1, is purely quantum mechanical. It describes each particle as being partly in the α (spin "up") state and the β (spin "down") state.

With specific choice of unit vectors: $\mathbf{b} = \mathbf{a}'$, $\mathbf{a} \cdot \mathbf{b} = \mathbf{b} \cdot \mathbf{b}' = \cos\theta$, the preceding relation becomes (see Problem 11.100)

(11.167) $S_{QM} = 2\cos\theta - \cos 2\theta \leq 1$

This inequality is violated over an interval of θ with maximum violation at $\theta = \pi/3$, where $S_{QM} = 3/2$. We may conclude that quantum mechanics is in conflict with assumptions employed in the derivation of Bell's inequality (11.165) (eg., the principle of locality). The proof of the existence of such conflicts with quantum mechanics is the essence of Bell's theorem.

We return now to the two-photon decay process, discussed at the start of this section. In applying Bell's analysis to this process, the unit vectors, $\mathbf{a}$, $\mathbf{b}$, introduced above, are associated with respective polarization axes of the two polarizers (see Fig. 11.17). In this representation relevant to Bell's analysis, the particle decay process depicted in Fig. 11.17 and the two-photon decay process depicted in Fig. 11.20, are effectively equivalent. Thus, one expects the two-photon decay process also to carry an inequality violation paralleling (11.167).

A generalization of Bell's inequality (11.165) was obtained by Clauser and Horn.[1]

---

[1] J. F. Clauser and M. A. Horn, *Phys. Rev.* **D 10**, 526 (1974). See also, L. E. Ballentine, *Quantum Mechanics*, Prentice-Hall, Englewood Cliffs, N.J. 1990, Chapter 20.

When applied to the two-photon decay process one finds

(11.168) $$3N(\chi) - N(3\chi) \leq 1$$

where $N(\chi)$, as given by (11.159), is interpreted as probability of coincidence. Substituting this value for $N(\chi)$ into the preceding equation indicates that the inequality is violated over angular intervals at $\chi \geq 0$ and $\chi \leq \pi$. Again, we are confronted with conflicts between assumptions leading to the inequality (11.168) and quantum mechanics.

However, it should be borne in mind that the preceding statements of inequality are idealized in the sense that they do not incorporate limitations of the apparatus of an actual experiment. Results to date, stemming from experiments performed in the laboratory, indicate that quantum mechanics is consistent.[1]

PROBLEMS

11.85 In actual S-G experiments, measurement of particle spin through direct observation of deflection of such particles runs into difficulty with the uncertainty relation if particle mass is too small.[2] Thus, in the original S-G experiment, electron spin was measured through observation of a beam of silver atoms. Such atoms contain an uncoupled electron in the 5s shell.

Consider a beam of particles of mass M and spin $\frac{1}{2}$, propagating in the $+x$ direction. The beam has cross section d. It interacts with an S-G apparatus whose field is in the z direction. Employing relevant uncertainty relations, show that the smallest uncertainty in normal displacements (Δz) grows large with decreasing mass M.

Answer

The uncertainty relation indicates that the minimum perpendicular component of momentum of a particle in a beam of cross section d is

$$p_z \simeq \frac{\hbar}{d}$$

Let the beam pass through the S-G apparatus in time t. Then the related spread in normal displacements of particles is given by

$$\Delta z \simeq \frac{p_z}{M} t = \frac{\hbar}{Md} t$$

We may view t as an interval over which energy states of the particles are measured. For the case

---

[1] A compilation of experimental results in this context is given by B. d'Espagnat, *Scientific American* **241**, 158 (Nov. 1979). See also, J. F. Clauser and A. Shimony, *Repts. Prog. Phys.* **41**, 1881, (1978); M. Redhead, *Incompleteness. Nonlocality, and Realism*, Clarendon Press, Oxford, 1987, Chapter 4.

[2] I am indebted to Norman Ramsey for bringing this problem to my attention and for reference to Pauli's discussion on this topic; see *Hand. der Physik* **XXIV**, 1st part, 83–272 (1933).

at hand, particles have two energy states and $\Delta E = E_+ - E_-$. With Problem 5.28 we then write

$$t \gtrsim \frac{\hbar}{\Delta E}$$

Substituting in the preceding equation gives

$$\Delta z \simeq \frac{\hbar t}{Md} \gtrsim \frac{\hbar^2}{Md\,\Delta E}$$

Thus we find that the smallest transverse uncertainty spread in the beam grows large with decreasing mass.

11.86 (a) How many spin states are there in the uncoupled representation for three electrons?

(b) Calling individual particle states $\alpha(1)$, $\beta(1)$, $\alpha(2)$, etc., write down these uncoupled spin states.

11.87 With the states found in Problem 11.86 used as basis states, employ a Clebsch Gordan expansion to find the normalized coupled spin states for three electrons. In listing these states, in addition to spin quantum numbers (s, m_s), designate, also $s_{1,2}$, the quantum number corresponding to the subcomponent spin, $\mathbf{s}_{1,2} = \mathbf{s}_1 + \mathbf{s}_2$.

Answer

With the notation

$$(\alpha\beta\beta) \equiv \alpha(1)\beta(2)\beta(3)$$

the eight coupled spin states are given by

$$|s_{12}, s, m_s\rangle = \left|1, \frac{3}{2}, \frac{3}{2}\right\rangle = (\alpha\alpha\alpha)$$

$$\left|1, \frac{3}{2}, \frac{1}{2}\right\rangle = \frac{1}{\sqrt{3}}[(\alpha\alpha\beta) + (\beta\alpha\alpha) + (\alpha\beta\alpha)]$$

$$\left|1, \frac{3}{2}, -\frac{1}{2}\right\rangle = \frac{1}{\sqrt{3}}[(\beta\beta\alpha) + (\alpha\beta\beta) + (\beta\alpha\beta)]$$

$$\left|1, \frac{3}{2}, -\frac{3}{2}\right\rangle = (\beta\beta\beta)$$

$$\left|1, \frac{1}{2}, \frac{1}{2}\right\rangle = \sqrt{\frac{2}{3}}(\alpha\alpha\beta) - \sqrt{\frac{1}{6}}[(\beta\alpha\alpha) + (\alpha\beta\alpha)]$$

$$\left|1, \frac{1}{2}, -\frac{1}{2}\right\rangle = \sqrt{\frac{1}{6}}[(\alpha\beta\beta) - (\beta\alpha\beta)] - \sqrt{\frac{2}{3}}(\beta\beta\alpha)$$

$$\left|0, \frac{1}{2}, \frac{1}{2}\right\rangle = \frac{1}{\sqrt{2}}[(\alpha\beta\alpha) - (\beta\alpha\alpha)]$$

$$\left|0, \frac{1}{2}, -\frac{1}{2}\right\rangle = \frac{1}{\sqrt{2}}[(\alpha\beta\beta) - (\beta\alpha\beta)]$$

11.88 An important theorem in quantum mechanics addresses coupled spin states for three electrons. The theorem states that a *coupled* spin state which is antisymmetric with respect to all three particles vanishes identically.[1] Another theorem states that if $u(1, 2, 3)$ is antisymmetric with respect to interchange of particles 1 and 2, then

$$w(1, 2, 3) = u(1, 2, 3) + u(2, 3, 1) + u(3, 1, 2)$$

is totally antisymmetric.

Consider the spin state $|0, \frac{1}{2}, \frac{1}{2}\rangle$ given in Problem 11.87, which is antisymmetric with respect to particles 1 and 2 but not 3. Show that construction of the totally antisymmetric function $w(1, 2, 3)$ for this case gives $w = 0$.

11.89 (a) Repeat Problem 11.86 for four electrons.

(b) What are the possible total s-values for this situation?

(c) Working in the notation of Problem 11.87, in the coupled representation, write down an antisymmetric spin state for four electrons. What variables are designated in the state you have written down? For which particles is your state antisymmetric?

11.90 In relativistic formulations one works with observables called *four-vectors*. For example, an event which occurs in a frame S at the coordinates x, y, z at the time t is represented by the four-vector $X = (x, y, z, ict)$. Let the frame S' be in relative motion with respect to S with velocity v parallel to the z axis. If at $t = 0$ the two frames are coincident, then the event X as observed in S' has components $X' = (x', y', z', ict')$ which are given by the *Lorentz transformation* (where $\beta \equiv v/c$ and $\gamma^2 \equiv 1 - \beta^2$)

$$X' = \hat{L}X, \qquad \hat{L} = \begin{pmatrix} 1 & 0 & 0 & 0 \\ 0 & 1 & 0 & 0 \\ 0 & 0 & \gamma & i\beta\gamma \\ 0 & 0 & -i\beta\gamma & \gamma \end{pmatrix}$$

(a) Show that $\hat{L}$ is orthogonal and that $\hat{L}$ preserves the inner product (i.e., $X_1 \cdot X_2 = X'_1 \cdot X'_2$).

(b) Show that the Lorentz transformation reduces to the *Galelian transformation*, $z' = z - vt$, $t' = t$, in the limit $\beta \ll 1$.

11.91 With reference to (11.140) it is generally the case that the commutator $[\hat{A}, \hat{B}]$ is equal to $i\hbar$ times the Poisson bracket considered as an operator.[2] For this property to be satisfied it is required that the Poisson bracket is symmetrized in its arguments. Show that

$$[x^2, p^2] = i\hbar[x^2, p^2]_{\text{Cl}}$$

Answer

First, we write

$$[x^2, p^2]_{\text{Cl}} = \tfrac{1}{2}([x^2, p^2] - [p^2, x^2]) = 2(xp + px)$$

[1] For further discussion see, R. L. Liboff, *Am. J. Phys. 52*, 561 (1984). This topic is returned to in Section 12.4.

[2] For further discussion, see D. Bohm, *Quantum Theory*, Prentice-Hall, New York, 1951.

where the symmetrized term on the right is considered as an operator. For the commutator we find

$$[x^2, p^2] = 2i\hbar(xp + px)$$

which is seen to agree with the given rule.

11.92 (a) Integrate Heisenberg's equation of motion (11.139a) for a free-particle Hamiltonian to obtain $\hat{q}(t)$ and $\hat{p}(t)$ as a function of $\hat{q}(0)$ and $\hat{p}(0)$.

(b) Show that in this case

$$[\hat{q}(t), \hat{q}(0)] = -\frac{i\hbar}{m}t$$

Answer (partial)

Inserting

$$\hat{H} = \frac{\hat{p}^2}{2m}$$

into (11.139a) and dropping primes on operators, we find

$$i\hbar\frac{d\hat{p}}{dt} = 0, \qquad \hat{p} = \hat{p}(0)$$

$$i\hbar\frac{d\hat{q}}{dt} = [\hat{q}, \hat{H}] = i\hbar\frac{\hat{p}}{m}$$

$$\hat{q}(t) = \hat{q}(0) + \frac{\hat{p}(0)t}{m}$$

11.93 Write down the Clebsch–Gordan expansion for two coupled electrons with $j = 2$, $m_j = 0, l = 1, s = 1$, in terms of the product states $|lm_l\rangle|sm_s\rangle$. Evaluate coefficients.

11.94 Show that the total angular momentum of two fermions is always an integer. (*Note:* This problem establishes that two coupled fermions constitute a *composite boson*.)

Answer

The total angular momentum operator of two fermions is

$$\hat{\mathbf{J}} = \hat{\mathbf{S}}_1 + \hat{\mathbf{S}}_2 + \hat{\mathbf{L}}$$

where **L** is the orbital angular momentum of the two-particle system. With $s_1 = n_1/2$ and $s_2 = n_2/2$, where n_1 and n_2 are odd integers, we find the total spin numbers

$$s = \left(\frac{n_1}{2} - \frac{n_2}{2}\right), \dots, \left(\frac{n_1}{2} + \frac{n_2}{2}\right)$$

which are all integers. Total j is then given by

$$j = l - s, \dots, l + s$$

which, for all s given above, are integers.

11.95 Employing the vector operator formula given in Problem 11.68, show that

$$\frac{\{\hat{\boldsymbol{\sigma}} \cdot [\hat{\mathbf{p}} - (e/c)\hat{\mathbf{A}}]\}^2}{2m} = \frac{[\hat{\mathbf{p}} - (e/c)\hat{\mathbf{A}}]^2}{2m} - \frac{e\hbar}{2mc} \hat{\boldsymbol{\sigma}} \cdot \hat{\mathbf{B}}$$

where $\hat{\mathbf{A}}$ is vector potential.

11.96 A one-dimensional harmonic oscillator is initially in the state $|\psi(0)\rangle = |1\rangle$ [in the notation of (7.61)]. The system is acted upon by an exponentially decaying potential in time. The total Hamiltonian is given by

$$\hat{H} = \hbar\omega_0(\hat{N} + \tfrac{1}{2}) + (\hat{a} + \hat{a}^\dagger)V_0 e^{-(t/\tau)}$$

where $V_0 \ll \hbar\omega_0$ and τ is a constant relaxation time.

    (a)   Working in the interaction picture, calculate $|\psi_I(t)\rangle$ to first order in $\lambda \equiv V_0/\hbar\omega_0$.

    (b)   From your answer to (a), obtain the wavefunction in the Schrödinger picture, $|\psi(t)\rangle$.

    (c)   What is the probability $P_{1\to n}$ that a transition to the state $|n\rangle$ has occurred after time t due to the perturbation?

Answers

    (a)   First note that

$$|\psi_I(0)\rangle = |\psi(0)\rangle = |1\rangle$$

and

$$\begin{aligned}
\hat{V}_I(t)|\psi_I(0)\rangle &= e^{it\hat{H}_0/\hbar}\hat{V}e^{-it\hat{H}_0/\hbar}|1\rangle \\
&= e^{it\hat{H}_0/\hbar}(\hat{a} + \hat{a}^\dagger)V_0 e^{-t/\tau}e^{-i3\omega_0 t/2}|1\rangle \\
&= e^{it\hat{H}_0/\hbar}(|0\rangle + \sqrt{2}|2\rangle)V_0 e^{-t/\tau}e^{-i3\omega_0 t/2} \\
&= V_0[e^{-(t/\tau)-i\omega_0 t}|0\rangle + \sqrt{2}e^{-(t/\tau)+i\omega_0 t}|2\rangle]
\end{aligned}$$

Substituting this expression into (11.145) and integrating, we obtain

$$|\psi_I(t)\rangle = |1\rangle + \frac{iV_0}{\hbar\omega_0}[\Lambda(t)|0\rangle + \Lambda^*(t)\sqrt{2}|2\rangle]$$

where

$$\Lambda(t) \equiv \frac{e^{-t\phi} - 1}{\phi/\omega_0}$$

$$\phi \equiv \frac{1}{\tau} + i\omega_0$$

    (b)   This answer is obtained by substituting the preceding expression in the inverse of (11.141). There results

$$|\psi(t)\rangle = \Lambda e^{-it\omega_0/2}|0\rangle + \Lambda^*e^{-it5\omega_0/2}\sqrt{2}|2\rangle$$

(c)   We may view the preceding result as an expansion of $|\psi(t)\rangle$ in the basis $\{|n\rangle\}$ and further consider the state $|\psi(t)\rangle$ as given in (b) to be normalized to $O(\lambda)$, where $\lambda \equiv \hbar\omega_0/V_0$. There results

$$P_{1 \to 1} = 1 + O(\lambda^2)$$

$$P_{1 \to 0} = |\langle 0|\psi(t)\rangle|^2 = \lambda^2|\Lambda|^2 + O(\lambda^3)$$

$$P_{1 \to 2} = |\langle 2|\psi(t)\rangle|^2 = 2\lambda^2|\Lambda|^2 + O(\lambda^3)$$

where

$$|\Lambda|^2 = \frac{1 - e^{-2t/\tau} - 2e^{-t/\tau}\cos\omega_0 t}{1 + (1/\omega_0\tau)^2}$$

Thus we find it is most probable for the oscillator to remain in the first excited state. The probability for a transition to the second excited state is twice as probable as that for a transition to the ground state.

11.97   Employing the spin-zero antisymmetric state (11.160), evaluate (11.161) to obtain (11.162). *Hint*: See Problem 11.68.

11.98   Show that $|P_{12}\rangle$ as given by (11.153) follows from (11.152) through the transformation (11.147–150).

11.99   Employing the *triangle inequality*

$$\|\psi_1 + \psi_2\| \leq \|\psi_1\| + \|\psi_2\|$$

and conditions (11.164b) establish Bell's inequality (11.165).

11.100   Derive (11.167) from (11.168). *Hint*: Recall the rules:
  (a)   If $|A| + |B| \leq n$, then $|A + B| \leq n$.
  (b)   If $|A| \leq n$, then $-n \leq A$.

CHAPTER 12

APPLICATION TO ATOMIC, MOLECULAR, SOLID-STATE, AND NUCLEAR PHYSICS. ELEMENTS OF QUANTUM STATISTICS

Having developed methods for addition of spin angular momentum in Chapter 11 and properties of the three-dimensional Hamiltonian in Chapter 10, these formalisms, together with the Pauli principle, are now applied to some practical problems. Symmetry requirements on the wavefunction of the helium atom imposed by the Pauli principle serve to couple electron spins. This coupling results in a separation of the spectra into singlet and triplet series for helium as well as other two-electron atoms. Symmetrization requirements stemming from the Pauli principle are also maintained in calculation of the binding of the hydrogen molecule. The relevance of Bose–Einstein condensation to superconductivity and superfluidity is described. Application of the Pauli principle is further noted in two examples. In the first of these, semiconductor theory, previously encountered in Chapter 8, is extended to extrinsic (impurity) conductivity, in which the Fermi–Dirac distribution comes into play. The second example addresses nuclear physics, and with concepts of the nucleon and isotopic spin, the totally antisymmetric ground state of the deuteron is constructed.

12.1   THE TOTAL ANGULAR MOMENTUM, J

In this section we consider the addition of spin and orbital angular momentum for atomic systems. As previously noted, the total angular momentum of a system (e.g., an atom or a single electron) which has both orbital angular momentum **L** and spin angular momentum **S** is called **J**.

(12.1) $$\hat{\mathbf{J}} = \hat{\mathbf{L}} + \hat{\mathbf{S}}$$

It has components

$$\hat{J}_x = \hat{L}_x + \hat{S}_x$$
$$\hat{J}_y = \hat{L}_y + \hat{S}_y$$
$$\hat{J}_z = \hat{L}_z + \hat{S}_z$$

Its square appears as

(12.2) $$\hat{J}^2 = \hat{L}^2 + \hat{S}^2 + 2\hat{\mathbf{L}} \cdot \hat{\mathbf{S}}$$

In obtaining this expression we have used the fact that **L** and **S** commute.

L-S Coupling

Individual electrons in an atom have both orbital and spin angular momentum. Among the lighter atoms, individual electrons' **L** vectors couple to give a resultant **L** and individual **S** vectors couple to give a resultant **S**. These two vectors then join to give a total angular momentum **J** (Fig. 12.1). This is called the *L-S* or *Russell–Saunders coupling scheme.*[1]

Eigenstates in this representation are simultaneous eigenstates of the four commuting operators

(12.3) $$\{\hat{J}^2, \hat{J}_z, \hat{L}^2, \hat{S}^2\}$$

There are six pairs of operators in this set which must be checked for commutability.

(i) $[\hat{L}^2 + \hat{S}^2 + 2\hat{\mathbf{L}} \cdot \hat{\mathbf{S}}, \hat{L}^2] = 0$ (iv) $[\hat{L}^2, \hat{J}_z] = [\hat{L}^2, \hat{L}_z + \hat{S}_z] = 0$

(ii) $[\hat{L}^2 + \hat{S}^2 + 2\hat{\mathbf{L}} \cdot \hat{\mathbf{S}}, \hat{S}^2] = 0$ (v) $[\hat{S}^2, \hat{J}_z] = [\hat{S}^2, \hat{L}_z + \hat{S}_z] = 0$

(iii) $[\hat{J}^2, \hat{J}_z] = 0$ (vi) $[\hat{L}^2, \hat{S}^2] = 0$

---

[1] This scheme is also relevant to "one- or two-electron atoms." More generally, in heavy elements with large Z, the spin-orbit coupling (Section 12.2) becomes large and serves to couple $\mathbf{L}_i$ and $\mathbf{S}_i$ vectors of individual electrons. giving resultant $\mathbf{J}_1$. $\mathbf{J}_2$. . . . . values. These individual electron $\mathbf{J}_i$ values then combine to give a resultant **J**. This coupling scheme is known as *j-j* coupling.

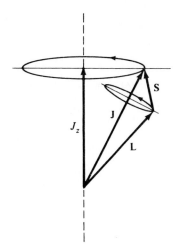

FIGURE 12.1   Schematic vector representation of the L-S scheme of angular momentum addition. J^2 and J_z are fixed, as are L^2 and S^2.

In (i), $\hat{L}^2$ commutes with all its components. In (ii), $\hat{S}^2$ commutes with all its components. The remaining relations are self-evident.

Eigenvalue equations related to the commuting operators (12.3) appear as

$$\hat{J}^2|jm_jls\rangle = \hbar^2 j(j+1)|jm_jls\rangle$$

$$\hat{J}_z|jm_jls\rangle = \hbar m_j|jm_jls\rangle$$

$$\hat{L}^2|jm_jls\rangle = \hbar^2 l(l+1)|jm_jls\rangle$$

$$\hat{S}^2|jm_jls\rangle = \hbar^2 s(s+1)|jm_jls\rangle$$

For a given value of J, m_j runs in integral steps from $-j$ to $+j$.

A very important operator that commutes with all four operators (12.3) is $\hat{\mathbf{L}} \cdot \hat{\mathbf{S}}$.

$$\hat{\mathbf{L}} \cdot \hat{\mathbf{S}}|jm_jls\rangle = \tfrac{1}{2}(\hat{J}^2 - \hat{L}^2 - \hat{S}^2)|jm_jls\rangle$$

$$= (\hbar^2/2)[j(j+1) - l(l+1) - s(s+1)]|jm_jls\rangle$$

Eigen-j-values and Term Notation

In the L-S representation, l and s are known. What are the possible j values corresponding to these values of l and s? Since **J** is the resultant of two angular momentum vectors, the rules of Section 9.4 apply. These rules indicate that j values run from a maximum of

$$j_{\max} = l + s$$

to a minimum of

$$j_{\min} = |l - s|$$

in integral steps.

$$(l + s) \geq j \geq |l - s|$$

$$j = l + s, l + s - 1, l + s - 2, \ldots, |l - s| + 1, |l - s|$$

For $s < l$, there are a total of $(2s + 1)$ j values. The number $(2s + 1)$ is called the *multiplicity*. In the section to follow, these different j values are shown to correspond to distinct energy values for the atom. Thus for a one-electron atom $(s = \frac{1}{2})$, a state of given l splits into the doublet corresponding to the two values

$$j = l + \tfrac{1}{2}, \qquad j = l - \tfrac{1}{2}$$

(Fig. 12.2). The *term notation* for these states is given by the following symbol

$$^{2s+1}\mathscr{L}_j$$

where $\mathscr{L}$ denotes the letter corresponding to the orbital angular momentum l value according to the following scheme:

l	0	1	2	3	4	5	6	7	8	9	10	$\cdots$
letter $(\mathscr{L})$	S	P	D	F	G	H	I	K	L	M	N	$\cdots$

The doublet P states of one-electron atoms are denoted by the terms

$$^2P_{1/2}, \, ^2P_{3/2}$$

The doublet F states $(l = 3)$ are denoted by

$$^2F_{7/2}, \, ^2F_{5/2}$$

In two-electron atoms, the resultant spin quantum number is either 0 or 1. These, we recall (Section 11.10), are the resultant s values corresponding to the addition of two $\frac{1}{2}$ spins. These two values of s give rise to two types of spectra (this is the case, for

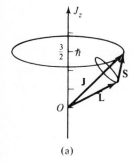

(a)

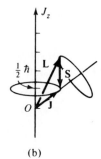

(b)

FIGURE 12.2 Diagrams depicting coupling of the L and S vectors of a single p electron, in the L-S scheme. The doublet contains two values of j.

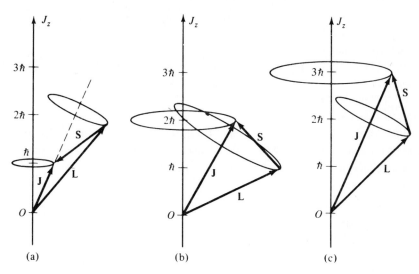

FIGURE 12.3 **Diagrams depicting coupling of L and S vectors for two electrons in an orbital D state and a spin-1 state. The resultant triplet of j values is**

$$j = 1, 2, 3$$

example, for He):

$$s = 0 \rightarrow \text{singlet series:} \quad {}^1S, {}^1P, {}^1D, \ldots$$

$$s = 1 \rightarrow \text{triplet series:} \quad {}^3S, {}^3P, {}^3D, \ldots$$

The j values of the ${}^3D$ states are $j = 1, 2, 3$. These correspond to the states

$$ {}^3D_1, {}^3D_2, {}^3D_3 $$

In general, any state with $l > 1$ becomes the triplet

$$ j = l + 1, l, l - 1 $$

(Fig. 12.3).

The multiplicity corresponding to the case $s > l$ is $2l + 1$. For example, if $s = \frac{3}{2}$ and $l = 1$, there are three j values: $j = \frac{5}{2}, \frac{3}{2}, \frac{1}{2}$. However, the notation for this state remains ${}^4P_{5/2, 3/2, 1/2}$, with 4 written for $2s + 1$, although in fact the multiplicity is $2l + 1$.

PROBLEMS

12.1 (a)   For given values of L and S, show that the four operators

$$ \{\hat{L}^2, \hat{L}_z, \hat{S}^2, \hat{S}_z\} $$

form a commuting set of observables. This representation is akin to the "uncoupled" representation discussed in Sections 9.4 and 11.9, while the L-S representation (12.3) compares to the "coupled" representation.

(b) Which of these operators are incompatible with those of (12.3)?

12.2 Show that

(a) $[\hat{J}^2, \hat{L}_z] = 2i\hbar e_z \cdot (\hat{L} \times \hat{S})$

(b) $[\hat{J}^2, \hat{S}_z] = 2i\hbar e_z \cdot (\hat{S} \times \hat{L})$

12.3 What are the multiplicities of the G ($l = 4$) and H ($l = 5$) states for the two spectral series related to a three-electron atom? What is the complete term notation for all these states?

12.4 What kind of terms can result from the following values of l and s?

(a) $l = 2, s = \frac{7}{2}$

(b) $l = 5, s = \frac{3}{2}$

(c) $l = 3, s = 3$

12.5 What are the l, s, j values and multiplicities of the following terms?

(a) $^3D_2$

(b) $^4P_{5/2}$

(c) $^2F_{7/2}$

(d) $^3G_3$

12.2 ONE-ELECTRON ATOMS

In this section we consider the manner in which the spin of the valence electron in one-electron atoms interacts with the shielded Coulomb field due to the nucleus and remaining electrons of the atom. One-electron atoms are better known as the alkali-metal atoms.[1] In such atoms all but one electron are in closed "shells" (to be discussed below). These "core" electrons, together with the nucleus, present a radial electric field to the outer valence electron (Fig. 12.4). Furthermore, the total orbital and spin angular momentum of a closed shell is zero, so that the angular momentum of the atom is determined by the valence electron.

Spin-Orbit Coupling

The interaction between the spin of the valence electron and the shielded Coulomb field arises from the orbital motion of this electron through the Coulomb field. When an observer moves with velocity **v** across the lines of a static electric field $\mathscr{E}$, special

---

[1] This analysis is also relevant to hydrogenic atoms (Section 10.6).

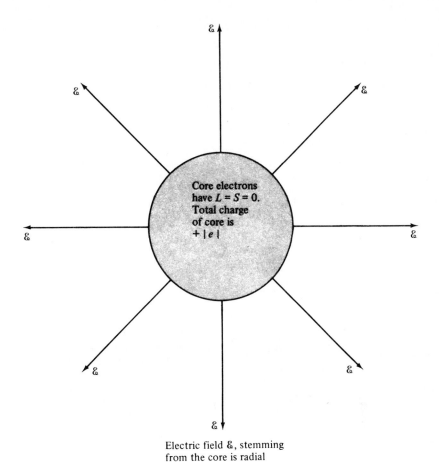

Electric field &, stemming
from the core is radial

FIGURE 12.4 **Properties of "one-electron" atoms.**

relativity reveals that in the frame of the observer, a magnetic field

$$\mathscr{B} = -\gamma\boldsymbol{\beta} \times \mathscr{E}$$

$$\boldsymbol{\beta} \equiv \frac{\mathbf{v}}{c}, \qquad \gamma^{-2} \equiv 1 - \beta^2$$

is detected (Fig. 12.5). Keeping terms to first order in β gives

$$\mathscr{B} = -\frac{\mathbf{v}}{c} \times \mathscr{E}$$

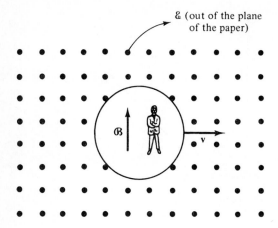

FIGURE 12.5   An observer in the (perfectly transparent) sphere which is moving with velocity v across the electric field $\mathscr{E}$ detects a magnetic field

$$\mathscr{B} = -\frac{\mathbf{v}}{c} \times \mathscr{E}$$

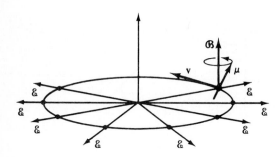

FIGURE 12.6   The magnetic moment $\boldsymbol{\mu}$ of the satellite electron sees a magnetic field due to its orbital motion across the radial, static Coulomb lines of force which emanate from the nucleus. The resulting torque on the moment produces a precession of the spin axis of the electron as shown.

It follows that (to this order) if an electron moves with momentum **p** across a field $\mathscr{E}$, the electron will feel a magnetic field[1]

(12.4)
$$\mathscr{B} = -\frac{\mathbf{p}}{mc} \times \mathscr{E}$$

This is the nature of the magnetic field with which the magnetic moment of the orbiting valence electron interacts (Fig. 12.6). The interaction energy between $\boldsymbol{\mu}$ and $\mathscr{B}$ is given by (11.89), modified by the Thomas factor, $\frac{1}{2}$. This correction factor represents an additional relativistic effect due to the acceleration of the electron.[2] Thus the interaction energy between the spin of the orbiting electron and the magnetic field (12.4)

---

[1] Relativistic momentum is $\mathbf{p} = \gamma m\mathbf{v}$, so that to terms of $O(\beta^2)$, $\mathbf{p} = m\mathbf{v}$. In these formulas m is the rest mass of an electron; $mc^2 = 0.511$ MeV.

[2] L. H. Thomas, *Nature* **117**, 514 (1926).

appears as

(12.5)
$$H' = -\frac{1}{2}\boldsymbol{\mu} \cdot \mathscr{B} = \frac{1}{2}\boldsymbol{\mu} \cdot \left(\frac{\mathbf{v}}{c} \times \mathscr{E}\right)$$

$$= -\frac{1}{2}\frac{\mu}{mc} \cdot (\mathscr{E} \times \mathbf{p})$$

In spherical coordinates, $\mathscr{E}$ has only a radial component

$$\mathscr{E}_r = -\frac{d}{dr}\Phi(r)$$

where $\Phi(r)$ is the static Coulomb potential (ergs/esu) seen by the valence electron. Substituting this expression for $\mathscr{E} = (\mathscr{E}_r, 0, 0)$ into (12.5) gives

$$H' = \frac{1}{2}\frac{1}{mc}\left[\frac{1}{r}\frac{d\Phi(r)}{dr}\right](\mathbf{r} \times \mathbf{p}) \cdot \boldsymbol{\mu}$$

Recalling the linear relation (11.86) between $\boldsymbol{\mu}$ and $\mathbf{S}$,

$$\boldsymbol{\mu} = \left(\frac{e}{mc}\right)\mathbf{S}$$

and rewriting $\mathbf{L}$ for $\mathbf{r} \times \mathbf{p}$ permits H' to be rewritten in terms of its operational equivalent.

(12.6)
$$\hat{H}' = \frac{e}{2m^2c^2}\left[\frac{1}{r}\frac{d\Phi}{dr}\right]\hat{\mathbf{L}} \cdot \hat{\mathbf{S}} \equiv f(r)\hat{\mathbf{L}} \cdot \hat{\mathbf{S}}$$

which serves to define the scalar function $f(r)$.

Approximate Wavefunction

The Hamiltonian of a typical one-electron atom, neglecting the L-S coupling term just discovered, appears as [recall (10.93)]

$$\hat{H}_0 = \frac{\hat{p}^2}{2m} + V(r) = \frac{\hat{p}_r^2}{2m} + \frac{\hat{L}^2}{2mr^2} + V(r)$$

where $V = e\Phi$, $\mathbf{L}$ is the orbital angular momentum of the valence electron, and p_r is its radial momentum. The eigenstates of $\hat{H}_0$ are hydrogenlike in structure. They are comprised of the eigenstates of $\hat{L}^2$ (spherical harmonics) and solutions to the radial equation [recall (10.95)]. Incorporating the spin-orbit interaction (12.6) gives the total Hamiltonian

$$\hat{H} = \hat{H}_0 + \hat{H}' = \frac{\hat{p}_r^2}{2m} + \frac{\hat{L}^2}{2mr^2} + V(r) + f(r)\hat{\mathbf{L}} \cdot \hat{\mathbf{S}}$$

Rewriting $\mathbf{L} \cdot \mathbf{S}$ in terms of J^2, $\hat{L}^2$, S^2 (12.2) permits this Hamiltonian to be rewritten

$$(12.7) \qquad \hat{H} = \hat{H}_0 + \frac{f(r)}{2} [\hat{J}^2 - \hat{L}^2 - \hat{S}^2]$$

In the preceding section we showed that $(\hat{J}^2, \hat{J}_z, \hat{L}^2, \hat{S}^2)$ comprise a set of commuting operators. Since these operators also commute with $\hat{H}_0$, approximate eigenstates of $\hat{H}$ may be taken to be of the form[1]

$$(12.8) \qquad |\varphi\rangle = |nl\rangle |jm_j ls\rangle$$

where $|nl\rangle$ represents the *radial* component of the eigenstates of $\hat{H}_0$.

$$\hat{H}_0 |nl\rangle = E_n |nl\rangle$$

For hydrogen, for example, $|nl\rangle$ are solutions to the radial equation (10.103), that is, weighted Laguerre polynomials (Table 10.3).

Substituting the product form (12.8) into the Schrödinger equation

$$\hat{H} |\varphi\rangle = E |\varphi\rangle$$

with $\hat{H}$ given by (12.7) gives

$$(12.9) \qquad \left\{ E_n + \frac{\hbar^2}{2} f(r) \left[j(j + 1) - l(l + 1) - \frac{3}{4} \right] \right\} |\varphi\rangle = E |\varphi\rangle$$

It follows that the product solutions (12.8) are *not* eigenstates of $\hat{H}$ (i.e., $\hat{H} |\varphi\rangle \neq$ constant $\times |\varphi\rangle$). But due to the fact that the spin-orbit correction to E_n is small compared to E_n, they do serve as approximate solutions. Approximate eigenvalues of $\hat{H}$ may then be obtained by constructing the expectation of $\hat{H}$ in these states.

$$(12.10) \qquad E_{nlj} = \langle \varphi | H | \varphi \rangle$$

$$= E_n + \frac{\hbar^2}{2} \left[j(j + 1) - l(l + 1) - \frac{3}{4} \right] \langle f(r) \rangle_{nl}$$

Since j can have two values $(l \pm \frac{1}{2})$ for a given value of l, it follows that an energy state of given l separates into a doublet when the spin-orbit interaction is "turned on." The two corresponding values of energy are

$$E_{nlj}^{(+)} \equiv E_{j = l + 1/2} = E_n + \frac{\hbar^2}{2} l \langle f \rangle_{nl}$$

$$(12.11)$$

$$E_{nlj}^{(-)} \equiv E_{j = l - 1/2} = E_n - \frac{\hbar^2}{2} (l + 1) \langle f \rangle_{nl}$$

---

[1] The states $|jm_j ls\rangle$ may be constructed from the product states $|lm_l\rangle |sm_s\rangle$ in a Clebsch-Gordan expansion (with $m_s + m_l = m_j$).

An estimate of $\langle f \rangle_{nl}$ may be obtained using hydrogen wavefunctions and assuming the Coulomb potential for V,

$$V = -\frac{Ze^2}{r}$$

where Z is an effective atomic number. (See Problem 12.13.) Substituting this potential into $f(r)$ as given in (12.6) gives

$$f(r) = \frac{1}{2m^2c^2}\frac{1}{r}\frac{dV}{dr} = \frac{Ze^2}{2m^2c^2}\frac{1}{r^3}$$

(12.12)
$$\langle f \rangle_{nl} = \frac{Ze^2}{2m^2c^2}\int_0^\infty \frac{|R_{nl}(r)|^2}{r^3}r^2\,dr$$

$$\frac{\hbar^2}{2n}\langle f \rangle_{nl} = \frac{(me^4Z^2/2\hbar^2n^2)^2}{mc^2(l + \frac{1}{2})(l + 1)l}$$

(where we have used the results of Problem 10.48).

Fine Structure of Hydrogen

Recalling the energy eigenvalues of hydrogen (10.110)

$$|E_n| = \frac{m(Ze^2)^2}{2\hbar^2n^2}$$

permits the correction factor (12.12) to be written

$$\frac{\hbar^2\langle f \rangle_{nl}}{|E_n|} = \frac{2n}{l(l + \frac{1}{2})(l + 1)}\frac{|E_n|}{mc^2} = \frac{(Z\alpha)^2}{n}\frac{1}{l(l + \frac{1}{2})(l + 1)}$$

where α is the *fine-structure constant* (see Problems 2.20 and 2.29),

$$\alpha = \frac{e^2}{\hbar c} = \frac{1}{137.037}, \qquad \alpha^2 = 5.33 \times 10^{-5}$$

Substituting these forms into the doublet energies (12.11) gives the values

(12.13) E_n

"spin up"
$j = l + \frac{1}{2}$
$$E_{nlj}{}^{(+)} = -|E_n|\left[1 - \frac{1}{(2l + 1)(l + 1)}\frac{(Z\alpha)^2}{n}\right]$$

$$E_{nlj}{}^{(-)} = -|E_n|\left[1 + \frac{1}{l(2l + 1)}\frac{(Z\alpha)^2}{n}\right]$$
"spin down"
$j = l - \frac{1}{2}$

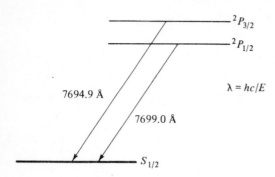

$$\lambda = hc/E$$

FIGURE 12.7   Wavelengths, in angstroms, corresponding to the transition from the lowest ^2P states to the ground state, for potassium.

(where we have written $E_n = -|E_n|$). Thus we see that the spin-orbit corrections to the "unperturbed" energies E_n are about 1 part in 10^5. The fact that these corrections are indeed small lends consistency to our original assumption that the product eigenstates (12.8) closely approximate the eigenstates of the total Hamiltonian $\hat{H}$ (12.7).

The two energies (12.13) correspond to the two possible orientations of **S** with respect to **L** (see Fig. 12.2). When the spin is "down," the magnetic moment of the electron ($\boldsymbol{\mu} \sim -\mathbf{S}$) is aligned with the magnetic field $\mathscr{B}$ (12.4) of the relative electron-nucleus motion. This is the configuration of minimum energy. Thus the correction increment corresponding to the "spin-down" case is smaller than that due to the "spin-up" case, in agreement with the expressions (12.13).

The spin-orbit interaction serves to remove the l degeneracy of the eigen-energies of hydrogenic atoms. If the spin-orbit interaction is neglected, energies are dependent only on the principal quantum number n and are independent of l (and m_l). In the L-S representation, $nljm_j$ (and $s = \frac{1}{2}$) are good quantum numbers, and the l degeneracy is removed. Degeneracy with respect to m_j, however, remains. Eigen-energies are dependent only on (n, l, j), as indicated by expression (12.10). For a given principal quantum number n, the orbital quantum number is restricted to the values $l = 0, 1, \dots, (n-1)$ [recall (10.113)], while for a given l, the total angular momentum quantum number j can take the two values $j = l \pm \frac{1}{2}$.

The partial energy-level diagram[1] for potassium, depicting the transition from the doublet ^2P states to the ground state, is shown in Fig. 12.7. The corresponding radiation lies in the near infrared.

The selection rules for dipole radiation developed in Section 10.7 are generalized to the following, for one-electron atoms.[2]

---

[1] It was in explanation of such doublet spectra that G. E. Uhlenbeck and S. Goudsmit first postulated the existence of electron spin. *Naturwiss.* **13**, 953 (1925), and *Nature* **117**, 264 (1926).

[2] Derivation of these selection rules may be found in R. H. Dicke and J. P. Wittke, *Introduction to Quantum Mechanics,* Addison-Wesley, Reading, Mass., 1960.

$$\Delta l = \pm 1$$

$$\Delta j = \pm 1, 0 \qquad \text{(but } j = 0 \to 0 \text{ is forbidden)}$$

$$\Delta m_j = \pm 1, 0$$

Δn is unrestricted

Photons are emitted only for transitions between states which obey these conditions.

Relativistic Corrections

The spin-orbit corrections to the energies of hydrogen are the same order of magnitude as the corrections due to the relativistic speed of the electron in its orbit. This small correction to the "unperturbed" energies E_n may also be obtained using the technique developed above to find the spin-orbit correction. The relativistic Hamiltonian for a particle of mass m moving in a potential field V is

$$H = (p^2 c^2 + m^2 c^4)^{1/2} - mc^2 + V$$

(12.14)
$$V = - \frac{Ze^2}{r}$$

If $p \ll mc$, then the radical may be expanded to obtain

$$H = \frac{p^2}{2m} - \frac{p^4}{8m^3 c^2} + \cdots + V$$

(12.15)
$$\hat{H} = \left(\frac{\hat{p}^2}{2m} + V \right) - \frac{\hat{p}^4}{8m^3 c^2} \equiv \hat{H}_0 + \hat{H}'$$

Again, as was done above, the correction to the eigenenergies E_n of $\hat{H}_0$ due to $\hat{H}'$ may be obtained by calculating $\langle H' \rangle$ using the hydrogen wavefunctions (10.112). There results

$$\langle H' \rangle = \langle nl | \hat{H}' | nl \rangle$$

$$= - \frac{1}{8m^3 c^2} \langle \hat{p}^4 \rangle_{nl}$$

To evaluate this expectation value, we recall that the eigenstates $|nl\rangle$ satisfy the equation

$$\hat{H}_0 | nl \rangle = E_n | nl \rangle$$

$$\left(\frac{\hat{p}^2}{2m} + V \right) | nl \rangle = E_n | nl \rangle$$

so that (writing $|nl\rangle \equiv |\varphi_{nl}\rangle$)

$$\hat{p}^2 |\varphi_{nl}\rangle = 2m(E_n - V)|\varphi_{nl}\rangle$$

Owing to the Hermiticity of $\hat{p}$,

$$\langle \varphi_{nl}|\hat{p}^2\hat{p}^2|\varphi_{nl}\rangle = \langle \hat{p}^2\varphi_{nl}|\hat{p}^2\varphi_{nl}\rangle$$

It follows that

$$\langle p^4\rangle_{nl} = \int_0^\infty [2m(E_n - V)\varphi_{nl}^*][2m(E_n - V)\varphi_{nl}]r^2 \, dr$$

Hence

$$\langle H'\rangle = -\frac{1}{2mc^2}(E_n^2 - 2E_n\langle V\rangle_{nl} + \langle V^2\rangle_{nl})$$

The integrals $\langle r^{-1}\rangle_{nl}$, $\langle r^{-2}\rangle_{nl}$ may be obtained using the results of Problem 10.48. There results

$$\langle H'\rangle = -|E_n|\frac{\alpha^2 Z^2}{4n^2}\left(\frac{8n}{2l+1} - 3\right)$$

As with the case of spin-orbit correction, we again find that $\langle H'\rangle$ is smaller than E_n by a factor of the order of α^2. Combining this correction with the expressions for the spin-orbit correction (12.13) gives the result

$$(12.16) \qquad E_{nlj}(\text{spin-orbit + rel.}) = -|E_n|\left[1 + \left(\frac{Z\alpha}{2n}\right)^2\left(\frac{4n}{j+\frac{1}{2}} - 3\right)\right] \qquad (j = l \pm \tfrac{1}{2})$$

This expression for the energies of hydrogen include the "fine-structure" corrections, due to spin-orbit and relativistic effects. The energies so found are in quite good agreement with observed hydrogen emission spectra.[1]

PROBLEMS

12.6 The magnetic field (12.4) due to the relative nucleus-electron motion may also be thought of as arising in the following way. If one "sits" on the electron, the nucleus is seen to move in orbital motion about this position. This nuclear orbit constitutes a current loop, which in turn generates a magnetic field. Calculate the value of this magnetic field for a given value of **L** and compare it to the value obtained from (12.4).

12.7 In quantum mechanics, when one says that the vector **J** is conserved, one means that for any state the system is in, the expectations of the three components of **J** are constant. This

---

[1] Calculation of higher-order effects may be found in L. I. Schiff, *Quantum Mechanics*, 3rd ed., McGraw-Hill, New York, 1968.

follows if these three operators all commute with the Hamiltonian. Show that for a one-electron atom with spin-orbit coupling, **L** and **S** are not conserved.

12.8 There is no spin-orbit interaction if an electron is in an S state ($l = 0$). Why?

12.9 What is the difference in energy between the two states of a doublet for a typical one-electron atom as a function of n and l?

12.10 What is the wavelength of a photon emitted by a typical one-electron atom when the valence electron undergoes a spin flip from the $2^2P_{3/2}$ to the $2^2P_{1/2}$ state? In this notation, 2 is the value of the principal quantum number n. Give your answer in terms of Z. According to the selection rules cited for dipole radiation, is such a transition allowed?

12.11 Make an estimate of the rotational kinetic energy, $L^2/2mr^2$, of an electron in a 2P state of hydrogen. What is the ratio of this energy to the rest-mass energy $mc^2 = 0.511$ MeV?

12.12 (a) What are the respective frequencies (Hz) emitted in the transitions (1) $2P_{1/2} \rightarrow 1S_{1/2}$, (2) $2P_{3/2} \rightarrow 1S_{1/2}$, for lithium?

(b) What is the percentage change between these frequencies?

12.13 In the theory developed in Section 12.2, the effect of the inner core electrons of an alkali metal atom on the energy spectrum was described by an effective atomic number. Owing to penetration of the valence electron wavefunction into the core, such a simple model proves insufficient. A more quantitative model which includes the effects of this *quantum defect* is described in the following example.

One assumes that the potential seen by the outer valence electron is of the form

$$V = \frac{-Z'e^2}{r}\left(1 + \frac{b}{r}\right)$$

This modified potential has the effect of making the force of attraction between the valence electron and the nucleus (of charge Ze) grow with penetration of the valence electron into the core. The deeper this penetration, the larger is the net positive charge "seen" by the valence electron. The effective nuclear charge $Z'e$ and displacement b may be chosen so as to give the best fit with observed spectral data.

(a) Show that the method used to solve for the energy levels of the hydrogen atom can be applied to this problem with only slight modifications to give energy levels of the form

$$E_{nl} = \frac{-Ze^2}{2a_0[n + D(l)]^2}$$

Here a_0 is the hydrogen Bohr radius and

$$D(l) \equiv \sqrt{(l + \tfrac{1}{2})^2 - \frac{2bZ'}{a_0}} - (l + \tfrac{1}{2})$$

represents the l-dependent quantum defect.

(b) How do these E_n energy states vary with increasing l? Give a physical explanation of this variation in terms of core penetration of the valence electron.

12.14 At sufficiently high temperatures, a diatomic dumbbell molecule may suffer vibrational modes of excitation above the normal rotational modes. Consider that the two atomic nuclei are bound through a central potential $V(r)$ which has a strong minimum at the separation $r = a$. At low temperatures the nuclei stay at this interparticle spacing and the effective Hamiltonian is

$$\hat{H} = \frac{\hat{L}^2}{2\mu a^2} + V(a)$$

where μ is reduced mass. At higher temperatures, the particles separate and the Hamiltonian becomes

$$\hat{H} = \frac{\hat{L}^2}{2\mu(a + \xi)^2} + V(a + \xi)$$

where ξ is the (small) radial deviation from the equilibrium separation, a. For the case that $\langle L^2/2\mu a^2 \rangle \ll (\xi/a)V(a)$:

(a) Obtain a form of $\hat{H}$ that is valid to order of $(\xi/a)^2$.

(b) By appropriate choice of product wavefunctions, obtain the eigenenergies E_{ln} of the Hamiltonian you have constructed.

(c) Argue the consistency of omitting the radial kinetic energy $\hat{p}_r^2/2\mu$ in the Hamiltonians written above.

[*Hint* [for part (a)]: Since $V(r)$ is minimum at $r = a$, it follows that $V'(a) = 0$.]

12.15 In Chapter 11 [Eq. (11.86)] it was noted that the magnetic moment of an electron due to its spin is

$$\boldsymbol{\mu}_S = \frac{e}{mc}\,\mathbf{S}$$

Consider that the electron moves in a circle with angular momentum $\mathbf{L}$. Show classically that the magnetic moment due to such orbital motion is

$$\boldsymbol{\mu}_L = \frac{e}{2mc}\,\mathbf{L}$$

[*Note:* Thus we see that the total magnetic moment of an atomic electron

$$\boldsymbol{\mu} = \frac{e}{2mc}\,(\mathbf{J} + \mathbf{S}) = -\frac{\mu_b}{\hbar}\,(\mathbf{J} + \mathbf{S})$$

may not in general be assumed to be parallel to its total angular momentum $\mathbf{J}$.]

12.16 Consider an atom whose electrons are L-S coupled so that good quantum numbers are $jlsm_j$ and eigenstates of the Hamiltonian $\hat{H}_0$ may be written $|jlsm_j\rangle$. In the presence of a uniform magnetic field $\mathscr{B}$, the Hamiltonian becomes

$$\hat{H} = \hat{H}_0 + \hat{H}'$$

$$\hat{H}' = -\hat{\boldsymbol{\mu}} \cdot \mathscr{B} = \frac{e}{2mc}\,(\hat{\mathbf{J}} + \hat{\mathbf{S}}) \cdot \mathscr{B}$$

where $\mathbf{J}$ and $\mathbf{S}$ are total and spin angular momenta, respectively, and e has been written for $|e|$.

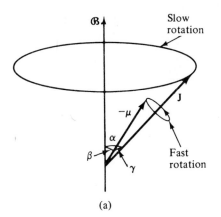 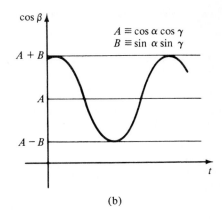

FIGURE 12.8   (a) In the presence of a weak $\mathscr{B}$ field, the precession of J about $\mathscr{B}$ is slow compared to that of μ about J. (See Problem 12.16.) (b) Variation of $\cos \beta$.

Before the magnetic field is turned on, L and S precess about J as depicted in Fig. 12.1. Consequently, $\mu = -(\mu_b/\hbar)(L + 2S)$ also precesses about J.

After the magnetic field is turned on, if it is sufficiently weak compared to the coupling between L and S, the ensuing precession of J about $\mathscr{B}$ is slow compared to that of μ about J, as depicted in Fig. 12.8a.

(a)   In the same limit show that time averages obey the relation

$$\langle \mu \cdot \mathscr{B} \rangle \simeq \left\langle \frac{(\mu \cdot J)(J \cdot \mathscr{B})}{J^2} \right\rangle$$

(b)   Assuming, as in the text, that eigenstates $|jlsm_j\rangle$ are still appropriate to the perturbed Hamiltonian $\hat{H}_0 + \hat{H}'$, show that an eigenenergy $E_{jls}{}^0$ of $\hat{H}_0$ splits into $2j + 1$ equally spaced levels

$$E_{jlsm_j} = E_{jls}{}^0 + \Delta E_{m_j}$$

$$\Delta E_{m_j} = \frac{\hbar\Omega}{2} g(jls)m_j \qquad (m_j = -j, \ldots, j)$$

where $\Omega/2$ is the Larmor frequency introduced in Section 11.9 and g is the *Landé g factor*

$$g(jls) = 1 + \frac{j(j + 1) + s(s + 1) - l(l + 1)}{2j(j + 1)}$$

also briefly discussed in Section 11.9. (*Note:* This splitting of lines due to the presence of the magnetic field is called the *Zeeman effect*. Note the inferred relation $\langle \mu_z \rangle = -g(\mu_b/\hbar)\langle J_z \rangle$.)

Answer (*partial*)

(a)   From the orientation of vectors shown in Fig. 12.8a we see that the relation to be established is correct provided that time averages satisfy

$$\overline{\cos \beta} \simeq \overline{\cos \gamma \cos \alpha}$$

Again in reference to the figure one finds that

$$\beta_{max} = \alpha + \gamma$$
$$\beta_{min} = \alpha - \gamma$$

The variation of $\cos \beta$ between these extremum values is very nearly harmonic (Fig. 12.8b), and forming the time average of $\cos \beta$ gives the desired result.

(b) With the given approximation, one obtains (show this)

$$\hat{H}' = \left(\frac{\Omega}{2}\right) \frac{\hat{J}^2 + \hat{\mathbf{J}} \cdot \hat{\mathbf{S}}}{J^2} \hat{J}_z$$

Expanding the vector $(\mathbf{J} - \mathbf{S})^2$ and forming the expectation $\langle jlsm_j | \hat{H}' | jsm_j \rangle$ gives the desired result.

12.17 Use the results of Problem 12.16 to obtain the Zeeman pattern of spectral lines which stem from the transition $^4F_{3/2} - {}^4D_{5/2}$.

12.18 In the classical formulation of the Zeeman effect, one views the orbital motion of an atomic electron as being perturbed by the imposed magnetic field, thereby altering frequencies of rotation. Assuming circular motion of unperturbed frequency ω_0, show classically that in the presence of a sufficiently weak magnetic field ($\Omega \ll \omega_0$), this frequency divides into the two lines

$$\omega_{\pm} = \omega_0 \pm \frac{\Omega}{2}$$

while radiation polarized parallel to $\mathscr{B}$ remains with frequency ω_0.

Answer

The atomic component of force maintaining circular motion before the magnetic field is applied may be written $m\omega_0^2 r$. If the magnetic field is normal to the plane of the orbit, the magnetic force is also along the radius and provided that the field is sufficiently weak, we may assume the perturbed motion to be slightly altered with small variation in frequency. Let the new frequency of rotation be ω. The total force may then be written

$$F = m\omega^2 r = m\omega_0^2 r + m\Omega\omega r$$

Solving for ω gives the roots

$$2\omega = \Omega \pm \sqrt{\Omega^2 + 4\omega_0^2}$$

The assumption of a weak magnetic field ($\Omega \ll 2\omega_0$) allows the radical to be expanded, giving the roots

$$\omega = \omega_0 \pm \frac{\Omega}{2}$$

Components of motion parallel to $\mathscr{B}$ are unaffected by $\mathscr{B}$ so that frequency ω_0 maintains for polarization parallel to $\mathscr{B}$.

12.19 In Problem 12.16 it was discovered that a magnetic field will split an eigenenergy corresponding to given *jls* values into $2j + 1$ levels

$$\Delta E_{m_j} = \left(\frac{\hbar\Omega}{2}\right) g m_j$$

Show that this proliferation of levels leads to the classical Zeeman splitting of a single frequency into three new lines (as demonstrated in Problem 12.18) in the event that the Landé *g* factors of both levels of the transition are the same. (*Hint:* Frequency displacements Δv from the original unperturbed frequency are given by

$$h \Delta v = \Delta E_{m_j'} - \Delta E_{m_j}$$

This, together with the selection rules $\Delta m_j = 0, \pm 1$, gives the desired result.)

12.3 THE PAULI PRINCIPLE

Indistinguishable Particles

The concept of symmetric and antisymmetric wavefunctions was encountered in Section 8.5. These wavefunctions are appropriate to systems containing identical particles. There is a very fundamental distinction between the quantum and classical descriptions of such systems. At the quantum mechanical level of description, identical particles are also *indistinguishable*. In the classical description of a system of identical particles, one may conceptually label such particles and follow their respective motion. This is impossible at the quantum level. There is no experimental result that distinguishes between two states obtained by exchange (interchange) of identical particles.

Consider a system that consists of two identical particles (e.g., electrons) moving in one dimension (x). Let x_1 be the coordinate of the first particle and x_2 be the coordinate of the second particle. Then

$$P_{12} \, dx_1 \, dx_2 = |\varphi(x_1, x_2)|^2 \, dx_1 \, dx_2$$

denotes the probability of finding particle 1 in the volume element dx_1 about the point x_1 and particle 2 in the volume element dx_2 about the point x_2 [recall (8.98)]. In this notation the first slot in the wavefunction $\varphi(\ ,\ )$ is reserved for the position of particle 1, while the second slot is reserved for the position of particle 2. Now if these two particles are truly indistinguishable, then it is impossible to discern between the two states:

$$\begin{pmatrix} \text{No. 1 at } x_1 \\ \text{No. 2 at } x_2 \end{pmatrix} \quad \text{and} \quad \begin{pmatrix} \text{No. 2 at } x_1 \\ \text{No. 1 at } x_2 \end{pmatrix}$$

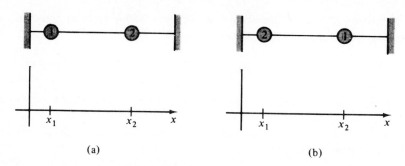

FIGURE 12.9 Two classically distinct configurations of two identical particles on a wire. The probability density $|\varphi(x_1, x_2)|^2$ pertains to configuration (a) and $|\varphi(x_2, x_1)|^2$ to configuration (b). In quantum mechanics, the identical particles 1 and 2 are also indistinguishable, so the probability densities associated with these configurations must be the same.

$$|\varphi(x_1, x_2)|^2 = |\varphi(x_2, x_1)|^2$$

It follows that the probability of finding these two configurations is the same (Fig. 12.9).

(12.17) $$|\varphi(x_1, x_2)|^2 = |\varphi(x_2, x_1)|^2$$

Only wavefunctions with this exchange-symmetry property are valid wavefunctions for a system of identical particles. Of these, it turns out experimentally that wavefunctions relevant to quantum mechanics fall into two categories: symmetric (φ_S) and antisymmetric (φ_A). These functions have the respective properties

$$\varphi_S(x_1, x_2) = \varphi_S(x_2, x_1)$$

$$\varphi_A(x_1, x_2) = -\varphi_A(x_2, x_1)$$

which both obey (12.17). For particles free to move in three dimensions, we write

(12.18) $$\varphi_S(\mathbf{r}_1, \mathbf{r}_2) = +\varphi_S(\mathbf{r}_2, \mathbf{r}_1)$$
$$\varphi_A(\mathbf{r}_1, \mathbf{r}_2) = -\varphi_A(\mathbf{r}_2, \mathbf{r}_1)$$

Let the Hamiltonian $\hat{H}$ describe a system that contains two identical particles: 1 and 2. If these two particles are truly indistinguishable, the Hamiltonian $\hat{H}$ must be symmetric with respect to the positions of these particles, that is,

$$\hat{H}(\mathbf{r}_1, \mathbf{r}_2) = \hat{H}(\mathbf{r}_2, \mathbf{r}_1)$$

Exchange Including Spin

In addition to space coordinates $\mathbf{r}$, a particle also has spin coordinates S. The state of a free particle, for example, may be given in terms of the eigenvalues of these commuting observables (e.g., $\hat{S}^2$, $\hat{S}_z$, $\hat{x}$, $\hat{y}$, $\hat{z}$). Thus, more generally, the Hamiltonian $\hat{H}$ must be symmetric with respect to spin as well as position coordinates of particles. This symmetry property for $\hat{H}$ appears as

$$(12.19) \qquad \hat{H}(\mathbf{r}_1, \mathbf{S}_1; \mathbf{r}_2, \mathbf{S}_2) = \hat{H}(\mathbf{r}_2, \mathbf{S}_2; \mathbf{r}_1, \mathbf{S}_1)$$

The properties (12.18) for φ_S and φ_A become

$$(12.20) \qquad \begin{aligned} \varphi_S(\mathbf{r}_1, \mathbf{S}_1; \mathbf{r}_2, \mathbf{S}_2) &= +\varphi_S(\mathbf{r}_2, \mathbf{S}_2; \mathbf{r}_1, \mathbf{S}_1) \\ \varphi_A(\mathbf{r}_1, \mathbf{S}_1; \mathbf{r}_2, \mathbf{S}_2) &= -\varphi_A(\mathbf{r}_2, \mathbf{S}_2; \mathbf{r}_1, \mathbf{S}_1) \end{aligned}$$

Again, as in the one-dimensional case (12.17), the probability densities associated with these wavefunctions are totally symmetric. Writing "1" for $(\mathbf{r}_1, \mathbf{S}_1)$, and "2" for $(\mathbf{r}_2, \mathbf{S}_2)$, this symmetry property appears as

$$|\varphi_S(1, 2)|^2 = |\varphi_S(2, 1)|^2$$
$$|\varphi_A(1, 2)|^2 = |\varphi_A(2, 1)|^2$$

These symmetry concepts are conveniently expressed in terms of the properties of the exchange operator $\hat{\mathfrak{X}}$ [recall (11.112) et seq.], which is defined by the equation

$$(12.21) \qquad \hat{\mathfrak{X}}\varphi(1, 2) = \varphi(2, 1)$$

The operator $\hat{\mathfrak{X}}$ has two eigenvalues:

$$\hat{\mathfrak{X}}\varphi_S(1, 2) = \varphi_S(2, 1) = +\varphi_S(1, 2)$$
$$\hat{\mathfrak{X}}\varphi_A(1, 2) = \varphi_A(2, 1) = -\varphi_A(1, 2)$$

The eigenfunction φ_S corresponds to the eigenvalue $+1$. It is even under particle exchange. The state φ_A corresponds to the eigenvalue -1. It is odd under particle exchange.

Owing to the exchange symmetry of the Hamiltonian (12.19), $\hat{H}$ commutes with $\hat{\mathfrak{X}}$.

$$(12.22) \qquad [\hat{\mathfrak{X}}, \hat{H}] = 0$$

It follows that $\hat{\mathfrak{X}}$ is a constant of the motion.

$$\frac{d}{dt}\langle\hat{\mathfrak{X}}\rangle = 0$$

If at time $t = 0$, $\varphi(0)$ is such that

$$\hat{\mathfrak{X}}\varphi(0) = +\varphi(0)$$

then the two-particle system remains with this property for all time. At time $t > 0$,

$$\hat{\mathfrak{X}}\varphi(t) = +\varphi(t)$$

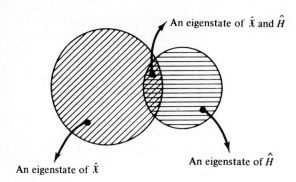

An eigenstate of $\hat{x}$ and $\hat{H}$

An eigenstate of $\hat{x}$

An eigenstate of $\hat{H}$

FIGURE 12.10 Venn diagram exhibiting the simultaneous eigenstates of the exchange operator $\hat{x}$ and the Hamiltonian $\hat{H}$.

Since $\hat{x}$ commutes with $\hat{H}$, it is possible to find simultaneous eigenstates of both these operators (Fig. 12.10). These common wavefunctions are the eigenstates appropriate to systems of identical particles. Furthermore, such eigenstates may be classified in terms of their symmetric or antisymmetric properties under particle exchange.

Bosons and Fermions

The constancy of $\langle \hat{x} \rangle$ is an immutable property of a system of identical particles. Owing to the permanence of this property, one may assume that it is a property of the particles themselves (as opposed to a property of the wavefunction). Particles characterized by the eigenvalue $+1$ of $\hat{x}$ are called *bosons*. The wavefunction for a system of bosons is symmetric (φ_S). Particles characterized by the eigenvalue -1 of $\hat{x}$ are called *fermions*. The wavefunction for fermions is antisymmetric (φ_A).

The characteristic of a particle that determines to which of these categories it belongs is given by the *spin* of the particle.[1] Bosons have integral spin, while fermions have half-integral spin. Electrons and neutrons are examples of fermions. Photons, π, and K mesons are examples of bosons.

Antisymmetric Wavefunctions

The Pauli principle is obeyed by fermions. It states, as described above, that the wavefunction for a system of identical fermions is antisymmetric. Consider a system of two fermions. They are in the state $\varphi_A(1, 2)$. This state has the property that

$$\varphi_A(\mathbf{r}_1, \mathbf{S}_1; \mathbf{r}_2, \mathbf{S}_2) = -\varphi_A(\mathbf{r}_2, \mathbf{S}_2; \mathbf{r}_1, \mathbf{S}_1)$$

---

[1] See Appendix B.

Consider that both particles have the coordinates $\mathbf{r}_1$ and $\mathbf{S}_1$. That is, the system is in the state, $\varphi_A(1, 1)$. This state has the property

$$\varphi_A(\mathbf{r}_1, \mathbf{S}_1; \mathbf{r}_1, \mathbf{S}_1) = -\varphi_A(\mathbf{r}_1, \mathbf{S}_1; \mathbf{r}_1, \mathbf{S}_1)$$

The only value of φ_A which has this property is

$$\varphi_A(\mathbf{r}_1, \mathbf{S}_1; \mathbf{r}_1, \mathbf{S}_1) = 0$$

There is zero probability of finding the particles at the same point in space, with the same value of spin. This is the essence of the Pauli principle: *two fermions cannot exist in the same quantum state.*[1]

Let us consider a problem: What is the wavefunction of two *free* electrons moving in 3-space? The Hamiltonian for this system is

(12.23)
$$\hat{H}(1, 2) = \frac{\hat{p}_1^{\,2}}{2m} + \frac{\hat{p}_2^{\,2}}{2m}$$

The space component wavefunctions of this Hamiltonian are the product states

$$\varphi_{\mathbf{k}_1}(\mathbf{r}_1)\varphi_{\mathbf{k}_2}(\mathbf{r}_2) = \frac{1}{(2\pi)^3} e^{i\mathbf{k}_1 \cdot \mathbf{r}_1} e^{i\mathbf{k}_2 \cdot \mathbf{r}_2}$$

$$E_{\mathbf{k}_1\mathbf{k}_2} = \frac{\hbar^2}{2m}(k_1^{\,2} + k_2^{\,2})$$

This same value of energy characterizes the exchange state,

$$\varphi_{\mathbf{k}_1}(\mathbf{r}_2)\varphi_{\mathbf{k}_2}(\mathbf{r}_1) = \frac{1}{(2\pi)^3} e^{i\mathbf{k}_1 \cdot \mathbf{r}_2} e^{i\mathbf{k}_2 \cdot \mathbf{r}_1}$$

From these two (degenerate) states one may form the symmetric and antisymmetric eigenstates

$$\varphi_S(\mathbf{r}_1, \mathbf{r}_2) = \frac{1}{\sqrt{2}}[\varphi_{\mathbf{k}_1}(\mathbf{r}_1)\varphi_{\mathbf{k}_2}(\mathbf{r}_2) + \varphi_{\mathbf{k}_1}(\mathbf{r}_2)\varphi_{\mathbf{k}_2}(\mathbf{r}_1)]$$

(12.24)
$$\varphi_A(\mathbf{r}_1, \mathbf{r}_2) = \frac{1}{\sqrt{2}}[\varphi_{\mathbf{k}_1}(\mathbf{r}_1)\varphi_{\mathbf{k}_2}(\mathbf{r}_2) - \varphi_{\mathbf{k}_1}(\mathbf{r}_2)\varphi_{\mathbf{k}_2}(\mathbf{r}_1)]$$

Since the two particles in this problem are fermions, wavefunctions for the system must be antisymmetric with respect to exchange of particle spin and position.

---

[1] This property of fermions is also called the *Pauli exclusion principle.*

Inasmuch as the Hamiltonian (12.23) does not contain the spin, it commutes with all spin functions. Thus, if ξ denotes a spin state for the two-electron system, then

$$\xi \varphi_S(\mathbf{r}_1, \mathbf{r}_2) \quad \text{or} \quad \xi \varphi_A(\mathbf{r}_1, \mathbf{r}_2)$$

are possible eigenstates of $\hat{H}$. In Section 11.10 we found that in the coupled representation, two spin-$\frac{1}{2}$ particles combine to give three symmetric ($s = 1$) states, $\xi_S^{(1)}$, $\xi_S^{(0)}$, and $\xi_S^{(-1)}$, and one antisymmetric ($s = 0$) state, $\xi_A^{(0)}$. Combining these with the space state (12.24), one obtains the four antisymmetric states

$$^1\chi_A(\mathbf{r}_1, \mathbf{S}_1; \mathbf{r}_2, \mathbf{S}_2) = \varphi_S(\mathbf{r}_1, \mathbf{r}_2)\xi_A(1, 2)$$

(12.25)
$$^3\chi_A(\mathbf{r}_1, \mathbf{S}_1; \mathbf{r}_2, \mathbf{S}_2) = \varphi_A(\mathbf{r}_1, \mathbf{r}_2)\begin{Bmatrix} \xi_S^{(1)}(1, 2) \\ \xi_S^{(0)}(1, 2) \\ \xi_S^{(-1)}(1, 2) \end{Bmatrix}$$

$$E_{\mathbf{k}_1\mathbf{k}_2} = \frac{\hbar^2}{2m}(k_1^2 + k_2^2)$$

Here we are using the simple rule that the product of a symmetric and antisymmetric state is antisymmetric.

This technique of incorporating spin states to ensure antisymmetry of a given state also applies to Hamiltonians that include interaction between particles, or interaction between particles and a central force field, but which are otherwise spin-independent. For such cases, the structure (12.25) of antisymmetric eigenstates is maintained. This concept finds direct application below in discussion of the helium atom. Symmetrization of the wavefunction for bosons will be applied in construction of the nuclear component eigenstates of the deuterium molecule.

PROBLEMS

12.20 Show that the two operators

$$\hat{\mathfrak{X}}_\pm = \frac{\hat{I} \pm \hat{\mathfrak{X}}}{\sqrt{2}}$$

have the projection property

$$\hat{\mathfrak{X}}_+ \varphi(1, 2) = \varphi_S(1, 2)$$
$$\hat{\mathfrak{X}}_- \varphi(1, 2) = \varphi_A(1, 2)$$

That is, $\hat{\mathfrak{X}}_+$ projects φ onto φ_S and $\hat{\mathfrak{X}}_-$ projects φ onto φ_A.

12.21 (a) Consider a two-particle system with relative radius vector $\mathbf{r} = \mathbf{r}_2 - \mathbf{r}_1$ (Figs. 10.8 and 10.9). Show that $\hat{\mathfrak{X}}\psi(\mathbf{r}) = \hat{P}\psi(\mathbf{r})$, where $\hat{P}$ is the parity operator introduced in Section 6.5 (see Problem 9.46).

(b)   Is the parity of a two-particle system a "good" quantum number in the energy representation?

(c)   Two particles interact under a central potential $V(r)$. It is known that the system is in the state $R_{43}(r)Y_3{}^0(\theta, \phi)$. What is the symmetry of the state?

Answer (partial)

(a) $\hat{x}\psi(\mathbf{r}) = \hat{x}\psi(\mathbf{r}_2 - \mathbf{r}_1) = \psi(\mathbf{r}_1 - \mathbf{r}_2)$

$\hat{P}\psi(\mathbf{r}_2 - \mathbf{r}_1) = \psi(-\mathbf{r}_2 - (-\mathbf{r}_1)) = \psi(\mathbf{r}_1 - \mathbf{r}_2)$

12.4   THE PERIODIC TABLE

Central Field Approximation

In this model it is assumed that each electron "sees" only the electrostatic field due to the nucleus and remaining electrons and that this combined field is spherically symmetric. Owing to this spherical symmetry, the operators

$$\hat{H}, \hat{L}^2, \hat{L}_z, \hat{S}_z$$

relevant to a single electron form a set of commuting operators so that the state of each such electron is specified in terms of the eigenvalues

$$n, l, m_l, m_s$$

These quantum numbers correspond to a wavefunction for the ith electron of the form

(12.26) $$\varphi_{nlm_lm_s}(\mathbf{r}_i, \mathbf{S}_i) = R_{nl}(r_i)Y_l{}^m(\theta_i, \phi_i)\xi_z{}^{\pm}(i)$$

$$\xi_z{}^{\pm} \equiv \alpha(i) \text{ or } \beta(i)$$

The Pauli principle precludes any two electrons being in the same state [i.e., having the same set of (n, l, m_l, m_s) values]. In our discussion of the hydrogen atom, we found that there are n^2 distinct states corresponding to a given value of n (10.114). When spin dependence is included in these states [e.g., (12.26)], this degeneracy is doubled. There are two values that m_s may assume for any set of values nlm_l. Thus, corresponding to any value of n, there are $2n^2$ functions of the form (12.26) which give the same energy for the ith electron.

This degeneracy rule, taken together with the Pauli principle, serves to explain the "shell structure" of the electronic configurations of the elements. As the atomic number Z increases, electrons fill the one-electron states of lowest energy first. For the lighter elements these are shells of lower n values. Within an n shell, for any given value of l there is an l subshell with $2(2l + 1)$ states, corresponding to two m_s values and

TABLE 12.1   Diagrammatic enumeration of states available in the first three atomic shells[a]

$m_l \rightarrow$	-2	-1	0	$+1$	$+2$	$\uparrow\downarrow$	Number of available states in each shell
M shell, $n = 3$	↑↓	↑↓	↑↓	↑↓	↑↓	2	$\left.\begin{array}{l}10\\6\\2\end{array}\right\}18 = 2 \times 3^2$
		↑↓	↑↓	↑↓		1	
			↑↓			0	
L shell, $n = 2$		↑↓	↑↓	↑↓		1	$\left.\begin{array}{l}6\\2\end{array}\right\}8 = 2 \times 2^2$
			↑↓			0	
K shell, $n = 1$			↑↓			0	$2 = 2 \times 1^2$

[a] Vertical arrows represent S_z values. The total orbital and spin angular momentum is zero for a closed shell.

$2l + 1$ values of m_l. Atoms with filled n shells have a total angular momentum and total spin of zero (see Table 12.1). Electrons exterior to these closed shells ("valence" electrons) determine the chemical properties of the atom. The "periodicity" of these properties owes to the fact that valence numbers repeat after shells become closed.

For $n = 1$, l can only be 0, but m_s may take on the two values, $\pm\frac{1}{2}$. Therefore, there can be, at most, only two electrons in the $n = 1$ "shell." (This is called the "K shell" in x-ray notation.) In the ground state of helium ($Z = 2$), the $n = 1$ shell is filled. The electronic configuration for this state is described by the notation $1s^2$. This is read: there are two (the exponent) electrons in the state with $n = 1$ and $l = 0$ (denoted by the letter s). The term notation for this state is $^1S_0$.

The electronic configuration of the ground state of lithium ($Z = 3$) is

$$1s^2 2s^1$$

There are two electrons with $n = 1$, $l = 0$ and one electron with $n = 2$, $l = 0$. The uncoupled spin gives a j value of $\frac{1}{2}$ with corresponding term notation, $^2S_{1/2}$. In beryllium ($Z = 4$) the ground-state configuration is $1s^2 2s^2$. All spins are paired and the ground state is given by the term $^1S_0$.

The electronic configurations and corresponding ground states for the first 36 elements are given in Table 12.2. Note that ground states follow the L-S coupling scheme. For example, for nitrogen ($Z = 7$), whose electronic configuration is

$$1s^2 2s^2 2p^3$$

the ground state is $^4S_{3/2}$. The three p electrons can have total spin values

$$s = \tfrac{1}{2}, \tfrac{3}{2}$$

TABLE 12.2 Distribution of electrons in the atoms from $Z = 1$ to $Z = 36$

X-ray notation			K	L		M			N				
Values of n, l			1,0	2,0	2,1	3,0	3,1	3,2	4,0	4,1	4,2	4,3	
Spectral notation			$1s$	$2s$	$2p$	$3s$	$3p$	$3d$	$4s$	$4p$	$4d$	$4f$	
Element	Atomic number Z	Ionization potential (eV)[a]											
H	1	13.595	1										$^2S_{1/2}$
He	2	24.481	2										$^1S_0$
Li	3	5.39	2	1									$^2S_{1/2}$
Be	4	9.32	2	2									$^1S_0$
B	5	8.296	2	2	1								$^2P_{1/2}$
C	6	11.256	2	2	2								$^3P_0$
N	7	14.53	2	2	3								$^4S_{3/2}$
O	8	13.614	2	2	4								$^3P_2$
F	9	17.418	2	2	5								$^2P_{3/2}$
Ne	10	21.559	2	2	6								$^1S_0$
Na	11	5.138	Neon configuration			1							$^2S_{1/2}$
Mg	12	7.644				2							$^1S_0$
Al	13	5.984				2	1						$^2P_{1/2}$
Si	14	8.149				2	2						$^3P_0$
P	15	10.484				2	3						$^4S_{3/2}$
S	16	10.357	10-electron core			2	4						$^3P_2$
Cl	17	13.01				2	5						$^2P_{3/2}$
Ar	18	15.755				2	6						$^1S_0$
K	19	4.339							1				$^2S_{1/2}$
Ca	20	6.111							2				$^1S_0$
Sc	21	6.54						1	2				$^2D_{3/2}$
Ti	22	6.82						2	2				$^3F_2$
V	23	6.74						3	2				$^4F_{3/2}$
Cr	24	6.764						5	1				$^7S_3$
Mn	25	7.432	Argon configuration					5	2				$^6S_{5/2}$
Fe	26	7.87						6	2				$^5D_4$
Co	27	7.86						7	2				$^4F_{9/2}$
Ni	28	7.633						8	2				$^3F_4$
Cu	29	7.724	18-electron core					10	1				$^2S_{1/2}$
Zn	30	9.391						10	2				$^1S_0$
Ga	31	6.00						10	2	1			$^2P_{1/2}$
Ge	32	7.88						10	2	2			$^3P_0$
As	33	9.81						10	2	3			$^4S_{3/2}$
Se	34	9.75						10	2	4			$^3P_2$
Br	35	11.84						10	2	5			$^2P_{3/2}$
Kr	36	13.996						10	2	6			$^1S_0$

TABLE 12.2 (*Continued*)

Electronic configurations for the alkali metals

Shell	K	L	M	N	O	P
Li	$1s^2$	$2s$				
Na	$1s^2$	$2s^22p^6$	$3s$			
K	$1s^2$	$2s^22p^6$	$3s^23p^6$	$4s$		
Rb	$1s^2$	$2s^22p^6$	$3s^23p^63d^{10}$	$4s^24p^6$	$5s$	
Cs	$1s^2$	$2s^22p^6$	$3s^23p^63d^{10}$	$4s^24p^64d^{10}$	$5s^25p^6$	$6s$

[a] Data obtained from *Handbook of Chemistry and Physics*, 56th ed., CRC Press, Cleveland, Ohio, 1976.

and total l values (see Section 9.5)

$$l = 0, 1, 2, 3$$

The corresponding j values are given by the sequence

$$j = |l + s|, \ldots, |l - s|$$

for all pairs of l, s values. The doublet ($s = \frac{1}{2}$) states thus obtained are

$$^2S_{1/2}, \, ^2P_{1/2, \, 3/2}, \, ^2D_{3/2, \, 5/2}, \, ^2F_{5/2, \, 7/2}$$

The quartet ($s = \frac{3}{2}$) states are

$$^4S_{3/2}, \, ^4P_{1/2, \, 3/2, \, 5/2}, \, ^4D_{1/2, \, 3/2, \, 5/2, \, 7/2}, \, ^4F_{3/2, \, 5/2, \, 7/2, \, 9/2}$$

Of these 19 possible states, the exclusion principle permits only the ^2D, ^2P, and ^4S states.

To understand this reduction in terms, consider the simpler case of the carbon atom which has two p electrons in its outer shell (with $n = 2$; see Table 12.2). Following the preceding construction, we find that the total l values, $l = 0, 1, 2$, and total s values, $s = 0, 1$, give the following terms:

$$^1S, \, ^1P, \, ^1D, \, ^3S, \, ^3P, \, ^3D$$

The allowed states for these two electrons must be antisymmetric under particle exchange. Wavefunctions have the product structure

(12.27) $$|n_1 n_2 jm_j ls\rangle = R_{\left[\begin{smallmatrix} S \\ A \end{smallmatrix}\right]}(1, 2)|jm_j ls\rangle$$

We assume for the moment that the two electrons are not "equivalent," i.e., that they have different principal quantum numbers, $n_1 \neq n_2$. The radial functions then have the form

(12.28)
$$R_{[^S_A]} = \frac{1}{\sqrt{2}}[R_1(n_1 l)R_2(n_2 l) \pm R_1(n_2 l)R_2(n_1 l)]$$

where the R functions are hydrogen-like wavefunctions. Angular momentum components $|jm_j ls\rangle$ are constructed as follows. In the Russell–Saunders coupling scheme total **L** adds to total **S** to give the atomic total angular momentum **J**. Wavefunctions are then given by the Clebsch–Gordan expansion,

(12.29)
$$|jm_j ls\rangle = \sum_{m_j = m_2 + m_s} C_{m_l m_s}|lm_l\rangle |sm_s\rangle$$

The $|lm_l\rangle$ and $|sm_s\rangle$ components as well are given by Clebsch–Gordan expansions, which for the case at hand are displayed in Tables 9.5 and 11.3, respectively. We see that orbital states of odd l are antisymmetric and those of even l are symmetric. All three $s = 1$ states are symmetric, whereas the singlet $s = 0$ state is antisymmetric. So we may construct the following list:

(12.30)

$$^3D: \quad l = 2(\text{sym}), \quad s = 1(\text{sym}) \rightarrow |jm_j\rangle_S \rightarrow R_A$$

$$^1P: \quad l = 1(\text{anti}), \quad s = 0(\text{anti}) \rightarrow |jm_j\rangle_S \rightarrow R_A$$

$$^3S: \quad l = 0(\text{sym}), \quad s = 1(\text{sym}) \rightarrow |jm_j\rangle_S \rightarrow R_A$$

$$^1D: \quad l = 2(\text{sym}), \quad s = 0(\text{anti}) \rightarrow |jm_j\rangle_A \rightarrow R_S$$

$$^3P: \quad l = 1(\text{anti}), \quad s = 1(\text{sym}) \rightarrow |jm_j\rangle_A \rightarrow R_S$$

$$^1S: \quad l = 0(\text{sym}), \quad s = 0(\text{anti}) \rightarrow |jm_j\rangle_A \rightarrow R_S$$

For the np^2 configuration under consideration, $n_1 = n_2 = n = 2$, so that $R_A = 0$. Only the terms which include the R_S factor survive. These are the ^1D, ^3P, and ^1S terms, which in all comprise 15 independent states corresponding to five energy levels. Of these, Hund's rules (Problem 12.25) determine the ground state. The first rule indicates that the ground state is among the $^3P_{0,1,2}$ states. Since there are only two electrons in the $2p$ shell of carbon, and this shell can accommodate six electrons, the third rule indicates that $^3P_0$ is the ground state of carbon. The same is true for the two-electron (np^2) atoms, Si and Ge. Remaining allowed atomic states for configurations up to the nd^{10} shell are shown in Table 12.2.

A note of caution is in order for three (or more) -electron atoms. For such cases spin states which are antisymmetric with respect to exchange of *any two* of the three particles do not exist (see Problems 11.86 et seq.). In this case, generalizations of the product form (12.27) comes into play.[1]

---

[1] For further discussion, see M. Weissbluth, *Atoms and Molecules*, Academic Press, New York, 1978. Recall also, Problem 11.88.

TABLE 12.3 States allowed by the exclusion principle in the L-S coupling scheme

config	singlet/doublet	(cont.)	triplet/quartet	quintet	sextet
ns^0	^1S				
ns^1	^2S				
ns^2	^1S				
np^0	^1S				
np^1	^2P				
np^2	$^1S, {}^1D$		^3P		
np^3	$^2P, {}^2D$		^4S		
np^4	$^1S, {}^1D$		^3P		
np^5	^2P				
np^6	^1S				
nd^0	^1S				
nd^1	^2D				
nd^2	$^1S, {}^1D, {}^1G$		$^3P, {}^3F$		
nd^3	$^2D, {}^2P, {}^2D, {}^2F, {}^2G, {}^2H$		$^4P, {}^4F$		
nd^4	$^1S, {}^1D, {}^1G, {}^1S, {}^1D, {}^1G, {}^1F, {}^1I$		$^3P, {}^3F, {}^3P, {}^3D, {}^3F, {}^3G, {}^3H$	^5D	
nd^5	$^2D, {}^2P, {}^2D, {}^2F, {}^2G, {}^2H, {}^2S, {}^2D, {}^2F, {}^2G, {}^2I$		$^4P, {}^4F, {}^4D, {}^4G$		^6S
nd^6	$^1S, {}^1D, {}^1G, {}^1S, {}^1D, {}^1G, {}^1F, {}^1I$		$^3P, {}^3F, {}^3P, {}^3D, {}^3F, {}^3G, {}^3H$	^5D	
nd^7	$^2D, {}^2P, {}^2D, {}^2F, {}^2G, {}^2H$		$^4P, {}^4F$		
nd^8	$^1S, {}^1D, {}^1G$		$^3P, {}^3F$		
nd^9	^2D				
nd^{10}	^1S				

The symmetry of the states shown in Table 12.3 about the midvalue of occupation in a given l shell is due to the following. Consider, for example, the p shell which can be occupied by six electrons. In what manner does the p^2 configuration resemble the p^4 configuration? The p^4 configuration may be viewed as one in which two holes occupy the p shell. This atomic configuration yields the same allowed states as does the p^2 configuration (see Problem 12.49). The same holds true for the d^3 and d^7 configurations, and so on.

A complete description of an atomic state involves the total L, S, and J of the atom in addition to the quantum numbers of individual electrons. The first of these is given by the term notation of the state (e.g., $^5F_{7/2}$), whereas the second is given by the electronic configuration (e.g. $1s^2 2s^2 2p^3$).

The periodic chart is shown in Table 12.4. The ground states of elements and outer electron shell configurations[1] are listed, as well as atomic numbers.

---

[1] These data were obtained from G. Baym, *Lectures on Quantum Mechanics*, W. A. Benjamin, New York, 1969; S. Fraga, J. Karwowski, and K. Saxena, *Handbook of Atomic Data*, Elsevier, New York, 1976; and H. Gray, *Electrons and Chemical Bonding*, W. A. Benjamin, New York, 1964.

TABLE 12.4 The periodic table

Group →
Period ↓

Main groups

Period	I	II	III	IV	V	VI	VII	VIII
1	H^{1} $1s^1$ $^2S_{1/2}$							He^{2} $1s^2$ $^1S_0$
2	Li^{3} $1s^2 2s^1$ $^2S_{1/2}$	Be^{4} $1s^2 2s^2$ $^1S_0$	B^{5} $2s^2 2p^1$ $^2P_{1/2}$	C^{6} $2s^2 2p^2$ $^3P_0$	N^{7} $2p^3$ $^4S_{3/2}$	O^{8} $2p^4$ $^3P_2$	F^{9} $2p^5$ $^2P_{3/2}$	Ne^{10} $2p^6$ $^1S_0$
3	Na^{11} $3s^1$ $^2S_{1/2}$	Mg^{12} $3s^2$ $^1S_0$	Al^{13} $3s^2 3p^1$ $^2P_{1/2}$	Si^{14} $3s^2 3p^2$ $^3P_0$	P^{15} $3p^3$ $^4S_{3/2}$	S^{16} $3p^4$ $^3P_2$	Cl^{17} $3p^5$ $^2P_{3/2}$	Ar^{18} $3s^2 3p^6$ $^1S_0$
4	K^{19} $4s^1$ $^2S_{1/2}$	Ca^{20} $4s^2$ $^1S_0$	Ga^{31} $4s^2 3d^{10} 4p^1$ $^2P_{1/2}$	Ge^{32} $3d^{10} 4p^2$ $^3P_0$	As^{33} $3d^{10} 4p^3$ $^4S_{3/2}$	Se^{34} $3d^{10} 4p^4$ $^3P_2$	Br^{35} $3d^{10} 4p^5$ $^2P_{3/2}$	Kr^{36} $4s^2 4p^6$ $^1S_0$
5	Rb^{37} $5s^1$ $^2S_{1/2}$	Sr^{38} $5s^2$ $^1S_0$	In^{49} $5s^2 4d^{10} 5p^1$ $^2P_{1/2}$	Sn^{50} $4d^{10} 5p^2$ $^3P_0$	Sb^{51} $4d^{10} 5p^3$ $^4S_{3/2}$	Te^{52} $4d^{10} 5p^4$ $^3P_2$	I^{53} $4d^{10} 5p^5$ $^2P_{3/2}$	Xe^{54} $5s^2 5p^6$ $^1S_0$
6	Cs^{55} $6s^1$ $^2S_{1/2}$	Ba^{56} $6s^2$ $^1S_0$	Tl^{81} $6s^2 5d^{10} 6p^1$ $^2P_{1/2}$	Pb^{82} $6p^2$ $^3P_0$	Bi^{83} $6p^3$ $^4S_{3/2}$	Po^{84} $6p^4$ $^3P_2$	At^{85} $6p^5$ $^2P_{3/2}$	Rn^{86} $6p^6$ $^1S_0$
7	Fr^{87} $7s^1$ $^2S_{1/2}$	Ra^{88} $7s^2$ $^1S_0$						

Transition elements

Period										
4	Sc^{21} $4s^2 3d^1$ $^2D_{3/2}$	Ti^{22} $4s^2 3d^2$ $^3F_2$	V^{23} $4s^2 3d^3$ $^4F_{3/2}$	Cr^{24} $4s^1 3d^5$ $^7S_3$	Mn^{25} $4s^2 3d^5$ $^6S_{5/2}$	Fe^{26} $4s^2 3d^6$ $^5D_4$	Co^{27} $4s^2 3d^7$ $^4F_{9/2}$	Ni^{28} $4s^2 3d^8$ $^3F_4$	Cu^{29} $4s^1 3d^{10}$ $^2S_{1/2}$	Zn^{30} $4s^2 3d^{10}$ $^1S_0$
5	Y^{39} $5s^2 4d^1$ $^2D_{3/2}$	Zr^{40} $5s^2 4d^2$ $^3F_2$	Nb^{41} $5s^1 4d^4$ $^6D_{1/2}$	Mo^{42} $5s^1 4d^5$ $^7S_3$	Tc^{43} $5s^2 4d^5$ $^6S_{5/2}$	Ru^{44} $5s^1 4d^7$ $^5F_5$	Rh^{45} $5s^1 4d^8$ $^4F_{9/2}$	Pd^{46} $4d^{10}$ $^1S_0$	Ag^{47} $5s^1 4d^{10}$ $^2S_{1/2}$	Cd^{48} $5s^2 4d^{10}$ $^1S_0$
6	La^{57} $6s^2 5d^1$ $^2D_{3/2}$ / Hf^{72} $6s^2 5d^2$ $^3F_2$	Ta^{73} $6s^2 5d^3$ $^4F_{3/2}$	W^{74} $6s^2 5d^4$ $^5D_0$	Re^{75} $6s^2 5d^5$ $^6S_{5/2}$	Os^{76} $6s^2 5d^6$ $^5D_4$	Ir^{77} $6s^2 5d^7$ $^4F_{9/2}$	Pt^{78} $6s^1 5d^9$ $^3D_3$	Au^{79} $6s^1 5d^{10}$ $^2S_{1/2}$	Hg^{80} $6s^2 5d^{10}$ $^1S_0$	
7	Ac^{89} $7s^2 6d^1$ $^2D_{3/2}$									

Rare earths[a]

Ce^{58} $6s^2 5d^1 4f^1$ $^3H_5$	Pr^{59} $6s^2 4f^3$ $^4I_{9/2}$	Nd^{60} $6s^2 4f^4$ $^5I_4$	Pm^{61} $6s^2 4f^5$ $^6H_{5/2}$	Sm^{62} $6s^2 4f^6$ $^7F_0$	Eu^{63} $6s^2 4f^7$ $^8S_{7/2}$	Gd^{64} $6s^2 5d^1 4f^7$ $^9D_2$	Tb^{65} $6s^2 5d^1 4f^8$	Dy^{66} $6s^2 4f^{10}$	Ho^{67} $6s^2 4f^{11}$	E^{68} $6s^2 4f^{12}$	Tm^{69} $6s^2 4f^{13}$ $^2F_{7/2}$	Yb^{70} $6s^2 4f^{14}$ $^1S_0$	Lu^{71} $6s^2 5d^1 4f^{14}$ $^2D_{3/2}$

Heavy elements[b]

Th^{90} $7s^2 6d^2$	Pa^{91} $6d^3$	U^{92} $6d^1 5f^3$ $^5L_6$	Np^{93} $5f^5$	Pu^{94} $5f^6$	Am^{95} $5f^7$ $^8S_{7/2}$	Cm^{96} $6d^1 5f^7$	Bk^{97} $5f^9$	Cf^{98} $5f^{10}$	E^{99} $5f^{11}$	Fm^{100} $5f^{12}$	Md^{101} $5f^{13}$

[a] With La^{57} included, this group is also called the *lanthanides*.
[b] With Ac^{89} included, this group is also called the *actinides*.

As stated above, chemical properties of elements are determined by the electron configuration in the unfilled shell. Atoms with similar valence electron configuration have nearly the same chemical properties. The properties of atoms in some of these groupings are described below.

The Alkali Metals. Group I

The alkali metals are the atoms with one valence electron: Li^3, Na^{11}, K^{19}, Rb^{37}, Cs^{55}, and Fr^{87}. The ground state of these elements is $^2S_{1/2}$. Ionization energy is low. The spectra of these "one-electron" elements resemble that of hydrogen. Valence is $+1$.

The Alkaline Earths. Group II

All the alkaline earths have two s electrons outside a closed p subshell. They are: Be^4, Mg^{12}, Ca^{20}, Sr^{38}, Ba^{56}, and Ra^{88}. The ground state is $^1S_0$. Ionization remains relatively small. Their valence is $+2$. When singly ionized, these atoms are known as "hydrogenic" ions; their spectra resemble that of hydrogen.

The Halogens. Group VII

These are the elements: F^9, Cl^{17}, Br^{35}, I^{53}, and At^{85}. They are all missing one electron in the outermost p subshell and therefore have a valence of -1. Halogens form stable molecules with the one-electron (alkali metal) atoms (e.g., NaCl) through ionic bonding.

The Noble Elements. Group VIII

The noble elements are also called the "rare gases" or the "inert elements." They are He^2, Ne^{10}, Ar^{18}, Kr^{36}, Xe^{54}, and Rn^{86}. Except for He, all these atoms have a completed outermost p subshell. The ground state of these elements is $^1S_0$. Total spin and orbital angular momentum are zero, so the atom has no magnetic moment. Ionization energy is large (see Table 12.2); electrical conductivity is low. Noble elements are chemically inert and have low boiling points.

The Transition Group

In the transition-group elements the incomplete $3d$ subshell is filled while 2 (or 1) electrons remain in the outer $4s$ subshell. The incomplete $3d$ subshell gives rise to magnetic properties. For example, the ground state of Cr^{24} is $^7S_3$. The spin of the atom is $s = 3$, which implies that the five $3d$ electrons and one $4s$ electron all have their spins aligned. These parallel spins contribute to the large magnetic moment of Cr. The

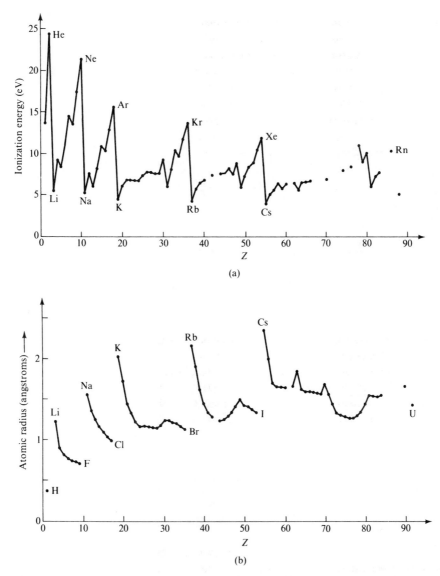

FIGURE 12.11 (a) First ionization energies versus atomic number Z. (b) Atomic radii versus atomic number. Recall conclusions of the Thomas–Fermi model concerning variation of atomic radius with increase in Z discussed in Section 10.8.

chemical properties of elements in the transition group are due primarily to the $4s$ electron(s).

The grouping of atoms according to their shell structure is further motivated by two very significant sets of observational and inferred data. These are, respectively, first ionization energies and radii of atoms. Thus, in Fig. 12.11a we see that binding in a given n shell grows roughly as the shell is completed, with one-electron atoms most weakly bound and noble elements most strongly bound. In Fig. 12.11b we note that the radii of atoms in a given n shell decrease roughly with increase in Z, with one-electron atoms having the largest radii and halogens the smallest radii. These properties become less valid at higher Z where outer electron shells of atoms in a given row of the periodic chart (Table 12.3) are not characterized by constant principal quantum number n.

PROBLEMS

12.22 Show that the ground-state electronic configuration of Cr^{24} does not violate the Pauli principle.

12.23 What are the possible states for the ground configuration of O^{16} which includes four p electrons in its outermost shell? Check that the $^3P_2$ ground state is included in your list.

12.24 Describe the energy band structure of the metal lithium. Specifically, indicate which electrons fill the valence band and which electrons contribute to the conduction band. How full is the conduction band? (The band theory of conduction was discussed previously in Section 8.4.)

12.25 Show that the ground states for the first three elements in the "neon configuration" ($Z = 11$ to 18) are consistent with *Hund's rules*:

   (a)   The lowest energy state is the LS multiplet with largest s value.

   (b)   When more than one value of l is associated with this maximum s value, the lowest energy state is the one with largest l value.

   (c)   For a given l subshell containing n_e electrons, in the lowest energy state the total angular momentum number j has the value $|l - s|$ for $n_e < N/2$ and $|l + s|$ for $n_e > N/2$, where N is the number of electrons in the completed subshell.

12.5   THE SLATER DETERMINANT

In the central field approximation, the Hamiltonian for an N-electron atom is written

$$(12.31) \qquad \hat{H}(1, 2, \ldots, N) = \hat{H}_1(1) + \hat{H}_2(2) + \cdots + \hat{H}_N(N)$$

In this notation the number 2 denotes the coordinates of the second electron. The eigenstates of the individual Hamiltonians are of the form (12.26). Calling the set of eigenvalues (n, l, m_l, m_s) of the ith electron v_i, these eigenstates obey the equations

(12.32)
$$\hat{H}(1)\varphi_{v_1}(1) = E_{v_1}\varphi_{v_1}(1)$$
$$\hat{H}(2)\varphi_{v_2}(2) = E_{v_2}\varphi_{v_2}(2)$$
$$\vdots$$
$$\hat{H}(N)v_{v_N}(N) = E_{v_N}\varphi_{v_N}(N)$$

In this notation the product eigenstates of $H(1, \ldots, N)$ appear as

(12.33)
$$\varphi_{(v_1,\ldots,v_N)}(1, 2, \ldots, N) = \varphi_{v_1}(1)\varphi_{v_2}(2) \cdots \varphi_{v_N}(N)$$

However, this function is not properly antisymmetric. If $\hat{\mathfrak{X}}_{1,3}$ denotes the exchange operation of the coordinates of electrons 1 and 3, the correct antisymmetric wavefunctions of $\hat{H}(1, \ldots, N)$ have the property

(12.34)
$$\hat{\mathfrak{X}}_{1,3}\,\varphi(1, 2, 3, \ldots, N)$$
$$= \varphi(3, 2, 1, \ldots, N) = -\varphi(1, 2, 3, \ldots, N)$$

The normalized wavefunction that obeys this rule (for all pairs of particles) and is an eigenstate of $\hat{H}(1, \ldots, N)$ is given by the *Slater determinant*,

(12.35)
$$\varphi_A(1, 2, \ldots, N) = \frac{1}{\sqrt{N!}} \begin{vmatrix} \varphi_{v_1}(1) & \varphi_{v_2}(1) & \cdots & \varphi_{v_N}(1) \\ \varphi_{v_1}(2) & \varphi_{v_2}(2) & \cdots & \varphi_{v_N}(2) \\ \vdots & \vdots & & \vdots \\ \varphi_{v_1}(N) & \varphi_{v_2}(N) & \cdots & \varphi_{v_N}(N) \end{vmatrix}$$

This determinant has four outstanding properties, which we discuss next.

Eigenvalues

It is an eigenstate of (12.31) with eigenvalue

(12.36)
$$E_{(v_1, v_2, \ldots, v_N)} = E_{v_1} + E_{v_2} + \cdots + E_{v_N}$$

The explicit form of φ_A appears as

(12.37)
$$\varphi_A = \frac{1}{\sqrt{N!}} \sum_{P(v_1, v_2, \ldots, v_N)} (-1)^{|P|}\varphi_{v_1}(1)\varphi_{v_2}(2) \cdots \varphi_{v_N}(N)$$

The sum is over all permutations P of the quantum indices $(v_1, v_2, \ldots, v_N)$. The symbol $|P|$ is zero or 1. It is zero if the permutation $P(v_1, \ldots, v_N)$ can be obtained from

$(v_1, \ldots, v_N)$ through an even number of exchanges of two indices. It is 1 if $P(v_1, \ldots, v_N)$ involves an odd number of exchanges. For example, the term

$$-\varphi_{v_4}(1)\varphi_{v_1}(2)\varphi_{v_2}(3)\varphi_{v_3}(4) \cdots \varphi_{v_N}(N)$$

corresponds to the permutation

$$P(v_1, \ldots, v_N) = v_4, v_1, v_2, v_3, \ldots, v_N$$

This sequence only involves permutation of the first four indices. To obtain $|P|$ we must rearrange the sequence $(4, 1, 2, 3)$ in the form of the original sequence $(1, 2, 3, 4)$ through exchanges of two integers only and count the minimum number of such exchanges which do the job.

$$(4, 1, 2, 3) \to (1, 4, 2, 3) \to (1, 2, 4, 3) \to (1, 2, 3, 4)$$

Three exchanges suffice so that $(-1)^{|P|} = -1$. We conclude that the preceding product wavefunction carries a minus sign, as written.

Each such N-particle product function is an eigenstate of $\hat{H}$ corresponding to the degenerate eigenenergy (12.36), whence the determinantal form (12.37), which is merely a linear combination of these product states, is also an eigenstate of $\hat{H}$ corresponding to the eigenenergy (12.36).

Orthonormality

The second property that the determinantal states (12.35) have is that they form an orthonormal sequence. That is,

$$\langle \varphi_A | \varphi_A \rangle = \frac{1}{N!} \sum_P \langle \varphi_{v_1}(1) \cdots \varphi_{v_N}(N) | \varphi_{v_1}(1) \cdots \varphi_{v_N}(N) \rangle$$

$$= 1$$

since there are $N!$ terms of unit value in the sum. Furthermore, owing to the orthogonality of single-particle eigenstates,

$$\langle \varphi_{A(v_1, \ldots, v_N)} | \varphi_{A(v_1', \ldots, v_N')} \rangle = 0$$

$$(v_1, \ldots, v_N) \neq (v_1', \ldots, v_N')$$

This establishes the orthogonality of these states.

Antisymmetry

The third property concerns the symmetry of φ_A. If $\hat{\mathfrak{x}}_{ij}$ denotes the exchange of the ith-particle coordinates with those of the jth particle, then

$$\hat{\mathfrak{x}}_{ij}\varphi_A(i, j) = \varphi_A(j, i) = -\varphi_A(i, j)$$

Exchanging particle coordinate numbers in φ_A, as given by (12.35), is effected by an exchange of two rows of the determinant. But a determinant changes sign under interchange of two rows. It follows that $\varphi_A(i, j) = -\varphi_A(j, i)$. For example, for the exchange of particles 1 and 2,

$$\hat{\mathfrak{x}}_{1,2} \begin{vmatrix} \varphi_{v_1}(1) & \varphi_{v_2}(1) & \cdots & \varphi_{v_N}(1) \\ \varphi_{v_1}(2) & \varphi_{v_2}(2) & \cdots & \varphi_{v_N}(2) \\ \vdots & \vdots & & \vdots \\ \varphi_{v_1}(N) & \varphi_{v_2}(N) & \cdots & \varphi_{v_N}(N) \end{vmatrix} = \begin{vmatrix} \varphi_{v_1}(2) & \varphi_{v_2}(2) & \cdots & \varphi_{v_N}(2) \\ \varphi_{v_1}(1) & \varphi_{v_2}(1) & \cdots & \varphi_{v_N}(1) \\ \vdots & \vdots & & \vdots \\ \varphi_{v_1}(N) & \varphi_{v_2}(N) & \cdots & \varphi_{v_N}(N) \end{vmatrix}$$

$$= - \begin{vmatrix} \varphi_{v_1}(1) & \varphi_{v_2}(1) & \cdots & \varphi_{v_N}(1) \\ \varphi_{v_1}(2) & \varphi_{v_2}(2) & \cdots & \varphi_{v_N}(2) \\ \vdots & \vdots & & \vdots \\ \varphi_{v_1}(N) & \varphi_{v_2}(N) & \cdots & \varphi_{v_N}(N) \end{vmatrix}$$

This property establishes the antisymmetry of the state φ_A.

Exclusion Principle

Finally, we note that if particle 2 has the same quantum numbers as particle 1, then $\varphi_A = 0$. This property also follows from the determinantal structure of φ_A: namely, if two particles are in the same eigenstate, then two columns of the determinant (12.31) are equal and φ_A vanishes. Thus φ_A written in the Slater determinant form is consistent with the Pauli exclusion principle.

PROBLEMS

12.26 Which purely determinantal properties related to the Slater determinant are involved in: (a) the antisymmetry of φ_A; (b) the Pauli exclusion principle?

12.27 Two spin-$\frac{1}{2}$ neutrons move in a two-dimensional box of edge length L and impenetrable walls. If the neutrons do not interact with each other, construct the antisymmetric determinantal spin and coordinate-dependent energy eigenstates for the system.

12.6 APPLICATION OF SYMMETRIZATION RULES
TO THE HELIUM ATOM

We have already found the Pauli principle to be an important rule in forming the periodic chart of the elements and in construction of properly antisymmetrized wavefunctions for atomic electrons in the central field approximation.

In this and the next section we will further demonstrate the important role played by symmetrization principles in analysis of two elemental systems in nature. The first of these is the helium atom, which has two outer electrons. Among other properties we will find how the Pauli principle influences the coupling between these electrons in the construction of properly antisymmetrized wavefunctions for the atom. The second example is the deuterium molecule, whose two deuteron nuclei are bosons. Consequently, the nuclear component of the molecular wavefunction must be symmetrized with respect to exchange of space and spin coordinates. Construction of such properly symmetrized states leads very simply to intensity rules for emission.

The Helium Atom

The Hamiltonian of helium, in a frame where the nucleus is at rest, is

$$(12.38) \qquad \hat{H} = \left(\frac{\hat{p}_1{}^2}{2m} - \frac{2e^2}{r_1} \right) + \left(\frac{\hat{p}_2{}^2}{2m} - \frac{2e^2}{r_2} \right) + \frac{e^2}{r_{12}} + \hat{H}_{SO}$$

$$= \hat{H}_0(1) + \hat{H}_0(2) + \hat{H}_{ES} + \hat{H}_{SO}$$

The last term, H_{SO}, is written for the spin-orbit interaction between the electrons and the nucleus, while H_{ES} is written for the electrostatic interaction, e^2/r_{12}, between the two electrons. The interelectron displacement is

$$r_{12} = |\mathbf{r}_2 - \mathbf{r}_1|$$

(Fig. 12.12). If the electrostatic as well as spin-orbit terms are neglected, $\hat{H}$ reduces to the sum of two hydrogenic Hamiltonians (each with $Z = 2$).

$$(12.39) \qquad \hat{H}_0(1, 2) = \hat{H}_0(1) + \hat{H}_0(2)$$

This Hamiltonian (as well as the total Hamiltonian) is symmetric with respect to the interchange of the two electrons

$$(12.40) \qquad [\hat{x}_{12}, \hat{H}_0(1, 2)] = 0$$

This merely reflects the indistinguishability of the two electrons. This property must be maintained in the eigenstates that we construct for $\hat{H}_0(1, 2)$. With the abbreviations

$$v_1 \equiv (n_1, l_1, m_{l_1}); \qquad v_2 \equiv (n_2, l_2, m_{l_2})$$

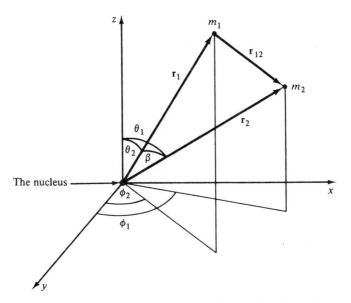

FIGURE 12.12 **The coordinates of the two electrons in helium. The six-dimensional volume element** $(d\mathbf{r}_1 d\mathbf{r}_2)$
is given by

$$d\mathbf{r}_1 \, d\mathbf{r}_2 = r_1{}^2 \, d\Omega_1 \, dr_1 \, r_2{}^2 \, d\Omega_2 \, dr_2 = r_1{}^2 r_2{}^2 \, dr_1 \, dr_2 \, d\cos\theta_1 \, d\cos\theta_2 \, d\phi_1 \, d\phi_2$$

The potential of interaction between electrons is given by

$$V = \frac{e^2}{r_{12}} = \frac{e^2}{\sqrt{r_1{}^2 + r_2{}^2 - 2r_1 r_2 \cos\beta}}$$

(See Fig. 9.16 for addition formulas connecting β **to** $\theta_1, \theta_2, \phi_1, \phi_2$.**)**

these symmetrized eigenstates of $H_0(1, 2)$ appear as[1]

(12.41) $$\varphi_{S, A}(\mathbf{r}_1, \mathbf{r}_2) = \frac{1}{\sqrt{2}} [\varphi_{v_1}(1)\varphi_{v_2}(2) \pm \varphi_{v_1}(2)\varphi_{v_2}(1)]$$

The plus sign gives a symmetric state, φ_S, while the minus sign gives an antisymmetric
state, φ_A. The energy eigenvalue corresponding to either of these states is

(12.42) $$E^{(0)}_{n_1, n_2} = -4\mathbb{R}\left(\frac{1}{n_1{}^2} + \frac{1}{n_2{}^2}\right) = -2mc^2\alpha^2\left(\frac{1}{n_1{}^2} + \frac{1}{n_2{}^2}\right)$$

---

[1] In the event that $v_1 = v_2$, then $\varphi_S(\mathbf{r}_1, \mathbf{r}_2) = \varphi_{v_1}(1)\varphi_{v_1}(2)$.

Separation of Multiplets Due to Spin Symmetry

Although the spin-orbit correction to the Hamiltonian of helium is small $[\Delta E_{SO}/E \sim \alpha^2$; see (12.13)] and may well be neglected in a first approximation, the spin of the electrons still has an important influence on the properties of helium. This occurs through a combination of Pauli antisymmetrization requirements on wavefunctions with respect to exchange of space and spin coordinates and the relatively large electrostatic interaction between electrons (see Problem 12.28). In what follows, first we will construct the properly antisymmetrized space-spin dependent eigenstates of $\hat{H}_0(1, 2)$. This immediately implies a coupling between the electron spins. Three states emerge with $s = 1$ (the triplet series) and one state emerges with $s = 0$ (the singlet series). When the electrostatic interaction e^2/r_{12} is brought into play, it is found that the triplet states all lie lower in energy than the singlet states. In this manner we will find that symmetry requirements couple the electron spins and electrostatic interaction separates the resulting singlet and triplet states.

Insofar as $\hat{H}_0(1, 2)$ does not contain the spin, spin-dependent eigenstates of $\hat{H}_0(1, 2)$ are quite simple to construct. If $\varphi(\mathbf{r}_1, \mathbf{r}_2)$ is a space-dependent eigenstate (12.41) of $H_0(1, 2)$, then following the procedure described in Section 12.3 for two free electrons, we find that the properly antisymmetrized wavefunctions are given by

$$^1\chi = \varphi_S(\mathbf{r}_1, \mathbf{r}_2)\xi_A(1, 2) \qquad (s = 0)$$

(12.43)

$$^3\chi = \varphi_A(\mathbf{r}_1, \mathbf{r}_2) \begin{cases} \xi_S^{(1)}(1, 2) \\ \xi_S^{(0)}(1, 2) \\ \xi_S^{(-1)}(1, 2) \end{cases} \qquad (s = 1)$$

The ξ-spin functions are listed in Table 11.3. We see how symmetrization requirements, together with the Pauli principle, effect a coupling between the spins of the two electrons in helium. In the triplet state ($s = 1$) the spins are aligned, whereas in the singlet state ($s = 0$) the spins are antialigned.

Electrostatic Interaction

To understand how the coupling augments the energies of helium, we recall the following property of symmetrized states (see Problem 8.32): namely, two particles in a symmetric state attract one another (in a statistical sense). It follows that the two electrons in the singlet $^1\chi$ state, which contains the symmetric φ_S state, are closer to each other than they are in the triplet $^3\chi$ state, which contains the antisymmetric φ_A state. Thus, owing to the positive repulsive energy of the electrostatic interaction, e^2/r_{12}, the triplet states lie lower in energy than do the singlet states (Fig. 12.13).

This is the mechanism behind Hund's first rule (see Problem 12.25)—that the total spin assumes the maximum value consistent with the Pauli principle. In this

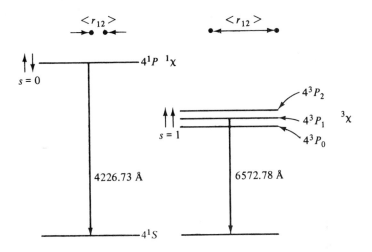

FIGURE 12.13   The $4P$ states of calcium, illustrating the fact that in two-electron atoms, triplet states lie lower than the corresponding singlet state. This is due to the fact that the average interelectron distance is smaller in the symmetric space state, φ_S, than in the antisymmetric space state, φ_A, thereby increasing the singlet electrostatic contribution to the total energy compared to the corresponding triplet contribution.

$$\langle {}^1\chi | H_{ES} | {}^1\chi \rangle > \langle {}^3\chi | H_{ES} | {}^3\chi \rangle$$

$$H_{ES} = + \frac{e^2}{r_{12}}$$

symmetric spin state, the space component of the wavefunction must be antisymmetric so that electrons are further removed from each other than in the corresponding symmetric space state.

Exchange and Coulomb Interaction Energies

Let us consider the space component eigenstates (12.41) of helium in more detail. As it turns out, the only states of helium that are of practical significance are those for which one of the two electrons is in its own ground state, with $n = 1, l = m_l = 0$. The reason for this is that it takes less energy to ionize a helium atom from the ground state than it does to raise both electrons to excited levels (Fig. 12.14). This means that one is more likely to find an He^+ ion (hydrogenic ion with $Z = 2$) than a helium atom with both electrons in excited states. It follows that the space-dependent states of helium atoms that exist under natural conditions are mostly of the form

$$(12.44) \qquad \varphi_{S,A}(\mathbf{r}_1, \mathbf{r}_2) = \frac{1}{\sqrt{2}} \left[\varphi_{100}(\mathbf{r}_1)\varphi_{nlm}(\mathbf{r}_2) \pm \varphi_{100}(\mathbf{r}_2)\varphi_{nlm}(\mathbf{r}_1) \right].$$

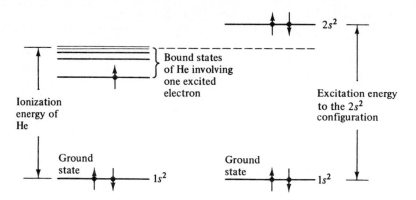

FIGURE 12.14 **The ionization energy of He is smaller than the energy of the lowest energy state of He with both electrons excited. It is possible for He in this excited state to decay to He$^+$ and a free electron.**

One may use these eigenstates to calculate the corrections to the eigenenergies of $\hat{H}_0$ due to the electrostatic interaction, e^2/r_{12}. As described above, we expect the triplet states to lie lower in energy than the singlet states.

One obtains

(12.45)
$$\left\langle \frac{e^2}{r_{12}} \right\rangle_{\text{singlet}} = \langle {}^1\chi | \left(\frac{e^2}{r_{12}} \right) | {}^1\chi \rangle$$

$$= A + B$$

where

$$A = \langle \varphi_{100}(1)\varphi_{nlm_l}(2) | \frac{e^2}{r_{12}} | \varphi_{100}(1)\varphi_{nlm_l}(2) \rangle$$

$$= \int d\mathbf{r}_1 \, d\mathbf{r}_2 |\varphi_{100}(1)|^2 |\varphi_{nlm_l}(2)|^2 \frac{e^2}{r_{12}}$$

(12.46)
$$B = \langle \varphi_{100}(1)\varphi_{nlm_l}(2) | \frac{e^2}{r_{12}} | \varphi_{100}(2)\varphi_{nlm_l}(1) \rangle$$

$$= \int d\mathbf{r}_1 \, d\mathbf{r}_2 \, \varphi^*_{100}(1)\varphi_{nlm_l}(1)\varphi_{100}(2)\varphi^*_{nlm_l}(2) \frac{e^2}{r_{12}}$$

The energy A is called the *Coulomb interaction energy*. It is akin to the classical interaction potential of two electron clouds with respective charge densities $e|\varphi_{100}(1)|^2$ and $e|\varphi_{nlm}(2)|^2$. The second term, B, has no counterpart in classical physics. It is called the *exchange interaction energy*.

In the triplet state the Coulomb interaction energy becomes

(12.47)
$$\left\langle \frac{e^2}{r_{12}} \right\rangle_{\text{triplet}} = \left\langle {}^3\chi \left| \left(\frac{e^2}{r_{12}} \right) \right| {}^3\chi \right\rangle$$

$$= A - B$$

For A and B positive this correction energy is smaller than the corresponding singlet correction energy, $A + B$ (see Problem 12.31).[1]

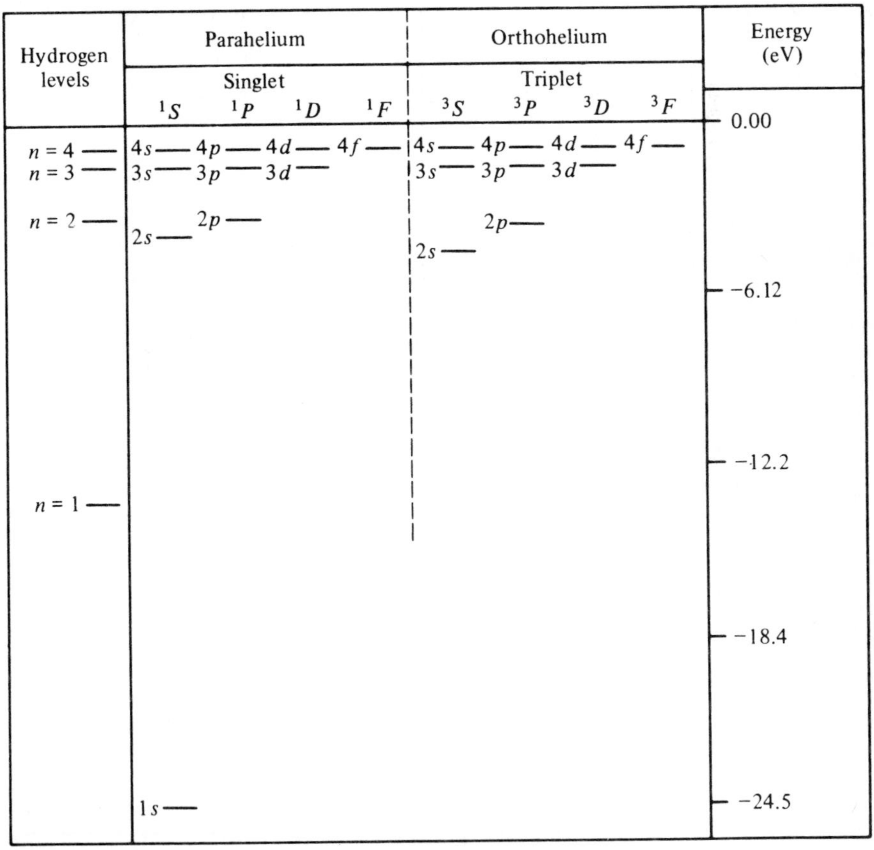

Hydrogen levels	Parahelium				Orthohelium				Energy (eV)
	Singlet				Triplet				
	^1S	^1P	^1D	^1F	^3S	^3P	^3D	^3F	0.00
$n = 4$	$4s$	$4p$	$4d$	$4f$	$4s$	$4p$	$4d$	$4f$	
$n = 3$	$3s$	$3p$	$3d$		$3s$	$3p$	$3d$		
$n = 2$	$2s$	$2p$				$2p$			
					$2s$				−6.12
									−12.2
$n = 1$									
									−18.4
	$1s$								−24.5

FIGURE 12.15 Energy levels of helium, illustrating the singlet and triplet series. The fine structure of the triplet levels is not shown. The energies of hydrogen appear at the left.

[1] The positivity of these integrals is demonstrated in J. L. Slater, *Quantum Theory of Atomic Structure*, Vol. 1, McGraw-Hill, New York, 1960, Appendix 19.

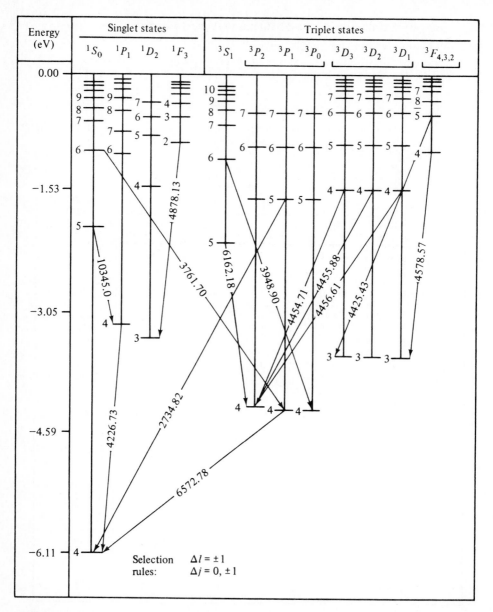

FIGURE 12.16 Energy levels of calcium, exhibiting the fine structure of the triplet states. A few typical transitions are also shown. Transition wavelengths are in angstroms. Principal quantum numbers appear at the left of levels.

Thus we find that the electrostatic interaction separates the singlet and triplet states of helium. Furthermore, when spin-orbit interaction is brought into play, different j values have slightly different energies in the triplet states. For a given value of l, the total angular momentum j number has the three values

$$j = l - 1, l, l + 1$$

which in turn give three distinct values for the spin-orbit coefficient $\langle \mathbf{L} \cdot \mathbf{S} \rangle$, thereby splitting l-levels into triplets.

Helium in antisymmetric spin states (singlet series) is called *parahelium*. Helium in symmetric spin states is called *orthohelium*. The distinct spectra associated with these different atomic configurations are shown in Fig. 12.15.

Similar descriptions apply to the heavier two-electron atoms. Their spectra also are observed to separate into singlet and triplet series. The corresponding energy-level diagram for Ca^{20} is shown in Fig. 12.16.

PROBLEMS

12.28 (a) What is the ground-state wavefunction for helium in the approximation $\hat{H}_{ES} = \hat{H}_{SO} = 0$ in (12.38)?

(b) What is the ground-state energy in this approximation?

(c) What is the correction to this ground-state energy due to the electrostatic interaction, e^2/r_{12}? What is the total ground-state energy obtained in this manner?

(d) In view of your answer to part (c), is the ground state of $\hat{H}_0(1, 2)$ a good guess for the ground state of $\hat{H}_0(1, 2) + \hat{H}_{ES}$?

Answers

(a) The ground state is $\varphi_S(\mathbf{r}_1, \mathbf{r}_2)$, with $n_1 = n_2 = 1$ and $l_1 = l_2 = 0$. This gives $(Z = 2)$

$$\varphi_S = \frac{Z^3}{\pi a_0{}^3} \exp\left(-\frac{r_1 + r_2}{a_0/Z}\right)$$

(b) $\quad E_{11}{}^{(0)} = -4\mathbb{R}(1 + 1) = -4mc^2\alpha^2 = -108.8 \text{ eV}$

(c) $$\Delta E = \langle \varphi_S | \frac{e^2}{r_{12}} | \varphi_S \rangle$$

$$= \iint |\varphi_S|^2 \frac{e^2}{r_{12}} d\mathbf{r}_1 \, d\mathbf{r}_2 = \frac{5e^2}{4a_0} = 34 \text{ eV}$$

$$d\mathbf{r}_1 = r_1{}^2 \, dr_1 \, d\Omega_1, \qquad d\mathbf{r}_2 = r_2{}^2 \, dr_2 \, d\Omega_2$$

$$r_{12}{}^2 = r_1{}^2 + r_2{}^2 - 2r_1r_2 \cos \beta$$

(See Figs. 9.9 and 12.12, and recall the generating function for Legendre polynomials given in Table 9.2.) This gives the corrected ground-state energy

$$E = E_{11}{}^{(0)} + \Delta E = -74.8 \text{ eV}$$

(d)   The fact that ΔE is not small compared to $E_{11}{}^{(0)}$ means that φ_S is not a good guess for the ground state of $\hat{H}_0 + \hat{H}_{ES}$. (*Note*: $E_{obs} = -78.98$ eV.)

12.29   Which of the following operators are diagonalized by the states ${}^1\chi$ and ${}^2\chi$ (12.43) relevant to helium?

$$\hat{L}^2, \hat{L}_z, \hat{J}^2, \hat{J}_z, \hat{S}^2, \hat{S}_z, \hat{H}_0, \hat{\mathfrak{X}}_{12}$$

12.30   The spin-orbit interaction in two-electron atoms gives three distinct energies in the triplet series. What are the values of $\langle \mathbf{L} \cdot \mathbf{S} \rangle$ if one electron is an s electron and the other is a d electron.

12.31   Show that the integrals A and B in (12.45) are both positive. [*Hint:* Recall the Fourier transform of the Coulomb potential,

$$\frac{1}{r_{12}} = \frac{1}{2\pi^2} \int \exp\left(i\mathbf{k} \cdot \mathbf{r}_{12}\right) \frac{d\mathbf{k}}{k^2}.\right]$$

12.32   A positron is the antiparticle of an electron. When in the presence of an electron, it may bind to the electron, forming a *positronium atom* that is unstable to positron–electron annihilation. Prior to annihilation, the energies and wavefunctions of the atom may be approximated by those of hydrogen with the Bohr radius replaced by $2a_0 = 2h^2/me^2$, owing to the change in reduced mass. What are the spin-dependent components of the wavefunctions of positronium? (*Note*: The annihilation time of the ${}^1S$ state of positronium for decay into two photons is $\simeq 1.2 \times 10^{-10}$ s. The ${}^3S$ state decays into three photons and lasts $\simeq 1.4 \times 10^{-7}$ s. The fact that positronium in the ${}^3S$ state must annihilate through the emission of three photons is due to the principle of charge conjugation. This principle states that electromagnetic interactions are invariant under change of all particles to their antiparticles.[1])

12.7   THE HYDROGEN AND DEUTERIUM MOLECULES

Exchange Binding

Another important area in which symmetrization requirements imposed by the spin of constituent particles plays a significant role is that of the theory of diatomic molecules. The simplest of these is the hydrogen molecule, H_2. The fact that the proton nuclei are extremely more massive than the electrons permits analysis of the molecule to be divided into two parts.[2] The first of these concerns the chemical binding between

---

[1] For a further discussion, see S. Gasiorowicz, *Quantum Physics*, Wiley, New York, 1974.

[2] This approximation is due to M. Born and J. Oppenheimer, *Ann. Physik* **84**, 457 (1927).

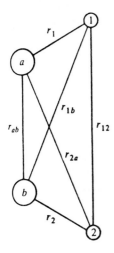

FIGURE 12.17   **Radial distances appropriate to the hydrogen molecule.**

the two atoms due to electron coupling. The second addresses the motion of the nuclei within the bound configuration.

The Hamiltonian of the molecule, neglecting all but electrostatic interaction, is (Fig. 12.17)

(12.48) $$\hat{H} = \hat{H}_{\text{atom } a} + \hat{H}_{\text{atom } b} + V_{ee} + V_{pp} + V_{ep} + \hat{T}_{\text{nuc}}$$

$$\hat{H}_{\text{atom } a} = \frac{\hat{p}_1{}^2}{2m} - \frac{e^2}{r_1} \qquad V_{ee} = +\frac{e^2}{r_{12}}$$

$$\hat{H}_{\text{atom } b} = \frac{\hat{p}_2{}^2}{2m} - \frac{e^2}{r_2} \qquad V_{pp} = +\frac{e^2}{r_{ab}}$$

$$\hat{T}_{\text{nuc}} = \frac{1}{2M}(\hat{p}_a{}^2 + \hat{p}_b{}^2) \qquad V_{ep} = -e^2\left(\frac{1}{r_{1b}} + \frac{1}{r_{2a}}\right)$$

The mass of an electron is m, nuclear mass is M, and r_1 is the distance from electron 1 to nucleus a. The repulsion between the two electrons is given by the positive potential e^2/r_{12}, while the repulsion between the nuclei is e^2/r_{ab}. The potential V_{ep} represents the cross attraction between electrons and protons. The kinetic energy of the two nuclei is written $\hat{T}_{\text{nuc}}$. Since $m/M \ll 1$, it is consistent to view the electrons as moving in the field of two fixed protons ($p_a{}^2 = p_b{}^2 = 0$, $r_{ab} = \text{constant}$).

To uncover the binding between the two atoms, one constructs the wavefunctions of $\hat{H}$, further neglecting the interaction $V_{ee} + V_{ep}$. The coupling between atoms which follows is then due primarily to antisymmetrization requirements imposed on these wavefunctions in accord with the Pauli principle. The residual Hamiltonian appears as

(12.49)
$$\hat{H} = \hat{H}_{\text{atom } a} + \hat{H}_{\text{atom } b}$$

with eigenstates

(12.50)
$$\varphi_{S, A} = \frac{1}{\sqrt{2}} [\varphi_{v_a}(\mathbf{r}_1)\varphi_{v_b}(\mathbf{r}_2) \pm \varphi_{v_a}(\mathbf{r}_2)\varphi_{v_b}(\mathbf{r}_1)]$$

Following previous notation [e.g., (12.41)] atomic eigenvalues have been written $v_{a, b}$. Spin-dependent states parallel those constructed for the helium atom (12.43) in that both problems address the antisymmetric states for two electrons. There results

(12.51)
$$^1\chi = \varphi_S(\mathbf{r}_1, \mathbf{r}_2)\xi_A$$
$$^3\chi = \varphi_A(\mathbf{r}_1, \mathbf{r}_2)\xi_S$$

Using these state functions, it is possible to calculate the expectation of the total potential of the hydrogen molecule contained in $\hat{H}$ given by (12.48).

(12.52)
$$\langle V \rangle = e^2 \left\langle \frac{1}{r_{12}} + \frac{1}{r_{ab}} - \frac{1}{r_1} - \frac{1}{r_2} - \frac{1}{r_{1b}} - \frac{1}{r_{2a}} \right\rangle$$

In calculating this average, r_1 and r_2 dependence is lost to integration, leaving only dependence on the internuclear distance r_{ab}. Thus we may write

$$\langle V \rangle = \bar{V}(r_{ab})$$

The resulting two curves for the triplet and singlet states are shown in Fig. 12.18. The potential of interaction is seen to have a minimum for the singlet state corresponding to antiparallel spins and symmetric space dependence as given by φ_S. Thus, binding of the atoms is possible in the singlet state. As discussed in the previous section, electrons in the state φ_S tend to occupy the same region of space. This common domain lies between the nuclei. At this location the electrons serve to attract each of the protons and bind the molecule. The same mechanism, we recall, is responsible for the triplet states of two-electron atoms lying lower in energy than the singlet states (see Fig. 12.13). However, in the present case the positive energy of repulsion between electrons is overbalanced by the negative energy of attraction of the protons toward the overlap domain. If r_{ab} is decreased beyond the minimum in $\bar{V}(r_{ab})$, the nuclear repulsion begins to overcome the binding afforded by the intermediary electrons and the atoms repel.

In the triplet state the antisymmetric wavefunction $\varphi_A(\mathbf{r}_1, \mathbf{r}_2)$ is appropriate, for which case $\varphi_A(\mathbf{r}_1, \mathbf{r}_1) = 0$, so that electrons do not tend to occupy the common domain between nuclei and there is no binding. This repulsion in the triplet state is

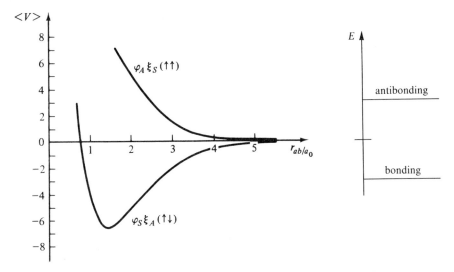

FIGURE 12.18 Expectation of the hydrogen molecule potential (in units of 10^{-12} erg) versus internuclear distance r_{ab} (in units of Bohr radii). Shown also is a schematic of the corresponding "bonding" and "antibonding" levels of the molecule.

evidenced by the monatonic increase of $\bar{V}(r_{ab})$ with decreasing internuclear distance, as shown in Fig. 12.18.

In the symmetric electronic state, electrons may be said to be "shared" by the two hydrogen atoms, thereby allowing each atom a completed $(1s^2)$ K shell. The bond so effected is called a *covalent bond*. This bonding is to be differentiated from that which couples, say, the NaCl molecule. In this case the sodium atom gives its isolated $3s^1$ electron to the vacancy in the $(3s^2 3p^5)$ M shell of the chlorine atom. In the resulting configuration, the positively charged sodium ion is knitted to the negatively charged chlorine ion in what is termed an *ionic bond*.

Symmetric States for the Nuclear Motion of D_2

Having discovered the nature of the binding of the H_2 molecule, we turn next to a discussion of the nuclear motion within this bound configuration. However, in that we wish to address the construction of symmetric states relevant to bosons, we will consider the isotope of hydrogen, deuterium. The deuterium atom has at its nucleus a deuteron that has spin 1 and is therefore a boson. The Hamiltonian (12.48) carries over to D_2 with the change that M becomes the mass of a deuteron instead of a proton. The mass ratio then becomes $M/m \sim 34{,}000$ and approximations introduced above for H_2 are even more appropriate to D_2.

In the bound configuration the deuterons move within the effective potential field $\overline{V}(r_{ab})$, and the Hamiltonian for the deuteron motion may be written

$$\hat{H}_{\text{nuc}} = \hat{T}_{\text{nuc}} + \overline{V}(r_{ab}) \tag{12.53}$$

The kinetic energy of the two deuterons $\hat{T}_{\text{nuc}}$ may be rewritten in terms of center of mass motion and motion relative to the center of mass [see (10.81)]. There results

$$\hat{H}_{\text{nuc}} - \hat{H}_{\text{CM}} = \frac{\hat{p}_r^{\,2}}{2\mu} + \frac{\hat{L}^2}{2\mu r^2} + \overline{V}(r) \tag{12.54}$$

The variable r is written for the interdeuteron distance r_{ab}, and $\mu = M/2$ is the reduced mass of the two deuterons. If $V(r)$ has its minimum at r_0, then near equilibrium one may write

$$\overline{V}(r) = \overline{V}(r_0) + \frac{1}{2}(r - r_0)^2 \left(\frac{\partial \overline{V}}{\partial r} \right)_{r_0} + \cdots \tag{12.55}$$

This parabolic potential gives rise to vibrational motion. At moderately low temperature these vibrational modes are "frozen in." That is, they are not excited and the deuterons assume the shape of a rigid dumbbell which is free only to rotate.[1] The Hamiltonian in this temperature domain reduces to the simple rotational form

$$\hat{H}_{\text{nuc}} = \frac{\hat{L}^2}{2I} \tag{12.56}$$

where $I = \mu r_0^{\,2}$ is the moment of inertia of the dumbbell two-deuteron system. Eigenstates of this purely rotational Hamiltonian are the spherical harmonics $|lm_l\rangle$ with corresponding eigenenergies (9.49)

$$E_l = \frac{\hbar^2 l(l + 1)}{2I}$$

The frequencies of emission due to transitions between rotational states (see Problem 9.6) lie in the infrared and are clearly distinguished from frequencies due to transitions in the electron states which lie in the ultraviolet–visible portion of the spectrum.

Spin-dependent eigenstates of $\hat{H}_{\text{nuc}}$ (12.56) are simply constructed in the product form

$$\chi_{\text{nuc}} = |lm_l\rangle \xi$$

---

[1] Characteristic rotational temperature for the D_2 molecule is $T = \hbar^2/2Ik_B = 44K$. Vibrational modes are excited at 4500 K and electron states are excited at temperatures several orders of magnitude larger. For further reference, see G. Herzberg, *Molecular Spectra and Structure*, Van Nostrand Reinhold, New York, 1950.

Since the deuteron has spin 1, it is a boson and χ_{nuc} must be properly symmetrized with respect to exchange of spin and space coordinates. The spin component ξ is composed of the nine states derived from the addition of two spin-1 particles. These states were previously constructed in Chapter 9 and are listed in Table 9.5. Of these nine states, six corresponding to $s = 0$ and $s = 2$ are symmetric with respect to exchange of spin coordinates, while the remaining three corresponding to $s = 1$ are antisymmetric. This separation of states is listed below.

Symmetric	Antisymmetric
$\hat{x}\lvert sm_s s_1 s_2\rangle = +\lvert sm_s s_1 s_2\rangle$	$\hat{x}\lvert sm_s s_1 s_2\rangle = -\lvert sm_s s_1 s_2\rangle$
$\left.\begin{array}{l}\lvert 2211\rangle \\ \lvert 2111\rangle \\ \lvert 2011\rangle \\ \lvert 2-111\rangle \\ \lvert 2-211\rangle\end{array}\right\} s = 2$ $\left.\begin{array}{l}\lvert 0011\rangle\end{array}\right\} s = 0$	$\left.\begin{array}{l}\lvert 1111\rangle \\ \lvert 1011\rangle \\ \lvert 1-111\rangle\end{array}\right\} s = 1$

In that the two deuterons form a dumbbell configuration, exchange of deuterons is equivalent to inversion through the origin. It follows that this operation is identical to the parity operation, so that the symmetry of a rotational state with angular momentum quantum number l is $(-1)^l$ (see Problem 12.21). Thus in order that the states χ_{nuc} be totally symmetric, they must be of the form

$$(12.57) \qquad \begin{aligned} {}^{6}\chi_{nuc} &= \lvert l_{even} m_l\rangle \xi_S \qquad \text{(ortho)} \\ {}^{3}\chi_{nuc} &= \lvert l_{odd} m_l\rangle \xi_A \qquad \text{(para)} \end{aligned}$$

For a given value of l, there are six χ_{nuc} states corresponding to l even and three states corresponding to l odd. As with the states relevant to helium, those containing an antisymmetric spin component are denoted as *para* states, while those containing a symmetric spin component are denoted as *ortho* states.[1]

Owing to the commutation property (12.22), the exchange operator $\hat{x}$ is a constant of the motion and one may expect transitions between states of different exchange symmetry to be forbidden. Assuming a uniform population of states, there are twice as many ortho states as there are para states. Thus in transitions between states of different angular momentum, radiation due to (even–even) decay is roughly twice as intense as that due to (odd–odd) decay.

A comparison of these properties of the nuclear wavefunctions for H_2 and D_2 is listed in Table 12.5. Since the proton nuclei of H_2 are fermions, the corresponding

---

[1] States of greater statistical weight carry the prefix *ortho*, whereas those of smaller statistical weight carry the prefix *para*.

TABLE 12.5 Properties of the nuclear component wavefunctions for D_2 and H_2

D_2

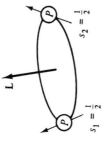

$s_1 = 1$ $s_2 = 1$ L

$\chi_{nuc} = |lm_l\rangle \xi$
χ_{nuc} is symmetric

s	Multiplicity $2s+1$	Symmetry of ξ_{spin}	Classification	l
2	5	Symmetric	ortho	even
1	3	Antisymmetric	para	odd
0	1	Symmetric	ortho	even

Intensity ratio for rotational transitions: odd-odd/even-even	1 : 2

State of lowest angular momentum: $^1S_0$(ortho)

H_2

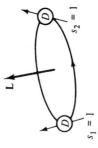

$s_1 = \frac{1}{2}$ $s_2 = \frac{1}{2}$ L

$\chi_{nuc} = |lm_l\rangle \xi$
χ_{nuc} is antisymmetric

s	Multiplicity $2s+1$	Symmetry of ξ_{spin}	Classification	l
1	3	Symmetric	ortho	odd
0	1	Antisymmetric	para	even

Intensity ratio for rotational transitions: odd-odd/even-even	3 : 1

State of lowest angular momentum: $^1S_0$(para)

χ_{nuc} function must be antisymmetrized with respect to exchange of space and spin coordinates. This reversal of symmetry requirements on χ_{nuc} results in nearly a complete reversal of intensity rules obtained above for the D_2 molecule. (These properties of H_2 are further discussed in Problems 12.37 and 12.38.)

Finally, we note that the χ_{nuc} wavefunction of lowest angular momentum for the D_2 molecule is the ortho $^1S_0$ state. In this configuration both orbital and spin angular momentum vanish ($s = l = 0$). The relative orientation of the spins of the deuterons in this state may be described as antiparallel, although in fact this $|001\uparrow\rangle$ state is the superposition

$$\uparrow\downarrow + \cdots + \downarrow\uparrow = \sqrt{\tfrac{1}{3}}|11\rangle_1|1, -1\rangle_2 - \sqrt{\tfrac{1}{3}}|10\rangle_1|10\rangle_2 + \sqrt{\tfrac{1}{3}}|1, -1\rangle_1|11\rangle_2$$

(The diagrammatic representation on the left is explained in Table 9.5.)

In this section we have found how symmetry requirements imposed by the spin of constituent particles strongly influence the physical properties of diatomic molecules. Antisymmetrization of the electron wavefunctions was found to give rise to exchange binding of the hydrogen molecule. Symmetry requirements on the wavefunctions for the two-boson deuteron system were found to give rise to intensity rules for molecular radiation. Symmetrization requirements will enter again in the last section of this chapter, wherein the quantum mechanical basis of superconductivity and superfluidity will be described.

PROBLEMS

12.33 It is found experimentally that hydrogen atoms with parallel electron spins repel in scattering from each other. What is the reason for this repulsion?

12.34 Using moment of inertia values relevant to the H_2 molecule, show that the frequency $\hbar/I$ lies in the infrared.

12.35 (a) What is the spring constant K for the vibrational coupling between the deuterons in the D_2 molecule in terms of the effective potential $\overline{V}(r_{ab})$ and the equilibrium radius r_0?

(b) Is there a coupling between the spin of two deuterons and their vibrational motion (in one dimension)? Explain your answer.

12.36 (a) In what manner are the following two physical phenomena related: (1) the triplet states of He lie lower than the singlet states; (2) H_2 is bound in the singlet electronic state.

(b) What is the radial probability distribution for the nuclei of either H_2 or D_2 in the $^1S_0$ state? Where are the electrons with respect to this distribution? In what temperature domain is your description appropriate?

12.37 The nuclei of the ordinary H_2 molecule are protons that have spin $\tfrac{1}{2}$ and are therefore fermions.

(a) What exchange symmetry must ψ_{nuc} have for the H_2 molecule?

(b) How many antisymmetric and symmetric spin states are there for two spin-$\tfrac{1}{2}$ particles? (See Table 11.2.)

(c)   What is the ratio of intensities of spectral lines due to transitions between even rotational states to that of lines due to transitions between odd rotational states? How does this ratio compare to that for the heavy hydrogen molecule D_2?

12.38   As discovered in Problem 12.37, the nuclear component of the wavefunction for the H_2 molecule must be antisymmetric. In that there are three symmetric, $s = 1$, triplet spin states and only one antisymmetric, $s = 0$, singlet spin state, the symmetric (or ortho) states are three times more prevalent than the antisymmetric (or para) states. At temperatures sufficiently high to populate the rotational levels, one expects to find molecules predominantly in odd rotational states. Describe qualitatively how this population of rotational and spin states changes with decrease in temperature. Specifically what rotational state should prevail near $0\,K$?

Answer

For $k_B T$ less than the energy between rotational levels, the molecules in an S rotational state cannot be excited out of that state. Owing to the Pauli principle, this symmetric rotational state must be accompanied by the antisymmetric (para) spin ($s = 0$, ^1S) state. Hydrogen at room temperature is a mixture of about 3:1 ortho to para molecules. However, near 20 K, the sample undergoes an ortho–para conversion. Beneath this temperature, molecules that are in the ^1S state become "frozen" in this state and the sample becomes comprised almost entirely of para molecules.

12.8   BRIEF DESCRIPTION OF QUANTUM MODELS FOR SUPERCONDUCTIVITY AND SUPERFLUIDITY

Bose–Einstein Condensation

The spin-statistics relation, which requires that fermions obey the Pauli exclusion principle and that bosons exist in totally symmetric states, has profound physical implications. We have seen that (Section 8.4) the mechanism of conduction in solids is intimately related to the fact that electrons, which have a spin of $\frac{1}{2}$, are fermions and therefore obey the Pauli exclusion principle.

Bosons (i.e., particles with integral spin values) have equally significant properties. Most interesting of these perhaps is the phenomenon of *Bose–Einstein condensation*. Since bosons do not obey an exclusion principle, a gas of such particles can conceivably be in a state in which all particles have the same momentum, same energy, and so on.

From kinetic theory we recall that the temperature T of a gas of particles is defined as[1]

$$(12.58) \qquad \frac{3}{2} k_B T = \frac{(\Delta p)^2}{2m} \equiv \frac{1}{2m} \langle (\mathbf{p} - \langle \mathbf{p} \rangle)^2 \rangle$$

---

[1] Recall also the thermodynamic identification, $T^{-1} = \partial S/\partial E$, where S is the entropy. This definition of temperature is more uniformly valid for low-temperature quantum systems than is (12.58). Thus, whereas the latter formula implies a finite temperature for a collection of fermions in the ground state, the thermodynamic relation gives the correct value, $T = 0$.

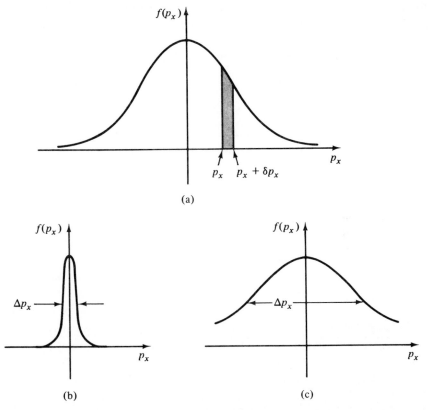

FIGURE 12.19   (a) The distribution function $f(p_x)$ **relevant to a gas of particles constrained to move in one dimension. The function** $f(p_x)$ **is such that the number of particles with momentum in the interval** p_x **to** $p_x + \delta p_x$ **is** $f(p_x)\,\delta p_x$ **(the shaded area). The total number of particles in the gas is**

$$N = \int_{-\infty}^{\infty} f(p_x)\,dp_x$$

The temperature of a gas is a measure of the mean-square deviation from the mean, $(\Delta p)^2$**, of momentum values of the particles in the gas. Thus a one-dimensional gas of particles with the distribution (b) is colder than the same gas of particles in the distribution (c).**

The temperature is proportional to the mean-square deviation of the momentum of the particles in the gas. It follows that at low temperatures, there is a small spread of momentum values away from the mean. A gas of particles which all have the same momentum has zero temperature, even though this momentum value may be large. In a frame moving with the gas, however, zero temperature means that all momenta are zero[1] (Fig. 12.19).

---

[1] For further discussion of the kinetic definition of temperature, see R. L. Liboff, *Introduction to the Theory of Kinetic Equations*, Wiley, New York, 1969; second printing, Krieger Publishing, Huntington, N. Y., 1979.

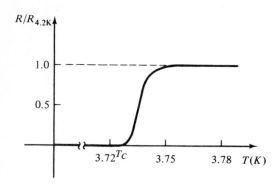

FIGURE 12.20 Resistance versus temperature for tin. The critical temperature for lead, $T = 7.18$ K, is higher than that of tin.

Suppose that we look at a box of bosons. The box is fixed in space. Lower the temperature. At zero temperature, momentum values drop to a minimal value which is consistent with the uncertainty principle (see Problem 12.40). This collapse of an aggregate of bosons to a collective ground state in which they all have the same minimal ground-state momentum eigenvalues is called *Bose–Einstein condensation*. The interesting properties of a system in such a condensed state may be related to the uncertainty principle. As the momentum of a particle in the system falls to lower values, its uncertainty in position grows. Loosely speaking, it is in many places at the same time. There are two well-established phenomena in nature which are related to Bose–Einstein condensation: *superconductivity* and *superfluidity*.

At very low temperatures (near 0 K), certain metals (e.g., tin and lead) become superconductors.[1] If a current is established in a superconducting loop, it maintains itself with zero loss. No potential difference is needed to keep the current flowing. The resistance drops to zero below a certain critical temperature, T_C (Fig. 12.20). Also, magnetic fields become completely excluded from a superconducting sample for temperatures below T_C (the *Meissner effect*; see Problem 12.39). For tin, $T_C = 3.73$ K, and for mercury, $T_C = 4.17$ K.

Cooper Pairs

It has been established[2] that below the critical temperature, interaction between electrons and the vibrational modes (phonons) of the positive ion lattice of the metal results in a diminution of the Coulomb repulsion between electrons. Phonons are quantized lattice vibrations. When averaged over many such phonon emissions and

[1] Superconductivity was discovered by K. H. Onnes in 1911.

[2] J. Bardeen, L. N. Cooper, and J. R. Schrieffer, *Phys. Rev.* **108**, 1175 (1957).

absorptions, at sufficiently low temperature, the effects of these deformations over-balance the Coulomb repulsion and yield a net attraction between electrons.[1] This attraction allows pairs of electrons to couple with spins antialigned so that each pair carries zero net spin. In that their spin values are zero, these *Cooper pairs* act like bosons. Beneath the critical temperature, these pairs collapse to a collective ground state, χ_G. This state is described as follows. Let $\zeta_i \equiv (\mathbf{r}_i, \mathbf{s}_i)$, where $\mathbf{r}_i$ and $\mathbf{s}_i$, respectively, are the space and spin coordinates of the ith conduction electron in the sample. The ground state is comprised of the appropriately symmetrized sum of products

$$\chi_G = \sum_{P(1,\,2,\,\ldots,\,N)} (\pm)^P \varphi(\zeta_1, \zeta_2) \varphi(\zeta_3, \zeta_4) \cdots \varphi(\zeta_{N-1}, \zeta_N)$$

relevant to an N-particle fermion system [compare with (12.33)]. The spin components of the bound two-particle states $\varphi(\zeta_{i-1}, \zeta_i)$ is the singlet spin-zero state, ζ_A (see Table 11.3). In this sense the ground state may be thought of as being comprised of bosons. However, the space component of these two-particle states is too widely spread to allow a consistent localized particle picture.[2] When Fourier analyzed, it is found that electron pairs in the preceding ground state have equal and opposite $\mathbf{k}$ vectors corresponding to zero total momentum.

Excited states in the material are separated from the ground state by an energy gap which at temperature 0 K is centered about the Fermi energy. When Cooper pairs are excited above the gap, they break up into two electrons that exhibit normal conduction. However, in the superconducting state, current flowing in a current loop has been observed to persist indefinitely. Such current is attributed to a collective motion of Cooper pairs in the superconducting ground state.

Superfluidity

A second example of Bose–Einstein condensation is superfluidity. In 1932 Kapitza[3] discovered that the viscosity of liquid helium drops dramatically beneath the λ point (2.19 K). This absence of viscous effects allows the helium to flow freely through capillaries with diameters as small as 100 Å.

At pressures less than 25 atm, helium is a liquid at 0 K. If the heat capacity of liquid helium is measured, a singularity is observed at about 2.2 K, which suggests a phase transition (Figure 12.21). The viscosity in the new phase (beneath T_λ) is essentially zero, while the thermal conductivity is very high. In the new phase, helium is a

[1] This attractive mechanism was first suggested by H. Frohlich, *Phys. Rev.* **79**, 845 (1950).

[2] For further discussion, see N. W. Ashcroft and N. D. Merman, *Solid State Physics*, Holt, Rinehart and Winston, New York, 1976, Chap. 34.

[3] P. L. Kapitza, *Nature* **141**, 74 (1932).

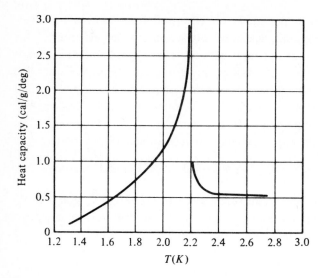

FIGURE 12.21 Heat capacity of liquid helium. The singularity near 2.19 K is evidence of a phase transition. The viscosity of the liquid above the transition temperature is similar to that of normal liquids, while below this temperature the viscosity is at least 10^6 times smaller than that above the transition.

superfluid. Below T_λ, helium is called *helium II*. Helium II is a mixture of superfluid and normal fluid.

The ground state of helium is $^1S_0$. The electron spins are antialigned with total spin zero. The orbital angular momentum is zero. The nucleus also has zero spin. The whole helium atom has zero angular momentum and is a boson. The superfluid component of helium II contains atoms condensed to the collective ground state.

Landau Theory

The first theoretical model related to superfluidity is due to L. Landau.[1] In this analysis the liquid interacts with the walls of the capillary through which it is flowing, via quantized vibrational modes of excitation generated in the liquid.

Consider that the superfluid is moving through the capillary with velocity **v**. In a frame moving with the fluid (in this frame the fluid is at rest) the capillary wall moves with velocity $-\mathbf{v}$ (Fig. 12.22). Owing to friction between the wall and the fluid, elementary excitations appear in the liquid. Let one such excitation be generated with momentum **p** and energy $\varepsilon(p)$. Transforming back to the frame where the tube is at rest (i.e., the lab frame), the energy of the liquid becomes

$$(12.59) \qquad E = \tfrac{1}{2}Mv^2 + [\varepsilon(p) + \mathbf{p} \cdot \mathbf{v}]$$

[1] L. Landau, *J. Phys.* **5**, 71 (1941); see also L. D. Landau and E. M. Lifshitz, *Statistical Physics*, Addison-Wesley, Reading, Mass., 1958.

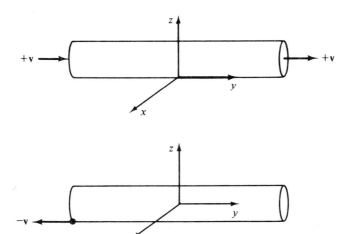

FIGURE 12.22   **(a) In the lab frame the capillary is fixed. Fluid moves with velocity $+\mathbf{v}$. (b) In the frame moving with the fluid, the fluid is at rest while the capillary moves with velocity $-\mathbf{v}$.**

where M is the mass of the liquid. The $\mathbf{p} \cdot \mathbf{v}$ term stems from Doppler shifting of the phonon frequency (see Problem 12.41). If no excitation is present, the energy of the fluid is $\frac{1}{2}Mv^2$. The presence of the phonon excitation causes this energy to change by the amount $[\varepsilon(p) + \mathbf{p} \cdot \mathbf{v}]$. Since the energy of the flowing liquid must decrease owing to such dissipative coupling,

$$\varepsilon(p) + \mathbf{p} \cdot \mathbf{v} < 0$$

This condition must be satisfied in order that an excitation appear in the liquid. Since

$$(\varepsilon + \mathbf{p} \cdot \mathbf{v}) \geq \varepsilon - pv$$

it follows that excitations have the property

(12.60)
$$v > \frac{\varepsilon(p)}{p}$$

We may conclude that an excitation of energy ε and momentum p cannot be created in a fluid moving past a wall with speed v unless the preceding inequality is satisfied. If ε/p has some minimum value greater than zero, then for small velocities of flow beneath this minimum, dissipative excitations will not appear in the liquid. That is, the liquid will exhibit superfluidity.

If the energy spectrum of excitations $\varepsilon(p)$ is plotted against p, then the minima of ε/p occur as those values of p where

(12.61)
$$\frac{d\varepsilon}{dp} = \frac{\varepsilon}{p}$$

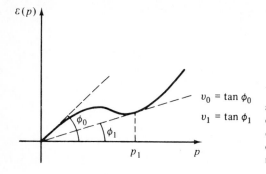

$v_0 = \tan \phi_0$

$v_1 = \tan \phi_1$

FIGURE 12.23 The minima of $\varepsilon(p)/p$ occur at values of p where a line drawn through the origin is tangent to $\varepsilon(p)$. For liquid helium this occurs at the origin and at p_1. Excitations at the origin are called phonons. Those at p_1 are called rotons.

that is, at points where a line drawn from the origin of the $p\varepsilon$ plane is tangent to the curve $\varepsilon(p)$ (see Problem 12.44).

The $\varepsilon(p)$ curve for liquid helium has been obtained by neutron scattering experiments.[1] It is sketched in Fig. 12.23. There are two values of p where ε/p is minimum, at $p = 0$ and $p = p_1$. The minimum at $p = 0$ is appropriate to temperatures near zero. At such temperatures superfluidity occurs for speeds

$$v < v_0 = \frac{d\varepsilon}{dp}\bigg|_{p=0}$$

For slightly larger temperatures the minimum at p_1 comes into play. Superfluidity occurs for this branch of excitations at fluid speeds

$$v < v_1 = \frac{d\varepsilon}{dp}\bigg|_{p=p_1} < v_0$$

The speed of phonons at $p = 0$ is that of sound in liquid helium at 0 K. The excitations at p_1 are called *rotons* corresponding, it is believed, to rotational motion of small clusters of helium atoms.[2]

The isotope of helium, He^3, has an unpaired neutron and is therefore a fermion. One would not expect this isotope to exhibit Bose–Einstein condensation. Recent experimental observation,[3] however, suggests the existence of a superfluid phase in this liquid as well. As with the case of superconductivity, such phenomena may be ascribed to a pairing process of fermions allowing for Bose–Einstein-like behavior.

[1] J. L. Yarnell et al., *Phys. Rev.* **113**, 1379, 1386 (1959).

[2] R. P. Feynman, *Phys. Rev.* **74**, 262 (1954); for further discussion of this topic, see D. L. Goodstein, *States of Matter*, Prentice-Hall, Englewood Cliffs, N.J., 1975.

[3] D. D. Osheroff, R. C. Richardson, and D. M. Lee, *Phys. Rev. Lett.* **28**, 885 (1972).

PROBLEMS

12.39 Consider a sphere of tin immersed in a uniform magnetic field at $T > T_C$. The finite conductivity of the tin permits the $\mathscr{B}$ field to penetrate. Inside the tin the field has the value $\mathscr{B}'$. If the conductivity becomes infinite as $T \to T_C$, what happens to $\mathscr{B}'$?

Answer

From Faraday's law

$$\frac{\partial \mathscr{B}}{\partial t} = -\nabla \times \mathscr{E} = -\frac{1}{\sigma} \nabla \times \mathbf{J}$$

we obtain that

$$\frac{\partial \mathscr{B}}{\partial t} = 0 \qquad \text{for } \sigma = \infty$$

so that $\mathscr{B}$ is constant in time and is *trapped* inside an ordinary conductor. This is not the case for a superconductor. At $T = T_C$, a magnetic field is excluded from a superconductor, so that $\mathscr{B}' = 0$ for the configuration given. In this sense a superconductor is said to have perfect diamagnetism. This effect is called the *Meissner effect*.

12.40 Consider a model of superconductivity where the critical temperature T_C describes the width of the Fermi sea (see Sections 2.3 and 8.4). Estimate the spatial spread (Δx) of the wavefunctions for such electrons in tin.

Answer

$$\frac{1}{2m} (\Delta p)^2 \simeq k_B T_C$$

$$(\Delta p)^2 \simeq m v_F \, \Delta p$$

where

$$m v_F^2 = 2E_F$$

Together with the uncertainty principle, this gives

$$\Delta x \geq \frac{h}{\Delta p} \simeq \frac{h v_F}{k_B T_C} \simeq 10^{-4} \text{ cm}$$

which is a macroscopic length.

12.41 An excitation has energy $\varepsilon = \hbar \omega$ and momentum $\mathbf{p} = \hbar \mathbf{k}$ in a frame S. Show that in a frame S' moving with velocity $\mathbf{v}$ with respect to S,

$$\varepsilon' = \varepsilon - \mathbf{p} \cdot \mathbf{v}$$

Answer

The energy of a photon of frequency ω in the S frame is

$$\varepsilon = \hbar \omega$$

In a frame moving with **v** with respect to S, the frequency is Doppler-shifted to the frequency

$$\omega' = \omega - \mathbf{k} \cdot \mathbf{v}$$

(If **k** is parallel to **v**, ω' decreases. If **k** is antiparallel to **v**, ω' increases.) Multiplying through by $\hbar$ gives the desired result.

12.42   (a)   Use the uncertainty relation to estimate the lowest temperature that a collection of bosons confined to a box of edge length L can have.

(b)   What temperature does this correspond to for helium in a box with edge length 1 cm?

12.43   A classical gas is said to be *degenerate* when thermodynamic properties of the system (equation of state, specific heat, conductivity, etc.) are governed by quantum statistics, as opposed to classical, *Boltzmann* statistics. A criterion that determines if a gas is degenerate may be obtained by comparing the mean interparticle distance $n^{-1/3}$ (n = particle number density) with the average de Broglie wavelength of particles.

(a)   What is this criterion in terms of n, T, and m, where T is the temperature of the gas and m is the mass of a particle?

(b)   Use this formula to estimate the temperature at which a *neutron star* of mass density 10^{14} g/cm^3 becomes degenerate.

12.44   Show that $\varepsilon(p)/p$ is minimum at those values of p for which

$$\frac{d\varepsilon}{dp} = \frac{\varepsilon}{p}$$

12.45   Show that superfluidity does not occur if excitations have the free-particle spectrum

$$\varepsilon = \frac{p^2}{2m}$$

[*Note:* If the liquid is a system of uncoupled bosons, one expects that excitations follow this spectrum. N. N. Bogoliubov[1] was the first to show that a gas of bosons with weak interactions has a spectrum of excitations $\varepsilon(p)$, which has a finite slope at $p = 0$ (such as sketched in Fig. 12.23).]

12.46   A certain Bose liquid has the excitation spectrum

$$\frac{\varepsilon(p)}{\varepsilon_0} = \tilde{p} \left[\frac{b^2}{3} + (\tilde{p} - b)^2 \right]$$

$$\tilde{p} \equiv \frac{p}{p_0}$$

where ε_0, p_0, and b are constants.

(a)   What are the maximum superfluid speeds for the phonon and roton branches of excitations, respectively?

(b)   What is the energy gap for the roton branch of excitations?

12.47   Give a qualitative explanation of the fact that superconductors are poor normal conductors.

[1] N. N. Bogoliubov, *J. Phys. USSR* **11**, 23 (1947).

Answer

Superconductivity is due to electron–phonon interactions. Metals with strong electron–phonon interaction will show large resistance at room temperature and therefore be poor conductors. On the other hand, strong electron–phonon interaction will raise the critical temperature beneath which superconductivity becomes evident. Such, for example, is the case for lead, which is a poor conductor but has one of the highest critical temperatures. On the other hand, superconductivity in gold and silver, which are very good conductors at room temperature and must therefore be typical of a weak electron–phonon interaction, proves difficult to exhibit.

12.48 When He II is constrained to flow in a circular channel, the circulation maintains itself with no dissipation. Let particles in the ground state have the wavefunction $\varphi = A \exp[i\phi(\mathbf{r})]$.

    (a)   The velocity field $\mathbf{u}$ of particles in this state is related to mass current[1] through the relation $\mathbf{J}_m = mA^2\mathbf{u}$. Show that $\mathbf{u} = (h/m)\nabla\phi$.

    (b)   What is the value of $\nabla \times \mathbf{u}$ for this flow?

    (c)   Show that the values of the *circulation* are restricted to the discrete quantum values

$$\oint \mathbf{u} \cdot d\mathbf{l} = na \qquad (n = 0, \pm 1, \pm 2, \ldots)$$

$$a \equiv \frac{h}{m}$$

The constant $a \simeq 10^{-3}$ cm²/s for He.

Answers

    (a) $\mathbf{J} = \hbar A^2 \, \nabla\phi$.

    (b)   In that $\mathbf{u}$ is the gradient of some function, $\nabla \times \mathbf{u} = 0$.

    (c)   Around the path of flow we have

$$\oint \mathbf{u} \cdot d\mathbf{l} = \frac{\hbar}{m} \oint \nabla\phi \cdot d\mathbf{l}$$

To ensure that the wavefunction is single-valued, change in ϕ about the closed loop is restricted to integral multiples of 2π.

12.49 The halogens, which comprise group VII of the periodic table, are characterized by the common property of missing one electron in the outermost p subshell. It is found that the ground states of these atoms are well described by the equivalent configuration of a single *hole* bound to an atom in an orbital p state. Using this model, obtain the possible ground states of a halogen atom. Check your answer with the ground states given in Table 12.3. The notion of a hole was introduced in Chapter 8 and is discussed further in the following section.

12.50 In the Heisenberg model for ferromagnetism, the Hamiltonian for an array of magnetic moments is given by

$$H = -\lambda \sum_i \sum_j \mathbf{u}_i \cdot \mathbf{u}_j - \sum_{i=1}^{N} \mathbf{u}_i \cdot \mathscr{B}$$

---

[1] Note the relation $\mathbf{J}_m = m\mathbf{J}$, where the particle current $\mathbf{J}$ is defined by (7.107).

where λ is a positive constant. The first sum is over "nearest neighbors," and the remaining sum extends over the N moments in the sample. The form of the first sum in this Hamiltonian presupposes that aligned magnetic moments are lower in energy than antialigned moments.

(a) Is this description consistent with classical physics?

(b) Offer a quantum mechanical explanation for the Heisenberg model, citing the Pauli principle appropriate to electrons.

12.51 (a) Referring to Fig. 12.15, give the "first" ionization energy of helium.

(b) Recalling that the remaining electron after ionization is left in the ground state of the $Z = 2$ atom, calculate the "second" ionization energy of helium. Compare your answer with the observed value, 54.4 eV.

12.52 The relativistic wave equation for bosons of rest mass m may be obtained by transforming the relation (see Problem 2.26)

$$E^2 = p^2c^2 + m^2c^4$$

through the identifications

$$E \rightarrow \hat{E} = i\hbar \frac{\partial}{\partial t}$$

$$\mathbf{p} \rightarrow \hat{\mathbf{p}} = -i\hbar \nabla$$

(a) Obtain the wave equation relevant to bosons of rest mass m. This equation is called the *Klein–Gordon* equation.

(b) What form does this equation assume for photons?

(c) Obtain a time-independent isotropic solution to the Klein–Gordon equation for bosons of finite rest mass. What characteristic decay length does this solution imply?

(d) What is this length for π mesons ($m_\pi \simeq 260m_e$)?

Answer

(a) $\nabla^2\psi - \dfrac{1}{c^2}\dfrac{\partial^2\psi}{\partial t^2} = \dfrac{m^2c^2}{\hbar^2}\psi$

(b) $\nabla^2\psi - \dfrac{1}{c^2}\dfrac{\partial^2\psi}{\partial t^2} = 0$

Equation (b) is appropriate to the propagation of electromagnetic fields in vacuum, for either the scalar or vector potential.

(c) Referring to Eq. (C.16), we find (for $r > 0$)

$$\psi(r) = \frac{Ae^{-r/a}}{r}$$

$$a \equiv \frac{\hbar}{mc}$$

(d) $a_\pi = \lambdabar_C/260 = 1.49 \times 10^{-13}$ cm $= 1.49$ fermi

12.53 A deuteron is comprised of a bound neutron and proton. Its angular momentum quantum number is $j = 1$. The spin-orbit component of the Hamiltonian is given by

$$\hat{H}' = -\frac{2\mu c^2}{\hbar^2} \hat{\mathbf{L}} \cdot \hat{\mathbf{S}}$$

where μ is the reduced mass of the two-particle system and c is the speed of light.

(a) Neutron and proton both have spin $\frac{1}{2}$. What are the possible values for the total spin quantum number s of the deuteron?

(b) Rewrite $\hat{H}'$ as a function of $\hat{\mathbf{J}}$, $\hat{\mathbf{L}}$, and $\hat{\mathbf{S}}$.

(c) What are the possible eigenvalues of $\hat{H}'$? Leave your answer in terms of μc^2.

12.54 The outer shell of neon in the ground state is in the $2p^6$ configuration. If one of these electrons is excited, one of four lines are observed in subsequent decay. Employing a model similar to that discussed in Problem 12.49, argue how these four lines might emerge from a coupling between the excited electron and the $2p^5$ unexcited electrons.

12.55 Lithium is a three-electron atom. Working in the central-field approximation and neglecting L-S coupling:

(a) Write down an appropriately symmetrized ground-state wavefunction, $\psi_G(1, 2, 3)$, for lithium with spin taken into account. Use the notation $\psi_{nlm}(i)\alpha(i)$ and $\psi_{nlm}(i)\beta(i)$ for single-electron wavefunctions. The variable i is electron number and runs from 1 to 3.

(b) What is the ground-state energy E_G of lithium corresponding to your answer in (a)?

(c) Your answer to (b) presumes that E_G is the eigenvalue of $\hat{H}$ corresponding to the eigenfunction ψ_G. That is,

$$\hat{H}\psi_G = E_G\psi_G$$

What is the explicit form of $\hat{H}(1, 2, 3)$, and what property of E_G allows it to be an eigenvalue of the extended wavefunction ψ_G?

(d) What is the answer to (b) if electrons have spin zero?

Answers (partial)

(a)

$$\psi_G(1, 2, 3) = \frac{1}{\sqrt{3!}} \begin{vmatrix} \psi_{100}(1)\alpha(1) & \psi_{100}(1)\beta(1) & \psi_{200}(1)\alpha(1) \\ \psi_{100}(2)\alpha(2) & \psi_{100}(2)\beta(2) & \psi_{200}(2)\alpha(2) \\ \psi_{100}(3)\alpha(3) & \psi_{100}(3)\beta(3) & \psi_{200}(3)\alpha(3) \end{vmatrix}$$

(b) $E_G = -Z^2 \mathbb{R}\left(\dfrac{2}{1^2} + \dfrac{1}{2^2}\right) = -\left(\dfrac{81}{4}\right)\mathbb{R}$

(c) E_G is degenerate, so that in the present instance, all six wavefunctions in $\psi_G(1, 2, 3)$ correspond to the same energy, $-9Z^2\mathbb{R}/4$.

12.9 IMPURITY SEMICONDUCTORS

The concepts of valence and conduction bands were described in Section 8.4 in reference to electrical conductivity in solids. Having developed the theory of atomic

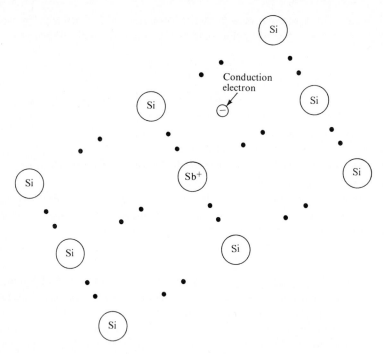

FIGURE 12.24 An antimony atom in a silicon crystal contributes a conduction electron.

structure in the early part of this chapter, we are now prepared to pursue the various aspects of impurity semiconductors. Semiconductors which conduct by virtue of impurities are said to be *extrinsic*. *Intrinsic* semiconductors, on the other hand, are not dependent on impurities for conduction.

Donor (*n*)-Type Impurities

Silicon, which is in the IVth group of the periodic chart, has four valence electrons ($3s^2 3p^2$).

Antimony lies in the Vth group of the periodic chart and has five valence electrons ($5s^2 5p^3$). Thus, if an impurity antimony atom replaces a silicon atom, four of the five valence electrons contribute to the bonding of the impurity atom to the silicon crystal. The fifth electron remains loosely bound to the parent antimony atom. A schematic two-dimensional illustration of this bonding is shown in Fig. 12.24. Under minor thermal agitation, the impurity atom is ionized and the loosely bound electron is excited to the conduction band. See Fig. 12.25. Once in the conduction band, the

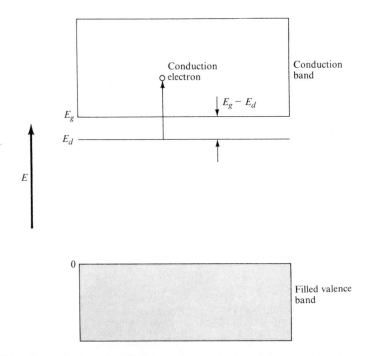

FIGURE 12.25 **Energy level E_d of an impurity antimony atom in a silicon crystal lies close to the bottom of the conduction band. The energy gap between valence and conduction bands is E_g.**

electron becomes a charge carrier and contributes to current. Since it is considerably more difficult to activate the valence electron of silicon to the conduction band, there will, in general, be a larger fraction of electrons in the conduction band from the impurity antimony component than from silicon atoms. This explains the experimentally observed large change in conduction for such crystals due to the presence of impurities.

Since conduction electrons are supplied by the antimony atoms, this form of impurity is called a *donor-type impurity*. Extrinsic semiconductors which contain a donor-type impurity are called *n-type semiconductors*. (Think of *n* for negative.)

Acceptor (*p*)-Type Impurities

Consider now that a silicon atom is replaced with an atom from group III of the periodic chart, such as indium. Since In has a valence of three ($5s^2 5p^1$), one of the four bonds to nearest neighbors will be missing an electron. See Fig. 12.26. This unsaturated indium bond effects an energy level slightly above the top of the valence band as shown

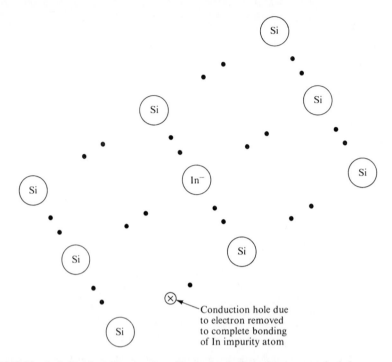

FIGURE 12.26 An indium impurity atom in a silicon crystal affects the presence of a hole.

in Fig. 12.27. Under thermal agitation an electron from the valence band of the host silicon crystal is excited to this level, leaving an unfilled level in the valence band. This unfilled level, which affects the presence of positive charge, is called a *hole*. If an electric field is applied, the hole will drift in the direction of the field and contribute to the current.

Since an electron is accepted by the impurity atom in the creation of a conducting hole, the impurity atom is called an *acceptor* and the semiconductor is said to be *p*-type. (Think of *p* for positive.)

Note that of the three atomic groups considered above, type IV atoms act as hosts to impurity atoms from type III (*p*-donor) and type V (*n*-donor) groups. Molecular semiconductors, on the other hand, are commonly comprised of type III and V atoms (e.g., GaAs, InP) and may be doped with impurity atoms from a number of other groups.[1]

---

[1] This topic is discussed further by J. C. Phillips, *Bonds and Bands in Semiconductors*, Academic Press, New York, 1973.

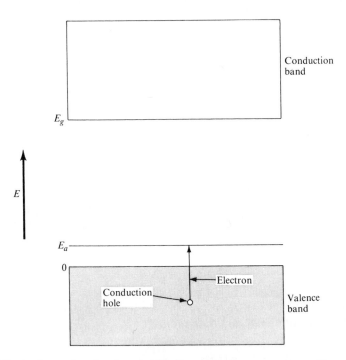

FIGURE 12.27 An electron from the valence band ionizes an impurity acceptor atom and leaves behind a hole.

For a semiconductor which contains both acceptor and donor impurities, one writes

$$\sigma = en\mu_n + ep\mu_p$$

for the conductivity, where μ denotes mobility and n, p denote electron and hole densities, respectively. (See Problem 8.27.) In germanium at room temperature (in MKS units),

$$\mu_n = 0.36 (m/s)/(V/m)$$

$$\mu_p = 0.17 (m/s)/(V/m)$$

General Results. Fermi–Dirac Distribution and Density of States

Consider a box of noninteracting electrons. The box has volume L^3 and is maintained at temperature T. Electrons may be found in any of a discrete infinity of states. Owing to the exclusion principle, not more than two electrons can be in any one energy state at any given time.

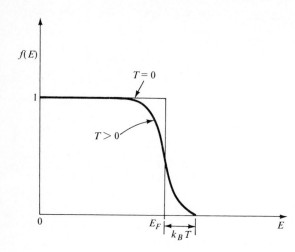

FIGURE 12.28 At 0 K all levels are oc-cupied up to the Fermi level. At finite tem-peratures, levels above E_F are occupied.

The *probability* that a given energy *level* at the value E is occupied is given by the *Fermi–Dirac* distribution relevant to fermions:[1, 2]

$$(12.62) \qquad f(E) = \frac{1}{e^{(E - E_F)/k_B T} + 1}$$

One may also interpret $f(E)$ as giving the fractional occupation of an electron in the energy level E. The Fermi energy E_F was discussed previously in Chapter 2.

At $T = 0$ K we see that $f(E) = 1$ for $E < E_F$ and $f(E) = 0$ for $E > E_F$. All levels up to E_F are occupied at zero temperature. See Figs. 2.5 and 12.28.

To obtain the density (number/cm^3) of electrons in the energy interval $(E, E + dE)$, one must multiply $f(E)$ by the density of states (number of states/energy − cm^3), which we have previously labeled $g(E)$ (see Section 8.8). It has the value

$$(12.63) \qquad g(E) = \frac{1}{2\pi^2} \left(\frac{2m^*}{\hbar^2} \right)^{3/2} E^{1/2}$$

Here we have written m^* for the effective mass introduced previously in Section 8.2. Thus the density of electrons with energy in the interval $(E, E + dE)$ is given by $f(E)g(E)\, dE$.

---

[1] Recall Section 12.3 and see Appendix A.
[2] The parameter E_F in (12.62) is the Fermi energy at 0 K. For $T > 0$ K, E_F in (12.62) is more correctly termed the chemical potential.

Electron Density

First we turn our attention to the density of electrons, n, in the conduction band. This is obtained by integrating the product $f(E)g(E)\, dE$ over conduction band energies. There results

(12.64)
$$n = \int_{\substack{\text{cond.}\\\text{band}}} f(E)g(E)\, dE$$

If we consider electrons in the conduction band to be free, then $g(E)$ is given by (12.62). Since energies in (12.64) are measured from the bottom of the conduction band, we write

(12.65)
$$g(E) = \frac{1}{2\pi^2}\left(\frac{2m_e^*}{\hbar^2}\right)^{3/2}(E - E_c)^{1/2}$$

where E_c denotes the energy at the bottom of the conduction band. Equation (12.65) reflects the property that there are no states available in the energy gap. With (12.62), we see that $f(E)$ decays rapidly with increasing E. This property permits the upper limit in (12.64) to be set equal to infinity. Combining results and setting

$$\beta \equiv \frac{1}{k_B T}$$

(12.64) becomes

(12.66)
$$n = \frac{1}{2\pi^2}\left(\frac{2m_e^*}{\hbar^2}\right)^{3/2}\int_{E_c}^{\infty}\frac{(E - E_c)^{1/2}\, dE}{e^{\beta(E - E_F)} + 1}$$

Here we have assumed that m_e^* is constant over the range of integration. Changing variables to

$$\eta = E - E_c$$

(12.66) becomes

(12.67)
$$n = \frac{1}{2\pi^2}\left(\frac{2m_e^*}{\hbar^2}\right)^{3/2}\int_0^{\infty}\frac{\eta^{1/2}\, d\eta}{e^{\beta\eta}e^{\beta(E_c - E_F)} + 1}$$

To simplify this integration, we assume that $E_c - E_F \gg k_B T$ or, equivalently, $\beta(E_c - E_F) \gg 1$. A material which satisfies this assumption is said to be *nondegen-*

erate.[1] If E_F lies in the conduction band, the material is said to be *degenerate.*

Given the preceding inequality, one may neglect the unity term in the denominator of (12.66) to obtain[2]

$$(12.68) \qquad n = \frac{1}{2\pi^2} \left(\frac{2m_e^*}{\hbar^2} \right)^{3/2} e^{-\beta(E_c - E_F)} \int_0^\infty \eta^{1/2} e^{-\beta\eta} \, d\eta$$

which gives

$$(12.69) \qquad n = N_c(T) e^{-(E_c - E_F)/k_B T}$$

where

$$(12.70) \qquad N_c(T) = \frac{1}{4} \left(\frac{2m_e^* k_B T}{\pi \hbar^2} \right)^{3/2}$$

represents an *effective density of states* in the vicinity of $E \simeq E_c$, i.e., near the bottom of the conduction band.

Hole Density

We turn next to a parallel calculation for density of holes in the valence band. A hole will be present at a given level providing an electron is missing from that level. So the Fermi distribution for holes becomes $1 - f(E)$. Thus the density of holes in the valence band is written

$$(12.71) \qquad p = \int_{\substack{\text{valence} \\ \text{band}}} g_p(E)[1 - f(E)] \, dE$$

When electrons in a partially filled band gain energy, they move upward in the band, leaving behind vacated states. Thus holes which gain energy move downward in their band. To incorporate this property and further stipulate that the zero of energy lies at the top of the valence band, the energy of a hole is written $E_v - E$, where E varies from E_v to $-\infty$. With this expression for energy substituted in the density of states

---

[1] That is, classical statistics applies. See Problems 2.47 and 12.43. At room temperature, $k_B T \simeq 0.025$ eV. A measure of $E_c - E_F$ is roughly given by E_g. Here are some typical values: $E_g = 1.14$ eV for Si, 1.25 eV for InP, and 1.4 eV for GaAs, for which cases $k_B T \ll E_g$.

[2] Here we recall $\int_0^\infty \eta^{1/2} e^{-\beta\eta} \, d\eta = \sqrt{\pi}/2\beta^{3/2}$.

(12.63), the preceding expression becomes[1]

$$(12.72) \qquad p = \frac{1}{2\pi^2}\left(\frac{2m_h^*}{\hbar^2}\right)^{3/2} \int_{-\infty}^{E_v} (E_v - E)^{1/2}\left[1 - \frac{1}{e^{\beta[E-E_F]} + 1}\right] dE$$

Following steps leading to (12.65), we introduce the variable $\eta \equiv (E_v - E)$ and again assume nondegeneracy, i.e., $\beta(E_F - E_v) \gg 1$. There results

$$(12.73) \qquad p = N_v(T)e^{-(E_F - E_v)/k_B T}$$

where

$$(12.74) \qquad N_v(T) = \frac{1}{4}\left(\frac{2m_h^* k_B T}{\pi\hbar^2}\right)^{3/2}$$

is the effective density of holes near the top of the valence band. Multiplying (12.69) and (12.73), we find

$$(12.75) \qquad np = \frac{1}{2}\left(\frac{k_B T}{\pi\hbar^2}\right)^3 (m_e^* m_h^*)^{3/2} e^{-E_g/k_B T}$$

This gives the important result that the product of n and p for a given semiconductor is constant at a given temperature. The key assumption in obtaining the preceding result is that the semiconductor is nondegenerate. Thus, with the rule (12.75), we see that introducing an impurity which would, say, increase n would diminish p, since the product np is constant. The generality of the result (12.75) follows from the observation that it is an application of the *law of mass action* to electron–hole dynamics.[2]

Intrinsic Fermi Level

For intrinsic semiconductors, for every electron in the conduction band, there is a hole present in the valence band. Thus we may set

$$(12.76) \qquad n_i = p_i$$

where the subscript i denotes intrinsic. Equating (12.65) to (12.73) gives

$$(m_e^*)^{3/2} e^{-(E_c - E_F)/k_B T} = (m_h^*)^{3/2} e^{-(E_F - E_v)/k_B T}$$

---

[1] For holes, m_h^* is given by (8.45) with a sign reversal: $m_h^* = -\hbar^2(\partial^2 E/\partial k^2)$. *This expression as well as (8.45) assumes a one-dimensional model.*

[2] That is, for the chemical reaction $A \rightleftharpoons B + C$, one may write $n_B n_C/n_A = f(T)$, where n_A denotes number density of A particles, etc., and $f(T)$ is an arbitrary function of temperature. Thus, with A denoting neutral atoms, B electrons, and C ionized atoms, one obtains the form of (12.76).

Solving for E_F gives the *intrinsic* Fermi energy:

Intrinsic

(12.77) $$E_F = \tfrac{1}{2}(E_v + E_c) + \tfrac{3}{4}k_B T \ln\left(\frac{m_h^*}{m_e^*}\right)$$

If the zero of energy is taken to be at E_v, then E_c becomes the value of the energy gap E_g, and (12.77) may be rewritten in the more concise form

(12.78) $$E_F = \tfrac{1}{2}E_g + \tfrac{3}{4}k_B T \ln\left(\frac{m_h^*}{m_e^*}\right)$$

Note in particular that energy values in semiconductor theory are not absolute but are relative to a given reference level. Thus in (12.78), the $E = 0$ value is set at the top of the valence band and E_F (for $m_h^* = m_e^*$) has the value of $\tfrac{1}{2}E_g$, whereas in (12.77) E_F is equal to the average of E_v and E_c. In all reference schemes E_F (for equal m^*) lies halfway between E_v and E_c. However, in practical usage, this is not the case. For example, for Ge, $m_h^*/m_e^* = 0.67$ and for Si, $m_h^*/m_e^* = 0.54$. In both cases E_F falls slowly with increasing temperatures.

Returning to the density of charge carriers, for intrinsic semiconductors we may write

(12.79) $$n_i p_i = n_i{}^2 = p_i{}^2$$

With (12.69) and (12.73) there results

Intrinsic

(12.80) $$\boxed{n_i{}^2 = N_c(T)N_v(T)e^{-E_g/k_B T}}$$

The effective density of states in the conduction band $N_c(T)$ is given by (12.70), and $N_v(T)$, corresponding to the valence band, is given by (12.74).

Extrinsic Parameters, Low Temperature

Consider a crystal doped with donor-type impurities. Let N_d represent the density of impurity donor atoms. Then if density of unionized donor atoms is N_{d_0}, density of ionized donor atoms is $N_d - N_{d_0}$. Charge neutrality may then be written

(12.81) $$n = p + (N_d - N_{d_0})$$

At relatively *low temperature* there are very few electron–hole pairs created, and the preceding equation becomes

(12.82) $$n \simeq (N_d - N_{do})$$

Thus, in *n*-type semiconductors, electrons are *majority carriers* and holes are *minority carriers*.

To calculate the Fermi energy for this case, first note that the density of neutral donors N_{do} is equal to the number of donor states per volume that are occupied. Thus we may write

(12.83) $$N_{do} = N_d f(E_d)$$

Substitution of (12.82) gives

(12.84) $$n = N_d \left[\frac{1}{1 + e^{\beta(E_F - E_d)}} \right]$$

Passing to the limit $k_B T \ll E_F - E_d$ gives

(12.85) $$n = N_d e^{-(E_F - E_d)/k_B T}$$

At low temperatures, $n < N_d$ so that $E_F > E_d$, and we may conclude that for the present case E_F lies between E_d and E_c. See Fig. 12.29. Equating this latter expression to our previously derived expression for n (12.69) gives

(12.86) $$N_d e^{-(E_F - E_d)/k_B T} = N_c e^{-(E_c - E_F)/k_B T}$$

from which we find

Low T

(12.87) $$E_F = \tfrac{1}{2}(E_c + E_d) + \tfrac{1}{2}k_B T \ln \left(\frac{N_d}{N_c} \right)$$

At $T = 0$, E_F is midway between E_c and E_d. As temperature rises, E_F falls since $N_c > N_d$.

Extrinsic Parameters, Intermediate and High Temperatures

At intermediate temperature nearly all donors are ionized. Assuming sufficiently large impurity density so that $N_d \gg n_i$, (12.81) gives

(12.88) $$n = N_d$$

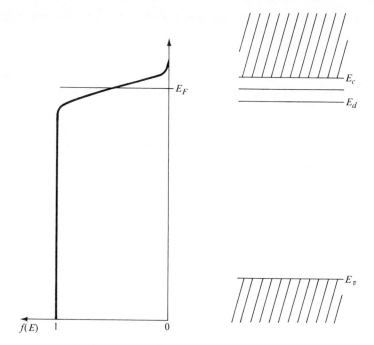

FIGURE 12.29 At low temperature, E_F lies between E_d and E_c for an n-type semiconductor. A similar diagram applies to p-type semiconductors for which, under similar conditions, E_F lies between E_a and E_v. At such temperatures, electrons are said to be "frozen out" of the conduction band (donor levels remain occupied). Likewise, holes are frozen out of the valence band (acceptor levels remain occupied with holes).

Again employing (12.69), together with the preceding equation, we find

(12.89)
$$N_d = N_c e^{-(E_c - E_F)/k_B T}$$

Solving for the Fermi energy gives

Int. T

(12.90)
$$E_F = E_c - k_B T \ln \frac{N_c}{N_d}$$

At sufficiently high temperature, electron–hole generation across the energy gap grows large compared with impurity contributions. In this limit the semiconductor behaves intrinsically and (12.77) et seq. apply.

For a semiconductor doped with acceptor-type impurities, charge neutrality is written

(12.91) $$p = n + (N_a - N_{a_0})$$

Again note that at low temperature there are very few electron–hole pairs, and the preceding relation becomes

(12.92) $$p = N_a - N_{a_0}$$

The right-hand side of this equation represents the density of ionized acceptor atoms. Each such ionized atom represents a filled acceptor state, and we may write

(12.93) $$N_a - N_{a_0} = N_a f(E_a)$$

Again we assume $\beta(E_a - E_F) \gg 1$. Substituting the resulting expression into (12.92) and recalling (12.73) gives

Low T

(12.94) $$p = N_a e^{-(E_a - E_F)/k_B T} = N_v e^{-(E_F - E_v)/k_B T}$$

(12.95) $$E_F = \tfrac{1}{2}(E_v + E_a) + \tfrac{1}{2}k_B T \ln\left(\frac{N_v}{N_a}\right)$$

Again, the Fermi level at $T = 0$ is midway between the top of the valence band and the acceptor levels. Since $N_v > N_a$, the Fermi level rises with increasing temperature. These effects are depicted in Fig. 12.30 for the case $m_h^* \gtrsim m_e^*$. A general compilation of results is presented in Table 12.6.

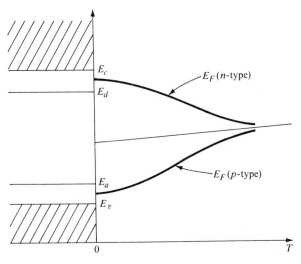

FIGURE 12.30 Change in E_F with temperature for n and p-type semiconductors (for $m_h^* \gtrsim m_e^*$).

TABLE 12.6 Semiconductor relations

General Results (Nondegenerate Crystal)

Conduction Band *Valence Band*

$$g_n(E) = \frac{1}{2\pi^2}\left(\frac{2m_e^*}{h^2}\right)^{3/2}(E - E_c)^{1/2} \qquad g_p(E) = \frac{1}{2\pi^2}\left(\frac{2m_h^*}{h^2}\right)^{3/2}(E_v - E)^{1/2}$$

$$n(T) = N_c e^{-(E_c - E_F)/k_B T} \qquad p(T) = N_v e^{-(E_F - E_v)/k_B T}$$

$$N_c = \frac{1}{4}\left(\frac{2m_e^* k_B T}{\pi h^2}\right)^{3/2} \qquad N_v = \frac{1}{4}\left(\frac{2m_h^* k_B T}{\pi h^2}\right)^{3/2}$$

$$np = \frac{1}{2}\left(\frac{k_B T}{\pi h^2}\right)^3 (m_e^* m_h^*)^{3/2} e^{-E_g/k_B T}$$

Intrinsic Semiconductor

$$E_F = \tfrac{1}{2}(E_v + E_c) + \tfrac{3}{4}k_B T \ln\left(\frac{m_h^*}{m_e^*}\right)$$

$$E_F = \tfrac{1}{2}E_g + \tfrac{3}{4}k_B T \ln\left(\frac{m_h^*}{m_e^*}\right)$$

Extrinsic, n-type

T_{low} $T_{\text{int}}{}^a$

$$n = N_d e^{-(E_F - E_d)/k_B T} = N_c e^{-(E_c - E_F)/k_B T} \qquad n = N_d = N_c e^{-(E_c - E_F)/k_B T}$$

$$E_F = \tfrac{1}{2}(E_c + E_d) + \tfrac{1}{2}k_B T \ln\left(\frac{N_d}{N_c}\right) \qquad E_F = E_c - k_B T \ln\left(\frac{N_c}{N_d}\right)$$

Extrinsic, p-type

T_{low} $T_{\text{int}}{}^a$

$$p = N_a e^{-(E_a - E_F)/k_B T} = N_v e^{-(E_F - E_v)/k_B T} \qquad p = N_a = N_v e^{-(E_F - E_v)/k_B T}$$

$$E_F = \tfrac{1}{2}(E_v + E_a) + \tfrac{1}{2}k_B T \ln\left(\frac{N_v}{N_a}\right) \qquad E_F = E_v + k_B T \ln\left(\frac{N_v}{N_a}\right)$$

^{a} Presumes large doping density.

Compensation

For semiconductors doped with both donors and acceptors, an effect may occur in which these impurity atoms compensate each other, thereby inhibiting impurity contribution to charge carrier density. Consider that both type of impurities are homogeneously distributed in the semiconductor. Then at low temperature it is possible that some electrons in the E_d level (see Fig. 12.30) "fall" to recombine with holes in the E_a level.

Suppose $N_d > N_a$ (n-type material). After compensation, some of the states at E_d are filled and some are empty. It follows that as $T \to 0$, $E_F \to E_d$ since this level represents the highest filled state. As temperature increases, the semiconductor enters the extrinsic range, with E_F given by (12.87). At higher temperature, the semiconductor enters the intrinsic range and (for $m_h^* = m_e^*$), E_F falls to $E_g/2$ [recall (12.78)].

PROBLEMS

12.56 Consider an *n*-type semiconductor.

   (a)   Give two relations in terms of density of ionized donors, $N_d - N_{d_0}$, and conduction electrons *n* which are obeyed in the T_{low} domain.

   (b)   Give two relations involving these variables which are obeyed in the T_{int} domain.

12.57 A *p*-type semiconductor has a hole density $p = 1.7 \times 10^{13}$ cm^{-3} at $T = 388$ K. If $E_g = 3.5$ eV and $m_e^* m_h^* = 0.25\ m_e^2$ at this temperature, what is the electron carrier concentration *n*?

12.58 The effective density of states in the conduction band of a semiconductor at $T = 370$ K is 6.1×10^{14} cm^{-3}. What is the value of m_e^*/m_e for this material at this temperature?

12.59 An intrinsic semiconductor has an energy gap of 1.03 eV with effective masses $m_e^* = 1.8 m_h^*$.

   (a)   What are the values of E_F at 300 K and 500 K?

   (b)   What are the charge carrier densities *n* and *p*, respectively, at these temperatures?

12.60 In the text we discussed Sb and In as impurity atoms in an Si host crystal. With reference to the periodic chart and an appropriate handbook for required additional data, choose another host element from the same group that Si lies in and two other impurity elements which would serve as donors and acceptors, respectively, for your host element. The device is to operate in the temperature interval 300 K $\leq T \leq$ 500 K.

12.61 Where does the Fermi energy of a conductor lie at 0 K with respect to valence and conduction bands? Your answer should include reference to the Fermi–Dirac distribution.

12.62 A one-dimensional model of a certain intrinsic semiconductor with lattice constant *a* has the following E versus k relations for the conduction and valence bands, respectively.

$$E_c(k) = 2E_0 - bE_0 \cos^2 \frac{ka}{2}$$

$$E_v(k) = bE_0 \cos^2 \frac{ka}{2}$$

The dimensionless parameter *b* is less than 1. At 0 K the conduction band is empty.

   (a)   Sketch these curves on the same E versus k graph in the reduced zone.

   (b)   Obtain expressions for the effective electron and hole masses, m_e^* and m_h^*, respectively. (See appropriate expressions in Chapter 8.)

   (c)   If $E_0 = 2.5$ eV, $b = \frac{1}{2}$, and $a = 1.5$ Å, obtain the value (in eV) of the Fermi energy for this semiconductor.

   (d)   At what frequency (Hz) would incident photons cause this semiconductor to conduct at 0 K?

12.63 A semiconductor is doped with donor impurity atoms of density N_d and acceptor atoms of density N_a. What is the equation of charge neutrality for this extrinsic semiconductor?

12.64 Electronegativity is a measure of the force required to remove the ionizing electron of negatively charged ions. Consider for example that an H atom in the ground state is negatively ionized, resulting in a $1s^2$ configuration.

(a) Repeating the calculation in the text relevant to helium, calculate the new ground-state energy of the H⁻ ion.

(b) Is the H⁻ a bound system?

(c) Repeat (a) for a one-electron atom with nuclear charge number Z. How does the energy you obtain depend on Z?

Answer (partial)

(a) Referring to Problem 12.28 and Table 10.3, we find

$$E = 2\mathbb{R} + \Delta E = -\frac{2e^2}{2a} + \frac{5e^2}{8a} < 0$$

Thus, to within the stated approximation, we find the ion to be a bound system.

12.65 Employing rules of electrostatics, show that the Coulomb interaction energy A (12.45) may be written as a single integral over electric field.

Answer

This integral is in the form

$$A = \iint d\mathbf{r}\, d\mathbf{r}'\, \frac{\rho(\mathbf{r})\rho(\mathbf{r}')}{|\mathbf{r} - \mathbf{r}'|}$$

where $\rho = e|\varphi|^2$ is charge density. Introducing the field (cgs)

$$\nabla \cdot \mathscr{E} = 4\pi\rho$$

permits the preceding integral to be rewritten

$$A = 2 \int \frac{|\mathscr{E}(\mathbf{r})|^2}{8\pi}\, d\mathbf{r} > 0$$

12.10 ELEMENTS OF NUCLEAR PHYSICS. THE DEUTERON AND ISOSPIN

The Pauli principle was applied earlier in this chapter in constructing properly symmetrized atomic and molecular states. In the preceding section we saw how this principle, through use of the Fermi–Dirac distribution, comes into play in describing election and hole properties in semiconductors.

In this concluding section of the present chapter we will find how the Pauli principle, through the notion of isotopic spin, comes into play in nuclear physics. This example addresses construction of the ground state of the deuteron.

The first section of the present discussion addresses the angular and spin components of the ground state of the deuteron. Recall that an estimate of the radial component of the ground state was obtained in Problem 10.30. In the present discourse, the angular component is obtained by employing conservation of parity and magnetic moment data. In this construction, noncentral, spin-dependent forces are encountered.

In the second component of the discussion, the notion of isotopic spin is developed, which, as noted above, permits quantum statistics to be applied, and the appropriately symmetrized ground state of the deuteron is obtained.

Magnetic Moment of the Deuteron

The magnetic moment of the deuteron is related to its angular momentum. In general, the angular momentum of a nucleus may be written as a sum of orbital and spin angular momenta of nuclear constituents, and we may write[1]

$$(12.96) \qquad\qquad \mathbf{J} = \mathbf{L} + \mathbf{S}$$

For the deuteron, $\mathbf{L}$ is attributed to rotational motion of the neutron and proton about each other, while $\mathbf{S}$ is the resultant of their spins.

As noted in Section 12.7, the spin of the deuteron is unity (i.e., $j = 1$). The spin of the neutron and proton are both $\frac{1}{2}$ so that allowed coupled spin values are $s = 0, 1$. The values of orbital angular momenta l which combine with these spin values to give $j = 1$ are listed below.

$$
\begin{array}{llll}
s = 1 & l = 0: & j = \underline{1} & {}^3S_1 \\
& l = 1: & j = 0, \underline{1}, 2 & {}^3P_1 \\
(12.97) & l = 2: & j = \underline{1}, 2, 3 & {}^3D_1 \\
s = 0 & l = 0: & j = 0 & \\
& l = 1: & j = \underline{1} & {}^1P_1
\end{array}
$$

Thus, the experimentally observed nuclear spin of unity for the deuteron in its ground state implies that the angular component of the wavefunction is some superposition of ${}^3S_1$, ${}^3P_1$, ${}^3D_1$, and ${}^1P_1$ states. We will now find how parity conservation properties of the Hamiltonian of the deuteron still further specify the composition of the ground-state wavefunction.

The Hamiltonian of the deuteron may be written $\hat{H}(\mathbf{r}_p, \mathbf{r}_n; \hat{\mathbf{S}}_p \hat{\mathbf{S}}_n)$, where the subscripts n and p denote neutron and proton, respectively. The parity operator $\hat{\mathbb{P}}$ (see Section 6.4) which reflects $\mathbf{r}_p$ and $\mathbf{r}_n$ through the origin leaves $\hat{H}$ unchanged.[2]

---

[1] Nuclear spin is often denoted by the symbol **I**.

[2] This statement is valid for the strong nuclear forces.

$$\hat{P}\hat{H}(\mathbf{r}_p, \mathbf{r}_n) = \hat{H}(-\mathbf{r}_p, -\mathbf{r}_n) = \hat{H}(\mathbf{r}_p, \mathbf{r}_n)$$

(12.98)

$$[\hat{P}, \hat{H}] = 0$$

It follows that the parity is a good quantum number and may be listed along with other quantum numbers in specifying the state of a deuteron. That is, states are of definite parity.

The combination of states $^3S_1$ and $^3D_1$ corresponds to even parity, while the combination $^3P_1$ and $^1P_1$ corresponds to odd parity (recall Problem 9.46). The specific combination which comprises the angular component of the ground-state wavefunction is obtained by finding the combination which gives the closest value to that of the observed magnetic moment of the deuteron,

$$\mu_d = 0.857\mu_N$$

The *nuclear magneton* μ_N has the value

$$\mu_N = \frac{\hbar e}{2M_p c} = 0.505 \times 10^{-23} \text{ erg gauss}^{-1}$$

The proton rest mass is M_p.

There are three distinct contributions to the deuteron's magnetic moment:

(12.99)

$$\boldsymbol{\mu}_d = \boldsymbol{\mu}_p + \boldsymbol{\mu}_n + \boldsymbol{\mu}_L$$

As with the electron, the proton and neutron magnetic moments are proportional to their respective spins. Experimentally, one obtains the values (noted previously in Section 11.9)

$$\boldsymbol{\mu}_p = g_p \left(\frac{\mu_N}{\hbar}\right) \mathbf{S}(p) \qquad g_p = 2(2.79)$$

$$\boldsymbol{\mu}_n = g_n \left(\frac{\mu_N}{\hbar}\right) \mathbf{S}(n) \qquad g_n = 2(-1.91)$$

The contribution to $\boldsymbol{\mu}_d$ due to orbital motion of the neutron and proton about each other is (see Problem 12.67)

$$\boldsymbol{\mu}_L = \tfrac{1}{2}\left(\frac{\mu_N}{\hbar}\right)\mathbf{L}$$

It follows that (12.99) may be rewritten

(12.100)

$$\boldsymbol{\mu}_d = \left(\frac{\mu_N}{\hbar}\right)[g_p\mathbf{S}(p) + g_n\mathbf{S}(n) + 0.50\mathbf{L}]$$

As stated in the previous chapter, it is conventional to measure the expectation of the z component of $\boldsymbol{\mu}_d$ in the state $m_j = j$ and to call this the magnetic moment of the deuteron μ_d.[1] We will calculate the contributions to this value from each of the four states listed in (12.97) and see which contribution of parity-conserving pairs of states ($^3P_1$ and $^1P_1$ or $^3S_1$ and $^3D_1$) gives a value closest to the experimentally observed value.

Introducing the spin operators

$$\mathbf{S} = \mathbf{S}(p) + \mathbf{S}(n)$$

$$\mathbf{S}^- = \mathbf{S}(p) - \mathbf{S}(n)$$

permits (12.100) to be rewritten

$$\boldsymbol{\mu}_d = \left(\frac{\mu_N}{\hbar}\right)\left[\left(\frac{g_p + g_n}{2}\right)\mathbf{S} + \left(\frac{g_p - g_n}{2}\right)\mathbf{S}^- + 0.50\mathbf{L}\right]$$

We are interested in the expectation of the z component of $\boldsymbol{\mu}_d$ in the three-triplet and one-singlet coupled spin states listed in Table 11.3. Calculation readily shows that $\langle S_z^- \rangle$ is zero in any of these four states (see Problem 12.66). Thus, for purposes of the said calculation, we are permitted to write

$$\boldsymbol{\mu}_d = \left(\frac{\mu_N}{\hbar}\right)[0.88\mathbf{S} + 0.50\mathbf{L}]$$

(12.101)

$$= \left(\frac{\mu_N}{\hbar}\right)[0.38\mathbf{S} + 0.50\mathbf{J}]$$

Construction of the Ground State

Following the description given in Problem (12.16) relevant to the Zeeman effect, we again consider that rotation of $\mathbf{J}$ about the z axis (direction, say, of a measuring $\mathbf{B}$ field) is slow compared with that of $\boldsymbol{\mu}$ about $\mathbf{J}$. With (12.101) this gives the form

$$\langle \mu_{d_z} \rangle = \left\langle \frac{(\boldsymbol{\mu}_d \cdot \mathbf{J})J_z}{J^2} \right\rangle$$

$$= \mu_N \left\{ 0.50 + 0.38\left[\frac{j(j + 1) + s(s + 1) - l(l + 1)}{2j(j + 1)}\right]\right\} m_j$$

---

[1] See also H. Von Buttlar, *Nuclear Physics*, Academic Press, New York, 1968.

Using this expression, one obtains the following contributions to $\langle \mu_d \rangle$ for the four states in question (setting $m_j = 1$ as indicated above):

$$\langle \mu_d(^3S_1) \rangle = \mu_N \left[0.50 + 0.38 \left(\frac{2+2}{2 \times 2} \right) \right] = 0.88 \mu_N$$

$$\langle \mu_d(^1P_1) \rangle = \mu_N \left[0.50 + 0.38 \left(\frac{2-2}{2 \times 2} \right) \right] = 0.50 \mu_N$$

(12.102)

$$\langle \mu_d(^3P_1) \rangle = \mu_N \left[0.50 + 0.38 \left(\frac{2+2-2}{2 \times 2} \right) \right] = 0.69 \mu_N$$

$$\langle \mu_d(^3D_1) \rangle = \mu_N \left[0.50 + 0.38 \left(\frac{2+2-2 \times 3}{2 \times 2} \right) \right] = 0.31 \mu_N$$

Since $\langle \mu_d(^3P_1) \rangle$ and $\langle \mu_d(^1P_1) \rangle$ are both less than $\langle \mu_d \rangle$, there is no combination of the form

$$\langle \mu_d \rangle = (a_{3P})^2 \langle \mu_d(^3P_1) \rangle + (a_{1P})^2 \langle \mu_d(^1P_1) \rangle$$
$$(a_{3P})^2 + (a_{1P})^2 = 1$$

which gives the observed value of $\langle \mu_d \rangle$. On the other hand, fitting $\langle \mu_d \rangle$ to the combination of even parity states,

$$\langle \mu_d \rangle = (a_S)^2 \langle \mu_d(^3S_1) \rangle + (a_D)^2 \langle \mu_d(^3D_1) \rangle = 0.86 \mu_N$$
$$(a_S)^2 + (a_D)^2 = 1$$

does allow the solution

(12.103) $(a_S)^2 = 0.96, \qquad (a_D)^2 = 0.04$

One may conclude that the ground state is a mixture of $\sqrt{0.96} \, ^3S_1$ and $\sqrt{0.04} \, ^3D_1$ states. That is,

(12.104) $|\psi_G \rangle = a_S |^3S_1 \rangle + a_D |^3D_1 \rangle$

The coordinate representation of this state is given by

(12.104a) $\langle \theta, \phi | \psi_G \rangle = a_S Y_0^{\,0}(\theta, \phi) + a_D Y_2^{\,m}(\theta, \phi)$

where the $Y_l^m(\theta, \phi)$ spherical harmonics are listed in Table 9.1.

As $a_D \ll a_S$, for purposes of calculating the magnetic moment of the deuteron, it suffices to take $|\psi_G \rangle$ to be entirely an S state. However, in evaluating the electric quadrupole moment, the D contribution to the ground state is necessary to obtain a finite result. Thus it is found that the superposition state (12.104) gives agreement with the experimentally observed quadrupole moment of the deuteron, $Q = 0.0027 \times 10^{-24}$ cm^2.[1] This moment is a measure of the deviation from spherical symmetry of the charge density of the deuteron. Furthermore, the sign $(+)$ of the quadrupole moment indicates that the charge distribution of the deuteron is prolate (resembling an egg) rather than oblate (resembling the earth).

Noncentral Forces

We note the following important fact about the neutron–proton interaction. Since the ground state of the deuteron is a superposition of states of different l, it is not an eigenstate of $\hat{L}^2$. If the interaction between these two particles were a central force,[2] then eigenstates of $\hat{H}$ would be common with eigenstates of $\hat{L}^2$, and $\hat{L}_z$ and would appear as in the form of (10.94). We conclude that the internucleon interaction is not a central-force field but rather is *spin-dependent*. That is, the potential of interaction between the particles is dependent on the relative orientation of the spins $\hat{\boldsymbol{\sigma}}_n$ and $\hat{\boldsymbol{\sigma}}_p$ of the neutron and proton, respectively, and the radius vector $\mathbf{r}$ separating them. An example of such a noncentral spin-dependent potential of interaction is given by the form[3]

$$(12.105) \qquad \hat{H}_S = f_S(r)\left[\frac{(\hat{\boldsymbol{\sigma}}_n \cdot \mathbf{r})(\hat{\boldsymbol{\sigma}}_p \cdot \mathbf{r})}{r^2} - \tfrac{1}{3}\hat{\boldsymbol{\sigma}}_n \cdot \hat{\boldsymbol{\sigma}}_p\right]$$

where $\hat{\boldsymbol{\sigma}}$ denotes the relevant Pauli spin matrices and $f_S(r)$ is a scalar function of r. Note that $\hat{H}_S$ preserves parity.

The fact that the deuteron has only one bound state (Problems 8.1 and 10.11) with parallel neutron–proton spin orientations (triplet state) is also evidence for the spin-dependent nature of the nuclear force. For if this were not the case, a bound state would also exist for the antiparallel case (singlet state). See Fig. 12.31.

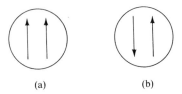

(a)   (b)

FIGURE 12.31   Relative possible orientations of the neutron and proton spins in the deuteron. The fact that the triplet spin state depicted by (a) is the ground state, whereas the singlet state depicted by (b) is not observed in nature, is evidence of the spin dependence of nuclear forces.

[1] This value takes electronic charge equal to one.

[2] Central potential was previously discussed in Section 10.5.

[3] H. Feshbach and J. Schwinger, *Phys. Rev.* **84**, 194 (1951). An assortment of forms for two-body nuclear interactions is presented in L. Eisenbud and E. P. Wigner, *Nuclear Structure*, Princeton University Press, Princeton, N. J., 1958.

Isotopic Spin

In nuclear physics, the neutron and proton are veiwed as two distinct states of a single particle called a *nucleon*. This point of view (due to Heisenberg, 1932) stems from the assumption that the nuclear force is far larger than the Coulomb force within the nucleus, or, equivalently, that the internucleon force is *charge-independent*.

These features are incorporated in a property called *isotopic spin*, or, equivalently, *isospin*. The operator corresponding to this property is written $\hat{I}$ and has three orthogonal components which obey the relation

$$(12.106) \qquad \hat{I}^2 = \hat{I}_1{}^2 + \hat{I}_2{}^2 + \hat{I}_3{}^2$$

These operators exist in isotopic spin space. Eigenvalues of the third component $\hat{I}_3$, relevant to a particle of charge Q (measured in units of electronic charge), are given by

$$(12.107) \qquad I_3 = Q - \tfrac{1}{2}$$

It follows that $I_3 = +\tfrac{1}{2}$ for a proton and $I_3 = -\tfrac{1}{2}$ for a neutron.

Let us ascertain the isospin states for a two-nucleon system. For this purpose we introduce the single-particle isospin states (see Fig. 12.32)

$$(12.108) \qquad \begin{aligned} \hat{I}^2|p\rangle &= \tfrac{3}{4}|p\rangle & \hat{I}^2|n\rangle &= \tfrac{3}{4}|n\rangle \\ \hat{I}^3|p\rangle &= \tfrac{1}{2}|p\rangle & \hat{I}^3|n\rangle &= \tfrac{1}{2}|n\rangle \end{aligned}$$

so that $|p\rangle$ and $|n\rangle$ correspond, respectively, to $I_3 = \tfrac{1}{2}$ and $I_3 = -\tfrac{1}{2}$. These states are written in analogy with the oridinary spin states α, β defined in Section 11.6. Thus, raising and lowering operators may be introduced, defined as

$$(12.109) \qquad \hat{I}_\pm = \hat{I}_1 \pm i\hat{I}_2$$

Applying these operators on the states $|p\rangle$ and $|n\rangle$ gives

$$(12.110) \qquad \begin{aligned} \hat{I}_+|p\rangle &= \hat{I}_-|n\rangle = 0 \\ \hat{I}_+|n\rangle &= |p\rangle, & \hat{I}_-|p\rangle &= |n\rangle \end{aligned}$$

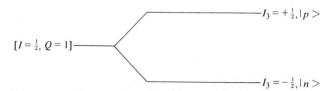

FIGURE 12.32 Nucleon with isospin $\tfrac{1}{2}$ and charge number 1 has two I_3 components corresponding to the eigenstates $|p\rangle$ and $|n\rangle$.

TABLE 12.7   Two-nucleon coupled isospin states

State	I	I_3
$\|pp\rangle_S = \|p\rangle_1 \|p\rangle_2$	1	1
$\|pn\rangle_S = \dfrac{1}{\sqrt{2}} [\|p\rangle_1 \|n\rangle_2 + \|p\rangle_2 \|n\rangle_1]$	1	0
$\|nn\rangle_S = \|n\rangle_1 \|n\rangle_2$	1	−1
$\|pn\rangle_A = \dfrac{1}{\sqrt{2}} [\|p\rangle_1 \|n\rangle_2 - \|p\rangle_2 \|n\rangle_1]$	0	0

With these results at hand, paralleling the construction of the coupled states shown in Table 11.3 gives the coupled isospin states listed in Table 12.7.

In the framework of isospin, as noted above, neutron and proton are viewed as separate states of a nucleon. As the nucleon we are considering has spin $\frac{1}{2}$, the coupled quantum states of a pair of such nucleons must be antisymmetric with respect to simultaneous interchange of particle coordinates, spins, and isospins.[1] Accordingly, the ground state of the deuteron is written

$$(12.111) \qquad |\bar{\psi}_G\rangle = |\psi_G\rangle \, |^3\chi\rangle \, |II_3\rangle$$

where the overbar on the left side indicates symmetrization. The component $|\psi_G\rangle$ represents the state (12.104) comprised of ^3S and ^3D states, with definite $j = 1$ and $s = 1$. The middle factor on the right side of (12.111) represents the triplet, symmetric, two-particle spin functions corresponding to spin 1 (see Table 11.3). Since these contributions are both symmetric states, the combined state is symmetric as well. It follows that $|II_3\rangle$ must be the antisymmetric state $|pn\rangle_A$.

Suppose this were not the case and that, instead, the proper isospin state were the symmetric $I = 1$ states. The Hamiltonian for our system is assumed to be rotationally invariant in isotopic spin space. Consequently, all three $I = 1$ states correspond to the same energy. This triplet includes the states $|pp\rangle_S$ or $|nn\rangle_S$ that correspond, respectively, to the diproton and dineutron which are not found in nature.

Thus, the appropriately symmetrized ground state of the deuteron (apart from the radial factor) is given by

$$(12.112) \qquad |\bar{\psi}_G\rangle = |\psi_G\rangle \, |^3\chi\rangle \, |pn\rangle_A$$

which is a common eigenstate of $\hat{J}^2, \hat{S}^2, \hat{S}_z, \hat{I}^2, \hat{I}_3$.

---

[1] I_3 is sometimes called the "charge variable" and is labeled I_ζ. For further discussion, see L. D. Landau, E. M. Lifshitz, and L. P. Pitaevskii, *Quantum Mechanics*, 3rd ed., Pergamon Press, New York, 1974.

Summary

The preceding result (12.112) for the ground state of the deuteron was based on the following concepts and properties: (1) Parity is conserved for the strong nuclear forces; (2) the value 1 for the spin of the deuteron; (3) values of the magnetic dipole and electric quadrupole moments of the deuteron. These properties gave rise to the even-parity, unsymmetrized component $|\psi_G\rangle$, (12.104).

Notions of the nucleon, charge-independent nuclear forces and isotopic spin were then introduced. With these concepts and the Pauli principle at hand, the isospin factor $|pn\rangle_A$ was obtained. When coupled with $|\psi_G\rangle$, this factor produced the appropriately symmetrized ground state (12.112).

It should be noted that until the notion of isotopic spin is introduced, there is no reason to incorporate the Pauli principle in the formalism, as neutron and proton are distinct particles. However, once isospin is introduced, neutron and proton become different states of the same particle, the nucleon. As this particle is a fermion, coupled states of such particles must be antisymmetric.

PROBLEMS

12.66 Show that $\langle S_z^-\rangle$ [defined following (12.100)] is zero for any of the four coupled spin states listed in Table 11.3.

12.67 Is the Pauli principle (neglecting isotopic spin) relevant to a proton–neutron system? Explain your answer.

12.68 Let the neutron and proton in a deuteron rotate about a common origin with combined orbital angular momentum **L**. Assuming circular motion, show classically that the magnetic moment due to this *orbital* motion is

$$\mu_L = \left(\frac{e}{4M_p c}\right)\mathbf{L} = \frac{1}{2}\left(\frac{\mu_N}{\hbar}\right)\mathbf{L}$$

where μ_N is the nuclear magnetic moment. Note that the orbital neutron motion contributes zero current.

12.69 Consider an ideal experiment in which a gas of deuterons is immersed in a steady $\mathscr{B}$ field of 10^3 gauss. Radiation of frequency ν is transmitted through the sample. At what frequency might you expect to see a diminishment in the transmission?

12.70 (a) What is the value of $\hat{\boldsymbol{\sigma}}_p \cdot \hat{\boldsymbol{\sigma}}_n \zeta$ for each of the four spin states ζ listed in Table 11.3?

(b) Let $f_C(r)$ and $f_S(r)$ represent central potentials. Is the interaction represented by the Hamiltonian

$$\hat{H} = f_C(r) + f_S(r)\hat{\boldsymbol{\sigma}}_p \cdot \hat{\boldsymbol{\sigma}}_n$$

a central potential? Explain your answer.

(c) What is $[\hat{H}, \hat{L}^2]$ for this Hamiltonian?

12.71 At which orientation of spins relative to interparticle radius, does the interaction $\hat{H}_S$ given by (12.105) lose **r** dependence?

12.72 A two-nucleon system is in a state with zero orbital angular momentum.

   (a) List the possible spin states for this system. (Each nucleon has spin $\frac{1}{2}$.)

   (b) List the possible isotopic spin states for this system.

   (c) Employing the preceding results, write down the appropriately symmetrized spin–isospin states for this system.

PERTURBATION THEORY

In this chapter perturbation techniques are described which serve to generate approximate solutions to the Schrödinger equation. Such solutions appear in the form of an expansion away from known, unperturbed values. A special procedure is developed for systems with degenerate eigenenergies. Application is made to problems in atomic physics and the problem of an electron in a periodic potential, encountered previously in Chapter 8. Harmonic perturbation theory is applied in a rederivation of the Planck radiation formula and the theory of the laser. The chapter concludes with a description of the interaction of a radiation field with an atom. Here we encounter the notion of oscillator strengths and the important Thomas–Reiche–Kuhn sum rule.

13.1 TIME-INDEPENDENT, NONDEGENERATE PERTURBATION THEORY

Approximate methods of solution to an assortment of problems in quantum mechanics were described previously in Chapter 7 (WKB analysis), Chapter 8 (LCAO approximation), Chapter 10 (Thomas–Fermi model), and Chapter 11 (interaction picture).

TABLE 13.1 **Examples of perturbation Hamiltonians**

Name	Description	Hamiltonian
L-S coupling	Coupling between orbital and spin angular momentum in a one-electron atom	$\hat{H} = \hat{H}_0 + f(r)\hat{\mathbf{L}} \cdot \hat{\mathbf{S}}$ $\hat{H}' = f(r)\hat{\mathbf{L}} \cdot \hat{\mathbf{S}}$ $\hat{H}_0 = \hat{p}^2/2m - e^2 Z/r$
Stark effect	One-electron atom in a constant, uniform electric field $\mathscr{E} = \mathbf{e}_z \mathscr{E}_0$	$\hat{H} = \hat{H}_0 + e\mathscr{E}_0 z$ $H' = e\mathscr{E}_0 z$ $\hat{H}_0 = (\hat{p}^2/2m) - e^2 Z/r$
Zeeman effect	One electron atom in a constant, uniform magnetic field $\mathscr{B}$	$\hat{H} = \hat{H}_0 + (e/2mc)\hat{\mathbf{J}} \cdot \mathscr{B}$ $\hat{H}' = (e/2mc)\hat{\mathbf{J}} \cdot \mathscr{B}$ $\hat{H}_0 = (\hat{p}^2/2m) - e^2 Z/r$
Anharmonic oscillator	Spring with nonlinear restoring force	$\hat{H} = \hat{H}_0 + \mathrm{K}'x^4$ $\hat{H}' = \mathrm{K}'x^4$ $\hat{H}_0 = (\hat{p}_x^2/2m) + \frac{1}{2}\mathrm{K}x^2$
Nearly free electron model	Electron in a periodic lattice	$\hat{H} = \hat{H}_0 + V(x)$ $V(x) = \sum_n V_n \exp\left[i(2\pi nx/a)\right]$ $\hat{H}_0 = \hat{p}_x^2/2m$

Our present concern lies in refinement of the approximation method described in Chapter 12, used in calculating both the ground-state wavefunction of the helium atom (Problem 12.28) and the fine-structure spectrum of hydrogen (12.13). In both these problems the Hamiltonian encountered was of the form

$$\hat{H} = \hat{H}_0 + \hat{H}'$$

This breakup of a Hamiltonian into a part $\hat{H}_0$, whose eigenfunctions are known, and an additional term $\hat{H}'$, which is in some sense small compared to $\hat{H}_0$, is typical of many practical problems encountered in quantum mechanics. The theory that seeks approximate eigenstates of the *total Hamiltonian* $\hat{H}$ is called *perturbation theory*. In the expression above, the Hamiltonian, $\hat{H}_0$, is called the *unperturbed Hamiltonian* while $\hat{H}'$ is called the *perturbation Hamiltonian*. Some typical perturbation problems are listed in Table 13.1.

The perturbation analysis we will develop in this chapter divides into three categories: (1) time-independent, nondegenerate; (2) time-independent, degenerate; (3) time-dependent. In the last category one investigates the time development of a system in a given state due to a perturbation on the system which is turned on at a given instant of time.

Smallness of the Perturbation

Perturbation theory begins with the assumption that the perturbation Hamiltonian, $\hat{H}'$, is in some sense small compared to the unperturbed Hamiltonian, $\hat{H}_0$. The criterion that establishes the smallness of $\hat{H}'$ compared to $\hat{H}_0$ will emerge in the course of the analysis. Another underlying assumption in perturbation theory is that the eigenstates and eigenenergies of the total Hamiltonian, $\hat{H}$, do not differ appreciably from those of the unperturbed Hamiltonian, $\hat{H}_0$. That is, suppose that $\{\varphi_n\}$ and $\{E_n\}$ are, respectively, the eigenstates and eigenenergies of the total Hamiltonian $\hat{H}$,

$$\hat{H}\varphi_n = (\hat{H}_0 + \hat{H}')\varphi_n = E_n\varphi_n$$

while $\{\varphi_n^{(0)}\}$ and $\{E_n^{(0)}\}$ are, respectively, the eigenstates and eigenenergies of the unperturbed Hamiltonian

$$\hat{H}_0\varphi_n^{(0)} = E_n^{(0)}\varphi_n^{(0)}$$

Then it is always possible to write

$$\varphi_n = \varphi_n^{(0)} + \Delta\varphi_n$$
$$E_n = E_n^{(0)} + \Delta E_n$$

where, owing to the smallness of $\hat{H}'$, $\Delta\varphi_n$ is a small correction to $\varphi_n^{(0)}$ and ΔE_n is a small correction to $E_n^{(0)}$.

To keep the smallness of $\hat{H}'$ in mind, we rewrite it as $\lambda\hat{H}'$, where λ is an infinitesimal parameter and is introduced for "bookkeeping" purposes only. The equation to which we seek a solution is of the form

(13.1) $$(\hat{H}_0 + \lambda\hat{H}')\varphi_n = E_n\varphi_n$$

The Perturbation Expansion

The eigenstates and eigenenergies of $\hat{H}_0$ are assumed known. Since $\varphi_n \to \varphi_n^{(0)}$ as $\lambda \to 0$, it is consistent to seek solution to (13.1) in the form of a series with $\varphi_n^{(0)}$ entering as the leading term. In similar manner, E_n is expanded, with $E_n^{(0)}$ entering as the leading term.

(13.2) $$\varphi_n = \varphi_n^{(0)} + \lambda\varphi_n^{(1)} + \lambda^2\varphi_n^{(2)} + \cdots$$
$$E_n = E_n^{(0)} + \lambda E_n^{(1)} + \lambda^2 E_n^{(2)} + \cdots$$

Substituting these expansions into (13.1) and arranging terms according to powers in λ gives

(13.3) $$[\hat{H}_0\varphi_n^{(0)} - E_n^{(0)}\varphi_n^{(0)}] + \lambda[\hat{H}_0\varphi_n^{(1)} + \hat{H}'\varphi_n^{(0)} - E_n^{(0)}\varphi_n^{(1)} - E_n^{(1)}\varphi_n^{(0)}]$$
$$+ \lambda^2[\hat{H}_0\varphi_n^{(2)} + \hat{H}'\varphi_n^{(1)} - E_n^{(0)}\varphi_n^{(2)} - E_n^{(1)}\varphi_n^{(1)} - E_n^{(2)}\varphi_n^{(0)}]$$
$$+ \cdots = 0$$

This equation is of the form

$$F^{(0)} + \lambda F^{(1)} + \lambda^2 F^{(2)} + \lambda^3 F^{(3)} + \cdots = 0$$

If this equation is to be true for *arbitrarily* small values of λ, then

$$F^{(0)} = F^{(1)} = F^{(2)} = \cdots = 0$$

In this manner (13.3) gives the coupled set of equations

(13.4)
- (a) $\hat{H}_0 \varphi_n^{(0)} = E_n^{(0)} \varphi_n^{(0)}$
- (b) $(\hat{H}_0 - E_n^{(0)})\varphi_n^{(1)} = (E_n^{(1)} - \hat{H}')\varphi_n^{(0)}$
- (c) $(\hat{H}_0 - E_n^{(0)})\varphi_n^{(2)} = (E_n^{(1)} - \hat{H}')\varphi_n^{(1)} + E_n^{(2)}\varphi_n^{(0)}$
- (d) $(\hat{H}_0 - E_n^{(0)})\varphi_n^{(3)} = (E_n^{(1)} - \hat{H}')\varphi_n^{(2)} + E_n^{(2)}\varphi_n^{(1)} + E_n^{(3)}\varphi_n^{(0)}$

$$\vdots$$

In the lowest approximation, (13.4a) returns the information that $\{\varphi_n^{(0)}\}$ and $\{E_n^{(0)}\}$ are, respectively, the eigenstates and eigenenergies of $\hat{H}_0$. The second (as well as all of the higher-order equations) has the following interesting property. The left-hand side of this equation remains the same under the replacement

$$\varphi_n^{(1)} \rightarrow \varphi_n^{(1)} + a\varphi_n^{(0)}$$

where a is an arbitrary constant. Suppose that one solves (13.4b) for $\varphi_n^{(1)}$ and $E_n^{(1)}$. Then $\varphi_n^{(1)} + a\varphi_n^{(0)}$; $E_n^{(1)}$ is also a solution. An extra constraint is needed to remove this arbitrary quality of solution. This constraint may be taken as follows.[1] We assume that all corrections to $\varphi_n^{(0)}$ in (13.2) are normal to $\varphi_n^{(0)}$.

(13.5) $\langle \varphi_n^{(s)} | \varphi_n^{(0)} \rangle = 0$ (for $s > 0$ and all n)

In Hilbert space this relation indicates that $\Delta\varphi_n$ is normal to $\varphi_n^{(0)}$. This condition will aid us in the construction of $\varphi_n^{(s)}$.

Returning to (13.4b) we note that $\hat{H}_0$ operates on $\varphi_n^{(1)}$ in this equation, which suggests that the solution may be obtainable through expansion of $\varphi_n^{(1)}$ in a superposition of the eigenstates of $\hat{H}_0$.

(13.6) $$|\varphi_n^{(1)}\rangle = \sum_i c_{ni} |\varphi_i^{(0)}\rangle$$

If this expansion is substituted into (13.4b), there results

$$(\hat{H}_0 - E_n^{(0)}) \sum_i c_{ni} |\varphi_i^{(0)}\rangle = (E_n^{(1)} - \hat{H}') |\varphi_n^{(0)}\rangle$$

---

[1] Another popular constraint is to construct $\varphi_n^{(s)}$ so that it is normalized. Both choices of constraint yield the same corrections to the energy, $\{E_n^{(s)}\}$, while the wavefunctions that emerge differ by at most a phase factor (see Section 4.1).

Multiplying from the left with $\langle \varphi_j^{(0)} |$ gives

(13.7)
$$(E_j^{(0)} - E_n^{(0)})c_{nj} + H'_{jn} = E_n^{(1)}\delta_{jn}$$

where H'_{jn} are the matrix elements of $\hat{H}'$ in the $\{\varphi_n^{(0)}\}$ representation

$$H'_{jn} \equiv \langle \varphi_j^{(0)} | \hat{H}' | \varphi_n^{(0)} \rangle$$

First-Order Corrections

With $j \neq n$, (13.7) gives the coefficients, $\{c_{nj}\}$, which when substituted into (13.6) gives the first-order correction to φ_n.

(13.8)
$$c_{ni} = \frac{H'_{in}}{E_n^{(0)} - E_i^{(0)}}$$

$$\varphi_n^{(1)} = \sum_{i \neq n} \frac{H'_{in}}{E_n^{(0)} - E_i^{(0)}} \varphi_i^{(0)} + c_{nn}\varphi_n^{(0)}$$

Here one assumes that all corrections, $\{\varphi_n^{(s)}\}$, lie in a Hilbert space that is spanned by the unperturbed wavefunctions, $\{\varphi_n^{(0)}\}$.

The coefficient c_{nn} is obtained from (13.5), which yields

$$c_{nn} = 0$$

With $j = n$, (13.7) gives the first-order corrections to the energy E_n.

(13.9)
$$E_n^{(1)} = H'_{nn}$$

These are the diagonal elements of $\hat{H}'$. Substituting (13.8) and (13.9) into (13.2) and setting $\lambda = 1$ gives

(13.10)
$$\boxed{\varphi_n = \varphi_n^{(0)} + \sum_{i \neq n} \frac{H'_{in}}{E_n^{(0)} - E_i^{(0)}} \varphi_i^{(0)}}$$

$$\boxed{E_n = E_n^{(0)} + H'_{nn}}$$

The first of these equations tells us that in order for the expansion (13.2) to make sense, the coefficients of expansion should be less than 1.

$$|H'_{in}| \ll |E_n^{(0)} - E_i^{(0)}|$$

The matrix elements of $\hat{H}'$ should be small compared to the difference between the corresponding unperturbed energy levels. In similar manner, the second equation in (13.10) reveals that

$$|H'_{nn}| \ll E_n^{(0)}$$

The diagonal elements of the perturbation Hamiltonian should be small compared to the corresponding unperturbed energy level.

Second-Order Corrections

To find the second-order correction to φ_n and E_n, we must solve (13.4c). Again we note that $\hat{H}_0$ operates on $\varphi_n^{(2)}$, and it is again advantageous to expand $\varphi_n^{(2)}$ in the eigenstates of $\hat{H}_0$.

(13.11)
$$\varphi_n^{(2)} = \sum_i d_{ni} \varphi_i^{(0)}$$

Substitution into (13.4c) gives

$$\sum_i E_i^{(0)} d_{ni} |\varphi_i^{(0)}\rangle + \hat{H}' |\varphi_n^{(1)}\rangle = E_n^{(0)} \sum_i d_{ni} |\varphi_i^{(0)}\rangle + E_n^{(1)} |\varphi_n^{(1)}\rangle$$

$$+ E_n^{(2)} |\varphi_n^{(0)}\rangle$$

Multiplying from the left with $\langle \varphi_j^{(0)}|$ gives

(13.12) $(E_j^{(0)} - E_n^{(0)}) d_{nj} + \langle \varphi_j^{(0)}|\hat{H}'|\varphi_n^{(1)}\rangle = E_n^{(2)} \delta_{jn} + E_n^{(1)} \langle \varphi_j^{(0)}|\varphi_n^{(1)}\rangle$

With $j = n$, this equation gives

$$E_n^{(2)} = \langle \varphi_n^{(0)}|\hat{H}'|\varphi_n^{(1)}\rangle$$

$$= \sum_{i \neq n} \langle \varphi_n^{(0)}| \frac{\hat{H}' H'_{in}}{E_n^{(0)} - E_i^{(0)}} |\varphi_i^{(0)}\rangle$$

$$= \sum_{i \neq n} \frac{H'_{ni} H'_{in}}{E_n^{(0)} - E_i^{(0)}}$$

Owing to the Hermiticity of $\hat{H}'$, this equation may be rewritten

(13.13)
$$E_n^{(2)} = \sum_{i \neq n} \frac{|H'_{ni}|^2}{E_n^{(0)} - E_i^{(0)}}$$

Note that in obtaining this result we have used the result that $c_{nn} = 0$. Substituting this expression for $E_n^{(2)}$ into (13.2) together with the expression for $E_n^{(1)}$ given by (13.9) gives the following second-order expression for E_n:

(13.14)
$$\boxed{E_n = E_n^{(0)} + H'_{nn} + \sum_{i \neq n} \frac{|H'_{ni}|^2}{E_n^{(0)} - E_i^{(0)}}}$$

To calculate the second-order corrections to the wavefunction φ_n, we must obtain the coefficients d_{ni} in (13.11). These are directly obtained from (13.12). With $n \neq j$ this equation gives

$$(E_n^{(0)} - E_j^{(0)})d_{nj} = \langle \varphi_j^{(0)}| \hat{H}' \sum_{k \neq n} \frac{H'_{kn}}{E_n^{(0)} - E_k^{(0)}} |\varphi_k^{(0)}\rangle$$

$$- H'_{nn}\langle \varphi_j^{(0)}| \sum_{k \neq n} \frac{H'_{kn}}{E_n^{(0)} - E_k^{(0)}} |\varphi_k^{(0)}\rangle$$

In the second sum, only the $k = j$ term survives the $\langle \varphi_j^{(0)}|\varphi_k^{(0)}\rangle$ inner product. All terms in the first term remain. There results

$$d_{nj} = \frac{1}{E_n^{(0)} - E_j^{(0)}} \left(\sum_{k \neq n} \frac{H'_{jk}H'_{kn}}{E_n^{(0)} - E_k^{(0)}} \right) - \frac{H'_{nn}H'_{jn}}{(E_n^{(0)} - E_j^{(0)})^2}$$

Again, owing to (13.5), one finds that

$$d_{nn} = 0$$

In this manner one obtains the following expression for φ_n, good to terms of second order in $\hat{H}'$.

(13.15)
$$\boxed{\varphi_n = \varphi_n^{(0)} + \sum_{i \neq n} \left[\frac{H'_{in}}{E_n^{(0)} - E_i^{(0)}} - \frac{H'_{nn}H'_{in}}{(E_n^{(0)} - E_i^{(0)})^2} \right. \\ \left. + \sum_{k \neq n} \frac{H'_{ik}H'_{kn}}{(E_n^{(0)} - E_i^{(0)})(E_n^{(0)} - E_k^{(0)})} \right] \varphi_i^{(0)}}$$

PROBLEMS

13.1 Calculate the first-order correction to $E_3^{(0)}$ for a particle in a one-dimensional box with walls at $x = 0$ and $x = L$ due to the following perturbations.
   (a) $H' = 10^{-3}E_1 x/L$
   (b) $H' = 10^{-3}E_1(x/L)^2$
   (c) $H' = 10^{-3}E_1 \sin(x/L)$

13.2 What is the eigenfunction φ_n for the same configuration as in Problem 13.1, to terms of second order, for the constant perturbation

$$H' = 10^{-3}E_1?$$

13.3 Calculate the eigenenergies of the anharmonic oscillator whose Hamiltonian is listed in Table 13.1, to first order in $\hat{H}'$.

Answer

In terms of raising and lowering operators, $(a^\dagger, a)$, the perturbation Hamiltonian appears as

$$\hat{H}' = Kx^4 = \frac{K}{4\beta^4}(\hat{a} + \hat{a}^\dagger)^4$$

The corrections to $E_n^{(0)}$ which we seek are given by

$$E_n^{(1)} = H'_{nn} = \langle n|\hat{H}'|n\rangle$$

The only terms in the expansion of $(\hat{a} + \hat{a}^\dagger)^4$ which give nonzero contributions are those which maintain the eigenvector $|n\rangle$. All other terms vanish because of orthogonality with $\langle n|$. Of the 16 terms in the expansion of $(\hat{a} + \hat{a}^\dagger)^4$, only six survive this orthogonality condition. The energy $E_n^{(1)}$ may be determined by a graphical analysis, according to which a diagram is associated with each integral that contributes to $\langle n|\hat{H}'|n\rangle$. The eigenvector $|n\rangle$ is represented by a dot drawn at the right of the diagram. Another dot drawn at the left of the $|n\rangle$ dot, but on the same horizontal, represents the eigenbra $\langle n|$. The creation operator, $\hat{a}^\dagger$, is represented by a diagonal arrow from the right and inclined upward at $\pi/4$, while the annihilation operator, $\hat{a}$, is an arrow from the right at $-\pi/4$. Thus the diagram related to $\langle n|\hat{a}\hat{a}^\dagger|n\rangle$ is

$$\langle n|\hat{a}\hat{a}^\dagger|n\rangle$$

The diagram that represents the fourth-order term, $\langle n|\hat{a}^\dagger\hat{a}\hat{a}^\dagger\hat{a}|n\rangle$, is

$$\langle n|\hat{a}^\dagger\hat{a}\hat{a}^\dagger\hat{a}|n\rangle$$

while the second-order term, $\langle n|\hat{a}^2|n\rangle$, is represented by

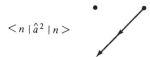

$$< n\,|\,\hat{a}^2\,|\,n >$$

Any sequence of arrows that do not join the two horizontal dots represents a zero contribution. Continuing in this manner, we find that, in all, there are 16 fourth-order diagrams. Of these, only six gave nonzero contributions. These six diagrams are:

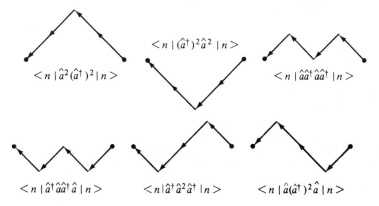

$$< n\,|\,(\hat{a}^\dagger)^2\hat{a}^2\,|\,n >$$

$$< n\,|\,\hat{a}^2(\hat{a}^\dagger)^2\,|\,n >$$

$$< n\,|\,\hat{a}\hat{a}^\dagger\hat{a}\hat{a}^\dagger\,|\,n >$$

$$< n\,|\,\hat{a}^\dagger\hat{a}\hat{a}^\dagger\hat{a}\,|\,n >$$

$$< n\,|\,\hat{a}^\dagger\hat{a}^2\hat{a}^\dagger\,|\,n >$$

$$< n\,|\,\hat{a}(\hat{a}^\dagger)^2\hat{a}\,|\,n >$$

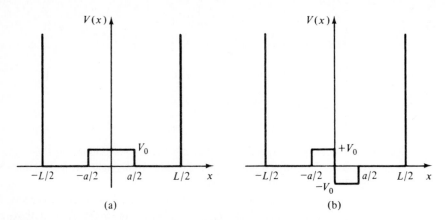

FIGURE 13.1   Potential configurations for Problems 13.4, 13.5, and 13.6.

Summing these contributions gives the desired result:

$$E_n^{(1)} = \frac{3K'}{4\beta^4}[2n(n+1)+1]$$

13.4   Consider a particle of mass m in a potential well shown in Fig. 13.1a. Suppose that the bump at the bottom of the well can be considered a small perturbation.

(a)   Calculate the corrected second eigenenergy and eigenfunction to first order in the perturbation.

(b)   What dimensionless ratio must be small compared to 1 in order for your approximation to be valid?

(c)   First-order corrected energies corresponding to even eigenstates are greater than energies corresponding to odd eigenstates. Why? (*Hint*: Note the behavior of eigenstates at the origin.)

13.5   (a)   Consider the perturbation bump shown in Fig. 13.1b. What are the first-order corrected eigenenergies of a particle of mass m confined to this well?

(b)   Calculate the eigenenergies of this configuration, in the domain $E \gg V_0$, using the WKB formula (7.191) and compare your results with those obtained in part (a).

13.6   Again consider the potential shown in Fig. 13.1a. What is the unperturbed ground state for:

(a)   Two identical bosons of mass m confined to the box.

(b)   Two identical fermions of mass m confined to the box.

(c)   What are the unperturbed ground-state energies E_S and E_A for these two cases?

(d)   Use first-order perturbation theory to obtain the new ground-state energies for these two cases.

Answers (partial)

(a) $\varphi_S(x_1, x_2) = \dfrac{2}{L} \cos\left(\dfrac{\pi x_1}{L}\right) \cos\left(\dfrac{\pi x_2}{L}\right)$

(b) $\varphi_A(x_1, x_2) = \dfrac{\sqrt{2}}{L}\left[\cos\left(\dfrac{\pi x_1}{L}\right)\sin\left(\dfrac{2\pi x_2}{L}\right) - \cos\left(\dfrac{\pi x_2}{L}\right)\sin\left(\dfrac{2\pi x_1}{L}\right)\right]$

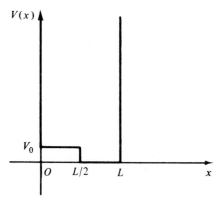

FIGURE 13.2   Potential configuration for Problem 13.7.

13.7   A particle of mass m is in an asymmetrical one-dimensional box, depicted in Fig. 13.2.

(a)   Use first-order perturbation theory to calculate the eigenenergies of the particle.

(b)   What are the first-order corrected wavefunctions?

(c)   If the particle is an electron, how do the frequencies emitted by the perturbed systems compare with those of the unperturbed system?

(d)   What smallness assumption is appropriate to V_0?

Answers

(a) $E_n = E_n^{(0)} + E_n^{(1)} = E_1 + \dfrac{V_0}{2}$

(b) $\varphi_n = \sqrt{\dfrac{2}{L}} \sin\left(\dfrac{n\pi x}{L}\right)\left[1 + \dfrac{V_0}{E_1} \sum_{l \neq n} \dfrac{l^2 n^2}{n^2 - l^2} \Lambda_{ln}\right]$

$\Lambda_{ln} \equiv \dfrac{1}{\pi}\left[\dfrac{\sin(n-l)\pi/2}{n-l} - \dfrac{\sin(n+l)\pi/2}{n+l}\right]$

(c)   They are the same.

(d) $V_0 \ll E_1$

13.8   Show that the matrix elements of the perturbation Hamiltonian $\hat{H}'$ obey the equality

$$\sum_m |H'_{nm}|^2 = (H'^2)_{nn}$$

Answer

$$\sum_m |H'_{nm}|^2 = \sum_m \langle n|\hat{H}'|m\rangle\langle m|\hat{H}'|n\rangle$$

$$= \langle n|\hat{H}'\hat{I}\hat{H}'|n\rangle = \langle n|\hat{H}'^2|n\rangle = (H'^2)_{nn}$$

Here we have recalled the relation

$$\hat{I} = \sum_m |m\rangle\langle m|$$

13.9   A hydrogen atom in its ground state is in a constant, uniform electric field that points in the z direction. The electric field polarizes the atom. Show that there is no change in the ground-state energy of the atom to terms of first order in the electric field. The interaction energy is

$$H' = e\mathcal{E}z = e\mathcal{E}r\cos\theta$$

13.10   The conditions are those of Problem 13.9. In calculating second-order corrections to the ground state, one must know the values of the matrix elements $\langle 100|H'|nlm\rangle$, where $|nlm\rangle$ denotes a hydrogen eigenstate. Show that these matrix elements vanish if $l \neq 1$.

13.11   Consider again, as in Problem 13.9, that a hydrogen atom is in a constant, uniform electric field $\mathcal{E}$ that points in the z direction. If α is the *polarizability* of hydrogen, then the change in the ground-state energy is $\frac{1}{2}\alpha\mathcal{E}^2$, so we may write

$$\tfrac{1}{2}\alpha\mathcal{E}^2 = \sum_{n>1} \frac{|H'_{1n}|^2}{E_n^{(0)} - E_1^{(0)}}, \qquad H'_{1n} \equiv \langle 100|H'|n10\rangle$$

(a)   Use the result of Problem 13.8 to show that

$$\tfrac{1}{2}\alpha\mathcal{E}^2 < \frac{(H'^2)_{11}}{E_2^{(0)} - E_1^{(0)}}$$

(b)   What maximum value for the polarizability of hydrogen does this inequality imply? How does this value compare with the correct value of α?

Answers

(a)
$$[E_n^{(0)} - E_1^{(0)}]_{\min} = E_2^{(0)} - E_1^{(0)}$$

so

$$\tfrac{1}{2}\alpha\mathcal{E}^2 < \sum_{n>1} \frac{|H'_{1n}|^2}{E_2^{(0)} - E_1^{(0)}} = \frac{1}{E_2^{(0)} - E_1^{(0)}} \sum_{n>1} |H'_{1n}|^2$$

Using the results of Problem 13.8, we obtain

$$\sum_{n=1}^{\infty} |H'_{1n}|^2 = |H'_{11}|^2 + \sum_{n>1}^{\infty} |H'_{1n}|^2 = (H'^2)_{11}$$

so that

$$\sum_{n>1} |H'_{1n}|^2 = (H'^2)_{11} - |H'_{11}|^2 < (H'^2)_{11}$$

hence

$$\tfrac{1}{2}\alpha\mathscr{E}^2 < \frac{1}{E_2^{(0)} - E_1^{(0)}} \sum_{n>1} |H'_{1n}|^2 < \frac{(H'^2)_{11}}{E_2^{(0)} - E_1^{(0)}}$$

(b)
$$\tfrac{1}{2}\alpha\mathscr{E}^2_{max} = \frac{4}{3\mathbb{R}} (H'^2)_{11}$$

$$= \frac{4}{3\mathbb{R}} e^2\mathscr{E}^2 \iiint |\varphi^*_{100}|^2 r^4 \cos^2\theta \, d\cos\theta \, d\phi \, dr$$

$$= \frac{8}{3} a_0{}^3\mathscr{E}^2$$

which gives

$$\alpha_{max} = \tfrac{16}{3}a_0{}^3$$

The more correct value is $\alpha = 9a_0{}^3/2$.

13.2   TIME-INDEPENDENT, DEGENERATE PERTURBATION THEORY

Again we consider a system whose Hamiltonian has the form

$$\hat{H} = \hat{H}_0 + \hat{H}'$$

where $\hat{H}'$ is a small perturbation about the unperturbed Hamiltonian, $\hat{H}_0$. In the present case, however, $\hat{H}_0$ has degenerate eigenstates. We have found previously (see Section 8.5) that degeneracy in quantum mechanics stems from symmetries inherent to the system at hand. Any distortion of such symmetry should therefore tend to remove the related degeneracy.

Suppose, for example, that the ground state of $\hat{H}_0$ is q-fold-degenerate. If the symmetry producing this degeneracy is destroyed by the perturbation $\hat{H}'$, the ground state $E_n^{(0)}$ separates into q distinct levels (Fig. 13.3). The primary aim of degenerate perturbation theory is to calculate these new energies. Suppose that we proceed as in the nondegenerate case described in the previous section and expand the first-order wavefunctions of $\hat{H}'$ in the eigenstates of $\hat{H}_0$ (13.6).

$$\varphi_n^{(1)} = \sum_i c_{ni} \varphi_i^{(0)}$$

The formula that emerges for the coefficients $\{c_{ni}\}$ is given by (13.8).

$$c_{ni} = \frac{H'_{in}}{E_n^{(0)} - E_i^{(0)}}$$

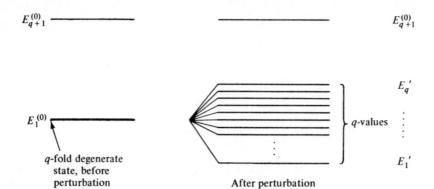

FIGURE 13.3 Perturbation causes a removal of degeneracy.

If $E_1^{(0)}$ is q-fold-degenerate, then

$$E_1^{(0)} = E_2^{(0)} = \cdots = E_q^{(0)}$$

and c_{ni} is infinite for $n, i \le q$. This situation is remedied by constructing a new set of basis functions from the set $\{\varphi_n^{(0)}\}$ which diagonalize the submatrix, H'_{in} (for $n, i \le q$). With the off-diagonal elements of H'_{in} vanishing, the corresponding singular c_{in} coefficients also vanish and we may proceed as in the nondegenerate case.

Diagonalization of the Submatrix

Thus the primary aim in degenerate perturbation theory is to diagonalize this submatrix of H'_{in}. As it turns out, the diagonal elements so constructed are the incremental energies, which when added to $E_1^{(0)}$ separate the q energies contained in the ground state.

Let the q functions that diagonalize H'_{in} ($i, n \le q$) be labeled $\bar{\varphi}_n$.

(13.16)
$$\bar{\varphi}_n = \sum_{i=1}^{q} a_{ni} \varphi_i^{(0)}$$

These linear combinations of the degenerate eigenstates $\{\varphi_i^{(0)}\}$ diagonalize H'_{in}, so

(13.17)
$$\langle \bar{\varphi}_n | \hat{H}' | \bar{\varphi}_p \rangle = H'_{np} \delta_{np} \qquad (n, p \le q)$$

These functions, when joined with the complementary set of non-degenerate states, $\{\varphi_i^{(0)}, i > q\}$, give the basis[1]

(13.18)
$$\mathfrak{B} = \{\bar{\varphi}_1, \bar{\varphi}_2, \ldots, \bar{\varphi}_q, \varphi_{q+1}^{(0)}, \varphi_{q+2}^{(0)}, \ldots\}$$

---

[1] The sequence (13.16) spans the same subspace of Hilbert space spanned by the degenerate states $\{\varphi_n^{(0)}\}$, $n \le q$. Thus the basis (13.18) spans the same Hilbert space spanned by the basis $\{\varphi_i^{(0)}\}$, $i \ge 1$.

The matrix of $\hat{H}'$ calculated in this basis appears as

(13.19)
$$\hat{H}' = \begin{pmatrix} H'_{11} & & & \vdots & H'_{1,q+1} & \cdots \\ & H'_{22} & \mathbf{0} & \vdots & & \\ \mathbf{0} & & \ddots & \vdots & & \\ & & & H'_{qq} & \vdots & \\ \hdashline H'_{q+1,1} & & & & & \\ \vdots & & & & & \end{pmatrix}$$

First-Order Energies

We will now show that the diagonal elements of the $q \times q$ submatrix of $\hat{H}'$ are the first-order energy corrections E_n' to $E_n^{(0)}$, $n \le q$. That is,

(13.20)
$$E_n' = \langle \bar{\varphi}_n | \hat{H}' | \bar{\varphi}_n \rangle = H'_{nn} \qquad (n \le q)$$

If these diagonal elements are mutually distinct, then the q-fold degeneracy of $\hat{H}_0$ is removed by the perturbation $\hat{H}'$. To establish the equality (13.20), we proceed as follows.

The Schrödinger equation for the total Hamiltonian appears as

$$\hat{H}\varphi_n = (\hat{H}_0 + \hat{H}')\varphi_n = E_n \varphi_n$$

The ground state of $\hat{H}_0$ is q-fold degenerate. Substituting

(13.21)
$$\left. \begin{array}{l} \varphi_n = \bar{\varphi}_n \\ E_n = E_n^{(0)} + E_n' \end{array} \right\} (n \le q)$$

into the Schrödinger equation gives

(13.22)
$$\hat{H}'\bar{\varphi}_n = E_n'\bar{\varphi}_n \qquad (n \le q)$$

Here we have recalled that $\bar{\varphi}_n$, being a linear combination of degenerate states, is itself a degenerate state (corresponding to the eigenvalue $E_1^{(0)}$). With the elements of $\{\bar{\varphi}_n\}$ (as well as those of $\{\varphi_n^{(0)}\}$) taken to comprise an orthogonal sequence,[1] one is able to identify (13.17) as being the matrix counterpart of the operator equation (13.22). This is so, provided that we set

$$E_n' = H'_{nn} \qquad (n \le q)$$

which again is the relation (13.20). This equality establishes the fact that the diagonal elements of the submatrix H'_{np} are the first-order corrections to the total Hamiltonian $\hat{H}$ (for $n \le q$).

---

[1] One may always construct an orthogonal set of functions from a given sequence of degenerate functions through the so-called Schmidt orthogonalization procedure. See Problem 13.53.

Let us now construct the new basis functions $\{\bar{\varphi}_n\}$ which diagonalize the said submatrix of $\hat{H}'$. These are given in terms of the a_{ni} coefficients in (13.16), which make $\{\bar{\varphi}_n\}$ obey the eigenvalue equation (13.22). Substituting the former into the latter gives

$$\hat{H}' \sum_{i=1}^{q} a_{ni} |\varphi_i^{(0)}\rangle = E_n' \sum_{i=1}^{q} a_{ni} |\varphi_i^{(0)}\rangle$$

Multiplying from the left with $\langle \varphi_p^{(0)}|$ gives

$$\sum_i a_{ni} H'_{pi} = E_n' \sum_i a_{ni} \delta_{pi} = E_n' a_{np} \qquad \text{(for fixed } n, p \leq q)$$

This equation may be rewritten as

(13.23)
$$\sum_{i=1}^{q} (H'_{pi} - E_n' \delta_{pi}) a_{ni} = 0 \qquad (n, p \leq q)$$

The coefficients $\{a_{ni}\}$ for a fixed value of n comprise the column vector representation of $\bar{\varphi}_n$ in the subbasis $\{\varphi_l^{(0)}, l \leq q\}$. Similarly,

$$H'_{pi} = \langle \varphi_p^{(0)} | \hat{H}' | \varphi_i^{(0)} \rangle$$

are the matrix elements of $\hat{H}'$ in this same basis.

The Secular Equation

For each value of n and p, (13.23) is one equation for E_n' and the q components $\{a_{ni}\}$. There are q such equations corresponding to the q values of p. For $n = 1$, for example, these equations appear as

(13.24)
$$\begin{vmatrix} H'_{11} - E_1' & H'_{12} & H'_{13} & \cdots & H'_{1q} \\ H'_{21} & H'_{22} - E_1' & H'_{23} & \cdots & H'_{2q} \\ \vdots & \cdot & \cdot & \cdot & \vdots \\ H'_{q1} & \cdot & \cdot & \cdot & \cdot \end{vmatrix} \begin{pmatrix} a_{11} \\ a_{12} \\ \vdots \\ a_{1q} \end{pmatrix} = 0$$

This is the matrix equivalent of (13.22) in the basis $\{\varphi_n^{(0)}, n \leq q\}$. Setting $n = 2$ in (13.23) generates (13.24), with the modifications that E_1' is replaced with E_2' and the column vector $\{a_{1i}\}$ is replaced by $\{a_{2i}\}$. As n runs from 1 to q, one obtains q such equations. The condition that there be a nontrivial solution $\{a_{ni}\}$ for any one of these q matrix equations is that the determinant of the coefficient matrix vanish, which gives the *secular equation*

(13.25)
$$\det |H'_{pi} - E_n' \delta_{pi}| = 0$$

This equation may be rewritten in a purely operational form,

$$\det |\hat{H}' - E_n' \hat{I}| = 0$$

The identity operator is $\hat{I}$ (or, equivalently, the $q \times q$ unit matrix). The q roots of the algebraic equation (13.25) are the eigenvalues of (13.22). They are the diagonal elements of the submatrix of $\hat{H}'$ depicted in (13.19). Substituting any value of E' so obtained, say E'_1, back into (13.24) permits one to solve for the coefficients $\{a_{1i}\}$. In similar manner, E'_2 permits calculation of $\{a_{2i}\}$, and so on. These coefficients, in turn, give the new basis functions $\{\bar{\varphi}_n\}$ in (13.18).

Using this new basis (13.18), the ambiguities due to the degeneracy of $\hat{H}_0$ are removed[1] and one may proceed with the analysis developed in the previous section for the nondegenerate case. For example, (13.2) appears as

$$\varphi_n = \bar{\varphi}_n \quad + \lambda \bar{\varphi}_n^{(1)} + \lambda^2 \bar{\varphi}_n^{(2)} + \cdots \quad n \le q$$

$$\varphi_n = \varphi_n^{(0)} + \lambda \varphi_n^{(1)} + \lambda^2 \varphi_n^{(2)} + \cdots \quad n > q$$

$$E_n = E_n^{(0)} + \lambda E_n'^{(1)} + \lambda^2 E_n'^{(2)} + \cdots \quad n \le q \quad (E_1^{(0)} = \cdots = E_q^{(0)})$$

$$E_n = E_n^{(0)} + \lambda E_n^{(1)} + \lambda^2 E_n^{(2)} + \cdots \quad n > q$$

$$E_n' = \langle \bar{\varphi}_n | \hat{H}' | \bar{\varphi}_n \rangle \quad n \le q$$

$$E_n^{(1)} = \langle \varphi_n^{(0)} | \hat{H}' | \varphi_n^{(0)} \rangle \quad n > q$$

An outline of this analysis is shown in Fig. 13.4. An important feature of degenerate perturbation theory is that solution of the matrix equation (13.24) gives (1) first-order corrections to the energy; and (2) corrected wavefunctions which, together with the nondegenerate states, serve as a proper basis for higher-order calculations.

Two-Dimensional Harmonic Oscillator

In this section we are primarily concerned with the degeneracy-removing property of the perturbation $\hat{H}'$. As a simple example of these procedures, consider the case of the two-dimensional harmonic oscillator whose Hamiltonian is

$$\hat{H}_0 = \frac{p_x^2 + p_y^2}{2m} + \frac{K}{2}(x^2 + y^2)$$

or, equivalently,

$$\hat{H}_0 = \hbar \omega_0 (\hat{a}^\dagger \hat{a} + \hat{b}^\dagger \hat{b} + 1)$$

where

$$x = \frac{1}{\sqrt{2}\beta}(a + a^\dagger) \qquad y = \frac{1}{\sqrt{2}\beta}(b + b^\dagger) \qquad \beta^2 = \frac{m\omega_0}{\hbar}$$

---

[1] It may be that first-order calculation does not remove the degeneracies of H_0. For example, this occurs if the off-diagonal elements of $\hat{H}'$ are zero. For such cases it becomes necessary to include higher-order terms to remove the degeneracy. For a discussion and problems relating to such *second-order degenerate perturbation theory*, see L. I. Schiff, *Quantum Mechanics*, 3rd ed., McGraw-Hill, New York, 1968.

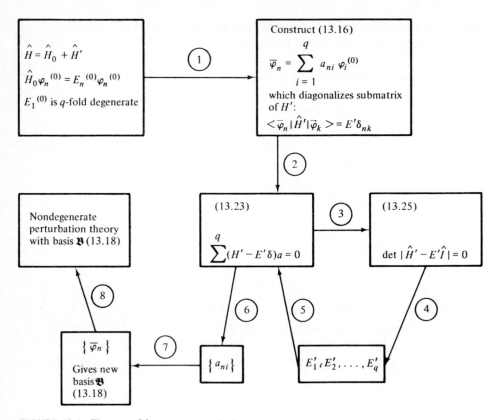

FIGURE 13.4 Elements of degenerate perturbation theory.

The eigenstates of $\hat{H}_0$ are the product forms (8.111)

$$\varphi_{np} \equiv \varphi_n(x)\varphi_p(y)$$

which we will label $|np\rangle$. The corresponding eigenenergy

$$E_{np} = \hbar\omega_0(n + p + 1)$$

is $(n + p + 1)$-fold degenerate. It follows that the energy

$$E_{10} = E_{01} = 2\hbar\omega_0$$

is two-fold degenerate, with corresponding eigenstates $|10\rangle$ and $|01\rangle$.

Let us apply the analysis developed above to determine how this energy separates due to the perturbing potential

$$H' = K'xy$$

Furthermore, we wish to find the two new wavefunctions that diagonalize $\hat{H}'$. These are given by the linear combinations (13.16)

$$\bar{\varphi}_1 = a\varphi_{10} + b\varphi_{01}$$
$$\bar{\varphi}_2 = a'\varphi_{10} + b'\varphi_{01}$$

The submatrix of $\hat{H}'$ in the basis $\{\varphi_{10}, \varphi_{01}\}$ appears as

$$\hat{H}' = K'\begin{pmatrix} \langle 10|xy|10\rangle & \langle 10|xy|01\rangle \\ \langle 01|xy|10\rangle & \langle 01|xy|01\rangle \end{pmatrix} = \mathbb{E}\begin{pmatrix} 0 & 1 \\ 1 & 0 \end{pmatrix}$$

$$\mathbb{E} = \frac{K'}{2\beta^2}$$

Consider, for example, the calculation of the (1, 2) elements of H'.

$$\langle 10|\hat{x}\hat{y}|01\rangle = \frac{1}{2\beta^2}\langle 10|\hat{a}\hat{b} + \hat{a}^\dagger\hat{b} + \hat{a}\hat{b}^\dagger + \hat{a}^\dagger\hat{b}^\dagger|01\rangle$$

$$= \frac{1}{2\beta^2}\langle 10|\hat{a}^\dagger\hat{b}|01\rangle = \frac{1}{2\beta^2}$$

With these values of the matrix elements of $\hat{H}'$, we are prepared to solve (13.25) for the incremental energies E'. This equation appears as

$$\begin{vmatrix} -E' & \mathbb{E} \\ \mathbb{E} & -E' \end{vmatrix} = 0$$

which has the solutions

$$E' = \pm\mathbb{E}$$

Thus we find that the perturbation separates the first excited state by the amount $2\mathbb{E}$.

(13.26)
$$E_{10} \left\langle \begin{array}{l} E_+ = E_{10} + \mathbb{E} \\ \\ E_- = E_{10} - \mathbb{E} \end{array} \right.$$

The corresponding new wavefunctions are obtained by substituting these values into the matrix equation (13.23), which for the present case takes the form

$$\begin{pmatrix} -E' & \mathbb{E} \\ \mathbb{E} & -E' \end{pmatrix}\begin{pmatrix} a \\ b \end{pmatrix} = 0$$

The two values of E' given above serve to determine two column vectors, (a, b) and (a', b'). The value $E' = \mathbb{E}$ gives $a = b$, while $E' = -\mathbb{E}$ gives $a' = -b'$. Thus we obtain for the new wavefunctions, $\bar{\varphi}_1$ and $\bar{\varphi}_2$, the values

$$E' = +\mathbb{E} \;\rightarrow\; \bar{\varphi}_1 = \frac{1}{\sqrt{2}}(\varphi_{10} + \varphi_{01})$$

(13.27)

$$E' = -\mathbb{E} \;\rightarrow\; \bar{\varphi}_2 = \frac{1}{\sqrt{2}}(\varphi_{10} - \varphi_{01})$$

These new wavefunctions serve to diagonalize the perturbation Hamiltonian H'.

PROBLEMS

13.12 How does the threefold-degenerate energy

$$E = 3\hbar\omega_0$$

of the two-dimensional harmonic oscillator separate due to the perturbation

$$H' = K'xy?$$

13.13 Consider a particle confined in a two-dimensional square well with faces at $x = 0, L$; $y = 0, L$ (see Section 8.5). The doubly degenerate eigenstates appear as

$$\varphi_{np}(x, y) = \frac{2}{L}\sin\left(\frac{n\pi x}{L}\right)\sin\left(\frac{p\pi y}{L}\right)$$

$$E_{np} = E_1(n^2 + p^2)$$

What do these energies become under the perturbation

$$H' = 10^{-3}E_1 \sin\left(\frac{\pi x}{L}\right)?$$

13.14 The eigenstates of a rotating dumbbell, with moment of inertia I,

$$E_l = \frac{\hbar^2 l(l + 1)}{2I}$$

are $(2l + 1)$-fold degenerate. In the event that the dumbbell is equally and oppositely charged at its ends, it becomes a dipole. The interaction energy between such a dipole and a constant, uniform electric field $\mathscr{E}$ is

$$\hat{H}' = -\mathbf{d}\cdot\mathscr{E} \qquad (\hat{H} = \hat{H}_0 - \mathbf{d}\cdot\mathscr{E})$$

The dipole moment of the dumbbell is $\mathbf{d}$. Show that to terms of first order, this perturbing potential *does not separate* the degenerate E_l eigenstates.

13.15 Consider again the dipole moment described in Problem 13.14. If both ends are equally charged, the rotating dipole constitutes a magnetic dipole. If the dipole has angular momentum **L**, the corresponding magnetic dipole moment is

$$\boldsymbol{\mu} = \frac{e}{2mc}\,\mathbf{L}$$

where e is the net charge of the dipole. The interaction energy between this magnetic dipole and a constant, uniform magnetic field $\mathscr{B}$ is

$$\hat{H}' = -\hat{\boldsymbol{\mu}}\cdot\mathscr{B} = -\frac{e}{2mc}\,\hat{\mathbf{L}}\cdot\mathscr{B} \qquad (\hat{H} = \hat{H}_0 - \hat{\boldsymbol{\mu}}\cdot\mathscr{B})$$

(a)   If $\mathscr{B}$ points in the z direction, show that $\hat{H}'$ separates the $(2l + 1)$-fold degenerate E_l energies of the rotating dipole.

(b)   Apply these results to one-electron atoms to find the splitting of the P states. (Neglect spin-orbit coupling.) (*Note*: This phenomenon is an example of the *Zeeman effect* discussed previously in Problems 12.15 et seq.)

13.3   THE STARK EFFECT

In Problem 13.9 we found that to within a first-order calculation, an electric field does not remove m_l degeneracy of states of definite orbital number l. However, as we shall now see, a similar calculation reveals that such a field will induce a partial separation of the n^2 degeneracy of eigenenergies related to one-electron atoms. This effect was first noticed in 1913 by Stark. He observed the splitting of the Balmer lines in a field of 100,000 V/cm. (The more readily observed Zeeman effect was first observed in 1897.)

The Hamiltonian of a one-electron atom in a constant, uniform electric field $\mathscr{E}$ which points in the z direction, neglecting spin, is

$$\hat{H} = \frac{\hat{p}_r^{\,2}}{2m} + \frac{\hat{L}^2}{2mr^2} - e\mathscr{E}z$$

$$= \hat{H}_0 + \hat{H}'$$

$$\hat{H}' = -e\mathscr{E}z = -e\mathscr{E}r\cos\theta$$

The eigenstates of the unperturbed Hamiltonian are n^2-fold degenerate. Let us consider how the perturbing electric field removes this degeneracy. Specifically, let us consider the fourfold degenerate $n = 2$ states. The related degenerate wavefunctions are, in the $|nlm\rangle$ notation,

$$|200\rangle, |211\rangle, |210\rangle, |21\text{-}1\rangle$$

To calculate the incremental changes in the energy E_2, we must solve the determinantal equation

$$0 = \begin{vmatrix} \langle 200|\hat{H}'|200\rangle - E' & \langle 200|\hat{H}'|211\rangle & \langle 200|\hat{H}'|210\rangle & \langle 200|\hat{H}'|21\text{-}1\rangle \\ \langle 211|\hat{H}'|200\rangle & \langle 211|\hat{H}'|211\rangle - E' & \langle 211|\hat{H}'|210\rangle & \langle 211|\hat{H}'|21\text{-}1\rangle \\ \langle 210|\hat{H}'|200\rangle & \langle 210|\hat{H}'|211\rangle & \langle 210|\hat{H}'|210\rangle - E' & \langle 210|\hat{H}'|21\text{-}1\rangle \\ \langle 21\text{-}1|\hat{H}'|200\rangle & \langle 21\text{-}1|\hat{H}'|211\rangle & \langle 21\text{-}1|\hat{H}'|210\rangle & \langle 21\text{-}1|\hat{H}'|21\text{-}1\rangle - E' \end{vmatrix}$$

(13.28)

Only two elements survive integration. All elements with different m_l numbers vanish by orthogonality of the $|nlm_l\rangle$ states. Equivalently, one says, "$\hat{H}'$ does not *connect* states of different m_l." Integration gives[1]

$$\langle 210|\hat{H}'|200\rangle = \langle 200|\hat{H}'|210\rangle$$

$$= -\frac{ea_0 \mathscr{E}}{32\pi} \int_0^\infty \rho^4 (2 - \rho)e^{-\rho}\, d\rho \int_{-1}^1 d\cos\theta \cos^2\theta \int_0^{2\pi} d\phi$$

$$= \frac{e\mathscr{E}h^2}{mZe} = -|e|3\mathscr{E}a_0 \equiv -\mathbb{E}$$

With these values inserted into the determinant above, (13.25) becomes

$$\begin{vmatrix} -E' & 0 & -\mathbb{E} & 0 \\ 0 & -E' & 0 & 0 \\ -\mathbb{E} & 0 & -E' & 0 \\ 0 & 0 & 0 & -E' \end{vmatrix} = 0$$

which has the four roots

(13.29)
$$E' = 0, 0, +\mathbb{E}, -\mathbb{E}$$
$$\mathbb{E} = 3|e|\mathscr{E}a_0$$

Thus we find that to terms of lowest order in the electric field $\mathscr{E}$, the degenerate $n = 2$ state separates into three states:

$$E_2 \begin{cases} E_2^{(0)} + \mathbb{E} \\ E_2^{(0)} \\ E_2^{(0)} - \mathbb{E} \end{cases}$$

To calculate the new $n = 2$ wavefunctions

$$\varphi = a|200\rangle + b|211\rangle + c|210\rangle + d|21\text{-}1\rangle$$

[1] The nondimensional radius ρ is defined in Table 10.3. See also Table 10.5.

we substitute the values (13.29) into the matrix equation

(13.30)
$$\begin{pmatrix} -E' & 0 & -\mathbb{E} & 0 \\ 0 & -E' & 0 & 0 \\ -\mathbb{E} & 0 & -E' & 0 \\ 0 & 0 & 0 & -E' \end{pmatrix} \begin{pmatrix} a \\ b \\ c \\ d \end{pmatrix} = 0$$

There results

(13.31)
$$E_2^+ = E_2^{(0)} + \mathbb{E} \quad \rightarrow \quad \varphi_+ = \frac{1}{\sqrt{2}}(|200\rangle - |210\rangle)$$

$$E_2^- = E_2^{(0)} - \mathbb{E} \quad \rightarrow \quad \varphi_- = \frac{1}{\sqrt{2}}(|200\rangle + |210\rangle)$$

$$E_2^0 = E_2^{(0)} \begin{cases} \varphi = |211\rangle \\ \varphi = |21\text{-}1\rangle \end{cases}$$

The perturbation mixes the $m = 0$ states, while the $m = 1, -1$ states are left degenerate. The values $\pm\mathbb{E}$ represent the average values of the interaction $\hat{H}'$ in the respective states, $\varphi_\pm$.

PROBLEMS

13.16 Show that the $n = 2$ matrix of the interaction Hamiltonian $\hat{H}'$, of the Stark effect, is diagonal in the basis (13.31).

13.17 What is the dipole moment of the hydrogen atom in the $\varphi_\pm$ states (13.31)?

Answer

The interaction energy between an electric dipole **d** and the electric field $\mathscr{E}$ is

$$H' = -\mathbf{d} \cdot \mathscr{E}$$

In the $\varphi_\pm$ states, the average value of H' is $E = \pm 3|e|a_0\,\mathscr{E}$. We may infer from this result that the magnitude of the dipole moment in the $\varphi_\pm$ states is $3|e|a_0$. The directions of these moments are parallel or antiparallel to the z axis (i.e., the direction of $\mathscr{E}$).

13.18 What is the charge density $q(r, \theta)$ of the hydrogen atom associated with the state φ_- (13.31)?

Answer

$$q(r, \theta) = e|\varphi_-|^2 = \frac{e}{16\pi a_0^3}\left[1 - \frac{r}{a_0}\sin^2\left(\frac{\theta}{2}\right)\right]^2 e^{-r/a_0}$$

13.19 Of the two states $\varphi_\pm$ in (13.31), φ_- is said to be more *stable* than φ_+. Why? Discuss your answer in light of the interaction energy, $-\mathbf{d} \cdot \mathscr{E}$.

13.4   THE NEARLY FREE ELECTRON MODEL

In this section we return to the problem of an electron in a periodic potential $V(x)$, discussed in Sections 8.2 and 12.9. Wavefunctions are in the Bloch form

$$\varphi(x) = u(x)e^{ikx}$$

where the periodic function $u(x)$ has the same period a as $V(x)$. Eigenenergies are functions of the crystal momentum wavenumber k. For a lattice of length L, the nearly continuous wavenumber k has the discrete values [see (8.42)]

$$k_j = \frac{j2\pi}{L}$$

We recall (see Problem 8.13) that in the high-energy domain ($E \gg V$), the energy spectrum reduces to the free-particle values $\hbar^2 k^2/2m$, or, equivalently,

$$E_{k_j} = \frac{\hbar^2 k_j^2}{2m} = \frac{j^2 h^2}{2mL^2}$$

which is doubly degenerate: $E_{k_j} = E_{-k_j}$.

We now wish to obtain an expression for the energy gap δE_n at the nth band edge. A band edge, we recall, is a break in the energy spectrum which occurs at the k values $k_j a = n\pi$.

In the present analysis the periodic potential is considered a small perturbation to the free-particle Hamiltonian

$$\hat{H}_0 = \frac{\hat{p}^2}{2m}$$

The electron is "nearly free," which is the same as saying that $E \gg V$.

Unperturbed eigenstates are normalized to the sample interval L.

$$\varphi_{k_j}^{(0)} = \frac{1}{\sqrt{L}} \exp\,(ik_j x)$$

With $k_j = j(2\pi/L)$ these functions comprise an orthonormal sequence, as may be seen as follows:

$$(13.32) \qquad \langle k_q | k_j \rangle = \frac{1}{L} \int_{-L/2}^{L/2} \exp[i(k_j - k_q)]\,dx = \frac{\sin\,[(k_j - k_q)L/2]}{(k_j - k_q)L/2}$$

$$= \frac{\sin\,(j - q)\pi}{(j - q)\pi} = \delta_{jq}$$

The Perturbation Potential

The perturbing potential is periodic and may be expanded in the Fourier series (see Problem 13.58).

$$(13.33) \qquad H' = V(x) = \sum_{n=-\infty}^{\infty} V_n \exp\left[i2\pi n\left(\frac{x}{a}\right)\right]$$

The zero energy line in the present analysis is set at the average of $V(x)$. This ensures that the dc component V_0 of V vanishes (see Fig. 13.5).

$$V_0 = \int_{-L/2}^{L/2} V(x)\,dx = 0$$

Application of first-order perturbation theory necessitates calculation of the matrix elements of H'.

$$(13.34) \qquad H'_{qj} = \langle k_q | \sum_n V_n \exp\left[i2\pi n\left(\frac{x}{a}\right)\right] | k_j \rangle$$

$$= \sum_n V_n \left\langle k_q \middle| k_j + \left(\frac{2\pi n}{a}\right) \right\rangle$$

$$= \sum_n V_n \, \delta_{k_q,\, k_j + (2\pi n/a)}$$

Substituting these matrix elements into the first equation of (13.10) gives the first-order corrected wavefunctions

$$(13.35) \qquad \varphi_{k_j} = \varphi_{k_j}^{(0)} + \frac{1}{\sqrt{L}} \sum_{n=-\infty}^{\infty} \frac{V_n \exp\{ix[k_j + (2\pi n/a)]\}}{E_{k_j}^{(0)} - E_{k_j+(2\pi n/a)}^{(0)}}$$

Calculation of the first-order corrected eigenenergies as given by the second equation in (13.10) gives

$$(13.36) \qquad E_{k_j} = E_{k_j}^{(0)} + \langle k_j | V | k_j \rangle = E_{k_j}^{(0)} + V_0$$

$$= E_{k_j}^{(0)} = \frac{\hbar^2 k_j^2}{2m}$$

To first order, the energy remains unperturbed. This free-particle spectrum was found in Chapter 8 to maintain in the domain $E \gg V$. However, in the present analysis there is explicit evidence which indicates that this result is invalid at the band edges of the energy spectrum. Namely, the summation in (13.35) for the first-order corrected wavefunctions becomes singular if the denominator of any term vanishes.

$$(13.37) \qquad E_{k_j}^{(0)} - E_{k_j+(2\pi n/a)}^{(0)} = -\frac{\hbar^2}{2m}\frac{4\pi n}{a^2}(k_j a + n\pi) = 0$$

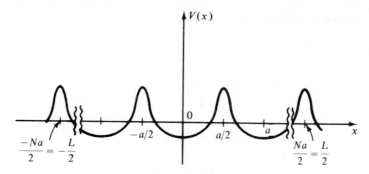

FIGURE 13.5 The zero in potential is chosen so as to eliminate V_0.

$$V_0 = \int_{-L/2}^{L/2} V(x)\, dx = 0$$

This singularity arises from the zeros of (13.37), corresponding to the degeneracy at the band edges, $k_j a + n\pi = 0$. To obtain correct energies at these values of k_j, one must use degenerate perturbation theory. As described in Section 13.2, the first step in this procedure is to construct a new basis $\{\bar{\varphi}\}$ that diagonalizes the relevant 2×2 submatrix of H'. This new subbasis $\{\bar{\varphi}\}$ is constructed from linear combinations of the degenerate portion of the unperturbed basis and we may write

(13.38)
$$\bar{\varphi} = \bar{a}e^{ik_jx} + \bar{b}e^{-ik_jx}$$

Diagonalization of H' in the basis (13.38) yields the matrix equation

(13.39)
$$\begin{pmatrix} H'_{k_jk_j} - E_{k_j}' & H'_{k_j,\,-k_j} \\ H'_{-k_j,\,k_j} & H'_{-k_j,\,-k_j} - E_{k_j}' \end{pmatrix}\begin{pmatrix} \bar{a} \\ \bar{b} \end{pmatrix} = 0$$

$$k_j = -\frac{n\pi}{a}$$

The matrix elements of H' are given by (13.34), from which we find

$$H'_{k_jk_j} = H'_{-k_j,\,-k_j} = V_0 = 0$$

Only the off-diagonal elements survive. Choosing the origin so that $V(x)$ is an even function gives

$$H'_{k_j,\,-k_j} = V_n = V_{-n} = H'_{-k_j,\,k_j}$$

The determinant equation (13.25) then becomes

$$\begin{vmatrix} -E_{k_j}' & V_n \\ V_n & -E_{k_j}' \end{vmatrix} = 0$$

which gives the roots

$$E_{k_j}' = \pm V_n$$

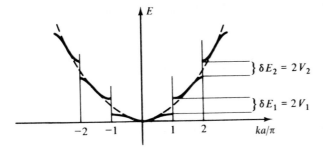

FIGURE 13.6 Energy gaps at the band edges in the nearly free electron model are given by twice the corresponding Fourier coefficient of the periodic potential.

Resubstituting into the matrix equation (13.39) gives two nondegenerate eigenstates, which written together with the first-order corrected energies appear as

$$E_{k_j}{}^{+} = E_{k_j}{}^{(0)} + V_n \qquad \bar{\varphi}_{+} = 2\bar{a} \sin k_j x$$
$$E_{k_j}{}^{-} = E_{k_j}{}^{(0)} - V_n \qquad \bar{\varphi}_{-} = 2\bar{a} \cos k_j x$$

These eigenstates, we recall, are the standing waves at the band edges previously obtained in Section 8.3 relevant to the Kronig–Penney potential. The present calculation affords an estimate (in the domain $E \gg V$) of the width of the energy gap at the band edges.

$$\delta E_n = E_{k_j}{}^{+} - E_{k_j}{}^{-} = 2|V_n| \qquad k_j a = \pm n\pi$$

Thus we find that the nth gap in the energy spectrum $E = E(k)$, at the values $k = \pm n\pi/d$, has width which is twice the nth Fourier coefficient of the potential $V(x)$ (Fig. 13.6).

In the nearly free electron model described above, conduction electrons are weakly bound to atoms of the lattice. This is opposite to the situation in the *tight-binding approximation* described in Problem 8.16 and in the LCAO analysis presented in Section 8.7.

PROBLEMS

13.20 Resketch the $E(k)$ curve shown in Fig. 13.6 for the case that the zeroth Fourier coefficient $V_0 > 0$.

13.21 A periodic potential has Fourier coefficients $V_n = V_1/n^2$. The width of the tenth energy gap is 0.031 eV. What is the value of V_1?

13.22 What are the energy gaps at the band edges for a particle in the periodic delta-function potential

$$V(x) = V_0 a \sum_{q=-\infty}^{\infty} \delta(x - qa)$$

defined over the entire x interval? [*Hint:* Consider the following delta-function representation

$$\delta(y) = \sum_{n=-\infty}^{\infty} \exp{(i2\pi n y)}$$

The right-hand side is periodic with period 1 and represents a delta function at $y = 0, \pm 1, \pm 2, \dots$. The left-hand side is a delta function only at the origin. The domain of validity of the equation may be extended to the whole y axis if one writes

$$\sum_{-\infty}^{\infty} \delta(y - m) = \sum_{-\infty}^{\infty} \exp{(i2\pi n y)}. \Bigg]$$

13.23 (a) Show that the first-order corrected wavefunction φ_k as given by (13.35) may be cast in the form of a Bloch wavefunction $\varphi_k = u(x)e^{ikx}$.

(b) Show that the expression you obtain for $u(x)$ has period a.

13.24 Estimate the energy gaps at the band edges for the Kronig–Penney potential of potential height V_0, well width a, and barrier width b.

Answer

$$\delta E_n \simeq \frac{2V_0}{n\pi} \sin\left(\frac{n\pi a}{a + b}\right)$$

13.25 Estimate the band-gap widths for the potential

$$V(x) = 2V_0 \cos\left(\frac{2\pi x}{a}\right)$$

Answer

Rewriting the cos term in terms of exponentials,

$$V(x) = 2V_0 \cos\left(\frac{2\pi x}{a}\right) = V_0 \left[\exp\left(\frac{i2\pi x}{a}\right) + \exp\left(\frac{-i2\pi x}{a}\right)\right]$$

reveals that this potential has only two nonvanishing Fourier coefficients. First-order perturbation theory implies the existence of the gaps at $k = \pm\pi/a$, of width $2V_0$. Higher-order perturbation theory would uncover energy gaps at subsequent band edges as well. This may be concluded directly by writing the Schrödinger equation for the given potential,[1]

$$\varphi_{xx} + \frac{2m}{\hbar^2}\left[E - 2V \cos\left(\frac{2\pi x}{a}\right)\right]\varphi = 0$$

This is a well-known equation in mathematical physics and is called the *Mathieu equation*. As with most such equations, it stems from writing the "wave equation," $(\nabla^2 + k^2)\varphi = 0$, in a particular orthogonal coordinate frame. For the Mathieu equation these are elliptic cylinder coordinates. Solutions of the equation are called *Mathieu functions* and have been studied in detail. These

---

[1] This case has been studied in detail by P. M. Morse, *Phys. Rev.* **35**, 1310 (1930). Here it is also shown that the energy bands approach the line spectrum of an infinitely deep well as the amplitude V_0 of the periodic potential grows infinitely large.

analyses reveal a sequence of intervals on the E axis in which solutions to the equation are unstable.[1]

13.26   At 0 K a certain semiconductor has its valence band filled and conduction band empty. The potential "seen" by electrons may be approximated by the function

$$V(x) = 2V_0 \cos\left(\frac{2\pi x}{a}\right)$$

$$V_0 = 0.1 \text{ eV}, \qquad a = 0.5 \text{ Å}$$

(a)   Assuming that the valence band is the band of lowest energy, estimate the width of the gap δE between the valence and conduction bands.

(b)   Setting $\delta E = k_B T$, estimate the temperature at which the sample will begin to conduct appreciably. ($k_B = 8.6 \times 10^{-5}$ eV/K.)

Answers

(a) $\delta E \simeq 2V_0 = 0.20$ eV

(b) $T \simeq 2300$ K

13.5   TIME-DEPENDENT PERTURBATION THEORY

Time-dependent perturbation theory addresses the following problem. Initially, the unperturbed system is in an eigenstate of $\hat{H}_0$. Then the perturbation, $\hat{H}'(t)$, is "turned on." What is the probability, after a time t, that transition to another state (of $\hat{H}_0$) occurs?

The total Hamiltonian for these problems is of the form

(13.40) $$\hat{H}(\mathbf{r}, t) = \hat{H}_0(\mathbf{r}) + \lambda \hat{H}'(\mathbf{r}, t)$$

where λ is again a parameter of smallness.

Let the time-dependent eigenstates of $\hat{H}_0$ be written

(13.41) $$\psi_n(\mathbf{r}, t) = \varphi_n(\mathbf{r})e^{-i\omega_n t}$$
$$\hat{H}_0 \varphi_n = E_n^{(0)}\varphi_n \equiv \hbar\omega_n \varphi_n$$

Suppose that at time $t > 0$, the system is in the state

(13.42) $$\psi(\mathbf{r}, t) = \sum_n c_n(t)\psi_n(\mathbf{r}, t)$$

---

[1] That is, series solutions in these domains do not converge. For further discussion of the properties of Mathieu functions, see P. M. Morse and H. Feshbach, *Methods of Theoretical Physics*, Chap. 5, McGraw-Hill, New York, 1953; also, E. T. Whittaker and G. N. Watson, *A Course of Modern Analysis*, Chap. 19, Cambridge University Press, New York, 1952.

Then, by the superposition principle, $|c_n|^2$ is the probability that measurement finds the system in the state ψ_n at the time t. Thus the primary aim of the present discussion is to calculate these coefficients. They are determined in the following manner. The wavefunction $\psi(\mathbf{r}, t)$ is a solution of

(13.43)
$$ih\frac{\partial \psi}{\partial t} = (\hat{H}_0 + \lambda \hat{H}')\psi$$

Substituting the expansion (13.42) into this equation and then operating from the left with $\int dr\, \psi_k^*(\mathbf{r}, t)$ gives

(13.44)
$$ih\frac{dc_k}{dt} = \lambda \sum_n \langle k|H'|n\rangle c_n$$

This is an infinite sequence of coupled equations for the coefficients $\{c_k(t)\}$. In the limit that $\lambda \to 0$, the c_k coefficients are all constant. It is therefore consistent to seek solution in the form

(13.45)
$$c_k(t) = c_k^{(0)} + \lambda c_k^{(1)}(t) + \lambda^2 c_k^{(2)}(t) + \cdots$$

Substituting this series into (13.44) and equating terms of equal powers in λ gives (with a dot denoting time differentiation and H'_{kn} written for the matrix elements of $\hat{H}'$)

$$ih\dot{c}_k^{(0)} = 0$$

$$ih\dot{c}_k^{(1)} = \sum_n H'_{kn} c_n^{(0)}$$

(13.46)
$$ih\dot{c}_k^{(2)} = \sum_n H'_{kn} c_n^{(1)}$$

$$\vdots$$

$$ih\dot{c}_k^{(s+1)} = \sum_n H'_{kn} c_n^{(s)}$$

The lowest-order equations for $c_k^{(0)}$ indicate that these coefficients are all constant in time. They are the initial values of the coefficients $\{c_k(t)\}$.

We now specialize to the problem in which it is known that initially the system is in a definite eigenstate of $\hat{H}_0$, say $\psi_l(\mathbf{r}, t)$. With (13.42) this implies that as $t \to -\infty$,

$$\psi(\mathbf{r}, t) \sim \psi_l(\mathbf{r}, t) = \sum_n \delta_{nl} \psi_n(\mathbf{r}, t)$$

(13.47)
$$c_n^{(0)}(-\infty) = \delta_{nl}$$

Note that we have taken "initially" to denote the time $t = -\infty$. Substituting this value into the second equation in (13.46) gives (dropping the superscripts 0 and 1)

(13.48)
$$ih\dot{c}_k(t) = \sum_n H'_{kn} c_n(-\infty) = H'_{kl}$$

For $n \neq l$, $c_n(-\infty) = 0$, so the first-order solution for $c_k(t)$ is given by

(13.49)
$$c_k(t) = \frac{1}{i\hbar} \int_{-\infty}^{t} H'_{kl}(\mathbf{r}, t') \, dt' \qquad (k \neq l)$$

If the time dependence of $\hat{H}'(\mathbf{r}, t)$ is factorable, then

$$\hat{H}'(\mathbf{r}, t) = \hat{H}'(\mathbf{r})f(t)$$

and the matrix elements of $\hat{H}$ become [with (13.41)]

(13.50)
$$\begin{aligned} H'_{kl}(t) &\equiv \langle \psi_k | H'(\mathbf{r}, t) | \psi_l \rangle = \langle \varphi_k | \mathbb{H}'(\mathbf{r}) | \varphi_l \rangle e^{i\omega_{kl}t} f(t) \\ &= \mathbb{H}'_{kl} e^{i\omega_{kl}t} f(t) \\ \hbar\omega_{kl} &\equiv \hbar(\omega_k - \omega_l) = E_k - E_l \end{aligned}$$

(Note that we have deleted the zero superscripts of E_k and E_l.) This gives the more explicit form of $c_k(t)$,

(13.51)
$$c_k(t) = \frac{\mathbb{H}'_{kl}}{i\hbar} \int_{-\infty}^{t} e^{i\omega_{kl}t'} f(t') \, dt'$$

These coefficients determine the effect of the perturbation on the initial state, ψ_l. The probability that the system has undergone a transition from this state to some other eigenstate of H_0, ψ_k, at the time t, is

(13.52)
$$P_{l \to k} = |c_k|^2 = \left| \frac{\mathbb{H}'_{kl}}{\hbar} \right|^2 \left| \int_{-\infty}^{t} e^{i\omega_{kl}t'} f(t') \, dt' \right|^2$$

Application of these results follows. The transition probability $P_{l \to k}$ is hereafter written P_{lk}.

Matrix elements in time-dependent perturbation theory are conventionally written with the initial state on the right and the final state on the left, as illustrated in the following symbolic form:

$$\langle \text{final state} \mid \text{interaction} \mid \text{initial state} \rangle$$

This convention is employed in the remainder of the text.

PROBLEMS

13.27 A system with discrete eigenstates $\{\varphi_n\}$ and eigenenergies $\{E_n\}$ is exposed to the perturbation

$$\hat{H}' = \hat{H}'(\mathbf{r}) \frac{e^{-t^2/\tau^2}}{\tau\sqrt{\pi}}$$

The perturbation is turned on at $t = -\infty$, when the unperturbed system is in its ground state, ψ_0. What is the probability that at $t = +\infty$ the system suffers a transition to the state ψ_k, $k > 0$?

Answer

To obtain the answer to this problem, we must calculate the time integral in (13.52). Writing ω for ω_{k0}, we have

$$I = \frac{1}{\sqrt{\pi}} \int_{-\infty}^{\infty} e^{i\omega t} e^{-t^2/\tau^2} \, d(t/\tau)$$

$$= \frac{1}{\sqrt{\pi}} \int_{-\infty}^{\infty} e^{i\bar{\omega}\xi - \xi^2} \, d\xi \qquad (\xi \equiv t/\tau, \bar{\omega} \equiv \tau\omega)$$

$$= \frac{1}{\sqrt{\pi}} e^{-\bar{\omega}^2/4} \int_{-\infty}^{\infty} \exp\left[-\left(\xi - \frac{i\bar{\omega}}{2}\right)^2\right] d\xi$$

$$= e^{-\bar{\omega}^2/4} = e^{-(E_0 - E_k)^2\tau^2/4\hbar^2}$$

Substituting this into (13.52) gives the desired result

$$P_{0k} = \left|\frac{H'_{k0}}{\hbar}\right|^2 e^{-(E_0 - E_k)^2\tau^2/2\hbar^2}$$

13.28   Consider that the unperturbed system in Problem 13.27 is a particle of mass m confined to a one-dimensional box of width L. The spatial factor in $H'(x, t)$ is

$$\hat{H}'(x) = \frac{10^{-4}\hat{p}_x^2}{2m}$$

What state does the perturbation leave the system in at $t = +\infty$?

13.6   HARMONIC PERTURBATION

Stimulated Emission

As a first application of the analysis above, we consider a perturbation that is switched on at $t = 0$ and is subsequently monochromatically harmonic in time. The perturbation acts on a system whose Hamiltonian is $\hat{H}_0$. If it is definitely known that the unperturbed system was in one of its own stationary states before the perturbation was applied ($t < 0$), in what state will measurement find the system after the perturbation has been turned on for t seconds? This problem is appropriate, for example, to an atom that interacts with a (weak) electromagnetic field. The explicit form of the perturbation $\hat{H}'$ is (with $\omega > 0$)

(13.53)
$$\hat{H}'(\mathbf{r}, t) = \begin{cases} 0 & t < 0 \\ 2\hat{H}'(\mathbf{r}) \cos \omega t & t \geq 0 \end{cases}$$

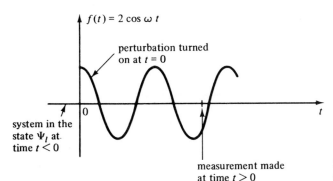

$$H' = \hat{H}'(\mathbf{r}) f(t)$$

FIGURE 13.7 Harmonic perturbation.

(see Fig. 13.7). Substituting this form into (13.51), with $f(t) = 2 \cos \omega t$, gives

$$c_k(t) = \frac{H'_{kl}}{i\hbar} \int_0^t e^{i\omega_{kl}t'}(e^{-i\omega t'} + e^{i\omega t'}) \, dt'$$

$$= -\frac{H'_{kl}}{\hbar} \left[\frac{e^{i(\omega_{kl} - \omega)t} - 1}{\omega_{kl} - \omega} + \frac{e^{i(\omega_{kl} + \omega)t} - 1}{\omega_{kl} + \omega} \right]$$

Employing the relation (see Problem 1.21)

$$e^{i\theta} - 1 = 2ie^{i\theta/2} \sin (\theta/2)$$

permits the equation above to be rewritten

(13.54) $$c_k(t) = -\frac{i2H'_{kl}}{\hbar} \left[\frac{e^{i(\omega_{kl} - \omega)t/2} \sin (\omega_{kl} - \omega)t/2}{\omega_{kl} - \omega} + \frac{e^{i(\omega_{kl} + \omega)t/2} \sin (\omega_{kl} + \omega)t/2}{\omega_{kl} + \omega} \right]$$

The dominant contributions to c_k come from the values $\omega \simeq \pm\omega_{kl}$. At these values, the moduli of the two bracketed terms in c_k, respectively, assume their maximum value, $t/2$. These *resonant* frequencies correspond to the energies

$$\omega \simeq +\omega_{kl} \longrightarrow E_k > E_l$$
$$\omega \simeq -\omega_{kl} \longrightarrow E_l > E_k$$

In the first case, the "final" energy E_k is larger than the "initial" energy E_l. The system absorbs energy and jumps to a higher energy level.

$$E_k = E_l + \hbar\omega$$

The energy absorbed, $\hbar\omega$, is that of a photon in the incident radiation field (Fig. 13.8).

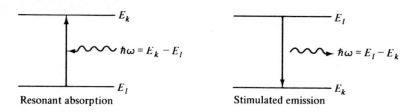

FIGURE 13.8 **Dominant transition processes due to harmonic perturbation.**

For the second case the perturbation induces a decay in energy

$$E_k = E_l - \hbar\omega$$

A photon of energy $\hbar\omega$ is radiated away from the system. This decay process is *stimulated* by a photon of the same frequency in the perturbation field.

Let us consider the case that the incident radiation field excites only higher energies, so that $\omega_{kl} > 0$. Under such conditions, the first term in (13.54) dominates and the probability that the perturbing field causes a transition to the kth state becomes

(13.55)
$$P_{lk} = |c_k|^2 = \frac{4|H'_{kl}|^2}{\hbar^2(\omega_{kl} - \omega)^2} \sin^2\left[\tfrac{1}{2}(\omega_{kl} - \omega)t\right]$$

The frequency dependence of this function is plotted in Fig. 13.9.

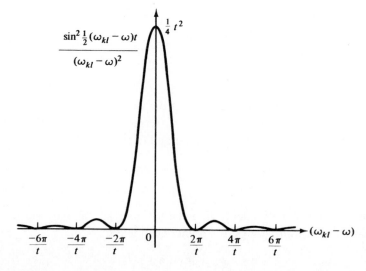

FIGURE 13.9 **Frequency dependence of probability of transition from the lth to the kth state at the time t due to harmonic perturbation of frequency ω.**

Energy-Time Uncertainty

From this sketch it is evident that states falling in the interval

$$(13.56) \qquad |\hbar\omega_{kl} - \hbar\omega| = |E_k - (E_l + \hbar\omega)| \lesssim \frac{2\pi\hbar}{t} \simeq \Delta E$$

have the greatest probability of being excited, after the perturbing field has acted for t seconds. If this perturbation is applied many times to an ensemble of independent, identical systems, all initially in the state ψ_l, then after a time t the energies excited among the members of the ensemble will lie primarily in the interval ΔE. Thus (13.56) gives the uncertainty in the values of energy observed. The final energies E_k are spread throughout the interval

$$\Delta E \simeq \frac{\hbar}{t}$$

$$(13.57) \qquad E_k \simeq (E_l + \hbar\omega) \pm \Delta E$$

$$E_{\text{final}} \simeq E_{\text{initial}} \pm \Delta E$$

Note that E_{initial} refers to the energy of the incident photon plus that of the system in its initial state. Our perturbation analysis returns the principle of conservation of energy, properly modified by the uncertainty relation, $\Delta E \simeq \hbar/t$.

Long-Time Evolution

The transition probability formula for absorption (13.55) and its counterpart for stimulated emission may be written together as

$$(13.58) \qquad P_{lk} = \frac{4|H'_{kl}|^2}{\hbar^2(\omega_{kl} \mp \omega)^2} \sin^2\left[\tfrac{1}{2}(\omega_{kl} \mp \omega)t\right]$$

where the $\mp$ signs refer to absorption and stimulated emission, respectively (see Fig. 13.8). This expression takes a convenient form in the long-time or, equivalently, high-frequency limit. This form follows from the delta-function representation

$$(13.59) \qquad \delta(\omega) = \frac{2}{\pi} \lim_{t \to \infty} \frac{\sin^2(\omega t/2)}{t\omega^2}$$

so that in the same limit,

$$(13.60) \qquad P_{lk} \to \frac{2\pi t}{\hbar^2} |H'_{kl}|^2 \delta(\omega_{kl} \mp \omega)$$

The corresponding transition probability *rate* appears as

$$(13.61) \qquad w_{lk} = \frac{2\pi}{\hbar^2} |\mathbb{H}'_{kl}|^2 \delta(\omega_{kl} \mp \omega)$$

In this or the formula above, the delta function expresses the fact that in the long-time limit, the Fourier transform of a monochromatic perturbation becomes sharply peaked about the frequency of perturbation. The system "sees" only a single frequency. Since the uncertainty in energy $\hbar/t$ vanishes in this limit, the argument of the delta function is also an expression of conservation of energy.

Short-Time Approximation

If a harmonic perturbation is applied to a system for a short-time interval such that $(\omega_{kl} - \omega)t \ll 1$, (13.55) may be expanded to yield

$$(13.62) \qquad P_{lk} = \frac{t^2 |\mathbb{H}'_{kl}|^2}{\hbar^2}$$

The related transition probability rate is

$$(13.63) \qquad w_{lk} = \frac{t |\mathbb{H}'_{kl}|^2}{\hbar^2}$$

At early times, the rate at which transitions to the kth state occur grows linearly with time.

The Golden Rule

In many problems of practical interest, the final excited states lie in a band of energies (Fig. 13.10). Such is the case, for example, for ionization or free-particle scattering states. Such states comprise a continuum. If the density of final states is $g(E_k)$, then

$$dN = g(E_k)\, dE_k$$

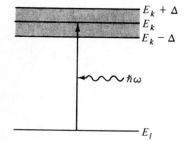

FIGURE 13.10 Resonant absorption to a band of energies. If the density of states about E_k is $g(E_k)$, then the probability rate of transition to the band is

$$\overline{w}_{lk} = \frac{2\pi}{\hbar} g(E_k) |\mathbb{H}'_{kl}|^2$$

is the number of energy states in the interval E_k to $E_k + dE_k$. The probability that a transition occurs to a state in a band of width 2Δ centered at E_k is

$$\bar{P}_{lk} = \int_{E_k - \Delta}^{E_k + \Delta} P_{lk} g(E_k') \, dE_k'$$

Inserting (13.55) gives

$$\bar{P}_{lk} = \int_{E_k - \Delta}^{E_k + \Delta} dE_k' g(E_k') \left| \frac{\mathsf{H}_{kl}'}{\hbar} \right|^2 \frac{\sin^2 \beta}{\beta^2 / t^2}$$

$$2\hbar\beta \equiv \hbar(\omega_{kl} - \omega)t = (E_k' - E_l - \hbar\omega)t$$

For fixed E_l, t and ω

$$dE_k' = \frac{2\hbar d\beta}{t}$$

and

$$\bar{P}_{lk} = \frac{2t}{\hbar} \int_{-\delta}^{+\delta} g(E_k') |\mathsf{H}_{kl}'|^2 \frac{\sin^2 \beta}{\beta^2} \, d\beta$$

where 2δ is the corresponding spread in β values. Owing to the rapid decay of the $\sin^2 \beta / \beta^2$ function (see Fig. 13.9), only a slight error is introduced in the expression above if we replace the interval $(-\delta, +\delta)$ by $(-\infty, +\infty)$. Furthermore, if we assume that g and H_{kl}' are slowly varying functions of E_k, they may be taken outside the integral. There results

$$\bar{P}_{lk} = \frac{2t}{\hbar} g(E_k) |\mathsf{H}_{kl}'|^2 \int_{-\infty}^{\infty} \frac{\sin^2 \beta}{\beta^2} \, d\beta$$

$$= t \left[\frac{2\pi}{\hbar} g(E_k) |\mathsf{H}_{kl}'|^2 \right]$$

The related transition probability rate is

(13.64) $$\bar{w}_{lk} = \frac{2\pi}{\hbar} g(E_k) |\mathsf{H}_{kl}'|^2$$

This formula was found to have such widespread application that Fermi[1] dubbed it "Golden Rule No. 2." It will be applied in Section 13.9 in study of atom–radiation interaction and in Chapter 14 in study of the Born approximation in the theory of scattering.

[1] E. Fermi, *Nuclear Physics*, University of Chicago Press, Chicago, 1950.

PROBLEMS

13.29 Show that if the perturbation field

$$\hat{H}' = 2\hat{H}'(\mathbf{r}) \cos \omega t$$

acts on a system initially in the lth state for a very long time, and $\omega \simeq \omega_{kl}$, then the only state that will be excited is the kth state. Interpret this result in terms of the Fourier decomposition (in time) of $\hat{H}'$.

13.30 The expression in the text obtained for $\bar{P}_{lk}$, the probability of transition to a band of states, is seen to be independent of the frequency ω of the perturbing field. In what approximation is this result valid?

13.31 What does the transition probability P_{lk} (13.55) become if the perturbing field is precisely "on resonance"; that is, if $\omega = \omega_{kl}$?

13.32 A polarized beam of current, J cm^{-2}/s, contains electrons with spins aligned with a steady magnetic field of magnitude $\mathscr{B}$ which points in the z direction. The beam propagates in the x direction. The wavefunction for an electron in the beam is

$$\psi = \frac{1}{(2\pi)^{1/2}} e^{i(kx - \omega t)} \binom{0}{1}$$

A monochromatic electromagnetic field of frequency

$$\hbar\omega = 2\mu_b \mathscr{B}$$

extends over a length of beam path, L cm long. A Stern–Gerlach analyzer is in the path of the beam at a point beyond the domain of the electromagnetic field. Its orientation is such that spins aligned with $\mathscr{B}$ are not deflected from the beam. If the beam moves with the speed v, what is the current of electrons with is scattered out of the beam by the S-G analyzer? Assume that the interaction between electrons in the bean and the electromagnetic wave is

$$\hat{H}' = -\boldsymbol{\mu} \cdot \mathscr{B}' \cos \omega t$$

where $\mathscr{B}'$ is the amplitude of magnetic field of the wave. The component of $\mathscr{B}'$ in the z direction is small compared to $\mathscr{B}$.

13.33 A one-dimensional harmonic oscillator of charge to mass ratio e/m and spring constant K is in its ground state. An oscillating uniform electric field

$$\mathscr{E}(t) = 2\mathscr{E} \cos \omega_0 t, \qquad \omega_0^2 = K/m$$

is applied for t seconds, parallel to the motion of the oscillator. What is the probability that the oscillator is excited to the nth state given that $(\omega_{n0} - \omega_0)t \ll 1$?

Answer

$$P_{0n} = \delta_{n,1} \left| \frac{e\mathscr{E}}{\sqrt{2\hbar\beta}} \right|^2 t^2, \qquad \beta^2 \equiv \frac{m\omega_0}{\hbar}$$

13.7 APPLICATION OF HARMONIC PERTURBATION THEORY

In this section we apply the transition probability formula (13.61) found above to Einstein's derivation of the Planck radiation formula (2.3) and to a brief qualitative description of the laser.

Einstein considered the equilibrium state between the walls of an enclosed cavity and the radiation field interior to the cavity. Atoms in the walls constantly exchange energy with the radiation field. The excited states of these atoms are very closely spaced and essentially comprise a continuum. Consider that the two energies E_l and E_k are representative of two states in the continuum (Fig. 13.11). Photons with energy $hv = E_k - E_l$ can be absorbed by atoms in the E_l state raising them to the E_k state. Atoms may decay from the E_k to the E_l state by *stimulated* or *spontaneous* emission. Stimulated emission was discussed in the preceding section. Spontaneous emission is related to the natural lifetime of the excited state and is more dependent on internal properties of the radiating system.

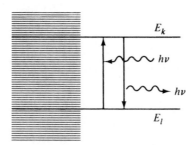

FIGURE 13.11 Energy states of closely packed atoms in the walls of a radiation cavity.

Einstein A and B Coefficients

The rate at which atoms in the E_k state decay by stimulated emission is proportional to the number of such atoms (N_k) and number of photons of frequency v in the radiation field. This number of photons is proportional to photon energy density, $u(v)$. The rate at which the atoms in the E_k state decay by spontaneous emission, on the other hand, is proportional only to the number of atoms N_k in the E_k state. Thus the total transition rate of decay of atoms with energy E_k to energies E_l is

(13.65) $$W_{kl} = [A_{kl} + B_{kl}u(v)]N_k = w_{kl}N_k$$

The transition probability rate per atom is written w_{kl}. The proportionality constants A and B are called *Einstein A and B coefficients*.

Atoms in the E_l state can be excited to the E_k state only by absorption of a photon of frequency $(E_k - E_l)/h$. Thus the rate of elevation of atoms in the E_l state to the E_k state is

(13.66)
$$W_{lk} = B_{lk} u(v) N_l = w_{lk} N_l$$

Planck Radiation Formula

In equilibrium, these rates must be equal.

$$W_{lk} = W_{kl}$$

Equivalently, we may write

(13.67)
$$\frac{N_l}{N_k} = \frac{A_{kl} + B_{kl} u(v)}{B_{lk} u(v)}$$

The ratio N_l/N_k may be obtained from elementary statistical mechanics. If the energy of an aggregate of atoms at the temperature T is partitioned such that N_1 of the atoms have energy E_1, N_2 have energy E_2, and so on, and the total energy of the aggregate is constant, then the most probable distribution of energies is given by the *Boltzmann distribution*.[1] In this distribution the number of atoms N_i with energy E_i is proportional to the Boltzmann factor, $\exp(-E_i/k_B T)$, where k_B is Boltzmann's constant. It follows that the ratio N_l/N_k has the value

$$\frac{N_l}{N_k} = e^{(E_k - E_l)/k_B T} = e^{hv/k_B T}$$

Substituting this into (13.67) gives

(13.68)
$$u(v) = \frac{A_{kl}}{B_{lk} e^{hv/k_B T} - B_{kl}}$$

The basic structure of the Planck radiation formula (2.3) follows if $B_{lk} = B_{kl}$. This equality may be obtained from the harmonic perturbation theory developed

---

[1] This distribution was employed previously in Problem 2.36. For a further discussion, see F. Reif, *Fundamentals of Statistical and Thermal Physics*, McGraw-Hill, New York, 1965.

above. To these ends we consider the interaction between a wave mode in the radiation field,

$$\mathcal{E} = \mathcal{E}_0 2 \cos \omega t$$

and an atom in the wall of the cavity. If $\mathbf{d}$ is the dipole moment of the atom, this interaction is

(13.69) $$H' = -\mathbf{d} \cdot \mathcal{E} = -\mathcal{E}_0 \cdot \mathbf{d} \, 2 \cos \omega t$$

which is identical to the perturbation (13.53). Substituting into (13.61) gives the transition probability rate for radiative absorption,

$$w_{lk} = \frac{2\pi}{\hbar^2} |\langle k | \mathcal{E}_0 \cdot \mathbf{d} | l \rangle|^2 \delta(\omega_{kl} - \omega)$$

If the electric field is isotropic ($\mathcal{E}_0$ is randomly oriented), then

$$|\langle \mathbf{d} \cdot \mathcal{E}_0 \rangle|^2 = \tfrac{1}{3} \langle \mathcal{E}_0^2 \rangle |\mathbf{d}_{ik}|^2$$

Furthermore, the energy density associated with this mode is

$$U(\omega) = \frac{1}{2\pi} \langle \mathcal{E}_0^2 \rangle$$

It follows that

$$w_{lk} = \frac{4\pi^2}{3\hbar^2} U(\omega) |\mathbf{d}_{ik}|^2 \delta(\omega_{kl} - \omega)$$

With $u(\omega) = U(\omega)\delta(\omega_{kl} - \omega)$, (1.66) may be rewritten

$$w_{lk} = B_{lk} \, U(\omega)\delta(\omega_{kl} - \omega)$$

Equating this value to the preceding expression for w_{lk} gives the desired result,

(13.70) $$B_{lk} = \frac{4\pi^2}{3\hbar^2} |\langle l | \mathbf{d} | k \rangle|^2$$

The square moduli of the matrix elements of $\mathbf{d}$ obey the relation

$$|\langle l | \mathbf{d} | k \rangle|^2 = \langle l | \mathbf{d} | k \rangle \cdot \langle k | \mathbf{d} | l \rangle$$
$$= \langle k | \mathbf{d} | l \rangle \cdot \langle l | \mathbf{d} | k \rangle = |\langle k | \mathbf{d} | l \rangle|^2$$

It follows that

$$B_{lk} = B_{kl}$$

and (13.68) reduces to the desired form,

$$(13.71) \qquad u(v) = \frac{A/B}{e^{hv/k_B T} - 1}$$

To obtain the ratio A/B we will use the correspondence principle. This rule stipulates that (13.71) should reduce to the classical Rayleigh–Jeans law (discussed in Section 2.2) in the limit $h \to 0$.

$$u_{RJ} = \frac{8\pi v^2}{c^3} k_B T$$

Expanding the exponential in (13.71) in this limit gives

$$\frac{A/B}{hv/k_B T} = \frac{8\pi v^2}{c^3} k_B T$$

from which we obtain the desired result (see Problem 13.52)

$$\frac{A}{B} = \frac{8\pi h v^3}{c^3}$$

Substituting this value into (13.71) gives the Planck formula, (2.3).

The Laser

The concepts of stimulated and spontaneous emission play an important role in the theory of the *laser*. A laser is a device for producing an intense beam of coherent, monochromatic light. A coherent beam may be defined as follows. Two or more collinear, unidirectional, monochromatic beams of electromagnetic radiation which propagate in the same region of space, and are in phase, form a coherent beam. In 1954, C. H. Townes[1] conceived of a process for the generation and amplification of such coherent radiation in the microwave domain. The device was called a *maser*. The term is an acronym for the words *m*icrowave *a*mplification by the *s*timulated *e*mission of *r*adiation. Shortly after, these concepts were extended to the optical region.[2] In this domain the corresponding device is called a *laser*.

[1] J. P. Gordon, H. J. Zeiger, and C. H. Townes, *Phys. Rev.* **99**, 1264 (1955).

[2] A. L. Schawlow and C. H. Townes, *Phys. Rev.* **112**, 1940 (1958). For further discussion and reference, see B. L. Lengyel, *Introduction to Laser Physics*, Wiley, New York, 1966; T. B. Melia, *An Introduction to Masers and Lasers*, Chapman & Hall, London, 1967.

Coherent Photons

The central principle in the realization of the laser is as follows. In constructing formula (13.65) for the transition rate from the E_k state to the E_l state, decay due to *stimulated* emission was taken to be proportional to the number of resonant photons present in the radiation field. Consider that a number of atoms in a gas are in the excited state E_k. Then when a photon of frequency

$$hv = E_k - E_l$$

falls on one of these atoms it stimulates the emission of another photon of the same frequency. These two photons, the emitted and incident, travel in phase in the same direction and combine coherently. A second atom, in the vicinity of the first and also excited to the E_k state, suddenly "sees" a duplication of resonant photons and is stimulated to emit another coherent photon, thereby adding to the intensity of the coherent radiation and amplifying it.

We found above that the matrix elements for radiative excitation, B_{lk}, and stimulated emission, B_{kl}, are equal [(13.70) et seq.]. To ensure that resonant photons stimulate decay to the E_l state more than they are absorbed in exciting the atom to the higher E_k state, the number of atoms, N_k, in the E_k state must outweigh the number of those in the lower E_l state, N_l. At any finite temperature, T, the ratio of these numbers of atoms, N_k/N_l, is given by the Boltzmann formula [preceding (13.68)].

$$\frac{N_k}{N_l} = e^{-hv/k_B T}$$

Population Inversion and Optical Pumping

The number of atoms in the higher E_k state decreases exponentially with energy difference, hv. To effect a *population inversion*, so that $N_k > N_l$, an outside source must be brought into play.[1] In *optical pumping*, N_k is increased by irradiation with light of frequency $v \geq (E_k - E_0)/h$, where E_0 is the ground state of the atom. Another technique for effecting a population inversion is by bombardment of electrons with energy $E > E_k - E_0$, thereby exciting atoms in the ground state to higher energy states. In a third technique, inelastic collisions between atoms in the ground state with those of a foreign gas which are in an excited state $E > E_k - E_0$ serve to populate the higher-energy states.

---

[1] Including the effects of degeneracy, the number of atoms in the nth state is given by $\bar{N}_n = g_n N_n$. The condition for growth of radiation then becomes $\bar{N}_k g_k > \bar{N}_j g_j$, which returns the inequality stated in the text. For further discussion, see A. Yariv, *Quantum Electronics*, Wiley, New York; 1968.

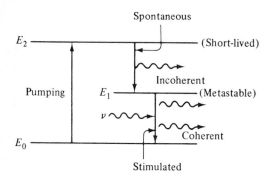

FIGURE 13.12   Schematic for the three-level laser.

Consider three atomic states: E_0, E_1, and E_2. The ground state is E_0, while E_2 is a short-lived state with lifetime of the order of 10 to 100 ns. The more stable ("metastable") E_1 state has a lifetime of the order of μs to ms. The upper E_2 state replenishes the metastable state through *spontaneous decay* [the A coefficient in (13.65)]. These randomly emitted photons comprise an incoherent radiation field. In Fig. 13.12 a scheme is depicted where the E_2 state is populated through pumping with an external source. When an atom in the metastable E_1 state decays to the E_0 state, neighboring atoms in the populated E_1 state decay, through stimulated emission to the ground state, and a coherent beam is generated.

Such a device is realized in an optical cavity of cylindrical shape with end mirrors positioned so that the cavity is tuned to the frequency mode $(E_1 - E_0)/h$. Photons propagate parallel to the axis of the tube. The radiation field is coherently amplified by stimulated emission $(E_1 \rightarrow E_0)$ and may be tapped through a small aperture on the axis, in one of the mirrors. The spontaneously emitted photons $(E_2 \rightarrow E_1)$ propagate in random directions and are dissipated in collisions with the walls.

PROBLEMS

13.34 Obtain a relation between the spontaneous emission coefficient A_{kl} and the dipole matrix element $|d_{kl}|$, analogous to (13.70).

13.35 The electric dipole moment of the ammonia molecule, NH_3, has magnitude $d = 1.47 \times 10^{-18}$ esu-cm. A beam of these molecules with dipoles polarized in the $+z$ direction enters a domain of electric field of strength 1.62×10^4 V/cm, which points in the $-z$ direction. The resulting interaction between the molecules and the field gives rise to coherent radiation. What is the frequency of this radiation (Hz)? (*Note:* Units of voltage in cgs are statV; 1 statV = 300 V.)

Answer

$$\nu = 24 \text{ GHz}$$

13.36 In deriving Planck's formula (see Problem 2.36 et seq.) for the density of photons in frequency interval v, $v + dv$, in a radiation field in equilibrium at the temperature T, in cubic volume $V = L^3$,

$$n(v)\, dv = \frac{8\pi v^2\, dv}{c^3}\, \frac{1}{e^{hv/k_BT} - 1}$$

one assumes the relation

$$n(v)\, dv = f_{BE}(v)g(v)\, dv$$

The term $g(v)\, dv$ is the density of available states in the said frequency interval, while f_{BE} is the *Bose–Einstein* factor, giving the average number of photons per state,

$$f_{BE}(v) = \frac{1}{e^{hv/k_BT} - 1}$$

Using the momentum-position uncertainty relation (appropriate to a three-dimensional box), derive the following expression for the *density of states* g, relevant to a radiation field:

$$g(v)\, dv = \frac{8\pi v^2\, dv}{c^3}$$

Answer

With $p = hv/c$, we look at Cartesian **p** space and note the volume of a spherical shell of radius p and thickness dp.

$$\text{volume of shell} = 4\pi p^2\, dp$$

All momenta in this shell correspond to nearly the same energy—and therefore the same frequency. Owing to the uncertainty principle for a particle confined to a three-dimensional box (see Problem 5.42), the smallest volume (a cell in **p** space) that may be specified with certainty to contain a state is

$$(\Delta p)^3_{\min} = \frac{(h/2)^3}{L^3} = \frac{(h/2)^3}{V}$$

It follows that the number of states in the dp shell is

$$\frac{\text{volume of shell}}{\text{volume of cell}} = \frac{4\pi p^2\, dp\, V}{(h/2)^3}$$

Finally, we note that one counts states only in the first quadrant of **p** space insofar as photons have only positive frequencies. Furthermore, to each such **p** state there are two photon states, corresponding to two possible polarizations. This gives

$$Vg(v)\, dv = \frac{1}{8} \times 2 \times \frac{4\pi V p^2\, dp}{h^3/8} = \frac{8\pi V p^2\, dp}{h^3}$$

The desired answer follows using the given relation between p and v for a photon.

13.37 (a) What are the Hamiltonian $\hat{H}_0$, eigenstates and eigenvalues of a collection of harmonic oscillators with natural frequencies: $\omega_1, \omega_2, \ldots$, in the number operator representation?

(b) The Hamiltonian constructed in part (a) represents a radiation field in *second quantization*. In this representation the radiation field is viewed as a collection of harmonic oscillators. A perturbation $\hat{H}'$ is applied to the radiation field such that

$$[\hat{H}', \hat{N}_1] = \frac{i\hbar}{\tau} \hat{N}_1$$

$$\hat{N}_1 \equiv \hat{a}_1^\dagger \hat{a}_1$$

The subscript 1 denotes the frequency ω_1. Show that $[\hat{H}_0, \hat{N}_1] = 0$. Then show that the perturbation $\hat{H}'$ represents a sink that diminishes the number of photons of frequency ω_1 exponentially with an *e*-folding time τ.

Answers

(a)
$$\hat{H}_0 = \sum_i \hbar\omega_i(\hat{a}_i^\dagger \hat{a}_i + \tfrac{1}{2})$$

Eigenstates take the form

$$|\psi_{n_1 n_2 \ldots}\rangle = |n_1, n_2, n_3, \ldots\rangle$$

The number of photons in the ω_i mode is n_i. These wavefunctions have the properties

$$\hat{a}_i|\ldots, n_i, \ldots\rangle = \sqrt{n_i}|\ldots, n_i - 1, \ldots\rangle$$

$$\hat{a}_i^\dagger|\ldots, n_i, \ldots\rangle = \sqrt{n_i + 1}|\ldots, n_i + 1, \ldots\rangle$$

It follows that

$$\hat{H}_0|n_1, n_2, \ldots\rangle = \sum_i \hbar\omega_i(n_i + \tfrac{1}{2})|n_1, n_2, \ldots\rangle$$

(b) The average number of photons of frequency ω_1 is

$$\langle N_1\rangle = \langle n_1, n_2, \ldots|\hat{N}_1|n_1, n_2, \ldots\rangle = n_1$$

The equation of motion for $\langle N_1\rangle$ is [recall (6.68)]

$$\frac{d\langle N_1\rangle}{dt} = \frac{i}{\hbar}\langle[\hat{H}, \hat{N}_1]\rangle = \frac{i}{\hbar}\langle[\hat{H}', \hat{N}_1]\rangle = -\frac{1}{\tau}\langle N_1\rangle$$

$$\frac{dn_1}{dt} = -\frac{1}{\tau}n_1 \qquad n_1 = n_1(0)e^{-t/\tau}$$

13.38 In Problem 13.37 we found that the Hamiltonian of an electromagnetic radiation field may be written[1]

$$\hat{H}_R = \sum_j \hbar\omega_j \hat{a}_j^\dagger \hat{a}_j$$

---

[1] Here we are omitting the infinite zero-point energy $\sum \tfrac{1}{2}\hbar\omega_j$.

Recalling that the frequency of a photon of momentum $\hbar\mathbf{k}$ is $\omega = ck$ [see (2.28)], it follows that the operator corresponding to the total momentum of the radiation field is

$$\hat{\mathbf{P}} = \sum_j \hbar\mathbf{k}_j \hat{a}_j^\dagger \hat{a}_j$$

Here we are assuming that the field is contained in a large cubical box with perfectly reflecting walls. Boundary conditions then imply a discrete sequence of wavenumber vectors $\{\mathbf{k}_i\}$. (See Problem 2.37.) Consider that a charged particle bound to a point within the radiation field vibrates along the z axis. Its Hamiltonian is

$$\hat{H}_P = \hbar\omega_0(\hat{a}^\dagger\hat{a} + \tfrac{1}{2})$$

(a) If the oscillator and the field are uncoupled from each other, what are the Hamiltonian and eigenstates of the composite system?

(b) Consider now that a small coupling exists between the particle and the field whose interaction energy is proportional to the scalar product of the displacement of the particle and the total momentum of the field, with coupling constant $\alpha\hbar\omega_0^2/mc^2$, where α is the fine-structure constant. Assuming (13.61) to be appropriate, calculate the probability rate that as a result of this interaction, the oscillator emits to the field a photon of energy $\hbar\omega_0$.

Answer

(a)
$$\hat{H}_{PF} = \hat{H}_P + \hat{H}_F = \hbar\omega_0(\hat{a}^\dagger\hat{a} + \tfrac{1}{2}) + \sum_j \hbar\omega_j \hat{a}_j^\dagger \hat{a}_j$$

$$|\psi\rangle = |\psi_P\rangle|\psi_F\rangle = |n; n_1, n_2, \ldots\rangle$$

$$\hat{H}_{PF}|\psi\rangle = \left[\hbar\omega_0(n + \tfrac{1}{2}) + \sum_j \hbar\omega_j n_j\right]|\psi\rangle$$

The notation is such that n_j denotes the number of photons with momentum $\hbar\mathbf{k}_j$.

(b) The interaction Hamiltonian is

$$\hat{H}' = \frac{\alpha\hbar\omega_0^2}{mc^2} \frac{1}{\sqrt{2\beta}} (\hat{a} + \hat{a}^\dagger) \sum_j \hbar k_{jz} \hat{a}_j^\dagger \hat{a}_j$$

where $\beta^2 \equiv m\omega_0/\hbar$ [recall (7.26)] and k_{jz} denotes the z component of $\mathbf{k}_j$. To apply (13.61) we must first calculate the matrix element of $\hat{H}'$ between the initial state

$$|\psi_i\rangle = |n; n_1, \ldots, n_k, \ldots\rangle, \qquad E_i = \hbar\omega_0(n + \tfrac{1}{2}) + \hbar\omega_0 n_k + \sum_{j \neq k} \hbar\omega_j n_j$$

and the final state

$$\langle\psi_f| = \langle n - 1; n_1, \ldots, n_k + 1, \ldots|, \qquad E_f = \hbar\omega_0(n - \tfrac{1}{2}) + \hbar\omega_0(n_k + 1) + \sum_{j \neq k} \hbar\omega_j n_j$$

Here n_k represents the number of photons with wavenumber ω_0/c. This choice of $\langle\psi_f|$ guarantees conservation of energy in the transition, as prescribed by the delta-function factor $\delta(E_i - E_f)$ in (13.61). Since $|\psi_i\rangle$ (as well as $\langle\psi_f|$) is an eigenvector of $\sum \hat{a}_j^\dagger \hat{a}_j$, it may be brought through the field component of $\hat{H}'$. When completed with the ket vector $\langle\psi_f|$, the inner product of the field component factors of the wavefunctions vanishes and we conclude that for the given perturbation,

$$\langle \psi_f | H' | \psi_i \rangle = H'_{if} = 0$$

The hypothetical field-particle coupling does not induce a transition in the state of the harmonic oscillator.

13.8 SELECTIVE PERTURBATIONS IN TIME

The Adiabatic Theorem

Let us now consider the case that $\hat{H}'$ is *adiabatically* turned on, which means that $\hat{H}'$ changes slowly in time (Fig. 13.13). Consequently, at any instant of time, the Hamiltonian may be treated as constant and an approximate solution can be obtained by regarding the Schrödinger equation as time-independent. This will be shown below.

The slowly changing quality of $\hat{H}'$ may be incorporated into the analysis through a parts integration of (13.51).

$$(13.72) \qquad c_k(t) = \frac{1}{i\hbar} \int_{-\infty}^{t} H'_{kl}(t') e^{i\omega_{kl}t'} dt' = -\frac{1}{\hbar\omega_{kl}} \int_{-\infty}^{t} H'_{kl}(t') \frac{\partial}{\partial t'} e^{i\omega_{kl}t'} dt'$$

$$= -\frac{1}{\hbar\omega_{kl}} \left\{ H'_{kl}(t) e^{i\omega_{kl}t} - \int_{-\infty}^{t} e^{i\omega_{kl}t'} \frac{\partial}{\partial t'} H'_{kl}(t') \, dt' \right\}$$

where we have set $H'_{kl}(t) \equiv H'_{kl} f(t)$. If $\hat{H}'(t)$ is slowly varying, the second term is small compared to the first and $c_k(t)$ is well approximated by

$$(13.73) \qquad c_k(t) \simeq -\frac{1}{\hbar\omega_{kl}} H'_{kl}(t) e^{i\omega_{kl}t} = -\frac{\langle k | \hat{H}' | l \rangle e^{i\omega_{kl}t}}{E_k^{(0)} - E_l^{(0)}}$$

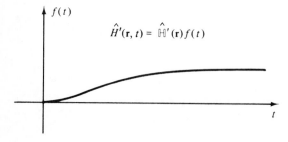

$$\hat{H}'(\mathbf{r}, t) = \hat{H}'(\mathbf{r}) f(t)$$

FIGURE 13.13 Adiabatic perturbation. The rise time, ω^{-1}, of the perturbation obeys the inequality

$$\omega \ll \omega_{kl}$$

where ω_{kl} are the natural frequencies of the unperturbed system.

Let us recall that the analysis leading to (13.49) presumes that the system is in the stationary state ψ_l at $t < 0$. To terms of first order in the perturbation H', the series (13.42) then appears as

$$\psi(\mathbf{r}, t) = \psi_l + \sum_{k \neq l} c_k \psi_k$$

As the perturbation is slowly varying, we write $\psi(\mathbf{r}, t) \simeq \exp(-i\omega_l t)\varphi(\mathbf{r})$. Substituting this expression into the preceding and employing (13.73) gives

$$(13.74) \qquad \varphi(\mathbf{r}) = \varphi_l + \sum_{k \neq l} \frac{H'_{kl}\varphi_k}{E_l^{(0)} - E_k^{(0)}}$$

This is precisely the first-order result which stationary (time-independent) perturbation theory gives, (13.10). But the solution (13.10) represents an eigenfunction of the new Hamiltonian, which in the present case is $\hat{H} = \hat{H}_0 + \hat{H}'(t)$. That is, the wavefunction (13.74) represents the lth eigenstate of $\hat{H}_0 + \hat{H}'(t)$, to first order in H'. [Note that we are writing $\hat{H}'(t)$ for the operator $\hat{H}'$ evaluated at the specific time t.] Since the system was originally in the lth state of $\hat{H}_0$, we may conclude the following. A system originally in the lth state of an unperturbed Hamiltonian will at the end of an adiabatic perturbation be found in the lth state of the new Hamiltonian. One says that the system "remains in the lth state."

Having established that φ as given by (13.74) is an eigenstate of the new Hamiltonian, it follows that expectation of energy in this state is the same as the eigenenergy of the state. The expectation of energy to first order in the perturbation Hamiltonian is easily calculated with the explicit form of φ as given by (13.74). There results

$$(13.75) \qquad E_l = E_l^{(0)} + H'_{ll}(t)$$

This is the lth eigenenergy of the Hamiltonian $\hat{H}_0 + \hat{H}'(t)$. So under an adiabatic perturbation, a system originally in the lth eigenstate of an unperturbed Hamiltonian will at time t be found in the lth eigenstate of the new Hamiltonian with the lth eigenenergy of the new Hamiltonian. With $\varphi_l(t)$ written for the time-independent wavefunction evaluated at the time t, we may write that under an adiabatic perturbation

$$(13.76a) \qquad \begin{aligned} \varphi_l(0) &\longrightarrow \varphi_l(t) \\ E_l(0) &\longrightarrow E_l(t) \end{aligned}$$

where

$$(13.76b) \qquad [\hat{H}_0 + \hat{H}'(t)]\varphi_l(t) = E_l(t)\varphi_l(t)$$

This equation is *not* a time-dependent equation. One first evaluates $\hat{H}'(t)$ and then finds the solution $\varphi_i(t)$. These results constitute the *adiabatic theorem*, first proved by Born and Fock in 1928.[1]

As an example of the use of this theorem, consider that a particle of mass m is in a one-dimensional box of edge length L. The width of the box is slowly increased to αL over a long interval lasting t seconds, where $\alpha > 1$. Let us calculate the new wavefunction and new energy of the system given that the particle initially was in the ground state. The adiabatic theorem tells us that the new state is the ground state of the new Hamiltonian.

$$\varphi_1 \rightarrow \varphi_1(t) = \sqrt{\frac{2}{\alpha L}} \sin\left(\frac{\pi x}{\alpha L}\right)$$

Here, as above, we are writing $\varphi(t)$ for the time-independent wavefunction evaluated at the time t. The new energy is the ground state of the new Hamiltonian.

$$E_1 \rightarrow E_1(t) = \frac{h^2}{8m(\alpha L)^2}$$

Note that in slowly expanding (as in the classical case), the particle loses energy to the receding walls in slowing down. The amount of work absorbed by the walls in the present example is

$$W = E_1 - E(t) = \frac{h^2}{8mL^2}\left(1 - \frac{1}{\alpha^2}\right)$$

Domain of Validity

Our conclusions regarding an adiabatic perturbation rest on neglecting the integral term in (13.72). Let us obtain a quantitative criterion which allows this term to be discarded. Let the perturbation be gradually turned on at $t = 0$. If $\hat{H}'$ changes slowly, then $\partial\hat{H}'/\partial t$ may be taken outside the integral term in (13.72) and we obtain

$$\left|\int_0^t e^{i\omega_{kl}t'}\frac{\partial H'_{kl}}{\partial t'}\,dt'\right| \simeq \left|\frac{\partial H'_{kl}}{\partial t}\right|\left|\int_0^t e^{i\omega_{kl}t'}\,dt'\right| \simeq \left|\frac{2}{\omega_{kl}}\frac{\partial H'_{kl}}{\partial t}\right|\left|\sin\left(\frac{\omega_{kl}t}{2}\right)\right|$$

It follows that the nonintegral term in (13.72) dominates provided that

(13.77)
$$|\omega_{kl}H'_{kl}| \gg 2\left|\frac{\partial H'_{kl}}{\partial t}\right|$$

[1] M. Born and V. Fock, *Z. Physik.* **51**, 165 (1928).

Thus a perturbation that undergoes a small fractional change in a typical period of the unperturbed system may be termed adiabatic.

Transition Probability

We have found that in the adiabatic limit, neglecting the integrated term in (13.72) leads to the system remaining in the initial state of the unperturbed Hamiltonian. What is the probability in this same limit that there is a transition out of the initial state? Suppose, for example, that an adiabatic perturbation is turned on at $t = 0$ and that the unperturbed system is originally in the lth state of the unperturbed Hamiltonian. The probability for a transition from the l to the k state is given by $|c_k|^2$. With (13.72) in the adiabatic limit, we obtain

$$
|c_k|^2 = \frac{1}{\hbar^2(\omega_{kl})^2} \left[|H'_{kl}|^2 + \frac{4}{(\omega_{kl})^2} \left| \frac{\partial}{\partial t} H'_{kl} \right|^2 \sin^2\left(\frac{\omega_{kl} t}{2}\right) \right.
$$

$$
\left. - \frac{2}{\omega_{kl}} \frac{\partial (H'_{kl})^2}{\partial t} \sin\left(\frac{\omega_{kl} t}{2}\right) \cos\left(\frac{\omega_{kl} t}{2}\right) \right]
$$

Over a long time interval, the last term averages to zero, whereas the $\sin^2$ term in the middle expression averages to $\frac{1}{2}$. As we have found above, the first term represents the probability that there is no transition out of the lth state. It follows that the probability that there is a transition out of the initial lth state to some kth state $(k \neq l)$ in the adiabatic limit is

(13.78)
$$
P_{lk} = \frac{2}{\hbar^2(\omega_{kl})^4} \left| \frac{\partial}{\partial t} H'_{kl} \right|^2
$$

Once more we find that there is zero probability of transition out of the lth state for sufficiently slowly changing perturbation.

An instructive laboratory example of the adiabatic theorem is given by elastic collision of molecules in a gas. During the collision, electron states are acted upon by intermolecular forces. This collisional interaction varies roughly as the intermolecular velocity, which, typically, is far smaller than mean atomic electron speeds. Thus, collisional forces which act on atomic states may be viewed as being adiabatically switched on. Consistent with this observation, one finds that whereas final rotational and vibrational molecular states are altered in the collision, atomic states suffer minimal change.[1]

---

[1] For further discussion, see R. D. Levine and R. B. Bernstein, *Molecular Reaction Dynamics and Chemical Reactivity*, Oxford, New York, 1987.

Sudden Perturbation

The next type of perturbation problem we wish to examine involves a sudden change in a parameter in $\hat{H}_0$. For example, suppose that the spring constant of a simple harmonic oscillator is suddenly doubled. If the oscillator is in its ground state before the perturbation, in what state is it after perturbation? For such problems it is presumed that the eigenstates of both Hamiltonians (i.e., before and after perturbation) are known. This, together with the assumption of an instantaneous change of $\hat{H}_0$, are basic to the *sudden approximation*.

Let us call the Hamiltonian before the change in parameter, $\hat{H}$, and the Hamiltonian after the change in parameter, $\hat{H}'$, so that

$$\hat{H}\varphi_n = E_n\varphi_n \qquad t < 0$$
$$\hat{H}'\varphi_n' = E_n'\varphi_n' \qquad t \geq 0$$

(Note that $\hat{H}'$ is now a total Hamiltonian.) Initially, the system is in an eigenstate ψ_l of $\hat{H}$.

$$\psi'(\mathbf{r}, 0) = \varphi_l(\mathbf{r})$$

At later times the system is in a superposition state of $\hat{H}'$.

$$\psi'(\mathbf{r}, t) = \sum_n c_n \varphi_n' e^{-i\omega_n' t}$$

Equating this function to its initial value at $t = 0$ gives

$$\varphi_l = \sum_n c_n \varphi_n'$$

so that

$$c_k = \langle \varphi_k' | \varphi_l \rangle$$

The probability that the sudden change from $\hat{H}$ to $\hat{H}'$ causes a transition from the lth state of $\hat{H}$ to the kth state of $\hat{H}'$ is

(13.79)
$$P_{lk} = |c_k|^2 = |\langle \varphi_k' | \varphi_l \rangle|^2$$

A criterion for the validity of this approximation scheme may be drawn from the preceding relation. As described above, let the sudden change in Hamiltonian occur at $t = 0$. For the relation (13.79) to have meaning it is necessary that the initial wavefunction $\varphi_l(\mathbf{r})$: (1) maintain its form at $t = 0^+$ and (2) lie in the Hilbert space spanned by the eigenstates of the new Hamiltonian. Thus, for a particle confined to a rigid-walled box, we may conclude that the approximation is inappropriate to sudden compression of the box but is appropriate to sudden expansion of the box.

To further demonstrate this approximation, let us apply the formalism to the latter case. Thus we consider a particle of mass m confined to a one-dimensional box which suffers a sudden expansion. With no loss in generality we take the box to be of unit width ($L = 1$). The particle is in the energy eigenstate φ_l at $t < 0$. At $t = 0$ the box is suddenly expanded to the edge length $\alpha > 1$ (recall Problem 5.11). With the preceding formalism we may establish the somewhat surprising result that although no work is done on the particle in the expansion, energy is not necessarily conserved.

The new Hamiltonian $\hat{H}'$ (after expansion) is given by (4.2,3) with L replaced by α. New basis wavefunctions are the eigenfunctions

$$\varphi'_k = \sqrt{\frac{2}{\alpha}} \sin\left(\frac{k\pi x}{\alpha}\right)$$

The probability for transition is given by

(13.80)

$$\sqrt{P_{lk}} = \langle \varphi_l | \varphi'_k \rangle = \frac{2}{\sqrt{\alpha}} \int_0^1 \sin l\pi x \sin\left(\frac{k\pi x}{\alpha}\right) dx$$

$$= \frac{1}{\sqrt{\alpha}} \left\{ \frac{\sin \pi[l - (k/\alpha)]}{\pi[l - (k/\alpha)]} - \frac{\sin \pi[l + (k/\alpha)]}{\pi[l + (k/\alpha)]} \right\}$$

Note that the integral in the preceding expression goes over $(0, 1)$ and not $(0, \alpha)$ because $\varphi_l(x)$ is zero over $(1, \alpha)$.

The transition which conserves energy occurs for the k value $\bar{k} = \alpha l$. With $A \equiv \hbar^2 \pi^2 / 2m$, we obtain

(13.81)

$$E'_k \bigg|_{k = \bar{k}} = A \frac{k^2}{\alpha^2} \bigg|_{k = \bar{k}} = A l^2 = E_l$$

At this value of k, (13.80) gives

(13.81a)

$$P_{l\bar{k}} = \frac{1}{\alpha^2} < 1$$

Thus, as stated, it is not certain that the transition conserves energy. However, it is simply shown that expectation of energy is conserved. To show this, we note that at $t = 0^+$ the wavefunction for the system at hand is given by

$$\bar{\varphi}_l = \varphi_l \qquad 0 \leq x \leq 1$$

$$\bar{\varphi}_l = 0 \qquad 1 < x \leq \alpha$$

It follows that

(13.82)

$$E_l = \langle E \rangle = \langle \varphi_l | \hat{H} | \varphi_l \rangle = \langle \bar{\varphi}_l | \hat{H}' | \bar{\varphi}_l \rangle$$

TABLE 13.2   Transition probabilities for time-dependent perturbations

1. $\hat{H}'$ is separable. $\hat{H}'(\mathbf{r}, t) = \hat{H}'(\mathbf{r}) f(t)$

$$P_{lk} = \frac{|H'_{kl}|^2}{\hbar^2} \left| \int_{-\infty}^{t} e^{i\omega_{kl} t'} f(t') \, dt' \right|^2$$

Harmonic perturbation, $f = 2\cos\omega t$, turned on at $t = 0$:

$$P_{lk} = \frac{|H'_{kl}|^2}{[\hbar(\omega_{kl} - \omega)/2]^2} \sin^2 [(\omega_{kl} - \omega)t/2]$$

DC perturbation $(\omega = 0, f = 1)$ turned on at $t = 0$:

$$P_{lk} = \frac{|H'_{kl}|^2 \sin^2 (\omega_{kl} t/2)}{(\hbar\omega_{kl}/2)^2}$$

Long-time or high-frequency limit:

$$P_{lk} = \frac{2\pi t |H'_{kl}|^2}{\hbar^2} \delta(\omega_{kl} - \omega)$$

$$w_{kl} = \frac{2\pi |H'_{kl}|^2}{\hbar^2} \delta(\omega_{kl} - \omega)$$

Short-time or low-frequency limit:

$$P_{lk} = \frac{t^2 |H'_{kl}|^2}{\hbar^2}$$

$$w_{kl} = \frac{t |H'_{kl}|^2}{\hbar^2}$$

Probability for transition to a band centered at E_k with $g|H'|$ slowly varying in energy:

$$\bar{P}_{lk} = \frac{2\pi t}{\hbar} g(E_k) |H'_{kl}|^2$$

$$\bar{w}_{kl} = \frac{2\pi}{\hbar} g(E_k) |H'_{kl}|^2$$

2. $\hat{H}' = \hat{H}'(\mathbf{r}, t)$ is slowly changing

Probability for transition out of the initial l state:

$$P_{lk} \simeq \frac{2}{\hbar^2 (\omega_{kl})^4} \left| \frac{\partial}{\partial t} H'_{kl} \right|^2$$

Adiabatic theorem:

$$P_{kl} \simeq 0 \quad (k \neq l)$$

3. $\hat{H}$ changes suddenly to $\hat{H}'$

Eigenstates of both $\hat{H}$ and $\hat{H}'$ are known.

$$P_{lk} = |\langle \varphi_k' | \varphi_l \rangle|^2$$

Note that the third term is defined over $(0, 1)$ and is the value of $\langle E \rangle$ at $t < 0$. The last term is defined over $(0, \alpha)$ and is appropriate to $t = 0^+$. We conclude that $\langle E \rangle$ is conserved in the expansion. Further developing (13.82) gives

(13.83)

$$E_l = \sum_k \langle \bar{\varphi}_l | \hat{H}' | \varphi_k' \rangle \langle \varphi_k' | \varphi_l \rangle$$

$$E_l = \sum_k P_{lk} E_k'$$

or, equivalently,

(13.84)
$$l^2 = \frac{1}{\alpha^2} \sum_{k=1}^{\infty} P_{lk} k^2$$

with P_{lk} given by (13.80).

This result is valid for arbitrary $\alpha > 1$, i.e., for irrational as well as rational values. In particular, note that if $k/\alpha \neq l$ for all k, then no direct transition conserves energy. However, with (13.83), expectation of energy is still conserved.

Expressions obtained above for the transition probability in various limiting cases are listed in Table 13.2.

PROBLEMS

13.39 A neutron in a rigid spherical well of radius $a = 0.1$ Å is in the ground state. The radius of the well is slowly decreased to $0.9a$.

(a) What is the energy and wavefunction of the neutron after the decrease in the well radius?

(b) How much work (in eV) is performed during the compression of the well?

13.40 A collection of $N_0 = 10^{13}$ independent electrons have spins polarized parallel to a uniform magnetic field that points in the z direction, of magnitude $\mathscr{B}_0$. A perturbation field is applied in the x direction of magnitude

$$\mathscr{B}'(t) = 10^{-3} \mathscr{B}_0 (1 - e^{-t/\tau}) \qquad (t \geq 0)$$

(a) Obtain a criterion involving τ which ensures that the perturbation is adiabatic.

(b) Given that $\Omega\tau = 10^2$ and $\mathscr{B}_0 = 10^4$ gauss, estimate the number of electrons ΔN that are thrown out of the ground spin state at $t = 10\tau$.

Answers

(a) $\tau \gg \Omega^{-1} = mc/e\,\mathscr{B}_0$

(b) A rough estimate is obtained from (13.78).

$$\Delta N \simeq 2N_0 \left| \frac{10^{-3} \mu_b \mathscr{B}_0}{\hbar\Omega} \right|^2 \left| \frac{e^{-t/\tau}}{\Omega\tau} \right|^2$$

13.41 A one-dimensional harmonic oscillator has its spring constant suddenly reduced by a factor of $\frac{1}{2}$. The oscillator is initially in its ground state. Show that the probability that the oscillator remains in the ground state is $P \simeq 0.98$.

13.42 A particle of mass m in a one-dimensional box of width L is in the third excited state. The width of the box is suddenly doubled. What is the probability that the particle drops to the ground state?

13.43   A one-dimensional harmonic oscillator in the ground state is acted upon by a uniform electric field

$$\mathcal{E}(t) = \frac{\mathcal{E}_0}{\sqrt{\pi}} \exp\left[-\left(\frac{t}{\tau}\right)^2\right]$$

switched on at $t = -\infty$. The field is parallel to the axis of the oscillator. What is the probability that the oscillator suffers a transition to its first excited state at $t = +\infty$? Are any other transitions possible?

13.44   Radioactive tritium, H^3, decays to light helium, He^{3+}, with the emission of an electron. (This electron quickly leaves the atom and may be ignored in the following calculation.) The effect of the β decay is to change the nuclear charge at $t = 0$ without effecting any change in the orbital electron. If the atom is initially in the ground state, what is the probability that the He^+ ion is left in the ground state after the decay?

13.45   A hydrogen atom in the ground state is placed in a uniform electric field in the z direction,

$$\mathcal{E} = \mathcal{E}_0 e^{-t/\tau}$$

which is turned on at $t = 0$. What is the probability that the atom is excited to the $2P$ state at $t \gg \tau$?

13.46   The perturbation

$$H' = \frac{A}{\tau\sqrt{\pi}} e^{i(\pi x/L)} e^{-t^2/\tau^2}$$

is applied to a particle of mass m in a one-dimensional box of width L at $t = -\infty$. At this time the particle is in the ground state. If $h/\tau \ll E_1$, in what state is it most probable that the particle will be at $t = +\infty$?

13.47   An electron in a one-dimensional potential well

$$V = \tfrac{1}{2}Kx^2$$

is immersed in a constant, uniform electric field of magnitude $\mathcal{E}$ which points in the x direction. The corresponding perturbation to the Hamiltonian is

$$H' = e\mathcal{E}x$$

(a)   Find the exact eigenenergies of the total Hamiltonian,

$$\hat{H} = \frac{\hat{p}^2}{2m} + \frac{1}{2}Kx^2 + e\mathcal{E}x$$

(see Problem 7.16). Discuss your findings with respect to the corresponding classical motion.

(b)   Show that the first-order corrections to the energy vanish. Then calculate the second-order corrections. Show that these agree with your answer to part (a), so that the second-order corrections give the complete solution for this problem.

Answers

    (a) Setting

$$\chi \equiv \frac{e\mathscr{E}}{K}$$

together with the transformation of variables

$$x' \equiv x + \chi$$

permits $\hat{H}$ to be rewritten

$$\hat{H} = \frac{p^2}{2m} + \frac{1}{2} K(x^2 + 2\chi x)$$

$$= \frac{p^2}{2m} + \frac{1}{2} Kx'^2 - \frac{1}{2} K\chi^2$$

$$= \hat{H}_0 - \mathbb{E}$$

Since

$$\mathbb{E} \equiv \frac{1}{2} K\chi^2 = \frac{e^2 \mathscr{E}^2}{2K}$$

is constant, the eigenenergies of $\hat{H}$ are simply

$$E_n = E_n{}^0 - \mathbb{E} = \hbar\omega_0 \left(n + \frac{1}{2}\right) - \frac{e^2 \mathscr{E}^2}{2K}$$

All levels are equally depressed by the constant energy $\mathbb{E}$. The new wavefunctions are

$$\varphi_n = \varphi_n(x') = \varphi_n(x + \chi)$$

The center of symmetry of these wavefunctions is at $x = -\chi$.

    In the corresponding classical problem, the potential of the electron in the presence of the uniform electric field is

$$V = \tfrac{1}{2}K(x + \chi)^2 - \mathbb{E}$$

This parabola is congruent to the original potential $Kx^2/2$. The new equilibrium at $x = -\chi$ occurs where the electric force is balanced by the spring force. This new potential minimum is lower than the original minimum by the amount $\mathbb{E}$. (Work must be done to move the electron, quasi-statically, from $x = -\chi$ to $x = 0$.) The classical analog of the quantum mechanical problem considered above involves an electric field that is very slowly (adiabatically) turned on. If the energy of the oscillator initially is $E^{(0)}$, what is it after the electric field is established? In classical mechanics, for such adiabatically changing harmonic motion, the ratio A/ω is constant (it is an *adiabatic invariant*). The amplitude of oscillation is A. Since the new potential well is congruent to the old well, ω is the same for both wells. This means that the amplitude of oscillation must also be preserved (during the adiabatic change). Owing again to congruency of the parabolas, this is

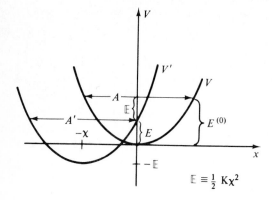

FIGURE 13.14 **The classical adiabatic change from a parabolic potential, V, to a congruent ($\omega_0 = \omega_0'$) parabolic potential, V', with a new center of symmetry, preserves amplitude A. We see that the amplitude of oscillation is preserved, $A = A'$, provided that $E = E^{(0)} - \mathbb{E}$. This result for the new energy of oscillation, E, is identical to the quantum mechanical result. (See Problem 13.47.)**

ensured (only) if the distance between $E^{(0)}$ and the bottom of the new well is the same as the distance between $E^{(0)}$ and the bottom of the original well. That is, if

$$E^{(0)} = E + \mathbb{E}$$

It follows that the new energy of the oscillator

$$E = E^{(0)} - \mathbb{E}$$

is depressed from the initial value, $E^{(0)}$, by the amount $\mathbb{E}$. This is identical to the quantum mechanical result (Fig. 13.14).

   (b)   The perturbation Hamiltonian may be rewritten

$$\hat{H} = \frac{e\mathscr{E}}{\sqrt{2}\,\beta}(\hat{a}^\dagger + \hat{a}) \equiv \mathscr{E}'(\hat{a}^\dagger + \hat{a})$$

$$\beta^2 \equiv \frac{m\omega_0}{h}, \qquad \mathscr{E}' \equiv \frac{e\mathscr{E}}{\sqrt{2}\,\beta}$$

It follows immediately that

$$\langle n|\hat{H}'|n\rangle = 0$$

so there is no first-order correction to $E_n^{(0)}$. To calculate the second-order corrections, we must evaluate the off-diagonal matrix elements of $\hat{H}'$.

$$\langle n|H'|l\rangle = \mathscr{E}'\langle n|\hat{a}^\dagger + \hat{a}|l\rangle$$

$$= \mathscr{E}'(\sqrt{l+1}\,\delta_{n,\,l+1} + \sqrt{l}\,\delta_{n,\,l-1})$$

Substituting this expression into (13.14) gives the desired result.

13.48   A system with discrete energy states and Hamiltonian $\hat{H}_0$ has the density operator $\hat{\rho}_0$, which is diagonal. Furthermore, $[\hat{\rho}_0, \hat{H}_0] = 0$, so $\hat{\rho}_0$ is constant in time. Show that after a perturbation $\hat{H}'$ is applied, the diagonal elements of $\hat{\rho}$ change according to the *Pauli equation*

$$\frac{\partial \rho_{nn}}{\partial t} = \sum_k w_{nk}(\rho_{kk} - \rho_{nn})$$

The transition rates w_{nk} are given by (13.63).

$$w_{nk} = \lim_{\Delta t \to 0} \frac{\Delta t}{\hbar^2} |H'_{nk}|^2$$

The density operator was discussed in Section 11.11.

Answer

For a short time interval after the perturbation $\hat{H}'$ is applied, we may expand $\hat{\rho}$ to obtain

$$\hat{\rho}(\Delta t) - \hat{\rho}(0) = \left(\frac{\partial \hat{\rho}}{\partial t}\right)_0 \Delta t + \left(\frac{\partial^2 \hat{\rho}}{\partial t^2}\right)_0 \frac{(\Delta t)^2}{2} + \cdots$$

Using the equation of motion (11.122) for $\hat{\rho}$ permits the last equation to be written, with $\hat{\rho}(0) = \hat{\rho}_0$,

$$\frac{\partial \hat{\rho}}{\partial t} = \lim_{\Delta t \to 0} \left[\frac{\hat{\rho}(\Delta t) - \hat{\rho}_0}{\Delta t}\right] = \lim_{\Delta t \to 0} \left\{\frac{1}{i\hbar}[\hat{H}, \hat{\rho}]_0 - \frac{\Delta t}{2\hbar^2}[\hat{H}, [\hat{H}, \hat{\rho}]]_0\right\}$$

The diagonal form of $\hat{\rho}_0$ leads to the following properties:

$$\langle n|[\hat{H}_0 + \hat{H}', \hat{\rho}_0]|n\rangle = 0$$

$$\langle n|[\hat{H}_0, [\hat{H}', \hat{\rho}_0]]|n\rangle = 0$$

Forming the diagonal elements of $\partial \hat{\rho}/\partial t$ then gives

$$\frac{\partial \rho_{nn}}{\partial t} = -\lim_{\Delta t \to 0} \frac{\Delta t}{2\hbar^2} \langle n|[\hat{H}', [\hat{H}', \hat{\rho}_0]]|n\rangle$$

The diagonal element on the right-hand side reduces to

$$\langle n|[\hat{H}', [\hat{H}', \hat{\rho}_0]]|n\rangle = 2\sum_k \{|H'_{nk}|^2 \rho_{nn}(0) - |H'_{nk}|^2 \rho_{kk}(0)\}$$

which when substituted into the preceding equation gives the desired result.

13.9 ATOM–RADIATION INTERACTION

Our first description of the interaction between an atom and a radiation field was given in terms of Einstein's derivation of the Planck radiation law (Section 13.7). We now wish to present a Hamiltonian formulation of this problem. The Hamiltonian of an electron in an electromagnetic field with vector potential **A** (previously introduced in

Section 10.4) and electric potential $V(\mathbf{r})$ is given by

(13.85)
$$\hat{H} = \frac{1}{2m}\left[\hat{\mathbf{p}} - \frac{e}{c}\mathbf{A}(\mathbf{r}, t)\right]^2 + V(\mathbf{r})$$

In the full quantum electrodynamic analysis of this problem, electrodynamic fields are quantized. The present analysis is termed a "semiclassical" description in that the vector potential $\mathbf{A}$ in (13.85) is taken as a classical field.[1]

Expanding the kinetic energy term in (13.85) gives

$$\hat{H} = \hat{H}^{(0)} + \hat{H}' + \hat{H}''$$

(13.86)
$$\hat{H}^{(0)} \equiv \frac{\hat{p}^2}{2m} + V; \qquad \hat{H}' = -\frac{e}{2mc}[\hat{\mathbf{p}} \cdot \mathbf{A} + \mathbf{A} \cdot \hat{\mathbf{p}}]$$

$$H'' = \frac{e^2}{2mc^2}A^2$$

We neglect $\hat{H}''$ and consider $\hat{H}'$ a perturbation term to the unperturbed atomic Hamiltonian $\hat{H}^{(0)}$.

The matrix element of $\hat{H}'$ between initial $|n\rangle$ and final $|n'\rangle$ states of $\hat{H}^{(0)}$ is given by

(13.87)
$$\langle n'|H'|n\rangle = -\frac{e}{2mc}\int \psi_{n'}^*(\hat{\mathbf{p}} \cdot \mathbf{A} + \mathbf{A} \cdot \hat{\mathbf{p}})\psi_n \, d\mathbf{r}$$

Note that the index n denotes a quantum state, not the principal quantum number. The first term in this integral may be written

(13.88)
$$\hat{\mathbf{p}} \cdot \mathbf{A}\psi_n = \frac{\hbar}{i}\nabla \cdot (\mathbf{A}\psi_n) + \frac{\hbar}{i}[\psi_n(\nabla \cdot \mathbf{A}) + (\mathbf{A} \cdot \nabla)\psi_n]$$

The gradient and curl qualities of the potentials which enter electrodynamics permit certain gauge conditions to be imposed. In the so-called *Coulomb gauge* one sets

$$\nabla \cdot \mathbf{A} = 0$$

so that (13.88) reduces to

$$\hat{\mathbf{p}} \cdot \mathbf{A}\psi_n = \frac{\hbar}{i}\mathbf{A} \cdot \nabla\psi_n$$

---

[1] For a more complete quantum analysis, see E. G. Harris, *A Pedestrian Approach to Quantum Field Theory*, Wiley, New York, 1972; A. S. Davydov, *Quantum Mechanics*, 2nd ed., Pergamon, Elmsford, N.Y., 1973.

Thus we may write (13.87) in the equivalent forms

(13.89a)
$$\langle n'|H'|n\rangle = -\frac{e}{mc}\int \psi_{n'}^{*}\mathbf{A}\cdot\hat{\mathbf{p}}\psi_{n}\,d\mathbf{r}$$

(13.89b)
$$\langle n'|H'|n\rangle = -\frac{e}{mc}\int \psi_{n'}^{*}\hat{\mathbf{p}}\cdot\mathbf{A}\psi_{n}\,d\mathbf{r}$$

A revealing insight into this analysis is gained if one associates the vector potential $\mathbf{A}$ with the wavefunction of a photon of energy $\hbar\omega$. This permits the preceding two matrix elements to be symbolically written

(13.90a)
$$\langle n'|\hat{H}'|n\rangle \to -\frac{e}{mc}\langle n';\hbar\omega|\hat{\mathbf{p}}|n\rangle \qquad \textit{(emission)}$$

(13.90b)
$$\langle n'|\hat{H}'|n\rangle \to -\frac{e}{mc}\langle n'|\hat{\mathbf{p}}|\hbar\omega; n\rangle \qquad \textit{(absorption)}$$

Thus, (13.90a) corresponds to the case where a photon is emitted by the system, whereas (13.90b) corresponds to the case of the absorption of a photon. Since these two integrals are equal, we conclude that probabilities of emission and absorption, at the same frequency, are equal. This is an example of the principle of *microscopic reversibility*.

To obtain a more explicit representation of the matrix elements (13.89), we consider the atom in interaction with a plane electromagnetic wave whose vector potential is of the form

(13.91)
$$\mathbf{A}(\mathbf{r}, t) = \mathbf{a}A_{0}\cos(\mathbf{k}\cdot\mathbf{r} - \omega t)$$

where $\mathbf{a}$ is a unit polarization vector ($|\mathbf{a}|^{2} = 1$) which is normal to the propagation vector $\mathbf{k}$. We wish to construct the amplitude A_{0} so that the corresponding wave carries one photon per unit volume. The time-averaged energy carried in a plane electromagnetic wave is

(13.92)
$$\langle U \rangle = \frac{1}{4\pi}\langle \mathscr{B}^{2} \rangle = \frac{1}{4\pi}\langle \mathscr{E}^{2} \rangle$$

With

(13.93)
$$\mathscr{B} = \nabla \times \mathbf{A}$$

and given the form (13.91), we find that

(13.94)
$$\langle U \rangle = \frac{1}{8\pi}k^{2}A_{0}^{2}$$

For one photon per unit volume we set

(13.95)
$$\frac{k^2 A_0{}^2}{8\pi} = \frac{\hbar\omega}{V}$$

which, with $\omega = ck$, gives

(13.96)
$$A_0{}^2 = \frac{8\pi\hbar c^2}{\omega V}$$

Henceforth we set the volume $V = 1$.

   In what follows, we will employ Fermi's golden rule (13.64). In formulating this relation it was found that for the time-dependent perturbation, $\cos\omega t$, the $\exp(-i\omega t)$ term was responsible for resonant absorption and the $\exp(+i\omega t)$ was responsible for decay. For consistent application of Fermi's rule, these observations must be incorporated into the present analysis. Thus, we first write (13.91) in the form

(13.97)
$$\mathbf{A} = \frac{A_0}{2}\,\mathbf{a}[e^{i(\mathbf{k}\cdot\mathbf{r}-\omega t)} + e^{-i(\mathbf{k}\cdot\mathbf{r}-\omega t)}]$$

which permits the identification

(13.98)
$$\mathbf{A}_\pm = \frac{A_0}{2}\,\mathbf{a}\,e^{\pm i(\mathbf{k}\cdot\mathbf{r}-\omega t)}$$

where A_+ corresponds to photon absorption and $\mathbf{A}_-$ to photon emission. So we may write

(13.99)
$$\mathbf{A}_\pm = c\left(\frac{2\pi\hbar}{\omega}\right)^{1/2}\mathbf{a}\,e^{\pm i(\mathbf{k}\cdot\mathbf{r}-\omega t)}$$

With

(13.100)
$$\hat{H}' = \hat{\mathbb{H}}_\pm(r)e^{\mp i\omega t}$$

and (13.89) we obtain

(13.101)
$$\langle n'|\hat{\mathbb{H}}_\pm|n\rangle = -\frac{e}{m}\left(\frac{2\pi\hbar}{\omega}\right)^{1/2}\langle n'|\mathbf{a}\cdot\hat{\mathbf{p}}e^{\pm i\mathbf{k}\cdot\mathbf{r}}|n\rangle$$

The Dipole Approximation

For typical atomic transitions $\lambda \gg a_0$, where the Bohr radius a_0 is a length characteristic of atomic dimensions. Thus, over the domain of integration in the preceding matrix element, $\mathbf{k}\cdot\mathbf{r} \ll 1$, and we may write

$$e^{i\mathbf{k}\cdot\mathbf{r}} = 1 + i\mathbf{k}\cdot\mathbf{r} + \cdots$$

In the *dipole approximation*, one sets $e^{i\mathbf{k}\cdot\mathbf{r}} = 1$. There results

$$(13.102) \qquad \langle n'|\mathbb{H}_{\pm}|n\rangle = -\frac{e}{m}\left(\frac{2\pi h}{\omega}\right)^{1/2} \langle n'|\mathbf{a}\cdot\hat{\mathbf{p}}|n\rangle$$

With the aid of the commutator relation (5.62) this matrix element may be transformed to one only involving $\mathbf{r}$ (see Problem 13.60):

$$(13.103) \qquad \hat{\mathbf{p}} = \frac{im}{\hbar}\,[\hat{H}^{(0)}, \mathbf{r}]$$

It follows that

$$(13.104) \qquad \langle n'|\mathbf{p}|n\rangle = \frac{im}{\hbar}\,\langle n'|H^{(0)}\mathbf{r} - \mathbf{r}H^{(0)}|n\rangle$$

Recalling that $|n'\rangle$ and $|n\rangle$ are eigenfunctions of the unperturbed Hamiltonian $\hat{H}^{(0)}$, we obtain

$$\langle n'|\mathbf{p}|n\rangle = -\frac{im}{\hbar}\,(E_n^{(0)} - E_{n'}^{(0)})\,\langle n'|\mathbf{r}|n\rangle$$

$$(13.105)$$

$$\langle n'|\mathbf{p}|n\rangle = -im\omega\langle n'|\mathbf{r}|n\rangle$$

Note that we are writing ω for $\omega_{nn'} > 0$. With these relations at hand, (13.102) may be rewritten more explicitly as

$$(13.106) \qquad \langle n'|\mathbb{H}_+|n, \mathbf{k}\rangle = \langle n', \mathbf{k}|\mathbb{H}_-|n\rangle = ie(2\pi\hbar\omega)^{1/2}\langle n', \mathbf{k}|\mathbf{a}\cdot\mathbf{r}|n\rangle$$

Here we have specified that in the absorption process, the initial state contains a photon of wavevector $\mathbf{k}$, whereas in the emission process, the final state contains a photon of wavevector $\mathbf{k}$. Again as in (13.90), we find that the related probabilities of these two processes are equal. Spontaneous decay may be associated with the matrix element $\langle n', \mathbf{k}|\mathbb{H}_-|n\rangle$.

Spontaneous Decay

As described previously, the Einstein A coefficient represents the probability rate for spontaneous decay. [See, for example, Eq. (13.65).] Ordinarily, one would suspect that such spontaneous decay occurs in the absence of any perturbation. However, in our earlier description of hydrogen (Section 10.6) we found atomic states to be stationary. So in reality, spontaneous decay must have a triggering mechanism. What is this mechanism?

The answer to this question stems from the observation that (as described in Problem 13.37) an electromagnetic field may be represented as a collection of harmonic oscillators. We have found previously that a harmonic oscillator always has a residual energy which is called its "zero-point energy." In like manner, no region of space is ever free of electromagnetic energy. It is such *vacuum fluctuations of electrodynamic fields* which are responsible for spontaneous decay.

Golden Rule Revisited

In describing spontaneous decay we concentrate on the emission matrix element $\langle n', \mathbf{k} | \mathbb{H}_- | n \rangle$, and consider that the emitted photon lies in the differential of solid angle $d\Omega$. The corresponding probable transition rate is given by Fermi's golden rule (13.64), which, in the present case, assumes the form

$$(13.107) \qquad dw_{nn'} = \sum_{\mathbf{a}_i} \frac{2\pi}{\hbar} |\mathbb{H}_{n'n}|^2 \bar{g}(E) \, d\Omega$$

Here $\mathbf{a}_i$ denotes possible photon polarization and $\bar{g}(E) \, d\Omega$ is the density of states of photons emitted in the solid angle $d\Omega$. This value of $\bar{g}(E)$ may be obtained from the value of $g(\nu)$ given in Problem 2.37 by setting

$$(13.108) \qquad V g(\nu) \, d\nu = V g(\omega) \, d\omega = 2 \left[\int \bar{g}(E) \, d\Omega \right] dE$$

The factor 2 in this equality accounts for photon polarizations. Setting the volume $V = 1$, we obtain

$$(13.109) \qquad \bar{g}(E) = \frac{E^2}{(2\pi\hbar c)^3} = \frac{\omega^2}{\hbar(2\pi c)^3}$$

Combining these results gives the following probability rate for spontaneous decay:

$$(13.110) \qquad dw_{nn'} = \sum_{\mathbf{a}_i} \frac{e^2 \omega^3}{2\pi\hbar c^3} |\langle n', \mathbf{k} | \mathbf{a}_i \cdot \mathbf{r} | n \rangle|^2 \, d\Omega$$

The summation over polarizations in (13.110) is performed as follows. First, we note that

$$(13.111a) \qquad \langle n', \mathbf{k} | \mathbf{a} \cdot \mathbf{r} | n \rangle = \mathbf{a} \cdot \langle n', \mathbf{k} | \mathbf{r} | n \rangle$$

We recall that in the dipole approximation the presence of $\mathbf{k}$ in the preceding matrix element is cosmetic. It merely reminds us that the element is relevant to spontaneous decay and includes a photon of wavevector $\mathbf{k}$ in the final state. It follows that the only vector in the matrix element on the right side of (13.111a) is $\mathbf{r}$. Consequently, this matrix element is likewise in the direction of $\mathbf{r}$ and we may conclude that the entire matrix element on the left side of (13.111) is proportional to $\mathbf{a} \cdot \mathbf{r}$. With this observation, the sum in (13.110) is evaluated as depicted in Fig. 13.15. The propagation vector of the

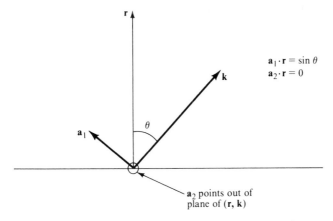

Figure 13.15   Configuration for summation over a_i in (13.110). The propagation vector k defines the direction of the differential of solid angle $d\Omega$.

emitted photon, **k**, makes an angle θ with the radius vector **r**, which is held fixed and taken as the polar axis. The unit polarization vectors $\mathbf{a}_1$, $\mathbf{a}_2$ are normal to each other as well as to the propagation vector **k**. In Fig. 13.15 the vectors **k** and **r** lie in the plane of the paper, as does the unit vector $\mathbf{a}_1$. It follows that the vector $\mathbf{a}_2$ points out of the plane of the paper, so that $\mathbf{a}_1 \cdot \hat{\mathbf{r}} = \sin\theta$, $\mathbf{a}_2 \cdot \hat{\mathbf{r}} = 0$ ($\hat{\mathbf{r}}$ is a unit vector). Thus we obtain

$$\sum_{\mathbf{a}_i} |\langle n', \mathbf{k}|\mathbf{a}_i \cdot \mathbf{r}|n\rangle|^2 = |\mathbf{a}_1 \cdot \langle n', \mathbf{k}|\mathbf{r}|n\rangle|^2 + |\mathbf{a}_2 \cdot \langle n', \mathbf{k}|\mathbf{r}|n\rangle|^2 = \sin^2\theta|\langle n', \mathbf{k}|\mathbf{r}|n\rangle|^2$$

Inserting this result into (13.110) gives the following probability rate for spontaneous decay:

$$(13.111b) \qquad dw_{nn'} = \frac{e^2\omega^3}{2\pi\hbar c^3} |\langle n', \mathbf{k}|\mathbf{r}|n\rangle|^2 \sin^2\theta \, d\Omega$$

So we reach the conclusion that the differential transition probability rate $dw_{nn'}$ depends only on the angle θ between **r** and **k**. Consequently, the subsequent integration over $d\Omega$ is independent of the azimuthal angle, and with **r** still held fixed, we obtain

$$(13.112) \qquad A_{nn'} = \int dw_{nn'} = \frac{4}{3}\frac{e^2\omega^3}{\hbar c^3} |\langle n'|\mathbf{r}|n\rangle|^2$$

Note that we are writing

$$|\langle n'|\mathbf{r}|n\rangle|^2 = |\langle n'|x|n\rangle|^2 + |\langle n'|y|n\rangle|^2 + |\langle n'|z|n\rangle|^2$$

The coefficient $A_{nn'}$ represents the probable rate of transition for spontaneous decay and may be identified with the Einstein A coefficient in (13.65). To obtain the corresponding expression for radiated power, one multiplies $A_{nn'}$ by $\hbar\omega$. There results

$$(13.113) \qquad\qquad P_{nn'} = \frac{4}{3}\frac{e^2\omega^4}{c^3}|\langle n'|\mathbf{r}|n\rangle|^2$$

which agrees with our previous result (10.125).

The Einstein B coefficient may be found from our previously derived relation (see Problem 13.52):

$$\frac{A}{B} = \frac{\hbar\omega^3}{\pi^2 c^3}$$

which, with (13.112), returns our earlier finding (13.70).

It should be kept in mind that the preceding results are relevant to weak fields and the assumption that $\lambda \gg a_0$.

To apply the above formulas to an atom with Z electrons, one makes the replacement

$$(13.114) \qquad\qquad \mathbf{r}_{nn'} = \langle n'|\mathbf{r}|n\rangle \rightarrow \langle n'|\sum_{i=1}^{Z}\mathbf{r}_i|n\rangle$$

in the preceding expressions for A and P. The sum in (13.114) runs over all the electrons in the atom. For such cases $e\mathbf{r}_{nn'}$ represents the total dipole matrix element of the atom.

The total probable rate of spontaneous decay of an atom in the nth energy state is given by

$$(13.115) \qquad\qquad A_n = \sum_{E_{n'}<E_n} A_{nn'}$$

The summation runs over all states of lower energy than E_n. The corresponding mean lifetime of the nth excited state is then given by

$$(13.116) \qquad\qquad \tau_n = \frac{1}{A_n}$$

Oscillator Strengths

An important parameter in radiation analysis is the oscillator strength, defined as the dimensionless form

$$(13.117) \qquad\qquad f_{nn'} = \frac{2m\omega}{3\hbar}|\langle n'|\mathbf{r}|n\rangle|^2$$

It follows that

(13.118)
$$A_{nn'} = \frac{2e\omega^2}{mc^3} f_{nn'}$$

(13.119)
$$P_{nn'} = \frac{2e^2\hbar\omega^3}{mc^3} f_{nn'}$$

Oscillator strengths obey the so-called Thomas–Reiche–Kuhn sum rule,

(13.120)
$$\sum_{n'} f_{nn'} = 1$$

from which, together with (13.117), it may be concluded that $f_{nn'} < 1$.

The sum rule (13.120) is simply derived with the aid of the basic commutator relation (5.61), according to which, in one dimension, one obtains

(13.121)
$$\langle n|[x, p_x]|n\rangle = i\hbar$$

Equivalently, we may write (see Problem 11.1)

(13.122)
$$\sum_{n'} [\langle n|x|n'\rangle\langle n'|p_x|n\rangle - \langle n|p_x|n'\rangle\langle n'|x|n\rangle] = i\hbar$$

With (13.105) we obtain

$$\sum_{n'} [im\omega_{n'n}|\langle n'|x|n\rangle|^2 - im\omega_{nn'}|\langle n'|x|n\rangle|^2] = i\hbar$$

$$\sum_{n'} \frac{2m\omega_{n'n}}{\hbar} |\langle n'|x|n\rangle|^2 = 1$$

The same result follows with x replaced by y or z. There results

(13.123)
$$\sum_{n'} \frac{2m\omega}{3\hbar} |\langle n'|\mathbf{r}|n\rangle|^2 = \sum_{n'} f_{nn'} = 1$$

For an atom with Z electrons, with (13.114) one finds

(13.124)
$$\sum_{n'} f_{nn'} = Z$$

Values of the oscillator strengths of hydrogen corresponding to the first four $nP - 1S$ transitions are listed below.[1]

n	2	3	4	5
f_{nP-1S}	0.416	0.079	0.029	0.014

[1] R. Loudon, *The Quantum Theory of Light*, Clarendon Press, Oxford, 1973.

We conclude this section with an estimate of the lifetime of the excited states of one-electron atoms.[1] With $|\mathbf{r}_{nn'}| \simeq a_0$, we rewrite $A_{nn'}$ (13.112) as

$$(13.125) \qquad A_{nn'} \simeq \frac{4}{3} \frac{e^2 \omega^3 a_0^2}{\hbar c^3}$$

Introducing the effective nuclear charge $\bar{Z}$,

$$(13.126) \qquad \hbar \omega \equiv \frac{\bar{Z} e^2}{a_0}$$

permits the preceding formula to be written

$$(13.127) \qquad A_{nn'} \simeq \tfrac{4}{3} \alpha^3 \omega \bar{Z}^2$$

where $\alpha = e^2/\hbar c$ is the fine-structure constant. With $\bar{Z} \simeq 1$, and $\hbar \omega \simeq R = \alpha^2 mc^2/2$, we find

$$(13.128) \qquad A_{nn'} \simeq \frac{\alpha^5}{\hbar} \frac{mc^2}{2} = \frac{\alpha^5 c}{2 \lambda_C}$$

In this expression, $\lambda_C = \hbar/mc$ is the Compton wavelength. Inverting $A_{nn'}$ gives the lifetime[2]

$$(13.129) \qquad \tau \simeq \alpha^{-5} \left(\frac{2 \lambda_C}{c} \right) \simeq 10^{-9} \text{ s} = 1\text{ns}$$

PROBLEMS

13.49 The radioactive isotope C^{11} decays through positron emission to B^{11-}. With the same assumptions holding as described in Problem 13.44, estimate the probability that B^{11-} is born in the ground state. (*Hint:* For your estimate, consult Problem 12.28 and pay attention to Z dependence of wavefunctions.)

Answer

As discussed in Problem 12.28, two-electron, ground-state wavefunctions (with $\hat{H}_{SO} = 0$) are given by

$$\varphi_Z(r_1, r_2) = \frac{1}{\pi a^3} \exp\left[\frac{-(r_1 + r_2)}{a} \right]$$

$$a \equiv \frac{a_0}{Z}$$

---

[1] More detailed discussions of these topics may be found in H. A. Bethe and E. E. Salpeter, *Quantum Mechanics of One- and Two-Electron Atoms*, Plenum, New York, 1977; F. Constantinescu and E. Magyari, *Problems in Quantum Mechanics*, Pergamon Press, Elmsford, N.Y. 1971; S. Flügge, *Practical Quantum Mechanics*, Springer, New York, 1974.

[2] Nanosecond (10^{-9}), picosecond (10^{-12}), and femtosecond (10^{-15}) measurements are presently commonplace in many laboratories. A review of these topics is given by P. W. Smith and A. M. Weiner, *IEEE Circuits and Devices* **4**, 3 (May 1988).

with normalization

$$\iint d\mathbf{r}_1 \, d\mathbf{r}_2 |\varphi_Z|^2 = (4\pi)^2 \int_0^\infty dr_1 r_1^2 \int_0^\infty dr_2 r_2^2 |\varphi_Z|^2 = 1$$

The transition probability to the ground state of B^{11-} is given by

$$\sqrt{P} = \langle \varphi_C | \varphi_B \rangle = \frac{Z_C^3 Z_B^3}{\pi^2 a_0^6} \iint \exp\left[\frac{-Z(r_1 + r_2)}{a_0}\right] d\mathbf{r}_1 \, d\mathbf{r}_2$$

where $Z \equiv Z_C + Z_B$. The preceding integral may be rewritten

$$\sqrt{P} = \frac{Z_C^3 Z_B^3}{\pi^2 a_0^6} I^2$$

$$I = a_0^3 \int_0^\infty dx \, x^2 e^{-Zx} = 2\left(\frac{a_0}{Z}\right)^3$$

There results

$$\sqrt{P} = \left(\frac{4 Z_B Z_C}{Z^2}\right)^3$$

With $Z_B = 5$, $Z_C = 6$, $Z = 11$, we obtain $P = 0.951$.

13.50 (a) The density matrix (in a representation where $\hat{S}^2$ and $\hat{S}_z$ are diagonal) describing a beam of spinning electrons has the value

$$\hat{\rho} = \begin{pmatrix} 1 & \frac{1}{2} \\ \frac{1}{2} & 0 \end{pmatrix}$$

at $t < 0$. What are the values of $\langle S_z \rangle$, $\langle S_x \rangle$, and $\langle S_y \rangle$ for an electron in the beam?

(b) The beam interacts with a field which is turned on at $t = 0$. The corresponding interaction Hamiltonian has the matrix

$$\hat{H} = \hbar\omega_0 \begin{pmatrix} 0 & 1 \\ 1 & 0 \end{pmatrix}$$

where ω_0 is a characteristic frequency. Estimate the value of the matrix $\hat{\rho}$ at the time Δt, where $0 < \Delta t \ll \omega_0^{-1}$.

(c) What is the value of $\langle S_y \rangle$ at this value of time Δt? [*Hint:* Recall (11.122).]

Answers

(a) $\langle S_z \rangle = \langle S_x \rangle = \hbar/2$, $\langle S_y \rangle = 0$.

(b) Set

$$\hat{\rho}(\Delta t) = \hat{\rho}(0) + \left(\frac{\partial \hat{\rho}}{\partial t}\right)_0 \Delta t + \cdots$$

$$= \hat{\rho}(0) + \frac{1}{i\hbar} [\hat{H}, \hat{\rho}]_0 \Delta t + \cdots$$

Employing the given value of $\hat{\rho}(0)$ gives

$$\hat{\rho}(\Delta t) = \begin{pmatrix} 1 & 0.5 \\ 0.5 & 0 \end{pmatrix} - i\omega_0 \Delta t \begin{pmatrix} 0 & -1 \\ 1 & 0 \end{pmatrix} = \begin{pmatrix} 1 & a \\ a^* & 0 \end{pmatrix}$$

where

$$a \equiv 0.5 + i\omega_0 \Delta t$$

(c)   At $t = \Delta t$

$$\langle S_y \rangle = Tr\hat{\rho}\hat{S}_y = Tr\left[\hat{\rho}\,\frac{i\hbar}{2}\begin{pmatrix} 0 & -1 \\ 1 & 0 \end{pmatrix}\right]$$

$$= \frac{i\hbar}{2} Tr\left[\begin{pmatrix} 1 & a \\ a^* & 0 \end{pmatrix}\begin{pmatrix} 0 & -1 \\ 1 & 0 \end{pmatrix}\right]$$

$$= \frac{i\hbar}{2}(a - a^*) = -\hbar\omega_0\,\Delta t$$

Thus, the perturbation causes a change in $\langle S_y \rangle$.

13.51   A one-dimensional harmonic oscillator of charge-to-mass ratio e/m, and spring constant K oscillates parallel to the x axis and is in its second excited state at $t < 0$, with energy

$$E_2 = \hbar\omega_0\left(2 + \frac{1}{2}\right)$$

An oscillating, uniform electric field

$$\mathscr{E}(t) = 2\mathscr{E}_0 \cos \omega_0 t$$

is turned on at $t = 0$, parallel to the motion of the oscillator.

(a)   What is the new Hamiltonian of the oscillator at $t > 0$?

(b)   What are the matrix elements H_{2n} for this system? You may leave your answer in terms of $\beta = \sqrt{m\omega_0/\hbar}$.

(c)   What are the probabilities P_{2n} that the oscillator undergoes a transition to the nth state at the end of t seconds? [Hint: Use reasonant harmonic perturbation theory and look at the resonance limit.]

(d)   Using your answer to part (c), offer a technique for 'pumping' a harmonic oscillator to higher states.

13.52   In our discussion of Planck's radiation law in Section 13.7, an expression for A/B was obtained as a function of frequency, v. What is the corresponding expression for A/B in terms of angular frequency, ω? [Hint: see Problem 2.43.]

13.53   Suppose φ_1 and φ_2 are two normalized eigenfunctions of an operator $\hat{A}$ with the same eigenvalue. Construct two new normalized, orthogonal functions, ψ_1 and ψ_2, which are linear combinations of φ_1 and φ_2. Offer a geometrical description of construction in the appropriate Hilbert space.

Answer (partial)

Let

$$\psi_1 = \varphi_1, \qquad \psi_2 = \alpha\varphi_1 + \beta\varphi_2$$

and

$$\int \varphi_1^* \varphi_2 \, d\mathbf{r} = K \neq 0$$

Orthogonality of ψ_1 and ψ_2 gives

$$\alpha + \beta K = 0$$

whereas normalization of ψ_2 gives

$$|\beta|^2(1 - |K|^2) = 1$$

There results

$$\beta K = -\alpha = \frac{K}{\sqrt{1 - |K|^2}}$$

which determines ψ_2. This construction gives the essentials of the Schmidt orthogonalization procedure.

13.54 (a) Show that the time-dependent Schrödinger equation may be written [compare with (11.144)]

$$\psi(t) = \psi(0) + \frac{1}{i\hbar} \int_0^t \hat{H}(t')\psi(t') \, dt'$$

(b) Show that for small $\hat{H}$ the preceding equation gives the series (Neumann–Liouville expansion) with $t > t'$, etc.,

$$\psi(t) = \left[1 + \frac{1}{i\hbar} \int_0^t dt' \, \hat{H}(t') + \frac{1}{(i\hbar)^2} \int_0^t dt' \int_0^{t'} dt'' \, \hat{H}(t')\hat{H}(t'') + \cdots \right]\psi(0)$$

(c) Show that the nth-order term in the preceding series may be written

$$\psi^{(n)}(t) = \frac{1}{n!} \frac{1}{(i\hbar)^n} \int_0^t \cdots \int_0^t dt_1 \cdots dt_n \, \hat{T}[\hat{H}(t_1) \cdots \hat{H}(t_n)]\psi(0)$$

where the time-ordering operator

$$\hat{T} f(t_a)g(t_b) = \begin{cases} f(t_a)g(t_b) & \text{for } t_a > t_b \\ g(t_b) f(t_a) & \text{for } t_b > t_a \end{cases}$$

13.55 A particle of mass m is confined to a one-dimensional partitioned box with walls at $(-L/2, 0, L/2)$. The Hamiltonian for this configuration is labeled $\hat{H}_0$. See Fig. 13.16a. One of two perturbations, $\hat{H}'_{a,b}$, are applied as shown in Fig. 13.16b, c. The total Hamiltonians for

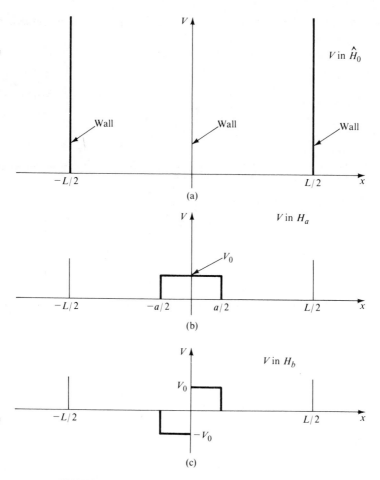

FIGURE 13.16 Potential configurations for Problem 13.55.

the particle are

$$\hat{H}_{a, b} = \hat{H}_0 + \hat{H}'_{a, b}$$

Let the perturbation potentials have magnitude $V_0 = 10^{-4} E_G$, where E_G is the ground state of $\hat{H}_0$, and let the normalized ground states of $\hat{H}_0$ be written $|r\rangle$ and $|l\rangle$ corresponding to the particle trapped in the right and left boxes, respectively.

(a)   What are the coordinate representations of $|r\rangle$ and $|l\rangle$?

(b)   What is the ground-state eigenenergy E_G of $\hat{H}_0$? What is the order of the degeneracy of E_G?

(c)   What is $\langle l|H_0|r\rangle$?

(d)   What are the values (zero or nonzero) of the commutators $[\hat{H}_0, \hat{\mathbb{P}}]$, $[\hat{H}_a, \hat{\mathbb{P}}]$, $[\hat{H}_b, \hat{\mathbb{P}}]$? Here we have written $\hat{\mathbb{P}}$ for the parity operator (6.90).

(e)   Construct two simultaneous eigenstates of $\hat{H}_0$ and $\hat{\mathbb{P}}$ in terms of $|l\rangle$ and $|r\rangle$. Label these states $|S\rangle$ (symmetric) and $|A\rangle$ (antisymmetric). What are $\hat{H}_0|S\rangle$, $\hat{H}_0|A\rangle$, $\mathbb{P}|S\rangle$, $\hat{\mathbb{P}}|A\rangle$?

(f)   Using degenerate perturbation theory, obtain the splitting of E_G due to $\hat{H}'_a$ and $\hat{H}'_b$, respectively. Do both $\hat{H}'_a$ and $\hat{H}'_b$ remove the degeneracy of E_G? Explain your answer. Call the new energies E_+, E_-. What effect does the nonsplitting perturbation have on the ground-state energy?

(g)   The trapped particle is an electron and $2a = L = 10$ Å. What are E_+ and E_- (in eV)?

(h)   Can you suggest an alternative way to evaluate $E_\pm$ which does not employ degenerate perturbation theory?

Answers (partial)

(a) $\langle l|x\rangle = 0$ on $(0, L/2)$ and $\langle r|x\rangle = 0$ on $(-L/2, 0)$.

(e)
$$|S\rangle = \sqrt{\frac{1}{2}}\,[|l\rangle + |r\rangle], \qquad |A\rangle = \sqrt{\frac{1}{2}}\,[|l\rangle - |r\rangle]$$

$$\hat{\mathbb{P}}|S\rangle = |S\rangle, \qquad \hat{\mathbb{P}}|A\rangle = -|A\rangle$$

$$\hat{H}_0|S\rangle = E_G|S\rangle, \qquad \hat{H}_0|A\rangle = E_G|A\rangle$$

(f)   We must examine the secular equation

$$\begin{vmatrix} \langle l|H'_{a,b}|l\rangle - E' & \langle l|H'_{a,b}|r\rangle \\ \langle r|H'_{a,b}|l\rangle & \langle r|H'_{a,b}|r\rangle - E' \end{vmatrix} = 0$$

There results

$$\mathscr{E}_a \equiv \langle l|H'_a|l\rangle = \langle r|H'_a|r\rangle$$

$$\mathscr{E}_b \equiv \langle l|H'_b|l\rangle = -\langle r|H'_b|r\rangle$$

$$\Delta\mathscr{E}_a \equiv \langle r|H'_a|l\rangle = \langle l|H'_a|r\rangle = 0$$

$$\Delta\mathscr{E}_b \equiv \langle r|H'_b|l\rangle = \langle r|H'_b|l\rangle = 0$$

Thus $\hat{H}'_b$ splits E_G but $\hat{H}'_a$ does not. The explanation of this result is that $\hat{H}'_a$ maintains the symmetry of $\hat{H}_0$. That is, for both $\hat{H}_0$ and $\hat{H}'_a$, there is no difference between right and left boxes. New energies (for $\hat{H}_b$) are given by

$$E_+ = E_G + \mathscr{E}_b$$

$$E_- = E_G - \mathscr{E}_b$$

13.56   An approximate form of the Hamiltonian of a positronium atom in the 1S state immersed in a weak magnetic field $\mathscr{B}$ is given by

$$\hat{H} = A\hat{\mathbf{S}}_1 \cdot \hat{\mathbf{S}}_2 + \boldsymbol{\Omega} \cdot (\hat{\mathbf{S}}_1 - \hat{\mathbf{S}}_2) \equiv H_0 + \lambda H'$$

$$\Omega = \frac{e\mathscr{B}}{\mu c}$$

where H' is the magnetic field term. The subscripts 1 and 2 denote electron and position, respectively, A is a constant, and μ is reduced mass.

(a)   Show that the coupled spin states ξ_S and ξ_A relevant to two spin $\frac{1}{2}$ particles are eigenstates of $\hat{H}_0$. What is the ground-state energy of $\hat{H}_0$ and which eigenstate does this correspond to? [See Table 11.3 and recall (11.111)]

(b)   Employing ξ_S and ξ_A as basis states, use first-order perturbation theory to obtain eigenvalues and eigenstates of $\hat{H}$. Assume that $\mathscr{B}$ is aligned with the z axis.

13.57   In Section 11.12 we encountered an approximation scheme centered about the interaction picture. Again consider the Hamiltonian given in Problem 11.96, which we now write in more symbolic form,

$$\hat{H} = \hat{H}_0 + \lambda \hat{V}(t)$$

Let the system be in the initial state

$$|\psi(0)\rangle = |n\rangle$$

$$\hat{H}_0|n\rangle = E_n|n\rangle$$

where, we recall $\hat{H}_0$ is appropriate to an harmonic oscillator with natural frequency ω_0.

(a)   Calculate the wavefunction $|\psi_I(t)\rangle$ to first order in λ.

(b)   For the problem at hand, show that

$$|\langle m|\psi(t)\rangle|^2 = |\langle m|\psi_I(t)\rangle|^2$$

(c)   Again, to $0(\lambda)$, obtain an integral expression for the time-dependent transition probabilities $P_{n\to m}$.

(d)   In the event that $\hat{V}$ is time-independent, show that your answer to (c) reduces to (13.55) corresponding to the DC perturbation, $\omega = 0$.

Answers

(a)
$$|\psi_I(t)\rangle = |n\rangle + \sum_m \frac{\lambda}{i\hbar} \int_0^t e^{i\omega_{mn}t} |m\rangle\langle n|\hat{V}(t')|n\rangle \, dt'$$

$$\hbar\omega_{mn} = E_m - E_n = \hbar\omega_0(m - n)$$

(b)   First, note that

$$|\psi(t)\rangle = |e^{-i\hat{H}_0 t/\hbar}\psi_I(t)\rangle$$

The desired equality follows since $\hat{H}_0|m\rangle = E_m|m\rangle$.

(c)   With the result (b), we write

$$P_{n\to m} = |\langle m|\psi_I(t)\rangle|^2$$

where, for $m \neq n$,

$$\langle m|\psi_I(t)\rangle = \frac{\lambda}{i\hbar} \int_0^t e^{i\omega_{mn}t'} \langle m|\hat{V}(t')|n\rangle \, dt'$$

(d) In the event that $\hat{V}$ is time-independent, integration of the preceding finding gives

$$\langle m|\psi_I(t)\rangle = \frac{V_{mn}}{\hbar\omega_{mn}}(1 - e^{i\omega_{mn}t})$$

$$V_{mn} \equiv \lambda\langle m|\hat{V}|n\rangle$$

which yields

$$P_{n\to m} = \left|\frac{2V_{mn}}{\hbar\omega_{mn}}\right|^2 \sin^2\frac{\omega_{mn}t}{2}$$

This result is seen to agree with (13.55) for the DC perturbation, $\omega = 0$.

13.58 Establish the following relations for the coefficients of the Fourier expansion (13.33) of the real potential function $V(x)$.

(a) $V_n^* = V_{-n}$.

(b) If $V(x)$ is even, then

$$V_n = V_{-n}$$

(c) If $V(x)$ is odd, then

$$V_n = -V_{-n}$$

(d) For $V(x)$ (even, odd) and with period $2a$, one writes

$$V(x)\binom{\text{even}}{\text{odd}} = \sum_{n=1}^{\infty}\left(\begin{array}{c} a_n \cos\dfrac{n\pi x}{a} \\[2mm] b_n \sin\dfrac{n\pi x}{a} \end{array}\right)$$

Show that (with $a_0 = 0$)

$$a_n = 2V_n = 2V_{-n}$$

$$b_n = 2iV_n = -2iV_{-n}$$

Answers (partial)

(a) As replacing n with $-n$ in the series (13.33) does not change the sum, we may set

$$\sum_{\forall n} V_n e^{i2\pi nx/a} = \sum_{\forall n} V_{-n} e^{-i2\pi nx/a}$$

For $V(x)$ real, $V(x) = V(x)^*$. With the preceding we then obtain

$$\sum_{\forall n} V_n^* e^{-i2\pi nx/a} = \sum_{\forall n} V_{-n} e^{-i2\pi nx/a}$$

whence $V_n^* = V_{-n}$.

(b) If $V(x)$ is even, then

$$\sum_{\forall n} V_n e^{i2\pi nx/a} = \sum_{\forall n} V_n e^{-i2\pi nx/a} = \sum_{\forall n} V_{-n} e^{i2\pi nx/a}$$

whence $V_n = V_{-n}$.

(c)   If $V(x)$ is odd, then

$$\sum_{\forall n} V_n e^{i2\pi nx/a} = -\sum_{\forall n} V_n e^{-i2\pi nx/a} = -\sum_{\forall n} V_{-n} e^{i2\pi nx/a}$$

whence $V_n = -V_{-n}$.

13.59   (a)   Evaluate the average dipole moment of hydrogen for the $3P \rightarrow 1S$ transition from the corresponding value of oscillator strength given in the text. Work in cgs units and state the dimensions of your answer.

(b)   Compare your answer with the classical estimate, $a \simeq ea_0$.

13.60   Establish the commutator relation (13.103).

Answer

Since **r** commutes with $V(\mathbf{r})$, the relation reduces to

$$\hat{\mathbf{p}} = \frac{im}{\hbar}\left[\frac{\hat{p}^2}{2m}, \mathbf{r}\right]$$

With reference to Problem 5.45, the preceding is rewritten

$$\hat{\mathbf{p}} = \frac{i}{2\hbar}\{\hat{\mathbf{p}}[\hat{\mathbf{p}}, \hat{\mathbf{r}}] + [\hat{\mathbf{p}}, \hat{\mathbf{r}}]\hat{\mathbf{p}}\}$$

$$\hat{\mathbf{p}} = -\frac{i}{2\hbar}\{\hat{\mathbf{p}}i\hbar\hat{I} + i\hbar\hat{I}\hat{\mathbf{p}}_n\}$$

where $\hat{I}$ is the identity operator,

$$\hat{\mathbf{p}}\hat{I} = \hat{I}\hat{\mathbf{p}} = \hat{\mathbf{p}}$$

which establishes the said relation.

13.61   A particle of mass m is in a three-dimensional, rigid-walled cubical box of edge length L. Edges of the box are aligned with the Cartesian axes, with one corner of the box at the origin.

(a)   Write down the normalized ground state for this configuration.

(b)   The face of the cube at $x = L$ is suddenly displaced to $x = 2L$. Obtain an expression for the probability that the particle remains in the ground state.

(c)   Given that the particle is an electron and that $L = 2$ Å, what is the numerical value of this probability?

13.62   A particle of mass m is confined to the interior of a rigid-walled spherical cavity of radius a. The particle is in the ground state. At $t = 0$ the radius of the sphere begins to expand according to

$$a(t) = a_0 e^{-(t/\tau)^2}$$

where $\tau \gg ma_0^2/\hbar$. What are the wavefunction and energy of the particle at $t = \tau/2$?

13.63   At a given instant of time an harmonic oscillator undergoes a sudden change in spring constant from K to K'. Show that for energy to be conserved in the accompanying transition,

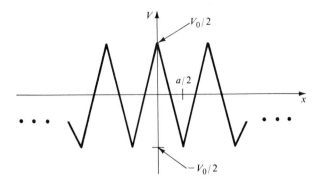

FIGURE 13.17   **See Problem 13.64.**

$\sqrt{K/K'}$ must be the ratio of two odd numbers.

13.64   The same conditions as in Problem 13.26 apply. However, now the periodic potential is as shown in Fig. 13.17.

If $V_0 = 1.5$ eV and $a = 2.3$ Å, at what energy (eV) will photons incident on the crystal cause it to conduct? [*Hint:* Before you start, decide on the period of $V(x)$.]

Answer

You should obtain

$$E_g = 2V_1 = \frac{4V_0}{\pi^2} = 0.61 \text{ eV}$$

13.65   In our analysis of the *nearly free electron model* we set
(a) $H'_{kj, kj} = H'_{-kj, -kj} = V_0 = 0$
(b) $H'_{kj, -kj} = V_n$
(c) $H'_{-kj, kj} = V_{-n}$
(d) $V_n = V_{-n}$
Establish the validity of these relations.

SCATTERING IN
THREE DIMENSIONS

In this concluding chapter an elementary description is offered of the quantum mechanical theory of scattering in three dimensions. Application of low-energy scattering is made to the Ramsauer effect, formerly encountered in Chapter 7, and scattering from a rigid sphere. The chapter continues with a discussion of the Born approximation. This important analysis permits certain scattering problems to be formulated in terms of harmonic perturbation theory developed previously in Chapter 13. The cross section of an atom interacting with a radiation field is obtained. For off-resonant incident phonons one encounters the line-shape factor. The chapter concludes with a description of the formal theory of scattering and derivation of the Lippmann–Schwinger equation in which the formalism of the interaction picture (Chapter 11) comes into play.

14.1 PARTIAL WAVES

The Rutherford Atom

One of the most fundamental tools of physics used for probing atomic and subatomic domains involves scattering of known particles from a sample of the element in question. Thus, for example, the description of an atom as being comprised of a positively charged central core of radius $\simeq 10^{-13}$ cm, with external satellite electrons, is due to scattering experiments performed by E. Rutherford in 1911. In these experiments α particles in an incident beam were deflected in passing through a thin metal foil. The prevalent model for an atom at the time was J. J. Thomson's "watermelon" model, in which negative electrons floated in a ball of positive charge. The relatively large angle suffered by a small fraction of the α particles in the incident beam in Rutherford's experiment was found to be inconsistent with Thomson's model of the atom. For it is easily shown that α particles, after passing through hundreds of such spheres of distributed charge, are deflected at most only by a few degrees. On the other hand the actual scattering data is consistent with an atomic model in which the positive charge is concentrated in a central core of small diameter. Large angle of scatter is then experienced by α particles which pass sufficiently close to the positive nucleus.

Scattering Cross Section

The typical configuration of a scattering experiment is shown in Fig. 14.1. A uniform monoenergetic beam of particles of known energy and current density J_{inc} (7.107) is incident on a target containing scattering centers. Such scattering centers might, for example, be the positive nuclei of atoms in a metal lattice. If the particles in the incident beam are, say, α particles, then when one such particle comes sufficiently close to one of the nuclei in the sample, it will be scattered. If the target sample is sufficiently thin, the probability of more than one such event for any particle in the incident beam is small and one may expect to obtain a valid description of the scattering data in terms of a single two-particle scattering event.

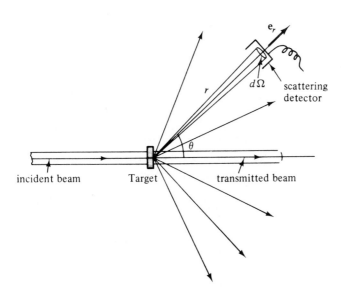

FIGURE 14.1 Scattering configuration.

Let the scattered current be $\mathbf{J}_{sc}$. Then the number of particles per unit time scattered through some surface element $d\mathbf{S}$ is $\mathbf{J}_{sc} \cdot d\mathbf{S}$. Let $d\mathbf{S}$ be at the radius $\mathbf{r}$ from the target. Then if $d\mathbf{\Omega}$ is the vector solid angle subtended by $d\mathbf{S}$ about the target origin, $d\mathbf{S} = r^2 \, d\mathbf{\Omega}$ (see Figs. 9.9 and 14.1). The vector solid angle $d\mathbf{\Omega}$ is in the direction of $\mathbf{e}_r$; that is, $d\mathbf{\Omega} = \mathbf{e}_r \, d\Omega$. It follows that

number of particles

passing through $d\mathbf{S} = dN = \mathbf{J}_{sc} \cdot d\mathbf{S} = r^2 \mathbf{J}_{sc} \cdot d\mathbf{\Omega}$

per second

Since the number of such scattered particles will grow with the incident current $\mathbf{J}_{inc}$, one may assume this number to be proportional to $\mathbf{J}_{inc}$ and can equate

(14.1) $$dN = r^2 \mathbf{J}_{sc} \cdot d\mathbf{\Omega} = \mathbf{J}_{inc} \, d\sigma$$

The proportionality factor $d\sigma$ is called the *differential scattering cross section* and has dimensions of cm^2. It may be interpreted as an obstructional area which the scatterer presents to the incident beam. Particles taken out of the incident beam by this obstructional area are scattered into $d\mathbf{\Omega}$. The *total scattering cross section* σ represents the obstructional area of scattering in all directions.

(14.2) $$\sigma = \int d\sigma = \int_{4\pi} \left(\frac{d\sigma}{d\Omega} \right) d\Omega$$

Scattering cross section has a classical counterpart. Classically, the total cross section seen by a uniform beam of point particles incident on a fixed rigid sphere of radius a is $\sigma = \pi a^2$. If the incident beam has current $\mathbf{J}_{inc}$, the number per second scattered out of the beam in all directions is $\pi a^2 \mathbf{J}_{inc}$.

The Scattering Amplitude

Returning to quantum mechanics, let the particles in the incident beam be independent of each other so that prior to interaction with the target a particle in the incident beam may be considered a *free particle*. If the z axis is taken to coincide with the axis of incidence, then a particle in the incident beam with momentum $\hbar\mathbf{k}$ and energy $\hbar^2 k^2 / 2m$ is in the planewave state,

(14.3) $$\varphi_{inc} = e^{ikz}$$

When this wave interacts with a scattering center, an outgoing scattered wave φ_{sc} is initiated. If the scattering is *isotropic* so that scattering into all directions (all 4π steradians of solid angle) is equally probable, we can expect the scattered wave φ_{sc} to be a spherically symmetric outgoing wave. The specific form of an isotropic outgoing wave was described previously [(10.65) and Problem 10.6].

$$\varphi_{\text{sc, iso}} = \frac{e^{ikr}}{r}$$

More often, however, the scattered wave is anisotropic. Anisotropy of the scattering component wavefunction φ_{sc} may be described by a modulation factor $f(\theta)$, and in general we write

(14.4)
$$\varphi_{\text{sc}} = \frac{f(\theta)e^{ikr}}{r}$$

The modulation $f(\theta)$ is called the *scattering amplitude* and will be shown to determine the differential scattering cross section $d\sigma$.

The number of particles scattered into $d\Omega$, which is in the direction of $\mathbf{e}_r$, is obtained from the radial component of $\mathbf{J}_{\text{sc}}$ [recall (7.107)]:

(14.5)
$$\mathbb{J}_{\text{sc},r} = \frac{\hbar}{2mi} \left(\varphi_{\text{sc}}^* \frac{\partial}{\partial r} \varphi_{\text{sc}} - \varphi_{\text{sc}} \frac{\partial}{\partial r} \varphi_{\text{sc}}^* \right)$$
$$= \frac{\hbar k}{mr^2} |f(\theta)|^2$$

Since the vector element of solid angle $d\mathbf{\Omega}$ is in direction $\mathbf{e}_r$, it follows that

$$r^2 \mathbf{J}_{\text{sc}} \cdot d\mathbf{\Omega} = r^2 \mathbb{J}_{\text{sc},r} \, d\Omega = \mathbb{J}_{\text{inc}} \, d\sigma$$

In that the current vector of the incident beam only has a z component with magnitude $\hbar k/m$ [see (7.113)], the preceding equation becomes

$$r^2 \mathbb{J}_{\text{sc},r} \, d\Omega = \frac{\hbar k}{m} \, d\sigma$$

Substituting (14.5) into this equation gives the desired relation,

(14.6)
$$\boxed{d\sigma = |f(\theta)|^2 \, d\Omega}$$

Thus the problem of determining $d\sigma$ is equivalent to constructing the scattering amplitude $f(\theta)$.

Owing to the rotational symmetry of the scattering configuration about the axis of the incident beam and the assumed radial quality of the interaction potential between incident particle and scatterer, the scattering cross section depends only on the scattering angle θ (and incident energy) and not on the azimuthal angle ϕ (see

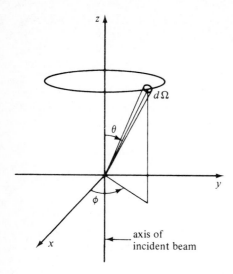

axis of
incident beam

FIGURE 14.2 The scattering cross section is independent of the azimuthal angle ϕ for central potentials of interaction $V(r)$.

Fig. 14.2). It follows that in integrating (14.6) over all directions, the integration over $d\phi$ may be done separately to obtain 2π. There results

$$(14.7) \qquad \sigma = \int d\sigma = 2\pi \int_0^\pi |f(\theta)|^2 \sin \theta \, d\theta$$

The total cross section is a simple integral over the square modulus of the scattering amplitude. Referring again to Fig. 14.2, we see that the same symmetry implies that $f(\theta)$ is an even function of θ or, equivalently, $f(\theta) = f(\cos \theta)$.

Partial-Wave Phase Shift

The form of the wavefunction·for the steady-state scattering configuration described above, at positions far removed from the scattering target, will contain a plane-wave incident component and an "outgoing" scattered component.

$$(14.8) \qquad \varphi(r, \theta) = e^{ikz} + \frac{f(\theta)e^{ikr}}{r} \qquad (r \to \infty)$$

(Fig. 14.3). The scattering amplitude is determined by matching (14.8) to the asymptotic form of the solution of the Schrödinger equation relevant to the configuration at hand. Such configuration includes a particle of mass m with known energy $\hbar^2k^2/2m$, interacting with a fixed scattering center through the central potential $V(r)$. The radial Schrödinger equation is given by (10.95).

$$(14.9) \qquad \left[\frac{1}{r}\frac{d^2}{dr^2}r - \frac{l(l + 1)}{r^2} + k^2 - \frac{2mV}{\hbar^2}\right]R_{kl}(r) = 0$$

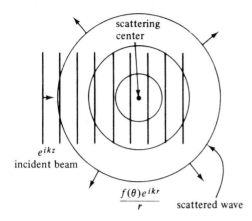

FIGURE 14.3 Incident plane wave and scattered outgoing spherical wave.

In the far field where $V(r)$ is rapidly approaching zero, one may expect the solution to this equation to be given approximately by the asymptotic form of the free-particle solution $j_l(kr)$ [see (10.55) and Table 10.1].

$$(14.10) \qquad R_{kl} \sim \frac{1}{kr} \sin\left(kr - \frac{l\pi}{2}\right)$$

Provided that $V(r)$ decreases faster than r^{-1}, this free-particle asymptotic form remains intact[1] save for a change in argument through a phase shift δ_l.

$$(14.11) \qquad R_{kl}{}^{\text{asm}} = \frac{1}{kr} \sin\left(kr - \frac{l\pi}{2} + \delta_l\right)$$

A superposition state comprised of these wavefunctions at fixed k has the form

$$(14.12) \qquad \varphi_k(r, \theta) = \sum_{l=0}^{\infty} C_l R_{kl}{}^{\text{asm}} P_l(\cos \theta)$$

The lth term in the sum is called the lth *partial wave* and δ_l is the phase shift that the partial wave incurs in scattering.

We must now match the asymptotic form of the general solution (14.12) to the form (14.8). With the expansion for exp (ikz) given in Problem 10.12, we obtain the asymptotic expression

$$e^{ikz} \sim \sum_{l=0}^{\infty} (2l + 1)i^l \frac{\sin(kr - l\pi/2)}{kr} P_l(\cos \theta)$$

---

[1] For example, the analysis is not valid for the Coulomb potential $V(r) = r^{-1}$. Proof of the validity of the stated criterion may be found in L. Landau and E. Lifshitz, *Quantum Mechanics*, Addison-Wesley, Reading, Mass., 1958.

The coefficients C_l and the scattering amptitude $f(\theta)$ are found from the matching equation

$$\sum_l C_l P_l(\cos\theta)\frac{\sin(kr - l\pi/2 + \delta_l)}{kr} = \sum_l (2l + 1)i^l P_l(\cos\theta)\frac{\sin(kr - l\pi/2)}{kr}$$
$$+ \frac{f(\theta)e^{ikr}}{r}$$

Expanding $f(\theta)$ in a series of Legendre polynomials, one obtains, after some trigonometric gymnastics,

$$C_l = i^l(2l + 1)\exp(i\delta_l)$$

(14.13)
$$f(\theta) = \frac{1}{k}\sum_{l=0}^{\infty}\frac{C_l}{i^l}\sin\delta_l P_l(\cos\theta)$$

The problem of calculating $d\sigma$ or, equivalently, $f(\theta)$ is reduced to one of constructing the phase shifts δ_l.

Two immediate results are evident: First, substituting the series (14.13) into (14.7) and taking advantage of the orthogonality of the $P_l(\cos\theta)$ polynomials, we obtain

(14.14)
$$\sigma = \frac{4\pi}{k^2}\sum_{l=0}^{\infty}(2l + 1)\sin^2\delta_l$$

The second result follows from setting $\theta = 0$ in (14.13), which yields

$$f(0) = \frac{1}{k}\sum_l(2l + 1)\cos\delta_l\sin\delta_l + \frac{i}{k}\sum_l(2l + 1)\sin^2\delta_l$$

Comparison with (14.14) reveals that

(14.15)
$$\sigma = \frac{4\pi}{k}\operatorname{Im}[f(0)]$$

This result is known as the *optical theorem*. It is a widely used relation connecting the forward scattering amplitude, $f(0)$, to the scattering in all directions, σ.[1]

Relative Magnitude of Phase Shifts

The problem of determining the partial wave phase shifts δ_l is often difficult. However, under certain conditions one may make simplifying assumptions which greatly facilitate calculation. In classical scattering one introduces the impact

---

[1] For inelastic scattering, (14.15) is still valid with σ replaced by the total cross section, $\sigma_T = \sigma_S + \sigma_A$, where σ_S is the elastic cross section and σ_A is the absorption cross section.

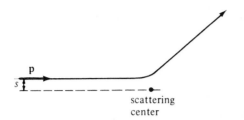

scattering
center

FIGURE 14.4 Classical trajectory and impact parameter s.

parameter. If L and p are the incident particle's angular momentum and linear momentum, respectively, then the impact parameter s is given by (see Fig. 14.4)

$$L = ps$$

Quite clearly, if the potential of interaction is appreciable only over the range r_0, then the interaction between incident particle and scatterer will be negligible for $s > r_0$. This criterion provides a useful rule of thumb applicable in quantum mechanics. With $L = \hbar\sqrt{l(l + 1)} \simeq \hbar l$ and $p = \hbar k$, interaction will be negligible if

(14.16) $$l > r_0 k$$

The incident energy is $\hbar^2 k^2/2m$.

Each partial wave in the superposition (14.12) represents a state of definite angular momentum. From (14.16) we can expect that partial waves with l values in excess of $r_0 k$ will suffer little or no shift in phase. In the corresponding expansion of the scattering amplitude $f(\theta)$ as given by (14.13) it follows that only those δ_l values will contribute for which $l < r_0 k$. For low-energy scattering with $kr_0 \ll 1$, only the $l = 0$ phase shift will differ appreciably from zero. When such is the case (14.13) reduces to

(14.17) $$f(\theta) = \frac{1}{k} e^{i\delta_0} \sin \delta_0$$

which is independent of θ. The scattering is isotropic and is called S-wave scattering. Only the S partial wave ($l = 0$) contributes to the scattering. In the opposite extreme of large incident energies, $kr_0 \gg 1$, we can expect all partial waves to suffer phase shifts and the cross section to be anisotropic.

PROBLEMS

14.1 From (14.1) we find that the number of particles scattered into the solid angle $d\Omega$ per second is

$$dN = \mathbb{J}\, d\sigma$$

or, equivalently,

$$\frac{dN}{\mathbb{J}} = \left(\frac{d\sigma}{d\Omega}\right) d\Omega$$

The Coulomb cross section for the scattering of a charged particle of energy E and charge q from a fixed charge Q is

$$\frac{d\sigma}{d\Omega} = \left(\frac{qQ}{4E}\right)^2 \frac{1}{\sin^4(\theta/2)}$$

(a)   What is the expression for the fraction of particles scattered into the differential cone $(\theta, \theta + d\theta)$ from a target comprised of Λ scattering centers per unit area?

(b)   Employ the expression you have obtained to find the fraction of α particles with incident energy 5 MeV which are scattered into a differential cone $(\theta, \theta + d\theta)$ at $\theta = \pi/2$, in passing through a gold sheet 1 μm thick.

Answers

(a)   If we assume that each particle in the incident beam sees only one scatterer and that there is a scattering event for each scatterer, then

$$\delta N = \Lambda \left(\frac{d\sigma}{d\Omega}\right) d\Omega$$

For the scattering into the cone $(\theta, \theta + d\theta)$

$$\delta N = \Lambda \int_0^{2\pi} d\phi \left(\frac{d\sigma}{d\Omega}\right) \sin\theta \, d\theta$$

$$= 2\pi\Lambda \left(\frac{d\sigma}{d\Omega}\right) \sin\theta \, d\theta$$

(b)   For a sheet of mass density ρ, thickness l, comprised of atoms with atomic mass A,

$$\Lambda = \frac{\rho N_0 l}{A}$$

where N_0 is Avogadro's number (N_0 atoms have mass A grams). For a gold foil l cm thick with $\rho = 19.3$ g/cm^3 and $A = 197$, we obtain $\Lambda = 5.9 \times 10^{22} \, l$ atoms/cm^2. For α particles of energy 5 MeV scattered by the nuclei of gold atoms, $qQ/E = e^2 \cdot 2 \times 79/E = 4.6 \times 10^{-12}$ cm. Thus we obtain

$$\delta N(\pi/2) = \frac{\pi}{2}\left(\frac{qQ}{E}\right)^2 \frac{\rho N_0 l}{A} d\theta \simeq 2 \times 10^{-4} \, d\theta$$

14.2   S-WAVE SCATTERING

Let us consider the configuration of a low-energy beam of point particles of mass m scattering from a finite spherical attractive well of depth V_0 and radius a.

(14.18)
$$V(r) = \begin{cases} -V_0 & \text{for } r < a \\ 0 & \text{for } r > a \end{cases}$$

If we assume that energies are sufficiently small that $ka \ll 1$, we need only look at the S-wave scattering. The corresponding Schrödinger equation is obtained from (14.9). Setting $l = 0$ and $u \equiv rR$ there results, for $r < a$,

$$\frac{d^2u_1}{dr^2} + k_1 u_1 = 0$$

$$\frac{\hbar^2 k_1^2}{2m} = E + V_0$$

The solution to this equation which corresponds to $R(r)$ remaining finite at $r = 0$ is

$$u_1 = A \sin k_1 r \qquad (r < a)$$

For $r > a$, $V = 0$ and we obtain the general solution

$$u_2 = B \sin (kr + \delta_0) \qquad (r > a)$$

$$\frac{\hbar^2 k^2}{2m} = E$$

Boundary conditions require continuity of $d \ln u/dr$ at $r = a$, which gives

(14.19) $$k_1 \cot k_1 a = k \cot (ka + \delta_0)$$

In that k_1 is finite, in the limit that k goes to zero,

$$\cot (ka + \delta_0) = \frac{k_1 \cot k_1 a}{k}$$

grows large so that $\sin (ka + \delta_0)$ grows small and we may set

$$\sin (ka + \delta_0) \simeq ka + \delta_0$$

Since $ka \ll 1$, this equation implies that $\delta_0 \ll 1$ as well. Under these conditions (14.19) reduces to

$$k_1 \cot k_1 a \simeq \frac{k}{ka + \delta_0}$$

or equivalently

$$\delta_0 = ka\left(\frac{\tan k_1 a}{k_1 a} - 1\right)$$

In that δ_0 is small, we may also set

$$\delta_0 \simeq \sin \delta_0 = ka\left(\frac{\tan k_1 a}{k_1 a} - 1\right)$$

We may now construct the scattering amplitude (14.17) and cross section (14.7).

(14.20)
$$\sigma = 4\pi a^2 \left(\frac{\tan k_1 a}{k_1 a} - 1 \right)^2$$

Two significant observations relevant to this study of attractive well scattering are discussed next.

S-Wave Resonances and Ramsauer Effect

First we note that when $k_1 a$ is an odd multiple of $(\pi/2)$, $\tan k_1 a$ is infinite and the cross section as given by (14.20) becomes singular. In that δ_0 is also infinite at these values of $k_1 a$, assumptions leading to (14.19) are violated and we must seek an alternative procedure to construct the cross section. Consider the relation (14.19), which assumes only that $ka \ll 1$. Let $k_1 a = n(\pi/2)$, where n is an odd number. At these values, (14.19) gives $\sin(\delta_0 + ka) = 1$, which with the condition $ka \ll 1$ yields $\sin \delta_0 \simeq 1$. Thus the maximum cross section at these *S-wave resonances* is

(14.21)
$$\sigma_{\max} = \frac{4\pi}{k^2}, \qquad k_1 a = n\left(\frac{\pi}{2}\right)$$

A more careful analysis pursued to higher angular momentum states, appropriate to larger incident energies, reveals corresponding resonances at $l = 1$, termed *P-wave resonances*, and so forth.

Whereas (14.20) suggests resonant scattering at odd multiples of $\pi/2$, it also indicates that the attractive scattering will become transparent to the incident beam at values of $k_1 a$ which satisfy the transcendental relation

$$\tan k_1 a = k_1 a$$

As noted in Section 7.8, such resonant transparency of an attractive well is experimentally corroborated in the scattering of low-energy electrons (~ 0.7 eV) by rare gas atoms and is termed the *Ramsauer effect*.

The Repulsive Sphere

The second observation related to our study of low-energy scattering by a scattering well is that merely changing the sign in the defining equations (14.18) produces the potential for a repulsive sphere of radius a. Solution for the corresponding scattering problem is effected by simply replacing k_1 by $i\kappa$, in the relations following (14.18). For the interior wavefunction we obtain

$$u_1 = A \sinh \kappa r \qquad (r < a)$$

$$\frac{\hbar^2 \kappa^2}{2m} = V_0 - E > 0$$

The exterior wavefunction u_2 maintains its sinusoidal dependence for $r > a$, as given in the equation preceding (14.19). Imposing boundary conditions at $r = a$ and, again assuming low-energy incident particles, we obtain the total scattering cross section,

$$(14.22) \qquad \sigma = 4\pi a^2 \left(\frac{\tanh \kappa a}{\kappa a} - 1 \right)^2$$

In the limit that $V_0 \to \infty$, the sphere becomes impenetrable and the total cross section reduces to

$$(14.23) \qquad \sigma = 4\pi a^2$$

In that this formula does not contain $\hbar$, our suspicion is that it is also appropriate to the classical domain. However, the obstructional area imposed by a rigid sphere of radius a to an incident beam of classical particles has the value πa^2, so the quantum cross section is larger than the classical one by a factor of 4. Although the cross section (14.23) does not contain Planck's constant, nevertheless one might still object to considering it a classical result in that it is relevant to the strictly nonclassical domain of large de Broglie wavelength. If a classical result is to be obtained, it should emerge in the limit of large incident energy, $ka \gg 1$. Such analysis, which includes the phase shifts of all waves,[1] again yields a cross section independent of $\hbar$, namely

$$\sigma = 2\pi a^2, \qquad ka \gg 1$$

which is still larger than the classical result. Thus the classical cross section does not emerge in the limit of large incident energy. This discrepancy may be ascribed to the sharp edge of the spherical potential barrier for the configuration at hand. Across the sharp potential step, dV/dx is infinite and it is impossible for the classical criterion (7.166) to be satisfied.

PROBLEMS

14.2   The scattering amplitude for a certain interaction is given by

$$f(\theta) = \frac{1}{k} (e^{ika} \sin ka + 3ie^{i2ka} \cos \theta)$$

where a is a characteristic length of the interaction potential and k is the wavenumber of incident particles.

(a)   What is the S-wave differential cross section for this interaction?

(b)   Suppose that the above scattering amplitude is appropriate to neutrons incident on a species of nuclear target. Let a beam of 1.3-eV neutrons with current 10^{14} cm^{-2} s^{-1} be incident on this target. What number of neutrons per second are scattered out of the beam into $4\pi \times 10^{-3}$ steradian about the forward direction?

---

[1] The calculation may be found in L. I. Schiff, *Quantum Mechanics*, 3rd ed., McGraw-Hill, New York, 1968.

14.3   Analysis of the scattering of particles of mass m and energy E from a fixed scattering center with characteristic length a finds the phase shifts

$$\delta_l = \sin^{-1}\left[\frac{(iak)^l}{\sqrt{(2l+1)l!}}\right]$$

(a)   Derive a closed expression for the total cross section as a function of incident energy E.
(b)   At what values of E does S-wave scattering give a good estimate of σ?

Answer (partial)

(a)

$$\sigma = \frac{4\pi\hbar^2}{2mE}\exp\left(\frac{-2mEa^2}{\hbar^2}\right)$$

14.3   CENTER-OF-MASS FRAME

In all of the preceding analysis, it has been assumed that the target particle remains fixed during the scattering process. This is the case if the mass of the target particle far exceeds that of the incident particle. More generally, however, the recoil motion of the target particle must be taken into account in any scattering analysis. Thus the general formulation of a scattering event involves two particles, of mass m_1 and m_2.

As described in Section 10.5, the motion of such two-particle systems may be described in terms of the motion of the center of mass and motion relative to the center of mass. The Hamiltonian of the relative motion (10.85) describes a single effective particle with reduced mass $\mu = m_1 m_2/(m_1 + m_2)$ at the radius $\mathbf{r} = \mathbf{r}_1 - \mathbf{r}_2$. This is the motion observed in a frame moving with the center of mass. So, in fact, in this center of mass frame, the scattering event may be described by a single particle of mass μ interacting with a potential $V(r)$ centered at a fixed origin. It follows that the preceding formulation of the cross section $\sigma(\theta)$ describing scattering from a fixed scattering center is appropriate to scattering in the center of mass frame. The only change is that the mass m of the incident particle is set equal to the reduced mass μ. In addition, we must note that the angle of deflection θ is measured in the center of mass frame. For example, in the expression (14.13) for the scattering amplitude, θ is the angle of scatter in the center of mass frame, which will henceforth be called θ_C. To obtain a relation between the scattering cross section $\sigma_L(\theta_L)$ in the frame of the experiment, or what is commonly called the *lab frame* and the cross section $\sigma_C(\theta_C)$ as measured in the center-of-mass frame, we note the following. The number of particles scattered into an element of solid angle in the lab frame $\mathbb{J}_{\text{inc}}(d\sigma_L/d\Omega_L)\,d\Omega_L$ is equal to the number scattered into the corresponding solid angle in the center of mass frame, $\mathbb{J}_{\text{inc}}(d\sigma_C/d\Omega_C)\,d\Omega_C$. This gives the equality

(14.24)
$$\frac{d\sigma_L}{d\cos\theta_L} = \frac{d\sigma_C}{d\cos\theta_C}\frac{d\cos\theta_C}{d\cos\theta_L}$$

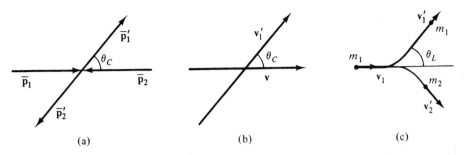

FIGURE 14.5   (a) In the center-of-mass frame, the total momentum is zero. (b) The relative velocity vector v rotates through the angle θ_C. (c) In the lab frame, m_2 is assumed to be at rest before collision.

The relation between $\cos \theta_C$ and $\cos \theta_L$ is obtained by examining the scattering in both frames. In transforming from one frame to the other, it is convenient to speak in terms of velocities. Such velocities are related to linear momentum through the prescription $\mathbf{v} = \hbar\mathbf{k}/m$. When one describes an "orbit" in this description, one has in mind a picture inferred by the direction of momentum $\mathbf{k}$ vectors. Thus, before collision, m_2 is at rest and the incident particle has velocity $\mathbf{v}_1 = \hbar\mathbf{k}/m_1$. After collision, m_1 is scattered through the angle θ_L.

The center of mass frame is characterized by the property that *total momentum in that frame is zero before and after collision* (Fig. 14.5). Letting barred variables denote values in the center of mass frame, and $\mathbf{v}$ the *relative velocity*,

$$\mathbf{v} \equiv \mathbf{v}_1 - \mathbf{v}_2$$

one obtains, for before the collision,

$$\bar{\mathbf{p}}_1 = m_1(\mathbf{v}_1 - \mathbf{v}_{CM}) = \mu\mathbf{v}$$

We may immediately conclude that

$$\bar{\mathbf{p}}_2 = -\mu\mathbf{v}$$

In a similar manner, after collision we write

$$\bar{\mathbf{p}}_1' = \mu\mathbf{v}'$$

$$\bar{\mathbf{p}}_2' = -\mu\mathbf{v}'$$

or, equivalently,

$$\bar{\mathbf{v}}_1' = \frac{\mu}{m_1}\mathbf{v}'$$

$$\bar{\mathbf{v}}_2' = -\frac{\mu}{m_2}\mathbf{v}'$$

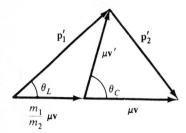

FIGURE 14.6 **Orientation of momentum and relative velocity for** $m_1/m_2 < 1$.

The corresponding relations in the lab frame are obtained by adding $\mathbf{v}_{CM}$ to the right-hand sides of these equations. Multiplying the resulting equations by m_1 and m_2, respectively, gives

$$\mathbf{p}_1' = \mu\mathbf{v}' + \frac{\mu}{m_2}\mathcal{P}$$

(14.25)
$$\mathbf{p}_2' = -\mu\mathbf{v}' + \frac{\mu}{m_1}\mathcal{P}$$

$$\mathcal{P} = \mathbf{p}_1 + \mathbf{p}_2$$

Since m_2 is at rest before scattering, $\mathbf{p}_1 = m_1\mathbf{v} = \mathcal{P}$. It follows that (14.25) may be rewritten

$$\mathbf{p}_1' = \mu\mathbf{v}' + \frac{m_1}{m_2}\mu\mathbf{v}$$

(14.26)
$$\mathbf{p}_2' = -\mu\mathbf{v}' + \mu\mathbf{v}$$

These vector equations imply the vector diagrams shown in Fig. 14.6. The desired relation between θ_L and θ_C is obtained by constructing $\tan \theta_L$ from the partial diagram shown in Fig. 14.7:

(14.27)
$$\tan \theta_L = \frac{\mu v' \sin \theta_C}{(m_1/m_2)\mu v + \mu v' \cos \theta_C}$$

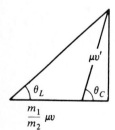

FIGURE 14.7 **Triangle used to obtain relation between** θ_L **and** θ_C.

Now H_{rel} is a conserved quantity throughout the scattering. Prior to, and after collision, H_{rel} is purely kinetic and has the respective values $\mu v^2/2$, $\mu v'^2/2$. It follows that the magnitude of the relative velocity is maintained in scattering

$$v = v'$$

Substituting this equality into (14.27) gives the desired relation,

$$(14.28) \qquad \tan \theta_L = \frac{\sin \theta_C}{\iota + \cos \theta_C} \qquad \left(\iota \equiv \frac{m_1}{m_2} \right)$$

This relation permits completion of (14.24):

$$(14.29) \qquad \frac{d\sigma_L}{d \cos \theta_L} = \frac{d\sigma_C}{d \cos \theta_C} \frac{(1 + \iota^2 + 2\iota \cos \theta_C)^{3/2}}{1 + \iota \cos \theta_C}$$

If the mass of the scatterer is very much larger than that of the incident particle, we may set $\iota = 0$ and the cross sections in both frames are equal. From (14.28) in this same extreme we obtain $\theta_L = \theta_C$.

In general, as (14.29) implies, scattering that is isotropic in the center of mass frame is not isotropic in the lab frame. For example, the isotropic cross section obtained for S-wave scattering (14.17),

$$\left(\frac{d\sigma}{d\Omega} \right)_C = |f(\theta)|^2 = \frac{\sin^2 \delta_0}{k^2}$$

when substituted in (14.29) yields [with (14.28)] an anisotropic cross section in the lab frame.

$$(14.30) \qquad \left(\frac{d\sigma}{d\Omega} \right)_L = \frac{\sin^2 \delta_0}{k^2} \frac{(1 + \iota^2 + 2\iota \cos \theta_C)^{3/2}}{1 + \iota \cos \theta_C}$$

Applications of results developed in this section appear in problems to follow. Whereas our primary example in the preceding analysis is relevant to low-energy scattering, where the potential of interaction plays a dominant role, the analysis to be developed in Section 14.4 addresses the case where the potential of interaction acts as a small perturbation on the incident plane-wave state. This analysis, known as the Born approximation, has many applications.

14.4 Assume that the differential cross section for a given interaction potential $d\sigma/d\Omega$ is isotropic in the center-of-mass frame. For mass ratio $\epsilon \ll 1$, what is the ratio of the differential cross section in the forward direction to that in the $\theta = \pi/2$ direction in the lab frame?

Answer

$$\frac{d\sigma(0)}{d\sigma(\pi/2)} = 1 + 2\epsilon$$

14.5 At what value of θ_C will the cross section vanish in the lab frame for S-wave scattering of two particles with mass ratio ϵ?

14.4 THE BORN APPROXIMATION

Harmonic perturbation theory, developed in Chapter 13, includes as a special case the example of a constant potential that has been turned on for t seconds. The perturbation Hamiltonian[1] is then given by (13.53) with $\omega = 0$. As was shown in Section 13.6, the theory of harmonic perturbation leads naturally to Fermi's formula (13.64) for cases where final states comprise a continuum. Such, of course, is the situation for scattering problems.

For these problems the perturbation Hamiltonian is the interaction potential, which is viewed as being "turned on" during the time that the incident particle is in the range of the potential. The incident particle enters the range of interaction with momentum $\hbar\mathbf{k}$ and leaves the range of interaction with momentum $\hbar\mathbf{k}'$. Such states of definite $\mathbf{k}$ before and after interaction correspond to plane-wave states (Fig. 14.8). Let us suppose that the scattering experiment is performed in a large cubical box of volume L^3. Normalized plane-wave states corresponding to $\mathbf{k}$ and $\mathbf{k}'$ are then given by

$$|\mathbf{k}\rangle = \frac{e^{i\mathbf{k}\cdot\mathbf{r}}}{L^{3/2}}$$

(14.31)

$$|\mathbf{k}'\rangle = \frac{e^{i\mathbf{k}'\cdot\mathbf{r}}}{L^{3/2}}$$

To apply Fermi's formula (13.64) for the rate of transition from the $\mathbf{k}$ to the $\mathbf{k}'$ state, caused by the perturbing potential $V(r)$,

(14.32)

$$\bar{w}_{\mathbf{k}\mathbf{k}'} = \frac{2\pi}{\hbar} g(E_{k'})|\langle\mathbf{k}'|V|\mathbf{k}\rangle|^2$$

---

[1] For the case $\omega = 0$, the factor 2 is deleted in (13.53).

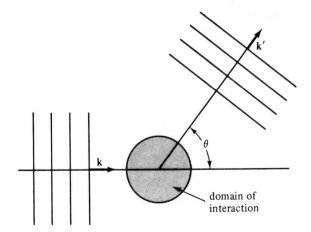

FIGURE 14.8   **In the Born approxima-tion, incident and scattered particles are in plane-wave states.**

domain of
interaction

we must know the density of final states $g(E_{k'})$. Having prescribed that the scattering experiment is performed in a large box of volume L^3, we may employ the expression for $g(E)$ as given by (8.141). Written in terms of final momentum $\hbar k'$, this expression becomes for nonspinning particles[1]

$$(14.33) \qquad\qquad g(E_{k'}) = \frac{mL^3 k'}{2\pi^2 \hbar^2}$$

Now we wish to use the rate formula (14.32) to obtain an expression for the differential scattering cross section $d\sigma$. This parameter was defined by (14.1) according to which the number of particles scattered into $d\Omega$ per second is $\mathbb{J}_{\text{inc}}\, d\sigma$. To relate the transition rate $\bar{w}_{\mathbf{kk'}}$ to $d\sigma$, we note that the incident plane wave $|\mathbf{k}\rangle$ given in (14.31) corresponds to an incident current

$$(14.34) \qquad\qquad \mathbf{J}_{\text{inc}} = \frac{\hbar \mathbf{k}}{mL^3}$$

In that $g(E_k)$ as given by (14.33) is isotropic in $\mathbf{k'}$, it represents the density of final $\mathbf{k'}$ states in all 4π solid angle. To select those scattered states that lie in the direction $d\Omega$ about $\mathbf{k'}$, we multiply g by the ratio $d\Omega/4\pi$. With $g(E_{k'})$ so augmented, $\bar{w}_{\mathbf{kk'}}$ then represents the rate at which particles of the incident flux (14.34) are scattered into $d\Omega$ in the direction of $\mathbf{k'}$. This rate is by definition the product $\mathbb{J}_{\text{inc}}\, d\sigma$. Thus we obtain the desired relation

$$\mathbb{J}_{\text{inc}}\, d\sigma = \frac{d\Omega}{4\pi}\, \bar{w}_{\mathbf{kk'}}$$

[1] The g factor in Problem 2.42 represents density of states per unit volume.

Inserting previous expressions, we obtain

$$(14.35) \qquad \frac{d\sigma}{d\Omega} = \left(\frac{mL^3}{2\pi\hbar^2}\right)^2 \frac{k'}{k} |\langle \mathbf{k}' | V | \mathbf{k} \rangle|^2$$

Since particles suffer no loss in energy in the scattering process, we may equate

$$k = k'$$

The Scattering Amplitude

Recalling (14.6), which relates $d\sigma$ to the scattering amplitude $f(\theta)$, and inserting the explicit forms (14.31) for incident and scattered states into (14.35), allows the identification (with a conventional minus sign)

$$(14.36) \qquad f(\theta) = -\frac{m}{2\pi\hbar^2} \int V(r) e^{i\mathbf{r} \cdot (\mathbf{k} - \mathbf{k}')} \, d\mathbf{r}$$

This formula for the scattering amplitude may be further simplified through the substitution

$$\mathbf{K} = \mathbf{k} - \mathbf{k}'$$

As is evident in Fig. 14.9a, owing to the equal magnitudes of $\mathbf{k}$ and $\mathbf{k}'$, we may set

$$K = 2k \sin\left(\frac{\theta}{2}\right)$$

where θ is the angle of scatter. With the differential volume of integration $d\mathbf{r}$ in (14.36) written in spherical coordinates and the polar axis taken to be coincident with $\mathbf{K}$ (Fig. 14.9b), we obtain

$$f(\theta) = -\frac{m}{2\pi\hbar^2} \int_0^{2\pi} d\bar{\phi} \int_0^{\pi} d\bar{\theta} \sin\bar{\theta} \int_0^{\infty} dr \, r^2 V(r) e^{iKr\cos\bar{\theta}}$$

$$= -\frac{m}{\hbar^2} \int_0^{\infty} dr \, r^2 V(r) \int_{-1}^{1} d\eta \, e^{iKr\eta}$$

Integrating over $\eta \equiv \cos\bar{\theta}$ gives

$$(14.37) \qquad \boxed{f(\theta) = -\frac{2m}{\hbar^2 K} \int_0^{\infty} dr \, r V(r) \sin Kr}$$

This expression for the scattering amplitude is called the *standard form of the Born approximation*.

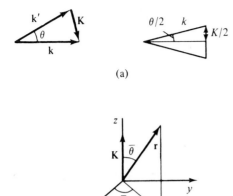

(a)

(b)

FIGURE 14.9 (a) Transformation $\mathbf{K} = \mathbf{k} - \mathbf{k}'$. (b) Spherical coordinate frame with $\mathbf{K}$ aligned with the polar axis.

In applying this formula for the scattering amplitude, one should keep in mind that it is derived on the basis of perturbation theory according to which the scattering potential should be small compared to the free-particle (unperturbed) Hamiltonian. This will be the case for sufficiently large incident energies or sufficiently weak strengths of potential.

The Shielded Coulomb Potential

Let us apply (14.37) to calculate the cross section for the shielded Coulomb potential,[1]

$$V(r) = - \frac{Ze^2 \exp{(-r/a)}}{r}$$

The exponential factor for $r > a$ acts to shield the bare Coulomb potential Ze^2/r between two particles with respective charges Ze and e. Thus beyond the range a, the potential is exponentially small. Within the range, $r < a$, the potential is essentially Coulombic. Substituting the shielded Coulomb potential into (14.37) gives

$$f(\theta) = \frac{2mZe^2}{\hbar^2 K} \int_0^\infty e^{-r/a} \sin Kr \, dr$$

$$= \frac{2mZe^2}{\hbar^2} \frac{1}{K^2 + (1/a)^2}, \qquad K = 2k \sin \left(\frac{\theta}{2} \right)$$

---

[1] This potential enters in three independent areas of physics, where it carries the following names: in plasma physics, the *Debye potential*; in high-energy physics, the *Yukawa potential*; in solid-state physics, the *Thomas–Fermi potential*.

The corresponding scattering cross section is obtained from (14.6).[1]

(14.38)
$$\frac{d\sigma}{d\Omega} = \frac{(2mZe^2/\hbar^2)^2}{[K^2 + (1/a)^2]^2}$$

In the limit of large incident energies $K^2 \gg a^{-2}$, the predominant contribution to $d\sigma$ is due to the bare Coulomb potential. The resulting cross section, employed previously in Problem 14.1, appears as

(14.39)
$$\frac{d\sigma}{d\Omega} = \left[\frac{Ze^2}{4(\hbar^2 k^2/2m) \sin^2 (\theta/2)} \right]^2$$

$$= \left(\frac{Ze^2}{4E} \right)^2 \frac{1}{\sin^4 (\theta/2)}, \qquad E = \frac{\hbar^2 k^2}{2m}$$

This is the precise expression for the Rutherford cross section for the scattering of a charged particle with charge e and mass m from a fixed charge Ze. Furthermore, the classical evaluation of the Rutherford cross section also gives (14.39), with $E = p^2/2m$.

PROBLEMS

14.6 Using the Born approximation, evaluate the differential scattering cross section for scattering of particles of mass m and incident energy E by the repulsive spherical well with potential

$$V(r) = \begin{cases} V_0, & 0 < r < a \\ 0, & r > a \end{cases}$$

Exhibit explicit E and θ dependence.

Answer

$$\frac{d\sigma}{d\Omega} = \left(\frac{2mV_0}{\hbar^2 K^3} \right)^2 (\sin Ka - Ka \cos Ka)^2$$

$$\hbar K = 2\sqrt{2mE} \sin (\theta/2)$$

14.7 Using the Born approximation, obtain an integral expression for the total cross section for scattering of particles of mass m from the attractive Gaussian potential

$$V(r) = -V_0 \exp\left[-\left(\frac{r}{a} \right)^2 \right]$$

[1] With m replaced by the reduced mass μ, (14.38) represents the cross section in the center-of-mass frame.

14.8 An important parameter in scattering theory is the scattering length a. This length is defined as the negative of the limiting value of the scattering amplitude as the energy of the incident particle goes to zero.

$$a = - \lim_{k \to 0} f(\theta)$$

(a)   For low-energy scattering and relatively small phase shift, show that

$$a = - \lim_{k \to 0} \frac{\delta_0}{k}$$

(b)   For the same conditions as in part (a), show that

$$\sigma = 4\pi a^2$$

(c)   What is the scattering length for point particles scattering from a rigid sphere of arbitrary radius $\bar{a}$?

14.5   ATOMIC-RADIATIVE ABSORPTION CROSS SECTION

Returning to our analysis of Section 13.9, again we consider a flux of photons incident on an atom, carrying one photon per unit volume. Since the photon moves with speed c, this gives an incident photon current

(14.40)
$$J_{\text{inc}} = c \times \frac{1 \text{ photon}}{\text{cm}^3} = c \left(\frac{\text{photons}}{\text{cm}^2 \text{ s}} \right)$$

There is a probability rate for each photon in the incident current to be absorbed by the atom. The principle of microscopic reversibility allows us to equate this probability rate for absorption to the corresponding probability rate of atomic decay.

With Fermi's golden rule (13.64) we may then write the probability rate for atomic absorption as

(14.41)
$$w_{nn'} = \frac{2\pi}{\hbar} |\langle n', \mathbf{k} | \mathbb{H}_- | n \rangle|^2 \frac{g(\omega)}{\hbar}$$

where we have made the replacement

$$g(E) = \frac{g(\omega)}{\hbar}$$

[See (13.111b) and recall that $g(E) = 8\pi\bar{g}(E)$ and that $V = 1$.] With (13.106), our preceding equation (14.41) becomes

(14.42)
$$w_{nn'} = \frac{2\pi}{\hbar} e^2 (2\pi\hbar\omega) \frac{g(\omega)}{\hbar} |\langle n' | \mathbf{a} \cdot \mathbf{r} | n \rangle|^2$$

If the polarization unit vector $\mathbf{a}$ of the incident field is randomly oriented, then as was previously demonstrated in our discussion of the Einstein B coefficient (Section 13.7), we may set

(14.43)
$$|\langle n'|\mathbf{a}\cdot\mathbf{r}|n\rangle|^2 = \tfrac{1}{3}r_{nn'}^2 = \frac{1}{3}\frac{d_{nn'}^2}{e^2}$$

Here we have reintroduced the dipole moment $\mathbf{d}$ (13.69).

With (14.43) placed into (14.42) we find

(14.44)
$$w_{nn'} = \frac{4}{3}\frac{\pi^2\omega d_{nn'}^2}{\hbar}g(\omega)$$

This is the probability rate for an atom to absorb a photon from an incident current carrying one photon per unit volume. Since the incident current in the present configuration is so normalized, $w_{nn'}$ represents the rate at which photons are absorbed by the atom from the incident beam. With our previous definition of cross section we may then write

(14.45)
$$J_{\text{inc}}\sigma_{nn'} = w_{nn'}$$

Note the dimensions:

$$J_{\text{inc}}\left(\frac{\text{photons}}{\text{cm}^2\,\text{s}}\right) \times \sigma\,(\text{cm}^2) = w_{nn'}\left(\frac{1}{\text{s}}\right)$$

With (14.40) and (14.45) we obtain the total cross section,

(14.46)
$$\sigma_{nn'} = \frac{4}{3}\frac{\pi^2\omega d_{nn'}^2}{\hbar c}g(\omega)$$

for resonant absorption at the incident frequency

$$\hbar\omega = E_n - E_{n'}$$

A more accurate description of this process includes the possibility of absorption of incident photons which are off-resonance. For such cases, with $\bar{\omega} \equiv \omega_{nn'}$, (14.46) is generalized to the form

(14.47)
$$\sigma(\bar{\omega}, \omega) = \frac{4}{3}\frac{\pi^2 d^2}{\hbar c}\bar{\omega}g(\bar{\omega}, \omega)$$

where we have set

$$g(\omega) \to g(\bar{\omega}, \omega)$$

The function $g(\bar{\omega}, \omega)$ is the so-called *line-shape factor*.

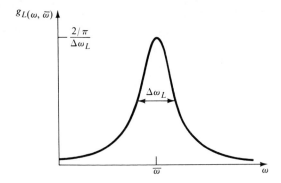

FIGURE 14.10   Lorentzian line-shape.

A realistic expression for $g(\bar{\omega}, \omega)$ which is appropriate to many line-broadening processes is the *Lorentzian line-shape factor*,

$$(14.48) \qquad g_L(\bar{\omega}, \omega) = \frac{1}{\pi} \frac{\Delta\omega_L/2}{(\bar{\omega} - \omega)^2 + (\Delta\omega_L/2)^2}$$

See Fig. 14.10.

The spreading of an absorption line is attributed to relaxation processes—such as, for example, the relaxation of excited atomic states incurred in atomic collisions. If τ represents the decay time for such processes, then one sets

$$\frac{\Delta\omega_L}{2} = \frac{1}{\tau}$$

In the idealized limit that these states last indefinitely, $\Delta\omega_L \to 0$, and $g_L(\bar{\omega}, \omega)$ becomes sharply peaked about $\omega = \bar{\omega}$. In this limit (14.47) becomes

$$(14.49) \qquad \sigma(\bar{\omega}, \omega) = \frac{4}{3} \frac{\pi^2 d^2}{\hbar c} \bar{\omega}\delta(\omega - \bar{\omega})$$

Here we have employed the delta function representation (C.9).

14.6 ELEMENTS OF FORMAL SCATTERING THEORY.
THE LIPPMANN–SCHWINGER EQUATION

In this concluding section we present a brief introduction to the formal theory of scattering, central to which is the Lippmann–Schwinger equation.[1] An elementary

---

[1] For further discussion, see E. Merzbacher, *Quantum Mechanics*, 2nd ed., Wiley, New York, 1970, Chap. 19.

derivation of this equation is presented based on the interaction picture described previously in Section 11.12. The Lippmann–Schwinger equation so derived appears in a form independent of representation. Writing this equation in coordinate representation is found to give an integral equation for scattered states, which in turn gives a general expression for the scattering amplitude. In the Born approximation this relation returns our previous expression for $f(\theta)$ given by (14.36).

We consider the Hamiltonian

$$(14.50) \qquad \hat{H} = \hat{H}_0 + \hat{V}e^{-\varepsilon|t|/\hbar}$$

where $\hat{H}_0$ is the free-particle Hamiltonian and the infinitesimal parameter ε has dimensions of energy. The interaction $\hat{V}$ is assumed to be independent of time. For small ε the exponential factor has the effect of "turning on" the interaction $\hat{V}$ in the interval about $t = 0$. We will also find that the presence of this factor insures convergence of integration in the derivation to follow (in both limits $t \to \pm \infty$).

Recall that the Schrödinger equation in the interaction picture has the integral form (11.144),[1]

$$(14.51) \qquad |\psi_I(t)\rangle = |\psi_I(t_0)\rangle + \frac{1}{i\hbar} \int_{t_0}^{t} \hat{V}_I(t')|\psi_I(t')\rangle \, dt'$$

where, we recall,

$$(14.52a) \qquad |\psi_I(t)\rangle = e^{it\hat{H}_0/\hbar}|\psi(t)\rangle$$

$$(14.52b) \qquad \hat{V}_I(t) = e^{it\hat{H}_0/\hbar}\hat{V}e^{-it\hat{H}_0/\hbar}$$

As we wish to apply (14.51) to scattering theory, we stipulate that $\hat{V}_I(t) \to 0$ in the limits $t \to \pm \infty$. At these asymptotic values, with the interaction vanishingly small, $|\psi_I(t)\rangle$ loses its time dependence, and we define

$$(14.53) \qquad |\varphi\rangle \equiv |\psi_I(\pm\infty)\rangle$$

which may be identified as free-particle states.

Rewriting (14.51) over the interval $t_0 = \pm \infty$, $t = 0$, and identifying the scattering states

$$(14.54) \qquad |\psi^{(\pm)}\rangle \equiv |\psi(0)\rangle$$

gives the equation

$$(14.55) \qquad |\psi^{(\pm)}\rangle = |\varphi\rangle + \frac{1}{i\hbar} \int_{\mp\infty}^{0} \hat{V}_I(t)|\psi_I(t)\rangle \, dt$$

---

[1] Ket notation is employed to obtain a relation independent of representations.

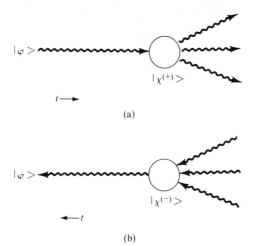

(a)

(b)

FIGURE 14.11 (a) "In" solution. (b) "Out" solution.

Note that with this choice of interval, $|\psi^{(\pm)}\rangle$, given by (14.54), represents scattered states in the domain of interaction. Furthermore, as $|\psi^{(+)}\rangle$ is relevant to the time interval $(-\infty \le t \le 0)$, we may identify it with incoming incident waves, commonly called the "in" solution. As $|\psi^{(-)}\rangle$ relates to the interval $(\infty, 0)$, it is the time-reversed state of $|\psi^{(+)}\rangle$ and is commonly called the "out" solution. See Fig. 14.11. In the limit $\varepsilon \to 0$ we take $|\psi^{(\pm)}\rangle$ to be an eigenstate of the total Hamiltonian and write

(14.56)
$$\hat{H}|\psi^{(\pm)}\rangle = E|\psi^{(\pm)}\rangle$$

Reduction of Interaction Integral

Consider the integrand in (14.55). With (14.52a, b) we write

(14.57)
$$\hat{V}_I(t)|\psi_I(t)\rangle = e^{i\hat{H}_0 t/\hbar} \hat{V} e^{-\varepsilon|t|/\hbar} e^{-i\hat{H}_0 t/\hbar} e^{i\hat{H}_0 t/\hbar} |\psi(t)\rangle$$
$$= e^{i\hat{H}_0 t/\hbar} \hat{V} e^{-\varepsilon|t|/\hbar} e^{-i\hat{H}t/\hbar} |\psi(0)\rangle$$

and note

$$|\psi(0)\rangle = |\psi_I(0)\rangle = |\psi^{(\pm)}\rangle$$

Substituting this identification into (14.57) followed by replacement of (14.57) into (14.55) gives

(14.58)
$$|\psi^{(\pm)}\rangle = |\varphi\rangle + \frac{1}{i\hbar} \int_{\mp\infty}^{0} e^{i\hat{H}_0 t/\hbar} \hat{V} e^{-\varepsilon|t|/\hbar} e^{-i\hat{H}t/\hbar} |\psi^{(\pm)}\rangle \, dt$$

With (14.56) the preceding becomes

(14.59)
$$|\psi^{(\pm)}\rangle = |\varphi\rangle + \frac{1}{i\hbar} \int_{\mp\infty}^{0} e^{i\hat{H}_0 t/\hbar} \hat{V} e^{-\varepsilon|t|/\hbar} e^{-iEt/\hbar} |\psi^{(\pm)}\rangle \, dt$$

For "in" solutions we encounter the integral

$$\hat{G}^{(+)} = \frac{1}{i\hbar} \int_{-\infty}^{0} dt \exp\left[\frac{-t(i\hat{H}_0 - iE + \varepsilon)}{\hbar}\right]$$

$$= \frac{1}{E - \hat{H}_0 + i\varepsilon}$$

Similarly,

$$\hat{G}^{(-)} = \frac{1}{E - \hat{H}_0 - i\varepsilon}$$

Note that without the presence of ε the integrals $\hat{G}^{(\pm)}$ do not converge. Substituting these expressions for $\hat{G}^{(\pm)}$ into (14.59) gives the *Lippmann–Schwinger equation*,

(14.60)
$$\boxed{|\psi^{(\pm)}\rangle = |\varphi\rangle + \frac{1}{E - \hat{H}_0 \pm i\varepsilon} |\psi^{(\pm)}\rangle}$$

which, as previously noted, is independent of specific representation.

Scattering Amplitude Revisited

In Problem 14.11 you are asked to show that the coordinate representation of the "in" solution to (14.60) assumes the form

(14.61)
$$\psi_\mathbf{k}^{(+)}(r) = \varphi_\mathbf{k}(r) - \frac{m}{2\pi\hbar^2} \int \frac{\exp(ik|\mathbf{r} - \mathbf{r}'|)}{|\mathbf{r} - \mathbf{r}'|} V(\mathbf{r}')\psi_\mathbf{k}{}^+(\mathbf{r}') \, d\mathbf{r}'$$

Here we have made the identifications

(14.62a)
$$\langle \mathbf{r} | \varphi \rangle = \varphi_\mathbf{k}(\mathbf{r})$$

(14.62b)
$$\langle \mathbf{r} | \psi^{(+)} \rangle = \psi_\mathbf{k}^{(+)}(\mathbf{r})$$

(14.62c)
$$\langle \mathbf{r} | \hat{V}\psi^{(+)} \rangle = V(\mathbf{r})\psi^{(+)}(\mathbf{r})$$

At large distances from the interaction domain we may write

$$\frac{1}{|\mathbf{r} - \mathbf{r}'|} \simeq \frac{1}{r} + \cdots$$

and

$$k|\mathbf{r} - \mathbf{r}'| = kr\left[1 + \left(\frac{r'}{r}\right)^2 - \frac{2\mathbf{r}\cdot\mathbf{r}'}{r^2}\right]^{1/2}$$

$$\simeq kr\left[1 - \frac{\mathbf{r}\cdot\mathbf{r}'}{r^2} + \cdots\right] = kr - \mathbf{k}'\cdot\mathbf{r}'$$

where we have set

$$\mathbf{k}' \equiv \frac{k\mathbf{r}}{r}$$

Substituting these expansions in (14.61) gives

(14.63)
$$\psi_{\mathbf{k}}^{(+)}(\mathbf{r}) = \varphi_{\mathbf{k}}(\mathbf{r}) - \frac{m}{2\pi\hbar^2}\frac{e^{ikr}}{r}\langle\varphi_{\mathbf{k}'}|V|\psi_{\mathbf{k}}^{(+)}\rangle$$

Comparison with (14.7) gives the following expression for the scattering amplitude:

(14.64)
$$f(\theta) = -\frac{m}{2\pi\hbar^2}\langle\varphi_{\mathbf{k}'}|V|\psi_{\mathbf{k}}^{(+)}\rangle$$

In the Born approximation

$$\psi_{\mathbf{k}}^{+} \to \varphi_{\mathbf{k}}$$

and (14.64) returns our previous finding (14.36). However, one should bear in mind that (14.63) is, more generally, an integral equation for $\psi_{\mathbf{k}}^{(+)}$, solution to which gives a more accurate expression for the scattering amplitude through (14.64).

PROBLEMS

14.9 A beam of photons at a given frequency propagates into a medium of atoms with density n (cm^{-3}). The cross section for absorption of photons at this frequency by the atoms is σ. If J is incident photon flux, then argue that the decrease in J in the distance dx due to absorption is

$$dJ = -\kappa J\,dx$$

where

$$\kappa = \mathbf{n}\sigma$$

14.10 A monochromatic beam of photons at frequency $\nu = 10^{14}$ Hz and intensity 1.4 keV/ cm^2 s is incident on a gas of atoms of density $n = 10^{18}$ cm^{-3}. At the given frequency the radiation is very near resonance of the atoms. The related transition dipole moment of atoms in the gas has magnitude $0.4a_0e$. At what distance (cm) into the gas will the intensity of the beam be e^{-1} times its starting value? (*Hint:* Use results of the preceding problem.)

14.11 Working in the coordinate representation, employ the Lippmann–Schwinger equation (14.60) to derive its coordinate representation (14.61).

Answer

Let us label

$$\hat{G}_\pm \equiv \lim_{\varepsilon \to 0} G^{(\pm)}$$

The Lippmann–Schwinger equation (14.60) may then be written

$$|\psi^{(\pm)}\rangle = |\varphi\rangle + \hat{G}_\pm \hat{V} |\psi^{(\pm)}\rangle$$

To obtain the coordinate representation of this equation, we operate on it from the left with $\langle \mathbf{r} |$ to obtain [see (14.62)]

$$\psi_{\mathbf{k}}^{(\pm)}(\mathbf{r}) = \varphi_{\mathbf{k}}(\mathbf{r}) + \langle \mathbf{r} | \hat{G}_\pm \hat{V} | \psi^{(\pm)} \rangle \equiv \varphi_{\mathbf{k}}(\mathbf{r}) + I_\pm$$

which serves to define the interaction term $I_\pm$. Developing this term, we obtain[1]

$$I_\pm = \int d\mathbf{r}' \int d\bar{\mathbf{k}} \, \langle \mathbf{r} | \bar{\mathbf{k}} \rangle \langle \bar{\mathbf{k}} | \hat{G}_\pm | \mathbf{r}' \rangle \langle \mathbf{r}' | \hat{V} | \psi^{(\pm)} \rangle$$

We recall[2]

$$\langle \mathbf{r} | \mathbf{k} \rangle = \frac{1}{(2\pi)^{3/4}} e^{i\mathbf{k} \cdot \mathbf{r}}$$

and

$$\hat{H}_0 | \mathbf{k} \rangle = \frac{\hbar^2 k^2}{2m} | k \rangle$$

whence

$$\langle \bar{\mathbf{k}} | \hat{G}_\pm | \mathbf{r}' \rangle = \lim \frac{1}{E - (\hbar^2 \bar{k}^2 / 2m) \pm i\varepsilon} \langle \bar{\mathbf{k}} | \mathbf{r}' \rangle$$

Thus we obtain

$$I_\pm = \frac{1}{(2\pi)^3} \int d\mathbf{r}' \int d\bar{\mathbf{k}} \, \frac{e^{i\bar{\mathbf{k}} \cdot (\mathbf{r} - \mathbf{r}')}}{E - (\hbar^2 \bar{k}^2 / 2m) \pm i\varepsilon} \langle \mathbf{r}' | \hat{V} | \psi^{(\pm)} \rangle$$

Consider the $\bar{\mathbf{k}}$ integration,

$$\Phi_\pm \equiv \int d\bar{\mathbf{k}} \, \frac{e^{i\bar{\mathbf{k}} \cdot \Delta \mathbf{r}}}{R_\pm}$$

---

[1] Here we employ the spectral resolution of unity (see Problem 11.1).

[2] Note that this form gives the proper normalization

$$\int d\mathbf{k} \, \langle \mathbf{r} | \mathbf{k} \rangle^* \langle \mathbf{k} | \mathbf{r}' \rangle = \delta(\mathbf{r} - \mathbf{r}')$$

See (C. 11).

where

$$\Delta \mathbf{r} \equiv \mathbf{r} - \mathbf{r}'$$

$$R_{\pm} = E - \frac{\hbar^2 \bar{k}^2}{2m} \pm i\varepsilon$$

With

$$\int d\bar{\mathbf{k}} = 2\pi \int_0^\infty \bar{k}^2 \, d\bar{k} \int_{-1}^1 d\mu$$

$$\mu \equiv \cos\theta = \frac{\bar{\mathbf{k}} \cdot \Delta \mathbf{r}}{|\bar{\mathbf{k}} \cdot \Delta \mathbf{r}|}$$

integration over μ gives

$$\int_{-1}^1 d\mu \, \exp i\bar{k} \, \Delta r \mu = \frac{1}{i\bar{k} \, \Delta r} \left[\exp(i\bar{k} \, \Delta r) - \exp(-i\bar{k} \, \Delta r) \right]$$

As $\bar{k}/R_{\pm}$ is an odd function of $\bar{k}$, we find

$$\Phi_{\pm} = 2\pi \int_0^\infty d\bar{k} \, \bar{k}^2 \frac{1}{i\bar{k} \, \Delta r} \frac{\left[\exp(i\bar{k} \, \Delta r) - \exp(-i\bar{k} \, \Delta r) \right]}{R_{\pm}}$$

$$= \frac{2\pi}{i \, \Delta r} \int_{-\infty}^\infty \frac{d\bar{k} \, \bar{k} e^{i\bar{k} \Delta r}}{R_{\pm}}$$

Next we set

$$E = \frac{\hbar^2 k^2}{2m}$$

which, by conservation of energy, is the same as the free-particle energy of the incident wave, $\varphi_{\mathbf{k}}(\mathbf{r})$. We obtain

$$\frac{2m}{\hbar^2} R_{\pm} = \left(k^2 - \bar{k}^2 \pm i \frac{2m\varepsilon}{\hbar^2} \right)$$

$$= (k - \bar{k} \pm i\bar{\varepsilon})(k + \bar{k} \pm i\bar{\varepsilon})$$

where $\bar{\varepsilon} \equiv m\varepsilon/\hbar^2 k$. Thus

$$\frac{\bar{k}}{R_{\pm}} = -\frac{2m}{\hbar^2} \left(\frac{1/2}{\bar{k} - k \mp i\bar{\varepsilon}} + \frac{1/2}{\bar{k} + k \pm i\bar{\varepsilon}} \right)$$

We are now prepared to integrate over $\bar{k}$ by contour integration. As the integrand contains the factor $\exp i\bar{k} \, \Delta r$, it must be closed in the domain $\text{Im } \bar{k} > 0$, that is, the upper half $\bar{k}$ plane. For Φ_{+} there is a pole at

$$\bar{k} = k + i\bar{\varepsilon}$$

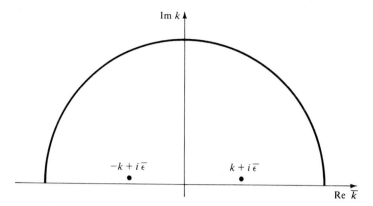

FIGURE 14.12   The contour for the integration in $\Phi_{\pm}$. The pole at $k + i\varepsilon'$ contributes to $\psi^{(+)}$ and the one at $-k + i\varepsilon'$ contributes to $\psi^{(-)}$.

whereas for Φ_- there is a pole at

$$\bar{k} = -k + i\bar{\varepsilon}$$

See Fig. 14.12.

Passing to the limit $\bar{\varepsilon} \to 0$, we obtain

$$\Phi_{\pm} = -\frac{2\pi}{i\,\Delta r}\frac{2m}{\hbar^2}\,2\pi i(\tfrac{1}{2})e^{\pm ik\,\Delta r}$$

$$= -\frac{4\pi^2 m}{\hbar^2}\frac{e^{\pm ik\,\Delta r}}{\Delta r}$$

whence

$$I_+ = -\frac{1}{(2\pi)^3}\frac{4\pi^2 m}{\hbar^2}\int d\mathbf{r}'\,\frac{e^{ik\,\Delta r}}{\Delta r}\,\langle\mathbf{r}'|\hat{V}|\psi^{(+)}\rangle$$

which when inserted in our starting relation

$$\psi_{\mathbf{k}}^{(+)} = \varphi_{\mathbf{k}}(r) + I_+$$

returns (14.61).

14.12   Is (14.61) a valid relation for inelastic collisions—for example, an ionizing collision in which an electron is emitted from an atom due to, say, electron scattering?

Answer

As the energy of the incident electron is not conserved (it loses energy in releasing the bound electron), one cannot equate energy of the scattered electron to its incident free-particle value, and the derivation in the preceding problem is invalid.

LIST OF SYMBOLS

a_0	Bohr radius
a	Scattering length
$\hat{a}, \hat{a}^\dagger$	Annihilation and creation operators
$\mathbf{A}$	Vector potential
$Ai(x)$	Airy function
A_{lk}, B_{lk}	Einstein A and B coefficients
$b(\mathbf{k})$	Momentum probability amplitude
$\mathscr{B}$	Magnetic field (also $\mathfrak{G}$ in figures)
$\mathfrak{B}$	Basis
c	Speed of light
$c_k(t)$	Transition probability amplitude
$C_{m_1 m_2}$	Clebsch–Gordon coefficient
$\mathscr{D}$	Coefficient matrix
$\hat{\mathscr{D}}$	Displacement operator
$\hat{D}$	Derivative operator $(\partial/\partial x)$
$\mathbf{d}$	Electric dipole moment
DC	Denotes constant in time
$\mathscr{E}$	Electric field (also $\mathscr{E}$ in figures)
$\mathbf{e}$	Unit vector
e	Charge
$\mathbb{E}$	Photoelectric energy
$\mathbb{E}$	Perturbation energy
E	Energy
E_F	Fermi energy
$f_{nn'}$	Oscillator strengths
$f(\theta)$	Scattering amplitude
g	Acceleration due to gravity; Landé g factor
$g(E), g(v)$	Density of states
$\bar{g}$	Degeneracy, Density of states
H	Hamiltonian
H'	Perturbation Hamiltonian
$\mathfrak{H}$	Hilbert space
$\mathscr{H}_n(\xi)$	Hermite polynomial
$\mathbb{H}$	r-dependent perturbation Hamiltonian

$h, \hbar$	Planck's constant
$\hat{I}$	Identity operator
j	Total angular momentum quantum number
$\mathbf{J}$	Spin, orbital, or total angular momentum
$\mathbf{J}$	Current density
k_B	Boltzmann's constant
k, κ	Wavenumbers
K^2	Eigenvector of $\hat{J}^2/\hbar^2$
l	Orbital angular momentum quantum number
$\mathbf{L}$	Orbital angular momentum
$\mathscr{L}$	Orbital-angular-momentum-term notation symbol
L	Edge length
$L_p^q(x)$	Laguerre polynomial
m	Mass
m^*	Effective mass
M	Mass of center of mass
$\mathscr{N}_E, \mathscr{N}_k$	Number of states in an energy band
N	Total number of particles
$\mathbf{p}$	Momentum
$\not{p}_r$	Unsymmetrized radial momentum
p_r	Radial momentum
P_{12}	Two-particle joint probability density
P	$P\,d\mathbf{r}$ is probability
P_r	$P_r\,dr$ is probability
$\bar{P}$	Nondimensional P_r
P	Radiated power
$\mathscr{P}$	Momentum of center of mass
P	Permutation operator
$|P|$	Order of permutation
$\mathbb{P}$	Parity operator
P_{lk}	Transition probability
$\bar{P}_{lk}$	Probability relevant to transition to an energy band
$P_l(\cos\theta)$	Legendre polynomial
$P_l^m(\cos\theta)$	Associated Legendre polynomial
$P_n^{\text{QM}}, P_n^{\text{CL}}$	Quantum and classical probability densities
q	Charge
q	Coordinate
r	Integral in WKB analysis
$\mathbf{r}$	Radius vector
$\mathscr{R}$	Radius to center of mass

R	Reflection coefficient
R	Radial wavefunction
$\bar{R}$	Nondimensionalized radial wavefunction
$\mathbb{R}$	Rydberg constant
$\hat{R}_{\delta\phi}$	Rotation operator
$\mathbf{S}$	Spin angular momentum
s	Spin angular momentum quantum number
T	Transmission coefficient
T	Temperature
T_C	Critical temperature
T	Kinetic energy
u	Radial wavefunction ($u = rR$)
$u(v)$	Energy density per frequency interval
U	Energy density
$\mathscr{U}$	Column vector
v_F	Fermi velocity
V	Potential energy
$\bar{V}$	Average potential
$\mathscr{V}$	Column vector
w_{lk}	Transition probability rate
$\bar{w}_{lk}$	Probability rate for transition to an energy band
W_{lk}	Total atomic transition rate
$\hat{\mathfrak{X}}$	Exchange operator
$Y_l{}^m(\theta, \phi)$	Spherical harmonic
α	Fine-structure constant
α	Polarizability
β	Harmonic oscillator wavenumber
β	Speed nondimensionalized with respect to the speed of light
α, β	Spin eigenstates
ε	Energy
ϵ	Mass ratio
η	Integral in WKB analysis
Γ	State of a system
$\Theta_l{}^m(\theta, \phi)$	Eigenfunction of $\hat{L}^2$
K	Spring constant
λ	Wavelength
λ	Parameter of smallness
Λ	Angular momentum parameter; scattering centers per unit area

μ	Reduced mass
μ	Chemical potential
$\boldsymbol{\mu}$	Magnetic moment
ν	Frequency
ξ	Nondimensional displacement
ξ	Spin state
ξ, η	Nondimensional wavenumbers
ρ	Nondimensional radius in hydrogen wavefunction
$\hat{\rho}$	Density matrix
$\rho(x)$	Particle density
σ	Stefan–Boltzmann constant
σ	Total scattering cross section
$\hat{\boldsymbol{\sigma}}$	Pauli spin operator
$d\sigma$	Differential scattering cross section
φ	Time-independent wavefunction
Φ	Work function
Φ	Electric potential
$\Phi_m(\phi)$	Eigenfunction of $\hat{L}_z$
χ_S, χ_A	Symmetric and antisymmetric wavefunctions
ψ	Wavefunction
ω	Angular frequency
Ω	Solid angle
$\Omega, \Omega/2$	Cyclotron and Larmor frequencies

Units

A	Ampere
cm, μm, m	Centimeter, micron, meter
C	Coulomb
s, ms, μs, ns	Second, millisecond, microsecond, nanosecond
V, eV, meV, keV, MeV	Volt, electron volt, milli electron volt, kilo electron volt, mega electron volt
W, kW, MW	Watt, kilowatt, megawatt

APPENDIXES

ADDITIONAL REMARKS ON THE $\hat{x}$ AND $\hat{p}$ REPRESENTATIONS

Let $|x'\rangle$ represent an eigenstate of $\hat{x}$. Then

(A.1)
$$\hat{x}|x'\rangle = x'|x'\rangle$$

These eigenstates obey the orthonormality condition

(A.2)
$$\langle x'|x\rangle = \delta(x - x')$$

The matrix elements of $\hat{x}$ are then given by

(A.3)
$$\langle x|\hat{x}|x'\rangle = x'\langle x|x'\rangle = x'\delta(x - x')$$

This is a continuous matrix with nonzero entries only on the diagonal $x = x'$.

As remarked in the text, summations over continuous matrices are replaced by integrations. For example, the multiplication of the matrix $\hat{x}$ by the column state vector $|\psi(x)\rangle$ gives

$$\int dx'|\psi(x')\rangle\langle x'|\hat{x}|x\rangle = x \int dx'|\psi(x')\rangle \, \delta(x' - x) = x|\psi(x)\rangle$$

In the x representation, $\hat{x}$ operating on a state has the effect of multiplying the state by the scalar x.

The projection of $|\psi\rangle$ onto the basis vector $|x'\rangle$ is the coordinate representation of $|\psi\rangle$.

(A.4)
$$\langle x'|\psi\rangle = \int \langle x'|x\rangle\langle x|\psi\rangle \, dx = \psi(x')$$

Here we have employed the spectral resolution of unity,

(A.5)
$$\hat{I} = \int |x\rangle \, dx\langle x|$$

Note in particular that the coordinate representation of $|x\rangle$ is the delta function, $\delta(x - x')$, as given by (A.2). This identification permits one to write the eigenvalue equation for $\hat{x}$ in the form (3.26).

If $|p\rangle$ represents an eigenstate of $\hat{p}$, then

(A.6)
$$\hat{p}|p'\rangle = p'|p'\rangle$$

The matrix of $\hat{p}$ in the coordinate representation is given by

(A.7)
$$\langle x|\hat{p}|x'\rangle = -i\hbar \frac{\partial}{\partial x} \delta(x - x')$$

This relation allows us to obtain an explicit form for the transfer matrix $\langle x|p\rangle$.

(A.8)
$$p\langle x|p\rangle = \langle x|\hat{p}|p\rangle = \int dx' \langle x|\hat{p}|x'\rangle\langle x'|p\rangle$$

$$= -i\hbar \int dx' \frac{\partial}{\partial x} \delta(x - x')\langle x'|p\rangle$$

$$= -i\hbar \frac{\partial}{\partial x} \langle x|p\rangle$$

The solution to this differential equation is

(A.9)
$$\langle x|p\rangle = \frac{1}{\sqrt{2\pi\hbar}} e^{ipx/\hbar}$$

The normalization ensures the unitarity of the continuous matrix $\langle x|p\rangle$. To see this, we first recall the condition for unitarity,

(A.10)
$$\int_{-\infty}^{\infty} \langle p|x\rangle^* \langle p|x'\rangle \, dp = \delta(x - x')$$

With the representation (A.9) for $\langle x|p\rangle$ and using the property $\langle p|x\rangle^* = \langle x|p\rangle$, we find that

(A.11) $$\text{LHS(A.10)} = \int_{-\infty}^{\infty} \langle x|p\rangle\langle p|x'\rangle \, dp = \frac{1}{2\pi\hbar} \int_{-\infty}^{\infty} e^{ipx/\hbar} e^{-ipx'/\hbar} \, dp$$

Setting $p/\hbar \equiv y$ reduces the right-hand side of the latter equation to

(A.12)
$$\frac{1}{2\pi} \int_{-\infty}^{\infty} e^{iy(x - x')} \, dy = \delta(x - x')$$

which establishes the unitarity of $\langle x|p\rangle$. Note that the projection (A.9) gives either the coordinate representation of the eigenstates of p or the momentum representation of the eigenstates of $\hat{x}$.

Let us see how the form (A.9) allows one to reconstruct the matrix for $\hat{p}$ as given by (A.7). In the p representation, we have

(A.13)
$$\langle p|\hat{p}|p'\rangle = p' \, \delta(p - p')$$

Using (A.9) together with the last equation gives

(A.14)
$$\langle x|\hat{p}|x'\rangle = \frac{1}{2\pi\hbar} \int_{-\infty}^{\infty} \int_{-\infty}^{\infty} dp \, dp' \, p' \, \delta(p - p')e^{ipx/\hbar}e^{-ip'x'/\hbar}$$

$$= \frac{1}{2\pi\hbar} \int_{-\infty}^{\infty} dp \, pe^{ip(x-x')/\hbar}$$

$$= -i\hbar \frac{\partial}{\partial x} \frac{1}{2\pi\hbar} \int_{-\infty}^{\infty} e^{ip(x-x')/\hbar} \, dp$$

$$= -i\hbar \frac{\partial}{\partial x} \delta(x - x')$$

which agrees with (A.7). We may use this relation to calculate the coordinate representation of $\hat{p}|\psi\rangle$.

(A.15)
$$\langle x|\hat{p}|\psi\rangle = \int_{-\infty}^{\infty} dx' \langle x|\hat{p}|x'\rangle\langle x'|\psi\rangle = -i\hbar \int_{-\infty}^{\infty} dx' \frac{\partial}{\partial x} \delta(x - x')\psi(x') = -i\hbar \frac{\partial}{\partial x} \psi(x)$$

This has the same effect as simply operating on the state ψ with the differential operator $-i\hbar \, \partial/\partial x$.

As a simple example of these concepts,[1] consider the problem of finding the matrix of $(\hat{x}\hat{p} - \hat{p}\hat{x})$ in the x representation. Let us first examine the term

$$\langle x|\hat{x}\hat{p}|x'\rangle = \iiint_{-\infty}^{\infty} dx'' \, dp' \, dp \, \langle x|\hat{x}|x''\rangle\langle x''|p'\rangle\langle p'|\hat{p}|p\rangle\langle p|x'\rangle$$

$$= \frac{1}{2\pi\hbar} \iint_{-\infty}^{\infty} dx'' \, dp \, x'' \, \delta(x - x'')pe^{ip(x''-x')/\hbar}$$

$$= \frac{x}{2\pi\hbar} \int_{-\infty}^{\infty} dp \, pe^{ip(x-x')/\hbar}$$

$$= \frac{x}{2\pi\hbar} \left(-i\hbar \frac{\partial}{\partial x}\right) \int_{-\infty}^{\infty} dp \, e^{ip(x-x')/\hbar}$$

$$= -i\hbar x \frac{\partial}{\partial x} \delta(x - x')$$

[1] For further development of these topics, see W. Louisell, *Radiation and Noise in Quantum Electronics*, McGraw-Hill, New York, 1964.

In like manner we find that

$$-\langle x|\hat{p}\hat{x}|x'\rangle = i\hbar x' \frac{\partial}{\partial x}\delta(x - x')$$

Combining these results gives

(A.16) $\langle x|\hat{x}\hat{p} - \hat{p}\hat{x}|x'\rangle = -i\hbar(x - x')\frac{\partial}{\partial x}\delta(x - x') = +i\hbar\,\delta(x - x')$

In concluding this discussion we note the following. Suppose that a complete set of commuting observables are diagonalized by the ket vectors $|\xi\rangle$. Then the coordinate and momentum representations of these states are $\langle x|\xi\rangle$ and $\langle p|\xi\rangle$, respectively. For example, consider the eigenvectors $|n\rangle$ that simultaneously diagonalize the number operator $\hat{N} = \hat{a}^\dagger\hat{a}$ and the Hamiltonian $\hat{H} = \hbar\omega_0(\hat{N} + \frac{1}{2})$, appropriate to the harmonic oscillator (Section 7.2).

(A.17) $$\begin{aligned}\hat{H}|n\rangle &= \hbar\omega_0(n + \tfrac{1}{2})|n\rangle \\ \hat{N}|n\rangle &= n|n\rangle\end{aligned}$$

No information is revealed by these equations other than the fact that $|n\rangle$ is an eigenvector of $\hat{H}$ and $\hat{N}$ with respective eigenvalues as shown. If, for example, one wishes the coordinate representation of these states, one must form the projections $\langle x|n\rangle$. These are the weighted Hermite polynomials (7.58).

In a similar vein the coordinate representations of the eigenvectors $|lm\rangle$ of the operators $\hat{L}^2$ and $\hat{L}_z$ are the projections $\langle\theta\phi|lm\rangle$ [i.e., the spherical harmonics, $Y_l^m(\theta, \phi)$].

As a further case of this formalism, consider the following example. We wish to show that the coordinate representation of the Schrödinger equation in abstract ket space

$$i\hbar\frac{\partial|\psi\rangle}{\partial t} = \hat{H}|\psi\rangle$$

is the standard relation (3.45).

Without loss in generality we work in one dimension. Multiplying the given relation from the left by $\langle x|$, we obtain

$$i\hbar\frac{\partial}{\partial t}\langle x|\psi\rangle = \langle x|\hat{H}|\psi\rangle = \int dx'\langle x|\hat{H}|x'\rangle\langle x'|\psi\rangle = \int dx'\,\delta(x - x')\hat{H}(x')\langle x'|\psi\rangle$$

where $\hat{H}(x)$ denotes $\hat{H}$ in the coordinate representation. The further identification

$$\langle x|\psi\rangle = \psi(x)$$

gives the desired relation.

SPIN AND STATISTICS

In this appendix we wish to offer a brief elementary outline of the argument connecting spin and the exclusion principle. As described in Chapter 12, particles with integral spin do not obey the exclusion principle, whereas those with half-integral spin do obey the exclusion principle.

The particle quality of a field may be described in second quantization, wherein, in accord with Problems 13.37 and 13.38, the state of the system is written $|n_1, n_2, \ldots\rangle$. In this notation n_i represents the number of particles in the ith state.

There are two prescriptions for the quantization of a field. The first is given by the Jordan–Wigner anticommutation rules (Problem 7.32),

$$\{\hat{a}_n, \hat{a}_m\} = \{\hat{a}_n{}^\dagger, \hat{a}_m{}^\dagger\} = 0$$

(B.1)

$$\{\hat{a}_n, \hat{a}_m{}^\dagger\} = \delta_{nm}$$

The second is given by the Bose commutation rules,

$$[\hat{a}_n, \hat{a}_m] = [\hat{a}_n{}^\dagger, \hat{a}_m{}^\dagger] = 0$$

(B.2)

$$[\hat{a}_n, \hat{a}_m{}^\dagger] = \delta_{nm}$$

As established in Problem 7.32, particles such as electrons, which obey the Jordan–Wigner anticommutation rules (B.1), exist in accordance with the exclusion principle. Number eigenvalues n_i are either zero or 1 ($n_i{}^2 = n_i$). Particles, such as photons, which obey the Bose commutation rules (B.2) do not adhere to the exclusion principle.

In his argument relating spin and exclusion, Pauli[1] imposed the following requirements on physical systems.

1. Let A be an observable pertaining to the space–time point $\mathbf{r}_1, t_1$, and let B be an observable at $\mathbf{r}_2, t_2$. Consequently, if $|\mathbf{r}_1 - \mathbf{r}_2|/|t_1 - t_2| > c$, then A and B commute. The rationale behind this stipulation is as follows. In that these space–time points are separated by speeds greater than that of light, relativity (or *causality*) specifies that measurement of A can in no way interfere with measurement of B. Equivalently, we may say that A and B commute.

---

[1] W. Pauli, *Phys. Rev.* **58**, 716 (1940). See also R. Streater and A. S. Wightman, *PCT, Spin and Statistics, and All That*, W. A. Benjamin, New York, 1964.

2. The total (relativistic) energy of the system is greater than or equal to zero. What Pauli then showed is that:

(a) Quantization of integral spin fields according to Jordan–Wigner anti-commutation rules (B.1), corresponding to exclusion, violates the first postulate.

(b) Quantization of half-integral spin fields according to Bose commutation quantization rules (B.2) violates the second postulate.

The distinction between half-integral spin fields and integral spin fields enters the argument through the manner in which these fields transform under a Lorentz transformation. The Lorentz transformation relates observation of properties (fields, mass, length, etc.) in one inertial frame to observation in another inertial frame of these same properties. The corresponding matrix is orthogonal (see Table 11.1) and represents a rotation in four-dimensional space. A somewhat similar distinction between integral and half-integral spin states evidenced under ordinary rotation of axes in 3-space, such as described in the discussion on the rotation operator in Section 11.5, is found to persist under Lorentz transformation of spin fields.[1]

Statistics

The property that particles have of either obeying or not obeying the exclusion principle has direct consequence in the distributions in energy that aggregates of particles have in equilibrium at a temperature T. Thus fermions (particles with half-integral spin) satisfy Fermi–Dirac statistics. A collection of such noninteracting particles at the temperature T has the energy distribution

$$f_{FD} = \frac{1}{e^{(E_i - E_F)/k_B T} + 1}$$

This expression gives the average number of particles per state at the energy E_i. The parameter E_F denotes the Fermi energy. At zero degrees Kelvin, no states of energy greater than E_F are occupied (see Fig. 2.5).

Bosons (particles with integral spin) satisfy Bose–Einstein statistics. A collection of noninteracting bosons at the temperature T has the energy distribution

$$f_{BE} = \frac{1}{e^{(E_i - \mu)/k_B T} - 1}$$

Here again, f represents the average number of particles per state at the energy E_i and μ is written for the chemical potential. This distribution appears in the Planck radiation formula (2.3) relevant to a photon gas in equilibrium at the temperature T, for which case $\mu = 0$.

---

[1] For further discussion of the distinction between integral spin and half-integral spin fields, see H. Yilmaz, *The Theory of Relativity and the Principles of Modern Physics*, Blaisdell, New York, 1965.

REPRESENTATIONS OF THE DELTA FUNCTION[1]

Cartesian Coordinates

$$(C.1) \qquad 2\pi\delta(x - x') = \int_{-\infty}^{\infty} e^{ik(x - x')} \, dk$$

$$(C.2) \qquad \pi\delta(x - x') = \int_{0}^{\infty} \cos k(x - x') \, dk$$

$(C.3)[2]$
$$2\pi\delta(x - x') = \sum_{-\infty}^{\infty} \exp\left[in(x - x')\right]$$

$(C.4)[2]$
$$2\pi\delta(x - x') = 1 + \sum_{1}^{\infty} 2 \cos n(x - x')$$

$$(C.5) \qquad \pi\delta(x - x') = \lim_{\eta \to \infty} \frac{\sin \eta(x - x')}{x - x'}$$

$$(C.6) \qquad \delta(x - x') = \lim_{\epsilon \to 0} \frac{e^{-(x - x')^2/\epsilon^2}}{\epsilon\sqrt{\pi}}$$

$$(C.7) \qquad \pi\delta(x - x') = \lim_{\eta \to \infty} \frac{1 - \cos \eta(x - x')}{\eta(x - x')^2}$$

$$(C.8) \qquad \pi\delta(x - x') = \lim_{\eta \to \infty} \frac{2 \sin^2 \left[\eta(x - x')/2\right]}{\eta(x - x')^2}$$

$$(C.9) \qquad \pi\delta(x - x') = \lim_{\epsilon \to 0} \frac{\epsilon}{(x - x')^2 + \epsilon^2} = \lim_{\epsilon \to 0} \mathrm{Im} \, \frac{1}{(x - x') - i\epsilon}$$

Let $\mathscr{H}_n(x)$ be the nth-order Hermite polynomial. Then

$$(C.10) \qquad \delta(x - x') = \sum_{n=0}^{\infty} \frac{1}{\sqrt{\pi} \, 2^n n!} \exp -\left(\frac{x^2 + x'^2}{2}\right) \mathscr{H}_n(x)\mathscr{H}_n(x')$$

---

[1] Additional properties of the delta function may be found in Problem 3.6.
[2] Domain of validity: $x' - \pi \le x \le x' + \pi$.

All the above representations obey the normalization

$$\int_{-\infty}^{\infty} \delta(x - x')\, dx' = 1$$

In three dimensions, with $\mathbf{r} = (x, y, z)$, one has

$$\delta(\mathbf{r} - \mathbf{r}') = \frac{1}{(2\pi)^3} \iiint_{-\infty}^{\infty} e^{i\mathbf{k}\cdot(\mathbf{r}-\mathbf{r}')}\, dk_x\, dk_y\, dk_z$$

(C.11)

$$\iiint_{-\infty}^{\infty} \delta(\mathbf{r} - \mathbf{r}')\, dx'\, dy'\, dz' = 1$$

Spherical Coordinates

Let $P_l(\mu)$ be the lth-order Legendre polynomial. Then

$$\delta(\mu - \mu') = \sum_{l=0}^{\infty} \frac{2l + 1}{2} P_l(\mu)P_l(\mu')$$

(C.12)

$$\int_{-1}^{1} \delta(\mu - \mu')\, d\mu' = 1$$

The delta function over solid angle may be written in terms of the $Y_l^m(\theta, \phi)$ spherical harmonics.

(C.13) $$\delta(\mathbf{\Omega} - \mathbf{\Omega}') = \frac{\delta(\theta - \theta')\delta(\phi - \phi')}{\sin\theta} = \sum_{l=0}^{\infty} \sum_{m=-l}^{l} [Y_l^m(\theta, \phi)]^* Y_l^m(\theta', \phi')$$

The directional coordinates of $\mathbf{\Omega}$ are θ and ϕ. Normalizations are given by

$$\int_0^{\pi} \delta(\theta - \theta')\, d\theta' = 1, \qquad \int_0^{2\pi} \delta(\phi - \phi')\, d\phi' = 1, \qquad \int_{4\pi} \delta(\mathbf{\Omega} - \mathbf{\Omega}')\, d\Omega = 1$$

$$d\Omega = \sin\theta\, d\theta\, d\phi$$

In three dimensions one obtains the representation

(C.14) $$\delta(\mathbf{r} - \mathbf{r}') = \delta(\mathbf{\Omega} - \mathbf{\Omega}') \frac{\delta(r - r')}{r^2}$$

$$= \frac{2}{\pi} \sum_l \sum_m [Y_l^m(\theta, \phi)]^* Y_l^m(\theta', \phi') \int_0^{\infty} j_l(kr)j_l(kr')k^2\, dk$$

where $j_l(kr)$ is the lth-order spherical Bessel function.

(C.15) $$\int_0^\infty j_l(kr)j_l(kr')k^2\ dk = \frac{\pi}{2r^2}\ \delta(r-r'),\qquad \int_0^\infty \delta(r-r')\ dr' = 1$$

We note also the differential relations

(C.16) $$(\nabla^2 + k^2)\frac{e^{ikr}}{r} = (\nabla^2 + k^2)\frac{\cos kr}{r}$$

$$= -4\pi\delta(\mathbf{r})$$

Cylindrical Coordinates

Let $J_m(x)$ be the mth integral-order Bessel function. Then

$$\frac{\delta(\rho-\rho')}{\rho} = \int_0^\infty J_m(k\rho)J_m(k\rho')k\ dk$$

(C.17)

$$\int_0^\infty \delta(\rho-\rho')\ d\rho = 1$$

With $k_j\rho_0$ denoting the zeros of $J_0(x)$, that is,

$$J_0(k_j\rho_0) = 0$$

one has the representation

$$\pi\rho_0^2\,\delta(\rho) = \sum_{j=1}^\infty \frac{J_0(k_j\rho)}{[J_1(k_j\rho_0)]^2}$$

(C.18)

$$\int_0^{\rho_0} 2\pi\,\delta(\rho)\rho\ d\rho = 1$$

Three other important normalizations of $J_m(x)$ are:

(C.19) $$k\int_0^\infty J_m(k\rho)\ d\rho = 1$$

(C.20) $$\int_0^\infty \frac{J_m(k\rho)}{\rho}\ d\rho = \frac{1}{m}\qquad (m>0)$$

(C.21) $$\int_0^{\rho_0} J_0(k_j\rho)J_0(k_l\rho)\ d\rho = \tfrac{1}{2}\rho_0^2 J_1^2(k_j\rho_0)\delta_{jl}$$

Green's Identities

First Identity

(C.22)
$$\int_V (\phi \nabla^2 \psi + \nabla \phi \cdot \nabla \psi) \, dV = \int_S (\phi \nabla \psi) \cdot d\mathbf{S}$$

Second Identity

(C.23)
$$\int_V (\phi \nabla^2 \psi - \psi \nabla^2 \phi) \, dV = \int_S (\phi \nabla \psi - \psi \nabla \phi) \cdot d\mathbf{S}$$

PHYSICAL CONSTANTS AND EQUIVALENCE RELATIONS

Velocity of light in vacuum	c	2.9979×10^8 m/s
		2.9979×10^{10} cm/s
Planck's constant	h	6.6261×10^{-34} J s
		6.6261×10^{-27} erg s
		4.1357×10^{-15} eV s
	$\hbar$	1.0546×10^{-34} J s
		1.0546×10^{-27} erg s
		6.5821×10^{-16} eV s
Avogadro's number	N_0	6.0221×10^{23} atoms/mole
Boltzmann's constant	k_B	1.3807×10^{-23} J/K
		1.3807×10^{-16} erg/K
		8.6174×10^{-5} eV/K
Room temperature		300 K $=0.02585$ eV
Gas constant	$R = N_0 k_B$	8.3145 J/mole K
		8.3145×10^7 erg/mole K
		1.9870 cal/mole K
Volume of 1 mole of perfect gas, at normal temperature and pressure		22.421 liters
Electron charge	e	1.6022×10^{-19} C
		4.8032×10^{-10} esu
Electron rest mass	m	9.1094×10^{-31} kg
		9.1094×10^{-28} g
		0.511 MeV
Electron classical radius	$r_0 = e^2/mc^2$	2.818×10^{-13} cm
Electron magnetic moment	μ_e	$1.001 \ \mu_b$
Fine-structure constant	$\alpha = e^2/\hbar c$	$7.297 \times 10^{-3} = 1/137.04$
Compton wavelength	$\lambda_C = h/mc$	2.426×10^{-10} cm
	$\lambdabar_C = \hbar/mc$	3.862×10^{-11} cm

Gravitational constant	G	6.672×10^{-8} dyne-cm^2/g^2
Proton rest mass	M_p	1.6726×10^{-27} kg
		1.6726×10^{-24} g
		1.0073 amu
		938.27 MeV
Neutron rest mass	M_n	1.675×10^{-27} kg
		1.675×10^{-24} g
		939.57 MeV
Bohr magneton	$\mu_b = \dfrac{eh}{2mc}$	9.27×10^{-21} erg gauss^{-1}
Ratio of proton mass to electron mass	$\dfrac{M_p}{m}$	1836.1
Charge-to-mass ratio of electron	$\dfrac{e}{m}$	1.7588×10^{11} C/kg 5.2730×10^{17} esu/g
Stephan–Boltzmann constant	$\sigma = \left(\dfrac{\pi^2}{60}\right)\left(\dfrac{k_B{}^4}{\hbar^3 c^2}\right)$	5.6697×10^{-5} erg cm^{-2} s^{-1} K^{-4}
Rydberg constant	$\mathbb{R} = \dfrac{me^4}{2\hbar^2}$	109,737.32 cm^{-1}
	$= \dfrac{e^2}{2a_0} = \dfrac{\alpha^2 mc^2}{2}$	13.61 eV
Bohr radius	$a_0 = \dfrac{\hbar^2}{me^2}$	0.52918 Å
Triple point of water		273.16 K
Atomic mass unit	1 amu	1.6605×10^{-24} g
		931.5 MeV
		Mass of $C^{12} \times \dfrac{1}{12}$
Wien's displacement law constant	$\lambda_{\max} T$	0.290 cm K

Useful Conversion Constants and Units

Constants, MKS units	ϵ_0	8.8542×10^{-12} farad/m
	μ_0	$4\pi \times 10^{-7}$ henry/m
Energy in wavenumbers		$E = hc\lambda^{-1}$
		$hc = 1.240 \times 10^{-4}$ eV cm
		$= 1.24$ eV μm

1 Electron volt	eV	10^{-6} MeV
		1.6022×10^{-12} erg
		1.6022×10^{-19} J $= [e, C]$ J
		3.829×10^{-20} Cal
		11,605 K
1 Coulomb		0.1 abcoulomb (emu)
		$(c/10)$ statcoulomb (esu)
1 emu		1 esu/c
1 weber per square		10^4 gauss
meter $= 1$ tesla		(1 gauss cm$^2 = 1$ tesla m^2)
1 Cal		4.186 J
		4.186×10^7 ergs

$k_B N_0$ $\qquad$ 2 Cal/K

$$\left[\frac{e\mathbf{B}}{mc}\right]_{esu} = \left[\frac{e\mathbf{B}}{m}\right]_{emu}$$

$$\left(\mathbf{p} - \frac{e}{c}\mathbf{A}\right)^2_{esu} = (\mathbf{p} - e\mathbf{A})^2_{emu}$$

INDEX

The letter "p" following a page number denotes a Problem.

Conflict Sociology

Toward an
Explanatory Science

Conflict Sociology

Toward an Explanatory Science

Randall Collins
Department of Sociology
University of California, San Diego
La Jolla, California

With a Contribution by Joan Annett

ACADEMIC PRESS New York San Francisco London

A Subsidiary of Harcourt Brace Jovanovich, Publishers

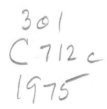

ACADEMIC PRESS, INC.
111 Fifth Avenue, New York, New York 10003

United Kingdom Edition published by
ACADEMIC PRESS, INC. (LONDON) LTD.
24/28 Oval Road, London NW1

Library of Congress Cataloging in Publication Data

Collins, Randall.
    Conflict sociology: toward an explanatory science.

    Bibliography: p.
    1.   Sociology.   2.   Social conflict.   I.   Title.
HM24.C65   301   74-5688
ISBN 0–12–181350–9 (Cloth)

ISBN 0–12–181352–5 (Paper)

Contents

111903

Preface

Sociology has been pursued for a number of motives. Many have sought in it practical benefits or ideological justifications – or perhaps only a pleasant career delving into some well-guarded academic niche. We have been at it on a large scale for quite some time now, and most of the results are disappointing. I make this judgment from the standpoint of yet another value: that a coherent, powerful, and verified set of explanatory ideas is one of the great things in the world, quite by itself. I do not assume that everyone ought to share this value; apparently too few in the social sciences have, for the promising advances of the past have given rise to little enough cumulative development. Yet practical application, for all the attention given it, has made little real headway, and the ideologists have settled nothing. I would maintain that neither of these enterprises will ever get anywhere, or even indicate where there is to go, until a true social science is made.

They are not the reasons for this book. It is motivated by a glimpse I have caught somewhere along the way of what some of the thinkers and researchers of past and present have accomplished. However incomplete and fragmented, there is a powerful science in the making. I have attempted to pull things together sufficiently to show where we are now.

This book focuses on conflict because I am attempting to be realistic, not because I happen to think conflict is good or bad. After reading this book, anyone who still judges explanatory concepts in terms of their value biases will not have grasped what it is about. Past theorists who have done most to remove our thinking from the murk of artificially imposed realities that populate our everyday worlds have found a guiding thread for explanation in the existence of plurality and conflict. Their lead is worth special emphasis right now, when there is so much potential for getting our science straight, and so many vestiges of utopian unreality burdening our habitual modes of analysis. Eventually, of course, there will no longer be such a thing as conflict sociology or any other label; there will be only sociology without adjectives.

Looking back, historians will see a great intellectual revolution in the twentieth century—the establishment of a true social science. Whether external observers will find it a desirable thing to happen to the world is open to question. The explorers of this frontier have only one real justification: the adventure itself. It is to this invisible community stretching across the years that this book is dedicated.

Acknowledgments

I am indebted to Arthur L. Stinchcombe for critical comments on the entire manuscript, and to Joseph R. Gusfield, Aaron Cicourel, Stanford M. Lyman, Herbert Isenberg, and Christine Chaillé for comments on various sections. Donald Pilcher kindly provided materials from his own research. Joan Annett, who coauthored Chapter 4, also provided assistance with other chapters. I have learned much from conversations and correspondence with Samuel W. Kaplan, Reinhard Bendix, Harold L. Wilensky, Alvin W. Gouldner, Bennett Berger, Warren O. Hagstrom, and Joseph Ben–David. What there is of value in this book is no doubt taken from them.

Why Is Sociology Not a Science?

Why is sociology not a successful science?

No one would dispute the general point. Certainly it is not, by comparison to physics, chemistry, or biology. Disagreement comes over the explanation. Some would argue that we are yet a young science and that we need more time, more research, or perhaps more scientific methods; others hold that sociology is not destined to be a science in that sense at all but, rather, an interpretive discipline, a part of the humanities or a political ideology.

Neither explanation is convincing. Society has been an object of thoughtful consideration at least as far back as Aristotle; sheer time alone is hardly the answer. Nor does rejecting the scientific ideal settle the issue. As an interpretive discipline, sociology is yet far short of the sophistication of history or literary criticism. In any case, it remains to be demonstrated that sociology cannot be a science.

My contention is that sociology can be a successful science and that it is well on the way to becoming so. The problem is to understand just what is involved in this. There are a number of different goals that have been set for sociology (and for the social sciences generally); it is necessary to understand that they differ and that somewhat divergent paths lead toward each of them. As things stand now, we have badly confused the explanatory, practical, ideological, and aesthetic aspects of the social sciences. Although we have made considerable progress in theorizing and in collecting data, the significance of what has been accomplished remains to be recognized.

Sociology is more advanced than we usually believe. The past four decades of research have brought considerable progress in key areas, and the elements of a powerful explanatory theory were laid down even before this, especially by Marx, Weber, and Durkheim. Subsequent advances in research and theoretical refinement have fleshed out these models to the point that we now have a solid framework for a scientific sociology. If this statement comes as a surprise, it is because other kinds of analyses (including ones springing from these same classic thinkers) have obscured the development of genuine explanation.

This may be seen by examining the different goals of sociology: generalized explanation, practicality, ideological evaluation, and aesthetic interpretation. I am not proposing that only one of these goals is legitimate, but it must be understood that they are distinct. It is the confusion among them which has prevented us from approaching any of them as closely as is already possible.

GENERALIZED EXPLANATION

The scientific ideal is to explain everything, and to do it by making causal statements which are ultimately based upon experience. The most powerful scientific theory is the one that can get the most explanatory mileage out of the most concise body of principles. Science is a way of finding the common principles that transcend particular situations, of extrapolating from things we know to things we do not, a way of seeing the novel as another arrangement of the familiar.

This aim is no different when applied to sociology than to physics, however different the empirical materials may be. Marx's observation that men's ideas are predictable from their economic positions is one of the great starting points of sociological science because it takes us far beyond Europe in the 1840s. With subsequent applications and critiques, it became both more refined and more powerful: Weber demonstrating that conflicting positions shape outlooks and behavior in the several realms of economics, politics, and community; Michels focusing in on the micro-class conflicts and resulting ideologies within single organizations; subsequent lines of research extending these principles to a variety of organizations and occupations, historical communities, and political structures, and even the internal structure of science itself. The principles change somewhat as we find more variations to explain (just as Einsteinian physics subsumes Newtonian as a special case), but it is precisely this capacity to give the conditions under which *some* things happen rather than *others* that is the essence of science. Science not only can but must continually seek out new applications and new tests of its explanatory power, even if they strain the existing framework of explanation. The theory that does not venture beyond the familiar will ultimately be overthrown by one that does, and even on its own home grounds. For the method of validating a theory—of showing that its explanations are true —ultimately depends on its capacity to act as an economical and coherent filter for our experience in the broadest sense.

The ideal of science is so popular that it would seem to require little defense. Indeed, it would seem that the opposite is more necessary—to defend other sorts of values that are ground under by a widespread and rather philistine adherence to the slogan of science. All this is admitted, and I shall do some of this defending shortly. But the scientific ideal does need to be defended after all. In the social sciences, in particular, there is a good deal of lip service to what is taken to be scientific method, but much less real science. It is for this reason that the cumulative advance of real sociological science is so difficult to see. If we can disentangle the false ideals from the true, much of the humanistic and dialectical critiques of science fall to a lower key: they are valid largely against pseudoscience rather than science itself.

The most vulgar error is to equate science with practicality, as if physics consisted of making atom bombs and television sets. It is true that advances in scientific knowledge sometimes give rise to practical inventions. For a considerable period in history, it has been more the reverse— practical inventions prodding science, where they have been related at all rather than completely independent. But even when they are connected, the two enterprises have distinct goals and ways of dealing with the world. Generalized explanation requires setting one's sights progressively wider, while practicality involves narrowing in to only those conditions in a specific situation which can be manipulated to achieve a given end. It is because of this difference in focus, as we shall see, that the heavy emphasis on hoped-for practicality in the social sciences has been such an obstacle to scientific advance.

A second error is to equate science with rationality in general. The positivist movement of the nineteenth century saw science as a comprehensive method for organizing life, for settling questions of what *ought to be* as well as what *is.* The fallacy here of confusing the ontological and the evaluative has been well-recognized in twentieth-century thought, but such recognition has been a long time in trickling down—not only to the layman for whom value judgments are routinely rationalized as part of the absolute order of things, but even to fairly sophisticated social and natural scientists. The failure to see what science is and what it is not—it is not a secular theology by which we are told what to do, but only a picture of the world which we must evaluate by our own lights—has been another major hindrance to the development of general explanatory sociology. We will return to this matter in more detail in the discussion of ideology.

A third error is to regard science as equivalent to precise measurement

and careful statement, especially in mathematical form. These are characteristics of science, especially in advanced stages of development; but,
important as they are, they are not the key. Science's most central characteristic is making *economical* yet *generalized, causal, empirical explanations.* Without this, there may be precision aplenty, but no science. This is
particularly important to understand in the social sciences, where emulation of the more superficial traits of physical science has been an actual
detriment to scientific advance. It has been detrimental because premature
emphasis on "hard data" has blinded us to the basic nature of what we
have to explain: dynamic, conscious, reality-creating human interaction.
It has also been detrimental because emphasis on measurement for its own
sake is most easily carried out under practical rather than theoretical
orientations; hence, the naïve cult of "hard science" and the exclusive
emphasis on practicality have been mutually reinforcing.

The basic method of explanatory science is that of controlled comparison. The essence of methodological empiricism—and that is what science
is—is to explain a phenomenon not by looking at it in isolation but by
comparison and contrast to other things. To understand a thing, we must
compare where it occurs with where it does not occur, and note the
difference in the accompanying conditions.[1] Durkheim understood this
well enough when he tried to demonstrate the conditions for social
solidarity by comparing it with situations in which solidarity breaks down
(for example, in suicide). The method of variation, however complex the
results flowing from it, is the basic method of science. The main hindrance
to the advance of sociology is that it has been insufficiently used. Practical,
ideological, and aesthetic orientations have focused attention elsewhere.
But what advances have occurred have come largely from its use.

The current opportunities for scientific advance in sociology are especially a matter of extending the explanatory principles we have already
derived. There is a great deal of empirical data now available which can
be used not merely for descriptive purposes but as so many cases for
comparative analysis of variations. In this enterprise, careful measurement

---

[1] What we mean by scientific explanation is to relate variations in one thing (or set of
elements) to variations in another thing (another set of elements) by means of a set of
principles that apply in a number of different situations. A "science" is the image of the
world that translates into a logically coherent body of such principles. The symbolism
(especially its verbal part) always contains some degree of unexplicated resonance with other
images, ideas, and experiences; contrary to some philosophies of science, this is not a limitation on the science, but the source of its fruitfulness in suggesting further applications and
of what sense of reality it can convey beyond narrowly defined experimental situations.
Science is a body of ideas that is both grounded in empirical variations and tightly interwov-
en into a systematic reality. Compare Stinchcombe (1968a: 15–56); Zelditch (1971); and
Chapter 9 in this volume.

is less important than making the attempt at comparative analysis. The comparative study aimed at formulating general causes of variations is more valuable for the advance of social science, *even if it is based on imprecise data,* than precise measurements that are not used for developing explanatory principles. The classic comparisons by Tocqueville and Weber, or more recent comparative studies such as those of Gerhard Lenski (1966) and Barrington Moore (1966), have done far more to set forth the principles of political and economic structure than any number of more precise investigations that lack either comparative scope or a serious theoretical perspective.

The initial testing of explanations is an uncertain adventure. There is no guarantee that one begins with the right concepts, that one has divided phenomena into categories that capture rather than obscure their natural lines of coherence and disjunction. Premature emphasis on precise measurement before we know what are the fruitful things to measure has often proved to be a waste of time. Having said this, I wish to avoid the romantic fallacy which arises in opposition to the cult of measurement. Measurement is not something to be dispensed with but, rather, to be used in its proper place. Once we have established the general lines of analysis along which explanation may fruitfully proceed, the task becomes one of tightening up the explanations—of seeing multiple variables where tentative formulations saw only one, of correcting misperceptions based on inaccurate data used in the heroic age of exploration. Once we are on the right explanatory track, the search for better measurements is one of the lines along which improvements in explanatory power take place.

As sociology stands now, its concern for precision in measurement operates more often as a hindrance to scientific advance than as an aid. It provides a lure for keen minds (and for comfortably mediocre ones, too) to attach themselves to problems below the level of genuine explanatory theory. It has kept us from seeing that the major elements of such an explanatory apparatus are lying about in the heritage of the last century of theorizing and the last half-century of empirical research. I am going to try to show that the fundamental ideas of Marx, Weber, Durkheim, and other classical theorists, when properly understood and fleshed out by empirical research on stratification, organizations, and personal interaction, constitute a core of reasonably well-validated explanation of how men think and act, which may be extended to account for a universal range of variations. If this theory or some other one that is genuinely explanatory can be established in sociology, then our problems of assigning a proper place to measurement would be resolved. Only from such a perspective will we be able to see which of our current efforts at quantification are fruitful lines of development and which are blind alleys.

The concept of science as an enterprise that proceeds by seeking the conditions for variations in phenomena gives a clear distinction between real explanation and pseudoexplanation. Much of what has passed for explanation in sociology is really the latter. To describe the "norms" governing a set of behaviors (e.g., to invoke the norm of recognizing priority, as sociologists of science have done, as a fundamental feature of behavior in the scientific community) may be a contribution to designating a phenomenon to be explained, but to use the suggested "norms" as an explanation of behavior is an empty gesture. It still leaves us with the problem of stating conditions under which things happen or do not happen (i.e., formulating principles whose explanatory power can extend to other situations). Without such a statement of conditions for variation, there is no proof that the explanation is right, that the way of conceptualizing the phenomenon captures its essential features.

The search for universals in sociology has taken us down the wrong road. For phenomena that are truly universal cannot be explained in any testable fashion, but only speculated about. Decades of inconclusive theorizing about the incest taboo or the nuclear family are cases in point. But even in such instances, we are not without explanatory resources, if only we orient ourselves to use them. Whether we see the universal or the variable aspects of things comes from the angle of vision, and it is possible to decompose abstract universals into more variable, concrete manifestations. We can look for the conditions of variation in the scope and strength of the incest taboo, for example, and move from unresolvable controversy over untestable (and ultimately meaningless) functional explanations to empirically testable explanations which reveal the incest taboo to be a phenomenon of power relations (see Chapter 5). In the same way, we can derive more powerful explanations of family sex-role structure than the supposedly universal necessity of socializing children. If the testable explanation of variations shows us conditions under which women are more and less strongly confined to the domestic role, we may avoid the shock of finding that, given specific conditions, the family as we know it should someday disappear.

Much of functional explanation in sociology, insofar as it is stated in terms of untestable universals, is really pseudoexplanation: in effect, an after-the-fact rationalization for whatever exists. The development of testable theory in sociology will make this kind of analysis an historical relic. At the same time, it is worth looking for what can be salvaged from functionalism. To the extent that statements about the interrelations among institutional arrangements (i.e., statements about what structures are compatible or incompatible with each other) are testable, they can and must be incorporated into sociological theory. The area of organizations (Chapter 6) is the principal area in which such incorporation seems likely, although under a perspective which is basically nonfunctionalist. As a universal system, functionalism has operated mainly as a mode of abstract description with claims to eventually becoming an explanation. As such, it has been yet another obstacle to the development of scientific sociology. It has hung on so long, despite its scant success in living up to its scientific pretenses, primarily because it satisfies ideological rather than scientific orientations.

Hypostatization versus Reductionism

The most serious strategic problem in building a scientific sociology is how to conceive of what we are explaining. Much of traditional sociology conceives of its subject matter as consisting of "society," "organizations," "classes," "communities," "roles," and "systems." None of these, in fact, is observable. All we can ever see are real people in real places, or the writings and artifacts they have made. Yet sociology has claimed its independence as a field of inquiry on the grounds that "society," and other invisible entities standing beyond particular individuals, are objective *things,* and that sociological theories are to be built on this level of analysis.

The rationale laid down by Comte and transmitted into modern sociology by Durkheim is that, although social relationships are carried out by individuals, they are not determined by the individuals making them up. Like a language, institutions are handed down from generation to generation, outlasting the individuals who come and go, and shaping their behavior while they are there. Any effort to explain behavior in terms of the individual alone commits the fallacy of psychological reductionism.

The canons of scientific explanation show why reductionism is a fallacy. To explain social institutions (e.g., the set of behaviors that make up the modern family) purely in terms of the individual does not sufficiently account for the behavior. We can still ask: Why does he receive this pattern of rewards and punishments? Or, more precisely: Why is it that the people involved give these rewards and punishments to each other? This drives us back to the social level, the structure of interaction among individuals—sometimes among a very large number of individuals, as in the network of economic relationships that help explain the structure of the modern nuclear family. All institutions are made up of individual behavior, but the behavior can be fully explained only by looking at the structure of relations at the group level.

Recent years have seen an upsurge of criticism of traditional sociological formulations from the vantage point of individualistic phenomenology. What may be loosely labeled as "ethnomethodology" can be characterized as a form of radical empiricism which refuses to accept the claims of prevailing theoretical abstractions and research methods to capture something which we may call "reality." This position draws American inspiration from Goffman and other researchers in the area of "deviance" who are concerned with how views of reality are socially constructed in concrete interactions. More broadly, the background is the whole of modern philosophical epistemology, treated as subject to be related directly to sociological research. The focus is on the actual talk, gesture, and presence of face-to-face contacts that make up the whole of sociology's subject matter—"the social"—and on the practical reasoning that goes on as individuals tacitly construct things that they take to be real. In this perspective, an "organization" is something that people construct as they go along, a topic of conversation and a taken-for-granted referent for making sense out of particular encounters.

In counterattack against the notion of reductionism, then, we find an onslaught on the error of reification or hypostatization. The phenomenologists do not stop at claiming independence for their level of analysis, but assert more broadly that sociology can only validly exist in terms of its own perspective. There seem to be a number of related critiques. First, the labeling theory of deviance, along with Goffman's (1959) work on the presentation of self in everyday interaction, has brought about a generalized recognition that the world is full of illusions perpetrated by men upon each other. Why not, then, regard "society," "organization," "classes," and so on, as so many more illusions, a kind of talk that men try to enforce on others they wish to control? Relatedly, work such as that of Douglas (1971a) and Cicourel (1968) on official governmental reports and statistics, point up the organizational and political biases which go into them; the "objective" world they claim to portray turns out, on closer examination, to be only the product of various social interests. More broadly, any version of "reality" must be understood as socially constructed, and these realities are plural. Setting one of them up as *the* reality is of a piece with the naïve belief of ordinary social actors; sociological sophistication must consist in directly confronting the plurality and making the *process* of reality construction its object of investigation.

A second aspect of social phenomenology emphasizes that immediate observational methods focusing on everyday life are the only valid sociological approaches. The immediate awareness of the sociologist—an extension of that of the phenomenologist—is the necessary and unavoidable starting point of all analysis; positivistic methods such as surveys, interviews, and use of official statistics are criticized as naïvely unaware of the cognitive constructions that have *already* gone into whatever they produce. Quantitative analysis is even more suspect, for it imposes its own structure upon social processes and hides its extraneous nature under the guise of objectivity. Following Alfred Schutz, social phenomenologists have been willing to venture only so far as to characterize the process of reality-constructing of social actors in general, including the process by which the flux of experience is reduced to abstract typifications; any methods or concepts that move farther afield have only the most tenuous existence as abstractions of abstractions, subject to so many contingencies in the process of creating them as to be little more than fictions. At best, one may investigate, in close observational detail, the process by which sociologists construct their data and the meanings they assign to it.

A third and most profound level of social phenomenology, however, renders even this problematic. That is the recognition, first emphasized by Garfinkel, that all social perception is constructed from inferences that take observables as indices of something else, and that imputations of meaning always include some features that are arbitrary to the specific situation. As Garfinkel (1967) shows in experiments in which people try to force others to say literally what they mean, people always reach the point of exasperation when they see that one must be willing to accept a tacit agreement (e.g., on the meaning of words in the particular situation) at some point. Applying this analysis to the imputation of meanings that a sociologist makes in the process of research—e.g., when he imposes his view of what an answer to a questionnaire means—leads to a view that sociological research is constructed, too, in ways that are not ultimately verifiable except to the degree that a community of scholars agrees to accept their own tacit criteria. The ideal of a literally objective science crumbles, and along with it the notion of "objective" entities like "social structure." Even the milder versions of social phenomenology, explained earlier, become suspect; the process of interpretation can never find solid ground, and the ethnomethodologist examining the positivist gives way to the hyperethnomethodologist examining the ethnomethodologist, and so on in infinite regress.

The complexities of these issues, together with a range of related research questions, have served to create yet another enclave within the fragmented world of sociology. For the sake of building a genuinely explanatory sociology, however, it is necessary to understand just how these positions may be reconciled with the main line of sociological development. Three points must be made: two of them against the extreme phenomenological positions on the criterion of truth and the use of sociological methods, and a third stressing the important contributions of phenomenology in grounding sociological science.

First, the extreme ethnomethodological attack on the possibility of science is an attack on a straw man. It assumes that science implies an objective world "out there" which we may approach via the proper research procedures, in a presuppositionless and unbiased manner. Thus, by showing the inevitable *activity* of reality-constructing that goes into any research, and especially the unspecifiable background understandings necessary, social phenomenology presumes to have destroyed the possibility of science. But this is an attack on vulgar positivism in its traditional form (which, to be sure, is the form that is handed down by many sociological methodologists). A more sophisticated philosophy of science, however, makes no such assumptions. The extreme claims of logical positivism in the 1920s and 1930s to reduce *all* statements to direct observational ones, foundered on the prior theoretical necessity of defending its own criteria of analysis. Since then, the philosophy of science has moved increasingly toward the position that empirical observations are always made within a prior conceptual scheme; in Kuhn's (1962) historical formulation, this stresses the *pragmatic* validity of reigning paradigms, as the basic context within which any judgments of "objectivity" can be made. "Truth" becomes a matter of strategic judgments about the coherence, at any point in time, of empirical procedures and theoretical models.

Presuppositionlessness, then, is not a part of scientific method at all, nor is a naïve correspondence theory of truth. The discovery of tacit, unexpressed groundings of particular statements is not an embarrassment, but only a particular feature of the coherence model and its relationship to the transverbal world in which we live.[2] The phenomenologists' assertion that objective science is impossible, therefore, holds only against a particularly naïve version of science; this aspect of phenomenology seems to be maintained by continually resuscitating its most vulgar opponents. It should be noted, moreover, that any sharp distinction between the realm of the natural sciences and the social world, *on these grounds,* is pointless. Natural science is no more objective, or presuppositionless, or lacking in tacit groundings than social science. The failing in social science has been on a substantive and not an epistemological level; natural science carries the same fundamental liabilities as social science, and there is nothing in principle to prevent the latter from carrying on to equal success in developing explanatory theory.[3]

---

[2] See Zelditch (1971) for a sophisticated statement within the positivist tradition.

[3] What about special problems of "meaningfulness" in the data social scientists deal with? I believe this is an artificial issue. That the sociologist must "get inside" the symbolic interpretations of his subjects to some degree, only adds a special range of materials to be interpreted and assessed—our criterion for success in such efforts remains their coherence with a more general explanatory model. Thus the epistemological issue here, despite its greater complexity in practical detail, is no different in principle from the basic epistemological problem of *all* scientific inquiry. If one *wants* to stress a special historical uniqueness or ambiguity in one's materials, that is a choice in angle of vision, which cannot be construed

Second, sociology cannot be successful on the microlevel alone, either conceptually or methodologically. Terms like "state," "social change," "economic structure," and "class structure," always refer us back empirically to the behavior of individuals at particular times and places. But the things that we are interested in when we use these terms are not just the behavior of a few individuals for a few minutes; we want to say something about the behavior of thousands or millions of people, over many days or many years. Sociology cannot give up this effort without cutting off its most interesting questions; and, without this long-range vision, it cannot properly interpret even momentary and small-scale situations. For the structure of any face-to-face interaction is influenced by an astronomical number of other encounters, in the sense that economic, political, ethnic, and other patterns all impose themselves. Microsociology by itself is doomed to be either extremely abstract, focusing on those processes that are found in *all* encounters, or extremely parochial, taking for universal those processes that are in fact limited to a particular social class in a particular time in history.

We cannot do without concepts that summarize the long-term and large-scale networks of interaction, the macrolevel of analysis. Nor are such concepts epistemologically objectionable, except under an unrealistic notion of knowledge. If our knowledge, in fact, proceeds by the gradual refinement of conceptual schemes to incorporate evidence more and more precisely and coherently, than macroconcepts are pragmatically justified to the extent that they enable us to move forward. The evidence is always dealt with in a summary form before it is subjected to more precise analysis; without the former, we have no way of knowing which of the potentially infinite set of detailed investigations would prove useful. This is not to say that any particular macrotheory is justified; some are more adequate to the evidence than others. It is not because organicist functionalism is macro that it is wrong, but because it fails to provide testable explanations.

The same logic tells us why observational methods by themselves cannot suffice for a sociology of any scope. We need to be able to summarize behavior over long periods of time; questionnaires, official reports, and summary historical accounts all serve their purpose in this way. The phenomenological critique reminds us of the biases of these materials, but the warning should be taken as a guide to their more realistic interpretation, not as an injunction against using them. Once again, the coherence

---

to exclude analysis of the same materials from a scientific angle. Moreover, the area that symbolic meanings has been claimed to occupy in sociology has been exaggerated, for all their importance. On these two points, see the section of this chapter on aesthetic orientations, and Chapter 3, respectively.

of all sources of evidence with our most extensive theoretical generalizations must be our standard of deliverance; we can have confidence in
surveys, historical accounts, and the like, precisely to the extent that their
results fit with observational data and with each other, as measured by
their contributions to a single body of theory (cf Verba, 1971).

Third, the phenomenological position gives us a crucial grounding for
explanatory theory and enables us to avoid errors of hypostatization. For
Durkheim's rejection of psychological reductionism—the recognition that
how people relate to each other in groups has an effect on what each of
them can do—does no more than establish the right context in which to
view individual behavior. As Homans has rightly emphasized, only real
people can do things.[4] "Structures" are a way of talking about the patterns
of what they do in groups. If we pay attention to what goes on around
us all the time, it is not hard to remember that "organizations," "classes,"
or "societies" never *do* anything. Any causal explanation must ultimately
come down to the actions of real individuals.

This needs reasserting at this time in history, because concepts deriving
from ideological language—a kind of drama in which reified abstractions
(hypostatizations) are the most prominent characters—are yet to be
purified from scientific sociology. The ideological rhetoric of "the organization's goals" must be translated into the aims of real people, usually
those who get power and personal glory from the work of subordinates.
"Paying one's debt to society" must be seen in the context of politicians,
judges, and policemen making their careers in an appeal to the support of
particular social groups whose material or status interests are being
upheld, usually in conflict against other group interests. Disengaging oneself from this sort of ideological hypostatization is not merely a matter of
taking the view of the underdog either. "Society is against me" is just a
grand way of saying that interaction with certain people is a losing struggle in terms of things that the particular individual wants. "A traitor to
one's class" or someone with "false consciousness" is usually someone
who pursues an individual interest that the speaker does not approve of.

These kinds of concepts have been built into our attempts at scientific
analysis, resulting in endless problems of defining formal and informal

---

[4] This is another recent effort at "bringing men back in," drawing on behaviorism. In
the work of Homans (1961, 1964), it tends to commit the reductionist error by ignoring the
causes of the variations in repetitive social behavior (what we call "organizations," "stratification," or other "structure"), while concentrating on showing that sociological evidence is
compatible with a view of the rewards and punishments individuals get. Blau's (1964)
exchange theory moves up to the right level of analysis, concerned at least in principle with
why actors give rewards and punishments as well as receive them. Traditional structural
sociology is thus under attack from both sociological camps, romanticist and positivist alike.

organization, defining the boundaries of societies, debating if white collar crime is "deviant," and so on. A phenomenological perspective not only remedies this but has positive benefits as well. It enables us to pinpoint just what social relationships we are explaining. A system of stratification becomes a complex network of individuals interacting, with nodes of solidarity and points of antipathy seen as variations to be explained. An organization is a particular sort of behavioral network, including some highly regularized encounters and some characteristic sorts of reality-defining conversations; again, these must be seen as explainable variations in interaction. Our perspective becomes inherently dynamic. "Social change" is not something that happens occasionally; it is, rather, the lack of change which is an illusion. The structure of power relations in an organization fluctuates minute by minute, although tending to return to a similar point. The formal–informal distinction largely disappears; both of these concepts refer to the kinds of relationships that are enacted as people happen to come into contact. Organizational structure must be seen as something that is continually recreated; its determinants must all flow through a set of pressures on individuals who go about enacting it.

There are a number of other examples of how a phenomenological perspective gives the necessary empirical underpinnings to a scientific sociology. Since these will be elaborated in later chapters, only the general principles need be mentioned here.

Phenomenological empiricism enables us to conceptualize institution-alized patterns of interaction in ways that allow us to see how ideals are imposed by those in power, rather than having to view the institution through the lens of the ideal and then struggling to correct it for distortion. Thus we can see the key place that behavior upholding exclusive sexual possession has in the complex of relations usually referred to as the "fami-ly." In a similar fashion, we can understand the importance of threatened violence in underpinning the ideals by which political men define the state. This approach also promises a grip on the problems in explaining distributions of goods: The analysis of "property" can go beyond reified conventional indicators of money value to a recognition of the power over behavior that underlies these indicators. Masses of wealth, for example, are really power positions in particular kinds of social networks of credit and influence. I shall try to indicate some consequences of this approach in Chapter 8. In a similar fashion, abstract counts in social mobility tables can be explained as they are resolved into movements through networks of personal acquaintanceship and organizational encounters.

Ultimately, a successful scientific sociology will have acquired a new vocabulary in which the misleading connotations of "organizations," "class," and so on, are remedied. This will happen as our explanations actually focus on the ways in which the maneuvers of large numbers of individuals go together in producing more or less enduring patterns. In the meantime, we cannot entirely dispense with the vocabulary we have. It operates as a necessary heuristic with which to summarize the myriad social encounters that are the concrete subject matter of such terms as "a decentralized organizational structure" or "an unequal distribution of wealth." The immediate problem is to find ways to remind ourselves of the sort of real particulars we are referring to whenever such a term is invoked.

The strategic problem in building a scientific sociology is to integrate macro- and microlevels of analysis. The microlevels must provide the detailed mechanisms through which the processes summarized on the macrolevels may be explained. Micro does not precede macro; progress goes on along both fronts, with each setting problems for the other and suggesting where the solutions lie. As we shall see, the phenomenological tradition has recently made a breakthrough on the level of individual cognitive mechanisms which goes a long way toward explaining some of the puzzles of macrosociology, puzzles associated with Weber's concept of legitimacy and Durkheim's concept of ritual-produced "collective consciences."

The justification for such an integration is pragmatic; only in this way can a satisfactory explanatory theory be created. The rest of this book will attempt to vindicate this claim. Conflict theory is the most appropriate vehicle in this respect as well as in others, because it fits well with the positive thrust of social phenomenology while situating it in an explicitly historical context. The conflict tradition of Machiavelli, Marx, and their successors has emphasized the social construction of subjective realities and the dramaturgical qualities of action, while viewing these as based upon an underlying world of historically conditioned material interests. In this perspective, social phenomenology flows into the mainstream of realistic sociological explanation.

PRACTICALITY

Practical research in sociology is aimed at helping achieve some concrete goals in the world. For a long time, the main or even sole avowed purpose of American sociology was to solve "social problems." These included various forms of "deviance" such as crime, delinquency, mental illness, alcoholism, divorce, and prostitution, as well as political and stratification issues such as poverty, social unrest, or racial discrimination.

Related work designed to make practical contributions, generally referred to as "applied sociology," involves research and consultation to aid industrial and organizational functioning, economic development, and planning or evaluating personal advertising, election campaigns, and military or paramilitary strategy.

This sort of work looms large in sociology. It is the only kind of research that most laymen can understand. No doubt the majority of sociologists who concentrate on it are committed to it personally, as well as find it the most convenient path to earning their bread and butter. This field also claims much attention because of controversies over which kinds of practical research (such as counterinsurgency research) should or should not be done. Related controversies arise over which interest groups define the terms in which social problems are studied, as in what is considered "deviant."

Important as these issues are from certain points of view, one consequence of our heavy concentration on practicality has gone unnoticed. As far as building a general explanatory theory is concerned, practical research has been a huge obstacle. It has channeled energies into the kinds of research that have tended to lead away from, rather than toward, general scientific explanations. And it has affected the ethos of the sociological community to the extent that the whole idea of a generalized scientific goal for the discipline is obscured. A great many sociologists fail to recognize that this could be a goal at all. The term "science" itself has been appropriated for the narrowly methodological aspect of practical research. It is no doubt for this reason, along with its use to avoid confronting the value judgments involved in such research, that the slogan of "science" is repugnant to many of the more intellectually sensitive members of the discipline.

There are some needless problems here. Practical and generalized scientific goals are not incompatible, but it is necessary to understand that they are distinct. Ideally, there may be no conflict between the two realms. Practical research involves collecting information about conditions in the world so that one knows the extent of a problem to be solved or the effects of a past policy, or in order to extrapolate the future within which he will operate. A further aspect of practical sociology may be called upon to produce policies that will reliably change one condition into another that someone desires. The first kind of practical research, descriptive data gathering, can be carried out entirely independently of the scientific enterprise, except perhaps for some borrowing of techniques. Market surveys or election polls go on more or less irrelevantly to the building of sociological science, and vice versa. The second kind of practical sociology, however, depends on some verified model of how the social world operates. In this area, a truly powerful explanatory science is the precondition for a successful policy planning.

This is not to say that a certain amount of reasonable advice may not now be given on an ad hoc basis because of intuitive feel or extensive familiarity with the dynamics of a particular case. But, in general, sociology is very far from being successful in policy matters. Prescriptions on such matters as crime, community development, or social mobility opportunities largely remain within the conventional wisdom of popular (especially liberal) political positions. Faith in schools, social work, and secularized Christian benevolence still generally defines the horizon even of the self-consciously antiestablishment wing of social problems sociologists. A truly realistic assessment of policy questions awaits a genuinely sophisticated, and scientifically tested, sociological science.

The achievement of such a science—and I am proposing that we can describe its outlines now, and should have it well along by the end of the century if we put ourselves to the task—will not usher in any golden age of applied sociology. Conflicts of interest over the goals for which policies might be stated will always remain a part of the surrounding world, not resolvable within the context of sociology at all. Nor would a truly scientific sociology in the hands of applied sociologists eliminate the maneuvering over power which goes on in any organization within which different factions attempt to bolster their own positions with allegedly neutral expertise. Even in a completely scientized future, one could expect that organizational politicians (who themselves might bear the titles of applied social scientists) will have occasion to produce just those kinds of research which support what they want to do.

But these interesting issues are not the main point here. I am concerned with what is necessary to achieve a scientific sociology, and this involves understanding which aspects of the current sociological science must be treated as obstacles to its attainment. Practical research is one of them, and for several reasons. Practical research has led us to concentrate more on collecting facts than on explaining them. Cumulative development has taken place mainly by refining methods, especially the statistical apparatus through which graduate students have been inducted into the field. We have accumulated a great deal of descriptive material on the distribution of wealth, amounts of social mobility, and the situations of racial and ethnic groups; relatively little has been done to explain and test explanations of these phenomena. Where explanatory work is done, it has generally been parochial and tied to the issues of the here and now.

This is apparent in the ways in which fields of research are divided. Race relations, crime, deviance, urban sociology, community development, are viewed as separate specialties—and separate courses in university catalogues—as if they were subjects which can be treated in their own right. But the fundamental explanatory theory that can make sense of the descriptive materials of each area must come from a more generalized area of explanation, stratification. Similarly, industrial sociology has a way of getting loose from organizational theory, and economic development from stratification and politics; even courses on war or conflict tend to be set in a realm that somehow manages to abstract itself away from more basic models. It is the narrow focus of a social problems orientation—the feeling that one should stick to the here and now of the particular problem—that prevents us from seeing how these materials are just one set of influences within a larger pattern of variations. The explanations of particulars, even if they are the only ones of practical interest, must take place by the process of comparison across the whole range of possibilities. What theory we have had has been largely social-psychological. This has made it possible to maintain the narrow practical focus and yet pretend to some broader explanation by confining this explanation to generalizations about the processes of individual socialization or behavior construction. Unfortunately, this fails to explain the larger framework within which these processes take place. It is not possible to explain poverty by focusing on the poor, whatever insight may be achieved into the psychology of poor people's lives. Why some people are poor is only one aspect of the same question as why some people are rich: a generalized explanation of the distribution of wealth is called for if one is to have a testable explanation of either particular. Similarly, race relations have been studied for decades in terms of the social psychology of discrimination and of being discriminated against; larger structural issues have been handled only in terms of a pseudoexplanatory model of stages of assimilation—in effect, little more than a framework for value judgments about what is desirable (Lyman, 1972).

The field of "deviance" has agonized over its self-imposed limitations, and now seems to be nearing a crisis. The opportunity model, white collar crime, crimes without victims, labeling theory—all of these are steps beyond the borders within which "deviance" can be taken as a self-confined issue. Recent work on the social construction of meanings points squarely to a conclusion: Different social groups use their power to enforce the standards they prefer. The next step clearly must be to abolish the field of deviance entirely, to link its materials with what is known of general explanations of stratification and politics. This means asking the question: What conditions determine the relative power of different social groups, the status ideals and material interests they uphold, and so forth?

Yet the field seems hesitant to make this step. It lingers on the threshold, turning out even more studies of police practices, judicial and correctional procedures, and sometimes historical descriptions of the interests that went into constructing official categories of official deviance. These practitioners seem satisfied to have made the point that social interests are involved and to draw out the policy implications for liberalizing laws. The bolder theoretical statements are elaborated in the direction of arguing a generalized political philosophy rather than of incorporating "deviance" into stratification and making it part of a general explanatory science. Thus, even our current sophistication about some practical issues has not yet resulted in our subordinating the practical orientation until its problems can be reconceptualized in a purely scientific framework.

IDEOLOGY

The division between practical and ideological orientations in sociology is not absolutely firm. Generally speaking, practical research takes the ends for granted and collects information on the extent of the problem or on how to bring about a desired change. Ideological work is concerned with the ends themselves. This can take the form of explicit debates over what is the best form of society, and evaluations of particular arrangements as good or bad. This was the exclusive emphasis in political science during the period when it concentrated on comparing constitutions and debating their merits. This activity still goes on in political science, supplemented by more scientifically explanatory orientation, although the normative concerns can still be seen in the latter sort of work and in political sociology. The ideological orientation can also operate implicitly. It affects what questions are asked, how information is treated, and where debates lead. I am especially concerned here with the latter—the more insidious effects of ideological concerns in hindering the development of scientific sociology.

Sociology developed in a context of political movements, and much of its theory and research has been oriented toward bolstering partisan positions. This is true for sociology's origins in Saint-Simon, Comte, Marx, and Spencer. It continues to have an effect in the modern era of empirical research, even where the ideal of separating scientific judgments from political judgments is explicitly upheld. A great deal of empirical work has stayed very close to the descriptive level, primarily because of the ideological concerns behind it.

Research on social mobility, for example, has concentrated on description rather than on explanation. We have been concerned to know whether mobility has been increasing or decreasing in America and whether America has more mobility than other countries. Our techniques of measurement have become progressively more sophisticated, but we have only now started to develop explanatory theory. The reason is that the research tradition did not arise for purposes of developing explanatory theory at all; it arose to contribute facts to the dispute that has gone on at least since the Progressive Era among liberal, conservative, and radical positions, the question of whether or not America is still (or ever was) the "Land of Opportunity." Once the research tradition is established (especially since the data on this question are pretty much in), the original ideological impetus can become muted, but its effects live on in the emphasis perpetuated by a succession of scholarly careers within the boundaries of well-marked descriptive and methodological specialties, even if the practitioners have forgotten what their predecessors were arguing about. Where the field does move toward explanation, it has been most interested in the *consequences* rather than the *causes* of mobility: for example, the concern for political extremism which blocked upward mobility or enforced downward mobility has long been feared to bring about.

Mobility research has thus been confined to the concerns of liberal faith in mobility as a formula for political stability. Explanatory work upon the causes of mobility has stayed close to the social psychology of the individual career upward, especially in school achievement. All this concern for the good little boy who studies hard has diverted attention from the causes of downward mobility, the lifetime sequence of jobs, or the structure of economic and political organization within which careers are shaped. The research focuses as closely as it can on the portion of reality that we wish to see: the success stories of individuals in American society, whether this is treated optimistically as something real, or critically as something whose obstacles are yet to be removed. In either form, the scope of what is to be explained is circumscribed. Even where empirical sociology has turned toward explanation, the marks of ideological concerns are evident in its limitations.

Similar interests can be seen in other areas of sociology. These go a long way toward explaining why there has been so little cumulation in sociology, despite its fifty-year history of self-consciously "scientific" research. Work on class cultures, for all its value in providing the basis for a genuinely powerful theory of behavior, has not been much used for such a purpose. Instead, the emphasis has been on polemics pro and con Marxist beliefs about class divisions and class consciousness; whether or not there is a self-sustaining culture of poverty; class limitations on full democratic participation; or a decline in the quality of life in the suburbs. The community power literature has been locked in a protracted debate over essentially descriptive issues; behind these, obviously enough, are commitments to upholding or attacking a liberal pluralist evaluation of America. The same can be said of the tradition following C. Wright Mills' (1956) description of the structure of national politics. In all of these areas and many more besides, the field has remained locked in a descriptive posture or in explanatory efforts that move only within the limits of ideological concerns. Thus we have political theories of extremism, or of the conditions for modern democracy, which have remained unconvincing because of the selective scope of the problems to be explained. Particular types of political movements cannot be explained in themselves, but only as one set of variations within a universe of variables. A valid theory of democracy will be achieved only as part of a general theory of political structures across the board, in which the categories are defined by the exigencies of generalized explanation, not simply by a favored type with all else treated as a residual category.[5]

One deleterious effect of our overriding ideological concerns, then, has been that description has crowded out explanation. Another has been the imposition of a mold over how far and in what directions explanation is to go. It has also operated to shape what we have usually thought of as theory, in directions that have led us away from, rather than toward, a generalized explanatory science. I have already touched on aspects of this in discussing the ways in which specific issues such as "social deviance" are conceptualized. Here I will mention two versions of "grand theory"—functionalism and ideological Marxism.

The functionalist effort to analyze human institutions as a system, despite its considerable sophistication in distinguishing levels of abstraction and in integrating multiple forms of interaction, has failed to pay off in a genuine explanatory theory. This failing is due to a commitment to certain political values, which can be seen in system theorists from Comte and Durkheim through Pareto and Parsons. The commitment is to political unity. Systems theory is, in effect, a political (usually nationalist) utopia;[6]

---

[5] The relationship of such dichotomous concepts to ideological issues is examined in Bendix and Berger (1959), and Bendix (1967).

[6] This is most apparent when systems theorists give empirical examples of social systems: they almost always name *states,* and all the groups and organizations within their territorial boundaries are treated as part of the "system" by virtue of that fact alone. Modern systems theory comes from a series of nineteenth-century liberal nationalists, especially Spencer, Durkheim, and Pareto, and reflects the rhetoric by which they tried to promote political loyalty to the national state. Parsons, Merton, and others have propagated the image of American value-consensus and of America's advanced evolutionary status in more or less the same terms, and have carried on the same subdued polemic against any suggestions that might legitimize ethnic pluralism, class conflict, or the notion that the state and its boundaries are essentially arbitrary. It might be noted that not all nationalists have held this position: Weber, for example, held a much more cynical view of politics and national unity,

hence the treatment of conflict as residual, of interest only insofar as it contributes to pluralist order. "Society" or "system" is hypostatized, made the referent around which theory is to be constructed.

I believe that the only viable path to a comprehensive explanatory sociology is a conflict perspective, in which solidarity of any particular sort is just one outcome among many to be explained, and by no means the most common. This has not been realized because "conflict" versus "order" perspectives have almost always been contrasted in evaluative terms as competing commitments to what is desirable. The choice has been posed as revolution versus stability or gradualist reform. I think it can be shown, without too much trouble, that the "system" is usually a myth; that everything does not affect everything else in an important way; and that the "needs of the system" is a way of expressing preferences for what a theorist believes is good, not a causal explanation of the way things actually happen. The conflict perspective, which grounds explanations in real people pursuing real interests, is a good deal more successful at realistic and testable explanation. The reason functionalist systems theories have been upheld in the light of the greater explanatory power of alternate assumptions comes down basically to ideological commitment. Not that systems theorists would not like to generate a real sociological science, but they wish to do it in a fashion that glorifies national (or occasionally international) unity. It is interested in explaining things that it desires, not the range of human behavior. Most of the latter falls into the realm of the undesirable, and hence is treated as a residual category. As Bennett Berger (1971: 240–241) has noted, the blandness of conservative analysis makes it look misleadingly detached, simply because it refrains from obvious polemics; the implicit values are still there.

Conflict theory is intrinsically more detached from value judgments than is systems theory. To be able to recognize competing interests as a matter of *fact*, without trying to squeeze some of them out of existence as unrealistic, deviant, or just plain evil, is the essence of a detached position. It is for this reason that I argue for conflict theory as the basis of a scientific sociology, precisely because it moves farthest from the implicit value judgments that underlie most other approaches. Conflict theorists have come in a variety of political shades, ranging from anarchists and revolutionary socialists through welfare-state liberals to conservative nationalists. They have hardly been averse to arguing for their political values, but it is not so difficult to separate their value judgments from their causal

---

perhaps because of his greater personal familiarity with the realities of politics. But then the German middle class, unlike the French, English, or American, has never really had an opportunity to indulge in sentimental self-justification of their political power.

analysis, and it is to the best of them—Max Weber, above all—that we owe the ideal of detachment from ideology in social science.[7]

But the conflict tradition has ideological problems of its own, and these must be faced if we are to build a true science out of it. Classical Marxism, which provides a starting point (along with Machiavelli, the German conquest theorists, and others) for conflict sociology, is a complex mixture of different themes. Marx's conflict model is situated in a larger functional system, with the emphasis placed on strains of transition from one stage of social development to another, and the functionalist utopia reserved for the end of history. This is another version of traditional organicist positivism, found most prominently in Marx's *economics*, and it is subject to the criticisms of all such models. Again, it is a question of separating evaluation from analysis if we are to pick out what is scientifically useful.

Modern Marxism (leaving aside the Stalinist orthodoxy of the Soviet bloc) has tended to emphasize yet a third aspect of Marx, the dialectical philosophy deriving from Hegel. Marx's scientific efforts are rejected because the hoped-for denouement of the capitalist stage has not materialized, while other political opportunities have opened up, both for revolutionists in the traditional autocracies and for parliamentary socialists in some Western democracies. Because both of the new routes called for political flexibility rather than economic or social determinism, the intellectual task has been primarily one of maintaining moral fervor and rationalizing political maneuvers, rather than revising Marxian conflict theory to make it scientifically more adequate. The historical nature and social conditioning of any claims to truth are taken as self-sufficient grounds for asserting that one should openly disavow any claims to rise above history and should commit oneself instead to intellectual work in the interests of the side one believes to be right.

As an effort to substitute simultaneously for science and religion, this dialectical approach is not entirely self-consistent. It usually takes for granted that one knows which side one ought to be on. Somehow history is supposed to make the underdogs of one period into the chosen people of the next, thus allowing one to be incorruptibly benevolent as well as certain of final victory. But none of this actually follows from a thoroughgoing historical relativism. In its own terms, we have no way of knowing what the future may bring, nor can we say unambiguously whose interests are the "right" ones in the conflicts of today. Even among the underdogs and their champions, there are reformers, revolutionists, and individualistic opportunists. There are conflicting demands for power, for autonomy, for material goods, for improvements in the aesthetic and the emotional

---

[7] Even within the revolutionary socialist tradition, it is possible to make this explicit separation, as has been eloquently argued by Genovese (1971).

quality of life. The dialectical position avoids these choices by sneaking Marxist positivism back in, once relativism has served its purposes in making a call for commitment.

This is not to say that intellectuals have no business making arguments of this sort. They are a form of political action (like much functionalist sociology) which we may consider good or bad depending on what political values we happen to have ourselves. The assumptions of the complacent liberal establishment in American sociology and in American politics are well worth shaking up, in my opinion, on both humanitarian and intellectual grounds. But to do a demolition job on the conventional beliefs shrouding the world does not put anything more reliable in their place; it merely opens up the possibilities. It is for this reason that the most powerful scientific developments of the conflict tradition have been made outside the ideological centers of continental Europe.

A totally relativistic position, whatever its political uses, can never be the basis for a science. Its very denial of the possibility of science cuts the ground from under itself. If there are no criteria for truth independent of political and moral interests, then any statement to this effect is itself dubious and need not be accepted. Even the theory of relativism takes some such standards as operative. Having seen this, we can move toward a pragmatic conception of theory building and theory testing which enables us to take full account of the fact that scientific theories are historically situated too, changing over the course of time. Our criteria for truth are in the method and the direction of this movement, not in the results of any particular stopping point along the way. This is missed because of the confusion among different areas of discourse. Presuppositionless science is an absurdity, but it is an error to suppose that *intellectual* presuppositions are necessarily the same as *political* or *moral* commitments. It is always possible to distinguish, if only one makes the effort, between statements of what is and statements of what we ought to do. The only necessity is that persons must wish to make the distinction. A scientific sociology can be built if we dedicate ourselves to making the effort to build it. This means choosing our concepts for their optimal explanatory adequacy rather than for their evaluative resonance, and setting our sights progressively outward to maximize the scope of causal explanation.

This is not to say that we *must do so.* That is a value judgement that can be accepted or rejected by the individual according to what he wishes to do with his life. For those who would like to do both things, to advance science and make a political contribution as well, the path will be particularly arduous. The separation of "facts" and "values," as Weber well recognized, is not a description of how things are in the sciences, but a commitment to a mental discipline that must be constantly enforced.

At this moment, the task in scientific-sociology-building is to free ourselves from the dead weight of ideological commitments implicitly molding our vision. No doubt we will be only partially successful in this. A sociology that is politically relevant is constantly in danger of adding these distortions at any time. We shall always need criticism in the light of the ideal of a totally explanatory theory.

AESTHETICS

There is a fourth orientation that obscures the path toward a scientific sociology. This may be called the "aesthetic," "interpretive," or "dramatic" approach. As with practical and evaluative orientations, it is not a question of the generalized explanatory orientation being right and the others wrong. All may be desirable activities. The problem is to understand that they are distinct so that we can see what is necessary and what is extraneous to achieving any one of them and particularly the most precarious goal, the establishment of a sociological science. In the case of the aesthetic orientation in sociology, the task of disentangling it from sociological science requires particular care, for there are crucial elements of explanatory science which have been preserved precisely by the aesthetically-oriented side of sociology.

The aesthetic orientation may be defined as follows: If one's prime purpose is not to win the agreement of the widest number of detached thinkers, nor to aid men in achieving the goals of practical economic or political activity, nor to persuade others to evaluate things as right or wrong so as to make them favor a particular moral community or political faction, but it is merely to produce an intellectual work the experience of which is a value in itself, one has an aesthetic orientation. Its slogans are beauty, form, style, drama, vision—or "truth" in a special sense suffused with these qualities.

Literary and art criticism provide a ready-made language for analyzing aesthetic orientations in sociology. Similar distinctions among types of products and techniques can be made in all of these realms; fiction is not the only vehicle for aesthetics. Different kinds of aesthetic experience are distinguished especially in terms of quality, however elusive any absolute standards may be. The simplest level of art attempts to entertain or divert by sensational action and sentimental appeals to emotion. At more complex levels—for which intellectual critics would reserve the term "literature"—there is a more complex response evoked. This is achieved through verbal style: from the timing and flow of thought and emotional response which make up the plot or structure, and from constructions in the form of character, theme, or meaning which result in a sense of having achieved a coherent (or "real") vision of the world.

Different critical theories offer different explanations of how these effects are achieved, and subscribe to different values in determining which are the most desirable (Bate, 1959; G. Watson, 1964). For our purposes here, it is enough to see some of the different sorts of means by which literature achieves its effects. Style, plot, and meaning all work in the same basic way: by calling forth a flow of emotional and intellectual responses from the reader which results in being a satisfying experience. For the trained and experienced reader, the experience will be a particular sort of complex response. Although its effects are varied, we can assume that good literature will avoid boredom and sentimentality—the calling forth of too simple or too often experienced intellectual and emotional responses. Aristotle and others have assigned a special value to literature which results in purifying one from the disabling emotions, leaving one on a plane of resolute and humane insight. Thus literature has sometimes been given the task of a kind of elevated psychotherapy or religious experience. Other critical theories have emphasized the sharing of the writer's creative enthusiasm. Still others have pointed to the kinds of meanings one can derive, to the vision that art produces. Art may be regarded as an illumination of our own world, either to show the universals involved in it (the classical emphasis) or to capture the uniqueness and ephemerality of a particular existence (the romantic emphasis). Of course, the world that is illuminated need not be our own but may be a different or a higher one.

From the various techniques that criticism has shown to be operating in art, there are two principles relevant to understanding the appeals of sociological writing. First, the techniques of literature must always work so as to evoke a series of responses in the reader or viewer. It is inherently a matter of sequence, of flow of response; and the great pitfall to be avoided is the short-circuiting of the response into boredom or triviality. It is for reasons of this order that ambiguity has been found to be so important in literature. The style, the flow of action, and the meanings should be rich and complex, whether this is achieved by the texture of Shakespearian imagery or by the cryptic language of symbolist poetry. In this sense, aesthetics is the enemy of the practical, the precise, and the transparently orderly, in which verbal textures are reduced to the simplest means to an ulterior end.

Second, literature is selective, and its criteria for selection come from the criteria for achieving literary effects. Of the great many things that could be said about any particular topic, literature chooses only a few.

Again, of course, all forms of discourse have their places at which they stop. For generalized explanatory science, the goal is in principle endless: a piece of scientific writing ends because it has reached the end of what can practically be said at that point of the research. For practicality, the information itself is only a means; a manual of useful principles or a catalogue of descriptive information is consulted without regard to the overall structure of the compilation. The criteria for selection is totally external, hence practical work can be formless in every respect except consultability. Moralistic ideological writing stands closest to literature here. The emphasis is on the response, and the criterion is whether or not one is persuaded, although this may have to be done over and over again.

In literature, a proper or a dramatic ending is especially crucial, since the emphasis is on the flow of experience and on the mood in which it leaves the audience. The sense of unity derived from great literature, the feeling of having seen a comprehensive vision, a completed experience, is virtually the hallmark of the genre. The guiding light of the literary writer, then, is when the sequence of thoughts and emotions in his reader can be fittingly resolved.

This is not to say that other orientations may not find their place in literature. Good literature often illuminates our view of reality in a sense that may be translated into scientific terms. It may describe accurately, its themes may reveal psychological, philosophical, or sociological generalizations. Moreover, a great deal of good literature has moralizing overtones, sometimes explicitly religious, sometimes a secularized version thereof, sometimes political. It may even convey some practical information, although most of this is incidental. What makes a work a literary experience, whatever else it may contain, is the way in which it uses its materials to achieve a complex flow of responses in the audience. Whatever literature may do intellectually or morally, it is defined by the aesthetic experience as an end in itself. Writing which has too little of this, too much of the other, fails to be literature, however well it may serve as a form of scientific reporting or philosophical argument, a sermon, or a political tract.

These considerations may be applied to sociology. On the simplest level, there are kinds of popular sociology obviously written for entertainment, which crowd the drugstore bookstands and show up on the bestseller lists. These generally correspond to popular literature in their sensationalism and their simplified picture of reality. There are also serious sociologists who write with an eye for style—Erving Goffman comes to mind, as well as various others mostly on the far side of the Atlantic. But what I am concerned with is neither of these more superficial sorts of aesthetics, but with the ways in which aesthetic concerns have molded the substance of serious sociology. This can best be seen through the rubrics of the "hard" versus "soft" debate in social science.

Underneath the various disputes over methodology, relevance, value judgments, and theoretical questions, the "hard" versus "soft" approaches can be seen as parts of two larger traditions. One is the positivist or utilitarian school. Born in England at the time of Hobbes and Locke, it takes physical science as its model, gradualist liberalism as its political faith, material progress as its practical concern, and religion and intuitionalism as its enemies. An opposition ideology crystallized around the turn of the nineteenth century, in the form of romanticism. Its strands have been politically and religiously diverse, embracing both iconoclasm and individualistic revolt as well as traditionalism and authoritarianism. Its unity comes from its intellectual themes. These have centered on a glorification of the aesthetic and an antipathy to the practical; a critique of the philistine character of modern society viewed as the concrete counterpart of the utilitarian ideology; a philosophy of subjectivity, consciousness, and creativity, in opposition to materialism and determinism; and an emphasis on the historical uniqueness of events and hence the necessity of a concrete and intuitive understanding of history. This position has been particularly strong among intellectuals in Germany and continental Europe generally. Elsewhere it has been limited mainly to literary circles.

This grand typology is not a history of modern thought in any sense. There have been many mixtures and crosscurrents among these camps. The geography is not so neat, although Continental philosophy, psychology, sociology, economics, and even some aspects of natural science (e.g., biology) have been dominated by the romanticist position; Anglo-American versions of these fields have been dominated by positivism—analytical philosophy, psychological behaviorism, and methodological quantification and organic system models in sociology and economics. But there have been important minority currents at various times: materialist positivism in Germany and Austria, idealist metaphysics in Britain, fluctuations between both camps in France. Many of the great sociological innovators combined elements from both sides: Marx, Freud, Weber, and Piaget are obvious examples.

In American sociology, the dominant schools have been positivist. This has meant an emulation of the physical sciences in methods and also in conceptions of sociological subject matter—attitudes, systems, "societies" treated as if they were things. This has not meant that positivist social science has actually been oriented in the right direction toward producing a genuine explanatory science. For the positivist approach—which must be understood as primarily a movement within philosophy and the social disciplines rather than within the natural sciences themselves[8] —has generally overlooked the analytical distinction between science and practicality and has ridden roughshod over questions of value conflict and remained oblivious to implicit constraints on the way problems are conceptualized. Positivism is basically a program for replacing politics and morality with technicism, in effect a value choice for marginal reform within the shell of any particular status quo.

At the same time, there has been a sizable antipositivist minority in American sociology. Part of it is underpinned by the position of sociology on the borderline between the sciences and the humanities. The influx of students—most of whom have the latter orientation, if any at all—or ties with the larger intellectual world have given a continual source of sympathy to the "soft" position.[9] More importantly, there are powerful intellectual traditions in American social science which come directly from the romanticist, idealist, and historicist schools of continental Europe. American universities have two foreign homes, one English and one German. American sociology in its formative period was a mixture of imports: of British utilitarian problem-solving and evolutionism, and of German historicism and neoidealism.

It is no simple matter to assess what these two traditions have contributed or how they have obstructed the development of generalized explanatory science. As I have argued, the positivist tradition has thrown up many obstacles, not the least of which has been its false pretense of constructing a science while actually doing something else. Nevertheless, it is the ideal of science that I am trying to advance here, and I have been harsh toward positivism because it has gotten in the way of its own promises. The case with romanticism is different. On the one hand, I am going to argue that crucial insights of idealism and historicism are absolutely essential for understanding the nature of reality about which our science is to be built. At the same time, these traditions have put up obstacles to building a science, certainly not as a betrayal of their own ideals but because those ideals are profoundly antiscientific. From the standpoint of building a scientific sociology, our attitude to these two schools must be different. Positivism needs to be purged. Romanticism needs to be borrowed from.

---

[8] There was also an explicit positivism in late nineteenth- and early twentieth-century physics, expressed most clearly by Mach. It was positivist in the narrower sense of emphasizing strict empirical observability and attacking what is considered to be idealism in the form of more abstract and systematic theorizing. With the victory of Einsteinian relativity, nuclear physics, and quantum mechanics, however, this form of positivism has declined, except among certain philosophers of science.

[9] This is part of what Gouldner (1970) refers to as the "infrastructure" of sociology, in his massive critique of the utilitarian (positivist) tradition in sociology.

The fundamental aim of the "soft" school is not science but aesthetics. Since art is very broad, it can include a great deal of insight into general causal principles and convey much concrete information about the world. It has room for moral and political exhortation, but also for questioning certainties on these scores; and it puts practicality in critical perspective. On all these points, the aesthetic tradition is greatly superior to positivism in flexibility, breadth, and sophistication. But when aesthetic sociology moves from criticism to statements of its own world view, its limitations become apparent. It cannot fill the place where a genuine science should stand; by the very nature of its basic concerns, it operates to evoke a particular sort of experience in the reader, not to lay out a comprehensive explanatory system.

This can be seen by examining three main versions of romanticist social science: the cultural critique of modern society; the interpretive social psychologies, from the symbolic interactionists to modern phenomenologists; and the historicists.

The Critique of Modernity

The tradition of cultural criticism can be summed up by such topics as "alienation," "mass society," and "popular culture," sometimes translated into more positivistic cover as "community mental health" or "environmental quality." These themes go back directly to the early romanticist movement, to the revolt of the artist against science and respectability which began with Rousseau and became a basic article of literary faith in the bohemian communities of the nineteenth century.[10] It is striking how little there is new about each successive exposition of the alienation of modern man or the quality of modern culture. To the extent that a systematic research tradition has risen from this in recent years, the results have also tended to have the same noncumulative quality. For the ordinary individual hardly suffers the way intellectuals do in contemplating him, and most research has simply thrown unwelcome facts into a discourse that is not a scientific debate but a lament.[11] Most of these theories, thus, can be viewed as species of literature and should be so judged. Much of it has too much bathos to be very good literature. To be sure, there are also sensitive observers like David Riesman,

---

[10] The romanticist tradition has passed on the old aristocracy's expressions of disdain for the industrial–bourgeois world, as artists attempted to make themselves into a new aristocracy. The process is described in Hauser (1951: Part IV), and Graña (1964).

[11] See, for example, the evidence presented in Blauner (1964). Even the best empirical work on alienation remains cut off from explanatory sociology in general; it is little recognized that alienation at work is the obverse of successful normative control, and that these results can be integrated with a more general theory of organizational power struggles. See Chapter 6 in this volume.

artistically aware of the complexities of such issues, capable of playing off philistine and romanticist simplifications against each other; but the orientation remains aesthetic.

My main point here is simply to illustrate how an aesthetic approach, of whatever quality, cuts off analysis to fit its own goals and is not a path forward to scientific sociology. At the same time, I do not want to leave the impression that such material is inevitably doomed to bathos unless handled with the genteel liberalism of a Riesman. There certainly is room for powerful literary statement—in nonfictional prose form—that criticizes modern society in the drastic tones of a Marcuse or an Ellul. From the point of view of our own best ideals, there is some devastating real-life comedy to be appreciated, tragedy to be experienced, and epic of rebellion to be formulated. But the standards of literary sociology can be pushed a great deal further than the melodrama we have so far been treated to. Among modern social scientists, only Levi–Strauss—but writing about the world of primitive tribesmen—has produced much of real aesthetic merit. The cultural critique of modern society needs an infusion of a new vision drawn from the best sociology of today, not from the worn-out images of two centuries ago.

Interpretive Social Psychology

Another sociological approach which falls into the category of romanticist social science is the analysis of the flow of individual experience. There are various types of such analysis: symbolic interactionism, social phenomenology and ethnomethodology, and sociological existentialism; all of them go back ultimately to the German idealist tradition. Some of these—e.g., Lyman and Scott's (1970) *Sociology of the Absurd*—explicitly set an aesthetic aim: to capture the pathos or drama of human existence, especially the existential tragedy of man imposing fragile meanings on a meaningless world. Others, such as ethnomethodology, are far from overtly humanistic in style or content. Ethnomethodology and phenomenology tend to exist in symbiosis with a counterimage of positivistic science which is a constant target of attack. A hidden rivalry to science comes to the surface in the ethnomethodologist' more positive aims, which is to purify research procedures. In this respect, ethnomethodology is reminiscent of a hypercritical positivist methodology taken as an end in itself. Some of the same rivalry is found in symbolic interactionism, which, although extremely critical of positivist formulations, often puts itself forward as the basis for a truer scientific explanation of social behavior.

Nevertheless, all of the interpretive sociologies seem to be basically organized around an aesthetic approach to their subject matter. The emphasis on the subjectivity of the human actor, the time-bound processes of behavior and experience, and the denial of external structure and of the scientific ideal of explanation—all of these leave the interpretive sociologist in the position of needing some criterion for organizing what he is going to say. The scientific ideal of progressively more powerful causal explanation is explicitly denied as a valid goal, although the ideal of "truly" scientific knowledge is sometimes smuggled back in to defend the critics against the charge that they are purely negativistic. The interpretive sociologists can also find an exterior purpose through their utility in attacking the political premises assumed in positivist treatments of "deviance" and "social problems" generally.

But the core of the appeal of Cooley and George Herbert Mead, Schutz and Merleau–Ponty, Garfinkel, Lyman and Scott, and others is the experience of insight that they give into social reality. The interpretive sociologies are truer to life than structuralist models. They attempt to capture the flow of our own experience just as it is and to demonstrate that all else is a myth built up from individual performances. Since science must be about reality, interpretive sociologists claim to be the real basis for any scientific sociology, in contrast to theories derived from reified ideals about structures, attitudes, norms, and values. But this promise (or threat) is never really lived up to, for the interpretive sociologists do not develop and expand a body of testable generalizations. The scope of application may expand, as symbolic interactionist interpretations have been given of various kinds of deviance, career mobility, ethnic relations and so on. But this is a matter of repeating the insight so that one sees how each of these is a processual phenomenon built up by the individual actors. This has been the case with the symbolic interactionism; ethnomethodology—insofar as it moves beyond purely philosophical discussions—often seems to tread the same path, with its own peculiar sophistication but substantively similar results.

This emphasis on an insight experience is the hallmark of an aesthetic orientation. Negatively, it is set off against the cold, uninsightful, and hence unreal constructions of positivist science which interpretive sociology constantly criticizes. Positively, its own claim to value is the experience it gives of showing how things really are, not through the cumulative results of explanatory research, but all at once with a vision of the universal processes of human experience. Like great literature (especially in the romanticist version), it captures the particularity of individual human existence, the pathos of its boundedness in time, the subjectivity that gives man his freedom but also cuts him off from fully encountering others. At the same time, it approaches the classical literary ideal in demonstrating the universality of these phenomena, giving the experience of recognizing the universal in the particular. It is the constant potential

for recreating these experiences in the reader that keeps interpretive sociology alive through endless repetitions, much as a Shakespearian play can
return for an infinite number of engagements.

The ambiguity of this position must be faced. On the one hand, interpretive sociologists are quite right that a science must be true to the basic
empirical reality it is trying to explain. A scientific sociology will have to
be built on an understanding of the enacted, subjectivistic, interpretive
nature of human experience. "Organizations," "societies," and so on must
be seen through and through as the ephemeral creations of real men in the
flux of constructing agreed-upon realities, which often are not really
agreed-upon but inalterably plural. But having performed both a critical
and a potentially reconstructive service for scientific sociology, the interpretive sociologists stop dead in their tracks. The aesthetic orientation—
which provided from the wealth of the romanticist–idealist tradition both
the critique of positivist social science and an alternative image of man—
becomes an obstacle to moving any further.

Even where it proposes to reorient the foundations of scientific sociology, ethnomethodology has jealously guarded its boundaries, refusing to
allow generalizations from outside its own precincts even in tentative
status. It has maintained an absolutist, nonpragmatist ideal of truth as a
stick that only itself escapes being beaten with—and this only because it
usually fails to attempt serious generalizations. Symbolic interactionism,
on the other hand, has lived in proximity with American positivism for
half a century now, and has accommodated by pretending more directly
to underpin a science of its own. Yet this has not come about, even with
several decades of empirical research informed by the symbolic interactionist tradition. It has described occupations, careers, and deviance (and
has contributed to the downfall of positivist theories in the latter area
based on projections of political values), but it has built no generalized
explanatory theory beyond a repetition of the same fundamentals. The
generalizing efforts of Hugh Dalziel Duncan (1962) and Anselm Strauss
(1971) point up the failures: a great deal of material is subsumed under the
model, but formulation never moves to testable generalizations about the
causes of variations in behavior.

It is the aesthetic orientation that brings them up short. For to state
causal generalizations in this context would be to introduce the cold,
unambiguous, and "external" qualities into human experience which the
subjective, interpretist approach tries to overcome. Despite its occasional
pretensions to develop a new science, interpretive sociology is bent on
protecting a romantic version of human experience from science. Ambiguity is of the essence in romantic literature, for it is in the reader's
responses to a complex of suggestions that its experience of drama is

found. The processual quality of our lives, in the same way, derives its drama from the ambiguity of our future at any given time. To limit that ambiguity with causal principles is to circumscribe our sense of freedom and to risk losing the dramatic experience that symbolic interactionists mean to convey.[12]

What kind of sociology one wishes to do, of course, is a value choice that everyone makes for himself. Explanatory, practical, ideological, aesthetic—there is no one type that can arbitrarily be called right and the others wrong, but they are different modes of discourse and there is no excuse for confusing them. My concern at this moment is for a successful scientific sociology. To build this, I have argued, we must free ourselves from practical and ideological orientations and their aftereffects, and from aesthetic concerns as well. In the last case, however, there are important things to be borrowed for scientific sociology. The aesthetic critique has been crucial in freeing ourselves from the practical and ideological projections that have propped up a pretended, an unsuccessful, positivist science. The interpretive sociologists give us a firmer basis on which to build.

But we cannot linger forever admiring the insight. In order to build explanatory sociology, we must work at building, recognizing all the tentativeness and pragmatics of our formulations but not using these as an excuse for abandoning the enterprise. Ironically, even the fate of the insight experience favored by an aesthetic orientation may depend on such advances, for the insights of interpretive sociology have a way of palling. The flash of recognition, which hopeful sociology teachers have sought for decades to make their students experience when they are brought to see the very world they live in as enacted and interpreted, has been all too ephemeral. The general idea of the social nature of our world has crept into the public world view. Symbolic interactionists, for all their self-image as proponents of freshness and life in sociology, are themselves largely responsible for the feeling that sociologists give other names to what everybody already knows. Ethnomethodology left to itself seems likely to suffer the same fate. It is only by producing explanations that give the conditions for why one thing happens rather than another that sociology can recover some of its magic. Though there are dangers here of turning sociology into a drab determinism, any scientific sociology built up from

---

[12] To take just one example: the compelling arguments of David Matza (1969) for the importance of consciousness, drift, and the ongoing self-shaping of experience come up against a limit of their own making; consciousness is invoked as an arbitrary realm beyond causality, without considering what kind of determinants might be located at this very level. An approach to causal explanation of the ongoing process of negotiating cognitive realities is presented in Chapter 3 in this volume.

an interpretive perspective will have built into it its own foil. It is the combination of determinism and freedom, after all, that constitutes the greatest art; advances in scientific sociology can move its aesthetics beyond a tired romanticism, hopefully into something greater.[13]

Historicism

There is yet another branch of aesthetic sociology. This is the version of sociology that emphasizes the historically unique qualities of what is being explained. Political structure and events, stratification, culture—indeed, any of the materials of sociology—exist in a unique configuration of particulars shaped by an idiosyncratic series of previous events. This again is a nineteenth-century tradition, transmitted into sociology especially through Max Weber and into anthropology through Bastian and Boas. It opposes simplistic "laws" of history and has done much to show how different societies can be from each other even though they have technical or economic structures in common. As we shall see, the sequence of struggles which makes up political history turns out to be one of crucial determinants of modern democracy or totalitarianism, of centralized or decentralized regimes, of the degree to which ethnic enclaves are preserved, and of the forms taken by mobility channels.

Like the interpretive sociologists, historicist sociology presents an ambiguous face to sociological science. On the one hand, it effectively criticizes poor scientific explanation and points to a range of empirical considerations. The long-term flow of history itself, and the crystallization of institutions that can perpetuate certain arrangements long after the other institutions that came into existence with them have passed away, must come into the body of any successful explanatory theory. If we are to have a real sociological science, it must be grounded in the lessons of history. But historicism sets itself against taking this step. Like symbolic interactionism, it takes the stance of a permanent opposition to generalizations. It is ready to criticize but not to construct. There is one major difference: more than the interpretive sociologies, historical sociology has provided our main body of insights which go beyond common sense. But these are limited to particular explanations of the importance of feudalism for modern democracy, the importance of ethnic migration and conflict for the structure of mobility channels in the United States, the importance of

---

[13] Richard Brown has pointed out to me that there are nonromanticist aesthetics, especially in contemporary Continental literary criticism, that do not fit the foregoing characterization. Whether they can maintain a link to a truly scientific sociology remains to be seen; so far, French structuralism connects all too readily with the noncausal systems models from which scientific sociology has been trying to disentangle itself.

Christianity for the original rise of capitalism. Larger generalizations are often regarded as impossible. The aim becomes a sociologically informed history, not a generalized sociology.

Our task is to decide how much the alleged impossibility of a transhistorical sociological science is an inherent limitation on what we can do and how much of this is due to a value choice in the direction of wishing to do something else with our intellectual efforts. If sociology is to become social history, then history itself must be examined.

History's task basically is to tell stories, to describe what happened. But its scope consists of an infinite number of facts, all the experience of everyone who has ever lived. Even limited by what records are available, the facts are myriad. How is this selection made? One guiding aim is ideological: history originally was written to glorify a king, a nation, a party, a religious faction. Debate between opposing points of view has made such history more accurate, more sophisticated in its interpretations. But ideological themes remain an important selective principle for setting the topics historians wish to study: assessing the causes of the Cold War, looking for continuities in foreign policy, and so on in many political and related topics.

A second orientation is aesthetic. Historical stories are sometimes put together for the effects they invoke in a reader; selection of what will enter into the narrative and in what order and emphasis is based essentially on the same criteria as literature. This is not intended as a disparagement of such history. Its task is to make drama out of true life, and one of the standards it sets itself is that of objective scholarship. Nor should we take sentimental melodrama—typical of "light history" like romantic biography—as the archetype of aesthetic history. These correspond to the more popularistic levels of literary art. Great history corresponds to great literature in its emphasis on the larger drama of many men across many years; it invokes the powerful effect of the ephemerality of events as well as their underlying continuity and their ironic breaks, which gives the value to serious literature. Scholarly truth is by no means neglected; in fact, the intellectual power to discern a verifiable pattern out of what has actually happened is one of the elements that adds to the experience of the reader. Explanatory scope and the excitement of intellectual discovery add to the aesthetic power of the historical narrative, just as they do in great literature. Aristotle, seeing history as mere uninspired description, called art truer than history; aesthetically inspired history reclaims a higher position.

There is also a sphere of history that can be called scholarly in a narrower sense. Ideological disputes over the causes of wars or the importance of class conflicts give rise to a community of researchers whose aims are defined by correcting the errors of their predecessors and achieving a more comprehensive and detailed explanation of all of the causes of a given event. Similarly, themes proposed by dramatically-oriented historians can also become material for specialists to revise. Methodological innovations take on an importance of their own in a competitive community of scholars whose members constantly need new fields to till. Social description in history—e.g., of the class structure of Colonial America or of the everyday life in ancient Athens—entered to fill such needs and, eventually, to become incorporated into the larger traditions of ideological and aesthetic interpretation. Or historical scholarship can take its guiding themes from the generalizing attempts of the social sciences, testing out, sometimes gleefully demolishing, sociological generalizations about mobility rates in nineteenth-century America, Marxist interpretations of politics, or Weberian interpretations of economic development.

The view that insists on the particularity of historical events as a complex sequence in time, then, reflects only one portion of historians' activity. Historians need a guiding concern by which to select their materials. In the smaller focus, this can be seen simply as an attempt to test out and improve on formulations of predecessors. But the themes of those predecessors must themselves have some aim, as must that of any contemporary historian with a claim to point the way to topics of new importance. These topics are sometimes ideological, sometimes aesthetic, sometimes scientific generalizations—usually treated as a foil—borrowed from other disciplines. The mutual criticism within the scholarly community tends to make these concerns interpenetrate each other.

It is not my purpose to assert the priority of one aim over another. Philosophers of science who claim that history is valid only to the extent that it conforms to a natural science model of tested explanation are simply imposing their value judgment on the kind of history they want written (cf. Weingartner, 1968; Dray, 1968). There are other criteria for how to organize a historical narrative into which the truthfulness of the facts and the validity of whatever general explanations are invoked are only some of the component parts. Description, for whatever purpose, is usually the strongest concern of historians; but it is equally true, in opposition to philosophers who defend the historicity of history, that history can and does contribute to generalized explanation and that many of its guiding concerns come from efforts to test such principles.

History can be many things at the same time. If one wishes to emphasize its complexity, the uniqueness of any particular portion of it, the drama of its transitory episodes and its long heritages, that is certainly valid. But one may also see it as a field in which general principles find their application and hence as a place for testing and developing sociological theory.

A generalized sociology founded on this perspective certainly does not dispense with history nor overrule the vision of particularity, uniqueness, sequence, conflict, and causative concatenation that it offers. What it does do is distinguish two levels of analysis. One is the vision of what concrete history is like; the other is the realm of analytical generalizations, causal principles that are testable across a multiple of particular instances. These will be found in the realm of organizations, stratifications, social psychology—principles which explain how people tend to act under given conditions. They make up the parts of an historical analysis. The whole, however, must be seen in its particularity, as a concatenation of individual causes. "Societies" do not exist except as ideological reifications. Hence we have no laws relating to them. What we do have is a complex of generally explainable processes occurring within government organizations and in particular communities.

The explanatory theory does not dispense with history, for the relationships among the various elements remain to be accounted for more concretely. Since we wish especially to account for one particular historical sequence—the present or near future of the people around us—such a vision is no hindrance to the immediate application of general laws to materials that concern us. The overall relationship between history and historical sociology and generalized scientific sociology is like that between evolutionary biology and geology, on the one hand, and the generalized theories of chemistry and physics, on the other. The latter provide the general explanations that go to make up any given state described by the former discipline, although the long-term sequences of natural history are themselves so complex as to require more particularistic explanation.

The positive insights of historicism, like those of the interpretive sociologies, define for us the basic nature of reality which a scientific sociology sets out to testably explain. In neither case should we take too seriously the accompanying claim that insightful interpretation is all that is possible. Once we see that the aesthetic commitment of the romanticist tradition is only one possible commitment among several, the path onward to a genuine science opens before us.

THE BASES OF SCIENTIFIC SOCIOLOGY

What assurance is there, if we set our minds straight on building an explanatory science rather than pursuing practical, ideological, or aesthetic aims, that we will succeed? The claim is not merely programmatic. I believe that a considerable number of advances have been made in theory and research. The task of putting them together into a generalized explanatory theory has simply been held back by the implicit dominance of nonscientific orientations. In particular, we know a great deal about stratifications and organizations; and these areas provide the core of a general sociology.

First of all, social class has proven to be the most important single variable in empirical research. Political attitudes, voting, organizational participation, religious behavior, styles of consumption, childrearing practices, media exposure, deviance patterns, world views—all of these and more show major social class differences. The theoretical significance of this has been lost in polemics pro or con the Marxist position. Clearly, behavior has multiple causes. A highly global view of class will not advance us in stating these precisely; neither will emphasizing the abstract proposition of multiple causality. The fruitful approach is to pick out those variables which are subsumed under the earlier, tentative uses of "class." The term, after all, derives from the operation of classification; all that "class" means in any realistic sense is that people who share certain common conditions in the ways they relate to others will tend to be alike in certain other ways. Early research efforts can only lump people together in very gross categories; once we see that this is on the right track, the next step is to make finer distinctions to improve the power of our explanations.

One set of meanings of "class" involves characteristics of occupations; the kind of work that one does is one major determinant of how one thinks and behaves. The task here is to show what it is about occupations that affects the individuals. As this progresses, our grosser categories of occupational classes will give way to a more precise set of occupational variations.

Another influence comes from the community that one lives in, the immediate milieu of people with whom one associates. This, too, has sometimes been subsumed under "class"; but, although it is affected by occupational position, it has organizing principles of its own. Some of these are in the structures of families and households; these living arrangements can make the occupational culture of some persons—typically those who dominate a family or household, the father or master as opposed to wives, children, or servants—carry over into the behavior and outlooks of others who are not directly in that occupational position. This shows up in survey research as the effect of parental occupation (and spouse's occupation) in addition to one's own occupation. In historical perspective, as the very form of household has changed, the set of such influences on the individual has altered.

There are other characteristics of community organization, or "status groups," as we somewhat vaguely refer to them. Religion and ethnicity are two important forms. These refer to membership in a group of persons who associate together, often around formalized community organizations, and who uphold a common culture. Again, these are affected by (and affect in turn) occupational position. An ethnic group is only the institutionalization of the kind of culture that grew up under the work (and community) conditions of a particular area, transplanted to a new context by migration. There is a good deal of evidence on how religion and ethnicity affect behavior independent of current occupation; these must all be included in our multiple causal model. The theory-building task now is to departicularize ethnicity and religion, to see them as concrete cases of more general processes. The problem is to show the conditions under which community cultures are institutionalized so as to survive as determinants of behavior in the context of a changed occupational situation. Ethnicity can thus be made part of a general explanatory theory instead of merely a matter for description. By far the greatest obstacle is to formulate the question in general terms, to break through our prevailing ideological concerns for Americanization or taking sides in ethnic conflicts. Once the first step is taken, the rest may not be so very difficult.

I am proposing, then, that we have the elements of a theory of social behavior and belief, based on a multiple causal model of stratification. Occupational situations, family structure, residential community organization, the institutionalization of various of these in religious and ethnic communities—these are the pieces of the puzzle, each to be stated in precise generalizations including their effects on each other. Other variables used in empirical research (education, friendship group, social organizations) can be fitted into this general schema of work organizations and cultural communities.

A second area that has made great advances is organizations. A unifying theory is emerging around the theme of power struggles and their outcomes. The elements of this theory have been developed in a number of separate areas. The degree of convergence has not yet been widely recognized, partly because of the localizing effects of ideological and practical concerns, partly because of overspecialization and fragmentation of research. Researchers in industrial relations have long known about informal group organization among workers and their tactics of struggle against superiors for controlling their own conditions of work. Blau (1955) and others have extended these findings to white-collar as well as blue-collar workers. Research on managers, from Barnard (1938) to Dalton (1959), has extended the picture of political maneuvering to higher organizational levels, and shown some of its tactics and conditions. Out of these various researches, Etzioni (1961), Crozier (1964) and others have formulated many of the principles of a general theory of control and conflict in organization.

Moreover, all of this converges with the longstanding tradition of power studies of political organizations. Weber's own models of patrimonial and bureaucratic organization, although frequently taken as no more than idealized taxonomies, actually contain a general theory of the conditions under which different kinds of power arrangements will be upheld or broken down. The implicit struggles between rulers and their organizational servants are explained here; from this springs the whole line of analysis of power struggles, ideology, and goal displacement in government bureaucracies, of which Mannheim (1940) and Selznick (1949) provided the classic works.

Research on membership-controlled, "democratic" organizations is a variant on the same themes. Michels' (1949) classic work on the conditions that allow political parties to be taken over by their elected leaders who become a self-interested and a self-perpetuating group, has been extended to a large variety of membership organizations, from unions and professional organizations to charities and social clubs. The contributions that this line of research makes to a general theory of power in organizations has been obscured by two things. One is the ideological context within which Michels wrote; hence, subsequent research and theory on the "Iron Law of Oligarchy" have been concerned to uphold or refute conclusions on the impossibility of democracy rather than to state more fully the variations in conditions that make organizational leaders more and less powerful vis-à-vis their followers. The other obstacle has been to treat democracy and voluntary organizations as separate issues, remote from power questions in other types of organizations. But in a more abstract sense, these are identical questions of power struggle in organizations that are officially controlled from the top and in those which are officially controlled from the bottom. In both bureaucratic theory and voluntary association theory, there is the same problem: how the official control is subverted by those who are supposed to be carrying out a task imposed by others. Moreover, given the differences in empirical arrangement of the variables, the conditions that determine just who wins what in these struggles of organizational levels—notably control over channels of communication—are much the same in both types of organization. Thus, both areas contribute to a generalized theory of organizational power.

The final area of convergence with general organizational theory comes from studies of occupations and professions. The major issue here has been to state the conditions under which an occupation becomes relatively free of vertical controls and becomes, instead, autonomously powerful through the organization of a professional community in which peers reserve the power to monopolize practice, train new recruits, judge performance, and control careers. The work of Hughes (1958) and Wilensky (1964) has moved to a general statement of these conditions. These have much to do with control over areas of uncertainty, particular states of the same general variables that have been found important in bureaucratic and voluntary-association power struggles. Professions can thus be seen as another set of arrangements explainable in terms of a general theory.

Organizational studies thus provide another well-documented area of explanatory development in sociology. It is probably the most advanced area in sociology in terms of conscious theoretical formulations. In addition to the general power theory just referred to, it has recently begun to state the general set of conditions for organizational structure. The work of Woodward (1965), Stinchcombe (1959) and others on the effects of technology and task arrangements are the most striking advances; classic works by Weber can be reincorporated in this light, as can historical works such as Chandler's (1962), and contemporary research on external organizational environments, notably by Thompson (1967). Moreover, as I shall try to show, structural theory and power theory mesh at crucial points, since organizational structure refers in large part to the network of attempted controls among organizational members.

Sociology, then, has a solid explanatory foundation in the area of stratification and organizations. These two are not unrelated, of course. Occupational positions, which are such a crucial element in stratification theory, are the very units of organization. Organizational power theory provides the key explanations of how individual behavior is shaped, and hence, of all the rest of stratification analysis. What we have is two different ways of focusing on a single reality. Organizational theory and stratification theory reinforce each other. In this perspective, the wider questions of stratification theory—the determinants of social mobility, of the distribution of power and wealth—can be seen as ways of enumerating some of the results of the struggle for occupational position. This enables us to solve the problem of hypostatization. After all, a set of statistics on mobility rates or financial distributions are not empirically observable human behaviors but are merely abstractions based on verbal reports about the results of lifetimes of behavior. The causal explanation of such distributions, then, will ultimately be established by focusing on the patterns of behavior of real people struggling for power in organizations and moving through networks of personal association.

What about the rest of sociology? Organizations and stratification are the explanatory core of the field. The rest are essentially empirical applications of their principles. Race relations is one particular type of stratification. Political sociology is another area of stratification—the causes and effects of the distribution of power—informed by the study of political organizations. Deviance, as the current line of researches have shown, is a matter of conflict among individuals and groups over cultural standards, material possessions, and power. A general causal explanation of the phenomena of this area will come essentially as an application of stratification theory; careers of individual "deviants" are a part of a more general theory of social mobility (in this case, mostly downward).

Virtually every area of sociology now isolated for special study, usually because of its special "social problems" interest, can be seen as an application of stratification, organization, or both. Specific institutional areas—education, religions, communities, occupations and professions, industrial sociology, the sociology of science, social movements, even the family—all take their explanatory principles from these more general formulations. Many of these simply refer to a particular type of organization. Even social demography, the study of population and its movements, may eventually merge with this body of explanatory principles. As of now, demography remains more a descriptive than an explanatory discipline, although it has been able to predict by extrapolation (not always successfully) even without general causal principles. What we now know of the effects of economic conditions and other effects of stratifications on births, deaths, and migration suggests that a more thorough grounding of demographic theory in stratification will be of use. At the same time, sheer numbers of persons and the geographic arrangements that affect their movements are conditions that can add independent explanatory power to the theories of stratification.

Social change, including such specialized topics as economic development, is essentially a matter of applying stratification and organizational principles to some long-term historical changes. As I have argued, there can be no general laws of history; our generalized explanations come from a more concrete level. As we get more global, the amount of concrete description of sequences and concatenations of causes becomes a greater and greater proportion of what we must include. Social change, then, is a kind of border area for generalized social theory.

There is another broad area of sociology that is not subsumed under stratification and organizations. This is the area of general theory inquiring into the principles of social interaction, or social psychology. As a very generalized and abstract inquiry, it crosscuts all the specific areas of sociological research. It is on a different level of analysis. There have been a number of important theoretical advances in this area which provide an underpinning for explanations of variations in behavior through the principles of stratification and organizations. The most fundamental breakthrough was made by Durkheim when he demonstrated that interaction is not simply a matter of cold-blooded bargaining, but that strong social ties are based on emotional bonds to which we attach moral ideals.

The significance of this breakthrough has been obscured by the way subsequent theorists have focused on Durkheim's functionalism and have used the analysis of moral bonds as an argument bolstering a reified conception of a "society" or "social system." What is necessary is to separate out the usable portion of Durkheim from the misleading part. It is true, as Talcott Parsons (1937) argued thirty-five years ago, that Durkheim and Weber converge in appreciating the crucial role that moral appeals play in social behavior. On other points, Durkheim and Weber were widely opposed: Durkheim as a functionalist deeply committed to the value of a nation-state and hence willing to reify his view of society to explain away conflict as pathological; Weber as an historicist and a conflict theorist deeply cynical about pretensions of social unity and aware of the competing interests behind any such claim. Parsons, in effect, emasculated Weber's theory by forcing it into an artificial convergence with Durkheim and with Pareto's reified social system. My proposal is just the opposite: The path forward to a general explanatory theory is to build on Weber's nominalist conflict approach to stratification and to organizations, and to treat any larger historical pattern as an historicist combination of these elements. Durkheim is to be borrowed from selectively in order to round out the theory at the point of a fundamental understanding of the emotional and cognitive dynamics of interpersonal interaction.

The path forward from Durkheim, then, is not to accept his overall conception of societies, but to understand what he shows about the nature of specific interactions. Particularly in his last work, *The Elementary Forms of the Religious Life*, Durkheim presented a powerful model of the ritual aspects of social behavior as the key to emotional solidarity and to our most fundamental conceptions of reality. Along these lines, Erving Goffman has shown how little "collective consciences" are created and recreated every-time people meet. Instead of the great balloon in the sky suggested by Parsonian interpretations of Durkheim, we see a plurality of moral realities populating the landscape. Goffman himself is not the most reliable guide on this point, as he concentrates on the most general conditions for social order and misses both the historical variations and the struggles for group domination that underlie ritual behavior. As in other cases, the task is to make use of the contributions of other researchers while disengaging them from orientations that are a hindrance to explanatory advance.

In the following chapters, I will draw on those elements of Goffman that transcend the functional model and the somewhat narrow historical period from which he generalizes a model of interpersonal relations. For all its limitations, Goffman's work situates itself precisely at the crossroads of all of the forms of analysis I shall be concerned with. Much of his work draws upon and summarizes the occupational and organizational literature in such a way as to show how power struggles are carried out through the manipulation of ritual encounters. In the same vein, many of Goffman's examples show how status communities—the associational side of stratification—also depend on ritual solidarity in the details of everyday social life.

There are several other approaches that resonate with Goffman's applied Durkheimianism, although on somewhat different wavelengths. On the one hand, there is the line of analysis stemming from Freud, concerning the dynamics of emotional interaction and especially the relations between what we are conscious of and what is unconscious. Freud's emphasis on childhood experience has been somewhat overdone, especially by followers who have adopted behaviorist psychology, or by functionalists looking for a stable resting place for society's value system. Modern versions of Freudian analysis, especially the Gestalt therapy of Perls, Hefferline, and Goodman (1951), have emphasized the contemporary quality of all causes of behavior—only the here and now, one's own body and cognitive states and the people around us exist; only these can have real effects. These approaches open the way to directly studying the emotional aspects of interaction, especially as bodily experience. Ultimately this line of analysis should tie in with an understanding of human physiology, and with animal ethology as well, linking the biological sciences with sociology.

There is another approach to interaction which extends our understanding in yet another direction. This is the detailed examination of human cognition carried out now under such labels as "ethnomethodology" and "phenomenology." Like modern Neo-Freudians and like Goffman, these emphasize how reality is constructed in an ongoing flow of social encounters. Behind this tradition lie such thinkers as George Herbert Mead, with their insights into the social nature of language and hence of thought. Other than a tendency to take a stance that rules out scientific contributions on principle, the weakness of this approach lies in its emphasis on purely cognitive problems. Its image of man is narrowed to that of a thinker, never a creature with a physical body and emotional responses. Its positive contributions include an understanding of the constructedness of social structure. This fits into a general conflict scheme, not only in guarding against the imposing of reified structures on an ongoing power struggle, but by showing some of the mechanisms of power in the manipulations of beliefs that must necessarily be taken for granted most of the time. Ethnomethodology and phenomenology also promise, through their application to sociolinguistics and childhood socialization, a new and powerful sociological theory of what have hitherto been thought of as questions of individual psychological functioning.

On the level of a general theory of interaction, then, there are also powerful lines of advance. They remain to be integrated. The emotional and cognitive aspects of interaction must be explained as part of a common framework, and their significance for the variable patterns of social interaction which make up the subject matter of sociology proper need to be shown. What follows is a tentative attempt to do so.

OVERVIEW OF CHAPTERS

The plan of the book is as follows. The center of a general explanatory theory is a model of stratification; this is attended to immediately in Chapter 2. The path pursued is a version of Weber, which incorporates the main thrust of the Marxian fundamentals but elaborates them in a direction that enables us to pick out the variations not only in formal occupational positions, but in the organization of domestic and associational networks, that shape individuals' outlooks and behaviors. The crucial dividing lines in the social structure are dominance relations in all of these spheres, and the resources and contacts that are attached to them. In keeping with my general program of linking structural and interactional levels of analysis, this chapter attempts to cast all structural variables in terms of actual differences in the experience of face-to-face encounters; Durkheim's theory of the conditions for rituals with varying emotional and cognitive effects is thus introduced to explain the basic variations in class and status group cultures. Throughout, the effort is to show that various forms of the ritualization of belief and behavior are intimately linked to the process of stratification; thus it should not be surprising that Weber's crosscultural description of the religious propensities of different classes should serve as prototype for class cultures in general.

Chapter 3 deals with the fundamental theory again, but from the perspective of microanalysis. The earlier part briefly reviews the history of various approaches to the sociology of face-to-face interaction, leading up to the argument that animal ethology and social phenomenology come together in pointing to the ritualized, verbally unformulated aspects of encounters as the underlying features that shape social relationships and that make it possible to construct shared symbolic realities. The remainder of the chapter presents a model of conversational interaction. It attempts to show that this most commonplace and ubiquitous social activity is the building-block of all the social networks, formal and informal alike, that we usually refer to as "structure"; and that what evidence is available on this detailed level of observation serves to fill in the minute-by-minute processes by which class and status-group relations, and their accompanying cultural outlooks, are continually negotiated.

The next two chapters elaborate particular aspects of the basic model. Chapter 4 continues the emphasis on the microlevel, but puts it into an historical perspective missing from most such analyses. Thus it presents a brief historical survey of the changing forms of ritualized encounters, which should serve to remind us again of how stratification variables affect interaction styles and are expressed through them. Chapter 5 applies the conflict theory of stratification to the interior of the family, showing via comparative evidence that variations in family organization and accompanying ideology can be explained in terms of a struggle for dominance among sexes and age groups which have differing personal and external resources. In the process, what traditionally has been referred to as "socialization" is more fruitfully redefined as a two-sided effort at control under varying circumstances of advantage and disadvantage.

Chapter 6 presents the other main sector of conflict theory: a fairly refined and formalized theory of organizations. The basic elements are once again familiar—interests and outlooks shaped by conflictful encounters; varying resources and tactics and their effects; and the structures that can be seen to result from these at any point in time. The attempt is to show that all kinds of organizations are the result of particular states of these same general variables. Chapter 7 goes on to apply this organizational model to politics; Chapter 9 applies it to the external and internal organization of intellectual communities, drawing especially on the sociology of science.

Chapter 8 attempts to show that some rather disjointed and theoretically underdeveloped areas—social mobility, the distribution of wealth, and deviance—can be formulated in terms of a single set of explanatory issues, and that the conflict theory of stratification and organizations provides a base for what generalizations may be offered.

Finally, Chapter 10 summarizes the historical development of sociology toward scientific status, assesses the existing level of documentation for the generalizations proposed here, and speculates about the future of the social sciences in relation to a successful scientific sociology.

Conflict theory is presented here semiformally. I have tried, as far as possible, to summarize the main causal statements in an intermittent list of numbered propositions. In some chapters, they are incorporated into the main body of the text, with the surrounding materials serving as commentary and evidence. This is done in Chapters 2 and 6, which present conflict theory in its most advanced forms, applied to stratification and organizations, respectively; and in Chapter 7, much of which is a straightforward extension of organizational conflict theory to politics. In the other chapters, the propositional summaries are added in appendices, reflecting the more tentative and hypothetical nature of conflict theory in those areas.

I have chosen this device for several reasons. It serves to emphasize that the goal of a scientific sociology must be some such network of definite and testable statements; although the propositions offered here may no doubt prove to be inadequate in various ways, it is preferable to have a clear formulation to be extended, refined, or attacked, than to be left with a purely conceptual argument or a vague mixture of empirical, causal, and programmatic issues. I do not wish the main theoretical claims to be missed or distorted, as they have been in the reception of many of their classical expositions, for lack of sufficient explicitness. Sociology has gone on too long in an unfocused state of disagreement about theories that are ill-remembered or -understood (where they are not intrinsically vague); I prefer to take the risk of being *proven* wrong, for that is the only path out of our present morass.

It will be apparent that many of the propositions are much better grounded than others, and that the effort to show their logical interconnections is often no more than a hint. This is the result of a deliberate choice, and I am aware of its liabilities. It seemed desirable to extend the propositional summary even to areas where conflict theory is very new, or the empirical materials are as yet sparse or hard to bear on the theory, to serve as a model for what must be done. As I have argued, theories advance from less to more precise and adequate formulations; the important step is to provide a genuinely causal model to work upon. By the same token, it seemed important at least to set in motion the effort to show which propositions are more fundamental and universal, and which are derivations or more limited additions. The only workable criterion for the truth of a scientific theory is the degree of coherence among its parts, both in the relations of propositions to evidence and in the internal relations of the theory itself. The more tightly the various propositions can be linked together, the more confidence we can have in any one of them; the empirical grounding of particular propositions provides corroborating evidence for other propositions that can be shown to be related species of the same larger generalizations. The strength of a theory is like the strength of a spiderweb: It is not only that many parts touch the external world, but that the inner links pass along this support from part to part and provide a foundation from which one may extend it to new areas with good hopes of success.

Having embarked on the path of formalizing conflict theory, both in terms of causal propositions and in terms of the inner coherence of these propositions, one opens up the issue of why the project has not been carried further. In principle, this is entirely just. The only reason I have left things in this mixed state, with some areas much more formalized than others, and none made as precise as desirable, is a practical one. Working up conflict theory to the point where we may see its application to the full range of sociology has been a very large task, and one requiring many strategic choices of organization and presentation; to push further in the direction of formalization would require yet another long working through of empirical and theoretical materials and other strategic restructurings. It seems desirable to reach a stopping point from which to survey the territory claimed, before setting off on the next stage.

This is an effort at synthesis. Most of its materials will be familiar to specialists. What I have tried to do is to put matters into causal generalizations with as wide a scope of application as possible. Wherever possible, the application is comparative and historical. Modern America makes up a small corner of the social universe, although it may be the corner of greatest interest to us. Buy we cannot really explain this narrow band of phenomena unless we know the conditions for the wider range. Without this perspective, the parochial interests of today will attract little more attention in a few years than those of the past we ignore today. We should never forget that the power of a science comes from taking the universe as its field. We will know we have a successful sociological science when the most remote societies and the most striking changes are no longer occasions for theoretical surprise.

Chapter 2

A Theory of Stratification

Any coherent theory of stratification is impressive. Stratification touches so many features of social life—wealth, politics, careers, families, clubs, communities, lifestyles—that any model tying these together is bound to occupy a prominent place in the conceptual landscape of sociology. This is true even if the theories do not really work. Social life has so far proven too complicated to be reduced to a single order. But even though the great theoretical models of stratification are failures, their hulks remain in full view, too massive to be dismantled, too central to be forgotten.

Two great rival systems, Marxism and functionalism, have been with us in articulate form for over a century now. A third model, Max Weber's, has been used primarily as an antisystem, a vantage point from which to survey the failings of the others. This has left us in something of an impasse in building a more powerful explanatory system. The classical Marxian model, for all the importance of the economic divisions on which it focuses, ultimately assumed a monocausal explanation for a multicausal world. The result has been either the untenable strategy of explaining all other conditions as correlates of economic ones, or the unfruitful one of leaving them unexplained entirely. The traditional Marxian model founders in the face of ethnic, racial, and religious divisions, political parties that do not coincide with economic groupings, organizational factions in the modern corporation and state, the complex linkages of friendship groups in the community, as well as the phenomenon of social mobility. These matters can, of course, be discussed in a Marxist perspective, but tend to be explained away rather than to be explained systematically.[1]

---

[1] Sophisticated versions of modern Marxism have dealt with such problems primarily by giving up the attempt at scientific explanation and retreating to a philosophical level of analysis. The Critical Sociology of the Frankfurt school, for example, has concentrated on retrieving the notion of rationality as an ethical and political imperative, and criticizing the establishment epistemologies of positivism and technocracy. In France, Marxism has been argued as converging or diverging with humanism, existentialism, and structuralism, but always in highly philosophical terms. There have been some efforts to revise Marxism as an

The penchant for intellectual symmetry, which has confined Marxist stratification theory to a set of clear-cut steps on a single ladder, can be found in its traditional rival as well. Functionalist theory from Saint–Simon and Comte through Sorokin, Davis, and Parsons envisions an order of social privileges based on an underlying dimension of social talent. Stratification is seen as a selective system for bringing the most able persons to positions of greatest service. This very abstract model can be interpreted in many ways, depending on what one believes to be the primary "needs" of a "society." The Durkheimian tradition, whose most prominent continuers are W. Lloyd Warner and Talcott Parsons, emphasizes the needs for ceremonial and cultural leadership in making a society an integrated moral community. Warner's (1949, 1959; Warner & Lunt, 1941) empirical studies thus focus on the hierarchy of cultural groups within a community. Coming to the study of American society from anthropological research on Australian aborigines, Warner describes the complexities of modern stratification as a kind of totem pole of tribes sitting one above another, occupying different rungs in the modern division of labor. As in Marx, we get a tidy picture of a set of coherent groups neatly bounded off from each other along a single dimension. The basis of categorization differs—economic categories in one case, cultural and sociable communities in the other; a unidimensional hierarchy cleanly cut by horizontal cleavages remains the same in both.

The inadequacies of such models have been apparent for some decades now. Lifestyle does not always neatly line up with occupational class, nor does ethnicity, political behavior, personal associations, or parental background, although these divergences are probably more striking in twentieth-century America than elsewhere. There have been several reactions. One has been to regard variations as the result of methodological impurities, to be overcome by treating the different variables as indicators of a single underlying stratification position. For a period during the 1940s and 1950s, followers of Marx and of Warner alike concentrated on building a composite index of SES (Socio-Economic Status).

This effort to salvage a unicausal model has tended to be supplanted by a second approach, roughly describable as pluralistic. The fact that different orders of stratification do not line up neatly becomes the center of attention, rather than an awkward failure of reality to fit theory. Max Weber's (1968: 926–939) tripartite model of class, status group, and power

---

explanatory science, notably by Touraine (1971). The most powerful development on Marx, however, is still the work of Max Weber, and efforts to deal with class conflicts in the world of modern organizations would be saved many problems if this line of development were taken seriously.

group ("party"), as independent although interacting forms of stratification, has become the new paradigm.

There are certain ideological overtones involved in Weber's popularity. The multiplicity of stratifying factors becomes a virtue rather than a vice because it can be used as a polemical critique of Marxism. The pluralist stance has preferred to sacrifice explanatory concerns; multiplicity is asserted to show that economic class is not all-powerful, but nothing definite is put in its place. An abstract assertion of multiple causality replaces the effort to state causes. Warner's version of functionalism, of course, goes down in the same barrage. What remains is a pluralist functionalism, made doubly abstract by both its functionalist and its pluralist unwillingness to come to grips with specific causal statements.

This is not the full story, of course. There have been explanatory efforts deriving from Weberian pluralism focusing on the study of incongruence effects. This is the effort to see if persons whose status varies on different dimensions are different from those whose positions are consistent. For example, it has been suggested that political extremism might be accounted for by persons whose ethnic or educational status was incongruent with their occupational class. Few such effects have actually held up under analysis, however, and this approach seems to be fading.

The strategic problem in theory building is to find the right guiding imagery. The complexities of the world cannot be comprehended all at once; what we need is a device for suggesting where order is to be found. A good theory gives a coherent vision within which research can elaborate complexities without having them overwhelm us. The problem is to choose an image that leads us to the key processes rather than one that obscures them. It is mainly in this respect that the existing theories fail us.

Our prominent images of stratification share the propensity to cloud our eyes with reifications. Stratification is seen as a ladder of success, as a hierarchy of geological layers, as a pyramid (or sometimes a set of ethnic pyramids side by side); but this is not what human society *looks like*. What it looks like, as anyone can verify by opening his eyes as he goes about his daily business, is nothing more than people in houses, buildings, automobiles, streets—some of whom give orders, get deference, hold material property, talk about particular subjects, and so on. No one has ever seen anything human that looks like a ladder or a pyramid, except perhaps in a high school variety show. What these images derive from, most likely, are the convenient ways we have for graphing statistics of occupation or wealth.

There is nothing inherently wrong in any use of metaphoric imagery. The question is whether or not it leads us forward. These images, I believe, have been misleading, for they have structured our thought in terms of relatively simple hierarchies when the problem is to understand very complex kinds of interpersonal relationships. They have led us to focus on a structure that is often not even there, as when class or status categories make us think in terms of distinct associational groups that no one has ever seen. And most importantly of all, these images operate as the ground for theory construction, directing our attention to causal agencies that do not exist: to hypostatized abstractions rather than to the behavior of real people.

This kind of Platonic illusion results not only from the concrete metaphors I have mentioned but also from more abstract notions of "system." The latter sort of image is generally used with a great display of sophistication in distinguishing the heuristic constructions of the theory from the multiplicity of social events to which it applies. Nevertheless, the guiding imagery is equally remote from the real behavior of observable human beings. It invokes a kind of ghostly machine of interacting parts, or perhaps a biological organism as diagrammed in the pages of a physiology textbook, or most recently a sort of heavenly computer with a perpetual motion program. I have already argued that such models mistake the degree of connectedness among the behavior of the real individuals who serve as the cogs in such machines. This imagery also shares a more general failing with the ladder–layer–pyramid metaphors used in more explicitly nominalistic models: It turns our attention from the real world of causal agents to a purely conceptual realm of categories.

Weber's tripartite model of class, status group, and party has been our best lead toward a multicausal model; but even this model has tended to confine our vision in misleading images. The terminology calls up not just one kind of ladder (or mineshaft or pyramid) but three, interpenetrating in some vaguely defined manner. Since groups, rather than individuals, are the units of analysis, we are forced into mental gymnastics to state the interrelations among them, to show that class is related to both status group and party, yet on occasion independent of these as well. It is not difficult to demonstrate both sides of this rather indefinite statement, and every undergraduate lecturer has his own favorite examples of overlap and independence. The problem arises in moving beyond this rendering of pluralism—which takes its punch primarily by contrast with a unicausal model like vulgar Marxism—to more precise, predictive statements. What is it about status groups that causes particular kinds of relationships with economic classes and so on? Once we begin to think in generalized causal terms, the group-relating-to-group model becomes sticky, because most of the causal statements relate to individuals rather than to the structure of the groups themselves.

Part of the problem has shown up in the debate over category boundaries: whether classes (or status groups or even parties) are divided off from each other by recognizable gaps or constitute a continuum. Most research on modern America has come to the latter conclusion; but if the boundaries are not always clear, what do we mean empirically when we refer to stratification in terms of the structure of group relations? Status groups in Weber's usage are explicitly associational communities. Given the importance of personal associations in influencing the individual, it is essential to retain some such group concept; but what does it mean to talk about groups that lack clear boundaries? How can we say how many status groups there are, as we surely must if we are to treat them as recognizably different milieux affecting behavior? There is a further problem which even the continuum model fails to make real: the fact that groups can be tighter or looser and that there are many unsociable individuals who do not happen to join clubs and attend dinner parties but who nevertheless bear a determinant place in the world of cultural ranking.

The Weberian approach is on the way to a solution, but it must be interpreted in terms of imagery that can capture far more than three dimensions of causality. Despite the foregoing imprecisions, almost any stratification categories—different measures of economic class (occupation, property, income) or of status (ethnicity, lifestyle, education, sociometric rating)—will yield important differences in the way people think and behave. This initial success comes from the fact that people experience very different sorts of social influences and are greatly affected by them; stratification is an important area because it latches onto the diversity of the world and to the basic processes of social causality. At the outset, almost any approach within this framework will yield results. The problem is to move forward to the set of categories that maximizes explanatory power.

A solution may be taken from the effort of phenomenological sociology to ground all concepts in the observables of everyday life. All social structure is problematic, quite possibly only a myth that people talk about or implicitly invoke when they encounter each other. This is a form of radical empiricism in which nothing is acceptable except as it is observed as people construct reality from minute to minute in everyday life. When the people who act out an organization are home asleep or chatting with their friends in the corridor, the organization at that moment does not exist. I believe that everything we have hitherto referred to as "structure," insofar as it really occurs and is not just one of those myths people fabricate, can be found in the real behavior of everyday life, primarily in repetitive encounters.

What I propose, then, is to treat the observable behavior of individuals in everyday life as the subject matter for a theory of stratification. This material is not the behavioral attributes of classes, institutions, or groups; these are only ways of referring either to sets of individuals who have things in common or to the behavior of individuals in associating with each other. Instead of trying to place individuals as members of certain groups, I should look for a set of influences on how each individual behaves, including what will make him associate with others in particular ways. Thus we can incorporate both the categorizing and the associational side of stratification theory.

What does explain an individual's behavior? Each person who has ever lived probably has a unique set of experiences; a complete list of influences might well prove infinite. Multiple causality should not be so multiple as to be unworkable. I believe that six or eight variables will give us a powerful approximation.

"Society" is just an abstract way of talking about people encountering each other. "Social determinism," then, means that individuals are influenced by other people they meet. If we want a realistic image of this, think of man as a smart, hairless monkey capable of infinitely subtle sorts of communication with his fellows and of constructing an invisible world inside his head by means of conversations with himself. Just what kinds of contact each monkey has, helps shape his subsequent thoughts and behavior.

The problem is to make a manageable classification of these contacts. But we are hardly starting from scratch here; it is only necessary to translate previously useful research variables into this frame of reference. How one relates to others at work is one crucial set of experiences; how one relates to others with whom one resides, plays, worships, and has sexual intercourse are another couple of sets. These two general realms correspond to the structures heretofore referred to as "class" and "status group." Weber's third major category, political party, refers to yet another range of crucial experiences, those in which men line up for potentially violent fights; these are the basic relationships of the political community.

These are rather general categories, and the task in improving explanatory power beyond current formulations is to state the major subtypes of encounters within each of these. Power relations, situations of giving and taking orders, seem to be the most important behavior-shaping experiences in the world of work. But there are a number of different types of power to be examined. Relations with subordinates, customers, clients, colleagues, rivals—setting these in theoretical order would add explanatory precision. Similarly, to state the variables involved in the influence of family, friendship, recreational, residential, religious, and political groupings on the individual would improve the explanatory power of the other realms of causality. Some efforts toward these refinements will be presented later.

There are two major advantages of this approach to stratification. One is that it enables us to make use of the precision which ethnomethodology proposes to provide. Sociology is to be grounded in the details of everyday experience. This makes it possible to see how ideas of reality are constructed and maintained in the moment-to-moment negotiations of our social encounters. This is particularly useful for a conflict theory that seeks to avoid the pitfalls of treating the reified ideology of dominant groups as the fundamental reality of group life. Ideas are constructed according to the power resources different people have as they encounter each other. The details of interpersonal negotiations show us the mechanisms by which social influence over mind and body operate. Power relations and their accompanying forms of reality construction can thus be understood as the key to the realm of "class" variables. In a similar fashion, the kinds of verbal and emotional negotiations among family members, sexual partners, friends, or community members can be seen as the explanatory links between status groups and the cultures they create and maintain. The social construction of reality can be treated in detail as the mechanism by which different sorts of associations influence the individual; and the kinds of subjective reality an individual constructs helps explain what associations he will choose to carry on.

The other advantage is to give us an approach to "social structure" which can incorporate any degree of complexity. Since "organizations" and "institutions" are only ways of talking about certain kinds of associations, we may comprehend a wide variety of occupational settings as relevant to explaining individual behavior. It is also possible to envision many different ways that the associational milieu can be organized, ranging from highly formalized to highly informal, relatively permanent or relatively fleeting, few or many, tightly integrated or loosely connecting isolated individuals. A close look, not only at modern but many traditional societies, reveals that social relations are quite complex; this conceptual strategy can help us avoid pouring them into a narrow conceptual mold. Kinds of associations can be treated as subjects to be explained in our multicausal model; their interrelations are neither fixed a priori nor left vague, but become subjects of testable causal propositions. As we shall see, it is possible to derive explanatory theories of power (as position in a network of bargaining relationships), social mobility (as movement among networks of occupational and sociable acquaintance), and wealth (as a result of both of the preceding) in this fashion, capable of treating matters of any complexity under these headings.

I have proposed an empirically realistic image of society as the encounters among a population of hairless monkeys with pictures in their heads. Those who prefer a more colorful imagery might try the following. Imagine the view of human society from the vantage point of an airplane. What we can observe are buildings, roads, vehicles, and—if our senses were keen enough—people moving back and forth and talking to each other. Quite literally, this is all there is; all of our explanations and all of our subjects to be explained must be grounded in such observations. "Social structure" could be brought into such a picture if we understand that men live by anticipating future encounters and remembering past ones. Structure is recurring sorts of encounters. An imaginary aerial time-lapse photograph, then, would render social structure as a set of light streaks showing the heaviness of social traffic. If we go on to imagine different colored streaks corresponding to the emotional quality of contacts—perhaps gray for purely formal relations, brown for organizational relations infused with more personal commitments, yellow for sociable relations, and red for close personal friendships—we would have an even more significant map. If Weberian and Durkheimian theories about what undergirds strong social structure are correct, we should have a picture of where strong cultural ties (in this phenomenological vision, pockets of strongly-believed reality construction) make structures strongest. Power, wealth, career possibilities —all of these will tend to flow around the places where strong sociable ties coincide with occupational ties.

This image does not include all the relevant contacts, of course. Written communications (also telephones and other long-distance technology) would have to be added on somehow. Thought is much better able to do this sort of thing than spatial pictures. In any case, it is not very likely that we will ever have a technique of this sort. But to the extent that an image is useful for guiding our theory, an ecological one seems to be best for the present task: to formulate a sociology that builds up complex interrelations from the empirical realities of everyday interaction.

THE BASICS OF CONFLICT THEORY

The level of interpersonal interaction is all-inclusive; by the same token, it is highly abstract. To reduce its myriad complexities to causal order requires theory on another level of analysis. The most fruitful tradition of explanatory theory is the conflict tradition, running from Machiavelli and Hobbes to Marx and Weber. If we abstract out its main causal propositions from extraneous political and philosophical doctrines, it looks like the following.

Machiavelli and Hobbes initiated the basic stance of cynical realism about human society. Individuals' behavior is explained in terms of their self-interests in a material world of threat and violence. Social order is seen as being founded on organized coercion. There is an ideological realm of belief (religion, law), and an underlying world of struggles over power; ideas and morals are not prior to interaction but are socially created, and serve the interests of parties to the conflict.

Marx added more specific determinants of the lines of division among conflicting interests, and indicated the material conditions that mobilize particular interests into action and that make it possible for them to articulate their ideas. He also added a theory of economic evolution which turns the wheels of this system toward a desired political outcome; but that is a part of Marx's work that lies largely outside his contributions to conflict sociology, and hence will receive no attention here. Put schematically, Marx's sociology states:

1. Historically, particular forms of property (slavery, feudal landholding, capital) are upheld by the coercive power of the state; hence classes formed by property divisions (slaves and slave-owners, serfs and lords, capitalists and workers) are the opposing agents in the struggle for political power—the underpinning of their means of livelihood.

2. Material contributions determine the extent to which social classes can organize effectively to fight for their interests; such conditions of mobilization are a set of intervening variables between class and political power.

3. Other material conditions—the means of mental production—determine which interests will be able to articulate their ideas and hence to dominate the ideological realm.

In all of these spheres, Marx was primarily interested in the determinants of political power, and only indirectly in what may be called a "theory of stratification." The same principles imply, however:

1. The material circumstances of making a living are the main determinant of one's style of life; since property relations are crucial for distinguishing ways of supporting oneself, class cultures and behaviors divide up along opposing lines of control over, or lack of, property.

2. The material conditions for mobilization as a coherent, intercommunicating group also vary among social classes; by implication, another major difference among class lifestyles stems from the differing organization of their communities and their differing experience with the means of social communication.

3. Classes differ in their control of the means of mental production; this produces yet another difference in class cultures—some are more articulated symbolically than others, and some have the symbolic structures of another class imposed upon them from outside.

These Marxian principles, with certain modifications, provide the basis for a conflict theory of stratification. Weber may be seen as developing this line of analysis: adding complexity to Marx's view of conflict, showing that the conditions involved in mobilization and "mental production" are analytically distinct from property, revising the fundamentals of conflict, and adding another major set of resources. Again making principles more explicit than they are in the original presentation, we may summarize Weber as showing several different forms of property conflict coexisting in the same society, and hence, by implication, the existence of multiple class divisions; elaborating the principles of organizational intercommunication and control in their own right, thereby adding a theory of organization and yet another sphere of interest conflict, this time intraorganizational factions; emphasizing that the violent coercion of the state is analytically prior to the economy, and thus transferring the center of attention to the control of the material means of violence.

Weber also opens up yet another area of resources in these struggles for control, what might be called the "means of emotional production." It is these that underlie the power of religion and make it an important ally of the state; that transform classes into status groups, and do the same to territorial communities under particular circumstances (ethnicity); and that make "legitimacy" a crucial focus for efforts at domination. Here, Weber comes to an insight parallel to those of Durkheim, Freud, and Nietzsche: not only that man is an animal with strong emotional desires and susceptibilities, but that particular forms of social interaction designed to arouse emotions operate to create strongly held beliefs and a sense of solidarity within the community constituted by participation in these rituals. I have put this formulation in a much more Durkheimian fashion than Weber himself, for Durkheim's analysis of rituals can be incorporated at this point to show the mechanisms by which emotional bonds are created. There involves especially the emotional contagion that results from physical copresense, the focusing of attention on a common object, and the coordination of common actions or gestures. To invoke Durkheim also enables me to bring in the work of Goffman (1956, 1967), which carries on his microlevel analysis of social rituals, with an emphasis on the materials and techniques of stage-setting that determine the effectiveness of appeals for emotional solidarity.

Durkheim and Goffman are to be seen as amplifying our knowledge of the mechanisms of emotional production, but within the framework of Weber's confict theory. For Weber retains a crucial emphasis: The creation of emotional solidarity does not supplant conflict, but is one of the main weapons used in conflict. Emotional rituals can be used for domination within a group or organization; they are a vehicle by which alliances are formed in the struggle against other groups; and they can be used to impose a hierarchy of status prestige in which some groups dominate others by providing an ideal to emulate under inferior conditions. Weber's theory of religion incorporates all of these aspects of domination through the manipulation of emotional solidarity, and thereby provides an archetype for the various forms of community stratification. Caste, ethnic group, feudal Estate *(Stand)*, educational–cultural group, or class "respectability" lines are all forms of stratified solidarities, depending on varying distributions of the resources for emotional production. The basic dynamics are captured in the hierarchy implicit in any religion between ritual leaders, ritual followers, and nonmembers of the community.

From this analytical version of Weber, incorporating the relevant principles of Marx, Durkheim, and Goffman, we can move into an explicit theory of stratification. It should be apparent that there are innumerable possible types of stratified societies; our aim is not to classify them, but to state the set of causal principles that go into various empirical combinations. Our emphasis is on the cutting tools of a theory, whatever the complexity of their application in the historical world.

For conflict theory, the basic insight is that human beings are sociable but conflict-prone animals. Why is there conflict? Above all else, there is conflict because violent coercion is always a potential resource, and it is a zero-sum sort. This does not imply anything about the inherence of drives to dominate; what we do know firmly is that being coerced is an intrinsically unpleasant experience, and hence that any use of coercion, even by a small minority, calls forth conflict in the form of antagonism to being dominated. Add to this the fact that coercive power, especially as represented in the state, can be used to bring one economic goods and emotional gratification—and to deny them to others—and we can see that the availability of coercion as a resource ramifies conflicts throughout the entire society. The simultaneous existence of emotional bases for solidarity—which may well be the basis of cooperation, as Durkheim emphasized—only adds group divisions and tactical resources to be used in these conflicts.

The same argument may be transposed into the realm of social phenomenology. Every individual maximizes his subjective status according to the resources available to him and to his rivals. This is a general principle that will make sense out of the variety of evidence. By this I

mean that one's subjective experience of reality is the nexus of social
motivation; that everyone constructs his own world with himself in it; but
this reality construction is done primarily by communication, real or
imaginary, with other people; and hence people hold the keys to each
other's identities. These propositions will come as no surprise to readers
of George Herbert Mead or Erving Goffman. Add to this an emphasis from
conflict theories: that each individual is basically pursuing his own interests and that there are many situations, notably ones where power is
involved, in which those interests are inherently antagonistic. The basic
argument, then, has three strands: that men live in self-constructed subjective worlds; that others pull many of the strings that control one's
subjective experience; and that there are frequent conflicts over control.
Life is basically a struggle for status in which no one can afford to be
oblivious to the power of others around him. If we assume that everyone
uses what resources are available to have others aid him in putting on the
best possible face under the circumstances, we have a guiding principle to
make sense out of the myriad variations of stratification.[2]

The general principles of conflict analysis may be applied to any empirical area. (*1*) Think through abstract formulations to a sample of the
typical real-life interactions involved. Think of people as animals maneuvering for advantage, susceptible to emotional appeals, but steering a
self-interested course toward satisfactions and away from dissatisfac-
tions. (*2*) Look for the material arrangements that affect interaction: the
physical places, the modes of communication, the supply of weapons,
devices for staging one's public impression, tools, and goods. Assess the
relative resources available to each individual: their potential for physical
coercion, their access to other persons with whom to negotiate, their
sexual attractiveness, their store of cultural devices for invoking emotional
solidarity, as well as the physical arrangements just mentioned. (*3*) Apply
the general hypothesis that inequalities in resources result in efforts by the
dominant party to take advantage of the situation; this need not involve
conscious calculation but a basic propensity of feeling one's way toward

---

[2] The proposition that individuals *maximize* their subjective status appears to contradict
March and Simon's (1958) organizational principle that men operate by *satisficing*—setting
minimal levels of payoff in each area of concern, and then troubleshooting where crises arise.
The contradiction is only apparent. Satisficing refers to a strategy for dealing with the *cognitive*
problem produced by inherent limits on the human capacity for processing information. The
principle of maximizing subjective status is a *motivational* principle, telling us what are the
goals of behavior. Any analysis of cognitive strategies is incomplete without some motivational principle such as the latter to tell us what are the purposes of action, and what areas
of concern are most emphasized. In other words, it is one thing to predict what goals someone
will pursue, another to predict what strategies he will use in pursuing them, given the
inability to see very far into the future or deal with very many things at once.

the areas of greatest immediate reward, like flowers turning to the light. Social structures are to be explained in terms of the behavior following from various lineups of resources, social change from shifts in resources resulting from previous conflicts. (4) Ideals and beliefs likewise are to be explained in terms of the interests which have the resources to make their viewpoint prevail. (5) Compare empirical cases; test hypotheses by looking for the conditions under which certain things occur versus the conditions under which other things occur. Think causally; look for generalizations. Be awake to multiple causes—the resources for conflict are complex.

Nowhere can these principles be better exemplified than on the materials of stratification. Especially in modern societies, we must separate out multiple spheres of social interaction and multiple causes in each one. These influences may be reduced to order through the principles of conflict theory. We can make a fair prediction of what sort of status shell each individual constructs around himself if we know how he deals with people in earning a living; how he gets along in the household in which he lives; how he relates to the population of the larger community, especially as determined by its political structures; and the ways in which he associates with friends and recreational companions. The conventional variables of survey research are all reflected in this list: occupation, parental occupation, education, ethnicity, age, and sex are cryptic references to how one's associations are structured at work, in the household, and in community and recreational groups. In each sphere, we look for the actual pattern of personal interaction, the resources available to persons in different positions, and how these affect the line of attack they take for furthering their personal status. The ideals and beliefs of persons in different positions thus emerge as personal ideologies, furthering their dominance or serving for their psychological protection.

I begin with occupational situations, as the most pervasively influential of all stratification variables. They are analyzed into several causal dimensions, elaborating a modified version of Marx, Weber, and Durkeim. Other stratified milieux are treated in terms of other resources for organizing social communities; here we find parallel applications of conflict principles as well as interaction with the occupational realm. The sum of these stratified milieux makes up the concrete social position of any individual.

OCCUPATIONAL INFLUENCES ON CLASS CULTURES

Occupations are the way people keep themselves alive. This is the reason for their fundamental importance. Occupations shape the differences among people, however, not merely by the fact that work is essential for survival, but because people relate to each other in different ways in this inescapable area of their lives. Occupations are the major basis of class cultures; these cultures, in turn, along with material resources for inter-communication, are the mechanisms that organize classes as communities, i.e., as kind of status group. The first process is dealt with here and the second takes up a later part of this chapter. The complexity of a system of class cultures depends on how many dimensions of difference we can locate among occupations. In order of importance, these are dominance relationships, position in a network of communication, and some additional variables, including the physical nature of the work and the amount of wealth it produces.

Dominance Relationships

Undoubtedly, the most crucial difference among work situations is the power relations involved (the ways that men give or take orders). Occupational classes are essentially power classes within the realm of work. In stating this, I am accepting Ralf Dahrendorf's (1959) modification of Marx. Marx took property as the power relation par excellence. The dividing line between possessors and nonpossessors of property marked the crucial breaks in the class structure; changes among different sorts of property—slaves, land, industrial capital—made the difference among historical eras. But, although property classes might be the sharpest social distinctions in certain periods, the twentieth century has shown that other types of power can be equally important. In capitalist societies, the salaried managerial employee has remained socially distinct from the manual worker, although a strictly Marxist interpretation would put both of them in the working class. In socialist countries where conventional property classes do not exist, the same sorts of social distinctions and conflicts of interest appear among various levels of the occupational hierarchy. As Dahrendorf points out, Marx mistook an historically limited form of power for power relations in general; his theory of class divisions and class conflict can be made useful for a wider range of situations if we seek its more abstract form.[3]

---

[3] This is not to say that Dahrendorf's (1959) position is completely satisfactory. Power organized as property, and power organized within a government or corporate structure, are not entirely equivalent. Men whose power depends on one of these forms are likely to be politically committed to maintaining it. The political differences among capitalists and socialists remain, even though the elites of both systems may have similar outlooks, in much the same way as holders of landed and industrial property have fought bitter political battles over whose organizational form should dominate. Dahrendorf's formulation is a product of the period of Cold War liberalism; he argued for decreasing international hostilities by focusing on those things that might be taken as structural convergence among all modern societies.

Max Weber (1968:53) defined power as the ability to secure compliance against someone's will to do otherwise. This is not the only possible use of the word "power," but it is the most useful one if we are looking for ways to explain people's outlooks. There is power like the engineer's over inanimate objects; there is power like the scholar's over ideas and words; there is the power of the planner to affect future events. But, since men encountering men is the whole observable referent of "social causation," a social power that will directly affect someone's behavior is that of a man giving orders to another. It affects the behavior of the man who gives orders, for he must take a certain bearing, think certain thoughts, and speak certain formulas. It affects the man who must listen to orders, even though he may not accept too many of them or carry them out, for he accepts at least one thing—to put up with standing before someone who is giving him orders and with deferring to him at least for the moment. One animal cows another to its heels: That is the archetypal situation of organizational life and the shaper of classes and cultures.

The situations in which authority is acted out are the key experiences of occupational life. Since one cannot avoid having an occupation or being cared for by someone who does, it influences everyone. On this basis, three main classes can be distinguished: those who take orders from few or none, but give orders to many; those who must defer to some people, but can command others; and those who are order-takers only. The readily understood continuum from upper class through middle class to working class corresponds to this dimension. This is especially clear if we note how the middle-class–working-class break is commonly assigned: not so much on the basis of the cleanliness of the work, or of the income derived from it; certainly not, today, on the basis of property distinctions; but on the basis of where one stands when orders are given.

Upper middle class and lower middle class correspond to relative positions within the middle group, based on the ratio of order-giving to order-taking. Lower class can be distinguished from working class as a marginal group who work only occasionally and at the most menial positions. Farmers and farm laborers can be fitted into this categorization at a variety of middle-class and working-class levels. Prosperous farmers are similar to other businessmen; tenant farmers and laborers are not unlike the urban working class, with differences attributable to the different

---

Ideological considerations aside, it is useful to retain both levels of analysis. Differences in power position, in whatever kind of organization, are the most fundamental determinant of mens' outlooks, and hence of where solidarity groups will form. Within the same general level of power, differences in the organizational basis of power—different forms of economic property or government organization—result in different political and ideological commitments. Men of power all resemble each other in general, but the specific source of power makes for some specific differences in political culture and creates definite political factions.

community structure rather than to occupational conditions per se. The power situation is similar, too, if one understands that the people who give orders are not necessarily all in the same organization and that one need not be an actual employee to be a subordinate; the small farmer or businessman meets the banker with much the same face as the foreman meets his supervisor. There are some differences too, of course. First, I want to show that the most powerful effects on a man's behavior are the sheer volume of occupational deference he gives and gets. Then I will show how some different types of situations at about the same class level can add variations on the pattern.

Dahrendorf's (1959) revision of Marx converges here with Weber's emphasis on power relations. It should be noted that this formulation brings us into the universe of Durkheimian sociology as well, at least in its Goffmanian variant. If the successful application of power is a matter of personal bearing (in which sanctions are implied but not called upon), Goffman's analysis of the ritual dramatization of status provides us with detailed evidence on the mechanism. In a sense, the apocryphal Weberian principle of the "means of emotional production" applies not only in the realm of community formation but in the heart of the occupational relationship. Hence, it happens that Weber's historical summary of the religious propensities of various classes epitomizes later evidence on class cultures.

Networks of Occupational Communication

Another dimension of occupational cultures comes from the sheer volume and diversity of personal contacts. The politician must see diverse audiences and the king receive the awe of crowds, whereas the tenant farmer and the servant rarely see outsiders, and the workman regularly deals with few besides his boss and a little-changing circle of friends and family. The greater cosmopolitanism of the higher occupational levels is one key to their outlooks. Cosmopolitanism is generally correlated with power because power is essentially the capacity to keep up relations with a fairly large number of persons in such a way as to draw others to back one up against whoever he happens to be with at the moment. But communications are also a separate variable, as we can see in the case of occupations that have greater contacts than power, such as salesmen, entertainers, intellectuals, and professionals generally. This variable accounts for horizontal variants within classes, and for their complex internal hierarchies (e.g., within professions or in the intellectual world) that stratify whole sectors over and above their actual order-giving power.

This dimension has its classic theoretical antecedants. Marx's (1963:

123–124) principle of class mobilization by differential control of the means of transportation and communication applies not only to politics but to the differentiation of class cultures themselves. Weber's extensions of this principle to the internal structure of organizations reinforces the implication, for organizational evidence not only documents the crucial distinctions in outlook and power derived from control over information and communications (Chapter 6 of this volume) but provides a look from a different angle at the *empirically* same phenomenon of occupational stratification. Durkheim's model of ritual interactions and their effects on the "collective conscience" provides a finer specification of the mechanisms involved. In the *Division of Labor in Society,* Durkheim shows that the content of social beliefs, and especially the pressure for group conformity and respect for symbols, varies with the intensity and diversity of social contacts. In *The Elementary Forms of the Religious Life,* Durkheim examines the mechanisms at the high-intensity end of the continuum and shows that the highly reified conception of collective symbols, and the intense loyalties to the immediate group, are produced by ceremonial interactions within a group of unchanging characters, in a situation of close physical proximity and highly concentrated attention. By abstraction, we can see that not only entire historical eras but particular occupational milieu vary along these dimensions and hence produce different sorts of cultural objects and personal loyalties. Weber's distinction between bureaucratic and patrimonial cultures captures this dimension, with its different centers of loyalty and standards of ethics; the bureaucratic and entrepreneurial sectors of the modern occupational world represent these variations across the dimension of class power.[4]

---

[4] Patrimonial organization, most characteristic of traditional societies, centers around families, patrons and their clients, and other personalistic networks. The emphasis is on traditional rituals that demonstrate the emotional bonds among men; the world is divided into those whom one can trust because of strongly legitimated personal connections, and the rest of the world from whom nothing is to be expected that cannot be exacted by cold-blooded bargaining or force. In modern bureaucratic organization, by contrast, personal ties are weaker, less ritualized, and emotionally demonstrative; in their place is the allegiance to a set of abstract rules and positions. The different class cultures in patrimonial and bureaucratic organizations are accordingly affected. Patrimonial elites are more ceremonious and personalistic. Bureaucratic elites emphasize a colder set of ideals.

The contrast is not merely an historical one. There are many elements of bureaucracy in premodern societies, notably in China; in Europe, bureaucracy gradually set in within the heart of the aristocracy, especially in France and Germany, around the seventeenth century. Patrimonial forms of organization exist in modern society as well, alongside and within bureaucracies. They are prominent in the entrepreneurial sector of modern business, especially in volatile areas like entertainment, construction, real estate, speculative finance, and organized crime, as well as in the politics of a complex, federated governmental system like

Wealth and Physical Demands

Besides the main variables of power and communications networks, occupations vary in additional ways that add to the explanation of class cultures and hence to their potential variety. One is the wealth produced and another is the kind of physical demands made. To insist on the importance of money as the main difference among social classes, of course, is vulgar Marxism. It is the organizational forms of power that produce the income that are crucial in determining basic distinctions in outlook. But money is important as one intervening link between occupational position and many aspects of lifestyle that set classes apart; as such it can have some independent effects. Income is not always commensurate with power. Some men make less or more than others of their power level. Power of position and power of money can be separate ways of controlling others, and hence have alternative or additive effects on one's outlook. Moreover, income can be saved, collected, or inherited so that an aspect of power can be passed on—and so preserve its accompanying culture— when its organizational basis is no longer present.

On the physical side, some work calls for more exertion than others; some is more dirty or more dangerous. These aspects tend to be correlated with power, since it can be used to force others to do the harder and more unpleasant labor. But physical demands do influence lifestyle, making the lower classes more immured to hardship and dirt, and allowing the upper to be more effete and fastidious. Physical demands also vary independently of class power, and help account for variations between more military and more pacific eras and occupations, and between rural and urban milieux.

With these variables in mind, let us proceed to a brief overview of the evidence on class cultures, concentrating on the main vertical distinctions among levels of power.

---

the United States. Weber (1958a: 57–58) caught the contrast between the two ways of doing business when he pointed out two kinds of business ethos throughout history. One has existed in all major societies: it emphasizes trickery, cleverness, and speculation aimed at making the greatest possible immediate profit. A second form is rationalistic, ascetic capitalism, which approaches business in a methodical and routinized fashion. Work and production are more ends in themselves, a way of life, rather than a means to get rich quick. In Weber's famous theory, capitalism developed in Europe precisely because business was dominated not merely by the entrepreneurial ethic, as in ancient and oriental societies, but by a sizable group holding the ascetic business ethic. The entrepreneurial type does not disappear once the modern economy is established, of course. He survives to skim the cream off of a system he could not have created.

A SUMMARY OF CLASS CULTURES

The "upper class" is a way of talking about people who command large numbers of men and defer very little to anyone. There are many ways people have gotten into this enviable position. Leaders of conquering armies, and their heirs, have been the commanders throughout most of history; some of them still survive on their landed estates in Latin America and Southeast Asia. Being at or near the head of a government is another way, and being one of the most powerful businessmen in an industrial society is yet another. In practice, business power has tended to mean involving oneself especially in financial matters; mere operating heads of firms have led a precarious existence independent of the financial community.

Max Weber (1968: 472–477) summarized the religious propensities of well-established elites of these sorts; it epitomizes their outlook in general (cf Baltzell, 1958; McArthur, 1955). The military nobility throughout history has upheld some code of honor, couched in whatever religious terms were convenient. Such classes have at all times used religions for political and military purposes, assimilating them to the ceremonious formalities of their relations with underlings generally. The more bureaucratic elites—the Chinese mandarins, the Catholic officials of the Middle Ages, the office-aristocracy of post-Renaissance Germany, France, and England—have been scrupulously orthodox, usually in some worldly religious philosophy, while maintaining that rites and traditions are good for domesticating the masses. Business elites fit into this pattern too, with certain variations depending essentially on just how elite business is in the context of the particular society. The merchant princes of Medieval Europe and of the Middle East have generally affected a worldly, ostentatious, ritualistic religion, not unlike the military and political elites. Something like this in its modern secular version can be found in the ritualistic social life of well-established industrial and financial upper classes which emerged by the twentieth century in England, the United States, and France. The significant exceptions are those businessmen, usually in the first generation of grand entrepreneurship, who have a more pious, inward and moralistic religion. But this seems to be the result of their middle-class origins rather than of their upper-class destination. Those who live most of their lives in the atmosphere of command tend to fall into the typical pattern.

What causes the upper-class outlook?—primarily, the experience of being continually in command. Getting deference is a matter of bearing; it depends on expecting obedience and treating disobedience as unthinkable. Upper-class assuredness, cool composure, unconscious arrogance is the result. The upper class are arbiters, they are the court of last resort, at least as long as they can convince other people of it. The resulting attitude is one of deliberateness and finality. The upper-class man is the one most committed to his organization, for he gets the most rewards from it, and it is in the organization's name that he gives orders and receives deference. Moreover, he *is* the organization more than anyone else; the network of authority that links his subordinates together would not be knit at all if he did not knit it, much as an army is destroyed not so much because of casualties but because the general cannot keep its pieces together. Whether the images of organization are religious, political, or secular, the upper class is the strongest believer in them. Their philosophies are, nevertheless, worldly in practice; they have nothing to gain from fundamental changes and no basic failures or lacks to make them humble or fanatical.[5]

All this follows from giving many orders and taking few. Some side effects come from the upper classes' social contacts and wealth. Upper-class persons are on top of the largest organizational networks of communications. This is more or less true by definition, since power over men is precisely this sort of manipulation of human networks. This means the upper-class man is necessarily sociable. At the same time, we have seen that he is awesome, self-important, deliberate, and dignified. The result is highly formalized codes of etiquette, ways in which potentates can deal with others without letting down the facades of their positions. The emphasis on forms for their own sake and the concern for tradition reinforce each other here. The possession of the largest available wealth—which usually goes along with high power, since power can procure almost anything—makes it possible for upper-class people to put on a very elaborate show of themselves. Expensive tastes in clothes, buildings, food, and other paraphernalia become part of the expected stage setting of power. Another side result is the ideal of generosity to those in need which leavens upper-class arrogance, at least in principle. This fits in with a paternalistic rationalization for their dominance, representing an investment in ceremonial leadership of the community which has comparatively low material costs and high returns in status, and additional insurance for their power.

The more consistently a person experiences situations of unmitigated deference, the more sharply this general outlook appears. Secure business tycoons, like hereditary monarchs, experience more of this in their lifetime

---

[5] Modern evidence show that the higher one's social status, the more one participates in organized religious activities, but the less emotionally absorbing it is for one (Demerath, 1965).

than insecure politicians in a competitive democracy. The more a man associates with others, whether family or friends who have experienced only the same situations, the stronger the culture. Weber (1968: 932–948) suggested that times of rapid change in power relations tend to break down upper-class culture into its basic elements—which are the sheer capacity to awe others, the energy to command, the self-identification with the organization's ideological reflection. With the passage of time, there appear the refined manners, the self-contained air, the elaboration of complex systems of mutual recognition through signals of material tastes.

A second major class consists of the functionaries, the middle classes who defer to some and exact deference from others. Actually the term "middle class" loosely covers a variety of situations. Just as "upper class", defined in terms of getting much deference and giving little, includes isolated country squires and industrial overlords of small towns as well as busy emperors and financiers, the "middle class" category contains men who are highly placed in administrative hierarchies, as well as clerks and supervisors at the very lowest level of command. We are dealing with a continuum here. At one end is the upper middle class, the functionaries who deal only with other functionaries, or the nominally independent larger businessmen and professionals who depend on good relations with bankers, clients, suppliers, and associates. At the other end is the lower middle class of first line supervisors who give orders only to men who give orders to no one at all—that is to say, those who face the sharpest class boundary in power relations.

The lower middle class is the most distinctive type. Weber (1968: 481–484) summarized its religious outlook as ascetic, moralistic, community-oriented, respectable, and hard working.[6] In the premodern societies Weber was reviewing, this class consisted primarily of independent artisans like the small shopkeepers and craftsmen among whom Christianity originated in the cities of the Roman Empire. The same general traits can be found in the modern petite bourgeoisie of minor clerical employees, small businessmen, and independent skilled workers. Weber explained the outlook mainly from the nature of the work situation in which success seems possible through constant self-discipline. No doubt the nearby bad example of working-class hedonism has something to do with the vehemence with which lower-middle-class persons drive themselves to keep up their respectability.

---

[6] See also Vidich and Bensman (1968), Gans (1967), and general summaries of class differences from community studies in Kahl (1957) and from survey studies in Glen and Alston (1968).

The causal conditions can be seen more clearly if we consider interpersonal relatons. The petit bourgeois has a stake in a system of organizational power, however tenuously. Particularly if he is a low-level employee of a larger organization, he has a reason to feel superior to at least some people, provided that he takes responsibility for his role. In compensation for the deference he must give to his superiors, for whom he carries out essentially menial and repetitive tasks, he can exact deference from a class of subordinates who have no power at all. But the last are outside the realm of power and hence have no reason to identify with it; the petit bourgeois thus has the hardest struggle in day-to-day class war. Least sure of his own authority and most pressed to stay on the top side of the sharpest division among power classes, he identifies with the values of the organization and of respectability and authority in the most rigid way. The rule-bound "bureaucratic personality" is the functionary with minor authority exercised for all it is worth, oblivious to the larger purposes of organizational coordination visible only to those with less specific responsibilities. Essentially the same pattern is found among small businessmen and minor professionals struggling to set themselves off from a clientele that may be only slightly poorer than themselves. The lesser range of contacts and more meager incomes of the lower middle class, compared to the upper class, accounts for its lack of cosmopolitanism, subtle manners, and refined tastes.

The higher levels of the middle class range in culture between the lower-middle class's rigid and tasteless respectability and the complacent gentility of the upper class (Seeley *et al,* 1956). In general, all of the middle class is set off from those below by their intermediate position in the communications network of the larger society; the working class makes up a set of little enclaves in the larger community; while the upper class occupies the central links. The lower middle class is just within the circle of these wider linkages; higher levels of the middle class become successively more organizationally conscious, more cosmopolitan, more involved in formalized sociability and community affairs. There also tends to be a continuum of tastes and manners from the crude pleasures and crude asceticism of the lower middle class through the progressively more expensive tastes and subtle manners that the upper middle class borrows from the elite.

Those who hold relatively steady jobs at the bottom of the economy have yet another occupational culture.[7] They are almost exclusively

---

[7] See Kahl (1957), Glen and Alston (1968), Gans (1962), Rainwater *et al.* (1962), and Weber (1968: 468–472, 484–486). There tends to be an antiworking-class bias in the literature, presented as it is by middle-class researchers; the data are usually presented so as to

subordinates. Since they do not give orders to anyone in the name of the organization, they do not identify with the organization. Often they identify against it, if only in diffuse apathy to the ideals put forth by their superiors. Working-class culture is localistic, cynical, and oriented toward the immediate present. Lacking an active position in the channels of organizational communication and recognizing that their superiors use control of information to justify and manipulate their dominance over them, workers see the world from an aggressively personal point of view.

The abstract rhetoric of their more cosmopolitan superiors is distrusted. The only accurate information is about what known individuals are doing. The viewpoint of workers is largely confined to what is physically present to themselves and their immediate circle of acquaintance. Thus, we find that workers tend to confine their social relations to their own family and groups of childhood friends. The middle-class cosmopolitanism expressed in joining political, social, and charitable organizations is largely absent in the working classes, as is the pattern of sociability in which strangers are invited to the home for dinner or parties found commonly in the upper- and upper-middle classes. Working-class values, like those of everyone else, emphasize the virtues of their own life situations: in this case, physical toughness, loyalty to friends, courage and wariness toward strangers and superiors. As Bennett Berger has remarked, values are always self-congratulatory.

The ethos is one that regards life as hard and unpredictable, with little long-range planning possible. One should be prepared to seize the opportunity to enjoy oneself when possible and to endure the inevitable periods of deprivation. In all these respects, industrial working-class culture is similar to the culture of peasants and farm laborers. Weber characterizes both groups, historically, as essentially worldly in their religious attitudes. Moralistic and ascetic religions have never been strong in either group, and the religions that do flourish are those that celebrate the periodic events of life with boisterous festivals. Religion, here, has a magic tone including faith-healing, future-predicting, luck-bringing, and other uses of religious ceremonies for worldly ends and emotional release. Rural pagan religions, with their annual festivals and fertility rites and the celebrations involved

---

characterize working people as parochial, authoritarian, uncultivated, materialistic, and lacking in moral virtues, self-discipline, and foresight. Without romanticizing manual workers, it should be pointed out that the same data can support value judgments of a very different sort: working people are more loyal to their friends, more physically courageous, more capable of enjoying themselves when the opportunity arises (instead of "delaying gratification"), less prone to abstract moralizing, and more realistic (although less informed) about the larger world than much of the middle class with its naïve faith in official definitions of work and politics.

in marriages and funeral wakes, have their equivalent in the industrial class's entertainment of violent (or at least highly active) sports and periodic drinking bouts punctuated with fights. Unlike the moralistic lower middle class, and also unlike the sophisticatedly public-image-conscious upper- and upper middle classes, the working-class male culture is explicitly interested in sex, with a rigid dual standard of male control over wives, sisters, and daughters, and free male access to prostitutes and unattached women. The emotional tone is relatively uninhibited, whether in fighting or celebrating. Work is regarded as an unavoidable evil.

Lower-class culture, finally, is built on the world view of persons with little stable attachments to the major organizations—transient workers at menial jobs, the chronically unemployed, beggars, outcasts, and derelicts (Roach and Gursslin, 1967; Liebow, 1967). Indeed, it has been widely debated whether or not the lower-class ethos can be referred to as a culture at all, since a prime feature of lower-class life is the lack of strong interpersonal ties and stable groups which might sustain and pass on a culture. If we avoid the terminological aspects of this problem, we can characterize the lower-class outlook. It is essentially amoral and individualistic, an attitude of every man for himself. Rules of honesty, suppression of violent impulses, or restrictions on extreme forms of self-indulgence such as alcoholism or narcotics addiction, have little or no force in social aggregates that do not form coherent groups. Weber's (1968: 486) historical survey reflects this best in the situation of slaves, whom he characterizes as religiously disinclined except for short-lived waves of chiliastic beliefs, of fantasies and emotional panics about the imminent destruction of the social order.

Clearly enough, these class cultures (upper class, middle class, working class, lower class) are ideal types, guideposts along a continuum. Even this is too simple, as there are several different dimensions along which people's occupational experiences can vary. We are dealing with individuals, each of whom may have his own situation; "classes" is just a convenient way of talking. The major occupational distinction involve the sheer amount of time spent in getting and giving deference, and the sheer amount of communicativeness involved. The former makes one dignified and arrogant, respectable and complaint, or cynical and defensive, depending on whether one mostly gives orders, takes and gives, or only takes. The latter makes one cosmopolitan and ceremonious, or localistic and unrefined. A third major variable is income: The more one has of it the more one is likely to be concerned about the refinements that can be bought with it. It is also important because it tends to determine who can associate with whom, and thus tends to knit together class cultures through associational groups.

One can conceive of various mixtures of these traits, along with further variations. For example, the more coercion there is involved in exacting deference, the more sharply the pattern of dignity, respectability, or defensiveness appears at the three main class levels, respectively. This explains some of the difference in tone between traditional societies with their omnipresent military force, and most industrial societies, in which overt coercion has generally declined.

Some Formal Principles

There are a number of different determinants of occupational class cultures. The available evidence, especially from community studies, tends to lump them together and to confuse them with other variables from the realm of status group organization. For the sake of clarity, I will state them here as formal propositions, beginning with some general postulates.

Postulates

I. Each individual constructs his own subjective reality.

II. Individual cognition is constructed from social communications.

III. Individuals have power over each other's subjective reality (from I and II).

IV. Each individual attempts to maximize his subjective status to the degree allowed by the resources available to himself and others he contacts.

V. Each individual values highest what he is best at, and attempts to act it out and communicate about it as much as possible.

VI. Each individual seeks social contacts which give him greatest subjective status, and avoids those in which he has lowest status (from III, IV, and V).

VII. Where individuals' resources differ, social contacts involve inequalities in power to define subjective reality.

VIII. Situations in which differential power is exercised, and withdrawal is not immediately possible, implicitly involve conflict (from IV and VI).

Propositions

1.0 Experiences of giving and taking orders are the main determinants of individual outlooks and behaviors.

1.1 The more one gives orders, the more he is proud, self-assured, formal, and identifies with the organizational ideals in whose name he justifies the orders.

1.2 The more one takes orders, the more he is subservient, fatalistic, alienated from organizational ideals, externally conforming, distrustful of others, concerned with extrinsic rewards, and amoral.

1.3 The more one interacts with others in egalitarian exchanges, the more he acts informal, friendly, and tends to accept others' ideals.

1.4 The more one *both* gives and takes orders, the more he combines both formality, self-assurance, and organizational identification with subservience and external conformity; he is little concerned with the long-range or abstract purposes of the organization (in whose name he is given orders), but strongly identifies with his own short-term order-giving rationale; he attempts to transform order-taking situations into orders that he passes on to others.

1.1, 1.2, and *1.4* are summaries of the occupational and organizational literature, giving us ideal types of upper-class, working-class, and middle-class attitudes. *1.3* is suggested by less explicit studies of middle-managers or professionals interacting among themselves; it is probably also borne out by cooperative activities in relatively unstratified tribal societies. Notice that a large number of different outcomes are possible for different individuals. Not only can one have different mixtures of order-giving and order-taking (thus giving us upper middle class, lower middle class, upper working class, etc.), with different amounts of egalitarian exchanges mixed in; but the sheer amount of *time* one spends doing these things can vary. (That is, both the proportion of order-giving and -taking, and the absolute amount of each, can vary.) Individuals like college students may have to take orders from their teachers at exam time, but the alienation is relatively slight because these situations happen so episodically. Where students have to do class drills daily, the amount of alienation is much greater.

1.5 The more coercion is used in backing up orders, the more accentuated are the effects in *1.1, 1.2,* and *1.4.*

This foreshadows a basic finding of organizational studies, to be presented in Chapter 6. Some other variations in enforcement procedures will also be given there, along with some organizational resources and problems that tend to make organizational authorities emphasize different kinds of controls. The egalitarian situation referred to in *1.3* is known in the organizational literature as a type of normative control. Another form of normative control is to offer opportunities to become an order-giver oneself:

1.6 The more one believes in the future possibility of being in a position as order-giver or order-taker, the more he takes on the attitudes of that position.

This is the well-known principle of anticipatory socialization (Merton, 1968: 316–325). It also works in retrospect; attitudes carry over from the past:

> *1.7* The more one remembers being in a position as order-giver or order-taker, the more he retains the attitudes of that period.

Since people tend to believe what is most pleasant for them (Postulate IV), downwardly mobile individuals retain old attitudes longer than upwardly mobile ones (Wilensky and Edwards, 1959). Eventually, we should be able to set a time period on these effects (all other influences, such as friendship ties, being equal). I would suggest two years is enough to assimilate anyone to any occupational change that involves taking on a new power position, even moving over the line of class deference from order-taker to order-giver; we might call this the "sell-out span". For shifting friends, the period of attitude change might be much shorter.

> *1.8* The more physical exertion and danger involved in the work (whether manual labor or fighting), the more one values toughness, courage, and action.

This proposition is based on the emphasis on physical toughness in working-class culture; but it is borne out also in upper-class culture in the era of the military aristocracy, and even in middle-class culture in organizations like the army. It also explains rural–urban and patrimonial–bureaucratic differences in the propensity to use violence (Gastil, 1971; Whitt *et al.,* 1972). The converse might be called the "effeteness principle." Both sides of it follow from Postulate V. Incidentally, situations of physical activity lead to a high value on toughness and hence a considerable use of coercion in backing up orders; the chain ends (via *1.5*) by accentuating the cultural differences given in *1.1, 1.2,* and *1.4,* as in the sharp class distinctions and powerful latent antagonisms of patrimonial societies.

> *2.0* The amount and structure of social communications make up a second set of determinants of individual outlooks and behaviors.

> *2.1 Mutual Surveillance.* The more one is in the physical presence of other people, the more he accepts the culture of the group and the more he expects precise conformity in others. Conversely, the less he is around others, the more his attitudes are explicitly individualistic and self-centered.

> *2.2 Cosmopolitanism.* The greater the diversity of communications one is involved in, the more he develops abstract, relativistic ideas and the habit of thinking in terms of long-range consequences. Conversely, the less the variety of communications, the more one thinks in terms of

particular persons and things, short-term contingencies, and an alien and uncontrollable world surrounding familiar local circles.

These two principles divide the Durkheimian notion (Durkheim, 1947) of social density into several variables.[8] The first has a great many ramifications, from differences in childrearing where surveillance is much or little available, to differences in the ethos of communities both within our own society and throughout history. It can crosscut all the authority variables (*1.1–1.8*), although part of the difference in occupational cultures are due to the lower experience of surveillance and the higher diversity of contacts at the higher occupational levels.[9]

It also helps explain the cultures of two occupational "classes" not given earlier. One of these is the lower-class culture of individuals who work only episodically, and that at the most menial levels of order-taking; their culture, accordingly, is an extreme form of amoral individualism, deriving from a combination of low authority, low surveillance, and low cosmopolitanism. The other occupational "class" consists of many artists, intellectuals, and other lone wolfs who relate to others as equals or even (with enough money or fame) as order-givers. Their culture is highly self-centered and nonconformist but couched in terms of (at least imaginary) control over the social world, a kind of creative megalomania. Intellectuals, whether personally isolated or not, are relatively high in terms of communications of a very complex sort, which thereby become a major standard of value for them (by Postulate V).[10] Thus intellectuals tend to be high on *2.2* but low on *2.1* (since intellectual pursuits usually require working alone), hence the culture of either arrogance or informality, individualism, and relativism.

High diversity of communications (*2.2*) can result from encountering or corresponding with a great many different kinds of people, which is the most common meaning of "cosmopolitan"; or from constantly being given new messages from the people one sees repeatedly. High-level executives

---

[8] For the sake of completeness, principles analogous to *1.6* and *1.7* should be added to *2.1, 2.2,* and *3.0,* since memory and anticipation effects may occur here too, although probably with relatively little additional explanatory power.

[9] Evidence that the closeness of surveillance and the diversity of contacts are both correlated with authority levels, and also contribute *independently* to class attitudes and behaviors, is found in Kohn and Schooler (1969), Pearlin and Kohn (1966), and Hagedorn and Labovitz (1968).

[10] If he is at all successful, an intellectual cannot really be too isolated. For getting into the center of a network of intensively felt, if slowly moving, communications is what intellectual success is all about. See Chapter 9 of this volume.

and members of the intellectual professions experience a great deal of the latter, even if they operate within a relatively homogeneous network of acquaintance. The effects are about the same in either case: the aspect of upper-middle-class and upper-class culture that makes men sophisticated, thinking in terms of abstractions and long-term consequences. This variable (*2.2*), together with *1.1*, helps explain why there have been a number of different upper-class cultures throughout history. The relatively isolated rural landowner or even the patrimonial king, for all the deference he got, operated in nothing like the communications network of the modern business executive or politician.

The difference between patrimonial and bureaucratic social organization can also be explained in terms of these variables. Patrimonial organizations have high surveillance but low diversity of contacts; the resulting outlook, which Weber (1968: 212–254) characterized as *traditional legitimacy,* emphasizes personal relationships, a purely local concern, and a surrounding world that is regarded as permanent and beyond human control.[11] Bureaucratic organizations tend to reduce direct surveillance and to increase diversity of communications, especially by adding written messages to personal contacts. Weber termed the legitimating world view *rational–legal:* the notion that arrangements are not only abstract and impersonal but deliberately enacted by human beings. In modern America, there are few truly patrimonial organizations left (especially with the separation of the work place from the family), but the sphere of small or entrepreneurial business provides a similar contrast to bureaucratic organizations. In the former, men are more likely to be authoritarian, inflexible, and unreceptive to change—differences that can be pinned down to the greater prevalence of diverse communications in bureaucratic organizations (Kohn, 1971).[12]

---

[11] This reified world view is what has been measured under the rubric of "authoritarian personality." See Gabennesch (1972), where it is interpreted as the result of a situation of low cosmopolitanism rather than as a personality trait. The effects of such conditions on *political* tolerance, however, have been misinterpreted with the usual antiworking-class bias of American sociologists. Many of the indicators of authoritarian attitudes refer to preferences for strict childrearing practices and sexual moralities, which tap stratification within the home and community rather than within the sphere of economic class conflicts. In the latter sphere, working-class people are more likely to have a conspiratorial view of the world than members of the higher classes; but, whereas the latter *prefer* inequality of political influence, it is the former who are more attached to the ideal of democratic pluralism (Form and Rytina, 1969). The effects of self-interest here are obvious.

[12] The differences are consistent across a great many measures, although the effects are relatively small. This seems to be due to the fact that for most workers, bureaucratic organizations *also* tend to have a rather high degree of surveillance. Kohn is characterizing lower-middle- and middle-class occupations especially; in those bureaucratic occupations with

Pursuing these variables further, we should be able to account for differences among a greater diversity of occupations than we usually capture in our conventional categories of class. Vocational interest scales, for example, show that the outlooks of occupations are lumped together as follows (Tyler, 1965: 198–199):

a. top-level business executive;

b. production manager;

c. middle-level bureaucrats (accountant, office man, purchasing agent, banker, mortician, pharmacist);

d. salesmen (salesmanager, real estate, life insurance);

e. political–social administrators (personnel manager, public administrator, vocational counselor, social worker, social science teacher, city school superintendent, minister, physical therapist);

f. verbal professions (advertising man, lawyer, author–journalist);

g. physical science professions (physicist, chemist, mathematician, engineer);

h. human science professions (artist, psychologist, architect, physician, psychiatrist, osteopath, dentist, veterinarian);

i. musicians;

j. lower technical or physical labor, and/or authoritarian (farmer, carpenter, printer, mathematics–science teacher, policeman, forest service, army officer, aviator).

The last, for example, seems to indicate the type of drill-master culture one would expect from being in the middle levels of line authority (*1.4*), doing physical or dangerous work (*1.8*), and operating in a situation of low surveillance (*2.1*) and low cosmopolitanism (*2.2*). Other similarities and differences among occupational groups should also be interpretable in terms of these variables (plus perhaps a few others).

3.0 Authority and social density are experienced in a number of different spheres of life: work, politics, home, sociable recreation, and moving about the geographic community. The individual's outlook is produced by the linear sum of all of these experiences.

We have been dealing with the effects of occupational settings. But people encounter each other in more places than at work, and a complete explanation must take all of these into account. There is one's home life; the groups of friends or coparticipants in whatever one does for entertainment; the people one encounters just because they are in the same piece

---

greater freedom (professional and higher managerial), we might expect more striking differences.

of territory where one lives, shops, plays, or travels to work. The latter make up the sphere that is sometimes called the "community," although obviously there can be a great many different communities in the same territory. It is easy to make up elaborate typologies of the different kinds of associations people can have with one another, but it is hard to make them inclusive without a lot of messy overlapping. I have implicitly included political organizations in the foregoing analysis of authority and communicative relationships, since political positions are occupations for some people. In a complete analysis, though, we would have to list political relationships separately, since they *can* differ from occupational experiences such as being simultaneously a store clerk and a county supervisor as opposed to being a store clerk in a society ruled by hereditary aristocrats.

If we want to explain people's behavior and beliefs, we must add up *all* the influences acting upon them: Weber's class, status, and party situations, and a lot more subcategories within these. Just listing all these possibilities would be a big task, and more confusing than otherwise at this point. The rest of the book is devoted to unraveling a number of these areas in some detail. Chapter 3 deals with the organization of friendship groups, along with looking at the whole process of stratification in a more fine-grained focus. Chapter 4 deals with the kind of deference relationships that exist in the larger community, and the conditions that have made them vary throughout history. Chapter 5 takes a close look at power relations within the family household, and Chapter 9 delves into the organization of the intellectual community.

In each of these areas, the foregoing set of hypotheses should apply. People within the household are divided into groups with different kinds of outlooks because of the ways in which they give and take orders and carry out communications among themselves and with outsiders. These are age and sex classes, which have a history of class warfare all their own. Similar things can be said about scientists or sociable acquaintances, although each area has its distinctive tone due to the kinds of resources that exist there. That is to say, all the principles (*1.1*, etc.) apply everywhere, but we do not always find empirical instances of certain combinations; and on both the structural and the microscopic level, a great many other principles need to be introduced to give a full explanation of the diversity of behavior. Here I will only sketch out the relationships among various spheres.

Occupational Effects on Status Group Membership

Occupational classes differ in their amounts and kinds of social participation in other realms (Laumann, 1966; Beshers, 1962). We know that the higher classes tend to have more friends, belong to more clubs and other formal organizations, and attend church more often; in premodern societies, they were usually more likely to be able to marry, and in modern societies there are still class variations in the age at which different classes marry. The relationships also tend to be endogamous in all spheres: people tend to marry into the same social class, join churches and clubs with members like themselves, and to have friends whose occupations are similar to their own.

The greater resources of the higher classes helps account for their wider social participation. In premodern agrarian societies, these differences were extreme, as only the upper classes had much opportunity to move about, while everyone else was usually confined to the household as servants or tied to the land as peasants. This is not to say that the lower classes are lonely individuals; on the contrary, they are usually in the presence of other people at all times. But it is a seldom-changing cast of acquaintances; sociable ties are very important to working-class people, although not necessarily very happy, precisely because one has less choice over who one is with. The higher classes, on the other hand, are more cosmopolitan in all societies, and it is those individuals who encounter the most people in their work who are the most socially active (Hagedorn and Labowitz, 1968).[13]

Class endogamy is produced by a network of links between occupation and association. (*1*) Occupations shape people's outlooks. Thus people are most comfortable in associating with others who share the same work situations: It gives them something to talk about and enough similarity of emotional commitments to join in the ritual participation that makes up so much of sociable ties. (*2*) People cannot be friends unless they can meet each other, and jobs bring people together primarily within the same occupational levels. It is probably for this reason that friends tend to be close not only in terms of occupational prestige (which is related to the number of people to whom one can give orders) but also in terms of sectors of the occupational world, including the split between bureaucratic and entrepreneurial organizations (Laumann and Guttman, 1966).[14] (*3*) Propin-

---

[13] These friendships within the upper or middle classes are relatively superficial (at least for men), as friends do not see each other often or devote much time to each other, or consider personal relationships very important compared to their careers (Kohn and Schooler, 1969; Booth, 1972).

[14] Caplow (1954: 54–56) presents some evidence that occupational prestige primarily reflects the cumulative number of persons one outranks in terms of order-giving. The prestige scale is also known to be highly correlated with income and education (Reiss, 1961: 83); these are correlated with order-giving ranks, although a precise multivariate analysis has not yet been done.

quity off the job is shaped by occupations too, via the intermediating link of money which puts people into class-divided neighborhoods and sends them to different vacation resorts (or none at all, if they are too poor), places of entertainment, and the like. (4) People become acquainted as friends of their friends, join clubs by sponsorship or invitation, and so on.

If most people in these chains have been influenced to select their friends on a class basis, the more remote links will be class-endogamous too. We should always remember, though, that the degree of associational endogamy is a matter of empirical variation; the foregoing contains a number of implicit contingencies that can bring people together with others in particular occupations that differ from their own. An advanced theory of occupational effects on associations will not predict just an overall correlation, but these variations as well.

Indirect Occupational Effects on Status Cultures

In a society with any fluidity of resources, a number of informal cross-class contacts are bound to occur among friends, relatives, or members of clubs, churches, or other voluntary associations. This means that individuals can be influenced by class cultures other than their own. The evidence for this is somewhat tangential, but we do know that whether or not one identifies himself as working class or middle class in modern America is based not only on having a manual or nonmanual occupation but also on the occupations of one's parents, friends, and neighbors (Hamilton, 1966; Hodge and Treiman, 1968a). One's propensity to belong to formal associations—the Masons, Elks, Knights of Columbus, and the like—also is affected by one's parents' social class (Hodge and Treiman, 1968b); and better-paid workers have more middle-class attitudes, probably because their money enables them to live in neighborhoods surrounded by middle-class people (Kohn, 1971).

When people associate across class lines, which class affects which? Probably this depends on the sheer numbers of contacts on each side, but it is also affected by social climbing. Everyone tries to imagine themselves in the highest prestige position, and there is evidence that people not only name higher class persons more often than lower as the friends they associate with, but actually do seek them out and pay attention to their opinions (see Chapter 3 in this volume). We should remember, though, that these effects occur only in freely chosen social relationships; work contacts between those who give orders and those who must listen to them usually results in little subjective influence.

Status Group Effects on Occupational Careers

The link between personal associations and occupational class can also run in the opposite direction. The effect of one's family ties on occupational attainment has been scrutinized in studies of social mobility. Such influences operate directly in most traditional societies; in modern America, the link is mainly indirect, through the provision of opportunities and motivation for education. But status group membership is not exhausted by one's family origins; it includes one's friends and associational members throughout one's life. This has been much less extensively studied, but the indications all point in the same direction: Careers of all sorts move through networks of acquaintanceship, and who one knows informally links together one's positions in the worlds of business, politics, science, and everything else (see Chapters 8 and 9 in this volume). The relationship between occupational and status group does not merely swing round in a big circle from generation to generation; it operates during every day of one's life, as one goes from work to home and back again, or even from the office to the water cooler. As sociology progresses, we should be hearing more and more about the interpenetration of these spheres.

Indigenous Organization of Status Communities

But who one associates with is not completely predictable from one's occupation, or even the indirect effect of the occupations of parents and other acquaintances. The same can be said about attitudes; these are influenced by other things besides occupational cultures. We need to fill in another side of a multicausal situation. People divide themselves into groups by age, sex, recreational interests, ethnicity, and education; each of these has special interests and creates an internal group stratification of its own, which shapes the individual's outlook.

The independence of these spheres is part of Weber's theory of status groups; it is only part of that theory, because, as we have just seen, occupational classes are one of the prime determinants of status group formation. Marx's conditions of class mobilization and Weber's means of emotional production, however, are not exhausted by their correlation with class; although they often operate as intervening variables, they may also operate to mobilize groups along the lines of other resources as well.

Age and Sex.   Wherever resources are available to allow people to go off by themselves, they tend to form special groupings of their own age and sex. Children of different ages, adolescents, unmarried young adults, the married middle-aged, and the elderly all tend to form distinct social circles in modern America and, to a degree, in many tribal societies where a rigidly enclosed household organization does not exist; where we find the latter, age-specific relationships are much more cramped. Some of the reasons will be dealt with in Chapters 3 and 5, but the main principle is clear: Different age groups have interests and resources of their own, which they can maximally enjoy in their own company, just as occupational classes tend to be attracted internally and repulsed externally. The same applies to males and females, especially as shaped by the kinds of sexual stratification found in that society. In modern America and Canada, for example, middle-class women are much more likely to participate in formal clubs and associations than in the more traditional household organizations found elsewhere in the world (Curtis, 1971). In the working class, where women have fewer resources, we find the traditional cliques of female relatives and neighbors who sharply separate themselves from the associational networks of their husbands, and view the world through a lens of traditional absolutist ideas (Nelson, 1966), as we would expect from *2.1* and *2.2* of the earlier discussion.

Recreational Groups. Friends are people who gather to enjoy themselves together; in a broader sense, the same category can be extended to companions at any game or entertainment, whether participant or spectator. Where people have relatively great leisure and disposable wealth—which, in modern affluent society, includes almost everyone to some degree—a considerable variety of entertainments spring up; a whole universe of little game worlds emerges. For many people, these become the most important realities, overshadowing work, politics, and family.

Modern America (and increasingly the rest of the affluent world as well) has become divided into football fans, golfers, bowlers, fishermen, bridge players, concert-goers, and movie fans. Such activities become major topics of conversation, the basis of friendships, the subjective worlds of greatest immediacy. This is not surprising in view of our premise that everyone tries to arrange things so as to maximize his own status. Games and entertainments, after all, are manufactured fantasy worlds for the display of specially cultivated skills. They are rigged precisely to show off whatever an individual is good at. To the extent that he has the resources, each individual gravitates toward that world where he shines brightest. He will participate where his skill stands out to best advantage; he will be a spectator where his knowledgeability or enthusiasm make him most esteemed. Social life becomes a series of leisure groups, each inflating its own particular bubble.

In principle, good bridge players or soccer stars or knowledgeable devotees of opera or rock music can come from any social class. Because of this and because entertainment has become the prime subjective reality, modern American class structure has become invisible to most people. As it happens, types of entertainment are not really so unrelated to class after all. Money, leisure, and occupational outlooks influence what one likes and what one can afford, so that contact sports are most popular in the working class, skiing and tennis in the higher classes, and sedentary and rule-encumbered games like bridge in the bureaucratic middle class (Clarke, 1956). The upper class still carries on the ostentatious ritual of supporting opera seasons precisely because they are expensive relics of historical elites; and the undignified rhythms of rural music are still popular in the lower classes (and among rebellious youth cults of the higher classes as well). Styles of entertainment tend to link together friends within class lines—which is, after all, where most friendships occur. But entertainment styles add additional structure to the social world, especially where resources are widespread. They make for additional distinctions *within* classes, and sometimes cut across class boundaries, especially in very popular entertainments like spectator sports. They also provide little worlds of stratification all their own, so that an individual's outlooks are determined not only by how much authority he gives or takes on his job but also on how much he dominates in these entertainment circles; being a star or a mediocre amateur athlete or culture buff can be an additional determinant of dominant or subordinant attitudes in any individual. Where these spheres take up much of people's time, the stratification of the entertainment worlds become increasingly important in shaping world views.

Ethnic Groups.   One other important variable is ethnicity. Sociologists have had a great deal of trouble in handling this in any theoretical fashion; most analysis has focused on the allegedly evolutionary process whereby ethnic distinctions disappear, not on explaining why ethnic groups became distinct in the first place and the conditions under which they remain so. Descriptive work is much better, and we have a great deal of information on how ethnic groups (at least in the United States) differ not only in their occupational attainments and community organization but in their values, styles of sociability, and emotional expression (Collins, 1975).

Ethnic groups are territorial groups that have been transplanted to another place and brought into contact with outsiders. This happens mainly because of migration or conquest. The main question in a general theory of ethnicity, then, is the conditions under which differences that existed for independent societies at one time and place become carried over to a multigroup situation somewhere else. The carryover is a matter of degree; obviously, such things as political independence tend to be lost immediately, and very often separate economic institutions as well. Other superficial differences—local cuisine, costumes, building materials—also tend to disappear in a new environment, as do language differences where a dominant group requires others to participate in their political or economic system.

The core of ethnic group identifications are the status community organizations just described: family, friends, church, clubs, and associations. Separate ethnic cultures and identities are carried on as long as people associate within their own ethnic community. Under what conditions are they maximally distinct? The most extreme cases we have are of two sorts: (*1*) Separate ethnic economies inhabit the same territory, as in Malaysia and Indonesia, so that there is no real superordination and little incentive to assimilate one way or the other; and (*2*) there are economically integrated societies with very sharp political and economic stratification along ethnic lines, in which cases the dominant group uses its monopoly of resources to strictly reserve the better positions for themselves, and hence, occupational cultures are made to reinforce ethnic distinctions. The control of political organization also serves to enforce deference relations between members of the different groups, further enhancing the distinctiveness of their cultural identities and sociable circles. This is the case of the Indian caste system, and the race lines found in the history of America and the European colonies in Africa and Asia.

Variations in ethnic distinctiveness depend on three main conditions: (*1*) the degree of participation in a common economy—the lower the participation, the sharper the ethnic distinctions; (*2*) the degree of monopolization of political and economic power by a dominant ethnic group; (*3*) and the extent to which ethnic cultures can be easily distinguished. The closer the emotional tones and rituals of social participation are to each other, the harder it is for people to tell each other apart, and the easier it is for them to associate informally.[15] Thus, the descendants of the various European peasant cultures in America have tended to assimilate together (usually within the bounds of a larger Catholic ethnic community), as have the offspring of the commercial small farmers of northern Europe (who make up the Protestant community historically); the greatest cultural antagonism has been between both of these groups (although they got along with each other none too well) and the carriers of the cultural tones of African tribal organization. Such differences in personal style themselves are carried over by the continuation of closest ethnic communities, and by work settings that fit these styles.

---

[15] Skin color is important in all of this only as a visible label; it makes it easier to act on distinctions, if the conditions just given hold.

Ethnic groups, then, are simply the results of a particular combination of very general stratification processes. Weber (1958b; 1968:385–398, 932–938) first sketched out the interrelations among occupational class, political power, and status communities along these lines: The sharpest distinction among stratified cultural communities occurs where initial cultural distinctions derive from different community forms, economic distinctions become superimposed, and the dominant group enjoys stable political resources. This was the basis of the legally defined estates of Medieval Europe; Bendix (1956) has shown how the continuation of such corporate distinctions in nineteenth-century Russia kept occupational classes maximally distinct as status communities, while the political shifts producing an ideology of individualism in England and America tended to reduce the salience of class-based status groups. The Indian caste system was built up by a long-term continuation of these conditions, building on even sharper cultural distinctions between horticultural tribes and civilized conquerors. Toward the other end of the continuum, where resources are more fluid and the population more generally mobilized, cultural communities are more ephemeral and tend to be based on short-term class relations, like the claims to family descent embodied in the Social Register in America.

In terms of stratification principles previously set forth, ethnicity derives originally from the cultures of communities with particular occupational, political, household, and recreational structures (variables *1.1–2.2,* operative in all of these realms). These produce the initial distinctions; migration or conquest shifts the lineup of variables affecting the mobilization of such communities vis-à-vis each other, because each originally geographically distinct group is drawn together by its possession of a common culture, and because its internal coherence is a powerful weapon to use in struggling with other groups over power and economic position. These noncultural goals of the conflict become mediated by cultural organization that need not in principle coincide with current class lines, and the economic and political antagonisms serve to reinforce these lines of associational inclusion and exclusion.

Education as Pseudoethnicity.   Education shows up in all the surveys as an independent contributor to cultural distinctions. This has mostly been taken as a brute fact, without explaining how it fits with the rest of the processes of stratification. We can see a more general pattern, however, if we treat it as a subcase of ethnic stratification—or, to be more precise, a subcase of the same processes that also produce ethnicity.

Education socializes people into a particular kind of culture, working best on those who already have acquired the general orientation in their families. Schools everywhere are established originally to pass on a particular form of religion or elite class culture, and are expanded in the interests of political indoctrination or ethnic hegemony. In these situations, education is nothing more than ethnic or class culture, although it can be taught to those who are not born into it. But long-standing and internally complex school systems bring about some goal displacement, changing the culture into something specifically scholastic; insofar as it goes on to provide the cultural identity for its graduates, it has an independent effect on class and status group cultures. We have been so concerned to determine whether or not schooling can provide social mobility apart from family origins that we fail to notice how the educated class itself is a kind of surrogate ethnic group, setting up job requirements in its own favor and discriminating against those who do not use its vocabulary and do not refer to the same literary classics or technicist ideals (Collins, 1971; Bourdieu and Passeron, 1970).

The same principles apply to the Confucian education of dynastic China, the Christian gentleman of traditional Europe, or the Communist theocracy of the Soviet states. The rhetoric of technocracy prominent in so many places today is not essentially different, except that it reflects much more bureaucratic school and work organizations, in which the legitimating ideology is influenced by middle-level specialists defending the autonomy of their positions; what one learns in school, even today, is not so much real technical skills (which are almost always learned on the job) as it is an esoteric rhetoric to keep outsiders at arms length (Collins, 1975).

A society with a large educational system, then, is different from other stratification systems only in how certain variables are arranged, and not in the basic processes of stratification. The interaction of status group cultures with occupational classes and political power is the main dynamic of stratification in all societies; whether status groups are organized around families, ethnic communities, or education is a set of variations on a common theme.

CONCLUSION

The possible influences on an individual's behavior are fairly diverse. In a society like modern America where political domination is loose and the number of different organizations and group settings very large, the range of contacts within one man's lifetime, or even within a few days, can be extraordinary. We should not be surprised that every individual has something unique about him. In other societies, the range of diversity has been smaller; perhaps in some cases, maybe of the future, the range may be even greater.

But our purpose here is not to marvel about diversity, but to explain behavior in terms of a combination of some relatively simple principles. The reader might try these on someone he knows to see how well that level of subjective reality is explained. An occupational career within the minor bureaucratic ranks of the army; a family background amid the prosperous small farmers of a Northern European village in a country still ruled by a landed aristocracy; the head of a mildly patriarchal household with a stay-at-home wife and submissively ambitious children; a respectable citizen of a small American town and of a complacently humdrum Protestant congregation; friendly contacts largely confined to a circle of kin, themselves from the same European communities and the same lower-middle-class occupations—these experiences shape the strongly held world view of this man of caution, with his respect for authority and his ambitious hopes for his children, inhabiting an ordered and complacent universe and threatened only by youthful cultures whose experiences he has never participated in. This describes the world of an old man. At any point in his life, of course, the set of influences might be somewhat different, and a closer look at these would show what moved him along from moment to moment. The example helps check out the theory in my own mind; an analogous set of explanatory rubrics makes sense of myself. To the extent that sociology becomes a powerful theory, it should illuminate more and more surely the lives we see around us.

What I have proposed is a subjective check on the validity and usefulness of this model. It shows where we can expect some of the payoffs of success in this direction, as well as a useful—perhaps indispensible—aid to theoretical development. Theories that do not make sense of our own everyday life cannot be very good theories. The main evidence used to build and test the theory, of course, is more systematic. I have indicated some of it, and where more may be found. Data gathering is hardly complete. For purposes of explicitly testing many of these theoretical propositions, much of it has hardly begun.

There is yet another way of validating the model. So far, I have said little or nothing about the determinants of power, politics, and policy, or about social mobility, the distribution of wealth, or the causes of social change. All of these are phenomena of stratification, which can be explained by applying the basic model just outlined. Just what political position one will take is a complex matter; class, community structure, ethnicity, all enter into it in an intricate way that will be described in Chapter 7. The mechanism through which power is obtained, irrespective of what it is used for, is more closely related to the conditions of class organization, for it is networks of communication, already introduced here, that are the basic resource of power. Social mobility is no simple subject, involving as it does the structure of formal organizations and their various linkages to family, friendship, and cultural groupings. The distribution of wealth follows essentially from the same considerations that can be used to explain social mobility; these topics will be handled together in Chapter 8.

The basic premises of the conflict approach are that everyone pursues his own best line of advantage according to resources available to him and to his competitors; and that social structures—whether formal organizations or informal acquaintances—are empirically nothing more than men meeting and communicating in certain ways. The outlooks men derive from their past contacts are the subjective side of their intentions about the future. Men are continually recreating social organization. Social change is what happens when the balance of resources slips one way or another so that the relations men negotiate over and over again come out in changed form. The general propositions put forth here are thus a basis for explaining social change. Every chapter in this book carries this out in some more detailed way. The next chapter takes a closer look at the same general principles from a different research perspective, the sociology of interpersonal encounters.

Chapter 3

Microsociology and Stratification

Microsociology deals with the details of human thought, emotion, and face-to-face interaction. This chapter will go over much of the same ground as the previous one, but from a more detailed perspective. The broad intention has already been announced: to ground our propositions about "structure" in the only material that is causally explanatory, the real experiences of individual human beings. Some general principles along these lines were used as the basis of the conflict theory of stratification just presented. This chapter will attempt to improve the precision of their coverage and to show how the two levels of sociological analysis may be united.

Microsociology itself is divided among competing approaches. The first task is to sketch out the major areas of theory and research and to relate them to each other. First is the perspective of animal ethology deriving from Darwin. I shall argue that Durkheim and Freud drew their power from this general insight: Man is an animal, albeit one with special faculties for incorporating the social world into the interior of the mind. In particular, I shall argue that Durkheim and his followers in microsociology—notably Goffman—provide our best basis for a comprehensive theory of man.

A second approach to microsociology comes from traditional philosophical concerns with thought and language. Recent phenomenological sociology has turned these into detailed research on human interaction; I shall try to show what has been accomplished here, as well as its limitations. These derive primarily from its continuing philosophical bias, its focus on man as a thinker rather than as a creature of emotions and activities. Social phenomenologists themselves have not been unaware of the difficulty; in fact, most of the recognized problems of the background traditions on which it draws—the philosophical phenomenology proceeding from Husserl, the linguistic philosophy of Wittgenstein and Austin that followed from the breakdown of early twentieth-century positivism, the pragmatism of Pierce and James which issued in symbolic interaction-ism—centered around precisely this feeling, that the linguistic symbol system within which philosophical discussion took place was not itself to be mistaken for the fundamentals of experience that it attempts to capture. But even in this recognition, these philosophers and their sociological descendants have continued to turn the problem around in philosophical terms, rather than taking the leap into the explanatory stance from a natural world in which human symbols arise. Thus, ethnomethodology tends to cling to its paradoxes, as the ordinary language analysts cling to their expressions of the limits and properties of language itself. The most recent developments, however, move from paradox to positive explana-tion in terms of a world of social action and nonverbal bases of communi-cation, and thus move toward linking up to the advances of animal ethology.

The bulk of this chapter applies what we have learned from animal ethology and from social phenomenology to the detailed foundations of stratification. The general assertion is that, if we can explain who will talk to whom and about what, we will have the centerpiece for a grounded theory of stratification and of social structure.

THE ETHOLOGY OF HUMAN ANIMALS

We have recently begun to rediscover that man is an animal. This takes us back to Darwin. Of course, we all know that man has a common ancestry with the apes; physical anthropology has established some dates for the various stages of divergence, and biochemistry has shown the basic similarity of the components of all living matter. What needs to be rees-tablished, however, is an immediate sense of the relevance of our biologi-cal presence as a source for explanatory principles. Darwin's initial wave of influence includes many of the crucial figures in modern social science, such as Freud, Durkheim, and Pavlov; the trouble is that these men's achievements have tended to lose Darwin's own sociological relevance in considerations on other levels of analysis.

Consider Darwin's own work, notably *The Expression of the Emotions in Man and Animals,* and that of his followers in the field of animal ethology. To be sure, some of the latter (and especially outside popularizers) have put forward foolish overgeneralizations about the biological inherency of man's propensity to violence, the forms of bourgeois marriage, or of the alleged tendency to form all-male groups. But without losing sight of the great cultural variability that particular historical conditions have brought about, we have things of considerable importance to incorporate from Darwin and ethology. First and foremost is the proposition that *society is subhuman*. Many animals live in groups just as we do. The distinctively human forms of cognition and communication are *built on top of* the preexisting capacity for social ties; they are not the basis of it. That is, in effect, the meaning of Durkheim's (1947) proof that rational contract does not produce society, but that it rests upon a prior solidarity. The collective conscience ultimately refers us to those elements of human shared cognitions that are subhuman and biological.[1]

A second consequence is methodological. Man is an animal; "human society" means a group of animals meeting each other, communicating through basic animal gestures for sex, parenthood, threat, submission, mutual support, and play. Our human capacity for symbolism only adds refinements to the basic gestures, however powerful they may be in their consequences for the number of persons who can live together and the complexity of the things they can do. The invisible world of organizations and institutions remains grounded in the behavior of human animals in real places; here is where the causal locus of sociological propositions must be. This is what I mean to convey in the previous chapter with the image of society as an aerial photograph.

More substantively, we are led to focus on the emotions. In the various strains of modern social science, this has been a surprisingly neglected subject, abjured alike by those who emphasize physical behavior and those who emphasize cognitive capacities. Human beings have a range of emotions that goes far beyond that of other animals. As we shall see, the additional ones are complex developments out of more basic ones shared with other species, in conjunction with man's unique cognitive development. The infant learns to laugh and to blush, for example—emotions that other species generally lack—at specific points in his cognitive development (Ambrose, 1960). From a sociological point of view, this meeting ground of mind and body is very important, for it is through emotional behavior that men exercise power, create religions and works of art, as well as enact bonds of solidarity among family and friends. Another way of seeing this is to recognize that *ideals* or *values* are emotionally charged ideas.

---

[1] Durkheim, of course, stressed that the collective conscience is a reality *sui generis*. He was arguing against the individualistic rationalism of the contract theorists, and against biological reductionists like Lombroso, Lapouge, and others who attempted to explain social phenomena by such features as the size of people's heads. My interpretation is consistent with Durkheim's themes that social interaction is a level of analysis in its own right, and that its most important features are the conditions of nonrational solidarity.

Durkheim's great contribution is to provide a method for explaining the variations in behavior and thought through the kinds of situations that influence emotions.

It is important to recognize, however, that some very serious misuses have been made of the Darwinian model. These are principally the result of using it as a macromodel for social structure rather than as a micromodel for the nature of human social bonds. The most foolish error has been to take the success of the Darwinian theory of evolution as a justification for the assertion that society is an organism characterized by homeostasis and the functional interdependence of parts, and changing by evolutionary differentiation. But the theory of evolution through natural selection has nothing to do with such images; it is as if Darwin had treated whole populations of trees, insects, or mammals as if they constituted huge (and invisible) organisms. The model of society-as-organism comes from embryology, not from evolutionary biology, and its use in sociology is simply a misapplied analogy.

The natural selection model also has been more narrowly applied, sometimes with greater disaster than otherwise. In addition to empirically erroneous theories about certain distinctions among human races, the nineteeth- and early twentieth-century Social Darwinists argued for the dominance of the upper classes and of the major imperialistic powers on the grounds of "survival of the fittest." This of course was largely a failure to distinguish value judgment from statements of fact. It rested on considerable vagueness about by just what mechanisms such "survival" (which was taken as equivalent to domination) came about, and just what characteristics could be specified as "fitness."

The same error lingers on in sophisticated modern versions of evolutionism. It is undoubtedly true that there is a natural selection among social institutions and the human populations that carry them; some societies have become much larger and more powerful than others, while some have disappeared entirely. The trouble is that the model is highly underdetermined. We do not know much about which developments will be "naturally selected" without specifying the conditions under which they happen, and there is no guarantee that these can be reduced to a tidy order or a sequence of stages. The evidence of biological evolution is that chance variations produce creatures that are shaped to survive in the ecological niches provided by other populations existing at the time. Later developments are usually more complex than earlier ones, but the simpler ones (like bacteria) can survive right along with the more complex. It is silly to speak of the later ones as being "better adapted." All the diverse creatures that survive at any one time are equally well-adapted. Indeed, highly complex and specialized creatures tend to have transient niches and die out when supporting conditions change.[2]

If we apply these considerations to the social world, we can see that the evolutionary model can be perfectly true in general but fail to explain anything in particular. It does not tell us anything about the course of history, for more complex and less complex forms of society can provide quite diverse niches for each other, and thus all will be able to survive at the same time. The fact that some societies conquer or culturally assimilate others (the modern model is for complex societies to do this to simpler ones, but the reverse has often been the case) is something that requires concrete explanatory principles of just the sort sociological specialists have tried to find. The error of self-styled evolutionary sociologists has been to assume there is more determinancy in their model than there actually is; their specific explanations are usually sneaked in as hidden value judgments about the superiority (and hence greater evolutionary advance) of their own society. If biological evidence teaches us anything, it is the advance of diversity rather than the dominance of a particular type. This throws sociology back on its own resources.

The value of the Darwinian perspective, as I have said, resides rather on the microlevel. Human beings are animals like many others; the basic variables that explain social interaction are provided in animal ethology. I have argued that Durkheim's proof that the basis of human solidarity lies in a prerational trust leads straight to this conclusion. Durkheim also gives us the means for understanding our distinctively human cognition, not as a radical break with the animal world but as a new faculty arising out of our animal heritage and fundamentally molded by it. In *The Elementary Forms of the Religious Life,* Durkheim proposed both a sociology of religion and a sociology of knowledge. These were further integrated around the fundamental questions of sociology, so that Durkheim could conclude that religion (and knowledge) are basically social and conversely, that society is basically religious. The key is Durkheim's rigidly empirical treatment of religion as behavior that evokes particularly strong ideas and emotions. "To do religion" is to create a mood and a belief, to call forth a Holy Spirit. How is this done?

It requires, first of all, *a group of people, concentrating their attention* in such a way as to generate a common mood and a common object of thought.

---

[2] A long-term overview of natural history on the earth would show that the various species have followed each other in response to climatic changes, as the earth has gradually cooled, released and condensed vapors, experienced volcanic eruptions and continental shifts, and so forth. Thus, many of the changes even in biological history, are not very "evolutionary" in the popular sense. Biological history is as much one-damn-thing-after-another as anything else, and human history is no exception.

There must be no individual concerns apart from the group, no side conversations or activities. The physical setting is structured to enhance this mood and to avoid distraction; hence, religious ceremonies are usually carried out in quiet and secluded buildings or in the remote countryside. The group's *actions are stereotyped* in some degree, whether these are ritual gestures or verbal formulas, and whether carried out in unison or by a leader holding the common attention of the congregation. The stereotyping may be seen as an aid to concentrating everyone's attention, evoking a single definition of reality and a mutually reinforced emotion. It is for this reason that strongly held religions require that gesture and formula be carried out precisely, and that errors, whether intentional or not, are believed to destroy the sacredness of the occasion. Finally, there is the *name or other symbol* for the spirit invoked; this makes the transient mood into a visitation of something more permanent and recallable.

Durkheim generalized from the behavior of Australian aborigines coming together periodically to celebrate the totems which are their basic religious (and social) reality. The same basic principles apply to the Christian church service, or even to secular equivalents like a flag salute. In every case, there is the group concentrating its attention, usually with the aid of stereotyped gestures and formulas, and thereby generating a strong emotion which suffuses the participants' thinking about the idea named or symbolized in the ceremony. It is apparent that the strength of the emotion can vary, depending on the conditions discussed in Chapter 2. It is where people are constantly in the presence of the same group (i.e., high surveillance and low cosmopolitanism) that beliefs are strongly reified and the stereotyped formulas are considered to be most sacred.

This is the general formula for producing not only religious ideas but moral ideas of all sorts. Since, in Durkheim's analysis, prerational moral solidarity is what underlies every successfully sustained social interaction, it is easy to see how Durkheim could argue that the periodic recurrence of such ceremonies is necessary to hold society together. The bonds that unite human beings are fundamentally the same that unite other animals. The only difference is that men have symbols that can invoke unseen realities and hence carry the past into the future. Religious and other moral ceremonies thus serve to attach the animal bonds of emotion to symbols which enable men to carry their solidarity in their heads even when they are not together physically.

Even more generally, Durkheim argues that such rituals are the basis of all cognition. That is, cognition is social and is based on the social organization of situations of communication. Durkheim's sociology of knowledge has been attacked on specific facts cited in *The Elementary Forms of the Religious Life* and in *Primitive Classification*, such as the interpretation of space and number as reflecting tribal organization (Worsley, 1956). Yet the fundamental insight is very strong. Our system of time, for example, is clearly social. Of course, there is an underlying physical reality of days, seasons, and years to which Durkheim did not pay much attention. But how men respond to this physical world is based on their social organization. Our calendar, for example, is organized around religious festivals—the Sundays which divide up the week; Christmas, Thanksgiving, and other holiday seasons; as well as their secular political equivalents. The point becomes even sharper if we compare the world of the Christian calendar with that of the Chinese or Arab calendars, in which men take for granted a basic framework of life which incorporates very different conceptions of human history and social organization. Even further removed from us are the calendarless tribal societies. The difference between the clock-conscious West, where the bureaucratic scheduling of daily activities enforces a reality of hours and minutes (or even seconds), and a loosely organized patrimonial or tribal society illustrates the general power of Durkheim's analysis (Hall, 1959: 128–145).

Durkheim's insight into the social nature of cognition is similar to that of Mead and also that of Freud. Clearly, the idea was brewing on the forefront of intellectual currents in the early years of the twentieth century. Durkheim's model of ritual behavior gives the most fruitful means for understanding how it operates; for language, the basis of human civilization, is fundamentally ritualistic in the sense just described. It requires a group of at least two persons. It focuses their attention by formalizing the gesture (in this case, the sound) so that its arbitrarily repetitive qualities are the essence of the experience. It calls forth a common emotional (we may say more generally, a perceptual–experiential) response, which it serves to select out and reinforce by mutual attention, and it provides a symbol (an idea) transcending the here and now of concrete experience and making it recallable in other contents. Human language, in other words, is the evocation of animal social bonds strengthened and focused by stereotyping and attached to the human capacity for symbolization.

Other animals, in fact, have language too. They make vocal noises, as well as gestures, that serve for communication: especially among herd or flock animals, as well as in bird and mammal familistic and mating signals, and the famous language of bees, wasps, and other social insects (Lorenz, 1966: 54–211; Thorpe, 1956; Lindauer, 1961). Moreover, animal language is strikingly similar to human rituals. It is especially ritualistic in the sense of stereotyped, social, and communicative rather than practical and goal-directed; it is emotionally charged (in a physiological sense) and expresses a mood that the individual tries to make others share by contagion, which is what the ritual is designed to do. The imitative alarm signals of birds or mammals provide a good example. All that animal language lacks, in comparison to human rituals, is a symbolic significance or *naming* quality that lifts its meaning to levels more remote from immediate experience.

This presents a puzzle. We usually assume that animal rituals are innate but that human rituals are learned (since they vary a great deal) and are carried on by convention. But the two are structurally very similar. Either human communications are more innate than we like to believe, or animal communicative stereotypes are more of a learned culture than we think.

Both of these possibilities seem to have some truth. The basic *elements* of human communicative rituals are innate: the gestures of bowing to show deference, playing, jeering, crying and responding to cries, and so forth (or perhaps the *truly* basic elements that underlie these). Verbal rituals are carried off because of emotional contagion, much in the way that chanting spreads in a football crowd or a hush falls upon a church service. As we shall see in greater detail while examining conversation, it is the nonverbal signals in which verbal formulas are embedded that holds human rituals together.

On the other hand, much of the animal communication that we observe is probably learned and passed on; animals may very well have a culture. There are regional differences in the language of bees, for example. The basic elements that are innate seem to be elaborated by learning and social negotiation. Apes that are raised without their mothers do not know how to play or mate; the observable training that adult lions and other mammals give their young in hunting techniques probably extends to communicative behaviors as well. Even the complex mating dances of certain birds or fish probably include a range of improvisation as they maneuver to present certain innate gestures.[3]

The language differences between humans and other animals are not so great. We only elaborate more on the arbitrary part of the dance built up around innate gestures. Our language is based on cries, pleas, snarls, demands, coos, and so forth, and talk is an effort to convey one's mood to someone else. It is not language at all that makes us distinctively human, but symbolization, which changes the meaning of our animal noises (cf. Lenneberg, 1964).

Piaget (1951, 1952, 1954; cf. Bruner, 1967) has shown how this capacity emerges in the mental forms underlying behavior and imagery of small children *before* they learn to talk; it is because they have this capacity for

---

[3] Lorenz (1966) argues strongly for the instinctual nature of crucial communicative signals, such as the built-in inhibitive response to deference gestures among killer animals. These types of responses certainly vary among species; what is innate and what is built up around innate elements still leaves a gap for social learning or negotiation (see Harlow, 1964).

evoking objects that are not present that they can learn language. Having done so, they have a much more flexible means of evoking symbols, namely sound, attached to ongoing social contacts which continually broaden the frame of reference. Early language behavior in children is especially ritualistic in the Durkheimian sense. But then, learning language is learning stereotyped verbal gestures, initially as elaboration of some highly repetitive kinds of social interactions with adults. Socialization *is* just this sort of ritual experience, as adults gradually negotiate the child into using their favorite verbal stereotypes.[4]

Subsequently, language becomes internalized and individualized so that the child comes to converse with himself (à la George Herbert Mead) in the same way that religious experience, once created in group situations, can be reinvoked in solitary prayers. As the more arbitrary categories of language become firmly established, more complex thoughts are possible, detached from the specific social context of action and emotion in which the components were originally learned. Sequences of utterances that are not stereotyped in overall patterns can emerge because the components have been so firmly stereotyped that their social meaningfulness is no longer in question.

Analogous changes have occurred historically. In societies with little division of labor, and hence little of what Durkheim (1954) called "organic solidarity," strong emotional trust had to be evoked to carry off most exchanges; otherwise, strangers encountered each other much the same way as mutually hostile animals might. This is one reason that stereotyped verbal formulas, as well as gestural manners, were rigidly adhered to; for example, why it was crucial to plead one's case in a Medieval law suit in precisely the rights words. Words had a stronger emotional significance because of the greater precariousness of the social bonds; in primitive religions (and in Medieval magic), to name a spirit was a serious and perhaps dangerous act. But, of course, to name someone is to know his social network and to figure in it oneself. Cursing—"God damn," or even "God's wounds" ("Zounds")—has its emotional effect where social structure is such as to make given relations precarious and in need of ritual behavior to uphold them.

In fact, it is likely that language and religion have a common origin. Historically, early speaking must have been ritualistic and regarded as rather sacred. Many abstract qualities seem to have come into language as names of gods. For example, the Greek god of war, Ares, is a shadowy

---

[4] Durkheim himself paid no attention to childhood socialization in his theory of cognition, but his general model (as interpreted here) fits readily enough. See Chapter 5 in this volume.

figure who appears only in the grim excitement of battle accompanied by his even less personified companions, Strife, Terror, and Panic (Hamilton, 1942:34). The rituals used for whipping up courage before a battle involved calling the name of the god who represented the mood they wanted. The same seems to have applied to Venus and Cupid, whose names were ways of talking about the moods of someone in love. The discovery of Socrates that there are such things as abstract ideas went along with his attack on the gods as mere names of more abstract qualities.

Durkheim's ideas have been developed along two different lines. One has been functional analysis. In this vein, we have Radcliffe–Brown's (1948) analysis of particular rituals in primitive society in terms of the social relationships they uphold and express, such as the way in which a funeral ceremony may serve to publicly express a loss in a group and to reknit the bonds of the remaining members. Goffman (1959, 1967) has made the most original application of this model by showing the dynamics of such secular rituals in modern everyday life. Radcliffe–Brown's transition rituals find their equivalent in every encounter where greetings and farewells are tailored to express the relationships to be sustained, and ritual proprieties are upheld and repaired according to general Durkheimian principles.

The weakness of this approach is its lack of concern for variations and their determining conditions. Durkheim himself was the first sociologist to emphasize that we can testably explain behavior by comparing the conditions under which it occurs with the conditions under which it does not occur. If one fails to do so, one is left in the position of rationalizing whatever happens to be the case; hence the conservative ideological bias (or at least complacent over estimation of the status quo) found in Durkheim's functionalist followers.

The other line of analysis focuses on different types of ritual order and on the different social conditions they reflect. This has been the path taken by Swanson (1962) in showing that different types of religion are correlated with different forms of social organization. In Chapters 4 and 7, I will attempt to show how particular kinds of moral ideas and particular kinds of religious and secular rituals may be explained by their historical setting.

There are also variations in the kinds of rituals performed within any given society, and in the degree of emphasis placed on them. Some general principles may be stated about their conditions. There are *rites of passage:* the social ceremonies of marriage, funerals, birth, puberty initiation, promotion, induction, and retirement, as well as the little greeting and departure rituals of everyday sociability. As the work of Radcliffe–Brown (1948), Goffman, (1971:62–187), and Young (1965) demonstrate, these are taken most seriously when the change is important for the individuals and especially for the group. Marriages are major celebrations in extended family peasant societies, much less important in the highly individualistic (and much divorced) modern middle class. School graduation becomes an "empty ritual" when it is only part of a long bureaucratic procession from kindergarten to postgraduate fellowship.

There are *emergency or danger rituals* of the sort Malinowski (1948) described as operative when Trobriand Islanders had to leave for fishing on the open seas. Modern equivalents include the ritualism of medical care as well as spontaneous religious invocations in time of danger: "Oh my God!" as a vestigal invocation of social support in a time of shock, or the irreligious soldier falling into prayer in the midst of combat. There are *celebration rituals* of both spontaneous and relatively deliberate sorts: victorious dancing and cheering, exalting, joking; eating and drinking, whose ritual character is expressed in the pressure for everyone to indulge regardless of how hungry or thirsty they may be; gift exchanges like those of Christmas, birthday, and Halloween, which serve to celebrate friendly relationships. At the more deliberate end are church services of thanksgiving and sociable parties in which the announced idea of having "a good time" becomes a guiding tool for directing emotional behavior (Goffman, 1961a: 66–79; cf. Mauss, 1967).

Appealing to different emotions are *deference rituals*. There are *asymmetrical* or *one-way deference rituals:* the ritual prostrating of subjects before a traditional monarch, soldiers standing at attention for the inspecting generals, or modern workers listening to the orders of their superiors (even if they do not carry them out). In polite bourgeois sociability, we find *mutual* or *two-way deference rituals* in which both persons treat the other with the respect due to objects of some sacredness (Goffman, 1959, 1967, 1971). (In Chapter 4, I will place this insight of Goffman's in historical perspective as characteristic of a particular class in a particular period.) And there are *collective deference rituals,* ranging from the awe-struck silence of a church congregation before the invocation of its high god, to the flag salutes, parades, and speeches of mass-participation politics (Durkheim, 1954). As we shall see in subsequent chapters, the degree of deference imposed (and the extent to which it is mixed with other kinds of ritual-based emotions) varies with specific power conditions. Weber's legitimacy types refer to particular sorts of ritual manipulations carried out by political leaders.

Finally, there are *punishment rituals.* Durkheim (1938) introduced this subject by arguing that executions, lynchings, and other acts of spontaneous outrage operate to recharge the sense of group solidarity. The offenses that call for such reactions, however, must occur against a preexisting state of moral solidarity or collective conscience. This can be seen from the fact that individuals respond with outrage when they themselves are not threatened (e.g., when someone else is raped or a racial barrier is transgressed, or when the offense itself is purely symbolic rather than practical —offenses of nudity, drugs, obsenity in polite conversation, or other breaches of ritual occasions). In a minor vein, cursing and other humiliation rituals use properties of everyday language and demeanor to call down the wrath of the symbolically organized group, usually for offenses against this group's symbolic solidarity or dominance. The extent and intensity with which punishment rituals are applied is based on the corresponding extent and kind of social organization.[5]

The general principles are these. Man is a social animal with a distinctively human capacity for symbolization of the unseen, amplified through the internalization of social language. But underlying this are the same kinds of emotional bonds found among animals: sexual, paternal, and related familistic responses; responses related to fighting, including mutual alarm signals, mutual support in attack and defense, as well as intragroup signals of threat and submission; and playing responses consisting especially of toned-down versions of fighting and struggling. Human beings seem to lack the extended innate sequences of complex gestures for mating preliminaries, such as those found among many birds and fish; or the ability to recognize members of their own immediate group by smell, as found among rats, wolves, and dogs; or the elaborate preset work coordinations found among bees and ants; as well as instinctual inhibition of aggression to others performing certain deference postures. Human beings are relatively nonspecific on these matters, much like the great apes. But the more basic patterns are there, and these seem to be nonverbal mechanisms that automatically call for alarm, clinging, and sexual arousal, or express dominance and submission without actual fighting. As Darwin (1965) points out, an animal ready to attack is erect, eyes staring, face expressionless (in contrast to the more terrified posture of a barking or roaring animal trying to scare another); the equally widespread deference posture is eyes lowered, head bowed, body slack. The same basic postures for expressing dominance and submission are found among human beings.[6] Cultural variations seem to enter in the degree to which these

---

[5] The classic analyses are Durkheim (1938, 1947). Durkheim's evolutionary bias is somewhat misleading as a picture of historical differences. There is much restitutive as well as punitive law in tribal societies, especially in regard to matters that are of little importance to members of the tribe (which may include sexual and violent offenses that we might consider more serious). See Chapter 4 in this volume.

[6] Darwin (1965) presents evidence that deference postures include the same basic elements in different cultures and that these are akin to the postures found among higher mammals. Darwin argues for similar conclusions in regard to expressions of grief, love, anger, contempt, surprise, and self-attention (although some of these are specific to humans, for

gestures are performed, not in the basic pattern. Deference in relatively more egalitarian America is done by averting one's eyes and nodding one's head, whereas the Chinese Mandarin demanded full prostration on the ground. The variations up around the same continuum.

Other rituals are built up around basic postures. Laughter seems to be a combination of cries of alarm broken up by recognition signals, with the joyous tone of the latter predominating and the level of affect amplified by the former. Play is built up along the same lines, using the patterns of mock fighting and overcoming incongruities with reassuring gestures of solidarity. Inclusion rituals, such as celebrations and rites of passage, may dramatize the period of pain and separation or move directly to gestures of solidarity; touching, handshaking, shoulder-clasping all appear to derive fairly directly from primal gestures of cuddling and huddling together, although different cultures vary in how broadly or subtly these are expressed. What ethologists glibly refer to as "pair-bonding" in fact seems to include many degrees of distance, negotiated by a complex sequence of gestures.

Once the basic rituals are laid down, they can be deliberately manipulated. Power in human society depends heavily on leaders being able to make raw coercion into ritual deference and especially being able to draw on spontaneous feelings of emotional solidarity by incorporating emergency, transition, and celebration rituals in the exercises of threat and punishment. It is for this reason that conflict not only tends to involve antagonistic groups performing solidarity rituals vis-à-vis each other, as Sorel (1961) observed, but conflicts *within* groups over control of rituals. The priest or other ritual leader gains personal power from his position in manipulating others' emotions and beliefs. Power struggles thus go through inevitable phases of deritualizing and reritualizing.

The specific factors that produce variations in ritual behavior are too numerous to be dealt with here. They will make up much of the remainder of this book. We have seen in the previous chapter that the core of both ethnic and class stratification involves solidarity groups with spontaneous feelings of recognition. A good deal of this solidarity derives from charac-

---

reasons that Darwin attributes to the physical structure of the human face and to other adaptive problems on the physical level). Birdwhistell (1970) has developed a refined notation system for describing nonverbal communications, to the extent of being able to distinguish 2500–10,000 microkinesic signals per second. The difficulty with such an approach is that the descriptive detail has been elaborated far beyond a theoretical model for dealing with it; beginning with larger types of social relationships and the rituals that underlie them seems to be a more likely path forward.

teristic kinds of rituals and the corresponding kinds and amounts of emotions that are expressed. Protestant culture takes on its peculiar emotional tone from its emphasis on dignified mutual deference rituals as well as from emotionless work; peasant cultures from their periodic celebrations and transition rituals. Other refinements along these lines will tell us what is spontaneously recognizable in Chinese or African styles of nonverbal behavior. In subsequent chapters, we will deal with the power relationships surrounding sexual behavior and the accompanying ideologies of kinship and love; the kinds of deference and demeanor among social classes in different historical periods; relationships of dominance and solidarity among parents, children, and other age groups; and the organization of violence and threat by the state.

One more corollary of the Darwinian perspective is worth mentioning here. Evolutionary biology uses the notion of the "ecological niche" to refer to the tendency for each species (and each member thereof) to fill the particular "slot" in the environment for which it is best suited. This means living off of resources that others have not touched, or that the species is better at competing than its rivals; this usually consists in living off of other living creatures. The underlying principle seems to apply to human beings as well. We live in a symbolic as well as a physical world; it is by means of these communications about invisible realities that we can coordinate our behavior in complex and distant ways. Each man takes up his ecological niche within this world as well, moving toward that slot in which he is best able to compete. The rewards involved are not only physical survival, comfort, and dominance but also the subjective equivalents of these. We live in a world of talk and of the internal talk of our own thoughts and emotions. As we shall see in a later section of this chapter, men seek out their best relative advantage in this symbolic world as well, even as they go about constantly recreating it.

HUMAN LANGUAGE AND CONSCIOUSNESS

It is strange that the insights of animal ethology still constitute so much of a frontier for social science. The reason this aspect of Darwin's work has been sidetracked seems to be because of other developments flowing very closely from it. If man became an animal for nineteenth-century thinkers, the burning question became to explain what made him such a special eruption of biological history. The answer had become well-known by the early twentieth century: Man is an animal with a capacity to use symbols that transcend the here and now. He is the time-binding animal who can learn from the past, imagine the future, and travel off into infinite realms of abstraction. All of this was taken to be the result of man's unique capacity for language. Man is the animal who talks.

The concern for language and symbolism has grown continuously since the turn of the twentieth century. In continental European philosophy, it fitted in with the neo–Kantian tradition of German idealism and especially with Dilthey's distinction between the cultural sciences and the natural sciences. The former was concerned with the historical unfolding of man's consciousness as expressed in his symbol systems of language, literature, and art. Cassirer's *The Philosophy of Symbolic Forms* and Suzanne Langer's *Philosophy in a New Key* brought this squarely into line with the anthropological problem of the distinctive nature of the human animal. In the orbit of English philosophy, analyses of the modes of expression in language, mathematics, and logic have virtually monopolized the field since the work of Moore and Russell at the turn of the century. A related emphasis developed in the social sciences. Physical anthropology gave special attention to the physiological evolution making speech possible, and anthropological linguistics laid claim to major importance through Whorf's theory that language differences mold cultures by controlling their members' perception of the world. In psychology, an approach to cognitive development via the study of language was initiated especially in Russia, later to become one of the more productive avenues of research in British and American psychology.

In sociology, the most powerful statement of the importance of the linguistic base of human cognition was made by George Herbert Mead. Mead drew upon the German tradition, especially upon the psychologist–philosopher–anthropologist Wundt, as well as on the pragmatism of Pierce and James and on Darwinian science, which Mead identified with the behaviorism then emerging in the twentieth century. Mead's innovation was to stress the social nature of human consciousness by showing how one's planned actions involved taking the stance of another and making an object out of one's own self. The social nature of human thought was already present in Dilthey (and even earlier in Herder and Hegel), who could point out that language and other aspects of culture are adopted from long generations of previous inventors. The individual can do no more than add a few modifications of his own to this culture through which he himself experiences his existence. Mead grounded all this in a model of how the individual man thinks: Thought is social and arises through the internalization of communication. Social interaction among human adults, then, can attain high complexities of mutual expectations through the process of projecting the self into the roles of others. Together with William I. Thomas' related idea that behavior is determined cognitively through the socially defined "reality" of a given situation, this laid the grounds for symbolic interactionism.

Mead and Thomas represent the first big impact of modern philosophy upon sociology, which took place in the years around World War I. A second influx has occurred in the last two decades. Once again, it proceeds both from Anglo-American and from German philosophy. The former stream, proceeding from Wittgenstein and Austin, has been less noticeable, but it underlies much of what is distinctive in Sacks' brand of ethnomethodology, as well as adding an important background for other versions of modern interpretive sociologies. The more spectacular wave came from the phenomenological philosophy which arose in Germany from around 1900 to 1930, a generation or two after the founders of American sociology had finished their studies abroad. The originating figure is Husserl, with his effort to eliminate the unknowable Kantian "thing-in-itself" by confining philosophy to a close analysis of the contents of consciousness. Husserl's own work was largely programmatic, and his particular version of "phenomenology," a method of "bracketing" types of experience to arrive at their essence, certainly does not monopolize the meaning of the term. In the wider sense of concentrating on the stream of consciousness as it actually presents itself, phenomenology has swept much of Continental philosophy (with the exception of the purely political philosophers) and psychology from Heidegger and the existentialists to the Gestalt psychologists and philosphers of the biological *umwelt.*

For sociology, the key figure is Alfred Schutz. Schutz applied the phenomenological perspective to a problem raised by Weber's concept of *Verstehen.* Weber stood firmly in the older neo–Kantian tradition, and both his sociological methodology and his conception of how human beings act are based on the idea that men pragmatically apply preconceived cognitions to make a meaningful world. Weber thus recognized "social structure" as constructed of subjective meanings imposed upon action, and restricted sociology to the analysis of meaningful behavior.[7] Schutz took up the task of describing exactly what the social world was like in the minute-to-minute perspective of the actor in everyday life. The approach is not unlike that of Heidegger and Sartre, but lacks their concentration on feelings, motivations, the meaningfulness of life in the face of death, and their metaphysical conclusions about *being* in general. Schutz immigrated to America in the late 1930s and taught for many years at the

---

[7] It should be noted, though, that Weber made this point most sharply in his contributions to an abstract methodological debate; in his actual analyses (1946), he stressed a world of material conditions and interests, which ideas only *interpreted.*

Not ideas, but material and ideal interests, directly govern men's conduct. Yet very frequently the "world images" that have been created by ideas have, like switchmen, determined the tracks along which action has been pushed by the dynamics of interest [p. 280].

New School for Social Research. There he influenced a former student of Talcott Parsons, Harold Garfinkel, from whom the current wave of phenomenological sociology developed.

Phenomenological or cognitive sociologists today comprise a very mixed group. No single label such as "ethnomethodological," "existentialist," or even "phenomenological" would be accepted by all. With some simplifications we can lay out the various positions on a continuum of "Left" and "Right," with the explicit recognition that these distinctions have nothing to do with politics in the larger sense.

Talcott Parsons himself can be placed on this continuum. His early works centered on a theory of action derived from Weber: a model of men making choices among alternative means toward various ends, in the context of others doing the same. This put Parsons in the voluntaristic and cognitive camp in opposition to the reductionist version of positivism. His later works became much more positivistic, with their emphasis on the systematic constraints on the behavior of individuals in a society and on the determinism of choice by values socialized in childhood.

In a sense, Talcott Parsons represents the far Right of the phenomenological world. Garfinkel is a critique from the Left. The basic questions are the same: the concern for how social order is possible, for the conditions of trust and distrust, and for the nature of human rationality. Garfinkel (1967) applies Schutz's phenomenological analysis (and many of the results Schutz arrived at by using this method) to these questions and evolved a quasi-experimental method of his own of disrupting normal social expectations in order to see what upholds them.

Garfinkel's conclusions are somewhat different from Parsons'. Social order when looked at closely becomes something of a myth which exists only because it is never questioned. Garfinkel finds that people take for granted that there is a rationale in what others do and say, and that they are able to carry on mutual activities not because they share common meanings but because they *do not question* whether or not this is the case. Parsons' social system, held together by value consensus, thus crumbles under close empirical examination. But the resulting interpretation is not very different. Both positions see human social behavior as carried off successfully because of underlying trust. Garfinkel simply transfers Parsons' value consensus to the properties of cognition per se: We have a social order because no one questions it, and this taboo on criticism is almost functionally given as necessary for practical reasoning in the world.

It is not surprising then that the ethnomethodologists split up into a Left and Right. The Left takes up the empirical criticism of "social order" and "meaning." It makes much of the "indexicality" of our expressions;

the fact that all statements contain ambiguous elements that are handled only by the participation of communicator and communicatee is a common unexplicated context. Following out Garfinkel's method of studying the assumptions made in categorizing social data, the ethnomethodological Left has moved toward the infinite regress of methodological self-questioning and to positions verging on solipsism (Blum, 1970, McHugh, 1970; Zimmerman and Pollner, 1970). In loose alliance with this ethnomethodological Left are the symbolic interactionists, waging their long-standing battle against positivistic determinism in favor of a cognitive approach and an emphasis on the pluralism of meanings (Blumer, 1969). Others have used ethnomethodology as a theoretical extension of the labeling approach to deviance (Douglas, 1971a); these are relatively unconcerned about the problems of methodological solipsism because their focus is on practical critiques of public policy rather than on formulating explanatory theory.

The Right is best defined by the work of Harvey Sacks. Sacks (1972, n.d.) pursues the close analysis of social interaction, usually as represented in tape-recorded conversations. The aim is to generate a logical system of underlying rules and categories which makes sense of the conversational behavior. These rules and categories presumably are known tacitly by the members themselves, who use them to communicate and to predict what will happen from moment to moment in a conversation. Sacks' concern, then, is basically to show in detail, by using his own member's knowledge, what it is that we know how to do when we converse. The answer he implies is that we know social structure, and that such knowledge is in fact what social structure is.

For example, Sacks analyzes telephone conversations to a suicide prevention center in terms of underlying categories of membership and the obligations involved in them. Talk revolves around terms like "family" and "friends" and around the implicit obligations to help involved in them. Thus, the potential suicide is seen as carrying on a rational argument over whether or not there actually is "no one to turn to" and whether or not he therefore should, logically, kill himself. In other works, Sacks is concerned with rules such as turn-taking in conversations and so on.

This analysis takes on resonance from similar developments in anthropology, such as componential analysis and Lévi-Straus' (1969) concern for the structure of systems of myth. Also parallel to it is Chomsky's formalization of linguistic structure. Sacks seems to be going for a "deep structure" of conversation (which for him seems to equal "society") of the same sort as Chomsky (1965) claims to have uncovered in syntax.

That this constitutes the phenomenological Right can be seen by its convergence with functionalism. Sacks does not explain why the rules or categories exist; the possibilities are left implicit that they are an inherent aspect of an underlying structure of the social world or else (as apparently in Parsons) of an ideal world which becomes realized in nature through human communication. As in Parsons, a high level of abstraction is maintained, and there is little concern for variations in language behavior. There is little recognition of conversational differences among social classes, ethnic groups, or historical periods; accordingly, there is no attempt at causal analysis which comparison of such variations would provide. The picture is of an idealized and static world of categories and rules; significantly, what is examined is the observer's conception of the logical structure of talk rather than emotions and behavior. In effect, we have a picture of the everyday ideologies of social structure as expressed by individuals, and not of the behavior that actually makes up their social relationships.[8]

Other ethnomethodological analyses of conversations fit more ambiguously on the Left–Right continuum. Schegloff (1972) and others are concerned with various kinds of unexamined categories within talk, such as what one means by *place* terms such as "here" and "where." This involves an emphasis on indexicality and hence on the situatedness and pluralism of the social universe. Other conversational analysis concerns itself with attempting to formulate universally applicable Sacks-like rules. A great deal is purely descriptive, reproducing great amounts of material from audio records with little effort to reduce it to more manageable generalizations. This line of ethnomethodological research now promises to join with more practically-oriented work in linguistics, such as that of Labov, concerned with topical issues such as demonstrating the rational structure of nonstandard English and the high linguistic skills involved in black culture.

Generally speaking, the phenomenological Left seems to be the most relevant for the development of explanatory sociology. This is because it incorporates the thrust of ethnomethodology as a movement of radical empiricism. Both the bag of abstractions that make up sociological theory and the data produced by our research machinery are measured against a strict standard for admission into the realm of reputable knowledge. The standard is the closely examined properties of the flow of human experience. In this perspective, abstract social systems without human actors are fantasies; highly technical questionnaire surveys which fail to examine

---

[8] Berger and Luckman (1967) are also part of the phenomenological Right, building up a rather idealized picture of the phenomenological world, but from a more philosophical stance.

the situation of question-answering in regard to the other reality-constructing activities of respondents' (and sociologists') lives are naïve. The ethnomethodological program is to take nothing for granted without close empirical examination; the prospect is to reconstruct sociology on a firm foundation.

The limitations of the phenomenological Left come from this radical empiricism and from its own philosophical heritage. Like other empiricist purifiers in sociology, ethnomethodologists have tended to become hung up on the method as an end in itself, and forget about the theory it was to aid in building. Overspecialization takes its toll as things are criticized for the sake of demolishing rather than of reconstructing. The solipsistic wing of ethnomethodology has tended to act as if there were no regularities at all in the way people affect each other, which might be called "social structure." The world dissolves into an ideational haze of free-floating reality constructors among whom sociologists and ethnomethodologists are just so many more fantasists.

This kind of philosophical game is hardly necessary for sociology. It depends on taking an absolutist conception of truth as the implicit platform from which one shoots down all epistimological pretensions. A more pragmatist version of knowledge (such as that found in Kuhn's (1962) version of the scientific enterprise) is adequate to get us under way again. The radical empiricist critique then takes on another light. It is not likely to be useful as an absolutist program, throwing out all existing theory and research and starting from phenomenological scratch; for when social phenomenologists get around to talking about social structure, they often tend to commit the same old fallacies of reifying their own cultural biases from which structural sociologists have spent so long disentangling themselves. Scientific validity lies in the direction of movement, not in a particular stopping point along the way. Experience is too complex to be understood all at once, and cruder efforts always proceed more precise approximations. Radical phenomenological empiricism could not have taken hold at this time unless sociology had already blocked out a frame for it to work on.

There is also the limitation of ethnomethodology's philosophical heritage: the tendency to emphasize thought to the exclusion of behavior and emotions. "Ethnomethodology," we may recall, means the ethnography of human methodologies for dealing with the practical world; it is sometimes referred to as "the study of practical reasoning." This takes man basically as thinker, and as actor only insofar as he thinks about what he does. The armchair tradition of German idealism lingers on here in ethnomethodological field research.

This may now be changing. Cicourel (1973) has demonstrated that

Chomsky's purely linguistic model of the structure of communication is an idealization of real conversation and that the latter depends on nonverbal kinds of interactions which situate the linguistic behavior.[9] By studying sign language, Cicourel highlights the *multimodal* nature of perception and cognition. Our usual auditory language is dominated by a mode in which communication is sequential, as words follow one another, and one must refer forward and backward in time to pull together meaning. But visual forms of communication give a more simultaneous pattern.[10] In attempting to translate from sign language to spoken language, some things are inevitably lost. And this is only a more apparent case of a crossmodal problem that is always with us. For we perceive in *all* modalities when we communicate with someone, but we usually identify ourselves with our verbal language or thoughts and hence lose some of what we "know." This translation problem is at the basis of Garfinkel's "indexicality," since it is the nonverbal elements that underlie and make possible any successful understanding. Ambiguities are resolved not because we verbally understand the nonverbal but because people nonverbally understand each other.

Cicourel thus moves from the verbal world of the classical ethnomethodologists to a two-level model: (*1*) a world of interpretive procedures that operate nonverbally, and in which individuals are able to perceive what others are doing in such a way as to sense a social structure; and (*2*) a world of verbal language constructions or surface rules (including ideas of norms, rules, social positions, and so on) which are applied to the various situations that have already been interpreted nonverbally.[11] The interpretive procedures are considered invariant, and presumably are biologically given; verbal communication and social norms rest upon them. The organization of memory, which upholds long-term social structure, presumably is based on a combination of both levels.

This moves us back toward the broader Darwinian perspective with which this chapter began. Human behavior and emotions thus should

---

[9] I am indebted to Paul Attewell (1974) for certain insights into this material.

[10] Written language of course is simultaneous, in the sense that all the words are there on the page at the same time. But our Western languages in particular are read off more or less like an auditory sequence. Ideographic scripts like Chinese are further down the continuum, with pictures at the other end. Perhaps this explains something of what is distinctive in Chinese character, regarded as a way of organizing one's perceptual and communicative relationships.

[11] This model is explicitly analogous to Chomsky's (1965) surface and deep structures in transformational grammar, although Cicourel is critical of the latter as not truly "deep" enough.

become crucial for linguistic communication, let alone social organization more generally. Cicourel's "cognitive sociology" does not yet take the step, lingering still in the idealistic tradition with its concern for thought rather than action; it seems to aim at putting a situational–interpretive model under the old normative order rather than seeing human animals struggling over physical and symbolic territory. What seems necessary is to press even further Left to challenge the sentimental notion that social behavior is inherently meaningful (i.e., based either on linguistic categories of thought or on an orderly nonverbal social structure beneath them). Meaning becomes, instead, something that each man tries to impose upon the world. Of all the interpretive sociologies, Lyman and Scott's (1970) *Sociology of the Absurd* alone manages to maintain this insight without falling into a purely philosophical position and missing the real conflicts in human behavior.

Ethnomethodological research shows that people impute various sorts of meanings to behavior *after they have done it;* which one they come up with depends generally on the situation of being questioned about it. On these grounds, symbolic interactionists are regarded as far too dogmatic about their beliefs in the symbolic guidance of behavior and in the ideal world of roles that it entails. The next step is to apply the same critical rigor to the cognitive bias of ethnomethodological and cognitive sociology itself. We have already seen that the talking animal has social ties below the level of verbal meaning, and his motives largely reside there as well. Meaning does not become quite an epiphenomenon (since it can react back on behavior), but it must be seen as a great deal more fragmentary and episodic than we have liked to believe.

AN EXPERIENTIAL FOUNDATION OF STRATIFICATION

What I intend to do here is apply the spirit of both the ethological and the phenomenological approaches to a major area of traditional structural sociology, the study of stratification. Everything that happens in the realm of society happens to some person, and it happens to him from minute to minute, making up the subjective flow of his life. I shall attempt to show how the detailed mechanisms of an explanatory theory of stratification works by looking at the commonest activity of everyday life—conversation.

My approach is in the general spirit of the ethnomethodological critique. It understands social structure as something enacted from moment to moment; reality as whatever people negotiate a belief in. This enables us to deal with stratification without imposing our preconceived value judgments in the form of conceptions of structure. We can deal with plural realities, with complex and disorderly interpersonal relations. Such an approach is essential for a sophisticated conflict theory. At the same time, it is obvious that I am not doing ethnomethodology in any strict sense. My analysis of conversation is from a special angle: to put a foundation under stratification in everyday life. This is different from ethnomethodological or sociolinguistic studies of conversation concerned with the close analysis of the most general properties of social interaction and with general questions of the nature and possibility of meaning and mind. These are important questions, and hopefully some day a unified sociology will forge the link between this abstract level of analysis and the more specific variations in behavior which make up such areas as stratification. Right now my task is to sketch the connections to show what even a loose application of the phenomenological approach can do for a theory of stratification. At this point, neither conceptual rigor nor close empirical documentation is as useful as a demonstration that the enterprise is fruitful for explanatory theory. But we are not exactly working in the dark. There is a fair amount of data, although collected for other purposes; it remains to be put in the proper framework.

The purely cognitive approach of ethnomethodology is not enough, of course; it needs to be situated in a real world of human animals, with their physical and emotional interests, and their surrounding material environment. We recapitulate once again the movement from Hegel to Marx, this time (hopefully) in a more balanced and empirically complex way. Darwin and his amplification in Durkheim give us the intersection of the two levels. It is the emotional and physical conditions that generate strongly believed-in symbols, and both of these together give us what regularities there are that we can call "social structure."

In the interactional perspective shared by ethology and phenomenology, social structure means two things. It means the real connections among men dealing with each other over and over again. An organization is a number of people who continually meet one another in the same ways; a status group is a network of acquaintanceship. This is the interactional side of social structure; there is also an ideational (or ideal) side. An organization is not only the supervisor talking with the foreman, the foreman with the workers, and so on; it is also a word, something these men talk about. It can be written down on paper and symbolized in pictures. Both sides of social structure, the interactional and the ideal, are socially constructed, but in different ways. The interaction structure is simply there, because it is what people do; the ideal side can be faked. Given the slipperiness of social symbols and the naïveté and sophistication with which they are used, it is not surprising that the ideal definitions of organizations are usually far afield from the realities of interaction. This does not make the ideal constructions irrelevant for sociology. They are important weapons and resources in the everyday maneuverings of men's lives, along with material resources. They are, in effect, the focal point of Durkheimian rituals. Weber's analysis of legitimacy points to the ways in which they are bases of power.

Social structure thus is made up of networks of acquaintanceship and the bubbles of subjective reality people inflate when they meet. Both of these are phenomena of conversation. Generally speaking, if people cannot talk to each other they cannot have much of a relationship. What they talk about determines what kind of a relationship they have. To describe who says what to whom would be to describe most of the social structure, especially the structure of power, acquaintanceship, and patterned world views that make up stratification. If we had a record of all the conversations that have ever taken place, we would have a more or less complete social history of the world—providing, of course, that we understood the physical behaviors involved as well.[12] To predict and explain who will say what to whom under what circumstances would be to have an extremely powerful sociological theory.

The subjectivistic sociologists have leaned in this direction for a long time. The symbolic interactionists' formulation was that behavior is determined by the definition of the situation. This is no doubt true, but it is not always very useful. The problem was to go beyond this very abstract statement encompassing all possible situations to a more manageable categorization of what different sorts of situations there are and how they developed. An approach to the different kinds of conversations men have makes this possible, for most human social situations are conversational situations. Each is generally defined by the previous conversations (if we stretch this to include their nonverbal encounters as well) the participants have had right up to the present moment. Their relations are summed up in what they remember about their past negotiations and how they construe their prospects for the future. In conversational terms, this comes down to what they expect to be able to say next to each other. The

---

[12] Much of the physical behavior involved in acting out an organization—making various things and shipping them from one part of a factory to another, for example—is really communicative, as far as the structure of the organization is concerned. It is the recognition that someone is *sending* these objects, and can be expected (given certain messages in the opposite direction) to send more, that makes this an organization rather than a series of accidents. In the same way, it is the *threat* of violence (and the aspect of its use that constitutes an implied further threat) that makes up the organization of an army or state. See Chapter 7 in this volume.

conversational approach thus captures one of the most attractive features of the situational paradigm: the time-bound quality of social motivation in which what happens at a given moment depends on what has gone just before. Life can be seen as a series of ongoing negotiations, and explanation is solidly rooted in the reality of little moments in time.

Types of Conversation

Most of what people talk about can be described as practical talk, ideological talk, intellectual discussion, entertainment talk, gossip, or personal talk. These six categories are neither exhaustive nor mutually exclusive. Perhaps a more thorough effort would improve them on both counts. I have made enough efforts in this direction, however, to satisfy myself that we can dissect conversation from a great many angles, and that the meaningful things that can be said about it are not easily exhausted in any one set of rubrics. These six are operable at least, in the sense that it is not difficult to identify instances; and they are fruitful in that specific hypotheses follow, both about the conditions under which different forms of each occur and about their consequences for group formation. The search for beautifully comprehensive sets of categories is hardly of the essence of explanatory science. The past half-century of typological sociologies should be enough to convince us of where it stands on the pole of scientific priorities. No doubt elegant categorization can be an absorbing game; what I am attempting here is something different.

Generally speaking, all talk is negotiation. Sustained talk requires the participation of at least two persons, either of whom can veto any subject or form of conversation by refusing to take part. Conversation creates subjective realities, and it also creates social relationships. What is being negotiated, then, includes the cognitive reality of the moment and the personal relationships among the conversationalists. Friendship relations often consist of the bargain: I'll provide a sympathetic audience for your recapitulation of the events of your recent experience, if you'll do the same for me. The existence of formal organizations also needs to be bargained out from moment to moment, in such implicit formulas as: I'll agree to give particular people the right to talk authoritatively about certain practical matters and will respond appropriately and talk authoritatively in my turn, if others will do the same and if I receive some material benefits for playing the game.

Much of the bargaining is tacit, as people negotiate what kind of a relationship they will have, and what kind of a reality they will enact. Not all of it is tacit or symbolic, however. Practical talk, at some point, includes overt negotiations about the material benefits of the deal. Sociable relationships can also include benefits that are not strictly conversational—

exchanges of gifts, ranging from dinner invitations and Christmas presents to loans and sexual favors. But even the nontacit and nonconversational things involved in social relationships are bargained about through some tacit and symbolic conversational maneuvers. As Goffman has shown, all social encounters have an expressive side, whatever else they are designed to do. What is expressed, in every case, is the capacity and willingness of the parties to carry on a particular type of relationship and a particular form of reality construction. In ethological terms, every human encounter picks up initially from ritually negotiating some emotional resonance between the human animals confronting one another.

Conversation, then, is negotiation on a number of levels. What the very abstract discussions of Goffman and the ethnomethodologists do not reveal, however, is the fact that people have different resources for carrying out particular sorts of negotiations. These resources can be traced to the forms of stratification discussed in Chapter 2. People with different stratification positions are differently equipped to carry out conversations, and hence to establish social relationships. There are a number of different things one might want from social encounters: power, material goods, information, admiration, entertainment, the emotional gratifications of belonging, the cognitive gratifications of experiencing a subjectively meaningful world. To get these, one needs resources, which tend to come from previous conversational encounters. Stratification is thus both cause and consequence of the different fates people experience in the world of conversations.

Practical Talk. One thing that people can talk about is practical matters. There are two main forms. *Work talk* is carried out for purposes extraneous to the talk process: for carrying out a task, getting compliance, material goods, or information. It occurs at work, in stores, in places of public transportation; it can also occur at home or in sociable situations, as in "Pass the salt" or "There's a phone call for you." It is the least ritualistic of all forms of talk, although it shares with all language at least a minimal emphasis on controlling side involvements, concentrating attention, and invoking common symbols. But the element of stereotyping is *least* tied to emotionally compelling animal postures, and the focus of attention is on the objects rather than the participants.

Shop talk is closely related. It is about the same matters as work talk, but with a different purpose: it is something to talk about rather than a way to accomplish some nonconversational end. The borderline is not always clear. Talking about work can be largely sociable if it takes place in a bar; it is virtually part of the work itself if it consists of asking advice in the midst of a job. Operationally, we can tell the difference by the increase in ritual elements as the talk becomes more sociable; the orienting gestures are toward each other rather than toward the topic, and there is more affect expressed.

Practical talk is extremely important in stratification. It makes up the social content of peoples' work roles; holding a particular kind of job means, more than anything else, being able to do certain kinds of practical talk. Organizations are the basis of economic and political stratification, and an organization empirically comes down to a number of conversations, especially reports.[13] The situations that tell us about organizational hierarchy are those in which subordinates report to a boss who has the right to begin and end communications, to ask questions that must be answered, and to give orders that must be assented to. The horizontal aspects of organization structure are given by other conversations in which information and requests are reported in a more voluntary fashion.

For a theory of stratification, there are two crucial points: Getting an occupational position depends on being able to do that kind of talk; being in an occupational position affects what kinds of conversations one can have off the job as well, and hence what friends one will have—in stratification terms, one's status group membership.

The conversation of work situations varies in a number of ways. First, there is the sheer amount of talk to be done. Manual labor requires very little, skilled labor somewhat more; administrative, business, and professional work are made up almost entirely of talk, especially if one includes written communication. Different class cultures thus vary in how talkative they are, ranging from the taciturnity of the isolated farmer to the continuous patter of the entrepreneur. Work situations lay out both networks of acquaintanceship and degrees of cosmopolitanism.

Then there is the content of different work talk. In effect, skilled mechanics are persons who know the names of different machine parts and can talk about how they function and do not function; the professional defines himself by his specialized realm of discourse, the clerical worker by his knowledge of forms, the businessman by his references to prices and costs. Each occupation is a little world of specialized language. Admittance requires learning the language.

As ethnomethodological studies indicate, this is not merely a matter of vocabulary, of picking up the argot of the trade. The structure of language is in its use, and much of the implicit grammar that makes terms meaningful is a matter of using them in the right context. The construction worker calling "Slab!" is Wittgenstein's (1953: 3–10) famous example; this is a meaningful sentence not to the formal grammarian but to the workers

---

[13] I implicitly include written communications such as memos, notices, letters, and records as part of organizational conversation.

themselves, because it fits into an ongoing physical activity of passing materials to build a wall. All occupations have their tacit conventions about communicating which the participants build up among themselves. The communications networks of occupational milieu are stitched together with threads invisible to the casual bystander.

The conversational conventions of a given occupation can vary in more arbitrary ways as well. Business and governmental relations, for example, are carried on with a good deal more conversational formality in Europe than in the United States. At the most formalized end of this continuum, certain rigid formulas of the sort we associate with the military—"All present and accounted for, sir"—make the relationships run off in the most impersonalized fashion possible. At the opposite end is the flexibility of conversational form, the lack of formal titles, and the tendency to mix sociable talk with business, that is referred to as "Americanization" when it penetrates practical affairs abroad. As we shall see in Chapter 4, these ritual forms have a history, emerging only at a particular time in America and in certain sectors more than others. Knowing the right degree of formality or informality of practical discourse is another prerequisite for admittance into a given occupational class in any time and place.

Another dimension of practical talk is the authority it conveys. Here practical talk merges with ideological or legitimacy talk, and becomes accordingly more ritualistic. Cultural differences among occupational classes are based especially on giving and talking orders; power is indexed by how often you report and how many people report to you, and by how much backtalk you give and take. Successfull authority depends on one's demeanor as an order-giver, a matter of conversational intonation, timing, vocabulary. This has not yet been investigated in great detail, but it is likely that the stuffy composure of the upper classes can be traced to the kinds of controls over the throat and face that are involved in talking as if one always has the last word. Correspondingly, the characteristic attitudes of the working classes, including the nonverbal styles that set classes apart without their even thinking about it, can be traced to the ways they hold themselves in conversation with superiors.[14]

There is also an informal kind of power, not recognized in explicit authority roles. This occurs when people possess knowledge that others need in order to carry out their work. The advice-giver, the information expert, is often the power behind the throne in many organizations. Blau (1960) has investigated in some detail the conversational interactions in-

---

[14] It is probably for these reasons that people can tell someone's social class with considerable accuracy from voice recordings alone (Shuy, 1970). Labov (1966) reports that social classes make distinctive sounds, but that ethnic differences are even sharper.

volved in giving advice. He finds a two-phase movement: first, the advice-giver demonstrates his expert status; then he shows off a weakness or two (often by joking at his own expense) in order to make himself approachable. The combination of these two moves yields popularity, and a kind of informal power.

Advice-giving is a kind of conversation that creates relationships of personal dependency within the more impersonal relationships of the work situation. There are other kinds of alliances and dependencies as well; the conversation that sustains them may be called *politicking*. As organizational studies show (Gross, 1953; Dalton, 1959), much of what goes on at work involves informal cliques. Every able administrator has his reliable contacts, his friends who can help him cut corners and short-circuit bureaucratic routine when he needs to. Clerical workers and manual laborers have tacit understandings among themselves over which rules need to be applied and which can be evaded. The exercise of authority often rests on informal understandings. As Crozier, (1964: 156–161) puts it, rules are most powerful in the hands of managers who can use their own discretion about when to enforce them. All of this tends to make organizations a network of personal alliances and understandings, counteralliances and antagonisms. Promotion, particularly in the higher administrative ranks, tends to occur through informal sponsorship, as juniors ride the coattails of their patron's success or failure (Martin and Strauss, 1956).

All of this is accomplished through a special form of conversation. Politicking involves a kind of work talk in which the tacit negotiations of practical alliances is paramount; its able practitioners must be adept at reading between the lines, noticing the offers that are being made without being spelled out. Talk involves many tacit elements; organizational politics adds yet another level of communicative subtlety.

Its degree and forms vary with the job. At the lower levels, politicking is least subtle. It mainly involves presenting a united front to the supervisor which appears to comply with his demands while giving the workers a breathing space in which to control their own work pace and free their subjective selves from a sense of external compulsion. In Goffman's (1959) famous metaphor, encounters with authority are frontstage, while the workers on their own are backstage. Here they can let down the masquerade and engage in "staging talk"—conversation about how the boss was fooled, how a member of the team almost blew its cover; jokes are made and anecdotes are told, creating an entire conversational culture of a particular backstage world.

Similar frontstages and backstages are found at other occupational levels. The more manipulative occupations, like salesmen, have the more closely guarded backstages, in which the staging talk reflects a relatively strong in-group tie. Professions, similarly, have their backstages in which mistakes are admitted and the secrets of self-presentation more frankly discussed; the more august their public image, the more heavily veiled these conversational enclaves. Since organizational authority, in general, involves manipulation of one's public self, conversational in-groups are generally split along authority lines. The higher executives with their keys to a private washroom, like the Chinese emperor in his forbidden city, illustrate the extreme of the same process.

Higher positions, in general, involve more talk; a specialized kind of argot (like all jobs); an authoritative stance toward underlings; an emphasis on conversational subtleties in negotiations among equals and near-equals; and great care in guarding conversational backstages. Indeed, in higher occupational positions, one spends proportionately more time acting on the frontstage, whereas the lower classes spend most of their time backstage in their own informality.

These conversational differences both create and sustain divisions among occupational cultures. To enter an occupation, one must either know how to converse properly or at least convince the gatekeepers that one has the rudiments upon which conversational refinements can be built. (Of course, low-ranking occupations, whose gatekeepers are in higher classes, have relatively less stringent requirements.) Moreover, being in an occupational milieu habituates one to a form of conversation that carries over off the job. Since status groups are made up of families and friends, and these in turn are put together by conversational relationships, the conversational forms learned at work affect the formation of status groups.

Sociable relations are essentially conversational relations: getting together for the sake of talking as an end in itself. A considerable amount of sociable talk is shop talk, and this, of course, varies in content with the social class. Businessmen at a party talking business, academics discussing research, farmers jawing about crops, workmen talking about auto repairs, all of these inhabit closed-off circles precisely because of the conversational ties that set them apart. Housewives occupy the other side of the room at a party, discussing childrearing and shopping, for the same reason. The isolation of the lower classes comes, at least in part, because their work (or lack of it) leaves them in a conversational vacuum:

> "You remember at the courthouse, Lonny's trial? You and the lawyer was talking in the hall? You remember? I just stood there listening, I didn't say a word. You know why? 'Cause I didn't even know what you was talking about. That's happened to me a lot."
> "Hell, you're nobody special. That happens to everybody. Nobody knows about everything. One man is a doctor, so he talks about surgery. Another man is a teacher, so he talks about books. But doctors and teachers don't know anything about concrete. You're a cement finisher and that's your specialty."
> "Maybe so, but when was the last time you saw anybody standing around talking about concrete?" [Liebow, 1967: 62]

Ideological or Legitimizing talk.   The most serious and impelling rituals are those which act out power or membership in the larger community. These are the rituals of politics and religion. The ties are of crucial importance, for the state and community always mark out who will be the object of potential violence and distrust, and shows of loyalty are motivated here at least implicitly by a concern for safety. At the same time, the organization of state and community is a resource that some men can use to dominate others. This is done by leading rituals of solidarity so as to identify the group with oneself and make loyalty to one seem to be loyalty to the other, and by special deference rituals that act out one's personal supremacy.

Ideological rituals, then, are the most solemn and impressive type, drawing as they do upon implicit threats against outsiders and the emotional security of insiders in a conflict situation. They are brought about by a combination of proud and humble gestures, and above all, complete seriousness. Ideological talk thus enacts the basic relationships that make up the organization of state and community, as well as the structure of other organizations (work, family) in which there is a structure of authority based on implicit dangers or threats of punishment or expulsion. Those who do most of this kind of talk acquire very different world views from those who merely listen or assent to it, and this accounts for a basic feature of class cultures.

As in the case of practical talk, official talk may be more or less formalized. The more elaborate stereotyped recitations, and the relatively greater ban on free improvisation, are what give traditional forms of legitimacy their distinctive flavor in everyday life. The modern forms of rational–legal authority, by contrast, differ both in the kinds of formulas uttered (more abstract in content, as well as secular rather than religious in metaphor) and in the amount of free improvisation allowed within the framework of bounding formulas. The decline of rigidly upheld formalization is part of the *entzauberung* of modern public life.[15] Such differences in communication style can be pursued in greater detail, of course: public life in Europe, China, Latin America, and the United States, for example, differ through various mixtures of these elements. A more precise characterization of the legitimating tones of different governments can be achieved along these lines.

---

[15] Bits of formalization remain to give us an idea of what more pervasive forms of traditionalism were like: "I hereby declare this session of Congress open" or "Dearly beloved we are gathered here together. . . ." Some of the international differences existing today are unsystematically sketched by Hall (1959).

Ideological talk can also make up a kind of conversation in the private realm. Men's conversations about politics or religion are one of the things that divide them into separate groups. As a general principle, politics and religion are matters of utmost seriousness, referring to a realm beyond the merely sociable, and the merely practical as well. They invoke ritually anchored emotional loyalties, often in conflict with other emotionalized factions. Conversational relations based on talking about politics or religion, then, are not likely to last very long unless the participants are generally in agreement. If they do agree, however, the conversation creates social ties because they are helping each other to invoke a symbolic reality of great emotional significance. Persons with strong political or religious views tend to talk about them with persons like themselves and to avoid persons with antagonistic views. The social word is thus divided into a series of mutually exclusive cults worshiping different gods, whether at the formal altars of churches and political meetings, or on the informal stage of private conversations. The extent to which one can participate in such conversations must considerably affect how strongly one believes in ideological values and formulas.

People vary in these respects. Different occupational classes tend to have different political values; insofar as these are subjects of conversation, they are drawn together within a class and set apart between classes.[16] There are also political factions within and across class lines, based on more specilized political interests, including ethnic, age, organizational, and other interest groups. All of the foregoing, however, is only potential until the individuals are actually mobilized into ideological conversation. The higher social classes are more mobilized because of their greater resources for communication, and their more active social lives; their ritual attachment to ideological values is correspondingly greater. A big conversational difference among social classes, then, is the sheer amount of talk about politics. This seems to be a staple of upper-class and upper-middle-class social life, at least among men; it is rare in working-class and lower-class circles. An analogous set of principles may be suggested for talk about religion.

Intellectual Discussion. Discussion is yet another kind of interaction. When people talk about books or ideas, the natural world or social events,

---

[16] Katz and Lazarsfeld (1955:278–283) and Lin (1967) show that most discussions are carried on among partners of the same sex, educational level, religion, age, and marital status. Where class boundaries are crossed, it is women who tend to name men as discussants, and lower-ranking persons to name higher-ranking ones instead of vice versa. The younger, though, do not tend to seek out the older in these samples.

purely from the point of view of establishing what is true, we have a type of conversational ritual that is different from any of the others. It is the talk itself that is important, not the practical results or the political membership that flows from it; it also differs from sociable talk (entertainment, gossip, personal) in focusing on the verbal part of the interaction, and playing down as far as possible the nonverbal and emotional appeals for solidarity. This not to say that there is no social organization in the realm of intellectual argument, but only that its defining rituals are peculiarly and narrowly verbal.

Defined in this way, pure intellectual discussion is rather rare. The world of science and scholarship gets very close to it; in the world of private sociability where intellectual matters are subjects of conversation, it becomes increasingly mixed with ideological appeals, or breaks down into heated argument which nonverbal appeals supplant verbal ones. Like most things sociologists study, conversation is often a mixture of types. But the pure type of intellectual discussion picks out some important tendencies; where the relatively emotionless and hyperprecise verbal style is found, we tend to find a special form of stratification.

How well one can keep up with a discussion depends on his sophistication. At the minimum, the boundary between outsiders and insiders is drawn between those who know something about a subject like idealistic philosophy, modern physics, or Russian novels, and those who do not know what is being talked about. But there are circles within circles; those who understand a subject better can carry on a conversation among themselves that less sophisticated bystanders must drop out of at some point. In an important sense, the backbone of stratification within a science or scholarly discipline is based on just such differences in conversational capabilities. The apex of any field is the conversational elite: those who can carry on to the last word, who can puncture the intellectual pretensions of others without raising voices and waving fists—in short, those who dominate the purely intellectual conversation, whether it goes on face-to-face or in the pages of books and journals. At the same time, this criterion enables us to distinguish horizontal factions who can argue with each other interminably because they appeal to different intellectual premises.

In the larger social arena, intellectual discussions reinforce some class boundaries and mark off some subcommunities of their own. Generally speaking, working-class persons rarely talk about abstract intellectual matters and (in the United States) are likely to jeer at those who do. The upper middle class, especially the sector of highly educated professionals, talk about such matters a great deal, and take much of their flat and controlled emotional tone from this activity. This is especially true in the academic sector, for whom discussion is a form of shop talk as well as a more general form of sociability. There is an expectation of general literacy in intellectual matters more broadly, as well as in particular academic specialties. The extent to which such an intellectual culture defines the sociable activities of the respectable classes as a whole varies with time and place. It is exceedingly strong in France and important in Europe generally; much less so in the United States. Such differences result from the histories that have shaped their stratification structures, especially in the degree of linkage between social classes, education, and positions of high occupational power; cultural differences play back upon and reinforce these structures.

Between the classes who spend much time displaying their intellectual cultivation and those who talk hardly at all about abstract matters, there is a range of intermediaries: the "middlebrow" world between "highbrow" and "lowbrow." This may be more precisely indexed by the amount of time spent on intellectual subjects, the level of precision with which they are discussed, the extent to which expert opinion is called upon to bolster one's own, and the amount of extraneous emotional appeals that are allowed in. Such differences may cut across class lines as well as follow along with them; they also define some of the distinct characteristics of particular ethnic cultures.[17] The causes and effects of such variations would fill an important slot in a comprehensive theory of stratification.

Like ideological talk, intellectual discussion is a vehicle for negotiating stronger personal ties than practical talk, but, at the same time, it creates greater dangers of conflict and requires more skill in interpersonal negotiations. Especially in the middle-brow range at which most discussion takes place, individuals have moral and aesthetic images of themselves invested in their views of the world and they dislike having their views challenged. Moreover, unless an individual has a strong desire for information (usually confined to the relatively specialized community of scholarship), he is not usually willing to let someone else show off his knowledge without acquiring some reciprocity. Discussion ties thus tend to be built up between individuals based on a tacit bargain of the form: I'll support you in your views and you'll support me in mine; or: I'll allow you to show off your knowledge if you'll do the same for me or allow some other reciprocity (such as letting me borrow the prestige of talking with you).

---

[17] American Jews seem to spend a great deal of conversation on intellectual discussions, judging from their expressed cultural interests; Anglo-Protestants are also relatively high (Greeley, 1969).

Not everyone has the desire or the skill to negotiate these implicit bargains, and hence discussions are often the shoal on which conversational relations break up. The rule of pre-World War II middle-class America, "Don't talk politics or religion," must have reflected a situation of much interclass contact in which strong personal ties were rather difficult to establish.

Entertainment Talk.   With conversation for enjoyment, we have left the serious realms and are now fully inside the world of sociability. The social relationships involved also change. Practical talk establishes relationships for a very limited purpose; at the most, alliances based on extrinsic rewards, not on personal liking. Similarly, ideological talk links together people as members of the community in which violence is organized. Intellectual discussions have much of the same quality, at least among serious discussants; it is less important who one talks to than where one arrives in the abstract realm of ideas. All of these kinds of conversation are fundamentally impersonal or universalistic.

Entertainment talk operates on a different level. It is talk for the sake of talking, as an enjoyment in itself. Here the talk is embedded in nonverbal rituals that are even more important than the words themselves, especially the signals of mock fighting and exaggerated gesture that make up animal play. People who participate in it to any substantial degree are friends of some sort. Friendship is the result of a relatively irreplaceable series of conversational exchanges; there is a great deal of subtle negotiation of just this sort involved for mutual entertainment talk to take place. As Suttles (1970) and others have observed, friendships are relationships in which people recognize each other as individuals. The main kinds of conversation that produce friendship, however, are not necessarily revelations of a disreputable self, as Suttles suggests, but can be any of the ritual exchanges involved in entertainment talk.

Personal relationships are built up out of a series of conversational exchanges that the participants feel are relatively unique and irreplaceable. It is successfully bargained conversational exchanges—mutual audiences for showing off, for example—that draws people together. Indeed, what we mean by a "friendship" is largely a history of mutually rewarding conversational exchanges. Such ties are particularly strong when the individuals find they seek each other eagerly to tell them their recent thoughts or information. "Hey, have you heard . . ." is a mark of this kind of conversational relationship. It usually develops on the basis of exchange about matters that the participants find morally or aesthetically important, or that constitute a realm of reality in which they vicariously participate with some commitment, such as politics, sports, or other subjects of which one can be a "fan." The participants help each other enact a larger and valued reality on the stage of their personal encounters.

Entertainment in conversation is talk that engenders feelings of enjoyment and conviviality, and thus is satisfying as an end in itself. It consists of anecdotes, jokes and witticism, bantering, put-ons, fantasies, and other forms of game-like contests. It is designed to generate emotions, but limits itself to the emotions of joy and laughter and expressly prohibits other emotions that might "spoil the fun." The side-involvements that are controlled here in the interest of the ritual are the serious ones of particality, ideology, or personal concerns (Goffman, 1961a; Simmel, 1950:40–57). There is also talk about entertainment—the discussions of movies, music, sports, actors, and athletes that make up so much of casual chitchat. This becomes an even bigger area if we include talk about things people consume or seek out—food, clothes, houses and gardens, restaurants, scenery, vacations. In some social circles, this entertainment talk makes up most of the content of conversations; in all classes, it is important.

Entertaining conversation depends on the mood established between talkers and listeners, and often on the quick and agile interchange of roles. The rules of humorous talk reside in the interplay between verbal and nonverbal levels, with the former often explicitly nonsensical while the meaning is redeemed in the tone of voice, posture, eye movements, or facial expressions (Roy, 1959; Coser, 1964; Labov, 1972a). The penalty for not learning the rules is to be laughed at or to be left out of a group experience. Entertaining talk also depends in considerable degree on common viewpoints and values among the participants. Jokes are often at the expense of some form of behavior that is described as incongruous in the light of higher values. This means that entertaining conversation can serve to exclude people as well as to include them. Joking is a way of being backstage from some other situation held up to ridicule: it creates a bond of cooperation because the participants must uphold at least a mild atmosphere of secrecy. Entertainment can be a risky form of conversation, and hence it requires some tacit negotiation to avoid the embarrassment of an unsuccessful attempt. For the same reasons, an entertaining conversation creates positive bonds of some strength. It tends to be uniquely specific to a certain combination of individuals, and thus serves to bind them together by their mutual and relatively irreplaceable contributions to happiness.

The differences in values expressed in entertainment talk are derived from life experiences, and hence reflect class and ethnic divisions. Since they serve as bases for negotiating personal relations, they tend to reinforce such distinctions. Research on personalities (Tyler, 1965:203) that people are distinguished more by things they dislike than by things they like. It is hardly too much to say that jokes are one of the main things that turns classes into exclusive status groups.

Similar stratification differences are reflected and reinforced in talk about entertainment. This is true because tastes are based on opportunities, and the richer and more mobile upper classes can afford to develop tastes and interests that poorer people cannot (Clarke, 1956). Being a gourmet, a polo player, or an art collector is not open to everyone. In a less restrictive sense, "highbrow" symphony music, theatre, and literature reflect both the greater financial resources of the upper classes, their leisure to train their tastes for complex matters, and their distinctive class attitudes and values. Working-class tastes are correspondingly molded toward action drama, simple melodic music, and cheap participant sports like bowling. Conversation between persons with different standards of entertainment is bound to lead to strains, even though the underlying class differences are never brought to consciousness. Persons who have had the most leisure and money to develop their tastes are likely to be better able to give a long and subtle defense of them; the implicit standards of evaulation involved in any taste thus become a battleground between persons with unequal conversational resources. The usual solution is to break off the relationship or keep it at a superficial level.[18] The same applies, of course, to different tastes in clothes, decor, personal grooming styles—all those superficial differences among social classes, based on their money and leisure to devote to such things.

All of this applies in a very general sense to styles of sociable conversation themselves. Social classes differ (as do historical periods) in the standards of what constitutes an acceptable conversational performance. Generally speaking, the higher social classes make sociable encounters more of a ritual in itself, emphasizing an idealized image of dignified, complacent enjoyment. Vocabulary, grammar, accent, use or nonuse of slang—all of these rules can vary in importance, with the more stringent requirements found in those classes attempting more strongly to idealize their everyday encounters (Labov, 1972b). Their conversations become an elaborate worship of the finicky gods of propriety, who turn out to be the conversationalists themselves. The relative emphasis on this kind of conversational style depends essentially on the inequality of power in the particular society. Some evidence on the sources of variation is presented in Chapter 4.

---

[18] The other possibility is for those with fewer resources to subordinate themselves to the consumption leaders and go to them for advice, as found to a certain extent among housewives in Katz and Lazarsfeld (1955). This social-climbing alternative to class conversational endogamy is probably found where there is considerable social mobility and people of different resource levels are physically adjacent, as in some housing developments.

Gossip. Gossip consists of talking about one's acquaintances. At any extensive level, it implicitly involves judging them, and ends by discussing their personal relationships. In this realm, one finds conversationalists keeping up with old friends; watching the progress of acquaintances for better or worse (with sadness or glee); telling stories and giving descriptions that reflect on their characters, abilities, and other personal qualities; relating the complexities of their acts and feelings toward each other. Gossip usually at least implicity involves the talker's evaluations, and certainly their dramatic sympathies or antipathies toward the events they unfold for each other on the stage which their acquaintances' lives provide for them. Gossipers reveal their own standards and personal relationships, and hence extensive gossip requires some careful negotiation to be successfully carried out. Moreover, most gossip is definitely backstage; one must usually be careful not to gossip about people in their hearing, or in the hearing of their friends, or sometimes of persons who are considered too distant to have the right to know about them.

Gossip is an informal and relaxed type of ritual exchange, much like personal "grooming talk." It has two qualities which conduce to fairly strong familiarity: a personal and relatively irreplaceable exchange, and the sharing of secrets. Gossip binds conversationalists together as spectators at a drama that few are privileged to witness; the more limited the number of persons who know all the actors, the stronger the bond the gossip can establish between conversationalists. In any case, those persons who have common acquaintances have the basis for conversational exchange that can bind them together more closely than those persons who have fewer acquaintances in common. Upper-class persons seem to spend much time gossiping about family ties, perhaps as an explicit effort to remain exclusive by requiring their friends to participate in a rather limiting conversation (Birmingham 1968: 8–9, 85–86). Also, the more that the spectators have themselves been parts of the drama they are recounting, or the stronger their emotional responses to these adventures, the more gossip becomes the sharing of personally meaningful experience. At this point, gossip shades over into the most familiar of the categories of sociable talk, personal narrations and concerns.

Class communities are based on differences in who is known, and in the values expressed in gossiping about them. Members of different occupations are involved in separate networks of acquaintanceship at work. One barrier to entering the status community of a social class different than one's own, then, is simply that one is left out of a lot of gossip. Who one knows in public or intellectual affairs sets further opportunities and barriers for negotiating conversational relations. The group tends to be closely knit to the degree that its members have mutual acquaintances they can sustain themselves by talking about. There is also the barrier resulting from different values people apply in discussing their acquaintances; successful gossip is not possible if people disagree too much about when to laugh and when to cry. Finally, there are different rules about how much gossip is allowed in polite conversation. In traditional upper classes with highly idealized manners, gossip of any degree of familiarily was taboo. On the other hand, gossip seems highly characteristic of tightly knit communities or cliques with little power or privacy, especially women in traditional marriages (Young and Willmott, 1962: 44–61; Nelson, 1966). Changes in such standards are one indicator of the way in which modern society and particular classes within it are distinctive.

Personal Talk.   Personal talk is about oneself. It includes several levels, from the relatively superficial mutual inquiries about health, family, and other biographical externals that go on as a greeting ritual among acquaintances, through a recount of the hopes and events of the day, to more intimate revelations of things ordinarily kept secret.

At the most remote level, we find the polite rituals of "small talk" that serve to act out the minimal concern for each other's personal fortunes that makes up a sociable acquaintanceship. Stronger bonds come from more intimate "grooming talk," to use Goffman's (1971) metaphor borrowed from the mutual lice-pickings of friendly monkeys.

Personal narrations can be described as gossip about oneself and one's feelings. The listener's reaction becomes itself a main concern. The conversation not only concerns fairly important emotions, but it begins to arouse such emotions and to define the relationship between the conversationalists in terms of them. The narrator may tell of an encounter at someone else's expense; if his listener joins in with appropriate derision, the relationship between them is cemented, and further strengthened if each joins in the initiative in putting down the common enemy. There is some risk when the narrator begins his story: If he is not joined in his feelings, he may at the least feel disappointed, and at worst have opened himself to a hostile counterattack. Shared ambitions, problems, and enjoyments can also generate a strong commitment insofar as the feelings that the individuals have about their experiences and projects are themselves quite strong. Strong bonds are generated by conversations that involve the negotiation of mutually supporting exchanges of strong personal feelings. Commitment is proportional to the costs in emotional response called for by the listener, and to the risks in self-revelation by the narrator.

Relationships established through this form of conversation are likely to be fairly unique. They are usually established in two-person conversation (tête-à-tête). The exchange need not be based entirely on mutual ego support. Some experiences strike one as exciting, significant, or important, and much of that feeling would be lost if it were not told to someone: "Wait till I tell Xenia about this!" Indeed much of the feeling at the time of the event itself may be anticipatory; it is the putting of the experience into words that creates the significance, and the internal audience in one's mind only foreshadows an imagined audience, itself based on the memory of a previous audience. Not any audience will do; the wrong listener destroys the value of the experience, so much may it hinge on the recounting. Some experiences, indeed, never take significant form at all for want of a conversation that may give inchoate daily events a retrospectively illuminating form.[20] Only the trained sensibilities of an experienced conversational team can give such a payoff. If this exchange is sure, the bond is strong.

Intimate conversation is the level of personal talk in which one lets another person into the privacy within which he prepares himself for and recuperates from forays into the world of ordinary sociability. Here conversation centers on feelings about others not ordinarily expressed or admitted in sociable talk: reactions to veiled or ignored insults; candid evaluations of acquaintances with whom one carries on pleasant or at least manageable relationships; admissions of anxieties, secrets, mistakes, put-ons, hopes, despairs, disreputable victories, and defeats. Intimacy is an arena for the expression of socially reprehensible feelings: unbounded egotism, subservience, silly delight, hatred, withdrawal, repose. The drama here goes backstage and concerns the way actors feel about roles that they do not always wish to play the way they are written, or sometimes that they do not want to play at all.

The narrator of intimacies risks a great deal. His listener must be willing to keep the confidence of the dressing room; the strongest assurance he can give is to make similar revelations. The exchange is then equal, not only in risks and in costs (listening with the appropriate show of support), but in payoffs to the participants (having an audience for these emotional expressions). On this level at least, Homans' (1950) theorum may indeed be true: The more people interact, the more they will like each other. Intimate exchange builds up quite strong bonds between individuals, since these bonds are based on a fairly unique exchange. This bond is usually solidified by a conversational exchange usually found only at the intimate level: the candid expression of *my* attitudes toward *you*. Ordi-

---

[20] Scott and Lyman (1968) describe various kinds of retrospective attribution of meanings, ranging from intimate to very distant and formal contexts.

nary sociable discourse limits this kind of honesty because most exchanges are based on mutual support of each other's ego-enhancing presentations of self. Intimate exchange allows people to be honest mirrors to each other because of the implicit rule of secrecy and of benevolent intent that has been negotiated. Moreover, the attitudes expressed toward each other when the intimate exchange is first reached are generally affectionate, a combination of the warmth and support called for in response to intimate revelations and a feeling of joy and gratitude for receiving such support. This analysis suggests that strongly affectionate emotions—close friendship or love—are generated by the dynamics of conversation reaching this level of intimacy, more than vice versa.

But intimate relationships, like sociable relationships in general, are not always very happy or very stable. Conversational dynamics produce a variety of emotions, but strong emotions of every sort seem to go along with strong interpersonal ties. Hostile relationships do not constitute a distinct level of closeness or distance, but can occur at any level of familiarity. Hostile talk usually grows out of nonhostile talk at that level: discussion may turn into heated dispute; entertaining banter may turn vicious; intimate revelations may be rejected or ridiculed. The varieties of hostile expression toward someone with whom one has not conversed before are limited to stereotyped and rather impersonal insults. As marriage well illustrates, a past conversational history is useful if the battleground is to be well-supplied with weaponry. Sociable exchanges probably generate more intense hostility than most practical disputes. Good turns and bad turns, gifts and thefts can create feelings, but the commitments of the individuals involved to affection or hatred are less strong than the commitment to carry through a relationship of love or hate shaped to the distinct contours of the individual personalities. As we know, most murders are commited by persons who know their victims personally.

Entrance into intimacy may be best described as moving behind the scenes of the usually guarded presentation of self to others. It is such shifts that make up the most dramatic periods of everyday life, which helps explain why the period of courtship is regarded with such anticipation and nostalgia. As we have seen, there are backstages within backstages; the shift from practical talk to politicking is also a movement into the dressing room. Intimacy is the most remote and heavily guarded backstage, and the fact that intimate talk is avoided in most conversations makes its occurence between a few individuals a particularly strong bond between them. The fact that such conversation takes place at all, perhaps serves to tie them together more strongly than the actual content of the exchange itself.

Overt sexual activity is also usually confined to a strongly guarded backstage which is supposed to remain unique to a particular couple.

Sexual relationships thus have many of the characteristics of conversational intimacy. Sexual intimacy is not necessarily emotionally and conversationally close, however; the combining of the two as an ideal is a feature of modern industrial societies, and especially Anglo-American. The American love ethic defines sexual relations as love relations, and this in practice means conversational intimacy. Sexual relations, at least ideally, require that a conversational negotiation of intimacy be reached first. The ideal is not always lived up to, but the formula "I love you" is widely accepted as at least a token of conversational intimacy.

It is not difficult to see the stratification processes involved in personal talk. Different social classes lead different sorts of lives; the kind of appreciative concern involved in sustained personal talk comes most easily to persons of common experience. Moreover, the high trust necessary for the more intimate kinds of relations tends to make equality an important precondition. This is not always the case, of course; in highly stratified societies, there are often institutionalized confidantes to the higher social classes, such as maids, secretaries, or paid companions. The bought time of the psychotherapist may be the modern equivalent—available, as before, primarily to the wealthier classes. In general, as we shall see in examining a market model of conversation, power and other external resources can be used as substitutes for conversational reciprocities. Patrimonial retinues can still be found in certain sectors of modern society, in which the leader can indulge in personal self-dramatization at great length before a captive audience which is not allowed a reciprocal indulgence. The conditions for such variations need to be spelled out.

Determinants of Conversation

How would we go about predicting who will talk to whom and what they will say to each other? Contrary to Chomsky's insistence that talking is an extraordinarily creative process, there is a lot of repetition, both in form and in specific content. The same verbal formulas are repeated millions of times. This is not to say that many conversations are exactly identical, but there is enough commonality for a sociological theory to begin to move. Once we get the general principles, more idiosyncratic utterances will begin to fall into the proper slots. After all, social structure is enacted largely through conversations; to the extent that sociology can explain anything at all, it is explaining talk, especially in the larger sense of ritual encounters.

Focusing on the conversational behavior of any one individual, we would want to know, first of all, *whom* he will talk to, which in turn depends on whom he meets. This is a matter of sheer physical *propinquity* at work, home, neighborhood, or recreation. It also depends on *deliberately seeking out* people at work, for shopping, with invitations to sociable occasions, and so forth. Both of these (especially the latter) somewhat depend upon what relationships have been previously negotiated, so that the individual wants to speak to certain people, or has to speak to others (such as those who have authority over him).

This leads us to the same variables that explain *what* people will say to each other. This depends on several things, including their *motivations*—what they want from social contacts—and their *resources*—the things that they can say (and do) to each other. Motivations are best handled in a market model of the various opportunities one chooses among in attempting to make the best conversational deal. Resources depend especially upon previous social encounters, the things people have said to one another (and to other people) before, which define their relationships up to that moment and provide ongoing worlds of socially constructed reality that they may use as a basis for further negotiation.

The best approach is a causal chain model, in which we attempt to state how what has gone before constrains what will happen at any one moment in time. Conversational negotiation has an emergent quality, so that there are important choice-points or gaps in the causal explanation; our task is to fill in the content that narrows them down to a given range of possibilities. Conversational relationships are constantly being renegotiated, with what happens at one point in time setting both the resources and the motivations for what is negotiated next.

Propinquity.   The sheer ecology of face-to-face contacts is the background condition for virtually all social structure. Organizations can be seen, first of all, as the network of people who are regularly accessible for that combination of practical and ideological–legitimizing talk that makes up the structure of authority; and second, as the horizontal network of practical talk that links together separate tasks into a division of labor. The structure of an organization means, above all, who is physically accessible to whom. What determines who will say what practical and official things to whom, then, is implicitly given in the part of organizational theory that gives the conditions for variations in structure and authority relations; although this material is rarely couched in terms of explaining talk, that is what we find if we take a close look at what "structure" refers to empirically. The result of how this communication is carried off at any one time is the structure of the organization at the next moment. Where the degrees of conversational propinquity change, the organizational structure changes with it; as we shall see in Chapter 6, the amount of face-to-face surveillance is a crucial determinant of what has gone on in organizations throughout history.

The patterning of propinquity at work has other effects besides structuring the formal side of organizations. Within work places, personal associations are limited by the lines of expressive performances that different "teams" put on for each other. Conversational relationships of much familiarity are unlikely to be built up across the customer–salesman line, the professional–client line, or the supervisor–supervisee line, when these groups commonly meet only across a "frontstage" and exclude each other from "backstage" settings. Since frontstage–backstage distinctions usually coincide with levels of authority in organizations, it is not surprising that friendships tend to be made within similar authority levels, as well as within particular horizontally distant sectors of the occupational world (Laumann and Guttman, 1966). Within levels, informal associations develop where individuals have schedules that allow them work breaks at the same times and places (Gross, 1953).

Neighborhoods are another major setting for conversational contacts. In housing projects, friendship patterns can be predicted by propinquity of houses or apartments, the sharing of common pathways and parking lots, the width of streets, occupying adjacent lawns, and by the routes of children at play (Festinger *et al.*, 1950; Whyte, 1956: 365–386; Priest and Sawyer, 1967). The tendency for marriages to be related to residential propinquity probably reflects the importance of common ecology on conversational associations (Katz and Hill, 1958). Schools are important places for associations because they provide a common ecology as well as a common culture. Since neighborhoods usually consist of dwellings of comparable price, wealth affects sociable associations by distributing people ecologically. The existence of different vacation resorts for rich and poor also contributes to associational groupings along wealth lines. Other class-related links are forged where people meet regularly at clubs, churches, places of recreation, sports, and voluntary associations. Personal contacts may also be structured ecologically on dimensions other than class; the fact that neighborhoods tend to be segregated by ethnicity, age, and marital status also tends to keep the pool of potential friends divided along these lines (Beshers, 1962). In contrast, the traditional household arrangement, which kept persons of different rank constantly in each other's company, produced a very different conversational market, where most exchanges were among unequals. This reminds us that the effects of propinquity must always be taken together with the effects of conversational resources and motivations.

Conversational Markets and Individual Motivation

Conversation is an exchange. It requires the cooperation of at least two persons to carry on talk, even if one's part of the bargain is only to listen. In the business of creating symbolic realities, no one can carry it alone.

Implicit negotiation goes on over whom to talk to and for how long, about what subjects, and under what terms of exchange.

This suggests a market model. Waller (1937), Davis (1941), and others have used a model of the sexual market to account for the pattern of exchanges involved in who marries whom. But sexual bargains are negotiated primarily through conversation. We may thus conceive of a larger conversational market, in which relationships of all sorts are bargained, among which sexual relationships are only one type (cf. Blau, 1964).

A basic premise is that each individual attempts to negotiate the best possible exchange available to him under the circumstances. But what is being exchanged? We can think of a number of things: material possessions (dinners, jobs, pay, etc.), power, practical aids, sex, satisfaction of curiosity, desires to be entertained, to play, to bolster a favorable subjective world view, to satisfy a variety of interests and activities, and more. There is no obvious way at this time to settle on a list of basic human motives. Nor would this be useful, since there is so much individual variation. But which things (and how much of each) people bargain for in conversations can be at least roughly accounted for in the market model itself.

That is, all of these exchanges may be regarded as ways of getting interpersonal status—resources through which one may optimize the amount of deference and support he gets from others. Even material goods, some of which have the further significance of keeping one alive, may be handled at least partially within this framework. For material goods are generally derived (and certainly are held onto) as a byproduct of getting deference from other people. Even their significance for livelihood may be essentially derivative of the process of social evaluation. What is considered an acceptable standard of living varies with the structure of deference, and even the minimal aim of staying alive is a matter of choice which depends ultimately on the amount of dignity with which one lives.

We can take it as a premise that everyone gravitates toward those things in which he can get the best deference deal. This depends on his opportunities, resources, native capacities, and competition from others. We may suppose that one's interests, abilities, and commitments are acquired in the process of pursuing the best market position. Whether one becomes an intellectual, an athlete, a musician, a dedicated businessman, a sexpot, or a righteous Christian, these commitments build up as one follows out a particular line of advantage, acquiring specialized skills and hence further opportunities that move one deeper into that socioecological niche.

This solution to the problem of basic values is essentially the same as that developed in economics. The search for an absolute standard of value was abandoned with the marginal utility concept elaborated in the late nineteenth-century by Walras and Marshall, in recognition of the subjective and continually shifting quality of the things men desire, and of the fact that *anything* someone desires, no matter how ephemeral it might seem, is just as real in its effects on the market as any other good. The economists' solution was a purely abstract definition of value in terms of its place in the system of exchange. This seems a reasonable procedure for sociology as well. The individual may be conceived of as a firm, buying and selling a variety of goods and services. How varied or how specialized his exchanges are depends on the condition of the market. The conversational market differs from the modern economist's market in one important respect: the absence of a conventionalized medium of exchange like money. The conversational market is mostly a barter system, dealing in some fairly personalized kinds of services. Individuals are more closely tied to particular conversational customers and suppliers than in a market based on money. There are variations on this score. Real money—or rather, the things one can buy with it—can enter in as a resource in conversational negotiations. More importantly, *social reputation* is a kind of medium of exchange which can transcend particular relationships, although obviously lacking the same convertability and precise calculability of money. When a model of conversational markets is more developed, we may be able to distinguish different sorts of conversational markets which arise as various systems of communicating reputations give greater or lesser approximations to an impersonal medium of exhange.

Conversational Resources and Conversational Costs

A consideration of the individual's conversational motivations is a consideration of the conversational market as he experiences it over time. The basic premise is that everyone moves toward the best available exchanges for creating his subjective status. But by no means is everyone able to get others to help him create a conversational world in which he can continuously show off, receive deference, and enjoy himself. Some persons must settle for lesser realities in which they are merely audiences and supporters of conversational heroes, or participants in tawdry worlds of minimal conversational interest. Where one stands depends on the resources that he brings to the market.

For people to talk depends on their having something to talk about. Such conversational resources come from a number of areas: sharing practical concerns; political, religious, or intellectual interests; participation in types of entertainment; similar tastes in consumption; similar ideals of conversational style, not the least of which is a common language; a set of mutual acquaintances; and similar standards by which to evaluate oneself, one's acquaintances, and the world in general. We have already seen how each of these tends to be distributed in such a way that different occupational classes tend to have the resources for closer conversational ties among themselves than with anyone else. These resources are a major intervening link between occupational class and the formation of status groups, and they also provide for status groups formed along lines of age, sex, ethnicity, and special political, intellectual, or entertainment interests. There is some evidence on the process of acquaintanceship which shows that strong relationships depend upon similar attitudes about politics, religion, the local community, and mutual acquaintances (Newcomb, 1961; Triandis, 1960).

There are other resources one can trade in conversation, consisting not of what one has to talk about but the role one is willing to play. One can make up for a lack of conversational resources by being a good listener, a sycophant, an entertainer or adviser waiting patiently for a call to perform, a pet, or a butt for jokes. These are essentially the resources of weakness. There are also the resources of strength, which tend to come from outside the realm of conversational exchange itself. Power is the most important resource for commanding the conversational deference of others, whether it is the toughest streetfighter cowing his gang, the patrimonial prince surrounded by his court, or the modern politician or movie producer with his retinue of advisers, lackeys, and hangers-on. Other nonconversational resources that can be exchanged for conversational relationships include money, whether it is used to buy dinners, drinks, gifts, or a country estate or a yacht for guests to visit; one's good offices in providing business, political, or social contacts; one's information or expertise; one's reputation; and one's personal attractiveness or sexual favors. Whether an individual using these nonconversational resources will be able to buy himself a leading position as conversational star on whose words everyone hands, or merely admittance to a more popular conversational circle, depends on how his conversational and nonconversational resources stack up against those that others bring to the market.

Which particular conversational trades, then, will be made? One might expect that trades among conversational equals, in which mutual ego support is the basic principle, are the most common. Each individual would thus get the fullest return on his conversational efforts, without paying the price of having to humiliate himself in a demeaning conversational role, or to feel the strains of distrust in his reality bubble that come from depending on the unspontaneous support of a hired audience. Upper-class persons would thus tend to associate among themselves, as would the middle classes, working classes, and so on. Other resources distributed by age, sex, ethnicity, personal attractiveness, or talents would produce corresponding forms of conversational endogamy.

But this may be confined to very modern sectors of industrial societies. A little thought shows that egalitarian conversational exchanges will occur only under particular conditions: where the participants' nonconversational resources (money, power, etc.) are roughly equal; their conversational resources (topics of interest, tastes, acquaintances to talk about, etc.) are also equal; they have the opportunity to associate among themselves; and there are no motivations or constraints to make them trade off their conversational resources (e.g., humbling themselves) to enter a higher material realm or their material resources to get deference from lower classes.

One feature of any conversational market is the external constraints on who is accessible to whom. In traditional societies built around militarily powerful households, gross inequalities in power dominated the local situation. The prince or magnate surrounded by his sycophants, and the domestic patriarch surrounded by his children and servants, were the inescapable rulers of interpersonal exchanges. Urban, commercial, and industrial transformations mobilized larger numbers of persons with some degree of independence from powerful masters, and gave them enough free conversational and nonconversational resources to strike out on their own level. It is in these situations that conversational markets tend to move toward equality of exchanges. One result has been that the different social classes have become more segmented in their daily lives, whereas in traditional societies they were closer together but interacting in very unequal relationships.

Differences in the relative extensity of choice in a conversational market can be found within modern society as well. The work place with few employees limits each of them in choosing his acquaintances; it seems to be for this reason that closer friendships are found in the larger workplaces (Lipset *et al.* 1956: 176–185). Similarly, small towns (or restricted urban ghettoes) have limited conversational markets, in which the dominant figures are inescapable. This helps account for the greater feeling of stratification in such communities, as well as for one aspect of provinciality noted at least as far back as Flaubert's *Madame Bovary:* the stifling quality of rural "friendliness," born of having to keep up conversational relationships with almost everyone if one is to have much sociable contact at all, but in circumstances that allow for little choice of conversational topics of real individual interest. Modern transportation, especially the private automobile, has helped transform this into a more individual-choice conversational market.

Lacking this freedom, as in boarding schools and other total institutions, the conversational market offers little flexibility, and hence one finds more traditional patterns: cliques with dominants, followers, and scapegoats who have nowhere else to turn (Goffman, 1961b: 48–70; Weinberg, 1967: 97–126). Since children are relatively immobile, and hence thrown in with a small group of acquaintances in school and neighborhood over whose composition they have little choice, it is not surprising that childhood social structure is more tightly stratified than adults', or that the transition from the petty rankings of adolescent culture to the more freely chosen sociable situations of adulthood comes to many persons as a liberation.[21]

Given the size of the potential market, how many relationships will an individual sustain? The person with the greater resources will be able to carry off more contacts. This seems to be empirically so, if one accepts higher social classes as having more conversational (as well as nonconversational) resources than lower, or men more than women (Booth, 1972). But there is a limit on how many relationships any one individual can have. Time and energy are scarce commodities, and conversational relationships require a certain amount of both if one is to build them up to the more satisfying levels. Moreover, many close relationships (notably marriage, but also very close friendships) call for fairly extensive commitments; once they are established, the individuals are in effect "off the market" as far as establishing such relationships with most other people.

A marginal utility principle of conversational payoffs may operate here. The easier it is for someone to get certain kinds of conversational exchanges, the less he is concerned about getting more payoffs of this sort. But some conversational relationships, especially the more personal ties, call for considerable long-term efforts and obligations. Persons who stand highest on the conversational market, then, should be expected to tie themselves down the least; their conversations will consist more of the superficialities of discussion, gossip, or the more public forms of discourse (e.g., shop talk). This in fact seems to be the case with upper- and upper-middle-class talk. Persons who stand low in conversational markets, on the other hand, are more willing to tie themselves to a few very long-term relationships. There is evidence for this on a class basis: Working-class persons tend to value lifetime friendships more than the higher classes, and to confine their sociable ties to childhood buddies; and women, who

---

[21] The childrens' restricted choices with interpersonal markets is a case of low cosmopolitanism, according to the principles (2.2) introduced in the previous chapter. The predicted consequences, an outlook of particularism and symbolic reification, is one aspect of the early phase of "socialization." This line of analysis is developed in the latter part of Chapter 5.

generally have fewer resources than men in the larger conversational markets, confine their social relationships more closely to their relatives than do men at all social class levels, but especially in the lower classes.[22]

In general, the particular kind of conversation carried out will depend on what it costs the participants, in relation to what their market position makes them willing to pay. Different sorts of conversation vary in how much secrecy or trust they require, and in the degree of personal irreplaceability. The more complex types of exchanges—work alliances based on implicit mutual understandings; styles of joking, gossiping, and personal narrations that hit off the emotional exchanges of a complex form of mutual support—are relatively unique to a particular combination of individuals. Talk about relatively well-known practical or public matters, sports or popular entertainments, and the more superficial forms of talk about the activities of self and others, on the other hand, can be carried off among large numbers of persons on relatively short acquaintance. Again, we have the principle that persons who stand highest in the conversational market will go in for the more volatile forms of exchange. Those who stand low will be willing to tie up their resources in a few long-term investments. But those who stand very lowest may lack the resources to be able to carry off such a strongly cemented exchange. We have at least a two-factor theory. The maximum investment in long-term exchanges occurs at the intersection of two distributions: the distribution of opportunities for negotiating large numbers of relationships, and the distribution of resources available for use in negotiation.

The Sequence of Conversational Negotiation

What people will say to each other at any moment depends on what has been said before. This is true on a number of levels of causality. For it is prior negotiation that defines all situations, including the structure of organizations as well as personal relationships; it is prior conversation that creates resources to be used in the present; and on the detailed level, the microsequencing of words and sentences has a complex but predictable structure of its own.

Situations are socially created, not given by God. What makes something a work situation or a shopping situation is the prior agreement among people to act out a particular task or to offer goods for sale. A

---

[22] Gans (1962:229–262), Booth (1972), and Kohn and Schooler (1969) show that working-class men are both more concerned than middle-class men about social relationships and less receptive to change. This fits with the finding that men in the higher social classes have considerable turnover in their relationships (Babchuck and Booth, 1966).

business corporation, a state, a store, a family, each of these is a reality defined by the arrangements previously negotiated among its participants. In some situations, you have negotiated the relationships personally; in others, you enter a situation which has an external reality for you because other people have already defined it. The store or the office already exists as a social organization for the newcomer walking in the door, because there is a group already on the premises with a conversational history of their own.

Of course, every organization is tacitly renegotiated all the time, if only by agreement not to disturb previous definitions of reality. By the same token, an organization can dissolve almost instantaneously, if most of the people involved decide, perhaps because of duress, emergency, or revolt, to stop enacting it. Between these two extremes, organizations are renegotiated in slightly changing forms all the time. Other relationships, such as an informal alliance among bureaucrats or businessmen, or a particular kind of exchange among friends, are more explicitly ephemeral. Lacking the conventionalizing support of a larger public network to help enact it, such relationships depend very closely on the ongoing flow of negotiations among the most immediate participants. For a friendship or an alliance, the reality of the relationship is summed up by the history of personal encounters, and nothing else.

These are the endpoints of a continuum. Personal relationships become institutionalized in various degrees; the most explicit form is a public announcement of marriage. A club with formal membership procedures has the same effect of making the relationships public and hence upheld by other persons besides those directly involved. This depends, of course, on there being more than two members. Simmel's (1950: 145–69) emphasis on the importance of the number three can be read in this light. To a lesser degree, formally announced moves, such as inviting someone to dinner, giving a gift or a kiss, or in some cultures, calling someone by his first name, explicitly mark a move toward particular relationships. Once established, the public reality of the relationship constrains the individuals to live up to it, or face the consequences of recognizing a formal break. The real flow of conversational exchanges can thus get out of phase with formally believed-in relationships.

What kind of conversation takes place, then, depends on conditions that produce an entire sequence of negotiations. There are two main possibilities. One can strike up a relationship out of the blue. This kind of self-introduction is the most difficult step to make, and hence is confined to the least binding sorts of exchanges: a brief shopping encounter, applying for a menial job, passing a sociable triviality on the street. Stronger kinds of exchanges are built up from self-introductions with great difficulty, or through a lengthy process of gradually escalating exchanges. It is for this reason that propinquity fosters relationships only where people can expect to see each other again regularly.

One can also build on intermediaries who have relationships with others. By being introduced, given a letter of recommendation or a formal credential, or invited to someone's home for a party, one can widen his network of acquaintanceship. In effect, the exchange of trust already negotiated by one's sponsor becomes surety for entering into relationships with strangers on a deeper level of commitment than otherwise. Entering into a formal organization is a form of sponsorship; once one occupies a job that entails practical relationships with others linked in a formal network, it can be used as a base from which to extend one's range of practical and other conversational encounters. Most work and political careers, as well as many sociable acquaintances, are built up via such intermediaries.

Occasions are socially defined by what one expects to happen in a given period of time spent with other people. They are, in effect, time contracts. Work situations are contracts to be "on the job"—that is, available for practical talk at certain times and places. One important dimension of authority, as Elliot Jacques (1956) has shown, is the freedom one has to set the specific conditions of his time contracts. Persons with more authority are usually committed for more of their time, but they get to set the specific times and places they will be available much more flexibly than lower-level employees, who typically are on duty for set hours and at set stations. The lower occupational classes thus occupy subjective worlds that seem massively bounded by external forces, experiencing a sharper contrast between the "in here" where they are confined, and the "world out there"—which, after all, consists of nothing more than the subjective world of other people, most of whom are just as closed in as themselves.

Political and religious occasions also consist of time contracts, periods set apart to do ideological talk with a given set of persons.

Social or sociable occasions fall into two types. There are *explicit invitations* to sociability: parties, dinners, luncheons, "coffees," visits, dates, receptions, and so on. At these, people enter into a contract to have sociable conversation for a given period of time. There are also *casual encounters* among persons who have defined each other as sociable acquaintances of some sort. This means that they have entered into a long-term contract, whether explicitly or by implication, to carry on sociable conversation whenever they meet, unless extenuating circumstances (practical or public) intervene. As Goffman and others who have studied such encounters have shown, there is usually considerable ambiguity about just how long a period of time the long-term contract implies for the individuals to maintain sociability each time they meet. Especially when the degree of relationships is being renegotiated due to changes in the resources and motives of the participants, there may be uneasy moments centering around when and how to close off such casual encounters.

Together, the conversational contracts involved in these two kinds of sociable situations—explicit invitations and casual encounters—make up the content of different sorts of sociable relationships. Thus in modern America we have: (*1*) strangers—people who do not know each other, and have no communication of any kind except for the "vehicular relationships" involved in walking down the street, driving a car, or otherwise being physically proximate (Goffman, 1971: 5–8); (*2*) enemies—persons who reject any communication, any effort to construct a common reality, except that of emnity itself; (*3*) formal associates—persons who carry on practical or public talk as part of a work, political, or religious occasion; (*4*) close associates—persons who have negotiated a special alliance to carry out some practical or public matter, conversations consisting of politicking, sharing backstages in practical or public settings; (*5*) acquaintances—persons who talk briefly about sociable matters when they encounter each other; (*6*) formal friends—persons who can carry on some extended sociable conversational exchanges; ritual obligations for some degree of "grooming talk" (greeting rituals including sympathetic inquiries into each other's personal affairs); some explicit invitations to sociable occasions, usually fairly nonexclusive ones (large rather than small parties), following fairly strict sequences of reciprocity; (*7*) personal friends—persons who have negotiated conversational ties of some degree of uniqueness and privacy; invitations to smaller and more exclusive parties and social occasions; visiting without formal invitations; less concern for reciprocity of invitations and visits; warm greeting rituals and extensive farewell rituals surrounding absences; (*8*) close friends, intimates—strong conversational bonds, including intimate or highly guarded backstage matters; no social barriers to physical proximity, hence invitations, visiting, and reciprocity are not issues.

This is a set of ideal types of relationships, giving a correspondence between two ways interpersonal relationships are socially defined: the set of occasions people may contract for, and the actual conversational interchanges that take place. These can get considerably out of key, as anybody can testify who has attended a party at which people are not motivated or prepared to do the conversation implied in the time contract for the occasion. This frequently seems to happen at extended family gatherings, or other situations where the motives for being together are extraneous to the actual conversational ties. This happens where there is a great deal of social climbing, organizational maneuvering (e.g., the pseudo–*Gemeinschaft* of office sociability), or purely sexual concerns (as on dates); or when the spouses of friends are dragged along to occasions that make the couple the unit for sociability. It is here that one finds the greatest amount of perfunctory "small talk": the drawing out of routine details of personal biographies, commentary on externals like the weather, the scenery, or regional differences, all carried out with dogged determination rather than spontaneous interest. Practical talk is another resource often dragged in to fill the gap in the contracted flow of sociability.

Things to talk about come from two sources: either stored-up experience can be recounted or the conversational history itself can generate new materials. Stored-up experience comprises most of the basic resources, already discussed, that determine whether or not persons will be able to negotiate a conversational bond. But many conversational ties also depend on being able to generate new materials as conversation goes along, in the form of jokes, games, plans, or new ideas. Merely recounting the events of one's life, listing one's acquaintances, repeating one's opinions—this use of materials seems to be characteristic of perfunctory conversations, in which one remark leads to another only with effort, not because they attach to a motivational thread that pulls conversation along of its own accord. Self-generating conversation, on the other hand, makes use of materials to establish a mood, which leads on to a fresh reality within the conversation itself. It may even go over the same ground repeatedly, in the way that jokes, mock fights, or teasing routines develop much of their punch from the sheer cumulative impact of verbal play. As professional comedians know, much of humor comes from a play on the absurdity of repetitions; remarks become funny once the right mood of relaxed expectancy is established, which would not even raise a smile out of context. Self-generating conversation in general, whether humorous or not, takes its appeal from these sorts of emotional rhythms, resembling nothing so much as a musical composition.

Conversational relationships must be continually renegotiated, and it is not surprising that there is always a certain amount of change. Conversational resources get used up; interests and concerns change as new life circumstances open up new opportunities and narrow down others. Multiple sociable bonds (or networks of practical alliances) continually introduce strains to balance different conversational realities. All of these things result in a gradual tearing down and building up of relationships, keeping the conversational markets open, to a degree. Just how the negotiation is going is best indexed by nonverbal, emotional behavior. Fun, contentment, anticipation indicate that the conversational resources are well enough matched to produce the desired ritual bonds. Embarrass-

ment (Goffman, 1967: 97–112) indicate the reverse. Boredom is not necessarily based on a desire for new experience (as nonboring conversation can
be quite repetitive), nor on social isolation (since one can be bored in
company). It arises from frustrated expectancies for a certain kind of
conversational exchange to which one has become committed, perhaps
because of a faulty time contract, or of attempts at conversational negotiation which one's market position will not support. As such, boredom may
serve as a powerful tool for dissecting the underlying dynamics of conversational relationships.

There is also a microlevel of conversational sequencing to be considered. What is said at one moment depends on what has been negotiated
the moment before, and each exchange is a matter of gambits offered or
declined, utterances that lead onward to further talk or close out the realm
of visible possibilities. What close analysis we have of conversational
sequences (Labov, 1972; Schegloff, 1968; J. Watson, 1958) has moved
somewhere between the realms of linguistics and sociology, attempting to
show that there is a logic of discourse that may not be displayed in
individual sentences but that links these pieces together because they are
embedded in the larger purposes (asking for information, challenging
someone's membership in the social network, picking a mock fight) of the
conversation. Thus, it becomes understandable that someone will answer
a question with another question on certain occasions, or answer a challenge with a ritual insult. Openings and closings have their expectable
structures, which make sense when one sees them as rituals for refurbish-
ing agreements that might have fallen (or might eventually fall) into decay
during the time of separation; accounts and excuses are predictable from
the motivations conversationalists have to overcome any disruptions in
their ongoing construction of conversational reality. In sociable talk, we
find that topics are chain-linked in a sequence whose logic may be found
on the underlying level of search for materials that will sustain a mutually
rewarding exchange.

What all of this tells us, above all, is that the broader structure of
conversational markets and negotiations is what makes the details of talk
meaningful. The narrower realm of grammar is rarely coherent in itself,
but depends on the larger structure of social interactions and nonverbal
rituals in which it is embedded. These results promise that the possibility
of being able to explain in considerable detail what people will say to each
other from one moment to the next is now coming into view.

How might this be done? We need to know what conversational markets an individual has been through and what contingencies he now faces.
These produce a set of subjective resources and possibilities with which
he presumably maneuvers to arrive at the best next conversational posi-

tion. More concretely, we should be able to predict generally what he will say if we know: (*1*) the structure of authority and social density he has experienced in the realms of work, home, and other areas of social participation; (*2*) what kinds of conversational relationships he has had recently with the people he has encountered, and especially with the person he is talking to now.

The first set of determinants (detailed in Chapter 2) gives us his general outlooks on the world: the type of demeanor he attempts to carry off with others, the realms of discourse and accompanying self-images that he values most highly. This, by itself, narrows down a great deal what sort of conversations he is likely to have.[23] The second set of determinants tells us what his market position is for carrying out particular kinds of conversation, and hence his motivation to talk about particular things. It also tells us what kinds of things have been negotiated with the person he now confronts; one might expect he will pick up the same general types of conversation (practical, ideological, intellectual discussion, entertainment, gossip, personal) as well as more specific topics and levels of familiarity within these.

Of course, personal relationships do not necessarily remain static, and new ones may be negotiated as any conversation goes along. On this microlevel, the same sort of principles apply: We must keep our eyes on the resources and motivation at any given moment in order to predict what will happen next. On this level, the important thing to pay attention to is the nonverbal, ritual side of the encounter. Once a particular kind of ritual begins (playing word games or teasing, testing ideological loyalty, enacting personal ties by grooming talk, etc.), it tends to set up an emotional tone that sustains itself for awhile (the time periods have not been much studied yet) and keeps producing appropriate verbal material. Where the resources and motivations are not available for successfully negotiating a ritual, the resulting tone tends to disrupt the whole episode. Beginnings and endings of ritual interchanges, then, block out much larger chunks of materials that are relatively predictable within those bounds. A good grasp of the principles of microsequencing, together with information on the individuals' store of verbal information, should enable us to predict conversations even on a fairly detailed level.[24]

---

[23] This set of conditions empirically refers to the same conversational interactions we consider in the second set; we only pick out different aspects.

[24] The methodological problems are considerable, but they are beginning to be surmounted by using tape recordings from natural settings. The problem is that we need much more long-term materials than has been collected so far. It is strategically appropriate to use one's own experience, in whatever form of notetaking or recording is available (as I have done with conversational notes over a period of some months for material that went into the

The Effects of Conversational Interaction

Conversations determine most of our social experience. We live in the shadow of conversations, when we are not in the midst. Our thoughts are a kind of internalized talk between very generalized speakers and listeners. This is true not only in the sense that children learn verbal thinking by learning to make silent conversation within their heads; the content of what we think about, every day of our lives, comes primarily from the conversations we have just had or expect to have. Thinking is mostly a pale reflection of our more vivid experiences of reality, and reality for every man is enacted primarily in the words that pass between himself and others.

International politics, for example, is real (if not necessarily accurately so) to upper-middle-class people because they talk about it daily; it is a haze to lower-class people who do not. The worlds of work organizations, intellectual topics, forms of entertainment, the lives of other people are all real objects in people's minds only insofar as they are talked about, and are forgotten when there is no longer anyone with whom to talk about them. Even our own lives are part of our conscious (i.e., verbal) reality to the extent that we have conversational relationships in which they can be talked about. What old-fashioned sociologists used to call "norms" are nothing more than whatever points of agreement there exist in and around conversations. Strong moral commitments, whether political, religious, or secular, arise from well-staged conversational reiteration of dramatic realities. As Durkheim (1938, 1954) argued, it is in social rituals that values are sustained; in modern society, it is not so much full-scale public ceremonies which do this as it is the minature enactments of everyday conversations. Commitments to a particular political or intellectual position depend on having conversational partners (or at least correspondents) with whom one can reaffirm their reality; lacking this, they fade within a period of months.

I have proposed that motivation is a matter of subjective orientation: The individual moves to optimize his status rewards according to his view of reality. Since each person's reality comes mostly from conversation, we can use the conversational model as a mechanism for capturing the dy-

---

analysis of this chapter), especially to set the general model, and whenever the important issues are how to structure such long-term materials. Although this kind of data collection has obvious problems of representativeness and bias, there is no real alternative right now if we want to be fruitful rather than merely precise. In the networks of communication that make up an organization, we may have fewer methodological problems, especially in such an organizational structure as a scientific community (Chapter 9 in this volume), inwhich the interchanges are relatively slow-moving and the major ones are available on paper.

namics of social motivation in some detail. The different kinds of conversational milieux, from a structural viewpoint, are what we mean by stratification. The individual's successive motivations are the subjective counterpart of his career—the results of the set of conversations leading up to the present and moving him in a particular direction into the future.

Organizations are networks of people carrying on certain regularized patterns of communication. An individual's work career is a series of different people he meets, a series of conversations on different subjects. What we mean by moving "up" is a matter of how much practical talking, order-giving, and order-taking he does, and in reference to how large a network of people. "Classes" are categories of persons similarly shaped by such experiences. Status groups are communities of sociable acquaintance, formed along class as well as ethnicity, age, and other more specialized lines of common interests, who are held together by conversational exchanges of a more intrinsically pleasurable sort. Sociable conversation, because of its greater emotional commitment and apparent spontaneity, is the basis for most strongly held beliefs about reality.

There are careers in the sociable realm as well as in work. This is what is usually referred to as "social climbing." It has more than a casual significance. Organizational and political careers are mediated by networks of sociable acquaintance. Families, friends, friends of friends—it is through such contacts that people learn of jobs, business opportunities, political support, allies in practical matters. Sociable relationships have a broader effect, even where there are gaps in networks of direct acquaintance (Kadushin, 1966, 1968). People can become sociable because they are able to create a common conversational reality. This is easiest to do if they already inhabit pretty much the same subjective world. The impersonal standards of selection that operate most strongly in modern organizations generally embody the beliefs of powerful groups about what is the desirable reality that recruits ought to help them enact. In effect, people who control organizational recruitment into their own ranks try to choose people they can expect to be friends with. A social class is not a group in which everyone knows everyone else, but it is a pool of potential friends.

The effects of conversational negotiations are broadly cumulative. On the subjective side, successful efforts to establish desired conversational relationships increase one's self-confidence and give an emotional boost, making future successes all that much more likely. Failures create a vicious spiral in the opposite direction. On the objective side, position in the conversational market improves or declines in cumulative spurts. This is reflected in sociable popularity or in success in practical or public careers. Of course, something must start a run, and shifts in external circumstances can slow one down or bring a reversal. Cumulative processes frequently develop along specialized dimensions, as one becomes more and more able at a particular type of interaction, whether it be political, intellectual, practical, sexual, athletic, or religious. Increasing confidence and commitment to such a specialty usually leads to increased ability to create satisfactory conversational resources within it. The price of extreme specialization, on the other hand, is to lose the resources and contacts that would allow moves in other directions.[25] To the extent that the distribution of resources allows such specialization, the society tends to become a set of mutually oblivious enclaves of conversational reality.

In a previous chapter, a visual image was introduced to remind us that stratification is enacted by real people in real places. This was the imaginary aerial time-lapse photograph, in which lines of interpersonal structure appeared as streaks of light marking heavily traveled social pathways.[26] Looked at in more detail, each individual might appear as a chemical atom surrounded by concentric rings of conversational relationships. Individuals are bound to each other like atoms sharing electrons in common orbits, ranging from the inner ring of intimacy to the outermost ring of limited practical relationships. The orbits are conversational situations that are jointly enacted. What we mean by an individual personality, hypostatized indications of which show up on personality tests, is based on the set of these relationships. Close sociable relationships make up the centers of maximum solidarity in the social structure, and these centers are linked with others through the intermediary of overlapping and successively more distant relationships with wider networks of individuals.

Social classes are broad groupings of potentially close ties, within the

---

[25] This seems to be found most in the upper middle class, whose associational boundaries are sharpest, work careers most specialized and regular, and attitudes most favor specialization rather than breadth (Laumann and Guttman, 1966; Kohn and Schooler, 1969; Wilensky, 1961; Landecker, 1960).

[26] Compare Leonard Bloomfield's statement (1933:46):

Imagine a huge chart with a dot for every speaker in the community and imagine that every time any speaker uttered a sentence, an arrow were drawn into the chart pointing from his dot to the dot representing each one of his hearers. At the end of a given period of time, says 70 years, that chart would show us the density of communication in the community. . . . We believe that the differences in communication are not only personal and individual but that the community is divided into various systems of subgroups, such that persons within a subgroup speak much more to each other than to persons outside their subgroup. . . . Subgroups are separated by lines of weakness in this net of social communications.

Evidence that class or caste lines, conversation contacts, and linguistic differences coincide is provided for India by Gumperz (1971), and references to less careful studies of the United States contained therein.

different power levels of the larger structures of practical relationships. This is what first strikes the eye in viewing social interaction from afar. A finer lens shows many other networks, some within class lines, some crisscrossing them. Depending on the particular society, we would be able to see groups formed along age and sex lines (male and female, married and unmarried, as well as more minute subdivisions corresponding to different levels within these conversational markets); groups upholding special religious, intellectual, and entertainment cults; as well as ethnic and racial lines of division. One task of general theory will be to formulate the conditions under which different varieties of networks form, making one historical society distinctive from another.

This is for later chapters. For the present, in order to make the link between conversation and stratification as real as possible, let us take a look at a descriptive summary of classes in the modern United States (drawing upon the materials given in Chapter 2), viewed as conversational groups.

Lower-class people have very few gatherings explicitly designed for sociable conversation, nor do they spend much time in formal organizational roles calling for practical talk. Conversation takes place in relatively unstructured encounters, often in hanging around public places. There is little privacy, hence fairly intimate matters can be discussed without the usual taboos of politeness. But for this very reason, these are not matters around which strong personal bonds are negotiated. There is widespread distrust, which prevents emotionally significant personal narrations. Bantering, joking, verbal contests make up much of the talk, at least in the black lower-class subculture. This seems to be suited for relationships in an atmosphere of distrust, since joking provides an ambiguous medium allowing much expression of hostility, and guards against wounds to the ego when social supports fail. Lower-class conversational styles, and the personalities and outlooks that derive from them, are a major part of the barrier that seals the lower class off from the more powerful networks of organized society.

Working-class people are less likely to congregate in public places like streets; they form sociable circles meeting habitually at home, in bars, or places of recreation. Men and women have separate groups, usually confined to a single generation. There are few explicitly sociable occasions based on formal invitation, except for family gatherings. The main exceptions are the larger celebrations surrounding traditional family events such as weddings. Conversation usually revolves around practical matters (autos, household repairs), sports and banter about personal prowess in the case of men, and household and family matters, and especially health, in the case of the women. Working-class people do not appear to talk very much, and sociability often consists of sitting together before the television. But this may be confined to certain ethnic groups; black Americans are much more sociable and verbal than other working-class cultures.[27]

Lower-middle-class people, in the more traditional sectors, tend to restrict their social contacts to relatively small groups of family and relatives. Situations such as visits to relatives are explicitly sociable, with much emphasis on ceremonial proprieties such as offering cake and coffee. Sociable talk is rigidly controlled unless the participants are quite intimate, with a strong emphasis on small talk (the weather, innocuous news about each other's doings). Among intimates, talk heavily emphasizes gossip within the circle of relatives.

The modern *middle-class* people who inhabit the large national bureaucracies, on the other hand, are explicitly sociable, especially in the post-World War II generation. There is considerable emphasis on daytime socializing: women's coffee klatches and bridge clubs, men's luncheon clubs and business contacts. In addition, there is much emphasis on belonging to community organizations and having their children involved in them, on participating in sports as sociable activities, and to a lesser degree on entertaining personal friends. The working-class social organization around one-generation, one-sex peer groups is displaced by a nuclear family society in which sociability is carried out in husband-and-wife teams, frequently with the children in attendance. This is the world of the PTA and the Little League. Conversation is explicitly affable. Serious

---

[27] Just what constitutes the difference between working- and middle-class cultures on the level of verbal practices is now becoming the subject of much empirical research and theoretical controversy; see Bernstein (1972), and Labov (1972c). Bernstein argues that the working class uses a restricted code of concrete utterances, the middle class an elaborated code with more abstract, context-independent and planful use of language. Labov counters that working-class language is just as logical and flexible (or perhaps more so) in its own contexts, particularly when one understands that discourse is a form of action in which words are only a part. The argument tends to be weighted around issues of ideological bias in the schools. Another caution is that Labov draws his evidence from black American children, from a culture which is far more verbal than most white working-class cultures. The argument takes on a somewhat different complexion if we regard verbal styles not as a matter of socialization, but as continually negotiated in particular sorts of situations. The finding that middle-class language deals with more abstract and long-range matters fits with Labov's contention that it is verbose and redundant rather than efficient; middle-class people spend much time in ideological and intellectual discussion, where the nonverbal side is stifled or made into a rigid stance characteristic of enacting authority. Working-class conversation seems to contain much more nonverbal ritual, with pithier phrases carried along by an active physical stance. This would be especially true of the verbal games which make up so much of the black culture Labov studied. Middle-class people engage in more purely verbal ritual; working-class people use language as part of more expressive nonverbal rituals. Which of these is considered more effective is primarily a matter of value judgments.

gossip and important personal narrations tend to be avoided on sociable occasions (although reserved for more intimate settings) in the interests of avoiding possible disruption. There is a well-developed flow of small talk, discussions (kept from becoming too "serious" in order to head off arguments) about local affairs, with practical matters filling in gaps in the front of sociability. It seems to be the strong normative pressures to carry on continuous sociable conversation of this sort that causes intellectuals to castigate the "conformity" of suburbia. Gossip, personal narrations, and intimate talk also occur in the private settings of the nuclear family and with close friends. Middle-class sociability thus contains a more differentiated range of relationships and situations than the sociability of any of the lower classes.

Upper-middle-class people tend to lead lives in the professions or higher managerial positions that cause work to spill over into sociable life. The backstage of the practical world, in which politicking is carried on, tends to merge with sociable occasions. Upper-middle-class people accordingly have a great deal of sociability, although much of it is sociable only in form. Sociable occasions tend to be somewhat impersonal, such as large cocktail parties or frequent dinner parties given on a basis of strict reciprocity. Conversation tends to involve much politicking, much shop-talk discussion. But the uppermiddle class is also infused with various cultural and intellectual hierarchies; hence, much talk is discussion of areas of cultivation and knowledge, public affairs, and whatever are considered the refined forms of entertainment at the moment. Gossip on such occasions usually concerns the backstages of work and public affairs, not expressly personal narratives. Because of their positions of control in occupational affairs, upper-middle-class persons are highly skilled at manipulating conversational forms, and can be less rigidly formalistic than the lower rungs of the middle classes. Even more so than middle-class talk, upper-middle-class conversations are highly differentiated, with a whole series of conversational front- and backstages to inhabit. Especially because of the vogue of psychoanalysis and more recent clinical psychologies, upper-middle-class people favor having some relationships in which they can speak about personal and intimate concerns. If necessary, they will pay a specialist $35 or so per hour to have an opportunity to do so.

Upper-class sociability tends to break up the husband-and-wife "couples society" found in the middle class into separate men's and women's sociable groups. There is a great deal of formalized sociability, especially for women: dances, debutante balls, "the season" at musical events, charity shows, teas, as well as formal parties and dinners, and entertaining guests at vacation homes. For men, there is a highly exclusive world of business meetings and clubs. Upper-class conversation places considerable emphasis on discussing the round of upper-class sociable activities themselves. There appears to be a distinctive form of upper-class gossip which consists of talking about family genealogies, which operates to exclude the nonelite. Among the unmarried younger set, there is the similar practice of exchanging common acquaintances at great length, called "playing 'Do you know ____?' " Talk about practical matters such as money is considered highly inappropriate for sociable settings, and is reserved for tightly guarded backstages of family councils and private men's clubs.

APPENDIX: SUMMARY OF CAUSAL PROPOSITIONS

It is of some value to put the main causal statements into a formal propositional summary. I do this not because I am convinced that all of these propositions are correct, complete, or necessary, or that any of them is precise, but simply as an effort to move in those directions. The lack of this sort of explicitness has been one of the reasons the classical theories have occasioned so little cumulative development and their causal statements have been so easily buried in their ideological or aesthetic implications. I have aimed at putting everything into causal or at least correlational form, and showing what general principles relate the more particular statements to each other logically. This enables us to work at theory-building empirically, to see just which part of the theory is already supported by evidence, as well as where evidence is lacking. Tying the whole together through general principles is also important, for this is what gives us the confidence that particular formulations are correct and that new ones can be derived. The name of the game, after all, is to be able to argue with maximal conciseness and power in the widest range of situations. What follows here is a sketch; the connecting links in particular are only a beginning.

Chapter 3 has been an attempt to ground the principles of stratification given in Chapter 2 in the dynamics of face-to-face encounters.

POSTULATES

IX. All animals have automatic emotional (hormonal, neural, postural, gestural) responses to certain gestures and sounds made by other animals.

X. The basic social ties among animals consist of the mutual arousal of alarm signals, recognition and affection signals, sexual arousal signals, antagonism signals, play signals, and of the asymetrical arousal of threat and deference signals.

XI. Human beings are animals, and human social ties are fundamentally based on automatically aroused emotional responses (from IX and X).

XII. Human beings have the capacity for symbolism (invoking the unseen past, future, or abstract notions representing combinations of many different experiences) by using images and/or sounds to represent them. (**N.B.** The capacity for symbolism is not equivalent to the ability to make certain sounds.)

XIII. *Cicourel's multimodality principle.* All human communications take place in several modalities (visual, aural, emotional) simultaneously.

XIV. *Garfinkel's indexicality principle.* Social interactions can be carried out smoothly to the extent that mutually accepted implications do not have to be verbally explicated (from XIII).

PROPOSITIONS

Now let us try to derive Durkheim's theory of ritual solidarity.

4.1 The longer human beings are physically copresent, the more likely automatic, mutually reinforcing nonverbal sequences are to appear, and the stronger the level of emotional arousal.

4.2 The greater the number of human beings who are physically copresent, the more intense the emotional arousal.

4.3 The greater the common focus of attention among physically copresent human beings, the more likely they are to experience a common emotional arousal or mood.

4.4 The more that people use stereotyped sequences of gestures and sounds, the greater the common focus of attention.

4.5 The more that people use stereotyped sequences of gestures and sounds, the more likely they are to experience a common mood (from *4.3* and *4.4*).

4.6 The stronger the emotional arousal, the more real and unquestioned the meanings of the symbols people think about during that experience.

4.7 Durkheim's social density principle. The longer people are physically copresent, and the more they focus their attention by stereotyped gestures and sounds, the more real and unquestioned are the meanings of the symbols people think about during that experience (from *4.1, 4.2, 4.3, 4.5,* and *4.6*).

This derives *2.1* (mutual surveillance) and *2.2* (cosmopolitanism) in at least a general sense. Obviously these are only sketches, not rigorous proofs. The reason two propositions were made out of this one is, first of all, because the empirical materials of class cultures based on different community structures seem to vary along two dimensions; and, second, because there are two main variables in *4.7*—the amount of physical copresence, and the amount of common attention (versus diversity of attention).

Notice that propositions *4.1–4.5* apply to any animals, and that *4.6* and *4.7* simply draw out the consequences for human symbolism. Notice also that Postulate II (individual cognition is constructed from social communications) is hardly primitive, but derives generally from *4.7*.

It is implicit that human beings can deliberately create rituals, using prior symbolic realities as devices for orienting their own thinking and communicating their intentions to others. Presumably, people will use rituals as they feel an advantage in them (cf. IV–VI). This enables us to formally derive certain propositions of Radcliffe–Brown, Goffman, Malinowski, and Durkheim.

4.8 The more the conditions for strong ritual experiences are met (given in *4.7*), providing they do not produce mutual antagonism or asymmetrical threat–deference arousals, the greater the interpersonal attachment and feeling of security (limitations from Postulates V and VIII).

4.81 The greater the previous ritual solidarity and symbol reification, the more painful a change.

4.82 The greater the stress, the more incentive to invoke ritual solidarities.

4.83 Malinowski's principle. The greater the physical danger, the greater the likelihood of invoking prior rituals of solidarity (from *4.8* and *4.82*).

Notice the interaction of this with *1.8* and *1.5*.

4.84 Radcliffe–Brown's principle of ritual transition. The stronger the ritual ties within a group, the more that entries and departures of individuals from participating in those rituals are handled by ritually enacting the transition (from *4.81* and *4.82*).

Radcliffe–Brown applies this to births, deaths, and marriages; Goffman to the greeting and departure ceremonies of everyday conversation.

4.85 Durkheim's principle of ritual punishment. The stronger the ritual ties within a group, the more that violations of ritual procedures are met by spontaneous outrage and by ritualized punishments (from *4.81* and *4.82*).

The deliberate use of rituals to bolster authority is treated in Chapter 4. Here it may be useful to define Weber's legitimacy types in terms of ritual interactions.

Traditional authority is a type of deference relationship in which the order-giver uses highly stereotyped gestures and verbal formulas, with the result that the symbols of authority are highly reified and emotionally compelling. (The result follows from *4.5* and *4.6;* cf. *2.2.*)

Rational–legal authority is a type of deference relationship in which the order-giver freely improvises much of the contents of communications, and little attention is paid to the surrounding postures, with the result that symbols are regarded as human enactments with little emotional compulsion. (The result follows from *4.5, 4.6,* and *2.2.*)

Charismatic authority is a type of deference relationship in which the order-giver produces a strong focus of attention purely by means of the contagiousness of his personal mood, based on the strength of his nonverbal communications, with the results that his symbols are received as personal revelations (cf. *4.3*).

It should go without saying that these are definitions of ideal types, which empirically may be found mixed in various proportions. These are modes of face-to-face encounters and may be highly ephemeral, unless work is put into arranging the conditions under which they occur. Among these are the conditions under which deference will actually be given to a particular person, which in most cases ultimately refers us back to the organization of physical threat. This is described in Chapters 6 and 7.

Predicting Conversation

Conversation is treated here as a kind of negotiation of ritual ties and of the cognitive realities they sustain.

5.0 Who will talk to whom about what is determined by the chances of physically encountering each other and by the previous conversations they have had with each other and with others. These previous conversations determine individual motivations by giving each person a sense of his market position, his favorable opportunities, sunk costs, and problems of extricating himself from announced intentions, as well as by shaping attitudes and creating resources for use in conversational negotiation. These sequential determinants operate both in the long run and in second-by-second microsequences; in the latter, the emotional dynamics of particular rituals are determining.

5.1 The more often the same people physically encounter each other, the greater the likelihood they will talk. (This is another version of *4.1*.)

5.11 The more that work relationships bring individuals repetitively into contact, the more likely they are to talk about subjects beyond the immediate content of work, *provided* that they are not in an authority relationship (limitation from VI).

5.12 The closer people live to each other, and the more their lines of movement cross, the more likely they are to talk.

5.13 The more that people take recreation in the same places, the more likely they are to talk.

An *organizational position* is a repetitive set of behaviors, sometimes including occupying a particular physical space and doing physical work, and always including certain repetitive practical and/or ideological conversations.

5.2 Occupants of organizational positions vary in the kinds of practical, ideological (and sometimes intellectual) talk they do, and hence in their outlooks and attachment.

Power is the experience of giving orders and receiving ritual deference. The amount of power is indexed by how many people defer to one (including the cumulative number of persons who defer to one's subordinates), and in turn how many others one must listen to authoritative talk from. Notice that I have defined power in terms of deference (D-power); the contrast with the notion of power as efficacy (E-power) is treated in Chapter 8.

5.21 The more power one has, the more one talks officially.

5.211 The more an occupational position involves negotiating personal contacts, the more its occupant talks.

5.212 The more often one talks, the more skilled one becomes at expressing the particular kind of content and accompanying mood.

5.213 (The *argot* principle.) The more experience one has in a particular kind of position, the more specialized and esoteric one's communications with surrounding positions becomes.

5.214 The more complex one's job experience, the more specialized and esoteric the communications with surrounding positions.

5.215 (The politicking principle.) The more that men who talk much encounter each other, the more subtle the negotiation that can be carried out among them (from *5.212–5.214*).

5.22 The more people one gives orders to and the more unconditional the obedience demanded, the more dignified and controlled one's non-verbal demeanor.

This follows in part from *4.6,* since order-giving is an effort to maintain a particular kind of strong mood by using certain stereotyped gestures. It is in effect a microlevel derivation of *1.1.* Propositions *1.2–1.4* can be similarly derived.

5.221 The more august one's public image, the more one guards backstage situations in which one's postures are relaxed.

5.23 The more one gives orders to others in the name of political, religious, organizational, or moral ideals, the more one is committed to those ideals (from V and VI).

5.231 The more one receives orders in the name of political, religious, organizational, or moral ideals, the more one is alienated from those ideals (from VI and VIII).

The foregoing propositions derive another aspect of *1.1* and *1.3.*

5.232 The more one talks about ideals with equals who agree with one, the more committed one is to those ideals.

This proposition refers to ideological talk off the job, as well as to the egalitarian work exchanges in *1.3.*

5.3 The more unique and irreplaceable a conversational exchange, the closer the personal tie among the individuals who can carry it out.

5.31 The more trust and secrecy involved in a conversational exchange (i.e., the more cost if the confidence is not kept), the closer the personal tie if it is successfully carried out.

5.32 The greater the penalty if an offer to do a conversational exchange were rejected, the closer the personal tie if it is not rejected.

5.33 The more that people share common outlooks deriving from their previous social experiences, the more they are able to engage in rituals of mutual entertainment by reenacting enjoyable experiences or playing conversational games.

5.331 The more that people share common outlooks, and the longer they have talked to each other, the more easily they can carry off conversational exchanges in which literal verbal meanings are untrue (e.g., joking).

5.34 The more that people know about common acquaintances, the more likely they are to have something to talk about.

5.341 The more that people know about common acquaintances, and the more they share common values, the more likely they are to enjoy gossiping.

5.35 The more that individuals have previously committed themselves to exchanges of personal concerns, the more they expect each other to carry on such exchanges.

5.351 The more that individuals have committed themselves to personal exchanges, the more anger or resentment when an exchange is not lived up to.

5.4 The more subjective status one receives from a conversation, the more one attempts to carry on that kind of conversation (from VI).

5.41 The more one encounters others of unequal status, the more the conversation is limited to superficial exchanges involving little interpersonal commitment (from *5.4*).

5.42 The more things people have to talk about with each other, the longer they can talk.

5.421 The more nonconversational resources one has (ability to give gifts, do favors, or confer status), the more conversational and nonconversational services can be demanded in return.

5.422 The more conversational resources one has (common practical concerns, common ideological interests, common tastes in entertainment and consumption, common acquaintances, and previous negotiations of trust), the more conversational and nonconversational services can be had in return.

5.43 The more conversational and nonconversational resources, the more one can choose with whom to converse and about what; conversely, the fewer the resources, the less choice.

5.431 The more people one encounters, the greater the control over with whom to converse and about what.

5.44 Conversational endogamy. The more the choice of conversational partners includes those who have equal conversational and nonconversational resources, the more likely conversational exchanges are to be carried out among equals, and the more likely exchanges are to be sociable and close.

5.45 Social climbing. The less closely the conversational and nonconversational resources of individuals are correlated, the more conversational exchanges occur among persons who are unequals in certain respects.

In combination with *5.41,* this implies that such conversations are likely to be full of superficial "small talk."

5.46 Marginal utility. The more opportunities for conversational relationships of a given type one has, the less motivation to seek more such relationships.

5.461 The more trust required in a conversation, the more effort it requires.

5.462 The more personal commitment in a conversational relationship, the more time and effort it requires, and the fewer other relationships it allows.

5.463 The more conversational resources one has, the less of them one invests in exchanges of high trust or intimacy (from *5.43, 5.46–5.462*).

The *definition of the situation* is what is remembered about previous conversational and nonconversational ritual negotiations among the persons present at a particular time and place, and what expectations they construct at that moment for future negotiations.

5.5 Publicly announced relationships. The more ritually an ideal symbolizing an interpersonal contract to carry on certain types of exchanges is announced to third parties, the more reified and constraining the ideal relationship becomes for the participants (from *4.7*).

5.51 The more ritually a conversational contract is announced between two persons, the more reified and constraining the ideal, but less so than if it is announced to third parties.

Conversation and Stratification

5.61 The closer people work together, the more likely they are to become friends (from *5.11, 5.34, 5.341,* and *5.42*).

5.62 The more similar people's occupational experiences, the more likely they are to become friends (from *5.2–5.231, 5.33, 5.41, 5.241,* and *5.44*).

5.63 The more similar their occupational positions, the more similar their incomes, and hence the more likely persons are to live in the same neighborhoods and have similar recreations, and to become friends (from *5.12* and *5.13*).

5.64 The more divergent the experiences of different age, sex, ethnic, or recreational interest groups, the more likely are friendships to be confined within these lines (from *5.3, 5.33,* and *5.341*).

Conversation and Thought

5.7 What an individual thinks about at any moment is determined by the conversations he has just had and looks forward to having.

Chapter 4

A Short History of Deference and Demeanor

Joan Annett and Randall Collins

> Many Gods have been done away with but the individual himself stubbornly remains as a deity of considerable importance. He walks with dignity and is the recipient of many little offerings. . . . Because of their status relative to his, some persons find him contaminating while others will find they contaminate him, in either case finding that they must treat him with ritual care.
>
> *Erving Goffman (1967: 95)*

Some of the most important developments of the Durkheimian tradition have come from W. Lloyd Warner and Erving Goffman. Warner linked the theory of ritual solidarity to stratification, showing that a modern community can be viewed from the same pith-helmeted perspective as a primitive society. Warner (1959) turned his attention to the ritual procedures of Protestant religion in Newburyport and its idealized support for the family structure; to the cemetery as a world of the dead continuous with the living, and a repository of sacred objects around which upper-class traditions could be maintained; and to the effects of patriotic ceremonies in reinforcing political domination. At the same time, Warner carried along the weaknesses of this anthropological approach: an excessive reliance on a few atypical informants ("the oldest member of the tribe"), as well as a functionalist bias that overstates the unity of the community structure and diminishes the importance of economic and political power. Although Warner's life work dealt mostly with stratification, he failed to forge tightly enough the link between Durkheimian theory and stratification, to show *which* groups are most attached to particular ceremonies and what advantages they derive from them in dominating other groups.

Goffman (1959, 1963, 1971) picked up his teacher's emphasis on the importance of rituals in modern society, while carrying us forward to new stratification-related materials and to the finer focus of analysis necessary to understand the full sense and significance of ritual activity in modern society. To Goffman we owe the insight that the long strings of obligatory rites that are normally associated with primitive societies have not disappeared: they remain on a less obtrusive scale in the ceremonies of day-to-day interaction. Man himself has become a little god, the *object* of ritual offerings as well as conveyer.

Goffman shows how the bourgeois emphasis on polite manners is a constant illustration of this principle. Deference is given to others in the form of a mutual exchange, allowing each individual the room to construct and uphold his own idealized image. Demeanor practices such as wearing the proper clothes, using the proper expressions and tone of voice are forms of respect given, as well as demanded, from others. A web of obligation and reciprocal contingencies is woven by the overlap between deference and demeanor: one is limited in the deference he can claim by the demeanor he is willing and capable of presenting. Social relationships are thus seen as rituals in which participants work together to uphold an idealized reality. Goffman here applies Durkheim's sociology of knowledge to everyday life.

Goffman's (1959) full-scale model of modern "interaction ritual," as he later termed it, is given in *The Presentation of Self in Everyday Life.* He uses a theatrical model to characterize the process by which individuals construct the world about them and their place in it: there is a frontstage of idealized performances and a backstage where preparation and recuperation occur. Throughout, he emphasizes the mutual aid between performers and audiences: the hostess is allowed to be gracious and unruffled; the dignitary's image is upheld by his courtiers.

Goffman's work is especially important because he incorporates the mainstream of empirical sociology, effecting a synthesis of Durkheimian theory with the research of the Chicago school. The frontstage–backstage model, for example, comes from a combination of two sources—studies of informal groups in work organizations, and of occupations attempting professionalization. In all of this, Goffman interprets the materials of organizational power relations in terms of the Durkheimian perspective of the ritual construction of moral realities. Rituals are thus found at the basis of class stratification; Goffman also deals with status groups when he shows that similar principles apply to the polite rituals of sociability and entertainment; visits, diplomatic receptions, grand balls, vacation fun, tea parties, and even the brief ritual greetings of acquaintances meeting on the street. This is an elaboration of the point that first Durkheim, and later Weber, were to appreciate: Raw coercion and material wealth must operate within a substrate of emotional ties and common constructions of subjective reality. This is what the "collective conscience" meant for Durkheim and "legitimacy" for Weber.[1]

Goffman's work is a powerful extension of this line of analysis, bringing it in line with contemporary phenomenological concerns with the everyday construction of reality. Goffman shows that moral solidarity and reality construction are microscopic and ephemeral things, not a property of some abstract system as a whole. Every time men come together, they must once again ritually reconstruct their ties. This does not prejudge the question of how much one set of encounters determines another; the linkage of situations that makes up institutions is an empirical matter to be investigated, not the straitjacket of an a priori given. In this way, the complexities of the historical view of social structure can be incorporated, as can the conflict view. Although Goffman himself does not take the step, his model shows us clearly enough the resources and maneuvers, the bargains and reciprocal contingencies, that determine who will be able to get away with what degree of power and prestige in the encounters of work and sociability. Legitimacy is not an absolute, but something to be continually bargained for.

The major weakness of Goffman's work is not so much its lingering functional bias as its obliviousness to historical context. Goffman tends to select those instances that illustrate most nicely the ritual character of life, hence his emphasis on the upper and upper middle classes in their most formalized period. He concentrates on diplomatic receptions and military parades, and attributes the refinements of hat-doffing and polite address to an amorphous "middle class" that is actually best represented by the British and Anglo-American higher classes of pre-World-War-I vintage. But great changes have occurred in the twentieth century, not to mention earlier. There has been a status revolution, at least in America about the time of World War II, in which the decline of old-fashioned deference and demeanor was a central part. David Riesman (1950) attributed this to the decline of the "inner-directed" personality and the rise of the "other-directed." This kind of popular psychoanalytical characterology will not do; Goffman's analytical method can give us a better grasp. The further revolution in casualness of the 1960s—variously attri-

---

[1] Durkheim and Weber do converge on this point, as Parsons (1937) observed; Parsons simply read the convergence at the wrong level of analysis, assimilating Weber's atomistic and historical conflict theory to Durkheim's (and Pareto's) idealized functional version of the social system. The reverse is more nearly appropriate.

buted to generational conflict, drugs, and politics—can be seen as yet another change in resources for defining personal status. Earlier revolutions in personal relationships—of the nineteenth-century bourgeoisie, of the seventeenth-century courtly aristocracy, and others—were perhaps even more important.

Goffman does not catch these because of his functionalist preoccupation with the highly general rather than the particular. But Durkheim himself made the crucial methodological point: Sociology as a science must be concerned with variations because it is only by comparing the conditions where something occurs with those where it does not can we reliably test and refine our explanations. Goffman's work begins from an observation that Durkheim explicitly situated in a historical context: Modern individual man is surrounded by a halo of respectability formerly reserved for the gods of the tribe. Durkheim's logic is clear: Gods represent social organization, and the densely packed social structure of tribal society is represented by the powerful and repressive presence of collective spirits. In a society with a high division of labor, the collective reflection of the social structure focuses on the individual himself.

Durkheim's formulation is a beginning, not the end of our research efforts. It is too casual an observation of modern society. It needs to be located explicitly in the class structure; at the time that Durkheim made it, the halo surrounded the respectable bourgeois but not the worker. "Modern" is always an historically relative term; 70 years later, we can see that much of the halo has dropped still further down the social scale, and changed its quality at the same time. The awesome demeanor of the bourgeois individual is not so common in the last few decades in America. The ritual forms that used to convey respect and the expectancy that it would be reciprocated (as the formal styles of addressing one another) now convey disrespect among many Americans, because intimacy and informality are what is desired. Physical trespasses, such as slapping another on the back and relating sexual jokes that comment on one's intimate behavior, become recognized signs of friendship. European and other societies have undergone changes of their own, although perhaps less striking ones. And Durkheim's observation has a past; in the nineteenth century the upper middle class acquired the halo formerly reserved for the aristocracy, but a new and distinctively more egalitarian halo. This in turn was prepared for by a transitional period in the courts of the absolutist states, themselves marking a status revolution from the patrimonial relations of medieval households. Behind this stretches still more history.

To explain deference rituals, we must look for the resources that enable persons to idealize themselves by controlling the stage-settings upon which social reality is enacted. These are always the subject of at least implicit conflict. The general principle is that individuals maximize their dominance, including the subjectively constructed realities of their social communications, according to the resources available to them. These include the resources to seek out favorable contacts and evade unfavorable ones, as well as to make others stay around as part of their scene even in subordinate roles.

Some principles of stratification were summarized in Chapter 2. There they were applied to occupational relationships (under which political relationships were temporarily subsumed),[2] showing the effects upon one's world view resulting from giving or taking orders, and from various amounts and kinds of social contacts and communications. These principles tell us, in a somewhat different guise, the conditions that produce variations in the strength of rituals, or in Durkheim's phrase, in the strength of the collective conscience. As argued in Chapter 3, a state of religious intensity is produced by a combination of: people being in each others' physical presence, i.e., a high state of mutual surveillance; people communicating together intensely by means of highly stereotyped gestures and symbols, equivalent to the principle that a high degree of communications produces conformity to the external forms prevalent in the community; a common focus of attention, without side involvements, equivalent to a low degree of communications diversity or cosmopolitanism, resulting in a strong identification with the local group and reification of its symbols. Substituting other values of these variables, we get different less formal, weaker, or more relativistic states of social consciousness. Cutting across these differences in social density are differences in authority; depending on the form of stratification, we find communal egalitarian rituals, asymetrical (subordination) rituals, and mutual deference rituals.

If we wish to explain the differences in rituals and related status beliefs throughout history, these are the relevant principles. But occupational and political relationships are only part of the explanation. The same principles apply in the realm of the home and the family, the residential community, and the circle of sociable acquaintances—all the things that make up the sphere that Weber called "status groups." These *independently* influence deference styles, particularly as they become organized distinctly from occupational relationships; in the industrial era, it is the form of community organization more than occupational relationships that ex-

---

[2] Property is yet a third form of power, which may be separate from occupation or political power although it is based upon them at least historically. A complete model of the determinants of class cultures should include this set of relationships in which one can also have experience at giving or taking orders in the realm of buying services and regulating the use of possessions.

plains a significant portion, perhaps the greatest portion, of deference and demeanor. But community organization is important throughout history, although the two realms have often been fused; for purposes of simplicity, it will be easier to outline the main historical shifts by concentrating on changes in the structure of household and community.

The same principles of stratification apply to household and community relationships as to occupational relationships. First, there are variations in the structure of authority. Within the household, the relations between men and women, adults and children, older and younger siblings have varied from extreme subordination to relative equality; and some households have had many servants and retainers, others few or none. Just as within the occupational world, the home can include several social classes with their own cultures: the ritually dignified and arrogant outlook of the masters, the repressed rebelliousness or sullen apathy of the order-takers, the narrow rule-following of the middle levels (such as higher servants or, often, women), or the informality of equals. Outside the household as well, stratification exists among friends and sociable acquaintances, depending on whether the distribution of resources enables people to single out their equals or forces them to associate with others who outweigh them in terms of wealth, power, or the capacity to shine in games, conversation, or sexual attractiveness.

Second, there are variations in social density: the degree to which people are in each other's presence and hence under each other's surveillance; and the diversity or cosmopolitanism of communication that goes on. These determine the degree of conformity or individualism, the amount of emphasis on formal rules of communication, and the degree to which outlooks are relativistic, abstract and subtle, or simple, reified, and localistic. These are affected by the ecology of the household and community. Where households are crowded and close together in an isolated village or manor, surveillance is high and cosmopolitanism low; the lone household of the farmer, the crowded urban tenement, the private apartment in a large city all vary in some respects among the several dimensions of social density. The form of household and family organization is one set of determinants of deference rituals; various kinds of rural and urban community structure is another. The technology of transportation and communication is a third, affecting the number and range of people one may come into contact with or escape from.

The different density and authority conditions interact, producing a large number of hypothetical combinations. We shall not attempt to construct a complete grid; many combinations probably do not occur empirically, and in any case a multifactor theory is more useful as a calculating device than as a typology-producing machine. But some of the major possibilities are worth looking at, as a guide to the complexities of deference and demeanor throughout world history.

High authority plus *high surveillance* leads to a culture of strong and inescapable subordination rituals. The head of a large patriarchal household was in this position: upholding his authority by constantly acting it out, demanding numerous petty acts of ritual deference that could not be evaded by his subordinates. With a *lower degree of surveillance,* however (as in the case of the head of a wealthy modern family), the collective conscience is much weaker and secular in form, deference rituals more perfunctory and easily evaded.

High authority and a *high degree of communicative cosmopolitanism* describes the situation of a sociable upper class or court aristocracy; it is here that we find the most rigidly formal codes of etiquette, and the elaboration of refined and esoteric tastes in consumption and conversation. *Low cosmopolitanism,* on the other hand, gives us the provincial tycoon or the boorish country lord.

Low authority and *high surveillance* describes the situation of the traditional servant class, or the subordinate members of a patriarchal household. The lower class here hardly has the opportunity to create a culture of its own, but emphasizes external obedience, making a show of assiduous compliance to the more observable demands. With *low surveillance,* we get more overt hostility, apathy, or attempts to escape.

Low authority and *high communicative cosmopolitanism* is a rather unusual combination, since access to a large network of communications is usually one basis of power; where it exists may be in the hustler mentality of the big-city lumpenproletariat or the servant with many jobs. *Low cosmopolitanism* is the more usual case; with the traditional peasantry or the modern working class (who have their own households, and hence are free of upper-class surveillance), the culture emphasizes local loyalties, a short-range view, an ethic of episodic hedonism, and little concern for abstract symbols or refined tastes.

Egalitarian relationships combined with *high surveillance* should produce a high degree of conformity to sociable exchange rituals, as exemplified both in tribal societies and in self-selected middle-class communities of the contemporary world. With a *high degree of cosmopolitanism,* the common culture becomes highly secular and relativistic; new sociable fads are constantly being invented, as in the middle-class youth culture of the twentieth century. With a *low degree of cosmopolitanism,* we would expect strong attachments to the reified symbols that represent the group's identity—hence the traditional conservatism of isolated, egalitarian communities.

A full explanation of the variety of status cultures found throughout history would have to add up occupational, political, household, sociable, and community relationships separately, as well as such variables as the degree of military force (*1.5*) and physical hardship (*1.8*) in each sphere. The overall equation for any particular case, then, would be rather lengthy. In many cases, this is simplified by the fact that most or all of these realms have been fused; in many societies, all of these activities are organized around the same kinship groups or household. There is a special theoretical significance about the cases where work is separated from the home, and where politics, public entertainment, or religion become carried out by separate groups meeting in separate places. For the more activities are separated from a single group, the lower the degree of surveillance over the individual, and the higher the degree of cosmopolitanism through diversity of communications. Thus, we expect a shift in the strength and content of collective consciousness on these grounds alone.[3]

Now let us see how well these principles are borne out by the main patterns of deference and demeanor throughout history.

TYPES OF SOCIETIES

The simplest way to capture the variety of human social organization is with a typological scheme. Gerhard Lenski (1966) provides a useful model based on different levels of technology. He distinguishes five main types.

1. Hunting-and-gathering societies produce scarcely enough food for the entire group to subsist; hence such societies are quite small, containing no more than a few hundred people, moving about over a very large territory.

2 and 3. Beyond this level are simple and then advanced horticultural societies. These have developed agricultural techniques, making possible a modest food surplus and, accordingly, a size of social unit extending up to small kingdoms of hundreds of thousands in advanced horticultural societies.

4. Agrarian societies are those in which the plow, metal working, and other technological advances make possible a quite considerable surplus. These societies are characterized by a tremendous degree of stratification, usually with a warrior aristocracy living off a conquered peasantry. These

---

[3] We should note that these spheres diverge more for some occupational classes than for others. In agrarian empires, the upper class is more cosmopolitan and pluralistically organized than peasants and servants; in industrial societies, the upper class has more overlap of family, work, and sociability than the middle class. The divergence is also much less for children and for women at all times.

societies also can have a fair degree of commerce, a government administration including at least some literate officials, as well as literate priests with their own administrative organization.

5. Finally, industrial societies are those in which inanimate energy sources make possible an extraordinary increase in the economic surplus and, hence, in the size and comlexity of society itself, as large numbers are freed from direct productive activities.

This scheme is a useful starting place, if we bear in mind certain limitations on it. A set of categories is simply a device for dividing up the complex materials of world history into more manageable portions. We should not forget this and reify these categories into a set of rungs on a ladder that mankind has climbed. To be sure, there has been a tendency for the types named later in this list to occur historically after the ones named earlier. However, all of these types of societies can exist at the same time; almost all have been found in the twentieth century. This means they modify each other by external contact. Later versions of hunting-and-gathering societies, such as the Australian aborigines found in the twentieth century, are almost certainly different from those found all over the world 30,000 years ago. It is not necessary for one group of people to go through all of the stages or to experience them in any particular order. Many of the important innovations in human history have occurred by confrontations between these different types, with entirely new forms of civilization arising by conquest, migration, or selective diffusion of techniques beyond the periphery of more complex societies then existing.

Nor can we assume that the historical record available to us gives pretty much all there is to know about possible social types under the technologies so far invented. Within the constraints set by a given kind of technology, it is possible to have considerable variation, especially given the interaction among societies just described. The societies existing around the world today differ largely because of their histories and the ways in which they are mixtures of peoples encapsulating in various degrees the cultures arising from previous conditions. Societies are not automatic melting pots; these differences may be maintained for extremely long periods, although under particular conditions they do amalgamate into more uniform cultures. Even this complexity is only a small sample of the variety of social organization possible. There is no reason to suppose that we have gone through all of the possible cases, even within given levels of technology. History is only a small sample of the combinations that can arise from the working out of particular circumstances. This does not mean that we cannot develop lawful generalizations about history, as extreme historicists suppose, but only that the laws that we can induct by careful examination of empirical conditions apply to underlying units of behavior within the combinations of elements that we see. The same kinds of conditions shuffled differently would no doubt produce a somewhat different set of societies. Our task, then, is to try to induct statements about the conditions that produce particular kinds of patterned behavior, in this case status behavior. Given these, and the capacity to extrapolate certain kinds of factual conditions, we should be able to state which kinds of status cultures will exist at a given time, including the future.

Simpler Societies

The simpler types of societies—hunting and gathering and simpler horticultural societies—are organized around small groups (Lenski, 1966: 94–141; Hoebel, 1961; Weber, 1968: 399–439; Durkheim, 1954). There are a variety of living arrangements. In some cases, there are families or small bands traveling together in search of food, such as the Eskimos or the Australian aborigines. In some places, these bands come together periodically, usually under the auspices of celebrating a religious festival, which defines the tribe or larger group as a unit. Elsewhere one may find settled villages, often organized around a kinship system. It is also possible to find rural neighborhoods in which the various families, although linked together in kinship networks, live at some distance from each other and have their own fields. Kinship systems are important in all of these societies, but not necessarily elaborate. Among the Eskimo, for example, the kinship system makes relatively few distinctions and places little emphasis upon them beyond the nuclear household itself. In other societies, the kinship system involves a complex set of relationships, with great emphasis placed on the precise nature of the ties with persons with whom one associates for various purposes, and on kinship categories designating with whom one can or cannot marry. In the more complex societies, one may find not only family groups of both the smaller and the extended kinds, as well as larger tribal organizations, but the larger tribe or clan may also be divided into subclans, clubs, or secret societies of warriors, usually grouped around some religious totem.

It is one thing to describe the actual behavior of people as they congregate together in particular kinds of groups. It is another to describe the ideal side, the verbal accounts that people offer to explain and justify what they are doing; these primarily take the form of religion. In primitive societies, the ideal world which is used so pervasively to explain the physical world is predominantly animistic. The physical world is permeated with spirits, which are used to explain and justify social events, making the world of nature into part of the society itself. As Durkheim pointed out, this is not merely a matter of the peculiar mentality of primitive man who makes mistakes about the nature of the external world. The religious system operates as a social medium of communication, classification, and symbolization. The most important spirits, and the rituals for propitiating them, serve to bring the group together and maintain it by participating in shared rituals. Religious reality-creating, in effect, is a way of maintaining personal relationships.

Deference and demeanor can be seen both in terms of the structure of the power and solidarity found in any society, and in the idealized beliefs by which men mark for themselves the boundaries of their groups and the relationships within them. Deference and demeanor here is found in markedly ritualized forms. Goffman's model of behavior in modern societies is an application of the observations that anthropologists had earlier made of highly ritualized behavior in these simpler societies. Social life is marked by taboos: behaviors, gestures, utterances that are not only forbidden but forbidden by supernatural sanctions. To eat the wrong kind of food or to eat the wrong combinations of foods can bring down the wrath of the surrounding spirits and destroy the prosperity of the tribe. Grooming rituals are also important: not only how one cuts one's nails and hair, but also what one does with the clippings. There are often tabooed places and tabooed objects.

In general, these societies exist under conditions that are conducive to emotionally compelling ritual realities. Social density is usually high, although members of hunting-and-gathering tribes may be very dispersed much of the time;[4] there is little privacy, and the individual is constantly face-to-face with the same people. Cosmopolitanism is low; contacts are uniform and familiar, and there is no written communication to compete with personal relationships. These are the conditions for a strong state of collective conscience, and the ideas reflecting group identity are reified in the form of spirits who live in the environment.

The particular form that rituals take depends on local conditions. In highly egalitarian hunting-and-gathering tribes, the margin of survival is relatively low. The Eskimos, for example, practiced not only infanticide but the exposure of the old, the sick, and the weak, as there was hardly enough for each active adult to support oneself, let alone others. It is not

---

[4] This implies that ritual solidarity is lower in most hunting-and-gathering societies than in horticultural societies, which indeed seems to be the case if we look at the diffuse religious systems and relatively weak constraints over individual behavior found in the former (Hoebel, 1961: 67–99). But the general form of collective consciousness is determined by the fact that cosmopolitanism is low and most contacts are within the family. Durkheim's evolutionary bias misled him into looking for the strongest collective conscience in the technologically most primitive societies. The lesson is to be aware that loose typologies are only a starting point, and that we must look at the precise set of conditions determining ritual behavior in each society.

surprising, then, that life is carried on in an atmosphere of fear, or that people should place special ritual attention on keeping oneself constantly alert and aware of dangers. But beyond this, taboos, like forms of religion, are important because, in being upheld by the group they operate to preserve the organization of the group. This is what individuals sense when they react strongly to someone else's violation of a ritual. In the simplest societies, there is virtually no stratification; the only role marked out from the others is that of shaman or sorcerer. These men are important essentially as manipulators of the emotional and ideal side of the world. They are correspondingly both looked up to and feared, and may be attacked by others when troubles occur that can be blamed on their magic. Sorcerers have a special role in upholding taboos and interpreting troubles as violations of them. We see here a form of power which becomes successively more important in societies with larger surplus.

In more complex horticultural societies, those which have developed enough surplus so that there can be stratification among their members, taboos are related to the system of stratification itself, or to particular subunits of memberships within the larger group. Just as in simpler societies a totem and its celebration can be the occasion for drawing together the entire tribe, in more complex and subdivided societies, such as the Comanches, totems are associated with specific groups like warrior clubs; membership within them depends on upholding particular rituals, which in turn create taboos that unknowing outsiders may violate. Among the Ashanti, there are a large number of taboos relating to what kinds of goods can be carried through the village, or what can be carried on one's head. These all relate to questions of deference, since there are village headmen who are given great religious significance and around whom an entire etiquette has developed. The reason that certain goods cannot be carried on one's head through the village, although explained in magical terms, essentially is because it is an insult to the village headman to hold oneself higher than he. Since rituals express stratified social relationships, it is not surprising to find that individuals attempt to manipulate them for their own advantage, as Forde (1962) has demonstrated in the mortuary rituals of the Yäko.

In simple societies, then, we find the forerunners of later forms of deference and demeanor, and indeed of the entire apparatus of status group stratification. We find versions of modern Rotary clubs with their ritualized foolishness, our sports and our cheerleaders, our dances and entertainments. Although the form of organization has changed, the mechanisms by which members of a group maintain their ties and set themselves apart from others are basically the same. Goffman's backstage–frontstage distinction can be applied not only to the modern theatre and the home performances of polite dinner parties and social gatherings; they are exemplified in even more striking form in the secret societies of tribal life in which initiation into the group is initiation into the backstage of masked dancers who put on ceremonies before the entire tribe. Like our clubs of today, these secret societies have political and social importance, for organized groups within a society always have more political resources than the unorganized mass. Similarly, grooming rituals—concern for hair, fingernails, shaving and its lack, wearing particular kinds of clothing—are obvious enough in our society, but found equally well in others. The main difference is that we have lost the magic or religious accounts given to them. But the underlying forces, the concern for exemplifying membership, particularly in subgroups of a special rank, has not changed.

Fortified Household Societies

Advanced horticultural and agrarian societies are larger societies with much greater economic surplus (Lenski, 1966: 142–296; Weber, 1968: 1006 –1109). Accordingly, they develop extensive stratification networks, ranging from principalities to very large empires. These political units come in a variety of types: relatively more or less centralized, with greater and lesser degrees of urbanization and of commercialization, greater and lesser degrees of bureaucratic government administration. By and large, what they have in common is the fact that people's lives are organized around fortified households. The larger political units are built up by households developing private armies. Alliances along major households are the primary basis of political empires. The emphasis on traditional inheritance of position is simply the result of this household structure. There is usually no distinction made between one's home and one's work place. The artisan works in his own home; if business expands, he takes in more apprentices, servants, and helpers, as well as uses his family as assistants. The same thing applies to a merchant, a farmer, or a prince. There is no police force or organized army except that built of such household units.

In such societies, the local village or tribal organization has usually been superceded as the larger political unit, with the rise of a cosmopolitan political hierarchy. Developments in military technology now make it possible for those who can afford horses, armor, or other extensive weaponry to constitute themselves as an elite fighting force. At the same time, the improved methods of agriculture and transportation make it possible to produce a much larger surplus than necessary to keep people alive. The ethos of simpler horticultural societies in which surpluses are relatively transient, and in which the great man is one who accumulates surpluses in order to give them away and acquire status by his largesse, has changed

into a situation of a much more permanent aristocracy that can monopolize the military means and live in considerable style by appropriating
most of the surplus wealth. This is the society of the manor house, the
lordly estate, the walled castle, or the urban patrician with his many
servants and retainers.

For such societies, explaining deference and demeanor becomes more
complex, for there are determinants operating on two levels. There are the
authority relations and the conditions of social density *within* the
household, and there are the variations in authority and social density in
relationships *outside* the household. For the most part, it is only the heads
of the more prominent households who participate in the larger sphere.
Because all of these societies are based on similar household organizations,
they all have similar kinds of deference systems in some respects. But the
external relationships can vary considerably, producing several different
upper-class cultures.

Within the household, we find a very powerful deference system organized around the household head. He is head of the family, surrogate
father to his servants and retainers, military leader, controller of the family
economy, and domestic priest. It is he who leads the family prayers and
ceremonies. In many such societies, there are special family deities; very
often these involve the worship of spirits of dead ancestors, as found
among the Romans, the Chinese, and the ancient Hindus. What such
manners do is to reinforce the authority of the family head by acting it
out over and over again many times each day. Modern analogues can be
found in the more traditional sectors of our societies, and in the memory
of perhaps most living adults. Take, for example, the still surviving
ceremony of having the family sit down together at the table. There are
ceremonial rules of the sort discussed in the previous chapter: controlling
side involvement by requiring that the whole family be there together,
that no one start before the entire group, and that no one leave before
everyone is through eating. This sets the stage for the performance and for
the leading figures in it. Rituals in which the head of the family says grace,
carves the roast, or serves the food, thus enact his authority in a way that
makes it taken for granted.

It is by such constant small repetitions that one is reminded of what
the authority relationships are, and by which the habit of giving and
taking orders and thinking about it through idealized notions is continually reconstituted. The ideal side, by which the father mediates for the
family with the spiritual realm, as in leading prayers, gives an emotional
sanction to his authority. In agrarian societies characterized by fortified
households, the authority of the household head is that much greater. One
does not speak to him until spoken to, and one greets him with elaborate,

quasi-religious formulas. These are what one may call extreme *asymmetrical* or *one-way deference rituals:* the bowings and scrapings, the ritual proprieties that underlings must give to the person of overwhelming power, by which he forces them to help in perpetuating his power by idealizing him.

Within the household, then, the usual conditions are a high degree of authority, powerful and inescapable surveillance, and, for most household members, a low degree of cosmopolitanism. The result is strongly upheld rituals and idealized beliefs surrounding household authority. Outside the household, though, political and religious organization can take a number of forms, ranging from the tightly imposed domination of one royal household over all the others, an egalitarian community of noble warriors, the complex hierarchy of bureaucrats and courtiers, or the cultivated communities of urban patricians.

Authority Relations Outside the Household

In a number of societies, the household of one great lord becomes elevated over all the others. Domestic deference to the "lord of lords" is demanded from everyone, including the heads of other large households; his personal gods become the gods of the entire society. In the earliest empires of Mesopotamia, Egypt, and China, and in the kingdoms of Africa and the Americas, there was little distinction between priests and rulers; the king himself was usually considered divine or a high priest. Here we have a case of tightly knit societies, with all power resources concentrated in the hands of the ruling house; surveillance was high and cosmopolitanism low, while authority relationships were even more unequal than within most patrimonial households. Such rulers lived in an atmosphere of continuous pomp and ceremony; it is here that (if technical conditions were favorable) we find art and architecture used almost exclusively to glorify the ruler–gods with grandiose monuments, such as pyramids, tombs, huge columns, and statues (Hauser, 1951: Vol. 1, 37–48, Vol. 2, 185–195). Similar styles and demeanor relations reappear wherever a powerful centralized kingship or theocracy can establish itself.

The extreme form of religiously bolstered legitimacy of an absolute ruler carried over to his officials' relations with the lower classes. The Chinese mandarins of the bureaucratic dynasties presented themselves in public as demigods (Weber, 1951: 128–145). They presided at the traditional religious rites which were maintained for the benefit of the lower classes, and took on as much of the traditional magic qualities as they could borrow. Commoners were required to bow to the ground before them, lest their eyes fall upon the mandarin as he was carried by in his sedan chair. Such officials lived in an atmosphere of great pomp and circumstance, surrounded by servants wearing elaborate costumes, and in stage settings of considerable beauty. It was by means of such settings and ritual deferences—enforced at the end of a spear—that the semisacred qualities of the mandarins were continually recreated. Similar procedures have been used elsewhere, most recently in fascist autocracies.

Warrior Aristocracies.   In other circumstances, the upper class's relationships outside their own households were a good deal more egalitarian. The culture of knightly chivalry is one such type, found in early Greece, India, and probably China before the establishment of centralized government, and in Medieval Japan, Persia, and Europe (McNeill, 1963: 185–268, 432–434, 588–590; Weber, 1968: 1070–1085). It arises where military technology favors the heavily armored individual knight rather than mass armies of infantry or horsemen, and most of the remaining populace are tied to the land as peasants. This gives us the familiar two-level society: the patrimonial household and its highly authoritarian and religious atmosphere within, and a larger military–political world outside, participated in only by the knights.

The knights were thus the recipients of highly ritualized deference from underlings, and evinced the corresponding attitudes of pride and paternalism. Among themselves, however, relationships were usually fairly equal. With a great deal of fighting going on, no one's position was very permanent, and notions of individual merit rather than family heredity were strong (Huizinga, 1954: 56–107). Ritual deference thus centered around membership rather than hierarchic position within the class; the important thing was to live as a knight and to treat other knights in the proper way. At the same time, men lived very public lives, whether in the crowded castles of their temporary lords or in the streets of the towns, and went everywhere with as great a retinue as possible (Holmes, 1964: 23, 36–41, 159–196). As knights traveled much and made cosmopolitan contacts, their manners became correspondingly sophisticated and elaborate.

The conditions shaping knightly culture, then, were relative equality among members of the ruling warrior class, along with absolute superiority over all other social classes; a highly public setting; a fair degree of cosmopolitanism (although limited usually by illiteracy); and, of course, a way of life involving considerable danger and physical exertion. A similar type of chivalrous culture emerged under these conditions everywhere. It involved a great deal of pride and ostentation, taking as much of the center of attention with bright clothes, weapons, banners, and accompanying servants as one could afford. A knight's behavior was designed to demonstrate his courage and superiority on every occasion; dueling, challenges, and oaths to undertake difficult tasks were the showy gestures most admired. At the same time, knights had an elaborate form of courtesy among themselves, emphasizing the keeping of one's word, spurning petty advantages, and arranging fights as if they were games with strict rules. This style was usually legitimized by reference to religion; knightly honor was considered a form of Christian virtue in Europe, and was closely related to the Zen Buddhist ideal of demeanor in Japan.[5]

The courteous ideals, of course, were not always lived up to. The knightly world, after all, was one of great contentiousness; the code of honor served as a standard by which one could take offense at others' slights, and also as a way of getting maximally showy stage-settings for one's battles. Courtesy was reserved for other knights; the failure of deference by commoners (even rich burghers) was punished with malicious cruelty. The code of etiquette served for very un-ideal purposes, but was upheld nonetheless for its usefulness in making the proper status display in this situation of individual competition and high social density. It died out with the coming of mass armies, both in Greek antiquity and in Renaissance Europe.

Courtiers and Bureaucrats. Where a powerful centralized kingdom is established, manned by civilian officials, a new upper-class deference culture emerges. This may be called the "style of courtly manners," found in the Chinese empire during its strong periods, in Persia and Islam at various times, and in Europe after about the fifteenth century. The military virtues were displaced by indoor forms of competition; conditions of public surveillance remained high, but sophistication and secularization increased with the addition of *literate* communications to the diversity of contacts, and rising material wealth made possible many refinements.

In the period when a centralized empire was emerging from the feudal fragmentation of China, religious advisers, oracles, and soothsayers took on great influence at court, and became the basis for an emerging bureaucracy which the emperor used as a counterpoise against the independent nobility (Weber, 1951; Franke, 1960). Gradually the class of hereditary nobles was eliminated, and noble rank depended upon achievement in the centrally ministered examination system which was used as a prerequisite for office-holding. At the same time, the empire was never rigidly controlled from the center. An amalgamation of interests developed among the gentry in various parts of the country, the wealthy landowners who alone could afford the leisure to study for the increasingly more elaborate

---

[5] Many of the legends of King Arthur and of Charlemagne emphasize the importance of keeping an oath in its precise words, even at great cost to oneself. This recalls the tribal religions and verbal duels in which precise formulas have magic significance; apparently there was a long period in which Nordic tribal culture became amalgamated with the Christian one imposed from above by ambitious rulers.

series of examinations. They managed to monopolize administrative offices and to administer the country for their own personal benefit.

Gentry control of this loosely bureaucratic system over many centuries led to progressive refinements in manners and tastes. The examination system became extended from a single exam to a long series, the content of which became more and more refined. It began with commentaries on Confucian classics and their moral precepts, and came more and more to emphasize the aesthetics of writing itself. The so-called "eight-legged" essay was judged on its merits in composition and calligraphy. The ideal of the gentry became a man who could afford to spend 20 or 30 years studying for the examinations and developing his appreciation of painting and poetry. The emphasis was on the idle, highly-cultivated man; status markings, such as wearing long fingernails to indicate remoteness from manual work, were used. The content of Chinese high culture thus developed great subtlety of tastes. Out of this situation came the high aesthetic standards of Chinese arts, cuisine, architecture, and furnishings. The Chinese high culture was one of aesthetic consumers. Admission into upper social ranks through the examination system, as well as through informal associations, depended upon one's cultivation.[6]

In Europe, the courts arose as collections of nobles and retainers gathered around powerful lords, partly for reasons of politics and hope of preferment; partly also because kings used this as a mechanism for controlling potential rivals, demanding that they spend at least part of their time at court and leave their children there.[7] With the decline of military careers, more sedentary entertainments replaced the rougher sports of the Middle Ages. "High culture" began to split off from the "low culture" of the people. Card games replaced dice; orchestral music replaced popular dances; the melodramatic religious plays and bawdy humor of the streets were replaced by the stylized plays of what we now call the "classical" tradition (e.g., Racine and Corneille). With the improvement in material

---

[6] In addition to this sophisticated worldly culture, there was a secondary upper-class culture. For those whose political fortunes were troubled, or simply for men jaded with worldly life, Chinese high culture had a recognized lifestyle expressed in the religion of Taoism. It meant giving up worldly goods (although apparently keeping enough to avoid hard manual labor), living in quiet seclusion, devoting oneself to nature and to meditation. It was in effect a Chinese version of the hippie ethic and other pastoral idylls that have been popular among affluent classes in the West.

[7] See Lewis (1957: 37–62, 195–213); Hauser (1951: Vol. 2, 47–53, 95–96, 156–157, 203–206); Stone (1967: 125–134, 303–341). In the more feudal period (Holmes, 1964: 85–91, 191, 223), material conditions were ruder, the knights spent their time hunting or fighting; but where political conditions draw them together in large courts, we find already descriptions (such as the following by a hostile priest) of "much vanity in spectacles, in empty conversations, in blandishing adulation, in detestable voluptuousness [P. 223]."

conditions came the introduction of tableware and refined table manners. No longer was it acceptable to eat out of a common bowl or to throw bones over one's shoulder. Standards of personal cleanliness gradually developed. Along with this came a new emphasis on polite conversational manners.

Courts became places of considerable enforced leisure. Large numbers of ladies and gentlemen were obliged to spend a great deal of time waiting upon the king, and developed means of whiling away the hours. Conversation became an art form with elaborate emphasis on politeness and topics of respectable entertainment. It is at this time that verbal taboos began to develop, especially taboos on the vulgar language of the people. Particularly at the court of Louis XIV in the seventeenth century, great emphasis was placed on elaborate mutual bowings and curtseying, flowery greeting and farewell rituals, and polite inquiries after one's health. The correct style of conversation kept politics and other business for the backstage, while entertaining conversation came to consist of *bon mots* and flowery compliments, clever discussions of personalities and love affairs. At Versailles, the mastery of trivial details of etiquette and an elaborate secret language of hints and catch-words could make or break careers.

This quasibureaucratized aristocracy began to create a new style of deference and demeanor. The earlier forms might be characterized as *asymmetrical deference rituals:* the elaborate prostrating of oneself before a Chinese mandarin; the ritual shunning of lower Indian castes by members of the higher; the patriarchal dominance of the traditional head of household. In the large courts, however, there developed what might be called *mutual deference rituals:* the system of greetings, bows, and curtsies upheld at both ends among the courtiers. It involved paying deference to each other; it was in effect an elaboration of deferential gestures that had previously been given by underlings to superiors. It now became a mark of courtly capacity to give these courtesies to others who were more or less one's equal, and to carry out battles over precedence by ironic manipulation of the arts of sociability. The mark of superiority was no longer the sheer arrogance of the man who expects others to defer completely; this even became a mark of boorishness, unsuited for the subtle interplays of courtly life. The new ideal became controlled and easy sociability.

Competitiveness had not disappeared, of course; it had only gone indoors. The centralized courts made minute rankings of precedence more visible, and the policy of the great kings was to encourage such rankings in order to keep the aristocracy under control of their own system of awards. The emphasis on heredity is not always characteristic of patrimonial societies per se, but reached its peak under the development of bureaucratic absolutism, with the elaboration of precise aristocratic rankings calculated upon the combination of titles in one's ancestry, as well as awards by the king. The concern for a noble style of life and knightly heredity became formalized because of the political maneuvers of a period when knights were already being replaced by the mundane officials of mass armies and centralized administrations.

Bourgeois and Clerical Elites

With the rise of networks of commerce, improved transportation and the development of literacy in the earlier states, priests tended to develop organizations of their own. These were usually closely allied with the prevailing military–political forces, but could begin to exert an independent influence. One of the main results was that they developed philosophical religions or theologies: abstract accounts of the nature of the universe (Weber, 1968: 439–467, 500–576; Parsons, 1966: 69–115). Within these accounts, the traditional animistic spirits, personified local political potentates, and imperial gods were seen either as illusions or as lesser forms of greater cosmic powers. Thus, we have a series of "philosophical breakthroughs" characterizing the world-historical societies: Buddhism and Brahminism in South Asia, Confucianism in China, as well as the philosophical systems of Hellenic society from which arose Christianity and Islam. In these religions, the spiritual world, the world of ideals, becomes finally separated from the physical world. This stands in contrast to the traditional religions, in which the spirit world permeates this world. Some important differences follow.

In traditional religion, there is no important difference between one's religious position and one's worldly position. One was in good grace with the spirits if he was prosperous, healthy, and victorious in war. If one fell on ill luck, it was interpreted as the result of spiritual transgressions. The transcendental religions created new possibilities. Good and evil became ends in themselves, separated from worldly success. It became possible for one to be a holy man, to be righteous or saved even though one was not prosperous. It now became possible to construct an ideal world in which to achieve salvation apart from one's worldly fortunes.

This could occur in two main ways: either by direct mystical experience, making one feel a part of the transcendental world and regard the physical world as an illusion or at least a lesser realm of existence; or by conceiving of a world of balance existing after death in which the good who suffer in this world are victorious, and the evil who prosper are finally punished. The latter is the predominant Christian and Moslem version; most of the oriental religions emphasize mystical salvation. Another version developed first in China. The religious wisdom did not so much emphasize a substantively real transcendent world, but rather an understanding of the basic principles of the universe, especially the principles by which things change constantly from one pole to the other. Confucians thus give a sophisticated philosophy of how to attain happiness by keeping oneself in tune with the underlying flow of events.

The creation of these new alternatives for pursuing the good life did not mean that men would abandon the usual struggle for political and social advantage. On the contrary, only a relatively small group of virtuosos would devote themselves to attaining strong mystical experiences or assurances of salvation. For the rest, religion continued to be a subject for political ideologies and alliances in the higher classes; in the lower classes, a means of momentary release; for those in the middle, a means of achieving a maximal amount of deference from others, since religions defined the proper standards of demeanor. For along with the more abstract accounts of the nature of the universe—the forces of Yin and Yang, the higher realms of Nirvana and the veils of illusion, the spiritual powers of an omnipotent and omniscient God presiding over the mundane world —came changes in the forms of daily deferential relationships and in the emotional tone of life.

The more abstract religions called for greater emotional self-control, for elaboration of a demeanor of relative inexpressiveness in this world as a counterpart to one's entry into the transcendental world. Whereas, in traditional religions, religious occasions were generally marked by group emotional outbursts, frenzied dancing, gluttonous eating, and an orgiastic emotional tone, the transcendental religions emphasized a frigid demeanor and great self-control. The ideal of peace and quiet, of eternal tranquility which characterizes their salvation worlds, was a reflection of the kind of demeanor preferred in this world.

This controlled demeanor arose from several sources. The transcendental religions arose in the cosmopolitan centers, although not always in the most powerful groups within them, and set themselves off in opposition to the traditional religions still being practiced, especially in the rural communities subjugated by the larger empires and in the urban lower classes. The orgiastic and worldly religions became negative instances of the low status behavior to be avoided. The frigid demeanor idealized by all of the transcendental religions takes its tone from the effort to show oneself superior.

On a microlevel, the emphasis on emotional control seems to have developed because the body had to be held rigid in order to concentrate one's nervous system on complex inner experiences and lengthy verbal and symbolic systems. In effect, to have a religion of The Book (or a religion of the philosophers) one had to concentrate to a tremendous degree on matters inside one's head and keep the rest of the body under control.

In the various societies that developed these transcendental religions, there also developed very refined interpersonal manners. Deference and demeanor in a Goffmanian sense moved from the highly demonstrative and theatrical activities of tribal societies into something that represents more closely the secular politeness Goffman describes for our own recent past. The great world cultures—Chinese, Japanese, Indian, Islamic, Hellenistic, European—developed complex forms of polite everyday behavior. These were based originally on the religious rituals that the members of the higher classes performed in their daily activities. The religious tone has persisted in many societies down into the 20th century, although belief in the transcendental world mentioned in the accounts may not have been very strong. The peculiar differences in tone which still make it difficult to penetrate comfortably into the personal life of other societies today are carryovers of these manners into the industrial era. Our own culture did not somehow separate itself from history as a "modern" form against which all the rest are "premodern"; we are an industrialized form of the particular rituals and emotional tone of *Christianity.*

It is the nonmilitary, nonofficial urban classes who developed these forms of deference and demeanor most fully. In structural terms, we are dealing with situations in which commercial or forensic virtues are emphasized over military ones, and the autocratic state is remote or relatively unimportant in determining one's status. There is a relatively egalitarian situation among "middle class" persons without the resources of fighting or government influence to set one above another. This made the problem of setting off oneself from the traditionalistic peasants and the shadier businessmen all the more acute. The main resource of the more cosmopolitan classes was their superior sophistication and openness to communications, their greater leisure, and their modest wealth. Their best deference strategy was to adopt a sophisticated inwardly oriented religion, or its secular equivalent.[8] Similar developments along these lines can be found in Indian, Islamic, Hellenistic, and Roman society, and in the independent cities of the European Middle Ages.

The Indian subcontinent was populated by an enormous diversity of ethically distinct tribes (Weber, 1958b; McNeill, 1963: 397–407). It was overrun early in its development by a number of (mostly small) conquest

---

[8] One way to raise one's status was to become a priest or a monk oneself. Presumably, this choice was limited to those whose special talents or opportunities lay in that direction, and whose worldly opportunities were less desirable. The kinds of hierarchies that arise among religious specialists in the transcendental religions, whether priests or holy men, are characterized by the same elaboration of subtle distinctions as laymen could derive from applying religious standards to their everyday manners.

states, thus creating agrarian forms with a peasantry and aristocracy, as well as commercial centers and urban commercial classes. But India developed no long-lasting united state, or even very stable local states. In a period of reaction against one imperial effort, the Brahmin religion came to dominance. It was a sophisticated version of earlier magical and ecstatic techniques, carried by a class of literate priests, who also developed themselves into wealthy landlords. They made their claims for prestige by emphasizing their ritual purity. A large number of food and cleanliness rules and taboos were elaborated, in such a fashion that only the wealthy could afford to keep them. The Brahmins also placed special emphasis on avoiding the orgiastic aspects of tribal religions. The theological doctrine of the transmigration of souls was essentially an ideology justifying the superiority of the Brahmins and maintaining the religious differences among the subordinate groups which gave the higher ones a basis for expressing their superiority. Although the response to this system might ideally be expected to be quietism and acceptance, it is not true that there were no efforts at social mobility in India. It is rather a matter of mobility taking place by whole groups rather than by individuals. Lower-ranked tribal groups (subcastes) could emulate the manners of Brahmins, taking on as many of their ritual taboos and purifications as they could afford. As a result of this system of cultural competition, a large number of subcastes emerged over the years. The basic form of the system of cultural competition in terms of a top reference group remained very stable, and survived the Islamic and British conquests.

Islam developed as a tribal war coalition, with a monotheistic religion as unifying agent, in an area of commerce between urban centers of the Middle East (McNeill, 1963: 461–483, 546–560). After the early phases of conquest, the Islamic empire soon fell apart into a patrimonial network and a variety of principalities. It contained within it a number of status cultures, all expressed in religious terms. At the wealthier courts, there developed sophisticated and aesthetically oriented manners not dissimilar to those found at the wealthy courts of the Orient and in Europe. In the rural areas, there developed a form of Islamic traditionalism expressed as religious legalism, led by the local prestige of the *ulema,* the local religious leaders and interpreters of the law. In the urban–commercial centers arose a variety of petty bourgeois pietism and puritanism. There also developed craft organizations, especially around orders of Sufi mystics, within which political and religious activities were carried on in secret hierarchies. On the ideal side, this hierarchy of masters was expressed as a series of initiations into secret doctrines and mystical techniques, but it also served as a means of bolstering an economic position and undergirding a respectable status claim, and even occasionally as a vehicle for political revolt.

Hellenistic and Roman society developed in yet another way (McNeill,

1963: 210–227, 277–321, 342–344; Fustel de Coulanges, 1864). Tribal war coalitions came together in the mountainous terrain of Italy and Greece to form cities, organized originally as military citadels. Out of their interaction developed new forms of urban and commercial societies. In the early period, great emphasis was placed upon the household organization of the family, bolstered by its domestic cults. Early Greek and Italian manners were thus based upon maintaining household rituals, especially those expressing filial piety. As the various cities began to amalgamate into political leagues and then into larger states, urban politics supplanted the household as the center of attention. The city itself was organized as a religious cult, which provided the vehicle for military alliance. As the military situation changed with the development of techniques of disciplined mass warfare, in which the manpower of massed citizenry became increasingly important, there were a series of revolutions. The clients, the lesser members of the great households, acquired political rights for themselves; eventually the earlier coalition of clan headmen became a larger democracy.

In this situation, approximately around the fifth century B.C., a new style of social manners appeared in Greece. With the emphasis on urban mass politics, men began to place great value on their oratorical skills; the profession of rhetors developed to teach men these manners. A new world of urban sociability developed. Friendship ties began to become more important than ties of kin; we begin to find ourselves in the world of dinner parties of the Socratic dialogues. Athletics, which had earlier been a form of preparation for warfare, became a mode of sociability as well (Hauser, 1951: Vol. 1, 81–100; Marrou, 1964: 21–94).

The Greek cities did not last long in their period of political glory. The semi-barbarian armies of Macedonia conquered them as a prelude to the empire-building of Alexander. This empire, however, relied heavily upon Greek troops and administrators. Greeks became a wealthy dominant class throughout the Mediterranean world. Their culture, with its heavy emphasis on verbal skills and consumer aesthetics, led to the rise of schools for keeping Greeks distinct from and superior to others, and education replaced earlier religious pieties as the fundamental mark of social distinction (Marrou, 1964: 137–164).

The Roman empire developed gradually on the eastern fringes of this Hellenistic empire. The original city was one of many such tribal coalitions; in the constant warfare of the Italian peninsula, the shuffle of chance brought Rome to the top. It too went through a series of internal political revolutions. As it began to take over the empires of the Carthaginians and then of the Hellenistic kings, however, the earlier relatively egalitarian society of fighting small farmers became more stratified. The conquered provinces abroad, as opposed to those in Italy, were no longer admitted into the state as coalition partners, but were divided up into estates for the more prominent Roman soldiers. Thus appeared a very wealthy upper class owning vast areas abroad. The latter period of Rome, after the wealthier classes had virtually destroyed the earlier democracy, was one of tremendous cultural conflict (Hauser, 1951: Vol. 1, 101–113; Murray, 1951: 76–165). Not only was there a cosmopolitan diversity of ethnic groups within the Empire, but the Roman population itself was split. On the one hand, there were the old upholders of the pious household virtues. On the other, there were the rich landowners emphasizing a new style of lavish display, waste, and debauchery, as described for example by Juvenal, a representative of the older modest householders. There also developed the moralistic cultures of the urban middle classes, Stoicism and Christianity, emphasizing transcendental beliefs as well as equality of dignity among all persons; these corresponded to a breakdown of the household system of loyalties in favor of a larger and more universalistic community. This universalism, however, tended to disappear with the disintegration of the empire.

With the revival of commerce in Medieval Europe, a pious bourgeois status culture came to dominate in the free cities of Italy, Germany, and the Netherlands. Relatively egalitarian patrician classes ruled the cities through guild organizations; social status within their ranks was primarily a matter of public charity, patronizing art in the churches, and giving to the poor (Hauser, 1951: Vol. 2, 245–267; Vol. 2, 16–40). (At the same time, they carried on strenous efforts to maintain control over the journeymen and other artisans who sometimes challenged their economic and political domination.) Once again, we find that conditions of relative equality within a dominant urban–commercial class results in status competition through displays of personal piety rather than ostentation.

The Community of Private Households

In the extensively commercialized and then industrialized societies of western Europe since the seventeenth century, the conditions for status display underwent a revolutionary shift. The large fortified household declined with the rise of the centralized bureaucratic state and its wresting into its own hands the concentration of armaments. The smaller household developed without the great numbers of retainers and the custom of hangers-on living in the households of the greater lords (Stone, 1967: 96–134; Ariès, 1962: 365–404). With the extension of commerce and manufacturing, as well as the expansion of the class of government administrators of medium rank, power came to be spread out through a larger class than previously. Men of moderate authority now lived under conditions where they were no longer at the beck and call and under the constant surveillance of their superiors.

This produced a significant shift toward secularization. The level of social density was drastically reduced for an important segment of the population; men were no longer in each other's presence so continuously, and one had more choice over the social stages on which to appear. At the same time, cosmopolitanism increased for the middle classes, with the elaboration of far-flung commercial networks, the rise of the press, and the diversification of perspectives deriving from the separation of work from the home all contributing to this result.

The moralistic status culture previously found among urban patricians now became the strategic self-image of a much larger group. The actual model of the eighteenth- and nineteenth-century bourgeoisie, however, was the courtly aristocracy rather than the Medieval patricians; for the latter style had largely disappeared in the Renaissance, as the great urban financers and merchants acquired autocratic power and transformed themselves into ostentatious military princes. The leisure culture of the court—the polite manners, the grand opera, and the theatre—thus became taken over by the bourgeoisie as soon as they could afford it.

It is here that we may locate Goffman's work most properly. The one-sided deference rituals of the traditional aristocracy had given way to two-sided rituals of mutual respect in courtly manners. Among the much larger and more egalitarian middle class, these manners became even more widespread and lost much of their previous aristocratic flourish. At the same time, they took on a stronger moral tone. They no longer represented the heroic ideals of the warrior aristocracy nor the aesthetic ideals of the leisure aristocracy, but rather an expression of moral relationships in everyday life. It is here that we see the application of Durkheim's idea that in "modern" (i.e., nineteenth-century) society, the individual becomes a little god surrounded by a halo of untouchability. Quite literally, the person becomes inviolate. The older forms of punishment, in which mutilation—hacking off a hand or an ear—were common, now become morally repugnant. Tortures, forced recantations, foul prisons, and degrading submissions all are no longer accepted, at least as public ideals. The individual walks through his daily routine accepting mild but strongly observed deferences in the form of greetings, bows, tippings of the hat, and polite conversational inquiries.

The home becomes a sacred arena of individual rights. No longer is it a place where strangers and underlings wait upon the pleasure of their superiors, but a place where equals gather for entertainment, or the family retires in its privacy. An elaborate system emerges for not encroaching upon other's newly found privacy. Calling cards, the practice of having certain afternoons at home to receive one's guests, the custom of salons and of dinner parties at which polite conversations are carried out, spread among the middle classes as wealth made this increasingly possible in the nineteenth century.

But matters did not reach a new equilibrium. On the contrary, the resources for putting on a proper status show were spreading lower and lower in the class structure with the increase in wealth, leisure, and mass-produced entertainment. The middle classes of the nineteenth century were under a terrific strain to set themselves off from their underlings as distinctly as the older aristocracy had been, whose former glory they now saw coming into reach. Ironically, in the wealthiest industrial societies, the resources for status display were spreading simultaneously with technical and organizational conditions that were destroying the older community settings in which display could be effective. The decline of authoritarian relationships in the household knocked an especially important prop from under the previous system of deference. The best place to follow this little drama is in the United States.

STATUS COMPETITION IN AMERICA

American society, from the middle of the 1800s to the 1950s, went through sweeping changes in the structure and style of its status communities. The outcome can be characterized as a general leveling effect on manners and customs, and a trend toward informality and familiarity in deference rituals. Goffman's (1967: 95) metaphor of the individual, as a little god owed ritual respect and freedom from physical handling, has disappeared in most circles; it certainly no longer acts as the dominant cultural ideal. In its place we find the "good fellow," an object of ritual disrespect and profanity, who openly courts familiarities such as back-slapping and nicknames to prove his acceptance by his crowd.

This change was hardly a smooth transition. As we would expect, most of the opposition came from the upper classes while the collapse of their influence was due to steady pressure from a rapidly expanding and mildly affluent urban middle class. But the pattern of conflict *between* these two groups was enormously complicated by changes *within* the upper class and the middle class as well; and, after World War II, we must add a third complicating variable, the public emergence of a working-class culture competing for status recognition on its own terms. A Goffmanian analysis of stratification as a reality constantly negotiated must emphasize that labels such as "upper class" and "middle class" are historically relative terms. The problem here, as it is in explaining and dating any historical change, is to convey the lag between structural changes which promote the development of new status groups with distinctive cultural forms within a given stratum of the stratification hierarchy, and the time it takes for these groups to gain the strength and assurance to challenge the ritual order defined by the established group.

The point of direct challenge is not often reached. The established groups can move to assimilate individuals and forestall contention, although assimilation can be a two-way street: the individuals who cross status lines can end up having greater influence on the lifestyle of the group whose boundary defenses they surmount than vice versa. This does not have to be a deliberate choice: the structural changes that bring new groups to the fore generally undermine the entire line of resources on which the old ritual order was based. If this later development is strong enough, confrontation is avoided altogether as the old order and the ideal image its rituals celebrate are removed to a corner of the stage where their activities can have little affect on the public at large.

As we shall see, the mid-eighteenth century presents a picture of a solidly entrenched upper class concentrated in the northern urban areas. Stow Persons (1966) has labeled this class a "gentry" elite, despite their urban location, to emphasize their quiet, industrious, and provincial manner. Not that they were overburdened with work: they just took themselves seriously and felt a duty to impress the lower classes with the evils of frivolity. Their cultural dominance appears to have depended as much on a lack of resources by other groups as on their own monopoly. Certainly the upper class Henry James describes in his novels about New York and Boston sensed no threat to their preeminence lurking on the horizon.

Yet one was there. Large-scale commerce and the railroad boom was already producing fabulous fortunes in the 1840s and 1850s. The term "millionaire" was coined in 1845 to describe such men as Astor, Harriman, Gould, Hill, and Vanderbilt (Wector, 1937:113). These *nouveaux riche* merchants and railroad barons, while capable of buying out the gentry elite many times over, were as yet too few and too lacking in cultural refinement to present a serious cultural challenge or a pressure for assimilation by the gentry. But they did begin efforts in that direction. In their attempt to play Society with a capital "S," they established a parallel culture, naturally emphasizing their greater economic resources by building expensive homes, dressing elaborately, and entertaining frequently and lavishly.

By the 1870s, this self-dubbed "fashionable elite" had grown sufficiently in numbers and refinement to be a force for the gentry to reckon with (Persons, 1966:88). A tacit agreement to assimilate the most respectable members of the fashionable circles was halfheartedly offered by the gentry. That the former would listen is evidence of the extent to which they were still uncertain of the status of their own standards and dominated by the ritual ideal of the old elite. Within another decade, however, as the second and third generations of the fashionable elite reached their prime, a victory of sorts was achieved: the gentry with their subdued manners and moral emphasis were not displaced so much as completely overshadowed by the energetic and extravangantly visible sociability of fashionable society.

During this period, in the decades around the turn of the twentieth century, the fashionable elite made a bid to establish themselves as an American aristocracy. Their attempt was doomed from the start. They had no political organization to define their status in legal terms, as the pre-Civil War Southern aristocracy had built on the institution of slavery or as their English counterparts had inherited the titular remnants of feudalism. They found themselves under attack on two fronts. On one hand, industrialization brought a new wave of even greater fortunes and status climbers. The fashionable elite were by nature more vulnerable to these new demands for assimilation than the old gentry elite had been to theirs. They themselves had defined upper-class membership by standards of material display; now they were in danger of being outbid for their own positions. The line they tried to impose between novel extravagance and vulgarity was too fine to withstand much pressure. Many of the industrialists knocking on the doors of fashionable society did manage to assume an aristocratic lifestyle, thereby gaining acceptance (although rarely without careful study and submission to sufficient degradation and initiation ceremonies). But others were still too closely tied to their occupations to break loose from the rituals of aggressive entrepreneurialship. Still others, faced with an upper class that had grown so rapidly and unevenly that it was breaking into competing factions, simply decided the traditional status game was not worth the effort. Not raised with any appreciation of "high culture," they pursued their own world of flashy entertainers and artists.

At the same time, the middle classes, who had only marginally been affected by the economic growth of the 1840s, were experiencing a significant rise in their standard of living. The reorganization of occupations and business enterprise by industrialization swelled their ranks and expanded their social horizons: the cultural gap between the upper and the middle class began to blur. The "new" middle class were still unable to dress and entertain lavishly, and lacked the education and leisure for purely aesthetic pursuits, but they formed their social clubs and held dinner parties in imitation of the upper classes.

The middle class itself was split between its more traditionally respectable members—the farmers, preachers, clerks, and eventually a placid group of bureaucratic functionaries—and its small-time entrepreneurs. Entrepreneurs, though never confined to a particular stratum, had long been associated with the more disreputable elements of the lower middle classes. But taking advantage of economic currents during the early twentieth century, they had gained a veneer of respectability commensurate with their more secure economic position. It was enough to move them into the middle-class status system, but they retained their distinctive lifestyle. They saw social position as an avenue to wealth, and wealth in turn could be translated into more prestige. Thus, they were more sociable and active in community affairs than their respectable neighbors; their tastes and lifestyle set the tone of middle-class culture far out of proportion to their numbers. We can expand this point as a general observation about the overweighted impact of the entrepreneurial style on American culture. Informality and familiarity may have been supported by structural changes, but it was the ubiquitous entrepreneur, as the most socially mobile individual (or, at least, the most energetic and voluble social climber) who carried this style with him as he made contact with all levels of American society; and it was entrepreneurs who developed and controlled the new and powerful instrument of cultural transmission, the mass media, who projected this image of the "typical" American in this nation and abroad.

These combined developments have signaled the disintegration of a uniform status hierarchy and the end of cultural dominance by a traditional upper-class elite. Faced with an increasingly reluctant or indifferent audience and the demise of their one reliable source of deference, the servant class, the fashionable elite found themselves infected by the informality and familiarity of these new groups whose paths they were increasingly crossing in the expanding public arenas of recreation, education, and civic activity. The stage-settings for the show of deference and demeanor had changed, and along with it the content of cultural stratification itself.

We have drawn upon contemporary novels as a primary data source for describing these transformations. This creates some problems of representativeness and accuracy, but novels provide one of the few sources available which depict social interaction in sufficient detail to reveal the deference rituals of the time. This is the best period for the purpose, for it was the heyday of the realistic social novel, and there were a great many authors who competed with each other in depicting manners.

The most serious problem is that not all groups produced authors or had interested observers from the outside to leave records of their lifestyles, nor publishing markets who wished to read of them. This is particularly true of the lower classes everywhere. Even Charles Dickens' celebrated observations about the grimmer sides of London in the mid-nineteenth century dwelt mainly on the unfortunate sunken members of the bourgeoisie. The respectable middle class likewise left few accounts of their lifestyles. The passing glimpses we catch from such novels as Theodore Dreiser's *Sister Carrie* leave little doubt that material for an exciting or amusing plot was not to be found in the monotonous routine of their lives. On the other hand, the groups on which we focus both left rich literary traditions. The upper class had their nineteenth-century "novels of manners" and we can trace the rising aspirations and fortunes of entrepreneurs in the more home-grown tradition of satires on the "democratic man" and American egalitarianism.

In general, the taken-for-granted assumptions of these novels are our most reliable sources of data, rather than their plots and incidents.[9] The "facts" of the size, location, and wealth of sociable groups, the physical settings of social activities and interaction, the basic units around which social activity is organized are relatively noncontroversial, and they tell us such things as when and where the family household is displaced by the large ball or the public nightclub as the locus of elite sociability. The description of social rituals, the topics and motives of conversation, and the general "atmosphere" of the culture found in these novels presents somewhat more questionable data, for authors may have been trying to impress their readers with a more novel or brilliant style than typically existed. Here we have to use our judgment, taking into account each author's perspective, comparing different novelists on the same materials, and testing out the implications with more conventional sources of data. In general, though, the authors of this period seemed to pride themselves on their detachment and their devotion to details and subtleties, and the cross-checks show a great deal of reliability.

The American Upper Class

Washington Square and *The Europeans* were written by Henry James in the 1880s about his class of origin in the 1840s and 1850s. We are struck by how relatively small, isolated, and poor the elite circles are in James' novels. We have indications from other sources that these impressions are accurate. The populations of even major cities such as Boston and New York were small by modern standards, and could easily be dominated by a single elite.[10] Visiting between cities with enough frequency to sustain

---

[9] This methodological strategy is discussed in Lowenthal (1957).

[10] The population of Boston in 1850, within the 3-mile radius of walking distance before

a network of social relations was clearly impossible as long as the horse and buggy remained the only form of transportation. European travel had not yet become a popular pastime. We can guess (particularly when we compare this period to the next) that the most important determinant of the size and the wealth of the gentry was that the pre-mass-industrialization economy of the time simply did not produce many families who could afford the material trappings of gentry culture, or who would have had enough leisure time to cultivate the tastes and activities that would bring them into contact with and make them acceptable to the established elite. At the same time, this economic structure produced *moderately* wealthy, rather than *extravagantly* wealthy families; their homes, servant staffs, dress, and entertainment do not evidence elaborate displays of material wealth.

Statistics based on contemporary government studies indicate that the average real per capita income increased 50% from 1815 to 1860; for laborers, the figure was 40%, and most of the increase went to those in a few skilled trades (Taylor, 1951: 296, 393; Martin, 1942: 396). A "typical" budget for a working family of five with an annual income of $538.44, published by the *New York Daily Tribune* in 1851, included no allotment "for amusements, for ice creams, his puddings, his trips on Sunday up or down the river in order to get some fresh air . . . to pay for pew rent in the church, to purchase books, musical instruments." Even this budget represented a standard of living out of the reach of most of the working class. A similar tabulation for a middle-class family earning $1,500 in 1857 reveals that the largest portion of their extra money went to better housing conditions. One significant luxury is noted—the middle class appears to have spent as much on servants as the working class did on rent. But there still was no allotment for amusement and entertaining. Moreover, the common working day for both groups was 10 hours; for laborers, this often included Saturdays. If the status game requires time and money, then most families before the Civil War had neither the wealth nor leisure to play.

The combined effect of these conditions was a social elite whose boundaries were secure, both from challenges by other groups for culture dominance and from large numbers of individuals demanding admittance, simply because of physical constraints. We see none of the elaborate rituals developed by less-isolated cultural groups for the introduction and evaluation of "unknowns." The Boston family in *The Europeans,* the Went-

---

the streetcars, was 208,000, having already undergone considerable growth in the previous decade. By 1900, the effective metropolis for railroad and streetcar commuters had a radius of 10 miles and contained 1,141,000 persons (S.B. Warner, 1962: Table 1).

worths, are quite surprised that their European cousin, Eugenia, sent her brother Felix to "announce" her, or that she considered the possibility of needing letters of introduction. To them, a blood relationship is a sufficient entrée. Kinship ties, as Felix and Eugenia readily admit to themselves, are not the only assets that accounted for their welcome by the Wentworths and by Boston society: they relied heavily on the "dazzling" effect created by their aristocratic manners and style of conversation. As Mr. Brand, one of the dullest members of the Wentworths' circle, comments in wonder: "Now I suppose that's what is called conversation . . . real conversation. . . . It must be quite the style that we have heard about, that we have read about—the style of conversation of Madame de Stäel, of Madame Récamier [James, 1959: 263]." Mr. Wentworth expressed most clearly the isolation and provincialism of his class in his reaction to Eugenia's fast-paced, witty, salon-styled conversation:

> She spoke, somehow, a different language. There was something strange in her words. He had a feeling that another man, in his place, would accommodate himself to her tone; would ask her questions and joke with her, reply to those pleasantries of her own which sometimes seemed startling as addressed to an uncle. But Mr. Wentworth could not do these things. He could not even bring himself to attempt to measure her position in the world. . . . He felt himself destitute of the materials for judgement [James, 1959: 256–266].

In an ironic twist, Eugenia and Felix are able to obscure the fact that they had come to live off their wealthy American relatives by making their cousins defer to their customs. When Eugenia refuses to participate in the customary rounds of visiting in the city, the Wentworths do not think her rude, only that she had tired of them.

Both money and breeding were considered necessary resources for those claiming elite status. The two generally came together, through family membership, which in turn kept these resources within the established elite. Small size and physical, economic, and social isolation had combined to create a tightly organized, inbred group. The households that comprised the social circle of the elite were bound by extensive kinship ties. The family was the basic channel for economic resources in that wealth was passed on through inheritances and through jobs dependent on family connections. The family was also the basic channel through which "good breeding" was acquired, and this was the major agency of socialization into gentry culture.

The social activities that made up gentry culture were organized around families operating as *total units:* Whatever the social status of the head of the family, the other members occupied the same status (Persons, 1968: 112; cf. Sennett, 1970). There was no evidence of age segregation in activities, once children reached 15 or 16 years. Although at dances, the parents and spinster aunts and uncles would retreat to the sideline while the younger people danced and acted out their rituals of flirtations and courtship, the older people were actively involved in pairing off couples by arranging introductions, whispering encouragements, and nodding approval. We notice that Morris Townsend, in *Washington Square* (1881), spent only enough time with Catherine at their initial meeting to dazzle her with his looks and "easy" conversation; he then devoted the rest of the evening to soliciting an invitation "to call" from Catherine's Aunt Penniman.

We can surmise that the rituals surrounding courtship did not provide the basis for an independent youth culture and an independent resource for status (as they have in the twentieth century) because, while young adults were ostensibly free to negotiate their own marriages, the young couples remained physically and emotionally within the orbit of their respective families until they were ready to start their own households. Their social contacts were restricted to whom they met through their family at family-dominated activities.

The family was still formally patriarchal. The moral traditions of the community still invested the father with considerable dignity and power. His own children had to "request interviews" with him to discuss personal affairs. Even here a chilly formality pertained. Yet this power was only a vestige of the era when fathers had absolute control over their wives and daughters. The implications of this change in family authority in terms of status negotiation become clear at the turn of the nineteenth century with the dominance of social life by wives as "hostesses" and in the mid-twentieth century by the dominance of youth culture.

Men and women were beginning to enter areas of activity outside the household—the men had their occupations, and women their charities—but full-blown, sex-divided subcultures were not apparent yet among the upper class of the 1840s and 50s. Men's occupations were important as a source of income but did not in themselves define circles of acquaintance-ship or bestow status: business friends, the ritual of the business luncheon, and men's professional and social clubs were not important until later in the century. Among the women, charitable institutions, asylums, hospitals, and aid societies were drawing them away from endless rounds of "at homes" and visits and needlework; but, as with men's occupations, charity works were undertaken *with* one's friends, not as a means of acquiring new ones. They formed no independent focus for sociable occasions or status displays.

The setting of social activity was almost exclusively within the home. The forms of entertainment were few: small dinner parties and large dinner dances were the only festive occasions mentioned by James; "at homes" and visiting were the most common forms of social interaction. Nowhere was the austerity and the provinciality of the gentry more apparent than at their sociable occasions. Their dances had none of the lavishness or resplendency of the grand balls of later decades; nothing but a hollow echo remained of the sophisticated conversations and intellectual pretensions of the European aristocratic salon, in the formal style of conversation and in passing references to literature and music.

Instead of intellectual achievement or material display, the gentry chose to distinguish themselves by their singular claim to moral superiority. Henry James was born in New York, but was educated and eventually took permanent refuge in Europe. He clearly scorned the moral preoccupations of the gentry class into which he was born, but even *he* accepted them as genuine. James described the effects of this moral preoccupation on his class of origin through the eyes of his European character, Felix:

> They are not gay ... they are sober; they are even severe ... they take things hard. I think there is something the matter with them; they have some melancholy memory or some depressing expectation. It's not the epicurean temperament. My uncle, Mr. Wentworth, is a tremendously high-toned old fellow; he looks as if he were undergoing martyrdom, not by fire, but by freezing [James, 1959: 236–237].

A rigid formality dominated their conversations and manners. But the formality was expressed in tight control over the emotions, in a dispassionate and calm appearance, to the extent we might call it a ritualization of the content, more than of the form, of expressions and actions. It was a formality that was to a large extent internalized—hence the distinction between genuine and pretentious, with the former the mark of the "true" gentlemen. The scene in which the Wentworths greet their European relatives for the first time has all the solemnity and propriety of any court presentation, yet is characterized more by what the Wentworths do not do, by their restraint, than by what they do. Emotions are expressed by the smallest, most controlled gestures: the eyebrows raised in sarcasm over Dr. Sloper's tone, or the expression of love in Lizzie Acton's blush and the incline of Clifford's head. Members seem quite adept at picking up these low-keyed communications; we see none of the elaborate debates of the later novels over what someone "meant" by his words or gestures. This again seems evidence of the closed, inbred nature of the gentry class.

Intimacy was constrained by this pattern of maintaining emotional control and formal conversational styles, thereby upholding the privacy of the individuals involved: theirs was a very frontstage culture. Friendships were measured by the frequency of visits, not by confidences exchanged. Goffman's little gods treating each other with dignity and ritual care are found here in a particularly solemn mood. Morris Townsend's future is destroyed by his failure to respect this principle in the presence of Catherine's father. Dr. Sloper condemns him with this indictment: "He is not what I call a gentleman; he has not the soul of one. He is extremely insinuating; but it's a vulgar nature. I saw it in a minute. He is altogether too familiar—*I hate familiarity* [James, 1959:52]."

Economic expansion in the 1840s and 1850s gave rise to individuals with fortunes that could easily challenge the material isolation of the gentry. William Dean Howells' (1971) character, Silas Lapham, is one of the first of this "new breed" we meet in American literature. Set in 1875, *The Rise of Silas Lapham* indicates that the gentry could manage to control these interlopers, although as much by the latter's ineptitude as by their own devise. It is mute tribute to the cultural dominance of the gentry, clipping coupons and living on a fraction of their opposition's income, that the *nouveaux riche* of this period were successfully denied equal status.

Lapham had earned his fortune in rural Vermont in the paint mining business. While not adverse to social climbing, he has no conception of how to break into the closed private world of Boston "Society":

> The fact that they lived in an unfashionable neighborhood was something they had never been made to feel to their disadvantage, and they had hardly known it till the summer before this story opens. . . . Lapham's idea of hospitality was to bring a heavy buying customer home to potluck; neither of them imagined dinners. . . . The girls had learned to dance at Papanti's, but they had not belonged to the private classes. They did not even know of them and a great gulf divided them from those that did. . . . They dressed for one another; they equipped their house for their own satisfaction; they lived richly to themselves not because they were selfish, but because they did not know how to do otherwise [Howells, 1971:118–120].

It is a young member (Thomas Corey of a gentry family, who quite by accident had met and fallen in love with one of Lapham's daughters) who engineers their introduction to Boston's high society. The dinner party arranged between the Coreys and the Laphams, which was to have served as the introduction, ends in disaster. All the etiquette books in the public library are unable to define the latest dress proprieties for the Laphams. Neither he nor his wife know the elaborate procedural rituals surrounding dinner parties—when to enter and leave rooms, with whom, and in what order. Lapham can only sit in silence as his companions for the evening make small-talk about novels he has never heard of, not to mention read, and gossip about other members of their group. The evening ends with Lapham getting drunk and dominating the conversation with his own "vulgar, bragging, uncouth nature [Howells, 1971:172]."

By the turn of the century, Lapham's counterparts were far more adept
at playing the social game, but then the context had changed considerably,
and structural changes greatly aided their attempt. In 1830, internal transportation was almost totally dependent on an elaborate but expensive and
slow system of rivers, canals, and turnpikes. The United States contained
only 73 miles of railroad track. By 1860, the total was 30,636 miles, and
in 1890, 492 million passengers were carried a total of 12 billion miles by
rail; in 1920, traffic reached a peak of 47 billion passenger miles (Taylor,
1951:79; Historical Statistics: Series Q66–67). One effect of the development of rail transport, of course, was the rise of the national industrial
economy and the development of large cities. A lesser-known impact of
the railroad was the effect of commuter trains on the character of urban
life. To the rich, it meant they could effect a peculiar combination of the
lifestyle and prestige of a landed aristocracy with the cosmopolitan diver-
sions of an urban elite, all the while maintaining personal control over
their urban-based corporations. In the 1890s, the thirty-mile commute
between Garrison-on-Hudson, where many of these wealthy magnates
built country estates, and New York City, where they constructed lavish
town houses, took less time than today. Edith Wharton's (1964) description of the rising aristocratic affectations of New York's fashionable elite
at the turn of the century corresponds directly with this increase in geographic mobility and the opportunities it presented for variety and choice
in sociable activities for the wealthy.

Improvements in sea transportation added another dimension for cultural pretensions by bringing America into the orbit of continental European culture. In the late 1840s, the Astor Place Opera House was built in
New York. As the name Astor indicates, this elite European pastime was
financed by the still embryonic fashionable elite. Ticket prices were astronomical and a strict dress code was imposed to assure the anxious *ton*
that they would not be associated with the vulgar spectacle of the Bowery
—a decidedly middle-class theatre. On opening night, a mob gathered to
protest the exclusiveness of the establishment and to "burn the damn den
of the aristocracy." Twenty persons in the crowd were shot and killed by
the police who were called in. A period of decline in opera attendance
followed, but it picked up rapidly in the 1880s; until World War I, a box
in the "Diamond Horse Shoe" at the Metropolitan Opera House remained
the "most luscious of social plums [Wector, 1937:463]."

Edith Wharton's (1964), *The House of Mirth,* describes in detail the hierarchy of social elites that grew up in New York under these conditions at
the beginning of the twentieth century. The *center* of elite culture had
shifted away from the low-keyed sociability of James' elite to the fast-
paced ("no one ever dined at home unless there was 'company'; a doorbell

perpetually ringing . . ."), resplendent, and highly formal and ritualistic culture of "fashionable society." Their lives revolved around rigidly defined "seasons": June to the Thanksgiving holidays were spent at country estates along the Hudson or at fashionable sea resorts such as Newport, and the other months were spent in New York.

Existing on the fringes of the fashionable world was a second group, with newly made fortunes but no social background or breeding, trying to gain entrance to that world. Farther down the social ladder, we see a third circle who had given up social climbing as boring and had "struck out on their own" into a newer mass culture. And finally, we see the "dimly lit region" of the "fashionable New York Hotel." This era seems to mark a real transition point away from the stable status structure of the early nineteenth century when classes were few, clearly defined, and readily distinguishable from one another in lifestyles, customs, activities, and manners.

Sustaining the proper *style* of conversation, including subtle gestures and expressions, now seemed more important to attaining or sustaining high status than the topic of conversation. Topics were generally gossip about the "social aspects" (i.e., fashions, architectural styles of homes, etc.) of various individuals. There were definite rules of politeness, familiarity, and display of interest that governed what status one wished to confer or extract. Because of their subtlety, knowledge of these rules from long exposure and use was an important resource used by the established members of fashionable society to recognize and exclude social climbers. But even those outsiders such as Rosedale, who had lived on the fringes of fashionable society long enough to imitate their manners, were not allowed to participate in such exchanges of familiarity as calling someone by his first name (as was done among established members of the fashionable circles), or to bring his hat and stick into the drawing room to add that touch of "elegant familiarity" to his appearance (Wharton, 1964:185). The fashionable elites clearly felt the threat to their solidarity and existence. Boundary maintenance was becoming a focus of considerable attention and not just an "instinctive" reaction.

We now see considerable contact between those who maintained the lifestyle of James' provincial gentry and the fashionable elites in terms of each knowing the members of each other's group, and occasionally coming together at weddings, dinner parties in private homes, and at "at homes." Contact seemed to be most curtailed by the failure of the gentry to enter into the new activities of fashionable society—the opera, the theatre, restaurants, week-long "house parties" at country estates, and traveling parties aboard private yachts and train cars—rather than from deliberate exclusion. The individuals who moved between the two circles (usually by reason of family connections or courtship) tried to conceal the gambling over bridge games, the lack of church attendance, and the flirtations of the fashionable circles. Sedate house parties were intentionally planned to promote marriages between heirs and heiresses and gentry families. This was tacitly accepted by the gentry as the only means of keeping money in the "right hands": there was clearly some sense of solidarity between these groups.

But in the end, as Wharton points out, the fashionable elite needed the input of *new* sources of money as much as the *nouveau riche* social climbers wanted social acceptance. The lifestyle of the fashionable society was simply too extravagant to be maintained by their own resources: Lily, one of the poorer members of her group, spends $9000 on clothes, jewels, and gambling debts in just six months; country estates, steam yachts, motor and horse carriages, large staffs of servants, palatial town houses on Fifth Avenue near Central Park had to be maintained; and entertainment often involved supporting dozens of guests for weeks, or even months, on trips abroad.

The only means of asserting their dominance that remained to the fashionable elite was to force those with "new money" to submit to an elaborate procedure or ritual of initiation. Rosedale, who has the double misfortune of being tainted by his entrepreneurial background and by his ethnic origins (Jewish) is a striking example:

> Already his wealth and the masterly use he had made of it were giving him an enviable prominence in the world of affairs, and placing Wall Street under obligations which only 5th Avenue could repay.... His name began appearing on municipal committees and charitable boards; ... his candidacy at one of the fashionable clubs was discussed with diminishing opposition. He had figured once or twice at the Trenor dinners, and had learned to speak with the right note of disdain of the big Van Osburgh crushes ... [Wharton, 1964:294].

Part of this elaborate procedure was the use of an impoverished member of the fashionable elite to act as a "sponsor" who had the contacts to arrange invitations to social functions, who could give advice on matters of etiquette, of dress, of where to go during what season, and who would receive "handsome presents" (i.e., money) in return.

The passage just quoted also points out the hierarchy of exclusiveness against which social activities were beginning to be ranked. The opera, the theatre, restaurants, large balls (the Van Osburgh crushes) were all large-scale, impersonal activities readily accessible to anyone who could afford the entrance price. Even those more private and exclusive activities of small dinner parties and house parties were carefully divided into those intended to fulfill obligations, and those reserved for intimate friends (i.e., *accepted* members of the social elite). It is in this sense that women, acting as hostesses who handled the invitations to these affairs, were the arbiters of social status.

It is remarkable to what extent the family had disappeared as the basic unit of sociable activity. Sexually mixed cliques, or circles of friends not necessarily including husband and wife as a couple, were predominant in daily activities. The husbands were absorbed in investment and speculation, riding the crest of the wave of business expansion. While financing and enjoying their wives' social successes, they seemed to prefer the seclusion of their clubs for a daily respite. It was left to their wives to organize daily sociable activities with what younger men (or momentarily free older ones) were available to "fight the boredom of leisure," and to organize the fancy-dress balls and elaborate dinner parties that gave this period the title "The Gilded Age." This was the period dominated by a handful of women—Mrs. William Astor, Mrs. Oliver Belmont (formerly Mrs. William K. Vanderbilt), the second Mrs. William K. Vanderbilt, Mrs. Ogden Mills, and Mrs. Ogden Goelet. Mrs. William Astor was the acknowledged queen; it was from the guest list to her grand ball of 1892 that Ward McAllister dictated the famous list of the Four Hundred to the society editors of the *New York Times* that effectively defined the families of the fashionable elite of that most fashionable of all cities [Wector, 1937:216).

But even as early as the 1890s, there appeared cracks in this wall of solidarity. The first apparently was in the person of Mrs. Stuyvesant Fish who "helped revolutionize the art of fashionable entertaining in America."

> With a smart, efficiently served fifty minute dinner she superceded the eight-course banquet with its fish and fowl, baroque confections, and half a dozen wines deemed necessary in the Brownstone Era . . . Mrs. Fish was apparently the first hostess of the upper reaches who sprinkled her invitation list freely with the names of amusing, attractive and talented people who had no social or financial claims—Mrs. Astor openly regarded her as a disintegrating force. Mrs. Fish offered her guests private theatricals with stars hired from Broadway, and in place of the old formal orchestra she introduced lighter brighter music by a small band. She was in fact the harbinger of the Jazz Age. . . . In accord with the new informality, everybody was called by his first name . . . and frequently there were name-callings of another sort. . . . Mrs. Fish cracked the impeccable dignity which the so-called Four Hundred enjoyed in the public eye, and prepared the way for its dissolution into small groups pursuing its own interests and amusements [Wector, 1937:341–343].

Parts of the upper class were clearly beginning to model their conventions more in line with the mass culture of the public sphere. By the end of the 1800s, although the group immediately below the fashionable elite remained captivated by its style and accolades, this trend was in full force. The circles of *nouveaux riche* illustrated by Wharton still participated in the more public activities of the established elite, but they openly preferred "a crowd they could really feel at home with." Formalities of social obligations, such as keeping appointments, maintaining visiting hours for "at homes," dressing to fit the occasion, are all ignored by the Gomers and their friends in *The House of Mirth.* Conversations become informal and familiar in style.

Even more disorganized and "socially irresponsible" are the fashionable hotel circle where hairdressers and manicurists are invited to stay for lunch. Wharton's heroine Lily Bart, a one-time member of the fashionable elite whose star has faded with her fortune, is hired to advance a member of this set socially. But Lily finds it impossible to separate Mrs. Hatch from her enthusiasms "culled from the stage, the newspapers, the fashion journals, and the gaudy world of sport [Wharton, 1964: 285]." Lily's dilemma is evidence of the growing multiplicity of elites, sharing the common denominator of money and leisure and crossing paths at some of the more public arenas of entertainment, yet living in completely different realities constructed by different tastes and customs. This transition was completed in the time F. Scott Fitzgerald was writing about the Jazz Age. The peculiar combination of familiarity and informality with impersonality, depicted in the famous party scene in *The Great Gatsby* (1925), where most of the guests do not even know what their host looks like, was the inevitable result.[11]

Entrepreneurs and the Middle Class

While the upper classes were lowering their tastes and standards, the middle classes were raising theirs. In the mid-nineteenth century it looked as if they had a long way to go; by the 1920s, it turned out not to be so far after all. Even as the fashionable elite in Wharton's novels were trying to crown themselves as an American aristocracy, the middle and working classes were turning their attention elsewhere, to forms of entertainment that emphasized their own cultural values.

Charles Dickens' *Martin Chuzzlewit* (1844), Herman Melville's *The Confidence Man* (1857), and Mark Twain's *Huckleberry Finn* (1884), describe middle- and lower-class culture before they were affected by industrialization and prosperity. All take as a major theme the style and activities of men who made their living by swindling the public—soliciting funds for

---

[11] This transformation in the basis of ritual precedence makes up the theme of all of F. Scott Fitzgerald's novels, with Hollywood coming more into the center of attention in his later works. See especially *The Beautiful and Dammed* (1922: 271–272).

nonexistent charities or selling nonexistent or worthless products—by creating elaborate disguises and fictions to gain the confidence of their audiences.[12]

While enterprising individuals undoubtedly exist in any historical time period among different levels of society, the pervasiveness (if the themes of these novels are trustworthy indicators) of hucksters and swindlers in nineteenth-century America can be explained by specific structural features. Most of the communities portrayed were newly established or caught in the tide of geographically mobile individuals and families that flooded the Midwest during this century of westward immigration. This meant that people were constantly forced into the company of strangers and into doing business with them. Despite the agricultural base and isolation of these communities, its members were oriented toward money exchange, speculation, and the market (Hofstadter, 1955:23–58). At the same time, lack of strong national, or even local, legal institutions made it virtually impossible to enforce formal contracts.

Feeding into this situation were the consequences of relative equality of conditions that existed particularly in the frontier regions. Melville was the most acutely aware of these authors that there was nothing particularly ennobling about democratization. By reducing material and social distinctions and by abolishing titles and legal distinctions necessary to build strong, relatively permanent status groups with distinctive life styles, manners, and customs, American-style democracy made deceit a constant possibility in human relations. Rituals of introduction and recognition (including specifics of demeanor such as dress, physical gestures, etc.) were easily faked and consequently meaningless among classes that did not have any standards or rituals for testing.

This opened the way for quick and observant individuals to fake status claims by simulating topics, styles, and manners of conversation. In fact, conversational skills were the major resources of the confidence man. We are left with the impression that the call of the dollar, even by disreputable means, prevailed, particularly since this kind of activity did not entail any loss in social status but was, on the contrary, an easy means of acquiring it within the transient boarding-house or hotel societies of the middle class that existed as pockets within the larger cities or as whole communities farther west.

Refraining from challenging the broker's claims about Eden (his west-

---

[12] This entrepreneurial style was also noted by Tocqueville in *Democracy in America,* where he took it as the dominant American culture of the 1830s, although he also noted the existence of a gentry group. Dickens verifies James' description of the latter group, depicting a polite, moralistic, and very status-conscious family who befriended his English hero, only to discover in shock that he is not part of the aristocracy but had come over in steerage.

ern land purchase) because to do so would have been "ungentlemanly and indecent," is the undoing of Martin Chuzzlewit, Dickens' English visitor. This is not to say that aristocratic values of personal honor as well as other upper-class pretensions such as dress, refinement, and elegance in manner were not affected by the confidence men, but these affectations were manipulated solely for the purpose of economic gain. It is only later, when these middle-class entrepreneurs develop steady and profitable careers, that they have the leisure and resources to pursue upper-class customs for purely status reasons. (In England, by contrast, the aristocratic and gentry classes not only engaged in business but were powerful enough to impose their cultural standards on commercial activity. Not only did business transactions, and that elusive area of business–social relations, maintain a more subdued, refined air, while being every bit as ruthless and opportunistic, but this situation was reflected in English law which gave much more weight to contract by "gentlemen's agreement".)

Several passengers aboard the riverboat Fidele in *The Confidence Man* placed their "confidence" and their money in the hands of one they judged to be a "gentleman" by the refinement of his dress and manner, his reserved yet respectfully-given attention, and his comfortable and prosperous air, only to find themselves deceived by a particularly clever actor. But here again, these individuals encouraged their own deception by their own eagerness to get something for little or nothing. Perhaps this latter factor provides partial explanation for the combination of extreme openness, volubility, and instant proffers of friendship among perfect strangers in both public (streets, coffee houses, boarding houses, ships, river boats, etc.) and private arenas within a society where deception and swindling were widely recognized as constant hazards.

Alongside these free-wheeling, fast-talking, hard-drinking members of the entrepreneurial middle and lowermiddle classes, there existed lower-middle-class cultural groups, especially farmers and preachers, that emphasized more traditional virtues of respectability and piousness. Most of the latter groups' time was spent in the home and within the family unit. Religious reading seemed to be the only acceptable daily leisure activity, but they did not seem to have much leisure time in the first place. Melville and Twain make it clear that the pious and hard-working were not immune to the deceptions of the confidence men, particularly those acting as subdued, respectful, and respectable clergymen soliciting funds for charities. Other than this, it is not clear what the relationship between these two branches of middle- and lower-middle-class culture were. Moral outrage characterized the reaction of the respectable class toward the entrepreneural class, but the former had neither the power nor the prestige to effectively sanction the latter.

While there undoubtedly existed a sizable group of families committed to ascetic living growing out of America's Puritan tradition, the fact that the church was the center of so much sociable activity in the frontier regions should not be allowed to inflate this figure. Martin Chuzzlewit's comment that lectures and sermons are to Americans what balls and concerts are to Englishmen has a larger meaning than is immediately apparent. As Twain seems to indicate, these activities were often the only form of entertainment available to people living in these isolated areas. Part of their susceptibility to the entertainment hoaxes of Twain's King and the Duke, we suspect, arose out of eagerness to do something new and exciting. The idea of church as entertainment seems a particularly accurate way to explain the circus-like atmosphere of the religious revival meeting Twain describes in *Huckleberry Finn*.

In the 1890s, America began to change from an urban society dominated by the private households and elaborate ceremonies of the upper class to the modern affluent society where everyone would be able to participate in some form of public entertainment and take advantage of widely available transportation, and where no one group would have the power to dictate cultural standards. Figures indicate that the mass flight to the suburbs dates back to the 1850s. In 1857, the Boston and Worcester Railroad alone reported carrying half a million passengers between Boston and stations no more than 10 miles away, to such middle-class bedroom communities as Dorchester, Milton, Dedham, Roxbury, Brookline, Brighton, Cambridge, Charleston, Sommerville, Chelsea, Lynn, and Salem (Taylor, 1951: 390–391). Even working men in the major urban areas benefited from transportation improvements with the development of horse-drawn street railways. The rich still had their private carriages; but in 1860, the longest horse-drawn railroad in Boston carried nearly 35 million passengers. The rapid spread of horse-car lines in Boston in the 1850s led to major population shifts, with large numbers of Irish workmen invading residential areas previously closed to them because of transportation difficulties.

The transportation revolution did far more than change residential patterns and mobility rates. Combined with technological advances in communications techniques, it lay at the bottom of much of the transformation in recreational habits from the 1890s on. The driving forces in this transformation were the ubiquitous American entrepreneurs. They took note of the rising affluence among the growing middle class and the working class as well, and the race for the consumer dollar was on. The fact that the field of communication was left wide open to commercial development is one of the most significant determinates of the direction American culture and patterns of status stratification have taken. Communications entrepreneurs were looking for customers, trying to create a demand where only potential existed: their offerings were geared directly to the tastes of the common man.

Daily newspapers in the early nineteenth century were relatively expensive and were intended to provide, in the words of one editor, "a complete history of the COMMERCE, POLITICS, and LITERATURE, of the times." The appeal clearly seemed to be geared to upper-class rather than mass circulation (Lee, 1937). But in the 1890s, innovations in printing techniques and the rise of chain newspapers brought the economies of standardization and size to the industry. Men like Scripps, Hearst, and Pulitzer now offered one- and two-cent papers to compete with the four-cent dailies. The whole new stratum of readers they found, however, was not attracted by the price alone but by the "sensational" journalism pioneered in Pulitzer's *World* and quickly followed by Hearst's *Morning Journal*. The "yellow press," as it was called, emphasized sex, crime, violence, and scandal "of the more general sort" which were said to appeal to the tastes of lower-class women, and extensive sports coverage to appeal to the working-class men. The popularity of the Hearst–Pulitzer style of sensationalism was confirmed by their skyrocketing circulations and the rapid collapse of many less adventuresome dailies.

The preoccupation with increasing profits (through advertising revenue as well as direct sales) by directly appealing to the interests and standards of the larger working and middle classes rather than the refined cultured and intellectual tastes of the upper class continues throughout the history of the media in the twentieth century. This same process has been at work in the changing arenas and character of public entertainment. Before the turn of the twentieth century, only the elite had much time and money for entertainment and recreation and, hence, controlled this second powerful means of defining the proper ritual reality and their place in it. The vaudeville and burlesque circuits had long brought troups of entertainers to small towns across the country, often providing the only contact these isolated communities had with the outside world. With the growth of urbanization, the "common man's opera" settled into permanent homes in the more sizeable metropolitan areas. But their tenure was rather brief. In the early 1900s, the first moving pictures were introduced. Extremely cheap and convenient, they could, quite literally, be shown in any abandoned drug store; and by 1907, there were over 5000 of these nickelodeans in operation. Jewish entrepreneurs, shut out of more respectable pursuits, had turned the motion picture industry into a 77-million-dollar industry by 1921. This was the era of the "million dollar movie palaces" built in obvious, if gaudy, imitation of the opera houses of the upper class and providing a combined bill of movies and vaudeville (Steiner, 1933: 108–109).

Movies were not just an isolated diversion from traditional sociability. The fairy tale world of the celluloid became part of a larger picture of amusement available to the more sociable members of the middle class— in the dance halls, road houses, and nightclubs that grew up almost overnight in the 1910s and 1920s (Steiner, 1933: 114–115). Much of the focus of public attention and curiosity shifted from the upper class to the glitter of Hollywood. This was spurred by new forms of transportation, which made it possible for most individuals to seek out their own settings for entertainment, free from controls by more traditional groups.

This was made possible especially by the private automobile. Automobiles were produced in America as early as 1893. At first they were regarded as play things for the rich. Woodrow Wilson, then president of Princeton, deplored the motor car as likely to stimulate socialism in the United States by inciting the poor to envy the rich. But in America, entrepreneurs like Henry Ford proved Wilson wrong by building an automobile for the common man; his mass-produced low-priced Model T was distributed in millions, as no such item of property had ever been distributed before. In 1910, the ratio of cars to people was 1 : 201. Ford's assembly line did not begin moving until 1913, but by 1920, the ratio had dropped phenomenally to 1 : 13, by 1930 to 1 : 5.3, and in 1969 to 1 : 2.6 (Rae, 1971: 43, 50).

The working class participated in these activities to the extent that they were able. Despite advances in union organization, between 1890 and 1920, the real income of wage laborers increased only 1%, but the normal work week in American industry between 1880 and 1930 was reduced approximately 20 hours (P. H. Douglas, 1930: 392; Steiner, 1933: 10). The "half-holiday" of Saturday afternoons and Sundays (once the religious objections of the respectable middle class had been overcome) became new markets for exploitation. Amusement parks and trolley parks existing at the city limits at the end of urban rail lines were popular among younger working-class couples. The masculine-oriented culture of the working-class sports occupied particular prominence in their sociable activities; with the shift toward a public center of attention, these became popular with the middle class as well.

The first professional baseball team was organized in Cincinnati in 1869; in 1886, a largely working-class crowd of 10,000 witnessed the opening game in New York. By the decade 1911–1920, World Series attendance averaged 173,000 and was to increase 41% in the next decade. Radio broadcasts, which at first had been rejected by club owners, became the true popularizer of the game. Boxing was another favorite of the working class, especially among the immigrant groups. The sport brought particularly heavy criticism from the upper class because of its "vulgarity," "brutality," and association with gambling: it was refused legal status in most states until the 1890s. But the criticism did not dent its popularity: in 1927, 120,000 saw the Dempsey–Tunney fight in Philadelphia, making it the largest single sports attraction of the times (Steiner, 1933: 6, 85–95).

Until after World War II, football was the only spectator sport to enjoy popularity among the upper class. This was because of its early and exclusive association with intercollegiate athletics. The game started as intercollegiate competition between the Ivy League schools. In 1889, the player names of the All-Americans were, with one exception, of Anglo–Saxon origins. After 1895, it was rare that at least one Irishman was not included. By 1927, ethnic-origin names dominated the list as the Anglo–Saxon names had earlier. This was paralleled by a shift of class identification with the game. In America, intercollegiate football became a symbolic battle ground for ethnic struggle. The "subway alumni" joined the legitimate alumni in rooting for "our boys" from Notre Dame. This element of class-ethnic competition had much to do with expanding the popularity of collegiate football to an audience with no other contact with higher education (Riesman and Denney, 1951). Colleges famous for their teams displaced the elite educational institutions in public esteem.

It is this new atmosphere of status made in the world of mass entertainment that we find in Sinclair Lewis' novel, *Babbitt* (1922), set in the early 1920s in the medium-sized city of Zenith (modeled after Cincinnati). Lewis is primarily describing the lifestyle of the middle class, middle-aged businessman in small businesses connected with consumer production and services, or high-level managers in medium-size firms. While there are decisive differences between the *activities* of Babbitt and his friends and the earlier confidence men (largely due to changed structural conditions such as greater community stability, the formal organization of commerce, increased prosperity which brought a desire for respectability as well as economic return, and the increased power and reach of the law), the similarities in *style* are striking, and we feel justified in placing them within the great American entrepreneurial tradition.

As reflected in *Babbitt,* we find the culture of the midwestern entrepreneurial middle class deliberately (i.e., ritualistically) informal and familiar. Sociable activity was dominated by the men and frequently took place outside the home within organizational contexts such as clubs and business conventions. Their business and social worlds were intertwined: success in one fed into and determined success in the other. Social relationships reflected this situation. "Friendships" were not just casually formed around shared positions in the occupational hierarchy but were actively sought among those whose ventures or contacts would be profitable to one's own business. Within this context, conversational skills were crucial in establishing one's position as a "good fellow" in order to be included in the inside business deals that were the payoff of social acceptance and provided the economic resources for further business and social success. In this world of hustling, fast-moving deals, and constantly shifting fortunes, conversations were dominated by rituals testing the continued prosperity (to see if one's audience was a desirable friend) and commitment (to see if he still wanted to be *your* friend) of a friend, and conveying the same. These gave rise to a distinctive conversational style of loudness, gregariousness, and physical and verbal intimacy.

Middle-class culture was a product of the fact that economic resources for putting on status displays were available for the first time to large segments of the population. But there was neither leisure nor money enough to support the individualistic, personalistic, and elegant style of the early upper classes. Middle-class culture was mass-produced culture; Lewis describes it as extremely conformist and monotonous. Its consumers lived in tract housing; bought mass-produced furnishings, assembly-line cars, readymade clothes; went to movies, listened to the radio, attended cultural lectures, or followed baseball; read newspapers, magazines, dime-novel Westerns, and love stories; and even planned all their dinner parties with the same menu served by the same caterer.

Important in terms of breaking down the exclusiveness of elite social circles was the fact that upper-middle-class and upper-class men were involved in many of the same organizations as the middle class. There remained a rigid stratification among the more purely social organizations such as the country clubs, and the social elites stayed away from business–social organizations such as the Elks, the Rotary Club, and the State Association of Real Estate Boards which were oriented toward the small business interests. But community political issues, often seen as moral concerns for business—such as unionization, the "immigrant problem," liberal political candidates—cut across status lines and forced businessmen of all classes to cooperate in the Chamber of Commerce, Zenith Booster's Club, political campaigning, and the Good Citizen's League (an "Americanization" group). Other centers of social life, such as the church, had begun to formally organize their activities to suit institutional purposes of increasing membership. Thus, most committees were open not only to those with high status but to those who would contribute time and effort to organizational goals. It was usually the middle class who were the working members of these organizations;[13] their payoff was civic

---

[13] Compare Davis *et al.* (1965: 78–79), and Lynd and Lynd (1929: 285–286, 301–308). The latter notes the proliferation of formally organized clubs and activities in the 1920s which replaced the more informally scheduled and family-centered sociability of the 1890s, and the

acclaim and the opportunity to mix publicly with the social elites, with the hope of being included in those few private activities, such as dinner and cocktail parties, which marked final admittance to elite status. No amount of civic contributions or recognition could break down status barriers unless they could be turned to economic profit. But once this was accomplished, the social barriers were mainly physical or material and could be crossed by buying a home in a more exclusive neighborhood and a chauffeured limousine. Manners and conversation were not so incompatible between the middle-class entrepreneur and the upper middle class in the provincial Midwest.

The upper middle classes lacked many of the signs and rituals that could have made their lifestyle distinct from and less accessible to the middle class. They dressed the same and also surrounded themselves with mass-produced goods. While the quality of their material possessions was higher, for the first time these goods were accessible to *anyone* with the purchase price. They had informalized many of their sociable activities. They also had adopted the public entertainment interests of the middle class, such as baseball and movies, which allowed for no privacy or separation of cultural groups. Their style and conversations also reflected a move toward more familiarity and informality.[14]

We hear no mention of relatives, maiden aunts, or extended family gatherings in *Babbitt*. Family life, such as there was, centered around the nuclear family unit. Even when not working at their businesses or in their organizations, men developed a series of all-male activities such as poker parties, fishing trips, business and club conventions, which isolated them from their families. Their children filled this gap with their own subculture.[15] There was a sharp associational break between young people still in school and men with jobs and families. Youth culture appeared to be simply a more flamboyant and intense (reflecting more leisure time) version of adult middle-class culture. Cars, baseball, and sports in general,

---

tendency for social clubs to be used for business purposes, thus corroborating Lewis' observations.

[14] The shift in both middle- and upper-class life from a household to a commercial focus of recreation is documented by a comparison of published biographies of the years 1880–1917 with surveys for the period 1918–1946 (Bossard and Ball, 1950). The ritualism of daily meals at home in the earlier period was often replaced by meals taken by family members apart and at restaurants; evening participation in family reading, prayer, or projects gave way to listening to the radio or the record player, and to commercial recreation and community activities outside the home; the Sunday drive in the automobile became an escape "rather than a horse-and-carriage open show of family pride to well-known neighbors [p. 103]."

[15] Compare Lynd and Lynd (1929), and Hollingshead (1949), the latter being a study made in 1941–1942.

and school organizations and fraternities were the centers of their interests and activities. There was also much more interaction between sexes, presumably reflecting courtship as a major preoccupation of this age group.

This relative fluidity of social boundaries at the higher levels and the widespread participation in a mass culture is borne out by a number of community studies made in the interwar period (Lynd and Lynd, 1929; Warner, 1949; Warner and Lunt, 1941). At the same time, the cultural and social gulf between middle and working classes was still relatively sharp; in certain parts of the country, such as New England and the South, a more traditional upper class continued to set itself off from the middle and upper middle classes. But even these surviving local elites could not dominate the community's attention; the forefront of the public scene was held by the ritualized familiarity of the middle class.

THE EVAPORATION OF DEFERENCE CULTURES

By the 1920s, the major transition in status cultures had already occurred in America, and subsequent developments merely extended the trend. Ritual deference is determined by authority relations, the degree of social density or surveillance, and the diversity of communications. All of these conditions differ markedly from those of the mid-nineteenth century, not to mention the earlier period of fortified household societies.

Authority relations have changed both at work and at home. The sphere of sedentary middle-class administrative, professional, and technical occupations has expanded, displacing the more clear-cut attitudes of order-givers and order-takers. Unionization has given much of the working class, if not a greater share of the wealth, at least shorter hours and more freedom from ritual deference-giving on the job. At home, the authority of patriarchs has declined with an increasing prevalence of working wives and peer-oriented children. Servants have disappeared with the coming of household appliances for cleaning, cooking, and lighting, which previously had been tedious chores that middle-class people avoided wherever possible, and with the general rise in wages reducing the numbers of those willing to undertake the ritually demeaning roles of household servants.[16] It has been nicely stated that the upper class has not

---

[16] The ratio of servants in the labor force to number of households went from .26 in 1850 to .04 in 1950, or from at least one servant for every middle and upper class household, to about one per ten such households. Along with this, the population per household dropped:

disappeared—it is only that the servant class has. This has made a considerable difference, because a great deal of the pomp and circumstance of the upper class could only be put on with a retinue of servants. The butler in the old-fashioned household was a key figure in presenting the distinguished play of his master and mistress' social engagements.

By the 1940s, the American working class had acquired its own automobiles, as had most middle-class youths and women; separate entertainment cultures were now fully possible, free from the surveillance of dominant classes. The result was an efflorescence of hotrod clubs and youth gangs, specialized sports cultures and ladies' bridge clubs. It was increasingly possible to escape from scenes in which one did not shine, and make an appearance only on stages of one's own choosing. The working-class flight to low-cost suburbs after World War II enhanced class (and racial) segregation.[17] Residence and work place became sharply distinguished for most people, especially with the decline of jobs as servants and of small family farms and stores. This reduction in the possibilities for close surveillance of individuals brought the obverse case of groups not being able to study and distinguish each other. Increases in the absolute (although not relative) level of wealth in the working and lower middle classes further reduced public surveillance by bringing a fairly respectable style of dress within reach of the great majority of the population. Studies in the 1950s (Form and Stone, 1957) showed that, although people still believed they could pick out the rich and the poor by obvious material artifacts (big cars, furs, jewels, versus dirty clothes, work uniforms, lunch pails), the distinguishing marks of the intervening middle classes were vague.

1790	5.79
1850	5.55
1900	4.76
1940	3.77
1970	3.17

(Historical Statistics, Series A255–257, D68, D457; Statistical Abstract, 1971, Table 44). The effect of the servant shortage on the upper class is described in Birmingham (1968: 14–16).

[17] This does not mean the disappearance of working-class culture, however, as the differences remain even in the suburbs (Berger, 1960; Gans, 1967). Interestingly, intellectual critics whose standards are those of the aristocracy-emulating upper class regard mass development suburbs with horror, apparently because they impinge upon the elite image of "country living." Critics usually fail to recognize that such suburbs are essentially horizontal versions of working-class tenements; the "little boxes made of ticky-tacky" are not filled with doctors and lawyers, and the impression that they are only illustrates how invisible social class has become.

Family- and home-centered activities declined, and social life moved to a wide choice of public settings—restaurants, bars, nightclubs, sports arenas, and movies. (The height of this public atmosphere was reached between the 1920s and the 1940s; since then, television has brought a new level of privatization.) Radio broadcasting began in the 1920s; by 1940, 91% of American families owned radios, and polls rated it America's favorite recreation (Bartlett, 1947). The introduction of television and cheap long-playing records further enhanced the popularity of the home as an entertainment center, but without reviving the surveillance of family-centered activities, since the mass-produced entertainments were highly voluntaristic in structure. Stage-settings that previously could be found only at collective festivals or scheduled performances could now be supplied on demand. At the same time (contrary to the outcries of elitist intellectuals), the older forms of fashionable entertainment did not disappear; libraries, book clubs, operas, and symphonies have actually spread, but as part of a diverse situation rather than the center of attention (Kluckhohn, 1958).

This disperson of resources for creating one's own ideal persona raised several possibilities. Many people took advantage of their increased resources to emulate the culture of the traditional upper classes. By the 1920s, piano lessons, attending concerts, and imitating old courtly manners at ladies' tea parties had spread to the lower middle class in America. By the 1930s, one could find working-class girls dressing themselves up at least once in their lives for their school graduation ball in the costumes of the courtly aristocracy of the eighteenth century. Detroit-made automobiles for this emulative market stressed size and grandeur reminiscent of the old limousines, reaching a zenith with the tailfins and enormous size of the 1950s models.

But status emulation could spread only so far without a reaction. After all, status-gaining techniques are relative and invidious things, designed to assert superiority over others. By the mid-twentieth century, there was little that remained exclusive about the externally visible side of elite culture, and a great many people had decided that there was little to be gained from it. Moreover, the extravagant and often boring elite culture had to compete with mass-consumption cultures geared to indigenous lower middle and working class tastes. The mass-circulation newspaper, with its comics and sensationalist news, led the way, followed by movies, sports, radio, and television. The devotees of elite culture, of course, did not yield hegemony without a fight, making initial efforts to control radio and television in the interests of "cultural uplift" and "public affairs" (now disguised as "education"), and episodically attempting to censor movies and rock 'n' roll music (the "payola" scandals of the 1950s), and to prohibit violent sports and gambling. But commercial competition favored those entertainments with the widest direct appeal. In the 1940s, it was the daytime soap operas that attracted the biggest radio audiences; in the evening, variety and comedy programs topped the list, with melodramatic serials next; symphonies and public forums were not found in the top 50 (Bartlett, 1947). The same pattern emerged in television. The commercialization of the media for a mass audience, beginning with mass-vulgar newspapers and carried on by the radio and television, eroded the monopoly the respectable elite once enjoyed over the means of propagating a public culture in their own idealized image.

The general trend has been the decline of the old pretensions, the evaporation of the old collective conscience and its supporting deference rituals. Changes in styles of clothing, manners, conversation, and entertainment have been apparent since at least the 1940s, when David Riesman (1950) seized upon them to lament the decline of an older ideal of personal dignity, while failing to see the stratification that had supported it.[18] But the observations themselves are valid: the disappearance of the old formal party with its obliging host and array of servants, along with the high theatricality of evening dress; the bowing, hat-tipping, and polite titles of address giving way to a pervasive nicknaming familiarity; the neglect of old rituals of standing when women and superiors entered a room, and of the elaborate handshakings and introductions; the gossipy sociability that replaced the stiff aloofness and clearly marked ritual barriers of the traditional gentry style and the urbane posturing of fashionable society. The informal style, to be sure, was nothing new; we find it in Sinclair Lewis describing the 1920s, and in the mid-nineteenth century among the more disreputable entrepreneurial classes. The 1940s merely elevated it to general cultural respectability, and broadcast it in the newer casual heroes of the movies and comics, with their wisecracks and their slang.[19] Even the old formality of "proper" grammar was on the

---

[18] Evidence from an empirical study of parties is given in Riesman (1960a, 1960b). Riesman summed up the findings with his characteristic value bias: "a shift toward more relaxed and more egalitarian norms of behavior in sociability. Monopolies of responsibility for the conduct of a party have tended to shift from the host to the guests . . . and acceptance of mediocre performance and even passive nonperformance has increased accordingly [Riesman and Watson, 1964: 239]."

[19] The entire shift is nicely illustrated in the history of the comics. These began in the mass-circulation newspapers of the 1890s with a cast of lower-class buffoons and ethnic stereotypes (the Katzenjammer Kids, Mutt and Jeff, Jiggs and Maggie, Barney Google). In the 1920s and 1930s, with the spread of comics to middle-class newspapers around the country came strips with idealized small-town characters (Little Orphan Annie, Dagwood Bumstead, Gasoline Alley, and Disney characters such as Donald Duck and Mickey Mouse). World War II shows an important break: its adventure heroes (like Terry and the Pirates, Steve Canyon,

wane, to the consternation of those who identified their own status pretentions with "standards" in the absolute.

What was happening was a decline in the conditions that supported strongly marked deference rituals. Strict authority relations affected far fewer people, and took up a much smaller space in their lives. Surveillance was reduced by transportation and wealth that made face-to-face contacts more voluntary than obligatory; the diversity of communicative contacts was generally increased. This meant a sharp decline in the intensity of collective conscience and group rituals; the content of public culture became more egalitarian, relativistic, and secular. This trend has been misleadingly characterized as "mass culture": as everyone acting, dressing, doing, and thinking the same things, and as the destruction of any kind of class distinctions. The dispersion of class and other cultural communities—who may or may not be doing the same thing—is a more accurate characterization. Individuals isolated at work, in their class-segregated neighborhoods, and by voluntary choice of recreation and entertainment, and whose main contact with the public is through the sterilized vision of the mass media, have little way of knowing what others are doing, let alone judging the status relationships among those outside their immediate reference group. The rituals that remain are primarily inclusion ceremonies among voluntarily assembled groups of equals.

The phenomenon has not been universal. Different cities, towns, and regions of America have their own ambience, ranging from the relatively intense collective respectability rituals of smaller towns, the closer ethnic communities and sharper social boundaries of the Northeast, the street culture of the ghettos, the patriarchal coerciveness of the rural South, to the extreme informality of the sprawling freeway cities of the far West. There is still a variety of class and ethnic cultures, anchored in different family styles and occupational milieux; these differences remain, although in a community structure that makes them into private enclaves rather than elevating a few into recipients of public deference.

Dick Tracy, Joe Palooka) are no longer portrayed as elegant gentlemen but as tough-talking, slang-using ordinary guys; at the same time, villains come to be portrayed as aristocrats and tycoons rather than as hoodlums. Humphrey Bogart, whose movie career was made at about the same time, played the same sort of working-class hero. The military mobilization of the working class during World War II must have had a good deal to do with popularizing its culture, although the trends appear even earlier in the realm of popular sports. Comics provide a peculiar mausoleum of these shifts, since they tend to preserve the styles and stereotypes of the period in which they began. A modern Sunday newspaper can display next to each other portraits of the boarding-house world of the 1890s, the small town life of the 1920s, the urban ambience of the 1940s, the suburban life of the 1950s, and the relativistic fantasies of the 1960s. See, for example, Perry and Aldridge (1967).

Nor is there any necessity for all industrial societies to take this form. We see certain elements of it in Europe, but a great deal more of the traditional deference styles remain, evidenced in the greater formality of dress and the sharpness of social barriers; the continuing importance of formal introductions and traditionally "correct" speech; the pervasive rituals of handshaking, rising, bowing, and otherwise showing mutual deference in the mid-nineteenth-century style; the continuing distinction in most languages between the formal and the familiar pronouns of address, which not only mark off sharp boundaries between intimates and others but allow superiors to treat subordinates with intrusive familiarity that is not allowed to be reciprocated (Brown and Gilman, 1960).[20]

A number of conditions support this greater ritualization in Europe. Public transportation outweighs the use of private autos; geographic mobility rates are low, and cities and towns retain their traditional high-density architecture; and the absolute level of wealth among the middle and lower classes is lower. All of these contribute to a greater degree of physical copresence and a higher condition of personal surveillance, resulting in an atmosphere more conducive to ritualization. There is a greater monopolization of the resources for economic and cultural domination, especially education and property ownership; and not least strikingly, the control of radio and television by governments who use them to impose traditional elite culture under the unexamined and supposedly class-free rubric of "good taste." Even in the Communist countries, an allegedly proletarian elite doggedly enforced nineteenth-century aristocratic and *haut bourgeois* tastes in music, art, and literature.

The totalitarian countries demonstrate in an extreme form what applies everywhere: The degree of surveillance and of cultural diversity can be either high or low in industrial societies, depending on a number of other variables—geography, absolute wealth, types of housing and transportation, and control over the mass media, all of which interact with the underlying system of political and economic stratification.

The form of status stratification found in the contemporary United States, then, is by no means an evolutionary stage to be gone through everywhere. We need to make our predictions from more specific conditions in each particular place. Nor are matters at an end in America. The 1960s witnessed yet another phase of the same development found here over the preceding century, with the rise of a new youth culture attacking

---

[20] The "thou" and "you" forms were used in Elizabethan English, but had disappeared by the nineteenth century (except perhaps in the most patrimonial rural areas), with the American colonies leading the way toward greater equality in titles of address (Main, 1965: 221–239).

many forms of ritual deference and demeanor that had been taken for granted. Since most of the changes we have been describing have cropped up first in the younger generation, this no doubt foreshadows changes in the adult world to come. These matters are considered in Appendix B.

APPENDIX A: SUMMARY OF CAUSAL PRINCIPLES

6.0 Deference and demeanor rituals are produced by a combination of social density conditions (*2.1, 2.2*) and authority relationships (*1.1– 1.4*).

6.1 The more unequal the power resources and the higher the surveillance, the more often acts of petty ritual deference are demanded (from *2.1* and *1.1*).

6.11 The more unequal the power resources and the lower the surveillance, the more perfunctory the compliance with deference rituals, and the greater the tendency for individuals to evade contact (from *2.1* and *1.3*).

6.2 The more unequal the power resources and the higher the diversity of communications, the more elaborate the deference rituals and the more complex the standards applied (from *2.2* and *1.1*).

6.21 The more unequal the power resources and the lower the diversity of communications combined with high surveillance, the more extensive the formality of conversational manners (from *1.1, 2.1,* and *2.2*).

6.22 The more unequal the power resources and the lower the diversity of communications, the more simplified the gestures of ritual interaction and the more emphasis on highly visible gestures of nonverbal deference.

6.3 The more equal the power resources and the higher the surveillance, the more conformity to rituals of group inclusion (from *1.3* and *2.1*).

6.31 The more equal the power resources and the higher the diversity of communications, the less emotionally compelling the inclusion rituals and the less reified and unchanging the attached symbols (from *1.3* and *2.2*).

6.4 The more dangerous the way of life and the more fear, the more use of emergency-related solidarity rituals, and the more severe the punishment for ritual violations (from *4.83* and *4.85*).

6.41 The more authority an individual gains from upholding rituals, the more he enforces rituals and ritual punishments and attempts to reify their symbolic rationale (from Postulate VI).

6.5 The more episodic the good fortune, the more ritualized the celebration (from *4.6* and the presumably heightened attention in such situations).

6.6 The more that internally egalitarian groups are stratified among themselves (i.e., externally ranked in relation to each other), the more important membership in or exclusion from a group is for individuals, and the more ritualized the marks of membership (from *1.3* and *4.84*).

6.61 The more conflict for precedence within a collegial group externally ranked above other groups (i.e., a group whose members are equal in the deference they receive from nonmembers), the more conflicts take place through ritual challenges and adherence to rules of contest, and the more precedence within the group is determined by adherence to ritual propriety expressed in codes of honor.

6.611 The more violent the conflict within such a group, the more the code of honor emphasizes physical prowess and games of danger (from *1.8*, *6.6*, and *6.61*).

6.62 The more peaceful the conflict within such a group, the more the code of honor emphasizes verbal and aesthetic skills (politeness, conversational skills, refined standards of consumption).

6.7 Wealth is used to produce the greatest amount of subjective status possible under given conditions of stratification.

6.71 The greater the wealth in a situation of highly concentrated resources for stratification, the more wealth is used for ostentatious dramatization of rank.

6.711 The greater the wealth in a situation of ranked collegial groups, the more wealth is used for subtle refinements of taste marking group membership and precedence (from *6.1*).

6.712 The greater the wealth in a situation of relatively dispersed resources for commanding deference, the more wealth is used for private entertainment and consumption and for escape from ritual encounters in which individuals receive the least deference (from Postulate VI).

6.72 The lower the wealth in a situation of highly concentrated resources for stratification, the more wealth is used to create status by gift-giving.

6.721 The lower the wealth in a situation of ranked collegial groups, the more ostentation is dishonored, and the more wealth is used for dramatizing pious morality by cleanliness, sedateness, and acts of public charity (from *5.23, 5.232,* and *6.61*).

6.722 The lower the wealth in a situation of relatively dispersed resources for commanding deference, the more wealth is used for episodic celebrations which create solidarity by gift-exchange within the private group of participants (from *6.5*).

6.81 The more efficient the technology of transportation and communication, the greater the potential diversity of communications and the lower the potential level of surveillance.

6.82 The more that political authority is separated from the household, the greater the potential diversity of communications and the lower the potential level of surveillance.

6.83 The more that work is separated from the household, the greater the potential diversity of communications and the lower the potential level of surveillance.

6.84 The greater the commercialization of communications (and the lower the political control over communications), the greater the potential diversity of communications.

6.85 The greater the availability of private transportation (rather than public transportation), the lower the potential level of surveillance.

This may be a convenient place to summarize the principles concerning ethnic group cultures presented in Chapter 2.

7.0 The economic, political, and community structures of the societies occupying particular territories shape habitual modes of ritual encounter and, hence, emotional expressions and symbolic attachments; the extent to which these are maintained or changed when territorial migration brings them into contact with other societies depends on the extent to which resources motivate, allow or require the continuation of ritually distinctive communities.

7.1 The lower the participation in a common economy, the more strongly ethnic distinctions are maintained.

7.2 The more economic and political power are monopolized by a dominant ethnic group, the more strongly ethnic distinctions are maintained.

7.3 The more distinctive the emotional tones and ritual postures and symbols of ethnic cultures, the more difficult it is for their members to associate informally across ethnic lines, and the more likely they are to have careers in distinctive occupational levels or sectors.

7.31 The more distinctive the economic, political, and community structures of societies coming into contact by territorial conquest or migration, the sharper the ritual barriers between them.

7.4 Weber's principle of status-group formation. The more stable the political resources, the more distinctive and ritually bounded the ethnic and class communities in that territory; the less stable the political resources, the more ephemeral the distinctions and the more the claims for status-group membership become based only on prior achievements of families or individuals.

APPENDIX B: YOUTH CULTURES AND DEFERENCE AND DEMEANOR

The important cultural shifts of twentieth-century America were all manifested first in prominent youth cultures. We should not overgeneralize these, for almost all were confined to an upper- or upper-middle-class minority. Not everyone was a flapper in the 1920s, a beatnik in the 1950s, or a hippie in the 1960s; most of them conformed closely enough to the organized school activities their elders had set up for them (Berger, 1971: 44–87). Minority movements have turned out to be important because others followed their lead, although in a less flamboyant manner, as resources spread down the class hierarchy.

The sexual emancipation of the post-Victorian era appeared before World War I in the upper-class colleges. Elite youth were the first to have their own cars, live away from home, and move in a national setting that took them out of the control of their parents. Women shared in these shifting resources, and hence a new form of sexual bargaining appeared, concentrating on short-term liaisons rather than long-term property only. The 1920s saw the style spreading to the provincial upper middle class, and later to groups further down the ladder, as transportation, the mass media, and other resources became common. In the same way, the suburban phenomenon of the 1940s and 1950s—the greater casualness, the withdrawal from the dignified status pretensions of the traditional bourgeois world into a semiprivate round of child-centered community activities, the world of coffee klatches and the Little League—was foreshadowed in the high schools of the 1920s and 1930s, with their heavy emphasis on clubs and sports, their fads and their sloppy-clothes styles (Lynd and Lynd, 1929: 211–224; Waller, 1932: 103–133; Hollingshead, 1949).

What was happening was that the changing distribution of resources for commanding deference was making the older models of aloof dignity useless. The shift caught on in the high schools first, not only because youth were more sensitized to newer conditions but because the schools themselves were one of the first places that the new kind of community organization showed up. Formerly an elite institution where aloofness was learned, school had become an unexclusive club containing the entire middle class and much of the working class as well. Being an expressive standout in an informal and egalitarian group thus became a better prospect than the traditional dignified *hauteur*—the very shift that Riesman and others would remark among the adults of the sprawling middle-class suburbs after World War II.

In the 1950s, many high schools acquired a heavily working-class composition, especially with the increasing urbanization of blacks. Schools were now mobilizing a working-class culture as well, and the middle-class reaction was a hue and cry about the juvenile delinquency menace (Trow, 1966; Cohen, 1955). The long-term consequences were more serious: the organization of the civil rights movement among black college students in the late 1950s and early 1960s, the increasing mobilization of a black revolt in the northern cities, as well as a widening crack in the middle-class cultural hegemony.

In the late 1960s, upper-middle-class college students and college dropouts launched an extreme attack on the traditional style of deference and demeanor (Yablonsky, 1968; Carey, 1968; Simmons and Winograd, 1966). The leaders rejected all the old standards of grooming and attire—shaving, haircuts, neckties, and other dignified clothes for men, high heels and makeup for women, and all the other marks by which the respectable classes advertised their ritual superiority. Old styles of polite sociability and public order went down under the same barrage; dignified postures were rejected in favor of sitting on the sidewalk and the floor; polite language taboos were punctured by deliberate use of "obscenity" (by both men and women) and working-class and black styles of grammar and slang; traditional middle-class courtship and parties, with their careful scheduling, polite small-talk, and serious discussions, were challenged by a new sociable ideal of extreme casualness, spur-of-the moment sociability and sex, and "tripping out" with the aid of psychedelic drugs into word games and perceptual fantasies.

Things had reached the point where the old demeanor style had very few payoffs to compensate for the effort and self restraint they demanded, even in their sociable forms, and the more creative upper-middle-class youth were the first to realize the implications. The affluence produced by high-level technology made for a great deal of leisure, much of it disguised in the form of bureaucratic job sinecures and a school system artificially lengthening with the ongoing spiral of rising educational attainment and rising educational certification for jobs. The public ideals justifying long years of passing artificial hurdles had become transparent, and political issues sparked off a ceremonial revolt that became a social movement for puncturing the entire legitimating structure of the old status system. Characteristically enough, it was the most secure group, the upper-middle-class students at the elite schools, who reacted to the new opportunities (Messer, 1969): What was the point in continuing a petty struggle for minor deferences that almost everyone could get, when one could become a new kind of leisure aristocracy by living for the present and making a joke out of the old pretensions? With the level of affluence, it had become possible to live without much work by spending only for consumption instead of for ostentation and by giving mutual aid in the form of communal living, hitchhiking, and a general ethos of sharing possessions.

The ethic of "do your own thing," the notion of "tripping"—whether on drugs, put-ons, fantasy games, or flamboyant political demonstrations —reflected the recognition that this was a world of plural world views and arbitrary formalities, and the fact that resources for reality-constructing were so widespread that no one really *could* impose his ideal upon others. Under these circumstances, one was better off not trying, and the new movement was geared precisely the other way, toward whatever advantages might be found in acting out plural realities. The early phase concentrated on the Sorelian joy of flaunting old pretensions and world views, and on the sense of a new egalitarian community emerging outside the traditional hierarchies of business, politics, school, and the military. The phase of initial enthusiasm was transitory, of course, as the avant garde hippie style carried by a relatively small elite was emulated in diluted form by a much larger group. As in the earlier movements, the significance of the extreme group is in its foreshadowing of a wider shift to changed circumstances; in this case, portending the more explicitly pluralistic, privatized, and deference-free society of the future.

The student movement of the 1960s was not only a conflict among age-groups over deference styles and political power (since the civil rights and antiwar movements were a revolt against the policies of a much older group of politicians and officials); it also represented shifting resources in some long-standing conflicts within the youth group itself. One of these was class-related. Youth culture in twentieth-century America had been heavily dominated by the upper and then the middle classes; it was these

groups who took the center of attention with their fraternities and sorori-
ties, clubs and activities, organized sports and cheering sections, and formal dances and sociable rituals of "big game weekends" and the like.
Under their shadow were a number of other groups: the grade-grubbing
career-oriented students attempting to move up from lower-middle- and
working-class (and often minority ethnic) backgrounds; and the working-
class youth ("the hoody element") who were staying in school in sufficient
numbers to make a cultural impact, beginning in the 1950s (Schwartz and
Morton, 1967; Clark and Trow, 1966). There were also several middle- and
upper-middle-class factions who received little public honor while all the
attention was going to the sports heroes and the apex of the dating hierarchy (Mead and Metraux, 1957; Coleman, 1961). These included the
serious intellectual students, those high on creativity (Getzels and Jackson,
1962) and interested in the arts, and the politicians and do-gooders. These
splits probably reflected different occupational sectors, with the latter
groups coming from (or aspiring to) intellectual professions and living in
urban settings, and the sociable leaders coming from the business class and
the suburbs.

The shifting social ecology of the 1960s broke down the monopoly on
ceremonial attention within the youth group. Not only high schools but
colleges had become unexclusive. The big universities had lost the sense
of a private club as the undergraduate culture of the sociable middle class
had become a rather small minority, and the large numbers of graduate
students dispersed the physical locus of the community, formerly the
dormitories and fraternity houses. The absolute numbers became large
enough to provide a critical mass for a plurality of scenes. Political demonstrations captured attention from the traditional collegiate culture, and
then the theatrical potential of the hippie style and the propaganda of
psychedelic rock music picked up momentum to become a full-blown
ideal of a counterculture.

Psychedelic rock music (often listened to while using drugs, whether
in private gatherings or massive public festivals) was clearly the ritual
center of the movement, and surveys of musical preferences give us the
best evidence that it had a primarily upper-middle-class composition
(Hirsch, 1970). The content of the music shifted sharply from the stereotyped sentimentalities of traditional "pop" music, to the explicitly iconoclastic and philosophical lyrics of rock music (Carey, 1969); singers began
to write their own lyrics, thereby becoming charismatic cult leaders in a
full religious sense. The phenomenon can be foreseen in earlier studies of
youth culture. Coleman (1961: 243) found that the fans of Elvis Presley
and black rock 'n' roll music did not dominate the high school group, but
came from the lower levels of popularity *within teenage* sociometric stratifi-

cation; the dominant teenagers came from solid middle-class families and emulated the clean-cut style of conventional pop stars like Pat Boone. Schwartz and Merton (1967) found in the early 1960s that it was dropouts from the ostentatiously competitive and sexually manipulative dating system that were the early Beatles fans. The conflict within the younger generation and the conflict between generations overlapped; the activities of the conventional youth culture were *sponsored* by adults, and thus the counterculture was revolting against a united front of both their enemies. In this sense, the revolts of recent years represent the rise of a real youth culture for the first time.

This counterculture is best understood in relation to styles of deference and demeanor, and these matters make most sense from the vantage point of the sociology of religion. Weber (1968: 468–486) shows that various social groups have their characteristic moral and ritual stances: the worldly magic of the working class, the pious moralism of the bureaucratic middle class, the genteel hedonism and cynical formalism of the upper classes. The collegiate culture of the early twentieth-century youth can be seen as a secularized version of the last, with the same ceremonial concerns for membership within a consciously elite, nonstriving group. But there have also been religious styles that reject rather than adapt to wordly arrangements of business and power: Weber (1968: 541–551) distinguishes two forms—an *ascetic* or moralizing effort to change oneself and the world to fit a religious ideal, and a *mystical* style of withdrawing from worldly routines to concentrate on subjective states of emotional experience.

The hippie movement fits the model of a mystical religion. It has the emphasis on inward experience, on the plurality of subjective realities, and the illusoriness of economic, political, and status ideals. It relegates rational knowledge to a lower form, below emotionally suffused vision. It emphasizes detachment from events and the freedom that comes from scorning worldly sanctions. Like Taoist and Zen Buddhist mysticism, it abandons conventional rules of conduct and idealizes the attitude of humor and paradox. Like all mystical movements, its social ideal is a community of immediate cosmic love. Like other mystical religions, it tends to come into conflict with its compatriots, the ascetic moralizers (in this case, the political radicals and reformers of the New Left), regarding them as too worldly and ambitious; they in turn regard the former as self-centered and amoral. Weber even noted that mysticism, more than ascetic movements, actually depends on worldly goods being provided without working for them; oriental mystics have typically come from aristocratic backgrounds (like the upper-middle-class hippies of today), and depend on gifts from the pious (begging, crash pads, communes).

The similarity is structural as well. The hippie world has no formal church hierarchy, but only a number of gurus or charismatic leaders (rock stars, psychotherapists and rap leaders, and, increasingly, oriental gurus themselves); mendicant monks who give up the worldly life (people who set themselves off by growing their hair very long, wear extreme clothes styles, and live on communes or by wandering); pious laymen who look up to them for spiritual style and support them with gifts (the large numbers of students, young professionals, and others who emulate the hip style, attend rock concerts, pick up hitchhikers, and generally identify with the lifestyle of the leaders); and, finally, the world of nonbelievers or the merely conventionally religious (including here the shifting mass of the old "collegiate" culture who affect some of the superficial traits of psychedelic culture). The counterculture, then, is no apocalypse, but a kind of secular (or not so secular) religion with a complex structure of inner and outer circles. The shades of difference from "weekend hippies" and temporary dropouts on through communards and gurus parallels the structure of religions of China, India, and the Middle East, in which members of the higher classes would make temporary visits to monasteries, or pilgrimages as mendicant monks. A rhythm of "dropping out" and "dropping in" is part of the resiliency of this form of organization. It seems to emerge in highly stable and relatively affluent societies that have reached a high level of cosmopolitanism and relativism in the upper classes.

In modern America, and the West more generally, we have reached a degree of cosmopolitanism and affluent leisure that includes an unprecedented proportion of the population. The material means of impression management have become dispersed; old methods of requiring deference become steadily less viable. The highly organized ideal which went along with bourgeois ambitions of the last few centuries gives way to the more relativistic and expressive concerns of a pluralistic situation. It is this culture that we have seen emerging in the youth movements of recent years.

Chapter 5

Stratification by Sex and Age

The family has always been regarded through a murk of sentimentality. In sociology, most of the emphasis has been on practical problems of marital adjustment and the like, couched in terms that reinforce our ordinary conceptions. But serious sociology has not been much better in this area. The sociology of family, kinship, and socialization has been the bastion of functionalism, framing its analysis against an ideal system in which men, women, and children all fit nicely in their places.

We are beginning to see that such idealizations are no more realistic in this area than anywhere else in sociology. The family is a structure of dominance like anything else, and we are beginning to see that it enforces a great deal of inequality. In sex roles, there has been considerable male dominance in areas ranging from who gets the best occupational and political positions and who does most of the menial household labor, to who controls the scheduling of sexual intercourse and who gets most of the orgasms. In terms of age, we are beginning to see a parallel structure of domination, in which adults not only control children's behavior but their interpretations of it, so that successful socialization means the thorough indoctrination of its recipients into believing in its necessity and legitimacy. Sociological theory is only starting to become liberated from the self-justifying viewpoints of dominant males and adults.

The advance of serious sociology in this area depends on taking advantage of this breakthrough in our preconceptions. But we need to go beyond the polemical reliance on favorite examples of oppression. After all, there is a two-way conflict, however unequal it may be in particular instances, and it is always possible to throw up counterexamples of the respects in which children dominate the suburban middle class, or women command ritual deference of their own. What we need is to state the general explanatory principles for variations in the phenomena of stratification by sex and age. From this point of view, exceptions become no longer embarrassments but opportunities for testing explanations.

Freud as a Conflict Theorist

The enduring value of Sigmund Freud is that he opened up a conflict model in both areas. Despite his great popularity, it is striking how little these major contributions have been appreciated. We have been diverted by Freud's more superficial appeals, and then disillusioned by his underlying weaknesses. We all know that Freud's psychological constellations of the Oedipus complex and the latency period are culture-bound. His hydraulic analogy of psychic forces now looks archaic, and some of his own descendants have rejected his emphasis on childhood causality of adult behavior and his retrospective method of psychotherapy. Many of Freud's formulations have turned out to be untestable interpretations, particularly where they depend on the notion of unconscious repression thrown in to account for any failure to turn up the hypothesized motivation.

Freud's strongest talent, in a way, was as a dramatic writer. His best papers read like Victorian detective stories: a mystery posed by a patient's visit, the marshalling of clues, the train of logical deduction, until finally the solution is revealed and the problem solved. It is not surprising, then, that Freud has been most popular in the interpretive rather than the scientific disciplines, among literary critics and biographical historians rather than experimental psychologists. Freudian theory has come to stand for far-fetched interpretations of symbols and cultish dogmatism about motives, for a set of moral and aesthetic judgments rather than causal explanations of the real world.

But Freud's own aims are not so lightweight. Subtracting his errors, his accomplishments remain not only historically important but still relevant, all the more so to the degree that they have not yet been properly made use of. More than anyone else, Freud saw the implications of Darwin's demonstration that man is an animal with physical appetites and instinctually aroused behaviors. He saw that the problem was to account for the distinctiveness of human consciousness, not as a new departure but as something emerging from a new combination of animal characteristics. Like Durkheim, he saw that emotional arousal is the basis of human social ties, and he went beyond Durkheim in seeing that such ties are always ambivalent, involving both attraction and domination. Like Mead, he saw that human consciousness is internalized from interaction with other individuals, and that it is built up in childhood by a process of gradually identifying with or taking the role of the other. He went beyond Mead in showing that conflict goes on here too: that social communications involve a struggle for control, and that the internalization of the dominant person's communications makes for intrapsychic conflict in the form of conscious and unconscious realms.

Freud saw, more clearly than anyone else, that conflict underlies the two main dimensions of family life: sexual and power struggles between males and females, and between adults and children. Unlike his academic interpreters, he did not try to put a respectable face on it: blind, selfish lust for sexual pleasure is a crucial motive in everyone, and the more inhibited striving for love is explained as emerging where social conflict requires that it be toned down or displaced. Aggression is treated in the same way: Freud's hypothesis is that the prime state of the organism is egocentric and self-assertive, and that morality and self-discipline are derivatives that appear as the result of external pressures. Freud gives us a perspective in which intrapsychic conflict reflects and internalizes social conflict, and social bonds are to be explained by the outcomes of the struggle for emotional gratification. Like Marxian conflict theory, there is the realm of real motives and material conditions, and a covering ideology or false consciousness that results from domination.

The iconoclastic thrust of Freud's theory has been a source of embarrassment to many of his followers, especially in academic circles, and much effort has gone into toning down just this aspect of it. Since so much else in Freud's long and speculative work was obviously to be jettisoned, it seemed reasonable to argue that the child's need for food or maternal care was more important than sexual drives; and the fact that Freud himself swallowed up aggressive drives into a metaphysical death instinct made it easy to get rid of both. Combined with Freud's own tendency to reduce everything to childhood events, the way was cleared for a view of docile, socialized man (and woman), neatly fitting into a network of social obligations and carrying the internal gyroscope of society's values in his head.

But Freud's original conflict model is more valuable now than what replaced it. Despite all the arguments that sex and aggression are not so important, the logic of Freud's model has never received an appropriate test. It implies that a great many human social arrangements can be explained by sexual and aggressive motivations, and that these are suppressed precisely to the degree that external conflicts overpower the individual. Ideal conceptions of virtue and social membership are to be explained as ideologies imposed by the structure of dominance. I propose to give this an historical–comparative test. The appropriate material to be explained, of course, is the structure of the family—relationships between men and women, adults and children.

This involves a number of departures from the various Freudian systems taken as wholes. In particular, the traditional emphasis on childhood as an enduring mold for the rest of life's patterns goes out; what I am concerned with is the way in which sexual and aggressive motives determine behavior between adult men and women in every encounter in their lives. This is in keeping with various contemporary offshoots of Freud, especially Fritz Perls' gestalt therapy.[1] The latter stresses that what is real is always the present moment in time, the here-and-now of *me*, in this *place*, feeling the sensations of this *body* and the verbal thoughts of this moment of *consciousness*. Taken in this way, Freud is entirely congruent with the other ingredients of conflict theory: with the animals in interaction model coming from Darwin and Durkheim, with the phenomenological emphasis on the immediacy of subjective interpretations at every point in time, with the effort to bring all idealizations down to the words and actions of real men and women as they go through the moment-by-moment business of constructing subjective reality in as favorable a form as they can get away with.

Even Freud's historical limitations have their use. Freud's work has generally been interpreted as a theory of individual psychological functioning. On this level of analysis, sociological and historical criticisms have been devastating, showing that sexual repression and its related family constellations are not universal, but only characteristic of Victorian Europe. But what may be a limitation on a psychological theory can prove fruitful in historical sociology. Freud's major discoveries—the biologically universal drives of sexuality and aggression, and the historically specific repression of these drives through an idealized moralism—thus become the keys to unlock the history of sexual stratification.

PART I: A THEORY OF SEXUAL STRATIFICATION

We may begin with three basic propositions.

1. All human beings have strong drives for sexual gratification. Such drives are very widespread among animals. But it is more than an animal-

---

[1] Perls' watchword is that only the *here and now* are real. The past and future are only ideas that we have now in the present; abstract ideas are always things passing our tongues or repeated within our heads at some particular time and place. This is a return to radical phenomenological empiricism, to the flow of experience as the only thing we really know. What is distinctively modern about it is the emphasis (as in the ethnomethodologists) on the *situatedness* of experience in time and space. The aim of gestalt therapy is to force the subject back on himself. Freudian therapy is criticized on the grounds that it encourages him to dwell in a realm of childhood memories or abstract self-examination instead of facing the present in which he is always acting. Perls' clinical concern is for each person to come to grips with his own postures in the world minute by minute; he must take responsibility for the fact that he himself is producing the behavior, for only in this way can he find the means for changing what he does not like. On the theoretical level, many of the exercises in sensory awareness proposed in Perls, Hefferline, and Goodman (1951) and elsewhere provide promising lines for empirical exploration of an updated version of Freud, simultaneously with a phenomenological approach to consciousness.

istic carryover among humans. In comparison to other species, human beings have the most pervasive sexual drives of all. Only humans, along with a few species of apes, copulate outside of a limited estrous period. Of all the animals, humans have by far the most developed capacities for sensual (not to speak of psychological) gratification from petting auxiliary to intercourse. As Morris (1967: 13–84) has argued, human beings are the sexiest of animals, and there are obvious evolutionary advantages to explain the fact.[2] Humans have the longest period of maturation and the greatest dependence of infants on their mothers; this pattern, which makes possible the uniquely human cognitive capacities, could only develop along with mechanisms to strongly attract adults and children and keep males and females cooperating together as a family economy during pregnancy and childrearing. Various physical peculiarities probably developed to maximize interpersonal attraction; this includes human's relatively bare skin, with its sensual potentialities; large female breasts (compared to those of monkeys and most other mammals), which serve both as erogenous zones and as salient sex identification for animals walking upright; facial hair in males, and pubic hair in both sexes, retained as visual sexual markers in the absence of smells, colored feathers, horns, or other distinctive markings found in other species;[3] and the capacity for unusually long intercourse (in the male) and for orgasm (in the female) compared to other species.[4]

It is not surprising, then, that virtually all known human societies are very active sexually, and that covert sexual activity flourishes where it is overtly suppressed. There are a good many instances, though, where females in particular are not very active sexually. Yet there are considerable variations comparatively, and the maximal female activeness gives us no reason to suppose that female sexual drives are any less than those of males; there is even some speculation that they may be greater. Rather, females appear to have been often more deeply and pervasively repressed sexually than men. To account for these variations, we need to invoke some further propositions.

2. Human beings all have the capacity for aggressive arousal, particu-

---

[2] See also Lorenz (1966: 144–211). Works of this sort by zoologists need to be treated with due caution whenever they get beyond their data into making human analogies while ignoring cultural variations; at this point, we tend to get cultural biases masqueraded as biological necessities.

[3] Thus, female apes have strong estrous smells and small breasts (Gough, 1971: 761).

[4] Morris (1967) argues that the uniquely strong female orgasm in humans is due to the vertical position of the vaginal canal when standing upright; hence, a prolonged orgasm is evolutionarily useful to keep the female lying down so that the sperm does not flow back out. For Freud's own theorizing about the effects of walking upright, see *Civilization and Its Discontents* (1961: 46–47).

larly in response to being coerced. This, too, is something we share with a number of other animals, especially other carnivores who live in groups. This is not to say, however, that we can use such a proposition all by itself to argue for the inevitability of war, or the particular forms human conflict may take. Such matters of historical variation in social organization must be explained by particular social conditions, not a universal drive. Nor are humans necessarily very aggressive much of the time. What this proposition tells us, though, is that humans are quite capable of coercing each other when the opportunity arises; and, just as importantly, that individuals will go to great lengths to avoid being coerced. This means that every society will have some kind of power situation, consisting of the way that violence and threats of violence are organized, and this will set the conditions within which men and women can pursue their other interests, including getting sexual gratification. The general conflict approach laid out in Chapter 2 here finds a more specific application.

3. Males, on the average, are bigger and stronger than females, in the human species. Women are also made physically vulnerable by bearing and caring for children. That is to say, resources for social domination are distributed unequally between the sexes in general (with individual variations contributing to further distinctions within the larger pattern). Following the conflict approach, we would then expect that persons take advantage of inequalities in resources, and that the recurrent behavior we call "structure," and the ideals used to justify it, reflect the underlying situation of power conflict.

The combination of these propositions means that, without considering other resources, men will generally be the sexual aggressors and women will be sexual prizes for men. Family organization, as stable forms of sexual possession, can be derived from conditions determining how violence is used. Political organization is the organization of violence, hence it is a major background variable here; when the political situation restricts personal violence and upholds a particular kind of economic situation, economic resources accruing to men and women can shift the balance of sexual power and, hence, the pattern of sexual behavior. These resources and their consequences will be considered later.

In its simplest form (i.e., no other resources being involved), superior male size determines the historically predominant pattern of male dominance. The historical pattern of greater female restraint on sexual drives can be explained in this way. Human beings have relatively strong sexual preoccupations compared to most other animals, but even so, no one person is sexually arousable all the time. Since members of the bigger sex can force themselves on the smaller sex, the former can satisfy their sexual drives at will, whereas the latter have sex forced upon them at times they may not want it. Unattractive males can force themselves on attractive females, but unattractive females can rarely do the reverse. Males thus become the sexual aggressors, and females generally adopt a defensive posture. The element of coercion is potentially present in every sexual encounter, and this has shaped the fundamental features of the woman's role. Sexual repression has been a basic female tactic in this situation of struggle among unequals in physical strength.

There are a number of alternative explanations. The traditional one is to argue that family patterns evolved out of functional necessity. Without them, the human species would have died out or evolved in another direction, without its capacity for cultural advance. This may be true in general, but evolutionary explanations of this sort are not specific enough to explain any of the variations. One might even regard the conflict theory proposed here as filling in the blank reserved for a mechanism of how this development has proceeded. We must also avoid retrospective overdetermination; explanation by natural selection requires that it be possible for things to die out, and we certainly cannot argue that any particular arrangements will prevail in the future, because human society as we now know it may otherwise disappear.

Another sort of sentimental explanation justifying the traditional family is that it is based fundamentally on the human need for love and affection; this need is particularly intense in modern society, in which particular individuals must rely on each other closely, but it operates in a more diffuse sense in traditional societies with their larger family networks. This is bolstered by the research of Harlow, Bowlby, and others, showing that both human beings and their closest animal relatives, the great apes, are severely damaged by lack of affectionate contact in infancy. But the "love-need" here is a very general need for physical contact and communicative response, and it can be satisfied in a great many forms, especially among adults. Coercive possession of other adults as sexual prizes would be one way to do so. Historical comparisons, too, show a great deal of variation in particular forms of contact and affection, coinciding with a great deal of coercion. What we call "love" in a more limited, ideal sense, occurs only under particular kinds of conditions, involving particular arrangements of conflict resources.

Another type of explanation accepts the existence of sexual domination, but argues that it is "natural," the product of generalized male superiority or special male tendencies to form all-male groups, which can then dominate the economic and political realms. This is a pseudoscientific form of argument; it provides no testable explanation, since it does not account for the conditions under which variations occur. It is essentially a way of elevating selected instances into a general rule.

Within the general realm of conflict explanations, there are other alternatives. The Marxian tradition generally holds that sexual domination is conditioned by particular economic conditions, and that its motivation is economic exploitation of women's labor. The first part of this is certainly true, and economic conditions play an important part as independent variables in the following model. But the initial situation of coercion among males and females is a crucial underlying condition, and the organization of the state plays an important role in determining the history of sexual stratification, interacting with economic conditions. Also, we must recognize that the fundamental motive is the desire for sexual gratification, rather than for labor per se; men have appropriated women primarily for their beds rather than their kitchens and fields, although they could certainly be pressed into service in the daytime too. Without this sexual complementarity on the genital level, it is hard to see why the whole apparatus of *specifically sexual* property and its surrounding ideals should have come about, why the family should have a structure independently of a sexual class domination. Females have been special prizes for men in ways that smaller males generally have not. There seems to be a long-standing prudishness in the Marxian tradition that has kept its theory of the family from developing to account for the full range of phenomena.[5]

Related models have emphasized that the appropriation of women is not for copulation but for procreation, especially to produce heirs as well as daughters to use as exchanges in cementing political alliances with other families. The usefulness of women as childbearers may provide an additional motive for males to use their superior resources for domination; obviously it does not explain those resources, and hence the fact of sexual stratification. Does this motive substitute for the desire for sexual pleasure in an explanatory theory; i.e., does it have testable empirical consequences? The evidence on sexual property (given later) indicates that it is intercourse, rather than childbearing, that is the focus on the struggle for sexual possession. In the same vein, the argument that women can produce daughters to use as exchange property for alliances raises the question of why men in other families want to accept exchanges of women in the first

---

[5] There is something peculiarly Victorian about the materialist approach shared by Marxism, classical (and neoclassical) economics, and utilitarian psychology as updated in behaviorism. Its effort to be realistic extends only to recognizing desires for food and economic goods, but not for power, sex, activity, social belonging, or other emotional or cognitive states. The Marxist theory would thus imply that sexual stratification cannot exist where there is no economic stratification, as in tribal societies without economic surplus, or socialist industrial societies.

place, which brings us back to the obvious role of lust. The production of male heirs, and of children as laborers, suggests an additional utility of possessing women; presumably, sexual property would be intensified—especially in those aspects that determine appropriation of *children*—where family inheritance is a major form of the organization of power, and family labor is crucial in the economy. By the same token, one would expect that concern over sexual access per se would remain an object of contention in all circumstances, and vary with resources for domination rather than motives for possession, if sexual pleasure is the fundamental feature on the motivational side.

Just how much female vulnerability through childbearing contributes to their subordination is a matter of debate. Prolonged pregnancy and infant dependency must have evolved together with superior male size and the rest of the complex of human physical characteristics; hence, it is difficult to compare their independent effects. But size would seem to be more important in that it makes coercion possible, and also enhances the vulnerability of childbearing because dominant males do not have to help women in caring for the children. In contemporary society, with the availability of both birth control and social organization for shifting child-care from individual mothers, this must be reduced to a minor factor; it can be invoked as a rationalization for sexual domination, but not as an explanation of it.[6]

The advantage of the size explanation is that it is amenable to empirical test. This can be done by comparing species in which the males are relatively bigger or smaller than females. Among the primates, it is the ground-dwelling baboons, macaques, and gorillas that are organized in large groups dominated by a small number of powerful males, and it is these species in which the males are considerably larger than the females (Gough, 1971). The tree-dwelling gibbons, orangutans, howler monkeys, and chimpanzees, on the other hand, have little male dominance and, for the most part, little stable sexual pairings (or "sexual property"); it is in these species that the males and females are similar in size. The tree-dwellers do much less fighting with other species, which probably accounts for the relative lack of male physical specialization; it is also easier for individual females to escape from males.[7] Patterns of male dominance

---

[6] The remaining biological explanation is the shape of the genitals. The erectile penis and the receptacle vagina seem to make it impossible for women to forcefully rape men, even if the size differences were reversed. But genital and size differences probably evolved together as part of the same complex. Given particular conditions of control upon force and economic property, however, it is quite plausible that women would be the courtship aggressors and controllers, short of rape.

[7] Aberle *et al.* (1963: 259–261) indicate the lack of instinctual restraints on sexual choice

are also found among other mammals, where the males are bigger than the females, as among lions and other predators. Cases of female dominance seem to occur mainly among insects, especially the highly collective societies of bees, ants, wasps, and termites, where a single large female (the "queen") lays most of the eggs and is assisted by large numbers of female workers; males are important only for fertilizing the queen, and live otherwise short and isolated lives. Relatively equal relationships (such as the geese families described by Lorenz (1966)) seem to occur mainly among birds, where sex is not much related to size. All of this is conditioned by another factor: whether the animals are capable of engaging in any cooperative activities that bring about permanent groups in which stratification can appear, or sexual contacts are confined to episodic fertilizing of self-maturing eggs, as among fish and reptiles.

In general, interspecies comparisons seem to bear out the importance of size in sexual dominance. This may also be tested among humans. The difference in strength between men and women is a matter of averages; some women are stronger than some men, and it is a testable proposition that in those cases where they encounter each other, the patterns of dominance will shift. However, since a number of resources go into domination —notably property, but also the prevailing ideology, which may be regarded as defining the way in which particular individuals can call on the larger group for support—this factor will not totally determine the situation. Big, wealthy women probably would have little trouble dominating smaller, poorer men, especially in situations where the larger society does not interfere. The effects of these economic and political–ideological conditions makes up the main part of the application of the conflict theory to historical variations.

Sexual Property

The basic feature of sexual stratification is the institution of sexual property, the relatively permanent claim to exclusive sexual rights over a particular person. The concept is parallel to that of economic property: Property is not the goods or lands or buildings themselves, but the social relationships that determine access to them. As such, it must be enforced by interpersonal threat and alliance, and it is subject to continual bargaining and redefinition. It also admits of many variants. The kinds and degrees of possession over land and other goods have varied a great deal throughout history; "property" is not an absolute, but an abstract label

---

among mammals generally, and the importance of individual size. It should be noted that chimpanzees, the species most similar to humans, live both on the ground and in trees.

for a large number of different arrangements in which certain individuals have rights of access and use, and certain others are denied these. Sexual property has a similar range of variations. With male dominance, the principal form of sexual property is male "ownership" of female sexuality; bilateral sexual property is a modern variant which arises with an independent bargaining position for women. Other variations in sharing sexual access, and limits on the kinds of sexual activities allowed, represent further complexities of sexual property systems.

Lévi–Strauss (1949; cf. K. Davis, 1949:175–194) has made the most sweeping use of the notion of sexual property, to explain the basic structure of kinship systems. Men taking permanent sexual possession of particular women constitutes the biological family; children are part of the family because they belong to the woman and hence to the owner of the woman. Within the family, we should note, the incest taboos reflect the facts of sexual property; indeed, they are the negative side of sexual property rights. The most serious kind of incest is mother–son incest, for this is a violation of the father's primary sexual property by his most immediate rival. Sibling incest violates the father's rights to dispose of the sexual property in his daughters as he sees fit. The father–daughter incest prohibition is hardest to explain in this fashion, although we may argue that the use of sexual exchange in the larger group may account for it. It is congruent with the weakness of the power situation to uphold the taboo that father–daughter incest is by far the most common form of nuclear family incest, especially if the mother is dead or absent (S. K. Weinberg, 1955).[8]

There are, of course, a number of alternative explanations of the incest taboo. The long-standing functional explanations, though, do not really qualify as causal models, since they merely argue for the consequences of incest taboos in forcing exogamy and, hence, in creating larger societal ties. A line of psychological explanation, that there is an inherent aversion among siblings reared together, has been recently revived as the result of evidence that children raised together in Israeli kibbutzim are not sexually interested in each other when they reach puberty (Talmon, 1964; Spiro, 1965). But the same evidence indicates that the adults exercise very strong controls over sexual activity; they give a spurious appearance of liberality because children bathe together and there are no taboos on nudity, but the effect is very like that of a nudist camp where strong efforts, reinforced by continual surveillance, are made to dissociate nudity from sexual arousal (M. S. Weinberg, 1965). Insofar as one can speak of a kibbutz-wide

---

[8] Conflict theory leads to the hypothesis that the father–daughter incest taboo should be strongest in matrilineal and especially matrilocal societies.

incest taboo, then, it appears to be backed up by the same kinds of power relationships that enforce sexual control in all families.

Again, this is a case where the problem becomes more tractable by treating it as a matter of variations. As Goody (1969:13–38), Young (1967), and others have pointed out, incest taboos are only one form of a larger class of controls over sexuality, along with rules on adultery and exogamy more generally. The strength and extensiveness of these controls depends on the form of social organization. The more wide-reaching taboos are found where extensive kinship organizations are the basis for strong economic and political ties, ranging on down to the vestigial nuclear family incest-taboo in modern industrial society. Examining the variations in strength of prohibitions indicates that they are based on the crucial power relations in the surrounding kinship network. Incest, adultery, and exogamy rules all reflect particular kinds of sexual property, and the power relations within which they are embedded.

As Lévi–Strauss argues, if sons cannot get women in their own families, they must get them elsewhere, and that means from those men who have women to spare—the fathers, brothers, or husbands who can give away their daughters, sisters, or wives. Lévi–Strauss applies Marcel Mauss' model of gift-exchange systems to marriage customs: They are sets of rules which guarantee that if a man gives away some of his women, he can get others back from other families. In this way, kinship networks develop as opposed to the biological family. Beneath variations in rules of descent, household locality, and marriage choice, kinship systems can be seen as based on sexual property and its related rules of sexual prohibition and exchange.

In modern societies, the pattern is overlaid with a complex set of moral injunctions supported by church and state, but beneath the ideological surface, similar forces operate. Marriage is fundamentally a socially enforced contract of sexual property, as indicated by the facts that marriage is usually not legal until sexually consummated, sexual assault within a marriage is not legally a rape, and the major traditional ground for divorce was sexual infidelity (Kanowitz, 1969). That the basis of sexual property rights is male violence is still demonstrated by the generally acknowledged dispensation of fathers and brothers to kill rapists of their daughters and sisters.

Variations on this theme may be found historically. At the precultural level, many primates are organized polygamously, with the strongest male surrounding himself with a retinue of females, and driving the smaller and weaker males away. This pattern is also found in some human societies, where political and social power enables dominant men to get more than their share of women.

Once women have been acquired for sexual purposes, they may also be used as menial servants. In primitive societies, women are generally the agricultural and handicraft workers, while men are the armed fighters and hunters. One may say generally that sexual stratification is the sole basis of social stratification in societies with a very low technological level. In a situation of social instability, women are often regarded as booty in war, and it is likely that the institution of slavery began with women and was later extended to men (Thompson, 1960).

Variations in Sexual Stratification

Male sexual property in women is the basic pattern of sexual stratification. Variations result from two factors: forms of social organization affecting the use of force, and those affecting the market positions of men and women. These two factors operate interdependently. Where force operates freely, the distribution of power among males determines the nature of sexual stratification quite straightforwardly, and women have no bargaining power of their own. In such a context, any market of sexual exchanges operates only as part of the system of bargaining among heads of families and is based on family resources of individual men and women. A market for personal sexual qualities and other personal resources can emerge only where the private use of force is limited by the state. Thus, the emergence of a personal sexual market, like that of an economic market, depends fundamentally on the emergence of a particular form of the organization of power. Hence, social structures determining the distribution of force and those producing individual resources for use on a sexual market must be treated together, as interrelated structural complexes.

As a general test, we may compare four historical cases. The major types are (1) those in which economic and political resources coincide in the same units of organization and (2) those in which they differ. We then have the further cases: (1a) little economic surplus and (consequently) little political stratification, hence females can be relatively little exploited as workers and have little value for political exchange; (1b) greater economic surplus, larger-scale political organization, but organized in the hands of the heads of fortified households; also (2a) bureaucratic state organized independently of households, making possible a widened market economy, eliminating much of one male weapon (monopoly on immediate domestic force) but leaving another (economic property); and (2b), the case where women participate in the market on their own, acquiring another resource (although not yet as much of it as men.)

Table 1 presents in summary form hypotheses about the effects of four main types of social structure on sexual stratification. It states the male and female resources made available in each situation, the resulting system of sexual roles, and the dominant sexual ideology. *Low-technology tribal*

TABLE 1
Types of Social Structure, Sexual Stratification, and Dominant Ideologies

Social Structure	Male and female resources	Sexual roles	Dominant ideology
1a Low-technology tribal society	Male: personal force, personal attractiveness. Female: personal attractiveness.	Limited male sexual property; limited female exploitation.	Incest taboos.
1b Fortified households in stratified society	Male: organized force; control of property. Female: upperclass women head lineage during inter-regnum of male line	Strongly enforced male sexual property; high female exploitation; women as exchange property in family alliances.	Male honor in controlling female chastity.
2a Private households in market economy, protected by centralized state	Male: control of income and property Female: personal attractiveness; domestic service; emotional support.	Sexual market of individual bargaining; bilateral sexual property in marriage.	Romantic love idea in courtship; idealized marriage bond.
2b Affluent market economy	Male: income and property; personal attractiveness; emotional support. Female: income and property; personal attractiveness; emotional support.	Multidimensional sexual market of individual bargaining.	Multiple ideologies.

societies are those in which the degree of economic productivity allows little stratification. *Fortified households in stratified society* refers to the typical preindustrial organization based on independent households; it corresponds to Weber's patriarchal and patrimonial forms of organization. *Private households in market economy* refers to the typical domestic organization in a society dominated by the bureaucratic state, where the work place is separated from the home. *Affluent market economy* refers to the development of the preceding type into a society of a high level of affluence and widespread nonmanual employment. The four types of social structure are ideal types, not an evolutionary sequence; combinations of these yield intermediate forms of stratification, combining elements from adjacent systems of sexual stratification. Historical reality is usually a mixture of forms and processes; we may characterize different historical situations (and the situations of different social classes in each period) by a particular weighting of resources in the struggle over sexual dominance.

1a. *Low-technology tribal societies.* These are societies in which the technology produces little or no economic surplus beyond that necessary to keep each producer alive, and which have little economic, political, or status stratification (Lenski, 1966: 94–141). Accordingly, sexual stratification can exist only in a mild form. Superior male force can be used to enforce sexual property rights (marriage, incest taboos), but women cannot be forced to do a disproportionate amount of the work, since all members of society must work to survive. Insofar as work is divided and leisure is possible, women appear to work longer and at the more menial tasks. Since there is little surplus and little economic and political stratification, which intermarriages occur makes little difference to affected families; where the economic system does not permit substantial brideprices or dowries and no families are powerful enough to be highly preferred for political alliances, there is little reason for daughters to be strongly controlled, since they are not used as property in a bargaining system. Thus, it is in low-technology tribal societies that most known norms favoring premarital sexual permissiveness are found.[10]

There are variations within this rather large category. The greater the economic surplus in such societies, the greater the tendency for male control over daughters to be asserted, and for women to do a larger proportion of the menial labor (Zelditch, 1964: 687). This category also includes the greatest variety of kinship systems; matrilineal and matrilocal types

---

[10] This may be established by comparing the data given by Murdock (1949:260–283), which deal almost exclusively with tribal societies and show widespread norms of sexual permissiveness, and by Goode (1963), which show strong restraints on sexual permissiveness in stratified agrarian societies, somewhat modified with industrialization.

are pretty much confined to horticultural and some hunting-and-gathering societies. Matrilineality results in rather different categories of sexual property being emphasized from those of patrilineality; matrilineality seems to improve the power position of women vis-à-vis their husbands, even in polygynous marriages (Goody, 1969: 13-38; Clignet, 1970). Matrilocality, which sometimes goes with matrilineality, seems to further improve the power of wives, since they are among their own relatives, whereas their husbands are visiting among outsiders (Gough, 1971: 768). It should be recalled, though, that matriarchy is a myth; even in matrilocal systems, it is the male kinsmen of the women who dominate.

1b. *Fortified households in stratified society.*   In most historical preindustrial societies, the basic social unit is the fortified household (Weber, 1968: 356–384, 1006–1069). The use of force is not monopolized by the state; economic and political organization usually coincides with the family community. Thus, the owner of a farm workshop, business, or political office not only makes his place of work or his official seat in his home; his own family helps in his work, as do family servants. All work subordinates are treated as servants (of higher or lower level) and are supported from the household economy. The family occurs in an intact form only around the heads of such establishments; servants generally do not have an active family of their own so much as they are attached to their master's family. In the absence of police or other peace-keeping forces, the household is an armed unit; its head is also its military commander. Such households may vary considerably in size, wealth, and power, from the court of a king or great lord, through the households of substantial merchants and financiers, and knightly manors, down to households of minor artisans and peasants. Stratified below the heads of even the smallest units, however, are nonhouseholders—propertyless workmen, laborers, and servants.

In this form of social organization, male sexual dominance is maximized. The concentration of force and of economic resources in the hands of household heads gives them virtually unopposable control. Where sharp inequality among households permits, an upper class may practice polygamy or concubinage, monopolizing more than their share of females. Correspondingly, men of the servant and laborer classes are sexually deprived, and may never be permitted to marry. Women are most exploited in such societies; they are likely to make up a considerable proportion of the slave class if there is one, as in ancient Greek and Roman society or in Arab society. Wives and daughters as well do most of the menial work, while men concentrate on military pursuits or leisure (Goode, 1963: 90, 141).

Male rights in sexual property are asserted most strongly in this type of society. Intrahousehold alliances carry much weight in a situation of general distrust and sporadic warfare; the giving of women in marriage is virtually the only gift-exchange system that can produce such ties regularly, and substantial dowries or brideprices usually add weight to the bargain. Women are thus important among the householding classes as exchange property, and hence are closely guarded so as not to lose their market value; the institutions of the harem, the veil, the duenna, and the chaperone are employed here.

On the ideological side, sexual property is regarded as a form of male honor. The honored man is he who is dominant over others, who protects and controls his own property, and who can conquer others' property. If a man has his sexual property stolen (like Menelaos in the *Iliad*), he is held up to scorn as a weakling unless he commits the violence necessary to recover his property. Women's behavior in itself is regarded as unimportant (thus, it makes little difference in the interpretation whether Helen was kidnapped by Paris or ran away with him), and it is assumed that a man and a woman will have sexual intercourse if left alone together.

In highly warlike societies like that of the Bedouin Arabs, the result is an overriding concern for adultery and the institution of extreme controls over women. The ideal of female chastity (including premarital virginity) is an aspect of male property rights and is regarded as enforceable only by males; women are commonly regarded as sexually amoral, unclean, and lacking in honor, and hence are to be controlled by force.[11] The practice of clitoridectomy among the Bedouin in order to reduce women's sexual drives is an extreme reflection of this belief (Goode, 1963: 147, 211).

The ideological pattern is based on the fact that women are used as sexual objects for the men who own them; they are to act as sexual creatures, although within the confines of a male property system. Total asceticism by women is not allowed. Hence, women have low status in the religious systems of societies of this type. In Brahminism, Islam, Jainism, Confucianism, and the official Roman cults, women are usually regarded as incapable of detaching themselves from the mundane world, and high religious status is reserved for men.[12]

---

[11] The literature of Medieval Arab culture shows an obsession with maintaining sexual property in the harems. The unabridged *Arabian Nights* consists largely of variations on the plot of men whose wives are unfaithful, usually with male slaves, and the same theme introduces Scheherazade in the opening episode.

[12] Christianity is a partial exception, for reasons discussed below. Buddhism in practice tends to give low status to women, although not in theory. Some of the heterodox Hindu and Buddhist–Taoist cults did explicitly include women as full members (Weber, 1968: 488–490).

Women can achieve power of their own in this system only as adjuncts to dominant men. Thus, the wife of a household head may derive some power over men servants in the household; in the case of a noblewoman, this can produce considerable deference. In the extreme case, a woman may exercise absolute authority as head of a household lineage during an interregnum in the male line. That this is an exceptional circumstance is proved by the fact that the general status of women does not improve during the reign of a queen; queens like Elizabeth I of England or Catherine the Great of Russia may combine a severe personal autocracy with the enforcement of traditional status of women in society.

In the lower ranks of fortified households, neither men nor women have much honor. Sexual permissiveness here is possible where opportunity permits, although women at the more attractive ages are likely to be monopolized (perhaps *sub rosa*) by their masters or masters' sons (Holmes, 1964: 114). In general, only the upper-class women have the leisure and wealth to make themselves sexually attractive; there may also be some genetic selection for attractiveness among upper-class women, as dominant males may select attractive women from the lower orders as mates. What is left is a true sexual underground, with neither stable opportunities for marriage, ideological restraints in the form of notions of sexual honor, nor physical attractiveness and personal leisure. Sexual activity among the underclasses of traditional society must have had much of the elements of the grotesque. This circumstance may have helped the growth of Puritanism among the lower middle class during the breakdown of traditional society.

2a. *Private households in a market economy.*   The basic structure of home life changes with the rise of the centralized bureaucratic state claiming a monopoly on the legitimate use of violence. A complex of interrelated changes occur: The grand household declines with the diminution of private armaments. The centralized state usually fosters expansion of commerce and industry, hence a proliferation of small shops and crafts enterprises, and large industrial establishments separated from the household. The bureaucratic agencies of the state provide further work places separate from the household. The result is that households become smaller and more private, consisting more exclusively of a single family. With the expansion of a market economy, more persons can afford households of their own; a private family-oriented middle class appears (Aries, 1962: 365–404; Stone, 1967: 96–134; Weber, 1968: 375–381, 956–1003).

In this situation, sexual roles also change. The use of force by men to control women diminishes as household armaments disappear and the state monopolizes violence, especially with the setting up of a police force to which appeal can be made in violent domestic disputes. Men remain heads of household and control its property; they monopolize all desirable occupations in state and economy as well. Women become at least potentially free to negotiate their own sexual relationships, but since their main resource is their sexuality, the emerging free marriage market is organized around male trades of economic and status resources for possession of a woman. In petit bourgeois families lacking servants, wives serve not only as sexual objects but as domestic labor as well. Where the crowded setting of a large household is replaced by the comparative solitude of a small one, the woman also can become an important source of companionship and emotional support. The woman's capacity to provide these things are her resources on the sexual market. In wealthier families, the woman's resources may also include her family's wealth and social status, although the general importance of interfamily alliances based on marriage diminishes as political and economic aid can be acquired from nonfamily organizations.

The ideology arising from this situation is that of romantic love, including a strong element of sexual repression. The most favorable female strategy, in a situation where men control the economic world, is to maximize her bargaining power by appearing both as attractive and as inaccessible as possible. Thus develops the ideal of femininity, in which sexuality is idealized and only indirectly hinted as an ultimate source of attraction, since sexuality must be reserved as a bargaining resource for the male wealth and income that can only be stably acquired through a marriage contract. As element of sexual repression is thus built into the situation in which men and women bargain with unequal goods.

In contrast to the male-supported female chastity norm of traditionalistic societies, the romantic sexual repression is upheld principally by the interests of women. A hierarchy of moral evaluation emerges among women, in which women who sell their favors for short-run rewards (prostitutes, "loose women") are dishonored; this moral code reflects female interests in confining sexuality to use as a bargaining resource only for marriage.

Within marriage itself, women can use their improved bargaining position to demand the extension of sexual property norms to the husband. Adultery becomes tabooed not only for women but for men. The strategy for the improvement of women's position both before and after marriage is the same: the idealization of sexuality, made possible by women's newly-freed bargaining position and greater protection from violence. Sexual bargaining now takes place by idealized gestures and symbolization rather than the frank negotiations of traditional parents or marriage brokers. The attractive (and the wealthy) woman, in particular, can demand much deference during courtship, including an outright ban on direct sexual advances and discussions. Sexuality is referred to only under its idealized aspect of spiritual devotion and aesthetic beauty, i.e., the romantic love ideal. Male sexual motives operate beneath the surface of polite manners, but they are forced out of official consciousness. It is in these social situations that sexuality becomes repressed in the sense that Freud observed it.

The romantic love ideal is thus a key weapon in the attempt of women to raise their subordinate position by taking advantage of a free market structure. Used in courtship, it creates male deference; after marriage, it expresses and reinforces women's attempt to control the sexual aggressiveness of their husbands toward themselves and toward other women. The idealized view of the marriage bond as a tie of mutual fidelity and devotion calls for absolute restriction of sexuality to marriage, thereby reinforcing the sexual bargaining power of the wife, since she is the only available sex object. Idealization further has the effect of reducing female subordination within marriage by sublimating aggressive male drives into mutual tenderness.

Goode (1959) has emphasized that this romantic love norm reflects increased needs for personal emotional support in a society of relatively isolated nuclear households. Both historical shifts from the large households of traditional society, and current shifts from the crowded, public living conditions of modern working-class life to a more affluent private homelife show an increase in the ideal and in mutual emotional support of spouses (Goode, 1963; Rainwater, 1964). This shift to greater interpersonal reliance is undoubtedly one source of the romantic love ideal; however, its classical features, notably the repression and idealization of sexuality, appear to derive only from the struggle over sexual domination. In this perspective, the needs of men for psychological support become important primarily as an additional bargaining point for women in attempting to improve their power position within the family. As the stratification theory predicts, women are a good deal more attached to the romantic ideal than are men (Burgess *et al.*, 1963: 368).

The first approximation to the private household market economy structure appeared in the cities of the Roman Empire, and it is here that the first major love ideal is found as well, contained within Christianity. Christianity had its origins among the small independent craftsmen, merchants, degraded landowners, and other petit bourgeois of this flourishing international economy (Weber, 1968: 481–484, 488–490). Its major innovations were the establishing of a community independent of family or ethnic ties and the admittance of women to full and equal membership.

The first feature Christianity shares with the other great world religions: Buddhism, Islam, and (to a lesser degree) Confucianism; the latter feature was developed nowhere else to a similar degree. Indeed, the appeal to women was a key in Christianity's rapid spread and eventual success over its rivals. The Christian community was united among its members by spiritual love and shared norms of asceticism; in a Durkheimian interpretation, the theological doctrine of mutual love between Christ and his followers reflects and sanctions this community bond. Christianity in this regard appears to be the adaptation of Oriental asceticism and spiritualism to an urban lower-middle-class community which could not escape into mystical contemplation in a tropical countryside.

The process by which Christianity arose cannot be reconstructed precisely. There is clearly an affinity of interests between the lower-middle-class women who were acquiring a sexual bargaining position during the shift from a society of armed households to the comparatively peaceful and highly commercial society of the Roman Empire, and the Christian priests attempting to spread an ascetic spiritual movement by acquiring new converts. A corollary of Christian church membership was the confining of sexuality to Christian marriage, with its idealized and desexualized view of the marriage bond. It is possible that the changed position of women in the Roman lower middle class had an important effect in shaping Christianity; it is also conceivable that the rise of the religion was important in raising women's status by giving them strong allies in the church. Probably influences operated in both directions.

Until the modern industrial era, religious organizations have been virtually the only specialized culture-transmitting institution; hence, all ideologies tended to take religious form, including the ideologies of sexual interest groups. The early Christian expression of the romantic love ideal has been obscured historically because of the decline of urban Rome and the reruralization of European society until approximately the fifteenth century A.D. The fortified household reappeared as the principal social structure; women's position reverted to traditional subordination. When private middle-class households do reemerge with the bureaucratic European state of the sixteenth and seventeenth centuries, the romantic love ideal, the idealized family ties, and the repression of male sexuality develop with it (Stone, 1967: 269–302; Rougemont, 1956: 49–139). The Victorian ethic of extreme prudery and sentimental idealization does not originate in the nineteenth century; it is an ideal of middle-class families, and especially of middle-class women, found as far back as the early stages of emancipation of middle-class households from the great households of patrimonial society. Its original ideological form was Christian, although the rise of secular culture (through the mass reading market and public education) has given it other cultural bases. Where the private middle-class household spreads throughout the modernizing world, the romantic–puritanical ideal seems likely to go with it.[13]

The great Victorian rise in women's status, then, must be attributed to their alliance with the church, since it was the latter, with its links to political legitimation and status stratification, that prevented men (at least of the middle class and above) from upholding their traditional rights by sheer force. Women also gained another ally in the modern state. The state's crucial characteristic is its effort to monopolize all force, and in this, women have been ready supporters. With the development of police forces, rapes could be punished by an impersonal agency, and women no longer needed to rely on the force of their sexual owners for protection against outsiders. The result has been that women have gained some measure of protection even from domestic violence. Laws against obscenity, pornography, and prostitution also reflect the combined women–church lobby in its effort to monopolize sexuality for marriage.

In general, women's greater religious attachment and religious conservatism (in Christian countries at least) (Fichter, 1952; Glock *et al.*, 1967: 41–59) reflects the continuing alliance between the interests of priests in promoting spiritualism and asceticism, and women in protecting themselves against subordination to male sexual aggression. Women's propensity to political conservatism (Lipset, 1960: 760) may be explained by their special reliance on the state to control private violence. "Law and order," for women, appears to have a special sexual meaning, as rhetoric about "crime in the streets" abundantly implies. The appearance of this conservatism, following the successful late-Victorian campaign for women's suffrage, is no anomaly, considering the nature of the battle for sexual liberation in that period. That the women's suffrage movement should overlap substantially with the temperance movement (Gusfield, 1963: 88–91) is in keeping with the conflict theory here presented; the prohibition of alcohol was an effort to eliminate a substance that made men uninhibited, as well as to destroy a masculine sanctuary, the saloon.

The campaign for women's suffrage was of no great significance in improving women's position. It split women into a more conservative Victorian camp who wanted to protect the idealized female role from such worldly involvements, and a more radical feminist movement by younger

---

[13] Goode (1963) shows that the pattern of free marriage and women's rights spreads even faster than economic development. This is because ideology can be an independent resource in power conflicts; world communications spread to the Third World even where Western economic institutions do not, and military and economic prestige makes the argument "Imitate the West!" a strong weapon for groups like women who have something to gain from change.

women brought up in the greater security that the previous generation had achieved. The granting of women's suffrage came relatively easily, because men correctly saw that it would make little political difference. Women's suffrage was not the beginning of something, but the end. It could have been successful only if the previous Victorian strategy had made men recognize a rough equality. The point can be readily seen if one imagines what the Romans or the Arabs would have done to a women's public demonstration: Nonviolent tactics work only in the context of a shared moral order.

The Victorian period, then, shows the first great rise of the position of women in world history. Prudery, religion, and romantic idealization gave women not only a status sphere of their own and a certain amount of control over men, but a great deal of minor deference in the form of polite gestures like hat-doffing, door-holding, and hand-kissing. But it was basically a strategy of only partly raising an inferior position within the existing order, as women remained confined to the roles of self-decoration and home services: The Southern belle who gets the most gallant compliments pays for it by being locked in the role of mindless femininity. Moreover, women's prudery tended to reinforce a separate male culture, in its own backstage guarded by norms of obscenity and other "man talk" not for the ears of ladies. It was this kind of male culture that kept men's occupational positions relatively impervious to female penetration. All this was reinforced by its links to class culture, so that "respectable" classes came to pride themselves on their prudery and their gallantry, which included protecting women from the realities of economic and political life.

The casual use of force is most prevalent in societies organized around fortified households. As we saw earlier, this type of social organization produces severe controls over female sexuality, but primarily as external restraints imposed by men. In the period of transition from the situation of patriarchal dominance to that of private households in a peaceful market economy, men's interests in controlling their women and women's interests in improving their position through an idealization of sexuality are likely to coincide in producing a maximal degree of puritanism. Idealization requires that women, although desired sexual objects on the courtship market, should be inaccessible to male assault; this allows women to exact deference and at least overt cooperation from prospective suitors in idealizing themselves. But in a situation where violence is still widespread, women must depend on men to protect them. Thus, the initial effort at idealization leads women to reinforce patriarchal efforts at female sexual restraint; the improvement in women's power position comes only from extending this restraint to the men themselves, in effect, getting men to enforce sexual restraint on each other and, hence, on themselves.

The period of the greatest idealization and sexual repression, then, is the transitional one during which the first great battle is fought by women to raise their status, using as resources both their new personal worth on a courtship market, and male ideologies and interests in female chastity surviving from traditional society. The heights of European sexual repressiveness, referred to popularly as "Victorianism," were not confined to the nineteenth century; sexual puritanism developed as far back as the fifteenth century wherever the fortified household was giving way to the newer middle-class home.[14] However, it was in the nineteenth century when the number of families first undergoing the transition was great enough to make sexual puritanism into a dominant public ideology

---

[14] Hysteria about witches appears to be closely related to the early period transitions in the structure of sexual stratification. Although belief in witches and magic is prominent in most nonbureaucractic societies, the great period of witch hunting was in Europe during the sixteenth and seventeenth centuries, the time when bureaucratic states were being consolidated and the household structure began to change. Cases of witchcraft fell into two main types, both of which involved the changing status of women. On the one hand, there were poor old women, spinsters or widows; essentially without resources, they occupied the lowest position in the stratification system, the lowest class within the female class (Thomas, 1971: 502–583). They lived mainly by begging and under the protection of the local church. But as society grew more cosmopolitan and mobile, local ties weakened, and along with them the sense of moral obligation. The great majority of all cases of witchcraft were ones in which an old woman cursed a wealthier man who had refused to give alms; he later suffered some misfortune and had the beggar publicly accused of witchcraft. These witch persecutions can be regarded as a form of sexual class warfare, where the dominant class, no longer hindered by a tightly organized community with its more extended kinship ties and moral obligations, was free to strike back against harassment from the weakest class's demands. The general upheaval of religious belief in this period of transition and political conflict and the accompanying atmosphere of hysteria provided a convenient ideological cover. At the same time, there is some evidence that many of the old women themselves believed in witchcraft, which may have served as a way of subjectively increasing their own status under circumstances that allowed greater personal freedom than before.

The other major cases involved unmarried girls or nuns (Demos, 1970). There would be an epidemic of hysterical seizures—uncontrollable laughing, giggling, dancing, and sexual gestures—together with stories of being bewitched and accusations against other persons, especially those who had authority over them. These cases are more clearly related to sexual property per se; they are a kind of revolt by sexually capable women against the adults who kept them controlled for use as exchange property. The nunnery, used as a kind of prison for the surplus females of the upper class, was an especially likely place for such revolts. This was made possible by the beginning of a shift in resources toward females, and the general atmosphere of ideological confusion and hysteria which gave them some opportunities for external support. The fact that the fantasy content of witchcraft was explicitly orgiastic makes these underlying motives very plausible. As in the case of the old women witches, a great deal of sexual frustration and resentment was there to be expressed: in the one case with outcomes generally favoring the men; in the other with at least temporary revenge for the women.

(Marcus, 1964: 77–160). As the new family structure came to prevail and the relatively peaceful middle-class social order became taken for granted, extreme sexual restraint disappeared.

Freud's discoveries about sexual repression and idealization thus grew out of a particular historical era, as he treated the casualties of the first major battle in the struggle for women's liberation. The pattern emerges in other places, however, whenever the same combination of conditions occurs. Families undergoing change from traditional rural settings to urban middle-class settings are likely to be the most puritanical, for the patriarchal organization tends to prevail in the countryside (especially in more backwoods or frontier areas), and hence, the initial battle for woman's liberation is still being fought. Societies that attempt massive modernization, such as the Soviet Union or Communist China, thus generally undergo periods of sexual repressiveness during the transition. Indeed, there may be an explicit alliance between radical politicians and women in bringing about a revolution both within the traditional male-dominated family and in the larger society; this has occurred most prominently in Communist China, and to a lesser degree in the Soviet Union.[15]

2b. *Affluent market economy.* A further shift in bargaining resources occurs with the attainment of a high level of affluence and the rise of widespread employment opportunities for women. Women become freed from parental homes to go to school and to work. This not only makes the sexual market freer, by reducing parental controls, but also gives women additional bargaining resources. To the extent that women have their own incomes, they are free to strike their bargains without economic compulsion; and their incomes may become a bargaining resource of their own.

But women's occupational position, even in advanced industrial societies, has remained an enclave subordinate to those of men (Epstein, 1970). Hence, the older sexual market in which female attractiveness tends to be traded for male economic prospects continues to operate (Elder, 1969; Waller, 1937). In the working class, however, a woman's earning capacity may be an important resource in establishing sexual relationships (although it may be balanced off by a freer use of male force) (Liebow, 1967: 137–160). In the educated middle class, women with qualifications for professional jobs can double a family income, and hence, represent consid-

---

[15] But a great deal of sexual inequality may still be found. The often-referred-to preponderance of women doctors in the Soviet Union becomes less significant when one considers that most Russian physicians are essentially medical aides or midwives (a much more efficient use of medical resources than in the West), and that most of the high-ranking specialists are men. Wilensky (1968a) shows that the labor force participation of women is much more strongly influenced by the level of industrialization than by political ideologies.

erable bargaining power. In general, the higher the relative income of a wife compared to her husband, the greater her power within the family (Zelditch, 1964: 707);[16] this circumstance no doubt gives a woman some bargaining power before marriage as well.

Although women are far from economically equal with men, it is now possible for a number of different things to be bargained: income resources as well as sexual attractiveness, social status, personal compatibility, deference, and emotional support. The greater freedom of women from economic dependence on men means that sexual bargains can be less concerned with marriage; dating can go on as a form of short-run bargaining, in which both men and women trade on their own attractiveness or capacity to entertain in return for sexual favors and/or being entertained. Where women bring economic resources of their own, they may concentrate on bargaining for sexual attractiveness on the part of men. The result is the rise, especially in youth culture, of the ideal of male sexual attractiveness (Walster *et al.,* 1966). A pure market based on ranking in terms of sexuality, in the sense discussed by Zetterberg (1966), thus becomes more prominent for both men and women, but as only one of the many sexual markets in existence.

But men are still the sexual aggressors in free courtship systems; men are much more motivated by sexual interests as a reason for marriage, whereas women emphasize romantic love, intimacy, and affection more highly than men (Burgess and Wallin, 1953: 669); exclusively male culture has a heavy component of sexual jokes, bragging of sexual conquests, pinups, and pornography, which have little or no equivalent among women (Polsky, 1967: 197); prostitution occurs almost exclusively among women, and male prostitutes are sex objects for male homosexuals, not for women (Kanowitz, 1969: 15–18); men are much more likely to masturbate, experience sexual arousal earlier in life, and are generally more active sexually than women (Kinsey and Gebhard, 1953: 422–427). Men act as the sexual aggressors in modern society, as in virtually all other societies.

This is partly because force is still available as a male resource even in the modern middle class. A 1954–1955 survey (Kirkpatrick and Kanin, 1957) of girls at a Midwestern college found that slightly more than half of them experienced sexual attacks, with the average number of reported experiences being six times in the year.[17] Paradoxically, the more serious

---

[16] Some studies show, though, that wives with high incomes are not necessarily more powerful, because they tend to be married to men with high incomes, who thereby tend to be dominant themselves (Centers *et al.,* 1971).

[17] Another study (Kanin, 1967) reported that 23% of a sample of midwestern male undergraduates admitted attempting rape on dates.

the attack—attempted intercourse versus petting below the waist versus petting above the waist—the *less* likely was the girl to report it to the authorities or even to talk about it with her girl friends. The reason becomes clear when we find that the more severe assaults usually came from steady boyfriends or fiancés. The closer to marriage, the more that sex is expected.[18]

Thus, a mild use of force is taken into account in the dating system; women generally allow themselves to be made subject to force only after a tentative bargain has been struck. The availability of male force simply adds another element to the bargaining situation, and generally requires women to take the role of the sexually pursued, and thus to attempt to enforce an ideal of some degrees of sexual inaccessibility except under the idealized bond of romantic love.

The dating system developed in the generation of youths in the wealthier classes just preceding World War I. F. Scott Fitzgerald was catapulted to fame in 1920 with his inside description in *This Side of Paradise* of this world in which kissing had become a game for its own sake. Fitzgerald (1956: 15) later wrote: "Among other young people the old standard prevailed until after the War, and a kiss meant that a proposal was expected, as young officers in strange cities sometimes discovered to their dismay." By 1937, Willard Waller had given his analysis of the "rating and dating complex" as a system in which marriageable young men and women tried to make erotic conquests and at the same time sorted themselves out into ranks of desirability, eventually culminating in marriage.

What had happened was that the family had lost control over the sexual behavior of its children. To some degree, this was the result of the decreasing economic importance of the family, with the decline of family enterprise (especially farming), and of the importance of inheritance in careers; without these interests, parents and children alike had less concern with each other. But especially, the change was due to urbanization, mass transportation, and the automobile, which freed youths from parental surveillance. Courtship no longer occurred at home visits and on the front porch swing, and began to take place on dates, at parties, dances, movies, and in parked cars. Parents did not give up entirely, of course. The college sorority was developed around the turn of the century by upper-middle-class parents in an effort to control the social class of their daughters' dates

---

[18] Conversely, rapes that are reported are almost always committed by strangers (Svalas-toga, 1962). We can surmise that many successful rapes are never mentioned because they are committed by boyfriends.

(J. Scott, 1965), and suburban curfews are a recent effort at making police into surrogate lower-middle-class parents.

But such efforts have only glided over the top of a new reality. The dating system resulted in a great deal more sexual permissiveness than the old courtship system. "Promiscuity" would hardly be the word for it, however, as unmarried girls built their own forms of sexual control. Essentially, girls learned to use sex to lure men into marriage. From evidence cited earlier, we can easily see (what every unmarried male knows) that males date primarily for sexual conquest, whereas women use sex as a way of attracting men to flatter, entertain, and eventually marry them. To the extent that there is more premarital intercourse, it is generally among couples who intend to be married (Reiss, 1960; Zelnick and Kantner, 1972). Sex is used progressively, as a bait that gets nibbled up bit by bit, with males paying for increasing sexual favors by increasing commitments—to going steady, being pinned, and getting engaged. What starts out as a seductive line on the part of the male gets transformed into a two-sided commitment of sexual property, as the ambiguous meanings of the word "love" come together between the male's and the female's competing viewpoints. The success of the system is shown in the early marriage uniquely characteristic of the twentieth-century United States, and in the very high rate of marriage among both males and females (over 95% are married at least once in their lifetimes). The system produces a variety of other bargains as well, from "career girls" who use sexual intercourse as a reward for dates who spend enough on entertaining them, to high school girls who trade kisses for a free movie and eventually permanent sexual rights for a marriage contract.

As a general principle, the more male violence is available in a sexual market, the more puritanical and sentimental the female ideology. Thus, working-class women are more puritanical than middle-class women, since male violence must be more continuously guarded against (Rainwater, 1964); it is among stably married working-class women that the distinction between respectable women and "loose" women is most strongly enunciated. As the peaceful market system becomes fully established, the degree of sexual restraint relaxes, and the feminine ideal becomes more overtly sensual and less sentimentalized. But sexual attractiveness can still be used as a bargaining resource, although in a more complex market than previously; it remains a resource only to the extent that it is not simply given away. Further shifts in sexual ideologies depend upon further equalization of the economic positions of women.

What of the future? There are new alternatives now opening up, but we can be quite sure that intrasexual stratification by erotic attractiveness will continue to exist. Even without current notions of sexual property, some persons will be more desirable than others, and accordingly a market will still operate to pair off persons on equal erotic levels. The form of bargaining could be a lot more open and egalitarian. Still, some women and men will be better able than others to attract the partners they want. Individuals will no doubt continue to sort themselves out into favorite partnerships, including long-term sexual bargains that amount to mutual sexual property. The hippie community shows much of this pattern, and couples living together are, for all practical purposes, acting as if they were married.

It still remains a question what marriages (or pseudomarriages) will be like: internally stratified or egalitarian? There are strong forces that will continue to produce fairly stable and conventional marriages, including male-dominated ones. The sexual market remains the key. Women who are highly attractive and not very talented or ambitious would no doubt find it to their greatest advantage to play the conventional female role, and make their worldly success through marriage or the dating market. Conversely, talented and active women would attempt to break the sex-role mold and find occupational success. Unattractive women might be more motivated in this direction initially, but attractive ones have additional resources for occupational success. Highly successful men will continue to use their economic and status resources in the sex market. Unattractive persons of both sexes (which includes almost everyone as they get old enough) may be motivated toward a lasting formal contract out of their weak market positions: they would have trouble getting anyone better, and need the companionship as well as the sex. The degree to which such marriages or living arrangements are dominated by the man or the woman will depend on the particular market relationships of the individuals involved. A highly secure and (erotically or economically) attractive person can profit from a "mismatch" to a less attractive person, by demanding greater subservience from him or her as the price of staying. The relative chances each has for making a better match elsewhere will determine the domestic balance of power.

The future, then, will hold a diversity of forms, ranging from the conventional dating and marriage systems of today, through short-term and long-term egalitarian liaisons in the hippie style and group living arrangements, and within these types a variety of shades of male and female dominance, based on individual market relationships. In this context, the organized women's liberation movement can have only a facilitating effect. As a political force, this barely developed movement has already had a sharp effect on abortion laws, and its future program for female emancipation will no doubt press demands for free public babysitting, and possibly a change in the definition of "legitimate" birth. A complete change throughout society would become more likely if women broke men's monopoly on the more powerful and lucrative jobs. But here women face the same problem as blacks and other discriminated-against minorities: Even if discrimination stops completely, the sheer cumulative disadvantage of starting out at the bottom makes it a very long haul to economic equality, and the vicious circle of upgrading educational requirements to match applicants' rising qualifications only makes the problem worse. But as economically disadvantaged workers, women will still face the choice of having independence in relative poverty, or entering marriages or other long-term sexual bargains in which they have no economic cards to play. Sexual stratification is not about to disappear, although its forms may become more complex.

Theoretical Implications

Crosscultural comparisons, then, indicate that the degre of subordination of women is determined by the distribution of coercive power and economic resources, and by the value of women as exchange property. The most favorable situations for women are found where they have the greatest economic resources (especially in horticultural societies where they produce most of the food, and in advanced market societies) and where political power is relatively low at the household level (either because there is too little surplus to support much stratification, or the bureaucratic state removes power to an impersonal level). The greatest male domination is found where political and economic differentiation are great and power is concentrated in the hands of the household head. Ideologies about sexual morality reflect the prevailing power situation. The same principles are borne out within societies; the most favorable position of women is found in social classes where overt use of force is minimized and female economic resources are maximized.

If sexual property is the basic phenomenon in this area, then by implication we have the basis of a conflict theory of family and kinship, and of other related behaviors. Kinship is a network of economic, legal, and status relationships extending along the lines of sexual property, its exchange, and appropriation of its offspring; household arrangements reflect different concentrations of resources among the individuals and families who participate in these exchanges. In effect, kinship tells us of the variations within the main types of sexual property outlined earlier. A related area, demography, may be seen as potentially falling into theoretical order as well.

Family and Kinship Structure

Family and kinship involve a rather large number of distinct variables. First, there are the ways in which descent is reckoned and inheritance of property patterned; the major forms have been matrilineal, patrilineal, and bilineal. These types may further differ within themselves as to how far the kinship links are traced—from siblings and parents to cousins, aunts, uncles, and more remote relatives—and in the kinds of sexual obligations and taboos, as well as economic and political ties, that go along with these relationships. Second, there are variations in living arrangements: matrilocal, patrilocal, and neolocal; there are the subtypes of stem and joint families, depending on how many of the married offspring continue to live with their parents. Other living arrangements may exist for particular age groups, such as dormitories for unmarried men or women. Third, there are variations in the quantitative organization of sexual property: monogamy, polygyny, and polyandry; and variations in its permanence—the frequency of divorce and of casual liaisons.

This would make a rather hopeless morass if one tried to construct a complete typological grid. If we approach the historical complexities of kinship from the point of view of the resources that produce sexual stratification, however, certain causal patterns become clear.

1. Lineage type is primarily determined by the distribution of economic resources among men and women. We know that matrilineal kinship is most prominent in horticultural societies and to a lesser degree in hunting-and-gathering societies; patrilineal kinship is found in virtually all pastoral and agrarian societies (Gough, 1971). Food gathering and hoe-culture are work that men could press women into doing while they themselves carried the weapons and did hunting, fighting, or just living as a leisured sexual class. Men were relatively dominant, but family life was organized around women who did the most productive work; hence, matrilineal families tended to predominate, in which property was passed on through the mother's household. More arduous hunting and fishing cultures, pastoralism, and plow agriculture (agrarian societies) placed the main economic goods in the hands of the men, hence patrilineal kinship. Bilateral kinship is widespread in highly bureaucratic and commercialized societies. Here property is a good deal more individualized and current income is more important than inheritance of land, dwellings, animals or food-stocks; hence, intergenerational transfers are treated in a more ad hoc fashion.

How far into one's network of potential relatives the ties are actually traced also depends on economic resources. The relationship has shown to be curvilinear to levels of economic productivity (Blumberg and Winch, 1972). In hunting-and-gathering societies where groups are relatively small, isolated, and self-sustaining, kinship ties make up the structure of the group but the group itself is little extended. In horticultural societies, kinship systems become larger and more complex along with the economic division of labor, reaching their height in advanced forms of primitive agriculture with supporting commerce. In the stratified agrarian empires, the trend reverses, with the dominant urban and aristocratic classes moving to more flexible forms of political negotiation of kinship; here the exchange of women to cement alliances becomes too valuable a resource to be tied down by tradition, and extended kinship ties become superseded in some degree by other means of organizing large-scale undertakings. In highly bureaucratic and especially industrial societies, the importance of family resources drops off sharply and lineage systems decline toward a nuclear family which episodically calls on nearer kinsmen for sociability and assistance. Variations within industrial societies are explainable by the same general principle: The larger joint and stem families survive where the family is an important holder of economic property (as a farm, store, or business); on a lesser scale, the extent of intrafamily sociability is related to class differences which determine whether or not individuals have considerable chances and resources for extrafamilial contacts (Winch and Greer, 1968; Parish and Schwartz, 1972).

As I have argued previously, the extent and manner of sexual obligations and taboos, whether we call them rules of incest, adultery, preferred marriage statuses, or exogamy, are part of the system of sexual property. The availability of economic and political resources to individual men and women determines these along the general lines indicated. Explaining the extent of economic and political obligations among kinsmen reduces to the same question, for what we mean by an extended kinship system is precisely the salience of certain kinship categories (out of all the potential family ties an individual might think of *at any time in history*) in terms of economic, political, and sexual obligations.

2. Living arrangements are somewhat subsidiary to the explanation of lineage structures. Matrilocal systems are almost always matrilineal; patrilineal ones are usually patrilocal; and neolocality is correlated with bilateral (and relatively unextended) kinship. The economic and political resources that determine lineage systems usually affect living arrangements, since one usually goes to live where the property is. But there are a good many variants; in particular, matrilineal systems often seem to have lost matrilocal residence (assuming the two were correlated originally) (Aberle, 1961). This exemplifies a more general historical process found in many areas: Part of an institutional structure will change over time and new conditions, but part will continue on. Lineage descent is the more enduring pattern because it has general political uses; it is an important means of maintaining ongoing alliances, whereas the locality of the home is more a matter of immediate convenience that may shift with more superficial changes in the balance of domestic power.

Special age-group living arrangements will be dealt with in considering age stratification.

3. Lineage systems already summarize a certain amount of concrete information about sexual property relations. Some additional variations are in terms of the numbers of men and women involved. Polyandry is extremely rare. It means dominant men sharing a woman among themselves; if the basic assumptions of the conflict theory are right, it would tend to produce a great deal of conflict within the dominant sexual class, since all are relatively capable of fighting. Thus, it should tend to disintegrate rapidly into an all-male hierarchy in which the most dominant males would appropriate all the women. Stable polyandry does occur where there is a shortage of women, and men are relatively unstratified among themselves, especially if we include here temporary forms of woman-sharing as among Eskimoes (who not only practice female infanticide but also depend on mutual hospitality on long, individual hunting trips alone), and shared prostitutes in isolated wilderness camps or military posts (Hoebel, 1961: 83–85).[19] All of these cases, though, are noted for the incidence of violent disputes about the possession of women.

Polygyny is related to class stratification; it occurs where some men are politically and economically superior to others to the degree that they can afford to possess several wives or concubines. It thus tends to coincide with considerable female subordination, since the conditions that make polygyny possible also produce rather severe sexual stratification. But sexual stratification is the result of several conditions; polygyny per se is not so important as the background conditions, especially whether the kinship system is matrilineal or patrilineal (Clignet, 1970).

Polygyny tends to disappear in urban commercial situations, partly as a result of the general push toward a simplified family structure. It is also the target of vehement attack by urban ideologies, including both Christianity and its secular offshoots, liberalism, socialism, and communism. This is not surprising, as these ideologies are themselves partly the result of an alliance between leaders of nonkinship communities and women attempting to raise themselves from the subordination of the traditional household. In the Third World today the prestige of "modernity" gives some support to women in their own power struggles. It is not clear, though, that polygyny or other complex family structures are necessarily functionally impractical in urban societies; they are found in varying degrees in contemporary African cities and in informal and perhaps epi-

---

[19] Polyandry as the basis for a kinship system would likely be stable only where females are dominant; this is borne out in the only species in which females are bigger, in the insect colonies.

sodic forms on the communal "dropout" fringes of Western societies. The relative absence in the West is primarily attributable to the fact that political power and dominant class cultures have been strongly in the hands of the proponents of Christian and post-Christian ideologies, resulting in strong sanctions to enforce monogamy. The relative political strength of these ideological factions, both in the West and throughout the world, is probably a better predicter of the incidence of nonmonogamous families than urbanism and industrialism per se.[20]

Demography.   The study of population, although a highly technical area, is still relatively poor in explanatory theory. Extrapolating trends has been an unsatisfactory substitute, leading to some embarrassing failures of prediction. How many children are born, however, is clearly a result of how men and women go about defending their interests in their sexual and family relations, and their status positions in the larger society. Sexual stratification is the appropriate general theory from which such consequences can be drawn.

The variations in parents' motivations to have children (and children of a particular sex) depend on: (*a*) the extent to which men or women have the power to control sexual intercourse and its outcomes (i.e., the question of whose motives are relevant); and (*b*) the extent to which children are desirable as family laborers or warriers, exchange property, or undesirable as drains on the family economy. Generally, men have controlled most sexual activity throughout history; women have been relatively more powerful in horticultural societies and certainly in advanced market economies. Whether women, left to their own decision, would want to have many or few children is not entirely clear, although they probably would establish a lower limit than men throughout most of history; in the recent period, the availability of opportunities for good careers outside the home would tend to dictate less desire for children.

Men's motives are especially relevant in agrarian and early bureaucratizing societies. In the former, there should generally be motivation for a high birth rate, especially among the upper classes for whom a large household is useful as a source of servants and exchange property. The lower ranks, though, may not possess a household of their own and are likely to have much smaller families than the upper classes. With the initial mobilization of the underlying population in early modernization,

---

[20] These two are linked, of course. Urbanism and industrialism mobilize groups that favor monogamy, as well as tend to undermine the advantages of complex families of all sorts. But these are not absolute matters, and a range of variations is functionally possible in industrial societies, as elsewhere. What determines *which* variations are allowed or enforced is a matter of the power of contending interest groups.

the traditional high status motivations become more realistic for a much larger group; hence the population bulge observed at this point. The urban commercial classes, though, began to shift toward the private household form and, hence, lose their motives for having many children. In certain extreme cases, such as nineteenth-century France where the shift toward widespread ownership of small commercial enterprises (including rural ones) occurred abruptly, the result may be a completely stable population.

These hypotheses hardly make up a completely explanatory system, but they may serve to indicate the relationships between a general theory of sexual stratification and the questions of demography. The latter, though, involve a number of other variables as well, including health conditions, migration, and birth control technology. The problem is to integrate the more ad hoc historical conditions into what systematic theory is possible (cf. Freeman, 1961–1962; Cowgill, 1963).

PART II: AGE STRATIFICATION AND CONFLICT

Age stratification is yet another area beginning to explode into ideological controversy. There has, of course, been a long-standing popular cult of childhood. Now educational reformers are beginning to formulate a theory of the embattled child against adult forces of repression. This has been reinforced by the recent youth movement, including its efforts to change child-rearing in a radically egalitarian fashion in communal living arrangements. The term "ageism" is beginning to enter our vocabulary.

This is a useful corrective to the dominant attitude, which has viewed childhood almost entirely from the point of view of an idealized adult society as the abiding reality into which children must of necessity fit. But it will not help to sentimentalize a rebellious obverse of that coin. There has always been two-sided conflict between age levels. As in the case of sexual stratification, the resources that different age groups have determine their powers over each other as well as their capacity for formulating ideologies and making them prevail as definitions of reality. To study the shifts in these resources is our best approach to understanding the dynamics of age stratification as well as giving us a new perspective on social psychology.

Ideologies of Socialization

Almost all of the social science literature on childhood is based on adult-centered values. In sociology, the socialization literature is pervaded by the traditional ideologies of functionalism. This argues from the ab-

stract functional necessity of having families in order for society to survive, to a justification of whatever particular forms of parental domination
happen to exist. Implicit value judgments in favor of whatever happens
to be the dominant ideal in adult society pass into the allegedly explanatory scheme. Thus, the argument is made for the necessity of passing on the
culture so that the society can perpetuate itself. An argument with somewhat greater empirical content is that socialization tames the little barbarians who enter the world as infants and makes it possible for them to
associate in the civilized world. The particular culture into which children
are socialized is assumed to be necessary in some general sense; which
culture it will be is not explained in this model.

   This kind of explanation does not locate real motives in individuals.
It does not explain why parents should want to socialize children except
by invoking the presumed effects of their own previous socialization.
Freud's version was that the child comes to identify with his parent's
superego, whose superego in turn came from identifying with the su-
peregos of their own parents, and so forth. Kardiner (1945) has drawn out
the logical implications of this model. If culture is carried on by child-
rearing patterns, then changes in the culture can only come about by
changes in child-rearing practices at some particular point. How it is
possible to break this circle so that history can occur remains a paradox.
Implicitly, it seems to be recognized that changes occur because new forms
of behavior emerge in adult society. Even a model like Kardiner's recognized that differences characteristic of nonindustrial and industrial societies can be negotiated as the individual passes from one to the other. This
obviously depends on introducing another proposition into the model:
that new forms of behavior can be produced among adults as well.

   The socialization school had gradually and grudgingly come to recognize that there is "socialization after childhood" without quite realizing
the implications of this admission. As longitudinal studies have become
available, we discover that in fact very considerable changes can occur in
adulthood, including even habitual modes of emotional expression considered to be highly transsituational ("personality traits") (Bayley, 1968).
Changes in adult life are possible because adults negotiate relationships
continually, taking advantage of resources as they become available. There
is no reason to suppose that there is any radical discontinuity between
adulthood and childhood in this respect. The "socialization" of the child,
even in infancy, is an interaction between both participants, and the
outcome is not simply an imposition of the parent's culture—or some
theorist's idealized version of it—upon the child, but a negotiated product
which can change as the resources available to the parties change (cf.
Wrong, 1961).

Psychologists, perhaps even more than sociologists, hold this ideological bias. This is true of the Freudians and their neo-Freudian successors, even though the former opened up the analysis of childhood from a conflict perspective. Freud based his model on a Darwinian perspective. Human beings, like other animals, are impelled by some very primitive motives: desires for food, comfort, sex; emotional responses set off by interaction with others such as anger, aggression, and fear. Freud also caught the nature of distinctively human consciousness as built upon interaction and the internalization of the symbolic modes of interaction into each individual. Thus, he captured the main components that make human behavior especially complex and self-reflexive, and gave an approach to internal conflicts within the individual by linking them to social conflicts over gratification. But Freud located this process as more or less completed in childhood. Although the two analytical levels—the fundamental drives and emotional responses, and the social relationships internalized in idealized form—continue to interact through life, Freud's theoretical position, based upon his therapeutic procedure, was to trace these back to childhood rather than to view them as components that can be invoked and modified at any time.

A neo-Freudian version of this became extremely popular in the social sciences from the 1930s through the 1950s (Parsons, 1958; Erikson, 1959; Dollard and Miller, 1950). This eliminated some of Freud's less tenable formuations such as the overriding importance of sex and aggression, or the speculative notion of the death instinct, focusing instead on the process of socialization as the internalization of society's norms through the child's identification with the parent. The potentialities for developing Freud's model in the direction of explaining emotional and nonverbal interaction, and its interplay with cognitive reality construction, at any point in life, were not developed, despite their resonance with Durkheim's more explicitly sociological perspective. Of course, this was true to Freud's own predilections. Freud was an apologist for the emotionally repressive form of self-discipline that Weber characterized as the "Protestant Ethic"; the more orthodox psychoanalysts' therapeutic ideal has been to adjust the individual's internal gyroscope so that he might carry out his bourgeois responsibilities without being crippled by anxiety or over-repression.

Reinforcement theorists represent a more simplified and more extreme version of the same biases. The model is drawn quite literally from studies of animal training. It is a model of socialization in which the control is completely on one side; the animal or child is rewarded or punished until the desired behavior is produced. The earliest formulations of this around the turn of the twentieth century reflected an unsophisticated acceptance of the attitude of the authoritarian parents of the traditional bourgeois family. B. F. Skinner gives a more up-to-date version; like the less prepossessing middle-class parents in mid-twentieth-century America, he discovered that punishment is an alienating form of control, and that control by reward and withholding reward is a more effective means. The possibility that the subjects of training might have any cognitive or emotional capacities of their own seems not to have occurred to the conformist Anglo–Protestants who have made up the behavioristic movement.

The weaknesses of reinforcement theory as a general model for human behavior (and for animal behavior as well) are apparent. The experimental studies of reinforcement are done in an artificially closed field: The rat can turn only right or left in the alley; he has only one lever to push for his food. Thus, the experimental situation defines out of existence the cognitive choices of shaping one's environment that are so crucial in any real-life situation. It is for the same reasons that the reinforcement models have hardly been applied to the emotions, which arise in situations of interaction and which comprise a mixture of physical and cognitive elements.

The reinforcement model also ignores interaction. It is quite true that rewards and punishments have effects upon behavior. In the interaction between two or more creatures, the crucial question, however, is *who* is rewarding or punishing *whom*? In reality, both parties attempt to do so simultaneously; the outcome cannot be understood in terms of an individual theory, but only by incorporating it into a model of the resources available which affect the bargaining or conflict that occurs. The Skinnerian model, predictably enough, has been applied to human behavior only in artificial situations: one-sided controls designed to draw out psychotics, or to train the rudiments of literacy into children. Skinner's utopian society is a bland totalitarianism. Sociological exchange theory moves toward rectifying this one-sidedness, but as yet has given a rather idealized, conflictless version of the process of interaction.

There is a third approach to child development, through studies of cognition and language. This approach rectifies the anticognitive bias of the behaviorists, but contains serious omissions of its own. Most importantly, it ignores motivations and social interactions. This is true even of the greatest cognitive psychologist, Piaget, whose work is based on experimental interactions with children; but (except in his early study of childhood games where he gives an overly rationalistic evolutionary interpretation of the development of morality) he ignores the effects of the actors upon each other, and particularly the ways in which adults negotiate children into their adult versions of communications that serve to construct what we take for reality. This takes a number of years, so that what appear to be a series of natural stages can be found, reflecting successive compromises between experimenter and subject as their resources and inducements change.

Probably the worst offenders in this respect are the "mental abilities" theorists, constructing abstract entities such as "IQ" without attending to the interactive properties of the situations in which they are measured or otherwise presumably manifest themselves. Research on "creativity" is a little better, although it tends to be conceived in the same hypostatized fashion; it has, however, paid attention to such conditions as social isolation in the background of subsequently creative persons, and the antagonism between them and adult disciplinarians such as teachers. What is left out is how language and cognition are negotiated, however unequally, between adults and children, as well as among children themselves.[21]

Recent phenomenological approaches have begun to rectify this situation (Cicourel, 1970b; McKay, 1973). The general perspective on the social construction of subjective reality (outlined in Chapter 3) helps us understand that "cognitive development" is a continual process of negotiation. Its main dynamic, given the concentration of control resources in the hands of adults, is the imposition of standardized verbal meanings upon the behaviors of the child. Gradually he is manipulated into living in a world of interpretive accounts rather than unreflective behaviors and emotions. In this fashion, particularly through the development of language and the successively more thorough involvement of the child in adult conversations, stereotyped formulas are stressed which single out particular experiences and bring the child to evoke and manipulate them on his own.

The Freudian perspective could be fruitfully reintegrated into a sophisticated interactive view of cognitive development. This can be seen by examining the Freudian model of the unconscious. This refers to behaviors for which actors cannot offer adequate verbal accounts, and which are not controlled in advance by conscious, i.e., verbal, formulations. (Such accounts can be generated in psychoanalytical therapy, which also attempts to create the habit of planning one's future actions by verbal control.) This can be better understood from the sociological viewpoint of the social construction of reality. Goffman (1959) shows that much social behavior is divided into "frontstage" situations in which idealized performances are

---

[21] Yet another viewpoint on language and intelligence is now emerging. Piaget (1967) has begun to stress that language and cognition are *not* equivalent; that symbolism develops in interaction with the environment, and that language does not modify its operational structure, or even mirror its operations very well; and that much cognition even in adults is nonlinguistic. This seems to be congruent with the arguments presented earlier in Chapter 3; even on this nonverbal level, though, social interaction may well be more important for cognition than Piaget has taken into account.

carried off, and "backstage" situations in which the actors prepare and recuperate. Thought is internalized verbal conversation; what we mean by "consciousness" (at least in the Freudian sense) are those times when our attention is verbally structured.

Time and again in everyday life, an idealized world is acted out; every conversation has this constructed quality, imposing verbal stereotypes upon the flux of sensory impressions. Thought is a kind of backstage from actual vocal conversations. The distinction between "conscious" and "unconscious," then, is not equivalent to "frontstage" and "backstage,"— although it is the *latter* kind of talk that makes up most of our internalized thinking—but to a second backstage behind these two, consisting of physical sensations and fragmentary ideas. What one can think about by oneself seems to depend upon what one can talk about with others. It is possible to have some detachment from the pretenses of everyday life because there are backstage conversations with others; it is with more intimate friends that one arrives at satisfactory verbal accounts for keeping tabs on emotional states that could not be expressed while pretending the deference, affability, courage, or other idealized attitude of the more formalized encounters. In this way, emotional reactions and physical behaviors are kept linked to the guiding control of verbal consciousness.

Idealized situations that have no corresponding conversational backstage, however, break this link. Behaviors and emotions that are never or inadequately talked about cannot be adequately formulated in verbal terms; hence, the tools are not available for an internal conversation in the form of thought. The most common unconscious materials—sexual and aggressive impulses, expressions of fear and disgust—are those which the social relationships of the time rigidly overrule in idealized encounters and even in more intimate settings; the verbal expression of the emotion is more taboo than its physical expression. The emotions go on causing overt breaks in the continuity of conscious behavior, where the social controls are weak; and displacements into physical symptoms, where the controls are strong. The human animal continues to react to others in emotional and physical ways, regardless of what verbal content is passing through the forefront of his attention. Only when the verbal consciousness is focused on the physical level are the two integrated, and verbally formulated plans (what we mean by "will" or conscious "decision") can be carried out without clashing with nonverbal reactions. The process of "recovering the unconscious," then, is not a matter of bringing back things that were once conscious and no longer are, but of providing workable verbal labels for areas of experience that have never been allowed verbal formulation.

Much of the dynamics of power and solidarity in human relationships seem to depend on "unconscious" forms of behavior in this sense. Much in the same way that political legitimacy is maintained by the maneuvers of those in power to make disobedience unthinkable—frightening even to contemplate—the type of authority of parents over children that we call "superego" operates by keeping fear-of-punishment reactions nonverbal, while the forefront of verbal consciousness is monopolized by the idealized structure of control. Not only are forms of behavior prohibited, but its very verbal definition is controlled in a manner favorable to those on top. Of course, children (and subordinate classes generally) are not lacking in at least some resources; hence, counterdefinitions of reality get formulated, if in fleeting and fragmented ways. These make up the rebellious accounts that psychoanalysts find their patients can produce along with repressed emotions and behaviors. More generally, much of the "taken-for-granted" nature of everyday life uncovered by the ethnomethodologists is not inherently so, but is repressed by existing power relations.

A combination of Freud's motivational model (but without its classical childhood bias, and allowing for a great many more historical and class variations) with the phenomenologists' interpretive analysis provides us a more comprehensive and powerful model, incorporating the general principles of microsociology. The special value of this approach is that it makes no assumptions about what is "best" for the hypostatized abstraction, Society. Whatever their ages, individuals maneuver for their best possible position; whatever results from this will end by being called "the way it is" by those who find their own advantage served by it. Developmental psychology becomes a study of age stratification.

Age Resources and Conflicts

Explanation may proceed as in analyzing sexual stratification. Each individual is assumed to maximize his advantages according to the resources that are available to him, in relation to the resources of those around him. There are three main resources tied to age differences.

1. *Time advantages.* Simply being there first gives an advantage. This is a property of physical interaction as much as anything else. Possession is nine-tenths of the law simply because it takes more effort to take something away from somebody than it does to sit on it. In age terms, this means that the older generation has already accumulated advantages which puts the younger at a disadvantage. The dynamics of power reinforce this situation. Power operates essentially as a self-fulfilling prophecy. The leader can use the organized behavior of the followers to enforce his controls upon each other, to the extent that they believe in his power. Power goes to the center of the network of communications, where men are able to define the behavior of peripheral groups to each other. There

is a time advantage here to those who start first in building the network of communication, whereas the later arrivals can seek to make their careers either by acquiescing or by trying to build up a counternetwork in a situation in which the self-fulfilling prophecy of power has massed most of the resources against them.

The socialization of children operates in essentially the same fashion. Adults have organized the pre-existing order against which the new arrivals have a very peripheral position. But changes do occur, and it is possible for extensive conflicts to happen. Nothing is accepted without some degree of bargaining. This is because there are other resources besides the advantage of time position.

2. *Size and strength.* Physical coercion is always a potential resource among human beings. Despite the sentimentality with which we now treat child-rearing, it remains a fact that physical punishment of children by adults is always the ultimate sanction in maintaining adult control. Related to this are the goods that adult strength provide, including food, shelter, and protection. Children of varying ages acquire countervailing resources of their own, thus shaping the conditions of conflct.

3. *Physical attractiveness.* Persons of different ages tend to differ systematically in their attractiveness. This singles out not only their sexually active years but also earlier ones. One of the resources of infants and small children is the direct pleasure they can give to adults as pets. As Freud was the first to systematically point out, the reverse is true as well: adults are a source of direct physical pleasure for children, thus providing bargaining resources on both sides.

Combining these principles, we get a schematic model of the different resources available to different age groups, and of the conditions that bring variations in them.

Most research on socialization has focused on the effects of different child-rearing techniques on children's subsequent personalities. Since "personality" is a way of summarizing recurrent behavior that from another point of view may be called a "culture," this is equivalent to studying the effects of various kinds of age stratification on society more generally. There are four main types of parental control techniques.[22]

Physical punishment tends to produce boys who fight back against parents and who are aggressive toward outsiders; it also tends to create a strong imitation of the father, authoritarian and ethnocentric attitudes, and a set of moral beliefs based on fear of punishment rather than internalization of moral standards.

---

[22] Evidence for the following is summarized in Whiting and Child (1953); Sears *et al.* (1957); Goode (1971); and Maccoby (1968).

Control by *shaming or ridicule* produces strong self-control, especially over one's public demeanor and emotional expressiveness, and an emphasis on conforming externally to the group's expectations.

Using *deprivation of love* as a threat or punishment produces strong internalization of the point of view of the parent. Moral standards are strongly felt as intrinsic obligations, regardless of the consequences. This usually goes along with considerable emotional inhibition, sexual repressiveness, romantic ideals, and under some conditions, high striving for achievement. Love deprivation is the only technique that produces strong feelings of guilt for trespasses.

Control by *rewards* in child-rearing has been given less systematic attention than the other forms, perhaps because it is so ubiquitous. It is found in ritual form in Christmas and birthday presents, not to mention in everyday interactions. Where it is the outstanding form of control (usually mentioned residually in studies focusing on other control techniques), it probably has several effects. Control by *material* rewards (e.g., candy, money) presumably leads children to comply overtly in order to get the reward, rather than because they internalize the parents' desires.[23] If the rewards are *social* (parents' attention, joining in play) the child tends to like the parents and to be sociable and expressive.

We should bear in mind that these are the strongest effects of pure types of control, and that all parents use several or all types. Their effects on children should be a mixture, depending on the proportion of each type used. Moreover, how long-lasting these effects are depends on the grown-up children continuing to find themselves in conditions that support their earlier personality traits. Even in childhood, the effects may be a good deal weaker, depending on how strictly the control technique is applied, and on the capacity of children to resist. In effect, there is a fifth pure type mixed with any of the others, consisting of the extent to which the child escapes from being controlled by the parent at all; this should reduce the effects of all other variables.

What determines which of these control types is used? This is not a unilateral decision. Parents' are constrained by the resources they have, the social conditions that make them desire certain things from their children, and by the resources that children have to put up resistance or strike out on paths of their own.

Parental Resources. Parents make greatest use of the techniques they can best afford. Control by material rewards is probably most common in

---

[23] This version of the "spoiled child" syndrome is inferrable from evidence on industrial relations, which show that control by rewards leads to behavior designed to manipulate the reward-giving situation rather than to produce work. See Chapter 6.

social classes that have relatively more of this than other resources. Similarly, playing with and paying attention to children depends on a relatively leisure situation, where parents have the time to devote at home, and hence seems to be largely a phenomenon of very recent decades in the West (although perhaps also in certain horticultural societies). Violence, on the other hand, is a cheap resource; this helps explain why it is used so much in the modern working class, which lacks most other resources of control, and in certain hard-pressed peasant cultures.

Control by shaming requires another resource: a high degree of surveillance, and especially the presence of an external audience of peers and adults. Thus, it is found especially in very high-density communities such as the agrarian societies of East Asia; its continued effects into adulthood, however, must be linked to the importance of the family throughout one's career.

Control by love deprivation requires very special conditions. Parents must give a great deal of individual attention to each child, and concern themselves with rather subtle symbolic communications. Thus, it is generally confined to small middle-class households of the bureaucratic era, where the mother has considerable direct contact with a few children. This is also the situation in which the romanticist–Victorian love ideal arises, as a vehicle for raising women's status and power. The children are, if not the main target of this ideal, at least the most accessible ones, and they are required to learn a form of self-restraint and respect for ideals and symbolic communications centered around the notion of "love."

On the theoretical level, there remains a puzzle in explaining just why love deprivation works so effectively, producing a form of control that then becomes independent of further sanctions. The Freudian model is that the child identifies with the parent, and thus acquires an internal parent in the form of a superego, who exerts continual surveillance and punishment in the form of guilt. But why does this kind of identification occur? From the point of view of reinforcement theory, the problem is to explain why one goes on complying with commands a long time after they occur, even though they cause him to forego rewards and undergo deprivations.[24] To argue that there is a strong innate need for maternal love does not solve the problem; the evidence on this need (for contact, being held by the mother) are from earliest infancy (Spitz, 1950; Bowlby, 1958; Harlow and Zimmerman, 1959), and does not explain why parents'

---

[24] This is the same logical problem as explaining some forms of neurosis. A sharp formulation is given in Mowrer (1950). Freud recognized the problem as unsolved in his last work in basic theory, *The Problem of Anxiety* (1936: 120).

manipulations of this need should result in long-term effects, even after the parent has become only a memory.[25]

The solution must consist in establishing a form of cognitive organization that interprets external rewards and punishments in relation to one's own actions and intentions. Piaget (1951, 1952, 1954) shows that it takes approximately 1½ years for the infant to develop a conception of the external world as consisting of objects that exist permanently outside of his own experience with them; only upon attaining this level (the end of the sensorimotor period) can he begin to symbolize unseen things for himself, begin to plan for contingencies, learn referential language, and engage in fantasy play. This also provides a new sense of autonomy in his actions in relation to other people.[26] But parents may continue some of the infant dependency by playing upon *earlier* cognitive developments. In the period between approximately 6 and 18 months, the child usually acquires an especial need, not for *general* adult contact, but for particular *communicative* interaction with his mother or nurse, recognized as an individual (Sullivan, 1947; Bowlby, 1958). This is related to a particular phase in the Piaget sequence, when the child has acquired a sense that he can *cause* events to occur in his perceptual field, even though it is not yet a world of permanent objects (Gouin-Decarie, 1962). This sense of *intentionality* as independent of the observed events seems to extend to familiar humans as well, so that the mother becomes a much more impressive force in the world than in the earlier period lacking completely in cognitive differentiation; it also makes possible a subtle form of communication concerning intentions rather than overt behaviors. But object permanence has not yet been established; mother's subjective intentions (and one's own) have become very important, but the permanent *existence* of the mother herself is still in doubt. Hence, this is a period of heightened subjective insecurity for the child, which empathic communications alone can alleviate.

With the attainment of object constancy in the middle of the second year, the child's special symbolic love-need can subside. Depending on the form of child-rearing, he may then acquire considerable egocentric autonomy, or be subjected to various external controls. I would suggest that

[25] Traditional learning theory has the problem in even more acute form, since it attempts to explain the need for the mother's love itself as a secondary reinforcement, generalized from the primary rewards she provides. But experimentally, secondary rewards have always been found to be weaker than primary ones, so it is hard to explain why the child should comply with demands involving primary deprivations in return for secondary rewards, even during childhood itself, not to mention later.

[26] Hence, the "terrible twos" of the developmental literature, the period in which the child tries out his new-found powers and resists outside controls.

the love-deprivation technique takes its power if the parents begin to play upon it before this time, keeping the child in a state of extended anxiety concerning the basic security of his cognitive world. The permanence of the external world would be built up through the matrix of language learning that is controlled by the mother, rather than on its own through direct manipulation of objects. The love-deprivation technique works so powerfully, to the extent of overriding subsequent rewards and punishments, because it affects *cognitive* development and, hence, screens all perceptions of the external world through an internal world in which a strong identification with a controlling parent is the basic prop. Freud's model of basic personality dynamics being laid down around early nursing, weaning, and toilet-training experiences can thus be given a broader interpretation in terms of cognitive structures. These can have long-term effects, although vastly different subsequent experience can modify them. What is crucial for a sociological psychology is the fact that there are special *vulnerabilities* at this age, which parents may exploit if they have the resources to devote to it; missing these opportunities, a strongly internally-controlled personality cannot be established later.

Parents' Motivation.   Which type of control technique parents use depends on their motivations as well as their resources. Children provide a number of possible uses for parents. Once into middle childhood, they can be made to do *work* around the household, in agriculture, or a family business. They may contribute to *family status* as sexual exchange-property, future adults whose prestige reflects on the family, or decorous members of the home. They may be sources of *companionship* and entertainment. Residually, they may be of *minor value* or outright *nuisances* tolerated only to the extent that they stay out of the way.

These produce rather different control techniques. Work can be got out of children with almost any technique (expect sociable rewards, which tend to set the tone of parent–child interaction as an expressive rather than an instrumental one). Love deprivation, though, is a high-cost, time-consuming form of control, and not much used in the kinds of agrarian or household craft-production situations in which children are of greatest labor value. Violence and ridicule are the most likely controls; the former where work is crude, the latter where it demands more assiduous application.[27]

Where the child's value is as a contributor to family status, violent punishment would be less appropriate, because of its alienating effects. Shaming (surveillance) and rewards are appropriate ways to secure compliance in making the proper show to outsiders; these are most used in

---

[27] Compare the analysis of work controls in Chapter 6.

patrimonial households, where children are important for interhousehold alliances, and contributors to the family fortunes. In the period when private bourgeois households were the locus for sociability, the important thing was for children to be decorous; love deprivation, which could be applied only in such a private household, was a technique achieving good results of this sort, especially in the area of inhibiting sexual advances and boisterous behavior.

Where children become an important source of direct gratification and entertainment for parents, the main form of control is by parents becoming companions and playmates of their children. This milder orientation has emerged in the affluent societies of the twentieth century, with the decline of status differentiation within the large middle class. In the anonymity of large urban situations where sociability depends to a large extent on the initiative of each adult couple, the attractiveness of children as playthings and friends gives them comparatively great bargaining power over their parents. Comparatively mild child-rearing systems may also be found in relatively unstratified horticultural tribes, in contrast to the stringent controls of militarily oriented tribal societies or agrarian societies with property organized in terms of the family.

Finally, children are of relatively little interest to adults in a number of situations. This includes both the modern lower class and much of the working class, where children are of no value for work or status, entertainment is mostly centered away from the home, and children are a drain on economic resources and scarce living space; and much of the upper middle class, pursuing profitable careers and cosmopolitan sociability. In the former, child control is especially by violence; in the latter, by obliviousness (or paid help in the form of servants and boarding schools). In hunting-and-gathering societies (like the Eskimo), infanticide is widely used when children are a burden. In the modern cases, the children are left primarily to the influence of the peer culture or their own inventions.[28]

Children's Resources.   But adult–child encounters are interactions, and the outcomes depend on the resources of all parties. The child's input is

---

[28] Mixtures of these conditions may be found within the same society. Thus, contemporary America includes upper-class adults concerned with controlling children as manpower for family holdings; professionally busy or highly sociable upper-middle-class parents paying little attention to their children, and harassed lower-class parents doing the same; small communities in which family respectability is still an important motivation for controlling children; vestiges of peasant authoritarianism in immigrant families, often combined with great pressure on children for achievement; and a nonstriving sector of the middle mass for whom children provide entertainment and companionship. It is in the most nonstriving group of today, the middle-class dropout youth culture, that the most egalitarian child-rearing is found (Berger and Hackett, 1973).

least in infancy and early childhood, when his only resources are his attractiveness to adults and his capacity for crying and making a nuisance of himself. But these are effective weapons only to the extent that social conditions motivate the parents to respond. The moral sanctions that give an idealized belief among adults about how small children ought to be treated vary with the interests of adults regarding their family as a unit vis-à-vis the larger society, and the conditions of outside surveillance.

Preadolescence and the Development of Long-Term Motivations.   Medium-sized children, approximately between the age of six and puberty, are big enough to be made to work at helping adults. They are still small enough so that all adults can coerce them, and they cannot successfully fend for themselves by running away. Children reaching this age tend to become too large for adults to easily pick them up; this means that children lose one of their resources (and one of their rewards), since holding small children is one of the ways adults receive direct physical pleasure from them. The prepubescent years, as well, tend to include periods of considerable physical awkwardness and unattractiveness, further diminishing children's resources. It is probably for this reason that children between the age of five and puberty are comparatively docile. Freud regarded these years as a "latency" period in terms of childhood sexuality. Since this sexuality manifests itself (or is identical with) the child's physical contacts, his decreasing physical suitability as a pet must bring a decline in related emotional concerns. Freud drew the erroneous conclusion that children at this age no longer have any sexual interests; to the extent that contacts are available, their sexual interests are simply transferred to their peers. (In the middle-class Victorian families Freud treated, however, children were not generally allowed to play outside the household, thus making his observations accurate but historically overgeneralized.)

But middle-sized children can acquire some new resources. They become capable of ranging about more widely, making it possible for autonomous childhood peer groups to appear. This reduces children's psychological dependence on adults. Such groups take on importance only where there is sufficient leisure for children to play, and where social ecology is such that large enough numbers of children live near each other and are not confined to the household. Childhood peer groups exist only in certain horticultural societies, and in industrial societies since the decline of child labor. The rise of schools has also been important, serving among other things as collective caretaking institutions to free parents (especially women) for other activities, and greatly facilitating the formation of childhood peer groups and the emancipation of children from their parents.

Boys in particular have more resources vis-à-vis adults because they are stronger and thus more capable of fighting back. It is probably for this reason that violent punishment is more often used against boys, whereas girls are subject to love-deprivation techniques. Violent punishment is especially used in families where all or most of the siblings are male, thus giving them the strength of numbers (Goode, 1971:629).

Children may also develop some important motives of their own at this age. Peer groups offer a field for achievement in terms of sports, fighting, or social popularity. It is also in these years that subsequent ambition and creativity are founded, and that intelligence scores begin to diverge fairly sharply.

High ambitiousness or achievement-orientation is related to child-rearing techniques, especially to some variant on the love-deprivation pattern. But it also depends on the opportunities that open up for the child himself, and the closing down of possibilities in other directions. Children become specialists in school achievement, machines, or small business ventures because of initial opportunities and the skills that develop with practice, plus growing specialization due to cumulative lack of motivation to try other areas (sociability, sports, etc.) where the initial opportunities were lacking and other children have got relatively ahead of them. McClelland (1953) has argued that this has an important internal component: that *activity* itself is pleasurable in a zone between total familiarity and complete unfamiliarity; one can develop a steadily self-reinforcing pattern of achievement activity by learning how to set tasks with a medium degree of challenge or risk, leading one into the graduated hierarchies of business, mechanics, or intellectual enterprise.

Intelligence apparently has some genetic component. But a considerable amount of variability in it develops over time, particularly with the cumulative effects of school success and resulting encouragement (Rehberg et al., 1970; Bayley 1968). Moreover, IQ measures predict school success, but very little else (Bajema, 1968); they do not predict occupational success, with schooling held constant, and they fluctuate considerably in adulthood. What is involved becomes clearer if we compare it to measures of creativity, which involves a very different personality type. High-creative persons are personally autonomous, antisocial, resistant to authority, and tend to be disliked by school teachers and other children. High-IQ persons, however, are the teachers' favorites, capable of focusing their attention and giving concentrated effort to the demands of formal systems of symbolization (Barron, 1957; Getzels and Jackson, 1962; Torrance, 1962). A large component of IQ, then, appears to be the capacity for highly disciplined, internalized compliance with social demands.[29] A great deal of what we take to be evidence for its heredity—the high correlations between scores of parents and children, and among siblings—may actually reflect high-compliance-producing socialization techniques.

Creativity arises under different social conditions (Cattell, 1963; MacKinnon, 1962; Roe, 1952; Weisberg and Springer, 1961; Smith, 1968). There seem to be two personality traits involved: a high degree of personal independence and desire to make one's own judgments; and a high motivation to succeed in art, science, or other innovative field. The first is found among successful artists and scientists, but also among persons with other careers who happen to score high on tests of novel thinking; the second tends to be confined to successful and aspiring artists and scientists. High cognitive independence seem to be related to a great deal of independence in early childhood, with opportunities for novel experience and activity, and parents who are highly competent but distant. Motivation to excel in a creative field comes later, especially through periods of solitude in middle childhood (because of illness, geographic or social estrangement), which are filled by reading or experimenting. The type of self-rewarding activity drive described by McClelland takes over and cumulative processes reinforce the specialization. Social conflict with others, because of this autonomy and its related iconoclasm, wears the groove ever deeper as time goes on.

Puberty and Organized Age Groups.   Large children, approximately from the age of puberty to the attainment of full physical stature in their later teens, develop considerable resources of their own. Depending on the individual, many reach the borderline at which they can no longer be physically coerced; all become better in some degree at fighting back or running away. They also begin to acquire full mobility and capacity to support themselves, depending on the type of economy. In addition, they become sexually active; in many cases, they are at or near their height of sexual attractiveness. Given this shift in the balance of resources, it is not surprising that the teenage years are ones of tremendous conflicts with adults to the extent that the latter do not allow adolescents to be inducted into at least the initial privileges of adult life, but attempt to maintain childhood controls over them.

---

[29] Compliance seems to be what is most highly valued in school situations; Torrance (1962) shows that in America, Germany, Australia, and India, creativity scores drop steadily the longer a child has been in school. See Inhelder (1969) for a critical history of the use of IQ testing in school administration.

Eisenstadt (1956) has given a comparative analysis of the conditions under which formally institutionalized age groups emerge. These are groups that are recognized by adult society as holding a particular place in relation to members of other ages, and are often given ceremonial and sometimes economic and political functions. Eisenstadt finds that these emerge where the family is an important organization in society, but other activities—economic, political, religious—go on in a larger group drawn from across the boundaries of various families. These are found among tribal societies in which adolescent boys, and sometimes adolescent girls as well, are formed into special clubs, sometimes with their own dormitories, and have a special ceremonial place in the tribe. This was also found in ancient Greek cities (themselves developing as coalitions or tribes); the age groups here were composed of young men organized as training groups for military service.

Such groups are also found in modern industrial societies. Particularly strongly institutionalized examples are the Communist youth organizations in the Soviet Union, and comparable organizations in the Israeli kibbutzim. These serve to bring the individual out of the family and make him a member of the very collectively organized adult society. A comparable kind of organization with a less society-wide emphasis is the debutante group in the industrial upper class, where the family is an important wealth-holding unit. The general thesis is validated by the negative cases: Institutionalized age groups are lacking where the family is either highly inclusive, or is organized so as to produce little strain of transition to the larger society. The former is the case of most peasant societies such as traditional China, where there is no need for the individual to move outside of family relationship to become a full member of society. The latter case includes the individualistic rather than collectivistic modern societies, although there are somewhat milder equivalents of adult-controlled youth groups in the form of school groups and adult-sponsored recreations.

Eisenstadt also gives the conditions under which informal, noninstitutionalized age groups emerge. These seem to emerge spontaneously by action on the part of the young age group itself, where the familistic structure is strong but autocratically run so that economic goods or sexual opportunities are monopolized by family heads and denied to the younger members. Eisenstadt's main case here is the traditional Irish peasantry, where economic conditions limited the supply of land, requiring the sons to delay their marriage and wait for their fathers' death to achieve independence. Similar conditions are found in some primitive tribes with thorough age-grading of positions, amounting to a formal gerontocracy. In these cases, semirebellious groups of youths emerge to carry on their own socializing apart from their elders, and tend to mutter against them.

Eisenstadt's functional analysis is that institutionalized age groups appear where they are necessary to link together the family structure with extrafamilial activities carried out by adults. If we put aside the mystifying agency by which the hypostatized society finds ways of fulfilling its functions, the material may be interpreted along the lines of a conflict theory. In the case of the firmly institutionalized age groups, the nonfamilistic organizations of adults deliberately organize the youth in order to coopt them into their system of power. They are able to do this because they have more resources, making use of the full weight of adult political, economic, and ceremonial power. This is most obviously the case in Eisenstadt's modern examples. The Young Communist organizations of the Soviet Union are part of the totalitarian apparatus of the state, and are deliberately manipulated to break children from the ties of their family and attach them to the political regime. Comparable institutions in modern Israel, ancient Athens, Sparta, or primitive societies were probably upheld in a similar fashion, as are elementary and secondary schools in the modern West.

Eisenstadt's nominally functionalist analysis of *informal* age groups tends to bear this out. These latter groups do not serve any "function" in the social structure in any more than the most vacuous sense; they are organized by a disgruntled age-class on their own as a form of rebellion. The rebellion does not get very far because the resources are so much concentrated in the hands of the adults. The adults do not institutionalize, or even really accept the age-group organization, since it serves no purposes of their own; they already have full control without coopting the youth. Eisenstadt attempts to deal with the most obviously nonfunctional youth groups by labelling them "deviant," but this only applies a value judgment from the point of view of those who retain power over youth who are well enough organized to rebel but not enough to make a change.

The Conflict Perspective

The purpose of this sketch is not so much to formulate a complete theory of age stratification, but to show that the entire area traditionally subsumed under "socialization" may be more fruitfully understood in terms of the conflict perspective. Just as traditional functional treatments of the family have ignored sexual stratification and obscured the real motives and resources of individuals that go into its variations, the traditional treatments of childhood have missed an entire range of variables and misunderstood the social dynamics of the phenomenon. In both areas, we have had to disentangle ourselves from concepts reflecting the ideals of dominant groups—men and adults, respectively. To press for a conflict theory of sex and age stratification must be an effort at a superior level of detachment, not simply a plea for the underdog; the aim of constructing the most powerful explanatory theory requires seeing the resources and interests of the conflict in their fullest light, not to elevate the ideals of one particular party into our standard of theoretical adequacy. Whatever our values in these conflicts, as sociologists we are pressing for a vision within which the arrangements of human society are not "natural" but caused in specific ways. The content of our consciousness and our culture is never a teleological cause of what produces it, but the end product at one point in time of a mixture of contending forces.

In the latter section of this chapter, I have pushed conflict theory far in the territory usually considered part of psychology. That is intentional; the traditional division of psychology and sociology has preserved the former in an abstract realm in which not only social conflict but most interpersonal influences on the individual's psychic structure have been slighted. I have used Freud as a vehicle to reclaim this sphere for sociology —appropriately enough, since Freud has been an outcast from the mainstream of academic psychology, and he represents the classic formulation of conflict theory on the individual sphere, focusing on just those disreputable phenomena that are missed by the idealizations of existing psychology. Together with the model of conversational negotiation sketched in Chapter 3, this section attempts to establish the basis of a more realistic sociological approach to individual psychic functioning.

On the more traditional sociological level, the theory of age stratification is only now beginning to develop. We have started to collect information on inequalities in wealth, power, and prestige among age groups (e.g., Riley *et al.,* 1972; Lansing and Kish, 1957). When this is cast in terms of historical and cross-societal comparisons, and related to the resources age groups possess—variations in physical vigor, sexual attractiveness, advantages of temporal priority—and to differing military and economic situations favoring certain of these, we will be in a better position to explain variations in such inequalities, and in movements of conflict. The revolts of youth, the political tendencies of the aged, the struggle over requirements of seniority, and the length of educational sequences will no longer be dismissed as perennial manifestations of "generational conflict," but regarded as variations occurring under particular conditions. Here, as elsewhere, the thrust of conflict theory is to remove explanations from the plane of ordinary rhetoric, and move toward a more satisfactory science.

APPENDIX A: SEXUAL MANNERS AND CULTURAL STRATIFICATION

Sexual behavior is part of the larger organization of status dominance in any society. Stratification is not only a matter of power and wealth but of group membership and individual impressiveness; and the latter are crucial resources in gaining and retaining the former. Under particular conditions, cultural stratification in the community can produce advantages and disadvantages independently of political and economic struggles. Where the family residence is also the place for sociability, work, and politics, autonomy is minimal and cultural acceptability is identical with economic and political status. Where economic and political activities are carried on outside the family (as happens to some degree in most urban societies or large empires) but the family home is the location for sociability, it becomes possible for one's status to be different in the two spheres. This means that one can attempt to compensate for occupational failure by sociable (i.e., cultural) success, or vice versa; it also means that one can pursue occupational careers by the roundabout means of sociable connections. In the case where sociability takes place in the home, ideals of sexual morality and family decorum are most important. If sociability, however, becomes split off from the family, different standards of cultural stratification come to the fore, emphasizing more hedonistic entertainment and personal display, and downgrading sexual propriety.

Mixtures of these three conditions can occur within the same society, but for purposes of exposition it is enough to treat the main types separately. Cultural stratification is relatively closely tied to economic and political stratification in tribal and fortified-household societies. It is in the periods of bureaucratic societies when the private family household is the center of sociable entertainment that sexual moralism is most important in cultural stratification. This is the "Victorian" period, broadly speaking, in which the respectable classes do not attend restaurants, bars, or other forms of public entertainment, and socializing is largely confined to formal calls and "at-homes." The content of this home culture was shaped by the Victorian revolt of middle-class women. Given the payoffs of making a good cultural showing, both men and women were motivated to put on the best possible appearance of polite gallantry and sexual decorum. The result was an era of excruciatingly precise manners and doggedly dull conversations, when men and women sat primly in drawing rooms talking about the weather, "good" books, and other distant and unimpeachable topics.

At other times, sociability was more usually organized outside the household. The sexual content of sociable culture depended on what type of sexual stratification prevailed. In the great aristocratic courts throughout history, and in the urban periods of the ancient world, women were still without economic resources of their own, and the patrimonial households usually still existed in the background; hence, the sexual culture was a mixture of *machismo,* whoring, and flirtatious gallantry. A man might achieve social eminence as a carousing brawler, or in more peaceful settings, as an elaborately mannered courtier. Opportunities for women, though, were usually limited. In post-Victorian societies, on the other hand, women had acquired some economic resources, and hence, autonomy of their own; especially for the unmarried, active participation in the sexual market became the new basis for sociable eminence. Escorting attractive women and showing the right kind of flirtatious display became a new cultural ideal.

The history of fashions in clothes and dancing exemplify these variations. Both of these are highly expressive and sociable things, although some of the variation in clothing styles can be accounted for by climatic conditions and economic productivity. The Victorian style was to wear a great many clothes, covering as much of the body as possible, which were almost never taken off in sociable or public situations. These served as marks of economic status, both as conspicuous consumption and idealization of the human figure. This is especially the case with the wearing of hats, which can be doffed and thus show deference in a highly stratified society. Later, when such hierarchial deference became less compulsory, mutual hat-doffing became a mark of "good breeding," and hence, of social exclusiveness in the higher classes. (Egalitarian religious sects of the late Middle Ages refused to remove their hats for this reason.) Customs of wearing or doffing hats indoors and especially in churches show the same hierarchial principles. For women, wearing heavy skirts, stays, bustles, and the like also served as a kind of armor against male advances, as well as a sort of long-distance enticement by accentuating the female figure and offering up minor revelations of ankles and laces while keeping their primary sexual characteristics shrouded in mystery.

Post-Victorian styles went through various phases. Generally, the trend for both sexes was for less clothes. Hats for American men have generally disappeared since the status revolution around World War II, and other pieces of traditional decoration such as neckties have subsequently declined for most middle-class sociability and even for work. Women's clothes have also grown more informal for most situations, including public appearances such as shopping. All this seems to be a result of a general decline in resources for commanding status deference in the old manner. *Sexual* styles added a separate set of variations. Women's clothing became lighter, shorter, and generally tighter; the overall shift was to a more direct sexual appeal. With the more flexible use of sexual favors as bargaining power, women had less need to armor themselves; also, the more active short-term sexual market (the dating system for unmarrieds, and a greater use of feminine flirtatiousness generally) escalated the amount of sexual advertisement.

Dance styles provide a parallel set of evidence. In the era of family-controlled sexual markets, dancing was almost always in large groups, whether the stately minuet of the court or more active peasant dances (of which the modern square dance is a descendant). The waltz created a scandal among traditionalists when it was introduced in the nineteenth century, because it involved partners holding each other; but it fitted the romantic individualism of the time, and the fast but stately pace kept its eroticism within strict bounds. The twentieth-century dating culture introduced dances that were overtly sexual: slow, clinging dances with sentimental music, or erotically rhythmic ones, influenced first by lower-class Latin and then African rhythms.

The type of clothing and dance style preferred can be explained by the form of sexual stratification of the time. These acquired more general social significance because sociable eminence had come to depend on a sex-market-oriented culture. Stratification by sexual attractiveness began to have some importance of its own—not just to catch a desirable mate, but in order to shine in the most prominent social circles. Respectable clothing styles came to require sexual display; by the 1930s, the "proper" female attire for public occasions had become high heels, tight girdle, and make-up, the accoutrement of prostitutes in the pre-World War I generations.

As the older elite styles have come within reach of the wider populace, elites have shifted to more casual clothing styles. But stratification by good looks has probably become even more important. Modern affluence has raised the general level of attractiveness through better diet, health, and leisure for exercise; the competition for personal attractiveness grows more salient as more people find they have opportunities in that direction. But class advantages are still strong (Sorokin, 1927: 215–279; Elder, 1969). The middle and upper classes are usually a good deal more attractive than working and lower classes; they eat better food (and hence avoid the obesity and complexion problems of cheaper carbohydrate diets), play more outdoor sports (although the value of sedentary occupations versus hard physical labor is not clear), as well as tend to pick the gene pool by marrying the more attractive women of the lower classes. Similar class distinctions in appearance seem to have existed earlier, but at a much lower average level of attractiveness. In the twelfth century in Europe, disfigured people were seen everywhere, and bad teeth and skin diseases were prevalent (Holmes, 1964: 225–227). In the sixteenth and seventeenth centuries in England, about a sixth of the population had pockmarks on their faces from smallpox, and the poor were particularly distinguishable by the results of undernourishment (Thomas, 1971: 6–7). Similar gulfs in personal appearance between rich and poor are striking today in countries like India.

There is probably also a psychosomatic advantage. Attractiveness is a social role as much as a state of appearance; learning to play it with self-confidence results in a self-fulfilling prophecy, while the opposite process leads to cumulative failure through awkwardness of posture, complexion, and the like. (This is a testable proposition; a self-fulfilling prophecy would tend to produce high correlations among complexion, posture, muscle tone, and other components of good looks.) The overriding subjective salience of personal attractiveness in mid-twentieth-century status cultures has tended to make the underlying class stratification invisible, and a great deal of class warfare is fought out today in terms of acne and obesity. Such differences in attractiveness, along with more subtle ones, also add internal complexity to status groupings within classes, and influence group cultures in ways that have hardly begun to be investigated.

APPENDIX B: SUMMARY OF CAUSAL PRINCIPLES

POSTULATES

XV. Human beings have strong drives for sexual gratification.
XVI. Human beings strongly resist being coerced.
XVII. Human males are usually larger and stronger than females.

PROPOSITIONS

8.0 The greater a sex's control over the means of violence and over material resources, the greater its control over sexual activities, and the more auxiliary services it can command from the subordinated sex (from XVI and *8.0*).

8.1 In the absence of other resources, the members of the larger sex are the sexual aggressors and control the scheduling of sexual activities, and members of the smaller sex attempt to avoid sexual access to avoid being coerced (from XVI and *8.0*).

8.11 The more subordinated the members of a sex are for sexual intercourse, the more subordinated they are in work and in deference relations.

8.12 The greater the concentration of political and material resources within one sex, the more permanently it holds rights of sexual access (sexual property) over members of the other sex (from *8.0*).

8.2 The greater the power of dominant individuals to appropriate others as sexual property, the stronger the taboo and the greater the outrage at violations of these property rights.

8.21 The more political, material, or status benefits resulting from the negotiation of intergroup ties through the exchange of sexual property (intermarriage), the greater the controls upon appropriation of sexual property by individuals other than those involved in the intergroup exchanges (i.e., taboos upon sibling incest, father–daughter incest, and certain categories of adultery).

8.22 The less stratified the political and material resources in a society, the less controls over extramarital sexual behavior (from *8.21*).

8.3 The lower the economic surplus in a society beyond that necessary to keep each producer alive, the less work can be forced upon the subordinate sex by the dominant sex.

8.4 The more economic resources are organized around the work of women, the more likely the kinship system is to be matrilineal.

8.41 The more economic resources are organized around the work of men, the more likely the kinship system is to be patrilineal.

8.42 The more material resources depend upon current income rather than inheritable property, the more likely the kinship system is to be bilateral.

8.5 The greater the lineage solidarity along female lines, the greater the power of women in the household.

8.51 The greater the lineage solidarity along male lines, the greater the power of men in the household.

8.52 The more that women are physically copresent with their own relatives, the greater the power of women in the household.

8.53 The more that men are physically copresent with their own relatives, the greater the power of men in the household.

8.6 The greater the concentration of force in the household (rather than at some other political level of organization), the greater the power of men over women in terms of menial labor, ritual deference, and standards of sexual morality (i.e., asymmetrical sexual property ideals) (from *8.0*).

8.61 The greater the monopolization of force by political agencies outside the household, the lower the power of men over women.

8.62 The more that women rely upon the state and church as their major resources against male dominance, the greater their propensity to political and religious conservatism.

8.7 The greater the economic and political value of the family as a unit of organization, the more remote the kinship ties that are counted and interacted with.

8.71 The more complex the political and economic organization sustained by families, the more remote the kinship ties that are counted and interacted with.

8.72 The more wealth and influence within a particular family, the more remote the kinship ties that are counted and interacted with.

8.73 The more complex the family organization, the more emphasis upon using women as exchange property for familial alliances, and the greater the control over sexual property (from *8.21* and *8.71*).

8.8 The more material property controlled by individuals of a given sex, the more the other sex must withold sexual access for use as a bargaining good in long-term economic exchanges (i.e., marriage arrangements).

8.81 The more that livelihood is derived from inheritable property (rather than current income), the more emphasis upon long-term sexual bargains, and the more sexual intercourse is confined to marriage.

8.82 The greater the potential male violence in the bargaining situation, the more puritanical and sentimental the female ideology.

8.83 The more puritanical and sentimental the female ideology, the more internal solidarity there is within sexes, and the less conversational equality among them, and the more they occupy different occupational spheres or levels of authority (from *5.221–5.231, 5.44,* and *5.64*).

8.9 The more sexually attractive the individual, and the greater the individual's economic resources in a situation of sexual bargaining excluding physical coercion, the more deference the individual may successfully command.

8.91 The more that material goods are derived from current income rather than inheritable property, and the more equally incomes are distributed, the less there is of an ideology of sexual repression and the more sexual bargaining takes place for short-term arrangements (from *8.9* and *8.81*).

Age Stratification

9.11 The greater the use of physical punishment by the dominant age group, the more counteraggression by the subordinated group, the more emphasis upon external conformity, and the more the subordinates will value the possession of power.

9.12 The more control by shaming, the more the subordinates emphasize self-control over visible demeanor, and conformity to group expectations.

9.13 The more control by threats of love deprivation, the more the subordinates identify with moral principles, inhibit their emotions and sexual impulses, and suffer subjective experiences of guilt for trespasses.

9.14 The more control by material rewards, the more the subordinates comply with visible demands closely tied to the rewards.

9.15 The more control by sociable rewards, the more the subordinates are sociable and expressive.

9.2 The dominant age group uses the control resources that are most readily available to it.

9.21 Other resources being equal, the more material wealth possessed by the dominant age group, the more emphasis upon material rewards as a means of control.

9.22 Other resources being equal, the more parental leisure, the more emphasis upon control by sociable rewards.

9.23 The fewer other resources are available, the more use of violence as a control.

9.231 The more violence is used among adults, the more adults use it as a means of control over children.

9.24 The more surveillance there is over children (i.e., the greater the numbers of persons who are habitually present), the more use of shaming as a control.

9.25 The more time parents have for each individual child, especially in early childhood, the more that threats of love deprivation may be used as a control.

9.3 The more work or status that children can provide for adults, the greater the emphasis upon controlling them.

9.31 The more crude the labor demanded from children, the more that control by violence will be used (from *9.11*).

9.32 The more easily inspectable the outcomes of work, the more control by shaming will be used (from *9.12*).

9.33 The less valuable the work, the less use of love deprivation (as a high-cost technique of control) (from *9.25*).

9.34 The more important the child is for family status, the more emphasis upon nonalienating controls.

9.341 The more emphasis upon demeanor styles as a basis of family status, the more emphasis upon control by shaming.

9.342 The more emphasis upon individual achievement outside the family as a basis of family status, the more emphasis upon control by love deprivation.

9.35 The more important children are as a source of sociable gratification for adults, compared to other sources, the more emphasis upon control by sociable rewards.

9.36 The more children are nuisances without redeeming rewards for parents, the less attention is paid to them and the more they are controlled by material rewards or violence.

9.41 The more attractive the child, the more successfully it can negotiate desired treatment from adults.

9.42 The larger the child, the more easily it can evade parental control.

9.43 The more mobilized the childhood peer group, the more independent children are from parental ideals.

9.431 The more mobilized the childhood peer group independently of adult control, and the more concentrated resources for power or wealth are in the hands of adults, the greater the age-group conflict.

9.432 The more emphasis an extrafamilial political, religious, or economic organization places upon limiting loyalty to families and family-based cultures, the more emphasis upon extrafamilial youth groups organized and controlled by adults.

9.5 The more change, geographic mobility, and violent conflict in a society, the less emphasis upon stratification by seniority.

Chapter 6

A Conflict Theory of Organizations

The field of organizations is the most advanced part of sociology. A wide variety of organizations and occupational positions have been re-searched to the point that isolated case studies no longer attract very much attention. We have moved on to comparisons among organizations, and to testing hypotheses about their determinants. Theoretical syntheses of various parts of the field have been appearing for over a decade now, grounded not only in general considerations but in empirical evidence. If there is one area of sociology where serious cumulative development has taken place, it is in organizations.

Yet the field as a whole seems isolated from the rest of sociology. This is so even though most of the other things sociologists study—stratification, politics, education, deviance, social change—are based on organizations or take place within them. Organizational analysis is the most fruitful way to causally explain phenomena in almost any field of sociology; we have been too much locked into a way of dividing up problems because of their practical and ideological implications to see this. Even within the field of organizations, there is too much of this fragmentation, for approximately the same reasons.

Fortunately, there is a guiding thread that pulls together not only the various subfields of organizational studies, but links them to the main questions of general sociology. This is the organizational sociology of Max Weber. This statement may come as a surprise, for there is nothing better known in the field of organizations, perhaps in all of sociology, than Weber's model of bureaucracy. It also happens that there is no more complete misunderstanding of a major sociological theory than the way Weber's organizational theory has been treated in American sociology. By comparison, Marx's theory of stratification has only been caricatured; Weber's theory of organizations has been stood on its head.

Weber's theory is regarded as dealing only with bureaucracy, and with a particularly rigid and idealized form at that. He has been taken as the historical apostle of managerial efficiency, presenting a type of organization with carefully geared and completely interchangeable parts, a kind of perfect organizational machine without any human flaws. What raises this above the ordinary run of traditional administrative science is Weber's historical vision: his long-term view of bureaucracy's inevitable advance, its capacity for routinizing charisma and gobbling up all human elements, the metaphysical pathos of organization developing purposes of its own and trapping even its controllers in its iron cage.

Weber has thus appealed to both the cult of efficiency and the cult of romantic alienation. Not surprisingly, he has become the object of attack by both of their opponents: those who believe an ideal bureaucracy is not so efficient, as well as those who wish to show that things are not so bad after all. This has tended to make Weber even less accessible to the empirical researches of recent years, since so many of these have been carried out in opposition to what has been taken to be Weber's theory. It is ironic and a little absurd that so much of the anti-bureaucratic development in sociology has simply been recapitulating from modern materials organizational processes that Weber identified 50 years ago from his own historical studies. Informal groups and personal ties were hardly discovered by Mayo and Barnard in the 1930s; Weber built an entire theory of personalistic organization and its dynamics under the rubrics of patrimonialism and charisma. The notion that bureaucracies are full of conflicts and maneuvers did not just spring up after the World War II; it is the very essence of Weber's analysis of bureaucracies and every other organizational type. Weber himself was the first to see that "professionalization" and the imposition of "expert" qualifications are the result of efforts toward collegial power and autonomy, and that they raise a sacred aura of legitimacy about many arbitrary procedures. Even the recent analysis of organizational structure in terms of technology, and the very different impulse to see organizations in phenomenological terms, were already begun by Weber in a sophisticated form.[1]

---

[1] These aspects of Weber are hardly hidden in his work. Much of them are explicit in his famous chapter on bureaucracy (in which most readers' attention seems to have focused on the first few pages), and certainly they can be gathered by carefully relating this chapter to the equally famous one on class, status, and party. The main source of confusion (other than the way various commentators have attempted to draw on Weber's authority for purposes that contrasted with his own, or to set him up as a straw man) comes from the way in which these two chapters (11 and a fragment of 9 of Part III, respectively) have been lifted out of their context in *Economy and Society* (1968; originally published as *Wirtschaft und Gesellschaft* in 1922), where they are part of a sequence of chapters on political domination (9 through 16). Immediately after the chapter on bureaucracy comes a counterpart chapter on patriarchalism and patrimonialism, with many examples of how the ideal types are used

Let us be clear about what is being argued. I am not concerned about questions of priority of the who-said-it-first variety. Historical figures are of purely historical interest, and should not be invoked to attempt to inhibit modern work. But in a peculiar sense, Weber is not really an historical figure yet. He might just as well have died in 1968 rather than 1920, because it is only in the latter year that his major work became available as a whole in English; given the state of international communications (and the particular ideological bent of Continental sociology), we are just now discovering a body of work that has the most serious implications for all of sociology. We are now in a position to pick up where Weber left off, after a gap of half a century. That means precisely what it says; it does not imply that we have to accept Weber's politics or his philosophy, or any aspect of his work that is not suitable for our further scientific advance. In my view, Weber's most serious shortcoming was that he limited the use of analytical tools to aiding historicist interpretations. But that is not a position we are stuck with, and we are free to make explicit causal and testable generalizations out of materials that Weber never developed to that point. If we do so, I believe we can see the outlines of a powerful theory of organizations that can not only encompass our research advances but tie the field directly to the heart of general sociology.

It is something of an exaggeration, of course, to talk as if Weber really only became transmitted to us a few years ago. Bits and pieces of his mode of analysis have been split off along the way; some have branched into autonomous traditions of their own, others have gradually been recapturing for us the central viewpoint on organizations in general. For example, there is the line of analysis made popular by Weber's protegé Robert Michels (1949; originally published 1911) (although it is not entirely clear who influenced whom), from which Lipset (1956), Selznick (1949, 1952), and others have developed a whole tradition of studying democracy and oligarchy in voluntary associations, and the political maneuvers underlying goal displacement in hierarchic organizations. Etzioni's (1961) synthesis of empirical studies around a model of control types explicitly brings Weber's stratification model back into the organizational field, serving to remind us of the intimate connection between these areas and of the power of a multidimensional conflict theory in both. Stinchcombe (1965) has

---

together to analyze concrete situations. It is this part of Weber's *magnum opus,* his sociology of political struggle and control (to which he devoted more pages than any other topic), that has been left largely untranslated until very recently. For a careful discussion of these matters and a reliable guide to this part of Weber's work, see Guenther Roth's (1968) introduction to *Economy and Society,* and his (1971) "Max Weber's Historical Approach and Historical Typology."

built on some of the implications of Weber's technological and political dimensions of organization, and Crozier's (1964) conflict theory has more than an echo of Weber's approach. And most explicitly of all, Bendix (1956, 1964) and Roth (Bendix and Roth, 1971) have brought Weber's full-sided approach to light and applied it to modern social change.

This can be seen as the core tradition in the sociology of organizations: not that it has been the dominant approach, but that it has the underlying theoretical unity and empirical power to pull the whole field together as an explanatory science. The fact that Weber carries forward the most sociological side of Marx gives this even greater resonance, and enables us to see a modern class-conflict model like that of Dahrendorf (1959) as part of the central pattern.

What does this tradition tell us? First of all, organizations are best understood as arenas for conflicting interests. Weber developed his organizational theory as part of the sociology of political struggle which makes up so much of his effort to isolate the building blocks of world history. Much of this concerns the sociology of armies, of the ways in which fighting forces can be put together and the contingencies that face conquerors attempting to administer a conquered territory. The basic problem is that everyone within an army is capable of pursuing his own interests in opposition to its head, and that many techniques of administration put resources in the hands of subordinates who proceed to undermine authority from above. Weber's model of bureaucracy has been pulled out of this context; bureaucracy is a peculiarly efficient form of organization *for political purposes* because it keeps the organization together. But it can be set up only if certain technical and social conditions are available; and it does not stop the struggle for control but only transfers it to a new ground, one on which the bureaucrats themselves tend to make their superiors into figureheads.

This suggests some of the basic elements of an organizational theory: individuals pursuing their own interests; sanctions they may use to gain compliance, and the administrative forms through which they are applied; the way in which particular kinds of tasks, attempted with particular technologies and in particular geographical situations, shape these conflicts and give the organizational network its particular changing shape over time. Let us look at each of these in a little more detail.

Weber constantly refers to the interests that individuals are pursuing as an explanation of organizational arrangements of all sorts. One place where he pulls these together is in his treatment of class, status group, and party, another piece that has become famous out of its original context, which was the sociology of politics. Classes are groups who pursue economic interests vis-à-vis opposing interests, of which Weber distinguishes a number of kinds—questions of paying for labor, questions of financial loans, and questions of buying and selling goods. If we want to predict which interest an individual will follow, we must locate his degree of investment in these organizational positions and in the ongoing coalitions that operate to enhance their different interests. But individuals are also interested in power and status, and there may be distinctive social networks involved in each of these: the state and its parties or factions in the one case, communities and especially religions in the other. Each of these is a realm of struggle too: The state is full of factions maneuvering for power, and the various parties themselves are made up of power positions with their own interests; status communities and religions struggle with each other for precedence, and within each of them is a form of status stratification that generates its own implicit conflicts and maneuvers.

Given the fact that everyone is concerned with all these goods at once—economic, political, and status—and the fact that business, state, and church can be closely connected or at least resemble each other, we seem to find ourselves in an endless game of classifying and cross-classifying, and wondering if there is not some simpler way that will yield causal principles instead of only ideal types. But I believe that the ideal types are a causal model in embryo. Each of the three dimensions—class, status group, party—refers us simultaneously to a type of organizational control and a locus of interest-group conflict. Amitai Etzioni (1961) has demonstrated that each of these corresponds to a type of control used in organizations, and that each has very different sorts of consequences. Controlling men by material rewards makes them maneuver for their own economic advantage, and hence, produces work only to the extent that rewards are closely tied to what they produce; controlling by coercion leads to alienation and an implicit effort to escape or counterattack; controlling by normative ideals requires ceremonial membership in a status community and results in identification with organizational ideals but also in high-level disputes over purposes. Knowing what degree of each control type is being used, then, tells us how much men will emphasize their economic, political, or status interests, as well as what kind of organizational structures are possible and what the dynamics of conflicts will be.

The structure of an organization may be regarded as a network for applying controls so that certain tasks can be carried out or at least attempted, since the outcomes are rarely just what was expected. Just what the structure will be depends on what the controls are, as well as on the techniques available for administering them. These include both material technology and the ideas and ritual procedures by which they are communicated. The mounted cavalry, mass-produced firearms, writing paper, the telegraph and telephone each has affected organizational controls by making possible a different arrangement of men with the resources to exercise coercion, watch over others' actions, or make reports. In this sense, Weber extends Marx's notion that material conditions determine social organization.

But there is always another level of determination: the content itself of the communicative interactions that make up an organization. What Weber meant by legitimacy is the way in which men influence each others' emotions and ideas. One of the things that makes this work, as Stinchcombe (1968a: 158–163) points out, is the fact that there are real sanctions in the background: The foreman has authority over his employees because he can always call on his superior to fire someone, the superior in turn can call on the police to eject him if he does not leave, and the policeman can call on other police and ultimately on the army to reinforce him against resistance. Where emotions and ideals enter is precisely in the act of communicating all these implications. The successful authority, after all, is he who does not have to carry out his threats; the most successful of all is the one who couches his orders in such a way that disobedience is unthinkable, and threats (or even rewards) are never raised to the level of explicit issues. This involves carrying out the proper ceremonial procedures; authority is a ritual action to present the proper image of reality and acquire the proper submission to it.[2]

On this point, Weber converges with Durkheim, especially in the version of ritual reality construction presented in Chapters 3 and 4. The connection can be seen most clearly in the work of Erving Goffman (1959), who draws most of his original material for ritual in everyday life from studies of workers confronting managers, and professionals confronting their clients. In each case, we have two sides to an encounter, each attempting to uphold its own authority or autonomy by presenting an idealized "frontstage" image, and accepting ceremonial compliance as a sufficient indicator of recognition of its position. This helps us to understand how conflicts of interest may be reconciled with the implicit cooperation involved in such ritual encounters, for ceremonies mark the borders to which authority extends, the negotiated dividing lines between group territories. It is the ceremonial violations that are called insubordination— refusing an order rather than failing to carry it out; and when shifts in authority (and hence, in organizational structure) occur, it is through ceremonial encounters in which particular, often trivial, matters are disputed that the borderline is redrawn.

Now consider what is involved in the notion of ideal types. They

---

[2] At the top level of control in a government, conveying such a feeling is the masters' highest priority for keeping the organization together; for soldiers themselves can be made to comply only as long as they are afraid of each other. See Chapter 7.

imply multiple causality; any one situation has several determinants operating within it. They also imply variables; ideal types mark off the end
points on a continuum, with progressively weaker effects as we proceed
down it toward the opposite type. These may be used for descriptive
purposes, but also for explanation. Concretely, this means that any organization is to be explained in terms of a number of variables. Coercion is
involved in almost every organization, but in differing degrees; it is most
explicit in the military arm of the state, implicit in any organization
(including both businesses and the family) that has property, since that
is ultimately upheld by the state; and it is lacking only in pure voluntary
associations without property (such as ephemeral sects, rather than
churches). Material rewards are a focus in most organizations, but especially in organizations that produce and distribute them and use them as
a primary form of bargaining with its members. And status ideals are
found in all organizations, expressed in the ceremonies through which the
core of organizational relationships are dramatized; these may focus either
on the coercive threats, the bargaining situation, or the sheer emotional
and symbolic qualities of the membership rituals themselves. Just what
the interest groups will be in an organization, how conflicts are carried out,
and where the negotiated borderlines will fall depends on the particular
mixture of these variables.

   This, in turn, depends on the material conditions available, and on the
prior network of organizations and their accompanying ideas. Weber distinguishes three main types of organizations: ad hoc groups based on
personal ties; more permanent organizations (referred to mainly as patrimonial) which acquire property and raise personal loyalties into a long-
term world view; and impersonal bureaucracies which organize property
around an abstract set of positions and treat individuals as temporary
actors filling permanent slots. Each has its own conditions and dynamics,
and its own form of legitimacy.

   1. *Ad hoc groups* are formed whenever individuals who have a temporary coincidence of interests come together. Their members' control over
each other depends on how strong their common interests are and how
long they last; their greatest strength is when a *charismatic* leader arises
whose emotional intensity focuses their concerns and galvanizes the group
around himself. Such a group, then, is held together to the degree that the
conditions are met for intense ceremonial encounters: members meeting
face-to-face, isolating themselves from outside distractions, focusing on
distinctive gestures, insignia, and verbal formulas. For such an organization to become very large, the personal ties must be extended into a

network; usually this is done by the disciples of the charismatic leader branching out to gather personal disciples of their own, and they in turn acquiring further disciples, with each individual participating periodically in ceremonies both in the lower- and the higher-level groups. This is the organizational form of religious and political sects, and of cultural movements more generally.

2. Where property is involved, however, even a personalistic organization takes on a different form. And property does become involved whenever men attempt to carry out any extensive economic production, or to organize a state that appropriates certain goods, territories, labor services, or autonomies by maintaining a stable threat of violence. The family also falls into this category, since it consists of permanent property rights over sexual access, children, labor services, and household goods. *Patrimonial organizations* are those which organize such property on a personal basis; particular individuals make demands or bargains with other particular individuals to share in administering the property. Loyalty is expressed in ceremonies that dramatize the notion of permanence and unquestioning personal ties; this involves the ideals of *traditional legitimacy*, with their emphasis on unshakable loyalty to particularized symbols representing the cultural identity of the family, ethnic group, or club. Patrimonial organizations are extended by adding more individuals into the network, expanding the domestic group, or having protegés acquire their own networks of protegés.

3. Patrimonial organizations develop to administer property through a personal network. *Bureaucracies* develop to administer it impersonally. Bureaucracy is designed to overcome certain problems of patrimonial chiefs: For the emphasis on unquestioned personal ties means that individuals do not distinguish between the organization's property and their personal use of it; the subordinate takes his grant of authority as a mandate to act on his master's behalf in all affairs; and the personal nature of ties means that subordinates of subordinates are loyal only to their immediate masters. The result is that patrimonial organizations cannot be very well-controlled much beyond the immediate sight of a master; every man is a petty chief when he is out of range of his superiors, appropriating as much power, status, and material goods as he can get away with; and when the geographic range becomes great enough, the organization collapses into feudalism or outright independence.

These are inevitable results when one tries to administer large-scale property through a personal network. But no other method is possible as long as households are the main base of operations, family ties are the most stable form of control, and techniques for standardizing, subdividing, and checking upon complex tasks at long distance are not available. Bureaucracy is an organizational counterattack against these problems. It involves separating the man from the job and the family from the man, creating a set of abstract rules and regulations that subdivide responsibility and give each man only part of it, provide for codifying performance standards and keeping formal records of what is done, fixing formal qualifications for putting men in positions and for paying and punishing them. The crucial encounters in this form of organization, then, are those which constantly reinforce the notion that men are only occupants of positions who are subject to records and formal rules; ritual deference is to the rules themselves and to the organization in the abstract, not to any particular individuals. The accompanying concept of legitimacy is *rational–legal:* the ideal that the organization is enacted by human rationality along general principles, not arbitrary loyalty to particular individuals sanctified by custom.

Such an organization can arise only under certain conditions. On the technical side, it requires the skills and implements for codifying orders and keeping records; the communications and transportation equipment for sending messages on rigidly enforced schedules; a means of production that enables large numbers of people to free themselves from agricultural production and live away from a household economy; and a money economy that enables rewards to be administered in abstract tokens instead of available goods. On the organizational side, it requires that men have developed forms of loyalty outside their family or local cultural group, and an accompanying ideal of abstract rules.

These latter conditions were always developed first in religious cults, beginning around charismatic leaders and then developing into universal churches which held property (and hence, had a permanent organization) but did not transmit it through the family (hence, recruiting new members outside of patrimonial groups). Most of the first approximations to bureaucracy—the Chinese and Egyptian empires, the Buddhist empire of Asoka, the Christian church and the centralized European kingdoms— were staffed by priests, both because they were literate administrators and because they carried the ideal of control by abstract rules and loyalty to an abstract purpose, represented in religious terms by a transcendent God. Christianity, whose God is closest to the ideal of a rational rule-giver, gave rise to and developed along with the most bureaucratic type of administration.[3] Later, secular bureaucracies substituted an educational

---

[3] The Greek rational philosophies, from which Christianity developed, spread among the administrative classes of the Hellenistic Empires and their Roman successor; with the

system as a means of training in the ideals of abstract rules and purposes.

If we recall that ad hoc groups, patrimonial organizations, and bureaucracies are all ideal types, we can see that mixtures of them are empirically possible, and in fact very common. Weber refers to the Chinese empire, for example, as a patrimonial bureaucracy because of its mixture of elements; the reasons for this mixture can be traced to certain conditions (the extent to which magical as well as universalistic elements were part of the Chinese religion, the agricultural emphasis, the willingness of the emperor to countenance strong familistic ties among his officials, and so forth), and the consequences for Chinese economic and cultural development can be drawn out. As Guenther Roth (1968) has pointed out, mixtures are prominent in the modern world as well. Not only are there strong patrimonial networks underlying the superficially bureaucratic organizations of Latin America and much of Asia today, but personalistic networks of various kinds are found in the bureaucracies of the West, including America.

These continue to exist for a very important reason. Organizations are arenas for struggle, and bureaucratic rules and task divisions are one means by which superiors attempt to exert control over their subordinates. One counter-response is for subordinates to use bureaucratic procedures as a barrier against direct surveillance from above. Superiors may counter by developing personal networks of information and private favors. But this has its dangers too—notably the difficulty of controlling at more than one remove in a personal network—and hence, administrators must be wary of how they use this technique. As Roth points out, modern organizations are a mixture of control strategies, with the successful organizational politicians using both formal rules and informal networks and personal emissaries in an attempt to gain maximum leverage.

Organizations, then, can be seen as power struggles along several dimensions and as using a number of tactics and devices, according to availability and the personal predilections of the individuals involved. It should not be surprising, given the instabilities of most forms of control, that organizations tend to change their structures and goals as well as their personnel. This has been considered extensively in the modern literature on goal displacement; the main difficulty has been that the official idea of the organization at one point in time (typically the ideal of those who set it up, or of some outside observer) has been reified, and changes described in moral tones, instead of looking at the real complex of interest groups who are always on the scene and among whom power has shifted over time. Thus Robert Michels (1949), who began this form of analysis, acted

thorough bureaucratization of Roman administration under the Antonines, Stoicism became almost an official ideology, supplanted by the adoption of Christianity as the state religion.

as if there were a movement from an ideal, which he attributed to a party organization, toward its opposite, without considering just which members of the organization were actually concerned with upholding this ideal.

But if we separate out the moral evaluations, we can see some crucial variables for explaining who wins what degree of control over the products of an organization. Michels focused on the channels of administration and communication. Subsequent research along these lines leads us to a number of variables that determine how power will be distributed: the existence of channels of contact with important outside groups, internal controls over areas of technical uncertainty, the spatial distribution of the tasks in relation to the technology of communications. Some of this is implicit in Weber's treatment of how geographically large empires administered without adequate technical resources lead to the fragmentation of authority. If we recall that organizational structure is actually the network of power relations, we can see the elements of a general theory of the variations in organizational structure.

I have argued that the Weberian tradition provides the core of a unified theory of organizations. Much of the material for this theory comes from rather different traditions. One of the most important of these is the neorationalist line of managerial theory from Chester Barnard (1938) and Herbert Simon (1947) to James D. Thompson (1967). This has attempted to combine an understanding of organizational politics with a notion of rational decision-making in conditions of limited cognitive capacities and imperfect information. This approach gives some useful analysis of different types of administrative techniques and of the technical contingencies that go into the structure of a productive organization. What limits this mode of analysis is primarily the point of view; it is designed as a practical guide for managers as well as a general theory, and thus tends to lose sight of the political self-interest of the managers themselves by speaking in terms of the organization as if it were an actor, when what is meant is what the rational manager does (or should do).[4] The Human Relations tradition

---

[4] Thus, Thompson (1967) couches many of his propositions in terms of what an organization will do "under norms of rationality." This enables him to ambiguously make propositions about what organizational structure will be like under certain conditions, while giving advice as to what the rational manager should do. Empirically, these "norms of rationality" seem to refer us to a particular kind of competitive situation, perhaps the adjustment of organizations to the pressures of an idealized capitalist market. Thus a number of propositions about organizational growth attributed to needs of "rationality" really seem to mean that capitalist competitive conditions (probably in the realm of financial credit above all) pressure organizations into expanding in certain directions—unless, of course, they fail or are bought out.

is somewhat less useful, although it pioneered the study of informal work groups; it constantly invokes a naïve ideal of how peaceful and harmonious an organization could be if only we could find the right way of giving informal groups a sense of belonging.[5] This work tends to be theoretically limited by its practical concerns, and biased against any extensive analysis of power conflicts. Of much greater value has been the Chicago tradition of studying occupations and professions (Hughes, 1958; Wilensky, 1956), culminating in Wilensky's (1964) model of the conditions that enable an occupation to gain the advantages of a professional form of organization. And finally, there is a good deal of recent work on the determinants of organizational structure. Some of it has focused rather narrowly on the somewhat popular question formulated in Parkinson's Law, the alleged tendency of the administrative sector to grow faster than the organization as a whole. More significant work has developed a number of structural types related to differences in tasks and technology (Stinchcombe, 1959; Woodward, 1965; Perrow, 1967).

What is striking is the extent to which a number of key variables reappear in the most powerfully developed areas. The importance of un-certainty (in contrast to routine) and access to information about it for power relations and organizational structure is one such variable, appear-ing in Wilensky's theory of professions, Stinchcombe, Thompson, and Perrow's analyses of organizational structure, Wilensky and Crozier's studies of power in bureaucracies, and the Michelsian tradition on organi-zational oligarchy; the importance of technology in shaping task struc-tures and administrative channels, found in Woodward and Thompson's structural models, Blau and Scott's model of environmental control, and in Weber's contrast of bureaucratic and patrimonial politics; and the polit-ical role of beliefs about reality, from Weber and Michels to Goffman and Simon.

What follows is an attempt to state some formal principles that encom-pass the results of these analyses in a unified model. The principles are meant to apply to all forms of organizations, not to a limited range of types only. Once we take this stance, it is apparent that the theory of professions and the theory of democracy in voluntary associations are variations on a common set of principles stating the conditions for an inner group to control a membership association. These principles are strikingly similar

---

[5] A summary of this tradition of research is given in Blau and Scott (1962) and in Caplow (1964). These books move beyond the Human Relations school to the point of building theories around the dilemmas of attempting to maximize both communications and control, or initiative and control, or other things desired by organizational masters. The implicit value judgment hangs on in the formulation of incompatibilities as dilemmas rather than as conflict among groups with differing interests.

to those that determine power within bureaucracies, and the reason becomes clear when we see that a membership association is like a bureaucracy upsidedown, with the control structure running in the opposite direction. Patrimonial and bureaucratic organizations can be characterized by different states of the same variables; even such extreme groups as families and small personal groups can be explained under a unified scheme. Centralized and decentralized forms cut across all of these, and are determined by some core variables. And it should go without saying that political, economic, and cultural organizations are all subject to a common analysis, which enables us to deal with political ideologies and religious ideals in a causal manner.

I shall begin with some propositions about control strategies and their outcomes, move on to administrative techniques, and attempt to show how certain of these can accomplish certain kinds of tasks but not others. The way in which controls are structured at any one time shapes the interests of the individuals who make up the organization, and it is the struggle to pursue their interests that leads to whatever is produced in the organization, what ideals and goals are expressed, and the changing network of relationships that make up the organizational structure. This model will then be applied to different kinds of organizations, to show how they explain the structure of hierarchic organizations of various kinds, of membership-controlled associations including political democracies, and of professions and pseudoprofessions.

ORGANIZATIONAL CONTROL

An organization is a network of interpersonal influences, and the most fundamental determinant of how men influence each other is the type of sanction they apply. The main types are coercive threats, material rewards, and loyalty to ideals.[6]

10.1 Coercion leads to strong efforts to avoid being coerced.

10.11 If resources for fighting back are available, the greater the coercion that is applied, the more counter-aggression is called forth.

10.12 If resources for fighting back are not available but opportunities to escape are, the greater the coercion that is applied, the greater the tendency to leave the situation.

10.13 If resources for fighting back and opportunities to escape are not

---

[6] Evidence relating to these forms of control is summarized in Etzioni (1961).

available, *or* if there are other strong incentives for staying in the situation (material rewards or potential power), the greater the coercion that is applied, the greater the tendency to comply with exactly those demands that are necessary to avoid being coerced.

10.14 If resources for fighting back and opportunities to escape are not available, and there are no strong positive incentives for staying, the greater the coercion applied, the greater the tendency to dull compliance and passive resistance.

We have already met some of these and related principles (Postulate XV and Propositions *8.2* and *9.11*) in considering patterns of sex and age stratification. More generally, they follow from the fundamental postulates that individuals seek to maximize their subjective status (IV), avoid contacts in which they have low status (VI), and engage in implicit conflict when they cannot withdraw from an unfavorable situation (VIII). Similar results have been found in psychological experiments with animal behavior; it is for this reason that B. F. Skinner (1961, 1969) has argued so strongly that rewards are much superior to punishments as an incentive for learning (by which he means control over behavior).

This is obviously a very powerful generalization, for the same results appear in sociological and historical descriptions of coercive organizations, such as prisons, armies, concentration camps, and of slave labor and premodern organizations generally. Throughout history, slaves, prisoners, and oppressed minorities have acquired the reputation of being dull, childish, irresponsible, and careless. Dominant classes have incorporated this into a self-justifying ideology. But the behavior results from the situation of being coerced without opportunity for rebellion or escape; being noncooperative is the only means left for retaining subjective dignity, and appearing stupid and irresponsible is the best cover to avoid being punished. Even in extreme cases of unequal resources, the fight for autonomy goes on, even if it must go underground. Given the fact that there is an element of coercion in any form of control in which someone can back up his orders by bringing undesirable consequences (including such milder forms of coercion as taking away rewards or the privileges of membership), we can see why the culture of order-taking classes (*1.2*) is always built around some degree of implicit rebellion against authority.

10.2 Control by material rewards leads to compliance to the extent that rewards are directly linked to the desired behavior.

10.21 The greater the emphasis on material rewards, the greater the acquisitive orientation on the part of the individuals being controlled, and the greater their effort to manipulate to their own advantage the situations in which performance is measured and rewarded.

Throughout most of history, large-scale organizations have operated principally by coercion. With the industrial revolution came a shift in control forms. A whole ideology sprang up to the effect that if people could now be controlled by their own self-interest they would be happy, free, and willing workers, and that productivity would be increased by an enormous degree (Bendix, 1956: 22–116). Adam Smith is the great ideologue of this point of view, which comprises a basic theme in classical liberal thought. It took some time to discover that it was not quite so ideal. Controlling people by material rewards does not produce the kind of alienation that coercing them does, but it produces only a very specific kind of motivation: the motivation to be acquisitive. If one tries to control people strictly by material rewards, one finds that they are willing to cooperate precisely to the extent to which they are paid. Workers motivated by money and given freedom from coercive controls are likely to form a trade union. At higher levels, this control form frees men to pursue their own careers rather than committing themselves to the organization.

These effects have been mitigated only by tying rewards closely to productivity, such as in piece-work or bonus systems. Even so, the evidence shows that people working on incentive systems in factories concentrate on manipulating the incentive system (Roy, 1952). The famous informal system of control over rates of work was first discovered in such situations; workers formed their own rates in order to keep management from setting the piece rate too low. Workers also manipulate the situation when a new job is set up and the foreman comes around to see what the average time is so he can set an incentive rate; experienced workers were able to perform with a great amount of diligence, but nevertheless not producing too much. At higher levels of authority, heavy emphasis on material incentives leads to a similar concentration on presenting a favorable image of accomplishments.

A third form of control is to acquire members who are intrinsically committed to the same goals as their superior. Such control by ideals (normative control) is the strongest kind of motivation. This was recognized by the human relations school of the 1930s and 1940s, which discovered how workers' informal controls undermine formal authority. Their program of trying to make the workers identify with the company, however, had little effect; it did not stop them from unionizing, striking, or trying to control their work pace.[7] Normative control is very good if it can

---

[7] See Wilensky and Wilensky (1951). Interestingly enough, the evidence that purports to show that workers increased their productivity simply because greater personal attention was given to them (the so-called "Hawthorne Effect") has been misinterpreted; the real

be had, but it is extremely hard to fake it by friendly gestures and speeches. There are some actual methods for establishing normative control, but they are more costly: giving members power or opportunities for power, or creating informal solidarity.

10.3 The more one gives orders in the name of an organization, the more one identifies with the organization (cf. *1.1* and *5.23*).

10.31 The more one has to prove to others one's loyalty in order to acquire a position of authority, the more one identifies with the organization.

At increasingly higher levels in an organization, it is increasingly more likely that its members will identify with their jobs and with the organization. One method of acquiring normative control, then, is to coopt members into responsible positions; a related method is to offer them a chance to be promoted. This is effective if they perceive they have a realistic chance, and even more so if the chances are real but uncertain. In the latter case, they have to prove their loyalty, resulting in a process of self-indoctrination.[8]

Why are men in power committed to the organization? Having power is a very considerable reward in itself. Even more, exercising power always involves some sort of exercise of ideals. That is essentially the same argument that Weber made about legitimacy. It has been widely misinterpreted, as if Weber were saying that organizations are based on consensus and that conflict is a state of deviance from the ideal. But in fact, the ideals and the conflicts are intertwined. Power itself gives rise to at least implicit conflict between those who have the power versus those who do not. Moreover, organizations are nothing but repetitive behavior within a group of people; hence, an individual can be powerful only through his continual influence over others, including ultimately the support of those who will coerce others to maintain his authority or property.

As Goffman (1959) has shown on the microlevel, persons enact their authority by idealizing themselves and reifying their positions and their organization. Their effort at reality-constructing is simultaneously an effort at maintaining power. Usually they have to rely on teamwork among the group in control, the managers, executives, or officials, which involves primarily a theatrical performance in preserving each other's

---

operative variables appear to have been changes in material incentives (A. Carey, 1967). In other words, the Hawthorne Effect is a myth.

[8] B. F. Skinner gives evidence for the parallel case of animal behavior rather than human attitudes; the greatest level of performance (in a closed situation, where pigeons were pecking for their food), occurred on an uncertain schedule of reinforcement.

idealized selves. In order to have power, one must have a considerable commitment to acting like a person in power. One can not have power oneself unless one claims that the organization itself is something important and ought to be respected. The result is recurrent self-indoctrination, every time an official speaks in the name of the organization. Among the managers who uphold the power, ideals are the common currency they use to hold themselves together as a team.

Weber's legitimacy types—traditional, charismatic, and rational–legal —are attempts to idealize particular control devices; in Goffmanian terms, they are attempts to define the reality of one's relationship to others as if it were an eternal and transcendent object rather than (as it actually is) a transitory thing to be continually reenacted. A charismatic leader is continually putting on a performance that idealizes himself or the power that he stands for. Prophets always create the god that gives them power; the better they are at recreating it in every *here and now,* the more power they have. Traditional leaders rely on creating a different sort of reality, the reality of the past, of the taken-for-granted, of an order that it is unthinkable to change, and of course, an order that calls for obedience to his person. Such a leader is engaged in recreating the reality of tradition which people otherwise would forget if it were not continually enacted and dramatized. Rational–legal authority, prominent in modern organizations, is also continuously enacted. Power that depends on rules and regulations requires that they must be enacted just for purposes of maintaining power and quite apart from whether or not the rules accomplish anything practical. The key ritual in a rational–legal government is the ritual of a court of law, a situation in which people are required to give great respect to a hypostatized world of legality enacted by judges and lawyers.

Weber's analysis of legitimacy, then, fits nicely with Goffman's description of how people create an ideal reality in everyday life. It is an attempt to control other people by defining oneself as a powerful person. To the extent that other people will back up each other, there will exist an organized system of power, and its leaders can rely on control of further sanctions, such as material rewards or coercion, which the group can command.

There are certain drawbacks for individuals trying to achieve normative control over others in this fashion. If an organization consists of a structure of power in which certain people control other people, the more normative control that is attempted by sharing power, the less an hierarchic organization one has.

Efforts at worker participation have not worked very well, because most organizations cannot be run efficiently by giving the workers in each location absolute power in deciding how the organization is going to be run.[9] Since managers have seldom been willing to give up their apparatus of coordination and planning, workers have merely been given the chance to give their opinions, which may or may not be taken into account. The result is that workers are no more happy or attached to the organization than before. Normative control cannot be faked, either by pep talks or by pretending to give widespread power, if one is not really willing to give up centralized controls.

The strongest commitment to a group and its ideals results from successful social rituals. As we have seen (from Propositions *2.1, 2.2,* and *4.7*), a high degree of physical copresence and a high focus of attention tend to produce a strong state of collective consciousness: highly reified ideas and a strong expectancy of conformity among members of the group. But the content of these ideas and behaviors can vary a great deal, depending on the authority relations. It those who participate in such rituals derive power from them, they become especially dedicated to the organizational ideals and forms (*1.1* and *6.1*); but those who are given orders in this highly ritualized way tend to emphasize perfunctory compliance to only the most observable demands (*6.2* and *6.23*). Strong attachment to the group *per se* requires that the rituals take place among equals (*1.3, 4.8,* and *5.232*).

So another path to normative control, besides offering power, is to make organizational members committed to each other as equals—that is to say, as friends rather than merely co-workers. There are a number of ways to do this.

10.4 The more the people recruited to an organization are already committed to each other, the more potential loyalty to the organization.

This is a traditional method for securing organizational loyalty; from the point of view of rational–legal values, it is called *nepotism.* It is the basic principle of patrimonial organizations: to recruit persons who already have strong ties because of the high degree of ritualization of the family, or who are personal friends or followers. The family is no longer very much used for this purpose in Western societies, partly because organizations are far too big to fill them in this way, and partly because families themselves are no longer very ritualized internally. But personal ties remain important, especially in all of the more "political" positions in the higher levels of

---

[9] Lammers (1967) summarizes evidence that workers' councils operate harmoniously and cut strike rates only if they stay on extraneous issues (safety, fringe benefits, working conditions), and avoid direct issues of what work is done and how; this is clear even though the author struggles to be as optimistic about the arrangement as possible. Conversely, happy workers do not necessarily mean an efficient organization.

organizations, and especially in "entrepreneurial" kinds of interorganizational networks in business, finance, government, and the professions.

10.5 The more conducive the conditions for creating personal friendships in an organization, the greater the potential loyalty.

10.51 Barnard's Principle. The more similar recruits are in cultural background, the more likely they are to become friends, and the greater the potential loyalty. (cf. *5.3, 5.33, 5.34, 5.42,* and *5.44*).

10.52 The more opportunities for informal gatherings on the job, the more likely members are to become friends (cf. *2.1* and *5.11*).

10.53 The greater the isolation of organizational members from outsiders, the greater the potential loyalty to the organization (cf. *2.2*).

10.54 The more celebration rituals among equals with similar experiences, the closer the personal ties and the more potential loyalty to the organization (cf. *4.8*).

There is a great deal of evidence for how these mechanisms of informal solidarity operate. The organizations with the highest degree of member commitment rely heavily upon them. Churches put their officials through a period of isolation from outsiders and intensive interaction among themselves; the more the aim is to commit the priest's or monk's loyalty to the organization itself rather than to a local congregation, the greater the emphasis on celibacy and communal living, which cut him off from any ties except those within the church. Utopian communities have been successful precisely to the degree that they have selected members from similar backgrounds, sharply isolated themselves from the rest of the world, and maintained a high level of mutual surveillance, and highly ritualized and emotion-producing (charismatic) group ceremonies (Kantner, 1968). Military officers acquire an *esprit de corps* by similar devices: training in a total institution that breaks their ties to the outside world (Dornbush, 1955), a form of communal living at the officers' mess emphasizing the equality within the off-duty brotherhood, and backed up by numerous ceremonial occasions (banquets, toasts, oaths, ritual games with strong notions of honor; cf. *6.6–6.62*).

Most organizations lack the resources to make members undergo these kinds of commitment-producing situations, but milder forms produce their own levels of normative control. Jobs that remove men from the ordinary schedule experience of work and give considerable autonomy on the job (such as policemen, printers) tend to have high interpersonal commitments and job attachment.[10] Jobs that allow much informal contact facilitate the formation of informal groups; those which keep men isolated by their physical settings or by rigid adherence to rule-bound encounters produce alienation both from each other and from the organization (Aiken and Hage, 1966; Blauner, 1964). Since managers are most concerned with loyalty at the higher levels of the organization, it is here that the strictest tests for cultural similarity are applied (Collins, 1971).

There is yet another way to produce strong ritual ties within a group: conflict against outsiders may be used to promote solidarity through participation in the rituals that accompany stress.

10.6 The more that members of an organization are aware of danger and hostility from another organization, the more loyalty to the organization (provided that there is not already greater hostility between groups *within* the organization) (cf. *4.82* and *4.83*).

10.61 The more exclusively an organization recruits from a cultural group (ethnic, racial, class) that is in conflict with another cultural group, and the more an organization emphasizes ceremonial tests of cultural similarity for membership, the greater the loyalty to the organization.

10.62 The more an organization emphasizes ceremonial deference by lower-ranking groups whose membership is temporary and assured of promotion, the more the loyalty to the organization after they have been promoted.

Some organizations are expressly built around conflict with outsiders. The leaders of an army, church, police force, or political party command greater loyalty to the organization, the more its members are oriented toward conflict with outsiders rather than toward the stresses of internal authority. This is easiest in time of war, electoral campaign, doctrinal conflict, and so forth; the army can relax much of its disciplinary procedures in combat precisely because the enemy serves to keep the troops in line (cf. Shils and Janowitz, 1948). When there is no real conflict going on, leaders may create similar effects by substituting ceremonial conflicts. Much of the solidarity-producing effect of political speech-making, for

---

[10] See Skolnick (1966: 52–65). Police organizations (in America) are one of the few in which lines of promotion are open from the bottom to the top, thus furthering normative commitment in the group, as well as its potential estrangement from outside control.

example, comes from jibes against enemies; the same holds for other organizations. A strong church needs its devil, and any organizational leader can enhance his power by making sure there are visible enemies or competitors against whom a minor conflict can be quickly escalated when internal stresses become more serious. The heavy stress on athletic contests in American schools (which hold the largest compulsory clientele in the world) is just one example of this form of ceremonial control (Bidwell, 1965: 978–984).

In this perspective, racial, ethnic, and other cultural discrimination falls into place, not as a peculiar anachronism in the world of modern ideals, but as a deliberate control device. This is not to say employers necessarily tell themselves that ethnic discrimination will be a useful policy; they simply maneuver for the situation that yields the greatest day-to-day advantages, and defend their policy in terms of whatever ceremonial ideals are publicly acceptable at the time. Studies of organizations practicing discrimination show that they receive considerable benefits in terms of the loyalty of their own employees, although there is a cost in the form of antagonism on the part of groups who are discriminated against (Hughes, 1949; Nosow, 1956). Ethnic splits within the working class have probably contributed to the slower pace of unionization in American history in comparison to Europe, and to its much lower political organization along class lines (Johnson, 1968).[11]

It is where ethnic groups become sufficiently mobilized to engage in political protest that a policy of discrimination is modified, more or less proportionately to the power each group can bring to bear. But even in a highly mobilized society, cultural criteria for employment are not likely to disappear; the very existence of ethnic tokenism or quotas tends to keep ethnic distinctions important. A high level of ethnic mobilization also tends to shift control emphasis toward other cultural devices that may function as surrogates. Educational requirements for employment are best explained in this manner, and the form they take is best predicted from the balance of advantages they give to cultural groups with varying degrees of power (Collins, 1971).

Hazing rituals is another form of producing solidarity through conflict. They are found especially in clubs and fraternities, military academies, and other schools for potential elites (Dornbush, 1955; Weinberg, 1967:

---

[11] Whether this has been a blessing to American businessmen is questionable. Formal organization of unions tends to reduce the intensity of conflict, and American labor history has been especially violent for this reason. Moreover, trade-union-based political parties have been notoriously unmilitant, again because of the conservatizing effect of formal organization.

97–126). The standard form is for the higher-classmen to be given the power to demand all sorts of ritually demeaning services from the lower. For the former, this enhances their attachment to the organization because they get to enjoy the exercise of power in a highly ritualized form, which makes for much emotional experience and strong reification of the symbols and ideals of the organization (*1.1* and *4.7*). For the lower-classmen, the main compensation is that the experience is a collective one, which draws the class together and usually gives them some opportunities for ritual rebellions of their own (*4.82*). They also have the status-saving knowledge that their very subservience is part of a transition that sets them off from outsiders, and that results in making them part of an elite group (*1.6*). Eventually they reach the other side of the fence, reinterpret the rituals by going through them again with different rewards, and end up with a strong sentimental attachment to the organization.

There are a number of ways, then, that personal ties of loyalty can be established within an organization. It should be borne in mind, however, that having people like each other is *not* tantamount to having them produce what their organizational leaders wish. Strong informal groups can be an increased *problem* for organizational control if they are motivated against the leader's policies, and that depends especially on whether they are essentially order-givers or order-takers. For this reason, the deliberate stress on creating informal ties is mostly confined to the higher levels of organizational authority.

It is apparent that any single form of control is rarely used alone. For the effects of mixtures, I would suggest the hypothesis:

10.7 The effects of coercion, material rewards, and normative controls used as mixtures are a linear combination of the weights of each.

Administrative Devices

Tactics of control involve not only basic sanctions, but the minute-to-minute business of issuing instructions and seeing how they are carried out. The two are complementary: Particular control devices work best with certain kinds of sanctions. Controls may be either direct or indirect, and several kinds of controls may be operating at the same time.[12]

11.1 Surveillance. The more closely a superior watches the behavior of his subordinates, the more closely they comply with the observable forms of behavior demanded. (cf. *2.1* and *6.2*).

---

[12] The following typology of administrative devices is modified from Simon (1947).

11.12 The closer the surveillance, the greater the ratio of supervisors to workers.

11.13 The closer the surveillance, the greater the alienation from long-term organizational ideals, and the lower the initiative of the workers.

Surveillance is alienating because it limits freedom of action; it creates a reified form of consciousness (*4.7*) that is highly mechanical and conservative. To the extent that surveillance is used to control not only workers but administrators as well, there is virtual doubling of the control hierarchy. The czarist government in Russia, with its heavy emphasis on mutual spying, was typical of many traditionalistic organizations in this respect; the Communist regime uses a similar technique with a party hierarchy and an administrative hierarchy to check on each other (Bendix, 1956: 341–433).

11.2 Efficiency criterion. The more emphasis on control by inspecting the products of work, the greater the worker's concern for the most visible outcomes.

11.21 The more emphasis on control by inspecting outcomes, the greater the emphasis upon controlling the situations of inspection and the form of the records.

Control by an efficiency criterion is much less alienating than surveillance. It requires that the outcomes be directly observable, and hence is often difficult to apply in the case of managers, since the actions of a particular individual are somewhat remote from the measurable outcomes of the group performance. The battle over piece rates previously described illustrates the type of maneuvering involved at the workers level. This is even more prominent among various staff divisions. In contemporary organizations, a typical conflict is between the controller's office and the other managers who are supposed to give the information on which they are to be checked (Simon *et al.,* 1954). Fights center around information and its interpretation. Records are often guarded or suppressed; two sets of records may be kept, one to be viewed by supervisors and outsiders and one essentially for internal consumption. Highly bureaucratized school guidance counsellors, for example, often keep one set of records for parents to see, and another recording their candid evaluations of students (Cicourel and Kitsuse, 1963: 122–130).

Another problem that arises from control by efficiency criteria is the fact that there are multiple criteria of efficiency. One might attempt to minimize cost, maximize speed, maximize safety, maximize perfection, or maximize volume. These tend to be incompatible with one another. The ideology that goes with the use of efficiency controls does not usually recognize that multiplicity. Those aspects that can be most easily quantified are the ones that become emphasized. Quality is the hardest to quantify, and hence is most often left out.

11.3 Rules. The more reliance on written or formally codified rules, the more standardized the behavior of organizational members.

11.31 The more reliance on rules, the less the authority of any individual, and the more impersonal the relationships among organizational members.

11.32 The more reliance on rules, the slower and less adaptive the behavior of organization members.

Control by rules is another form of direct control. It is not an alternative to the others, but a way of formalizing either surveillance or standards of efficiency. One of its effects is a proliferation of the administrative side of the organization, as it requires a good deal of written communications and record-keeping. Other effects are somewhat mixed. Having written formal rules makes the control structure of the organization more public and objective; it reduces uncertainty about what the supervisors will do. It also appears fair in the sense that written statements of abstract generalizations have an idealized tone lacking in any particular ruling about some individual case. One of the effects of control by rules is a reduction in the power of the supervisors. Both sides in organizational conflicts can be interested in expanding rules, depending on what the particular rules are.

The ideology that goes along with this form of control is that of a perfectly running organization with a set of rules to cover everything, so that one would have merely to plug the people into it. This is an impossible ideal. The main problem is that the rules cannot cover all details, especially in areas of uncertainty. The more that rules are elaborated to cover all contingencies, the more effort must be exerted in trying to locate the proper rule to apply, as illustrated by the enormous inefficiency of the army. Among British dockworkers, a "rules strike"—following all regulations precisely—is used to create the effects of a crippling slowdown. Almost any task has to be carried out with a certain degree of human initiative: To the extent that one proliferates the rules to cover everything, one creates a situation that reduces efficiency, or in which people ignore the rules most of the time (Turner, 1947).

In organizations with elaborate rules, they are often kept as a form of control in reserve. Bureaucracies typically have two realities: a detailed rule book which is thrown aside most of the time but is available in case somebody wishes to assert his authority or cover himself in a crisis by pulling out the rule book to decide who exactly is to blame for something and who is not to blame. Members keep elaborate written records just to cover themselves (as in the army or other governmental agencies), but operate informally with a different understanding of what those records mean. Power thus tends to be dispersed within the organization toward those individuals who are best protected from outsiders by rules and regulations which they can apply or not apply at their own discretion (Crozier, 1964: 160–161).

There are two main types of indirect control: control over information and control over the environment.

11.4 Information control. The more exclusively an individual controls information about areas of uncertainty in an organization, the greater his power to control others' behavior.

11.41 The greater the consequences the resolution of an uncertainty has for an individual, or the larger the network of individuals affected by the resolution of an uncertainty, the greater the power of whoever controls information about that area of uncertainty.

Information is anything that defines reality. Since organizations are invisible—they are sets of beliefs and rules and accepted ways of acting which large numbers of people carry around in their heads and sometimes write down on paper—information is central in determining what the organization *is.* Covert power goes to whoever is in a strategic position to control information that somebody else will act on, whether or not he has formal authority over others.

This does not apply only to explicitly defined "experts." The pattern of organizational interaction is crucial in determining which information will carry the most influence. Wilensky (1956) discovered that the influentials in labor unions were lawyers who had contacts outside the organization, and who dealt with some area of uncertainty to which they alone had access.[13] Since they were the only ones with the expert contact with the courts, for example, they could define legal reality for the leaders of the union, who would tend to follow their proposals even though they lacked official authority. Similarly, Michels (1969) demonstrated that the leaders of a democratic association tend to take control away from the general membership by virtue of their control of the administrative machinery. In the folklore of radical politics, "He who controls the mimeograph machine controls the party." Members can act only on what they believe the situation is, and whoever defines the situation indirectly controls what they do. It is in that sense that secretaries can often have

---

[13] See also Wilensky (1968b) for an application of this principle to intelligence-gathering agencies in government and business.

considerable power within organizations, precisely because they control lines of communication, with their guarding of the telephone lines and their opening of the mail and deciding what is junk mail and what has to be acted upon by their bosses (Mechanic, 1962).

Information control can operate even in a coercive situation like a prison. In general, prisons are controlled by the old cons. This may sound odd, since all the force is in the hands of the guards and the warden. But like any high administrator, the warden is considerably removed from the places at which things actually happen, and acts only upon third-hand reports that come up to him. He cannot usually remedy this by direct inspection, because the underlings protect their autonomy by putting on an idealized performance for his benefit. Nor do prison guards do much except keep order. If the prisoners wished to riot, they could probably get away with it for a while; it could be put down, but the guards would just as soon not have it happen. So there is a certain amount of power that the prisoners have, just in terms of whether or not they are going to give their compliance. Usually the prisoners are controlled by a certain number of old cons, who control things primarily because they are the sources of information. In a situation of severely limited mobility, those few prisoners with greater privileges become the centers for rumor networks, which in turn gives them great influence in stirring up or calming down dissension. McCleary (1960) gives an example of a prison reform in which a new warden decided to change the emphasis from punishment to rehabilitation. By relaxing restraints, he challenged the power of the old cons who benefited from the old system because they monopolized information within it. The result was a series of rumors during the transition period, which resulted in a riot.

Crozier (1964: 145–208) states more generally that power is based on areas of uncertainty. In the French organizations he studied, engineers and maintenance men had power because they were the only ones in a highly routinized operation who dealt with the unexpected, in the form of machine breakdowns. Technical change, Crozier maintains, tends to routinize processes, so that the power of experts is constantly self-liquidating. In stagnant, noncompetitive organizations like the French monopoly Crozier studied, the vested interest of engineers was in maintaining unreliable techniques. In a more competitive situation where changes may be continual, technical experts are interested in innovation, as they are continually pushing toward the frontier areas in which they have the maximal power.

Information control also underlies legitimacy in organizations. Legitimacy is based on successfully defining one's position as both powerful and just. Men will be accepted as legitimate leaders as long as they can maintain control of the channels of information. Charismatic leaders have the most unstable position in this regard, since they must continually project an image of spontaneous inspiration and hide the machinery of control. At the most fundamental level, power is based on a self-fulfilling prophecy: If somebody becomes defined as powerful, other people will then defer to him, which in turn gives him more levers by which to control other people; because he is defined as powerful, he is powerful. On the other hand, if he starts becoming defined as nonpowerful (as when the rumor gets around that a politician's power is slipping), other people stop deferring to him, stop trying to make deals with him, which means that he actually does have less power, which tends to make his image slip still further, and the result is a vicious spiral in the other direction (see Schelling, 1963: 89–92; Bendix, 1968).

11.5 Environmental control. The fewer the physical alternatives within easy reach of an organizational member, the more likely he is to concentrate on the task to which he is assigned.

11.51 The more dispersed individuals with common work experiences are, the less likely they are to formulate a strongly held common outlook, and the less power they have. (That is partly the converse of *2.1.*)

Environmental control is a notion that arose from studying modern industrial organizations, but can be generalized further (Blau and Scott, 1962: 176–183). There are two main types of environmental control. The first is epitomized by an assembly line, or by trench warfare. In both of these situations, the worker is physically gripped to his job. The structure of the physical world around him gives him very few alternatives. The other kind of environmental control consists in physical controls over communications. The dispersion of a work force to many separate places gives an environmental control advantage to the authorities, since the workers cannot easily communicate among themselves; conversely, large-scale work settings reduce this environmental control. This type of control was characteristic of most premodern organizations, in which households were the main unit, and the servants had little opportunity to communicate with servants in other households.

Tasks, Sanctions, and Administrative Devices

There are systematic compatibilities and incompatibilities among types of sanctions, administrative devices, and kinds of tasks that an organization can do (Etzioni, 1961; D. Warren, 1968). The kinds of tasks that people can do in an organization may be placed on a continuum, from those requiring high initiative to those requiring low initiative. Low-initiative tasks include simple manual labor, like shoveling coal or loading trucks. Requiring somewhat more initiative are routine operating of machines or routine clerical jobs. Above that are more skilled or complex manual jobs such as jobs of craftsmen and administrative jobs that involve checking up on other people; these require more initiative because the outcomes are less certain. At the level of high initiative are planning and decision-making, since by its very nature one cannot predict in advance what a plan is going to be.

12.1 The lower the initiative required in doing a task, and the more visible the activity, the more successfully it can be accomplished using a high degree of coercion administered with surveillance, rules, or environmental control.

12.11 The more coercion is used, the more surveillance or environmental control is needed to prevent escape, rebellion, and organizational disintegration.

Organizations that use coercion produce high alienation. Workable control devices are close surveillance and strict environmental control; rules may be elaborated rigidly if surveillance is available to enforce them. Thus, coercive organizations are often "total institutions" bounded by walls which prevent escape and which prevent members from communicating with the outside except insofar as the authorities control the communication. Since members are alienated, authorities act to remove their resources for organizing as a conflict group and to control their actions as closely as possible. The tasks that can be carried out under these controls are limited to ones of low initiative; even here the productivity—as evidenced by forced labor camps or slavery—is very low.

To the degree that an elite uses high coercion but lacks the appropriate administrative controls, there will be an enormous number of cliques and conflicts in the organization.[14] Members will not be very committed to the organization, but defensively concerned for themselves. This is the case in most patrimonial political regimes with their emphasis on force, the "off with their heads" mentality; these are sites of constant intrigues among competing cliques, and of what in modern terms would be called "corruption," the appropriation of organizational property for private ends.

---

[14] See the description of Soviet industry under Stalin in Berliner (1957). Traditional regimes are described in Frykenberg (1968) and Wertheim (1968).

12.2 The more standardized and predictable the products, the more successfully tasks can be accomplished by emphasizing material rewards, administered by an efficiency criterion.

Control by material rewards fosters an acquisitive orientation. Members are not alienated from the organization, but their interests in it are dependent on the extent to which it rewards them. Control can be administered less directly, with less surveillance and more efficiency criteria. Instead of checking their actions, one checks what they produce. This leads to emphasizing products that are easily measurable; remunerative control operates best where there is a highly standardized output. Tasks can be in the middle range of the continuum of initiative beyond the level of sheer physical compliance, but still within the range of predictable outcomes. Such an organization will have some emphasis on rules, especially stating the terms of productivity and reward in a precise fashion; it is unlikely to be rigid about procedural rules, however, to the extent that the enterprise is concerned about its profits.

12.3 The more initiative required in a task, and the less predictable or visible the outcomes, the more its successful accomplishment depends upon strong normative control.

12.31 The greater the emphasis on normative control, the greater the reliance on indirect administrative devices, especially control over information.

12.32 The more exclusive the emphasis on normative control, the more likely there are to be intense factional conflicts at policy-making levels.

If one can institute commitment to ideals, tasks that require high initiative can be carried out. Within any organization, there is a range of tasks. Those that require the most initiative are the highest administrative jobs, where decisions are made and plans formulated; this includes decisions about how to control others. The nearer to the top of an organization, the more emphasis on normative control. The exercise of power itself tends to create this form of self-indoctrination, the more so to the extent that subordinates are also controlled in this fashion, by offering mobility chances, by dispersed power, or by ceremonial procedures. Professions fit into the same category; they are occupations carrying out tasks that require very high initiative if they are to be done at all well. Organizations hiring professionals thus must give them a considerable degree of autonomy, depending on just how far the operations depart from routine accountability.

Organizational leaders who attempt to use normative control, however, face a continual series of dilemmas. Certain tasks cannot be done without some degree of normative commitment; but attempting to get that commitment is likely to undermine the leader's own power over others. The two main ways of getting normative commitment, after all, are giving away some of one's power, or fostering highly mobilized egalitarian groups within the organization. Both of these are very dangerous, especially since men who internalize ideals come to consider themselves the best judges of what the organization should do to uphold them. Conflicts are quite frequent within the most idealistic organizations, such as religious and political movements or intellectual organizations such as universities, precisely because the members are both highly mobilized and highly motivated to identify the interests of the organization with their own personal accomplishments.[15] Astute organizational politicians, then, always attempt to mix normative incentives with material rewards and perhaps subtle coercive threats, since the latter can be administered much more routinely and kept under more stable control. In every instance, efforts at control have their problems, and organizations are always the site of political maneuvers in some form.

ORGANIZATIONAL STRUCTURE

There is some advantage in reminding ourselves periodically of what an organization looks like phenomenologically. Literally and empirically, there is no organization except in the actual behavior of real people at some moment in time; the organization *is* whatever they do and think and say. It is all too easy to reify the shorthand by which we conveniently lump together the actions of many people over long periods of time. Formulations about organizational structure tend to be statements to the effect that an organization facing certain contingencies in the environment or certain problems of coordination or numbers of members will react (especially if it is being rational) by taking a certain form. Strictly speaking, this cannot be the cause of an organizational structure. The "organization" is only people attempting to get certain things for themselves and using other people as a means; what many such statements mean *empirically* is that the leaders or owners of an organization, in trying to get their

---

[15] Even where normative control is effectively maintained, usually by emphasizing strict equality and pervasive group rituals, there is a price: the organization is not likely to be very efficient. The biggest weakness of utopian communities has been economic productivity, and even the Israeli kibbutzim have survived primarily as ideological showpieces and military outposts subsidized from outside (Caplow, 1964: 291–316).

subordinates to do certain things, will end in arranging them in a certain way. Even this is not quite accurate, because "organizational structure" is only a way of referring to how people behave repetitively toward each other; no one man can unilaterally decide how large numbers of people will interact, and any pattern is the result of bargaining among many parties. They may be very unequal in their resources, to be sure; but what patterns emerge is to be *causally* understood as the result of a struggle over who will do what among people whose very inequality gives them different aims as to how they want the others to behave.

This approach has the advantage of bringing theories of organizational structure back in touch with the core of sociological analysis, with its concern for rituals and interests, day-to-day struggles and individual cognitions. It also keeps us close to earth, since we must always think our abstract terminology through to real empirical variables, and our convenient quantitative measures into the full behavior of human beings minute by minute and day by day. At the same time, we can avoid going overboard in the opposite direction, reacting against some over-abstract or over-static conceptions of organization to take the stand that structure is a myth and that only a minute (or philosophical) analysis of an individual's consciousness is realistic. As we have seen, it is apparently very tempting for phenomenologists to ignore behavior in favor of cognition, and to set such lofty epistemological ideals as to overlook the simple fact that people tend to repeat some very important interactions over and over again, even over a period of years. Terms like "bureaucratization" and "centralization" have very striking referents in the flow of everyday experience; the canon of reducing everything to precisely these flows of experience is useful to make us identify just what kinds of behavior we are talking about, not to badger us into ignoring these long-term and repetitive patterns.

The task, then, is to understand structure as the behavior and cognitions of real people. An organizational hierarchy is centralized if there are long chains of people who report one to another in one direction and who pass on orders in the other; it is decentralized if people pass along few orders and reports along the chain, and themselves initiate actions at many points.[16] Put this way, we can obviously think of several different degrees

---

[16] Both March and Simon (1958) and Stinchcombe (1967) describe organizational structure as a nested set of plans for action. The most basic plans give the main lines of command; subsidiary plans tell what specific actions various individuals should perform. Decentralization here means that many plans are left to be filled in by subordinates. Of course, this is structure from the point of view of organizational masters; the actual structure of reports may turn out to be somewhat different. In either form, it is hard to summarize all this in an organizational chart, and the popularity of such charts is one reason why we tend to have

and types of centralization and decentralization; it is one of the advantages of this approach that we can increase the precision of our coverage to any desired level. Another advantage is that we can dispose of some murky arguments about reductionism and reification. All organizational structure is the behavior of individuals, and it is caused by the behavior of individuals; it also happens that people react to what each other are doing, and hence, many crucial elements of explanation relate to the *pattern* of their interaction, and not just to the properties of individual motivation and cognition. Phenomenological and psychological principles are important elements, but they must be worked into the social scene as it actually unfolds.

In what follows, I will sketch the outlines of a comprehensive theory of organizational structure. There are a great many things to be explained: the numbers and sequences in which people give orders and advice to each other, the kind of physical work people do and how they pass along the fruits of their labors through different kinds of networks, the amount of written reports and instructions, the kinds of deference styles, and the degree to which conversational content is made up of abstract rules or expressions of personal obligation. Strategically, there is nothing to be done but to plunge into what seems like the central part of some very complex materials. Any trained social scientist, after all, can show how complicated things are; what we want is a way to simplify them to a few crucial formulas. Hopefully, if we discover the right ones, we will have the whole complex world back in any degree of detail desired, but with a device for locating quickly its connections from any point to another. I believe that the key lies in the area of power. The following pursues its structural principles in somewhat summary terms; but they are meant to be summaries of real-life behavior and ones which can be unpacked with greater precision as we go along.

Weber's model of military–political administration is a useful place to begin, for it is our only treatment of organizational structure in terms of the contingencies of power. It may be stated schematically as follows:

13.0 Centralization and decentralization. The greater the concentration of control resources, the greater the centralization of authority.

13.1 The smaller the size of the group, the fewer the links in a chain of command.

13.11 The fewer the links, the greater the surveillance by the leader

---

a fallacious image of over-centralized control; the little boxes and lines convey too few dimensions of behavior.

over his subordinates, and the more closely the subordinates comply with the observable behavior demanded (by *11.1*).

13.2 The less geographically dispersed the group, the greater the surveillance by the leader, and the more closely members comply with the observable behavior demanded. (by *11.1*).

13.3 The greater the concentration of material resources in the hands of the leader (food, land, equipment, weapons), the greater the centralization of authority.

This is a way of explaining the rigid control in small patriarchal household administrations, and the implied converse: the tendency for patriarchal regimes to become decentralized into patrimonialism and then into feudalism as they became larger, conquer more territory, and rely on local sources of supplies.[17] But these relationships are affected by another set of determinants as well. Size and geographical dispersion can be offset by shifting from controls emphasizing surveillance and personal loyalties to those which check on results and foster impersonal ties, and by technical control of material resources at a distance.

13.4 The greater the use of written rules and standardized reports, the more central control can be maintained in a situation of large size and geographical dispersion.

13.41 The greater the use of written rules, the less personal loyalty to immediate superiors, and the less subversion of authority through the chain of command (by *11.3* and *11.31*).

13.42 The greater the use of written rules, the more authority can be subdivided into specific responsibilities, and the less the personal authority of any individual.

13.43 The greater the use of standardized written reports, the greater the members' concern for precise, visible outcomes.

13.5 The more efficient the technology of transportation and communication, the more central control can be maintained in a situation of large size and geographical dispersion.

13.51 The more efficient the transportation, the more centralized con-

---

[17] A more precise version of this theory would have to add a proposition on the time periods involved in this shift toward decentralization and disintegration; there is a lag which seems to be due to the tendency for personal loyalties established under strongly centralized conditions to persist for a while when conditions change. Possibly such lags could be predicted from the principles of interpersonal negotiation suggested in Chapter 3 earlier.

trol can be maintained over the supply of food, equipment, and other material necessities.

13.52 The more efficient the transportation, the more easily may inspections be made, and the more often may officials be transferred to new positions (thereby reducing personal ties).

13.53 The more efficient the communications, the more frequent and detailed orders and reports are possible.[18]

13.54 The more efficient the communications, the more easily may an official call on others for aid in disputes with others, and hence, the more power to those in the most central position in the network of communications.

This casts Weber's model of the effects of bureaucratization into control forms. We might note, in passing, that a money economy is a control resource because it enhances central control of material goods, but does it as a highly standardized, impersonal, yet flexible communications channel (Stinchcombe, 1968a; Parsons, 1964: 349–350). Strong governments are concerned with maintaining a reliable monetary system, not primarily to provide services to the rest of the society but to make it possible to appropriate local goods while maintaining centralized control.

Put a different way, Weber's model also gives us the conditions for another characteristic of organizational structure: the quality of interpersonal relationships, expressed in terms of personalistic (patrimonial) or bureaucratic styles.

14.0 Personalism and bureaucratization. The degree of bureaucratization is determined by the availability of technical and organizational resources, together with a situation motivating the individuals involved to rely on impersonal means of control.

14.1 The greater the availability of skills and materials for written records and communications, the higher the potential bureaucratization.

14.2 The more efficient the technology of transportation and communications, the higher the potential bureaucratization.

14.3 The greater the availability of religious or other cultural organizations (e.g., schools) that commit individuals to each other independently of their ties to families, local communities, or other previous solidarities, the greater the potential for bureaucratization.

---

[18] Compare the modern case of enhanced central control in police departments resulting from the introduction of radio communications (Bordua and Reiss, 1966).

14.4 The larger the size and the greater the geographic dispersion of the organization, the more its leaders are motivated to use bureaucratic control techniques.

14.5 The greater the threat from insubordinate local officials, or from outside organizations that have already bureaucratized, the more an organizational leader is motivated to bureaucratize.[19]

14.6 The weaker the leader's resources for personalistic control, the more he is motivated to bureaucratize.

The last point refers to the tendency for rulers without strong family ties (such as usurpers) or local ethnic attachments, or the capacity to provide the ceremonial ethos of traditional modes of legitimacy, to prefer bureaucratic means of control (historically, by favoring priesthoods; Weber 1968: 1160). There is one other, rather different path, to bureaucratization. We have so far dealt with hierarchies controlled from above, and their leaders' strategies for maintaining power. But as Weber (1968: 983–985) noted, governments that are built up from below, as democratic coalitions of equals, also tend to emphasize impersonal rules and to separate the man from the position that he temporarily occupies. Similarly, rebellions that attempt to replace arbitrary authority by democratic participation also take a constitutional—that is to say rule-bound—form.

14.7 The greater the pressure to disperse control equally and prevent personal long-term appropriation of authority, the greater the emphasis on bureaucratization.

Put this way, Weber's model applies most directly to military and political organizations, and especially to the transition from personalistic to bureaucratic organizations that went along with the decline of the household economy and improvements in transportation and communication. But the analytical model is of much wider use. Even in the modern era, some organizations are more personalistic and decentralized than others; even though the material and communicative resources are available for centralized bureaucracies, they are not always used for this purpose. I think Weber's model provides the core of a general explanation. The degree of visibility of tasks is one key variable to be elaborated. This is not done in Weber's own treatment, except for geographical disperson, simply because the main tasks of a military–political administration are fairly simple: They all come down to whether or not the leader can gain compliance in the use of force. An army or a government, after all, has no

---

[19] Bendix (1967) argues that the latter has been the primary mechanism of "modernization" in recent centuries, especially in terms of the military threat posed by bureaucratized states for nonbureaucratized ones.

standard of efficiency imposed upon it except to dominate; even its competition with others simply concerns the territory of effective physical threat. A government can fail at anything else it attempts to do, as long as it keeps its military apparatus together. When we extend the Weberian model to business, medical, or other kinds of organizations, however, the variations in task visibility and uncertainty become crucial determinants of structure, along with those already isolated.[20]

Stinchcombe (1959) has shown that certain types of modern production are organized in a decentralized and personalistic fashion. The construction industry, for example, is made up of a loose coalition of small organizational units, with considerable local autonomy and relatively personalistic relationships within each. The situation is reminiscent of Weber's model of feudalism. In construction, the task calls for many intermittent and highly skilled operations following in a sequence. That is to say, the skills are such that control by direct surveillance or by rules will not give the desired outcomes; considerable reliance must be placed on the workers using their own judgment, and management confines itself to inspecting outcomes. This gives a great deal of power to the workers themselves; to keep their loyalty, the lower-level bosses rely on normative control, allowing the formation of an egalitarian and informal group instead of imposing strict or impersonal controls. This makes the workers even more powerful vis-à-vis central managers; the latter, then, must give their units considerable autonomy, even to the extent of making them only intermittent members of the organization through subcontracting. In general terms, a high level of task uncertainty, and the absence of reciprocal dependence of work units on each other's operations, gives very few power resources to a centralized hierarchy and results in a decentralized structure.[21]

Woodward's (1965) comparative study of different industrial types provides evidence for a general formulation. She considers three main kinds of production:

---

[20] Another variable suggested in Weber's model: The kind of ethnic or cultural organization of the surrounding community will determine the viability of personalistic strategies for controlling organizations.

[21] This control interpretation of structure could be challenged by the functional argument that it is too expensive to maintain centrally administered units of specialists who are needed only intermittently. But large enough economies of scale would offset that expense; the problem is to explain why the construction industry has not become dominated by a few financial giants, as in most other areas of modern business. The answer involves the fact that the work relations themselves have already put power resources in the hands of the lower levels, which the workers use to actively foster a decentralized structure that enhances their personal autonomy. One attractive possibility that workers frequently follow is to split off and start one's own business.

1. *Unit production* involves making one unique item at a time (as in aerospace or other highly innovative areas, movie making, custom-made clothes or furniture, etc.) Here the structure is relatively flat and informal; most workers are highly skilled; managers have few subordinates and do much mutual consultation rather than order-giving; and there is little distinction between technical staff and line authority.

2. *Mass production* involves making many small, routine items and then assembling them (as in automobiles and other machinery). The resulting structure is pyramidal, with many unskilled and semiskilled workers, and a tall, narrowing hierarchy of managers who coordinate the various areas. Relationships are formal and bureaucratic, with many rules, reports, and much record-keeping paperwork; staff and line positions are clearly divided; and there is much conflict among both different sectors and different levels.

3. *Process or continuous flow production* involves taking the same materials through a series of operations and mixtures (as in oil refining, chemicals, or food processing). Here, much of the unskilled labor and a great deal of the coordination is done by the organization of the machinery; the core of the organization consists of a few skilled workers who check the process and trouble-shoot difficulties. Most of the personnel in such an organization are likely to be concentrated in auxiliary white-collar units, concerned with research and planning, sales, and other external operations. The hierarchy has few problems of coordinating production or dealing with reciprocal adjustments; hence, it can run in a fairly standardized bureaucratic fashion, but under low pressure and with a low level of conflict.

Woodward was concerned primarily with the tasks involved in industrial production, although she took as dependent variables the structure of the entire organization. It should be borne in mind that the precise effect of the variables should be traced in the specific part of the organization to which they apply; the organization as a whole may have rather different tasks in different sectors, each with a corresponding divisional structure, and the overall structure is made up from a combination of these different units. For example, the computer departments within a highly routinized government agency are rather different from the rest of the organization, showing the same kind of informal relationships and mutual consultations among middle-level authorities and skilled workers that is characteristic of the unit production type (Meyer, 1968).[22]

---

[22] Computer divisions are often involved in a series of complex and idiosyncratic problems of programming and trouble-shooting. The evidence presented by Meyer also warns us to look beyond formal titles to actual interactions; the many "managers" in these units

The principles are by no means confined to industrial production; similar problems of coordination and environmental uncertainty are found in various white-collar operations, and similar structural consequences can be observed. Scientific research generally takes the form of unit production (although, as we shall see in Chapter 9, different kinds of research vary somewhat on this score). Sales and procurement activities can have various structural consequences, depending on external and internal contingencies. Purely coordinative activities (such as those provided by banks, insurance companies, and telephone services or many governmental services and controls such as welfare administration) derive their structures from a fairly routine set of independent activities at the lower level, with coordination problems confined to long-range dispensation of resources by higher-level officials.

It is useful to label a fourth type, in addition to the three Woodward described.

4. *Pooled production* (Thompson, 1967: 54–64). This is the case where an organization carries out many separate operations (e.g., a government bureau handing out welfare checks or driver's licenses, or an accounting office auditing a series of separate operations). Here, coordination needs are low; it is only necessary that the various workers be supplied and checked upon, and that the organization as a whole be kept viable by maintaining an overall level of profitability or political influence. Such organizations can be highly bureaucratized, with great emphasis on standardized rules and fairly lengthy hierarchies, precisely because there is little need for close surveillance and mutual consultation in order to maintain control.[23] A parallel situation may be found in the interrelations among units (whatever their internal tasks may be) that are related only by pooling. In this case, the overall structure will be centralized around a bureaucratic main office, from which staff units impose standardized rules, and line authority branches out into a number of geographic or functional subdivisions.[24]

On this level of overall structure among units, the opposite case occurs where a number of different divisions are *reciprocally* related; each is an

---

do not mean a tall hierarchy, since most of their supervisory responsibilities are nominal and their main activity consists of mutual consultation.

[23] For example, the public personnel agencies and the government finance departments investigated by Blau (1968).

[24] The same is true of organizations with simple sequential interdependence, where one unit makes parts or carries out operations that can be stockpiled for the next unit to use (Thompson, 1967: 15–21, 54–78).

uncertain environment for each other, since their actions have to be continuously coordinated. This gives division managers mutual power over
each other (since they control areas of uncertainty and cannot be coordinated by rules); the result will be a high level of conflict, unless these
interdependent divisions are put together into an autonomous structure,
with internal relationships coordinated by mutual consultation. Units that
depend on each other have to be grouped together so that they can work
out their political problems by themselves; and if they face special environments that give them unique sources of uncertainty (and hence, of
autonomy), central control cannot be maintained by means of formal rules
applied from above, but must give way to a decentralized structure that
is made to account only for overall outcomes.[25]

It should be apparent that there can be a great many combinations of
structural contingencies, and few organizations are exactly alike. Global
typologies are useful mainly for assembling the evidence so that we can
see some of the general principles operating; a more advanced theory picks
out multiple variables that apply to the many parts of an organization, and
add up to its overall observed structure. For each task of the organization
in making a physical product or dealing with outsiders, we must look at
the predictability of the operation and the resulting constraints upon what
controls can be successfully applied. We must also pay attention to the
kind of coordination that is needed, and to what kinds of situations for
political maneuvers will result.[26] The same principles apply again for
interrelations among larger units of the organization. Let us see if we can
generate the foregoing empirical results:

15.1 The more unique the product or unpredictable the problems of
the task, the less reliance on rules and the greater the decentralization
of authority.

---

[25] This has happened to large American corporations (e.g., General Motors, Sears and
Roebuck, Dupont) which have expanded into diverse environments, while producing in such
a way as to require a high degree of intercoordination between procurement, production, and
sales for each of a number of different products or different geographical areas. The functional divisions for each of these operations centralized at headquarters had to be split up and
regrouped in autonomous production or geographical divisions (Chandler, 1962, as analyzed
in Thompson, 1967: 73–78).

[26] Perrow (1967) gives an analysis and extension of Woodward's findings that is similar
to what follows. He distinguished two aspects of technology: the number of exceptions that
occur in the work, and the routineness of the search procedures used when exceptions occur.
The first is the same as *15.1* and *15.11*; the second seems to refer to the level of managerial
coordination, but I have chosen to conceptualize it in terms of political interdependence
rather than the way in which exceptions are followed up.

This follows from *11.32* (rules are slow and unadaptive) and *12.3* (the greater the unpredictability, the more reliance on normative control if one wishes to achieve results). The main ways to achieve normative control, in turn, are to share out authority (and reduce authority from above), and to foster information and personal relationships. Conversely:

> *15.11* The more standardized and predictable the tasks, the greater the emphasis on rules and written reports, and the greater the centralization of authority, in organizations of any size and complexity.[27]

This follows from *12.2* (standardized products may be successfully produced by inspecting results and giving out material rewards according to fixed criteria) and *13* (the more control resources in the hands of the organizational leaders rather than the subordinates, the more centralized authority they will be able to wield). The provision that the organization be large and complex follows from *14.4* and *13.11* (in smaller or simpler organizations, the leaders can rely more directly upon surveillance, but substitute rules when surveillance becomes more difficult).

> *15.2* The more an individual linked to another by reciprocal interdependence also faces an unpredictable external task environment, the less he can be controlled by externally imposed standards.

This is to say, the technical managers in a unit production operation have to be allowed to work out their mutual relations without interference from above, if anything is to be accomplished (Thompson, 1967: 54–61). The various departmental managers in a mass-production operation, however, may need to coordinate their schedules, but the greater predictability of each operation means that their consultations can be subjected to stricter guidelines by higher-level authorities who, in effect, order and control the consultations.

> *15.22* The more independent power among the parties to reciprocal negotiations, the more likely they are to establish informal ties, and the less amenable they are to control from above.

This gives another reason why power is decentralized in organizations carrying out unique, complex, and unpredictable tasks. There is a high level of informal contact, which tends to make co-workers into friends (*10.52*); the result is more personal loyalty to horizontal rather than vertical ties. It is at this level that men are most mobilized for concerted action. An organizational leader who wants to maintain control over such an operation has only one course of action: to become part of the informal

---

[27] For some additional evidence, see Hall and Tittle (1966); Harvey (1968).

group by joining directly in the technical activities themselves. It is for this reason that unit production organizations tend to have a relatively flat hierarchy, or else take the form of a feudal coalition between autonomous units and a remote administrative center that handles routine operations (such as distribution and sales) (cf. Hirsch, 1972).

15.3 The greater the proportion of tasks in an organization that are predictable, the more emphasis on specialized staff attached to its central headquarters to formulate and apply standardized rules across divisions (Thompson, 1967: 60–61).

15.4 The more sequential interdependence (i.e., the more the operations of one individual or unit depend on the completion of activities by another, but not vice versa), the more these activities will be grouped under the single authority of a higher manager (Thompson, 1967: 60).

15.41 The more that activities are linked by sequential interdependence, the more covert power is in the hands of those who control the earlier activities in the chain.

15.42 The more different the activities to be linked sequentially, the more levels of the hierarchy of command.

If we assume that no one likes being controlled (Postulate VIII), it is likely that there will be several conflicts in any situation of sequential interdependence: those later in the chain chafing against those earlier in it, and those earlier in it (who have covert power) against the manager at the next higher level who has covert power to settle lower-level disputes. Any manager of such a situation will have limits to how much he can handle, and different parts or branches of a very long or complex chain will have to be put under different managers. Larger segments of the chain can be coordinated only by higher-level managers. The total number of levels should be determined by the political resources of each manager to control his subordinates (who have both resources and conflicts of their own) and by the number of activities to be linked.

Any large-scale mass-production operation, then, will have a complex administrative hierarchy (*15.42*), a great emphasis on rules (*15.11*), and considerable emphasis on staff divisions to administer the rules (*15.3*). This gives us the complex bureaucracy that has received so much attention in the classic case studies. Its use of fairly alienating and inflexible means of control, the many levels of hierarchy through which orders are passed, and the long chains of sequential interdependence all are conducive to many slip-ups that are difficult to trace, and to some rather bitter conflicts. Moreover, it is likely that there will be some degree of reciprocal interdependence, as different members of the organization may be able to solve their own performance problems by adjustments at either end (e.g., a scheduling problem can be resolved by increasing production at one place or slowing down at another). All of this means that men will be motivated to enhance their own positions by informal bargaining, although it must be done covertly and by violating official rules (Dalton, 1959).

15.5 The more that reciprocal interdependence develops within a complex bureaucratic hierarchy, the greater the split between formal and informal channels of influence, and the more conflict among staff and line units and among different levels and divisions.

This has something of the quality of a vicious circle; the more that informal bargaining is used as a way of meeting official requirements of performance, the more that men become reciprocally interdependent, and the greater the split between formal and informal structure.

This situation can be approached from a different angle as well. The stronger the political motives for maintaining control from above, the more bureaucratization is emphasized, even where it is not appropriate to the task situation. This results both in greater difficulty in reaching output targets, and in a greater emphasis on covert bargaining to avoid being penalized, as in the most politicized periods in Soviet industry (Berliner, 1957).

15.6 The greater the environmental control over the work process, the less need for personal surveillance or normative appeals, and the lower the control emphasis within management (by *11.5*).

15.61 The greater the environmental control over the sequencing of work operations, the lower the coordination needs, and the less power and informal bargaining among production managers (by *11.4*).

15.62 The greater the environmental control, the more predictable the work process, and the greater the emphasis on rules and written reports (by *11.5* and *15.11*).

15.63 The greater the environmental control, the lower the dispersion of power and the greater the centralization of authority (by *15.61* and *15.11*).

Process production organizations have a high degree of environmental control, at least within the production division itself. The core of the organization is thus a centralized and smooth-running bureaucracy, although auxiliary divisions—research and development, sales, and so forth —may add complexity to the upper structure.

15.7 The more an organization is made up of pooled production activities (those which are carried out drawing upon the same organizational resources and contributing their inputs to the same pool, but which are not dependent upon each other), the lower the power of managers in relation to each other.

15.71 The more pooled activities, the less horizontal coalitions among managers, and the greater the power of centralized staff administering standardized rules.

15.72 The more pooled activities, the fewer the organizational conflicts, and the more levels of hierarchy are tolerated by organizational leaders between themselves and the production activities.

Principles *15.7–15.72* are essentially the obverse of *15.2, 15.22,* and *15.5.* Pooled activities, such as those of many financial, service, and governmental agencies, make for another kind of ideal–typical bureaucracy. Especially when the production tasks themselves are highly routine, the problems of control and coordination are minimal, and the hierarchy is correspondingly formalized.

Not all organizations with pooled activites are necessarily so placid, however. Where the activities themselves involve subtle or unpredictable products (as in certain kinds of schools), the bureaucratized procedures are more closely confined to the administrative hierarchy rather than to the production activities (e.g., teaching advanced subjects). If the different production units have outside sources of support (their own property, income, clients, or legal autonomies, as is the case both in many modern universities and in many government agencies within a federated system), the unit leaders have considerable autonomy vis-à-vis the central administration. The more such resources there are, the more political maneuvering there is over how the pooled resources of the organization are to be distributed.

With all this behind us, it is apparent that the question of the ratio of administrative to production personnel is a relatively superficial one. The quantitative shape of the organization is the result of many variables, and the same ratios can be produced in a number of different ways. Process production organizations, for example, tend to have a high administrative ratio, not because there is a great deal of coordination to be done but because the technology of production eliminates not only coordination problems but most of the unskilled work force as well. Organizations that emphasize coercive controls need a large hierarchy to exercise surveillance. Organizations facing highly unpredictable tasks give the appearance of a large administrative ratio because the technical workers are given nominal supervisory rank, while supervisors immerse themselves in technical responsibilities. Most research on administrative ratios in relation to size has concentrated on pooled production organizations carrying out routine white-collar tasks (especially school districts, public personnel or finance agencies); in these, size itself is the prime determinant, producing a diminishing ratio (at an accelerating rate of diminution) because economies of scale in supervisory capacity apparently increase faster than the dispersion and complexity of operations (Blau, 1970; Meyer, 1972; Hendershot and James, 1972). But this may not be the case in organizations with a greater level of interdependence among their operations; we do know that where size increases technical complexity in reciprocally interdependent organizations (or perhaps politically decentralized organizations), a larger administrative component results (Anderson and Warkov, 1961). In either case, what we are dealing with is one of many structural outcomes of power struggles in situations of varying resources.

Membership-Controlled Organizations

Our analysis has been couched in terms of hierarchic organizations, controlled from the top. Membership organizations are those which are controlled from the bottom: Control is formally democratic and all members are the ultimate decision-makers. The typical image of a hierarchic organization is a pyramid; its typical control problems arise from the fact that the man who is in charge is the most remote from the work the lower members are actually doing, especially since the lower and middle levels attempt to guard information which would reflect unfavorably on them. The typical membership-controlled organization is also a pyramid, except that it is upside down. Formally, it has many bosses and few workers, the former being the elected officials and their staff, who represent the membership, speak for them, negotiate for them, present their view, or carry out services for them. Except for the relative numbers, the control problems are the same as in hierarchic organizations. The hierarchic chief has problems exerting control over what the workers are doing because there are so many links between, and also because the latter are on the spot and have the power that can be derived from the uncertainty of the tasks. The elected officials of a membership-controlled association are the "workers" in that organization, and benefit from a similar, although usually more extreme, displacement of power.

The "structure" of an organization refers primarily to the network of power relationships among its members. The basic theory of power in membership associations, Michels' (1949) Iron Law of Oligarchy, is a theory of organizational structure and its changes over time. The focus is on the factors that diffuse power from its legitimate center; formally, this is the Weberian theory of centralization and decentralization, applied to a type of organization in which the numbers and resources are distributed rather differently from the hierarchic organizations we have been considering.

The model stated by Michels for political parties has been found applicable to labor unions, professional associations, private clubs, and legislative bodies (Lipset *et al.,* 1956; Selznick, 1952; Tannenbaum, 1965; Schlesinger, 1965). Membership associations generally have a rather small number of decision-makers, compared to the total membership. The decision-makers tend to form a single, informal group or clique; they usually maintain secrecy about internal discussions not only against outsiders to the organization but against members too. Whatever disagreements there are among the leaders tend to be voiced in private, and their conclusions presented as a united front. They tend to have a very long tenure in office; although democratic forms are maintained, there is a strong tendency for the same people to be elected over and over again. Generally, they are selected from above rather than projected from below, so the new men entering the leadership group tend to be sponsored by those already at the top. As a result of all this, Michels argues, the leaders of voluntary organizations usually are not easily influenced by the membership.

The goals of leaders are likely to differ from those of their followers, and in a predictable direction. The leaders are more interested in "realistic" considerations of maintaining the organization and gaining personal power, and less concerned with the public ideals that attract outsiders to the organization or that were the reasons for its foundation. Since there is commonly a split between the interests of leaders and followers, most such organizations have implicit internal conflicts. In these conflicts the leaders are much more likely to win than the dissenters; they often have the capacity to mobilize the support of the usually apathetic majority of members, or at least to cut off resources from their opponents. Their control over administrative resources gives them the decisive advantage. In Michels' words, "Who says organization, says oligarchy."

It has been noted by several observers that Michels' argument is rather over-determined in a dramatic sense while under-determined in terms of causal explanation (Gouldner, 1955). If organizations are so rigidly oligarchic, why is it that revolts keep breaking out from time to time to be put down? And since they can break out, it is presumably by an extension of the same underlying conditions that they succeed at particular times. Michels' model in its classical form is too concerned with denouncing a state of affairs to state clearly the conditions that produce a range of variations. There is a second source of distortion in the model: Michels argued as if the leaders of an association always tended to become more conservative than their members, and to sell out the people's interests. In many cases, though, the evidence indicates that organizational leaders are much more ideologically committed and extreme than the rank and file (as in American political parties), and the latter often are interested in an organization for purely personal and sociable reasons (as in many churches, charities, and unions) (McClosky *et al.,* 1960). The real reference point of Michels and many other critics has been intellectuals like themselves, who are usually much more committed to organizational ideals than the practicing politicians. Moreover, sometimes organizations become taken over by a leadership that is very militant indeed, far more so than either the membership or the conventional leaders; this is usually when the intellectuals win control. The "People," then, is a term that must be seen in the context of a battle over legitimate control of a membership association, and one that intellectuals out of office are adept at cloaking themselves in.

This is not to dismiss Michels' cynical insights, as some complacent thinkers have done, or to turn them into an argument against criticizing the status quo. Properly understood in terms of variations, Michels' model points us to the mechanisms by which politics is carried out. Since democratic forms of community and national governments are a type of large membership-controlled association, Michels gives us a general key to community and national power structure, which can move us beyond merely describing what goes on, to explaining why particular sorts of things happen. The fact that intellectuals are the main carriers of ideals does not make those ideals irrelevant, but simply allows us to be realistic about who wants what and their chances of getting what proportion of it. Recasting Michels' model in terms of variables and their causes, then, not only makes it possible to test it and integrate it with more general organizational theory; what it loses in rhetorical power is made up in the capacity to locate just which of our values (whatever they may be) are being violated, and to see what needs to be done to realize them.

There are two kinds of dependent variables to be explained: the degree of oligarchy in an organization, and the kinds of policies of its leaders. The "degree of oligarchy" refers to such things as how widely people participate in organizational government, how closely elections are contested, how much turnover there is in office (or in behind-the-scenes cliques that exercise control), and how open the organization is to debate among competing factions. In general, we are concerned here with how close the organization is to the formal ideals of equal dignity and participation among members: Is the structure close to the democratic ideal of purely temporary and shared authority, or has the structure shifted into a hierarchy in which deference is monopolized by a small group? The second question, what the leaders do, is logically independent of the answer to the first question; leaders who are tightly in control of their offices may be highly militant, idealistic, corrupt, or conservative. I shall try to explain why their policies diverge along three dimensions: (*1*) how much the leaders pursue their material self-interests—in the extreme form, what we call "corruption"; (*2*) how much they devote themselves to the conservative interests of organizational members—maintaining organizational services on a routine level, appealing to members' concerns for sociability, pushing neither the members nor themselves to work hard for realizing organizational ideals in the world; and (*3*) how much they pursue militant, idealistic policies—whether attempting a revolution, pushing for sharp political reforms or changes, organizing strikes with strong demands, proselytizing to convert unbelievers or spread a gospel of moral behavior —in short, an activist orientation that does not hesitate to conflict with a recalcitrant world.

Determinants of Oligarchy.   The power of the leadership group to maintain itself in office, set its own policies, select new recruits, and win out against challenges, comes especially from its resource of informality, in contrast to the formality of mass meetings. One of the things that will produce formality is the attempt to be fair to large numbers of people. If many people wish to speak, fairness requires some impersonal means such as a list of speakers. Procedural rules become more involved, the larger the meeting and the more disagreement there is within it (*14.7*). But, as we have seen from the discussion of rules, formality is not an efficient way to get things done (*11.32*). A smaller group, such as the executive committee, has an advantage precisely because of its size, which allows greater informality. This enables them to be more efficient both in carrying out the policy of the membership and in pursuing their own concerns. For the same reason, informal cliques develop within hierarchic organizations; these are allowed or encouraged by higher authorities concerned with results, although, once in existence, the cliques can subvert formal controls to their own purposes. In membership associations, reformers periodically attempt to bring the leadership group back under mass control by enforcing open meetings and formal procedures; the leadership reacts by taking their informal deliberations to a new backstage.

As in a hierarchic organization, the more complex the tasks and the greater the volume of activity, the more it is divided among specialists. If those tasks themselves involve a high degree of uncertainty (*15.1*) or there is much mutual coordination to be done, as there is when there is much internal bargaining to be carried out among factions (*15.2*), the result is a high degree of informality among the men involved, and considerable power vis-à-vis outsiders to their cliques. This structural transformation is parallel to the decentralization of authority in hierarchies, except in this case it shifts from the mass membership to their agents.

The leadership's advantages include their access to information and other administrative resources. The leaders of the organization are those who are in contact with the outside world. The strike committee of a union or student political movement, or the elected officials of a democratic state, control the most recent information, especially in areas of political negotiation. Such information is often guarded with a shield of secrecy, rationalized in the name of efficiency or of danger from enemies. Its main effect is to protect the informality of the leadership group and to give it a secure monopoly over areas of uncertainty. Leaders thus have an important indirect control, since they can define what the problems are in such a way as to make their policy seem the only one that is rational.

Administrative resources include both control of communications within the organization and material resources. Such resources—an office, a treasury, a telephone, the ability to make long distance calls, a car, the ability to pay for plane tickets—can be used either to carry out the official tasks of the organization, or in a control fight against challengers to the leadership. Such resources make the leaders highly mobile, and enhance their internal communications and give them the capacity to mount an internal political campaign. They can also use them for rewards with which to coopt dissident individuals. Leaders can head off mass dissatisfaction by allowing it to become organized, and then attempting to bring its leaders into their own circle. This is done by giving some material rewards and personal deference of power; they are simultaneously bribed and made to share the public responsibilities of power, thus cutting them off from their followers, without the autonomy to set their own policies (cf. Selznick, 1949).

The leaders control information, so they can define reality to their members. They have control of finances and staff assistance. Very often they have control of time in the sense that they are likely to be the only full-time workers in the organization. They are likely to have the resource of secrecy. They are likely to have a monopoly on the skills or talents involving public speaking and organizational administration. Those resources reinforce the informal organization among the leadership because their resources depend on mutual cooperation. This is particularly true in the case of information control and secrecy, which are probably the basic resources. Other material advantages reinforce these, and provide a motivation for individual leaders to keep internal jealousies within certain ground rules.

Tactics of fending off challengers are standard. Leaders are in the position of calling the membership together to decide on issues, by organizing meetings and agendas recognizing speakers and calling votes at opportune times. Many challenges are headed off by packed meetings or partial publicity; much routine control over political parties is maintained by "open" meetings not generally known about by the mass membership. Leaders also make use of the primitive democratic notion of the "people's will"—a hypostatized concept of the group as an indivisible entity standing above its individual members (Clegg, 1951). When a conflict arises between the leaders and a dissident section of the membership, each claims to represent the people, or the organization, as a whole. The leaders of the organization have an advantage in this contest to identify themselves with the Durkheimian collectivity, since they are the officially selected ritual leaders; they have many occasions for identifying themselves publicly as representing the people, whereas the dissidents are clearly rebelling against the ostensible unity of the group. Moreover, the dissident group is usually a minority, since most of the membership is apathetic, and only a minority can organize itself. The leaders, of course, are a faction too, but are better able to hide the fact—the more so, in fact, when there is organized opposition than when there is merely diffuse dissatisfaction.

Variations in the degree of oligarchy are determined by the same conditions that determine the decentralization of authority in hierarchic organizations.

16.1 The larger a membership association, the greater the tendency to oligarchy.

16.2 The more dispersed the members, the greater the tendency to oligarchy.

Unless mitigated by other factors, large size is conducive to oligarchy because it increases the value of administrative resources (Marcus, 1966). Coordination problems are enhanced; surveillance is reduced; more administrative resources are put in the hands of the leadership, while challengers have greater organizational problems to overcome in making themselves heard. It is for these sorts of reasons that Rousseau believed democracy to be possible only in a group small enough to meet face-to-face, no doubt with the example of the ancient city-states in mind.

Dispersion has similar effects. Members who are spread out have few opportunities to communicate among themselves and to maintain coherent organizational ties. In labor unions, it is those which concentrate their members in few large workshops, or which otherwise foster a strong occupational community, that have the most democratic structures; corruption is most common in those in which the men are scattered in many work places—painters, teamsters, longshoremen—and the union officials alone maintain communications among them (Lipset *et al.,* 1956; Raphael, 1967; Bell, 1961). Similar effects seem to operate in politics; it is in rural areas that we find the most heavily entrenched, single-party regimes with their "safe" seats and powerful resistance to reform.

We are dealing here with the Weberian principles in regard to size and dispersion (*13.11* and *13.2*). In the case of hierarchic organizations, leaders may counter the diffusion of power by bureaucratization; similar efforts in membership associations are one source of pressures toward elaboration of constitutional safeguards. But the effect is somewhat like the development of bureaucracy in a situation of unpredictable tasks (*15.5*): In a situation of any political complexity, the result is a widening gap between formal rules and the informal groups that run things.

16.3 The greater the division of material and administrative resources, the lower the degree of oligarchy.

It has been observed in a number of contexts that democratic participation is related to the existence of autonomous organizational resources (Edelstein, 1967; Schlesinger, 1965; Duverger, 1954). In unions, democracy is higher where there is local control over material resources and appointed positions. The mass-based European parties tend to be highly oligarchic, because the central hierarchy controls party newspapers, summer camps, credit unions, and other bureaucratic positions; by contrast, American parties are concatenations of local committees which depend on intermittent fund-raising and office-holding for material resources. Party politics in the United States is thus more participatory in the sense of a competition among a number of local oligarchies rather than control by one central one.

In community politics, we find that it is the more ethnically and economically diversified cities that have a wider degree of participation in political decisions (T. Clark, 1968; Banfield and Wilson, 1964). The more that community groups based on class or ethnic lines are mobilized as distinct communities, the more they have the internal cohesion to act as agents of political conflict, and the more closely contested are political issues (Gamson, 1966). It is organizations rather than individuals that are crucial, both in shaping interests and providing the resources with which to engage in political activity (Banfield, 1961).[28] Presumably, the more

---

[28] Perrucci and Pilisuk (1970) and Turk (1972) present evidence that interorganizational links are even more important for the way in which community issues are resolved.

independent organizations are of each other, the greater the pluralism of power; where they are linked by reciprocal or sequential exchanges, they tend to make a resource bloc and contribute to oligarchy.

The mere existence of numerous large organizations in a community, though, does not guarantee that there will be democratic participation on all issues. If there is only one type of organization (or interdependent network of organizations) interested in a particular issue, effective opposition will not exist (Lieberson, 1971). In these areas (which tend to include foreign policy) the resources are not dispersed, and oligarchy is the result, or at best unequal conflict between groups with very different capacities for mobilization. It should also be noted that the sheer diversity of interests and resources in a complex situation like a large city with a network of overlapping governmental jurisdictions tends to make organizations or factions reciprocally interdependent; the result, as we can predict from our theory of organizational structure (*15.2*), is the emergence of an inner circle of power-brokers, whose position astride the channels of informal communications makes them difficult to displace and unaccountable to formal democratic controls (Banfield, 1961).

16.4 The greater the gap between the rewards of office and the alternative career possibilities of a leader, the greater the tendency to oligarchy.

If the organization does not have paid leaders, the incentives for staying in office, and the tendency to oligarchy, are lower than if leadership positions are paid. Paid union leaders, especially if they have risen from the ranks of the union, face an enormous loss of status if they lose their job as the union leader, and thus have strong motives for using whatever techniques are necessary to keep themselves in. This is more prominent in liberal parties than in conservative ones; conservative politicians are more likely to have lucrative opportunities elsewhere, whereas liberal ones are much more dependent on getting some sort of political job, and hence, they must keep their political ties strong.[29] The extreme case is where the motives for staying in office become a matter of life and death. In the politics of violent *coups* and purges the leader at any one time must

---

[29] I know of no general comparative studies of corruption, but the historical record seems to point up a pattern of the recurrent fortune-seeking of popular liberal leaders, opposed by righteous conservatives of the traditional property-owning classes. We find this in Roman history, with the liberal dynasty from Marius to the Caesars on one side, and conservative reformers like Sulla on the other; and the corrupt immigrant-based parties of nineteenth- and twentieth-century America against the older wealth of the civic reformers. A similar pattern seems to occur in Latin American politics (Huntington, 1968: 59–71).

continue to ensure his position against vengeance, as in escalating terrors of twentieth-century dictatorships.

16.5 The more frequently crises arise in the external environment which affect the immediate interests of the membership, the lower the degree of oligarchy.

It appears to be the onset of major crises—wars or military defeats, economic depressions, natural disasters—that bring about many of what changes do occur in highly oligarchic organizations (Craig and Gross, 1970). These events are a substitute for organizational resources in mobilizing the membership. Even so, this factor is never the single determinant of outcomes; the direction in which shifts will occur depends on which groups have the most resources to channel their aroused numbers into effective pressure.

16.6 The older an organization, other things being equal, the greater the tendency to oligarchy.

This is a hypothesis added in an effort to fill in the time-sequences implicit in Michels' model (cf. Marcus, 1966; Craig and Gross, 1970). Possibly it needs to be cast in a more periodic form. The initial act of setting up an organization (usually in the context of some larger mobilizing event) is almost always the high point of democratic participation; subsequent crises may be the points from which we must count, perhaps with additive effects subsiding over time. Why should there be time effects? Crisis periods bring about high face-to-face participation, and set the conditions for highly ritualized commitment to a set of abstract ideals (*4.7* and *4.82*). With time, this strong state of collective conscience declines, and the building of networks of informal bargaining transfers loyalty among the inner core to more realistic and pragmatic beliefs about the organization (*2.2, 5.214,* and *5.215*). The more complex the internal bargaining, the less room for maneuver any individual finds, the less venturesome the organization, and the fewer new resources coming in, thus further consolidating the existing networks. Unless there is a sufficiently severe external crisis, the organization reaches a point, which Selznick (1957) describes in the case of hierarchic organizations, at which it becomes highly inflexible and scarcely amenable to official control. The time periods involved doubtless depend upon the size of the organization and the nature of its task environment.

Determinants of Leaders' Policy. Values are determined by social position, and the leaders of organizations view the world differently from ordinary members because they derive different things from it. Giving orders or deriving public status from speaking in the name of an organization tends to make one identify with it (*1.1, 5.23,* and *5.221*), and concern oneself with maintaining it. In this sense, there are class positions within a democratic organization with distinct cultures, just as in the stratification deriving from hierarchic organizations. In general, the heads of all organizations, even if some of them represent unskilled workers, tend to share something of a common outlook, and it is for this reason that the formal organizing of conflicts over long periods of time tends to moderate the conflicts (Dahrendorf, 1959).

But leaders can identify with their organizations in a number of ways, and pursue a variety of paths. They may attempt to maximize their personal material gains; they may conservatively protect the existence of the organization which gives them a position, in implicit alliance with the members interested in the most routine services; or they may become intoxicated with their own ideals and seek the glory of spreading their program against external enemies or an indifferent world. The evidence on these alternatives is less systematic than for some of the previous propositions, but a number of hypotheses are suggested by historical examples and the logic of our general theory of stratification.

17.1 The more political bargaining a leader engages in within or across organizations, the greater his tendency to make compromises and the lower his idealism and militancy.

17.12 The more isolated a leader is from bargaining exchanges with outside organizations, and the more homogeneous his contacts with his organization, the greater his idealism and the lower his tendency to compromise.

This pair of propositions follows generally from the conditions that produce strong or weak attachments to ideals (*2.2* and *4.7*).

17.2 The greater the material resources of an organization (especially permanent property or stable arrangements from producing income), the greater its leader's tendency to conservatism.

17.21 The greater the material resources of an organization, and the lower the standard of living of the organization's members, the greater the leader's tendency to pursue his own material interests.

Highly successful political organizations or unions, or churches and recreational associations with a great deal of property, may be expected to have conservative leaders. If the leaders are especially motivated to retain their positions and avoid falling back into the ranks of the followers (*16.4*), they are likely to become even more estranged from them and to become corrupt. On the other hand, new organizations and weak ones such as political sects are likely to be low on resources, internally homogeneous, and externally in no position to bargain; it is their leaders (and members) who have the purest ideological commitments, especially if they are recruited largely from intellectuals. It is recognition and success that lower their militancy, whether this takes place after a successful revolution, or only by the smaller success of finding a stable niche in the world of political bargaining.

17.3 The more active participation of the membership is needed to carry out an organization's program, the more closely the leaders will adhere to the immediate desires of the members.

This is equivalent to saying that where organizational tasks are nonspecialized, there is a flat organizational structure; the same is the case where resources are concentrated in the hands of the legitimate authorities (in a democracy, the members as a whole) rather than their agents (*13.0*). Active mass movements which depend on members to turn out for demonstration, battles, or proselytizing are at one end of the continuum, at which leaders must be very close to the mood of followers if the organization is to exist. Political parties are at the median; they need the voters, campaign workers, and contributors, but they need them intermittently at election times. Trade unions need their members even less, except when the union is first being organized; once compulsory check-off of union dues is established and the leaders have a secure material basis, leaders become free to pursue their own interests. A striking example of the latter is the case of the political stands of clergymen. The more episcopal the form of church organization (where property is controlled and appointments made by the church hierarchy), the more free the clergy are to make unpopular political stands on moral issues; the more congregational control, the closer to the habits of his church members the clergyman remains (Wood, 1970).[30]

One other variable increases a leader's need for active support: the extent to which competing organizations exist which can draw his followers away from him by appealing more directly to their interests. This, of course, is operative only to the extent that the organization itself does not

---

[30] Wood's study concerns the stands of American clergymen on racial issues in the 1960s. Such recent policy stands pose a general question; in modern history, the large property-owning churches have tended to support conservative parties, usually in opposition to anticlerical liberals. Obviously there are several variables operating. There are several factions within the Catholic church today, for example, the more conservative property-managers and perhaps an intellectual contingent which has recently acquired greater organizational resources. The role of seminaries and the greater isolation of the administrative center from laymen produced by the ecumenical movement have had a parallel effect in promoting militancy among Protestant clergy.

command sufficient permanent resources of its own. The total level of appeal to members' interests is affected by a range of values on several variables, as is the case with all of the effects considered here.

Professional Communities

One more area fits closely into our general theory of organizations. Professions are not an isolated field, but a particular kind of membership-controlled association. At the same time, professionals more often than not work in hierarchic organizations (and not only in recent years, as the oldest professions, military and clerical, are essentially organizational); we find ourselves dealing with a more extreme version of ordinary organizational themes, a case of horizontally interdependent groups cutting across formal lines of authority and even extending beyond organizational boundaries. There is a second connection between the two types of organization structure: Professions have a basic equality among their members, like any democratic organization, but resources are distributed among them in such a way that an internal hierarchy emerges, with powerful controls over individual careers by an inner circle. Professions thus give us a chance to apply some of the ramifications of the Iron Law of Oligarchy, while at the same time they illustrate once again the conditions that disperse formal authority in hierarchic organizations.[31]

Professions are occupations which form highly self-conscious and self-regulating colleague groups. At the extreme, a profession has acquired the exclusive jurisdiction to practice a particular skill; to admit new practitioners; to train the practitioners; and to judge whether or not the skills are carried out properly (Wilensky, 1964). Laymen—that is, nonmembers of the community—are excluded from these rights, even if they hold hierarchically superior positions in organizations. Like all occupations and organizations, professions have an ideological cover. Theirs is in the form of an ideal of skill, impartiality, and altruism surrounding the exercise of their practice. Unlike other occupations, however, strong professions are relatively successful in getting wide public acceptance of their ideology. As a result, professions have high autonomy and power on the job; accordingly, their members enjoy their work the most. They also benefit by high prestige and wealth.

---

[31] Because of the similarities between hierarchic organizations and professions, it is not surprising that professions of some kinds are not so incompatible with bureaucratic employment as once was believed. The congruence is especially easy in the case of decentralized organizations or weak professions. See Hall (1968). The idea of incongruity came from a failure to see that struggles for power and autonomy go on in all organizations, and only becomes obvious where an occupation is highly mobilized to express its own point of view.

There is no point in debating a definitional dividing line between which occupations are or are not professions. There is a continuum of occupations with greater or lesser characteristics of being a profession. Given the rewards of a strong professional organization, it is not surprising that all occupations tend to push in this direction to the extent that their resources permit.

What are these resources? That is, how do men that band together as a collegial group acquire power over individual practitioners pursuing their own private interests? In the case of established professions, there are two main forms of control: (*1*) The state enforces their monopoly through licensing procedures so that coercive power may be used against individuals; (*2*) and the collegial profession operates as a validating group mediating the material rewards of their members by determining individual's careers. Men become successful in a profession because their colleagues claim the exclusive right to judge their behavior expertly, and recommend them to each other and to outsiders.

More fundamentally, what determines how much of these mechanisms a particular occupational group will be able to use? The first, the coercive power of the state, depends on the political resources of the group. Trade unions may be analyzed within the same universe of variables. They are organizations of workers who have successfully created an organization to serve their interests in conflicting with employers. They are operators of private government controlling their members through their voluntary adherence or through the use of coercion, as in physical intimidation of strike-breakers. Given enough political influence of workers in the larger state, the same ends can be attained through a closed shop backed by the coercion of the government itself.

But not all professions depend on coercive power. Mature sciences may be regarded as one of the strongest of all professions, without any licensing or malpractice laws whatsoever. This is because the power of the collegial group to validate the behavior of its members is overwhelming and crucial. Scientists' careers depend upon convincing others within the community that they have made a contribution to the ongoing development of knowledge within that group. The product, knowledge, is an intrinsically social thing; it is created in the very fact of communicating it, since it is the willingness of the collegial group to propagate the ideas of their members that validates them as scientific (see Chapter 9 later).

In general, the strong professions depend on one degree or another of the latter form of control. Many groups, due to various circumstances, are able to get the power of the state to monopolize positions for them. Castes, guilds, ethnic groups, and class-based status groups (estates) have managed to do this at various times throughout history. The empirical difference is in the degree to which the occupational group can maintain its autonomous privileges even when the resources for controlling the state are more widely dispersed. What we mean by "strong" professions are those which are based essentially on the second type of power, the capacity of the colleague group to validate the expertise and thus mediate the careers to its members, because of some intrinsic properties of the tasks they are carrying out. Given an organization based upon this capacity, a profession is usually then in a better position to get the power of the state on their side. The precise degree of the state influence, of course, depends on the particular political history and lineup of conflicting groups in that society (Collins, 1975).

Strong professional groups emerge around particular kinds of tasks and technologies for carrying them out. These are tasks that fit at the extreme end of the continuum just outlined, requiring high initiative and commitment on the part of the worker if they are to be carried out well. Professions arise out of situations where normative control is the only means by which bosses or clients can get a job done (12.3). Normative control is acquired by allowing practitioners great autonomy to carry out their tasks, as well as to organize themselves in such a way as to determine the normative commitment of the individual to the ideals of the group. High pay is also generally needed.

Such high initiative tasks have two main dimensions here. As Wilensky (1964) points out, the skill must be applied in an area which is relatively unpredictable, otherwise the client or boss can judge by results and thus exercise control without giving a great deal of autonomy. At the same time, the skill must be definite enough so that actual results can be produced and so that the practitioners themselves are capable of actually teaching those skills to new members. Only in the latter case do the practitioners have the intrinsic power to control each other: by deciding who will be trained and admitted, and by judging according to their esoteric criteria whether or not the practitioner is acting reasonably in ambiguous cases. If there is no relatively definite and teachable skill, or there is one that depends upon individual gifts or intuition, a colleague group is relatively weak. It is for this reason, for example, that business administrators or social workers cannot form intrinsically strong collegial organizations.

A second dimension also applies. The power of the practitioner vis-à-vis the client, and accordingly the attractiveness of the collegial organization for the practitioners, depends on how highly valued the task is by laymen. The strongest professions occur in the case of the skills which are of the right degree of effectiveness and ambiguity, but also ones that provide highly desired services. At the extreme, we have those skills which are called for in an emergency. In the case of physical hurt, imminent danger, conflict or psychic stress, those who provide services are in a strong position vis-à-vis their clients. It is not absolutely essential that the practitioners be able to deal with the emergency in a practical sense. In fact, as we have seen in discussing the previous variable, if they are completely successful, then the emergency becomes much less so, in effect a matter of routine. Where the practitioner is able to provide a sense of security against the unknown that nevertheless continues to remain at least somewhat unpredictable, he will be in the strongest position. This requires that the practitioners depend on each other; they are engaged in a form of ritual reality-creating, of emotional manipulation the object of which is primarily to convince others that they are a pillar of strength in the storm (although at the same time, they must make sure that there still continues to be a storm). Such ritual self-assurance depends upon the closure of the group of practitioners in supporting each other's claims. It is thus that the self-fulfilling prophecy of reassurance can be carried out.

The professional organization has an ideological side, as part of a complex form of exercising power both vis-à-vis outsiders and within the community itself. As we have seen, power is always organized around legitimating ideals. These usually conceal the actual sanctions involved by which power is maintained, but they are isometric with the structure of sanctions as well. The legitimating ideals of professions emphasize altruism, service, and disinterestedness to the highest degree of any type of organization. This partly expresses the collegial group's felt dependence upon each other; and in a Durkheimian sense, the group is symbolized by its gods—in this case, the dynamic gods of their services. Interaction with clients is also important. The legitimating ideals are enacted in actual minute-by-minute contact with outsiders as a highly polished performance in the Goffmanian sense. The basis of the professional's power vis-à-vis the layman in his capacity to continually present himself, at least when he is with them, as the expert. In many professions, the tone of benevolence is crucial here. This is found particularly in those professions that deal with highly emotional emergency situations, in which the layman is at the mercy of an expert who claims neither perfect nor predictable remedies, and does not allow the layman to judge the success. Such practitioners are subject to a great deal of potential hostility, and the strong service ideals seem to be a defense against this. This does not mean that doctors or lawyers or priests are actually oblivious to personal self-interest —far from it. But the colleague group itself has an interest in disciplining at least those of its membership who most flagrantly upset the image of benevolence and thus endanger the entire group.

We have a set of three types of resources which determine an occupation's chance of becoming a powerful collegial group.

18.1 The more successful trained practitioners are in comparison to untrained practitioners in providing a service in areas of uncertainty, the higher the potential power of the occupational group over clients and over its own members (by *11.4*).

18.2 The more valuable are the services to clients or employees, the greater the potential power of the occupational group controlling those services.

18.21 The more emotionally upsetting the situation of the client, the more the power of the occupational group that reduces anxiety.

18.211 The more emotionally upsetting the situation of the client, the more power a service occupation may achieve by invoking ritual solidarity (by *4.82*).

18.212 The more an occupation's power depends on reducing anxiety through ritual manipulations, the more dependent its members are upon each other to maintain the sanctity of rituals, and the more they punish each other for ritual violations (from *6.41*).

18.22 The more emotionally upsetting the situation of the client, and the more powerful that occupational group, the more the group emphasis on an altruistic service ideal.

18.221 The greater the emphasis on an altruistic public image, the more closely guarded and secretive the actual practices of the occupation (from *5.221*).

18.3 The greater the political resources of an occupational group, the more likely it is to achieve a coercive monopoly over practice.

18.31 The more cohesive an occupational group, the greater its political resources.

18.32 The greater the wealth of its clients, and the more valuable to them a type of service, the greater the wealth of the occupational group that controls those services.

18.321 The greater the wealth of an occupational group, the greater its political resources and the more likely it is to achieve a coercive monopoly over practice (from *18.3* and *18.32*).

Various occupations in different historical social settings can be characterized by a particular combination of these resources which determines its degree of professionalization. Some of the major types are the following.

Task skills with verifiable products but complex means of attaining them give rise to technically oriented professions: modern medicine, engineering, dentistry, applied scientific researchers. Where conditions for the optimal degree of skill complexity and outcome uncertainty are met, these professions enjoy great autonomy, income, and prestige. As the outcomes become increasingly predictable, however, employers or clients can control the practitioners by simpler economic means. This continuum shades off in the direction of engineers, technicians, and skilled mechanics.

The second type of skills are ritual or organizational. That is, the nature of the problems that clients face are not independent of the existence of the group of practitioners who in effect create the problems, or at any rate perpetuate or exacerbate them in order to maintain the demand for their services. This is clearly the case with lawyers. As one can see from a comparative historical analysis, the relative degree to which laymen need the services of a lawyer depends upon the political history of a government structure; the great power of lawyers in common-law countries, for example, results from their historical success in placing themselves as officials mediating between the government and citizens, and elaborating a complex and intuitive system of interpretation which the colleague group itself can monopolize (Collins, 1975). Under other political conditions such as the more centralized bureaucracies of the continent, lawyers have much less collegial power and are assimilated as members of government bureaucracies. Similarly, priests purveying ritual services to assuage emotionally disturbing occasions of life have organized themselves throughout history to maximize the layman's sense of emotional dependence upon them. The rise and fall of the priesthood as a profession is related to both political organization in general and to a degree to which emotionally disturbing emergencies of life have been taken over by other professions, such as medicine, law, and psychiatry.

Teaching also qualifies as a profession, offering mostly organizational and ritual skills; these are important for members of society to learn to the degree that teachers have been able to advance their interests in making school important parts of careers. The degree of professionalization of such ritual–organizational occupations depends on the historical specifics of the conflicts between them and other groups. Before the scientific developments which took place after 1870, medicine fell into this category; it was essentially a combination of priestcraft and fraud, since there were virtually no known cures and doctors' success depended on their being able to give inspiring performances and being lucky to be around when spontaneous cures occurred. Their wealth and prestige depended upon the degree to which they were able to attach themselves to the lifestyle of the upper classes and to use these connections to get political sanction for their monopoly.

Finally, there is a category of pseudoprofessions, based on monopoly-created esoteric knowledge. The purest examples here are politicians. As we have seen earlier in our analysis of the Iron Law of Oligarchy, the control of communications and the capacity to define reality to the members of the organizations are the keys to the politicians' power. They form a somewhat collegial group because their power depends upon their mutual cooperation in reality-creating. The features that Michels distinguished for a powerful political oligarchy parallel those of a successful profession: a monopoly on practice, admitting new members by cooptation, and claiming the exclusive right to judge themselves the feats of the layman. The major underlying resource is similar in both cases: the control of an area of uncertainty. But in the case of politicians, the uncertainty about information is largely due to the efforts of the politicians themselves to maintain a monopoly upon it. The degree of collegial autonomy of such groups depends on the variables given earlier to form variations in the Iron Law of Oligarchy; these have to do with the extent to which resources are split up among competing elites or are dispersed among the members generally. A similar set of considerations applies to the power of policy-advising experts in bureaucracies, as Wilensky (1968b) has shown in his analysis of the information-gathering occupations.

These concrete examples are given for purposes of illustration. In reality, every occupation in every particular society and point in historical time has its somewhat unique combination of resources on all of these dimensions. The relative professionalization of an occupation is not an all-or-nothing thing but shifts continuously as varying resources feed into an ongoing struggle of individuals to increase their power, prestige, and autonomy at work. The theory of professions is only a subcase of the more general theory of stratification.

CONCLUSIONS

Organizations are an intimate part of general sociology. Since so much of what we study under other rubrics is the same empirical materials as the behaviors that make up organizational structure, it is hardly surprising that the theoretical principles should dovetail. The general principles of conflict and the formation of interests by available resources are at the center of organizational analysis; so is the importance of personal associations for generating ideals and commitments, and the ubiquitous effort to structure interpersonal ties so as to maximize autonomy and subjective status.

Organization theory is only the more carefully studied side of general sociology, where the complex causes of things are beginning to be separated into more precise bodies of principles. Politics begins to fall into greater order when we avoid the usual ideological questions and concrete descriptions, and focus on the organizational weapons over which men struggle. The broad outlines of the state can be treated as a question of military and governmental administration; democratic politics in party, community, legislature, or national government can be explained as a special version of general organizational processes.

It is in organizations that social classes acquire their distinctive outlooks; class cultures are the effects of organizational positions. The theory of professions gives us the conditions for variations in the strength of occupational communities with horizontal bonds and a collective identity. The similarity to the general principles of status group formation is not accidental; most strong professions are in fact status groups within the upper middle or upper classes. The same resources that give them their high mobilization and their ability to ritualize their encounters and defend an idealized image, produce simultaneously their professional organization, their class position, and their status culture.

I have not attempted to link organizational principles in any detail to small friendship groups or to the structures of sex and age stratification that make up the family, but there is no reason why this extension cannot be made. Most of the things that friends exchange in conversation come from their organizational experiences; it is likely that the more permanent personal relationships are based on strong links to organizational networks, and that ephemeral ones come from more volatile structures. It should be possible to deal with sexual property, age-related controls, and the other variables that go into family structure as an even more direct application of organizational principles. Families and kinship networks, after all, are organizations, and we have touched upon certain kinds of them in considering patrimonialism. A more thorough effort should enable us to reduce all organizational phenomena to a single body of explanatory principles.

Chapter 7

State, Economy, and Ideology

Politics is a subject that we know a great deal about, without seeming at all close to a real explanatory science. The reason is that, of all the fields of social science, politics has been studied with the least detachment from value judgments. The effort to be value-free has not gone very deep, extending primarily to techniques of data handling and a ban on intemperate language, while the topics selected and the concepts in which they are analyzed have stayed very close to ordinary political discourse. Political sociologists have tried to do politics while doing research on it, and the result has been that the knowledge accumulated has not made up much of a systematic explanation.

A great deal of political sociology has consisted of descriptive research on topical questions: the effort to show that the upper class or political insiders do or do not control national or community politics, that extremists of the Left and/or Right are or are not irrational maniacs, or that international politics is or is not imperialistic. Such controversies often come down to matters of degree, of the variations which occur under specific conditions. But they are rarely resolved in this way, because the underlying controversies are about definitions which reflect very different value judgments. How much influence amounts to "upper-class control," or how much competition of influentials amounts to "pluralism" cannot be settled empirically at all, because men may differ on what they think is *desirable*. Furthermore, political sociologists rarely wish to bring their value judgments out into the open, because it is an effective weapon in political argument to assume that political goodness is transcendent and its recognition incumbent upon everyone; then one can take one's opponents to task for being morally obtuse as well as empirically blind to the ways in which the world or its opponents fall short of the ideal. This is the tacit side of most sociological controversy over politics; the journals enforce a neutral appearance by concentrating on the way data are presented, but most men react to articles by their implications for bolstering Left or Center (the Right hardly existing in American scholarship today) rather than what they contribute to a general explanation of politics.

This dynamic is no less in evidence where the task is making generalizations rather than descriptions. Many scholars would like to make a science of politics, but they aim for a science that resonates with the kind of imagery favored by their political faction. The playing-the-game-by-the-rules liberals like to concentrate on voting and legislative behavior, and on attacking what falls outside this as pathological; the militant Left looks for clandestine influences; the nationalists and the admirers of power reify the state into a system of inputs and outputs standing above partisan viewpoints and conceptualize political change as an evolution toward the apex represented by their own country; the would-be government advisers project an image of a new world of technocracy run by experts like themselves. Even where political sociology is not written for immediate partisan advantage, it tends to reflect deeply engrained rhetorical stances.

Take for example S. N. Eisenstadt's (1963) effort to derive general sociological laws for a wide range of historical states. This attempts to show the conditions under which centralized bureaucratic empires arose in the preindustrial world, survived, and either changed into modern states or retrogressed into more traditional forms. The aim is an admirable one: to use the logic of scientific analysis to formulate and test out generalizations by comparing the variations in as many historical cases as possible. The limitations, though, are built into the conceptual framework. Eisenstadt picks out some of the important processes, but his analysis of them needs to be viewed with considerable detachment; for everything is subsumed within a reified organic system in which rulers solve "integrative problems" for the populace and the latter provide "resources" to raise the "generalized power" of the "system." The conditions under which such empires arise are seen as a combination of free-floating resources (a market economy, nonlocal religious organizations, urban social classes) and autonomous nontraditional goals on the part of the rulers. The same conditions must continue in order for empires to be preserved. The ruler cannot rigidify the class system too much without destroying the economic basis that supports him, or (for the same reason) allow too many privileges to the aristocracy or the bureaucracy, nor can he undermine the universalism of the church without reducing his power back to the level of a traditional monarch. On the other hand, the further development of free-floating economic and political resources and of religious universalism can undermine his power in the opposite direction, leading to modernization and democracy.

If one constantly translates the functional terminology, with its implicit adulation of authority and its references to abstract processes of "problem-solving," into the real actions of individuals, the basic variables are important enough. Would-be rulers can build empires only upon certain economic and organizational resources. Eisenstadt's notion of autonomous nontraditional goals is an obscure way of saying that an ambitious man must be able to set himself above the masses—there cannot be a democracy like the commercial city-states of Greece—, one who is willing to free himself from aristocratic or ethnic loyalties and build his *own* bureaucracy, preferably employing priests of a universalistic church. But what is really going on does not come through easily in functionalist writers of this sort; Eisenstadt is a good representative of the bended-knee type of political sociologist, who must always talk about powerful men as if they were more than mere mortals winning out in the struggle with others. The habit of talking about politics as a system is a carry-over of the ancient requirement of speaking of the ruler in awe-struck abstractions, and sociologists of this political persuasion cannot come too close to the realities of power, even if the men involved are long dead.

If is apparent from the evidence referred to that there is no real "need" for a ruler, and he only emerges when the conditions of political struggle within the most mobilized strata favor centralization. But this is the kind of explanation, along with the effects of wars and conquests, that does not fit into the functional image of politics as the organic development of a System. The model also leads to serious errors in explaining those cases (all of them in European history) in which there is some degree of public participation in politics. With his unidimensional (if not unidirectional) model of evolutionary change, Eisenstadt treats this as an advanced level of a functional development, failing to see that independent law courts, parliaments, and the like are not part of the functional system of bureaucratic empires at all, but carry-overs of tribal and feudal arrangements. Like other modern admirers of power, Eisenstadt takes democracy far too much for granted, unwilling to see the fragmentation of power that it represents and the long struggles by which it survives.

What political sociology needs above all else is a resolute effort to establish a base of explanation above the topics and concepts of political controversy. Descriptive studies need to be subordinated to generalizations, and isolated areas of research need to be related to each other. Only in this way can we get off the narrow tracks laid by ideological interests, and collect a set of concepts for their usefulness in causal explanation rather than for their evaluative resonance. Voting studies in contemporary societies cease to be an alternative world view to that proposed by studies of covert community power, when both are treated as material for a set of generalizations about the way in which political interests are shaped and the conditions that mobilize them to influence the state. From such a perspective, new investigations are suggested not merely by the stage of ideological debate or the possibilities of methodological virtuosity, but by the dynamics of controversy within a genuinely scientific community.

The theoretical viewpoint on which such a science can be constructed is pretty certainly not functionalism or any other "systems" image; these are vestiges of the traditional rhetoric of political legitimation. Counter-systems like classical Marxism will not do either, for the same underlying reasons. Some version of historical conflict theory is the most promising lead, because of its empirical realism, its insistence on treating individual attitudes and organizational arrangements as a dynamic complex, and its wide comparative view of political variations. Once we establish such a viewpoint, a surprising amount of material begins to fall into place; as in so much of sociology, we know a great many things about politics without seeing what it all adds up to.

Most political sociologists probably will not wish to give up their commitments to political argument, and we can expect the same covert themes to be with us for a long time. The effort to build a truly scientific political theory will likely be confined to a minority, and then not without ambivalences of their own. If we want such a science, we must be continuously aware of this fact. Political struggle does not go away just because men try to study it in a detached way, and it is only by recognizing ongoing ideological controversies that we have any chance to keep our science above them. The work of the last two decades is ample proof that ideological positions masquerading under the pose of value freedom are even more serious threats to explanation than are overt ideologies. Only by accepting ideological controversy as a legitimate part of political sociology can we keep it in the open, and thus try to keep fundamental theory clear. There should also be a benefit in the opposite direction. Such values as democracy are hard to support by reasoned action if our theories about the conditions that uphold them are biased by the rhetoric of political argument, as is so manifestly the case today. To the extent that we can be realistic about the conditions for political arrangements, we will not bury our heads in the sand when it comes to defending them. In this sense, I hope that this scientific work may contribute to whoever else has political values similar to my own.

THE ORGANIZATION OF VIOLENCE

If we try to avoid reifying some ideal, we find that what we mean by the state is the way in which violence is organized. The state consists of those people who have the guns or the other weapons and are prepared to use them; in the version of political organization found in the modern world, they claim monopoly on this use. The state *is,* above all, the army and the police, and if these groups did not have weapons we would not have a state in the classical sense. This is a type of definition much disputed by those who like to believe that the state is a kind of grade-school assembly in which people get together to operate for their common good. Of course this is one of the things that a state can do; *if* its members are organized in such a fashion that there is a stalemate over the use of violence, so that no particular faction can control it for its own ends, then all the members could possibly use the state to carry out projects of mutual benefit. The extent to which this happens is grossly overestimated by the rhetoric of functionalism.

Like all matters of definition, this can be settled arbitrarily. What we shall deal with here are the ways in which violence has been organized in society. The justification for doing this is pragmatic. I believe it can be shown that in this fashion we can deal with all questions that might arise about politics. Who will fight or threaten whom and who will win what? This should cover the ground in which we might be interested. One might object that "crime"—private violence—would thus fit into the definition of state. I am quite willing to accept this. The only difference is in the extent of the regular organization of violence; it is only because we use the rhetoric of those that have the most stable threat that we make a distinction between crime and politics. It is only in societies in which there is one very large apparatus of violence which makes private violence seem trivial by comparison, that we can see a difference. In nonbureaucratic societies, in fact, private violence and politics are virtually identical.

This approach can handle economic conflicts as well. Economic issues, after all, involve some aspect of property rights, whether over money, labor, land, buildings, machinery, or persons. The property, of course, is not the physical goods themselves, but rather a set of rights stating what individuals can do and other individuals cannot do. Land is property not because of some metaphysical relationship between the owner and the earth, but because he has organized his relationship to his fellow men so that he can step on it and they cannot without his permission, and he can draw on the violence of the state to enforce his claim. The system of credit and the relationships of employer–employee are similarly based on social arrangements in which the ultimate sanction is the state. Economic conflicts occur within a social organization in which organized violence can be regularly called out or appealed to as a threat; various groups try to call on the violent coercion of the state to enforce their particular interest. Status group politics, whether involving religions or secular ideologies, similarly makes use of the violence of the state to enforce certain kinds of rituals and ban others.

Politics, in this approach, involves both outright warfare and coercive threats. Most of what we refer to as politics in the internal (but not external) organization of the modern state is a remote version of the latter. Elections, bargaining within parliamentary bodies, and bargaining within political parties, are maneuvers to create a coalition to gain control of the apparatus of the state; much politics does not involve actual violence but consists of maneuvering around the organization that controls the violence. With this influence, a group can use the state to implicitly coerce whatever outcomes are desired, whether they involve wealth (ranging from who is taxed and regulated to how money is spent and who is employed), legal regulation of organizations and occupations, support or prohibition of rituals and status displays, and the like.

A general theory of politics, then, is an explanation of how violence is organized, the structure of the state; of what determines the interests people have in influencing or controlling the policy of the state, or molding its structure; and of what determines just how much particular interest groups actually win in the process. It becomes a matter of explaining what the structure of the state will be, and what its policies will be. Both of these are seen as the results of contending groups, whose interests are shaped by their social positions, and whose degree of success is proportional to their resources for engaging in political struggle.

Such a theory draws upon most of the other branches of sociology discussed in this book. The state is a kind or organization, and its structure is determined by the variables dealt with in Chapter 6. The conditions for various interests that people might want the state to support—and these include economic, power, and status concerns—are contained in a general theory of stratification, such as that given in Chapter 2 and expanded in Chapters 4, 5, and 8. The resources that enable men to win influence over the state include both formal organizational arrangements (systems of property, networks of communication, coordination of weaponry) and the emotionally founded beliefs that make up systems of religion and status.

A theory of politics must deal with all of this together. Obviously, the problem is where to begin. A useful strategy is to start with the broadest outlines, and then fill in. If a new inhabited planet were discovered, what would we wish to know about its history to predict its politics? Of course, we must turn to our own history to get a clue. But it is a good mental device to keep us from seeing our history as inevitable instead of a sample of outcomes from a much larger range of possibilities. It is equivalent to asking: What kinds of principles would have enabled us at some time in the past to predict the history (especially the political history) of Rome,

China, or the United States? The fundamental circumstances would be wars. These provide the bases of states, of their growth and disintegration, as well as the particular shapes those states take. The general theory of political organization, then, must begin by understanding the conditions for particular kinds of military organization, and for the success and failure of each in maintaining a stable organization of threatened violence.

Appendix A gives a brief overview of the relevant historical materials. What laws of political motion would predict such events? Certainly we could not predict the details. It does not seem at all necessary that Rome should be the peripheral society to conquer the Hellenistic world; it could just as well have been the rival Italian city of Veii that emerged from the marchlands of the Hellenistic world to take advantage of the power stalemate among the successor states of Alexander. There are also many possible turning points which could have determined very different lines of historical development. These have been especially the outcomes of particular battles and campaigns, determining conquests which have cut off or opened up new possibilities of development. Max Weber, among others, thought that the Greek defeat of the Persians was a crucial event, making possible the whole development of European history. Another important turning point occurred when a Chinese emperor in the fifteenth century decided to destroy his fleet, at the time clearly the greatest in the world, and to cut back on commercial development. This effort to maintain traditional gentry power against new forms of social organization probably prevented further Chinese economic innovations, although they already had a commanding position in technological advances—advances which were diffused to the West and provided the basis for development there. Certainly world history would be different if the Chinese had been the first to industrialize. A nuclear war in the twentieth century destroying the Western nations would be another important turning point.

Our aim is not to predict long-term sequences followed by all societies, nor a global evolution through a series of stages. History is a concatenation of changing structures impinging upon each other from without. I shall try to summarize first what we know about the structure of military organization—the core of the state—beginning with the technologies of economic production, of weaponry and administration. From there we move to the tactics of political control within the state, including both ideological (mainly religious) appeals and the use of civilian administration, and to the factions and maneuvers which arise within a complex state. The conditions producing various forms of democracy are considered next. Finally, I consider what the overall flow of historical change looks like, and offer some comments on the problem of political prediction.

TECHNOLOGY AND MILITARY ORGANIZATION

19.0 The technology of economic production limits the potential size and structure of political organization.

19.1 The greater the surplus of goods produced per worker in primary production beyond what is necessary to stay alive, the greater the potential for political organization and political inequality.

19.11 The greater the reliance on *hunting and gathering* or *primitive fishing techniques,* the less likely a society's size is to exceed a few hundred persons ranging over a wide territory and assembling only periodically, and the closer its members will be to political equality.

19.12 The greater the use of *primitive horticultural* or *pastoral techniques,* the more likely a society's size is to be between a few hundred and a few thousand, moving periodically with the exhaustion of crop or pasture land, and the more likely it is to be ruled by councils of adult males, sometimes in conjunction with particular ceremonial leaders.

19.13 The greater the use of *advanced horticultural techniques* (especially metal tools and irrigation), the more likely a society's size is to be between 10,000 to a few million, spread over a defined territory as large as several hundred thousand square miles, and the more likely it is to be ruled by a small upper class of priests or military aristocrats living off of tribute.

19.14 The greater the use of *agrarian techniques* (especially metal plows and wheeled carts pulled by animal power), the more likely a society is to be ruled by an aristocracy of warriors, priests, or officials which claims permanent ownership of land and attaches agricultural labor to it, and organizing an effective political network over a population of 10,000 through many millions, and over territories of several hundred thousand through several million square miles.

19.15 The greater the use of *industrial techniques* (especially machinery run by inanimate energy sources), the more likely there is to be intensive political administration of large and densely populated territories, with upper limits of at least hundreds of millions of persons, and intercontinental territories.

These are empirical generalizations (Lenski, 1966). Their rationale is fairly straightforward. If there is little or no economic wealth beyond what is needed to keep its producers alive, it is impossible for some men to permanently dominate others, for their supporters will soon die. Increasingly powerful forms of productive technology make it possible for there to be a ruling class that lives by military dominance, religious charisma, or political manipulation, or some combination of these. The more intensive the productive techniques, the more densely populated a territory can become, and at a certain point this concentration can move out of agricultural areas and into cities. In either case, the concentration of the population makes it easier to unite them under a single political regime. This is particularly true where the technology changes from one that rapidly exhausts the land and requires frequent migrations, to one in which the same land can be worked more or less permanently; only in the latter case is it worth putting much stress on landed property, and only then is it possible for the ruling class to tie men to the land as serfs or habitual tenants. Where this can be done, the political power of the upper class increases enormously, for reasons that we shall examine.

Industrial technology has some rather complicated effects; this is why I have stated *19.15* so abstractly. It increases productivity and puts great resources in the hands of whoever can marshal them, and in this sense has a potential for state control over very large territories and populations. At the same time, it tends to increase the complexity of administration and thus disperse power resources into at least part of the hierarchy; and it mobilizes much of the population in large cities and long-distance transportation networks, making more people into political forces to be reckoned with. Industrial societies can have a variety of political forms, depending on how these forces balance out; the varieties of modern democracy and totalitarianism give an indication of what there is to explain. Nor are industrial societies necessarily large, as a glance around the world today will show. Productive technology determines the possibilities, but more directly, political forces determine which of them are realized.

Since the state is the organization of violence, this means looking for the conditions that affect the size and organization of armies.[1] The types of *economic production* just considered are one factor, since a larger surplus population means a potentially bigger army. (But rulers may well try to avoid mobilizing the maximum size army if they can avoid it.) The kind of *weapons* available is another technological variable, distinct from productive technology. There are variations in the technology and

---

[1] My main source here is Weber (1968: 901–1368). McNeill (1963) strikingly demonstrates how subsequent historical studies bear out much of Weber's leads from 60 years ago; of course, economic, legal, political, and religious history were already highly sophisticated studies by that time, and Weber's analytical talents had substantial materials to work on. Another convenient compilation of information is Eisenstadt (1963), treated with the cautions just noted.

organization for *supplying* and *controlling* armies, which interact with the other conditions to determine whether the military skeleton of the state will be large or small, centralized or decentralized, collegial or hierarchic. The conditions for democracy are one variant of these outcomes.

20.1 The greater the *economic surplus,* the greater the potential size of armies proportional to the population.

20.11 The greater the *population density* supported by the productive technology, the greater the potential absolute size of armies.

20.2 Weapons. The organization of armies is determined by the expense of weapons, and whether they are wielded individually or by a group.

20.21 The more reliance on *expensive, individually operated weapons,* the more fighting is monopolized by an aristocracy of independent Knights, and the greater the stratification of the society.

20.211 The more reliance on *cheap, individually operated weapons,* the more of the able-bodied population may participate in fighting, and the greater the democracy and decentralization of the society.

20.22 The more reliance on *cheap weapons operated by a group,* the more the interdependence among democratic equals, and the higher the level of potential conflict within the group.

20.221 The more a group of this sort is *actually engaged in fighting,* the greater its cohesion; the longer it is in a situation of peace (either after victory or defeat), the greater its tendency to split up.

20.222 The more reliance on cheap weapons operated in a group, the greater the tendency of leaders to seek new areas for conquest (by *20.221*).

20.23 The more reliance on *expensive weapons operated by a group,* the more an army takes the form of a central command hierarchy and a subordinate group of common soldiers.

Many of the shifts throughout history have been associated with changes in military technology. Expensive individual weapons—the first bronze swords and spears, the war chariot, the heavily armored horseman—tended to alternate with cheaper weapons such as iron spears, horsemounted archery, and mass-produced rifles and pistols. Expensive group-operated weapons are the mark of modern warfare, from artillery and armored vehicles to war planes and rockets, but they have their preindustrial counterparts in the form of catapults, warships, and especially fortresses. Inexpensive group-operated weaponry is the most complex category, since virtually any weapons may gain in power by being used together in a concerted group; the extreme cases I have in mind, though, are the long-boats of the Vikings and the galleys of ancient Mediterranean warfare (operated by oar as well as sail, but fighting largely hand-to-hand), and the Greek phalanx of spearmen. As always, we must think in terms of continuums instead of absolute contrasts, and of possible mixtures of the two dimensions.

Individual weapons are expensive, of course, only in relation to a given type of economy. But a similar pattern emerged on very different occasions (see Appendix A). Pastoral societies split into a warrior aristocracy and a class of commoners around 1500 B.C. with the spread of expensive bronze weapons, recapitulating an earlier development in the horticultural march-lands of Mesopotamia; the same thing happened again in advanced hor-ticultural societies with the technique of horseback riding around 900 B.C., and again after A.D. 200 in agrarian societies with the development of very heavily armored knights, first in Persia and later spreading to Europe. When the most effective military techniques of the time can be afforded only by a small group who wield them as individual warriors, the result is a rather small military class lording it over the rest of the population. By contrast, the development of cheaper weapons led to more massive armies, and the revival of participatory democracy. Thus, iron weapons appearing around 1200 B.C., and the later domestication of horses and the perfection of horseback archery on the plains of Asia, both led to a leveling of social differences and the revival of the tribal-council form of organi-zation, with only temporary authority for war leaders.

So far, I have concentrated on whether a society is divided into two classes of men or only one, ignoring for the moment the relations between this society and another one it may conquer. I have also put aside the question of how the military group is organized within itself: whether there is a collegial–democratic group of equals, or a hierarchy of com-manders and subordinates. This is determined by a number of variables, including how the army is supplied, what kind of economic organization there is in the territories they conquer, and the size and geographical dispersion of their conquered populations. We catch another glimpse here of Weber's theory of centralization and decentralization of organizations.

There is one variable that fits that model deriving from the type of military technology itself: the extent to which weapons are operated in-dividually or by a group. The extreme here is where weapons *technically* need to be operated by a group or they cannot be used at all (i.e., a fortress or an artillery battery); in this case, the army becomes a clear and perma-nent hierarchy of command, under the control of whoever holds posses-sion of the material resources. The more expensive these resources, the more stratification results from their existence. At the other extreme, where all warriors outfit themselves individually, there are only temporary coalitions without strongly imposed discipline, made up of individuals who guard their autonomy and fight among themselves when there is no external enemy.

The case of cheap weapons used in concert is a mixed one. It includes the coalition of men who arm themselves but come together to use a common weapon (such as a boat), and all other self-armed coalitions which depend for effectiveness on their cooperation (of which the phalanx is an outstanding example). In a sense, even individual knights who charge together in battle have some degree of interdependence as a unit, although they are highly capable of fending for themselves. These variations are best understood as combinations of different resources.

20.24 The greater the *proportion of their weapons supplied by the members themselves,* the more temporary and less disciplined the war coalition; the greater the proportion of weapons supplied for the group by an individual, the greater his discipline over the group, and the more permanent the organization.

20.241 The greater the *advantages of fighting as a group* (as compared to fighting as individuals), the greater the commitment to the group, and the more power for whoever provides its weapons used in common.

The first proposition means that knights charging together as a battle line are less committed to the group and its leader than Vikings in a longboat, and that a Greek phalanx is more democratic than an artillery brigade. The second means that a phalanx is more a permanent organization than a band of nomadic horsemen, even though the members are self-supplied in both cases; for in the former, their fighting superiority comes precisely from their greater use of disciplined tactics. It is not surprising, then, that the phalanx (as developed by the Macedonians and the Romans) eventually became the basis of bureaucratic army hierarchies and centralized empires (even though it originally had been the main support of civic democracy). The less disciplined form, on the other hand, tended to hold together only in times of battle, and any extended period of peace would destroy it. Nomadic coalitions would break apart if their leaders did not produce a continuous string of victories; ironically, a successful conquest had the same effect over a longer time span, as the warriors would reassert their claims to equality by making their shares of the booty into autonomous feudal regimes. It is for this reason that the leaders of relatively democratic self-equipped armies are so aggressive, and that the most warlike societies have fitted this description: the feudal Japanese (compared to the more bureaucratic Chinese), the nomads of central Asia (Mongols,

Turks, and others) and of Arabia, the Vikings, the Greeks and Italians in their period of ancient democracy, and the European feudal nobility.

The provision and use of weapons is only one determinant of military structure. Another set are the ways in which troops are fed and supplied during a campaign, and how they are rewarded when a war is won.

20.3 Supplies. The more centrally administered the army's supplies, the greater the power of the commander and the more permanent the unified organization.

20.31 The more an army is supplied by *individual soldiers* requisitioning their own supplies as they travel, or by individually trading booty for supplies in nonhostile areas, the more decentralized the army organization.

20.32 The more an army is supplied by *individuals* or *groups of individuals carrying their own supplies with them,* the more control is decentralized to the individual or group level.

20.33 The more an army is supplied by the *central command requisitioning or carrying stores which it periodically distributes to its troops,* the more centralized the control over the entire army.

20.34 The more an army is supplied by the *central command periodically providing money for unit commanders or individual troops to buy their own supplies,* the more control is a subject of conflict between central and local levels.

We have a continuum: The army living off the land or by trading booty for supplies gives the most power to the individual soldiers, and the centrally supplied army gives the most power to the commander. The other cases are intermediate. Where each type will be used depends on the resources and motives of the individuals involved. Armies from hunting-and-gathering or horticultural societies (which may well have acquired more advanced weapons by diffusion) have little or no supplies to take with them, and their alternatives are limited to raiding close to a home territory or living off what they can conquer. Since these societies tend to be relatively democratic, the motives of the troops themselves would probably favor such an arrangement in order to keep it that way. The leader wishing to prolong his temporary war-time power would probably adopt the tactic of acting as agent for the group in requisitioning supplies, distributing booty, and trading booty with nonhostile people.

Individuals or small groups carrying their own supplies are found primarily in the military forces of pastoral and maritime societies. Pastoralists are most able to take their supplies with them by moving their herds; but since herding is a group enterprise, the individual is strongly dependent upon the group, if not on the central commander. It is in pastoral armies (such as those of the early Islamic conquests) that the family or tribe is the strongest unit of control.

The central administration of supplies requires that the level of productivity, transportation, and communications be relatively high. Hence, such armies are confined to advanced agrarian, irrigation, or industrial societies—and within the more primitive of these types, to a river valley heartland where transportation is good. But the potential for supporting a centralized army may not be fully actualized; that depends on the motives of the individuals involved, and particularly on the struggle for control between the central government and local commanders. Where the latter are politically powerful, the government may only pass out funds with which each unit commander buys his own supplies, thus giving much effective control to the local level. This was the procedure in the British army, for example, until the middle of the nineteenth century; it was part and parcel of an entrepreneurial system in which regimental commands were bought and sold to individuals, who might hope to make a profit on the concession. (Where the army chief dispersed money for supplies directly to the soldiers themselves, as in the Roman legions, power tended to remain more in the hands of the general rather than those of his captains.) A key aspect of the centralization of power in Prussia and France several centuries earlier was to eliminate this regimental system in favor of centrally administered supplies; this marked a turn toward absolutism that had been found much earlier in some of the armies of the Middle East and the Orient.

20.4 Administering conquered territories. The form of a state depends on the way in which the dominating army is supported from its wealth.

20.41 The more the *individual soldiers of a conquering army are allowed to collect booty* and leave for home or new conquests, the less of a permanent and centrally controlled government will be established.

20.42 The more the *army commander collects booty to redistribute to his troops,* the more the army will hold together as a governing and governable unit until the booty is actually distributed.

20.43 The more an army is rewarded by *assigning conquered land to individuals,* the more likely the territorial regime is to be feudal decentralization.

20.44 The more an army is supported by *quartering troops in permanent garrisons supported by local tribute,* the more likely the territorial regime is to be moderately centralized, with political control the subject of struggle between garrison commanders and the central government.

20.45 The more an army is supported by *quartering troops in garrisons supplied by a central administration separately collecting taxes,* the more likely the territorial regime is to be permanent and centrally controlled.

I have proposed five main types. The first two are booty systems; used by themselves, they allow no permanent administration of a conquered territory at all, although the second may hold the army together until it gets home. Such conquests are little more than raids, although they may well destroy a previous government (as in the case of the Germanic attacks on the Roman Empire, or the early Viking attacks on the Kingdoms of northern Europe). The third is the classic road to feudal disintegration; it places a new government upon the conquered territory, but under a very weak form of centralized control. The last two are methods of holding together a centralized state; the degree to which a leader succeeds in this is very much a function of the administrative resources available, although his personal capacity for political maneuver may tip the balances, especially in the lower range where resources are more scarce. In particular, his ability to rotate commanders of local garrisons and to mobilize the army under his own command in periodic campaigns may serve well to combat the disintegrative possibilities of the more decentralized supply system.

When is each type likely to be used? The booty system is found where the conquerors themselves are organized in a highly individualistic manner, or the conquering territory contains no resources that can be permanently administered. Fighting among primitive horticultural societies is likely to be booty raids, for the latter reason; the very individualistic nature of the early Germanic and Viking armies made for a similar effect, even though there was more permanent wealth in the conquered territories, if they could have administered it.

The kind of weapons used is one determinant of how individually or collectively the army is organized, and thus influences these choices. The army based on heavily armored and self-equipped knights, as we have seen, is itself an unstable coalition whose chief is regarded only as the first among equals. Where the territory conquered has an agrarian economy (or where one can be easily introduced), there is the basis for permanent appropriation of land that will be quickly seized upon by the knights. Such expensive equipment requires that each knight have a considerable number of commoners to support him; both the technical problems of supply and the wishes of the knights themselves combine to pressure the chief into distributing land, usually to his regret. Where warfare depends on expensive, group-operated weapons, however, the position of the leader is much stronger. The sea-based power of the Byzantine emperors, for example, made for a stable pattern of collectively supplied garrisons; the

Great Wall of the Chinese state allowed a similar concentration of troops.

The economy of the conquered territory is equally important. An agrarian economy, where land can be continually reused, is suitable for appropriation by individuals. An irrigation economy, howver, is inherently more centrally controlled; it makes it easier for the king to supply the troops himself, as well as more difficult for individuals to supply themselves independently of the king. Similarly, an urban commercial economy (such as the ancient trading cities, as well as modern industrialism) cannot be so easily broken into pieces of landed property. The latter two types, then, are conducive to organization around garrisons rather than landed estates. Hence, we find more centralized armies in the Egyptian, Mesopotamian, and Chinese river valleys, and in the commercial cities upon which the Greek, Roman, and Islamic empires rested at their heights; the agrarian economies of plains of Persia and northern Europe are the sites of classic knightly feudalism.

The extent to which the garrison form is conducive to genuine central control depends upon the balance of resources between local commanders and the ruler. The more locally supplied the troops, especially if the tribute is in kind rather than money, the greater the power of the local commander. Thus, the city garrisons of the Islamic conquest tended to become the seats of independent states. The ruler is much more secure if he himself supplies the garrisons from his own resources. This requires an efficient tax system on a national basis, together with efficient means of transportation. Such centralized control is easiest in a highly commercialized and especially an industrial society, and hence, it is such societies that have had the most powerful governments.

Industrialization by itself is not a sufficient explanation of centralized control; it only provides resources for the struggle over control. In the modern world, the boundaries and structures of states continue to depend on how military force is organized, including both the army and the police. Effective political decentralization depends on the local supply of armed forces. The reality of what state and local power there is in the United States depends ultimately on the remaining powers over the use of force, as one may appreciate by comparing the structure of power in a state like Russia or France which has only a nationally organized police. (The relative preeminence of the United States government over the individual states is ultimately a matter of the superiority of the national army over the states' police and militia, as the various school integration cases of recent decades illustrates.) The extent to which contemporary China may be lapsing into decentralized control depends on how autonomous the regional army commanders can make their supply sources; similar considerations hold for the effective centralization of power in the states of Latin

America and Asia. What applies to internal autonomies applies to external coalitions as well: Important as common markets and other forms of international cooperation may be, they do not become a real government until acquiring an army supplied and controlled by themselves. It was the creation of a centrally administered army after 1871 that effectively united Germany, not the customs union of 35 years earlier. Without this coercive power, the European Common Market or the United Nations is not a state, and its decisions cannot be securely enforced and hence acquire the legitimacy that goes along with the power to coerce consent.

IDEOLOGY, RITUAL, AND CONTROL

The foregoing has been an expansion of Weber's (1968: 956–1110) organizational theory of centralization and decentralization, applied to the military structure that is the skeleton of every state. The basic principle has been that the centralization of authority is proportional to the concentration of control resources (*13.0*); but as the size of the army and its geographical dispersion increases, greater resources are needed to maintain control (*13.1–13.5*). For an army, such resources are of several types: the kinds of weapons used (*20.2–20.241*), the means of supply (*20.3–20.34*), and the means of administering the wealth of a conquered territory (*20.4–20.45*). The control over material resources (*13.3*) is a composite of these variables, with various mixtures being conducive to a proportional degree of centralized control. For example, the practice of dividing conquered land among the victorious soldiers tends to disintegrate the state; this did not happen to the territories of the Roman Republic, even though it rewarded its soldiers in this way, because the technique of warfare was highly collective. The combination of the two conditions did produce an intermediate form of decentralization, however, in the form of mass democracy among the soldier-citizens. It was with the shift to long-term enlistments and centrally administered pay that the power of the generals began to increase, resulting in a trend toward assigning lands only as large commercial estates, and culminating in the overthrow of Roman democracy.

Efficient techniques of transportation and communications, including a system of money and the use of writing, are crucial aids in maintaining central control of military resources (*13.4–13.5*). When they are available, it becomes possible to reduce the personal ties which are so threatening to the state, operating even at a distance (*14.1–14.2, 14.4–14.5*). All of this is easier in particular economic and geographical situations: Irrigation systems in river valleys both centralize material resources and provide an easy means of transportation and hence of control; agrarian economies are conducive to feudalism under some weapons and supply systems, but favor centralized armies if combined with an urban commercial network and a money economy; and industrial societies are high on all of the centralized control variables (*19.13–19.15*).

But it has been necessary to put all this in rather guarded generalizations, statements of potential rather than of necessity. For men do not give up their autonomous dignity without a fight. The more primitive societies are relatively democratic, and tend to remain so even when the diffusion of weapons from the more productive civilizations enables them to form war coalitions and embark on careers of conquest. The leader of such a coalition may wish to make himself a permanent king, but his followers are not likely to give him much power if they can avoid it. The same applies to the more aristocratic forms of decentralization; a coalition of self-equipped charioteers or mounted knights will put its weight on the side of forms of weapons, supplies, and division of spoils that maintain their feudal autonomies. New technology and economic resources for centralized control become available from time to time, but they are not automatically applied. The men supporting the older forms may be strong enough to maintain them for some time; in some cases, as in the Greek city-states of the fifth century B.C., it may be with the ultimate result of turning over military dominance to a peripheral party (Macedonia) which organizes along more powerful lines; in others, such as the long period of often successful resistance to centralized administration by the British nobility, with no great disaster (and some unforeseen benefits).[2]

Establishing centralized control, then, required astute political maneuvering and organization-building on the part of the would-be ruler. He had to have not only the technical and material resources, but the skill to weld them into a new framework. In particular, the leader of a temporary war coalition had to do two things. He had to reorganize his followers so that they would become loyal to him, and especially so that he could use them to enforce his commands *upon each other;* and he had to take over the administration of supplies and the collection of taxes himself, if possible by means of a separate, nonmilitary hierarchy of officials under his command alone. Both of these were primarily matters of religious organization.

---

[2] This resistance to military innovations may also come from the central authority. Both the Japanese and Chinese rulers forbade naval development at times when it could have been highly successful. The reason was not pacifism, but the realistic fear that military lords would build up distant bases of their own in the absence of efficient means of long-distance control, and eventually threaten the ruler himself (Weber, 1968: 912).

To invoke religion in this context sounds incongruous to modern ears. Religion seems to be a matter of other worldly concerns and personal moralities, or at best of social status and personal demeanor. But religion and politics have been closely connected until very recently, and even in the modern era there are important interactions and analogies. In tribal societies, and onward through advanced horticultural ones, political and religious authority are usually vested in the same man. Where kings exist, they are often considered divine gods themselves, or their priests. The establishment of military coalitions such as the Greek cities was a matter of establishing a joint cult. In agrarian societies, in the commercial societies growing up in their interstices, and in the cosmopolitan empires built up from conquests across a number of types and cultural identities, religious organization usually grows up separately from political organization, but it maintains important connections; governments draw upon it for ideological legitimation and for administrative organization, and priests in turn play politics in the hope of state support or even theocratic (strictly speaking, hierocratic) power. This is just as true of mystical religions like Buddhism, and moralistic ones like Christianity, as it is of more overtly militaristic Islam or secular Confucianism. It is hard to imagine what world history would be like without the Islamic conquests, the Christianization of Europe, the wars of the Reformation or the comparable Sunni-Shi'ite struggles in Islam. Even secular modern politics would be unrecognizable without the moralistic evangelism of Jacobinism, Socialism, Communism, and anti-Communism.

The principles of religious interaction are important for politics in two senses, one corresponding to the exercise of power itself, and the other to the more limited problem of establishing a bureaucratic civil service to administer military resources.

The power of the state is ultimately the power to use violence. The coercive sanction can override all others; even though it causes intense alienation and dangerous rebellion, its quality as the ultimate thing men wish to avoid gives its wielders the power to shape the rest of the social order. Used as a means of appropriating material property, it becomes the basis for control by rewards; operating as a threat that makes men dependent upon one another, it underlies the most intense emotional bonds (cf. *4.83* and *4.85*).

But violence is never very effective except as exercised by a social coalition. Except in a very small group, one individual cannot coerce many others, even if he is bigger or holds most of the weapons. Organization is the crucial factor. This can be seen by examining an extreme case. A dictator controls, not because he can personally coerce every single member in his regime but because his army can do so. Furthermore, he controls his army, not by his personal might but because he is in a position to use the power of his group against any individual who rebels against him. Even if the entire group is antagonistic to him (which may well be the case whenever autocratic and coercive systems of power exist), the organization makes it risky for any individual to rebel. The dictator reigns by organizing matters so that his followers watch each other and are afraid to take the lead in acting against him. Schelling (1963: 53–118) has given this a formal basis in his model of tacit coordination. In effect, power creates a self-fulfilling prophecy—men who are powerful are so because others believe they are powerful. It is this belief, in turn, which makes subordinates maintain the sanctions that keep their rulers powerful.

This kind of organization around self-fulfilling prophecies operates within politics at all levels. It explains the power of the dictator and also gives the conditions under which he will be undermined, for this occurs when the general belief in the reality of his power begins to crumble. It is for this reason that symbolic acts are crucial, for they are tests of whether or not the whole system of mutual fear will hold up. Similar principles apply, although on a lower level of fear, in electoral politics; the influence of a political boss is based on the fact that he is constantly being sought for support by a network of those who owe him favors, which he can return only because of the continued belief of others in his potency (Banfield, 1961). Liddell–Hart (1954) has shown that violent battles operate on the same principles. A battle is lost not so much because an army is decimated but because its organizational basis falls apart. This occurs when various sections of the army conclude that the other parts of the organization will not support them, which in turn makes it reasonable to save themselves; once such a reign of panic begins, a self-fulfilling prophecy operates in the negative direction. The maneuvering of warfare consists to a very large degree in trying to produce a feeling in the opposing army that it is crumbling before a superior organization, since disciplined military force can destroy a much larger undisciplined group.

Politics of all sorts, then, involves the effort to define oneself as the powerful center in an interpersonal network, and especially in terms of the part of the network that wields the weapons of violence. This maneuvering may involve some actual battles, but these tend to be dramatic shows, exemplary punishments, and tests to show which way the wind is blowing. It is in this sense that politics is clearly in the same sociological universe as religion. Men's beliefs about power are as crucial as the actual exercise of coercion itself, and the minute-by-minute tactics of organizing power involve not only fighting and organizing supplies but putting on ritual shows in an effort to influence solidarity and deference among supporters and enemies alike.

Weber referred to this aspect of power as "legitimacy." It involves two related levels of maneuver. On one level, the ruler must maintain the impression that he is fully able and willing to call upon others to coerce any dissident (see Stinchcombe, 1968a: 158–163). The loyalty of one section of troops is ensured by the threat from the others. This is one reason why improved transportation and communications helps extend the range and intensity of a ruler's power; they enable him to bring up supporting troops to any spot, or—since this is a struggle played more in gestures than in battles—simply to call upon others for assurances of support. Providing he can place himself at the center of communications, then, the ruler's position is secure (an application of *11.4*).

Now, all this display of threat and loyalty results in a distinctive kind of organization on the cognitive and emotional level. If a ruler has the material and communications resources well in hand, the words and symbols by which men refer to this state of affairs take on the solidity of physical objects. The government, although it is only a network of men reacting to each other, becomes a real thing in their minds, and its leader becomes more than an ordinary man. If the conditions for social rituals are of the right kind, he becomes a god or a priest, reflecting the awe of his coercive power and the impenetrableness of his network of mutual threat. In this sense, Durkheim's god is a collective conscience of a particular sort: It represents the political society standing over and out of the control of most individuals, but sanctifying the individual who manipulates the network for his own benefit.

I have taken the extreme cases for the sake of clarity. Weber's notion of legitimacy by no means implies that all governments are equally secure, or that men do not challenge the system of belief and ritual, or that there are not a great many different degrees and types of legitimacy attached to different ways in which political resources are distributed. It certainly does not mean that violent coercion and material resources are unnecessary or subsidiary to emotions and beliefs; legitimacy is nothing more than a shorthand way of communicating about the organization of these things, the ultimate sanctions in human behavior. But physical resources do not automatically flow into power, and it makes all the difference both in individual careers and in the long-run changes of human history just how men manipulate beliefs and emotions.

Throughout most of history, political power and the coalitions formed around it have been reflected on the ideological side in religious beliefs. With the separation of church from household and state, religions have both changed and become more complicated in their relations with politics. In commercialized, cosmopolitan, or industrialized societies, governments and political factions have even organized themselves in secular terms, and religions have fallen into the background of politics. But modern ideologies are variants of the same basic set of conditions, new forms appropriate to modern conditions of the same appeals for moral solidarity and for obedience to the organization stretching beyond individuals that make up the social essence of religion. What follows is an attempt to state some general conditions that produce the main types of ideology.

21.1 Ideals of loyalty and morality reflect the social unit to which the individual must orient for his physical safety and support.

21.11 The more *coercive power and material resources are located in the family or household,* the more the tendency for religious ceremonies to take place within it and to glorify loyalty to it, spirits or gods to be identified with the family–household, and the ceremonies and supernatural sanctions to bolster the authority of its head.

21.12 The more *power and resources are located in the joint activities of a tribe or community,* the greater the tendency for ceremonies to take place involving the whole community, spirits or ideals to represent the community, and both to bolster the authority of community leaders and loyalty to its members.

21.13 The more *power and resources are located in a dominant household ruling over other households,* the greater the tendency for religious ceremonies to take place in that household, invoking gods which watch over or are represented in the household and which bolster the authority of that household over the others.

21.14 The more *power and material resources are located in specially formed associations* (whether bands of warriors, monasteries, or other associations), the greater the tendency for ceremonies to take place within the association, invoking gods or ideals that are identified with it, and bolstering loyalty to the group and its leaders.

21.15 The more *power and material resources are located diffusely or remotely in a cosmopolitan community,* the greater the tendency for ceremonies to take place within specially designed organizations admitting members regardless of background, the invoking of gods or ideals that are believed to be universally accessible to all persons, and representing an ideal of moral behavior toward everyone.

This is a cryptic summary of Weber's religious sociology (1968: 399–467). Most types should be familiar from Chapter 4: the family- and comunity-based religions; the household gods of a king, raised to a second level of religious splendor over the spirits of subordinate households (and sometimes represented as a divine king himself); and the universal churches arising within urban centers of cosmopolitan civilizations. I have added another form of free-brotherhood, the military band or monastery, since it acquires autonomous ideological force of its own when it operates as a self-sufficient living group.

Principles *21.1–21.15* tell us where an individual's loyalty will lie. Given these units, what will be its gods or ideals?

> *21.2* The kind of spirit or ideal reflects the nature of the social relationships within the ceremonially united group.

> *21.21* The greater the *equality within the ceremonially united group,* the more likely the religion is to emphasize mass participation rituals, signs of membership, and the ideal of group brotherhood (from *6.3*).

> *21.22* The greater the *hierarchy within the ceremonially united group,* the more likely the religion is to emphasize awe before a powerful god or spirit, ceremonies carried out by a special staff, and confining most members to deferential gestures and spectatorship (from *6.1* and *6.23*).

> *21.221 Swanson's Principle.* The more *levels of hierarchy within a ceremonially united group,* the more likely the religion is to admit only a single and highly powerful god.

We have met these general principles before, in considering the determinants of deference rituals and claims for status in different types of societies. Together with the typology just given, they tell us where the main forms of political legitimation are found. Thus, the egalitarian tribal group or warrior band operating in a self-sufficient manner is a moral entity unto itself; its rituals and spiritual guardians not only serve to whip up loyalty and enthusiasm within the group, and to present a face to frighten outsiders, but operate as the testing ground for internal politics. Anyone who wishes to make himself preeminent in the group must challenge the rituals through which the membership asserts their equality; the difficulty in doing this is precisely what is meant by the power of tradition, not some superstition lurking in men's minds.

The head of an authoritarian household, on the other hand, is in a good position to expand his power outside it, for he is already the beneficiary of rituals of personal deference and awe. If he can raise his household to dominance in a patrimonial system, his household gods can become the basis for a ceremonial system through which he can enforce the loyalty of an entire society. Swanson's (1962) Principle, tested upon a sample mainly of horticultural societies of different kinds, refers primarily to the contrast between such a dominant household presiding over a large number of subordinate units, and lesser structures of domination. It also reminds us by implication that a king does not acquire his legitimation automatically and without a struggle. For the lesser units have their gods and spirits too, and it takes some maneuvering on the ceremonial and theological as well as the military and political levels before the ruler's symbols eclipse all others. That this is not always successful is shown by the imperfect correlation between political hierarchy and theological omnipotence. It is likely, too, that rumblings on the theological front foreshadow efforts at more substantial rebellion.

If we remember that secular ideologies operate in most respects like religious ones, some modern applications become apparent. The tactics of political maneuver are very much the same; ideological currents are straws in the wind by which men decide how to line up on the winning side, and symbolic issues and confrontations are the stuff of at least the public, ceremonial side of politics. Election campaigns and the routine business of presenting the government's viewpoint are directed at moral issues above all, appeals to ritual solidarity in the name of the ideals representing group membership. Speeches invoking the enemy threat and the nation's honor are only one form of this; everyday political discourse constantly harps on themes of justice and service versus injustice and selfishness. This occurs no matter what the substance of the issues, whether it is taxes or political patronage, office-seeking or a status crusade. The key events in political life, accordingly, are the dramatic ones: scandals in which someone is made an outcast to the moral community which politicians are constantly referring to, or emergencies in which one can show heroic commitment to the ceremonial ideals, even over trivialties. The skilled politician is he who orients himself most successfully to the demands of self-dramatization (with control over unfavorable information operating as an important background resource); it is for this reason that successful politicians always seem so diabolical to the partisans of rival causes.

The correlation between the degree of political hierarchy or equality and the contents of ideological symbols helps explain some striking variations in the modern world. Why do some societies (the Communist states, as well as the Fascist and other highly centralized conservative authoritarian states) exhibit such a cult of personality around their leaders, with mammoth placards carried in parades, and pictures in homes, and the constant ritual reference to their names? It certainly does not follow from the content of Communist doctrine, although it does from some conservative ones; moreover, particular leaders in nonauthoritarian regimes have also acquired enthusiastic followings, but without the official and ceremonial quality. The answer would seem to be that the cult of personality is a secular equivalent to a hierarchic religion; it is found where the regime is in fact highly centralized, and the ceremonial procedures which are so prevalent in all politics are concentrated around one man. Whatever his personal charisma, the results are the same: He becomes the ritual center of the regime, whether official doctrine requires or contradicts this procedure.[3]

21.3 The goals furthered by rituals vary with the extent to which recruitment to ceremonial participation is voluntary or follows automatically from membership in the group.

21.31 The more *automatically ceremonial participation follows from group membership,* the more likely are the ceremonies to foster particularistic loyalty to the gods or symbols that represent that organization, and the more likely are the gods or spirits to be invoked for material well-being (cf. *1.1, 1.2,* and *1.4*).

21.32 The more *voluntarily members are recruited into ceremonial participation,* and the more sharply the group is distinguished from organizations whose primary purpose is the pursuit of material aims, the more likely are the ceremonies to foster dedication to universalistic goals achievable by all, and consisting of inner states of morality or salvation apart from material well-being (cf. *6.72*).

Throughout most of history, religions have been ceremonial means of bringing good crops, warding off sickness, assuring victory in war, and other material benefits. Looked at behaviorally, it is easy to see what was going on. These were the ceremonial procedures or organizations through which men made a living or exercised power, and they were invoked to tighten the group bonds and give emotional tonic for the practical tasks and difficulties the group faced. This is not to say there was no moral content to traditional religion; whenever there are ceremonial tests of people's loyalty to a group, that is a moral matter, as Durkheim so well recognized. But the object of morality was strictly circumscribed. Killing, lying, stealing, and the like were morally proscribed, and mutual aid called for, only within the particular group itself. Such religions served to knit together the group at the same time that it made it an enemy of everyone else; outside its bounds there were only alien gods with whose worshippers one had no ceremonial communion.

The universalistic world religions changed all this, precisely because they were based on a new organizational form. Their universalism of morality was the direct counterpart of their universalism of recruitment, for they linked together people in a new ceremonial community, from

---

[3] The highly concentrated nature of ceremony in Communist regimes explains why intellectuals are so severely controlled there. The ritual order is a very tight one, with no allowance for cosmopolitan relativism; hence, even minor symbolic threats, if left unchallenged, portend a serious crack in the structure of legitimacy.

whatever families, tribes, or societies they might come. Such religions were much more universalistic than the national cults arising from royal households, for they set their membership aims in terms of the entire world. Such religions, understandably enough, appeared in cosmopolitan centers when the ebb and flow of rival states had disenchanted the worshippers of national gods, and the amount of migration and the growth of a complex economy brought men increasingly into dependence upon strangers with whom they had no ceremonial ties.

The universalism and the moralism of these religions were part of the same pattern. The success of such religious entrepreneurs depended on their offering to set up freely chosen and newly created communities within which men might trust one another enough to carry on honest dealings, whether or not they shared the same traditional ties. By the same token, this organization specializing in the cement of community-building per se had to distinguish itself from traditional religions with their worldly orientations bolstering group boundaries; it was precisely these boundaries that were to be overcome, and hence, magical (selfish) aims were downgraded. The trademark was the *act* of solidarity itself, not whatever it might be used for, hence the religious concern for enlightment or salvation as an experience of value *in itself.* The result of performing religious acts was always a *religious* reward, although it might come now, in a future condition of the perfect society of brotherhood, or after death.[4]

Universalistic religions received the most attention in Weber's political sociology, both because they were the main vehicle for legitimating rulers with truly independent powers over their troops, and because they provided an organizational basis for civilian administration. On the ideological and ceremonial side, the difficulty standing in the way of a personally ambitious army leader was the network of particularistic (and even tribally egalitarian) relationships that left him dependent upon troops that he did not fully control. All of this was reinforced by a set of traditional ceremonies and related religious ideas; for him to violate these was a sure sign that he was intent on usurping the others' powers. What the ambitious man needed whenever his conquests stretched far enough to make his position in the traditional manner tenuous was another religion with superior claims that he could put in place of the old ceremonies. Then he would have a ceremonial means of expressing loyalty at a distance, to a

---

[4] This discussion has concentrated on the moral side of universalistic religions, and virtually ignored their mystical side. The latter has somewhat different appeals and dynamics, and is certainly stressed more in some universalistic religions than in others. But both mysticism and moralism create impersonally recruited communities, and hence, their political and organizational effects are pretty much the same, at least for the purposes of this discussion.

remote but awesome power that everyone would know well enough stood in the same place as himself. As a weapon of legitimation for a huge and intensely administered empire, a universalistic religion was a superior means of symbolic communications and supplier of loyalties.

Along with this went the actual administrative work that the church could do for him. A key problem was maintaining centralized control over army supplies; this could best be done if taxation were put on a regular basis and taken out of the hands of the army. Even better would be to put the entire administration of the regime on a thoroughly bureaucratic basis, undercutting personal cliques by subdividing authority and transferring men from one abstract position to another, and controlling by formal rules and written reports. This required both literate officials who could devote themselves to this nonmilitary work, and even more importantly, the concept of such an impersonal organization and such a body of general laws. These were supplied by the priests. Their organization was the prototype of a bureaucracy, with members recruited into an impersonal community loyal only to their unworldly goals. Its abstract god or spiritual reality negated loyalty to particular individuals, and thus expressed the very ideal a ruler would find useful; the notion of the lawfulness of the lower world dependent upon transcendent powers above gave the model for rational rules and regulations, often exemplified in the general regulations of the church itself.

Virtually all of the great bureaucratic and quasibureaucratic empires were centralized by a priestly administration. The Christian bureaucracies of late Medieval Europe are only the latest example. The only premodern empire to conquer all of India, Asoka's rather short-lived one, attempted a totalitarian administration by using Buddhist priests. The stability of the Chinese empire began with the displacement of patrimonial warlords by a civil service of Confucians. The long-lived Byzantine empire was a Caesaro–papist regime; its Greek predecessors in the Eastern Mediterranean were also upheld by quasireligious bureaucrats, especially the followers of Stoicism and other universalistic philosophies, and with their generalized cultural loyalties established in the Hellenic schools.[5] It was Stoic and Christian administrators who bureaucratized the Roman empire.

It should be borne in mind that there is a certain amount of variation within this range. Some of these states had much more thoroughly imper-

---

[5] Notice that schools are the organizational equivalent of a universalistic church. Both serve to separate men from their family and ethnic ties and to inculcate the notion of membership in a new community by virtue of a shared abstract culture alone; thus secular schools are highly suitable for recruiting loyal bureaucrats.

sonalistic administrations and more universalistic religions than others. The centralized states of Egypt and Mesopotamia were administered for considerable periods by priests, but the religions were nationalistic rather than universal; the irrigation economies and river transportation provided compensating resources for centralization. The Chinese empire had strong mixtures of patrimonialism, and in some respects, Confucianism was not very rationalistically bureaucratic. Such variations, though, tend to prove the general correlation; they also remind us that political centralization is due to multiple conditions, bureaucratization being only one of them. Moreover, religions themselves can shift in their degree of bureaucraticism; Confucianism probably grew more rationalistic over the centuries as it became more and more strongly identified with government; Islam's universalistic potentials slid back into a form of nationalism as the great war coalition it had mobilized began to fragment into its ethnic and tribal elements; and the mass conversion of the German and Nordic tribes to Christianity, motivated by their leaders' hopes of making themselves into civilized emperors, tended to reduce Medieval Christianity to the level of a magic cult. A centralized bureaucratic regime could take hold only where favored by a number of conditions—in economic and military arrangements as well as the availability of priestly allies.

The distinction between particularistic and universalistic ideologies has its application in modern politics as well. The more closely political activities are based on organizations used for mundane purposes, the more particularistic the thrust of the ideological arguments and the more materialistic the goals. That is a way of saying that political factions that arise out of and use the organizational resources of businesses and banks, trade unions, or social clubs will have ideologies that identify the moral community with their particular organizational interests, and define their goals in terms of worldly goods. Where political factions are organized, on the other hand, around special-purpose organizations based on voluntary recruitment and setting themselves off from the material interests of other organizations, they tend to argue in terms of what is right for the entire world, and of actions which are to be judged in and of themselves (perhaps as contributing to an ultimate utopia) rather than by their material consequences. This is the case with political organizations based upon universalistic cultural organizations (churches, student politics, and above all, the politics of seminarians), and of the more cosmopolitan and intellectual professions generally. It also includes parties that organize themselves most like a religion, shunning ties with other organizations and claiming moral self-sufficiency, above all the radical parties of Left and Right. In contrast, there are the conservative and mildly liberal parties embedded in the class and status group organization of existing society, maneuvering for practical advantage and making moral claims primarily as a means of mustering their own constituencies.

21.4 The secular or transcendental nature of ideological beliefs is determined by the degree of cosmopolitan contacts within the group's experience.

21.41 The *lower the diversity of communications* experienced by a group, the more likely it is to reify its symbols and regard them as existing in a spiritual world transcending physical reality (from *2.2*).

21.42 The *higher the diversity of communications,* the more likely the group is to regard its symbols as human creations referring to goals that can be achieved in the physical world.

What we mean empirically by a transcendental or religious belief, as opposed to a secular one, is a difference between the way in which people talk about certain ideas and symbols. To refer to the name of a god one believes in is to say a word that does not fit with ordinary discourse; it is not part of the empirical world at all, and cannot be discussed in the same terms. This is so even though the god supposedly may have very decided preferences for worldly arrangements, and may interfere in the world, undergird it, or make appearances in it. Religious language always emphasizes the quality of going beyond things that have an empirical referent, and reserves the most *powerful* reality for just those qualities that cannot be talked about.

The most important change in political ideologies of the last few hundred years was the decline of religious and the rise of secular ones. Instead of proclaiming their power in terms of the spirits they propiated or the gods that supported them, rulers came to claim that they represented the nation, the people, and/or certain fundamental rights or principles for the correct administration of human affairs, whether these might be order, property, freedom, equality, or the common good. This shift still allowed for both fervent and complacent, moralistic and materialistic, hierarchic and egalitarian kinds of ideologies. What it did change was the form of political maneuvering. Politicians now had to explain their policies at great length (although not necessarily with any great accuracy), instead of merely leading religious rituals. The essence of the change was that politics became much more verbal, and speeches and debates became the major ceremonial maneuvers. The problems of maintaining and challenging political power thus came to center around making plausible arguments, including such background features as restricting speakers to be heard and topics to be raised, and controlling information.

Universalistic religions are one major step toward this kind of secular ideology, since they are usually religions of the book and the doctrine, religions of writing and theology as well as of symbols and ceremonies. The sermon is the antecedent of the political speech, and the universal ideals of religious morality are the basis of secular ideals of justice and right. The Republican ideals of Jacobin revolutionaries in eighteenth- and early nineteenth-century Europe tended to be carried in religious secret societies like the Masons (paralleling the populist political movements in Medieval Islam organized by Sufi cults); socialism (which has always been popular among liberal Protestant theologians) and Communism are secular extensions of the evangelical and egalitarian forms of Christianity.

Universalistic religions arise under the conditions of cosmopolitan societies, as in the ancient commercial cities amidst competing empires, where the diversity of contacts generated awareness of the transience of human arrangements and the ideal of a universal order. Hence, they are in a sense already more secular than traditional religions, with their multiple taboos and many gods whose actions cannot be explained. The universalistic religions do not completely eliminate reified symbols and pockets of transcendent reality, but they gather them all up in one lump and leave the rest of the world in its mundane clarity. The supernatural is sharply distinguished from the natural, and the latter becomes the deterministic realm of Karma, Yin and Yang, or God's Law. Mundane events are explainable in principle (just as they can be administered by bureaucratic rules), and only the center of the religious (and political) system is exempted from accountability.

What keeps the cosmopolitan conditions that produce universalistic religions from going all the way to producing purely secular ideologies? The answer, I think, must be in the fact that there are a number of different realms of social experience, so that cosmopolitanism in some areas may coexist with a low diversity of communications in others. In particular, there is the family, which is highly conducive to reified symbols because individuals are born into it helpless and thus with an initial sense of the arbitrariness and uncontrollability of power. Childrens' earlier world views are highly reified, and the longer and more extensive the control of the family over the individual, the more likely such world views are to be prolonged in religious beliefs sanctifying family structure. Universalistic religions, although they set up ceremonial communities outside of the bounds of families, usually exist within a world in which families are still important economically and socially (if not always politically). Universalistic religions appear in complex societies that have a cosmopolitan political order outside the family, but with an underpinning for symbolic reifications in the continued importance of the family unit for cultural training and social status. The universalistic religions usually make an explicit alliance with family authority, regarding family ties as among the most sacred moral obligations; this is most obvious in Confucianism and Christianity, and in the quasi-universalism of Brahminism and Judaism.

What gives rise to thoroughly secular ideologies is a social situation that not only is cosmopolitan on the public side, but has moved private status from the arena of the home to a world of freely chosen social gatherings. This occurs especially in affluent urban situations where public entertainment and mass media of communications are developed (as we have seen in considering the history of deference and demeanor). Of special importance is the rise of schools for teaching a literate culture, for these remove the individual from the family at an early age, and put him into a specialized status community of other students. Schools also provide the organizational basis for intellectuals to detach themselves from religious commitments and develop the logic of extended argument in its own right. Thus, the first really secular culture emerged in the Hellenistic cities, in conjunction with an educational system that had taken over from families the marks of cultural status (Marrou, 1956). The other famous period of secularization occurred in Europe from the sixteenth century onward, and again was associated with the rise of independent intellectuals and education forming a status culture that transcended particularistic loyalties and appealed to classes that were becoming more and more cosmopolitan. Other flurries of secularization popped up in the histories of China, Islam, and elsewhere, always in conjunction with cosmopolitan urban centers and their intellectuals and situations of public status competition. The spread of Western secularization around the world over the last hundred years has repeated the pattern, taking hold in the urban schools and under the leadership of intellectuals.

The political impact of this shift has been that states and politicians have had to shift their claims for legitimation from churches to schools. Schools replaced churches as sources of bureaucratic officials, but also as places where the battle over subjects' loyalties was carried out. The imposition of mass compulsory education can be understood only in this light: for schools may not be very effective in imparting technical knowledge, but they do serve well enough in drilling students into loyalty to the state. Thus, we find compulsory education beginning in the eighteenth century under Prussian absolutism and its imitators, and in the twentieth century it is used as the primary means of indoctrination and control by the Communist states. The more a state can depend on religious legitimacy and traditional organizational ties, on the other hand, the less its rulers feel the need to depend on massive state education, as evidenced by Britain up through recent decades, as well as by the more traditionalistic parts of the world.[6] The shift to educational legitimation is not without its dangers. Schools serve to mobilize students and intellectuals in a particularly moralistic and volatile setting, although this is much more true of certain types of schools than others (Ben–David and Collins, 1967). New ideologies challenging existing authority have arisen mainly in these places. Students and intellectuals by themselves, like priests in traditional societies, are not necessarily a very serious threat to a regime, as long as it maintains firm military control, and all the more so when other social groups are stably coerced or bargained with. But when these supports are threatened, such as when intellectuals associate themselves with a mass movement of dissatisfaction with the regime, they become a serious threat to the regime's legitimacy; and if this coincides with a break in military control—in the event of severe economic crisis, a split within the elite, or a defeat in war—the result is likely to be revolution.

21.5 The intensity of commitment to an ideology depends on the social density of interaction and the level of physical threat.

21.51 The more that *members of a group are in each others' physical presence,* the more emotionally intense its ceremonies, the more committed to the group are its members, and the more vehemently they punish members who violate its ritual requirements (from *2.1, 4.7,* and *4.85*).

21.52 The more that *members of a group are threatened by violence,* either from its own hierarchy or from outsiders, the more intense the commitment to the group and its symbols (from *4.83* and *6.4*).

The high-intensity ideologies, whether secular or religious, are found where men are continually together physically. This is the case in the men's house of tribal warrior communities and in the traditional household, in isolated military camps, monasteries, and modern cadre political parties. The totalitarian regimes maintain an atmosphere of total commitment by frequent parades and rallys, and by keeping individuals in the midst of group activities with the same companions. On a small scale, the same atmosphere is created in the school or college which can mobilize its students in mass athletics and cheering sections, assemblies,

---

[6] The massive educational system of the United States, much of it compulsory, is the result of several pressures. One of them was the effort of the original Anglo-Protestant settlers to maintain cultural preeminence over a massive influx of alien immigrants. Once such a structure was established, pressures from below for social mobility operated to further expand the system, especially at the noncompulsory higher levels. As usual, we are dealing with multiple determinants. In any case, it is apparent that compulsory public education is more effective in inculcating unthinking political loyalty than it is in teaching academic knowledge (Hess and Torney, 1967; Collins, 1971).

and group activities.[7] At the other extreme, it is situations of privacy that reduce ideological intensity, hence the generally lower key of political consciousness where modern affluence gives individuals more room to themselves.

Ideological intensity is also affected by conditions that make men emotionally aroused. The most important of these for politics is physical threat, for it is when people are in danger (including when they are challenged to fight) that they turn to each other for ritual support (*4.83*). The obvious case is the increase in national solidarity in time of war, or of party loyalty during an electoral campaign. But the same result may be brought about if the threat is internal to the group. Durkheim (1938) observed that attending an execution or other ritual punishment enhances solidarity within the group, and this is so whether the group is egalitarian or highly stratified. In either case, the threat is in the air, and people fall back on ritual self-discipline to avoid having it fall on them. Since political power is ultimately a matter of a leader being able to scare his followers into disciplining each other, rituals with an implicit threat to their participants are very prominent in politics. The danger of dealing with sacred objects, long noted by students of religion, is probably due everywhere to the atmosphere of political coercion (but sometimes also dangers in the physical world) with which rituals are associated.

Since threats and resulting ritual solidarity are so effective in politics, it is not surprising that much political maneuvering is an effort to invoke, seek out, or perpetuate threats. The revolutionary or wartime leader can prolong his dominance to the degree that he maintains an atmosphere of simulated wartime mobilization; hence, cold wars may often be expected after hot ones, and violent revolutions are conducive to postrevolutionary authoritarianism, especially in proportion to the degree of counterrevolutionary threat. That is to say, the degree of ritual solidarity is not determined only by the current threat at any one time. Politicians with the resources for controlling information and the skills to effectively dramatize conflict can prolong the power that comes from an emergency, or generate that power by making future conflicts seem imminent. This tactic can be used to enhance various kinds of economic, political, or status interests. The extent to which it is used depends on the resources available for carrying it off, and on the apparent chances of success in following this path as opposed to alternative ones. Both of these conditions are higher the more centrally the military figures in a state.

---

[7] Hence the tendency to political conservatism among athletes. This is probably most intense in the team sports with the most militaristic discipline, and especially in amateur sports, which resemble religious rituals most because players are supposed to be motivated by loyalty to team and school rather than individualistic goals like money.

The Internal Politics of Complex States

When the military structure of the state becomes supplemented by a civilian administration, political maneuvering becomes more complex, and it becomes possible for a number of interests to try to use the state's coercive threat for their own interests without engaging in actual warfare. This is particularly so where (as is usually the case) such states are built around urban centers and commercial networks, and where churches are organized outside of households. This does not mean that external wars and internal fighting become irrelevant, but that they are supplemented by maneuvers for political influence on the part of soldiers, aristocrats, bureaucrats, priests, landowners, merchants, manufacturers, and workers. Internal politics is added to external, and peaceful coalitions form around military ones.

Marx began this kind of analysis by picking out classes of interests, which he attempted to link to positions in the structure of material resources, especially the means of economic production. From this starting point, Weber, Michels, and subsequent sociologists have elaborated on the varieties of positions that shape political interests. Weber's distinction among economic class, cultural status group, and political party (organization or faction) is a famous way of indicating some of these complexities. But Weber's model can be confusing if we do not realize that *everyone* has all three kinds of interests. Everyone is concerned about material wealth in some degree, just to stay alive; churches need their supplies just as much as businesses, and the state both needs material resources and is a means to material comfort for politicians, soldiers, and administrators. Everyone would like to be given as much status honor as possible, to have the community and its rituals and ideals centered around himself and the things he is good at. And everyone is concerned with power, since all wish to avoid being coerced and, if it comes to a choice, to be on the side doing the coercing.

This is not to say that everyone does equally well on all dimensions, or that they take the same routes toward getting what desirable things they can. The differences are exactly what we mean by stratification and organizational structure. Some people in some historical circumstances are in a position to maximize honor, wealth, and status simultaneously; others have had better opportunities in some spheres than others, and some have had low resources on all three. Weber's model is useful for giving us some guidelines among the myriad interests people may pursue and for suggesting how they mesh into coalitions that struggle over control of the state.

Economic Class Interests.   Coercion is the ultimate sanction in establishing property, whether over land, goods, or labor rights; given the interest of any centralized ruler in maintaining a system of economic communications (for military resources if no other reasons) the state will be interested in upholding and legally regulating currency, credit, and business contracts. Hence, every possible economic interest is a potential political interest, since the state can be a great aid or a deadly enemy. Conflicts among economic rivals are most quickly politicized, for fear of the other side acquiring an advantage by reaching the state first.

There are a great many such conflicts, and I will only list them here.[8] Since all property is enforceable by coercion, *owners* wish the state to protect them, whether from isolated robbers or expropriation by the propertyless classes; they also wish to be protected against factions in the state itself who threaten expropriation, whether totally, or only partially through taxation. This is a collective interest of all owners. They also may use the state to enhance their property, by giving them gifts, monopolies and other forms of protection of commercial property, conquering land, and so forth. Here we shade into more individualistic property interests, as particular businessmen and landowners struggle for political advantage over each other.

Employers and *employees* attempt to use the state to regulate the terms of their relationships, whether using the state to enforce slavery or serfdom, to break unionizing efforts, or to guarantee work conditions, union control, and wage levels. An extreme form of state-enforced advantage for employees is one in which the client is an unorganized public, and the workers are given a guild or professional monopoly. The basic structure of labor relations upheld by the government is a collective interest among at least large sections of employers and employees; where particular working groups acquire strong monopolies, it becomes largely a matter of individualistic factions struggling for their own advantage.[9]

Debtors and *creditors* are another set of interests that can emerge out of a situation of property, whether in the case of land or other goods being rented out, or of money (abstract credit) being lent. There are collective interests among debtors (of each kind) for easing the terms of their debts with the aid of the state, whether by fixing rates of rent or interest, manipulating currency, or outright repudiation. (In the case of rent,

---

[8] This analytical typology is derived from Weber (1968: 927–932). Wiley (1967) has recently revived it to show how class conflicts have indeed been prominent in American history, especially on the dimensions of creditor–lender (the agrarian politics of the nineteenth century), and seller–consumer (the ghetto riots of the 1960s).

[9] Notice that what enables someone to become an employer is possession of property, with which he pays or otherwise motivates other people to work for him. Thus, the interests of owner and employer tend to coincide in the same individual.

repudiation amounts to expropriation.) On the other side, creditors have political interests in the opposite direction. There are also individualistic interests here, as lenders may compete with each other over favorable opportunities.

Sellers and *consumers* are a final set of variants on property interests. In addition to their general interests in protection as property-owners, sellers push for a favorable state attitude in the realm of prices and competition, in opposition to consumers' interests in keeping prices down and quality up. There are a number of different strategies that sellers may pursue: depending on the situation, they may prefer guaranteed high prices or freedom from regulation; government-enforced or -countenanced price fixing, or unlimited competition. The more particular determinants need to be worked out there.[10] In general, this is an area in which there may be a great deal of struggle for individual advantage from the state, rather than collective actions versus consumers—particularly since the latter are rarely organized as a coherent group.

Power Interests. Everyone is concerned with his direct position vis-à-vis the state, just in terms of whether he is subject to arbitrary coercion, or guaranteed certain autonomies or a share in the power. The broadest power interests are over matters of citizenship, franchise, or independence; where these issues become dramatized and the relevant groups are mobilized for action, the result can be the bitterest and most fervent of political conflicts (as one would expect from *10.1*). There are also highly organized political groups within and around the state itself: military factions, maneuvering for power both among themselves and against civilian administrators and politicians; different branches of government administration (as in modern interagency conflicts); battles among executive, legislative, and judiciary powers; disputes between centralized and decentralized centers of organization; and struggles both between political parties and within them (such as Michels brought up in his Iron Law of Oligarchy). Each of these organizational situations offer its occupants a chance for moving up in individual careers, or enhancing the power of the position (and thus hopefully its wealth and prestige). Whichever strategy the individuals involved may choose, such power struggles tend to be relatively particularistic, unlike the broad collective interests involved in the basic forms of

---

[10] These probably have to do with how concentrated resources are among buyers and sellers in each case, and whether or not the most powerful individuals feel they can achieve monopoly by driving others out of business rather than by using a seller's cartel. Compare agricultural products with industrial ones, or industry in nineteenth-century America with French or German industry.

property, or the fundamental political questions of citizenship and of state boundaries.

Status Interests.   Groups that are mobilized as communities, and thereby have been able to carry on rituals and create commitments to a set of ideals and symbols, are all at least implicitly concerned with politics. At a minimum, they need to protect themselves against annihilation or prohibition by the wielders of organized violence; at a maximum, they can use the state to support their community and broadcast its ideals, give its members special privileges, exhalt them over others, or even destroy rival communities and impose their own by force. Since all men create their subjective realities through a set of symbols upheld by social rituals, there is always the temptation to make one's reality absolute by prohibiting any rituals that reduce its believability, or punishing deviants from ritual practices. We have seen a great deal of this already in the historical struggles over styles of deference and demeanor and in struggles to impose state religions; the same interests survive today in the form of ethnic groups and the pseudoethnic cultures produced by education.

Which interests will be in operation at a given time, and what determines who will win what? Marx's abstract assertion that history is the record of class conflicts can be taken as true. But as we can see, there are a great many kinds of "classes"—arising from economic, political, and status organization alike—and the conflicts at any moment do not reduce to the simple division popularized by vulgar Marxism. Marx's own solution was to stress the long-term structural changes producing and mobilizing interests, and to concentrate on predicting when the great collective interests would override individualistic striving and bring about revolutionary confrontations. This model is on the right track, if we guard against the naïve evolutionism that underlies Marx's own historical theory. We would like to predict *all* the political conflicts, not just a few, and we wish to be able to account for a number of divergent historical paths. Obviously, this will be a rather complex task. Here I will give only a few general hypotheses and illustrations.

The most important thing to bear in mind is that everyone is interested in power, material goods, and status simultaneously. The way in which the last combines with the others is particularly important. Ethnic groups or religious communities happen to be ways that people can group together to struggle for wealth and power as well as subjective status. As such, they are prototypes of almost *all* of the effective groupings to influence the state; for as we have seen in discussing ideology, leading a ritual community is the only way in which power can be stably organized. The major route to power, whether one wants it primarily for protecting an economic interest, for the sheer enjoyment of political position, or for glorifying one's ideals, is by making one's network of threats and rewards into a ritual community claiming to state the moral ideals and the symbols of reality which are incumbent upon everyone. This is what I believe Marx recognized in the notion that the ruling class has the higher class consciousness, that it is a *Klasse für sich* as well as *an sich* (for modern evidence, see Mann, 1970).

Now let us attempt some general principles.

22.0 Marx's mobilization principle. The political power of an interest group is proportional to its degree of mobilization.

22.1 The greater the material resources controlled by an organizational position, the more likely political influence can be exercised on behalf of that position's interests (cf. *20.1–20.4*).

22.11 The greater the wealth of an individual, the greater his bargaining resources for political support.

22.12 The greater the resources for transportation available to an individual, the more he is able to participate in political bargaining.

22.121 The more sharply resources for transportation and communications are restricted to one class of interests, the more politics consists only of individualistic disputes within the existing system of property and organization.

22.122 The larger and more geographically dispersed the population of a state, the more resources for transportation and communications are needed to participate in politics, and the more effect the distribution of transportation and communications has on the distribution of political influence (from *16.1–16.2*).

Marx (1969: 61–62, 123–124; Marx and Engels, 1947: 27–43; 1959: 7–20) proposed that it was not sheer numbers that counted in politics, but effective mobilization. The reason that a tiny feudal aristocracy could control great masses of peasants was because only the former had the resources to move about and take part in the military coalitions and court intrigues; the latter had only the most local contacts, and hence, their numbers equaled only an artificial unity, "much as potatoes in a sack form a sack of potatoes." The material conditions of agrarian society made the aristocracy the ruling class; the shift to industrialism brought a shift in power because the business networks of the bourgeoisie made them even more mobilized than the aristocracy. The appearance of the working class on the political scene was produced the same way: The large factories of advancing industrialism brought the workers together so that their numbers could count, and thus the capitalist system was to supply its own gravediggers.

The general model is correct. We must bear in mind, though, that there are several forms of mobilization. The possession of wealth (or, in the case of military politics, suitable weapons (*20.2*)) is a mobilizing resource. To bring these to bear requires transportation and communications (just as in the problems of controlling any organization (*13.5–13.54*)), all the more so when the state covers a large population or territory. The population in primitive horticultural and in pastoral tribes was generally mobilized, hence a high degree of political participation. In agrarian societies with a large-scale ruling network, most of the population lacked the resources to have any influence on the remote centers of power, and local uprisings could be easily crushed by bringing in outside forces. Industrialism again shifted the resources, most especially in the areas of population density and availability of mass communications, mobilizing both an urban bourgeoisie, and more sporadically, the workers. Above all, the fortified household was broken down, and the disappearance of the barriers that had isolated servants from each other allowed the mobilization of previously fragmented interests. As subsequent research has shown, it is where workers are together in large numbers, and less interspersed among other classes, that they are most engaged in conflict along class lines.[11]

22.2 The greater the resources for organizing itself as a status group, the greater the political influence of a collection of individuals.

22.21 The higher the level of intercommunication and the rate of physical contacts among a collection of individuals, the more likely they are to formulate distinctive ceremonies and symbols and hence to become a status community (from *4.7*).

One way in which the more mobilized interests are able to maintain their political influence is by acquiring a sense of community that goes beyond their particular disputes. This happens because the sheer volume of common contacts and the experience of giving orders tends to generate a distinctive outlook; the more one participates in this network, the more intensely does its symbols become one's reality and one's focus of loyalty (*21.5*). This collective consciousness of the elite is sophisticated rather than reified (from *2.2*), but it is a collective outlook nonetheless, especially

---

[11] See Portes (1971); Kerr and Siegel (1954). Paige (1971) shows that participation in the American ghetto riots of the 1960s was highest among highly mobilized blacks; Burstein (1972) shows that for all classes, location within social networks is the main determinant of political participation, and that class, age, and sex are related mainly because they affect the chances of mobilization.

in opposition to the reified symbolic commitments of the noncosmopolitans whom they dominate. It is a commitment to a life of manipulating the symbols that others believe in, and guarding the backstage of what they are doing (*18.212* and *18.221*). This gives the members of the elite a form of mutual interdependence that potentially opposing groups rarely possess, and helps translate their resources for greater mobilization into a stable monopoly on power. The same thing is true to the extent that the mobilized group has ethnic, religious, educational, or other cultural unity. Opposing class and power interests become more threatening when they are culturally unified as well. This, of course, is implied in the theory of pluralism, with its focus on the opposite case, in which groups opposed to a system of power are too fragmented ethnically or otherwise to mobilize effective opposition (Gusfield, 1962).

22.3 The greater the resources an interest group has for leading and shaping the community rituals of the relatively less mobilized interests, the less the chance of countermobilization.

22.31 The more a community is organized around a single compulsory ceremonial center (a household religion, a state church, a compulsory school system), the greater the power for the interests who lead or influence the ceremonies.

22.32 The greater the control by an interest group over information about political events, the greater its political influence.

22.321 The degree of diversity of viewpoints from which political information is publicly broadcast is determined by the diversity of interests with effective political influence.

In agrarian societies, one source of the power of the heads of fortified households is their position as ritual leaders, demanding continued shows of ceremonial loyalty by their subordinates. The servants were not only prevented from forming an autonomous community of their own but were forced to acknowledge their masters' version of cultural reality. The state church surrounding the ruler had a similar effect. Information on politics, in that age of meager communications, was likewise monopolized by the upper classes. In industrial societies, it is much more difficult to impose ceremonial participation in a single system upon everyone; the totalitarian states do this only with a great commitment of resources. All governments, though, make use of a milder version in the form of education, patriotic ceremonies whenever crowds are gathered, and so forth. Which interests can influence the content of these depends on their particular resources. The cheapest and most direct form of political influence of this sort is by pressure upon the mass media (Breed, 1955, and Tuchman, 1972, indicate some of these mechanisms). This is hard to assess precisely in democracies because of the highly partisan nature of complaints, and the fact that there usually is some variation in news slants. The only way to get beyond the particular subjects of controversy is by dealing with general causal principles. I offer the following: The range of what is considered "respectable opinion" will vary with the range of effective political interests with an immediate chance of gaining state power. The ruling ideas of every period are those of the groups who control the means of intellectual production, and the existence of competing ideas depends on what competing controls there happen to be.

22.4 The more that the mobilized individuals share a collective interest, the more likely the state is to uphold that interest.

Collective interests, like upholding or destroying a system of property, obviously command much more agreement than individual interests within a class, such as seeking franchises, monopolies, or appointments. Collective interests are easily translated into joint action by their mobilized partisans; for individual interests to be translated into influence requires either that an individual have enough personal resources to be effective, or his participation in a system of individual bargaining, which thereby gives power to political brokers.

22.41 The more highly individuals are interdependent in a system of markets or communications, the higher the general level of mobilization for political conflict within that community.

Being highly interdependent (reciprocally or sequentially interdependent, in the terminology of Chapter 6) is by no means the same thing as having a common interest. Which economic interests are most active politically, for example, depends on which ones are directly exposed to the market. The manor or hacienda system in agriculture worked by peasants producing only for themselves and their aristocratic lords kept the peasants from experiencing the fluctuations of the market, and hence kept them uninterested in the politics of the larger world (Stinchcombe, 1961). A system of family-owned farms producing for the market, on the other hand, has highly mobilized farm politics (as in the history of the United States), responding to fluctuations in prices and credit. Tenants of absentee landlords (like those of prerevolutionary France and China) are in the most extreme case, for they take the brunt of market fluctuations while paying constant sums to the landowners who are protected from risk; here the class interests are not only highly mobilized but have an immediate target, and hence, crises can produce peasant revolutions aimed at expropriation. The same kinds of principles probably apply to different kinds of urban business and labor interests, including professions.

22.5 The greater the crisis, the more interests are mobilized for political action, and the greater the potential for a change in government policy or structure.

Crises are events that threaten a great many interests, and motivate people to make the extra effort to mobilize themselves even if it is ordinarily too costly to do so. It is for this reason that crises are so important in political change. Ordinarily, the advantages of the best-mobilized groups are such that only imperceptible advantages seem to follow from considerable efforts on the part of groups that have fewer organizational resources. Crises change the nature of politics because they require negotiation among many more groups than usual, and in an emotional atmosphere conducive to rapid realignments in power. Hence, it is not surprising that so much of political history occurs in episodes, or that men who want sharp changes look for crises in which they can seize their moment.

The analytical problem is that crises mobilize very diverse interests. What gives them any coherence, so that the result of crisis is usually a restructuring rather than anarchy? I believe the answer is:

22.51 The more severe the crisis, the greater the tendency to polarize into two-sided conflicts, between the supporters of existing ceremonial order and its opponents.

The crucial device in maintaining control over the state, as we have seen, is the collective ceremony of deference to its symbols and, hence, an implicit show of one's loyalty to the wielder of those symbols. The key challenge to power, then, is to destroy the legitimacy of the head of state. Whatever the myriad interests of those who want to influence the state in their direction, the first question in a crisis always comes to whether or not to uphold the ceremonial structure of the moment.[12]

This gives us the most characteristic factions found in any crisis. There are the *conservatives,* the upholders of existing power. There are the *ultraconservatives,* whose attitude is: "The men in power are incompetent or soft or corrupt and are selling out the true ideals, but we will support them for the time being in order to defeat those who would destroy our order." There are the *liberal conservatives,* whose attitude is: "I certainly don't agree with everyone who supports this order, and I think we may be too harsh on its opponents, but I will support it until the crisis is over and we are assured there still is an order." On the other side there are the *radicals,* who

---

[12] A wonderful illustration of how the two-sided ceremonial crisis can hide a tremendous diversity of mutually antagonistic interests is Barrington Moore's (1966: 70–92) treatment of the French Revolution of 1789–1795.

take the position: "Break down the existing ritual order, eliminate its supporters, and institute a new order with its own tests of loyalty." Finally, in either camp, there are the *conciliators*, who argue characteristically: (*a.* to the conservatives) "I'm not one of 'them,' you understand, but you had better avoid worse trouble by making some concessions to me if you don't want to have to deal with those others"; (*b.* to the radicals) "Don't risk a defeat and lose everything, but take what you can get for now." The rhetoric of the "lesser evil" is thus of the essence of political discourse, on all sides.

I have avoided using terms like "Left" and "Right," because the same dynamic seems to occur if the radical threat is from the Right, or in the maneuverings which take place after the radicals have taken power. In most cases, of course, someone nearer the center wins. It should also be borne in mind that the crises do not have to grow to revolutionary proportions for this alignment to take place; what *22.51* says is that this alignment increasingly overrides complex maneuverings among factions with the increasing intensity of conflict, which may range from a heated legislative session to a general insurrection.

Now what determines which interests will assume what position in this continuum?

22.52 The greater the similarity among the ceremonial ideals of particular factions, the more they will coalesce in a common position in political crises (cf. *1.1–1.4, 2.1–2.2, 21.1–21.52*).

22.521 The greater the interpersonal contacts among leaders of particular factions, the more likely they are to coalesce in political crises.

22.53 The greater the compatibility of individualistic interests, the more likely they are to coalesce in a political crisis.

22.54 The more collective interests are shared by mobilized groups, the more likely they are to coalesce in a political crisis.

22.541 The more severe the crisis (that is, the greater the threat to the political order), the more likely groups are to coalesce along the lines of collective interests.

The major collective interests center around a system of property or the two great issues in political structure, the extent of the franchise and the borders of the state. The latter two issues are often stated in ethnic or religious terms, since they involve questions of just what communities are going to be subordinated or included in the state and its ceremonial order. Most major conflicts seem to form along those lines. Marx correctly singled out the other key line of division, since property is the collective interest underlying most other more limited economic conflicts. Where two or all three of these lines of collective interest coincide (as in the case of alien property owners with monopoly on political rights), the situation is revolutionary in the fullest sense. Where the collective interests are more limited or divergent on different lines, a sweeping change is less likely. In these cases, concessions are often made to particular subfactions, usually the ones that are least threatening to the existent holders of power.

What causes crises? Often there are a combination of contributing conditions: Economic depressions, crop failures and famines, runaway inflation all have a mobilizing effect, but almost always as facilitating rather than overriding. (Hence, it is not necessarily the worst economic crises that produce the greatest political uprising.) The problem of succession after the death of the ruler is important where the system is highly centralized and especially where it is despotic. Extended conflicts within the ruling group are especially important, for they crack the stable structure of mutual fear that makes for automatic obedience. Internal uprisings that dramatically challenge the ceremonial order by making it appear either weak or more arbitrary than its ideal image have sometimes been catalysts to an already polarized situation among the more mobilized groups. Wars are the most thoroughly mobilizing events. When there is a dramatic-appearing external threat, a war tends to strengthen the power of whoever leads the army and benefits from being the center of ceremonial attention; if the system of weapons depends on massive initiative at the lower levels, however, wars tend to mobilize interests that have not been represented at the center. Most important of all, defeat in war has been the trigger for the most striking changes (including the French commune of 1871, the revolts and revolutions in Germany and eastern Europe in 1918–1919, Russia in 1917 and 1905, and China in the Japanese invasion); clearly enough, it is when the military core of the state breaks down that it becomes easiest to reconstruct in a new way.[13]

Which interests win in a crisis?

22.6 The *more mobilized an interest group for organized political action,* the more likely it is to take command of the state in a crisis.

In many cases, the most mobilized group is a conservative one, which was already tied to state power.[14] Just because there is a crisis and the

---

[13] If one adds regimes imposed or aided from without, there are the cases of Japan, Germany, and Eastern Europe after 1945, France in 1940 and 1945, and others.

[14] In the Fascist successes in Italy and Germany, the crisis consisted of escalating vio-

opponents of a regime are mobilized, does not mean that they inevitably win. In most cases they do not. But sometimes the regime is shaken and a new faction takes over. This faction is always the one that is best organized for the purpose; it is not at all necessarily the most radical group, and it is never an episodically mobilized group that may have done most of the fighting (or the demonstrating or voting, as the case may be). Those who make a revolution are rarely the ones who benefit from it, certainly not in terms of holding power.[15] Thus, the French revolutions from 1789 through 1871 were usually fought by the *sansculottes* but brought power to the bourgeoisie; the Russian revolutions of 1917 brought to power successively the liberal aristocracy and the radical intelligensia; and the ghetto riots of America in the 1960s brought the greatest benefits to middle-class blacks.

The group that represents the radically disaffected, and that benefits from their episodes of mobilization, are themselves always part of the highly mobilized upper or upper middle classes, but a particular faction within them. Almost always they are intellectuals, priests, or other men who depend only very indirectly on the existing organization of power and property.

22.61 The less closely attached a mobilized group is to managing or defending the major institutions of property and power, the more likely they are to seek political influence as the representatives of the disaffected and unmobilized groups.

Finally, we should note that the first group to come to power in a crisis does not necessarily stay there long enough to reap any benefits or make any permanent changes.

22.7 The sooner a new ruling group is able to end a state of political crisis, the more likely it is to stay in power.

Just what resources and strategies are needed to end a crisis depends on what kind of crisis it is. The state of war that provoked the first Russian revolution of 1917, or the atmosphere of chaotic internal conflict that brought down the French regime in 1789, was not ended by the groups

---

lence between Leftists and radical anti-Leftists, eventually polarizing the entire populace; the conservative center finally gave power to the ultraconservatives, as the group which drama-tized itself as best organized to deal with the crisis—defined as a threat from the Left (O'Lessker, 1968). A milder version of this dynamic, without actual street fighting, took place in the United States in the late 1940s through the 1950s.

[15] Guerilla wars, rather than urban uprisings or electoral revolts, appear to be the exception here.

that were in the position to take over first; hence, there were further revolutions until a group came to power with the resources to take the steps to restore order on a new footing. Obviously more propositions are needed here.

Marx (1906–1909) laid great stress on economic crises arising from technological unemployment and resulting overproduction, accompanied by drastic credit fluctuations.[16] The archaic labor theory of value upon which he based his reasoning has made it the subject of much justifiable criticism. But the same general conclusions can be reinstated by a neo-Keynesian logic of the sort used by Baran (1957) and Sweezey (1946). The trend of technological development in a market economy certainly *can* produce such crises. The political consequences, however, are more varied than classical Marxism supposed. An economic crisis of this sort does tend to mobilize a dissatisfied population, and to put power in the hands of those who are able to resolve the crisis. But there are a number of possible resolutions: piecemeal efforts at redistribution (such as by minimum wages, shorter hours, or the hidden welfare of government employment or education); Fascist state regulation leaving much power in the hands of private businessmen and traditional landowners; the liberal mixed welfare state; as well as a takeover by the ideological followers of Marx. The fact that some of these are more temporary resolutions than others makes them nonetheless empirically possible. What Marx's model of capitalist economic crises tells us, then, is not which particular political regimes will win out, but what sorts of recurrent political issues will be found in capitalist societies. How they will be resolved is explainable in terms of the social and ideological organization of interest groups.

THE STRUCTURE OF DEMOCRACY

This interest-mobilization theory applies to many different forms of politics. I will conclude by attempting to show how it can explain not only the policy of a government but its structure—in this case, the structure of democracy.

The core of democracy is a collegial structure of authority, in which an assembly of equals meets periodically to make decisions on state policies, long-term principles (laws), or the delegation of executive authority.

---

[16] Not all politically mobilizing economic crises are of this sort, however; bread riots caused by wartime disruption or crop failure have been even more prominent in political history. Marx was less interested in these because he sought a long-term prediction which followed from inevitable principles of development.

This may take a great many different forms, ranging from mass voting to legislative assemblies of notables. Such collegial structures can include a large or small proportion of the population, and they can be mixed with various degrees of hierarchic administration, as when a college of notables advises a king, or a mass plebiscite elects (or affirms) the ruler of a bureaucratic administration. As usual, we are dealing with a range of variations, and the task is to explain how strong the collegial control is.[17]

There are a number of historical examples. Primitive horticultural societies and pastoral societies are generally collegial in government, with the leader a first among equals; this is related to the nature of military organization (*20.211*) and the small scale of affairs (*16.1–16.2*). The ancient city democracies of Greece, Italy, and possibly northern India and elsewhere, were war coalitions of tribes, which gradually increased their scope of citizenship from the tribal heads to all military participants as the nature of warfare came to stress mass-participation (but self-equipped) armies (*20.211* and *20.24*). The cities of Medieval Europe (but not those of the Middle East or the Orient) were ruled for a period by assemblies of the leading merchants; this seems to have been facilitated by the feudal fragmentation of the countryside and the defensability of civic fortifications, and by a general carry-over of tribal traditions in the form of collegially administered laws. Such carry-overs were important also in giving European feudalism—the decentralization of power among individually armed knights—a collegial form not found elsewhere in the world, with the knights regarding themselves as a corporate body with rights and autonomies preserved from the king, and the power to withhold assent to his collective projects. Finally, we have the various forms of modern democracy, growing out of traditional European institutions (such

---

[17] This may seem an unconventional definition of democracy; the term more commonly implies, among other things, mass participation in political life, a wide franchise. The important thing in any definition is that it should be clear, it should not include any hidden dimensions, and it should be fruitful in pointing us to crucial variables. The usual definition of democracy is weak on all of these counts. Above all, it points to mass participation while ignoring the crucial difference in what kind of structure of power—autocratic or shared—there is to participate in. What Bendix (1964) calls "plebiscitarian" rule—the affirmation of an autocratic ruler by a show of mass support—is structurally different from a collegial form in which power is split among some community of equals. The things that make democracy seem attractive—safeguards against arbitrary rule, effective representation of multiple interests, genuinely shared power—are results of collegial structures, although these have often been taken for granted by political theorists familiar only with the last two centuries of Western European politics, when issues centered around widening the franchise within such institutions. My choice of a definition is on strategic grounds, that it points us to crucial explanatory variables. As we shall see later, the degree of participation can be treated as a further set of variations within more basic patterns.

as the aristocratic parliaments and the representative assemblies) but expanded to include much wider or more continuous participation in government.

It should be apparent that the collegial form is not linked to any one form of economic organization; all types, including modern industrialism, have both more collegial and more despotic types. Nor is there any smooth trend; the record in any part of the world has its ups and downs. What determines how much collegial power there will be? We have already met the most general principles in considering organizations and the Iron Law of Oligarchy (especially *16.3*): Collegial power can be upheld where the administrative resources are divided approximately evenly among separate but interdependent groups. Notice that the formulation is congruent with the principles of mobilization in politics: It is the mobilized groups which must have autonomous resources, not the population in general; and the degree of resources needed increases with the size and geographical scope of government (*22.122*). Tribal democracy, tribal coalitions, and the dispersion of powers in the military situation of medieval Europe—*together* with the availability of institutional forms for creating horizontal coalitions to resist hierarchical authority—are fairly direct applications of these principles.

The last case suggests an important historical element. Organizational forms deriving from past conditions may be maintained even when those conditions change, if the interest groups that benefit from them have more resources than the groups opposed. The European knights and burghers alike resisted the imposition of patrimonial hierarchies because they had collegial forms still in use from the days of Nordic tribal organization (and more remotely, from the tribal coalitions that created the Mediterranean cities whose traditions were institutionalized in certain ideas of Roman law). The newer ways these forms were used were not the same as the earlier ways, but it is because conflicting groups fought to maintain or change the forms they had that these types existed. In this way, the Medieval estates and the civic corporations survived, even when amalgamated into the centralized kingdoms that emerged when control resources shifted toward the king and his bureaucracy.

The various fortunes of different parts of early modern Europe—the centralized despotisms of Spain, Russia, and Eastern Europe generally, the parliamentary regime in Britain (and its even more decentralized transplants in North America), the unstable mixed of centralization and collegial control in France—were due to the struggles which enabled rulers to destroy aristocratic and civic autonomies in some places, while victory went to the other side elsewhere, and in yet other cases the battle was indecisive.[18] Often the resistance was overt and military. English parliamentarism survived through revolts of the barons and through a civil war begun by the minor aristocracy; the basis of modern French democracy, unstable as it is, was laid by a last-ditch revolt of the aristocracy against the encroachments of royal bureaucracy, leading to a revival of parliamentary government from which the bourgeoisie were ultimately to benefit. Everywhere, centralization has proceeded as the resources for bureaucratization have become available; modern collegial forms have intruded on this through crucial episodes of revolt.

The coming of pervasive commercialism and then industrialism once again changed the balance of political forces, creating new interests and mobilizing new groups, as well as giving new opportunities to the older ones. Barrington Moore (1966) has made the most extensive comparison of the political structure emerging from different types of confrontations. There are five main interest groups involved: the central government and its bureaucrats; the landowning aristocracy; the peasants; the bourgeois industrialists and merchants; and the industrial workers.

The *bourgeoisie* is one fulcrum of the system, for it is the major group that prefers democracy. Presumably this is because it is the form of government which they can most easily influence; in an urban commercial situation, they are the most mobilized class, operating a business system that is essentially a network of communications tying them together and linking them with the government via the system of banking. They also benefit from the underlying necessity of protecting their interests if economic crises are to be avoided; hence, they can rule indirectly, without having to engage in much coercive manipulation of the government. They may not always support democracy, however; if the threat of socialism is strong (because of a revolutionary working class or strong pressure from government bureaucrats), they act to protect the private property that gives them their autonomous base of power, wealth, and honor, by allying with a fascist regime. Fascism is used here as a generic term for a modern, highly mobilized version of authoritarianism under the command of military and landowning classes;[19] by becoming a junior partner to such a

---

[18] The classic analysis, concentrating on the growth of royal bureaucracy in France and the survival of the decentralized institutions of the nobility in England is Alexis de Tocqueville's (1955; originally published 1856). Huntington (1966) extends this to America; Rosenberg (1958) shows how bureaucratic absolutism extended itself in Prussia, and provides an introductory overview of all the European states. Another classic treatment is Hintze (1962; originally published 1900–1920), which shows how European constitutionalism was linked to the legal traditions of feudalism, and behind that, to the tribal forms that made feudalism in Europe distinctively collegial.

[19] Again, we encounter a definitional problem. It has been argued that "fascism" is an

regime, the bourgeoisie must give up much honor and power to other classes, but it retains its property. Hence, fascism is preferred to socialism, which would eliminate it completely as an autonomous class.

The *landowners* can go two ways. They favor fascism as a means of retaining their old honor and privileges against a new order dominated by the bourgeoisie, by their former peasants, or by urban workers; but they may also favor democracy if they are the mobilized class which benefits from it, and their main opponent is the central government.[20] Which way they go depends on how commercialization reaches their agricultural property. (*1*) If the landowners go directly into the market by selling the produce their peasants traditionally have given up to them, the result is likely to be fascism; the landowners become tied to the state in order to maintain and increase their traditional hold over the peasants. In this variant, the landowners are a highly mobilized class, and the peasants are a very impotent one, tied down under the close attention of their masters and simultaneously under the distant surveillance of the state. This is the formula for successful fascism. (*2*) Where the landowners are absentee rentiers, however, their fascist tendencies are less effective, for the peasants are mobilized by the market, and the rentiers become increasingly isolated (Stinchcombe, 1961). (*3*) The other possibility is that the landowners themselves go into modern capitalist farming, eliminating both peasants and renters and working the land with hired labor. This makes them rural capitalists and brings them into amalgamation with the urban bourgeoisie, and thus more likely to favor democracy.

The *central government* and its bureaucrats have two main preferences.

---

historically very specific phenomenon, consisting of secular, nontraditionalistic, militant anticommunism, and hence applying only to Germany and Italy (and perhaps the *Action Francaise*) of the early twentieth century; or again, that fascism should be sharply distinguished, as a regime which mobilizes the population for mass ceremonial support, from traditional authoritarianism which aims at keeping the populace depoliticized. I have retained Moore's usage, however, because it points to a crucial variable underlying these several types of authoritarianism: the underlying interests of traditionally privileged classes in retaining their powers, especially over the labor force. Whether the methods they use involve mass mobilization or depoliticization, or whether their legitimating ideals are traditional or newly created, and just whom they see as their ideological enemies, are variations on a common set of conflicts; and given the fact that industrial societies have intrinsically more mobilized and secular populations than traditional ones, all efforts at preserving traditional privileges in industrial societies must have certain things in common. That it is possible to deal with variants within a common rubric called "fascism" is demonstrated by Lipset (1960: 127–179).

[20] Strictly speaking, this is not Moore's position, which concentrates on economic relationships, and tends to assume that the aristocracy per se is always the enemy of democracy. I have broadened his model to include the Tocquevillian case explained earlier.

The first is for traditional authoritarianism (fascism), which keeps the prestige of the old ruling classes high; this is especially likely where the military sector of the bureaucracy is drawn from the landowning aristocracy which is commercializing agriculture by intensifying traditional labor discipline. The second choice is for socialism, because it gives great power to the bureaucracy, and because it is preferable to bourgeois democracy, which puts the center of honor and power outside its own ranks. This choice is likely where bureaucrats are drawn from an educational system rather than from the aristocracy per se (by *22.61*).

The *peasants* are a largely unmobilized group, and tend to be pawns of other forces. Their revolt against absentee landlords, though, can provide the shock troops for socialism organized by would-be government officials (as in China and Russia), or for bourgeois democracy (as in the French revolution). If they succeed in expropriating the land and becoming small commercial farmers, they become a section of the bourgeoisie; as its least mobilized group, though, they benefit least from democracy and hence are the first to join the fascist coalition against the threat of socialism and its accompanying ideology of working-class honor.

The *workers* are the most consistently antifascist element, since their most immediate interests are in freeing themselves from authoritarian labor controls. Whether they opt for socialist centralization or democracy seems to depend on several conditions (which Moore does not treat). Probably the most important one is the extent to which the bourgeoisie has already made an alliance with the traditional state, and thus enforces labor discipline with the aid of authoritarian controls (Bendix, 1956). Where this happens (as in Prussia, Russia, and elsewhere where the bourgeoisie is largely alien and called in by special arrangement with the government), the workers' aim is to destroy the bourgeoisie and to take over the state, hence socialism. Where the bourgeoisie is more independent and relies on individualistic market incentives, working-class politics is aimed at more particular trade-union interests and at acquiring the franchise in a democracy.

Thus, the early transformation of the British (and American) landowners to capitalist enterprise, and the weakness of the centralized coalitions, both favored democracy. In Germany and Japan, modernization was carried out by enforcing traditional labor discipline, hence an alliance of military aristocracy and the state; the industrialists joined the coalition (or grew out of it, in the case of Japan) to enforce the same powers over unruly workers. In China and Russia, the aristocracy had become officials of the bureaucratic state long before commercialization became advanced; hence, there was neither a basis for collegial democracy, nor interests to support it, for the bourgeoisie was weak and alien and peasant revolts against rentiers were easily transformed by the intelligentsia (trained to become state officials) into the catalyst for socialism. In France, conditions were a mixture of those found in China and in England, with a bureaucratic government and absentee landlords, but also a native bourgeoisie and a tradition of collegial power; hence, the peasant revolt eliminated one supporter of fascism—the aristocracy—and left French politics in an uneasy seesaw between socialist and antisocialist tendencies.

The model may be applied more generally. Democratic forms have been imposed from time to time in societies where they did not develop; how long they last, and how collegially the power is actually exercised, depends on the organization of interest groups favoring one form or another. Germany in the 1920s and Japan in the early twentieth century both went through periods of parliamentary democracy imitated from abroad; neither had any structural supports, and the former collapsed in the struggle between socialist and fascist forces, while the latter reverted to fascism under minimal provocation.[21] Similarly, the superficial imitation of European parliamentary forms in nineteenth-century Latin America resulted in a version of overt or thinly disguised fascism since that time, as the main interests continued to be an alliance between the military government and landowners producing for markets by means of traditional labor discipline, with the bourgeoisie joining as a junior partner in opposition to Leftist workers and intellectuals. The situation in modern Greece, Spain, Iran, southeast Asia, and elsewhere can likely be explained by a combination of the same variables.

The effect of commercialization and industrialization per se has not been to foster or oppose collegial forms of government, but to create problems for interest groups which would then have to choose the forms they wanted as a tool. The only direct effect found everywhere has not been on the form of government, but on the degree of participation in politics. Industrialism has everywhere mobilized the population (*22.11* and *22.12*), so that there have been more participants in the struggle for power, whether by expanding the scope of collegial institutions (e.g., by widening the franchise) or by providing more material for bureaucratic control (e.g., in totalitarian mass ceremony) (Bendix, 1964: 74–104). The changes in several variables need to be tracked. If we are to explain the ebb and flow of democracy in the societies of today, the main focus must be on the interests that find their chances of prevailing over others best served by a particular structure of the state.

---

[21] The strength of the democratic forms imposed by their conquerors after World War II should depend on whether or not these forces have been changed, most particularly, by the elimination of the old landed classes.

THE PROCESS OF HISTORICAL CHANGE

To return to our original question: How much of history could we predict if this theory were fully developed? A great deal, I think, depending on what set of facts we had to begin with and how much detail we wanted to derive. But there is one major limitation. History is no simple developmental sequence. Societies are rarely very pure types, because of the selective diffusion of techniques and religions and because external conquests and resulting mixtures are so very important. There is no way to avoid the ramifications of this fact, which are bound to be large precisely because military power is *the* basis of the state. Still, it is possible to say something systematic about diffusion and conquest, and some principles are suggested in Appendix B to this chapter.

We are left with this position. It is possible to make lawful generalizations about every aspect of political conflict and historical change. A general model is that the network of habitual interactions, as human animals spread themselves across the landscape seeking natural goods and advantageous relations with their fellows, shapes both men's interests and the outlooks and the resources that will enable them to struggle to further them. Knowing men's positions sufficiently well, we can predict what they will attempt to do next and what the likelihood is of their succeeding. But all of this has a very time-bound quality. Their "positions," after all, are not physical things (except in the sense in which they are located in a territory amidst physical objects) but the habitual interchanges that they have negotiated in the past and labeled with ideas to which they have a greater or lesser emotional commitment. These arrangements are mostly very volatile, especially in the realms of power and ideology; not that they change very often, but when they do, it is with the extraordinary eruption of a war or revolution or a religious crusade.

As historical sociologists, we are in the position of being able to explain the likelihood of events from one moment to the next. But when the chain grows very long, the branching out of cumulative possibilities becomes very great. And when eruptions occur—even predictable ones—just where they will end is often impossible to say. Since the variety of history is very great, no simple typology is of much use, except as an analytical tool for identifying *aspects* of things which are explainable. But the overall patterns are certainly bedazzling.

The great transcendental religions, for example, over the centuries formed complex mixtures of tribal, patrimonial, and bureaucratic societies in Confucianism, Buddhism, Brahminism, Islam, and Christianity, in a way that can be accounted for but could not, in my opinion, have been predicted for the long run. And this is not just a curiosity, for as Weber showed, the very possibility of modern industrialism, with its extraordinary ramifications, was made possible by the structural transformations of Christianity. Weber's argument is fairly complex, and I can scarcely sketch it here, but it can be pointed out that Weber completely rejected the dream of an evolutionist sequence.[22]

Weber's economic theory is easiest to understand if we compare it to Marx's version of social change as proceeding through a series of economic changes, each of which is determined by a series of technological inventions. Ironically, Marx is much better at explaining the consequences of economic change than he is its causes. For Weber demonstrates that mass production technology using inanimate energy sources—the essence of industrialism—is not feasible unless things can actually be mass produced; and that requires large-scale markets, a free-floating labor force responding to market conditions, and a credit and legal system to facilitate the movement of resources. But these are very far from the typical conditions of the agrarian societies prevailing in the richest lands prior to industrialism. To break out of the usual complex of fortified households, patrimonialism and political despotism took a very unusual combination of circumstances: beginning with the peculiar way in which Christianity provided a moralistic universal church with a this-worldly yet less than completely politicized status system, on through the nature of late Medieval bureaucracy and legal systems, and to the carry-overs of Nordic tribalism that fought off royal despotism in England and left the path open for a competitive class to fight out its status battles in economic entrepreneurship.

Certainly there was nothing inevitable about industrialism, although, once unleashed, it spread by selective diffusion and enforced conquest, just like other powerful developments of previous history. And just as complex political, cultural, and economic mixtures have been the earlier results of major innovations, the modern world has seen its own range of new combinations. Certainly Russia, Japan, France, and the United States give us a variety of types, not to mention the part of the world that combines a rather small industrial sector with a larger, no longer traditional, one. The evasive argument that things are eventually becoming alike

---

[22] For those who know only Weber's early essay, *The Protestant Ethic and the Spirit of Capitalism,* it should be made clear that Weber in his mature work was more interested in the distinguishing organizational features of Christianity in world perspective than in its Protestant variants, and in the way in which religion meshed with political and legal structure, more than its motivating effects on individuals. His full analysis is contained in *Economy and Society* and in the comparative studies of China, India, and Judaism. For a summary, see Weber (1961: 207–270), or even more briefly, Collins and Makowsky (1972: 111–117).

is almost certainly wrong. Some things become alike, but new combinations and divergencies keep appearing. The world is no more homogeneous now than it was 100 years ago, and there is no reason to expect it will be any more so 100 years hence; it will certainly be different from what it is now, but it will have new diversities of its own. The abstract argument that industrialism produces certain patterns does not enlighten much in a stratified world in which some states are rapidly depleting their natural resources and others lack the resources to begin with.

If we wish to predict politics, we will have to remain in the realm of the possible. A highly developed theory, combined with enough descriptive facts about a point in time, will enable us to predict the next steps with some degree of accuracy. Moving very far along that chain of events becomes increasingly problematic, and there is no simple pattern of possibilities to help us. The world of conflict is a complex one, and will remain so even when we make a science of its elementary processes.

APPENDIX A: THE ORGANIZATION OF VIOLENCE IN WORLD HISTORY[23]

A brief review of earth history would begin about a million years ago with the rise of various protohuman species about the same time as the rise of other mammalian species. The most advanced of these protohumans displaced the others, possibly amalgamating with them, out-competing them for ecological resources, or killing them. There is a long period, perhaps several hundred thousand years, of small roving bands of human beings, especially in the warm areas of the globe but gradually spreading out after the recession of the last ice cap. These paleolithic cultures lived by hunting and gathering, using chipped stone tools. Since

---

[23] This section draws heavily on William McNeill (1963), *The Rise of the West*. McNeill's is far and away the best presentation of world history that we have, avoiding both the straitjacket of cyclical models found in Toynbee, Sorokin, and Spengler, and the almost inescapable tendency of narrative histories to overemphasize the Middle-East-to-Mediter-ranean-to-Western-Europe sequence. McNeill's is virtually the first truly *world* history, and it is only in this perspective that the last millennium of Western Europe really falls into place, and the crucial importance of external contacts, conquests, and cultural diffusion among different geographical areas comes clearly into focus. It should be noted that this general orientation—as well as the willingness McNeill shows to treat economics, politics, religion, health, and other matters as contributions to historical change in their own rights, without forcing them into a harmonious picture of each time period—is not entirely new. It was the hallmark of the German historical school of the late nineteenth century, and it is from this school that Weber drew his main orientations. For an example of the earlier version, see Hintze (1968).

there was virtually no surplus beyond what could be gathered to keep each member alive, there was no stratification to speak of. Tribes were small, certainly fewer than 100 members, and the only ceremonial organization was probably some form of animal-oriented religion, centering around the hunt. In the years between 8000 and 4500 B.C., there developed a transitional mesolithic culture, in which bows, arrows, and canoes were developed, as well as fishing techniques.

Around 6500 B.C. came the neolithic revolution: the invention of agriculture. In its most primitive form this was the slash-and-burn method in which trees were killed and the soft soil underneath prepared with digging sticks, or later with hoes. This form of agriculture was limited to the woods and hillsides with soft soils. Along with it went the development of pottery, cloth, and polished stone tools. Communities became semisettled, although they would move from time to time as the soil was exhausted by this method of agriculture. Religion came to center around ceremonies for producing agricultural fertility. Since agriculture was women's work, such societies tended to be centered around female-oriented religions, and matrilineal organization of the family was probably prominent. The family societies themselves could now be somewhat larger, supporting possibly several thousands in an organized community, usually ruled by a tribal council of males. This type of organization developed about 6500 B.C. in the Middle East, spread to Europe and Central Asia by about 4000 B.C., and to China and Mexico by 2500 B.C. Along with it developed the domestication of animals, and a second specialized farming society, the pastoral, gradually split off. Herdsmen probably amalgamated with paleolithic hunters, preserving a wandering way of life and a somewhat more military orientation.

Around 3000 B.C. in the Middle East, agriculture was further developed with the invention of the plow pulled by oxen. This made possible not only larger areas of cultivation per man, but better fertility of the soil due to deeper plowing, and it also made it possible for civilization to move down from the hills and onto plainsland which previously had been too tough for hoe culture. Since plowing was a male pursuit, agriculture was once more taken over by the men, and religion became more male-dominated once again. This period also saw the beginning of metallurgy, beginning with relatively soft copper metal. Plow culture spread to Europe by about 1500 and reached China by around 350 B.C.

Around the same time, the technique of irrigation was developed in the large river valleys of Mesopotamia. This made possible a tremendous increase in the development of agriculture; enough surplus could be produced so that a substantial class of nonfarmers could appear. Between 3900 and 3000 B.C., cities arose in Mesopotamia under an aristocracy of priests who owned the fields. Other technical developments took place, including the wheeled cart, bronze weapons and armor, writing (used first for temple records and later for government administration), and sailing ships. This made possible the first classes of professional soldiers and administrators, as well as specialized artisans and merchants. This is the transition referred to as the "rise of civilization."

Settlement of the areas lying between the cities gradually led to clashing boundaries and intercity warfare. In this manner, the first kingdoms arose. The dynamics of subsequent history center around military conquest.

Even prior to this, the diffusion of culture had become an important source of social change. Moreover, the various technical inventions did not diffuse together. Metal, wood, wool, and other items not found in the river valleys were brought by expeditions from the hinterlands. In this way, trade gradually grew up and various elements of civilized culture spread. Already in the earlier period, a complex mixture known to archeologists as the "megalithic culture" had spread throughout Europe, carried especially by seafaring traders, who adopted the techniques of sailing from the Middle East. They developed a far-spread network of priests, probably using ceremonial techniques that promised life after death, adapted from the more advanced societies. (These cults produced the famous rock formations such as those found at Stonehenge.) Other trading cultures with a more military orientation are found in the Eastern Mediterranean, such as the Minoans of Crete.

With the development of crude military techniques in the civilized centers, however, the barbarian imitations of the ciilization took on an increasingly military tinge. As barbarians became dependent on trade relations with the civilized center for luxuries as well as food, the center and the periphery entered into a lasting symbiosis. Political distractions in the center which broke the supply of trade could provoke movements for raiding and conquest, as could any military weakness in the wealthy part of their world.

Around 2400 B.C., an outlying city in Mesopotamia created an empire extending throughout the area, demanding tribute from the various cities. This was followed by a cycle that was to occur many times over throughout history, in which empires would become established and then disintegrate over a period of centuries, fall into a state of internal warfare among internal factions, and be conquered by barbarians from outside the area. This happened at least six times in the Middle East between 2400 and 1500 B.C. In each case, the tribal coalition of conquerors would divide the land among themselves. The leader would attempt to put himself in the place of the deposed king, using the administrative techniques of the civilization; this would lead to conflict with the aristocrats, and eventually to disintegration of the state.

A similar sequence is found elsewhere. Around 3100 B.C., the techniques of irrigation diffused to the Nile. The valley was well-protected by deserts from most outside incursions, and had particularly good navigation both to the north and the south. Because of the ease of transportation, quite soon after the development of irrigation, a centralized state was created without a phase of independent city-states. Again, the initial empire was ruled by priests, although probably originating in a conquest by pastoral tribesmen from the south. Egypt also went through a series of disintegrations and reconquests, beginning around 2400 B.C.

For some 4000 years, subsequent history becomes an almost monotonous cycle of barbarian conquests giving rise to new civilizations which disintegrate and are conquered again. Most of the major technical inventions increasing economic productivity were made before this period; they merely diffused to more and more areas. The technological inventions which did occur were primarily in the form of weapons and in transportation and communication techniques, and their significance was primarily military and political. Thus we have in schematic form:

1. The Mesopotamian city-states were overrun by a series of conquests from pastoral tribes beyond their borders between 2400 and 1500 B.C.

2. The development of the two-wheeled war chariot led to a series of highly mobile bronze-age barbarian armies. These were the Indo-European or Arian conquerors who overran Egypt and Mesopotamia in the years around 1500 B.C., completely destroyed a more primitive irrigation culture of the Indus valley around the same time, overran the neolithic tribes of Europe, and destroyed the Megalithic priest culture as well as the Minoans of Crete. The technique of chariot warfare reached China by around 1300 B.C., enabling pastoralist warriors to conquer the neolithic farmers in the river valleys and create the first stratified society of China.

3. The development of cheap iron weapons led to more massive armies, and to a wave of barbarian invasions breaking up the successor states to the bronze-age conquests around 1200 to 1100 B.C. The Egyptian empire which had dominated the Middle East gave way to Assyrian domination from Mesopotamia.

4. Around 900 B.C. the technique of riding horseback was developed, making possible wide-ranging pastoralism in Central Asia. Combined with a technique of shooting arrows from horseback, this innovation led to a new series of conquests. The Persian empire developed in this manner,

overrunning Egypt and Mesopotamia between 550 and 350 B.C. Another branch of the same technical development carried Celtic horsemen into Europe between 800 amd 600 B.C., conquering the previous inhabitants.

5. Yet another military development occurred beyond the borders of the Persian empire. In Greece, relatively democratic societies of small farmers adopted the techniques of heavy armor and created a mass unified infantry, as well as a large fleet, which was able to defeat the Persians and then embark on a program of conquest and colonization around the Mediterranean. Many Greek cities, notably Athens, also developed specialized commercial agriculture, exporting olive oil and wine, and relying on food from abroad. A stalemate among the Greek cities, however, allowed a number of more barbarian peripheral states to the north to imitate their military techniques and become the major military power in the area. This produced the Macedonian conquests and Alexander's short-lived empire in the Middle East. This is turn was displaced by yet another semicivilized state arising further to the west, imitating the military techniques of the Greek colonies. This was the Romans, who rose to dominance in a free-for-all among various powers in the western Mediterranean and extended their empire to take over the Hellenistic empire, now locked in conflict among its fragments.

6. As the new empire became highly stratified internally and caught up in domestic political conflicts, new barbarian military power grew up beyond its borders on both ends of the empire. Central Asian nomads conquered Persia and Mesopotamia to the east. In the west, the Germans, who had developed a mixture of pastoral and agricultural economies by around A.D. 100, began to be used as Roman mercenaries against other German tribes; they eventually inherited the western sector of the empire after nomadic Huns destroyed Rome between A.D. 375 and 450. A later wave of less-civilized German tribes destroyed these successor kingdoms, and gave rise, between A.D. 500 and 800, to the Frankish kingdoms which briefly unified Europe.

7. These Germanic societies took over the civilized administrative techniques of the Romans primarily through the adoption of Christianity. They were broken up in turn between A.D. 800 and 1000 by Scandinavian pirates from the north adopting their military technology. At the end of this period of conquest, the Vikings themselves began to create kingdoms of their own, including those of Russia and England. After this, a new form of heavy armor spread, and a manor economy developed, giving a new permanence to European military structure.

8. Similar cycles were taking place on the other side of the world. In

Northern China around 1000 B.C., a dynasty that had managed to unite all of the currently civilized territory was overthrown by barbarians, who produced a new and more centralized dynasty. The latter in turn disintegrated under new nomad invasions around 770 B.C. A border state adapting horseback tactics from the nomads created the Han state around 200 B.C. This disintegrated some 400 years later under barbarian pressure from the outside, as well as internal conflicts and peasant revolts, somewhat paralleling the disintegration of Rome. Civilization spread into south China as well as to peripheral areas. There followed a succession of nomadic conquests, most of which conquered only north China, usually followed by a reassertion of centralized control emanating from the south. This cycle occurred some five times up to the twentieth century.

9. In the Middle East, the now settled nomads of Persia developed a technique of heavily armored knights on horses who were able to defeat the faster but more lightly protected nomadic bowmen. This is turn gave rise to an economic system of feudal landowners to support the armored knights, a system which subsequently spread to Western Europe and produced its own period of knighthood. As commerce carried civilization into the Arabian peninsula, and chronic conflicts at the civilized centers to the north created a considerable political weakness, Islam arose and united the various nomadic tribes of the south in a war coalition which, between A.D. 620 and 700, conquered all of Persia, Mesopotamia, Egypt, most of North Africa, and Spain. The Islamic Empire then disintegrated into a series of civil wars and accompanying religious heresies. Turkish nomads from Central Asia then rejuvenated the Islamic states, first as mercenaries, then as conquerors. The Mongol conquest, which in the thirteenth century briefly took control of an area stretching from China to Northern India to Persia, put a temporary halt to Turkish expansion. But after the disintegration of the Mongols, the Turks took control of the Islamic states. This produced a second wave of Islamic expansion, extending into the Balkans as well as India and Indonesia.

10. This Islamic expansion, however, was met headlong by a wave of European expansion that began around the fifteenth and sixteenth centuries with developments of techniques of navigation and firearms spreading from China. These techniques had spread to the Islamic states as well, but the superior bureaucratic organization as well as greater commercial development of Europe led to their faster adaption. The first great sea powers were Spain and Portugal, succeeded by England and Holland, who were able to take military advantage of their commercialization. On the continent, France and Prussia put together bureaucracies to support their military machines; in Russia after the decline of Mongol power, the state of

Moscow in the north began a steady expansion at the expense of the Turks.

11. Cycles similar to those which began in Mesopotamia were developing in the Americas between A.D. 700 and 1600. Irrigation states developed in Peru, and there were a succession of priest-controlled states in Mexico. These, however, were completely destroyed by the expansion of European power in the sixteenth century.

12. The development of commerce in Europe took place in a situation of politicaL decentralization resulting from a relatively recent incursion of barbarian tribal forms into a conquered civilization; this spurred a new period of technological invention culminating in the industrial revolution of the nineteenth century. Several hundred years earlier, the European states achieved great military power by making use of commercial wealth in bureaucratic forms of administration. Strikingly enough, the balance between European states was such that no empire could hold its grounds for very long before being opposed by the others. Thus, Spanish power of the sixteenth century was destroyed in the seventeenth by a coalition to the north; French power of the seventeenth and eighteenth centuries, culminating in the brief interlude of the Napoleonic empire, was destroyed by the coalition of the rest of Europe. English power in the nineteenth century was challenged by the Germans in the twentieth; the brief period of German expansion was checked by the economically prosperous outliers of Europe, America and Russia. The overseas empires of the leading European states were lost as the result of conflicts among themselves at home. The aftereffects were to set in motion once again the process of emulation by the less advanced periphery, of which Japan was the first to catch up, prompted by the usual military considerations.

APPENDIX B: POLITICAL GEOGRAPHY AND POLITICAL TIME

Geography interacts with technology and organizational styles in affecting the type and location of states, and in shaping the wars that are so crucial for political history. A comprehensive theory of the state would include geographical conditions, along with the organization of military forces and administrative machinery, and would add some time limitations on the various types of political stability and change. I offer the following principles as an illustration of how this part of the theory might proceed, without attempting to justify them or tie them systematically to the rest of this chapter.[24] They are left unnumbered to emphasize their tentative status.

Location of Technological Types

Hunting-and-gathering societies originate in warmer zones and spread to colder climates under pressures of competition and with the development of appropriate technological modifications.

Slash-and-burn horticultural societies locate in the wooded areas of the warmer parts of the globe.

Irrigation agriculture locates in river valleys with appropriate patterns of flooding.

Plow agriculture makes possible a move to the plains and deciduous forest areas of the temperate zones.

Pastoral societies develop in symbiosis with horticultural or agricultural societies, occupying adjacent grassland steppes, mountains, or semideserts. With improved transportation techniques and the existence of agricultural societies adjacent to grasslands at several points, pastoral societies become much more geographically mobile.

Commercial networks emerge where there are waterways (long navigable rivers, inland seas, island collections) linking a number of territories at least some of which are settled by agriculturalists, and especially containing different types of natural resources (mineral deposits, woods, grazing areas, agricultural land of different climates).

Industrial societies develop and have their centers in large state-organized territories in the deciduous forest temperate zones, and expand into mountains and the conifer forests of cold zones in search of mineral supplies.

Diffusion

Selected items are more likely to diffuse than are entire cultural patterns. Techniques of transportation (e.g., ships, domestication of animals for hauling and riding) and weapons (e.g., metal arms, techniques of massed warfare, firearms) diffuse fastest. Advanced agricultural techniques (e.g., plows) diffuse faster to tribal societies than does the complex of urban civilization that can be built upon their productivity.

Universalistic religions diffuse as military leaders of tribal coalitions

---

[24] The major bodies of information on which I have drawn are McNeill (1963), Shepherd (1964), and McEvedy (1961).

acquire the technical resources to establish a permanent state organization; literacy diffuses with universalistic religions.

Conquests

Isolated areas suitable for horticulture or fishing, but inaccessible under existing transportation techniques to communities of pastoralism, irrigation, or plow agriculture, develop little or no state organization, and develop highly elaborated ethnic identities in the form of tribal or intertribal (e.g., caste) cultures. I have in mind here isolated islands and tropical rain forests, especially if divided from other climatic zones by mountain ranges or deserts (cf. south India, sub-Saharan Africa, Oceania).

Sparsely populated areas in temperate zones containing hunting-and-gathering, horticultural, or diffused agricultural tribes develop no states with stable boundaries until population density limits movement and increases fighting over resources, and/or high dependence upon trade or booty relationships with adjacent civilized areas leads to the adoption of civilized techniques of military organization.

The effort of leaders of tribal coalitions to establish permanent state power through the adoption of a universalistic religion results in a period of ferocious conquest (as a means of holding together the tribal coalition against efforts to reassert equality).

Grazing territories adjacent to wooded areas in warm zones are conducive to raids by pastoral warriors upon horticulturalists, and eventually to stratified quasitribal societies.

Grazing territories adjacent to irrigable river valleys are conducive to pastoral conquests and the establishment of highly unified urban-based states.

Once a heartland (the territory that is militarily accessible from a self-sufficient economic center; Stinchcombe, 1968a: 216–230) is unified by a single civilized state (characterized by urban administration, a money economy, and literacy):

    a.   The extension of agriculture to militarily accessible adjacent lands results in the extension of the heartland.

    b.   The lack of cultivatable adjacent lands results in merely defensive military expeditions against pastoralists, and occasional pastoral conquests (if attacks coincide with a period of internal struggle).

    c.   Isolation from other heartlands under existing conditions of transportation and military supply leads to the consolidation of civilian rather than military administration power and an accompanying religion without an activist orientation, and a highly consolidated ethnic identity around a national religious culture (cf. early dynastic Egypt, China).

d. Conditions of geographic propinquity and/or technology, making possible military contact among several such heartlands, results in a highly military emphasis—many wars and constant military threat, conquests, and nativist reactions, a high level of commerce, and universalistic but highly activist or militaristic religions (e.g., the Middle East after 1500 B.C.).

e. The diffusion of agricultural techniques to adjacent lands that are *not* militarily accessible on a permanent basis because of geographical barriers (mountains, deserts, lack of connecting rivers) results in the growth of outlying tribal populations and their increasing military power, internal conflicts, and sudden military outbursts against civilized territory.

Where there are a number of at least minimally civilized states (those with the core of permanent army under centralized control, with some literate administrators and nominal adherence to a universalistic religion) adjacent to each other, there is a high level of warfare and realignment of boundaries. The outcomes depend upon:

a. the army chief acquiring stable control over an economic heartland (usually a river valley or a seaport);

b. battles to deprive adjacent opponents of their heartlands, these battles usually taking place in the territory accessible to both sides (e.g., the areas between river valleys in France, Germany–Poland, Poland–Russia where there are no intervening geographical obstacles);

c. the advantage of the peripheral state (whose heartland is militarily threatened from only one direction—or not at all, if it is economically less attractive), as compared to the tendency of stalemate among states facing several potential enemies simultaneously.

State borders will extend beyond natural barriers (mountains, ocean straits) if the heartland state behind the barrier has more population, economic, and military resources than the nearest heartland state beyond it; natural barriers become state boundaries when an equal level of military organization is achieved on each side.

Where there are no natural barriers among heartlands, and equally powerful states form in each, the intervening territory is the site of constant tension (e.g., the French–German border; cf. the German–Polish border, where the heartland resources have rarely been equal).

Where highly bureaucratic states based on commercial or industrial economies conquer

    a. hunting-and-gathering societies, the result is cultural destruction and extreme depopulation of the latter;

    b. horticultural tribes, the result is sharp ethnic–racial stratification;

    c. advanced horticultural states (e.g., the Incas), the result is total destruction of the state structure;

    d. agrarian societies with quasibureaucratic states, the result is a long period of political countermobilization culminating in active movements or wars of national liberation.

Time Periods

Wars produce very rapid changes in borders, the more so the less bureaucratic the states involved; in the case of conquests by tribal coalitions, very large empires can be established in 3–20 years; the same empires may disintegrate in a war period of equal length.

Conquests by pastoral tribes over settled agricultural areas have generally led to patrimonial tributary kingdoms. Such empires have sometimes extended over a very large territory, but have been quite unstable. The waves of barbaric conquerors of Egypt, Mesopotamia, and China have dominated 50 to 100 years before disintegrating in the face of native reaction. A similar time period can be found in the case of the Macedonian Empire of Alexander, the Mongol Empire, the Seljuk Turks, the Islamic States, the German kingdoms succeeding to the Roman Empire, and the Frankish Empire of Charlemagne. A succession of short-lived patrimonial states has made up most of the history of India. The larger empires on the Indian subcontinent, the Mauryas and the Guptas, created large states which lasted only 100 years before beginning to disintegrate.

Feudal–patrimonial states based on the military technology of heavily armed knights have generally split up into a system of local lords, with a resulting condition of chronic instability which might last for 500 years. This was the case in China before the unification of the Han Dynasty around 200 B.C., and in Medieval Europe after the decline of Frankish power, and in the Sassanid Kingdom of Persia.

Quasibureaucratic empires organized around a centralized administration and an officialdom collecting taxes directly for the state are more durable, especially in the early periods of their development when rival states with congruent levels of organization have not developed. The first Egyptian Empire lasted some 750 years, and the smaller Minoan Dynasty of Crete lasted perhaps some 750 years. Later, with the development or massive rival states in the Middle East, Egypt went through a series of 150-year cycles of foreign conquests. Expansive states based on the incursion of mass barbarian armies into a quasibureaucratic civilization resulted in empires lasting from 300 to 400 years, as in the case of the Assyrians, the middle Egyptian kingdom, and the Persians. In China, the bureaucracy organized after the long feudal period went through a series of five centralized dynasties lasting approximately 250 years each, but interspersed with periods of approximately 100 to 150 years of dissolution and barbarian conquest.

The most stable states are the most highly bureaucratized preindustrial states. Rome, after a period of some 250 years as a distinctively urban tribal coalition operating as a tributary empire, transformed itself into a centralized bureaucracy which was able to hold the Western Mediterranean for some 500 years. The eastern branch centered in Byzantium lasted for about 1500 years, based primarily on sea power. This holds the world's record to date. The bureaucratized Turkish Mameluke Kingdom in Egypt lasted some 350 years, and the Ottomans, successors to the Byzantine control, set up a state that has lasted some 600 years.

The development of modern bureaucratic organization based upon firearms, elimination of patrimonial elements, and subdivision of authority has produced a number of highly stable states. The English state has now lasted some 900 years in unified form, although perhaps the first 500 years of its unity were due primarily to its island isolation while the centralized administration developed. France, Spain, and Russia have also had centralized and rather highly bureaucratized states for some 500 years. The German and Italian states, battered apart in the Middle Ages as the result of their central position in the struggle to reestablish a reunified European Empire while more cohesive states grew up on the periphery, have been unified only within the last 100 years. The former of these has, of course, been split since 1945 between the contending powers of East and West.

With the exception of this last case, modern bureaucracy has been probably the most stable of all state forms, at least against external military break-up, despite internal changes in regime. Their overseas colonial empires have been a great deal less stable, the Spanish lasting some 300 years, the English 200 years, the French and Dutch less. These have been broken up primarily by conflicts among contending bureaucratized European states themselves. Conquest states established among the bureaucratic societies themselves have been notably unstable; Napoleon's and Hitler's European empires each lasted less than 15 years. It appears that modern bureaucracies, far from eliminating the traditional solidarities of ethnic groups, have given them an organizational basis making them much more permanent and resistant to amalgamation by external conquest than previous patrimonial forms.

Chapter 8

Wealth and Social Mobility

The first step in sophistication about social inequality is to realize that it comes in two different kinds. There is the distribution of wealth, and there is the distribution of opportunities to acquire it. One of the first to recognize this was Tocqueville, in his study of America. Inequality, in the form of a distinction between rich and poor, may coexist with equality if everyone has a chance to become rich. This distinction has been much emphasized in American sociology, in opposition to radical polemics which concentrate on the inequality of distribution and ignore opportunities for mobility. Mainstream American sociology has tended the other way, ignoring anything but opportunity structures and beliefs about them.

It is important to retain the distinction in an explanatory theory. Both the structure of privilege and the structure of opportunities are factors that influence individuals' behaviors and outlooks. But a higher level of sophistication requires our seeing that the two are not independent of each other, but are interrelated in important ways. This is especially important to grasp if we are trying to state the principles explaining the distributions of both wealth and mobility. Approaches to both areas derive from the same underlying conceptual models, and both have experienced the same sorts of problems in developing explanatory theory.

One such problem has been the overwhelming emphasis on description rather than explanation. We have had a great deal of research on how much poverty there is, and how much wealth the upper class holds; there has been some research on trends in the distribution of wealth, guided by hopeful or critical perspectives. Research on social mobility has also revolved around descriptive questions. The original problem was whether or not the golden age of opportunity is over; then it became a question of whether or not the golden age ever existed. Much emphasis has been given to technical methods of measurement.

We have only recently begun to adopt the goal of stating the general conditions under which there are differing degrees of inequality of each kind. Research on the distribution of wealth has consisted largely in descriptive exposés and defenses, with the major exception of Gerhard Lenski's (1966) comparative historical theory. Mobility research in the last decade has become concerned with a version of explanation, but has concentrated primarily upon individual movement within a given social structure. We have thus arrived at a sequence of variables which explains a large proportion of individual variation in educational attainment. Education in turn is moderately correlated with occupational attainment in modern societies; and we know a little about the sequence of jobs.

There has been too little recognition as yet that societies—even modern societies—do not all have the same structure. Some have much larger educational systems than others; the United States has a gigantic, unitary, and rapidly expanding system; other industrial societies have small and class-divided systems. Other societies are not organized around educational systems at all. A general theory of social mobility must apply to all societies, and should state the conditions under which men move in different ways and degrees: in tribal societies, horticultural and agrarian states, industrial societies, and the various types within these categories, as well as combinations of them. It is silly to suppose that the sequence of variables leading up to educational attainment is a general theory of social mobility. Even within our own society, a theory of educational attainment needs to be fitted into the testable explanation of how and why education becomes linked with particular jobs; and it needs to be supplemented by principles that explain the variations in mobility which occur after education has been completed. More generally, we need to show the principles of movement within different types of social structure. Analysis of the factors leading up to a particular kind of achievement in our society is not irrelevant to this, but must be interpreted as an instance of more general principles. Research on educational attainment tells us something about general processes of motivation and social selection, which may fit into a number of different contexts provided we have an overview to integrate them.

Why the parochial narrowness of social mobility research? The reasons are basically ideological. In the United States, schools are an article of liberal faith as a panacea for social problems. Our sociological theory, only one step removed from the social problem phase, has concentrated on that part of social reality which fits the vision of the world as a place in which hard-working students make good, and which makes sociologists look like they are producing practical suggestions on how to make everyone fit the accepted mold. What is ignored is the larger structural context which explains why and where this particular kind of educational attainment may be important. Without this, the practical proposals are of little value;

expanding the school system or raising educational attainment while ignoring the larger context of stratification into which the schools are integrated does not eliminate inequality, but simply shifts it to a different level of educational attainment (Collins, 1971a; Spady, 1967).

Research on wealth and social mobility share the same obstacle to theory-building: ideological concerns limiting analysis to description and to a set of explanatory ideas which focuses on easily manipulable and hopefully effective reforms. The two areas are also linked on a more abstract theoretical level. The traditional model for dealing with both of them has been functionalist. This is a version of neoclassical economics in which social placement and social rewards are conceived of as the results of a market in which individual skills and effort meet social demands. This model has operated as metatheory rather than as an actual explanatory theory; only through the efforts of its opponents has it been cast into testable form. But the neoclassical market model does give us some leads, not so much because of the way it fits reality, but because of the ways it does not. As we shall see, obstacles to the working of a market whose outcomes can be abstractly computed gives us the rudiments of an explanatory solution.

Both areas share another problem. They are both extremely complex areas of explanation, probably the most difficult within sociology. Several levels of analysis are involved—the individual, the institutional, and a very summary level of rates of change—and these need to be interrelated. Both research specialization and ideological concepts have made it difficult to understand this problem of levels of analysis. As in most areas in social science, the studies of social mobility and the distribution of wealth began with problems framed by ordinary language. The reality, however, is a good deal more complex than the simplified terms we ordinarily use to refer to it. We must be aware, then, of the way in which words that mark out an area for investigation do not suffice to explain it. Rebuilding in this area is at least partly a matter of refining our vocabulary.

Our problem involves understanding how the distribution of wealth and opportunities for social mobility are interrelated; to cut ourselves free from concepts loaded with value judgments which have framed our analyses; and to establish the relationship between the individual, institutional, and rates-of-change of analysis. Once we have reorganized the fields this way, we can see that yet another sociological specialization falls within its framework. This is the area of deviance. Here again we have an area heavily loaded with ideological and practical concerns. The field has been organized around solving the problems of crime, juvenile delinquency, alcoholism, drugs, mental illness, vice, and other ills. As in most such practical areas, research has been heavily descriptive. Such description of course can stand by itself, since it gets effects by shocking its audience into recognizing the existence of a problem, or in a more liberal vein, by attempting to contribute to alleviating the problem by a sympathetic portrait of the deviant. Most of the theoretical work has concentrated on social psychological explanations of the behavior of the individual deviant; it has taken for granted the institutional framework within which he lives.

More recently, there have been attempts on the general theoretical level to break out of this framework. The labeling theory and related phenomenological approaches have attempted to show that the categories of deviance themselves are socially constructed by interested groups. Hence, it is necessary to look at the entire social structure, not only the deviant but the normality-upholders and deviance-punishers. This work is moving in the right direction. What is necessary now is to go beyond the philosophical level of general statements, and beyond phenomenological microstudies of the procedures of the police, the courts, and so on, to state the general explanatory principles for variations in the construction of deviance. What is necessary, in other words, is to connect this social problem area with the mainstream of explanatory sociology, and to relate individual social psychology to explanations of social institutions. To be precise: Theories of stratification, organization, and deviance are fundamentally unitary. To explain how a social group is stratified, including what status ideals are upheld, is by implication to state what will be considered deviant. Questions of power, group struggle, and the organization of status communities are exactly the same processes that are involved in deviance; the latter only shows the negative side of the coin. I propose to treat the area of deviance, then, in relation to social mobility. Detaching ourselves from value judgments, deviant "careers" are generally a form of downward mobility, usually with the same sort of cumulative processes (in this case cumulative disadvantage) that are found in other careers. If we can state any explanatory principles about social mobility in general, it should be possible to apply them to the particular case of deviance. This is attempted in Appendix B.

THE DISTRIBUTION OF WEALTH: MARKET MODELS

It would seem that the distribution of wealth is an economic question to be settled by economic theory. Unfortunately, this is the sort of question that neoclassical economics fails to handle very well. Modern economics presents an imposingly sophisticated model, but it is weak on the empirical side. It tells us little to explain the actual distribution of wealth among the individuals, and is not very usable in dealing with the distribution of wealth among nations, providing a workable explanation of economic development, or even predicting the actual movements of prices in the world of the modern corporation (Heilbroner, 1970).

This is because neoclassical economics is built upon an ideal of an abstract market operating according to certain laws of exchange. Sophisticated tools have been developed for stating different aspects of supply and demand in this market. Notions of elasticities, the marginal utility method for talking about economic value, and techniques for graphing relationships make for an economic theory susceptible to precise mathematical treatment. Certainly it is greatly superior in this respect to the older classical system and its heretical variants, all of which operated with absolute notions of value, presumably grounded in the empirical world. But modern economics remains a resolutely ideal system, anchored in two kinds of value preferences: scholars' preferences for intellectual clarity and mathematical formalization; plus an underlying ideological belief in the value of a market system. This leaves modern economics in a position of having no systematic explanations of nonmarket systems, whether of the preindustrial or modern state-directed forms. Even within societies with some version of a free market, departures from the market model are treated as a kind of residual category. The system of concepts provides no real way of grasping such variations.

We find, for example, a precise description of what "monopoly" can mean; but this analysis has done little toward answering the larger questions of the conditions under which differing degrees of monopoly occur, and of their effects on stratification (Schumpeter, 1954: 1150–1152). Most of the treatment of monopoly is like sociological arguments over power structure. The most sophisticated work has been done in showing that a perfect monopoly is hardly possible (because it must involve perfect inelasticity of substitution, etc.); the upshot is to argue that Marxists' and other critics' use of the term is incorrect. But that is as far as it goes.

Neoclassical economics is unable or unwilling to deal with the basic realities as seen from the point of view of a conflict perspective. It avoids questions of violence, including the coercive power of the state which underlies property and makes the economic system possible; as well as the politics of stratification within economic organizations and in the surrounding cultural groupings of class, race, and ethnicity within which such organizations are embedded. Neoclassical economists prefer to deal with a gentlemanly world of exchanges in good faith, where goods and services are traded according to agreements and calculations. Despite the material entities that make up their subject matter, the world of neoclassical economics is one in which every man is a mathematician, and material goods and physical services are not really things in time and space, but abstract tokens that men move around in their heads. Recall our earlier analysis (in Chapter 5) of the unrealism of utilitarian psychology. It, too, pretends to be about hardheaded material interests, but actually gives an ideal view of man as a rational thinker whose interests involve neither human solidarity nor social domination.

This accounts for the persistent tone of unreality in the best economic work. Economics is truly a "worldly philosophy" in which the ideal media of exchange are reified and the real social and material world is mostly ignored. What one misses is the fact that people are trying to survive and dominate each other, and that techniques of economic production and relationships of exchange are the means for doing this.

Despite its self-satisfied claims for an advanced scientific status among the social sciences, modern economic theory remains analytically equivalent to nineteenth-century sociology before Durkheim and Weber. In this respect, the analytical apparatus of economics has not really changed a great deal since Alfred Marshall (Schumpeter, 1954: 1140–1147). Keynes' famous work consisted of drawing new practical applications from it; there has also been much refinement of its analytical tools, and it has been applied to new sources of data since the rise of national statistical systems beginning in the 1920s. For all its sophistication, the neoclassical revolution in economics begun by Walras, Jevons, and Menger, and consummated by Marshall, is a matter of giving precision to the ideal, rather than coming to closer grips with the real world. What is still needed is to place the exchange model within the appropriate social context.

This is especially relevant for sociology because most of what has passed for sociological theories of wealth has come from the old economic vision. (This leaves aside for the moment the great exception of the work of Gerhard Lenski, who applies a very different perspective and constitutes our main empirical explanatory achievement in this area.) The classical theorists—Schumpeter (1951), Sorokin (1927), Davis and Moore (1945)—all present the same basic model. They conceive of a market for services that "society" needs. Those persons who provide the most needed services, and/or the services that are most difficult to supply (because of the effort of training oneself, a short supply of innate talent, and so on) receive the highest rewards. As Davis and Moore put it: Stratification is a mechanism to insure that the most important positions are filled by the most capable persons.

This theory, especially in the Davis–Moore version, has been extensively but inconclusively debated. (The Schumpeter model, however, is the most realistic of the three, since it expressly confines itself to "ethnically homogeneous" societies, implying that other factors besides market ones can enter into stratification by wealth.) The concern has been mainly with whether or not stratification is functional. Some theorists have asserted the dysfunctional consequences of stratification; others have raised the question of whether or not the egalitarian society is possible. The debators seem to have little recognized that this is mostly to talk about value judgments. The theoretical critiques have not moved much beyond pointing out that the existence of inequality of opportunities, and of resources for conflict, limit the market model, and thus disparage the eufunctional view of stratification. Critics of the functional theory of stratification tend to take a utopian tone in opposition to the functionalists' fatalism, asserting that things do not necessarily have to be that way.

Within this tradition, there has been little attempt to state and to test principles giving variations in types of stratification. This is not surprising, since the underlying neoclassical economic model is weak in explaining wages, except in a very abstract manner. One can state in very general terms that all prices are set interactively within the system by the supplies and demands for various products and services; the returns of each worker and each seller are determined by interaction with all the others (Chamberlain, 1958: 339–378; Mincer, 1970). There has also been some attention to the practical questions, usually treated in an ad hoc fashion: for example, whether or not unions seem to raise wages above nonunion workers.

But there is a more basic problem. Following through the pure market model leads us to a startling conclusion: The system must tend toward perfect equality in the distribution of wealth (L. Thomas, 1956: 92ff).[1] In a situation where labor is totally free to move wherever it wishes, jobs that pay high wages tend to attract a surplus of workers, which in turn will lead to a decline in their income. Jobs paying low wages tend to produce the opposite effect. Wherever jobs pay above or below the average, processes are set in motion through labor mobility which eventually bring wages back into line with all the others.

What about external conditions such as the inherent attractiveness of work? Some kinds of work are intrinsically more rewarding because they give power, status, clean working conditions, or interesting activities. But these added incentives should tend to produce high surplus labor conditions in those jobs, hence relatively lower pay to compensate for the added advantages. On the other hand, unattractive work must offer higher wages to attract the labor force. We must conclude, then, that in the perfect market situation, garbage collectors would receive very high pay and political leaders and business executives very low.

---

[1] The argument derives from H. F. Clark (1931) and ultimately from Taussig (1911, Vol. II, Chapter 47).

There is another condition in the model, of course: the possibility of short supplies of talent. But shortages in an innate sense probably do not exist, although they were much stressed by old-fashioned and quasiracist thinkers like Sorokin. It is not necessary to assert dogmatically that there are no innate differences among individuals; the question is whether or not the distribution of talent for socially demanded services is sufficiently broad so as to exceed demand in any particular occupation. Lawrence Thomas (1956: 285–310) has presented evidence to show that it is. In those few areas where we do have some evidence of truly innate specialized talent—primarily in art and music—the demand for services, or the organization of these markets, has usually been such that highly talented individuals do not necessarily achieve high wages.

The functionalist argument thus is mostly a value judgment: It is the usual practice of assuming that whatever is, is good, and trying to justify it, instead of stating the conditions under which particular arrangements exist. The notion of what society "needs most" is a value judgment, not an explanatory concept. It does not "need" any particular arrangement. It simply gets whatever the existing social forces produce. Indeed "society" is only a way of talking, in a single word, about the way people behave; we should not let ordinary language mislead us about empirical reality. There is no objective "need" for a particular kind of government leader; we simply get whoever triumphs, and of course he will proceed to claim that his kinds of qualities are what is needed. Logically, this kind of analysis is a tautology; since he is part of the situation, of course he is "needed" by it, otherwise it would be a different situation and someone else would be "needed."

We can salvage something important from the market model, however. A perfect market tends to produce perfect equality of incomes; hence, it is the *restraints* on the market that account for inequality in wealth. For example, the high pay of medical doctors today can be explained by the fact that they have the resources to limit their supply of services by limiting training, licensing, and so forth.[2] Similarly, Lawrence Thomas (1956: 389–403) has attempted to show that it is educational requirements for employment that are primarily responsible for barriers between different occupational labor markets and which thus enforce inequalities in incomes in modern societies.

Let us reconsider another verbal notion that often gets us into trouble: the concept of a "position." A "position" is an abstraction, not a visible thing. We think of it as a noun because we are used to summary symbols like the little blocks in an organizational chart, the categories in a statisti-

---

[2] This is demonstrated empirically in the case of dentists by Friedman and Kuznets (1945).

cal graph, or even the pithy little word which functions as a noun in sentences. But in reality, like so many things in the territory of sociology, it should be a verb, not a noun. A "position" is a way of talking about behavior: people treating each other in certain ways, giving and taking orders, doing work, occupying rooms, chairs, desks, and machines. The justification for using a noun here is simply that behavior repeats itself so that the same people give orders over again, the same people receive them, and so on. This lends some empirical basis on which to hang the hypostatizations of everyday discourse.

Now, the "position" is a *restraint* on the labor market. A man holds a position if he is not simply paid for each thing that he does but is paid only at the end of a period of time. His position approaches truly noun-like solidity if he achieves provisions for tenure, so that he cannot be removed from his rewards and privileges at all, regardless of what he does (almost). This behavioral pattern, of course, is a matter of degree. Property is an extreme form of a hypostatized position. Thus, we can have property in work positions. These were particularly prominent in premodern governments; owning one's own business is a typical current example. Land ownership, too, is a form of behavioral tenure. Property, if we think about it, is obviously not any particular "thing" itself, but the power to use it and to keep other people off it.

Differing degrees of tenure on the more short-term side of the continuum are found in the distinction between salaries and wages (Stinchcombe, 1965: 162–163). In the first, one is paid on a long-term basis by the fortnight or the month, or sometimes by the entire year. In the latter, work is calculated by the hour or minute, and paid daily or weekly. At the extreme of this continuum are minute-by-minute exchanges such as transactions in a store or a shoeshine parlor. From our previous discussion of organizational controls (Chapter 6), we can recognize here a dimension of administrative control structure, ranging from close surveillance at one end to occasional checking of results at the other. Unions, professions, and other forms of occupational membership organization surround these relationships with other forms of autonomy and security.

Relatively tenured positions, then, are one kind of restraint on the labor market, and produce some of the observed inequality. Partly, this will depend on whether the positions command large or small resources. Another form of restraint on the market comes from the status communities which operate as a basis for assigning jobs. Well-documented examples include discrimination against blacks, peasants, rural people, and women, and at higher class levels, positive preferences on the basis of educational culture and ethnicity (Collins, 1971: 1008–1012).

Market theory can help us when we detach ourselves from the ideological bias in which it has been enshrouded, by showing the necessity of a conflict model as its basis. We can calculate the effects of supply and demand conditions and hence the degree of pressure toward equality of incomes within the social context which limits the market. This type of analysis helps explain why social mobility theory (of which the Davis–Moore model is a version also) is at the basis of a theory of the distribution of wealth. More precisely, both theories of social mobility and the distribution of wealth must derive from the same underlying model.

This reconceptualization does fit one emphasis of the original market model. This is the tone of cynical realism in Davis–Moore and Schumpeter not found in many of their critics. They recognize that life is inherently conflictful, and that people who struggle harder than others tend to get more of the goods. This is a better basis for explanatory theory than pious liberal hopes that somehow eliminating certain external advantages or previously accumulated wealth will make everyone equal so that no further conflict will occur.

What we need is to rigidly apply the fact–value distinction. We must avoid the fallacy of assuming that the results of the strong and energetic getting to the top are necessarily good for everyone else. Wealth and mobility are the results of effort and struggle; this applies also to static periods in which all the effort is concentrated in keeping resources neatly monopolized. Men's efforts in these struggles are not necessarily aimed at producing what others want; the main effort, in fact, usually goes into keeping one's opponents from being able to compete on even terms with oneself. This is easiest to see in politics. Politicians rarely produce what anyone else wants, but they do make a great effort to feed and glorify themselves. They may well be more "talented" than others in the sense of being able to get themselves onto the top. Less-talented people, if by chance they fell into those positions, would not necessarily do worse for the population but would probably be less capable of defending themselves. And, of course, let us not overlook the fact that there is a great deal of luck involved in success and failure. This is one point which functional theories (and self-serving ideologies) almost always ignore. The same may be said about success in business, where for every stock-market millionaire there are millions of losers, and industrial tycoons seem to share the merit primarily of happening along at the right time.

The market model may now be seen as a special case of a more general conflict model. Where men's efforts produce increased wealth that may be distributed to others, this is as a by-product of their own self-interest. The conditions under which this happens need to be specified. The idealized market situation seems to operate precisely to the extent that there is a balance of conflicting resources so that others can demand that equal exchanges be carried out.

Lenski's Conflict Theory

In fact, the only empirically validated model we have is a conflict theory. This is Gerhard Lenski's (1966). Its basic principles are: Men largely follow their own self-interests; technology produces and determines the total wealth in a society, and also the means for dominating others; and the combination of variations in total wealth and in means of domination determines its distribution.

Different societal types are classified according to their main types of technology. Arrayed in order of how much wealth they can produce, there are five main types: hunting-and-gathering societies; primitive and advanced horticultural societies; agrarian societies; and industrial societies. We have already met these types in Chapters 4 and 7, where they provided us with a starting place for comparative analysis of status stratification and of politics. Essentially, Lenski's model proposes that politics determines the distribution of wealth. These five types, in the cited order, successively allow more production, greater population density, and also progressively greater surplus. This frees men from agricultural production for crafts, and for warfare and administration.

Therefore, the greater the technological productivity, the greater the potential inequality. As more is produced, the elite can get more and more of it while keeping the underlying population alive in order to support them. This holds in intersocietal comparisons on through agrarian societies (urban literate civilizations supplied by animal-powered plow agriculture). In industrial society, however, the curve reverses and there is a shift toward less inequality than in agrarian societies. The main reason again is political. Industrialization mobilizes an increased portion of the population. As we have seen in Chapter 7, the movement of population into cities, and the organization of work in factories and offices rather than households leads to a shift in the concentration of political resources. The result is mass politics (whether in democratic or totalitarian form) and the elites are forced to give material concessions to the masses. This is possible, among other reasons, because the total wealth is now so great that the upper class may be still better situated materially (if not in terms of status and power) while giving up proportionately more of the total.

Continuing this line of argument, we can see that the same process occurs on a smaller scale within modern organizations. The existence of many areas of uncertainty in work processes organized around a complex technology leads to a greater distribution of power within the work force itself, especially at the middle levels. Redistribution of wealth occurs because of processes on two levels: political concessions in the larger society (shifts in tax policies away from those characteristic of the predatory state; state suffrance or protection of trade unions; welfare services; government sinecures) and the internal politics of organizations, which forces some redistribution of the organizational profits in order to keep control of the work force.

Lenski's model not only fits the conflict theory of politics and of organizations, but is testable on at least a general level. Lenski himself has demonstrated the differences in the distribution of wealth in the five historical types using historical, anthropological, and contemporary evidence. Other cross-sectional data has further tested parts of the model on the mixed industrial, agricultural, and horticultural societies existing today. That greater equality in the distribution of wealth is found today in the more industrial societies is well-documented.[3]

Lenski's model does have certain weaknesses. Once we are within industrial societies, it leaves us without specific guidance. If we assumed that the trend toward greater equality continued, we would be empirically in error; among purely industrial societies, there is no evidence that the more technologically advanced are more egalitarian. For the 1950s and 1960s, Germany, France, and the Netherlands show especially high concentration of incomes; Israel and Norway are relatively the most egalitarian, with the United States in the middle, along with Britain and Sweden. Nor do time sequences show any general trend. In the United States, for example, there was an *increase* in the concentration of *income* from 1910 to 1929, followed by an abrupt decline through the World War II period, and a more or less steady state since then. The concentration of *property* wealth, on the other hand, has increased since World War II. For Europe in the postwar period, there is little change.[4]

---

[3] Kuznets (1963), Andic and Peacock (1961), Russett (1964), and other sources, as analyzed in Pilcher (1972). The strongest evidence deals with the distribution of property, although the concentration of land ownership is mildly and inversely correlated with industrialization. Pilcher's analysis shows that income shifts are specific to particular sectors of the population. Strikingly, the *poorest fifth* of the population is somewhat better off relatively in the *non*industrialized societies. The principal decline in concentration is in the top fifth, and especially the top tenth, of the population; the main beneficiaries of redistribution are the upper middle sectors, especially the second, third, and fourth deciles. This is congruent with the power-resource model; industrialization means a shift from a tiny elite (on the order of less than 10% of the population) vis-à-vis a largely undifferentiated mass of peasants, toward an urban organizational complex manned by a highly mobilized middle (and especially upper middle) class of functionaries, professionals, and entrepreneurs.

[4] Pilcher (1972: 38–50); Kolko (1962: 14). The pattern again is that what changes do occur are mostly a matter of redistributing relative shares of wealth from the upper class to the

Such evidence does not test an assumption of Lenski's theory, however, but a popular belief. Its problem, rather, is a lack of specificity. Lenski's categories are useful for broadly establishing the importance of his variables. What we need is a more refined model, especially for explaining the variations in the distribution of wealth found within modern industrial societies. Lenski's model of political conflict within the context of available technological resources provides some good leads, but we need a more refined mechanism.

There is also a weakness in dealing with preindustrial societies. Lenski's broad types do not give us a model of social change. History is a matter of fluctuations and mixtures rather than trends. As we have seen, much of the dynamics of history has come from the external relationships among geographically contiguous but typologically different societies, and the innovations that have resulted from *partial* diffusion of their cultures. Lenski has pointed to the overwhelming importance of politics in premodern societies. This needs to be elaborated to show the diversity of politics and the elements of market relationships that have operated in some of these societies, which help account for the larger variety of historical reality.

A general model of the distribution of wealth in industrial society needs to incorporate the effects of market conditions within a political context. An industrial economy can operate only by mass production, which in turn requires mass markets. Some redistribution of the wealth among the population (or else, massive government expenditure) is necessary whenever inequality becomes too great, just in order to keep the system going.[5] By the same logic, some degree of equality of distribution was necessary in order to initiate an industrial economy in the first place. Industrialization could not easily arise in highly stratified and politically centralized societies because of the lack of a mass market. This explains the historical fact that industrialization arose not in an archetypal agrarian society, but in mixed types combining tribal and agrarian elements. It was the relative equality of corporate (rather than purely hierarchic) society in Europe, and especially in England, that provided a basis for a mass market. Our problem, then, is to develop a model that can deal with the nonlinear aspects of history as the most crucial explanatory points.

---

upper middle class. Clearly, welfare systems, which operate primarily in shifting token amounts to the lowest groups, have little effect compared to the much more massive distribution of wealth in wages, salaries, and property.

[5] For a general discussion of this aspect of the Keynesian problem, see Baran and Sweezey (1966: 27–86); Bensman and Vidich (1971: 5–31).

The Marxian Model

Both the neoclassical model and Lenski's conflict theory come down to the same problem: how to relate political and technological contexts to the economic market. It is this complex that is at the heart of the Marxian model (Marx, 1906–09; Mandel, 1968).

Its weaknesses are well-known. It is derived from classical economics and still uses an absolutist labor theory of value. In addition to this is the heavy Hegelian manipulation of abstractions, that makes applying it to empirical problems (e.g., how prices are set) a much clumsier operation than using neoclassical marginalist economics. Some of its major empirical predictions are weak. Marx correctly foresaw trends toward monopoly in market economies; but they have not moved toward absolute monopoly. We have seen the destruction of the environment by blind economic interests, but not entirely and not without some instances of remedy and compensation within some version of the market system. His prediction of growing inequality in industrial society is not generally true. The impoverishment of workers has not happened, nor has there necessarily been an increase in relative inequality. The situation has rather been one of fluctuations, to which neither Marx nor his ideological opponents have been reliable explanatory guides. Unemployment has been a continual problem, but the great mass army of the unemployed that was to bring about the collapse of capitalism has not appeared (although it has been approached on occasion).

The underlying weakness is in the labor theory of value. Marx assumed that all wealth is produced by labor. Machines, in this model, do not increase the supply of wealth. It takes labor to make machines; a well-functioning market makes everything sell for the cost it took to produce it; hence, the introduction of labor-saving machinery can only produce as much wealth (measured in hours of labor power) as it took to make them. Profits from innovation come only temporarily to the few capitalists who get the jump on the others. Once their competitors introduce the same machinery, profits are reduced to cost level, and the system goes into a crisis.[6]

We are in the atmosphere of the classical economics of the turn of the nineteenth century. The basic weakness, from our point of view, is the lack of a concept of economic growth, the implicit notion that the total amount of wealth in the world is static. Any profits must come out of someone's hide. Marx expands on a problem already inherent in Adam

---

[6] A similar dynamic of pressures for innovation is presented in Schumpeter (1934), although with very different analytical apparatus and value judgments.

Smith: Where do profits come from? Since investors always move toward those opportunities where profits are being made, they increase competition and eventually bring profits in that sector down toward the average of the entire economy. We have seen the same problem before in the difficulty of the market model in explaining inequality of wages. Classical economists like Malthus and Ricardo solved these issues by invoking population increase; neoclassical economics solved it by discarding the absolute concept of value, concentrating only on a relative measure of exchange. Marx, of course, had ideological preferences for holding to the classical labor theory, since it provides an argument for the intrinsic fairness of socialism.

This theoretical construct is what makes Marx's predictions go wrong. He argued that technological advance must gradually reduce the amount of labor used in production and thus cut into the basis of profits. Capitalist competition for labor-saving machinery leads to a falling rate of profit. The result is a series of crises, business failures, and the advance of monopolization and the eventual collapse of capitalism. Another source of error is Marx's tendency to argue, especially in his economic writings, for a monocausal model in which politics depends upon economics. But this is secondary; if Marxian economics were true, the political alternatives at the crucial turning points of history would be entirely constrained by economic conditions, even though political factors could be regarded as hypothetically independent.

Marx's conception of labor as the basis of value is essentially a value judgment. Of course, definitions are arbitrary; the only question is whether or not the use of a particular word enables one to single out aspects of reality and talk about them in the most useful way. What we wish to know is how much wealth (goods and services) there is in a particular society, and who gets what part of them. It is obvious that machines (and particular administrative arrangements) can produce a great deal more or less of these, even if there is still a question of how to distribute them. Human labor is not the only source of energy. Machines can be regarded as mechanical slaves who can be exploited even more than men (i.e., one can get more out of them than one has put in); in fact, that is precisely for what they are designed.

What can we salvage from Marx's theory? In a historical sense, he is the great originator of modern conflict theory: the vision of society as people pursuing their own interests, concerned about survival and dominance, whose ideas tend to be ideological self-justifications. It is the basis for subsequent insights along these lines, including complexities that Marx did not foresee. We can also learn from his realism, which enables us to avoid certain reifications common to other sociological perspectives. This is particularly true in the economic realm, where we must constantly remind ourselves that money is only a symbolic medium, that the important economic realities are physical things: goods, machines, buildings, land—and the human behavior that appropriates them and uses them. The distribution of wealth, then, means who lives in what house, uses what land, has what clothes, food, and cars, and gives orders to what servants and employees. It is in this sense that the labor theory of value is meaningful: It reminds us to look at what people are doing, rather than at the abstractions they use to talk about (and obscure) their actions.

This realism is particularly important if we are to have an explanatory theory of property. Political struggle is the basis for both creating and maintaining property. This is easiest to see in the case of land. All titles to land existing today are historically from someone's military appropriation of it. The same applies to manmade artifacts. Employers own equipment and hence can set terms in bargaining with laborers who need jobs, because the violence of the state is organized to uphold their property rights. This leads us directly to a Weberian emphasis. We will see that politics is the crucial stage-setting for all economic activity. Political power can operate in several directions: It can uphold particular kinds of landed and capital interests; it can also support various kinds of workers' monopolies and interests, whether in the forms of unions, professions, or immigrant policies, and so forth. Approached without ideological preconceptions, an examination of the political appropriation of controls over land, artifacts, and social positions is the key to a realistic theory of wealth.

The reason that some people are rich in a market society, then, is because they own things that are crucial resources in producing goods and services, and thus control other people's job opportunities. Property is a form of organizational power. Similarly, money and credit are also power. Money in the bank has no use except that it is used as bargaining power (this was a central feature in Schumpeter's dynamic theory of the economy). Wealth is a form of power; explanation must deal with the different ways in which it is organized. Money in the bank is only a convenient way of talking about and calculating this in quantitative terms; it can be misleading, if exclusive concentration on sums of money hides important variations in the social organization of power. A crucial distinction is the degree to which an economy is organized around labor or machine production. In the latter, we can find a society like our own in which many people are rich in things that are easy to produce, like television sets and cars, but may be poor in labor services like exterminating rats or building well-made houses.

We are thus oriented toward an analysis of power struggles, particularly those involving the control of machines. In modern society, this especially concerns technological unemployment and its ramifications. Debate over this question has been heavily weighted in an ideological direction; most effort has been concerned to show descriptively the (actual or potential) amount of technological unemployment, or the extent to which technological advances can produce new jobs. From a more detached perspective, we can see a multifactor situation. New jobs do not appear or disappear automatically with technological change but under particular conditions. The larger distribution of economic and political power is the crucial background here.

Social Mobility Research

Before attempting to state the general principles of a theory of the distribution of wealth, it is worthwhile to survey one more related area. This is the field of social mobility research. It has recently begun to disentangle itself from ideological biases and to reach a level of sophistication in dealing with very complex matters that make it a good indicator of the potential scientific status of sociology. From it we can gain some precision for our general theory, while at the same time the conflict analysis of wealth provides a crucial lead to understanding social mobility.

Earlier work on mobility rates was largely descriptive. It attempted to answer such questions as: Was the rate of mobility in America declining? Did the United States have higher rates of mobility than the rest of the world? In answering these questions (the answers to which seemed to be "no," with various qualifications), debate came to focus on the issue of how to measure mobility rates. The occupational structure has been constantly changing in modern society; the agricultural sector has been declining and the white-collar sector growing. Therefore, the argument went, some mobility was "forced" by changes in the occupational structure: Some sons of farmers or workers would have to move up to fill the new white-collar positions. This was referred to as "structural" or "forced" mobility; if it could be subtracted out, we would have a measure of "free" or "exchange" mobility. It was only the latter that was felt to represent a truly "open" system—the ideal against which America was being measured.

Accordingly, indices of "pure" mobility were constructed, usually by calculating the proportions of sons from each class background level who would be expected to fill each of the different occupational categories purely on a basis of chance. For example, whether or not farmers are a declining group, their sons should be expected to fill professional, managerial, clerical, manual, and farm occupations according to their proportions in the population at the time of the survey; if professions comprise 10% of the occupations, 10% of the farmers' sons should be in professions. The expected numbers of sons of each background in each destination are calculated, and these figures are compared with the actual numbers in these positions. The resulting index summarizing these figures gives the mobility occurring independent of structural changes. Debate went on for a time over the question of whether or not the values of proposed indices were actually unaffected by shifts in occupational distribution; the earlier observed–expected type of index was criticized on this ground and Yule's Q proposed as a proper measure (Yasuda, 1964).

The sons-replacing-fathers model also implies that differential fertility among social classes may also account for some mobility. If the upper classes have fewer sons than necessary to replace themselves, and the lower classes have more than enough, then some upward mobility may be expected to come about to fill the vacuum. Variations in amounts of mobility thus might be predicted by differential class fertility ratios (Sibley, 1942; Kahl, 1957).

O. D. Duncan (1966) has presented a powerful critique of the idea that intergenerational replacement correctly represents the demographic facts expressed in a mobility table. Mobility is conventionally studied by taking a sample of men ("sons") and asking them the occupations of their fathers. But as Duncan points out, the distribution of occupations among the fathers of such a sample does *not* represent the occupational structure at a previous point in time. In fact, since sons are of different ages (perhaps separated by as much as forty years) and are born at different periods in their father's lives, the occupational distribution of fathers is drawn from a period of several decades. Instead of replacing their fathers, many sons are in the labor force with them at the same time: Some of the fathers of 1930 were still working in 1960 and some of the sons in 1960 were also working in 1930. Duncan thus points out that one cannot estimate from such a table how much mobility is "required" by demographic differentials among classes, or by structural shifts in occupations, since the earlier structure (the distribution of fathers) is not actually "replaced" by the sons in such a sample.

The importance of this critique is to demonstrate that fertility patterns, changes in occupational distributions, and processes of social mobility are independent, although interacting (Matras, 1967). Any particular ratio between class fertilities—relatively high or relatively low upper-class fertility—can be compatible with either high or low mobility in the population. Shifts in occupational distribution can offset any such apparent pressures, as when "extra" upper-class sons increase the size of the upper occupational category, or "missing" sons decrease it. Changes in population size can have a similar effect: In a growing population—where most fathers have more than one surviving son—sons find their places by occupying positions that did not exist before, not by displacing someone else.

Changes in occupational distribution are not a direct cause of mobility either. Since sons usually join their fathers in the work force instead of replacing them, occupational shifts can be seen as *results* of mobility, but not necessarily *causes* of it. Even with a large-scale process of occupational change like industrialization, population shifts could bring about shifts in occupational distribution without "forcing" movement. In a growing population, for example, the *proportion* of farmers can decrease without an *absolute* decrease in the number of farmers, and hence, without any necessity for farmers' sons to leave the farm. Indeed, increases in the number of industrial workers or in white-collar classes could conceivably occur simply because of population growth in those categories.[7] Again, we see that any given change in occupational structure or in population can be compatible with a wide range of mobility rates, and that the notion of "forced" mobility on either score alone is untenable.[8]

Duncan's critique marks a turning point in the analysis of social mobility. The older mobility table is now being left behind as a focus for research, although new samples are still reported in this manner for descriptive purposes. Wilensky (1966) has argued that the incomparability of samples makes comparisons among different mobility structures extremely hazardous, and that causal explanations of mobility are not the most fruitful path at this point. Instead, he proposes that research concentrate not on intergenerational shifts but on the lifetime sequence of career and jobs, and on the effects of these on the attitudes and behaviors of individuals. Duncan has proposed giving up mobility tables in favor of the analysis of correlations between fathers' and sons' occupational positions, arrayed across the wide range of occupations ranked by occupational status scale rather than by grosser categories of white-collar and blue-

---

[7] There are indications that the industrial labor force in England was initially created in this way rather than by depopulation of the countryside (Habakkuk and Postan, 1965: Vol. 6, 604–605, 611–612). An example of a proportional but not absolute decrease in farmers is found in America, 1900–1920 (Historical Statistics, Series D 86).

[8] Arthur L. Stinchcombe points out (private communication) that "forced mobility" can be calculated from a model of job vacancies including several variables, cast in terms of single age cohorts rather than the conventional mobility table, with its mixture of cohorts. The problem is to state "the determinants of the mean distance a cohort will have traveled from their mean social origin when they entered the labor force." The difficulty with formulations of this type, however, is that the analysis is necessarily retrospective; we can say something about the effects of vacancies only because we already know what those vacancies turn out to be. As we shall see later, vacancy models are a step toward the solution, but a successfully predictive causal theory needs to be built up from a perspective in which the organizational structure of the future is itself something to be explained by variations in the behavior of the individuals who will act it out.

collar occupations. This has several advantages. It enables us to answer the question of how "open" or "closed" the opportunity system is, in one particular respect—that is, in terms of how important the father's occupation is as a determinant of the son's occupation—while avoiding problems of statistical interpretation associated with earlier measures of "pure" mobility. It also makes it possible to introduce other variables which determine the individual's career, thus leading to the elaboration of the model of career mobility. Duncan (Blau and Duncan, 1967: 165–177) has shown how the use of path coefficients makes for mathematically powerful models of this sort.

A rival school has sprung up around a second mathematical technique, the use of stochastic models, particularly Markov chains. These researchers regard mobility as a probabilistic rather than as a deterministic process, and attempt to show that the observed data on mobility can be generated by mathematical models embodying particular probabilistic assumptions about the chances for individuals moving from time to time and from point to point. Representatives of this group (White, 1970a: 281–282) have argued that the path coefficient models are not realistically causal, since they simply associate a series of variables with categories of individuals rather than viewing their movement through a network of organizations.[9] Most of the stochastic models, to be sure, are also hypothetical rather than realistic, although, in one case, Harrison White has shown that movement within organizations (the careers of ministers in several American churches) can be represented by stochastic models. The weaknesses of this mode of analysis so far have been its failure to incorporate the kinds of variables that have been demonstrated to be important for individual mobility,[10] and to generalize it beyond the closed organizational systems with which White has worked.

There are still several problems that need to be surmounted. An explanatory theory should be completely general; it should predict outcomes in all possible cases. Most mobility research has concentrated on industrial societies and especially on the United States. In principle, explanatory theory should apply equally well to Imperial China, sixth-century India, ninth-century Islam, ancient Greece, or primitive tribes. In such a theory, the causal variables should be formulated so as to be applicable to all cases.

---

[9] White (1970a) also criticizes the notion of distinguishing "structural" (or "forced") mobility as the antithesis of "exchange" (or "pure") mobility, since all job moves, in his view, require that there be a job vacancy. Exchange mobility, he argues, is not a residual, but should be replaced by a broadened version of structural mobility which distinguishes various types of vacancy chains.

[10] Such a model is proposed in Spilerman (1972), applied to geographical rather than occupational mobility.

Modern societies would be distinguished (if in fact they are) by particular states of these variables. Dividing societies into "caste" or "ascribed" systems on the one hand and "free mobility" or "achievement" systems on the other is far too crude, since mobility occurs in all societies, although in different amounts and forms.[11] This sort of contrast reflects the predominantly ideological concerns underlying our heritage of social mobility theorizing. This heritage goes back to the ideologies of the period of the French Revolution, which contrasted "modern" meritocracy against the rigidly hereditary old regime. Like all ideological contrasts, it overstated the case on both sides, and it has tended to lock us into a simplistic view of the causes of social mobility.

Even more crucial has been the failure to relate social mobility research to the rest of sociology. But it merely takes a different perspective on the same processes. A mobility survey asks individuals to summarize, in a few words, their life's activities in relation to the life activities of their parents. Our explanation of social mobility ultimately rests upon the same factors that explain men's behavior in organizations, professions, politics, and status communities. But the problem is not simply that the overspecialization of social mobility research has cut it off from the rest of sociology. To the extent that there have been guiding explanatory principles in social mobility research, they have tended to come from our traditional sociological models of society as a functional system. Thus, the "forced" mobility model that Duncan criticized is essentially a version of functional theory applied to industrialization. Other aspects of mobility research—for example, the interpretation of correlations between education and achievement in modern societies—are similarly interpreted in terms of the alleged functional needs of society.

How can a more sophisticated conflict theory, which avoids reifying the social structure and sees it instead as the ongoing construction of individuals and groups attempting to maximize their interests, help explain social mobility? Duncan's critique provides a useful starting place, although the full implications of his analysis have not yet been drawn out. One of these is that *social change and social mobility are substantively the same phenomenon.* Sons rarely replace their fathers; especially as the population grows they create new positions that did not exist before. Conversely, as population shrinks, old positions simply disappear. The occupational distribution is the result of the mobility in the lives of the men in any given survey, not the cause of their mobility. Conventional interpretations of

---

[11] Indeed, the popular assumptions may be the reverse of what is empirically demonstrable, at least when comparing the structure of industrial and horticultural societies (Kelley, 1972).

"forced mobility" commit the same sort of error that functionalists make in general—they take a given state of affairs as the cause of the individual processes that make it up. Seeing this in a proper causal perspective means that our understanding of social history becomes relevant to a causal explanation of mobility. The analysis of stratification presented elsewhere in this book in terms of a struggle among groups with different resources to control or expand organization positions, can now be fitted onto an understanding of what is implied in mobility statistics. In place of an imminent development of the occupational structure that has been mistakenly read into mobility tables, we come to look for the conditions that allow groups to change the occupational structure or prevent changes in it. Economic development is one such result.

In this light, we are in a position to incorporate the leads provided by analysts of the distribution of wealth. Analytically, explaining the distribution of wealth and explaining social mobility is a matter of asking slightly different questions about the same processes. In both areas, we need to understand that we are dealing with several levels of analysis. In the social mobility field, this is reflected in the relationships among *path analysis* models of the factors that determine individual careers, *stochastic models* of how vacancies are filled in the social structure, and *rates of mobility,* which tell us something about the distribution of resources that make up social structure. These are on different levels of analysis. We will make greater progress toward a comprehensive theory once we understand that no one level is the right level of analysis, nor is any particular measure of mobility right or wrong. They simply tell us about different aspects of what is happening in society. Once we achieve a comprehensive theoretical model of these basic processes, we will be better able to understand what the different measures tell us, and how they interrelate.

The same questions can be asked in terms of either wealth or occupational position. We can ask three sorts of questions:

1. *The determination of individual careers:* Who will become rich or poor, or who will attain what occupational position? In practice, this means describing the characteristics of individuals who become successful, by dividing them into various categories to be compared.

2. *The rate of movements during people's careers:* How many people become rich; how many people will become successful? This essentially asks how much changing about goes on in a society over a given period, whether over an entire lifetime, parts of a lifetime, particular decades or centuries and so on.

3. *Determinants of the overall structure:* How is wealth distributed among

various groups; how are occupational positions distinguished from each other in terms of their powers and privileges?

The three levels of analysis are interrelated. The third, the structural level, is in a sense basic to the other two, for it determines who can be successful and what kinds of rates of movement are possible. Different types of persons (characterized in terms of variables reflecting their particular career experiences) will be successful in different types of social structure. At the same time, the individual level of analysis is also crucial; as we have seen throughout this book, social structure is nothing more than the behavior of individuals. The struggle to build organizations or to gain control of them is a matter of individual behaviors and their historical consequences. If we could fully describe all individual careers, they would add up to a complete description of the social structure. Social mobility rates (for which analogies in terms of wealth have not been much developed) are analytically more superficial, and should be derivable by special calculations if we can answer the first two types of questions.

Which level to analyze first, the structural or the individual, is essentially a strategic matter. There is no point in arguing which of these is more fundamental. Our problem is simply one of looking for the most convenient device for exposition. I propose that if we can answer the listed questions in the following order—(*a*) the determinants of overall structure, (*b*) the determinants of individual success, and (*c*) the determinants of overall rates—we will have discovered all of the things about social mobility and wealth in which we are interested.

TOWARD A THEORY OF WEALTH AND MOBILITY

Social Structure and Wealth

What determines the organizational structure of a society? Essentially, this is the same question of social history that we have asked throughout this book, particularly in Chapters 4 and 7. Technological developments are the most important background, as they bear both on economic productivity and on the structure of political organization. These affect social mobility, since they shape the organizational channels within which careers can take place, and the different kinds of positions which determine what success means in a particular society. The effects of these major variables in broad outline has been demonstrated for the distribution of wealth by Lenski. The whole analysis of political sociology given in the previous chapter is a crucial background both for the possible forms and types of social mobility and for the distribution of wealth.

Let us look at the social structure in more detail, beginning with the distribution of wealth. This is another way of talking about social structure—or at least one aspect of it—since wealth refers to people's behavior in appropriating material goods, and in giving and demanding services from each other.

It is analytically useful, first of all, to list the range of factors that determine the *total amount* of wealth in a society, and then to analyze the processes by which it is distributed *among individuals.*

The total goods and services that people consume or use for any given time period is the result of a combination of seven factors: (*1*) natural resources, including the land and its climate; supplies of metals, oil, coal, and other inanimate power sources; natural transportation advantages or barriers, such as rivers and mountains; on the negative side, the degree to which climatic conditions pose special needs for shelter, clothing, and food; (*2*) the number of people working; (*3*) the duration and intensity of work; (*4*) the technological lore available for use; (*5*) the store of past improvements and goods that are available for use, including improved lands, buildings, and machines, and stores of goods; (*6*) the effectiveness of the economic division of labor; (*7*) from all of which must be subtracted the amount destroyed in that particular time period by warfare, natural disaster, or deterioration. This summary is useful primarily for emphasizing a point that is implicit in Marx and made explicit more recently by Fourastié (1960). The fact that the total amount of wealth is determined by a number of conditions, including ones that do not reflect the current social structure, means that the distribution of wealth depends not only on the organization of exchange but on a sheer possession of resources that are physically present in some places but not in others. As Fourastié argued, redistributing wealth in terms of money will make little real difference if the ownership of basic goods and resources and their rate of production is not affected; redistributing money in a society with a shortage of housing simply raises the prices of housing rather than automatically producing better housing for all. In the same fashion, we are reminded that some countries are wealthier than others primarily because of the natural resources they possess; the sheer military control of territory is thus a crucial determinant of relative wealth.

The *distribution* of wealth is a function of the total amount of wealth and of the organization of power in that particular society. What units we use to mark off the population for measuring the distribution of wealth among them is an arbitrary matter. This can be decided heuristically in terms of what data is available, and what divisions are most fruitful for explanatory advance. Ultimately, of course, a successful theory should be able to explain the distribution among any units whatsoever (fifths, tenths, one-hundredths, men, women, age groups, ethnic groups, occupations, geographical groups, and so on).

As a general principle, wealth is based on power, both as it directly operates to appropriate goods and services, and as it indirectly operates to limit access to markets. The more individuals or organizations competing in a market, the more equality of distribution will be found within it. To the extent that labor markets are kept separate by barriers to mobility among them (what Clark Kerr (1964) called the "Balkanization" of labor markets), there will be more inequality among the different sectors. The same applies to organizations. The most favored sectors of the organizational world—those which have captured the greatest natural resources and advantages—get the greatest amount of wealth, and have an advantage when dealing with other organizations in a market situation.

Markets are those relationships beyond the realm of direct power, where exchange must take place; an explanatory theory of wealth must show the power conditions that bound the market. This can be pursued on two levels. First, on the overall level of the entire society and the interorganizational relationships within it: Here we are concerned with the direct political power of the state, and with the political organization of the media of exchange in society-wide markets (that is, with the system of money, credit and banking). This will make clear which organizations get what proportion of the total wealth. On a second level of analysis, we may be concerned with the internal politics of particular organizations or sectors of the labor market; this explains how the sector's wealth is subdivided.

A first determinant, then, is the organization of the state. Does the organization of violence directly control economic distribution or does it allow a market to some degree? The first alternative describes a major tendency in most preindustrial societies, as well as in the socialist variant of industrialism. In these cases, the distribution of property and income should be closely correlated with the distribution of political power. The more centralized the apparatus of the state—the greater the concentration of organizational power in one unit—the greater the concentration of wealth and income. Conversely, decentralization in the form of feudal fragmentation or federalism should widen the distribution of wealth. But whether the state structure is centralized or decentralized, there is a separate question of the number of people who participate in it. As this number becomes greater, the wealth of the state is more shared out. Examples of the latter would be the early periods of popular tribute-collecting warfare states such as those of ancient Greece and Italy. Similarly, in industrial state-controlled economies, the state officials may acquire larger or smaller amounts of the wealth individually, depending on the degree of participation of the population in the state.

If the state does not more or less completely control the economy, however, but merely confines itself to upholding the system of property, the variables affecting income and those affecting property will differ. We may expect, in general, that the extent of the political participation in such a state will primarily affect the distribution of property rather than the distribution of income. This should particularly affect the distribution of land, the extent of licensing monopolies for exploiting economic opportunities, and the system of taxation. Income will be affected, but indirectly.[12]

In states allowing a market, the determinants of the wealth of various organizations and organizational sectors will be of another sort. As Stinchcombe (1968b) has pointed out, different industries appropriate different amounts of wealth. Organizations tend to continue to use the technology and administrative forms that characterized them during the period in which they were founded. Older and less effective forms of organizations are those in sectors of which the initial advantages in resources, transportation, and so on, have diminished relative to other organizations, thus bringing them a relatively smaller proportion of the total wealth. But since we are talking about organizations operating within a market, we must tie the fate of organizations with different resources to their position in this market in general.

The fate of any organization is not determined by itself alone but by its competitive position in relation to the rest of the organizations in the market. This operates in two ways. First of all, they are competing for customers, and thus any organization can be displaced by another one that can cut into its market with a new product or form of organization. Even more importantly, the solvency of an organization depends upon its financial position. If we recall that money is not a solid good but a token of credit, a claim upon others' behavior, we can see that the monetary system reflects the constant conflict for control over each others' behavior which characterizes society. An organization must be continually able to control to a sufficient degree the behavior of those around it if it is to survive. This is made most explicit by Schumpeter (1934), who pointed out that, in a competitive situation, the organization that is not moving ahead cannot be standing still but must be falling behind. This is reflected in monetary

---

[12] Evidence for Britain, for example, shows that the richest tenth of the population received most (about 75%) of its income from property, and that no other group received any significant income from property (Meade, 1964; analyzed in Pilcher, 1972: 59–60). Since most of the variations in inequality observed in the industrial world are based on the relative standing of this upper class, political conditions affecting property ownership may turn out to be of major explanatory importance. The greatest inequality of incomes is found in Germany and France (Pilcher, 1972), where private business cartels are given strong government support.

terms: The organizations with the greatest amount of credit, reflecting their real power, will be solvent, whereas the sheer sum of money (or even gold) in the bank belonging to a stagnant, and thus relatively weaker, organization progressively becomes worth less. Eventually, such organizations go bankrupt or are bought out by stronger ones; not because their own organization is necessarily weak in itself, but because of the competitive relationships which determine the value of their resources in the overall market.

What determines which organizations will dominate? The general principle applies: The organization that can subject others to the conditions of the market while protecting itself from the market will do best. We must recognize here that the credit system is essentially a political one. This means not only that banks are generally regulated by the state to some degree; there is also an *internal politics* of banking. The best protection, then, comes from organizations that make the proper "political" alliances, or transform themselves into "political" powers within the financial realm. There are several means by which this can be done. The history of business enterprise in America shows that the most successful industries are those that amalgamate relatively early with the largest banking interests (including insurance companies).[13] The major industrial blocs, which we can chart by interlocking boards of directorates, are always organized around banks as well as major industries. A related tactic, pointed out by Berle (1959: 27–58), is for organizations to generate internally as much capital as possible, through such devices as employee stockholding, management of insurance and pension plans, and so forth. The point here is that the control of ready sources of cash at any one point in time is crucial for maintaining the political position of the organization in the system of credit, and thus for keeping it solvent; this is a precondition for continual control in the hands of a particular group. In a few cases, families (such as the Duponts) are able to provide this function if they are large and internally cohesive (Chandler, 1962: 63–67, 75–80; Lundberg, 1968: 155–294). Foundations serve an analogous function as financial reserves for families and for nonfamily organizational elites as well; even though they are set up to "give away" great blocks of money, the shares are usually politically available to maintain stock holding-control of the company for the family's allies, and to maintain its financial solvency (Lundberg, 1968: 465–531).

All these effects take place over time. The extent to which a given degree of political organization is necessary to maintain organizational

---

[13] See Chandler (1962: 25–28, 42–45, 480–481); for current evidence, Levine (1972).

control essentially depends on how well the other companies are organized. Eventually, we should be able to work out time sequences for these shifts in concentration, depending on various conditions. The question may be put in the form: What enables a company to drive others out of business? Which companies will survive in the competition and which will not? Sheer size would seem to be a crucial factor, resulting in a series of advantages for the company: economies of scale in production and distribution; the capacity to produce a greater diversity of fairly cheap products, thus bolstering itself against shifts in any particular market; better servicing of standardized parts; the capacity to undercut the prices of competitors in periods of price war; a critical mass in public fame and thus in customer loyalty; a critical mass in credit and thus in "political" influence within the credit system; a critical mass in political influence in the state through direct ties with politicians, and also by the creation of a vested interest in the survival of their jobs among large numbers of employees. The organizations which become vertically integrated the earliest—those that secure their most needed and most vulnerable sources of supply or markets by incorporating them within the organization and thereby removing them from the system of market competition—increase all of the cited advantages (Chandler, 1962: 1–21). The opposite tactic is also useful in particular cases: to externalize through franchise arrangements those areas (usually distribution or servicing) that are most erratic or costly, thus placing the risk upon smaller entrepreneurs (Macauly, 1966; cf. Thompson, 1967: 25–38). In general, political power within the credit system goes to the most secure investments, including the security of being able to innovate while having others take the risk. As in politics in general, political influence in the credit system is a self-fulfilling prophecy. The hypotheses about the political system advanced in the previous chapter may be applied more precisely to this particular political subsystem.

Within organizations, we deal with a different sphere of causality. The economic good that is distributed here is primarily income rather than property (the more stable form of wealth). The greatest incomes go to those who have the greatest political power within the organization.

Stinchcombe (1965: 164–167) gives a convenient typology of organizational situations.

1. *Families,* as economic organizations, are characteristic especially in the preindustrial societies; in modern industrial societies, they are found primarily in agriculture, small-scale retail trade, and in a few crafts. Within these organizations, the distribution of wealth follows the distribution

of power within the family (as indicated in Chapter 5). Usually this means that the head of the family appropriates all of the income and distributes it to its members on a purely ad hoc basis.

2. *Bureaucracy* is a generic term to refer to organizations which have become separated from the family. The conditions for the rise of bureaucracy have been given in Chapter 6. They include literacy, a particular type of community organization, and various conditions of size and technology. Under particular conditions, families may maintain their control at the top of the organization. Within the organization, however, there is a complex political situation.

In analyzing organizations, we must distinguish between two types of power: *deference power* (D-power) and *efficacy power* (E-power). The distinction is particularly important under conditions in which subordinates carry out tasks that require a great deal of initiative or whose outcomes cannot be specified in advance. The heads of such organizations have relatively little direct control over what the organization actually does or produces. D-power, however, is not necessarily dependent upon E-power; it is the former—the mere formal authority position—that seems to be the main determinant of relative incomes within the organization. That is because an organizational official may not be able to control much of what his subordinates do; but the payment of his salary is a clear and obvious form of behavior, put in simple quantitative terms, that cannot easily be evaded. The correlation between income and authority is amply demonstrated in the personnel literature, where it is shown that serious conflicts emerge whenever pay structure gets out of line with the structure of *formal* authority (Strauss and Sayles: 1960: 581–611).[14] (There is also a covert form of rewards in organizations, sometimes semiofficially referred to as "perquisites"—the use of the organization's equipment for one's private use. This is highly correlated with organizational power, and may be a primary form of the distribution of wealth within nominally socialist countries.)

Internally, then, an organization may be regarded as a struggle over power, of which the most important consequence is how much everyone

---

[14] Roberts (1959) shows that the salaries of top executives are directly related to the gross income of the organization, rather than to its profitability. Lenski (1966: 354–356) bolsters this with comparisons of governmental and military organizations; the general principle seems to be that executive pay is based on the total amount of wealth available to be appropriated, and the degree to which the executives have the power to set their salaries themselves. The same big companies with well-paid executives also provide the highest pay for middle management (Editors of Fortune, 1956: 97–135), and for manual labor (Reynolds, 1951). The moral is: If you want a big income, work for an autonomous enterprise with large assets (but not necessarily large profits).

gets paid. Stinchcombe distinguishes two main groups within large bureaucracies: (2a) *salaried employees* and (3) *unionized manual laborers.* It should be noted that the positions themselves are the result of power struggle. As men set up organizations, they usually find it convenient to divide its members into two groups: a group to whom they make concessions of power in order to acquire allies in controlling the others; and a group they attempt to keep as powerless as possible in order to keep them in a sheer market situation, and thus keep down the expense of paying them. The distinction between salaries and wages reflects this division.

Just where this distinction is drawn, and whether it is drawn or not, depends on the power resources of various groups and the historical patterns that they produce. Where one ethnic group owns most of the property and can hire its own members as an administrative staff while hiring alien ethnic groups or a group of a very distinct class culture on the wage market basis, the wage–salary distinction is strongly drawn. Sex has been another ground for keeping labor pools separate. Medical practice, for example, is divided between highly authoritative, largely male professionals carrying on the pretensions of earlier gentry lifestyle, and women for whom the menial tasks are reserved; the institutionalization of power interests and the cumulative advantages they produce thus keeps the medical field from being organized in other technically feasible ways, such as the creation of a much larger hierarchy of medical aides and specialists with only limited authority. One prominent organization in which the distinction between the wage-earning group and the salary group is not made is the American police, where the contingencies of ethnic politics have resulted in a single unified career hierarchy (Johnson, 1973).

Within organizations, as elsewhere, the groups that receive the greatest incomes are those which are capable of removing themselves from the market while keeping their subordinates (who compete with them for a share of the organization's total income) as much as possible in the market situation. The organizational structure itself provides many of these resources: the cumulative advantages of power and status control of communications channels, and superior group cohesiveness in particular ranks. It is for this reason that the higher ranking persons receive more income. This is enhanced to the degree that they have external bases of cohesion. A strongly organized ethnic hierarchy, or long-standing class status distinctions in the surrounding community, provides an anchor for keeping different organizational positions in the hands of limited groups.

The power of the state is another resource. Where political conditions allow one ethnic group to dominate, it can be used directly to enforce a caste or estate system of occupational opportunity, such as those found in the Ancien Regime, or in parts of America more or less up until the present. In a somewhat more covert form, favored groups use the coercive powers of the state to monopolize positions (or at least limit the extent of the market) by having it uphold licensing requirements. The formalization of the professions in America came with a period of intense ethnic conflict; the earlier Anglo–Protestant settlers managed to have their cultural qualifications officially protected by the state in the form of licensing laws prescribing education in their traditional culture (Collins, 1975).

Conflicts occur not only between the administrative class and the working class, but within each of these. A major struggle in industrial societies today goes on within the administrative sector, among professional groups attempting to extend their monopolies, and administrators pursuing more individualistic careers. At the working-class level, conflict goes on among ethnic and racial groups, particularly with the use of trade unions as an organizing device. Racial segregation and control over immigration policy are also tactics by which workers attempt to control the market situation for themselves.

Stinchcombe's other categories reflect different forms of organization within the manual labor force. *Craft unions* (4) are those which are organized around a highly monopolizable technical skill which gives them an intrinsic organizational power. They tend to develop in decentralized situations. The employers are in more a market situation than the employees; accordingly, crafts workers enjoy relatively high incomes. The interpersonal nature of the market situation, and the communications network by which information about jobs is passed along, tends to bolster the position of family and friendship networks and to make craft unions the most nepotistical and racially segregated of all manual occupations (Greer, 1959; Meyers, 1946). *Organized labor employed in large bureaucracies* (3) is in a relatively weaker position because of their exposure to a more open market. This varies according to the extent to which political parties in that society are organized around trade unions and thus put governmental pressure on the employers. In general, industrial workers are much more subject to employment according to short-term demand for products and thus enjoy lower wages than crafts workers. Finally, there is an unorganized sector of the manual labor force consisting of *casual employment* (5), especially in service jobs. Here the worker is entirely in a market situation, usually with little internal organization to bring political pressure to bear on improving its position or putting a floor under the other market in terms of minimum wages. Here employment is most episodic; power is most heavily in the hands of the employer; and the relative portion of the organizational income going to the employee is lowest.[15]

---

[15] Another explanation is that these are usually retail service organizations with low

Calculating the various forces that produce the overall distribution of wealth in a society, then, is a rather complex business. Explanatory theory can move forward on several fronts. The easier of these is to test particular propositions: those explaining the distribution of wealth among various industrial sectors, and the conditions determining the distribution of wealth within those sectors. The overall distribution of wealth within a society must be added up from these various processes. The distribution of *property* seems to be determined primarily by the overall political organization of the society and the "financial politics" connecting organizational sectors. The overall distribution of *income* within the population must be calculated from both the intersectoral and the internal organizational conditions. That is, we look at the different industrial sectors and the conditions producing the total amount of wealth in each of these; then we look at the conditions dividing each of those sectors into different occupational strata, and at the conditions for relative power among these. Adding up the total amount of income going to groups within each of these, then adding across the entire society, should give us the observed overall income distribution.

Social Structure and Mobility

The same framework may be applied to the analysis of social mobility. The overall political structure of society sets the conditions for social mobility in at least two senses. First, it determines what kinds of positions there are to rise or fall into. It also determines the total amount of movement; this is dependent upon whether political resources are highly restricted, resulting in a static political situation, or widely enough distributed to result in a possibility of considerable conflict and thus change. It is this political change—the rise and fall of kings and advisors, priests and bureaucrats, or of whole organizations—that makes up one large component of social mobility throughout history.

Where the political structure allows the existence of market organizations, we have another sphere of competition. Again we have the same two kinds of conditions effecting mobility: how much vertical differentiation of positions there is; and how resources are distributed among organizations so that there is a high or low rate of change in organizational fortunes, or in the creation of new organizations. One form of the latter, of course, is what we refer to as "economic development."

We have here two aspects of mobility rates which can be tested by available evidence: the limits set on mobility by the total number of

gross income, therefore less of a total pie to be carved. This would be the obverse of the case noted in footnote 14.

positions of various kinds; and the production of mobility by the distribution of conflict resources producing organizational change.

It is known that the proportions of different kinds of positions limits the mobility among them. If there is a relatively small number of upper-class positions, for example, then only a small proportion of the lower ranks can possibly move up, even if these positions are not taken by upper-class sons. Conversely, a large upper-class stratum allows opportunities for more persons of lower-class background to move up. Marsh (1963), who first put forward this argument, showed that comparable amounts of mobility could be found in Imperial China and in the modern United States if one focused only on elite occupational structures of approximately the same size in both countries. In a comparative analysis of ten modern societies, Marsh demonstrated that the amount of mobility into elite positions was correlated with (and presumably determined by) the size of the elite strata.

Other evidence tends to corroborate the importance of those "opportunity structures." A comparative study of six American cities (Lane, 1968) found higher mobility rates in the larger cities; and it is larger cities that have a more diversified occupational structure, with larger elite occupational sectors. Relatedly, a comparison of the ease of mobility into elite levels in the United States, England, and Denmark (Beshers and Laumann, 1967) found that the rates vary directly with the size of the elite structure. Harrison White (1970a), using stochastic models of vacancies for ministers in three large American religious denominations, found that the rate of mobility (here confined to a segment of middle- and upper-middle-class positions) can be predicted from the occurrence of vacancies plus certain characteristics of the rate at which they are filled.

This finding emerges even from a study designed to test a functional theory. Kingsley Davis (1962) has argued that industrialization requires that men be chosen for jobs on the basis of their individual talents, whereas nonindustrial societies can survive if men merely inherit their fathers' positions. Cutright (1968) has attempted to test this hypothesis by comparing the pure mobility rates (measured by Yule's Q) of thirteen European and North American societies. Mobility samples were from the 1950s and early 1960s, and were found to be the most strongly correlated with the level of industrialization and with the level of communications (literacy rates, newspapers and telephones per capita) in 1930. From these results, Cutright concluded that Davis' theory was supported.

Actually, these results do not clearly test Davis' theory, which would involve showing that pure mobility rates *result in* economic development, rather than are *preceded by* it. But there is another question of interpretation. Yule's Q, the measure of "pure mobility" used, does not rule out the effects of "opportunity" structure. It is designed to separate out the effects of *changing* occupational structures, but is also affected by the *static* size of occupational categories.[16] Industrial development (measured in Cutright's study by energy per capita) is a good index of the proportionate development of the higher occupational categories; so are measures of communications (directly implying something about the number of white-collar positions).[17] The most plausible interpretation of Cutright's data is that such mobility which is not directly an aspect of changing economic structures is related to the proportions of opportunities at the higher levels. Marsh's "opportunity structures" (a renaming of his term "demand structures") thus appear to be the most successful explanation of differences in mobility rates.

We should not hypostatize "opportunity structures," however. Struggle between social groups is what creates and sustains organizational positions. It is not simply a case of how much water can be poured into a big can or a small can, as the *size* of the cans can change. The reason a small aristocracy stays small is because it attempts to keep others out, and because it has the resources to carry it off. The reason that modern professions grow larger (or fail to) is because of the power of social groups to force their way in or to keep others out.[18] Even bureaucratic organizations,

---

[16] $Q = \dfrac{ad - bc}{ad + bc}$, where a, b, c, and d are the cell frequencies in a fourfold table. When any one of the cells is empty, $Q = \pm 1.0$, indicating perfect inheritance or disinheritance. A Q of zero supposedly indicates that occupational destinations are independent of occupational origins. But except for extreme cases where cells are actually empty, the relative size of the categories influences the value of Q. Thus, taking the simple case where the occupational distribution is the same in both generations, and lower-class sons move up into 40% of the available upper-class positions, Q depends on the ratio between lower-class and upper-class positions:

When 50% of the jobs are upper-class, $Q = .38$;
when 30% of the jobs are upper-class, $Q = .76$;
when 20% of the jobs are upper-class, $Q = .86$;
when 10% of the jobs are upper-class, $Q = .89$;

Even where the lower-class sons get half of the upper-class positions, Q is zero only when the two categories are equal in size:

When 30% of the jobs are upper-class, $Q = .57$;
when 10% of the jobs are upper-class, $Q = .89$.

[17] Communications and energy consumption are highly correlated ($r = .90$) in Cutright's sample.

[18] In the United States, the unusually massive development of the school system provided many middle-class jobs of its own, as well as facilitating the expansion of other credentialed positions in American society; this resulted from dispersed organizational resources, especially the political decentralization of the United States, the ethnic pluralism of its immigrants, and from ensuing conflicts (Collins, 1975). In a preindustrial society like

which we tend to think of as expanding purely according to economic demand for their services, may be understood in this light.

"Economic development" is not a *deus ex machina* we can invoke retrospectively to explain social structures. If certain positions are not created, a certain degree of economic development simply does not occur; the explanation must focus on the actual causes, which means especially the resources that groups have to limit or to create positions. These resources are largely carried over from the past; the previous efforts of groups to better themselves and control others sets the opportunities and limits for persons living at any particular time. The process of industrialization includes such mutually reinforcing occurrences as the creation of markets which provide opportunites to expand positions for manufacturers, administrators, professionals, and others. In effect, a high level of continual conflict with shifting outcomes is a major cause of industrial development, allowing groups to acquire resources for themselves which also enable them to act as markets for others. Conversely, a tightly controlled situation of group dominance prevents such changes, and hence discourages economic development.

Economic development may thus be incorporated into such a struggle model of social mobility. This is not meant to give the impression that economic development is the only or necessary result of all struggles which result in creating new positions. Government employment, for example, is heavily determined by the struggle of various groups for patronage and varies a great deal historically according to changes in group power.[19] The causes of government employment have been obscured by an overriding concern for its Keynesian effects in offsetting technological unemployment. It should be recognized that "economic necessity" is actually a very loose thing—depending, first of all, on the commitment of powerful groups to have a certain kind of economy, and secondarily, on the relative power of different interests which prefer to increase aggregate demand by government employment, by laws forcing private employers to redistribute wealth (e.g., via shorter hours), or by direct redistribution to various social classes. Similarly, the amount of "fat" that modern oligopolistic corporations have at various ranks may also be seen as the result of group struggles; the proportions of staff and managerial positions

England, dispersed political resources resulted in the expansion of upper-class positions in the legal profession, the clergy, and the military. Another way that new positions could be created for younger sons of the aristocracy was conquest and colonization.

[19] The introduction of parliamentary democracy into poor and highly unequal countries has usually expanded government sinecures for political patronage. This was the case in Italy after 1871, as well as in Latin America in the twentieth century (Soares, 1966).

must be understood in terms of the power of these groups themselves to defend and expand their own ranks.

There are some indications that structural change in the occupational distribution is a major correlate of variations in mobility rates.[20] The usual explanatory model, reflected in the term "forced" mobility, is that the observed changes *cause* the shifts that make them up. But this is retrospective causality, committing the sort of error that is basic to functional analysis. Putting the matter more realistically, we can say that most of the variation in mobility is associated with, results in, change in the occupational structure. Like the "opportunity structures" explanation of mobility rates, measures of "structural mobility" tell us something about the concentration of resources among contending groups for limiting or expanding desirable positions. The validity of this interpretation, of course, depends on its coherence with the results of research on the same social processes from other more detailed points of view.

Yet another aspect of mobility is internal to organizations. Here we apply the same variables as those explaining the distribution of wealth. Internal organizational resources determine the differentiation of positions within organizations, the barriers to movement among them. That is, it is the policy of groups with organizational resources to attempt to close off their position to others in order to keep their incomes high by making them less subject to market conditions. The extent to which this happens seems to depend on certain external conditions. Strong ethnic group domination in the larger society is one condition. A high degree of professionalization and of unionization are others. Introduction of educational requirements for employment is yet another.

Systematic evidence is available only on the last; other possibilities still need to be tested by comparative evidence. We know from American evidence that the rise of educational attainment in the population has gone along with and probably produced a shift toward higher educational requirements for employment (Collins, 1971) and that, as a result, there has been no shift in the rate of social mobility during the twentieth century, as far as we can measure it (Blau and Duncan, 1967: 110). For African evidence, Kelley and Perlman (1971, 1972) have shown that the introduction of an educational system into a society that previously lacked one resulted in a decline in the mobility rate; the degree of decline could

---

[20] Matras (1961), drawing upon data on mobility across the manual–nonmanual line from Bendix and Lipset (1959: Table 2.1 and Figure 2). Matras shows that changes in occupational distribution and class fertility differences together "account for" 20–50% of the total mobility in different samples, but that the residual, "exchange mobility," is virtually the same in all samples (as a proportion of the total careers, mobile and nonmobile).

be directly attributed to the effects of education on mobility.[21] Education, as the control of cultural resources, is apparently more stably monopolizable by elites than political or economic resources. But then, that is what it was designed for.

Determinants of Individual Careers

The two questions—which individuals become rich and which individuals become occupationally successful—are essentially the same. The latter has been much more studied, so we shall have to extrapolate from occupational mobility to the study of economic careers. These questions are analytically subordinate to the previous analysis of historical structure. History shows a variety of different types of competition. Periods of military competition in agrarian and horticultural societies favored those individuals who had the resources (including the conditions to produce the proper motivation) to become successful warriors. Other historical conditions have favored bureaucrats, priests, businessmen, scientists, engineers, or politicians of particular sorts. The analysis of individual careers and their determinants, then, must always be put into historical context. We should be careful about over-generalizing results; the role of education and the variables leading to it in modern America are not universal. Once we understand the historical varieties, we can begin to see just which processes on the individual level *are* universal.

One of these certainly must be the role of chance. The use of stochastic models shows us how far we can go in explaining careers as random movement within a given structure. The same result is found from biographical perspectives; success in politics and in business seems to depend to a great degree upon happening along at the right time. One might cite, for example, the career of Winston Churchill, whose rhetorical skills made him a popular leader during wartime, but whose peacetime career was relatively mediocre. Such diverse figures as Lenin and Lincoln demonstrate the same principle. In business, it appears that the careers of many major executives (and their resulting fortunes) result from being around at the proper time. Townsend of Chrysler Corporation, for example, seems to have made his career by being on the spot when his organization finally came out of a business depression that had removed the previous management; McNamara at Ford made his career by a show of technical expertise during a period in which Ford Corporation, after the death of its founder, was drastically in need of organizational renovation. A similar role of

---

[21] Kelley (1972) proposes an explanatory theory parallel at many points to the model advanced here; cf. Bourdieu and Passeron (1970).

chance in the careers of entertainers is well-known. The subject of careers, of course, has been the object of so much folklore that this mechanism has been overlooked. Most accounts have concentrated on how dedicated and hard-working the successful individuals have been. What is missing is the fact that there are a great many people competing, and the proportion of those who succeed is relatively small. Those who work at it longest and hardest, and those who begin with the greatest resources, are most likely to triumph over the others; but even within this group, the role of chance seems very great.

Chance conditions are also important in relating particular kinds of skills to the kinds of wealth available. The skills that would make a man a successful preacher were of much greater importance in eighteenth-century America than in the twentieth century, with a relative decline in clerical pay. Similarly, academic talents and connections were of greater importance in America in the 1950s and 1960s, during a period of expansion in the educational system and a resulting rise in the pay scale, than before or (no doubt) after.

More generally, what we need to formulate are the total organizational conditions that make particular types of individuals successful. There are some revealing cases in business history: the creative inventor Edison, whose development of the electric lighting system brought profit not to himself but to J. P. Morgan, who financed and organized the operation; or Henry Durant, a free-wheeling entrepreneur who organized General Motors Corporation only to be squeezed out in a period of bureaucratic consolidation, with the G. M. fortunes going to Alfred P. Sloan, the organizational politician. Which individual characteristics will be important, then, depends on the overall distribution of organizational resources at the time. These determine both the possibilities for individual movement and the total amount of rewards in wealth and power that are available to them.

A convenient mathematical tool for summarizing the effects of variables on individual mobility is path analysis. By means of this technique, a set of correlations bearing on occupational attainment is arrayed as a series of processes in time. Most such studies have concentrated on those conditions leading to educational attainment. Some recent analysis has begun to focus on the period after education, showing how the sequence of first and subsequent jobs leads to the present job. Explanatory models have been primarily social–psychological, concentrating on the ways in which ability, motivation, family settings, and encouragement by others affects school attainment (Sewell *et al.,* 1970; for a general summary, see Collins, 1975). But education and family background together explain only about 40% of the variance in early occupational attainment, and these effects decline for later jobs.[22]

What explains the remaining variation in occupational achievement? The most promising approach is to examine the sequence of jobs each individual holds. One job is relatively close to the next in the occupational prestige hierarchy; sampled over periods of about five years, each job explains about 40–50% of the variance in the prestige of the next (Featherman, 1971; Kelley, 1973). One conclusion we can draw is that, from the point of view of the individual, there is much random movement or emergent causality; his own past determines his future only within short time periods, and even there only partially. The explanatory problem, however, can also be treated as one of levels of analysis; we may do better by referring to a structural level of analysis.

This is particularly a matter of understanding the variety of structures that an individual can move through. Some of this can be seen by studies that record the entire lifetime job histories of a group of individuals rather than sample their jobs at a few points in time. At least in modern America, individuals pass through quite a few jobs in their lives, including considerable shifting among major occupational sectors and across the manual–nonmanual line. Some persons, though, get into much more orderly careers, especially in the professions and in certain bureaucratic jobs (Wilensky, 1961, 1966; Gusfield, 1961). We could say more about the determinants of occupational mobility if we decompose our implicit notion of a single structure into a number of different types of career channels—including both tighter and looser ones—and incorporate the surrounding social–cultural milieux as well.

The conflict model proposes that careers take place within an ongoing struggle of cultural groups to control positions by imposing their standards upon selection. Success comes to individuals who fit into the culture of those who hold the resources to control old positions or create new ones. A number of studies (Dalton, 1951; Coates and Pellegrin, 1957; O. Hall, 1946) have shown that careers are made by moving through networks of acquaintanceship—including both informal cliques and formal club memberships—which in turn are facilitated by cultural similarity.[23] We also know that men are more likely to stay in a particular type of job if it is filled by others from backgrounds like their own (Blau, 1965).

---

[22] The effects of education on *income* are delayed into midcareer, but are a good deal smaller than its effects on occupational prestige level (Featherman, 1971; and a reanalysis of the same data corrected for measurement error, Kelley, 1973).

[23] Manual workers also find their jobs primarily through information supplied by acquaintances (Reynolds, 1951; Sheppard and Beliksky, 1966).

We might expect to predict careers better if we could incorporate more detailed information on the cultural similarity between respondents and those who control positions. Another way of approaching this would be to predict occupational moves from the milieux of friends that individuals have at various times. There are some promising leads along these lines. Through the research of Laumann and his colleagues (1966, 1969), we have begun to acquire a picture of the friendship structures of American society. Laumann's data show that the likelihood of friendship is distributed along several dimensions: In one set of researches, we find that it falls within ethnoreligious lines, along broad underlying dimensions of cultural and economic similarity among denominations. Another set of researches show that friendships correspond to three dimensions in the occupational structure: one related to hierarchy of power and prestige; a second corresponding roughly to entrepreneurial and bureaucratic types of jobs; and a third as yet undefined. This evidence suggests that mobility could be better predicted by taking into account the different *dimensions* along which one may move in the occupational structure through the informal milieux of cultural similarity and friendship links.

In effect, this means refining our measures of what we mean by "mobility." What do we mean by saying that someone has "moved"? Usually, this implies more than that a person is no longer in exactly the same job as his father (or himself at an earlier point). We group jobs into broad levels that we consider similar, and we propose a dimension along which these categories are arrayed. Thus, if we have five categories—professional and managerial, clerical, skilled labor, semiskilled and unskilled labor—we usually place them in a hierarchy and describe a move from unskilled to professional as *"more mobility"* than from unskilled to clerical.

It is obvious that our explanatory efforts will work better if we have arrayed the categories in a way that corresponds to the actual difficulties of moving among them than if we have them in an "unnatural" order. But this implies that a good deal of *implicit explanation* has gone into constructing the categories in the first place. It follows that explanation can be improved by refining the dimensions, especially on the lines of Laumann's model of ethnic and occupational sectors of friendship milieux.[24] This, in turn, brings us again to focus on the *structures* within which mobility takes place, as a crucial element in explanation.

The line of analysis concerning the causal links among jobs and their surrounding milieux will probably be found to be more powerful and more universal than the current fashion of concentrating on education.

---

[24] Beshers and Laumann (1967) derive measures of the social distance of occupations inductively from mobility data on different societies, but confine their analysis to a single hierarchy of five occupational levels.

The reasons for the relationship between school and subsequent occupational attainment are hardly self-evident. The evidence best fits the interpretation that education is important, not for providing technical skill but for membership in a cultural group which controls access to particular jobs (Collins, 1971). This may be the traditional culture of a class or ethnic group; or it may be a new culture created in the schools, with feedback from the organizations to which they are linked; or it may be some combination of both. In any case, education will be found to be important in occupational mobility where it provides cultural entrée into dominant groups, operating as a modern bureaucratic equivalent of ethnicity. Stating the conditions under which either reflective or autonomous educational cultures are important, of course, brings us back to the historical–structural level of analysis. Social mobility theory necessarily must swing in this circle.

Mobility Rates and Links between Levels of Analysis

The general strategy of conflict theory is to look for the organizational resources that distinguish conflicting groups and enable them to compete with each other. The model is not rigidly deterministic because the outcome of the conflicts at any given time sets the stage for a new alignment of resources at a given time. However, we can predict the likely form of the next structure, calculate which individuals will change their positions, and an even more complex task, predict rates of movement. There has been relatively little research on mobility rates in terms of wealth, although perhaps these might be constructed by invoking some of the known correlations between occupation and wealth. The following analysis will concentrate again on occupational mobility.

"Mobility rate" is an abstract term referring to a number of different measures of change in social position within a given community. They can vary in the specific time period covered. At one end, we can be concerned with changes in position in a relatively short period or in movements during an individual's lifetime; at the other, we may concentrate on movement across generations (which is equivalent to the relationship between the family and the individual). The latter sort of analysis has been more popular, as it continues on an essentially ideological issue—family influence on success being considered an unfair advantage. More recent studies show that the influence of family is *now* pretty much confined to its influence on education; it remains to be shown through historical comparative analysis the conditions under which family influence is more or less pervasive in other respects.

Further variety comes from the different populations who may be studied: the mobility rate within an entire society, within a particular community, a particular occupation, a particular industrial sector, or even within an organization. Each of these will probably give us a somewhat different rate, even within the same society. The complete explanatory theory should be capable of giving the conditions predicting any of them. The core of such a theory, though, will be found on a more fundamental level of analysis; mobility rates are a kind of derivation that we can make if we have the proper theoretical tools.

Most of what is known about the determinants of mobility rates has been discussed already, where rates were taken as indicators of the organization of the social structure. It is apparent that all three of our levels of mobility analysis—structural, individual, and rates—are interrelated. Overall rates and individual careers are affected by the same underlying processes. They are simply different measures abstracting out different aspects of the same phenomenon; each sort of information can be used to make inferences about the other. Analyses of overall mobility *rates* in a population will tell us something about conditions affecting *who* will be able to move. I have argued that measures of "opportunity structures" reflect the line-up of resources that elite and subelite groups have for restricting or expanding elite positions. Conversely, if we knew all the determinants of individual mobility (and how these resources were distributed in the population), we should be able to construct a composite mobility rate. Work on stochastic models has begun, in a rudimentary fashion, to link together these two levels. The approach has been to make certain probablistic assumptions about individuals in a sample, and then to see if the observed mobility rate can be reconstructed on this basis.

Thus far, the debate has concentrated on what modifications in a conventional Markov chain must be made to account for the observed data. The Markov model assumes that the probabilities of moving are the same for all individuals and that they remain the same for each move—i.e., that each move by an individual is unaffected by his previous moves. The problem is that conventional Markov chains predict more mobility than is actually observed; that is, they fail to account for the density along the main diagonal in mobility tables, reflecting the tendency for occupational inheritance. In order to account for this fact, assumptions have been modified in two different ways. One form of analysis introduces the notion of cumulative inertia: that an individual's probability of staying in a position increases the longer he has been in it (McGinnis, 1968). A second tack is to assume that the population is divided into two groups, "stayers" and "movers," who have different probabilities of movement (McFarland, 1970; White, 1970b). Neither of these two modified models has, as yet, predicted observed data patterns very well. The cumulative inertia model thus far has been tested only on geographic mobility, not occupational attainment; in any case, its long-term implications appear to be that movement eventually slows down and stops entirely—raising the questions of how, under the assumptions of this model, any mobility could originate in the first place. The stayers–movers model seems more plausible, but it has been possible to fit it to data only by ad hoc restrictions to particular social classes, chosen after the fact in particular samples (White, 1970b). Moreover, when applied to intergenerational mobility, its assumptions seem to require dividing *families* into stayers and movers, suggesting a genetic interpretation of mobility of the sort that, on other grounds, again fails to be very plausible.

Stochastic models, then, have so far been mainly of hypothetical interest for a realistic mobility theory. Nevertheless, the mode of analysis may be potentially useful in several ways. It lets us think, although as yet in a very abstract and unspecified fashion, about constructing an overall mobility rate out of individual moves. In this sense, it marks a goal toward which mobility analysis may eventually move. In the interim, efforts to see what is involved in bridging this gap may help bring to light the implications of our current models. The effort to construct stochastic models, incorporating what is known on other levels of analysis, may lead to the refinement of these models.

The ideal stochastic model will no doubt be a rather complex one. It will have to incorporate the independent effects of population shifts, vacancies within existing organizational structure, and the resources that different generational cohorts, class, and ethnic groups have in struggling for positions. It will have to avoid the retrospective fallacy that comes from treating a given change in the organizational structure (labeled "economic development" or "technological change") as if it were a cause rather than a consequence of structural shifts and hence of "vacancies." Positions must be constantly upheld if they are to last; the vacancy model must become more sophisticated to understand that what it is dealing with are stable resources that make the social structure relatively stable in certain areas, and that explaining the degrees of instability is crucial in predicting careers. A more appropriate *general* model of mobility would be a matchmaking model (White, 1970a: 193–215), with stable vacancies as a special case to be derived from it under particular conditions. The retrospective fallacy of making things seem a simple replacement process is easy to fall into when dealing with data from the past. We will be sure of the power of our theory only if it can predict mobility rates in the future, starting from conditions in the present.

APPENDIX A: SUMMARY OF CAUSAL PRINCIPLES

23.0 Total wealth. The total of goods and services of a society during a given time period is the result of the amount of natural resources that are appropriated in its territory; the size of the population working; how long and how hard they work; the efficiency of the technology used; the efficiency of the economic division of labor; and the store of past goods and improvements; *minus* the amount of goods and improvements that are destroyed during that period.

23.1 Direct state distribution. If the state directly distributes wealth, the distribution of property and income correlates with the distribution of political power (cf. *19.0*).

23.11 Lenski's principle. The greater the concentration of political resources, the greater the concentration of wealth.

23.12 The greater the portion of the population which effectively participates in state power, the greater the equality of wealth (cf. *22.0* and *16.3*).

23.2 State and property. If the state merely upholds a system of property and allows income to be distributed by a private market, the main effect of political power is upon property wealth (land, taxation, expropriation, monopolies) rather than upon income.

23.21 The greater the concentration of political resources, the greater the concentration of property wealth.

23.22 The greater the portion of the population which effectively participates in state power, the greater the equality of property wealth.

23.3 Wealth of organizational sectors. In a private market economy, the greater an organizational sector's advantages, compared to other sectors, in wealth of natural resources appropriated, in ease of transportation, and in availability of productive and administrative techniques, the greater the proportion of the total wealth appropriated by that sector.

23.31 The greater the resources of an organization for attracting customers, compared to its competitors, the greater the proportion of the total wealth it appropriates.

23.32 The more an organization can protect itself from experiencing competitive market conditions while its competitors are exposed to the market, the greater the proportion of the wealth it appropriates.

23.33 The greater the credit resources an organization can call upon compared to its rivals, the greater the proportion of the total wealth it appropriates in the long run.

23.331 The earlier an organization amalgamates with its largest potential sources of financial credit, compared to its competitors, the greater the proportion of the total wealth it appropriates in the long run.

23.332 The closer the links between an organization and its sources of financial credit, the greater the proportion of the wealth it appropriates in the long run.

23.34 The larger the size of an organization, the greater its potential for diversification and the lower its dependence upon particular market sectors, and the more secure its financial credit.

23.341 The larger the size of an organization relative to its competitors, the greater its "political" influence in the community of financial credit, and with the state (cf. *22.0*).

23.35 The earlier an organization secures its most vulnerable markets or suppliers (by vertical integration) compared to its competitors, the less it is subject to market fluctuations in prices, the higher its influence in the credit community, and the greater the proportion of the wealth it appropriates.

23.351 The more an organization externalizes the least profitable risks (through franchising or other arrangements), the greater the proportion of the total wealth it appropriates.

23.4 Distribution of wealth within organizations. The greater the power of a group to command deference within an organization, the greater the proportion of the organization's wealth it appropriates.

23.41 Where power to command deference diverges from power to affect organizational production, incomes are correlated with deference power rather than efficacy power.

23.42 The more a group is able to protect itself from experiencing the fluctuations of the labor market by controlling admissions into its own ranks, the greater the proportion of organizational income it appropriates; the more a group is directly exposed to short-term fluctuations of the labor market, the lower the proportion of wealth it receives (cf. *23.32*).

23.43 The longer the terms of appointment and periods of payment, the greater the proportion of the organizational income appropriated (from *23.42*).

23.44 The more sharply the surrounding community is divided by internally cohesive and culturally distinctive groups, the more likely organizational positions will be divided into sharply different levels of tenure and power, and the more unequal the distribution of wealth within organizations (cf. *22.2* and *22.52*).

23.45 The greater the political influence over the state held by a cultural community, the more likely it is to monopolize desirable organizational positions by means of licensing or other credential systems (cf. *18.3*).

23.46 The more intrinsically monopolizable the training in a technical skill, and the more autonomous its exercise in practice, the more likely are its practitioners to be protected from labor market fluctuations, and the higher the proportion of income they receive (cf. *18.1*).

23.47 The more that employees are protected from experiencing market fluctuations by union organization, the greater the proportion of organizational income they receive.

23.48 The more episodic and casual the employment, the lower the power of the worker to set terms of employment, and the lower the proportion of organizational income he receives.

24.0 Social mobility. The amount of social mobility is determined by the distribution of resources for monopolizing resources as permanent positions and organizations, and for creating new positions and organizations.

24.1 The more widely organizational resources are distributed, the higher the rate of social mobility.

24.11 The more unstable organizational resources are in a society, the higher the rate of social mobility.

24.111 The more concentrated the political resources within a society, the lower the rate of political change and the less the mobility in political careers.

24.112 The more concentrated the economic resources within a society, the lower the rate of economic change and the less the mobility within economic careers.

24.12 The greater the diversity of positions that have been established at any point in time in a society, the greater the potential social mobility (by definition).

24.121 Opportunity structures. The larger the proportion of positions out of the total which are in a given category (e.g., elite positions), the greater the potential social mobility into those positions.

24.2 Mobility within organizations. The more concentrated the power resources within an organization, the less the mobility within the organization.

24.21 The more sharply the surrounding community is divided by internally cohesive and culturally distinct groups (e.g., ethnic groups), the lower the rate of social mobility (cf. *23.44*).

24.22 The higher the degree of professionalization within an organization, the lower the rate of social mobility (cf. *23.46*).

24.23 The higher the degree of unionization within an organization, the lower the rate of social mobility (cf. *23.47*).

24.24 The sharper the gap in educational requirements among different levels of organizational tenure and authority, the lower the mobility within the organization.

24.3 Individual success. The closer an individual is in the network of personal acquaintanceship to resource positions which fall vacant or are creatable in conflicts or other transitions, the greater that individual's chance of personal success (cf. *22.6*).

24.31 The closer that organizational positions are to each other in the network of personal acquaintanceship, the easier it is for individuals to move between them.

24.32 The more closely an individual's style of ritual interaction is to that of the persons who control positions of authority, the more likely he is to acquire such a position (cf. *5.2–5.232*).

APPENDIX B: DEVIANCE AND SOCIAL MOBILITY

When removed from the sphere of value judgments within which it is usually handled, the study of "deviance" is essentially a matter of explaining careers. A good deal of it emphasizes particular kinds of failures (i.e., downward mobility), particularly processes of cumulative failure produced by being labeled by official agencies. This rubric also includes particular kinds of forbidden success. Other areas of "deviance" concern particular realms of work and recreation that are not regarded as legitimate from the point of view of dominant classes. The latter is properly a matter for the theory of occupational classes and cultural communities, but may be handled here. The connection between the materials of deviance and conflict theory is obvious. A conflict theory of social mobility may be applied toward explaining various types of "deviance." Conversely, this area of study helps us to understand the complexity of some of the phenomenon of stratification.

Social mobility has been handled on three levels of analysis: The *structure of relationships* among groups and organizations can be explained by the historical sequence of conflicts among groups with varying resources; *individual careers* can be explained by the resources that individuals acquire by movement through particular parts of a given social structure and by the subjective reflections within the individual of those experiences; and *statistical rates* of social mobility can be explained by combinations of individual and structural factors. Applied to deviance, this helps us reconcile various types of theorizing about deviance. Labeling theory and related approaches emphasizing the social construction of deviance operate on a structural level and are a subcategory of the historical organization of stratification. The social–psychological models which have been so prominent in explaining individual deviance are like explanations of individual mobility. Finally, rates of deviance are quite analogous to rates of social mobility; in fact, they are rates of particular types of mobility, usually downward. The main difference is that, whereas social mobility rates have been most commonly studied intergenerationally, the study of deviant mobility has concentrated on lifetime mobility.

It is convenient to divide "deviance" into violent and nonviolent offenses. In the first category, we deal with crimes of violence; this will include crimes against property, since property is directly constituted by the potential coercion of the state. Combining these enables us to handle those aspects of deviance which arise from political and economic stratification. Nonviolent "deviance" includes all of the "crimes without victims" as well as offenses against other cultural standards that are punished by disapproval, incarceration, or treatment (such as for mental illness). These are not regarded as threats to the structure of economic and political power but to the status structure of the community and the rituals that uphold it.

VIOLENCE

Violence may be part of an occupational career or it may be an amateur sideline. Violence as a basis for careers is by no means confined to "crime"; it is also the basis for military careers and hence undergirds all states and systems of property. Explaining the extent of the use of violence in a society (the structural question) or the rates of success or failure in using violence (the individual and rate questions) is thus a matter that has already been treated in our discussion of political sociology. "Crime," as a specialized type of violent career, essentially refers to violence that is not successful at legitimizing itself: It is violence not by those who are all-powerful in the society, but by the underdog; it is violence by those who have not made the necessary alliances, particularly with religious organizations, to legitimate themselves.

Where there is no stably organized state, the category of crime is not generally applicable (Wertheim, 1968; cf. Roth, 1968). If it is applied, it is usually by outsiders who mistakenly regard the situation as equivalent to the private use of violence in their modern state. This is the situation of warring patrimonial clans, especially in rural areas of Asia, Latin America, and historically in the mountainous regions of Europe or North Africa. Such arrangements, of course, may be imported into a bureaucratic state, as in the continuation of patrimonial politics in the form of the Mafia in the United States, or the cliques of landowning families in modern Latin America.

Focusing on "crime" in a modern sense thus tells us something of the existence of multiple centers of political power within the modern state. There is considerable stratification within the world of crime itself. As in the legitimate world, stratification is based on those with the greatest amount of organization and the most stable resources. At the top of the world of crime are the organizations with private armies supplying the most regularized section of illegal activities: primarily the provision of illegal services such as gambling and drugs. The existence of "big business" in the crime world depends on the existence of cultural stratification making certain consumer goods and services illegal. As has been frequently pointed out, there is a symbiosis between "racketeers" and moralistic politicians and community leaders (Schelling, 1967). The racketeers exist primarily as a private government offering protection (and also making threats) to illegal services; hence, intermittent police raids against gamblers (and, at a lower level, against prostitutes) are a means by which the covert governments (the racketeers and the pimps) keep themselves in the position of a necessary protection, for which they are paid much of the profits of illegal business.

At a lower level in the crime world are the individual entrepreneurs staging robberies. This is a much riskier business, involving direct use of violence, whereas big business in crime prefers to use covert threat. The situation is analogous to that in the sphere of politics where states with concentrated coercive resources do much less fighting internally than those in which the resources are more widely dispersed.

There is also a thin borderline between big business crime and legitimate business which uses the official state to enforce its claims. Racketeers, where possible, invest their profits in legitimate business, usually of the entrepreneurial and speculative type in which their skills of manipulation make it possible to occasionally make use of their private armies to enforce illegal tactics. From the other direction, legitimate business also tends to verge on crime. As Sutherland (1949; see also Fuller, 1962; Dalton, 1959) has shown, legitimate business executives break the law when opportunities are favorable; their superior power keeps these offenses from being often prosecuted, and even then within a context in which they are treated as civil rather than criminal offenses. Offenses ranging from price fixing and embezzlement to appropriating company or government property thus are found widely in one degree or another. The extent to which these are labeled as "criminal" depends ultimately on political conditions. Openly conspiratorial cartels are not only legal but supported by the government in France, Germany, and Japan. At the opposite extreme, any effort to make private profits is a criminal offense in Communist countries.

Individual careers in crime, and composite crime rates, then, depend on the political context. It determines what, in effect, the potential criminal "positions" are. Higher crime rate thus will be found in a society in which more activities are called "crimes." This is equivalent to the opportunity-structure explanation of social mobility. On the individual level, which persons will be successful or unsuccessful in crime depends on a complex of factors. Alternative paths for success are one aspect of this; where legitimate careers are blocked and resources are available for careers in crime, individuals would be expected to move in that direction. The prominence of Italian-Americans in organized crime, for example, is related to the coincidence of several historical factors: the arrival of large numbers of European immigrants from peasant backgrounds who demanded cultural services that the dominant Anglo–Protestant American society made illegal; the availability of a patrimonial form of military organization that could be applied to protecting such services; and the relatively late arrival of the Italians in comparison to other ethnic groups (e.g., the Irish) who had acquired control of legitimate channels of political and related economic mobility (Bell, 1961: 127–150; Cressey, 1969). Similarly, the urbanization of American blacks in the mid-twentieth century was a precondition for a large market for gambling and narcotics in the black community; racial discrimination and lack of economic resources provided that many of the best career opportunities for this group have been in crime.

The organization of the criminal world itself places other limits on which individuals will be successful. Again, as Sutherland (1937) has pointed out, there is only room for a certain number of robbers, burglars, or providers of illegal services without cutting into each others' profits. At the big business level of crime, private armies are used to monopolize particular areas. It is in the most profitable areas that one finds the most stringent requirements for career entrance; in America, these are based on ethnic membership and family ties. This is analogous to the more stringent controls over entry into legitimate careers at the higher levels. Even at the lower levels, becoming a successful burglar depends on learning the techniques, primarily by a form of apprenticeship to successful criminals. If the existing group has control over its new members, those who fail to become integrated with the rest of the criminal community and serve the proper apprenticeship will be much more likely to be unsuccessful.

Considerable attention has been paid to the effects of conviction and imprisonment on careers in crime. Labeling theory has argued that these processes change the self-image of the individual into that of a full-fledged criminal; the alienating structure of incarceration reinforces his identification with the inmate culture and thus provides a means for socializing novices into the world of crime; employers' use of criminal records in their hiring practices keeps ex-criminals in menial occupations and thus makes opportunities in crime seem relatively more promising than legitimate careers (Glaser, 1964; Melchercik, 1960).[25] We find a vicious circle which is analogous to the processes of cumulative advantage that make up successful careers: sources of social support for reinforcing individual identity; accumulating resources and contacts in a particular direction, while cutting off ties to other areas and hence other opportunities.

The community in which criminals operate includes not only other criminals but also the police. A great deal of the recidivism cycle depends not only on the ineffectiveness of prison as rehabilitation, but also on police practices. Contrary to the impression given by television shows, the majority of crimes are, in fact, unsolved. The police maintain what record of success they do, primarily by maintaining surveillance on a group of known (i.e., priorly convicted) criminals (Skolnick, 1966: 112–181). Given the prevalence of narcotics use within the criminal community, police searches usually turn up incriminating evidence of some sort and criminals can be bargained into implicating others or confessing to a series of crimes in return for a relatively light sentence. There is a symbiosis between the police and the criminal world at the lower levels, analogous to the symbiosis between police and racketeers at the higher.

There are multiple factors producing criminal careers. The structure of the law-enforcement and corrections world, and the attitudes of "legiti-

---

[25] For a general review of the conflict approach, see A. T. Turk (1966); Quinney (1970); Douglas (1971b).

mate" society toward labeled criminals helps reinforce criminal careers; on the other hand, stratification within the criminal world places pressure in the opposite direction to keep the less successful members out. It is for the latter reason that the labeling process and the alienating aspects of incarceration do not always produce criminal careers. This can be seen particularly in the case of amateur "crimes"—private violence with no intention of making it a means of livelihood. Much of such "amateur violence" is found among youth gangs, usually organized on an ethnic basis; some of the conditions for their prevalence are given in the chapter on age stratification. The other main form of private violence is domestic; indeed, the most common type of murder occurs among relatives.[26] The causes here may be found in our analysis of sexual stratification and the family structure.

NONVIOLENT DEVIANCE

Nonviolent deviance falls into four main categories: sexual offenses, age offenses, illegal forms of consumption, and mental illness.

Sexual deviance includes homosexuality, adultery and incest, prostitution, exhibitionism, perversions, pornography, and obscenity, as well as giving illegitimate birth. As we have seen in analyzing sexual stratification and the family structure, these forms of deviance are merely the obverse of a given structure of power and status. Incest and adultery rules, as shown in comparative analysis, depend upon the structure of sexual property. Similarly, standards about pornography and obscenity are due to alliances between factions in domestic power struggles (particularly women in a particular stage of struggle against patriarchal dominance) and religious leaders, politicians, and other social groups attempting to uphold a given form of status stratification. The analysis given in Chapter 5, then, explains the structural conditions that set up the possibility of deviance. Individual careers in deviance here would seem to be the product of opportunity structures. Careers in homosexuality would seem to be based on the subjectively evaluated opportunities for status in either realm of sexual activity. The same probably applies to adultery and to the consumption of pornography (the last of which, in fact, is extremely common).[27]

A second category of deviance is specifically age-related: it consists of

---

[26] Of all homicides in America, 40–50% occur within the family (Goode, 1971: 631).

[27] *Report of the Commission on Obscenity and Pornography* (1970: 144–145). The core of opposition to sexual display seems to be the older, the more religious, the more rural, and the less

being of a certain age and doing things that adults can do. The category of "juvenile delinquency" is little more than an indicator of the existence of age stratification in an authoritarian system. Most of the offenses for which juveniles are punished consist primarily of symbolically overstepping their role and claiming adult status: smoking, drinking, sexual activities, or simple insubordination against authority. Vandalism seems to be a typically juvenile offense, and can be seen as a symbolic counterattack (Cohen, 1955). It seems to be produced in cases where youths have at least enough organizational resources to be able to strike back symbolically against the property of the oppressors. It is most of all in this area that strictures against the usual interpretations of deviance statistics hold most strongly: Juvenile offenses are extremely widespread, and the degree to which they are entered in the official statistics seems to be a matter of relative powerlessness of low-status juveniles from being apprehended or publicly punished (Cicourel, 1968; Douglas, 1971a: 42–132).

Illegal forms of consumption include alcohol, drugs, gambling, and certain kinds of violent sports. Just which of these are made illegal varies a very great deal throughout history. Prohibitions are rationalizated in terms of an idealized version of self-discipline, will power, or respectability, which reflects the status group interests of those who dominate the community. Often these are related to conflict among ethnic groups. The best structural study we have of any area of deviance is that by Gusfield (1963), who showed that prohibition of alcohol in the United States was a last-ditch effort of rural protestant Americans to maintain their cultural hegemony in the period when America was becoming an urban society numerically dominated by immigrants from peasant backgrounds.

Individual "careers" in deviance of this type depend on the historical conditions that make particular kinds of consumption deviant. Labeling theory on a social psychological level seems to be of relatively little value here, for "deviants" in the consumption sphere are rarely apprehended (Warren and Johnson, 1972). In many cases, there is a distinctive "deviant" identity—which might be better described as a "rebellious" identity —in which the members of the community of gamblers, drug users, or drinkers share a ritual camaraderie, reinforced by the latent dangerousness of their activities, the implied trust among those who participate in them, and the presence of a common enemy. These, we will recall, are the conditions given in Chapter 3 for strong states of ritual solidarity. The extent to which individuals participate in particular "deviant" consumer cultures, then, can be explained in terms of their alternative market opportunities for status and pleasure in various realms. The solidarity of the deviant group itself is one such motivator which, in particular cases such

---

socially active, especially women: all those who are lowest in terms of sexual activity and attractiveness, and whose greatest subjective status comes from attempting to devalue sexuality.

as the psychedelic drug culture of recent years among middle-class youth, may be a major incentive, compared to the lack of corresponding festiveness or solidarity among the officially sanctioned consumption cultures (Carey, 1968). There are also many isolated individuals who privately consume illegal services or substances; for them, the balance of rewards and opportunities in different directions must be calculated somewhat differently.

It is in the realm of consumer "deviance" that "moral entrepreneurs" are of greatest importance (H. Becker, 1963). The concern of dominant status groups to maintain their standards of what is deviant, would appear to be strongest in cases where the former has enjoyed control of a relatively small and tight community. Where the conditions for easy surveillance are lacking, as in large urban and suburban situations, there is relatively little direct status reward for puritanical vigilantes. However, certain consumption activities may continue to be officially deviant long past the strength of the status group that enforced them, in the same way that other institutions maintain themselves beyond the historical conditions that gave rise to them. This occurs to the extent that there are formal organizations that provide careers for deviance enforcers. In the realm of consumer deviance, this consists basically of the special police forces whose *raison d'etre* is the continuing existence of officially defined forms of deviant behavior. Theoretical explanation of consumer deviance in the contemporary urban society, then, must draw heavily upon the contingencies of organizational politics.

Finally, mental illness may be regarded as a residual category of deviance (Scheff, 1966). Into it can be fitted all those forms of behavior in which individuals do not meet the expectations of the vast majority of those around them, especially the acute form in which they are no longer able to participate in any of the organized forms of society. This includes the inability to participate coherently even in other organized deviant careers such as crime, sexual offenses, or illegal consumption. They are the outcasts of preindustrial societies, where their behavior is treated primarily in religious terms. We have seen in Chapter 5 how, under particular conditions, such individuals might become the object of purges. In the most part, however, they would simply be regarded as occupying a religiously peculiar state. In secular industrial societies, a number of occupations have expanded to make careers out of dealing with these individuals. This ranges at one extreme from incarceration or treatment (usually by physiological controls) of individuals more or less totally incapable of functioning, to the other extreme, which includes clinicians who pursue a lucrative practice among the upper middle class by popularizing various secular theories of psychic health or neurosis, and by offering various forms of companionship, recreation, or conversation for pay.

Given this variety of possible "mental illness" roles, individual careers in and out of these positions must be seen as a very complex thing. The general model applicable is that individuals follow the best opportunities available to them for subjective satisfaction. In the case of psychosis, it is reasonably clear that the individuals involved are subject to a great deal of stress (including stress that may be partially or totally determined by inherent physiological conditions). Under these conditions, it may be most advantageous to retreat into a totally subjective world and to ignore all external requirements. The possible relationships between creativity and mental illness has been much discussed. Since our verbal thought is socially constructed, both by being constructed out of internalized conversations, and reinforced by being expressed in actual social contacts, creativity is a matter of departure from the accepted social realities at any given time. The extent to which individuals are successfully able to make such departures and return, either to carry on ordinary social transactions or to transform them, depends on their psychic resources and their social opportunities. Under highly unfavorable external conditions, as R. D. Laing (1967; cf. Szasz, 1967) seems to imply, most creativity may be found among those who find that their best tactic is to withdraw from ordinary reality.

Suicide also seems to be appropriately handled under the rubric of "mental illness." It fits the definition given here by being an extreme form of nonparticipation. This is not to imply that any particular psychological theory can explain suicide. It is a form of behavior which again must be determined by the balance of opportunities that individuals have for trying various alternatives.[28] What is particularly interesting from a sociological point of view, however, is the structural question: What makes suicide into a form of deviance? It is striking that this question has almost never been raised, considering the classic status of Durkheim's treatment of suicide. But Durkheim was not interested in explaining suicide, but in using it as a test of his theory of conditions for social solidarity. But why should people become upset by suicide? It is a peculiarly Durkheimian question in the sense that someone else's death is rarely of practical significance to particular individuals; concern about suicide, then, is an indicator of the state of the collective conscience.

The best approach here, as for all such general questions, is comparative: In what societies or social groups is suicide most approved or disapproved of? Fragmentary evidence suggests that it is most approved of in situations of military honor, where death is an everyday occurrence; individual courage is highly valued; and great value is placed upon being able to control at least the details of one's destiny if not the broad outlines.

---

[28] For a general survey, see Douglas (1967).

The basic principle may be that suicide is generally an affront to others' construction of social reality. The ordinary world view almost everywhere is one in which men are continually convincing themselves that the world is worth living in. Thus, suicides can be accommodated among military officer corps or feudal nobilities (particularly when the status community is not organized in a form that upholds a moralistic religion like Christianity), since the suicide reinforces the accepted view of reality rather than shatters it. It is likely that disapproval of suicide has grown in modern society, reaching a culmination in the contemporary world, with its attempt to repress all evidences of death as much as possible (Blauner, 1966). At the opposite extreme, we find Eskimo societies where death is so common and the level of hardship is such that individuals contribute most to the group by killing themselves when they can no longer uphold their share of the burden. The repression of death in modern society also makes suicide a weapon in interpersonal struggles, since it provides the suicide with a means of punishing others.

Nonviolent deviance depends on the extent to which particular activities are defined as deviant, which in turn depends on the struggle of cultural groups for status advantage, and on specialized organizations who perpetuate these distinctions in order to maintain their power. Definitions of deviance change along with historical shifts in the organization of status communities, but are slowed by the institutionalization of organizational careers for deviance-punishers. It is probably for this reason that most change in definitions of deviance has occurred, not by changing the laws but rather by widespread disobedience to unpopular laws, to the point where they become too difficult to enforce. This has been the case, for example, with the effort to prohibit alcohol as well as with various regulations dealing with sexual offenses. It seems likely to be the case in the area of drugs and gambling. Since status groups and their conflicts, however, are not going to disappear in the future, we may expect that new forms of deviance will probably be invented or old ones revived.[29] The conditions for these could be predicted from applying a comprehensive theory of stratification and politics.

---

[29] The prohibition of smoking is a good candidate for manufacturing a huge deviance culture in the future. The politics of drugs in general seems likely to be central, with constant technological innovation (which has already produced, during the twentieth century, strong narcotics and psychedelics, as well as tranquilizers, amphetimines, and barbiturates). Categories of drug deviance will be the product of interactions among a number of interest groups: pharmacists and physicians with economic and status motives for monopolization; career interests of enforcement agency officials; various occupational and community groups with status interests in maintaining particular standards of demeanor; and politicians who play upon mixtures of such interests and act as brokers of pluralistic ignorance by which widespread consumer interests may be kept suppressed.

Chapter 9

The Organization of the Intellectual World

Explanatory sociology can give the conditions for our thought as well as our actions. We have already seen a good deal of this in the determinants of conversation, the world views of various occupations and status communities, religions and political cults. In a sense, the social construction of reality is a key to all of sociology. It is time to turn the enterprise upon ourselves, to locate our own theorizing in the same sociological universe as the behavior of the other people we analyze. Developments in this direction have been made by the sociology of science; from this base, we can understand the social structure of the intellectual world more generally.

What is science? There are various philosophical definitions. The idealist tradition has emphasized the construction of a *coherent thought-world* in which empirical observations take on meaning by being integrated into an overarching body of concepts and principles. The empiricist tradition places its emphasis at the other end, viewing science as a *summary of empirical observations* in the form of correlations among variables; exactness of observation may also be stressed. Both of these approaches define science in terms of *both* empirical reality and general ideas, but differ in the priority given to each. The empiricist tradition conceives of a real world out there which we gradually come to know in more and more detail. The sophistication of the last 200 years has moved it to give up the notion that our ideas might be made to correspond more and more closely to external reality; crude realism has been replaced in the empiricist camp by an emphasis on the techniques of measurement themselves. The idealists, on the other hand, regard "reality" as fundamentally shaped in our experience by the prior ideas we have of it; the very meaning of empirical observations depends on the idea model into which they are fitted, and "reality" is gradually constructed as thought progresses.

It is easier to follow these analyses with concrete examples, and, of course, philosophers of science probably have some favorite examples in mind. Everyone would agree that Newton's laws of mechanics are science, although we can disagree about whether they are essentially summaries of precise observations, or an idea model of the universe. Darwin's theory of evolution, Maxwell's electromagnetic equations, Einstein's general theory of relativity, Mendeleev's periodic table of chemical elements, the quantum theory of matter, or the Crick–Watson model of DNA structure are other examples that would be generally accepted; some would better support the notion of mental images, others the notion of observational summaries.

Dealing with the question concretely and historically in this way casts a different light on it. For "science" is continually changing; none of the examples just given is the last word on its subject. Maxwell's equations, giving up the attempt at a visualizable model in favor of heuristic principles for relating observations to each other, provided the background for a new effort by Einstein to construct a more sophisticated model of the universe. Einstein, in turn, has begun to be superceded by newer heuristic principles deriving from subsequent research, with perhaps yet another reality image to be created in the future. What is true today in physics (or in any science) is almost certainly different from what will be true in 50 years. Some elements will carry over, but the interpretation placed upon them will differ.

This makes the quest for a philosophical definition of science necessarily inconclusive. The various definitions seem to reflect favorite examples; from a sociological perspective, they look like ideologies that groups of scientists at particular times would favor in order to justify their work and disparage others. But if even the examples of the "great" works of science admit this historical ambiguity, what of the lesser or more controversial works? The Ptolemaic system is as much a part of the history of science as is the Copernican system. Science is full of erroneous and half-true models of reality, and of trivial accounts of empirical observations that may or may not be incorporated into a larger corpus of work and thus glorified with the label "truth."

Philosophical discussions of science, like most nonsociological thought about social activities, consists of the analysis of hypostatizations. An ideal entity called "science" is abstracted out of historical reality, against which someone else's philosophical ideal is measured and found wanting. To be strictly empirical, however, there is no such thing as "science" *as a noun,* except in the sense of a word that men apply to their ongoing activities. What we mean empirically by "science" are the activities of certain men: observing, manipulating, building apparatus, reading, writing, talking, thinking. The crux of their activities is their communication among themselves, as they argue about what statements are to be accepted. Much of the arguments are carried on in writing, often with a specialized symbol system; this helps give the impression that there is a *thing* called "science," since books and papers are things, although they are actually only instruments for carrying on the discussion.

I have not yet explained just what kinds of arguments are science and which are not; as we shall see, the differences depend on just how one wins an argument. Science differs from ideology, for example, because in the latter an argument prevails when its opponents are coerced into being quiet, whereas scientific arguments are those which exclude considerations of coercion and loyalty to interests of anyone but the community of arguers themselves.[1] There are different kinds of argument, some of which we can call "scientific," "ideological," "practical," or "entertainment." Scientific arguments may be briefly characterized as those carried on within self-contained intellectual communities. But different intellectual communities may also differ as to what "science" *is.* The standards of argument were different in Newton's day and in Einstein's, not to mention Aristotle's. There is no point in arguing which is the "true" definition of science, since all definitions are arbitrary; we can only choose on the grounds of which ones help us the most in making a coherent and acceptable argument. Recognition of this principle, in fact, is one of the major differences between Newton's day and Einstein's. Of course, I shall be arguing for a particular definition of science myself. I have chosen one on heuristic grounds which allows us to see historical variations as arising under particular conditions.

The matter, then, can be treated sociologically. Science is a human activity: a particular kind of verbal argument. "Science" in the hypostatized sense is simply the body of words that are accepted at any given time as true (that is, they admit of no further argument). It is the body of words that is repeated over and over, with minor variations in phrasing, when authoritative talk is done (as in university lectures).

The approach of the sociology of science (and, more generally, of the sociology of knowledge) has been objected to on several grounds. The objections are couched in terms of both empiricist and idealist philosophies. (*1*) Empiricists argue that science, unlike other forms of thought, escapes social determination in any important sense, because it is a method for investigating reality—what really exists "out there"; it is based on the

---

[1] These are matters of definition and hence cannot be rejected by showing instances where "scientists" coerce others into agreement. I am simply proposing a way of using words so that we can characterize different kinds of behavior, and then look for the conditions under which they occur.

facts, quite apart from men's perception of them. "I refute it thus!" said Dr. Johnson, kicking a boulder; it is not only an attack on Berkeleyan idealism but on its sociological equivalents. (*2*) The idealists take their stand on the autonomous logic of ideas; the history of science is determined, not only by empirical discovery, but by the progressive development of conceptual systems in following out their logical implications.

These defenses of the autonomy of science were more recently provoked by the development of the Marxist sociology of science (e.g., Bernal, 1939). In Britain during the 1930s and 1940s, there was a heated debate in response to Marxist contentions that science is a human activity; that it is socially grounded; that, like other human activities, it is based on class interests; and, therefore, that science ought to be directed toward the needs of the proletariat rather than toward the needs of the capitalist state or the personal interests of privileged intellectuals. This came in the atmosphere of Stalin's purges of his intellectual opposition, and the elevation of Lysenko's repudiation of Mendelian genetics into state dogma in the U.S.S.R. The response was correspondingly heated, with men like Popper (1945) broadening the counterattack to include all "sociologism" as an aspect of the totalitarian state. The debate has quieted down since then, although the themes linger on. Humanistic historians continue to use them to express their dislike of sociologists, and specialists in the history of science defend their autonomy against wider generalizations. The non-Marxist sociology of knowledge, stemming from Scheler and Mannheim (and behind them the idealistic tradition), generally skirted the issue, exempting science from their efforts at a thoroughgoing social relativism.

But except for ideological points, there is nothing to be gained by this sort of compartmentalization. Marxist sociology of science is only one version, an early and crude effort to begin sociological explanation. Its reductionist errors can be avoided in a more sophisticated multicausal model of organizations; the very autonomy of the intellectual community that the anti-Marxists wish to preserve may be understood in sociological terms.

What can the sociology of science explain? Obviously it cannot confine itself to merely describing scientists' behavior if it has any pretensions to becoming a science itself. This means it must state some conditions under which scientists will do and say certain kinds of things rather than others.[2]

---

[2] Purely descriptive sociology can serve ideological purposes, however; much functionalist sociology of science has been devoted to ideal descriptions of scientific norms, and descriptive research attempting to defend the fairness and functionality of scientific organization against criticism from Old and New Left. One might call this "cold war sociology of science."

But how much is explainable? The Marxists' most extreme claim to explain the contents of scientific thought virtually eliminates the objective world entirely, thus undermining their own materialism.[3] In contrast, the sociologist Joseph Ben–David (1971) confines himself to a much more modest claim, exempting the contents of science from sociological explanation and concentrating on the social conditions under which the scientific role is institutionalized. Thus, one can attempt to explain on the basis of comparative analysis why science appeared in Ancient Greece but relatively little in China; why in Renaissance Europe more than in Medieval Islam; why world leadership in science passed from France in the late eighteenth century to Germany in the nineteenth and to the United States in the twentieth. With closer analysis, we should be able to specify the reasons why one man rather than another was in a position to make certain discoveries and whether he gets the credit for them.

But the distinction between explaining background conditions and explaining content is not so easy to maintain—at least, it must be drawn only much later in our analysis. To state when and where science will exist in an historical analysis is to give conditions under which particular kinds of ideas are formulated and believed in. Why, for example, did philosophical systems hold sway in German biology during the first part of the nineteenth century while British and French biologists catalogued empirical observations? The answer would go some way toward telling us why ecological and ethological studies are mainly the product of German biologists in this century while Anglo–French biology has gone in the direction of biochemical analysis and mathematical genetics. Why did Descartes' cosmology prevail over Newton's in France during much of the eighteenth century? Why is mid-twentieth century experimental psychology divided into three mutually antagonistic and mutually oblivious world camps: the American, with its emphasis on behavioral experiments and animal studies; the French, with its emphasis on cognitive operations; and the Russian, with its emphasis on electrophysiology?

The sociology of science has made an approach to these matters and is not likely to stop now. One might object: Are not these merely cases of explaining where some societies are backward in a field of science? This is easy to say about German *naturphilosophie* or the French national bias toward Descartes. Even on this level, however, the question is hardly a trivial one, and it is of some value to state the conditions under which the

---

[3] This shift seems to have come about in response to the idealist philosophies of knowledge of Scheler and others in the 1920s; the result is to bring Marxism back to its Hegelian heritage in a fairly pure form. This, of course, has been a general trend in Marxism in the twentieth century.

most powerful scientific arguments are resisted. But, in many cases, it is not so simple. One or another of the three psychologies of today may eventually produce a completely accepted paradigm; it is even more likely that elements of all three will be incorporated when psychology becomes a unified science. But this may take a very long time; in the meanwhile, the sociology of science has some rather pointed relevance. The same could be said about different national emphases in biology today, not to mention the scholarly fields where international divergencies are famous: philosophy, anthropology, sociology, and, to a degree, history.[4]

The sociological explanation of the content of science, then, has a number of possibilities: It can attempt to explain the acceptance of outmoded science, as well as that of local variants on the frontiers of exploration. Nor is this simply a matter of dealing with the "preparadigm" stage. In some sense, all theories are limited; we have too much retrospective contempt for half-workable theories of the past which were replaced by slightly more workable theories. Copernicus' model of the universe was preferable to Ptolemy's in systematically accounting for observations in some ways but not in others. Kepler's, Brahe's, and Descartes' models all had advantages and disadvantages. So did Newton's, although we could only know some of them from the perspective of two centuries later. The line between explaining errors, half-truths, and truths, then, is never completely firm, and we may hope to say a good deal about why scientific beliefs have moved through their particular history.

There is even good reason to believe that there are many possible scientific histories. This sounds heretical because a fundamental scientific belief is antirelativistic: that agreement can eventually be reached in any argument carried on in the scientific manner. But even if one assumes this is true, it does not follow that different countries more or less cut off from communication with each other will produce the same science (or differ only in that one is behind the other). As a concrete example, we may refer to differences in Chinese and Western medical theories, just now coming into a common realm of investigation; or the diverse direction taken by modern chemistry in China and the West since the 1940s. If we discover another planet inhabited by civilized beings, will we expect it to have produced the same science? We might believe there would be some things in common, but would it have the same history, with the same errors, the same partial syntheses, the same "normal science" stopping points along the way? This seems highly unlikely, and it is useful to approach our own history with this thought experiment in mind.

---

[4] The last two diverge among nations most prominently in their localistic subject matters.

This does not mean that the sociology of science aims to be all-encompassing. The logic of the development of ideas, the recalcitrant facticity of experience, the place of human will or blind chance, all enter into the explanations of what men think. But we will be able to assign an appropriate place to each of these only after we have subtracted out what can be explained by social conditions. Perhaps there are some inherent "evolutionary barriers" in the development of science; it has been claimed that understanding the operation of the heart was a key to all future advance in physiology, as the question of gravity was for astronomy and physics, and the development of mathematical analysis was for many sciences. But this can be settled only when comparative sociological analysis has been pushed just as far as it can get results.

Our analysis can be taken deeper. There is no avoiding the implications of the fact that sociology has generated a general explanation for human consciousness. On the microlevel, the cognitive sociology stemming from Durkheim, Mead, and Garfinkel fits into a set of structural conditions first explored by Marx and Weber. Science is included in these general formulations. Empirically, it is organized by a set of communications, as are ultimately all other forms of human belief. What we mean by "true" or "real" always reduces to what social communications will be accepted. When we refer to an "external" world, we are using words, and what we believe "objectively" about that world consists of the words we can agree to use in describing it. The external world is not simply "there"; anything that we perceive *and* that we can persuade someone else to accept as "knowledge," is socially constructed through the concepts that we use to communicate about it. Moreover, nonspoken thought is mostly internalized communication; and the part of our subjective experience that we usually consider most "real" is the part we can verbally think about rather than only feel. Thus, social determinism enters the very core of our experience.

Our reality is always historically relative, depending as it does upon the concepts we have with which to isolate chunks of experience. Whether we see the thunder god's lightning bolts or a flash of static electricity depends on an accepted body of social ideas; so does seeing the sun move across the sky or the earth turn away from the sun. The history of the social sciences provides some of the most striking examples, for it is only by becoming aware of the nature of hypostatizations like "society" that we can make truly operable explanatory principles. And, of course, the recognition that there is such a thing as hypostatization is socially conditioned, and occurs only at particular places in history.

We gradually acquire our objective reality as we go along. This, I take it, is the basic value of Hegel's analysis of the "thing-in-itself," obfuscated as it has been by his own ideological interests and those of his successors. Our final test of objectivity is the transpersonality of observations: The criterion is the capacity of separate observers to send back and forth the same messages about their own experiences. Our confidence about what is truly there in the world (including other observers and our own selves) is something that develops historically through a gradual refining and coordinating of communications. We find ways to settle certain major arguments, even though debate over various points continues.

The sociological approach to science may also be justified heuristically. It is self-validating to the extent that it works. It gives us a view of science that we can accept as "reality," precisely to the extent that it can settle arguments by stating the coherence among the widest range of reports. As in the general ethnomethodological program of reducing all sociological statements to the second-by-second details of human experience, I am trying to state with at least a little precision just what the activities are that make up "doing science" in all its variations.

The sociological critique can be turned upon itself and give a basis for itself. It is a matter of finding the right sequence of abstract words that summarize many more particular statements (i.e., ones that use less abstract concepts), which in turn may summarize the general properties of even more particular statements; ultimately, these must summarize statements about particular people's moments of experience. Thus, one argument can subsume (and draw power from) a number of other arguments. The aim is to make the most "powerful" statements possible, in the sense of general statements that recapitulate the salient points of more particular statements. A successful theory is a way of settling arguments by appealing to other arguments, including ones that are irrefutable because they refer to statements about operations (logical or experiential) that everyone must agree to. It may be powerful in the sense of both recapitulating earlier arguments and of formulating subsequent, more specialized, arguments that are found to be successful in commanding agreement. (This is a way of saying that a good theory has further applications for research or practice.) This is the general model of a successful scientific argument; it can be applied to sociological arguments, or—as I am attempting to do here—to the social activity of science itself.

What I wish to apply are the explanatory principles of stratification and of organization developed throughout this book. In science as elsewhere, we can derive the most powerful explanations by taking these postulates: Individuals follow the path of maximizing their subjective status according to the resources available to themselves and their rivals: there are frequent points of conflict, and how resources are brought to bear upon them determines the repeatable patterns of behavior (structure) at any given time, as well as set up the social factions for subsequent conflict (and hence, change); and ideas about reality are weapons in the struggle for dominance and are to be explained by the interests of individuals with the resources to uphold them.

This approach enables us to understand much about science that is missed by naïvely assuming that an ideal version of science is its true structure. Ideals about science (including the philosophical notions given at the beginning of this chapter) are upheld by real men in particular situations where these ideals offer advantages; the empiricist and idealist theories, for example, seem to reflect the status interests of particular situations for researchers and theorists, respectively. Ideals are statements about what particular men would like; they are efforts to influence behavior. We do not leave them out of sociological analysis; they are among the things to be explained, but they should not blind us to the real behaviors that go on. The conditions under which behavior corresponds more and less closely to particular ideals need to be specified.

All the valid points that have been advanced against sociological determinism can be incorporated into a sophisticated sociology of science. A realistic view can recognize the existence of autonomous communities of intellectuals, of a recalcitrant world of facts (the nature of which, however, needs to be socially formulated), and of the inherent logic of the unfolding of scientific ideas. But these are not self-evident matters; the very objectiveness of reality, the very inherency of logical development are matters that can be deeply understood only by developing a successful model of the argumentative community that makes up science. Only by understanding the ideals of science in terms of the real interests, behaviors, and conflicts of individuals do we go beyond reified abstractions and move closer to a truer picture of scientific reality.

As in other areas of sociology, the sociology of science itself is quite a mixed bag. Much of the older theorizing has been devoted to upholding a saintly image of scientists. It gave an idealized code of truthfulness, altruism, and freedom from national bias and personal ambition, and explained away scientists' battles over recognition as evidence that scientists are only concerned that everyone be given his just due. The drift since then has been toward a more naturalistic approach, looking at scientists' behavior instead of observers' beliefs about norms. This has included several versions of "neosaintly" science, in which scientists politely exchange gifts of information (Hagstrom, 1965; Storer, 1966). As we press on toward principles that explain variations in what scientists actually do and produce, the field necessarily becomes even more realistic, and the most powerful leads are those that recognize science as an intensely competitive business and look for material resources, organizational structures, and numbers of competitiors, as bases of explanation.

Hagstrom (1965) has gathered evidence for the competitive nature of science from interviews with scientists showing variations in concern about being anticipated in a discovery, variations that are related to the structure of competitive conditions in the field. Hagstrom has also summarized the evidence for scientific competitiveness from statements and writings from scientists themselves, and has advanced the supporting argument that scientists work only on problems that have not yet been solved by others, hence scientists appear to be concerned with originality rather than with creativity alone. We may also add the fact that scientific advances are generally resisted by the old guard whose paradigm is overthrown; scientific theories win general acceptance only because their opponents die off.[5] Most importantly, the validity of the competitive model is borne out by its usefulness in explaining variations in innovation and productivity. That all this demonstration of scientific competition should be necessary, of course, tells us something about the role of ideology in science; for any honest participant in the academic world can tell how intensely competitive a world it is.

The intellectual world, in fact, seems to be made up of highly argumentative people. To the extent that they listen to others, it is because they must. This occurs where the group has evolved a set of standards for admission to the argument. If we look at science from the point of view of conversational analysis, the importance of competition in an explanatory theory of science becomes clear. "Science" after all, like all our hypostatized sociological entities—"organization," "class," "state"—consists empirically of the repetitive behavior and thought of many men; "science" is an abstract term summarizing millions of minutes of thought and communication. The basic question of the sociology of conversations is who will talk to whom, for how long, and about what. This applies to science quite precisely. For what is considered "truth" are those ideas that are repeated and drawn upon in the talk throughout the scientific community. A scientist's career consists of getting himself as much as possible into the center of the conversation, by having as many people as possible listen to him and talk about his ideas. The peculiar form of social control in science is based upon the capacity of scientists to validate each other's

---

[5] French science did not accept Newton's cosmology for over a half century; English scientists resisted Lavoisier's chemistry for an entire generation; relativity theory and quantum physics converted a younger generation rather than the older.

ideas and determine their success by the very act of communicating about them.

A realistic image of science, in fact, would be an open plain with men scattered throughout it, shouting: "Listen to me! Listen to me!" The polite formulations of the sociology of science take scientific communication as an exchange of gifts of information. But men are interested in such gifts mainly to the extent that it helps them to formulate statements that will cause others to listen to them. The more fundamental process is a competition for attention. Scholars send out preprints politely asking for comments to which they rarely attend; more realistically, it is apparent that everyone is trying to advertise himself as much as possible. The feeling that priority ought to be recognized is hardly a noble disinterest but something that appears only when scientific competition and division of labor reach a certain point so that men's fates are too much in each other's hands (or words) for them to risk offending each other—unless (and this states the rudiments of the conditions under which variations occur) the power resources are unequal enough so that one faction can ignore another's claims to priority.

What we are looking for, then, are explanatory principles stating the conditions under which men can get others' attention. There are a variety of strategies and advantages: being on the field earliest and longest; saying the most original things or those that interest the greatest numbers of listeners; talking to a selected audience; picking arguments with others who are better known; mentioning other people's names and ideas (since everyone likes to hear himself talked about); opening up new topics for others to follow. The political aspect of this is obvious, and some of the strategies for making alliances and using new resources were made explicit by scientific organizers like Francis Bacon, who knew governmental politics extremely well. The history of science abounds with men who pursued ambitious careers in both spheres, including Copernicus, Galileo, Newton, and Lavoisier; the gentlemanly Charles Darwin, the favorite of the saintly theories of science, is hardly a typical figure.[6]

----

[6] Newton was an unpleasant and egocentric man who criticized all contemporary scientists and praised only the ancient prophets and philosophers (see Manuel, 1968; Koyré, 1965). When successful, he became a highly autocratic president of the Royal Society, acting harshly toward his English rivals and even depriving them of power to publish their own works. He arranged a supposedly impartial inquiry by the Royal Society which awarded him priority over Leibniz in the discovery of the calculus. His wordly ambition makes it possible to understand several of his enterprises that modern commentators interested only in science have thought to be peculiar aberrations, such as Newton's powerful religious interests which led him to spend the greatest portion of his time attempting to prove the credibility of the Old Testament prophecies. Newton himself considered this to be his most important work and, in the public context of the time, it was, since controversies over the validity of religion

If we can state the conditions for men's successes or failures in the competition to get their ideas heard, we will be able to explain everything in traditional analysis of science, and more. A theory of career mobility is empirically equivalent to stating a theory of stratification or social structure. Since the structure of scientific stratification or organization is causally related to the structure of accepted scientific ideas (in the same way that ideologies are related to stratification and organization), this is also equivalent to explaining the social basis of scientific ideas. We can state the entire history of a science in terms of the career successes and failures of the men who practiced the science, providing we draw the net wide enough to capture more than a few of the most talked-about figures.

The principles of stratification and organization can thus be applied to intellectual communities. The major theoretical analyses of the nature of science have come very close to explicitly stating the organizational structure of scientific communities; this is particularly true of the work of Hagstrom and of Kuhn. We can now see that science is structured internally as a kind of profession, but containing a particular kind of hierarchical organization and emphasizing competition more than most. Similarly, the best historical sociology of science, especially the work of Ben–David, has been concerned with the place of scientists within the organizations of the surrounding society. Both internally and externally, a sociological analysis of science can proceed via the analysis of organizations.

---

and interpretations of scripture were violent political issues in the seventeenth century. Nor should it be considered a lapse from Newton's career ambitions that he gave up science entirely when given the opportunity to become Master of the Royal Mint; this was an office that he worked long and hard to achieve, more by manipulating personal and political connections than by his scientific eminence.

Most other prominent scientists have been involved in ambitious political careers as well as the pursuit of eminence within science; often the two projects coincided. Copernicus and Galileo were prominent in the dangerous business of Catholic Church reform. Bacon was not only the most astute organizational politician of science but also pursued a dangerous career that made him, for a while, Chancellor of England. Boyle, like Newton, was heavily involved in religious politics and spent much effort defending religious dogmas. Leibniz was a machinating diplomat as well as a cosmopolitan man of science. One of the most ambitious of all was Lavoisier, who devoted most of his time to a political career within the royal French government, which cost him his life during the Revolution. As a young man, he declared his intention to revolutionize chemistry and make it into a science, and then proceeded to do just that, including a standard nomenclature to embody his ideas throughout the field. He is also noted for having several times appropriated the discoveries of others as his own, and for believing that he had brought the field of chemistry to a conclusion through his own work (Butterfield, 1962: 217–221). The leaders of more recent science have been mostly university professors, but remain extraordinarily hard-working and ambitious men, such as Helmholtz, Wundt, Virchow, and Einstein; it is notable that these involved themselves prominently in politics when the opportunity presented itself.

EXTERNAL ORGANIZATIONAL ROLES

Intellectuals have occupied four main kinds of organizational positions: political roles, practical roles, leisure entertainment roles, and teaching roles.

1. *Political roles* are those in which intellectuals make careers of defending the legitimacy of their organization and attacking the legitimacy of competitors. These include high officials in governments or churches, office-holding politicians, leaders of political parties or movements, publicists, and journalists.

2. *Practical roles* are those in which intellectuals work for a customer, client, or boss, to achieve some practical result. This category includes most professions; doctors, builders, and engineers have been the main practical intellectuals through most of history, with modern staff positions proliferating with the growth of government and business bureaucracies.

3. *Leisure entertainment roles* include the positions of individuals who belong to a leisure class and produce intellectual work in order to amuse themselves, as well as intellectuals who make a living by producing entertainment for patrons or a mass market.

4. *Teaching roles* are positions that deal with communicating knowledge to specialized, more or less full-time students; this always involves some degree of cumulating and assessing previous work, and thus some degree of scholarship. In modern universities, this aspect of preparing materials for teaching has expanded to the point where some men are full-time specialists in it.

In all of these cases, we are dealing with intellectuals in occupational positions; in the third (leisure entertainment) we are also dealing with the organization of status communities. Applying stratification theory (Chapter 2), we can predict that intellectuals in varied positions will have different goals and outlooks in their work.

All *political roles* have some intellectual content. By their very nature, they are the cosmopolitan positions in the communication channels of social organization; and they involve defending religious or ideological orthodoxy as a basis of legitimacy of their power. Political leaders in any kind of complex administrative organization are a kind of professional talker or writer. It is not surprising, then, that they have sometimes contributed to intellectual developments in their writings or doctrines, particularly before the rise of specialized intellectual roles. But their intellectual work has a particular kind of content. Political roles result in ideological production: thought that is primarily concerned with values (argument about what is right or wrong according to moral and political criteria) plus arguments about what policy ought to be carried out. This kind of thought is usually concerned with the orthodoxy of particular ideas, from the standpoint of either maintaining a system of power or attacking it in order to set up another one.

Ben–David (1971: 28-30) provides corroborating evidence in his analysis of why science is relatively ephemeral in premodern societies. In Mesopotamia, China, and ancient Greece, scientific ideas were sometimes expressed; but the positions for intellectuals were those of priests, leaders of religious and political cults, or educated administrators and lawyers. Hence, the scientific ideas were always subordinated to the political or religious interests of the intellectuals, and confined only to particular transitional periods when an entire political cosmology was being challenged.

Some evidence is also available from a comparative analysis of the histories of six social sciences—history, economics, psychology, anthropology, sociology, and political science—in Europe and America since the Middle Ages.[7] The sample of social thinkers is derived by taking all the figures named in a number of histories of each discipline. The productions of intellectuals in political roles were primarily history, social philosophy, and economics. History writing by absolutist church and state officials was generally parochial self-justification of their institutions. The work of politicians sometimes shows capacity for intellectual generalization but tends to be primarily normative. Economic work is largely propagation of doctrine rather than technical theory or research. But there are some exceptions: the scholarly work of Tocqueville, the economics of Beccaria, Turgot, Marx, and Böhm–Bawerk, and, to a degree, the sociology of Comte.

Practical roles give rise to practical intellectual production: concrete proposals for action plus collections of facts. These roles do not give rise to general theories but to collections of recipes (for diagnosing and curing diseases, building bridges, fighting inflation, and so on). Practical roles are empirically oriented, not necessarily in order to test theories with facts but, rather, because a collection of facts is of use in diagnosing a situation so that a practical recipe may be applied to it. Facts are thus organized in the form of a catalogue ordered simply by an indexing method for retriev-

---

[7] Material is drawn from Barnes (1963); Becker and Barnes (1961); Ben–David and Collins (1966) and references cited therein; Bendix (1943); Bernard and Bernard (1943); Easton (1968); Lazarsfeld (1961); Oberschall (1965); Penniman (1952); Schumpeter (1954).

ing information. Typical practical information takes the form of a cook-book or a telephone directory.

Ben–David (1971: 23–27, 77–78) shows that, throughout history, practical roles have resulted in the accumulation of empirical facts but rarely of generalized explanatory principles. In architecture and engineering, the accumulated knowledge for the most part was not even committed to writing, but passed through apprenticeship as practical precepts. Similarly, in medicine, descriptive information and catalogues of supposed remedies were developed empirically by practitioners whose theories consisted of religiously based dogmas. The main contribution of practical work to generalized knowledge has come from a few occasions in astronomy when it was necessary to produce a more general model in order to make practical predictions, arrange calendars, or aid navigation. Once the practical task was accomplished, the theory tended to deteriorate or be forgotten; it provided only a stepping stone to a practical catalogue of information. Thus, astrology arose *after* the highest achievements of ancient astronomical science were supplanted by a magical and religious theory oriented toward practical results. In modern periods, heavy emphasis on practicality has also inhibited the development of general science; the British Royal Society founded in the mid-seventeenth century, despite its commitment to empirical experimentation, contributed almost nothing to the development of a cumulative scientific community for 200 years, precisely because it was narrowly oriented toward practical applications.

In the study of the six social sciences, practical intellectuals produced almost entirely descriptive catalogues. These included men in positions as staff recordkeepers and consultants, members of investigating committees of reform groups, and practicing physicians and hospital administrators. The work of practical economists proceeded for several centuries without issuing an economic theory. Similarly, the "moral statisticians" of the 1830s produced no theoretical generalizations, despite Quetelet's publicizing efforts on their behalf. The same can be said of the voluminous statistical information collected by twentieth-century government agencies, as well as by the British and American investigating commissions and the German *Verein für Sozialpolitik*. Practically employed French and British psychologists produced psychometric tests but little psychological theory. Most of the practicing physicians concerned themselves with the techniques of psychiatric treatment rather than with the development of theory; Freud and related theorists are the great exception. Other exceptions are the nineteenth-century British and French physicians who did anthropological research. But is seems more reasonable that their work was not part of their practical role but, rather, a hobby and hence fit under the entertainment role.

Another form of evidence comes from studies of the distinctive communication patterns in practical and scholarly disciplines (Price, 1969). In the world of basic science, publications refer to previous papers quite extensively, whereas most of the literature used by technologists contains few references to other literature. The success of a scientific paper can be measured by the extent to which it is cited in the scientific literature, whereas patents are rarely cited at all. As Price notes, science is organized around a highly publicity-conscious communications network drawing heavily on its archives; technologists draw on catalogues and tend to be secretive about the intellectual aspect of their work, reckoning success in terms of sales of products. The distinction between practical and basic research is usually quite clear, and the two communities are linked only indirectly. Work on basic research rarely pays off in practical applications to the organization that supports it; basic research becomes translated into practical payoffs only by being incorporated into standard theoretical textbooks which in turn provide the fundamental solution upon which applied researchers draw for their problems (Sherwin and Isenson, 1967). The lag between basic research advances and their practical applications is often a matter of several decades.

Leisure entertainment roles give rise to aesthetic productions. The emphasis is on pleasure rather than on the truth or falsity of the ideas, their practical use, or even their ideological purity in fitting prevailing values. Whatever standards are invoked derive from the particular status claims of the group that is being entertained, which, under certain conditions, may be the group of artists themselves.

The organization of intellectual activity in the Italian Renaissance was primarily of this sort (Ben–David, 1971: 59–65, 80–82). Intellectuals and artists gathered around wealthy patrons; as the aesthetic culture became a new status ideal, academies tended to become private clubs for the aristocracy. At all times, their interests were primarily literary and theatrical. Some scientific experiments were done and scientific ideas passed around, but primarily as a matter of amusement; serious development in science was carried out mainly by members (such as Galileo) who were tied to the universities. Similarly, the seventeenth-century French *Académie des Sciences,* although it pointedly avoided the practical emphasis characteristic of the British Royal Society, was initially unproductive in science because it emphasized heavily literary and entertainment orientations. Ben–David's analysis is that science was adopted only to the extent that groups attempting to acquire dominant status found it useful to use its intellectual orientations for marking its break with the prior establishment. Science enjoyed patronage in Italy by a new urban aristocracy attempting to live in a grand style and set itself apart from the earlier ethos of pious middle-class merchants; and by the grand courts of northern Europe, which were attempting to reduce the formally independent aristocracy to a courtier class attached to the leisure entertainment of the court. Once the transitional period was past, scientific ideas were largely dropped since it was no longer necessary to set off oneself from earlier religious orthodoxies.

Such leisure situations, in fact, tend to bring forth literary and artistic production rather than science or scholarship. The most successful approaches to the sociology of art and literature have been to link them to the status cultures of their audiences (Schücking, 1966; Hauser, 1951). In art, a number of styles recur throughout history. In absolutist kingdoms, grand, monumental, and austere art and architecture were favored; art was part of the stage-setting for awesome monarchs. Art produced for a traditional tightly knit middle-class community was smaller and less expensive; in content, it tended to be more piously religious, or in secular periods, to stress moral and domestic virtues and scenes. The military aristocracy, where it could afford it, preferred grandiose action-oriented heroic art; the art of periods in which courtiers gathered around great monarchs tended to be more complex, decorative, and erotic. Mixtures of social conditions explain mixed art styles.

The shift for artists from a patronage basis to a situation in which they produced for a large anonymous market tended to shift the evaluative audience to the community of artists themselves. It is under these conditions that there arose the ideology of "art for art's sake," and an emphasis on innovation and a departure from tradition, developing intellectual themes initiated within the artistic community itself. The same general principles hold for literature. Societies in which religious organization monopolizes intellectual production limits literary work to religious themes. Military aristocrats and idle courtiers patronized stylized heroic drama and poetry on the one hand and light erotic comedy on the other. The lower classes have always had their melodramas and burlesque comedies. Again, mixtures of social conditions can produce mixed literary styles; the great appeal of Elizabethan drama seems to be a product of courtly and popular audiences combined in the same theater (Hauser, 1951: vol. 2, 148–172).

The rise of a market for books large enough to support intellectuals apart from patronage relationships had the effect of making the internal organization of the literary community more important than external audiences (Graña, 1964; cf. Bourdieu, 1971). To be sure, the rise in the nineteenth century of a large middle-class audience with much time for being entertained was instrumental in the rise of the novel; its moralizing themes and its emphasis on the psychology of their characters' private lives is a piece with the moralism of middle-class culture, and particularly with the largely female nature of its audiences. The other theme of modern literature since the rise of the independent market has been the self-glorification of the artist and his alienation from the audience that does not sufficiently appreciate him. This appears to follow from the conditions under which writers and would-be writers form relatively self-contained communities in major centers, such as Paris and London, creating their own status system. "Poets are the unacknowledged legislators of the world," declared Shelley. Literary intellectuals, as the group with the greatest concentration of resources for creating reality out of sheer imagi-nation, thus create their own ideology, and have struggled bitterly against counterideologies that deny their peculiar claims to aristocratic status. The extraordinary innovativeness and detachment from prevailing definitions of reality that characterizes modern literature, particularly in the least of market-oriented segments (poetry, as well as the drama ever since its displacement from popular audiences with the rise of the mass media in the twentieth century) have been produced almost entirely by the compe-tition for status within the intellectual community itself.

To the extent that scientific and social science work has been done in leisure entertainment roles, it has a special stance. In the history of the social sciences, the work of leisure-consumer-oriented intellectuals is gen-erally unspecialized, combining many different fields. The market-orient-ed historians of antiquity and the Renaissance were often poets as well, and their history emphasized literary quality and dramatic or scandalous themes rather than scholarly accuracy. Such historians and philosophers in the eighteenth and nineteenth centuries had higher standards, but relied almost entirely on the works of academics rather than on their own re-search; the genre has lost virtually all intellectual esteem in the twentieth century by comparison to scholarly standards. Leisure gentry, doing intel-lectual work as a hobby, have been more research-oriented. In the case of nineteenth-century travelers and excavators, they did important pioneer-ing work in anthropology, although tending to compile atheoretical collec-tions of curiosities and to pursue a popularistic (and theoretically fruitless) search for the ultimate origins of various institutions.

In science, the gentleman amateur has been most prominent in descrip-tive zoology and botany, and in the exploratory work in geography and geology. Charles Darwin is the great exception, the gentleman amateur who is theoretically oriented in his research. As we shall see in examining stratification within the scientific community, there are systematic reasons for this peculiar case.

Teaching roles, under special conditions, can give rise to an orientation toward knowledge for its own sake: scholarly production with its own independent standards of truth. This distinguishes the production of scholarly roles from that of all other intellectual roles: Leisure entertainment roles have no such standards; practical roles have a purely pragmatic standard of truth or falsity, whether desired practical results are produced or not; and ideological roles, although they claim to deal with standards sometimes referred to as "truth," are essentially concerned with moral and political orthodoxy.

This is not to say that all teaching positions are oriented toward the production of ideas for their own sake. In fact, most schools throughout history have been oriented to ideological or practical purposes, sometimes to teach a particular status culture. Religious doctrine, law, and practical subjects such as astronomy and medicine have been the traditional curricula; if the orientation toward pure knowledge is created, it is because something has happened to subvert the original purpose of the organization. If knowledge is to be codified and taught so that it can be used to train lawyers, theologians, or doctors of medicine, then to be concerned about knowledge for its own sake is to take the means for the end. That is precisely what happened with the rise first of philosophy and then of science.

There are certain conditions under which this occurs. First, the educational system must become large enough so that groups of teachers and students may form a community oriented inwardly toward themselves, and thus develop a set of ideals of its own. A second factor is the relative autonomy of this group from outside control. The size of the Medieval university community in the twelfth century made it possible for teachers to be concerned with philosophy for its own sake, resulting in the efflorescence of Medieval philosophy which laid the basis for modern science. The situation in the Islamic countries was parallel (McNeill, 1963: 553–557). In Islam as in Christendom, schools grew up in the major cities around teachers of theology and law. Yet intellectual development in Islam came to a halt at the same time that European schools were undergoing a tremendous intellectual transformation, even though Islamic schools became increasingly large and well-supported financially. The key factor is the autonomy of the schools. In Europe, the political situation left a considerable degree of internal autonomy, whereas the intellectual life of Islam was destroyed by an orthodox religious reaction against the worldliness of the court circles and the Islamic scholars. Thus, in addition to the size of a system, the political conditions that give it autonomy are crucial in determining whether or not a scholarly goal displacement will occur and sustain itself.

A third organizational feature appears important in the rise of the autonomous scholarly orientation: the degree to which the school system is internally differentiated. Scholarly goal displacement takes root first in the subjects most insulated from outside pressure, namely, the preliminary subjects designed to train the student for his final practical or ideological studies; the maximum insulation occurs at the level of those teachers whose main job is to train other teachers. Thus, we find the scholarly goal displacement arising in some of the teaching–training schools of Greek antiquity (notably Aristotle's), which grew up as offshoots of religious–political cults (Marron, 1956: 256–295); and in the philosophical faculty of the Medieval universities, which prepared students for the higher faculties of theology and law (Ben–David, 1971: 47–54). The modern revolution in university scholarship began in the faculties of the eighteenth-century German universities training teachers for the newly created public school system (Bruford, 1935: 248–268; Schnabel, 1959; vol. 1, 408–457).

These shifts, like all organizational goal displacement, are the result of changing resources among different factions within the organizations. The ideal of truth as standing apart from political and other claims is the ideology of the group of scholars taking advantage of the power situation to claim autonomy and status for themselves. We can thus explain the conditions under which such an ideal arises quite apart from whether we ourselves happen to believe in it.

Of course, not all educational systems are equally productive intellectually. There are many examples of stagnation where schools merely propagate received beliefs in the spirit that all knowledge has already been accumulated and needs only to be taught. This happened to the Hellenic schools after 200 B.C. and to the European universities after the fourteenth century, lasting in the French and English universities until around 1870. The conditions for these periods of decline or stagnation are the obverse of the conditions under which educational systems lead to new knowledge. Stagnation occurs with the decline in the autonomy of these schools from outside political or religious interests. This is particularly the case where the teachers' audiences are students whose interests themselves are the practical, ideological, or status concerns of the outside world; the scholarly goal displacement occurs primarily when teachers can teach those who also aspire to be teachers or researchers at the highest level and in an expanding situation. Thus, secondary school teachers pass on knowledge rather than create it. Similarly, the decay of the Medieval European universities was due especially to the rise of the residential colleges providing cultural training for the sons of the genteel families (Aries, 1962: 137–175).[8]

---

[8] The decline was particularly severe in England, where Oxford and Cambridge lost

Throughout history, the periods during which teaching institutions have been relatively autonomous have been crucial for development of science and scholarship. Of course, some important work has occurred outside of teaching situations. In ancient Mesopotamia and Greece, as well as in sixteenth- and seventeenth-century Europe, scientific advances have been produced by men in practical, political, or entertainment roles. But as Ben–David (1971: 71–87) shows, scientific productivity resting upon these positions is relatively ephemeral.[9] The various sciences and fields of scholarship go into a long, steady period of development only with the rise of autonomous teaching institutions, especially in Germany after 1800 and, to a degree, in the technical schools of France somewhat earlier.

This is found in the history of the social sciences as well. The German university professors created scholarly history and experimental psychology; the Scottish professor, Adam Smith, turned the practical economics of his day into an economic system, and the move of economics into the universities in large scale after 1860 produced the marginalist revolution in economics; attempts at scholarly sociology and political science are largely the product of university professors; anthropology emerges from the phase of romantic explorations and curiosity collecting through its incorporation into the university. A striking example of the independent scholarly orientation is shown by the contrast between the amateurs who propagated racist doctrines in the nineteenth century and the German university anthropologists who disputed them on the non-political grounds that nonspecialists should not generalize about these complex matters (and thus confuse linguistic and biological evidence).[10]

---

control of training in theology, law, and medicine and became exclusively undergraduate institutions; the revival come only after the civil service reform in the late nineteenth century, and the foundation of new universities in the latter nineteenth century, brought a demand for training new university teachers and hence opportunities to pursue advanced training. In France, the professional faculties continued to exist alongside the residential colleges; but in the political maneuvers between the papacy and the rising French national bureaucracy, they became identified with the conservative side, and were displaced intellectually as the king set up new institutions for training engineers and civil servants.

[9] This analysis would stand out even more clearly if Ben–David did not overestimate the degree to which science was firmly institutionalized in its own right in seventeenth-century England; its subsequent decline after Boyle and Newton is more straightforwardly explained as a typically ephemeral scientist movement (i.e., an intellectual movement tied to a political upheaval).

[10] The main exceptions to the hypothesis are the German professors of *Staatswissenschaft,* who cling to practical fact-collecting and the preaching of policy doctrines and resist technical economic theory; the American political scientists, who up to World War II were almost exclusively normative; a strong emphasis on social problems in American sociology from the

It should be noted that these hypotheses have been cast in terms of ideal types. In reality, men can be influenced by several sorts of conditions, since many of them may have held different positions throughout their careers, or participated in intellectual dialogue among men from various kinds of positions. This helps us explain why there are exceptions to these hypotheses in their simpler form. There are many important hybrids in the history of science. Copernicus and Kepler worked outside the universities, but were trained there and communicated within intellectual networks that were centered around them. Galileo was a university professor during part of his career, although he took advantage of the opportuntities for support given by noblemen of his time. Although these men tended to be antagonists of the university teachings of their day (the Aristotelian philosophy), without the anchorage of the university intellectual community, they would probably have created no science. It was the content of the university intellectual role they were disputing, not the existence of the autonomous role itself. In medical science, where the university conservatives were less powerful, it is researchers at the University of Padua (the university of Copernicus and Galileo as well as Vesalius and Harvey) that produce all of the important advances in this period.

In the social sciences, scholarly advances by men in nonuniversity roles seem to be explained by these types of role hybrids. They take the scholarly community as their reference group; hence, the result of aspirations for academic careers combine this with the empirical materials provided by their nonacademic work roles. Marx, Comte, Leplay, and Freud all fit this model: they are all university-trained, and all of them sought academic careers but were barred from them—Marx, for his youthful political connections; Comte and Leplay, by the structure of the university system before the French Third Republic; Freud, partly by racial prejudice and partly by the stiffness of academic competition. The importance of the role hybrid for intellectual innovation becomes clear when we compare these men with their strictly academic and strictly nonacademic counterparts: compare Marx with the nonempirical German philosophers on the one hand, and with the nonscholarly radicals whom he overshadowed.[11]

1890s through the present; and post-World War II German and French sociologists, who produce largely political philosophy. Explanation would seem to come from their additional ties to practical and political organizations; the German political economists and the American political scientists, for example, situated within the law faculties and taking part in government administration.

[11] Role hybrids, in the opposite direction, are important for incorporating new materials into the academic role. Cases of nonacademics who eventually enter the academic world include Tylor, the first general theorist among British gentlemen-anthropologists, and Bastian, his German counterpart; and Walras and Pareto, the ex-engineers who mathematized

Teachers in autonomously structured university roles, then, are the main basis for an orientation toward standards of truth independent of practical and ideological and aesthetic concerns, and to the related ideal of research for the sole purpose of enlarging this body of truth. The universities may be influenced by other considerations, and the scientific ideal may be found outside of them at times, but it is the favorable structure of teaching organizations that is the anchor for autonomous science throughout history.

THE INTERNAL STRUCTURE

Sciences not only rest on other organizations, but each scientific discipline can be seen as an invisible organization in itself. Of course, all organizations—businesses, schools, government, and so on—are in a sense invisible, in that they consist of a network of communications among persons. They are not identical with the buildings and other physical artifacts that the organization may possess, but rather are a set of social relationships that enable people to appropriate and create these artifacts. This conception of organizations moves us in the direction of being able to see how a scientific discipline may be conceived of as an organization. We know further that it is not necessary for organizations to be based on coercion or on material reward, nor is it necessary to have an exact list of the members of an organization or a formal organization chart. Many voluntary associations, for example, lack the control structure of coercion or material reward. Patrimonial organizations, that is, most large-scale premodern organizations, lack the carefully delineated positions of a formal organizational chart and any clear notion of the boundaries of the organization. It is arguable further that even formally institutionalized organizations, such as modern business corporations, have boundaries that are not necessarily analytically clear. The question of whether or not a franchise is part of a gasoline company is really moot and depends on one's analytical purposes. In a similar fashion, we find that many groups of organizations may be treated for analytical purposes as one large organization. The propositions of organizational theory concerning power relationships, interest groups, conflicts, and tactics and their outcomes need not have any reference to organizational boundaries.

Why, then, should a scientific discipline be regarded as an organi-

---

economics. Adam Smith, who first systematized a general economic theory, also had a mixed career. The general model of role hybrids was introduced by Ben–David (1960a).

zation, given the possibility of doing so? An organization consists essentially of a network of relatively stable relationships among individuals. Wherever we find a set of positions that interact with each other in a regular way so that we find a regular division of labor and relatively stable forms of influence or control, we have an organization. Scientific disciplines are organizations in this sense.

The work of Thomas Kuhn (1962) makes this clear. Kuhn shows that any scientific discipline has what may be called a "paradigm," a generally accepted model of what the universe under investigation is like, as well as what methods may be fruitfully used in investigating it. Kuhn's distinction between normal science, in which paradigms operate strongly, and extraordinary periods of scientific revolution, in which one paradigm breaks down or is replaced by another, does not limit this generalization, but merely shows its variability. Such periods of revolution are extraordinary precisely because they disrupt the normal functioning of the organization. What Kuhn refers to as a paradigm is not merely an intellectual entity, a set of ideas, but is apparently both based on a social consensus and maintains an organized set of social relationships. The paradigm presents what Kuhn refers to as a number of "puzzles," that is, subsidiary problems witn the theory which remains to be investigated, and suggests the procedures to be used in solving them. The paradigm provides for the discipline an organization that is basically social, unifying its various members around a common enterprise. This is not to say that all sciences are equally well-organized. Indeed, Kuhn begins his discussion of the subject by noting the difference between the social sciences and the natural sciences in regard to consensus on theories and methods. But even fields that, by comparison to the highly advanced sciences, appear to lack a strong paradigm nevertheless can be seen upon examination to have some degree of organization in which people work on tasks in a similar fashion and adhere to at least a certain range of theories. Thus, our problem is to array the various disciplines along a continuum of forms of organizational control, and the task of organization theory here is to explain these differences.

We should remember, however, that organizations are also somewhat of a myth. The official unity of most business, political, or educational organizations is an ideal to be manipulated on appropriate occasions, while actually their members spend most of their time pursuing their own concerns, oblivious to each other. Where an organization is strongly held together, it is because the resources for control are distributed in a particular way. We shall be concerned with the degrees of unity or disunity of sciences as matters to be explained. Nor should we assume that this distinction corresponds exactly to mature sciences versus those in the preparadigm stage; many of the former, like highly rule-bound bureaucracies, are carried on by individual specialists paying little attention to each other or to any hierarchy of theorists.

We are in a good position to avoid the hypostatization error when studying sciences, because they are the most thoroughly knowable of organizations. If an organization is a network of communications, in science alone do we have most of the major communications recorded and readily inspectable, in the form of publications and citations. To put it another way: Scientific organizations can be studied in detail because they communicate very slowly, on a scale of months, years, and even centuries, compared to the minutes, hours, and days for communication and response in most other organizations. Sciences may thus be a convenient testing-ground for propositions in general organization theory.

Each scientist's work depends on both his predecessors and his contemporaries. He depends on those before him in the sense that, to have any value, his work must augment some tradition. Thus, the work of others both suggests the problems on which he will work, or the state of knowledge that he must advance, and gives him a community to which he can make a contribution. Further, the individual must be concerned about specialization, about working on things that are not precisely the same as someone else's, if he is to achieve any recognition for originality. There is also the problem of synthesis, since what is produced by a number of specialized individuals may very well not be recognized as a contribution to knowledge unless it can be brought together to form a part of a greater work through its relationship to other pieces of research.

There are three analytically distinct bases of power within an intellectual division of labor:

1. *Information.* The individual researcher depends on others to tell him what work has been done and by what methods, and what questions remain, so that an ongoing body of knowledge will be advanced (cf. Hagstrom, 1965; Stores, 1966). Power here can vary from one extreme at which all active scientists engage in equal exchanges of information, to the other extreme at which a theorist dictates all the research tasks for others. The conditions for variations in this power will be discussed later.

2. *Validation and recognition.* The individual normally depends on others in the intellectual community to validate his contributions as a legitimate form of knowledge, thus bestowing a status upon him according to the assessed value of his contribution. This is done as part of decisions about communication: The choice of whether or not to accept something as part of the corpus of knowledge is the same act as deciding whether or not to

pass it on, either by publishing it in a journal or book, or by making a reference to it in one's own writings. In particular, the specialized research-er is likely to have no audience, and hence to achieve no rewards from others, except as his work coordinates with the research of others to comprise a larger generalization. The work of a particular botanist with his description of a single species, or the chemist analyzing a particular chemi-cal reaction, may be of little significance by itself; hence, it is required that it be communicated to others, in particular to some scientist who periodi-cally synthesizes numbers of these pieces of research and thereby gives all of them meaning in contribution to a larger piece of knowledge. Power here rests with the gatekeepers of communication: in the more basic sense, the theoretical synthesizers of research; in a highly institutionalized sys-tem, the journal editors and publishers' consultants.

3. *Material resources.* Some intellectual work requires a great deal of research equipment, transportation, or paid assistance; other work, much less. Material resources are almost always involved in communicating one's work, whether in the form of direct verbal access, letters, journals, archives, libraries, or book publishing. And resources are needed for liveli-hood, especially in a fashion that allows ample time for intellectual work, whether in the form of a research-related job, a sinecure, or a patronage arrangement.

All of these material resources tend to be mediated by the intellectual community. That is to say, the process of *validation and recognition* funda-mentally constitutes the intellectual community, and gives it power over its members. In this respect, it is a form of strong profession (Chapter 6), in which the individual's reputation with laymen depends on the collec-tive recommendation of his peers, who alone have entered into the esoteric argumentation that makes up the internal content of the field. It is the world of laymen, generally speaking, which provides the material re-sources; but each intellectual, like the physician who must depend on his peers to validate his title to practice, can acquire these resources only through his reputation within the intellectual community at large.

The sources of material resources have been dealt with in the previous section, as external conditions for science. Teaching positions are especial-ly important, because they are the source of material resources which can be most stably controlled by the internal organization of the profession. They also provide primary and regular channels of information as well as initial validation in intellectual careers, and hence combine all three bases of control within the internal structure of science. Especially to the extent that professors mediate the material aspects of their students' careers, via access to research equipment, publication, and job placement, those professors' *ideas* are assured of continuous intellectual development. It is no doubt for this reason that, despite the availability of more distant communications via formal publication, the main schools of intellectual development so often depend on professor–student links (e.g., Ben–David and Collins, 1966), and that various intellectual factions and specialities are identified with particular universities and networks of job patronage (as is so apparent in contemporary sociology).

The internal structure of a science, then, is the network of communication and argumentation per se, whatever the external resources it may mediate. These resources need not all be of the same kind; political, practical, entertainment, and academic audiences and organizations may all be involved—more than likely, bolstering various factions within the intellectual field of discourse. The propensities already noted for given external roles to influence particular kinds of intellectual production are not static, but feed into an ongoing argument for intellectual dominance.

The individual intellectual, then, has a position to defend within the intellectual community, as well as vis-à-vis outsiders. It begins with his initiation into the field, as he acquires intellectual sponsors who provide him with information, access to material resources, and, above all, validation in the form of introduction to other intellectuals as having a legitimate claim on their attention. His intellectual career, as it proceeds, is a matter of ongoing choices and negotiations vis-à-vis his intellectual compatriots. He must decide what lines of argumentation and related research are likely to be well-received; what groups to ally himself with, by recognizing their work and building upon it; as well as what groups to ignore or attack. His initial contacts and sponsors lead to others, with possible choices for new alliances and conflicts. He must decide where he stands in the internal politics of the field; he must also make a personal decision as to whether his own chances of success are best as a theoretical leader or a specialized follower, as an innovator of a new specialty or of an overarching theory; and he must decide within what particular area of work his competitive chances seem best. Intellectual history, as it is usually written, is only the most visible side of this process; paradigms represent particular alliances, and scientific revolutions as well as periods of normal science are the reflection of political processes operating in the careers of the individuals involved.

The social contingencies affecting these developments constitute the sociological determinants of the course of "knowledge." These will be dealt with later, in the form of applications of the conflict theory of organizations and of social mobility. First, let us examine a concrete historical case.

The Internal Politics of Seventeenth-Century Physics

An example of an emerging organization structure in science may be found by examining the positions of Descartes and Newton in seventeenth-century physics. There is a retrospective fallacy to be avoided in this kind of historical analysis. We often assume things could only have gone the way they did, by concentrating on the few men whose ideas have subsequently become popular. We ignore deep-seated conflict among contending positions, the predecessors who failed to carry the opposition, and the periods of resistance to what we now think must be the truth—at least for that period. We can obtain a more realistic view of history by asking the sociological questions: What determines who is going to get credit for the famous discoveries? What ideas will be adequate to hold sway under the conditions of the time?[12]

The usual interpretation in terms of the history of ideas emphasizes a line leading from Copernicus, who attacked the Ptolemaic system and substituted a heliocentric one; to Kepler, who discovered the elliptical orbits of the planets and formulated their movements in terms of mathematical laws; to Galileo, who developed a new mechanics in mathematical form; to Newton, who constructed a universal system including both astronomical and terristical physics. What this misses is not only the well-known opposition that Copernicus and Galileo faced, but the fact that this was not entirely from the conservatives; that Kepler's work was scarcely known at all, until resurrected by Newton; that the Ptolemaic system, along with its Aristotelean base, was solidly displaced, not by any of these figures but by the physical system of Descartes, now almost completely forgotten; and that Newton himself was not really accepted in the major international community of scientists for many years, until his work had been reinterpreted and taken out of the intellectual context within which he meant it to apply. Still less was this simply a matter of practical inventions or Puritan viewpoints arising to challenge a dogmatic tradition. The Aristotelean orthodoxy was a rational—and by its own standards, a scientific—one; whatever the external influences, the task of overthrowing it was one of rearranging the structure of an already internally well-structured intellectual community. The opposition was not unitary but split among contending factions, above all the esoteric tradition of mathematical mystics and the religiously radical mechanists. A new scientific orthodoxy could arise only after a multisided conflict.

The framework of knowledge within which seventeenth-century

---

[12] Background for the following sketch is drawn from Mason (1962: 127–207); Butterfield (1962: 13–48, 67–170); E. T. Bell (1937: 35–130); Manuel (1968); Kearney (1971).

science appeared was developed in the Medieval universities during the previous four centuries. Philosophical inquiry had become independent of theology, resulting in some speculation in physics and astronomy. After an initial period of controversy, the ancient science transmitted by Aristotle became a rigid paradigm. Toward the seventeenth century, however, its position was weakened, not only because of intellectual shortcomings but because of a decline in its upholders' monopoly over material resources for intellectual life. Medieval universities had been politically powerful, providing the theologians and many of the bishops and popes at a time when the cosmopolitan Catholic Church was the most important force in fragmented Europe. In the fifteenth and sixteenth centuries, power shifted away from the universities with the rise of the national monarchies and their secular administrations, and ambitious intellectuals sought their careers elsewhere. The remaining university professors became part of the conservative faction loyal to the old regime of papal power, which in turn used political means to enforce Aristoteleanism as a church dogma. Despite the decline, the universities held the one intellectual system—Aristotle's—that could pretend to a full and coherent scientific explanation of the universe, one that incorporated not only logic and metaphysics, but ordered the empirical materials of physics and astronomy, biology and medicine. The universities held the single comprehensive paradigm; against this, particular factual anomalies could not be given much weight in comparison to the empirical and logical successes of the entire system.

The rise of important social groups outside the old church establishment—the Italian and German patricians, the royal bureaucracies of England, France, and Sweden, and elsewhere—are the social basis which led to the Reformation. This is not a matter of straightforward correlation of particular social groups with Protestant and Catholic divisions. There were many alliances and compromises, especially since the Reformation struggles took place over almost two centuries, and political and military events meant that both conservative and critical groups could be found in Protestant England and Germany and in Catholic France, Spain, and Italy. In a time of political ferment, political factions in several places looked for intellectual support to challenge established powers and their intellectual supporters in the church and the traditionalistic universities. Thus, resources for intellectual activity became dispersed, and the stage set for increased intellectual controversy.

A number of other conditions also obtained in the intellectual field of the time. Calendar reform was needed because of the disparity between the older chronology and the actual cycle of the seasons. The development of clocks and measuring instruments provided tools; and practical problems, especially in navigation, provided impetus for collecting new knowledge. The importation from China of the compass and the astrolabe provided new instruments with which to work; the development of fire-arms, also resulting from Chinese influence, provoked other practical investigations. The development of printing (yet another Chinese invention) improved the conditions for intellectual communication. The importation of Arabic (originally Indian) numerals, with their greater utility in commerce than the old Roman numerals, also took place at this time. With the general improvement of international communications, much of ancient thought was rediscovered, including the Neoplatonic and Hermetic mystical philosophies.

But it is important to note that the opportunities provided by the various conditions might very well have been missed. For if practical matters of calendar reform and improvement of navigation had been settled, investigation might well have ceased; the old knowledge was quite capable of being formed into a new orthodoxy, and, as we have seen, the opportunities for secular patronage and political relevance were likely to lead to political ideologies or the unsystematic trivia of entertainment rather than to a purely scientific explanatory system. It was only by the maneuvers of talented men of the time, seizing the opportunities for a sphere of purely intellectual argument within which they could excel, that real scientific advance came out of this situation.

Let us look now at the internal factions within astronomy in the sixteenth and seventeenth centuries.

First of all, there was the orthodox university community expounding the theories of Aristotle and Ptolemy. They saw the universe as a series of concentric crystal spheres, with the stars and planets attached to them and the earth at the center. To account for the movement of the heavenly bodies, 86 such spheres were necessary, with a number of smaller epicycles rotating on the surface of some of them in order to account for all the movements. It should be emphasized that there is nothing particularly Christian about this model; it came to be identified with church orthodoxy only because it was upheld in the conservative universities. But it was a theory with considerable intellectual appeal; it ordered most of the empirical observations, and it was based on a broad set of principles that synthesized knowledge in *all* areas of inquiry.

In the fifteenth century, a group of astronomers centering on Nuremberg (Müller with his assistants, Walther and Dürer), using improved techniques (mechanical clocks), amassed a body of observation of a higher degree of accuracy than had existed ever before. Müller and Walther were interested in practical matters—reforming the calendar and publishing nautical almanacs. But their data were made use of by the theorist Copernicus, who emphasized its discrepancies with the Ptolemaic model and proposed a new synthesis. Copernicus was part of the liberal faction within the Catholic Church, and was trained at the University of Padua, the most unorthodox university of its time. Copernicus' heliocentric system provided an easier way to calculate the orbits of planets and thus made better predictions than the Ptolemaic system. It also reduced the number of spheres from 86 to 34. But his system had major flaws. He proposed uniform circular orbits for the heavenly bodies, although this did not accurately fit the observations of the time. He also proposed a major contradiction for the physics of his day: to explain how the earth—the only heavy body within the system as conceived at the time—could move around the sun, particularly since theories of gravitational attraction depended upon the earth being the center of the universe. It was for this reason that Ptolemy had rejected the heliocentric model. Copernicus' bias was, in fact, aesthetic. Like the other Neoplatonists, whose ideas were popular in the courtly circles of Italy, he wanted a geometrically beautiful model. His strongest attacks on Ptolemy came from the messiness of its system and its lack of perfectly spherical orbits; his underlying criteria came from Hermetic mysticism.

Thus, long before religious and political controversy had made astronomical theories into political theories, there were good scientific reasons within the intellectual community at the time not to accept Copernicus. The best-equipped and best-assisted astronomer of the sixteenth century was Tycho Brahe, a Danish nobleman who received royal patronage in Germany, primarily to work on practical matters. Like most empirical specialists, he was not interested in theory but adopted a modified Copernican hypothesis solely in order to help order his observations. In this model, the planets go around the sun, but the sun and its planets swing around the earth, thus providing a compromise between the older and new models.

Brahe's assistant and successor, Kepler, was educated by the Copernicans and had wider ties to the community of Hermetic mysticism. He synthesized his theoretical approaches with Tycho's, thus discovering the elliptical orbits of the planets (and reducing the number of orbits to seven) and stating mathematical laws of planetary motion. These were not generally accepted, however, for two reasons: Kepler's ideas are buried in the midst of a great deal of other mystical analysis that appealed only to members of that particular faction; and, from the orthodox standpoint, his physics remained a serious problem.

No revolution could be carried out in astronomy without a revolution in physics. Experimental mechanics began in the fourteenth and fifteenth centuries as an alternative to the Aristotelian doctrines of inertia which held sway in the universities. Experiments with weights, ballistics, and falling bodies were begun by practical engineers such as DaVinci, Tartaglia, and Stevin. These were extended and synthesized in theoretical form by Galileo, who was at various times in his career both a university professor and a practical engineer and thus was able to state a theory of inertia in the general and mathematical form. Galileo became the head of a new and theoretically based school of mechanics, with disciples concentrated in Pisa and Florence. A different attack on scholastic physics came from experiments with magnets which had been carried out at least as far back as the thirteenth century. The English physician, Gilbert, produced a synthesis of his material in 1600. It contained little original work but provided a theoretical rubric and suggested various further applications, for example, that the earth as a whole might be a magnet. Gilbert may be regarded as a member of the scientific movement headed by Bacon; in contrast to the mechanical school of Galileo, this group emphasized mystical and magical traditions.

Yet another crucial development came in mathematics. Algebra in Europe began with the importation of Arabic numerals in the twelfth century. It was developed outside the universities, since geometry ruled academic mathematics. Such practically oriented individuals as Tartaglia, Viete, and Hariot developed algebraic notations and equations, although their work was limited by their practical outlook. It was finally synthesized and cast in the form of a generalized method by Descartes, whose university training placed him in a position to combine algebra with geometry. Descartes, indeed, was the prime mover of the first successful antiestablishment to challenge the Aristoteleans. Only with the publication of his physical system in 1644 was there a coherent new paradigm for science as a whole, which not only upheld the Copernican model in astronomy but grounded it in a consistent materialist physics that differed systematically from Aristotle's on all important points. Descartes thus managed to create a united front of all of the main empirical critics of traditional astronomy and physics, and the new discoveries in physiology, together with an underlying philosophy which justified a radical separation of science from religious questions. Descartes' innovations in mathematics gave an added attraction, for by bringing the new practical techniques of algebra into contact with the intellectually elite traditions of geometry, he opened up an entire new field for mathematical investigation.

Descartes' materialist synthesis gradually brought together the leading thinkers on the continent, and around him developed the basis of modern secular scientific establishment. But it did not command completely universal assent, even among intellectuals outside the conservative universities. One reason was its religious stance, which offended the mystical side of the scientific community; a second was its vagueness in relation to astronomical data. The materialists in effect were hidden atheists paying lip service to a god that their systems could just as well work without. These men were radicals within Catholic countries where the university orthodoxy had become increasingly rigid through alliance with the counter-Reformation. Thus, the reformers gave up on the church, and attempted to develop an entirely nonspiritual interpretation of the universe. Such were the followers of Galileo in Italy, and of Descartes in France and Holland. Descartes' vortex universe was filled with solid matter allowing no room for mystical causes. The Cartesian system was also weak in one respect that the mystical tradition was not: It proposed to explain planetary motion purely physically, in terms of direct contact among vortices of particles, and ignored the purely mathematical possibilities of gravitational attraction at a distance. Thus, Descartes was left with the crude Copernican orbits, oblivious to the greater sophistication already achieved by Kepler in purely mathematical fashion.

In England, however, religion was riding high and attracting the ambitious intellectuals to a purified Protestant form which had broken its ties with the orthodox universities. Around 1600, Francis Bacon united the main factions of English intellectual life, constructing an alliance between practical experimenters and the intellectual concerns of courtier mysticism (Rossi, 1968). Bacon was particularly attracted to the religious and aesthetic tone of the latter, but wanted to make it less elitist and philosophical. Baconian influence was strong in the 1640s through the 1660s, and it was his followers who founded the Royal Society (Purver, 1967). Interests of the time tended to give this a heavy practical emphasis in which there was a great deal of experimentation but little real scientific accomplishment.

But the influence of the religious wars then raging, both on the continent and within England itself, brought a reemphasis on the religious mystical side of the Baconian tradition. In particular, the British scientists around the Royal Society were involved in a counterattack against the apparently atheistic materialism of the French–Dutch school of Descartes, Huygens, Gassendi, Borelli, and the rest of Galileo's followers. For this reason, Kepler was given new prominence in England, and his mathematical laws of motion were elevated to the basis of a new attempted system which would overthrow the Cartesian model of the universe.

Borelli (university professor at Pisa) had also attempted to apply Galileo's mechanics to explain planetary orbits; a similar approach was taken by Huygens in Holland. But these attempted to incorporate this work within Descartes' materialistic mode. In England, a different tack was taken by Hooke, Wren, Halley, and Newton. In what may be regarded as a battle for high-level control of the scientific community, Newton defeated his rivals because he alone was a university professor able to draw on developments in mathematics; because he was least sidetracked in practical interests; and because his intense involvement in the Hermetic tradition made him most strongly oriented toward a purely mathematical model and most aware of the significance of Kepler's contributions. His invention of the calculus, which gave him leadership in the mathematical community, was the strategic base for his convincing synthesis of astronomical data and the leading principles of mechanics and magnetism in 1688. It should be noted that Hooke, Wren, and Halley had by 1679 deduced the key principle, the inverse square law, from experiments in mechanics combined with Kepler's laws, but they did not thereby convince the scientific community of their theory. Halley himself (who was an astronomer) was not convinced of the proof until the astronomical observations could be well-accounted-for, and this depended on Newton's command of mathematics. The conflict over priority between Newton and Hooke, when viewed in this light, may be seen as the rhetoric in terms of which domination over the organizational structure of science was to be legitimated. As in most such cases, priority was generally given to the man who most clearly dominated the organization, whether or not sheer temporal priority was necessarily his.

The aftermath must be understood in terms of the political situation of science. Newton became accepted in England as a national hero. French thinkers, however, resisted his model and stuck to Descartes' through the mid-eighteenth century. They could give a serious scientific reason for this resistance: Newton's mystical concept of gravity that acted at a distance did not fit with the prevailing conceptions of physics.

In a sense, the ultimate victory lies with the French. Science was not really institutionalized in the universities in England. The Royal Society, after a great deal of scientifically fruitless experimentation, degenerated into a honorific group with virtually no scientific membership at all. Newton's accomplishments again became a part of the ongoing enterprise of scientific investigation only late in the eighteenth century. The religious controversies that earlier had split the French and the English died down and the new French *écoles* provided a basis for teaching and investigation in mathematics and physics. It was here that Laplace incorporated Newton's laws into a more up-to-date development in mathematics and physics and gave it a basis in a unified structure of modern science. In the process, Newton was incorporated firmly into the mechanical model and came to be regarded as the discoverer of the clockwork universe. Ironically, it is only when enough time had elapsed so that his own religious intentions were forgotten, and the occultist community that supported them had disappeared as a serious rival to the materialists, that Newton was accepted.

What sociological conclusions may be drawn from the example?

First, we see illustrated the action of internal factions within the science world—in this case, the Aristoteleans, the materialists, and the mystics. Intellectual dominance went to the one that produced the most convincing *general* paradigm incorporating the broad sum of knowledge of the time. The Aristoteleans thus did not collapse until a generally satisfactory alternative paradigm was made available by Descartes; the Gilbert–Kepler–Newton mystical school became part of accepted knowledge only when it was reinterpreted to harmonize with the broader materialist paradigm. The existence of such internal factions also explains an otherwise paradoxical phenomenon: the recurrent pattern of rediscovery of retrospectively significant theories that had been ignored at the time. We might call this the "Gregor Mendel phenomenon," after its most famous example; but intellectual history is full of such rediscoveries—Spinoza, Hume, and Vico among them. Here we find Kepler, above all, rescued from oblivion by Newton; and earlier, Kepler himself developing the discredited ideas of Copernicus, and prevailing personally upon Galileo to give them a public defense. It is *within* a faction that such rediscoveries were made; Newton was specially sensitized to Kepler because they both belonged to the same esoteric faction, just as Kepler earlier had taken Copernicus seriously, when no one else did, because of their common membership in the Hermetic cult.

A second conclusion concerns the role of conflict in intellectual innovation. It is because of the independent resource bases sustaining the different factions—the Aristoteleans and the university and church establishment; the mystics with their appeal to the courtiers and religious-reforming cabals; the materialists with their links to practical innovators on one hand, and to nationalist, anticlerical political factions on the other—that each was strongly motivated to innovate. (This holds true even of the Aristoteleans, who produced considerable work in physiology at this time.) With the decline of competing paradigms at the end of the eighteenth century, by comparison, the level of scientific innovativeness fell off sharply, especially in England, where the Newtonians held undisputed sway. By themselves, the mystics were likely to have elaborated numerological and antiquarian games for the entertainment of their courtier audiences; mechanism would likely have remained a narrowly practical doctrine, without application to the intellectual world at large. Conflict was also important in bringing about crucial alliances: Descartes bolstered the materialist position by strategically incorporating a version of Copernicus' astronomy (without its mysticism) and Harvey's physiology (without its Aristoteleanism); Galileo adapted Copernicus, to bolster and generalize his attack on the Aristoteleans; Newton borrowed strategically from

Galileo's mechanism and Descartes' mathematical innovations to bolster his extension of Kepler's numerological universe.

Third, we are forcefully confronted with the fact that the ideas that dominate at a particular time do so because of their "political" relevance to the intellectual lineup of forces *at that time,* whatever the later verdict may be. Cartesian science *was* the majority consensus in the seventeenth century, and neither Kepler nor Newton commanded such *intellectual* influence in their day, whatever we may make of the retrospective merits of their respective ideas. The ideas that are able to sustain the most effective alliance are the ones that will be considered the firmest "knowledge"; the longest-lasting paradigm is one which serves to bind together an internal organization of the intellectual community with a hold on long-lasting material resources to support scientific investigation, communication, and careers. The same is no doubt true of the intellectual paradigms of today.

In terms of individual success, it is by attaching oneself to a dominant coalition that one achieves scientific fame, and that credit will be given *only* during that time that the coalition is in power—whether concurrently or retrospectively. *Within* a coalition, those individuals dominate who are most instrumental in organizing the coalition. Particular technical discoveries or empirical results (such as those of the mathematical predecessors of Descartes), or even general principles (the inverse square law as worked out by Newton's local rivals), are swallowed up as they are appropriated by the man who *organizes* a larger coalition. Scientific credit, in other words, goes above all for political accomplishment within the intellectual community.

Finally, there is much due to the role of chance. Retrospectively, we can see that the mystical tradition was better suited than the materialist (and the Aristotelean) to provide a convincing model of the universe, because of its mathematical emphasis and its willingness to deal with "occult" causes operating at a distance. If that and certain other factual conditions could have been known in advance, it might perhaps have been predicted that England would be the place where the astronomical system would finally be fully developed, because there the Aristotelean hold over the universities was weak (due to religious politics), and the form of opposition favored yet another kind of religious emphasis. If the sociology of science is successful, we will eventually be able to formulate principles that could tell us with some precision what organizational and political conditions favor particular intellectual emphases. Thus, part of the prediction might be made, given a range of empirical cases to plug into our sociological principles; but unless we know in advance the nature of the phenomenon being studied by our various scientists, we can predict only their approaches, not their successes. The same applies *a fortiori* to the viewpoint of the scientists themselves; in the seventeenth century, we see them walking backward into the future.

DETERMINANTS OF SOCIAL STRUCTURE IN SCIENCE

On the structural level, the explanatory problem is to systematically account for variations in the organization of different intellectual communities. The main distinction developed in the literature is between the "hard sciences" and the "soft." Kuhn (1962) began his analysis by considering the differences between the social sciences and the physical sciences. The latter are organized around a paradigm that provides for cumulative development through routine puzzle-solving; the former are much more diffusely organized and relatively noncumulative because they lack a paradigm. This distinction has been borne out by quantitative evidence. The hard sciences present more of their research in tables; they communicate primarily through short papers rather than by books; they are more impersonal and give references to other writers by their initials and last name rather than full names (Storer, 1967). Price's (1970) research on publications networks also bears out this distinction. The physical sciences are much more closely organized around a research front. In physics and biochemistry, for example, 60% to 70% of references are to publications within the last five years. At the opposite extreme, in the humanities, 10% to 20% of the references are to publications of the last five years. The social sciences tend to cluster around 40% references to the last five years. Price (1970) argues that the core of a hard science is an "invisible college" of researchers working on the most advanced materials, and communicating largely among themselves via unpublished papers. This has been borne out by Hagstrom's study of preprints, which are distributed primarily within the specialty; Hagstrom (1967) also shows that specialties tend to be relatively small, containing less than 50 members. Price also indicates the organization of science into closely knit schools, from the frequency with which particular papers are cited and the existence of long chains of papers citing each other, patterns that are not found in the humanities or in technology.

But the distinction between "hard" and "soft" does not characterize the organizational structure well enough. It should not be identified necessarily with the intellectual effectiveness or advance of the field. Thus, for example, psychology and sociology both have relatively high indices (around 45% references to the last five years), whereas mathematics articles are clustered around 30% (Price, 1970). There is also some evidence that the organization of the advanced hard sciences are not necessarily very tight. A particular specialty as a whole does not consist of a continually and thoroughly intercommunicating group, since its members tend to come and go; although a few stay usually at the center of what communications networks there are (Mullins, 1968, 1972). There is also a good deal of evidence that the larger community is split by deliberate efforts at secrecy, especially in the highly active and competitive areas (Hagstrom, 1967; Gaston, 1970). We also know that a sizable number of publications in the hard sciences are rarely referred to or or not referred to at all, and thus do not effectively enter into the communications network (Cole, 1970; Price, 1967). This is very striking in mathematics, where there are a very large number of separate specialties, and little unifying "theory" (Hagstrom, 1965: 227–235).

Thus, we find that the paradigm–nonparadigm distinction does not entirely coincide with hard versus soft, nor with the degree of intellectual advance in the field. These complications, of course, are already foreshadowed in Kuhn, where the tightly organized paradigm sciences are shown to have periods of crisis when the paradigms break down. It is apparent that we are dealing with a multidimensional situation.

Explanatory principles may be invoked from the sociology of organizations (Chapter 6, 13.0–15.7). There are a number of determinants of organizational structure: especially the degree of uncertainty of the task outcomes; the degree to which there are problems of coordination based on the need to bring together the results of many separate tasks, work places, or large numbers of workers; and the availability of communication technology, including writing, money, and the hardware of communication and transportation. Combining these schematically:

(A) *High task uncertainty* together with *low coordination problems* results in informal "crafts" types of organizations.

(B) *High task uncertainty* combined with *high coordination problems* results in collegial or professional types of organization.

(C) *Low task uncertainty* combined with *low coordination needs* results in a simple regularized bureaucracy.

(D) *Low task uncertainty* combined with *high coordination needs* results in a complex and conflictual bureaucracy.

These are the outcomes if advanced communications resources are available (as described by Weber; cf. 13.4–13.5). Without these resources:

(A) *High uncertainty–low coordination* results in much the same situation.

(B) *High uncertainty* and *high coordination problems* results in patrimonial organizations with a tendency to split when coordination problems become too great.

(C) *Low uncertainty* and *low coordination needs* results in small autocratic "patriarchal" organizations.

(D) *Low task uncertainty* and *high coordination needs* results in fragmented or feudal types of patrimonial systems.

These hypotheses may be applied to the organization of the sciences.

Coordination problems in the intellectual world must be seen from the point of view of the individuals affected. What conditions make intellectuals rely heavily or little upon their colleagues in order to have successful careers? There are four main variables:

1. the extent to which intellectuals communicate directly to external audiences rather than communicating only with other intellectual specialists;

2. the numbers of intellectuals competing in the field;

3. the degree to which intellectual activity depends upon material resources—research equipment, paid positions to allow time for research, expensive publication—which are widely available or controlled by a small group;

4. the scope of the problems attempted—ranging from simple questions that can be attacked by an individual, to those which require that large numbers of different investigations be related to each other.

The degree of *task uncertainty* in intellectual work in general is high compared to other kinds of work. But granted that scientific and intellectual communities will be at the far end of this continuum (and hence rely essentially upon normative control), we may characterize differences among intellectual communities. Task uncertainty is highest in frontier areas of research or in crisis periods, and is lowest where the field operates in terms of "normal science" within a well-defined and successful paradigm. In organizational terms, this is equivalent to the degree of environmental control, since a paradigm essentially marks out a definite intellectual environment, and the ease of formulating a paradigm is based upon the aspect of the world that is being investigated. Paradigms are easier to establish in some areas than in others; as for example, in the early phase of astronomy. This is also the case in some very advanced fields, where the basic nature of the phenomena is well-outlined and research tasks consist of filling in the details. Thus, there is relatively low uncertainty in the work of large chemical or biochemical laboratories pursuing relatively routine questions of detailed puzzle-solving and using routinized technology. The same may be found in fields that are not intellectually advanced but have distinguished a number of detailed areas of research with a standardized methodology; this is the case, for example, in a great deal of psychological experimentation with animals, and in certain areas of sociology. At the opposite extreme are those areas in which environmental control is least because intellectual work consists entirely of cognitive creativity. This is the case in mathematics and philosophy, where the world to be explored is created by the practitioners.

We must be careful to recognize that any particular science or intellectual discipline can be affected by several variables. Instead of attempting to characterize each one once and for all, we must recognize that fields can vary at different times in their histories as their paradigms and problems, equipment, number, and external organizational bases change. Moreover, even at the same time, different parts of the discipline can be organized in a different fashion and thus have a different ethos. The same, of course, may hold for differences between the sciences internationally. The following, then, merely characterizes particular phases of the history of the sciences for purposes of illustrating the hypotheses.

(A) *High task uncertainty* in science refers to sciences that are in a period of crisis, operating on a frontier area, and dealing with materials of great difficulty or requiring considerable subjective creativity. All of these conditions lead to relatively high autonomy for those individuals actually carrying out the tasks. Where a resultingly weak structure of coordination and control is combined with *low coordination problems*—in which individual intellectuals depend relatively little on each other for success—the result is an informal and unspecialized type of community. This low need for coordination can result from a number of different factors. One is the case in which intellectuals are close to external audiences, as when they deal with political or religious ideas, aesthetics, or practical matters, rather than with purely scholarly concerns. This is the position of the ancient philosophical cults, and continues to be the case with art and literature throughout history. Another is that in which there are relatively small numbers of intellectuals; especially when this condition is combined with ready availability of research equipment, it generally results in a loose organization of the intellectual community. As in the early periods of the sciences and the social sciences, the work that is produced tends to consist of diffuse interpretive schemes in which a particular argument depends on the persuasiveness of the whole philosophy rather than upon precise demonstration.

Individual figures loom large and their followers are personal disciples who carry on their entire system.

If we view this as a characterization of one end of the continuum, we can see that it is relevant for characterizing some of the relative differences within the modern intellectual world. Grand philosophical systems carried on by famous thinkers with devoted disciples are found especially in the university systems in modern Europe, where a few men teach entire fields. This also helps explain the more philosophical orientation of the European sciences. In a different sphere, we may explain the extreme disunity of modern mathematics in terms of the low need for coordination in terms of research resources; particularly in the United States, where university positions are widely available to support mathematicians, and the expenses of research often consist of no more than paper and pencil, there is little to unify the field.

(B)   *High task uncertainty* combined with a *high need for coordination* results in organization as an interdependent profession. There is heavy emphasis on the collective ethos and its "professional ideals." Individuals become more dependent on each other's professional evaluation to make their careers, although the organizational structure of the field still remains relatively unsystematic and the tone relatively egalitarian. This can come about in fields that are relatively remote from external audiences' interests and which have large numbers within the field; the former is often a result of the latter.

Since recognition in intellectual work depends on originality, increases in numbers raise the level of competition. This, in turn, tends to move science toward specialization, empirical methods, and logical rigor. Speculative combination and recombination of ideas can be carried on indefinitely; science did not arise because philosophy was exhausted. But competition is limited by the very structure of speculative thought. It is difficult for large numbers of persons to work at the same time in a speculative field; they all tend to tackle the central problems of the field since there are no clear criteria to indicate the potential contributions of less obvious problems. Everyone tends to be a system builder. Moreover, there is no good way to measure the relative success of their various speculative attempts; there are only a few recognizable ranks of achievement: "great," "good," "mediocre," and "valueless" would probably exhaust the list. The introduction of empirical methods and rigorous logic (especially in a mathematical form) has several advantages for the group of competitors as a whole. First, empirical fields lend themselves more readily to specialization, thus increased numbers of men can work on empirical materials without such great concern about duplicating the work of others. Empirical methods make it possible to distinguish among many closely related studies which vary slightly the different aspects of the situation in the course of investigating a problem.

Empirical methods provide much more easily inspected criteria of accomplishment than do speculative methods; hence, empirical methods are favored by the new men in a field as a means of combating entrenched and privileged elders (cf. Storer, 1967). After all, the conservatives in a speculative field already have what is perhaps the only solid sign of merit—their success in gaining formal recognition, such as the university chair; the newcomers can only compete among themselves in acquiring favor with the conservatives, or appeal to the easily discountable and nebulous force of "public opinion," unless there are clear standards by which the superiority of their new ideas may be demonstrated. It is perhaps for this reason that, as Kuhn noted, quantitative accuracy, once achieved, is never given up in scientific revolutions, no matter what other methods and concepts are given up in order to try a new paradigm. Kuhn (1961: 128–146) suggests that quantitative methods are retained because of their usefulness in settling disputes.

Empirical methods allow for a considerable expansion of the number of rungs on the ladder of scientific accomplishment. The order of the day becomes specialization rather than system-building, and there is room for the man who contributes a modest empirical study, as well as for the theorist who synthesizes diverse findings. This situation is of great advantage to the mass of men of medium talent who would otherwise be unable to attract any attention at all.

Increased numbers of men in the social and humanistic disciplines, and the consequent increase in competition within them, results in a similar trend toward objective and easily inspectable standards of scholarship: an outpour of increasingly specialized publications characterized by "objective" research techniques, such as mathematization in the social sciences and the emphasis on historical exactitude or analysis of technique in literary scholarship, usually at the expense of concern for humanistic meanings.

Thus, the scope of problems studied tends to diminish under the sheer weight of numbers. The difference in numbers of philosophers or psychologists in the United States versus Europe in the twentieth century may account for much of their difference in approach. Both large numbers and a large scope of problems undertaken increase the coordination problems within a discipline. The number seems to be a more autonomous variable and thus tends to reduce the scope of problems considered.

High coordination needs are also produced where expensive research equipment is involved and is monopolized by a relatively small group, or where full-time positions are controlled in a very hierarchical educational or patronage system. This results either in a severe limitation on the number of individuals who can enter the field or in a high degree of stratification with a few men controlling the resources and determining the career chances of the large number of aspirants. This became the case, for example, in the sciences at the German universities by the turn of the twentieth century. As Max Weber (1946: 129–156) noted, the sciences had become relatively autocratic due to the concentration of the means of intellectual production. The resulting organization was like a conservative guild with a relatively low rate of innovation. This structure, however, also meant a high degree of insulation of the scientific elite from the external pressures of practical interests, and a relatively unified intellectual structure. This helps account for national differences in mathematics, for example. The German universities have been the home of most of the efforts during the nineteenth and early twentieth centuries to systematically relate all mathematics in a coherent body of abstractions; the French Bourbaki group was the most recent to attempt this, whereas mathematics in the United States is much more diffuse and anomic, and directed toward particular problems. The same may apply to modern research in astronomy, centered very heavily around a few enormously expensive instruments. Without the current concentration of power over resources in the astronomical community, the recent discoveries of the vastness and the diversity of the universe might well be expected to result in a great deal more theoretic pluralism than is the case. The Velikovsky affair (in which astronomers acted peremptorily to deny research equipment to an outsider wishing to test an astronomical hypothesis) illustrates the interdependence of monopolizable research technology and theoretical consensus.

(c)   *Low task uncertainty* is found in sciences with a highly successful paradigm vis-à-vis their problems, or especially in those parts of sciences that operate with highly routinized research technology. Combined with *low coordination problems* the general result is comparable to a smoothly functioning bureaucracy. The ethos emphasizes routine and the seemingly immutable nature of the science. Low coordination problems can be produced by a number of factors. While the intellectuals are close to external audiences, we have a trivial case where they are simply teaching some form of dogma. Other possibilities are small numbers of men in the field, especially attempting minor routine problems; here again the matter is trivial. More interesting is the case in which modern sciences have readily available resources and are pursuing a highly successful paradigm. This is comparable to Joan Woodward's (1965; cf. Stinchcombe, 1959) analysis of process production organizations in which complex tasks may be carried out while coordination is provided automatically by the organization of the technology.[13] This would seem to be the case in modern chemical research. Parallel to Woodward's findings, there should be a high consciousness of the organization in the sense that everyone is aware of the scientific enterprise in which they are engaged. Since it is a well-operating network of communications in highly impersonal form, such an organization may operate with low coordination problems, even with large numbers, simply by enhancing the degree of bureaucratic specialization within it.

An especially interesting case are disciplines that have some very routine sectors of research but no overall guiding paradigm. That is, they are characterized by low task uncertainty in particular specialties but very high uncertainty at the theoretical levels of which fields are to be coordinated. This would seem to be the case in psychology and sociology today, as well as in biology before the Darwinian revolution. The ethos of the two different sectors of the field differs sharply; hence, we might expect considerable conflict between them.

The foregoing analysis has been given in terms of bureaucratic organization where the resources for standardized communication (analogous to Weber's writing, money, and so on) are widely available. Within the sciences, the equivalent of a writing or monetary system is a highly standardized, precise, easily communicable symbol system, appropriate for the materials at hand. Where this is lacking, we may expect, analogously to organizations without prerequisites for bureaucratic communication, to find patrimonial organizations. Relationships are more personal, consisting of masters with their disciples; networks fragment easily when they grow beyond a given size. This seems to characterize ancient science rather well. It also has modern applications. In the various social sciences that have not achieved an appropriate symbol system (a form of mathematics or a language that precisely captures the matters they need to explain), we find a type of patrimonial organization. Sociology, psychology, and anthropology thus are divided into various schools following after well-known leaders. This occurs even when the outward appearances are

---

[13] Crozier's (1964) analysis of the ethos of highly routine bureaucracies is also applicable to the corresponding type of scientific community.

"bureaucratic," as there are some highly routine areas of research being pursued. In keeping with this ethos, the heroes and disciples may be organized around particular methodologies and specialties rather than theories.

(D)   Finally, we have the case of *low task uncertainty* and *high coordination needs.* This is the situation that Woodward finds to produce a complex and conflictual bureaucracy. The most important cases here are those in which there is a well-working paradigm or set of subparadigms, or a large number of researchers in the fields, and/or highly monopolizable research resources. This seems to be the case in such fields as modern physics during their periods of high innovation. As is the case in complex bureaucracies, there is a heavy emphasis on rules and tremendous consciousness of the organization, but disagreement at local levels over how matters are to be interpreted. There is a tendency toward the defense of specialties, and there are conflicts and maneuvers because these specialities are periodically reorganized in middle- and higher-range syntheses. In bureaucracies of this sort, as studied by Woodward and others, the organization is constantly in turmoil: Specialists have considerable scope for autonomous operation and the sheer size of the administrative network of coordination enhances this; on the other hand, the specialties themselves cannot be successful without at least periodically being coordinated with the others. This seems to fit the description of large innovative modern sciences, with their division into a fluctuating research front of rapid communications, a periphery of isolated specialists in the routine areas, and a large middle ground in which men move back and forth.

Determinants of Individual Careers

We may also ask the question of who will become successful within science. As in our analysis of social mobility generally, an individual's chance for success depends on the structural organization of the scientific community at the time in which he begins. His career consists of the sequence of contacts he makes with others, both personally and through printed communications; success here is primarily a matter of fame, of being much referred to by others. As we have seen in analyzing the case of Newton, success means bringing oneself into a central place in the organization of science; the greatest fame goes to those who are most instrumental in reorganizing fields in a major way. It is for this reason, it has been much noted, that most discoveries in science are independently made by a number of different researchers. Discovery is not simply a statement of something about the external world, but the fitting into the ongoing structure of science at the time and the building of a new organization of science around it so that the discovery will be referred to. The problem of scientific careers, then, in a social sense is a political one as in other forms of social mobility. Generally, it would appear that success is produced by a combination of great personal ambition, the sheer luck of being in an organizational situation ripe for transformation, plus movement through the most favorable network of contacts. It is striking that most scientific revolutions have been carried out "within the citadel." Copernicus, Harvey, Galileo, all derive from the central vantage point of Padua; Newton at Cambridge is at the center of the national scientific community from the very beginning; so is Lavoisier in Paris; Darwin acquires the most favorable sponsorship in his college days and is in close contact with Lyell and the other leaders of British science; Thompson, Rutherford, and Bohr and his followers all branch off from a small group at the Cavendish laboratory.

The kind of tactics necessary in order to become successful in science depends on the structure of the field at the particular time. Fields with strong theoretical paradigms are thereby highly stratified in terms of intellectual power. Fields lacking strong paradigms may be much more egalitarian. If individuals follow their own interests, those in different positions within a field will wish the field to develop in different directions. Generally, in a relatively decentralized field, the interests of most of its rank-and-file members will be to maintain the status quo, to keep the organization decentralized and thus relatively egalitarian. The development of any theory that would change this situation thus involves a power struggle of a particular sort.

This may be illustrated by the development of Darwinian biology (Eiseley, 1958: 117–204). Eighteenth- and early nineteenth-century biology was relatively decentralized, carried out primarily by individual explorers hoping to discover new species. The community was unified mainly by Linneaus' classification system, which ensured a way of rewarding individual contributions, usually by memorializing the discoverer's name by eponymy. Any theory that unified this descriptive material around a dynamic explanatory mechanism would threaten the equality of these gentlemen-scientists, by making certain lines of work—those bearing on strategic points of synthesis—more important than others. Darwin's ambitious task was "political" as much as strictly intellectual. All of the elements of his theory were available in the scientific community during Darwin's youth, and indeed the major mechanism of evolution was publicly proposed by Matthew (1831) and by Chambers (1844). These theories were rejected, not simply by external religious interests, but by the scientific community of biologists itself, which was hypercritical of the loose manner in which the evidence was presented in support of the theory (especially by Chambers, who was a journalist, not a biologist). Darwin's long delay until 1858 in publishing his theory was due not to his concern with religious critics but to his efforts to convince the biological community by painstaking synthesis of the empirical evidence. As in the case of Newton, we see that intellectual influence eventually went to the man who dealt with the organizational realities of power in the field; rather than merely stating ideas of potential merit, he actually demonstrated how the middle and lower-middle level syntheses could be done, thus challenging his more ambitious or professionally minded colleagues to follow or be left behind.

The furor that followed Darwin's publication, then, was only superficially a matter of science versus religion; it also involved a shift in the organization of power within biology. It may be suggested that the length of time before a theory is accepted within its field reflects the organizational structure existing prior to its proposal. The years of controversy following Darwin's *Origin of Species* may be contrasted with the three weeks it took for Pauling to capitulate to Crick and Watson's solution for DNA structure (Watson, 1969: 138–141), and with the decades preceding even partial acceptance of major models in the social sciences, such as Weber's and Freud's, as guides for research. The length of delay appears to reflect the extent to which subsections of the field are already organized in a "bureaucratic" fashion, so that supporting evidence may be clearly and quickly marshalled for a major innovation.

Social mobility research within science has concentrated primarily upon movement among different university positions (Hagstrom & Hargens, 1968; Hargens and Hagstrom, 1967). As Hagstrom points out, the weakness of the field is the lack of an explanatory theory that may be tested. There is also the problem of the base of the analysis. It is of relatively little interest to compare the scientific success of individuals from various family backgrounds, since families are not part of the organization of science at all (except where the parents themselves are scientists). As in the analysis of social mobility generally, a great deal of implicit explanation goes into how the set of categories are constructed. "Position" within the scientific community, strictly speaking, is primarily a matter of having one's ideas recognized and referred to by a great many people rather than by a few or by no one. The criterion of success, then, ought to be the extent to which one's work is cited. (In fields like modern American physics, however, this seems to be correlated reasonably well with the eminence of the university position one obtains.) As in social mobility generally, the possibilities of success are limited by the structure of scientific organization at a given time, and by the possibilities for transforming it. The most successful career in science, for example, would be that which would supply a paradigm for a very large number of fields. In a situation where there is a high degree of relatively autonomous specialization, the height of the career is limited, but there may be possibilities for more men to make a success.

Explanations of mobility in science can apply hypotheses from mobility studies in general (Chapter 8), and must pay attention to the same sorts of problems. Some of these are particularly severe in science, since we measure success in terms of having one's ideas talked about. It is quite possible for people to have the best part of their careers after they are dead. What seems most plausible is a model in which we compare men who have different contacts with teachers and colleagues, looking for the conditions that move some of them into the more central positions in communications networks as they change over time. It is likely that in science even more so than in mobility generally, where moves are calculated on the basis of where one begins (one's teachers), moves are not made very far. Conditions for variations in *intellectual* career movement need to be stated. These will probably turn out to be equivalent, on a different level of analysis, to explaining the conditions for historical changes within the organization of the sciences.[14]

Rates of Mobility in Science

As in mobility studies generally, rates of mobility in science are equivalent to a sum of a number of careers measured against some initial point in the organization. They are equivalent to studying how the organizational structure changes as the individuals within it are replaced. In science, a career consists of becoming recognized by others, becoming the center of a communications network. Rates of mobility are equivalent to the rate of intellectual innovation within the field. The problem is different from general mobility studies only in that men continue to be part of the communications structure after they are dead, as long as their ideas continue to be talked about. We are talking, then, about the rate of mobility in men's ideas—which is equivalent to the rate of innovation. Any measure of rates of mobility, then, must compare the degree to which new ideas displace old ones. Price's (1970) measure of the proportion of new

---

[14] The material provided by Mullins (1972) shows the central importance of one man (Max Delbrück) in organizing a new community, gaining organizational resources for it, proselytizing new members, holding together its communications networks, and staying on while others came and went. The final result was the discovery of DNA structure and a revolutionary new model of the nature of living matter, to which Delbrück contributed few experimental or theoretical leads of his own, but only an aggressive personality and an overwhelming strategic commitment that the next great advance in science would be made in this area. He also happened to have previously been a member of the group around Niels Bohr in Copenhagen, which had revolutionized twentieth-century physics.

to old citations is an appropriate model; to be made precise for our purposes here, however, it should be made in terms of citations to particular men rather than to particular papers. Assuming that these two indices are correlated, however, we already know that there is a much higher rate of mobility (judged in terms of the half-life of the archives) in the physical sciences than in the humanities, with the social sciences somewhere in between.

There are a number of different questions that can be asked about mobility. The most interesting questions are those which enable us to state the conditions under which there will be a high or low degree of scientific innovation in a given community. Innovation seems to be related to two factors: the number of competitors and the degree of competition between competing paradigms. We know that the proportion of world productivity in a science is correlated with the sheer amount of funds spent on science and the number of scientists in a particular country (Price, 1971). But sheer numbers and funds do not necessarily indicate effective competition; in the period since 1950, the relatively small British, German, and French scientific communities have won Nobel Prizes and produced scientific papers considerably in excess of their numbers and funding. The large American and Japanese scientific communities are of average productivity and the very large Russian community is extremely inefficient.[15]

---

[15] In the period between 1950 and 1965, of the six largest scientific communities in the world, the United States had approximately 41% of the scientific personnel, won 54% of the Nobel Prizes, and produced 32% of the chemistry papers and 37% of the physics papers; the U.S.S.R. had approximately 38% of the personnel, won 9% of the Nobel Prizes, produced 24% of the chemistry papers and 20% of the physics papers; Japan had approximately 10% of the personnel, won 1% of the Nobel Prizes, produced 10% of both chemistry and physics papers; Britain had 5% of the personnel, won 20% of the Nobel prizes, produced 19% of the chemistry papers and 18% of the physics papers; Germany had 3% of the personnel, won 9% of the Nobel Prizes, produced 10% of the chemistry papers and 7% of the physics papers; France had 3% of the personnel, won 5% of the Nobel Prizes, produced 6% of the chemistry papers and 7% of the physics papers. The approximate ratios of productivity to numbers are shown in Table A. Figures are computed from data presented in Ben–David (1971: 174).

TABLE A

	Nobel Prizes	Chemistry Papers	Physics Papers
Britain	4.4	3.6	3.4
Germany	3.0	3.3	2.5
France	1.6	2.0	2.5
United States	1.3	.8	.9
Japan	.1	1.0	1.0
U.S.S.R.	.2	.6	.5

Situations of maximally intense competition foster the greatest innovation as long as the rewards are not centrally controlled by a single university or other organization. Major innovations occur where intellectual communities with independent bases come into contact. The origins of experimental psychology in the nineteenth-century German universities may be explained by the mobility of young scientists from the overcrowded field of physiology into the large but stagnant field of philosophy (Ben–David and Collins, 1966). The methods of natural sciences were retained by the men who changed fields, both for the prestige and the proven effectiveness of these methods; the application to the materials of philosophy created a new scientific field of experimental psychology. Such structurally induced "role hybridizations" may also be found in the origins of such fields as bacteriology and psychoanalysis (Ben–David, 1960a). In the modern American universities, the field of molecular biology appears to have arisen as a result of the movement of physicists into virology, bringing with them physical techniques and problems and replacing traditional biological concerns and methods.[16] One of the important modern developments, the "age of biology," seems to be the result of a continuing and large-scale movement of scientists trained in physical methods into the "underdeveloped areas" of biology.

The conditions producing competition may increase productivity within a field as well as give rise to new fields. The medical sciences in Germany were more productive than those in Britain and France in the nineteenth century because of the structure of the academic labor market in Germany (Ben–David, 1960b). The decentralized German university system made it necessary for German universities to give added inducement to lure important scientists away from other universities; among the most important of these inducements was the establishment of separate chairs for the scientist's preferred specialties. As a result, specialization and innovation in the German medical sciences were encouraged. Structural conditions affect the competitive situation in other ways; hierarchic laboratory organizations, as compared to decentralized departmental organizations, limit scientific productivity in the field by limiting the career chances of the younger scientists (Ben–David and Zloczower, 1962).

As in explaining social mobility in general, then, the degree of mo-

---

[16] Mullins' (1972) data show the central importance of the physicist Delbrück and the early influx of physics students into this area. Delbrück (one of the later students of Niles Bohr, whose group had accomplished the revolution in modern physics) decided early in his career that the great developments in physics were over for the time, and chose other areas which he felt would yield to the more advanced approach. Mullins' own analysis, however, concentrates only on the later phase in the development of this new research community.

nopolization of competitive resources is crucial in determining mobility rates within science.

Conclusions

Ultimately, we may be able to use the sociology of science for predicting certain aspects about the development of the sciences and other intellectual disciplines. Some of this will be attempted in Chapter 10. It should be noted, however, that the organization analysis of the structure of science gives an important place to environmental control. In science, this means that the structure of the intellectual community depends, at least in part, upon the nature of the environment that is being investigated. Since this is an exogenous factor, the sociology of science can hope for partial explanations only. If this were not true, science as a method for investigating that part of our reality that is not determined solely by preconceptions (even though it must operate through these preconceptions in order to investigate it) would not be possible.

APPENDIX: SUMMARY OF CAUSAL PRINCIPLES

25.0 Ideas about reality are weapons in the struggle for dominance over channels of communications, and can be explained by the interests of individuals with the resources to uphold them.

EXTERNAL ROLES

25.1 The more that intellectuals occupy political positions in state or church, the more that their intellectual productions consist of arguments over value judgments and policies.

25.11 Under these conditions, intellectual innovations occur primarily during periods when the political cosmology of a weakened organization is attacked by adherents to a rival organization with increasing resources.

25.2 The more that intellectuals occupy practical positions producing directly for clients, the more their intellectual productions consist of particularistic proposals for action, and catalogues of facts.

25.3 The more that intellectuals work to entertain a leisure audience (including themselves), the more likely are their intellectual productions to emphasize standards of pleasurable response through striking innovations, stylistic elegance, or factual curiosities, and to appeal to the status interests of the groups being entertained.

25.31 Under these circumstances, scientific work is supported only when its procedures are dramatically visible to observers and when it is a novelty.

25.4 The more that teachers are situated in a large community of teachers and students which is relatively autonomous from outside control, the more likely they are to emphasize knowledge as general principles, of value in themselves.

25.41 The greater the internal differentiation of a school system, the greater the emphasis upon knowledge as general principles, of value in themselves.

25.42 The more that teachers are engaged in training other potential teachers (especially in nonpractical or preliminary subjects), the more they are insulated from outsiders' definitions of their intellectual products, and the more they emphasize knowledge as general principles, and to treat these as being of value in themselves.

25.43 The more that intellectuals' careers are mediated by the approval of their own community, the more they emphasize the ideal of scientific or scholarly truth as existing independently of political, religious, or practical claims.

ORGANIZATIONAL STRUCTURE OF INTELLECTUAL COMMUNITIES

26.1 The higher the uncertainty of intellectual problems and the lower the coordination problems with the intellectual community, the more likely the community is to take the form of an informal "crafts" organization, with unspecialized intellectuals producing diffuse interpretative schemes (from *15.1* and *15.2*).

26.2 The higher the uncertainty of intellectual problems and the higher the coordination problems within the intellectual community, the more likely the community is to take a collegial–professional form, with high consciousness of professional ideals and strong but particularistic controls over admissions to the community (from *15.1* and *15.2*).

26.3 The lower the uncertainty of intellectual problems and the lower the coordination problems within the intellectual community, the more likely the community is to take the form of a regularized bureaucracy, with a high degree of consciousness over what the science has achieved (as its seemingly immutable store of knowledge), and a high emphasis upon strict methodological controls (from *15.1* and *15.2*).

26.4 The lower the uncertainty of intellectual problems, and the higher the coordination problems within the intellectual community, the more likely the community is to take the form of a conflictual and complex bureaucracy, with high emphasis on methodological rules, a high consciousness of the achievements of the organization, a high rate of conflict over the interpretation put upon specialized work when incorporated into larger syntheses, and a sharp split between routinized and innovative areas.

26.51 The more that intellectuals communicate with outsiders to the intellectual community, the lower the coordination problems within the community.

26.52 The greater the numbers within an intellectual community, the higher the coordination problems.

26.53 The greater the dependence upon expensive or large-scale material resources for research or publication, the higher the coordination problems within an intellectual community.

26.54 The more that intellectual problems require a large number of investigations to be related, the higher the coordination problems within the intellectual community.

26.61 The more there is a workable intellectual paradigm of scientific research, the lower the uncertainty of intellectual problems.

26.62 The greater the regularity of research practices, and the more suitable the research for detailed methodological sequences, the lower the uncertainty of intellectual problems.

26.63 The higher the element of subjective creativity in intellectual work, the higher the uncertainty of intellectual problems.

26.7 The greater the lack of precise symbols appropriate for communicating the materials of an area of research (mathematics or a refined verbal system), the more that the intellectual community is organized by personal relationships between masters, disciples, and allies (from *14.0–14.2*).

26.8 The larger the numbers in an intellectual community with a high level of coordination problems, the greater the emphasis upon empiricism and precision of observation and argument.

26.81 The larger the numbers in an intellectual community with a high level of coordination problems, the narrower the scope of problems investigated by each individual.

26.9 The more concentrated the resources for research or publication within an intellectual community, the greater emphasis upon producing unified theories.

Individual success in an intellectual career consists of being much referred to by other members of the community, and by being in the center of the communications network.

27.1 The closer an individual is to the network of personal associations through which the largest amount of intellectual communications flow, the more likely he is to innovate significantly and to become referred to widely throughout the community (cf. *24.3–24.32*).

27.2 The greater the transformation of the organization of an intellectual field brought about by an individual's communications, the more he is recognized.

27.3 The more the organization of an intellectual community is decentralized with individuals fitting many discoveries of equal status into a unifying classification scheme, the more resistance there is to a synthesizing theory that makes some areas of research more crucial than others.

27.31 In this situation, the more an individual theorist must produce detailed empirical demonstrations in order to convince others, the more slowly are theoretical controversies resolved.

27.32 The more an intellectual field is organized around a hierarchy of subfields connected by theoretical syntheses, the faster the ramifications of a general theory or a crucial piece of research can be assessed, and the more quickly controversies are resolved and contributions recognized.

Rates of mobility in science are the rates of replacement of the men's ideas which are being communicated about, and hence are equivalent to *rates of innovation* (cf. *24.111* and 24.112).

27.4 The more dispersed the material resources for scientific research, the more intense the competition and the higher the level of scientific innovation (and rate of intellectual mobility) (cf. *24.2*).

27.41 The more monopolized the resources for career support, the lower the rate of intellectual innovation and mobility.

27.42 Ben–David's role-hybridization principle. The more movement of intellectuals among independently organized fields or external settings, the higher the rate of scientific innovation, provided that the prestige of the internal orientation to the intellectual community is higher than the prestige of external orientations.

Chapter 10

Toward a Social Science

Where do we stand now?

I have tried to summarize developments from the point of view of what kind of science is emerging. But the enterprise of sociology has been pursued from a number of angles. Relatively few of its practitioners have been concerned with producing generalizable principles like those of physics or chemistry. Hence, it has been necessary to pick out those elements that move toward a scientific theory from the mass of materials that surround them. Because of this extremely mixed character of the social sciences, what progress there is toward a comprehensive theory has not been much recognized.

The task of creating a social science has been much more difficult for the conceptual side than in the natural sciences. The latter, of course, also had their origins largely in efforts to find practical principles. But it has been more difficult to distinguish practicality from general theory in the social sciences. This is because the material with which we deal consists of our own constructed realities: the organization of our collective behavior, and the labels we ordinarily place upon it. The necessary detachment does not come easily here. In fact, most "practical" sociology has consisted of efforts to construct a particular type of reality through the categories and focus of descriptive studies, and of efforts to manipulate other people's political beliefs and behaviors. Much work in the areas of "deviance," politics, social mobility, social problems, and even organizations, has consisted of arguments in favor of a particular preference for organizing other people's behavior, passed off in the guise of a theory. The distinction between facts and values—which is equivalent to a thoroughgoing distinction between means and ends—is most elusive in the social sciences. But without it, no real scientific generalizations are possible, even though to insist upon it is to relinquish one of the major weapons that intellectuals can use in the world: an argument in favor of one's particular desires about social order disguised as a theory.

Because of the limited and clear type of ends possible of being brought about in the physical world, the natural sciences found it much easier to maintain the distinction between practical ends and the general principles that might serve as means to these ends. The social sciences have been much more caught up in descriptive constructions of ideal social realities. Ideology in a more general sense has plagued the development of sociology in a way that it almost never has in the natural sciences. The realm of general theory, apart from self-conscious applications of it to social problems, has been permeated with efforts to extol a particular kind of idealized social order. Functionalists from Comte to Parsons are the best examples of this. This attitude has crippled historical and comparative sociology with an evolutionist interpretation. What most of this boils down to is a form of national self-congratulation; for what evolutionist ever places his own society anywhere but at the peak of human advance? Such theories have been used as excuses for imperialism and racism, as well as more subtle forms of justifying cultural domination by whatever group one happens to belong to; the latter enterprise goes on still. Against this, radical theorists have attempted to set up counterutopias, and have explicitly given up the enterprise of general theory-constructing because of its abuses at the hands of the proponents of the status quo.

In addition to this, aesthetic concerns have been mixed into the social sciences to a degree that the natural sciences have never experienced. Ever since the rise of a scientific community of physicists in the seventeenth century, it has been difficult to romanticize the heavens, and nature worship in great part waned after the establishment of scientific biology and geology in the mid-nineteenth century. But the social world is our prime target for romance by the entire literary profession, and given the inherent attractiveness of their work for intellectuals, as well as the striking insights about society it has preserved against the more distortedly ideological and technical practitioners of social science, it is not surprising that this orientation has been hard to disentangle.

In other words, we have been heavily shrouded in outright illusions and led astray by all manner of temptations. Some of these I, for one, am loath to give up completely, particularly the aesthetic tradition which breathes life into what otherwise is often a tedious and ungraceful scholastic enterprise. But sociologists have thought (as opposed to observed) too much in their capacity as members of societies, perpetuating popular illusions in theoretical forms instead of detaching themselves from them. Further intellectual advances depend on a severe effort at theory construction by its own standards.

It is possible to detach oneself from the social pressures of conventionally organized illusions blocking the way to a social science. The sociology of science itself shows how this is possible. The social basis of any science is a community of scholars, relatively autonomous from the concerns of the outside world. The criteria of objectivity and of verifiability of individually arrived-at results arise in situations of institutionalized competition. This is a situation in which intellectuals struggle to make their views prevail, without recourse to anything but intellectual resources. Such communities can exist only within a larger social structure which rewards the intellectual leaders in more conventional ways with careers, money, and prestige; but the intellectual criteria for discourse within the community are upheld to the extent that men must convince others within the community in order to attain a position of eminence within it. The autonomy of the community is always a matter of relative degree. Resources for carrying out research and for propagating ideas do accrue to particular members of the community, and orient it in various directions, as we have seen in Chapter 9. As in most things, this is a matter of degree: There is usually some pure competition plus some extraneous resources giving advantage to particular factions. When the extraneous conditions become all-powerful, then the intellectual community and its internal standards of discourse collapse and there is no science.

Taking this model in terms of variables gives us a perspective on the rise of sociology. Important developments have taken place under conditions fostering all manner of nonscientific productions as well. Most ideas about society throughout history have been part and parcel of traditional ideologies. The rise of autonomous communities of natural scientists in the seventeenth century created a focal point for socially oriented intellectuals. These achieved a new degree of autonomy in the wealthier commercial societies of the later eighteenth century. Economists, whose interests were close to the practical concerns of the day, were the first to achieve a relatively sizable specialized intellectual community and develop coherent internal standards of their own. History also began to become a community with at least some methodological independence from previous political and religious apologists. Accounts from the newly discovered Americas and from the South Pacific began to create the basis for the profession of anthropology, and to bring in comparative evidence which gave European intellectuals the beginnings of detachment from their own societies. Philosophy as well achieved new standards of sophistication, especially in epistemology. The result was an enthusiastic effort, especially among the *philosophes* of the French Enlightenment, to construct a scientific view of society. These efforts were enormously biased by the concern to establish the notion of reasonable innovation against the claims of traditional ideology, but a beginning was made.

In the nineteenth century, the basis for an intellectual community expanded considerably. With greater wealth and an expanding literate class, careers for writers became possible. The expansion of the universities and the development of public school systems provided positions for teachers. The church lost its previous near-monopoly over intellectual careers, and a variety of political movements in the volatile conditions of the time gave intellectuals considerable opportunities for breaking old ideologies and occasionally aspiring to great careers in carrying their new principles into practice.

Some accomplishments of the older sections of the intellectual community were crucial in laying the groundwork for a self-conscious effort at a social science. History became well-established in the early part of the nineteenth century, and the amount of comparative material available from traveler–anthropologists grew as well. The successes of the natural sciences in moving from mechanics and astronomy into the fields of thermodynamics, electricity, chemistry, physiology, and biology, not to mention practical developments popularly associated with the name of science, gave great prestige to the idea of creating still other sciences. The result was positivism, a movement of admiration for the natural sciences within social science fields. Much of this imitation was fruitless. Quetelet's "social physics," for example, produced a short-lived sensation based on the discovery of the year-in and year-out regularity of certain statistical phenomena, but no explanatory principles were produced. Social-problems-oriented reformers began to collect data, but made little use of it for theory construction. The more conservative wing of would-be sociology was somewhat more concerned with generalization but cast it primarily in the form of hastily thought-through and nationally biased racial ideas.

Somewhat more serious scientific efforts were produced by those who made use of the historical and comparative materials then emerging. But Comte's effort to be scientific was amalgamated with his own political biases through his borrowing of an inappropriate analogy from physiology. Other evolutionary thinkers similarly cloaked ideological self-glorification in scientific form. The contribution of these men was mainly to make explicit the idea of a social science and to provide a focal point for a community dedicated to building it. Beneath Comte's idealizations, one important idea emerged: the importance of ceremonial and emotional ties in holding institutions together, not only among the primitives but in a transmuted form in modern societies as well.

The greatest of the nineteenth-century social thinkers were two men who were opposed to most political trends then current. Possibly their aristocratic ties gave them a basis for detachment in the bourgeois era: Tocqueville, the self-conscious defender of the feudal aristocracy against modern society; and Marx, whose family had been assimilated into the urban patriciate of the Rhineland. Their views were based upon comparative and historical perspectives, which they penetratingly extended to analyzing the societies of their own day and to exerting an effort to understand the mechanisms of changes then in progress. Both began to formulate the principles of conflict and the organizational conditions determining its outcomes. Both broke with the naïve rationalistic views that mistook explicit policy for the guiding determinants of history; both endeavored to understand how the conscious choices of many individuals resulted in outcomes beyond the ken of each. Marx, in particular, gave special emphasis to placing ideological justifications and the web of human illusions generally in their place in a realistic model of human behavior.

By the turn of the twentieth century, the intellectual resources for establishing a serious social science were being developed. Especially on the continent of Europe, several vigorous university systems provided a favorable setting. History had become a mature field, so that, by the time of Weber, what were for Marx only sketchy materials about world societies had become solidly elaborated. Various explorers and colonial administrators by now had provided a very considerable amount of comparative materials about the variety of societies existing around the world; anthropology, although lacking much theoretical basis, had the empirical materials in hand to begin to build general statements. Economics, the oldest of the social sciences, had acquired a more sophisticated analytical framework by the 1870s, with the work of Menger, Walras, and Jevons. Psychology had begun to break away from philosophy through the application of experimental methods taken from natural science. And not least, Marxism had begun for the first time to attract a respectable intellectual following, primarily in Germany, including some penetration into the intellectual community surrounding the universities. It was still not quite out of the underground, but conventional academics at least on the Continent began to take serious consideration of it, as the establishment of a firmly based Social Democratic party (and its party press) gave it some external resources.

The time was ripe for theoretical synthesis, and a number of these were forthcoming from the generation of Weber, Toennies, Michels, Durkheim, Freud, Simmel, Cooley, and Mead. Weber, for example, was a talented individual who happened to live at just this time; a man with an extensive background in the various historical fields recently developed to maturity in the German university system. His own political concerns, plus the challenge of Marxist ideas and the attraction of setting a new science on its feet, motivated his comparative efforts. The result was twofold: There appeared the first truly scholarly model for showing the dynamics and the complexity of world history, which Weber organized around the theme of showing the preconditions for the rise of capitalism; and the analytical tools required provided the elements of a complex theory of stratification, organizations, politics, community structure, and religion.

Weber was simply the individual who took greatest advantage of the resources of the time. The general trend can be seen in other efforts in this direction: Toennies, who was also influenced by Marxism and drew on available historical and comparative materials for his orientation, but whose formulations are a good deal more philosophical than Weber's; Michels, who approached the history of politics with the detachment of an organizational perspective; Troeltsch, following a path similar to Weber's, but with a narrower focus; Sorel, introducing sociological detachment into the study of highly emotional political movements; as well as earlier figures such as Nietzsche, who prefigured in a more sketchy way many of Weber's themes (and Freud's as well).

The development of a considerable body of comparative material on more primitive societies was first fully exploited by Durkheim. (His earliest inspiration here probably came, however, from his studies with the historian Fustel de Coulanges, who provided a comparative analysis within the narrower scope of the tribal and religious organization of the city states of ancient Greece and Rome.) Durkheim was the first to seriously use the comparative method correctly in the scientific sense; that is, to compare variations in order to state the conditions under which they occur. Previous work in this area had simply arranged comparative material in an evolutionary sequence, like those of Comte or Spencer; or to search for origins, like the work of Tyler on religion. Durkheim was the first to realize that the comparative method could be used to get at basic mechanisms: general principles which would cast light on how social order is constructed at any time or place. His first work, *The Division of Labor,* although cast in the evolutionary framework borrowed from Comte and Spencer, is nevertheless concerned with formulating the basic principles of social organization and its variations. In *Suicide,* Durkheim attempted to test some of these generalizations about the basis of social solidarity, and he subsequently returned to anthropological materials in an effort to formulate more precisely how forms of social interaction are constructed and how they affect human ideas.

Durkheim's formulations of the social construction of reality were paralleled elsewhere. Freud, coming from quite a different background—a training in medical sciences—had opportunity, because of the vicissitudes of his career, to apply the scientific perspective to questions of the human psyche. The result was at least a rudimentary understanding of the relationship between the individual human as animal and the social bonds of cognition and emotion that came to interact within it. The development of a sophisticated neo-Kantian philosophy, and its grappling with the historicity of ideas, provided the background for similar insights on the part of Simmel, Mead, and Cooley, although cast in a somewhat more idealistic framework.

By World War I, sociology was on its way to becoming a serious science. From this peak, however, a rapid decline set in. The history of the decades following World War I is very baleful. In Europe, the achievements of the previous generation was largely obscured by the extreme politicization of the 1920s and 1930s. In Germany, the ferociousness of the conflict between radical Left and Right, growing already in the 1920s, can been seen in the sociological work of this period: the consistently applied political undertone in Mannheim, and the increasingly less empirical and more philosophical work of the Frankfurt school. The rise of the Nazi regime had a devastating effect on sociology in both Germany and France, completely obliterating it in the former, and leaving France in a situation where the highly engagé existentialist philosophy of resistance held sway. The establishment of Russian Communism had a similarly debilitating effect on Marxist sociology, elevating it to the rigidities of state dogma, and focusing on factional questions the attention of those interested in Marxism in the outside world.

Under these circumstances, the United States assumed the leadership of sociology, in large degree by default. England hardly accepted it all until the 1960s, since sociology there was identified with a narrow tradition of social problems with little intellectual content, and hence was not considered academically respectable. This same kind of sociology, however, was most welcome in the American universities. This very large and continually expanding university system, already the largest in the world in the late nineteenth century, never placed any great emphasis on the quality of intellectual standards for the vast majority of its students. It has been favorable to any number of practically oriented fields of little intellectual content, such as home economics, social work, education, business administration, and social-problems-oriented sociology. America thus provided the material resources for a large sociological community of very mixed quality.

Some intellectual guidance, however, was provided by Europe. In the half century before World War I, in fact, study in Germany was almost mandatory for aspiring American academics in all fields. The earlier imports into American sociology had come from Comte and Spencer, whose domestic reflections were Ward and Sumner. Then interest shifted toward Germany. Such men as Small, Thomas, Park, Cooley, and Mead journeyed to Germany, coming into contact with the psychologist Wundt, the historical economist Schmoller, as well as Simmel, Dilthey, and the neo-Kantians. What was brought into America was mostly social psychology. Given the emphasis on social problems in the United States, comparative and historical materials were of less interest. The empirical research arising first at the University of Chicago in the years following World War I concerned immediate social problems: crime, deviance, the assimilation of immigrants, race conflict. For the do-gooding atmosphere of the time, a voluntaristic social psychology was an adequate intellectual basis. The more serious consequences of applying structural theories of stratification and political conflict were generally avoided.

In the 1930s, more recent European theoretical sociology began to be imported into the United States. Talcott Parsons was quite instrumental in this, translating Weber's *Protestant Ethic* in 1930, and the first section of *Economy and Society* in the late 1940s. This view of Weber, however, was very one-sided, suppressing all of the conflict elements in it. Parsons presented the European theorists primarily in a functionalist system framework derived from Durkheim and Pareto. Freud also entered the mainstream of American intellectual life, and was soon assimilated to a functionalist model as providing the mechanism of socialization which holds society together.

By the 1940s, empirical research had begun to move beyond a purely social-problems framework and to amalgamate with general theory. Lloyd Warner applied the Durkheimian anthropological perspective to American communities, and along with more radically oriented work like that of Robert Lynd, began to map out the structure of social classes. Political sociology began to develop with the rise of voting surveys. Organizational sociology began from practically oriented work, such as that of Mayo and Barnard, emanating from the Harvard Business School. The Chicago school's research continued, so that by the 1950s a tremendous amount of information had been accumulated, describing organizations and occupations of various types and including a great many studies of deviance. Survey and observational materials became available on the range of class and ethnic differences.

What is striking in all this activity is the small extent to which explanatory theory of the turn of the century was expanded upon. Much empirical research was carried out oblivious to theory of any kind. Some of it, particularly in the area of organizations, gradually evolved explanatory principles of its own. Much research has been carried out in an entirely descriptive fashion, with serious intellectual reflection upon it confined to methodological questions. What self-consciously general theory there was has been mostly dressed-up versions of ideology and empty scholastic categorizing and conceptualizing. This also holds true for most "middle-range" theory which attempted to bridge the gap to empirical materials, and especially role theory, a way of conceptualizing rather than stating the conditions under which variations in behavior happen. Role theory seems to have been an effort to strike an alliance between the Ivy League functionalists and the Chicago social psychologists, mostly at the price of losing the dynamic qualities of the latter.

The more serious theory that has been available, and especially that which has been carried out in conjunction with empirical research, has been widely misunderstood. Lloyd Warner's attempts to apply Durkheimian analysis of rituals to his community studies has been largely overlooked in ideologically based attacks on the implications of his work, or in narrowly technical efforts to improve the methodology. Similarly, Goffman's efforts to extend the Durkheimian paradigm to the analysis of everyday life and to make it the building block of stratification and organizational authority has been missed in the usual reactions to his work. Weber's conflict theory has been slow to filter into American sociological consciousness, as have any but the more vulgar interpretations of Marx.

Why the foot-dragging on building a sociological science? Most of the empirical work has been carried out with practical interests in mind, for which description has seemed adequate. One might also cite the generally low intellectual level of American universities in comparison to the European ones where the major theories orginated, and the narrow and inadequate way in which the history of theory has been taught. The fragmentation of fields in the large-scale American system has also been of considerable importance. Unlike the natural sciences, sociology has grown very large before it has acquired a basic explanatory paradigm. To establish this, it is necessary to cut across what have become a great number of self-contained specialities: to see the processes of stratification, organizations, and social psychology in the light of comparative and historical perspectives which provide the necessary variations in them and make possible an adequate grasp on the mechanisms involved. The very success of sociology in accumulating information in a number of specialities has tended to obscure the development of general principles applicable in many areas.

Another important reason has been the primacy of ideology among those who consider themselves general theorists. Thus, the powerful thrust of pre-World War I European social science, of which Weber is the greatest representative, has been almost entirely obscured for many decades in America by political pressures. The great anti-Communist fear beginning with the depression of the 1930s, and the Fascist threat abroad along with milder domestic versions of it which were prominent through the 1950s and have not entirely disappeared since then, put American sociologists largely in the posture of showing their patriotism in their theories. Marx was mentioned only as a devil to be spurned; conflict theories such as those of Weber and Michels have been buried under timid reassurances that they do not apply here. This has gradually been remedied for at least a small core of scholars by the general depoliticization of the late 1950s. Comparative and historical scholarship, which is always so crucial theoretically, received greater attention as a by-product of America's new overseas policy interests. The subsequent revival of the Left in the civil rights and peace movements opened the way for incorporating the Marxist tradition more directly, although it has tended to diminish the scholarly focus in favor of ideological concerns.

Nevertheless, in recent years we have reached a position in which it is possible for scholars to pick and choose from among the accomplishments of the great theorists, relatively unhindered by ideological orthodoxy. The very balance in which we find ourselves, between the young Left and the old Cold Warriors, gives us room for maneuvering on purely scholarly grounds. The advance of empirical research is crucially important in this. For over a decade now, the field of organizations has been in a state of maturity and has received a number of theoretical syntheses. Research on stratification has become increasingly powerful, as has work on some aspects of politics. Some researchers here have been instrumental in taking from the Weberian and Marxist traditions some of the more important explanatory notions and extending them. Even the field of deviance has made some more general contributions, with the vogue of labeling theory bringing about increased interest in more fundamental questions of the social construction of reality, and providing a foothold in American social psychology for the philosophical sophistication of European phenomenology and language analysis.

The current situation is very mixed. Certainly ideological and practical interests of various sorts get a great deal of attention in both theory and research. With our very large numbers, the situation is as fragmented as it ever has been, and various theoretical orientations set themselves off with their own conceptual and methodological shibboleths. Nevertheless, for those who wish to see it, there are some powerful generalizations emerging. I have tried in this book to identify them along with the main bodies of evidence that support them. As a unifying device, I have used a conflict paradigm taken from Weber and Marx, integrated on the microlevel with a model of man having a distinctively *social* cognitive and emotional organization, drawn from the traditions of Darwin, Durkheim, and the German subjective idealists. These, after all, are the major traditions of sociological theory—with the omission of organicist evolutionism, which I do not believe has contributed very much except confusion.

Although the propositions presented in this book are only tentatively formulated and far from precise, I have used them as a way of showing some of what we do know. There are quite a few theoretical statements in sociology that are causal and empirically testable, not merely conceptualizations, trivialities, hidden value judgments, or tautologies. These statements are not only compatible with each other, but form an interconnected structure, with some deriving from others, and some fundamental postulates underlying the whole. I have suggested where only some of these connections are. Obviously, formal derivations of the propositions in this book would take a great deal more work, and the effort to do so would doubtless change the substance of the theory in various ways. This process always exposes multiple causes glossed over by ambiguous terms, and it probably would uncover some major inconsistencies; if so, these are precisely the points at which major theoretical advances may be expected.

The more elaborate and best-grounded part of this propositional scheme comes from the Weberian and Marxist tradition in stratification, politics, and organizations, fleshed out by much research in all of these and by some substantial developments in explanatory synthesis in the latter. I have tried to demonstrate that the tradition of microstudies (both phenomenological and Durkheimian) can be handled in the same way—organized into causal propositions and constructed from basic postulates; and, especially important, that this microlevel focusing closely on interaction and subjective experience is not only not incompatible with a structural level of analysis, but can be made into a basis for the causal principles involved in the latter. Structure *is* nothing more than men interacting and attempting to negotiate an advantageous subjective reality; it can be dealt with on another level of analysis simply because the interactions take place among many people and over long periods of time, and individuals take account of this background (not to mention its artifacts in the form of the material conditions surrounding them) when they meet face to face. The subjective idealist, materialist, and structural levels of analysis are not only not rivals, but none of them is an adequate explanation without the others. I have attempted to show this in concrete terms, not simply in abstractions.

This core of explanatory theory ramifies directly into the problems of most sociological specialties, with certain exceptions. Certainly some areas that have been rather isolated, such as the family and childhood, can be handled with the explanatory principles of stratification and organizations. In other areas, such as education, religion, medicine, politics, this is hardly anything new; in others, like science and culture, the viewpoint has been creeping in. It is true that certain areas of stratification itself—notably social mobility theory—have been sharply isolated conceptually from the main line of causal theory, even though the phenomena studied are *exactly* the same ones as in other areas; the main problem in this area is one of understanding how our measures summarize the results of a great many complex processes.

The major exception that comes to mind is demography, pursuing issues that seem only remotely connected with the main explanatory bulk of sociology. But I suspect that it will achieve the predictive power it desires only by forging this in much the same way that social mobility studies must if they are to achieve much generality. The conceptual problem is very similar—that of attempting to predict a single measure resulting from many complex behaviors. For demography, the crucial causes surely must be related to status stratification and sexual stratification, and perhaps age stratification as well. Another highly isolated area is experimental social psychology. This tends to suffer from a lack of clarity about levels of analysis, and especially from failing to see what structural questions it is relevant to. The result is often a kind of artificial seeking after research problems—as in looking for something relevant to say about small-group sociometric structure; with a better conceptual integration into the rest of the field, it might say something useful about the dynamics of microstratification. The same problems and possibilities seem to hold in the newly burgeoning area of conversational analysis.

What about the state of the evidence supporting the core propositions? The major difficulties here are not methodological, at least in the narrow sense. In many areas, the data are still relatively sketchy, and most can stand improvement. But contrary to the attitudes of those who make their living by trying to impress outsiders with their white lab coats, the data for some time now have not been the major problem in the development of a sociological science. The main obstacle, we have repeatedly seen, is the lack of a theoretical attitude toward the use of the data for elaborating causal principles which explain variations. In fact, the data are reasonably good in those areas in which the major theoretical generalizations are based. Historical data, which have provided an important basis for the comparative analysis of stratification, have become quite sophisticated. Survey research as well has developed quite a serious methodological standard; its equally serious limitations derive mainly from the static view of reality as seen through the narrow peepholes of questions, which results when it is *interpreted* in such a way as to rule out the dynamics of human behavior and thought. Similarly, organizations and communities have been quite well-studied by participant observation techniques, as indicated by the consistency of results for *theoretical* principles.

The phenomenological critique tends to share an overestimation of its methodological importance with opposing methodologists on the other side of the fence. Its main value is its conceptual reconstruction: the avoidance of hypostatization, the understanding of the negotiation of everyday reality, and particularly the need to ground everything in what happens among individuals minute by minute. Multiple methods are needed, and the main problem is first to overcome mutual biases among their practitioners and to arrive at a standpoint from which to integrate them. Minute description will not help us very much in explaining patterns of behavior involving large numbers of people over long periods of time. But in every case, personal observation should be added to the other methods as much as possible. Sociologists ought to be skeptical of any theory that is not true to their own experiences. We are often so busy being laboratory scientists that we do not open our eyes on the way home. It is this attitude that has stuck us with so many reified abstractions for so many years.

The movement toward close observation of the dynamics of interaction is crucial for the emergence of scientific sociology. But in most areas, once the conceptual lessons are digested, the methodology itself will not radically change things. The more minute data about just what people do in politics, or just what happens in a career that adds up to social mobility, will not change the causal generalizations that already can be made on the grounds of more summary methods. In some areas, the theory-building task primarily involves recognizing just what is being requested. The causal agents in sociology are always people behaving in relation to each other. Our terminology glibly reduces millions of encounters over many years when we speak of a "centralized state" or a "family," or cite the numbers that make up a "social mobility rate" or represent a "distribution of wealth."

In these areas, our explanations will be very complex, and necessarily historically situated. For example, the various processes that go into making a social mobility rate or a trend in the population size involve a great many causes in the form of human acts. It is not that general theoretical principles do not apply, but that they apply to small parts of the "phenomenon" that we measure in some simply summary fashion. The problem is to add them together. This is analogous to many problems in the natural sciences. Attempting to explain a social mobility rate is like attempting to predict the time and place of an earthquake from the knowledge of a number of physical and chemical laws, plus descriptive information of relevant factual conditions. What we need in these areas, in addition to the fundamental causal principles, are calculating techniques for bringing them together and into congruence with our superficially simple labels covering these complex phenomena. This is the place where mathematical models will fit most usefully. The main hindrance to their successful development so far has been their lack of integration—or often even concern—for the relevant causal principles. This has also been the case in much of the study of politics, social mobility, the distribution of wealth, and in demography.

Let us look briefly at the evidence for the various propositions in this book. I have indicated the supporting studies in some places but not in others; at some points, they are rather good and readily accessible; elsewhere they are more scattered or tentative.

The propositions about class cultures are quite strongly supported from a number of directions: many participant observation studies and surveys, as well as historical summaries (such as Weber's on class-based religions), cohere very well. Studies of organizational positions and of occupations (which are really the same thing) bolster this from a different angle.

Evidence on the effects of types of community organization on status group cultures is less precise. The problem is not so much that data does not exist but that it is rarely adequately distinguished from the effects of occupational position. Thus, community studies often label their data as "ethnic" or "class" more or less arbitrarily, without attempting the comparative procedures necessary to sort these out. But there is some survey data on ethnic and religious differences cross-classified with class differences; some anthropological and other crosscultural data that can be combined to make a comparative test; as well as historical materials on religion, manners, and household organization that give some confidence that the suggested principles refer to a real pattern. Community studies and some surveys have focused on informal interaction patterns, especially on friendship ties and voluntary associations, thereby supplying some of the structural linkages. And above all, Goffman's observational work on the tactics of face-to-face interaction increases the theoretical coherence of the area, by linking status groups with the mechanisms of individual cognitive and emotional organization. With this diversity of supports, I am confident that the general model of status group organization and its variations is on the right track.

The microlevel has been closely studied in some areas but not in others. Rituals have been studied for quite a long time by anthropologists, although for the most part without paying enough attention to the mechanisms of emotional influence and cognitive reality construction. This last topic now gets a great deal of philosophical attention, and some empirical treatment, especially in close studies of conversational interaction. (This is probably the closest any behavior has yet been studied in a naturalistic situation.) Such work is beginning to narrow in on cognitive mechanisms in a way that the abstract interpretations of symbolic interactionism never did. Its use as a basis for the processes of interpersonal negotiation that make up the building blocks of stratification, organizations, and politics, however, is very tentative. My propositions in this area are mainly hypotheses, although there is some relevant data (most of it not microobservational at all) on such matters as similarities of attitude among friends. The closest we have come to making this crucial link in any systematic research has been in the work on ethnic or class differences in language and nonverbal styles.

The propositions about sexual stratification and about family structure are reasonably well-founded. The data necessarily has been largely comparative and historical in nature, since the major variations have been very spread out in space and time. Anthropologists and historians have only now begun to look at much of this in the detail it deserves, and with the method of analytical comparison in mind. Some contemporary survey material on conjugal power structure also aids. The effort to remove the theory of family–kinship structure from its holistic evolutionary framework is only beginning on the theoretical side, although there has been good command over the comparative materials for quite some time; too much of the American survey material has been devoted to combating the notion that modern people do not see their relatives anymore, but it too is moving out of this vestige of the evolutionist debates. All in all, I think the family stratification model is consistent with a wide range of evidence, although the complexities are just coming into focus.

In the area of children, the most serious problem has been to break out of the strait jacket in which value judgments have confined theoretical concepts, limiting us to only a few dimensions of interaction. Many of the formal propositions given here have been culled from the psychological literature, and reinterpreted into the theory of conflict and control used elsewhere in this book. The fit is reasonably good, as far as it goes. The propositions about the economic and status positions of children are related to some suggestions in the historical and community literature, but this is a largely undeveloped area.

Organizations have the longest coherent research traditions of any area in sociology. They have been heavily studied by participant observation and informal interview, to the point where it is hard to think of a modern organization (or occupation) that has not been treated. More recently, a considerable amount of survey research has been done, as well as comparative studies of organizational samples ranging into the hundreds. The evidence for most of the propositions offered here is very good indeed, especially for those concerning control types and their consequences. Organizational structure has been adequately (i.e., comparatively) studied only within the last decade (with the very important exception of Weber's historical comparisons), but I am fairly confident about the generalizations here too. The coherence of principles from some very wide-ranging historical studies, from recent industrial and administrative comparisons and case studies, and from the comparative treatment of membership associations and of professions is striking, and raises confidence in a whole complex of interrelated ideas. If there is one place where things are really coming together, this is it.

Politics has been heavily studied, and with a variety of methods. But much of the work has been extremely fragmented, overly topical, and riding on strong ideological undertones. I have not done much with survey studies of voting and attitudes, partly because it is not the core of the field, and also because the earlier treatment of stratification (especially the model of multiple influences on individual outlooks) is the more general case relevant here. Doubtless this evidence could be brought in; the earlier work of Lipset, for example, would be quite coherent with the general model of interest mobilization. Most propositions deal with the structure of the state, and with the dynamics of struggle over its control. Here, the basic model is bolstered by the organizational evidence presented in the previous chapter. There is also a great deal of meticulous historical research which has gone into the basic comparative generalizations, beginning with the mid-nineteenth-century historians like Tocqueville and Fustel de Coulanges in France, and the even more deliberately institutional research in Germany from Gierke onward, which went in the syntheses of Hintze and Weber. (Marx also fits into these traditions, at an early point.) Recent work in these traditions, especially that of Reinhard Bendix and of Barrington Moore, has explicitly used the logic of comparative analysis. In my opinion, there is a highly refined model in these works (which is implicit already in Weber), although I have not provided the considerable exegesis that would be necessary to prove it.

On the topic of religion and ideology, the themes of many historical sociologists resonate strongly with certain fundamental principles articulated by Marx and Durkheim and given a comparative summary by Weber. Important segments of this have been bolstered very systematically by Swanson's comparative studies.

Most of the evidence for the core theory of politics is historical (naturally enough, given the time spans involved), although comparative surveys are of some use on particular points. The smaller-scale topics of community and national power are still heavily entwined in a polemical context; but comparative data has been becoming available, especially focusing on organizational and interorganizational materials. Its general coherence with organizational theory (especially the theory of membership associations) suggests that this may be fitted into systematic explanations before too long. Finally, the area of geopolitics which underlies the fundamental processes of state structure is another underdeveloped field, at least from the point of view of sociological research. There are abundant leads in the historical literature, and no doubt the discipline of geography has much to offer; but I do not know of any systematic studies that would support the propositions suggested here.

Social mobility and the distribution of wealth both have been researched extensively, but for the most part without real explanatory theory in mind, or at best very parochial ones. The major exception, of course, is Lenski's massive comparative structural study of the distribution of wealth. The propositions offered here extrapolate from that, from pieces of evidence on pay in organizational hierarchies, and from various theoretical suggestions that cohere with the general model of stratification and organizations. In social mobility, my main efforts have been to sort out what the theoretical questions are and to try to link some traditions of mobility research with the main body of sociological theory. There are pieces of evidence on the structure of education in the system of stratification that validate the general orientation expressed here, but the meager body of formal propositions are offered mainly as an index of how far there is to go. Similarly, my effort to bring deviance studies back into structural sociology is mainly conceptualization; other work along these general lines is proceeding with greater precision.

Finally, the propositions about the structure of science and the intellectual world generally are based on several types of evidence: general historical studies, some systematic comparisons of detailed historical materials, and some modern studies of citations, productivity, and other aspects of communications networks. The propositions relating organizational theory to structure in different sciences are extrapolations. Systematic evidence in this general direction seems to be mounting, but certainly no one would claim consensus in this field as of now.

In short, certain areas of stratification and organizations have detailed propositions supported by a good deal of evidence. Comparative materials on many aspects of politics and family organization make the propositions given here seen quite secure in general, although needing support in detail. Materials on status group organization, on microsociology, and on intellectual structure are highly suggestive, and cohere well with the main core of theory; the links between microsociology and stratification are mostly hypothetical, although formally definite. Wealth and social mobility are most loosely connected, with the major problem being to bring theory to bear on the evidence in a sufficiently systematic way.

What is likely to happen next (insofar as enough sociologists pay attention to building explanatory science rather than pursuing topical issues or technical refinements) is a great deal more change. Whatever can be stated now is only a way-station along the road, or, more precisely, an indication of where a network of paths coincide to form a highway. Scientific theories develop as earlier formulations are found to be too simple. Subsequent developments are always in the direction of comprehending greater complexities and getting a tighter grasp on particulars, eventually culminating in a more powerful overall perspective. We have seen this happen over and over again in the natural sciences, from Galileo and Kepler to Newton to Maxwell and Planck to Einstein and through the astronomical and infinitesimal studies of today pushing on to even more sophisticated theories.

Sociology is on its way. In the diffuse fashion characteristic of large twentieth-century fields, we are passing slowly into becoming a science.

THE AGE OF SOCIOLOGY

Historically, sociology split off from the other social disciplines over the last 200 years, in many cases for arbitrary reasons. Anthropology became separate primarily for practical differences between a community of scholars traveling abroad and those working at home. The more respectable and established fields of history and political science have been separate, among other reasons, because of the popularity of the parochial social-problems approach in sociology. Psychology and sociology have developed very far from their common border at social psychology. Economics and sociology, once united in political economy, became separated as their ideologies diverged and specialized research techniques developed first in the former.

Sociology has long been considered a junior partner intellectually, and sociologists have imitated both the methods and the concepts of other fields. These relationships are based largely on misunderstanding. With the exception of history and some areas of economics, most of the other social sciences are rather weak in substance. Especially when used as an integrated unit, the core of serious sociological theory is a good deal more powerful than anything the other disciplines have to offer. And sociological theory is in a particularly strategic position, for (unlike economic theory) its applications are very broad. As such, sociology holds the key to most of the possible breakthroughs in the social sciences. Interdisciplinary influences will have to flow the other way; the age of *scientific* social science will have to be the age of sociology. The end product can be nothing less than a general social science, with various branches.

One group of fields has considerable overlap with sociology. Anthropology is virtually identical with sociology in its scope of explanatory interests. Social anthropology sets for itself exactly the same problems as sociology, but for the most part has confined itself to primitive societies while sociologists have concentrated on industrial ones. Both groups have moved increasingly into studying peasant societies, or those mixtures of peasant, tribal, and industrial societies that are studied under the rubric of "modernization". As the old tribes die out, or are further transformed by contact with civilization, the two fields merge empirically as well as conceptually. Sociological theory is also anthropological theory; the same battles to become free of functionalist reifications and overly particularistic concerns goes on in each.[1] Other areas of anthropology are relevant to a general explanatory science as well. Archeology is continuous with the rest of historical sociology; physical anthropology should become more closely integrated as sociology takes account of man's physical capacities in its explanatory models. Much of anthropology, to be sure, has its particular descriptive bent, but so does much of sociology. Even the subfield of anthropological linguistics must become more closely related to the rest of the social sciences, as linguistics itself has become an increasingly central field on the microsociological level.

Political science is another field that can be increasingly merged into a general social science, as a subfield dealing with political phenomena. We have seen that general explanations of politics draw on mainstream sociological areas of stratification and organizations; the process of politics fits well into general considerations of the organization of human groups and the manipulation of cognitive reality constructions. What is left to provide a distinct focus for political science are two things. One is traditional descriptive work on the politics, laws, and agencies of various states; political science here operates as a practical field, acquainting prospective

---

[1] It is typical of the propensity of functionalists to rationalize existing structures, however arbitrary, that Parsons and Kluckhohn formally sanctioned the distinction between anthropology and sociology as one between the study of culture and the study of society—as if either could be explained without the other. Compare the recent viewpoint of an anthropologist (Goody, 1969: 9): "The distinction between sociology and social anthropology is basically a xenophobic one. Sociology is the study of complex societies, social anthropology of simple ones; sociology of Euro-Americans (with brief excursions into Asia), anthropology of non-Europeans; sociology of whites, anthropology of coloreds. As non-European societies become increasingly complex, so the distinction becomes increasingly meaningless as far as field work is concerned."

politicians, lawyers, and travelers with information on politics at particular times and places. There is also the traditional area of political philosophy. These discourses on what is better and worse in the realm of political organization will not disappear, as these questions will be of perennial interest. Hopefully, once the causal principles that explain empirical variations become clearer and better known, the value judgments involved in political philosophy will become less confused with explanatory analysis, and we will have fewer efforts to impose on the world the former in the guise of the latter. Political arguments can become more sophisticated, although scarcely more terminable.

One portion of economics also belongs in the realm of sociological science. This includes developmental economics; as we have seen already from Weber's work, the initial rise of the capitalist system, whose idealized workings constitute the subject matter for neoclassical economics, needs to be explained within the larger perspective of sociology. Its background explanation includes the entire organization of society, especially the structure of politics and stratification. Moreover, since the idealized economic system is actually situated within a political and social structure that determines how freely the market will operate, the applications of economics and its prognoses of the future always require an understanding of the surrounding conditions. This especially holds true for efforts to explain the course of economic development (or its lack) in the Third World, not to mention the future of the first two worlds. I have tried to suggest that the leading generalizations will come from the development of political sociology—of which a revived form of political economy plays an important part—rather than from the narrower field of contemporary economics per se. Similarly, the seemingly economic question of the distribution of wealth must be handled by invoking sociological explanations of stratification.

Perhaps even the general economic system can be related to more fundamental theory in sociology. Our analysis of the interpersonal dynamics that make up the basis of social structure involves patterns of bargaining, and markets for personal relationships. There is a basis here for bringing the precision of economic conceptualization into microsociology. It is likely that the mode of economic analysis handed down to us, with successive modification, from the eighteenth-century rationalists will find its place as a particular subtype of human behavior, subject to particular limiting conditions. Where this rationalist model is technically elaborated, of course, it supplies economics with its identity as an independent discipline, organized especially around applied purposes, including amassing descriptive information and working out particular projections of practical interest. More successful predictions, though, probably depend upon at least some economists following the route chart-ed by Weber, himself a former economist.

Geography is an area, generally overlooked in most catalogues of the social science, which has a part to play in a unified theory. We have seen this particularly in the analysis of politics. Analysis of economic stratifica-tion as well must contain geographic considerations. A truly successful social science must integrate the heretofore mainly descriptive informa-tion of geography in terms of more systematic causal generalizations.

History is another area with tremendous overlap with scientific soci-ology. There is no danger that it will lose its distinctive identity, as the guiding principles are somewhat different. History's task is to tell a de-scriptive story, although explanation is always involved. Much of history has been increasingly concerned with whether or not its particular expla-nations are valid on more general grounds. Thus, within its own ranks, the field of history has moved toward the concerns of social science. Increas-ingly, interpretations of history based on implicit or explicit notions about the kinds of effects that individuals, institutions, and social movements can have, are subject to validation in terms of generalizations from social science. Indeed, it has been this movement within history, as well as the interest in history on the part of sociologists and political scientists, that has constructed so much of the foundation for the emerging science of sociology. It should go without saying here that history overlaps with anthropology; the artificial distinction between societies with history and the so-called "timeless peoples" is now pretty much a thing of the past for serious thinkers. The social sciences and history promise to come increasingly closer in the future. With its more humanistic focus, however, history will probably maintain a distinct attractiveness of its own, which social science, operating in a more laboratory atmosphere, may progres-sively lose.

In several other fields, less immediately close to sociology in their current formulations, there are potential revolutions in store when scien-tific sociology makes a full impact. These include psychology, linguistics, and philosophy. Psychology has been making self-conscious efforts to be a science for about a century now, ever since Wilhelm Wundt transferred from physiology to philosophy and decided to make a laboratory science of the latter's concerns for mental states. Nevertheless, the initial period of enthusiasm for the development of a successful psychological science has long since passed. The basic slogans have had considerable appeal: Wundt's effort to analyze the mind into its constituent components; later, the behaviorists' promise of a learning theory that would show how behavior is shaped by experience. Neither of these have come to much. Wundt's system, current at the turn of the twentieth century, and the learning theories dating from the 1930s, are equally incomplete, equally inadequate. Despite a great deal of research, further theoretical progress along these lines seems unlikely.

Of all the social sciences, psychology is externally the most scientific. It includes the most individuals literally wearing white laboratory coats, carrying out series of well-defined experiments, each checking variables overlooked in previous trials. The field has a good grasp as well on the scientific idea of generating causal explantion through the analysis of variable conditions. Yet this work has gone on, very intensively over the last half century now, without striking results. The problem is primarily one of a misplaced focus of analysis for the explanatory aims involved: to explain human behavior (and cognition). Most research has been done on the individual, in an effort to generate a learning theory based on the effects of rewards and punishments, or sometimes on an information-processing mechanism, within the individual himself. The reason this has been so ineffective in generating explanations of important aspects of human behavior is that such individual mechanisms as can be isolated are only building blocks within a larger context of human interaction. Most human behavior is organized by cognition; and, as Mead and Durkheim were among the first to demonstrate, the variables of subjective reality-construction are to be found in the interactive situation itself. But their formulations have been followed up almost entirely within sociology, and their ideas are virtually unknown within psychology itself. Even the work on cognitive development stemming from Piaget, the great achievement of twentieth-century psychology, has been limited by its lack of attention to interaction. Similarly, the field of animal studies, so much investigated by psychologists, cannot really be handled without taking the interactive context quite seriously. The tiny field of animal ethology has done more for the development of general social science than have the thousands or millions of laboratory animals sent down alleyways or set before levers all by themselves. The basis here was laid many years ago by Darwin, but the peculiar concerns of psychologists have left this almost entirely to sociologists, including those in the Durkheimian tradition. I have tried to show how the analysis of animal social interaction is crucial in microsociology. The key is the analysis of social motivations and emotions, themselves scarcely handled within psychology, primarily because they are so much phenomena of interaction. Among human beings, of course, the cognitive component based on social interaction further permeates motivations and emotion, making them even more remote from the ordinary concerns of psychology.

This is not to overlook the field entitled "social psychology," as practiced by psychologists. This has consisted primarily of experiments in a social context concerned to verify some principle of individual behavior: Thus, social psychology has been organized around such topics as the

cognitive choices an individual might make according to dissonance theory or balance theory, or the influence of group pressures on attitude formation. The notion that consciousness is inherently a shared and
continuously constructed entity, although supported by some famous
psychological experiments, has not been followed up by psychologists
but, rather, by phenomenological sociologists. Thus linguistics, which in
its recent movement toward general explanatory principles has leaned
toward psychology, should increasingly find that sociology is more relevant. The ethnomethodologists have already taken a hand in this area, and
promise to transform it profoundly.

What should happen, I would suggest, is a considerable redefinition of
the field of psychology. Its anticognitive bias in America is a carryover of
its struggle, over half a century ago, to free itself from departments of
philosophy; this has left it in a peculiarly narrow and untenable position
with respect to undergirding a general theory of human behavior. Its
recent interests in cognitive development lead increasingly in the direction
of interaction; ethnomethodologists here are providing a framework within which a great deal of developmental work, including that in psycholin-
guistics, is being reformulated.

The area of psychology that has shown the steadiest progress is physiological psychology. Eventually, psychology as it now exists will be com-
pressed between sociology and biology, a narrowing front between these
two fields that may eventually disappear. The emerging vision is one of
man as a biological organism whose workings we are coming to understand at a chemical and electrical level. But we are also discovering that
the individual as a unit of analysis is too limiting. The proper focus is on
the collection of such physical entities in a much larger space, emitting
signals and making contacts that profoundly affect each other. It is along
this line that we will arrive at a firmer understanding of the great myster-
ies: the nature of "consciousness," the relation between "mind" and
"matter." These terms already are becoming much too simple given the
larger perspectives of Durkheim and Mead and the developments of
phenomenological sociology on the one hand, and our increasing understanding of neural networks, chemical reactions, and avant-garde research
on biofeedback on the other.

A successful thrust along these lines should have far-reaching reper-
cussions within philosophy as well. Our increasing understanding of the
social biological nature of cognition cannot but give rise to an entire new
perspective on traditional problems of epistemology. Many of the traditional difficulties from Descartes to Hume and Kant result from the way
the very questions were formulated: the rationalistic biases of individualistic psychology, with is hypostatized understanding of the nature of

mind. Kant's a priori categories of understanding are better understood with the lessons of Mead or Durkheim in mind. Twentieth-century philosophy, particularly under Wittgenstein's lead, began to clear up the difficulties in the way these questions were formulated, but was confined mainly to a critique, lacking the empirical techniques and sociological perspective from which to construct a better system. Certain areas of phenomenological sociology now set themselves almost exclusively to handle epistemological problems with the tools of empirical analysis.

The field of ethics as well cannot but benefit from the development of a social science. This development, as I have demonstrated, cannot be carried out without a firm understanding of the distinction between value judgments and logical and descriptive statements. Ethics is always an area of the ultimately arbitrary, but concerns itself with drawing out the consequences from these choice points, or tracing courses of action back to them. With the aid of social science, ethics can move beyond its conventional middle-class Christian biases built into some notion of rationality, interest, or the concept of "good" itself, to a far more sophisticated view of the choices that confront us.

The applied fields, finally, should also be changed. For the most part, they will have to become less pretentious. Certainly, they will have to fight themselves free of the quackery that prevails now, as value choices are hidden under displays of technical jargon or collections of facts—which never, in fact, speak for themselves. A truly serious applied social science cannot attempt to hide value choices, but must make them explicit, to break through the web of hypostatizations that hide policy-makers from their own conclusions. Hopefully, this will have some democratizing effects upon politics, if its extreme relativism does not stoke an atmosphere of panicky cynicism. Such clichés as "quality of life," "equality of opportunity," "peace with honor," "the public interest," and the like will no longer pass in serious intellectual discourse as if they meant but one thing and did not involve the interests of particular groups against others.

The professional applied social scientist who wishes to build on the lessons of the best science will have a problem of self-discipline more serious than any profession has ever experienced. Certainly, simple notions like setting up a Council of Social Science to tell the government what ought to be done about "national problems" can only serve to mystify, if the advisors do not make it their first order of business to indicate the plurality of value judgments and interests applicable to any policy decision. The elitist notion that high-level information ought to be reserved for the highest political authorities is one on which men of democratic values might well turn their backs. Instead, they might try spreading social science sophistication, puncturing hypostatizations, and revealing the stage techniques by which leaders manipulate followers—all of these are services that social science can perform for a client that is too often forgotten, the public at large.

Needless to say, particular social problems as defined today will be viewed quite differently when we understand the interests that uphold particular conditions as "problems." Psychiatry has continued throughout its brief history to hide value judgments in favor of a particular style of life in its none-too-well-validated theories; it needs particularly to be raised to the professional mark. This is not to say that any one kind of therapy ought to be banned, but only that practitioners ought to be aware of the value choices implicit in them: the authoritarianism of behavior therapy, the anarchism of many sensitivity-training approaches, the bourgeois respectability of conventional psychoanalysis. General practitioners here, or perhaps all social scientists, can serve a useful purpose in educating the public to the variety of techniques that they, at their own choice, might call upon.

The world of the future will be both strikingly different and, on a deeper level, abidingly similar to the one we have now. Precise predictions are foolish, considering all the turning points that can occur. As we have seen in political sociology, wars are quite crucial in this respect. Stratification itself by no means disappears in a high-technology world; the resources only become distributed somewhat differently. Ideologies and conflicts continue; polarization and escalation can move societies off in very divergent pathways. Nevertheless, certain things might be foreshadowed, primarily by projecting the likely future of the sciences themselves. Improvements in medical science are likely to produce a considerable period of gerontocracy. New techniques of birth control, including new forms of artificial insemination, may very well change in a drastic fashion the resources holding together the family unit. New techniques of health, amounting perhaps even to rejuvenation, as well as chemical and electrical techniques for altering consciousness, are likely to produce people with capacities entirely different from today's. Given the way stratification operates, it is more than likely that such developments will become resources for particular groups who will use them to strengthen their position against the have-nots.

The development of the social sciences may have a crucial role to play here. I do not believe that its effects will be primarily in the direction of providing techniques for manipulating others, although certainly some possibility exists. If it does, it will probably come especially from pseudoscience, using descriptive information as a direct means of political control, more or less in the same way that traditional autocracies used dossiers and surveillance. Theoretical and explanatory developments in the social science, it seems to me, can have their greatest impact in making us aware of the plurality of realities, the multiplicity of interests, and the tricks used to impose one reality upon others. Here sociology may have a liberating effect. Illusions, after all, are primarily on the side of the oppressors. With new technological developments providing new resources for the ongoing conflict between man and man, this widespread sophistication could be of considerable importance.

Despite much romantic hankering after the past, and its use as an ideal with which to flay the trends of the present, has been a long slow progress, at least in the world of the intellect. For all its ups and downs, its effects over the last few centuries have been increasingly libertarian. The social science of the future thus may have something to contribute to free our minds still further from illusions, and to make it possible for all people to share the gifts that have hitherto been reserved for the very aggressive and the very lucky.

References

Aberle, D.
1961 Matrilineal descent in cross-cultural perspective. In *Matrilineal kinship,* edited by D. Schneider and E. K. Gough. Berkeley: Univ. of California Press.
1963 The incest taboo and the mating patterns of animals. *American Anthropologist* **65**: 259–261.

Aiken, M., and J. Hage
1966 Organizational alienation: A comparative analysis. *American Sociological Review* **31**: 497–507.

Ambrose, J. A.
1960 *The Smiling response in early human infancy.* Ph.D. thesis, Univ. of London.

Anderson, T. R., and S. Warkov
1961 Organizational size and functional complexity: A study of administration in hospitals. *American Sociological Review* **26**: 23–28.

Andic, B., and B. Peacock
1961 The international distribution of income, 1949 and 1957. *Journal of the Royal Statistical Society* **124**: 208–218.

Ariès, P.
1962 *Centuries of childhood.* New York: Random House.

Attewell, P.
1974 Ethnomethodology since Garfinkel. *Theory and Society* **1**: 179–210.

Babchuck, N., and A. Booth
1969 Voluntary association membership: A longitudinal analysis. *American Sociological Review* **34**: 31–45.

Bajema, C. J.
1968 Interrelations among intellectual ability, educational attainment, and occupational achievement. *Sociology of Education* **41**: 317–319.

Baltzell, E. D.
1958 *An American business aristocracy.* New York: Macmillan.

Banfield, E. C.
1958 *The moral basis of a backward society.* New York: Free Press.
1961 *Political influence.* New York: Free Press.

Banfield, E. C., and J. Q. Wilson
1964 *City politics.* Cambridge, Massachusetts: Harvard Univ. Press.

Baran, P. A.
1957 *The political economy of growth.* New York: Monthly Review Press.

Baran, P. A., and P. M. Sweezey
1966 *Monopoly capital.* New York: Monthly Review Press.

Barnard, C. I.
1938 *The functions of the executive.* Cambridge, Massachusetts: Harvard Univ. Press.
Barnes, H. E.
1963 *A history of historical writing.* New York: Dover.
Barron, F.
1957 Originality in relation to personality and intellect. *Journal of Personality* 25: 730–742.
Bartlett, K.
1947 The social impact of the radio. *Annals of the American Academy of Political and Social Science* 251: 89–92.
Bate, W. J.
1959 *Prefaces to criticism.* Garden City, New York: Doubleday.
Bayley, N.
1968 Behavioral correlates of mental growth, birth to 36 years. *American Psychologist* 23: 1–17.
Becker, G.
1957 *The economics of discrimination.* Chicago: Univ. of Chicago Press.
Becker, H. S.
1963 *Outsiders.* New York: Free Press.
Becker, H., and H. E. Barnes
1961 *Social thought from lore to science.* New York: Dover.
Bell, D.
1961 *The end of ideology.* New York: Free Press.
Bell, E. T.
1937 *Men of mathematics.* New York: Simon and Schuster.
Ben-David, J.
1960a Roles and innovations in medicine. *American Journal of Sociology* 65: 557–568.
1960b Scientific productivity and academic organization in nineteenth century medicine. *American Sociological Review* 25: 828–843.
1971 *The scientist's role in society.* Englewood Cliffs, New Jersey: Prentice-Hall.
Ben-David, J., and R. Collins
1966 Social factors in the origins of a new science: The case of psychology. *American Sociological Review* 31: 451–465.
1967 A comparative study of academic freedom and student politics. In *Student politics,* edited by S. M. Lipset. New York: Basic Books.
Ben-David, J., and A. Zloczower
1962 Universities and academic systems in modern societies. *European Journal of Sociology* 3: 45–85.
Bendix, R.
1943 *The rise and acceptance of German sociology.* Univ. of Chicago, M. A. thesis.
1956 *Work and authority in industry.* New York: Wiley.
1964 *Nation-building and citizenship.* New York: Wiley.
1967 Tradition and modernity reconsidered. *Comparative Studies in Society and History* 9: 292–346.
1968 Reflections on charismatic leadership. In *State and society,* edited by R. Bendix et. al., Boston: Little, Brown.
Bendix, R., and B. Berger
1959 Images of society and problems of concept formation in sociology. In *Symposium on sociological theory,* edited by L. Gross. New York: Harper.
Bendix, R., and S. M. Lipset
1959 *Social mobility in industrial society.* Berkeley: Univ. of California Press.

Bendix, R., and G. Roth
  1971 *Scholarship and partisanship: Essays on Max Weber.* Berkeley: Univ. of California Press.
Bensman, J., and A. Vidich
  1971 *The new American society.* New York: Quadrangle Books.
Berger, B.
  1960 *Working-class suburb.* Berkeley: Univ. of California Press.
  1971 *Looking for America.* Englewood Cliffs, New Jersey: Prentice-Hall.
Berger, B., and B. M. Hackett
  1973 On the decline of age-grading in rural hippie communes. Unpublished paper.
Berger, P., and T. Luckman
  1966 *The social construction of reality.* Garden City, New York: Doubleday.
Berle, A. A.
  1959 *Power without property.* New York: Harcourt.
Berliner, J. S.
  1957 *Factory and manager in the U.S.S.R.* Cambridge, Massachusetts: Harvard Univ. Press.
Bernal, J. D.
  1939 *The social functions of science.* London: Routledge and Kegan Paul.
Bernstein, B.
  1972 Social class, language, and socialization. In *Language and social context,* edited by Pier Paolo Giglioli. Baltimore: Penguin Books.
Bernard, L. L., and J. Bernard
  1943 *The origins of American sociology.* New York: Thomas Y. Crowell.
Beshers, J. S.
  1962 *Urban social structure.* New York: Free Press.
Beshers, J. S., and E. O. Laumann
  1967 Social distance: A network approach. *American Sociological Review* **32**: 225–236.
Bidwell, C. E.
  1965 The school as a formal organization. In *Handbook of organizations,* edited by J. G. March. Chicago: Rand McNally.
Birdwhistell, R. L.
  1970 *Kinesics and context: Essays on body motion communication.* Philadelphia: Univ. of Pennsylvania Press.
Birmingham, S.
  1968 *The right people: A portrait of the American social establishment.* Boston: Little, Brown.
Blau, P. M.
  1955 *The dynamics of bureaucracy.* Chicago: Univ. of Chicago Press.
  1960 A theory of social integration. *American Journal of Sociology.* **65**: 545–556.
  1964 *Exchange and power in social life.* New York: Wiley.
  1965 The flow of occupational supply and recruitment. *American Sociological Review* **30**: 457–490.
  1968 The hierarchy of authority in organizations. *American Journal of Sociology* **73**: 453–467.
  1970 A formal theory of differentiation in organizations. *American Sociological Review* **35**: 201–218.
Blau, P. M., and O. D. Duncan
  1967 *The American occupational structure.* New York: Wiley.
Blau, P. M., and W. R. Scott
  1962 *Formal organizations.* San Francisco: Chandler.
Blauner, R.
  1964 *Alienation and freedom.* Chicago: Univ. of Chicago Press.
  1966 Death and social structure. *Psychiatry* **29**: 378–394.

Bloomfield, L.
1933 *Language.* New York: Holt.
Blum, A.
1970 Theorizing. In *Understanding everyday life,* edited by J. D. Douglas. Chicago: Aldine.
Blumberg, R. L., and R. F. Winch
1972 Societal complexity and familial complexity: Evidence for the curvilinear hypothesis. *American Journal of Sociology* **77**: 898–920.
Blumer, H.
1969 *Symbolic interactionism.* Englewood Cliffs, New Jersey: Prentice-Hall.
Bordua, D. J., and A. J. Reiss
1966 Command, control, and charisma: reflections of police bureaucracy. *American Journal of sociology* **72**: 68–76.
Booth, A.
1972 Sex and social participation. *American Sociological Review* **37**: 183–192.
Bossard, J. H. S., and E. S. Ball
1950 *Ritual in Family Living.* Philadelphia: Univ. of Pennsylvania Press.
Bourdieu, P.
1971 Intellectual field and creative project. In *Knowledge and control,* edited by M. F. D. Young. London: Macmillan.
Bourdieu, P., and J. C. Passeron
1970 *La reproduction: Eléments pour une théorie du système d'enseignement.* Paris: Editions de Minuit.
Bowlby, J.
1958 The nature of the child's tie to his mother. *International Journal of Psychoanalysis* **39**: 350–373.
Breed, W.
1955 Social control in the newsroom. *Social Forces* **33**: 326–335.
Brown, R., and R. Gilman
1960 The pronouns of power and solidarity. In *Style in language,* edited by T. A. Sebeok. Cambridge, Massachusetts: M.I.T. Press.
Bruford, W. H.
1935 *Germany in the eighteenth century.* New York: Cambridge Univ. Press.
Bruner, J. S.
1967 *Studies in cognitive growth.* New York: Wiley.
Burgess, E. W., H. Locke and M. Thomas
1963 *The family.* New York: American Book Company.
Burgess, E. W., and P. Wallin
1953 *Engagement and marriage.* Chicago: Lippincott.
Burstein, P.
1972 Social structure and individual political participation in five countries. *American Journal of Sociology* **77**: 1087–1110.
Butterfield, H.
1962 *The origins of modern science.* New York: Collier-Macmillan.
Caplow, T.
1954 *The Sociology of Work.* Minneapolis: Univ. of Minnesota Press.
1964 *Principles of organization.* New York: Harcourt.
Carey, A.
1967 The Hawthorne studies: A radical criticism. *American Sociological Review* **32**: 403–417.
Carey, J. T.
1968 *The college drug scene.* Englewood Cliffs, New Jersey: Prentice-Hall.
1969 Changing courtship patterns in the popular song. *American Journal of Sociology* **74**: 720–731.

Cassirer, E.
1953 *The philosophy of symbolic forms.* Originally published 1923.

Cattell, R. B.
1963 Personality and motivation of the researcher from measurements of contemporaries and from biography. In *Scientific creativity,* edited by C. W. Taylor and F. Barron. New York: Wiley.

Centers, R., B. H. Raven, and A. Rodrigues
1971 Conjugal power structure: A reexamination. *American Sociological Review* **36**: 264–278.

Chamberlain, N. W.
1958 *Labor.* New York: McGraw-Hill.

Chandler, A. D.
1962 *Strategy and structure.* Cambridge, Massachusetts: M.I.T. Press.

Chomsky, N.
1965 *Aspects of the theory of syntax.* Cambridge, Massachusetts: M.I.T. Press.

Cicourel, A. V.
1968 *The social organization of juvenile justice.* New York: Wiley.
1970a Basic and normative rules. In *Recent sociology number 2,* edited by H. P. Dreitzel. New York: Macmillan.
1970b The acquisition of social structure. In *Understanding everyday life,* edited by J. D. Douglas. Chicago: Aldine.
1973 *Cognitive Sociology.* Baltimore: Penguin Books.

Cicourel, A. V., and Kitsuse, J.
1963 *The educational decision-makers.* Indianapolis: Bobbs-Merrill.

Clark, B. R., and M. A. Trow
1966 The organizational context. In *College peer groups,* edited by T. M. Newcomb and E. K. Wilson. Chicago: Aldine.

Clark, H. F.
1931 *Economic theory and correct occupational distribution.* New York: Teachers College Press.

Clark, T. N.
1968 Community structure, decision-making, budget expenditures, and urban renewal in 51 American communities. *American Sociological Review* **33**: 576–593.

Clarke, A. C.
1956 The use of leisure and its relation to levels of occupational prestige. *American Sociological Review* **21**: 301–307.

Clegg, H. A.
1951 *Industrial democracy and nationalization.* Oxford: Blackwell.

Clignet, R.
1970 *Many wives, many powers: Authority and power in polygynous families.* Evanston, Illinois: Northwestern Univ. Press.

Coates, C. H., and R. J. Pellegrin
1957 Executives and supervisors: Informal factors in differential bureaucratic promotion. *Administrative Science Quarterly* **2**: 201–215.

Cohen, A. K.
1955 *Delinquent boys: The culture of the gang.* Glencoe, Illinois: Free Press.

Cole, J. R.
1970 Patterns of intellectual influence in scientific research. *Sociology of Education* **43**: 377–403.

Coleman, J. S.
1961 *The Adolescent Society.* New York: Free Press.

Collins, R.
1971 Functional and conflict theories of educational stratification. *American Sociological Review* **36**: 1002–1019.

1975 *The religion of America: An analysis of stratification and ideology.*
Collins, R., and M. Makowsky
1972 *The discovery of society.* New York: Random House.
Commission on Obscenity and Pornography
1970 *Report.* New York: Random House.
Coser, R. L.
1964 Laughter among colleagues. *Psychiatry* **23**: 81–95.
Cowgill, D. O.
1963 Transition theory as general population theory. *Social Forces* **41**: 270–274.
Craig, J. G., and E. Gross
1970 The forum theory of organizational democracy: Structural guarantees as time-related variables. *American Sociological Review* **35**: 19–33.
Cressey, D. R.
1969 *Theft of the nation: The structure and operation of organized crime in America.* New York: Harper.
Crozier, M.
1964 *The bureaucratic phenomenon.* Chicago: Univ. of Chicago Press.
Curtis, J.
1971 Voluntary association joining: A cross-national comparative note. *American Sociological Review* **36**: 872–880.
Cutright, P.
1968 Occupational inheritance: A cross-national analysis. *American Journal of Sociology* **73**: 400–416.
Dahrendorf, R.
1959 *Class and class conflict in industrial society.* Stanford, California: Stanford Univ. Press.
Dalton, M.
1951 Informal factors in career achievement. *American Journal of Sociology* **56**: 407–415.
1959 *Men who manage.* New York: Wiley.
Darwin, C.
1965 *The expression of the emotions in man and animals.* Chicago: Univ. of Chicago Press (Originally published 1872).
Davis, A., B. B. Gardner, and M. R. Gardner
1965 *Deep South.* Chicago: Univ. of Chicago Press (Originally published 1941).
Davis, K.
1941 Intermarriage in caste societies. *American Anthropologist* **43**: 376–395.
1949 *Human society.* New York: Macmillan.
1962 The role of class mobility in economic development. *Population Review* **6**: 67–73.
Davis, K., and W. E. Moore
1945 Some principles of stratification. *American Sociological Review* **10**: 242–249.
Demos, J.
1970 Underlying themes in the witchdraft of seventeenth-century New England. *America Historical Review* **75**: 1311–1326.
Demerath, N. J. III
1965 *Social class in American Protestantism.* Chicago: Rand McNally.
Dollard, J., and N. E. Miller
1950 *Personality and psychotherapy.* New York: McGraw-Hill.
Dornbush, S. M.
1955 The military academy as an assimilating institution. *Social Forces* **33**: 316–321.
Douglas, J. D.
1967 *The social meanings of suicide.* Princeton, New Jersey: Princeton Univ. Press.
1971a *American social order: Social rules in a pluralistic society.* New York: Free Press.
1971b *Crime and justice in American society.* Indianapolis; Bobbs-Merrill.

Douglas, P. H.
1930 *Real wages in the United States 1890–1926.* New York: Houghton.
Dray, W. H.
1968 Explaining "What" in History. In *Readings in the philosophy of the social sciences,* edited by M. Brodbeck. New York: Macmillan.
Duncan, H. D.
1962 *Communication and social order.* New York: Bedminister Press.
Duncan, O. D.
1966 Methodological issues in the analysis of social mobility. In *Social structure and mobility in economic development,* edited by N. J. Smelser and S. M. Lipset. Chicago: Aldine.
Durkheim, E.
1938 *The rules of the sociological method.* Chicago: Univ. of Chicago Press (Originally published 1895).
1947 *The division of labor in society.* Glencoe, Illinois: Free Press (Originally published 1893).
1954 *The elementary forms of the religious life.* Glencoe, Illinois Free Press (Originally published 1915).
Durkheim, E., and M. Mauss
1963 *Primitive classification.* Chicago: Univ. of Chicago Press (Originally published 1903).
Duverger, M.
1954 *Political Parties.* New York: Wiley.
Easton, D.
1968 Political science. In *International encyclopedia of the social sciences,* edited by D. L. Sills. New York: Macmillan.
Edelstein, J. D.
1967 An organizational theory of union democracy. *American Sociological Review* **32**: 19–31.
Editors of Fortune
1956 *The executive life.* Garden City, New York: Doubleday.
Eiseley, L.
1958 *Darwin's century.* Garden City, New York: Doubleday.
Eisenstadt, S. N.
1956 *From generation to generation.* Glencoe, Illinois: Free Press.
1963 *The political system of empires.* New York: Free Press.
Elder, G. H. Jr.
1969 Appearance and education in marriage mobility. *American Sociological Review* **34**: 519–533.
Epstein, C. F.
1970 *Woman's Place.* Berkeley: Univ. of California Press.
Erikson, E. H.
1959 *Identity and the life cycle.* Psychological Issues, Monograph 1.
Etzioni, A.
1961 *A comparative analysis of complex organizations.* New York: Free Press.
Featherman, D. L.
1971 A research note: A social structural model for the socioeconomic career. *American Journal of Sociology* **77**: 293–304.
Festinger, L. S. Schachter, and K. Back
1950 *Social pressures in informal groups.* New York: Harper.
Fichter, J. H.
1952 The profile of Catholic religious life. *American Journal of Sociology* **58**: 145–149.
Fitzgerald, F. S.
1922 *The Beautiful and Damned.* New York: Scribner's.

1956 Echoes of the jazz age. In *The Crack Up*, edited by E. Wilson. New York: New Directions (Originally published 1931).

Forde, D.
1962 Death and succession: An analysis of Yako mortuary rituals. In *Essays on the ritual of social relations*, edited by M. Gluckman. Manchester, England: Manchester Univ. Press.

Form, W. H., and J. Rytina
1969 Ideological beliefs on the distribution of power in the United States. *American Sociological Review* **34**: 19–30.

Form, W. H., and G. Stone
1957 Urbanism, anonymity, and status symbolism. *American Journal of Sociology* **62**: 504–514.

Fourastiè, J.
1960 *The causes of wealth.* New York: Free Press.

Franke, W.
1960 *The reform and abolition of the traditional Chinese examination system.* Cambridge, Massachusetts: Harvard Univ. Press.

Freeman, R.
1961–62 The sociology of human fertility: A trend report and a bibliography. *Current Sociology:* 10–11.

Freud, S.
1936 *The Problem of Anxiety.* New York: Norton.
1961 *Civilization and its discontents.* New York: Norton (Originally published 1930).

Friedman, M., and S. Kuznets
1945 *Income from independent professional practice.* New York: National Bureau of Economic Research.

Frykenberg, R. E.
1968 Traditional processes of power in South India. In *State and society,* edited by R. Bendix et. al. Boston: Little, Brown

Fuller, J. G.
1962 *The gentlemen conspirators.* New York: Grove Press.

Fustel de Coulanges, N. D.
n.d. *The ancient city.* Garden City, New York: Doubleday (Originally published 1864).

Gabennesch, H.
1972 Authoritarianism as world view. *American Journal of Sociology* **77**: 857–875.

Gamson, W. A.
1966 Rancorous conflict in community politics. *American Sociological Review* **31**: 71–80.

Gans, H. J.
1962 *The urban villagers.* New York: Free Press.
1967 *The Levittowners.* New York: Random House.

Garfinkel, H.
1967 *Studies in ethnomethodology.* Englewood Cliffs, New Jersey: Prentice-Hall.

Gastil, R. D.
1971 Homicide and a regional culture of violence. *American Sociological Review* **36**: 412–426.

Gaston, J. C.
1970 Competition and secrecy in science: Some comparisons between American and British scientists. Paper delivered at the Annual Meeting of the American Sociological Association, Washington, D.C.

Genovese, E. D.
1971 On being a socialist and a historian. In *The red and the black: Marxian explorations in Southern and Afro-American History.* New York: Random House.

Getzels, J. W., and P. W. Jackson

1962 *Creativity and intelligence.* New York: Wiley.

Glaser, D.

1964 *The Effectiveness of a prison and parole system.* Indianapolis: Bobbs-Merrill.

Glen, N. D., and J. P. Alston,

1968 Cultural distances among occupational categories. *American Sociological Review* **33**: 365–382.

Glock, C. Y., B. Ringer, and E. R. Babbie

1967 *To comfort and to challenge.* Berkeley: Univ. of California Press.

Goffman, Erving

1959 *The presentation of self in everyday life.* Garden City, New York: Doubleday.

1961a Fun in games. In *Encounters.* Indianapolis: Bobbs-Merrill.

1961b *Asylums.* Garden City, New York: Doubleday.

1963 *Behavior in public places.* New York: Free Press.

1967 *Interaction ritual.* Garden City, New York: Doubleday.

1971 *Relations in public.* New York: Basic Books.

Goode, W. J.

1959 The theoretical importance of love. *American Sociological Review* **24**: 38–47.

1963 *World revolution and family patterns.* New York: Free Press.

1971 Force and violence in the family. *Journal of Marriage and the Family* **33**: 624–636.

Goody, J.

1969 A comparative approach to incest and adultery. In *Comparative studies in kinship.* Stanford, California: Stanford Univ. Press.

Gough, K.

1971 The origin of the family. *Journal of Marriage and the Family* **33**: 760–771.

Gouin-Decarie, T.

1962 *Intelligence et affectivité chez le jeune enfant.* Neuchatel: Delachaux et Niestle.

Gouldner, A. W.

1955 Metaphysical pathos and the theory of bureaucracy. *American Politician Science Review* **49**: 496–507.

1970 *The coming crisis in Western sociology.* New York: Basic Books.

Graña, C.

1964 *Bohemian versus bourgeois.* New York: Basic Books.

Greeley, A. M.

1969 A note on political and social differences among ethnic college graduates. *Sociology of Education* **42**: 98–103.

Greer, S. A.

1959 *Last man in: Racial access to union power.* New York: Free Press.

Gross, E.

1953 Some functional consequences of primary controls in formal work organizations. *American Sociological Review* **18**: 368–373.

Gumperz, J. J.

1971 Dialect difference and social structure in a North Indian village. In *Language in social groups.* Stanford, California: Stanford Univ. Press.

Gusfield, J. R.

1961 Occupational roles and forms of enterprise. *American Journal of Sociology* **66**: 571–580.

1962 Mass society and extremist politics. *American Sociological Review* **27**: 19–30.

1963 *Symbolic crusade: Status politics and the American temperance movement.* Urbana: Univ. of Illinois Press.

Habakkuk, H. J., and M. Postan
  1965 *The Cambridge economic history of Europe,* volume VI. New York: Cambridge Univ. Press.
Hagedorn, R., and S. Labovitz
  1968 Participation in community associations by occupation. *American Sociological Review* 33: 272–283.
Hagstrom, W. O.
  1965 *The scientific community.* New York: Basic Books.
  1967 Competition and secrecy in science. Report to the National Science Foundation.
Hagstrom, W. O., and L. Hargens
  1968 Mobility theory in the sociology of science. Paper presented at the Cornell Conference on Human Mobility, Ithaca, New York.
Hall, E. T.
  1959 *The silent language.* Garden City, New York: Doubleday.
Hall, O.
  1946 The informal organization of the medical profession. *Canadian Journal of Economic and Political Science* 12: 30–44.
Hall, R. H.
  1968 Professionalization and bureaucratization. *American Sociological Review* 33: 90–103.
Hall, R. H., and C. R. Tittle
  1966 A note on bureaucracy and its "correlates." *American Journal of Sociology* 72: 267–272.
Hamilton, E.
  1942 *Mythology.* Boston: Little, Brown.
Hamilton, R. F.
  1966 The marginal middle class: A reconsideration. *American Sociological Review* 31: 192–199.
Hargens, L. L., and W. O. Hagstrom
  1967 Sponsored and contest mobility of American academic scientists. *Sociology of Education* 40: 24–38.
Harlow, H. F.
  1964 A behavioral approach to psychoanalytic theory. In *Science and Psychoanalysis,* volume VII, edited by J. H. Masserman. New York: Grune and Stratton.
Harlow, H. F., and R. R. Zimmerman
  1959 Affectional responses in the infant monkey. *Science* 130: 421–431.
Harvey, E.
  1968 Technology and the structure of organizations. *American Sociological Review* 33: 247–258.
Hauser, A.
  1951 *The social history of art.* 4 vols. New York: Knopf.
Heilbroner, R. L.
  1970 *Between capitalism and socialism.* New York: Random House.
Hendershot, G. E., and T. F. James
  1972 Size and growth as determinants of administrative-production ratios in organization. *American Sociological Review* 37: 149–154.
Hess, R. D., and J. V. Torney
  1967 *The development of political attitudes in children.* Chicago: Aldine.
Hintze, O.
  1962 *Staat and Verfassung.* Goettingen: Vandenhoeck and Ruprecht (Originally published 1900–1920).
  1968 The state in historical perspective. In *State and Society,* edited by R. Bendix et al. Boston: Little, Brown (Originally published 1897).
Hirsch, P. M.
  1970 Sociological approaches to the pop music phenomenon. *American Behavioral Scientist*

1972 Processing fads and fashions; an organization-set analysis of cultural industry systems. *American Journal of Sociology* **77**: 639–659.

Historical Statistics of the United States.

Washington: Government Printing Office.

Hodge, R. W., and D. J. Treiman

1968a Class identification in the United States. *American Journal of Sociology* **73**: 535–547.

1968b Social participation and social status. *American Sociological Review* **33**: 722–739.

Hoebel, E. A.

1961 *The law of primitive man.* Cambridge, Massachusetts: Harvard Univ. Press.

Hofstadter, R.

1955 *The age of reform.* New York: Random House.

Hollingshead, A. B.

1949 *Elmtown's youth.* New York: Wiley.

Holmes, U. T., Jr.

1964 *Daily living in the twelfth century.* Madison: Univ. of Wisconsin Press.

Homans, G. C.

1950 *The human group.* New York: Harcourt.

1961 *Social behavior: Its elementary forms.* New York: Harcourt.

1964 Bringing men back in. *American Sociological Review* **29**: 809–818.

Howells, W. D.

1971 *The rise of Silas Lapham.* Bloomington: Indiana Univ. Press (Originally published 1885).

Hughes, E. C.

1949 Queries concerning industry and society growing out of the study of ethnic relations in industry. *American Sociological Review* **14**: 211–220.

1958 *Men and their work.* Glencoe: Free Press.

Huizinga, J.

1954 *The waning of the middle ages.* Garden City, New York: Doubleday.

Huntington, S. P.

1966 Political modernization: America vs. Europe. *World Politics* **18**: 378–414.

1968 *Political order in changing societies.* New Haven, Connecticut: Yale Univ. Press.

Inhelder, B.

1969 *The diagnosis of reasoning in the mentally retarded.* New York: John Day.

Jacques, E.

1956 *The measurement of responsibility.* London: Tavistock.

James, H.

1959 *The Europeans.* New York: Dell (Originally published 1878).

Johnson, B. C.

1968 The democratic mirage: Notes towards a theory of American politics. *Berkeley Journal of Sociology* **13**: 104–143.

1973 Discretionary justice and racial domination. Unpublished dissertation, Univ. of California, Berkeley.

Kadushin, C.

1966 The friends and supporters of psychotherapy: On social circles in urban life. *American Sociological Review* **31**: 786–802.

1968 Power, influence, and social circles: A new methodology for studying opinion makers. *American Sociological Review* **33**: 685–699.

Kahl, J. A.

1957 *The American class structure.* New York: Holt.

Kanin, E.

1967 Reference groups and sex conduct norm violation. *Sociological Quarterly* **8**: 495–504.

Kanowitz, L.
1969 *Women and the law.* Albuquerque: Univ. of New Mexico Press.
Kantner, R. M.
1968 Commitment and social organization: A study of commitment mechanisms in utopian communities. *American Sociological Review* 33: 499–517.
Kardiner, A.
1945 *The psychological frontiers of society.* New York: Columbia Univ. Press.
Katz, A. M., and R. Hill
1958 Residential propinquity and marital selection: A review of theory, method, and fact. *Marriage and Family Living* 20: 27–35.
Katz, E., and P. F. Lazarsfeld
1955 *Personal influence.* Glencoe, Illinois: Free Press.
Kearney, H.
1971 *Science and change, 1500–1700.* New York: McGraw-Hill.
Kelley, J. L.
1972 A resource theory of social mobility: Modernization and mobility. Paper read at the Annual Meeting of the American Sociological Association, New Orleans.
1973 Causal chain models for the socioeconomic career. *American Sociological Review* 38: 481–493.
Kelley, J. L., and M. L. Perlman
1971 Social mobility in Toro: Some preliminary results from western Uganda. *Economic Development and Cultural Change* 19: 204–221.
1972 Social mobility in traditional society: The Toro of Uganda. Paper read at the Annual Meeting of the American Sociological Association, New Orleans.
Kerr, C.
1964 *Labor and management in industrial society.* New York: Doubleday.
Kerr, C., and A. Siegel
1954 The interindustry propensity to strike: An international comparison. In *Industrial conflict,* edited by A. Kornhauser. New York: McGraw-Hill.
Kinsey, A. C., and P. H. Gebhard
1953 *Sexual behavior in the human female.* Philadelphia: Saunders.
Kinsey, A. C., W. B. Pomeroy, and C. W. Martin
1948 *Sexual behavior in the human male.* Philadelphia: Saunders.
Kirkpatrick, C., and E. Kanin
1957 Male sex aggression on a university campus. *American Sociological Review* 22: 52–58.
Kluckhohn, C.
1958 Have there been discernable shifts in American values during the past generation? In *The American style,* edited by Elting E. Morison. New York: Harper.
Kohn, M. L.
1971 Bureaucratic man: A portrait and an interpretation. *American Sociological Review* 36: 461–474.
Kohn, M. L., and C. Schooler
1969 Class, occupation, and orientation. *American Sociological Review* 34: 659–678.
Kolko, G.
1962 *Wealth and power in America.* New York: Praeger.
Koyré, A.
1965 *Newtonian studies.* Cambridge, Massachusetts: Harvard Univ. Press.
Kuhn, T. S.
1961 The function of measurement in modern physical science. In *Quantification: A history*

of the meaning of measurement in the natural and social sciences, edited by Harry Woolf. Indianapolis: Bobbs-Merrill.

1962 *The structure of scientific revolutions.* Chicago: Univ. of Chicago Press.

Kuznets, S.

1963 Quantitative aspects of economic growth of nations: Distribution of income by size. *Economic Development and Cultural Change* **11**: supplement.

Labov, W.

1966 *The social stratification of English in New York City.* Washington, D.C.: Center for Applied Linguistics.

1972a Rules for ritual insults. In *Studies in social interaction,* edited by D. Sudnow. New York: Free Press.

1972b The study of language in its social context. In *Language and social context,* edited by P. P. Giglioli. Baltimore: Penguin Books.

1972c The logic of non-standard English. In *Language and social context,* edited by P. P. Giglioli. Baltimore: Penguin Books.

Laing, R. D.

1967 *The politics of experience.* New York: Pantheon.

Lammers, C. J.

1967 Power and participation in decision-making in formal organizations. *American Journal of Sociology* **73**: 201–216.

Landecker, W. S.

1960 Class boundaries. *American Sociological Review* **25**: 868–877.

Lane, A.

1968 Occupational mobility in six cities. *American Sociological Review* **33**: 740–749.

Langer, S.

1942 *Philosophy in a new key.* Cambridge, Massachusetts: Howard Univ. Press.

Lansing, J. B., and L. Kish

1957 Family life cycle as an independent variable. *American Sociological Review* **22**: 512–519.

Laumann, E. O.

1966 *Prestige and association in an urban community.* Indianapolis: Bobbs-Merrill.

1969 The social structure of religious and ethnoreligious groups in a metropolitan community. *American Sociological Review* **34**: 182–197.

Laumann, E. O., and L. Guttman

1966 The relative associational contiguity of occupations in an urban setting. *American Sociological Review* **31**: 169–178.

Lazarsfeld, P. F.

1961 Notes on the history of quantification in sociology. In *Quantification: A history of the meaning of measurement in the natural and social sciences,* edited by Harry Woolf. Indianapolis: Bobbs-Merrill.

Lee, A. M.

1937 *The daily newspaper in America.* New York: Macmillan.

Lenneberg, E. H.

1964 A biological perspective on language. In *New directions in the study of language,* edited by E. H. Lenneberg. Cambridge, Massachusetts: M.I.T. Press.

Lenski, G. E.

1966 *Power and privilege: A theory of social stratification.* New York: McGraw-Hill.

Levine, J. H.

1972 The sphere of influence. *American Sociological Review* **37**: 14–27.

Lévi-Strauss, C.

1949 *Les structures élémentaires de la parenté.* Paris: Presses Universitaires de France.

1969 *The raw and the cooked.* New York: Harper.

Lewis, W. H.

1957 *The splendid century: Life in the France of Louis XIV.* Garden City, New York: Doubleday.

Liddell-Hart, B. H.

1954 *Strategy.* New York: Praeger.

Lieberson, S.

1971 An empirical study of military-industrial linkages. *American Journal of Sociology* 76: 562–584.

Liebow, E.

1967 *Tally's corner.* Boston: Little, Brown.

Lindauer, M.

1961 *Communication among social bees.* Cambridge, Massachusetts: Harvard Univ. Press.

Linn, E. L.

1967 Social stratification of discussions about local affairs. *American Journal of Sociology* 72: 660–668.

Lipset, S. M.

1960 *Political man.* Garden City, New York: Doubleday.

Lipset, S. M., M. A. Trow, and J. S. Coleman

1956 *Union democracy.* Garden City, New York: Doubleday.

Lorenz, K.

1966 *On aggression.* New York: Harcourt.

Lowenthal, L.

1957 *Literature and the image of man.* Boston: Beacon Press.

Lundberg, F.

1968 *The rich and the super-rich.* New York: Lyle Stuart.

Lyman, S. M.

1972 *The Black American in sociological thought.* New York: Putnam.

Lyman, S. M., and M. B. Scott

1970 *A sociology of the absurd.* New York: Appleton.

Lynd, R. S., and H. M. Lynd

1929 *Middletown.* New York: Harcourt.

Macauly, S.

1966 *Law and the balance of power: The automobile manufacturers and their dealers.* New York: Russell Sage Foundation.

Maccoby, E.

1968 The development of moral values and behavior in childhood. In *Socialization and society,* edited by J. A. Clausen. Boston: Little, Brown.

MacKay, R.

1973 Conceptions of children and models of socialization. In *Recent sociology number 5,* edited by H. P. Dreitzel. New York: Macmillan.

MacKinnon, D.

1962 The nature and nurture of creative talent. *American Psychologist* 17: 484–495.

Main, J. T.

1965 *The social structure of revolutionary America.* Princeton, New Jersey: Princeton Univ. Press.

Malinowski, B.

1948 *Magic, science, and religion.* Garden City, New York: Doubleday.

Mandel, E.

1968 *Marxist economic theory.* London: Merlin.

Mann, M.

1970 The social cohesion of liberal democracy. *American Sociological Review* 35: 423–439.

Mannheim, K.
1940 *Man and society in an age of reconstruction.* New York: Harcourt. Originally published 1935.
Manuel, F. E.
1968 *A portrait of Isaac Newton.* Cambridge, Massachusetts: Harvard Univ. Press.
March, J. G., and H. A. Simon
1958 *Organizations.* New York: Wiley.
Marcus, P. M.
1966 Union conventions and executive boards: A formal analysis of organizational structure. *American Sociological Review* **31**: 61–70.
Marcus, S.
1964 *The other Victorians: A study of sexuality and pornography in mid-nineteenth century England.* New York: Basic Books.
Marrou, H. I.
1964 *A history of education in antiquity.* New York: New American Library.
Marsh, R. M.
1963 Values, demand, and social mobility. *American Sociological Review* **28**: 567–575.
Martin, E.
1942 *The standard of living in 1860.* Chicago: Univ. of Chicago Press.
Martin, N. H., and A. Strauss
1956 Patterns of mobility within industrial organizations. *Journal of Business* **29**: 101–110.
Marx, K.
1906-09 *Capital.* Chicago: C. H. Kerr (Originally published 1867-1894).
1963 *The eighteenth brumaire of Louis Napoleon.* New York: International Publishers (Originally published 1852).
Marx, K., and F. Engels
1947 *The German ideology.* New York: International Publishers (Originally written 1846).
1959 Manifesto of the Communist Party. In *Marx and Engels: Basic writings on politics and philosophy,* edited by L. S. Feuer. Garden City, New York: Doubleday (Originally published 1848).
Mason, S. F.
1962 *A history of the sciences.* New York: Collier-Macmillan.
Matras, J.
1961 Differential fertility, intergenerational occupational mobility, and change in occupational distribution: Some elementary interrelationships. *Population studies* **15**: 193–194.
1967 Social mobility and social structure: Some insights from the linear model. *American Sociological Review* **32**: 608–614.
Matza, D.
1969 *Becoming deviant.* Englewood Cliffs, New Jersey: Prentice-Hall.
Mauss, M.
1967 *The gift.* New York: Norton.
McArthur, C.
1955 Personality differences between middle and upper classes. *Journal of Abnormal and Social Psychology* **50**: 247–254.
McCleary, R.
1960 Communication patterns as bases of systems of authority and power. *Theoretical studies in social organization of the prison.* New York: Social Science Research Council, Pamphlet 15.
McClelland, D. C.
1953 *The achievement motive.* New York: Appleton.
McClosky, H., P. Hoffman, and R. O'Hara
1960 Issue conflict and consensus among party leaders and followers. *American Political Science Review* **54**: 406–427.

McEvedy, C.
1961 *The Penguin atlas of medieval history.* Baltimore: Penguin Books.
McFarland, D. D.
1970 Intragenerational social mobility as a Markov process. *American Sociological Review* **35**: 463–476.
McGinnis, R.
1968 A stochastic model of social mobility. *American Sociological Review* **33**: 712–722.
McHugh, P.
1970 On the failure of positivism. In *Understanding everyday life,* edited by J. D. Douglas. Chicago Aldine.
McNeill, W.
1963 *The rise of the West: A history of the human community.* Chicago: Univ. of Chicago Press.
Mead, M., and R. Metraux
1957 The image of the scientist among high school students. *Science* **126**: 384–390.
Meade, J. E.
1964 *Efficiency, equality, and the ownership of property.* London: Allen and Unwin.
Mechanic, D.
1962 Sources of power of lower participants in complex organizations. *Administrative Science Quarterly* **7**: 349–364.
Melchercik, J.
1960 Employment problems of former offenders. *National Probation and Parole Association Journal* **6**: 138–145.
Merton, R. K.
1968 *Social theory and social structure.* New York: Free Press.
Messer, M.
1969 The predictive value of marijuana use: A note to researchers of student culture. *Sociology of Education* **42**: 87–91.
Meyer, M. W.
1968 Automation and bureaucratic structure. *American Journal of Sociology* **74**: 256–264.
1972 Size and the structure of organizations: A causal analysis. *American Sociological Review* **37**: 434–440.
Michels, R.
1949 *Political Parties.* Glencoe, Illinois: Free Press (Originally published 1911).
Mills, C. W.
1956 *The Power Elite.* New York: Oxford University Press.
Mincer, J.
1970 The distribution of labor incomes: A survey. *Journal of Economic Literature* **8**: 1–26.
Moore, B., Jr.
1966 *Social origins of dictatorship and democracy: Lord and peasant in the making of the modern world.* Boston: Beacon Press.
Morris, D.
1967 *The naked ape.* New York: McGraw-Hill.
Moskos, C. C. Jr., and W. Bell
1967 Emerging nations and ideologies of American social scientists. *American Sociologist* **2**: 67–72.
Mowrer, O. H.
1950 Learning theory and the neurotic paradox. In *Learning theory and personality dynamics.* New York: Ronald Press.
Mullins, N. C.
1968 The distribution of social and cultural properties in informal communication networks among biological scientists. *American Sociological Review* **33**: 786–797.

1972 The development of a scientific specialty: The phage group and the origins of molecular biology. *Minerva* **10**: 51–82.

Murdock, G. P.

1949 *Social structure.* New York: Macmillan.

Murray, G.

1951 *Five stages of Greek religion.* Boston: Beacon Press.

Myers, R. R.

1946 Interpersonal relations in the building industry. *Applied Anthropology* **5**: 1–7.

Nelson, J. I.

1966 Clique contacts and family orientations. *American Sociological Review* **31**: 663–672.

Newcomb, T. M.

1961 *The acquaintance process.* New York: Holt.

Nosow, S.

1956 Labor distribution and the normative system. *Social Forces* **30**: 25–33.

Oberschall, A.

1965 *Empirical social research in Germany 1848–1914.* The Hague: Mouton.

O'Lessker, K.

1968 Who voted for Hitler? A new look at the class basis of Naziism. *American Journal of Sociology* **74**: 63–69.

Paige, J. M.

1971 Political orientation and riot participation. *American Sociological Review* **36**: 810–819.

Parish, W. L., and M. Schwartz

1972 Household complexity in nineteenth century France. *American Sociological Review* **37**: 154–173.

Parsons, T.

1937 *The Structure of social action.* New York: McGraw-Hill.

1958 Social structure and the development of personality. *Psychiatry* **21**: 321–340.

1964 Evolutionary universals in society. *American Sociological Review* **29**: 339–354.

1966 *Societies: Comparative and evolutionary perspectives.* Englewood Cliffs, New Jersey: Prentice-Hall.

Pearlin, L. I., and M. L. Kohn

1966 Social class, occupation, and parental values: A cross-national study. *American Sociological Review* **31**: 466–479.

Penniman, T. K.

1952 *A hundred years of anthropology.* London: Duckworth.

Perls, F. S., R. Hefferline, and P. Goodman

1951 *Gestalt therapy.* New York: Julian Press.

Perrow, C.

1967 A framework for the comparative analysis of organizations. *American Sociological Review* **32**: 194–208.

Perrucci, R., and M. Pilisuk

1970 Leaders and ruling elites: The interorganizational bases of community power. *American Sociological Review* **35**: 1040–1056.

Perry, G., and A. Aldridge

1967 *The Penguin book of comics.* Baltimore: Penguin Books.

Persons, S.

1966 The origins of the Gentry. In *Essays on history and literature,* edited by R. H. Bremner. Columbia: Ohio Univ. Press.

Piaget, J.

1951 *Play, dreams and imitation in childhood.* New York: Norton.

1952 *The origins of intelligence in children.* New York: International Universities Press.

1954 *The construction of reality in the child.* New York: Basic Books.

1967 *Six psychological studies.* New York: Random House.

Pilcher, D. M.

1972 A comparative study of income inequality in selected industrialized nations. Univ. of California, San Diego, unpublished paper.

Polsky, N.

1967 *Hustlers, beats, and others.* Chicago: Aldine.

Popper, K.

1945 *The open society and its enemies.* London: Routledge.

Portes, A.

1971 Political primitivism, differential socialization, and lower-class leftist radicalism. *American Sociological Review* **36**: 820–834.

Price, D. K. de S.

1967 Networks of scientific papers. *Science* **149**: 510–515.

1969 The structure of publication in science and technology. In *Factors in the transfer of technology,* edited by W. H. Gruber and D. R. Marquis. Cambridge, Massachusetts: M.I.T. Press.

1970 Differences between scientific and technological and non-scientific scholarly communities. Paper presented at the Seventh World Congress of Sociology, Varna, Bulgaria.

1971 Measuring the size of science. Cited in *The scientist's role in society,* by J. Ben-David. Englewood Cliffs, New Jersey: Prentice-Hall.

Priest, R. F., and J. Sawyer

1967 Proximity and peership: Bases of balance in interpersonal attraction. *American Journal of Sociology* **72**: 633–649.

Purver, M.

1967 *The Royal Society.* Cambridge, Massachusetts: M.I.T. Press.

Quinney, R.

1970 *The social reality of crime.* Boston: Little, Brown.

Radcliffe-Brown, A. R.

1948 *The Andaman islanders.* New York: Free Press.

Rae, J.

1971 *The road and car in American life.* Cambridge, Massachusetts: M.I.T. Press.

Rainwater, L.

1964 Sexual life and interpersonal intimacy: Class patterns. *Journal of Marriage and the Family* **26**: 457–466.

Rainwater, L., R. P. Coleman, and G. Handel

1962 *Workingman's wife.* New York: Macfadden.

Raphael, E. E.

1967 The Anderson-Warkov hypothesis in local unions: A comparative study. *American Sociological Review* **32**: 768–776.

Rehberg, R. A., W. E. Shafer, and J. Sinclair

1970 Toward a temporal sequence of adolescent achievement variables. *American Sociological Review* **35**: 34–48.

Reiss, A. J.

1961 *Occupations and social status.* New York: Free Press.

Reiss, I. L.

1960 *Premarital sexual standards in America.* New York: Free Press.

Reynolds, L.

1951 *The structure of labor markets.* New York: Harper.

Riesman, D.

1950 *The lonely crowd: A study in the changing American character.* New Haven, Connecticut: Yale Univ. Press.

Riesman, D., and R. Denney

1951 Football in America: A study in cultural diffusion. *American Quarterly* **3**: 309–325.

Riesman, D., R. J. Potter, and J. Watson

1960a Sociability, permissiveness, and equality: A preliminary formulation. *Psychiatry* **23**: 323–340.

1960b The vanishing host. *Human Organization* **19**: 17–27.

Riesman, D., and J. Watson

1964 The sociability project. In *Sociologists at work,* edited by P. E. Hammond. New York: Basic Books.

Riley, M. W., M. E. Johnson, and A. Foner

1972 *A sociology of age stratification.* New York: Russell Sage Foundation.

Roach, J. L., and O. R. Gursslin

1967 An evaluation of the concept "culture of poverty." *Social Forces* **45**: 384–392.

Roberts, D. R.

1959 *Executive compensation.* New York: Free Press.

Rodman, H.

1967 Marital power in France, Germany, Yugoslavia, and the United States. *Journal of Marriage and the Family* **29**: 320–324.

Roe, A.

1952 A psychologist examines 64 eminent scientists. *Scientific American* **187**: 21–25.

Rosenberg, H.

1958 *Bureaucracy, aristocracy, and autocracy.* Cambridge, Massachusetts: Harvard Univ. Press.

Rossi, P.

1968 *Francis Bacon: From magic to science.* London: Routledge.

Roth, G.

1968 Personal rulership, patrimonialism, and empire-building in the new states. In *State and society,* edited by R. Bendix. Boston: Little, Brown.

1971 Max Weber's historical approach and historical typology. In *Comparative methods in sociology,* edited by I. Vallier. Berkeley: Univ. of California Press.

Rougemont, D. de

1956 *Love in the western world.* New York: Pantheon.

Roy, D.

1952 Quota restriction and goldbricking in a machine shop. *American Journal of Sociology* **57**: 427–442.

1959 Banana time. *Human organization* **18**: 158–168.

Russett, B.

1964 *World handbook of political and social indicators.* New Haven, Connecticut: Yale Univ. Press.

Sacks, H.

1972 An initial investigation of the usability of conversational data for doing sociology. In *Studies in social interaction,* edited by D. Sudnow. New York: Free Press.

n.d. Mimeographed lecture notes. Univ. of California, Irvine.

Scheff, T. J.

1966 *Being mentally ill: A sociological theory.* Chicago: Aldine.

Schegloff, E. E.

1968 Sequencing in conversational openings. *American Anthropologist* **70**: 1075–1095.

1972 Notes on a conversational practice: Formulating place. in *Studies in social interaction,* edited by D. Sudnow. New York: Free Press.

Schelling, T. C.

1963 *The strategy of conflict.* Cambridge, Massachusetts: Harvard Univ. Press.

1967 Economic analysis of organized crime. In *Task force report: Organized crime.* Washington, D.C.: Government Printing Office.

Schücking, L. I.

1966 *The sociology of literary taste.* Chicago: Univ. of Chicago Press.

Schlesinger, J. A.

1965 Political party organization. In *Handbook of organizations,* edited by J. G. March. Chicago: Rand McNally.

Schnabel, F.

1959 *Deutsche Geschichte im Neunzehnten Jahrhundert,* Vol. I. Freiberg: Verlag Herder.

Schumpeter, J. A.

1934 *The Theory of Economic Development.* Cambridge, Massachusetts: Harvard Univ. Press.

1951 *Social classes in an ethnically homogeneous environment.* New York: Augustus M. Kelley (Originally published 1911).

1954 *History of economic analysis.* New York: Oxford Univ. Press.

Schwartz, G., and D. Merton

1967 The language of adolescence: An anthropological approach to youth culture. *American Journal of Sociology* **72**: 453–468.

Scott, J. F.

1965 The American college sorority: Its role in class and ethnic endogamy. *American Sociological Review* **30**: 514–527.

Scott, M. B., and S. M. Lyman

1968 Accounts. *American Sociological Review* **33**: 46–62.

Sears, R. R., E. Maccoby, and H. Levin

1957 *Patterns of child rearing.* Evanston, Illinois: Row, Peterson.

Seeley, J. R., R. A. Sim, and E. W. Loosley

1956 *Crestwood heights.* New York: Basic Books.

Sewell, W. H., A. O. Haller, and G. W. Ohlendorf

1970 The educational and early occupational attainment process: replication and revision. *American Sociological Review* **35**: 1014–1027.

Selznick, P.

1949 *TVA and the grass roots.* Berkeley: Univ. of California Press.

1952 *The organizational weapon: A study of Bolshevik strategy and tactics.* New York: McGraw-Hill.

1957 *Leadership in administration.* Evanston, Illinois: Row, Peterson.

Sennett, R.

1970 *Families against the city.* Cambridge, Massachusetts: Harvard Univ. Press.

Shepherd, W. R.

1964 *Historical atlas.* 9th edition. New York: Barnes and Noble.

Shepperd, H. L., and A. H. Beliksky

1966 *The job hunt.* Baltimore: Johns Hopkins Press.

Sherwin, C. W., and R. S. Isenson

1967 Project hindsight. *Science* **156**: 1571–1577.

Shils, E. A., and M. Janowitz

1948 Cohesion and disintegration in the Wehrmacht in World War II. *Public Opinion Quarterly* **12**: 280–315.

Shuy, R. W.

1970 The sociolinguists and urban language problems. in *Language and poverty,* edited by F. Williams. Chicago: Markham.

Sibley, E.

1942 Some demographic clues to stratification. *American Sociological Review* 7: 322–330.

Simmel, G.

1950 *The sociology of Georg Simmel,* edited by K. Wolff. Glencoe, Illinois: Free Press (Originally published 1908).

Simmons, J. I., and B. Winograd

1966 *It's happening: A portrait of the youth scene today.* Santa Barbara, California: Marc–Laird.

Simon, H. A.

1947 *Administrative behavior.* New York: Macmillan.

Simon, H. A., H. Guetzkow, G. Kozmetsky, and G. Tyndall

1954 *Centralization versus decentralization in organizing the controller's department.* New York: Controllership Foundation.

Skinner, B. F.

1961 *Cumulative record.* New York: Appleton.

1969 *Contingencies of reinforcement.* New York: Appleton.

Skolnick, J. H.

1966 *Justice without trial.* New York: Wiley.

Smith, M. B.

1968 Competence and socialization. In *Socialization and society,* edited by J. A. Clausen. Boston: Little, Brown.

Soares, G. A. D.

1966 Economic development and class structure. In *Class, status, and power,* second edition, edited by R. Bendix and S. M. Lipset. New York: Free Press.

Sorel, G.

1961 *Reflections on violence.* New York: Collier-Macmillan (Originally published 1908).

Sorokin, P. A.

1927 *Social mobility.* New York: Harper.

Spady, W. G.

1967 Educational mobility and access: Growth and paradoxes. *American Journal of Sociology* 72: 273–286.

Speier, M.

1972 Some conversational problems for interactional analysis. In *Studies in social interaction,* edited by D. Sudnow. New York: Free Press.

Spilerman, S.

1972 The analysis of mobility processes by the introduction of independent variables into a Markov chain. *American Sociological Review* 37: 277–294.

Spiro, M.

1965 *Children of the Kibbutz.* New York: Schocken Books.

Spitz, R.

1950 Relevancy of direct infant observation. *Psychoanalytic Study of the Child* 5: 66–73.

Statistical Abstract of the United States

1971 Washington D.C.: Government Printing Office.

Steiner, J.

1933 *Americans at play.* New York: McGraw-Hill.

Stinchcombe, A. L.

1959 Bureaucratic and craft administration of production. *Administrative Science Quarterly* 2: 137–158.

1961 Agricultural enterprise and rural class relations. *American Journal of Sociology* 67: 165–176.

1965 Social structure and organizations. In *Handbook of organizations,* edited by J. G. March. Chicago: Rand McNally.

1967 Formal organizations. In *Sociology: An introduction,* edited by N. J. Smelser. New York: Wiley.

1968a *Constructing social theories.* New York: Harcourt.

1968b The structure of stratification systems. In *International encyclopedia of the social sciences,* edited by D. L. Sills. New York: Macmillan.

Stone, L.
1967 *The crisis of the Aristocracy 1558–1641.* New York: Oxford Univ. Press.

Storer, N. W.
1966 *The social system of science.* New York: Holt.
1967 The hard sciences and the soft. *Bulletin of the Medical Library Association* **55**: 75–84.

Strauss, A. L.
1971 *The contexts of social mobility.* Chicago: Aldine.

Strauss, G., and L. R. Sayles
1960 *Personnel.* Englewood Cliffs, New Jersey: Prentice-Hall.

Sullivan, H. S.
1947 *Conceptions of modern psychiatry.* Washington: William Alanson White Foundation.

Suttles, G. D.
1970 Friendship as a social institution. In *Social relationships,* edited by G. J. McCall. Chicago: Aldine.

Sutherland, E. H.
1937 *The professional thief.* Chicago: Univ. of Chicago Press.
1949 *White collar crime.* New York: Dryden.

Svalastoga, K.
1962 Rape and social structure. *Pacific Sociological Review* **5**: 48–53.

Swanson, G. E.
1962 *The birth of the gods.* Ann Arbor: Univ. of Michigan Press.

Szasz, T.
1967 *The myth of mental illness.* New York: Dell.

Sweezey, P. M.
1946 *The theory of capitalist development.* London: Oxford Univ. Press.

Talmon, Y.
1964 Mate selection in collective settlements. *American Sociological Review* **29**: 491–508.

Tannenbaum, A. S.
1965 Unions. In *Handbook of organizations,* edited by J. G. March. Chicago: Rand McNally.

Taussig, F. W.
1911 *Principles of economics.* New York: Macmillan.

Taylor, G. R.
1951 *The transportation revolution 1815–1860.* New York: Holt.

Thomas, K.
1971 *Religion and the decline of magic.* New York: Scribner's.

Thomas, L.
1956 *The occupational structure and education.* Englewood Cliffs, New Jersey: Prentice-Hall.

Thompson, J. D.
1967 *Organizations in action.* New York: McGraw-Hill.

Thompson, E. A.
1960 Slavery in early Germany. In *Slavery in classical antiquity,* edited by M. I. Finley. Cambridge: W. Heffer.

Thorpe, W. H.
1956 *Learning and instinct in animals.* Cambridge, Massachusetts: Harvard Univ. Press.
Tocqueville, A. de
1955 *The old regime and the French Revolution.* Garden City, New York: Doubleday (Originally published 1856).
Torrance, E. P.
1962 Cultural discontinuities and the development of originality in thinking. *Exceptional Child* **29**: 2–13.
Touraine, A.
1971 *The post-industrial society.* New York: Random House.
Triandis, H. C., and L. M. Triandis
1960 Race, social class, religion and nationality as determinants of social distance. *Journal of Abnormal and Social Psychology* **61**: 110–118.
Trow, M. A.
1966 The second transformation of American secondary education. In *Class, status, and power,* edited by R. Bendix and S. M. Lipset. New York: Free Press.
Tuchman, G.
1972 Objectivity as strategic ritual: An examination of newsmen's notions of objectivity. *American Journal of Sociology* **77**: 660–679.
Turk, A. T.
1966 Conflict and criminality. *American Sociological Review* **31**: 338–352.
Turk, H.
1972 Interorganizational networks in urban society: initial perspectives and comparative research. *American Sociological Review* **35**: 1–18.
Turner, R. H.
1947 The Navy disbursing officer as a bureaucrat. *American Sociological Review* **12**: 342–348.
Tyler, L. E.
1965 *The psychology of human differences.* New York: Appleton.
Verba, S.
1971 Cross-national survey research: The problem of credibility. In *Comparative methods in sociology,* edited by I. Vallier. Berkeley: Univ. of California Press.
Vidich, A. J., and J. Bensman
1968 *Small town in mass society.* revised edition. Princeton, New Jersey: Princeton Univ. Press.
Waller, W.
1932 *The sociology of teaching.* New York: Russell and Russell.
1937 The rating and dating complex. *American Sociological Review* **2**: 727–734.
Walster, E., V. Aronson, D. Abrahams, and L. Rotterman
1966 Importance of physical attractiveness in dating behavior. *Journal of Personality and Social Psychology* **4**: 508–516.
Warner, S. B., Jr.
1962 *Streetcar suburbs: The process of growth in Boston 1870-1900.* Cambridge, Massachusetts: Harvard Univ. Press.
Warner, W. L., and P. Lunt
1941 *The social life of a modern community.* New Haven, Connecticut: Yale Univ. Press.
Warner, W. L.
1949 *Democracy in Jonesville.* New York: Harper.
1959 *The Living and the Dead.* New Haven, Connecticut: Yale Univ. Press.
Warren, C. A. B., and J. M. Johnson
1972 A critique of labeling theory from the phenomenological perspective. In *Theoretical perspectives on deviance,* edited by R. A. Scott and J. D. Douglas. New York: Basic Books.

Warren, D. I.
1968 Power, visibility, and conformity in formal organizations. *American Sociological Review* **33**: 951–970.

Watson, G.
1964 *The literary critics.* Baltimore: Penguin Books.

Watson, J. D.
1969 *The double helix.* New York: New American Library.

Watson, J.
1958 A formal analysis of sociable interaction. *Sociometry* **21**: 269–280.

Weber, M.
1946 *From Max Weber: Essays in sociology,* edited by H. H. Gerth and C. W. Mills. New York: Oxford Univ. Press.
1951 *The Religion of China.* Glencoe, Illinois: Free Press (Originally published 1916).
1958a *The Protestant ethic and the spirit of capitalism.* New York: Scribner's (Originally published 1904–1905).
1958b *The religion of India.* Glencoe, Illinois: Free Press (Originally published 1916–1917).
1961 *General economic history.* New York: Collier-Macmillan (Originally published 1924).
1968 *Economy and Society.* New York: Bedminster Press (Originally published 1922).

Wector, D.
1937 *The saga of American society.* New York: Scribner's.

Weinberg, I.
1967 *The English public school.* New York: Atherton Press.

Weinberg, S. K.
1955 *Incest behavior.* New York: Citadel Press.

Weinberg, M. S.
1965 Sexual modesty, social meanings, and the nudist camp. *Social Problems* **12**: 311–318.

Weingartner, R. H.
1968 The quarrel about historical explanation. In *Readings in the philosophy of the social sciences,* edited by M. Brodbeck. New York: Macmillan.

Weisberg, P. S., and K. J. Springer
1961 Environmental factors in creative functioning. *Archives of General Psychiatry* **5**: 554–564.

Wertheim, W. F.
1968 Sociological aspects of corruption in Southeast Asia. In *State and Society,* edited by R. Bendix et al. Boston: Little, Brown.

Wharton, E.
1964 *The house of mirth.* New York: New American Library (Originally published 1905).

Whiting, J., and I. Child
1953 *Child training and personality.* New Haven, Connecticut: Yale Univ. Press.

White, H. C.
1970a *Chains of opportunity: System models of mobility in organizations.* Cambridge, Massachusetts: Harvard Univ. Press.
1970b Stayers and Movers. *American Journal of Sociology* **76**: 307–324.

Whitt, H. P., C. C. Gordon, and J. R. Hofley
1972 Religion, economic development, and lethal aggression. *American Sociological Review* **37**: 193–201.

Whyte, W. H.
1956 *The organization man.* Garden City, New York: Doubleday.

Wilensky, H. L.
1956 *Intellectuals in labor unions.* Glencoe, Illinois: Free Press.
1961 Orderly careers and social participation. *American Sociological Review* **26**: 521–539.
1964 The professionalization of everyone? *American Journal of Sociology* **70**: 137–158.

1966 Measures and effects of mobility. In *Social structure and mobility in economic development*, edited by N. J. Smelser, and S. M. Lipset. Chicago: Aldine.

1968a Women's work: Economic growth, ideology, structure. *Industrial Relations* 7: 235–248.

1968b *Organizational intelligence.* New York: Basic Books.

Wilensky, H. L., and H. Edwards

1959 The skidder: Ideological adjustments of downwardly mobile workers. *American Sociological Review* 24: 215–231.

Wilensky, J. L., and H. L. Wilensky

1951 Personnel counseling: The Hawthorne case. *American Journal of Sociology* 57: 265–280.

Winch, R. F., and S. A. Greer

1968 Urbanism, ethnicity, and extended familism. *Journal of Marriage and the Family* 30: 40–45. *Journal of Sociology* 57: 265–280.

Wittgenstein, L.

1953 *Philosophical investigations.* New York: Macmillan.

Wiley, N.

1967 America's unique class politics: The interplay of the labor, credit, and commodity markets. *American Sociological Review* 32: 529–540.

Wolfe, T.

1968 *The pump house gang.* New York: Farrar, Straus, and Giroux.

Wood, J. R.

1970 Authority and controversial policy: The churches and civil rights. *American Sociological Review* 35: 1057–1068.

Woodward, J.

1965 *Industrial organization.* London: Oxford Univ. Press.

Worsley, P.

1956 Durkheim's theory of knowledge. *Sociological Review* 4: 47–61.

Wrong, D. H.

1961 The oversocialized conception of man. *American Sociological Review* 26: 184–193.

Yablonsky, L.

1968 *The hippie trip.* New York: Pegasus.

Yasuda, S.

1964 A methodological inquiry into social mobility. *American Sociological Review* 29: 16–23.

Young, F. W.

1965 *Initiation ceremonies: A cross-cultural study of status dramatization.* Indianapolis: Bobbs-Merrill.

1967 Incest taboos and social solidarity. *American Journal of Sociology* 72: 589–600.

Young, M., and P. Willmott

1962 *Family and kinship in East London.* Baltimore: Penguin Books.

Zelditch, M. Jr.

1964 Family, marriage, and kinship. In *Handbook of modern sociology,* edited by R. E. L. Faris. Chicago: Rand McNally.

1971 Intelligible comparisons. In *Comparative methods in sociology,* edited by I. Vallier. Berkeley: Univ. of California Press.

Zelnick, M., and J. Kantner

1972 Sexuality, contraception, and pregnancy among pre-adult females in the United States. In *Demographic and social aspects of population growth,* volume I. Washington, D.C.: Government Printing Office.

Zetterberg, H. L.

1966 The secret ranking. *Journal of Marriage and the Family* 27: 134–142.

Zimmerman, D. R., and M. Pollner

1970 The everday world as a phenomenon. In *Understanding everday life,* edited by J. D. Douglas. Chicago: Aldine.

Index of Casual Principles

Index

A	7
B	8
C	9
D	0
E	1
F	2
G	3
H	4
I	5
J	6